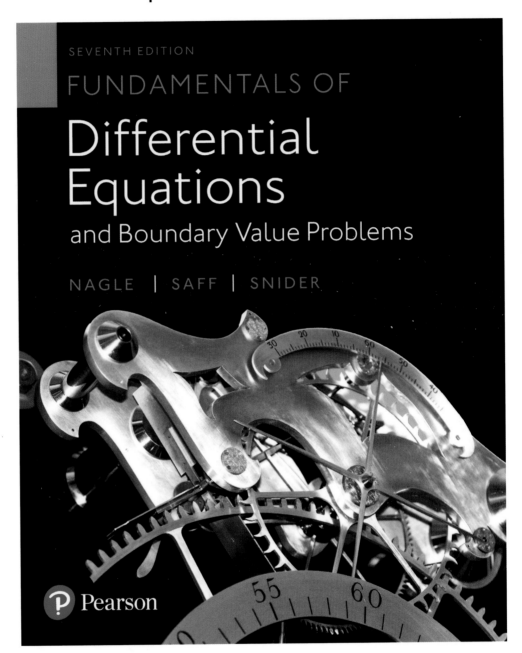

SEVENTH EDITION

FUNDAMENTALS OF
Differential
Equations

and Boundary Value Problems

NAGLE | SAFF | SNIDER

Pearson

ISBN-13: 978-0-321-97718-2
ISBN-10: 0-321-97718-1

EAN

MyMathLab®: Support You Need, When You Need It

MyMathLab is the world's leading online program in mathematics that has helped millions of students succeed in their math courses. Take advantage of the resources it provides.

Just-in-time help

MyMathLab's interactive exercises mirror those in the textbook but are programmed to allow you unlimited practice, leading to mastery. Most exercises include learning aids such as "Help Me Solve This," "View an Example," and "Tutorial Video," and they offer helpful feedback when you enter incorrect answers.

Tutorial Video support

Instructional videos narrated by the author cover key examples from the text and can conveniently be played on any mobile device. These videos are especially helpful if you miss a class or just need another explanation.

Interactive eText

The Pearson eText gives you access to your textbook anytime, anywhere. In addition to letting you take notes, highlight, and bookmark, the Pearson eText offers interactive links throughout, so you can watch videos as you read.

SEVENTH EDITION

Fundamentals of Differential Equations and Boundary Value Problems

R. Kent Nagle

Edward B. Saff

Vanderbilt University

Arthur David Snider

University of South Florida

Director, Portfolio Management: Deirdre Lynch
Executive Editor: Jeff Weidenaar
Editorial Assistant: Jennifer Snyder
Content Producer: Patty Bergin
Managing Producer: Karen Wernholm
Media Producer: Erin Carreiro
MathXL Content Manager: Kristina Evans
Product Marketing Manager: Yvonne Vannatta
Field Marketing Manager: Evan St. Cyr
Marketing Assistant: Jennifer Myers
Senior Author Support/Technology Specialist: Joe Vetere
Rights and Permissions Project Manager: Gina Cheselka
Manufacturing Buyer: Carol Melville, LSC Communications
Associate Director of Design: Blair Brown
Composition: Cenveo
Text Design, Production Coordination,
Composition, and Illustrations: Cenveo
Cover Design: Cenveo
Cover Image: Donald J. Saff

Cover is detail of Martin Burgess' "Clock B" that was constructed to demonstrate the efficacy of John Harrison's (1693-1776) science. Begun in 1974, and completed by the Charles Frodsham & Co. Ltd. of London in 2011, Clock B was placed on time trial at the Royal Observatory, Greenwich, England where it remained within 5/8 seconds in 100 days, and officially dubbed "the world's most accurate pendulum clock operating in free air." The clock exhibits oscillation consistent with van der Pol's nonlinear differential equation with an amplitude that minimizes external perturbation effects.

http://burgessclockb.com/
Photo courtesy of: Charles Frodsham & Co. Ltd.
Cover photo courtesy of: Donald Saff

Library of Congress Cataloging-in-Publication Data

Names: Nagle, R. Kent. | Saff, E. B., 1944- | Snider, Arthur David, 1940-
Title: Fundamentals of differential equations and boundary value problems.
Description: Seventh edition / R. Kent Nagle, Edward B. Saff, Vanderbilt University, Arthur David Snider, University of South Florida. | Boston : Pearson, [2018] | Includes index.
Identifiers: LCCN 2016030692| ISBN 9780321977106 (hardcover) | ISBN 0321977106 (hardcover)
Subjects: LCSH: Differential equations—Textbooks. | Boundary value problems—Textbooks.
Classification: LCC QA371 .N243 2018 | DDC 515/.35—dc23
LC record available at https://lccn.loc.gov/2016030692

1 16

www.pearsonhighered.com

Student Edition ISBN13: 978-0-321-97710-6
Student Edition ISBN10: 0-321-97710-6

Dedicated to R. Kent Nagle

He has left his imprint not only on these pages but upon all who knew him. He was that rare mathematician who could effectively communicate at all levels, imparting his love for the subject with the same ease to undergraduates, graduates, precollege students, public school teachers, and his colleagues at the University of South Florida.

Kent was at peace in life—a peace that emanated from the depth of his understanding of the human condition and the strength of his beliefs in the institutions of family, religion, and education. He was a research mathematician, an accomplished author, a Sunday school teacher, and a devoted husband and father.

Kent was also my dear friend and my jogging partner who has left me behind still struggling to keep pace with his high ideals.

E. B. Saff

Contents

CHAPTER 3 Mathematical Models and Numerical Methods Involving First-Order Equations

CHAPTER 4 Linear Second-Order Equations

CHAPTER 5 Introduction to Systems and Phase Plane Analysis

CHAPTER 6 Theory of Higher-Order Linear Differential Equations

CHAPTER 7 Laplace Transforms

CHAPTER 8 Series Solutions of Differential Equations

CHAPTER 9 Matrix Methods for Linear Systems

CHAPTER 10 Partial Differential Equations

CHAPTER 11 Eigenvalue Problems and Sturm–Liouville Equations

CHAPTER 12　**Stability of Autonomous Systems**

CHAPTER 13 Existence and Uniqueness Theory

13.1 Introduction: Successive Approximations 783

13.2 Picard's Existence and Uniqueness Theorem 789

13.3 Existence of Solutions of Linear Equations 797

13.4 Continuous Dependence of Solutions 802

Chapter Summary 809

Review Problems for Chapter 13 810

Technical Writing Exercises for Chapter 13 810

Appendices

A. *Review of Integration Techniques* A-1

B. *Newton's Method* A-9

C. *Simpson's Rule* A-11

D. *Cramer's Rule* A-13

E. *Method of Least Squares* A-14

F. *Runge–Kutta Procedure for n Equations* A-16

G. *Software for Analyzing Differential Equations* A-17

Answers to Odd-Numbered Problems B-1

Index I-1

Preface

Our Goal

Fundamentals of Differential Equations and Boundary Value Problems is designed to serve the needs of a one- or two-semester course in basic theory as well as applications of differential equations. For this purpose, we have augmented our shorter text by adding Chapter 11, Eigenvalue Problems and Sturm–Liouville Equations; Chapter 12, Stability of Autonomous Systems; and Chapter 13, Existence and Uniqueness Theory. We have striven to create a flexible text that affords the instructor substantial latitude in designing a syllabus (sample syllabi are provided in this preface), in course emphasis (theory, methodology, applications, and numerical methods), and in using commercially available computer software.

New to This Edition

- This text now features a MyMathLab course with approximately 850 algorithmic online homework exercises, tutorial videos, and the complete eText. Please see the "Technology and Supplements" section below for more details.

- In the Laplace Transforms Chapter (7), the treatments of discontinuous and periodic functions are now divided into two sections that are more appropriate for 50 minute lectures: Section 7.6 "Transforms of Discontinuous Functions" (page 383) and Section 7.7 "Transforms of Periodic and Power Functions" (page 392).

- New examples have been added dealing with variation of parameters, Laplace transforms, the Gamma function, and eigenvectors (among others).

- For students interested in bio-mathematics, in addition to the updated Section 5.5, a new section "Neurons and the FitzHugh–Nagumo Equations" has been added to Chapter 12 (page 770). It outlines the (Nobel prize-winning) work of Hodgkin and Huxley in modeling the electrochemical generation of action potentials inside axons and guides the reader through computer phase plane simulations of the remarkable responses produced by various stimula.

- New problems added to exercise sets deal with such topics as axon gating variables and oscillations of a helium-filled balloon on a cord. Additionally, novel problems accompany the new projects, focusing on economic models, disease control, synchronization, signal propagation, and phase plane analyses of neural responses. We have also added a set of Review Problems for Chapter 1 (page 29).

- Several pedagogical changes were made including amplification of the distinction between phase plane solutions and actual trajectories in Chapter 5, incorporation of matrix and Jacobian formulations for autonomous systems, and revised discussion of limit cycles in Chapter 12.

- A new appendix lists commercial software and freeware for direction fields, phase portraits, and numerical methods for solving differential equations. (Appendix G, page A-17.)

- "The 2014–2015 Ebola Epidemic" is a new Project in Chapter 5 that describes a system of differential equations for modeling for the spread of the disease in West Africa. The model incorporates such features as contact tracing, number of contacts, likelihood of infection, and efficacy of isolation. See Project F, page 314.

- A new project in Chapter 4 called "Gravity Train" invites the reader to utilize differential equations in the design of an underground tunnel from Moscow to St. Petersburg, Russia, using gravity for propulsion. See Project H, page 240.

- A new project in Chapter 1 called "Applications to Economics" deals with models for an agrarian economy as well as the growth of capital. See Project C, page 35.

- Phase-locked loops constitute the theme of a new project in Chapter 5 that utilizes differential equations to analyze a technique for measuring or matching high frequency radio oscillations. See Project G, page 317.

- A new project in Chapter 10 broadens the analysis of the wave and heat equations to explore the telegrapher's and cable equations. See Project E, page 637.

Prerequisites

While some universities make linear algebra a prerequisite for differential equations, many schools (especially engineering) only require calculus. With this in mind, we have designed the text so that only Chapter 6 (Theory of Higher-Order Linear Differential Equations) and Chapter 9 (Matrix Methods for Linear Systems) require more than high school level linear algebra. Moreover, Chapter 9 contains review sections on matrices and vectors as well as specific references for the deeper results used from the theory of linear algebra. We have also written Chapter 5 so as to give an introduction to systems of differential equations—including methods of solving, phase plane analysis, applications, numerical procedures, and Poincaré maps—that does not require a background in linear algebra.

Sample Syllabi

As a rough guide in designing a two-semester syllabus related to this text, we provide two samples that can be used for a sequence of two 15-week courses that meet three hours per week. The first emphasizes applications and computations including phase plane analysis; the second is designed for courses that place more emphasis on theory. Chapters 1, 2, and 4 provide the core for any first course. The rest of the chapters are, for the most part, independent of each other. For students with a background in linear algebra, the instructor may prefer to replace Chapter 7 (Laplace Transforms) or Chapter 8 (Series Solutions of Differential Equations) with sections from Chapter 9 (Matrix Methods for Linear Systems).

Semester I	Methods, Computations, and Applications	Theory and Methods	Semester II	Both Courses
Week	**Sections**	**Sections**	**Week**	**Sections**
1	1.1, 1.2, 1.3	1.1, 1.2, 1.3	1	9.1, 9.2, 9.3
2	1.4, 2.2	1.4, 2.2, 2.3	2	9.4, 9.5
3	2.3, 2.4, 3.2	2.4, 2.6	3	9.6, 9.7, 9.8
4	3.4, 3.5, 3.6	3.2, 3.4, 4.1	4	12.1, 12.2
5	3.7, 4.1	4.2, 4.3, 4.4	5	12.3, 12.4
6	4.2, 4.3, 4.4	4.5, 4.6	6	12.5, 12.6
7	4.5, 4.6, 4.7	4.7, 5.1, 5.2	7	12.7, 10.2
8	4.8, 4.9	5.3, 5.4	8	10.3, 10.4
9	4.10, 5.1, 5.2	5.5, 6.1	9	10.5, 10.6
10	5.3, 5.4, 5.5	6.2, 6.3, 6.4	10	10.7, 11.2
11	5.6, 5.7, 7.2	7.2, 7.3, 7.4	11	11.3, 11.4
12	7.3, 7.4, 7.5	7.5, 7.6	12	11.5, 11.6
13	7.6, 7.7, 7.8	7.7, 7.8, 8.1	13	11.7, 11.8
14	8.1, 8.2, 8.3	8.2, 8.3, 8.4	14	13.1, 13.2
15	8.4, 8.6	8.5, 8.6	15	13.3, 13.4

Retained Features

Flexible Organization Most of the material is modular in nature to allow for various course configurations and emphasis (theory, applications and techniques, and concepts).

Optional Use of Computer Software The availability of computer packages such as Mathcad®, Mathematica®, MATLAB®, and Maple™ provides an opportunity for the student to conduct numerical experiments and tackle realistic applications that give additional insights into the subject. Consequently, we have inserted several exercises and projects throughout the text that are designed for the student to employ available software in phase plane analysis, eigenvalue computations, and the numerical solutions of various equations.

Review of Integration In response to the perception that many of today's students' skills in integration have gotten rusty by the time they enter a differential equations course, we have included an appendix offering a quick review of the basic methods for integrating functions analytically.

Choice of Applications Because of syllabus constraints, some courses will have little or no time for sections (such as those in Chapters 3 and 5) that exclusively deal with applications. Therefore, we have made the sections in these chapters independent of each other. To afford the instructor even greater flexibility, we have built in a variety of applications in the exercises for the theoretical sections. In addition, we have included many projects that deal with such applications.

Projects At the end of each chapter are projects relating to the material covered in the chapter. Several of them have been contributed by distinguished researchers. A project might involve a more challenging application, delve deeper into the theory, or introduce more advanced topics in differential equations. Although these projects can be tackled by an individual student, classroom testing has shown that working in groups lends a valuable added dimension to the learning experience. Indeed, it simulates the interactions that take place in the professional arena.

Technical Writing Exercises
Communication skills are, of course, an essential aspect of professional activities. Yet few texts provide opportunities for the reader to develop these skills. Thus, we have added at the end of most chapters a set of clearly marked technical writing exercises that invite students to make documented responses to questions dealing with the concepts in the chapter. In so doing, students are encouraged to make comparisons between various methods and to present examples that support their analysis.

Historical Footnotes
Throughout the text historical footnotes are set off by colored daggers (†). These footnotes typically provide the name of the person who developed the technique, the date, and the context of the original research.

Motivating Problem
Most chapters begin with a discussion of a problem from physics or engineering that motivates the topic presented and illustrates the methodology.

Chapter Summary and Review Problems
All of the main chapters contain a set of review problems along with a synopsis of the major concepts presented.

Computer Graphics
Most of the figures in the text were generated via computer. Computer graphics not only ensure greater accuracy in the illustrations, they demonstrate the use of numerical experimentation in studying the behavior of solutions.

Proofs
While more pragmatic students may balk at proofs, most instructors regard these justifications as an essential ingredient in a textbook on differential equations. As with any text at this level, certain details in the proofs must be omitted. When this occurs, we flag the instance and refer readers either to a problem in the exercises or to another text. For convenience, the end of a proof is marked by the symbol ◆.

Linear Theory
We have developed the theory of linear differential equations in a gradual manner. In Chapter 4 (Linear Second-Order Equations) we first present the basic theory for linear second-order equations with constant coefficients and discuss various techniques for solving these equations. Section 4.7 surveys the extension of these ideas to variable-coefficient second-order equations. A more general and detailed discussion of linear differential equations is given in Chapter 6 (Theory of Higher-Order Linear Differential Equations). For a beginning course emphasizing methods of solution, the presentation in Chapter 4 may be sufficient and Chapter 6 can be skipped.

Numerical Algorithms
Several numerical methods for approximating solutions to differential equations are presented along with program outlines that are easily implemented on a computer. These methods are introduced early in the text so that teachers and/or students can use them for numerical experimentation and for tackling complicated applications. Where appropriate we direct the student to software packages or web-based applets for implementation of these algorithms.

Exercises
An abundance of exercises are graduated in difficulty from straightforward, routine problems to more challenging ones. Deeper theoretical questions, along with applications, usually occur toward the end of the exercise sets. Throughout the text we have included problems and projects that require the use of a calculator or computer. These exercises are denoted by the symbol 🖥.

Laplace Transforms
We provide a detailed chapter on Laplace transforms (Chapter 7) since this is a recurring topic for engineers. Our treatment emphasizes discontinuous forcing terms and includes a section on the Dirac delta function.

Power Series | Power series solutions is a topic that occasionally causes student anxiety. Possibly, this is due to inadequate preparation in calculus where the more subtle subject of convergent series is (frequently) covered at a rapid pace. Our solution has been to provide a graceful initiation into the theory of power series solutions with an exposition of Taylor polynomial approximants to solutions, deferring the sophisticated issues of convergence to later sections. Unlike many texts, ours provides an extensive section on the method of Frobenius (Section 8.6) as well as a section on finding a second linearly independent solution. While we have given considerable space to power series solutions, we have also taken great care to accommodate the instructor who only wishes to give a basic introduction to the topic. An introduction to solving differential equations using power series and the method of Frobenius can be accomplished by covering the materials in Sections 8.1, 8.2, 8.3, and 8.6.

Partial Differential Equations | An introduction to this subject is provided in Chapter 10, which covers the method of separation of variables, Fourier series, the heat equation, the wave equation, and Laplace's equation. Examples in two and three dimensions are included.

Phase Plane | Chapter 5 describes how qualitative information for two-dimensional systems can be gleaned about the solutions to intractable autonomous equations by observing their direction fields and critical points on the phase plane. With the assistance of suitable software, this approach provides a refreshing, almost recreational alternative to the traditional analytic methodology as we discuss applications in nonlinear mechanics, ecosystems, and epidemiology.

Vibrations | Motivation for Chapter 4 on linear differential equations is provided in an introductory section describing the mass–spring oscillator. We exploit the reader's familiarity with common vibratory motions to anticipate the exposition of the theoretical and analytical aspects of linear equations. Not only does this model provide an anchor for the discourse on constant-coefficient equations, but a liberal interpretation of its features enables us to predict the qualitative behavior of variable-coefficient and nonlinear equations as well.

Review of Algebraic Equations and Matrices | The chapter on matrix methods for linear systems (Chapter 9) begins with two (optional) introductory sections reviewing the theory of linear algebraic systems and matrix algebra.

Technology and Supplements

MyMathLab® Online Course (access code required) Built around Pearson's best-selling content, MyMathLab is an online homework, tutorial, and assessment program designed to work with this text to engage students and improve results. MyMathLab can be successfully implemented in any classroom environment—lab-based, hybrid, fully online, or traditional.

MyMathLab's online homework offers students immediate feedback and tutorial assistance that motivates them to do more, which means they retain more knowledge and improve their test scores. Used by more than 37 million students worldwide, MyMathLab delivers consistent, measurable gains in student learning outcomes, retention, and subsequent course success. Visit **www.mymathlab.com/results** to learn more.

Learning and
Teaching Tools

- **Exercises with immediate feedback**—Nearly 850 assignable exercises are based on the textbook exercises and regenerate algorithmically to give students unlimited opportunity for practice and mastery. MyMathLab provides helpful feedback when students enter incorrect answers and includes optional learning aids including Help Me Solve This, View an Example, videos, and an eText.

- **Learning Catalytics™** is a student response tool that uses students' smartphones, tablets, or laptops to engage them in more interactive tasks and thinking. Learning Catalytics fosters student engagement and peer-to-peer learning with real-time analytics.

- **Instructional videos** are available as learning aids within exercises and for self-study within the Multimedia Library. The Guide to Video-Based Assignments makes it easy to assign videos for homework by showing which MyMathLab exercises correspond to each video.
- **The complete eText** is available to students through their MyMathLab courses for the lifetime of the edition, giving students unlimited access to the eText within any course using that edition of the textbook.

- **Accessibility** and achievement go hand in hand. MyMathLab is compatible with the JAWS screen reader, and enables multiple-choice and free-response problem types to be read and interacted with via keyboard controls and math notation input. MyMathLab also works with screen enlargers, including ZoomText, MAGic, and SuperNova. And, all MyMathLab videos have closed-captioning. More information is available at **mymathlab.com/accessibility**.
- **A comprehensive gradebook** with enhanced reporting functionality allows you to efficiently manage your course.
 - **The Reporting Dashboard** provides insight to view, analyze, and report learning outcomes. Student performance data is presented at the class, section, and program levels in an accessible, visual manner so you'll have the information you need to keep your students on track.

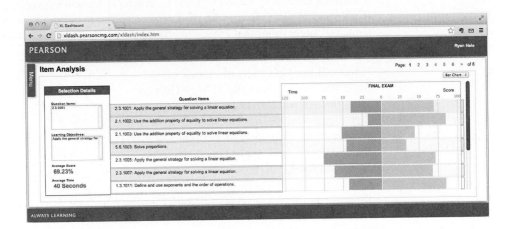

 - **Item Analysis** tracks class-wide understanding of particular exercises so you can refine your class lectures or adjust the course/department syllabus. Just-in-time teaching has never been easier!

MyMathLab comes from an experienced partner with educational expertise and an eye on the future. Whether you are just getting started with MyMathLab, or have a question along the way, we're here to help you learn about our technologies and how to incorporate them into your course. To learn more about how MyMathLab helps students succeed, visit **mymathlab.com** or contact your Pearson rep.

MathXL® is the homework and assessment engine that runs MyMathLab. (MyMathLab is MathXL plus a learning management system.) MathXL access codes are also an option.

Student's Solutions Manual	ISBN- 10: 0321977211 I ISBN-13: 9780321977212 Contains complete worked-out solutions to odd-numbered exercises, providing students with an excellent study tool. Available in print and for download within MyMathLab.
Instructor's Solutions Manual (downloadable)	ISBN-10: 0134659244 I ISBN-13: 9780134659244 Contains answers to all even-numbered exercises, detailed solutions to the even-numbered problems in several of the main chapters, and additional projects. Available for download in the Pearson Instructor Resource Center www.pearsonhighered.com/irc as well as within MyMathLab.
MATLAB, Maple, and Mathematica Manuals (downloadable)	By Thomas W. Polaski (Winthrop University), Bruno Welfert (Arizona State University), and Maurino Bautista (Rochester Institute of Technology), respectively. These manuals contain a collection of instructor tips, worksheets, and projects to aid instructors in integrating computer algebra systems into their courses. Complete manuals are available for instructor download within MyMathLab. Student worksheets and projects available for download within MyMathLab.

Acknowledgments

The staging of this text involved considerable behind-the-scenes activity. We thank, first of all, Philip Crooke, Glenn Webb (Vanderbilt University), and Joanna Wares (University of Richmond) who have continued to provide novel biomathematical projects, as well as Greg Huffman (Vanderbilt University) for his project on economics. We also want to thank Frank Glaser (California State Polytechnic University, Pomona) for many of the historical footnotes. We are indebted to Herbert E. Rauch (Lockheed Research Laboratory) for help with Section 3.3 on heating and cooling of buildings, Project B in Chapter 3 on aquaculture, and other application problems. Our appreciation goes to Richard H. Elderkin (Pomona College), Jerrold Marsden (California Institute of Technology), T. G. Proctor (Clemson University), and Philip W. Schaefer (University of Tennessee), who read and reread the manuscript for the original text, making numerous suggestions that greatly improved our work. Thanks also to the following reviewers of this and previous editions:

*Miklos Bona, University of Florida
Amin Boumenir, University of West Georgia
Mark Brittenham, University of Nebraska
*Jennifer Bryan, Oklahoma Christian University
Weiming Cao, University of Texas at San Antonio
Richard Carmichael, Wake Forest University
*Kwai-lee Chui, University of Florida
Karen Clark, The College of New Jersey
*Shaozhong Deng, University of North Carolina Charlotte
Patrick Dowling, Miami University
*Chris Fuller, Cumberland University
Sanford Geraci, Northern Virginia
David S. Gilliam, Texas Tech University at Lubbock
Scott Gordon, State University of West Georgia
*Irvin Hentzel, Iowa State University
*Mimi Rasky, Southwestern College
Richard Rubin, Florida International University
John Sylvester, University of Washington at Seattle
Steven Taliaferro, Texas A&M University at College Station
Michael M. Tom, Louisiana State University
Shu-Yi Tu, University of Michigan,
Flint Klaus Volpert, Villanova University
*Glenn F. Webb, Vanderbilt University

E. B. Saff, A. D. Snider

*Denotes reviewers of the current edition

Introduction

1.1 Background

In the sciences and engineering, mathematical models are developed to aid in the understanding of physical phenomena. These models often yield an equation that contains some derivatives of an unknown function. Such an equation is called a **differential equation.** Two examples of models developed in calculus are the free fall of a body and the decay of a radioactive substance.

In the case of free fall, an object is released from a certain height above the ground and falls under the force of gravity.[†] Newton's second law, which states that an object's mass times its acceleration equals the total force acting on it, can be applied to the falling object. This leads to the equation (see Figure 1.1)

$$m\frac{d^2h}{dt^2} = -mg\,,$$

where m is the mass of the object, h is the height above the ground, d^2h/dt^2 is its acceleration, g is the (constant) gravitational acceleration, and $-mg$ is the force due to gravity. This is a differential equation containing the second derivative of the unknown height h as a function of time.

Fortunately, the above equation is easy to solve for h. All we have to do is divide by m and integrate twice with respect to t. That is,

$$\frac{d^2h}{dt^2} = -g\,,$$

so

$$\frac{dh}{dt} = -gt + c_1$$

and

$$h = h(t) = \frac{-gt^2}{2} + c_1 t + c_2\,.$$

Figure 1.1 Apple in free fall

[†]We are assuming here that gravity is the *only* force acting on the object and that this force is constant. More general models would take into account other forces, such as air resistance.

We will see that the constants of integration, c_1 and c_2, are determined if we know the *initial* height and the *initial* velocity of the object. We then have a formula for the height of the object at time t.

In the case of radioactive decay (Figure 1.2), we begin from the premise that the rate of decay is proportional to the amount of radioactive substance present. This leads to the equation

$$\frac{dA}{dt} = -kA, \qquad k > 0,$$

where $A(>0)$ is the unknown amount of radioactive substance present at time t and k is the proportionality constant. To solve this differential equation, we rewrite it in the form

$$\frac{1}{A}dA = -k\,dt$$

and integrate to obtain

$$\int \frac{1}{A}dA = \int -k\,dt$$

$$\ln A + C_1 = -kt + C_2\,.$$

Solving for A yields

$$A = A(t) = e^{\ln A} = e^{-kt}\,e^{C_2-C_1} = Ce^{-kt},$$

where C is the combination of integration constants $e^{C_2-C_1}$. The value of C, as we will see later, is determined if the *initial* amount of radioactive substance is given. We then have a formula for the amount of radioactive substance at any future time t.

Even though the above examples were easily solved by methods learned in calculus, they do give us some insight into the study of differential equations in general. First, notice that the solution of a differential equation is a *function*, like $h(t)$ or $A(t)$, not merely a number. Second, integration[†] is an important tool in solving differential equations (not surprisingly!). Third, we cannot expect to get a unique solution to a differential equation, since there will be arbitrary "constants of integration." The second derivative d^2h/dt^2 in the free-fall equation gave rise to two constants, c_1 and c_2, and the first derivative in the decay equation gave rise, ultimately, to one constant, C.

Whenever a mathematical model involves the **rate of change** of one variable with respect to another, a differential equation is apt to appear. Unfortunately, in contrast to the examples for free fall and radioactive decay, the differential equation may be very complicated and difficult to analyze.

Figure 1.2 Radioactive decay

[†]For a review of integration techniques, see Appendix A.

Figure 1.3 Schematic for a series RLC circuit

Differential equations arise in a variety of subject areas, including not only the physical sciences but also such diverse fields as economics, medicine, psychology, and operations research. We now list a few specific examples.

1. In banking practice, if $P(t)$ is the number of dollars in a savings bank account that pays a yearly interest rate of $r\%$ *compounded continuously*, then P satisfies the differential equation

(1) $$\frac{dP}{dt} = \frac{r}{100}P\ ,\quad t \text{ in years.}$$

2. A classic application of differential equations is found in the study of an electric circuit consisting of a resistor, an inductor, and a capacitor driven by an electromotive force (see Figure 1.3). Here an application of Kirchhoff's laws[†] leads to the equation

(2) $$L\frac{d^2q}{dt^2} + R\frac{dq}{dt} + \frac{1}{C}q = E(t)\ ,$$

 where L is the inductance, R is the resistance, C is the capacitance, $E(t)$ is the electromotive force, $q(t)$ is the charge on the capacitor, and t is the time.

3. In psychology, one model of the learning of a task involves the equation

(3) $$\frac{dy/dt}{y^{3/2}(1-y)^{3/2}} = \frac{2p}{\sqrt{n}}\ .$$

 Here the variable y represents the learner's skill level as a function of time t. The constants p and n depend on the individual learner and the nature of the task.

4. In the study of vibrating strings and the propagation of waves, we find the *partial* differential equation

(4) $$\frac{\partial^2 u}{\partial t^2} - c^2\frac{\partial^2 u}{\partial x^2} = 0\ ,^{‡}$$

 where t represents time, x the location along the string, c the wave speed, and u the displacement of the string, *which is a function of time and location.*

[†]We will discuss Kirchhoff's laws in Section 3.5.

[‡]*Historical Footnote:* This partial differential equation was first discovered by Jean le Rond d'Alembert (1717–1783) in 1747.

To begin our study of differential equations, we need some common terminology. If an equation involves the derivative of one variable with respect to another, then the former is called a **dependent variable** and the latter an **independent variable.** Thus, in the equation

$$(5) \qquad \frac{d^2x}{dt^2} + a\frac{dx}{dt} + kx = 0 \,,$$

t is the independent variable and x is the dependent variable. We refer to a and k as **coefficients** in equation (5). In the equation

$$(6) \qquad \frac{\partial u}{\partial x} - \frac{\partial u}{\partial y} = x - 2y \,,$$

x and y are independent variables and u is the dependent variable.

A differential equation involving only ordinary derivatives with respect to a single independent variable is called an **ordinary differential equation.** A differential equation involving partial derivatives with respect to more than one independent variable is a **partial differential equation.** Equation (5) is an ordinary differential equation, and equation (6) is a partial differential equation.

The **order** of a differential equation is the order of the highest-order derivatives present in the equation. Equation (5) is a second-order equation because d^2x/dt^2 is the highest-order derivative present. Equation (6) is a first-order equation because only first-order partial derivatives occur.

It will be useful to classify ordinary differential equations as being either linear or nonlinear. Remember that lines (in two dimensions) and planes (in three dimensions) are especially easy to visualize, when compared to nonlinear objects such as cubic curves or quadric surfaces. For example, all the points on a line can be found if we know just two of them. Correspondingly, *linear* differential equations are more amenable to solution than nonlinear ones. Observe that the equations for lines $ax + by = c$ and planes $ax + by + cz = d$ have the feature that the variables appear in *additive combinations of their first powers only.* By analogy a **linear differential equation** is one in which the dependent variable y and its derivatives appear in additive combinations of their first powers.

More precisely, a differential equation is **linear** if it has the format

$$(7) \qquad a_n(x)\frac{d^ny}{dx^n} + a_{n-1}(x)\frac{d^{n-1}y}{dx^{n-1}} + \cdots + a_1(x)\frac{dy}{dx} + a_0(x)y = F(x) \,,$$

where $a_n(x), a_{n-1}(x), \ldots, a_0(x)$ and $F(x)$ depend only on the independent variable x. The additive combinations are permitted to have multipliers (coefficients) that depend on x; no restrictions are made on the nature of this x-dependence. If an ordinary differential equation is not linear, then we call it **nonlinear.** For example,

$$\frac{d^2y}{dx^2} + y^3 = 0$$

is a nonlinear second-order ordinary differential equation because of the y^3 term, whereas

$$t^3\frac{dx}{dt} = t^3 + x$$

is linear (despite the t^3 terms). The equation

$$\frac{d^2y}{dx^2} - y\frac{dy}{dx} = \cos x$$

is nonlinear because of the $y\,dy/dx$ term.

Although the majority of equations one is likely to encounter in practice fall into the *nonlinear* category, knowing how to deal with the simpler linear equations is an important first step (just as tangent lines help our understanding of complicated curves by providing local approximations).

1.1 EXERCISES

In Problems 1–12, a differential equation is given along with the field or problem area in which it arises. Classify each as an ordinary differential equation (ODE) or a partial differential equation (PDE), give the order, and indicate the independent and dependent variables. If the equation is an ordinary differential equation, indicate whether the equation is linear or nonlinear.

1. $5\dfrac{d^2x}{dt^2} + 4\dfrac{dx}{dt} + 9x = 2\cos 3t$

(mechanical vibrations, electrical circuits, seismology)

2. $\dfrac{d^2y}{dx^2} - 2x\dfrac{dy}{dx} + 2y = 0$

(Hermite's equation, quantum-mechanical harmonic oscillator)

3. $\dfrac{dy}{dx} = \dfrac{y(2 - 3x)}{x(1 - 3y)}$

(competition between two species, ecology)

4. $\dfrac{\partial^2 u}{\partial x^2} + \dfrac{\partial^2 u}{\partial y^2} = 0$

(Laplace's equation, potential theory, electricity, heat, aerodynamics)

5. $y\left[1 + \left(\dfrac{dy}{dx}\right)^2\right] = C$, where C is a constant

(brachistochrone problem,[†] calculus of variations)

6. $\dfrac{dx}{dt} = k(4 - x)(1 - x)$, where k is a constant

(chemical reaction rates)

7. $\dfrac{dp}{dt} = kp(P - p)$, where k and P are constants

(logistic curve, epidemiology, economics)

8. $\sqrt{1 - y}\,\dfrac{d^2y}{dx^2} + 2x\dfrac{dy}{dx} = 0$

(Kidder's equation, flow of gases through a porous medium)

9. $x\dfrac{d^2y}{dx^2} + \dfrac{dy}{dx} + xy = 0$

(aerodynamics, stress analysis)

10. $8\dfrac{d^4y}{dx^4} = x(1 - x)$

(deflection of beams)

11. $\dfrac{\partial N}{\partial t} = \dfrac{\partial^2 N}{\partial r^2} + \dfrac{1}{r}\dfrac{\partial N}{\partial r} + kN$, where k is a constant

(nuclear fission)

12. $\dfrac{d^2y}{dx^2} - 0.1(1 - y^2)\dfrac{dy}{dx} + 9y = 0$

(van der Pol's equation, triode vacuum tube)

In Problems 13–16, write a differential equation that fits the physical description.

13. The rate of change of the population p of bacteria at time t is proportional to the population at time t.

14. The velocity at time t of a particle moving along a straight line is proportional to the fourth power of its position x.

15. The rate of change in the temperature T of coffee at time t is proportional to the difference between the temperature M of the air at time t and the temperature of the coffee at time t.

16. The rate of change of the mass A of salt at time t is proportional to the square of the mass of salt present at time t.

17. Drag Race. Two drivers, Alison and Kevin, are participating in a drag race. Beginning from a standing start, they each proceed with a constant acceleration. Alison covers the last $1/4$ of the distance in 3 seconds, whereas Kevin covers the last $1/3$ of the distance in 4 seconds. Who wins and by how much time?

[†]*Historical Footnote:* In 1630 Galileo formulated the brachistochrone problem ($\beta\rho\acute\alpha\chi\iota\sigma\tau\sigma\varsigma$ = shortest, $\chi\rho\acute o\nu o\varsigma$ = time), that is, to determine a path down which a particle will fall from one given point to another in the shortest time. It was reposed by John Bernoulli in 1696 and solved by him the following year.

1.2 Solutions and Initial Value Problems

An nth-order ordinary differential equation is an equality relating the independent variable to the nth derivative (and usually lower-order derivatives as well) of the dependent variable. Examples are

$$x^2\frac{d^2y}{dx^2} + x\frac{dy}{dx} + y = x^3 \text{ (second-order, } x \text{ independent, } y \text{ dependent)}$$

$$\sqrt{1 - \left(\frac{d^2y}{dt^2}\right)} - y = 0 \text{ (second-order, } t \text{ independent, } y \text{ dependent)}$$

$$\frac{d^4x}{dt^4} = xt \text{ (fourth-order, } t \text{ independent, } x \text{ dependent).}$$

Thus, a general form for an nth-order equation with x independent, y dependent, can be expressed as

(1) $$F\left(x, y, \frac{dy}{dx}, \ldots, \frac{d^ny}{dx^n}\right) = 0,$$

where F is a function that depends on x, y, and the derivatives of y up to order n; that is, on x, $y, \ldots, d^ny/dx^n$. We assume that the equation holds for all x in an open interval I ($a < x < b$, where a or b could be infinite). In many cases we can isolate the highest-order term d^ny/dx^n and write equation (1) as

(2) $$\frac{d^ny}{dx^n} = f\left(x, y, \frac{dy}{dx}, \ldots, \frac{d^{n-1}y}{dx^{n-1}}\right),$$

which is often preferable to (1) for theoretical and computational purposes.

Explicit Solution

> **Definition 1.** A function $\phi(x)$ that when substituted for y in equation (1) [or (2)] satisfies the equation for all x in the interval I is called an **explicit solution** to the equation on I.

Example 1 Show that $\phi(x) = x^2 - x^{-1}$ is an explicit solution to the linear equation

(3) $$\frac{d^2y}{dx^2} - \frac{2}{x^2}y = 0,$$

but $\psi(x) = x^3$ is not.

Solution The functions $\phi(x) = x^2 - x^{-1}$, $\phi'(x) = 2x + x^{-2}$, and $\phi''(x) = 2 - 2x^{-3}$ are defined for all $x \neq 0$. Substitution of $\phi(x)$ for y in equation (3) gives

$$(2 - 2x^{-3}) - \frac{2}{x^2}(x^2 - x^{-1}) = (2 - 2x^{-3}) - (2 - 2x^{-3}) = 0.$$

Since this is valid for any $x \neq 0$, the function $\phi(x) = x^2 - x^{-1}$ is an explicit solution to (3) on $(-\infty, 0)$ and also on $(0, \infty)$.

For $\psi(x) = x^3$ we have $\psi'(x) = 3x^2$, $\psi''(x) = 6x$, and substitution into (3) gives

$$6x - \frac{2}{x^2}x^3 = 4x = 0,$$

which is valid only at the point $x = 0$ and not on an interval. Hence $\psi(x)$ is not a solution. ◆

Example 2 Show that for *any* choice of the constants c_1 and c_2, the function

$$\phi(x) = c_1 e^{-x} + c_2 e^{2x}$$

is an explicit solution to the linear equation

(4) $y'' - y' - 2y = 0$.

Solution We compute $\phi'(x) = -c_1 e^{-x} + 2c_2 e^{2x}$ and $\phi''(x) = c_1 e^{-x} + 4c_2 e^{2x}$. Substitution of ϕ, ϕ', and ϕ'' for y, y', and y'' in equation (4) yields

$$\left(c_1 e^{-x} + 4c_2 e^{2x}\right) - \left(-c_1 e^{-x} + 2c_2 e^{2x}\right) - 2\left(c_1 e^{-x} + c_2 e^{2x}\right)$$
$$= \left(c_1 + c_1 - 2c_1\right)e^{-x} + \left(4c_2 - 2c_2 - 2c_2\right)e^{2x} = 0.$$

Since equality holds for all x in $(-\infty, \infty)$, then $\phi(x) = c_1 e^{-x} + c_2 e^{2x}$ is an explicit solution to (4) on the interval $(-\infty, \infty)$ for any choice of the constants c_1 and c_2. ◆

As we will see in Chapter 2, the methods for solving differential equations do not always yield an explicit solution for the equation. We may have to settle for a solution that is defined implicitly. Consider the following example.

Example 3 Show that the relation

(5) $y^2 - x^3 + 8 = 0$

implicitly defines a solution to the nonlinear equation

(6) $\dfrac{dy}{dx} = \dfrac{3x^2}{2y}$

on the interval $(2, \infty)$.

Solution When we solve (5) for y, we obtain $y = \pm\sqrt{x^3 - 8}$. Let's try $\phi(x) = \sqrt{x^3 - 8}$ to see if it is an explicit solution. Since $d\phi/dx = 3x^2/\left(2\sqrt{x^3 - 8}\right)$, both ϕ and $d\phi/dx$ are defined on $(2, \infty)$. Substituting them into (6) yields

$$\frac{3x^2}{2\sqrt{x^3 - 8}} = \frac{3x^2}{2\left(\sqrt{x^3 - 8}\right)},$$

which is indeed valid for all x in $(2, \infty)$. [You can check that $\psi(x) = -\sqrt{x^3 - 8}$ is also an explicit solution to (6).] ◆

Implicit Solution

Definition 2. A relation $G(x, y) = 0$ is said to be an **implicit solution** to equation (1) on the interval I if it defines one or more explicit solutions on I.

Example 4 Show that

(7) $x + y + e^{xy} = 0$

is an implicit solution to the nonlinear equation

(8) $\left(1 + xe^{xy}\right)\dfrac{dy}{dx} + 1 + ye^{xy} = 0$.

Solution First, we observe that we are unable to solve (7) directly for y in terms of x alone. However, for (7) to hold, we realize that any change in x requires a change in y, so we expect the relation (7) to define implicitly at least one function $y(x)$. This is difficult to show directly but can be rigorously verified using the **implicit function theorem**[†] of advanced calculus, which guarantees that such a function $y(x)$ exists that is also differentiable (see Problem 30).

Once we know that y is a differentiable function of x, we can use the technique of implicit differentiation. Indeed, from (7) we obtain on differentiating with respect to x and applying the product and chain rules,

$$\frac{d}{dx}\left(x + y + e^{xy}\right) = 1 + \frac{dy}{dx} + e^{xy}\left(y + x\frac{dy}{dx}\right) = 0$$

or

$$\left(1 + xe^{xy}\right)\frac{dy}{dx} + 1 + ye^{xy} = 0,$$

which is identical to the differential equation (8). Thus, relation (7) is an implicit solution on some interval guaranteed by the implicit function theorem. ◆

Example 5 Verify that for every constant C the relation $4x^2 - y^2 = C$ is an implicit solution to

(9) $y\dfrac{dy}{dx} - 4x = 0$.

Graph the solution curves for $C = 0, \pm 1, \pm 4$. (We call the collection of all such solutions a *one-parameter family of solutions*.)

Solution When we implicitly differentiate the equation $4x^2 - y^2 = C$ with respect to x, we find

$$8x - 2y\frac{dy}{dx} = 0,$$

[†]See *Vector Calculus*, 6th ed, by J. E. Marsden and A. J. Tromba (Freeman, San Francisco, 2013).

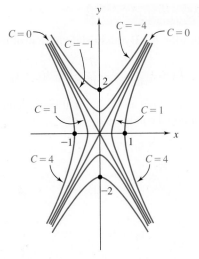

Figure 1.4 Implicit solutions $4x^2 - y^2 = C$

which is equivalent to (9). In Figure 1.4 we have sketched the implicit solutions for $C = 0, \pm 1, \pm 4$. The curves are hyperbolas with common asymptotes $y = \pm 2x$. Notice that the implicit solution curves (with C arbitrary) fill the entire plane and are nonintersecting for $C \neq 0$. For $C = 0$, the implicit solution gives rise to the two explicit solutions $y = 2x$ and $y = -2x$, both of which pass through the origin. ◆

For brevity we hereafter use the term *solution* to mean either an explicit or an implicit solution.

In the beginning of Section 1.1, we saw that the solution of the *second*-order free-fall equation invoked two arbitrary constants of integration c_1, c_2:

$$h(t) = \frac{-gt^2}{2} + c_1 t + c_2 ,$$

whereas the solution of the *first*-order radioactive decay equation contained a single constant C:

$$A(t) = Ce^{-kt} .$$

It is clear that integration of the simple *fourth*-order equation

$$\frac{d^4 y}{dx^4} = 0$$

brings in four undetermined constants:

$$y(x) = c_1 x^3 + c_2 x^2 + c_3 x + c_4 .$$

It will be shown later in the text that in general the methods for solving *n*th-order differential equations evoke *n* arbitrary constants. In most cases, we will be able to evaluate these constants if we know *n* initial values $y(x_0), y'(x_0), \ldots, y^{(n-1)}(x_0)$.

Initial Value Problem

Definition 3. By an **initial value problem** for an nth-order differential equation

$$F\left(x, y, \frac{dy}{dx}, \ldots, \frac{d^n y}{dx^n}\right) = 0,$$

we mean: Find a solution to the differential equation on an interval I that satisfies at x_0 the n initial conditions

$$y(x_0) = y_0,$$

$$\frac{dy}{dx}(x_0) = y_1,$$

$$\vdots$$

$$\frac{d^{n-1}y}{dx^{n-1}}(x_0) = y_{n-1},$$

where $x_0 \in I$ and $y_0, y_1, \ldots, y_{n-1}$ are given constants.

In the case of a first-order equation, the initial conditions reduce to the single requirement

$$y(x_0) = y_0,$$

and in the case of a second-order equation, the initial conditions have the form

$$y(x_0) = y_0, \qquad \frac{dy}{dx}(x_0) = y_1.$$

The terminology *initial conditions* comes from mechanics, where the independent variable x represents *time* and is customarily symbolized as t. Then if t_0 is the starting time, $y(t_0) = y_0$ represents the initial location of an object and $y'(t_0)$ gives its initial velocity.

Example 6 Show that $\phi(x) = \sin x - \cos x$ is a solution to the initial value problem

$$(10) \qquad \frac{d^2 y}{dx^2} + y = 0; \qquad y(0) = -1, \qquad \frac{dy}{dx}(0) = 1.$$

Solution Observe that $\phi(x) = \sin x - \cos x$, $d\phi/dx = \cos x + \sin x$, and $d^2\phi/dx^2 = -\sin x + \cos x$ are all defined on $(-\infty, \infty)$. Substituting into the differential equation gives

$$(-\sin x + \cos x) + (\sin x - \cos x) = 0,$$

which holds for all $x \in (-\infty, \infty)$. Hence, $\phi(x)$ is a solution to the differential equation in (10) on $(-\infty, \infty)$. When we check the initial conditions, we find

$$\phi(0) = \sin 0 - \cos 0 = -1,$$

$$\frac{d\phi}{dx}(0) = \cos 0 + \sin 0 = 1,$$

which meets the requirements of (10). Therefore, $\phi(x)$ is a solution to the given initial value problem. ◆

Example 7 As shown in Example 2, the function $\phi(x) = c_1e^{-x} + c_2e^{2x}$ is a solution to

$$\frac{d^2y}{dx^2} - \frac{dy}{dx} - 2y = 0$$

for any choice of the constants c_1 and c_2. Determine c_1 and c_2 so that the initial conditions

$$y(0) = 2 \quad \text{and} \quad \frac{dy}{dx}(0) = -3$$

are satisfied.

Solution To determine the constants c_1 and c_2, we first compute $d\phi/dx$ to get $d\phi/dx = -c_1e^{-x} + 2c_2e^{2x}$. Substituting in our initial conditions gives the following system of equations:

$$\begin{cases} \phi(0) \quad = c_1e^0 + c_2e^0 = 2 \ , \\ \dfrac{d\phi}{dx}(0) = -c_1e^0 + 2c_2e^0 = -3 \ , \end{cases} \quad \text{or} \quad \begin{cases} c_1 + c_2 \quad = 2 \ , \\ -c_1 + 2c_2 = -3 \ . \end{cases}$$

Adding the last two equations yields $3c_2 = -1$, so $c_2 = -1/3$. Since $c_1 + c_2 = 2$, we find $c_1 = 7/3$. Hence, the solution to the initial value problem is $\phi(x) = (7/3)e^{-x} - (1/3)e^{2x}$. ◆

We now state an existence and uniqueness theorem for first-order initial value problems. We presume the differential equation has been cast into the format

$$\frac{dy}{dx} = f(x, y) \ .$$

Of course, the right-hand side, $f(x, y)$, must be well defined at the starting value x_0 for x and at the stipulated initial value $y_0 = y(x_0)$ for y. The hypotheses of the theorem, moreover, require *continuity* of both f and $\partial f/\partial y$ for x in some interval $a < x < b$ containing x_0, and for y in some interval $c < y < d$ containing y_0. Notice that the set of points in the xy-plane that satisfy $a < x < b$ and $c < y < d$ constitutes a *rectangle*. Figure 1.5 on page 12 depicts this "rectangle of continuity" with the initial point (x_0, y_0) in its interior and a sketch of a portion of the solution curve contained therein.

Existence and Uniqueness of Solution

Theorem 1. Consider the initial value problem

$$\frac{dy}{dx} = f(x, y) \ , \qquad y(x_0) = y_0 \ .$$

If f and $\partial f/\partial y$ are continuous functions in some rectangle

$$R = \{(x, y): a < x < b, c < y < d\}$$

that contains the point (x_0, y_0), then the initial value problem has a unique solution $\phi(x)$ in some interval $x_0 - \delta < x < x_0 + \delta$, where δ is a positive number.[†]

[†]We remark that the continuity of f alone in such a rectangle is enough to guarantee the existence of a solution to the initial value problem in some open interval containing x_0, but uniqueness may not hold (see Example 9).

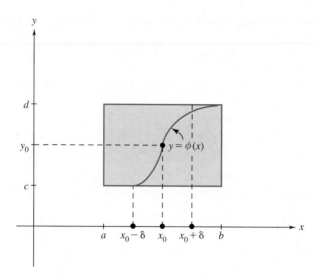

Figure 1.5 Layout for the existence–uniqueness theorem

The preceding theorem tells us two things. First, when an equation satisfies the hypotheses of Theorem 1, we are assured that a solution to the initial value problem exists. Naturally, it is desirable to know whether the equation we are trying to solve actually has a solution before we spend too much time trying to solve it. Second, when the hypotheses are satisfied, there is a **unique** solution to the initial value problem. This uniqueness tells us that if we can find a solution, then it is the *only* solution for the initial value problem. Graphically, the theorem says that there is only one solution curve that passes through the point (x_0, y_0). In other words, for this first-order equation, two solutions cannot cross anywhere in the rectangle. Notice that the existence and uniqueness of the solution holds only in *some* neighborhood $(x_0 - \delta, x_0 + \delta)$. Unfortunately, the theorem does not tell us the span (2δ) of this neighborhood (merely that it is not zero). Problem 18 elaborates on this feature.

Problem 19 gives an example of an equation with no solution. Problem 29 displays an initial value problem for which the solution is not unique. Of course, the hypotheses of Theorem 1 are not met for these cases.

When initial value problems are used to model physical phenomena, many practitioners tacitly presume the conclusions of Theorem 1 to be valid. Indeed, for the initial value problem to be a reasonable model, we certainly expect it to have a solution, since physically "something does happen." Moreover, the solution should be unique in those cases when repetition of the experiment under identical conditions yields the same results.[†]

The proof of Theorem 1 involves converting the initial value problem into an integral equation and then using Picard's method to generate a sequence of successive approximations that converge to the solution. The conversion to an integral equation and Picard's method are discussed in Project A at the end of this chapter. A detailed discussion and proof of the theorem are given in Chapter 13.[‡]

[†]At least this is the case when we are considering a deterministic model, as opposed to a probabilistic model.

[‡]All references to Chapters 11–13 refer to the expanded text, *Fundamentals of Differential Equations and Boundary Value Problems,* 7th ed.

Example 8 For the initial value problem

$$(11) \qquad 3\frac{dy}{dx} = x^2 - xy^3 \,, \qquad y(1) = 6 \,,$$

does Theorem 1 imply the existence of a unique solution?

Solution Dividing by 3 to conform to the statement of the theorem, we identify $f(x, y)$ as $(x^2 - xy^3)/3$ and $\partial f/\partial y$ as $-xy^2$. Both of these functions are continuous in any rectangle containing the point $(1, 6)$, so the hypotheses of Theorem 1 are satisfied. It then follows from the theorem that the initial value problem (11) has a unique solution in an interval about $x = 1$ of the form $(1 - \delta, 1 + \delta)$, where δ is some positive number. ◆

Example 9 For the initial value problem

$$(12) \qquad \frac{dy}{dx} = 3y^{2/3} \,, \qquad y(2) = 0 \,,$$

does Theorem 1 imply the existence of a unique solution?

Solution Here $f(x, y) = 3y^{2/3}$ and $\partial f/\partial y = 2y^{-1/3}$. Unfortunately $\partial f/\partial y$ is not continuous or even defined when $y = 0$. Consequently, there is no rectangle containing $(2, 0)$ in which both f and $\partial f/\partial y$ are continuous. Because the hypotheses of Theorem 1 do not hold, we cannot use Theorem 1 to determine whether the initial value problem does or does not have a unique solution. It turns out that this initial value problem has more than one solution. We refer you to Problem 29 and Project G of Chapter 2 for the details. ◆

In Example 9 suppose the initial condition is changed to $y(2) = 1$. Then, since f and $\partial f/\partial y$ are continuous in any rectangle that contains the point $(2, 1)$ but does not intersect the x-axis—say, $R = \{(x, y): 0 < x < 10, 0 < y < 5\}$—it follows from Theorem 1 that this *new* initial value problem has a unique solution in some interval about $x = 2$.

1.2 EXERCISES

1. **(a)** Show that $\phi(x) = x^2$ is an explicit solution to

$$x\frac{dy}{dx} = 2y$$

on the interval $(-\infty, \infty)$.

 (b) Show that $\phi(x) = e^x - x$ is an explicit solution to

$$\frac{dy}{dx} + y^2 = e^{2x} + (1 - 2x)e^x + x^2 - 1$$

on the interval $(-\infty, \infty)$.

 (c) Show that $\phi(x) = x^2 - x^{-1}$ is an explicit solution to $x^2 d^2y/dx^2 = 2y$ on the interval $(0, \infty)$.

2. **(a)** Show that $y^2 + x - 3 = 0$ is an implicit solution to $dy/dx = -1/(2y)$ on the interval $(-\infty, 3)$.

 (b) Show that $xy^3 - xy^3 \sin x = 1$ is an implicit solution to

$$\frac{dy}{dx} = \frac{(x \cos x + \sin x - 1)y}{3(x - x \sin x)}$$

on the interval $(0, \pi/2)$.

In Problems 3–8, determine whether the given function is a solution to the given differential equation.

3. $y = \sin x + x^2 \,,$ $\dfrac{d^2y}{dx^2} + y = x^2 + 2$

4. $x = 2 \cos t - 3 \sin t \,,$ $x'' + x = 0$

5. $\theta = 2e^{3t} - e^{2t} \,,$ $\dfrac{d^2\theta}{dt^2} - \theta\dfrac{d\theta}{dt} + 3\theta = -2e^{2t}$

6. $x = \cos 2t$, $\qquad\qquad \dfrac{dx}{dt} + tx = \sin 2t$

7. $y = e^{2x} - 3e^{-x}$, $\qquad \dfrac{d^2y}{dx^2} - \dfrac{dy}{dx} - 2y = 0$

8. $y = 3\sin 2x + e^{-x}$, $\qquad y'' + 4y = 5e^{-x}$

In Problems 9–13, determine whether the given relation is an implicit solution to the given differential equation. Assume that the relationship does define y implicitly as a function of x and use implicit differentiation.

9. $x^2 + y^2 = 4$, $\qquad\qquad \dfrac{dy}{dx} = \dfrac{x}{y}$

10. $y - \ln y = x^2 + 1$, $\qquad \dfrac{dy}{dx} = \dfrac{2xy}{y-1}$

11. $e^{xy} + y = x - 1$, $\qquad \dfrac{dy}{dx} = \dfrac{e^{-xy} - y}{e^{-xy} + x}$

12. $x^2 - \sin(x+y) = 1$, $\qquad \dfrac{dy}{dx} = 2x\sec(x+y) - 1$

13. $\sin y + xy - x^3 = 2$,

$$y'' = \dfrac{6xy' + (y')^3\sin y - 2(y')^2}{3x^2 - y}$$

14. Show that $\phi(x) = c_1\sin x + c_2\cos x$ is a solution to $d^2y/dx^2 + y = 0$ for any choice of the constants c_1 and c_2. Thus, $c_1\sin x + c_2\cos x$ is a two-parameter family of solutions to the differential equation.

15. Verify that $\phi(x) = 2/(1 - ce^x)$, where c is an arbitrary constant, is a one-parameter family of solutions to

$$\dfrac{dy}{dx} = \dfrac{y(y-2)}{2}.$$

Graph the solution curves corresponding to $c = 0$, ± 1, ± 2 using the same coordinate axes.

16. Verify that $x^2 + cy^2 = 1$, where c is an arbitrary nonzero constant, is a one-parameter family of implicit solutions to

$$\dfrac{dy}{dx} = \dfrac{xy}{x^2 - 1}$$

and graph several of the solution curves using the same coordinate axes.

17. Show that $\phi(x) = Ce^{3x} + 1$ is a solution to $dy/dx - 3y = -3$ for any choice of the constant C. Thus, $Ce^{3x} + 1$ is a one-parameter family of solutions to the differential equation. Graph several of the solution curves using the same coordinate axes.

18. Let $c > 0$. Show that the function $\phi(x) = (c^2 - x^2)^{-1}$ is a solution to the initial value problem $dy/dx = 2xy^2$, $y(0) = 1/c^2$, on the interval $-c < x < c$. Note that this solution becomes unbounded as x approaches $\pm c$. Thus, the solution exists on the interval $(-\delta, \delta)$ with $\delta = c$, but not for larger δ. This illustrates that in Theorem 1 the existence interval can be quite small (if c is small)

or quite large (if c is large). Notice also that there is no clue from the equation $dy/dx = 2xy^2$ itself, or from the initial value, that the solution will "blow up" at $x = \pm c$.

19. Show that the equation $(dy/dx)^2 + y^2 + 4 = 0$ has no (real-valued) solution.

20. Determine for which values of m the function $\phi(x) = e^{mx}$ is a solution to the given equation.

(a) $\dfrac{d^2y}{dx^2} + 6\dfrac{dy}{dx} + 5y = 0$

(b) $\dfrac{d^3y}{dx^3} + 3\dfrac{d^2y}{dx^2} + 2\dfrac{dy}{dx} = 0$

21. Determine for which values of m the function $\phi(x) = x^m$ is a solution to the given equation.

(a) $3x^2\dfrac{d^2y}{dx^2} + 11x\dfrac{dy}{dx} - 3y = 0$

(b) $x^2\dfrac{d^2y}{dx^2} - x\dfrac{dy}{dx} - 5y = 0$

22. Verify that the function $\phi(x) = c_1e^x + c_2e^{-2x}$ is a solution to the linear equation

$$\dfrac{d^2y}{dx^2} + \dfrac{dy}{dx} - 2y = 0$$

for any choice of the constants c_1 and c_2. Determine c_1 and c_2 so that each of the following initial conditions is satisfied.

(a) $y(0) = 2$, $\quad y'(0) = 1$
(b) $y(1) = 1$, $\quad y'(1) = 0$

In Problems 23–28, determine whether Theorem 1 implies that the given initial value problem has a unique solution.

23. $\dfrac{dy}{dx} = y^4 - x^4$, $\qquad y(0) = 7$

24. $\dfrac{dy}{dt} - ty = \sin^2 t$, $\qquad y(\pi) = 5$

25. $3x\dfrac{dx}{dt} + 4t = 0$, $\qquad x(2) = -\pi$

26. $\dfrac{dx}{dt} + \cos x = \sin t$, $\qquad x(\pi) = 0$

27. $y\dfrac{dy}{dx} = x$, $\qquad y(1) = 0$

28. $\dfrac{dy}{dx} = 3x - \sqrt[3]{y - 1}$, $\qquad y(2) = 1$

29. (a) For the initial value problem (12) of Example 9, show that $\phi_1(x) \equiv 0$ and $\phi_2(x) = (x-2)^3$ are solutions. Hence, this initial value problem has multiple solutions. (See also Project G in Chapter 2.)
(b) Does the initial value problem $y' = 3y^{2/3}$, $y(0) = 10^{-7}$, have a unique solution in a neighborhood of $x = 0$?

30. Implicit Function Theorem. *Let $G(x, y)$ have continuous first partial derivatives in the rectangle $R = \{(x, y) : a < x < b, c < y < d\}$ containing the point (x_0, y_0). If $G(x_0, y_0) = 0$ and the partial derivative $G_y(x_0, y_0) \neq 0$, then there exists a differentiable function $y = \phi(x)$, defined in some interval $I = (x_0 - \delta, x_0 + \delta)$, that satisfies $G(x, \phi(x)) = 0$ for all $x \in I$.*

The implicit function theorem gives conditions under which the relationship $G(x, y) = 0$ defines y implicitly as a function of x. Use the implicit function theorem to show that the relationship $x + y + e^{xy} = 0$, given in Example 4, defines y implicitly as a function of x near the point $(0, -1)$.

31. Consider the equation of Example 5,

$$(13) \qquad y\frac{dy}{dx} - 4x = 0 \,.$$

(a) Does Theorem 1 imply the existence of a unique solution to (13) that satisfies $y(x_0) = 0$?

(b) Show that when $x_0 \neq 0$, equation (13) can't possibly have a solution in a neighborhood of $x = x_0$ that satisfies $y(x_0) = 0$.

(c) Show that there are two distinct solutions to (13) satisfying $y(0) = 0$ (see Figure 1.4 on page 9).

1.3 Direction Fields

The existence and uniqueness theorem discussed in Section 1.2 certainly has great value, but it stops short of telling us anything about the *nature* of the solution to a differential equation. For practical reasons we may need to know the value of the solution at a certain point, or the intervals where the solution is increasing, or the points where the solution attains a maximum value. Certainly, knowing an explicit representation (a formula) for the solution would be a considerable help in answering these questions. However, for many of the differential equations that we are likely to encounter in real-world applications, it will be impossible to find such a formula. Moreover, even if we are lucky enough to obtain an implicit solution, using this relationship to determine an explicit form may be difficult. Thus, we must rely on other methods to analyze or approximate the solution.

One technique that is useful in visualizing (graphing) the solutions to a first-order differential equation is to sketch the direction field for the equation. To describe this method, we need to make a general observation. Namely, a first-order equation

$$\frac{dy}{dx} = f(x, y)$$

specifies a slope at each point in the xy-plane where f is defined. In other words, it gives the direction that a graph of a solution to the equation must have at each point. Consider, for example, the equation

$$(1) \qquad \frac{dy}{dx} = x^2 - y \,.$$

The graph of a solution to (1) that passes through the point $(-2, 1)$ must have slope $(-2)^2 - 1 = 3$ at that point, and a solution through $(-1, 1)$ has zero slope at that point.

A plot of short line segments drawn at various points in the xy-plane showing the slope of the solution curve there is called a **direction field** for the differential equation. Because the direction field gives the "flow of solutions," it facilitates the drawing of any particular solution (such as the solution to an initial value problem). In Figure 1.6(a) on page 16 we have sketched the direction field for equation (1) and in Figure 1.6(b) we have drawn several solution curves in color.

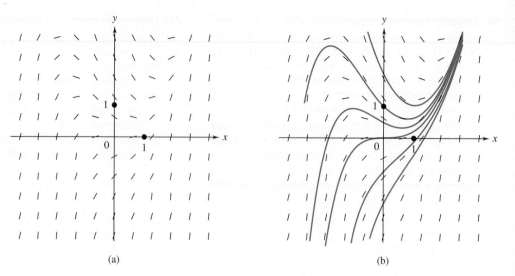

(a) (b)

Figure 1.6 (a) Direction field for $dy/dx = x^2 - y$ (b) Solutions to $dy/dx = x^2 - y$

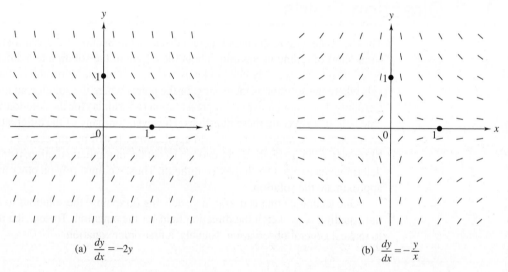

(a) $\dfrac{dy}{dx} = -2y$ (b) $\dfrac{dy}{dx} = -\dfrac{y}{x}$

Figure 1.7 (a) Direction field for $dy/dx = -2y$ (b) Direction field for $dy/dx = -y/x$

Some other interesting direction field patterns are displayed in Figure 1.7. Depicted in Figure 1.7(a) is the pattern for the radioactive decay equation $dy/dx = -2y$ (recall that in Section 1.1 we analyzed this equation in the form $dA/dt = -kA$). From the flow patterns, we can see that all solutions tend asymptotically to the positive x-axis as x gets larger. In other words, any material decaying according to this law eventually dwindles to practically nothing. This is consistent with the solution formula we derived earlier,

$$A = Ce^{-kt}, \quad \text{or} \quad y = Ce^{-2x}.$$

From the direction field in Figure 1.7(b), we can anticipate that all solutions to $dy/dx = -y/x$ also approach the x-axis as x approaches infinity (plus or minus infinity, in

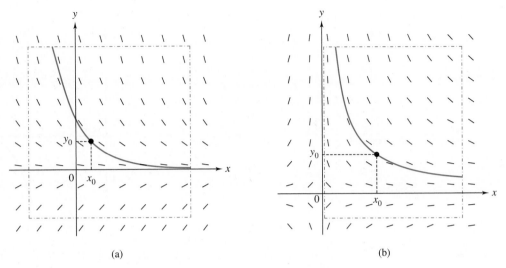

(a) (b)

Figure 1.8 (a) A solution for $dy/dx = -2y$ (b) A solution for $dy/dx = -y/x$

fact). But more interesting is the observation that no solution can make it across the y-axis; $|y(x)|$ "blows up" as x goes to zero from either direction. Exception: On close examination, it appears the function $y(x) \equiv 0$ might just make it through this barrier. As a matter of fact, in Problem 19 you are invited to show that the solutions to this differential equation are given by $y = C/x$, with C an arbitrary constant. So they do diverge at $x = 0$, unless $C = 0$.

Let's interpret the existence–uniqueness theorem of Section 1.2 for these direction fields. For Figure 1.7(a), where $dy/dx = f(x, y) = -2y$, we select a starting point x_0 and an initial value $y(x_0) = y_0$, as in Figure 1.8(a). Because the right-hand side $f(x, y) = -2y$ is continuously differentiable for all x and y, we can enclose any initial point (x_0, y_0) in a "rectangle of continuity." We conclude that the equation has one and only one solution curve passing through (x_0, y_0), as depicted in the figure.

For the equation

$$\frac{dy}{dx} = f(x, y) = -\frac{y}{x},$$

the right-hand side $f(x, y) = -y/x$ does not meet the continuity conditions when $x = 0$ (i.e., for points on the y-axis). However, for any *nonzero* starting value x_0 and *any* initial value $y(x_0) = y_0$, we *can* enclose (x_0, y_0) in a rectangle of continuity that excludes the y-axis, as in Figure 1.8(b). Thus, we can be assured of one and only one solution curve passing through such a point.

The direction field for the equation

$$\frac{dy}{dx} = 3y^{2/3}$$

is intriguing because Example 9 of Section 1.2 showed that the hypotheses of Theorem 1 do *not* hold in any rectangle enclosing the initial point $x_0 = 2$, $y_0 = 0$. Indeed, Problem 29 of that section demonstrated the violation of uniqueness by exhibiting *two* solutions, $y(x) \equiv 0$ and $y(x) = (x - 2)^3$, passing through $(2, 0)$. Figure 1.9(a) on page 18 displays this direction field, and Figure 1.9(b) demonstrates how both solution curves can successfully "negotiate" this flow pattern.

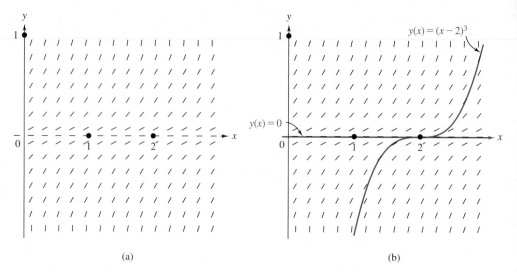

Figure 1.9 (a) Direction field for $dy/dx = 3y^{2/3}$ (b) Solutions for $dy/dx = 3y^{2/3}$, $y(2) = 0$

Clearly, a sketch of the direction field of a first-order differential equation can be helpful in visualizing the solutions. However, such a sketch is not sufficient to enable us to trace, unambiguously, the solution curve passing through a given initial point (x_0, y_0). If we tried to trace one of the solution curves in Figure 1.6(b) on page 16, for example, we could easily "slip" over to an adjacent curve. For nonunique situations like that in Figure 1.9(b), as one negotiates the flow along the curve $y = (x - 2)^3$ and reaches the inflection point, one cannot decide whether to turn or to (literally) go off on the tangent $(y = 0)$.

Example 1 The *logistic equation* for the population p (in thousands) at time t of a certain species is given by

(2) $$\frac{dp}{dt} = p(2 - p) .$$

(Of course, p is nonnegative. The interpretation of the terms in the logistic equation is discussed in Section 3.2.) From the direction field sketched in Figure 1.10 on page 19, answer the following:

(a) If the initial population is 3000 [that is, $p(0) = 3$], what can you say about the limiting population $\lim_{t \to +\infty} p(t)$?

(b) Can a population of 1000 ever decline to 500?

(c) Can a population of 1000 ever increase to 3000?

Solution (a) The direction field indicates that all solution curves [other than $p(t) \equiv 0$] will approach the horizontal line $p = 2$ as $t \to +\infty$; that is, this line is an asymptote for all positive solutions. Thus, $\lim_{t \to +\infty} p(t) = 2$.

(b) The direction field further shows that populations greater than 2000 will steadily decrease, whereas those less than 2000 will increase. In particular, a population of 1000 can never decline to 500.

Figure 1.10 Direction field for logistic equation

(c) As mentioned in part (b), a population of 1000 will increase with time. But the direction field indicates it can never reach 2000 or any larger value; i.e., the solution curve cannot cross the line $p = 2$. Indeed, the constant function $p(t) \equiv 2$ is a solution to equation (2), and the uniqueness part of Theorem 1, page 11, precludes intersections of solution curves. ◆

Notice that the direction field in Figure 1.10 has the nice feature that the slopes do not depend on t; that is, the slopes are the same along each horizontal line. The same is true for Figures 1.8(a) and 1.9. This is the key property of so-called **autonomous equations** $y' = f(y)$, where the right-hand side is a function of the dependent variable only. Project B, page 33, investigates such equations in more detail.

Hand sketching the direction field for a differential equation is often tedious. Fortunately, several software programs have been developed to obviate this task[†]. When hand sketching is necessary, however, the **method of isoclines** can be helpful in reducing the work.

The Method of Isoclines

An **isocline** for the differential equation

$$y' = f(x, y)$$

is a set of points in the xy-plane where all the solutions have the same slope dy/dx; thus, it is a level curve for the function $f(x, y)$. For example, if

(3) $\qquad y' = f(x, y) = x + y ,$

the isoclines are simply the curves (straight lines) $x + y = c$ or $y = -x + c$. Here c is an arbitrary constant. But c can be interpreted as the numerical value of the slope dy/dx of every solution curve as it crosses the isocline. (Note that c is not the slope of the isocline itself; the latter is, obviously, -1.) Figure 1.11(a) on page 20 depicts the isoclines for equation (3).

[†]Appendix G describes various web sites and commercial software that sketch direction fields and automate most of the differential equation algorithms discussed in this book.

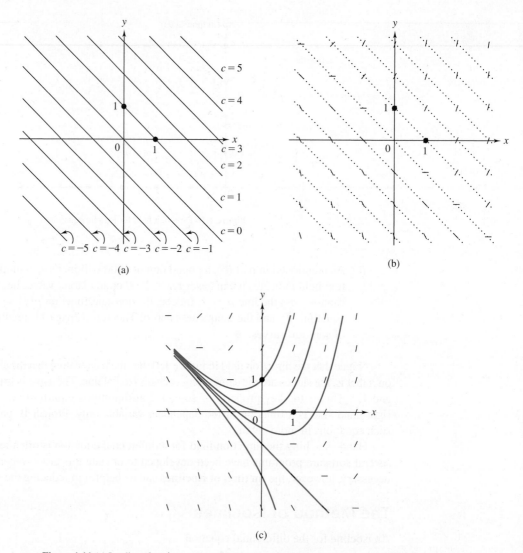

Figure 1.11 (a) Isoclines for $y' = x + y$ (b) Direction field for $y' = x + y$ (c) Solutions to $y' = x + y$

To implement the method of isoclines for sketching direction fields, we draw hash marks with slope c along the isocline $f(x, y) = c$ for a few selected values of c. If we then erase the underlying isocline curves, the hash marks constitute a part of the direction field for the differential equation. Figure 1.11(b) depicts this process for the isoclines shown in Figure 1.11(a), and Figure 1.11(c) displays some solution curves.

Remark. The isoclines themselves are not always straight lines. For equation (1) at the beginning of this section (page 15), they are parabolas $x^2 - y = c$. When the isocline curves are complicated, this method is not practical.

1.3 EXERCISES

1. The direction field for $dy/dx = 4x/y$ is shown in Figure 1.12.
 (a) Verify that the straight lines $y = \pm 2x$ are solution curves, provided $x \neq 0$.
 (b) Sketch the solution curve with initial condition $y(0) = 2$.
 (c) Sketch the solution curve with initial condition $y(2) = 1$.
 (d) What can you say about the behavior of the above solutions as $x \to +\infty$? How about $x \to -\infty$?

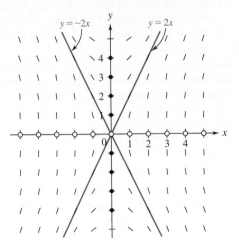

Figure 1.12 Direction field for $dy/dx = 4x/y$

2. The direction field for $dy/dx = 2x + y$ is shown in Figure 1.13.
 (a) Sketch the solution curve that passes through $(0, -2)$. From this sketch, write the equation for the solution.

Figure 1.13 Direction field for $dy/dx = 2x + y$

 (b) Sketch the solution curve that passes through $(-1, 3)$.
 (c) What can you say about the solution in part (b) as $x \to +\infty$? How about $x \to -\infty$?

3. A model for the velocity v at time t of a certain object falling under the influence of gravity in a viscous medium is given by the equation

$$\frac{dv}{dt} = 1 - \frac{v}{8}.$$

From the direction field shown in Figure 1.14, sketch the solutions with the initial conditions $v(0) = 5, 8,$ and 15. Why is the value $v = 8$ called the "terminal velocity"?

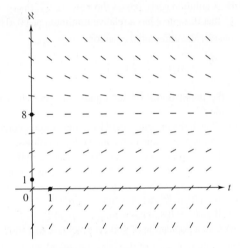

Figure 1.14 Direction field for $\dfrac{dv}{dt} = 1 - \dfrac{v}{8}$

4. If the viscous force in Problem 3 is nonlinear, a possible model would be provided by the differential equation

$$\frac{dv}{dt} = 1 - \frac{v^3}{8}.$$

Redraw the direction field in Figure 1.14 to incorporate this v^3 dependence. Sketch the solutions with initial conditions $v(0) = 0, 1, 2, 3$. What is the terminal velocity in this case?

5. The logistic equation for the population (in thousands) of a certain species is given by

$$\frac{dp}{dt} = 3p - 2p^2.$$

 (a) Sketch the direction field by using either a computer software package or the method of isoclines.

(b) If the initial population is 3000 [that is, $p(0) = 3$], what can you say about the limiting population $\lim_{t \to +\infty} p(t)$?

(c) If $p(0) = 0.8$, what is $\lim_{t \to +\infty} p(t)$?

(d) Can a population of 2000 ever decline to 800?

6. Consider the differential equation

$$\frac{dy}{dx} = x + \sin y.$$

(a) A solution curve passes through the point $(1, \pi/2)$. What is its slope at this point?

(b) Argue that every solution curve is increasing for $x > 1$.

(c) Show that the second derivative of every solution satisfies

$$\frac{d^2y}{dx^2} = 1 + x \cos y + \frac{1}{2} \sin 2y.$$

(d) A solution curve passes through $(0, 0)$. Prove that this curve has a relative minimum at $(0, 0)$.

7. Consider the differential equation

$$\frac{dp}{dt} = p(p - 1)(2 - p)$$

for the population p (in thousands) of a certain species at time t.

(a) Sketch the direction field by using either a computer software package or the method of isoclines.

(b) If the initial population is 4000 [that is, $p(0) = 4$], what can you say about the limiting population $\lim_{t \to +\infty} p(t)$?

(c) If $p(0) = 1.7$, what is $\lim_{t \to +\infty} p(t)$?

(d) If $p(0) = 0.8$, what is $\lim_{t \to +\infty} p(t)$?

(e) Can a population of 900 ever increase to 1100?

8. The motion of a set of particles moving along the x-axis is governed by the differential equation

$$\frac{dx}{dt} = t^3 - x^3,$$

where $x(t)$ denotes the position at time t of the particle.

(a) If a particle is located at $x = 1$ when $t = 2$, what is its velocity at this time?

(b) Show that the acceleration of a particle is given by

$$\frac{d^2x}{dt^2} = 3t^2 - 3t^3x^2 + 3x^5.$$

(c) If a particle is located at $x = 2$ when $t = 2.5$, can it reach the location $x = 1$ at any later time? [*Hint*: $t^3 - x^3 = (t - x)(t^2 + xt + x^2)$.]

9. Let $\phi(x)$ denote the solution to the initial value problem

$$\frac{dy}{dx} = x - y, \qquad y(0) = 1.$$

(a) Show that $\phi''(x) = 1 - \phi'(x) = 1 - x + \phi(x)$.

(b) Argue that the graph of ϕ is decreasing for x near zero and that as x increases from zero, $\phi(x)$ decreases until it crosses the line $y = x$, where its derivative is zero.

(c) Let x^* be the abscissa of the point where the solution curve $y = \phi(x)$ crosses the line $y = x$. Consider the sign of $\phi''(x^*)$ and argue that ϕ has a relative minimum at x^*.

(d) What can you say about the graph of $y = \phi(x)$ for $x > x^*$?

(e) Verify that $y = x - 1$ is a solution to $dy/dx = x - y$ and explain why the graph of $\phi(x)$ always stays above the line $y = x - 1$.

(f) Sketch the direction field for $dy/dx = x - y$ by using the method of isoclines or a computer software package.

(g) Sketch the solution $y = \phi(x)$ using the direction field in part (f).

10. Use a computer software package to sketch the direction field for the following differential equations. Sketch some of the solution curves.

(a) $dy/dx = \sin x$

(b) $dy/dx = \sin y$

(c) $dy/dx = \sin x \sin y$

(d) $dy/dx = x^2 + 2y^2$

(e) $dy/dx = x^2 - 2y^2$

In Problems 11–16, draw the isoclines with their direction markers and sketch several solution curves, including the curve satisfying the given initial conditions.

11. $dy/dx = -x/y$, $y(0) = 4$

12. $dy/dx = y$, $y(0) = 1$

13. $dy/dx = 2x$, $y(0) = -1$

14. $dy/dx = x/y$, $y(0) = -1$

15. $dy/dx = 2x^2 - y$, $y(0) = 0$

16. $dy/dx = x + 2y$, $y(0) = 1$

17. From a sketch of the direction field, what can one say about the behavior as x approaches $+\infty$ of a solution to the following?

$$\frac{dy}{dx} = 3 - y + \frac{1}{x}$$

18. From a sketch of the direction field, what can one say about the behavior as x approaches $+\infty$ of a solution to the following?

$$\frac{dy}{dx} = -y$$

19. By rewriting the differential equation $dy/dx = -y/x$ in the form

$$\frac{1}{y}dy = \frac{-1}{x}dx$$

integrate both sides to obtain the solution $y = C/x$ for an arbitrary constant C.

20. A bar magnet is often modeled as a magnetic dipole with one end labeled the north pole N and the opposite end labeled the south pole S. The magnetic field for the magnetic dipole is symmetric with respect to rotation about the axis passing lengthwise through the center of the bar. Hence we can study the magnetic field by restricting ourselves to a plane with the bar magnet centered on the x-axis.

For a point P that is located a distance r from the origin, where r is much greater than the length of the magnet, the **magnetic field lines** satisfy the differential equation

$$(4) \qquad \frac{dy}{dx} = \frac{3xy}{2x^2 - y^2}$$

and the **equipotential lines** satisfy the equation

$$(5) \qquad \frac{dy}{dx} = \frac{y^2 - 2x^2}{3xy} .$$

(a) Show that the two families of curves are perpendicular where they intersect. [*Hint:* Consider the slopes of the tangent lines of the two curves at a point of intersection.]

(b) Sketch the direction field for equation (4) for $-5 \le x \le 5$, $-5 \le y \le 5$. You can use a software package to generate the direction field or use the method of isoclines. The direction field should remind you of the experiment where iron filings are sprinkled on a sheet of paper that is held above a bar magnet. The iron filings correspond to the hash marks.

(c) Use the direction field found in part (b) to help sketch the magnetic field lines that are solutions to (4).

(d) Apply the statement of part (a) to the curves in part (c) to sketch the equipotential lines that are solutions to (5). The magnetic field lines and the equipotential lines are examples of *orthogonal trajectories*. (See Problem 32 in Exercises 2.4, page 65.)[†]

1.4 The Approximation Method of Euler

Euler's method (or the tangent-line method) is a procedure for constructing approximate solutions to an initial value problem for a first-order differential equation

$$(1) \qquad y' = f(x, y) , \qquad y(x_0) = y_0 .$$

It could be described as a "mechanical" or "computerized" implementation of the informal procedure for hand sketching the solution curve from a picture of the direction field. As such, we will see that it remains subject to the failing that it may skip across solution curves. However, under fairly general conditions, iterations of the procedure do converge to true solutions.

The method is illustrated in Figure 1.15 on page 24. Starting at the initial point (x_0, y_0), we follow the straight line with slope $f(x_0, y_0)$, the tangent line, for some distance to the point (x_1, y_1). Then we reset the slope to the value $f(x_1, y_1)$ and follow this line to (x_2, y_2). In this way we construct polygonal (broken line) approximations to the solution. As we take smaller spacings between points (and thus employ more points), we may expect to converge to the true solution.

To be more precise, assume that the initial value problem (1) has a unique solution $\phi(x)$ in some interval centered at x_0. Let h be a fixed positive number (called the *step size*) and consider the equally spaced points[‡]

$$x_n := x_0 + nh , \qquad n = 0, 1, 2, \ldots .$$

[†] Equations (4) and (5) can be solved using the method for homogeneous equations in Section 2.6 (see Exercises 2.6, Problem 46).

[‡] The symbol := means "is defined to be."

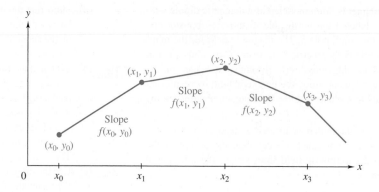

Figure 1.15 Polygonal-line approximation given by Euler's method

The construction of values y_n that approximate the solution values $\phi(x_n)$ proceeds as follows. At the point (x_0, y_0), the slope of the solution to (1) is given by $dy/dx = f(x_0, y_0)$. Hence, the tangent line to the solution curve at the initial point (x_0, y_0) is

$$y = y_0 + (x - x_0)f(x_0, y_0) \ .$$

Using this tangent line to approximate $\phi(x)$, we find that for the point $x_1 = x_0 + h$

$$\phi(x_1) \approx y_1 := y_0 + h f(x_0, y_0) \ .$$

Next, starting at the point (x_1, y_1), we construct the line with slope given by the direction field at the point (x_1, y_1)—that is, with slope equal to $f(x_1, y_1)$. If we follow this line[†] [namely, $y = y_1 + (x - x_1)f(x_1, y_1)$] in stepping from x_1 to $x_2 = x_1 + h$, we arrive at the approximation

$$\phi(x_2) \approx y_2 := y_1 + h f(x_1, y_1) \ .$$

Repeating the process (as illustrated in Figure 1.15), we get

$$\phi(x_3) \approx y_3 := y_2 + h f(x_2, y_2) \ ,$$
$$\phi(x_4) \approx y_4 := y_3 + h f(x_3, y_3) \ , \qquad \text{etc.}$$

This simple procedure is **Euler's method** and can be summarized by the recursive formulas

(2) $x_{n+1} = x_n + h \ ,$

(3) $y_{n+1} = y_n + h f(x_n, y_n) \ , \qquad n = 0, 1, 2, \ldots .$

[†]Because y_1 is an *approximation* to $\phi(x_1)$, we cannot assert that this line is tangent to the solution curve $y = \phi(x)$.

Example 1 Use Euler's method with step size $h = 0.1$ to approximate the solution to the initial value problem

(4) $y' = x\sqrt{y}, \qquad y(1) = 4$

at the points $x = 1.1, 1.2, 1.3, 1.4,$ and 1.5.

Solution Here $x_0 = 1, y_0 = 4, h = 0.1,$ and $f(x, y) = x\sqrt{y}$. Thus, the recursive formula (3) for y_n is

$$y_{n+1} = y_n + h f(x_n, y_n) = y_n + (0.1) x_n \sqrt{y_n} \, .$$

Substituting $n = 0$, we get

$$x_1 = x_0 + 0.1 = 1 + 0.1 = 1.1 \, ,$$
$$y_1 = y_0 + (0.1) x_0 \sqrt{y_0} = 4 + (0.1)(1)\sqrt{4} = 4.2 \, .$$

Putting $n = 1$ yields

$$x_2 = x_1 + 0.1 = 1.1 + 0.1 = 1.2 \, ,$$
$$y_2 = y_1 + (0.1) x_1 \sqrt{y_1} = 4.2 + (0.1)(1.1)\sqrt{4.2} \approx 4.42543 \, .$$

Continuing in this manner, we obtain the results listed in Table 1.1. For comparison we have included the exact value (to five decimal places) of the solution $\phi(x) = (x^2 + 7)^2/16$ to (4), which can be obtained using separation of variables (see Section 2.2). As one might expect, the approximation deteriorates as x moves farther away from 1. ◆

TABLE 1.1	Computations for $y' = x\sqrt{y}$, $y(1) = 4$		
n	x_n	Euler's Method	Exact Value
0	1	4	4
1	1.1	4.2	4.21276
2	1.2	4.42543	4.45210
3	1.3	4.67787	4.71976
4	1.4	4.95904	5.01760
5	1.5	5.27081	5.34766

Given the initial value problem (1) and a *specific* point x, how can Euler's method be used to approximate $\phi(x)$? Starting at x_0, we can take one giant step that lands on x, or we can take several smaller steps to arrive at x. If we wish to take N steps, then we set $h = (x - x_0)/N$ so that the step size h and the number of steps N are related in a specific way. For example, if $x_0 = 1.5$ and we wish to approximate $\phi(2)$ using 10 steps, then we would take $h = (2 - 1.5)/10 = 0.05$. It is expected that the more steps we take, the better will be the approximation. (But keep in mind that more steps mean more computations and hence greater accumulated roundoff error.)

Example 2 Use Euler's method to find approximations to the solution of the initial value problem

(5) $y' = y$, $y(0) = 1$

at $x = 1$, taking 1, 2, 4, 8, and 16 steps.

Remark. Observe that the solution to (5) is just $\phi(x) = e^x$, so Euler's method will generate algebraic approximations to the transcendental number $e = 2.71828. \dots$

Solution Here $f(x, y) = y$, $x_0 = 0$, and $y_0 = 1$. The recursive formula for Euler's method is

$$y_{n+1} = y_n + h y_n = (1 + h) y_n \,.$$

To obtain approximations at $x = 1$ with N steps, we take the step size $h = 1/N$. For $N = 1$, we have

$$\phi(1) \approx y_1 = (1 + 1)(1) = 2 \,.$$

For $N = 2$, $\phi(x_2) = \phi(1) \approx y_2$. In this case we get

$$y_1 = (1 + 0.5)(1) = 1.5 \,,$$
$$\phi(1) \approx y_2 = (1 + 0.5)(1.5) = 2.25 \,.$$

For $N = 4$, $\phi(x_4) = \phi(1) \approx y_4$, where

$$y_1 = (1 + 0.25)(1) = 1.25 \,,$$
$$y_2 = (1 + 0.25)(1.25) = 1.5625 \,,$$
$$y_3 = (1 + 0.25)(1.5625) = 1.95313 \,,$$
$$\phi(1) \approx y_4 = (1 + 0.25)(1.95313) = 2.44141 \,.$$

(In the above computations, we have rounded to five decimal places.) Similarly, taking $N = 8$ and 16, we obtain even better estimates for $\phi(1)$. These approximations are shown in Table 1.2. For comparison, Figure 1.16 on page 27 displays the polygonal-line approximations to e^x using Euler's method with $h = 1/4$ ($N = 4$) and $h = 1/8$ ($N = 8$). Notice that the smaller step size yields the better approximation. ◆

TABLE 1.2	Euler's Method for $y' = y$, $y(0) = 1$	
N	h	Approximation for $\phi(1) = e$
1	1.0	2.0
2	0.5	2.25
4	0.25	2.44141
8	0.125	2.56578
16	0.0625	2.63793

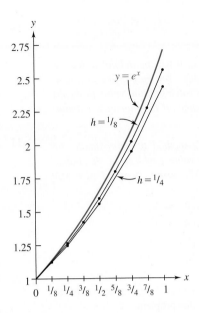

Figure 1.16 Approximations of e^x using Euler's method with $h = 1/4$ and $1/8$

How good (or bad) is Euler's method? In judging a numerical scheme, we must begin with two fundamental questions. Does the method converge? And, if so, what is the rate of convergence? These important issues are discussed in Section 3.6, where improvements in Euler's method are introduced (see also Problems 12 and 13 of this section).

Example 3 Suppose $v(t)$ satisfies the initial value problem

$$\frac{dv}{dt} = -3 - 2v^2, \quad v(0) = 2 .$$

By experimenting with Euler's method, determine to within one decimal place (± 0.1) the value of $v(0.2)$ and the time it will take $v(t)$ to reach zero.

Solution Determining rigorous estimates of the accuracy of the answers obtained by Euler's method can be quite a challenging problem. The common practice is to repeatedly approximate $v(0.2)$ and the zero crossing, using smaller and smaller values of h, until the digits of the computed values stabilize at the required accuracy level. For this example, Euler's algorithm yields the following values:

$h = 0.1$	$v(0.2) \approx 0.4380$	$v(0.3) \approx 0.0996$	$v(0.4) \approx -0.2024$
$h = 0.05$	$v(0.2) \approx 0.6036$	$v(0.35) \approx 0.0935$	$v(0.4) \approx -0.0574$
$h = 0.025$	$v(0.2) \approx 0.6659$	$v(0.375) \approx 0.0750$	$v(0.4) \approx -0.0003$
$h = 0.0125$	$v(0.2) \approx 0.6938$		
$h = 0.00625$	$v(0.2) \approx 0.7071$		

Acknowledging the remote possibility that finer values of h might reveal aberrations, we state with reasonable confidence that $v(0.2) = 0.7 \pm 0.1$. The Intermediate Value Theorem would imply that $v(t_0) = 0$ at some time t_0 satisfying $0.375 < t_0 < 0.4$ if the computations were perfect; they clearly provide evidence that $t_0 = 0.4 \pm 0.1$. ◆

1.4 EXERCISES

 In many of the problems below, it will be helpful to have a calculator or computer available.[†] You may also find it convenient to write a program for solving initial value problems using Euler's method. (Remember, all trigonometric calculations are done in radians.)

In Problems 1–4, use Euler's method to approximate the solution to the given initial value problem at the points $x = 0.1, 0.2, 0.3, 0.4$, and 0.5, using steps of size 0.1 ($h = 0.1$).

1. $dy/dx = -x/y$, $y(0) = 4$
2. $dy/dx = y(2 - y)$, $y(0) = 3$
3. $dy/dx = x + y$, $y(0) = 1$
4. $dy/dx = x/y$, $y(0) = -1$

5. Use Euler's method with step size $h = 0.1$ to approximate the solution to the initial value problem
$$y' = x - y^2, \qquad y(1) = 0$$
at the points $x = 1.1, 1.2, 1.3, 1.4$, and 1.5.

6. Use Euler's method with step size $h = 0.2$ to approximate the solution to the initial value problem
$$y' = \frac{1}{x}(y^2 + y), \qquad y(1) = 1$$
at the points $x = 1.2, 1.4, 1.6$, and 1.8.

7. Use Euler's method to find approximations to the solution of the initial value problem
$$y' = 1 - \sin y, \qquad y(0) = 0$$
at $x = \pi$, taking 1, 2, 4, and 8 steps.

8. Use Euler's method to find approximations to the solution of the initial value problem
$$\frac{dx}{dt} = 1 + t \sin(tx), \qquad x(0) = 0$$
at $t = 1$, taking 1, 2, 4, and 8 steps.

9. Use Euler's method with $h = 0.1$ to approximate the solution to the initial value problem
$$y' = \frac{1}{x^2} - \frac{y}{x} - y^2, \qquad y(1) = -1$$
on the interval $1 \le x \le 2$. Compare these approximations with the actual solution $y = -1/x$ (verify!) by graphing the polygonal-line approximation and the actual solution on the same coordinate system.

10. Use the strategy of Example 3 to find a value of h for Euler's method such that $y(1)$ is approximated to within ± 0.01, if $y(x)$ satisfies the initial value problem
$$y' = x - y, \qquad y(0) = 0.$$

Also find, to within ± 0.05, the value of x_0 such that $y(x_0) = 0.2$. Compare your answers with those given by the actual solution $y = e^{-x} + x - 1$ (verify!). Graph the polygonal-line approximation and the actual solution on the same coordinate system.

11. Use the strategy of Example 3 to find a value of h for Euler's method such that $x(1)$ is approximated to within ± 0.01, if $x(t)$ satisfies the initial value problem
$$\frac{dx}{dt} = 1 + x^2, \qquad x(0) = 0.$$

Also find, to within ± 0.02, the value of t_0 such that $x(t_0) = 1$. Compare your answers with those given by the actual solution $x = \tan t$ (verify!).

12. In Example 2 we approximated the transcendental number e by using Euler's method to solve the initial value problem
$$y' = y, \qquad y(0) = 1.$$

Show that the Euler approximation y_n obtained by using the step size $1/n$ is given by the formula
$$y_n = \left(1 + \frac{1}{n}\right)^n, \qquad n = 1, 2, \ldots$$

Recall from calculus that
$$\lim_{n \to \infty} \left(1 + \frac{1}{n}\right)^n = e,$$

and hence Euler's method converges (theoretically) to the correct value.

13. Prove that the "rate of convergence" for Euler's method in Problem 12 is comparable to $1/n$ by showing that
$$\lim_{n \to \infty} \frac{e - y_n}{1/n} = \frac{e}{2}.$$

[*Hint*: Use L'Hôpital's rule and the Maclaurin expansion for $\ln(1 + t)$.]

14. Use Euler's method with the spacings $h = 0.5, 0.1, 0.05, 0.01$ to approximate the solution to the initial value problem
$$y' = 2xy^2, \qquad y(0) = 1$$

on the interval $0 \le x \le 2$. (The explanation for the erratic results lies in Problem 18 of Exercises 1.2.)

Heat Exchange. *There are basically two mechanisms by which a physical body exchanges heat with its environment. The* **contact** *heat transfer across the body's surface is driven by the difference in the body's temperature and*

[†]Appendix G describes various web sites and commercial software that sketch direction fields and automate most of the differential equation algorithms discussed in this book.

*that of the environment; this is known as Newton's law of cooling. However, heat transfer also occurs due to **thermal radiation**, which according to Stefan's law of radiation is governed by the difference of the fourth powers of these temperatures. In most cases one of these modes dominates the other. Problems 15 and 16 invite you to simulate each mode numerically for a given set of initial conditions.*

15. Newton's Law of Cooling. Newton's law of cooling states that the rate of change in the temperature $T(t)$ of a body is proportional to the difference between the temperature of the medium $M(t)$ and the temperature of the body. That is,

$$\frac{dT}{dt} = K[M(t) - T(t)],$$

where K is a constant. Let $K = 0.04 \ (\text{min})^{-1}$ and the temperature of the medium be constant, $M(t) \equiv 293$ kelvins. If the body is initially at 360 kelvins, use Euler's method

with $h = 3.0$ min to approximate the temperature of the body after

(a) 30 minutes.
(b) 60 minutes.

16. Stefan's Law of Radiation. Stefan's law of radiation states that the rate of change in temperature of a body at $T(t)$ kelvins in a medium at $M(t)$ kelvins is proportional to $M^4 - T^4$. That is,

$$\frac{dT}{dt} = K(M(t)^4 - T(t)^4),$$

where K is a constant. Let $K = 2.9 \times 10^{-10} \ (\text{min})^{-1}$ and assume that the medium temperature is constant, $M(t) \equiv 293$ kelvins. If $T(0) = 360$ kelvins, use Euler's method with $h = 3.0$ min to approximate the temperature of the body after

(a) 30 minutes.
(b) 60 minutes.

Chapter 1 Summary

In this chapter we introduced some basic terminology for differential equations. The **order** of a differential equation is the order of the highest derivative present. The subject of this text is **ordinary** differential equations, which involve derivatives with respect to a single independent variable. Such equations are classified as **linear** or **nonlinear.**

An **explicit solution** of a differential equation is a function of the independent variable that satisfies the equation on some interval. An **implicit solution** is a relation between the dependent and independent variables that implicitly defines a function that is an explicit solution. A differential equation typically has infinitely many solutions. In contrast, some theorems ensure that a unique solution exists for certain **initial value problems** in which one must find a solution to the differential equation that also satisfies given initial conditions. For an nth-order equation, these conditions refer to the values of the solution and its first $n - 1$ derivatives at some point.

Even if one is not successful in finding explicit solutions to a differential equation, several techniques can be used to help analyze the solutions. One such method for first-order equations views the differential equation $dy/dx = f(x, y)$ as specifying directions (slopes) at points on the plane. The conglomerate of such slopes is the **direction field** for the equation. Knowing the "flow of solutions" is helpful in sketching the solution to an initial value problem. Furthermore, carrying out this method algebraically leads to numerical approximations to the desired solution. This numerical process is called **Euler's method.**

REVIEW PROBLEMS FOR CHAPTER 1

In Problems 1–6, identify the independent variable, dependent variable, and determine whether the equation is linear or nonlinear.

1. $5\dfrac{dx}{dt} + 5x^2 + 3 = 0$

2. $3r - \cos\theta \dfrac{dr}{d\theta} = \sin\theta$

3. $y^3\dfrac{d^2x}{dy^2} + 3x - \dfrac{8}{y-1} = 0$

4. $t^3 \left(\dfrac{d^3x}{dt^3} \right)^2 + 3t - x = 0$

5. $6 - t^3 \dfrac{d^2v}{dt^2} + 3v - \ln t = \dfrac{dv}{dt}$

6. $x \dfrac{d^2y}{dx^2} + 3x - \dfrac{dy}{dx} = e^{3x}$

In Problems 7–13, decide whether the statement made is True or False.

7. The function $x(t) = t^{-3} \sin t + 4t^{-3}$ is a solution to
$t^3 \dfrac{dx}{dt} = \cos t - 3t^2 x$.

8. The function $y(x) = -\dfrac{1}{3}(x+1)$ is a solution to

$\dfrac{dy}{dx} = \dfrac{y-1}{x+3}.$

9. The relation $x^2 + y^3 - e^y = 1$ is an implicit solution to
$\dfrac{dy}{dx} = \dfrac{e^y - 2x}{3y^2}.$

10. The relation $\sin y + e^y = x^6 - x^2 + x + 1$ is an implicit solution to

$\dfrac{dy}{dx} = \dfrac{6x^5 - 2x + 1}{\cos y + e^y}.$

11. The initial value problem $dx/dt = 3x^{2/3}$, $x(0) = 1$ has a unique solution in some open interval around $t = 0$.

12. The initial value problem $dy/dx = 3y^{2/3}$, $y(2) = 10^{-100}$ has a unique solution in some open interval around $x = 2$.

13. The solution to the initial value problem $dy/dx = (x-2)(y-3)^2$, $y(0) = 0$, will always be less than 3; that is, $y(x) < 3$ for $x \geq 0$.

14. Use Euler's method with step size $h = 1/2$ to approximate the solution to the initial value problem $dy/dx = x - y^2$, $y(1) = 2$, at $x = 2$.

15. If Euler's method with step size $h = 1/n$, where n is a positive integer, is used to approximate the solution to the initial value problem $dy/dx + y = 0$, $y(0) = 1$, at $x = 1$, what formula (expressed as a function of n) do you obtain for the approximation of $y(1)$? What is the exact value of $y(1)$?

16. Using the method of isoclines sketch the direction field for $y = -4x/y$.

17. The direction field for the equation

$$\dfrac{dp}{dt} = p(2p-1)(p-3)/9,$$

where p is the population (in thousands) at time t of a certain species, is plotted in Figure 1.17.

(a) If the initial population is 2500, what can you say about the limiting population as $t \to \infty$?

(b) If the initial population is 4000, will it ever decrease to 3500?

(c) If $p(1) = 0.3$, what is $\lim_{t \to \infty} p(t)$?

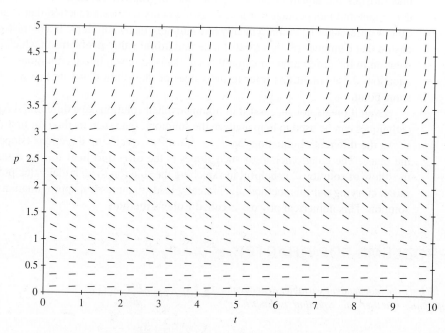

Figure 1.17 Direction Field for Problem 17

TECHNICAL WRITING EXERCISES FOR CHAPTER 1

1. Select four fields (for example, astronomy, geology, biology, and economics) and for each field discuss a situation in which differential equations are used to solve a problem. Select examples that are not covered in Section 1.1.

2. Compare the different types of solutions discussed in this chapter—explicit, implicit, graphical, and numerical. What are advantages and disadvantages of each?

Projects for Chapter 1

A Picard's Method

The initial value problem

$$(1) \qquad y'(x) = f(x, y), \qquad y(x_0) = y_0$$

can be rewritten as an **integral** equation. This is obtained by integrating both sides of (1) with respect to x from $x = x_0$ to $x = x_1$:

$$(2) \qquad \int_{x_0}^{x_1} y'(x)\,dx = y(x_1) - y(x_0) = \int_{x_0}^{x_1} f(x, y(x))\,dx .$$

Substituting $y(x_0) = y_0$ and solving for $y(x_1)$ gives

$$(3) \qquad y(x_1) = y_0 + \int_{x_0}^{x_1} f(x, y(x))\,dx .$$

If we use t instead of x as the variable of integration, we can let $x = x_1$ be the upper limit of integration. Equation (3) then becomes

$$(4) \qquad y(x) = y_0 + \int_{x_0}^{x} f(t, y(t))\,dt .$$

Equation (4) can be used to generate successive approximations of a solution to (1). Let the function $\phi_0(x)$ be an initial guess or approximation of a solution to (1). Then a new approximation function is given by

$$\phi_1(x) := y_0 + \int_{x_0}^{x} f(t, \phi_0(t))\,dt ,$$

where we have replaced $y(t)$ by the approximation $\phi_0(t)$ in the argument of f. In a similar fashion, we can use $\phi_1(x)$ to generate a new approximation $\phi_2(x)$, and so on. In general, we obtain the $(n + 1)$st approximation from the relation

$$(5) \qquad \phi_{n+1}(x) := y_0 + \int_{x_0}^{x} f(t, \phi_n(t))\,dt .$$

This procedure is called **Picard's method.**[†] Under certain assumptions on f and $\phi_0(x)$, the sequence $\{\phi_n(x)\}$ is known to converge to a solution to (1). These assumptions and the proof of convergence are given in Chapter 13.[††]

[†]*Historical Footnote:* This approximation method is a by-product of the famous Picard–Lindelöf existence theorem formulated at the end of the 19th century.

[††]All references to Chapters 11–13 refer to the expanded text, *Fundamentals of Differential Equations and Boundary Value Problems*, 7th ed.

Without further information about the solution to (1), it is common practice to take $\phi_0(x) \equiv y_0$.

(a) Use Picard's method with $\phi_0(x) \equiv 1$ to obtain the next four successive approximations of the solution to

(6) $y'(x) = y(x), \quad y(0) = 1$.

Show that these approximations are just the partial sums of the Maclaurin series for the actual solution e^x.

(b) Use Picard's method with $\phi_0(x) \equiv 0$ to obtain the next three successive approximations of the solution to the nonlinear problem

(7) $y'(x) = 3x - [y(x)]^2, \quad y(0) = 0$.

Graph these approximations for $0 \leq x \leq 1$.

(c) In Problem 29 in Exercises 1.2, we showed that the initial value problem

(8) $y'(x) = 3[y(x)]^{2/3}, \quad y(2) = 0$

does not have a unique solution. Show that Picard's method beginning with $\phi_0(x) \equiv 0$ converges to the solution $y(x) \equiv 0$, whereas Picard's method beginning with $\phi_0(x) = x - 2$ converges to the second solution $y(x) = (x-2)^3$. [*Hint:* For the guess $\phi_0(x) = x - 2$, show that $\phi_n(x)$ has the form $c_n(x-2)^{r_n}$, where $c_n \to 1$ and $r_n \to 3$ as $n \to \infty$.]

B The Phase Line

Sketching the direction field of a differential equation $dy/dt = f(t, y)$ is particularly easy when the equation is **autonomous**—that is, the independent variable t does not appear explicitly:

(9) $\dfrac{dy}{dt} = f(y)$.

In Figure 1.18(a) on page 34 the graph exhibits the direction field for

$y' = -A(y - y_1)(y - y_2)(y - y_3)^2$

with $A > 0$, and some solutions are sketched. Note the following properties of the graphs and explain how they follow from the fact that the equation is autonomous:

(a) The slopes in the direction field are all identical along horizontal lines.
(b) New solutions can be generated from old ones by time shifting [i.e., replacing $y(t)$ with $y(t - t_0)$.]

From observation (a) it follows that the entire direction field can be described by a single direction "line," as in Figure 1.18(b).

Of particular interest for autonomous equations are the constant, or **equilibrium,** solutions $y(t) = y_i$, $i = 1, 2, 3$. The equilibrium $y = y_1$ is called a *stable equilibrium,* or *sink,* because the neighboring solutions are attracted to it as $t \to \infty$. Equilibria that repel neighboring solutions, like $y = y_2$, are known as *sources*; all other equilibria are called *nodes,* illustrated by $y = y_3$. Sources and nodes are unstable equilibria.

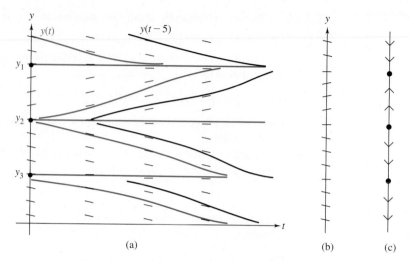

Figure 1.18 Direction field and solutions for an autonomous equation

(c) Describe how equilibria are characterized by the zeros of the function $f(y)$ in equation (1) and how the sink–source–node distinction can be decided on the basis of the *signs* of $f(y)$ on either side of its zeros.

Therefore, the simple **phase line** depicted in Figure 1.18(c), which indicates with dots and arrows only the zeros and signs of $f(y)$, is sufficient to describe the nature of the equilibrium solutions for an autonomous equation.

(d) Sketch the phase line for $y' = (y - 1)(y - 2)(y - 3)$ and state the nature of its equilibria.

(e) Use the phase line for $y' = -(y - 1)^{5/3}(y - 2)^2(y - 3)$ to predict the asymptotic behavior as $t \to \infty$ of the solution satisfying $y(0) = 2.1$.

(f) Sketch the phase line for $y' = y \sin y$ and state the nature of its equilibria.

(g) Sketch the phase lines for $y' = y \sin y + 0.1$ and $y' = y \sin y - 0.1$. Discuss the effect of the small perturbation ± 0.1 on the equilibria.

The splitting of the equilibrium at $y = 0$ that you observed in part (g) is an illustration of what is known as *bifurcation*. The following problem provides a dramatic illustration of the effects of bifurcation, in the context of a herd-management situation.

(h) When the *logistic model*, to be discussed in Section 3.2, is applied to the existing data for the alligator population on the grounds of Kennedy Space Center in Florida, the following differential equation is derived:

$$y' = -\frac{y(y - 1500)}{3200}.$$

Here $y(t)$ is the population and time t is measured in years. If hunters were allowed to thin the population at a rate of s alligators per year, the equation would be modified to

$$y' = -\frac{y(y - 1500)}{3200} - s.$$

Draw the phase lines for $s = 0, 50, 100, 125, 150, 175,$ and 200. Discuss the significance of the equilibria. Note the bifurcation at $s = 175$; should a depletion rate near 175 be avoided?

C Applications to Economics

Courtesy of Greg Huffman, Vanderbilt University

"The more, the better" is a description that applies to many things, particularly in economics. Any increase in wealth results in an improved lifestyle (country music and fairy tales notwithstanding). But once you have become a billionaire, earning another million dollars is not as gratifying as earning the first million was. So a realistic "benefit" function is likely to take the shape shown in Figure 1.19. It is monotonically increasing - its slope is always positive - but the slope decreases; economists say it has diminishing *marginal* benefit.

The benefit function $y(x)$ has the properties $y(x) \geq 0$, $y'(x) > 0$, $y''(x) < 0$ for positive x. (It is *concave*.) Such a function is $y(x) = x^a$ with $0 < a < 1$, since $y'(x) = ax^{a-1} > 0$ and $y''(x) = a(a-1)x^{a-2} < 0$. Below we give two examples where concave functions arise in the field of economics.

Agrarian Economy As a model of an agrarian economy, where a population of size N consumes food produced from the land, the total amount F of food available should increase with the amount L of land available per person, in the manner of a diminishing marginal benefit. It should also increase with the number N of people available to work the land. This is commonly modeled by the formula

$$(10) \quad F = A\left(\frac{L}{N}\right)^a N,$$

where $0 < a < 1$ and the constant A is a parameter measuring the effectiveness of the farming technology. Assume the survival birth rate increases with the amount of food available per person, so if D denotes the death rate (assumed constant), the rate of change of the population might be modeled by the differential equation

$$(11) \quad \frac{dN}{dt} = \left[Q\left(\frac{F}{N}\right)^b - D\right]N,$$

where $0 < b < 1$ and Q measures the effectiveness of the fertility technology. (Note the diminishing marginal benefit mode in (11) as well as (10).)

(a) Insert (10) into (11) and use phase line analysis (Project B) to investigate the (nonzero) equilibrium population level. Is the equilibrium stable? How does this equilibrium population depend on the technology parameters A and Q?

(b) Show that when the farming technology improves, the equilibrium population increases so as to maintain the same level of food consumed per person F/N.

(c) Show that when the fertility technology improves, everyone's dinner portions F/N diminish.

Figure 1.19 Increasing concave function ($y = x^{1/2}$)

Figure 1.20 Curves $y = Dk$ and $y = (Bk^a - c_m)/3$ for 3 different values of c_m

Growth of Capital Let k be the amount of capital invested in a country's industries at time t months. The investment yields a monthly return of Bk^a units of capital, where the constant B is a measure of the country's technological prowess and a is a constant satisfying $0 < a < 1$. Part of this output, which is denoted by c_m, is needed for subsistence (consumption), part of the net output $Bk^a - c_m$ is reinvested (i.e., added to the capital accumulation) and the remainder is used for additional consumption. Assume that the subsistence consumption per month remains constant and that one-third of the net output is reinvested each month. If the capital depreciates at a constant rate D, then a simple model for the rate of change in capital per month is the following:

(12) $$\frac{dk}{dt} = \frac{1}{3}\left[Bk^a - c_m \right] - Dk .$$

More precisely, since the consumption cannot exceed the output, we obtain

(13) $$\frac{dk}{dt} = \frac{1}{3}\max\left\{ Bk^a - c_m, 0 \right\} - Dk .$$

(d) Typical curves for the terms in the right-hand member of (13) are displayed in Figure 1.20 for three different values of c_m. Identify the equilibria in each case, sketch the phase line (see Project B), and determine the stability of the equilibria (are they sources, sinks or neither?).

(e) For the situation depicted by the third curve in Figure 1.20, set $a = 1/2$ and express the equilibrium levels for k as functions of B and D. How do the equilibrium levels behave when B is increased, or when D is increased?

D Taylor Series Method

Euler's method is based on the fact that the tangent line gives a good *local* approximation for the function. But why restrict ourselves to linear approximants when higher-degree polynomial approximants are available? For example, we can use the **Taylor polynomial** of degree n about $x = x_0$, which is defined by

$$P_n(x) := y(x_0) + y'(x_0)(x - x_0) + \frac{y''(x_0)}{2!}(x - x_0)^2 + \cdots + \frac{y^{(n)}(x_0)}{n!}(x - x_0)^n .$$

This polynomial is the nth partial sum of the Taylor series representation

$$\sum_{n=0}^{\infty} \frac{y^{(k)}(x_0)}{k!}(x - x_0)^k .$$

To determine the Taylor series for the solution $\phi(x)$ to the initial value problem

$$dy/dx = f(x, y), \qquad y(x_0) = y_0,$$

we need only determine the values of the derivatives of ϕ (assuming they exist) at x_0; that is, $\phi(x_0), \phi'(x_0), \ldots$. The initial condition gives the first value $\phi(x_0) = y_0$. Using the equation $y' = f(x, y)$, we find $\phi'(x_0) = f(x_0, y_0)$. To determine $\phi''(x_0)$, we differentiate the equation $y' = f(x, y)$ implicitly with respect to x to obtain

$$y'' = \frac{\partial f}{\partial x} + \frac{\partial f}{\partial y}\frac{dy}{dx} = \frac{\partial f}{\partial x} + \frac{\partial f}{\partial y}f$$

and thereby we can compute $\phi''(x_0)$.

(a) Compute the Taylor polynomials of degree 4 for the solutions to the given initial value problems. Use these Taylor polynomials to approximate the solution at $x = 1$.

 (i) $\dfrac{dy}{dx} = x - 2y; \quad y(0) = 1$. (ii) $\dfrac{dy}{dx} = y(2 - y); \quad y(0) = 4$.

(b) Compare the use of Euler's method with that of the Taylor series to approximate the solution $\phi(x)$ to the initial value problem

$$\frac{dy}{dx} + y = \cos x - \sin x, \qquad y(0) = 2.$$

 Do this by completing Table 1.3. Give the approximations for $\phi(1)$ and $\phi(3)$ to the nearest thousandth. Verify that $\phi(x) = \cos x + e^{-x}$ and use this formula together with a calculator or tables to find the exact values of $\phi(x)$ to the nearest thousandth. Finally, decide which of the first four methods in Table 1.3 will yield the closest approximation to $\phi(10)$ and give the reasons for your choice. (Remember that the computation of trigonometric functions must be done in the radian mode.)

(c) Compute the Taylor polynomial of degree 6 for the solution to the Airy equation

$$\frac{d^2y}{dx^2} = xy$$

 with the initial conditions $y(0) = 1, y'(0) = 0$. Do you see how, in general, the Taylor series method for an nth-order differential equation will employ each of the n initial conditions mentioned in Definition 3, Section 1.2?

TABLE 1.3		
Method	**Approximation of $\phi(1)$**	**Approximation of $\phi(3)$**
Euler's method using steps of size 0.1		
Euler's method using steps of size 0.01		
Taylor polynomial of degree 2		
Taylor polynomial of degree 5		
Exact value of $\phi(x)$ to nearest thousandth		

CHAPTER 2

First-Order Differential Equations

2.1 Introduction: Motion of a Falling Body

> **An object falls through the air toward Earth. Assuming that the only forces acting on the object are gravity and air resistance, determine the velocity of the object as a function of time.**

Newton's second law states that force is equal to mass times acceleration. We can express this by the equation

$$m\frac{dv}{dt} = F,$$

where F represents the total force on the object, m is the mass of the object, and dv/dt is the acceleration, expressed as the derivative of velocity with respect to time. It will be convenient in the future to define v as positive when it is directed *downward* (as opposed to the analysis in Section 1.1).

Near Earth's surface, the force due to gravity is just the weight of the objects and is also directed downward. This force can be expressed by mg, where g is the acceleration due to gravity. No general law precisely models the air resistance acting on the object, since this force seems to depend on the velocity of the object, the density of the air, and the shape of the object, among other things. However, in some instances air resistance can be reasonably represented by $-bv$, where b is a positive constant depending on the density of the air and the shape of the object. We use the negative sign because air resistance is a force that opposes the motion. The forces acting on the object are depicted in Figure 2.1 on page 39. (Note that we have generalized the free-fall model in Section 1.1 by including air resistance.)

Applying Newton's law, we obtain the first-order differential equation

$$(1) \qquad m\frac{dv}{dt} = mg - bv.$$

To solve this equation, we exploit a technique called **separation of variables,** which was used to analyze the radioactive decay model in Section 1.1 and will be developed in full detail in Section 2.2. Treating dv and dt as differentials, we rewrite equation (1) so as to isolate the variables v and t on opposite sides of the equation:

$$\frac{dv}{mg - bv} = \frac{dt}{m}.$$

(Hence, the nomenclature "separation of variables.")

Figure 2.1 Forces on falling object

Next we integrate the separated equation

(2) $$\int \frac{dv}{mg - bv} = \int \frac{dt}{m}$$

and derive

(3) $$-\frac{1}{b} \ln|mg - bv| = \frac{t}{m} + c \, .$$

Therefore,

$$|mg - bv| = e^{-bc}e^{-bt/m}$$

or

$$mg - bv = Ae^{-bt/m} \, ,$$

where the new constant A has magnitude e^{-bc} and the same sign ($\pm$) as $(mg - bv)$. Solving for v, we obtain

(4) $$v = \frac{mg}{b} - \frac{A}{b}e^{-bt/m} \, ,$$

which is called a **general solution** to the differential equation because, as we will see in Section 2.3, every solution to (1) can be expressed in the form given in (4).

In a specific case, we would be given the values of m, g, and b. To determine the constant A in the general solution, we can use the initial velocity of the object v_0. That is, we solve the **initial value problem**

$$m\frac{dv}{dt} = mg - bv \, , \qquad\qquad v(0) = v_0 \, .$$

Substituting $v = v_0$ and $t = 0$ into the general solution to the differential equation, we can solve for A. With this value for A, the solution to the initial value problem is

(5) $$v = \frac{mg}{b} + \left(v_0 - \frac{mg}{b} \right)e^{-bt/m} \, .$$

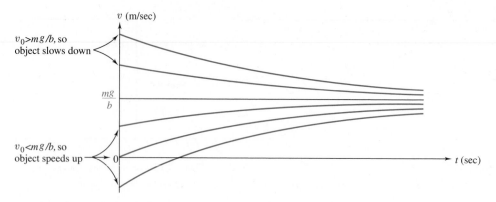

Figure 2.2 Graph of $v(t)$ for six different initial velocities v_0. (g = 9.8 m/sec^2, m/b = 5 sec)

The preceding formula gives the velocity of the object falling through the air as a function of time if the initial velocity of the object is v_0. In Figure 2.2 we have sketched the graph of $v(t)$ for various values of v_0. It appears from Figure 2.2 that the velocity $v(t)$ approaches *mg/b regardless of the initial velocity* v_0. [This is easy to see from formula (5) by letting $t \rightarrow +\infty$.] The constant mg/b is referred to as the **limiting** or **terminal velocity** of the object.

From this model for a falling body, we can make certain observations. Because $e^{-bt/m}$ rapidly tends to zero, the velocity is approximately the weight, mg, divided by the coefficient of air resistance, b. Thus, in the presence of air resistance, the heavier the object, the faster it will fall, assuming shapes and sizes are the same. Also, when air resistance is lessened (b is made smaller), the object will fall faster. These observations certainly agree with our experience.

Many other physical problems,[†] when formulated mathematically, lead to first-order differential equations or initial value problems. Several of these are discussed in Chapter 3. In this chapter we learn how to recognize and obtain solutions for some special types of first-order equations. We begin by studying separable equations, then linear equations, and then exact equations. The methods for solving these are the most basic. In the last two sections, we illustrate how devices such as integrating factors, substitutions, and transformations can be used to transform certain equations into either separable, exact, or linear equations that we can solve. Through our discussion of these special types of equations, you will gain insight into the behavior of solutions to more general equations and the possible difficulties in finding these solutions.

A word of warning is in order: In solving differential equations, integration plays an essential role. In particular, the separable equations in Section 2.2 always entail integration, as demonstrated in equations (2) and (3) above. For your convenience, Appendix A reviews three standard techniques for integrating the functions encountered in this text.

[†]The physical problem just discussed has other mathematical models. For example, one could take into account the variations in the gravitational field of Earth and the more general equations for air resistance.

2.2 Separable Equations

A simple class of first-order differential equations that can be solved using integration is the class of **separable equations.** These are equations

(1) $$\frac{dy}{dx} = f(x, y) \,,$$

that can be rewritten to isolate the variables x and y (together with their differentials dx and dy) on opposite sides of the equation, as in

$$h(y)\, dy = g(x)\, dx \,.$$

So the original right-hand side $f(x, y)$ must have the factored form

$$f(x, y) = g(x) \cdot \frac{1}{h(y)} \,.$$

More formally, we write $p(y) = 1/h(y)$ and present the following definition.

Separable Equation

Definition 1. If the right-hand side of the equation

$$\frac{dy}{dx} = f(x, y)$$

can be expressed as a function $g(x)$ that depends only on x times a function $p(y)$ that depends only on y, then the differential equation is called **separable.**[†]

In other words, a first-order equation is separable if it can be written in the form

$$\frac{dy}{dx} = g(x)p(y) \,.$$

For example, the equation

$$\frac{dy}{dx} = \frac{2x + xy}{y^2 + 1}$$

is separable, since (if one is sufficiently alert to detect the factorization)

$$\frac{2x + xy}{y^2 + 1} = x\frac{2 + y}{y^2 + 1} = g(x)p(y) \,.$$

However, the equation

$$\frac{dy}{dx} = 1 + xy$$

admits no such factorization of the right-hand side and so is not separable.

 Informally speaking, one solves separable equations by performing the separation and then integrating each side.

[†]*Historical Footnote:* A procedure for solving separable equations was discovered implicitly by Gottfried Leibniz in 1691. The explicit technique called separation of variables was formalized by John Bernoulli in 1694.

Method for Solving Separable Equations

To solve the equation

$$(2) \qquad \frac{dy}{dx} = g(x)p(y)$$

multiply by dx and by $h(y) := 1/p(y)$ to obtain

$$h(y)\,dy = g(x)\,dx\,.$$

Then integrate both sides:

$$\int h(y)\,dy = \int g(x)\,dx\,,$$

$$(3) \qquad H(y) = G(x) + C\,,$$

where we have merged the two constants of integration into a single symbol C. The last equation gives an implicit solution to the differential equation.

Caution: Constant functions $y \equiv c$ such that $p(c) = 0$ are also solutions to (2), but will not be included in (3) (see remarks on page 45).

We will look at the mathematical justification of this "streamlined" procedure shortly, but first we study some examples.

Example 1　Solve the nonlinear equation

$$\frac{dy}{dx} = \frac{x-5}{y^2}\,.$$

Solution　Following the streamlined approach, we separate the variables and rewrite the equation in the form

$$y^2\,dy = (x-5)\,dx\,.$$

Integrating, we have

$$\int y^2\,dy = \int (x-5)\,dx$$

$$\frac{y^3}{3} = \frac{x^2}{2} - 5x + C\,,$$

and solving this last equation for y gives

$$y = \left(\frac{3x^2}{2} - 15x + 3C\right)^{1/3}.$$

Since C is a constant of integration that can be any real number, $3C$ can also be any real number. Replacing $3C$ by the single symbol K, we then have

$$y = \left(\frac{3x^2}{2} - 15x + K\right)^{1/3}.$$

If we wish to abide by the custom of letting C represent an arbitrary constant, we can go one step further and use C instead of K in the final answer. This solution family is graphed in Figure 2.3 on page 43. ◆

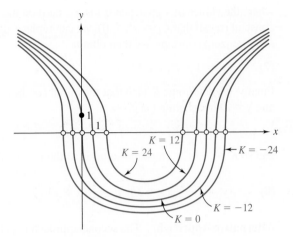

Figure 2.3 Family of solutions for Example 1[†]

As Example 1 attests, separable equations are among the easiest to solve. However, the procedure does require a facility for computing integrals. Many of the procedures to be discussed in the text also require a familiarity with the techniques of integration. For this reason we have provided a review of integration methods in Appendix A and a brief table of integrals at the back of the book.

Example 2 Solve the initial value problem

(4) $$\frac{dy}{dx} = \frac{y-1}{x+3}, \qquad y(-1) = 0.$$

Solution Separating the variables and integrating gives

$$\frac{dy}{y-1} = \frac{dx}{x+3},$$

$$\int \frac{dy}{y-1} = \int \frac{dx}{x+3},$$

(5) $$\ln|y-1| = \ln|x+3| + C.$$

At this point, we can either solve for y explicitly (retaining the constant C) or use the initial condition to determine C and then solve explicitly for y. Let's try the first approach.

Exponentiating equation (5), we have

$$e^{\ln|y-1|} = e^{\ln|x+3|+C} = e^C e^{\ln|x+3|},$$

(6) $$|y-1| = e^C |x+3| = C_1 |x+3|,$$

where $C_1 := e^C$.[‡] Now, depending on the values of y, we have $|y-1| = \pm(y-1)$; and similarly, $|x+3| = \pm(x+3)$. Thus, (6) can be written as

$$y-1 = \pm C_1(x+3) \quad \text{or} \quad y = 1 \pm C_1(x+3),$$

[†]The gaps in the curves reflect the fact that in the original differential equation, y appears in the denominator, so that $y = 0$ must be excluded.

[‡]Recall that the symbol $:=$ means "is defined to be."

where the choice of sign depends (as we said) on the values of x and y. Because C_1 is a *positive* constant (recall that $C_1 = e^C > 0$), we can replace $\pm C_1$ by K, where K now represents an *arbitrary* nonzero constant. We then obtain

(7) $y = 1 + K(x+3)$.

Finally, we determine K such that the initial condition $y(-1) = 0$ is satisfied. Putting $x = -1$ and $y = 0$ in equation (7) gives

$$0 = 1 + K(-1+3) = 1 + 2K ,$$

and so $K = -1/2$. Thus, the solution to the initial value problem is

(8) $y = 1 - \dfrac{1}{2}(x+3) = -\dfrac{1}{2}(x+1)$.

Alternative Approach. The second approach is to first set $x = -1$ and $y = 0$ in equation (5) and solve for C. In this case, we obtain

$$\ln|0-1| = \ln|-1+3| + C ,$$
$$0 = \ln 1 = \ln 2 + C ,$$

and so $C = -\ln 2$. Thus, from (5), the solution y is given implicitly by

$$\ln(1-y) = \ln(x+3) - \ln 2 .$$

Here we have replaced $|y-1|$ by $1-y$ and $|x+3|$ by $x+3$, since we are interested in x and y near the initial values $x = -1$, $y = 0$ (for such values, $y - 1 < 0$ and $x + 3 > 0$). Solving for y, we find

$$\ln(1-y) = \ln(x+3) - \ln 2 = \ln\left(\frac{x+3}{2}\right),$$
$$1 - y = \frac{x+3}{2},$$
$$y = 1 - \frac{1}{2}(x+3) = -\frac{1}{2}(x+1),$$

which agrees with the solution (8) found by the first method. ◆

Example 3 Solve the nonlinear equation

(9) $\dfrac{dy}{dx} = \dfrac{6x^5 - 2x + 1}{\cos y + e^y}$.

Solution Separating variables and integrating, we find

$$(\cos y + e^y)\, dy = (6x^5 - 2x + 1)\, dx ,$$
$$\int (\cos y + e^y)\, dy = \int (6x^5 - 2x + 1)\, dx ,$$
$$\sin y + e^y = x^6 - x^2 + x + C .$$

At this point, we reach an impasse. We would like to solve for y explicitly, but we cannot. This is often the case in solving nonlinear first-order equations. Consequently, when we say "solve the equation," we must on occasion be content if only an implicit form of the solution has been found. ◆

The separation of variables technique, as well as several other techniques discussed in this book, entails rewriting a differential equation by performing certain algebraic operations on it. "Rewriting $dy/dx = g(x)p(y)$ as $h(y)\,dy = g(x)\,dx$" amounts to dividing both sides by $p(y)$. You may recall from your algebra days that doing this can be treacherous. For example, the equation $x(x-2) = 4(x-2)$ has two solutions: $x = 2$ and $x = 4$. But if we "rewrite" the equation as $x = 4$ by dividing both sides by $(x-2)$, we lose track of the root $x = 2$. Thus, we should record the zeros of $(x-2)$ itself before dividing by this factor.

By the same token we must take note of the zeros of $p(y)$ in the separable equation $dy/dx = g(x)p(y)$ prior to dividing. After all, if (say) $g(x)p(y) = (x-2)^2(y-13)$, then observe that the constant function $y(x) \equiv 13$ solves the differential equation $dy/dx = g(x)p(y)$:

$$\frac{dy}{dx} = \frac{d(13)}{dx} = 0\,,$$

$$g(x)p(y) = (x-2)^2(13-13) = 0\,.$$

Indeed, in solving the equation of Example 2,

$$\frac{dy}{dx} = \frac{y-1}{x+3}\,,$$

we obtained $y = 1 + K(x+3)$ as the set of solutions, where K was a *nonzero* constant (since K replaced $\pm\, e^C$). But notice that the constant function $y \equiv 1$ (which in this case corresponds to $K = 0$) is also a solution to the differential equation. The reason we lost this solution can be traced back to a division by $y-1$ in the separation process. (See Problem 30 for an example of where a solution is lost and cannot be retrieved by setting the constant $K = 0$.)

Formal Justification of Method

We close this section by reviewing the separation of variables procedure in a more rigorous framework. The original differential equation (2) is rewritten in the form

$$(10) \qquad h(y)\frac{dy}{dx} = g(x)\,,$$

where $h(y) := 1/p(y)$. Letting $H(y)$ and $G(x)$ denote antiderivatives (indefinite integrals) of $h(y)$ and $g(x)$, respectively—that is,

$$H'(y) = h(y)\,, \qquad G'(x) = g(x)\,,$$

we recast equation (10) as

$$H'(y)\frac{dy}{dx} = G'(x)\,.$$

By the chain rule for differentiation, the left-hand side is the derivative of the composite function $H\big(y(x)\big)$:

$$\frac{d}{dx}H\big(y(x)\big) = H'\big(y(x)\big)\frac{dy}{dx}\,.$$

Thus, if $y(x)$ is a solution to equation (2), then $H\big(y(x)\big)$ and $G(x)$ are two functions of x that have the same derivative. Therefore, they differ by a constant:

$$(11) \qquad H\big(y(x)\big) = G(x) + C\,.$$

Equation (11) agrees with equation (3), which was derived informally, and we have thus verified that the latter can be used to construct implicit solutions.

2.2 EXERCISES

In Problems 1–6, determine whether the given differential equation is separable.

1. $\dfrac{dy}{dx} - \sin(x+y) = 0$

2. $\dfrac{dy}{dx} = 4y^2 - 3y + 1$

3. $\dfrac{ds}{dt} = t \ln(s^{2t}) + 8t^2$

4. $\dfrac{dy}{dx} = \dfrac{ye^{x+y}}{x^2 + 2}$

5. $(xy^2 + 3y^2)\, dy - 2x\, dx = 0$

6. $s^2 + \dfrac{ds}{dt} = \dfrac{s+1}{st}$

In Problems 7–16, solve the equation.

7. $x\dfrac{dy}{dx} = \dfrac{1}{y^3}$

8. $\dfrac{dx}{dt} = 3xt^2$

9. $\dfrac{dx}{dt} = \dfrac{t}{xe^{t+2x}}$

10. $\dfrac{dy}{dx} = \dfrac{x}{y^2\sqrt{1+x}}$

11. $x\dfrac{dv}{dx} = \dfrac{1 - 4v^2}{3v}$

12. $\dfrac{dy}{dx} = \dfrac{\sec^2 y}{1 + x^2}$

13. $\dfrac{dy}{dx} = 3x^2(1 + y^2)^{3/2}$

14. $\dfrac{dx}{dt} - x^3 = x$

15. $(x + xy^2)\, dx + e^{x^2} y\, dy = 0$

16. $y^{-1}\, dy + ye^{\cos x} \sin x\, dx = 0$

In Problems 17–26, solve the initial value problem.

17. $\dfrac{dy}{dx} = (1 + y^2)\tan x\,,\qquad y(0) = \sqrt{3}$

18. $y' = x^3(1 - y)\,,\qquad y(0) = 3$

19. $\dfrac{1}{2}\dfrac{dy}{dx} = \sqrt{y+1}\,\cos x\,,\qquad y(\pi) = 0$

20. $x^2\dfrac{dy}{dx} = \dfrac{4x^2 - x - 2}{(x+1)(y+1)}\,,\qquad y(1) = 1$

21. $\dfrac{1}{\theta}\dfrac{dy}{d\theta} = \dfrac{y\sin\theta}{y^2 + 1}\,,\qquad y(\pi) = 1$

22. $x^2\, dx + 2y\, dy = 0\,,\qquad y(0) = 2$

23. $\dfrac{dy}{dt} = 2t\cos^2 y\,,\qquad y(0) = \pi/4$

24. $\dfrac{dy}{dx} = 8x^3 e^{-2y}\,,\qquad y(1) = 0$

25. $\dfrac{dy}{dx} = x^2(1 + y)\,,\qquad y(0) = 3$

26. $\sqrt{y}\, dx + (1 + x)\, dy = 0\,,\qquad y(0) = 1$

27. Solutions Not Expressible in Terms of Elementary Functions. As discussed in calculus, certain indefinite integrals (antiderivatives) such as $\int e^{x^2} dx$ cannot be expressed in finite terms using elementary functions. When such an integral is encountered while solving a

differential equation, it is often helpful to use definite integration (integrals with variable upper limit). For example, consider the initial value problem

$$\dfrac{dy}{dx} = e^{x^2} y^2\,,\qquad y(2) = 1\,.$$

The differential equation separates if we divide by y^2 and multiply by dx. We integrate the separated equation from $x = 2$ to $x = x_1$ and find

$$\int_{x=2}^{x=x_1} e^{x^2}\, dx = \int_{x=2}^{x=x_1} \dfrac{dy}{y^2}$$
$$= -\dfrac{1}{y}\Big|_{x=2}^{x=x_1}$$
$$= -\dfrac{1}{y(x_1)} + \dfrac{1}{y(2)}\,.$$

If we let t be the variable of integration and replace x_1 by x and $y(2)$ by 1, then we can express the solution to the initial value problem by

$$y(x) = \left(1 - \int_2^x e^{t^2}\, dt\right)^{-1}\,.$$

Use definite integration to find an explicit solution to the initial value problems in parts (a)–(c).

(a) $dy/dx = e^{x^2}\,,\qquad y(0) = 0$

(b) $dy/dx = e^{x^2} y^{-2}\,,\qquad y(0) = 1$

(c) $dy/dx = \sqrt{1 + \sin x}\,(1 + y^2)\,,\qquad y(0) = 1$

 (d) Use a numerical integration algorithm (such as Simpson's rule, described in Appendix C) to approximate the solution to part (b) at $x = 0.5$ to three decimal places.

28. Sketch the solution to the initial value problem

$$\dfrac{dy}{dt} = 2y - 2yt\,,\qquad y(0) = 3$$

and determine its maximum value.

29. Uniqueness Questions. In Chapter 1 we indicated that in applications most *initial value problems* will have a unique solution. In fact, the existence of unique solutions was so important that we stated an existence and uniqueness theorem, Theorem 1, page 11. The method for separable equations can give us a solution, but it may not give us all the solutions (also see Problem 30). To illustrate this, consider the equation $dy/dx = y^{1/3}$.

(a) Use the method of separation of variables to show that

$$y = \left(\dfrac{2x}{3} + C\right)^{3/2}$$

is a solution.

(b) Show that the initial value problem $dy/dx = y^{1/3}$ with $y(0) = 0$ is satisfied for $C = 0$ by $y = (2x/3)^{3/2}$ for $x \geq 0$.

(c) Now show that the constant function $y \equiv 0$ also satisfies the initial value problem given in part (b). Hence, this initial value problem does not have a unique solution.

(d) Finally, show that the conditions of Theorem 1 on page 11 are not satisfied.

(The solution $y \equiv 0$ was lost because of the division by zero in the separation process.)

30. As stated in this section, the separation of equation (2) on page 42 requires division by $p(y)$, and this may disguise the fact that the roots of the equation $p(y) = 0$ are actually constant solutions to the differential equation.

(a) To explore this further, separate the equation

$$\frac{dy}{dx} = (x-3)(y+1)^{2/3}$$

to derive the solution,

$$y = -1 + (x^2/6 - x + C)^3 .$$

(b) Show that $y \equiv -1$ satisfies the original equation $dy/dx = (x-3)(y+1)^{2/3}$.

(c) Show that there is no choice of the constant C that will make the solution in part (a) yield the solution $y \equiv -1$. Thus, we lost the solution $y \equiv -1$ when we divided by $(y+1)^{2/3}$.

31. Interval of Definition. By looking at an initial value problem $dy/dx = f(x, y)$ with $y(x_0) = y_0$, it is not always possible to determine the domain of the solution $y(x)$ or the interval over which the function $y(x)$ satisfies the differential equation.

(a) Solve the equation $dy/dx = xy^3$.

(b) Give explicitly the solutions to the initial value problem with $y(0) = 1; y(0) = 1/2; y(0) = 2$.

(c) Determine the domains of the solutions in part (b).

(d) As found in part (c), the domains of the solutions depend on the initial conditions. For the initial value problem $dy/dx = xy^3$ with $y(0) = a, a > 0$, show that as a approaches zero from the right the domain approaches the whole real line $(-\infty, \infty)$ and as a approaches $+\infty$ the domain shrinks to a single point.

(e) Sketch the solutions to the initial value problem $dy/dx = xy^3$ with $y(0) = a$ for $a = \pm 1/2, \pm 1$, and ± 2.

32. Analyze the solution $y = \phi(x)$ to the initial value problem

$$\frac{dy}{dx} = y^2 - 3y + 2 , \qquad y(0) = 1.5$$

using approximation methods and then compare with its exact form as follows.

(a) Sketch the direction field of the differential equation and use it to guess the value of $\lim_{x \to \infty} \phi(x)$.

(b) Use Euler's method with a step size of 0.1 to find an approximation of $\phi(1)$.

(c) Find a formula for $\phi(x)$ and graph $\phi(x)$ on the direction field from part (a).

(d) What is the exact value of $\phi(1)$? Compare with your approximation in part (b).

(e) Using the exact solution obtained in part (c), determine $\lim_{x \to \infty} \phi(x)$ and compare with your guess in part (a).

33. Mixing. Suppose a brine containing 0.3 kilogram (kg) of salt per liter (L) runs into a tank initially filled with 400 L of water containing 2 kg of salt. If the brine enters at 10 L/min, the mixture is kept uniform by stirring, and the mixture flows out at the same rate. Find the mass of salt in the tank after 10 min (see Figure 2.4). [*Hint:* Let A denote the number of kilograms of salt in the tank at t min after the process begins and use the fact that

rate of increase in A = rate of input $-$ rate of exit.

A further discussion of mixing problems is given in Section 3.2.]

10 L/min
0.3 kg/L

$A(t)$

400 L

$A(0) = 2$ kg

10 L/min

Figure 2.4 Schematic representation of a mixing problem

34. Newton's Law of Cooling. According to Newton's law of cooling, if an object at temperature T is immersed in a medium having the constant temperature M, then the rate of change of T is proportional to the difference of temperature $M - T$. This gives the differential equation

$$dT/dt = k(M - T) .$$

(a) Solve the differential equation for T.

(b) A thermometer reading 100°F is placed in a medium having a constant temperature of 70°F. After 6 min, the thermometer reads 80°F. What is the reading after 20 min?

(Further applications of Newton's law of cooling appear in Section 3.3.)

35. Blood plasma is stored at 40°F. Before the plasma can be used, it must be at 90°F. When the plasma is placed in an oven at 120°F, it takes 45 min for the plasma to warm to 90°F. Assume Newton's law of cooling (Problem 34) applies. How long will it take for the plasma to warm to 90°F if the oven temperature is set at **(a)** 100°F, **(b)** 140°F, and **(c)** 80°F?

36. A pot of boiling water at 100°C is removed from a stove and covered at time $t = 0$ and left to cool in the kitchen. After 5 min, the water temperature has decreased to 80°C, and another 5 min later it has dropped to 65°C. Assuming Newton's law of cooling (Problem 34) applies, determine the (constant) temperature of the kitchen.

37. Compound Interest. If $P(t)$ is the amount of dollars in a savings bank account that pays a yearly interest rate of $r\%$ *compounded continuously,* then

$$\frac{dP}{dt} = \frac{r}{100}P, \qquad t \text{ in years.}$$

Assume the interest is 5% annually, $P(0) = \$1000$, and no monies are withdrawn.

(a) How much will be in the account after 2 yr?
(b) When will the account reach $4000?
(c) If $1000 is added to the account every 12 months, how much will be in the account after $3\frac{1}{2}$ yr?

38. Free Fall. In Section 2.1, we discussed a model for an object falling toward Earth. Assuming that only air resistance and gravity are acting on the object, we found that the velocity v must satisfy the equation

$$m\frac{dv}{dt} = mg - bv,$$

where m is the mass, g is the acceleration due to gravity, and $b > 0$ is a constant (see Figure 2.1). If $m = 100$ kg, $g = 9.8$ m/sec^2, $b = 5$ kg/sec, and $v(0) = 10$ m/sec, solve for $v(t)$. What is the limiting (i.e., terminal) velocity of the object?

39. Grand Prix Race. Driver A had been leading archrival B for a while by a *steady* 3 miles. Only 2 miles from the finish, driver A ran out of gas and decelerated thereafter at a rate proportional to the square of his remaining speed. One mile later, driver A's speed was exactly halved. If driver B's speed remained constant, who won the race?

40. The *atmospheric pressure* (force per unit area) on a surface at an altitude z is due to the weight of the column of air situated above the surface. Therefore, the drop in air pressure p between the top and bottom of a cylindrical volume element of height Δz and cross-section area A equals the weight of the air enclosed (density ρ times volume $V = A\Delta z$ times gravity g), per unit area:

$$p(z + \Delta z) - p(z) = -\frac{\rho(z)(A\Delta z)g}{A} = -\rho(z)g\Delta z.$$

Let $\Delta z \to 0$ to derive the differential equation $dp/dz = -\rho g$. To analyze this further we must postulate a formula that relates pressure and density. The *perfect gas law* relates pressure, volume, mass m, and absolute temperature T according to $pV = mRT/M$, where R is the universal gas constant and M is the molar mass (i.e., the mass of one mole) of the air. Therefore, density and pressure are related by $\rho := m/V = Mp/RT$.

(a) Derive the equation $\dfrac{dp}{dz} = -\dfrac{Mg}{RT}p$ and solve it for the "isothermal" case where T is constant to obtain the *barometric pressure equation* $p(z) = p(z_0)\exp[-Mg(z - z_0)/RT]$.

(b) If the temperature also varies with altitude $T = T(z)$, derive the solution

$$p(z) = p(z_0)\exp\left\{-\frac{Mg}{R}\int_{z_0}^{z}\frac{d\zeta}{T(\zeta)}\right\}.$$

(c) Suppose an engineer measures the barometric pressure at the top of a building to be 99,000 Pa (pascals), and 101,000 Pa at the base $(z = z_0)$. If the absolute temperature varies as $T(z) = 288 - 0.0065(z - z_0)$, determine the height of the building. Take $R = 8.31$ N-m/mol-K, $M = 0.029$ kg/mol, and $g = 9.8$ m/sec^2. (An amusing story concerning this problem can be found at http://www.snopes.com/college/exam/barometer.asp)

2.3 Linear Equations

A type of first-order differential equation that occurs frequently in applications is the linear equation. Recall from Section 1.1 that a **linear first-order equation** is an equation that can be expressed in the form

$$(1) \qquad a_1(x)\frac{dy}{dx} + a_0(x)y = b(x),$$

where $a_1(x)$, $a_0(x)$, and $b(x)$ depend only on the independent variable x, not on y.
For example, the equation

$$x^2\sin x - (\cos x)y = (\sin x)\frac{dy}{dx}$$

is linear, because it can be rewritten in the form

$$(\sin x)\frac{dy}{dx} + (\cos x)y = x^2 \sin x .$$

However, the equation

$$y\frac{dy}{dx} + (\sin x)y^3 = e^x + 1$$

is not linear; it cannot be put in the form of equation (1) due to the presence of the y^3 and $y\, dy/dx$ terms.

There are two situations for which the solution of a linear differential equation is quite immediate. The first arises if the coefficient $a_0(x)$ is identically zero, for then equation (1) reduces to

$$(2) \qquad a_1(x)\frac{dy}{dx} = b(x) ,$$

which is equivalent to

$$y(x) = \int \frac{b(x)}{a_1(x)}dx + C$$

[as long as $a_1(x)$ is not zero].

The second is less trivial. Note that if $a_0(x)$ happens to equal the derivative of $a_1(x)$—that is, $a_0(x) = a_1'(x)$—then the two terms on the left-hand side of equation (1) simply comprise the derivative of the product $a_1(x)y$:

$$a_1(x)y' + a_0(x)y = a_1(x)y' + a_1'(x)y = \frac{d}{dx}[a_1(x)y] .$$

Therefore equation (1) becomes

$$(3) \qquad \frac{d}{dx}[a_1(x)y] = b(x)$$

and the solution is again elementary:

$$a_1(x)y = \int b(x)\, dx + C ,$$

$$y(x) = \frac{1}{a_1(x)}\left[\int b(x)\, dx + C\right] .$$

One can seldom rewrite a linear differential equation so that it reduces to a form as simple as (2). However, the form (3) can be achieved through multiplication of the original equation (1) by a well-chosen function $\mu(x)$. Such a function $\mu(x)$ is then called an "integrating factor" for equation (1). The easiest way to see this is first to divide the original equation (1) by $a_1(x)$ and put it into **standard form**

$$(4) \qquad \frac{dy}{dx} + P(x)y = Q(x) ,$$

where $P(x) = a_0(x)/a_1(x)$ and $Q(x) = b(x)/a_1(x)$.

Next we wish to determine $\mu(x)$ so that the left-hand side of the multiplied equation

$$(5) \qquad \mu(x)\frac{dy}{dx} + \mu(x)P(x)y = \mu(x)Q(x)$$

is just the derivative of the product $\mu(x)y$:

$$\mu(x)\frac{dy}{dx} + \mu(x)P(x)y = \frac{d}{dx}[\mu(x)y] = \mu(x)\frac{dy}{dx} + \mu'(x)y .$$

Clearly, this requires that μ satisfy

(6) $\mu' = \mu P .$

To find such a function, we recognize that equation (6) is a separable differential equation, which we can write as $(1/\mu)\,d\mu = P(x)dx$. Integrating both sides gives

(7) $\mu(x) = e^{\int P(x)\,dx} .$

With this choice[†] for $\mu(x)$, equation (5) becomes

$$\frac{d}{dx}[\mu(x)y] = \mu(x)Q(x) ,$$

which has the solution

(8) $y(x) = \dfrac{1}{\mu(x)}\left[\displaystyle\int \mu(x)Q(x)\,dx + C\right] .$

Here C is an arbitrary constant, so (8) gives a one-parameter family of solutions to (4). This form is known as the **general solution** to (4).

We can summarize the method for solving linear equations as follows.

Method for Solving Linear Equations

(a) Write the equation in the standard form

$$\frac{dy}{dx} + P(x)y = Q(x) .$$

(b) Calculate the integrating factor $\mu(x)$ by the formula

$$\mu(x) = \exp\left[\int P(x)dx\right] .$$

(c) Multiply the equation in standard form by $\mu(x)$ and, recalling that the left-hand side is just $\frac{d}{dx}[\mu(x)y]$, obtain

$$\underbrace{\mu(x)\frac{dy}{dx} + P(x)\mu(x)y}_{\dfrac{d}{dx}[\mu(x)y]} = \mu(x)Q(x) ,$$

$$\frac{d}{dx}[\mu(x)y] = \mu(x)Q(x) .$$

(d) Integrate the last equation and solve for y by dividing by $\mu(x)$ to obtain (8).

[†]Any choice of the integration constant in $\int P(x)\,dx$ will produce a suitable $\mu(x)$.

Example 1 Find the general solution to

(9) $$\frac{1}{x}\frac{dy}{dx} - \frac{2y}{x^2} = x\cos x, \qquad x > 0.$$

Solution To put this linear equation in standard form, we multiply by x to obtain

(10) $$\frac{dy}{dx} - \frac{2}{x}y = x^2\cos x.$$

Here $P(x) = -2/x$, so

$$\int P(x)\,dx = \int \frac{-2}{x}\,dx = -2\ln|x|.$$

Thus, an integrating factor is

$$\mu(x) = e^{-2\ln|x|} = e^{\ln(x^{-2})} = x^{-2}.$$

Multiplying equation (10) by $\mu(x)$ yields

$$\underbrace{x^{-2}\frac{dy}{dx} - 2x^{-3}y}_{} = \cos x,$$

$$\frac{d}{dx}(x^{-2}y) = \cos x.$$

We now integrate both sides and solve for y to find

$$x^{-2}y = \int \cos x\,dx = \sin x + C$$

(11) $$y = x^2\sin x + Cx^2.$$

It is easily checked that this solution is valid for all $x > 0$. In Figure 2.5 we have sketched solutions for various values of the constant C in (11). ◆

Figure 2.5 Graph of $y = x^2\sin x + Cx^2$ for five values of the constant C

In the next example, we encounter a linear equation that arises in the study of the radioactive decay of an isotope.

Example 2 A rock contains two radioactive isotopes, RA_1 and RA_2, that belong to the same radioactive series; that is, RA_1 decays into RA_2, which then decays into stable atoms. Assume that the rate at which RA_1 decays into RA_2 is $50e^{-10t}$ kg/sec. Because the rate of decay of RA_2 is proportional to the mass $y(t)$ of RA_2 present, the rate of change in RA_2 is

$$\frac{dy}{dt} = \text{rate of creation} - \text{rate of decay},$$

(12) $$\frac{dy}{dt} = 50e^{-10t} - ky,$$

where $k > 0$ is the decay constant. If $k = 2/\text{sec}$ and initially $y(0) = 40$ kg, find the mass $y(t)$ of RA_2 for $t \geq 0$.

Solution Equation (12) is linear, so we begin by writing it in standard form

(13) $$\frac{dy}{dt} + 2y = 50e^{-10t}, \qquad y(0) = 40,$$

where we have substituted $k = 2$ and displayed the initial condition. We now see that $P(t) = 2$, so $\int P(t)\,dt = \int 2\,dt = 2t$. Thus, an integrating factor is $\mu(t) = e^{2t}$. Multiplying equation (13) by $\mu(t)$ yields

$$\underbrace{e^{2t}\frac{dy}{dt} + 2e^{2t}y}_{} = 50e^{-10t+2t} = 50e^{-8t},$$

$$\frac{d}{dt}(e^{2t}y) \ \ = 50e^{-8t}.$$

Integrating both sides and solving for y, we find

$$e^{2t}y = -\frac{25}{4}e^{-8t} + C,$$

$$y = -\frac{25}{4}e^{-10t} + Ce^{-2t}.$$

Substituting $t = 0$ and $y(0) = 40$ gives

$$40 = -\frac{25}{4}e^0 + Ce^0 = -\frac{25}{4} + C,$$

so $C = 40 + 25/4 = 185/4$. Thus, the mass $y(t)$ of RA_2 at time t is given by

(14) $$y(t) = \left(\frac{185}{4}\right)e^{-2t} - \left(\frac{25}{4}\right)e^{-10t}, \qquad t \geq 0. \ \blacklozenge$$

Example 3 For the initial value problem

$$y' + y = \sqrt{1 + \cos^2 x}, \qquad y(1) = 4,$$

find the value of $y(2)$.

Solution The integrating factor for the differential equation is, from equation (7),

$$\mu(x) = e^{\int 1\,dx} = e^x.$$

The general solution form (8) thus reads

$$y(x) = e^{-x}\left(\int e^x \sqrt{1 + \cos^2 x}\, dx + C \right).$$

However, this indefinite integral cannot be expressed in finite terms with elementary functions (recall a similar situation in Problem 27 of Exercises 2.2). Because we *can* use numerical algorithms such as Simpson's rule (Appendix C) to perform *definite* integration, we revert to the form (5), which in this case reads

$$\frac{d}{dx}(e^x y) = e^x \sqrt{1 + \cos^2 x},$$

and take the definite integral from the initial value $x = 1$ to the desired value $x = 2$:

$$e^x y \Big|_{x=1}^{x=2} = e^2 y(2) - e^1 y(1) = \int_{x=1}^{x=2} e^x \sqrt{1 + \cos^2 x}\, dx.$$

Inserting the given value of $y(1)$ and solving, we express

$$y(2) = e^{-2+1}(4) + e^{-2} \int_1^2 e^x \sqrt{1 + \cos^2 x}\, dx.$$

Using Simpson's rule, we find that the definite integral is approximately 4.841, so

$$y(2) \approx 4e^{-1} + 4.841 e^{-2} \approx 2.127. \quad \blacklozenge$$

In Example 3 we had no difficulty expressing the integral for the integrating factor $\mu(x) = e^{\int 1\, dx} = e^x$. Clearly, situations will arise where this integral, too, cannot be expressed with elementary functions. In such cases we must again resort to a numerical procedure such as Euler's method (Section 1.4) or to a "nested loop" implementation of Simpson's rule. You are invited to explore such a possibility in Problem 27.

Because we have established explicit formulas for the solutions to *linear* first-order differential equations, we get as a dividend a direct proof of the following theorem.

Existence and Uniqueness of Solution

Theorem 1. If $P(x)$ and $Q(x)$ are continuous on an interval (a, b) that contains the point x_0, then for any choice of initial value y_0, there exists a unique solution $y(x)$ on (a, b) to the initial value problem

(15) $$\frac{dy}{dx} + P(x)y = Q(x), \qquad y(x_0) = y_0.$$

In fact, the solution is given by (8) for a suitable value of C.

The essentials of the proof of Theorem 1 are contained in the deliberations leading to equation (8); Problem 34 provides the details. This theorem differs from Theorem 1 on page 11 in that for the *linear* initial value problem (15), we have the existence and uniqueness of the solution on the *whole* interval (a, b), rather than on some smaller unspecified interval about x_0.

The theory of linear differential equations is an important branch of mathematics not only because these equations occur in applications but also because of the elegant structure associated with them. For example, first-order linear equations always have a general solution given by equation (8). Some further properties of first-order linear equations are described in Problems 28 and 36. Higher-order linear equations are treated in Chapters 4, 6, and 8.

2.3 EXERCISES

In Problems 1–6, determine whether the given equation is separable, linear, neither, or both.

1. $x^2 \dfrac{dy}{dx} + \sin x - y = 0$

2. $\dfrac{dx}{dt} + xt = e^x$

3. $(t^2 + 1)\dfrac{dy}{dt} = yt - y$

4. $3t = e^t \dfrac{dy}{dt} + y \ln t$

5. $x \dfrac{dx}{dt} + t^2 x = \sin t$

6. $3r = \dfrac{dr}{d\theta} - \theta^3$

In Problems 7–16, obtain the general solution to the equation.

7. $\dfrac{dy}{dx} - y - e^{3x} = 0$

8. $\dfrac{dy}{dx} = \dfrac{y}{x} + 2x + 1$

9. $\dfrac{dr}{d\theta} + r \tan \theta = \sec \theta$

10. $x \dfrac{dy}{dx} + 2y = x^{-3}$

11. $(t + y + 1)\, dt - dy = 0$

12. $\dfrac{dy}{dx} = x^2 e^{-4x} - 4y$

13. $y \dfrac{dx}{dy} + 2x = 5y^3$

14. $x \dfrac{dy}{dx} + 3(y + x^2) = \dfrac{\sin x}{x}$

15. $(x^2 + 1)\dfrac{dy}{dx} + xy - x = 0$

16. $(1 - x^2)\dfrac{dy}{dx} - x^2 y = (1 + x)\sqrt{1 - x^2}$

In Problems 17–22, solve the initial value problem.

17. $\dfrac{dy}{dx} - \dfrac{y}{x} = xe^x$, $\quad y(1) = e - 1$

18. $\dfrac{dy}{dx} + 4y - e^{-x} = 0$, $\quad y(0) = \dfrac{4}{3}$

19. $t^2 \dfrac{dx}{dt} + 3tx = t^4 \ln t + 1$, $\quad x(1) = 0$

20. $\dfrac{dy}{dx} + \dfrac{3y}{x} + 2 = 3x$, $\quad y(1) = 1$

21. $(\cos x)\dfrac{dy}{dx} + y \sin x = 2x \cos^2 x$,

$y\left(\dfrac{\pi}{4}\right) = \dfrac{-15\sqrt{2}\pi^2}{32}$

22. $(\sin x)\dfrac{dy}{dx} + y \cos x = x \sin x$, $\quad y\left(\dfrac{\pi}{2}\right) = 2$

23. Radioactive Decay. In Example 2 assume that the rate at which RA_1 decays into RA_2 is $40e^{-20t}$ kg/sec and the decay constant for RA_2 is $k = 5$/sec. Find the mass $y(t)$ of RA_2 for $t \geq 0$ if initially $y(0) = 10$ kg.

24. In Example 2 the decay constant for isotope RA_1 was 10/sec, which expresses itself in the exponent of the rate term $50e^{-10t}$ kg/sec. When the decay constant for RA_2 is $k = 2$/sec, we see that in formula (14) for y the term $(185/4)e^{-2t}$ eventually dominates (has greater magnitude for t large).

(a) Redo Example 2 taking $k = 20$/sec. Now which term in the solution eventually dominates?

(b) Redo Example 2 taking $k = 10$/sec.

25. (a) Using definite integration, show that the solution to the initial value problem

$$\dfrac{dy}{dx} + 2xy = 1, \quad y(2) = 1,$$

can be expressed as

$$y(x) = e^{-x^2}\left(e^4 + \int_2^x e^{t^2}\, dt\right).$$

 (b) Use numerical integration (such as Simpson's rule, Appendix C) to approximate the solution at $x = 3$.

26. Use numerical integration (such as Simpson's rule, Appendix C) to approximate the solution, at $x = 1$, to the initial value problem

$$\dfrac{dy}{dx} + \dfrac{\sin 2x}{2(1 + \sin^2 x)} y = 1, \quad y(0) = 0.$$

Ensure your approximation is accurate to three decimal places.

27. Consider the initial value problem

$$\dfrac{dy}{dx} + \sqrt{1 + \sin^2 x}\, y = x, \quad y(0) = 2.$$

(a) Using definite integration, show that the integrating factor for the differential equation can be written as

$$\mu(x) = \exp\left(\int_0^x \sqrt{1 + \sin^2 t}\, dt\right)$$

and that the solution to the initial value problem is

$$y(x) = \frac{1}{\mu(x)} \int_0^x \mu(s)\, s\, ds + \frac{2}{\mu(x)}.$$

 (b) Obtain an approximation to the solution at $x = 1$ by using numerical integration (such as Simpson's rule, Appendix C) in a nested loop to estimate values of $\mu(x)$ and, thereby, the value of

$$\int_0^1 \mu(s) s\, ds.$$

[*Hint:* First, use Simpson's rule to approximate $\mu(x)$ at $x = 0.1, 0.2, \ldots , 1$. Then use these values and apply Simpson's rule again to approximate $\int_0^1 \mu(s)\, s\, ds.$]

 (c) Use Euler's method (Section 1.4) to approximate the solution at $x = 1$, with step sizes $h = 0.1$ and 0.05.

[A direct comparison of the merits of the two numerical schemes in parts (b) and (c) is very complicated, since it should take into account the number of functional evaluations in each algorithm as well as the inherent accuracies.]

28. Constant Multiples of Solutions.

(a) Show that $y = e^{-x}$ is a solution of the linear equation

(16) $$\frac{dy}{dx} + y = 0\, ,$$

and $y = x^{-1}$ is a solution of the nonlinear equation

(17) $$\frac{dy}{dx} + y^2 = 0\, .$$

(b) Show that for any constant C, the function Ce^{-x} is a solution of equation (16), while Cx^{-1} is a solution of equation (17) only when $C = 0$ or 1.

(c) Show that for any linear equation of the form

$$\frac{dy}{dx} + P(x)y = 0\, ,$$

if $\hat{y}(x)$ is a solution, then for any constant C the function $C\hat{y}(x)$ is also a solution.

29. Use your ingenuity to solve the equation

$$\frac{dy}{dx} = \frac{1}{e^{4y} + 2x}.$$

[*Hint:* The roles of the independent and dependent variables may be reversed.]

30. Bernoulli Equations. The equation

(18) $$\frac{dy}{dx} + 2y = xy^{-2}$$

is an example of a Bernoulli equation. (Further discussion of Bernoulli equations is in Section 2.6.)

(a) Show that the substitution $v = y^3$ reduces equation (18) to the equation

(19) $$\frac{dv}{dx} + 6v = 3x\, .$$

(b) Solve equation (19) for v. Then make the substitution $v = y^3$ to obtain the solution to equation (18).

31. Discontinuous Coefficients. As we will see in Chapter 3, occasions arise when the coefficient $P(x)$ in a linear equation fails to be continuous because of jump discontinuities. Fortunately, we may still obtain a "reasonable" solution. For example, consider the initial value problem

$$\frac{dy}{dx} + P(x)y = x\, , \qquad y(0) = 1\, ,$$

where

$$P(x) := \begin{cases} 1\, , & 0 \le x \le 2\, , \\ 3\, , & x > 2\, . \end{cases}$$

(a) Find the general solution for $0 \le x \le 2$.

(b) Choose the constant in the solution of part (a) so that the initial condition is satisfied.

(c) Find the general solution for $x > 2$.

(d) Now choose the constant in the general solution from part (c) so that the solution from part (b) and the solution from part (c) agree at $x = 2$. By patching the two solutions together, we can obtain a continuous function that satisfies the differential equation except at $x = 2$, where its derivative is undefined.

(e) Sketch the graph of the solution from $x = 0$ to $x = 5$.

32. Discontinuous Forcing Terms. There are occasions when the forcing term $Q(x)$ in a linear equation fails to be continuous because of jump discontinuities. Fortunately, we may still obtain a reasonable solution imitating the procedure discussed in Problem 31. Use this procedure to find the continuous solution to the initial value problem.

$$\frac{dy}{dx} + 2y = Q(x)\, , \qquad y(0) = 0\, ,$$

where

$$Q(x) := \begin{cases} 2\, , & 0 \le x \le 3\, , \\ -2\, , & x > 3\, . \end{cases}$$

Sketch the graph of the solution from $x = 0$ to $x = 7$.

33. Singular Points. Those values of x for which $P(x)$ in equation (4) is not defined are called **singular points** of the equation. For example, $x = 0$ is a singular point of the equation $xy' + 2y = 3x$, since when the equation is written in the standard form, $y' + (2/x)y = 3$, we see that $P(x) = 2/x$ is not defined at $x = 0$. On an interval containing a singular point, the questions of the existence and uniqueness of a solution are left unanswered, since Theorem 1 on page 53 does not apply. To show the possible behavior of solutions near a singular point, consider the following equations.

(a) Show that $xy' + 2y = 3x$ has only one solution defined at $x = 0$. Then show that the initial value problem for this equation with initial condition $y(0) = y_0$ has a unique solution when $y_0 = 0$ and no solution when $y_0 \ne 0$.

(b) Show that $xy' - 2y = 3x$ has an infinite number of solutions defined at $x = 0$. Then show that the initial value problem for this equation with initial condition $y(0) = 0$ has an infinite number of solutions.

34. **Existence and Uniqueness.** Under the assumptions of Theorem 1, we will prove that equation (8) gives a solution to equation (4) on (a, b). We can then choose the constant C in equation (8) so that the initial value problem (15) is solved.

 (a) Show that since $P(x)$ is continuous on (a, b), then $\mu(x)$ defined in (7) is a positive, continuous function satisfying $d\mu/dx = P(x)\mu(x)$ on (a, b).

 (b) Since

$$\frac{d}{dx} \int \mu(x) Q(x) dx = \mu(x) Q(x),$$

 verify that y given in equation (8) satisfies equation (4) by differentiating both sides of equation (8).

 (c) Show that when we let $\int \mu(x) Q(x) \, dx$ be the antiderivative whose value at x_0 is 0 (i.e., $\int_{x_0}^{x} \mu(t) Q(t) \, dt$) and choose C to be $y_0 \mu(x_0)$, the initial condition $y(x_0) = y_0$ is satisfied.

 (d) Start with the assumption that $y(x)$ is a solution to the initial value problem (15) and argue that the discussion leading to equation (8) implies that $y(x)$ must obey equation (8). Then argue that the initial condition in (15) determines the constant C uniquely.

35. **Mixing.** Suppose a brine containing 0.2 kg of salt per liter runs into a tank initially filled with 500 L of water containing 5 kg of salt. The brine enters the tank at a rate of 5 L/min. The mixture, kept uniform by stirring, is flowing out at the rate of 5 L/min (see Figure 2.6).

Figure 2.6 Mixing problem with equal flow rates

 (a) Find the concentration, in kilograms per liter, of salt in the tank after 10 min. [*Hint:* Let A denote the number of kilograms of salt in the tank at t minutes after the process begins and use the fact that

 rate of increase in A = rate of input − rate of exit.

 A further discussion of mixing problems is given in Section 3.2.]

 (b) After 10 min, a leak develops in the tank and an additional liter per minute of mixture flows out of the tank (see Figure 2.7). What will be the concentration, in kilograms per liter, of salt in the tank 20 min after the leak develops? [*Hint:* Use the method discussed in Problems 31 and 32.]

Figure 2.7 Mixing problem with unequal flow rates

36. **Variation of Parameters.** Here is another procedure for solving linear equations that is particularly useful for higher-order linear equations. This method is called **variation of parameters.** It is based on the idea that just by knowing the *form* of the solution, we can substitute into the given equation and solve for any unknowns. Here we illustrate the method for first-order equations (see Sections 4.6 and 6.4 for the generalization to higher-order equations).

 (a) Show that the general solution to

 (20) $$\frac{dy}{dx} + P(x)y = Q(x)$$

 has the form

 $$y(x) = Cy_h(x) + y_p(x),$$

 where y_h ($\neq 0$) is a solution to equation (20) when $Q(x) \equiv 0$, C is a constant, and $y_p(x) = v(x)y_h(x)$ for a suitable function $v(x)$. [*Hint:* Show that we can take $y_h = \mu^{-1}(x)$ and then use equation (8).]

 We can in fact determine the unknown function y_h by solving a separable equation. Then direct substitution of vy_h in the original equation will give a simple equation that can be solved for v.

 Use this procedure to find the general solution to

 (21) $$\frac{dy}{dx} + \frac{3}{x}y = x^2, \qquad x > 0,$$

 by completing the following steps:

 (b) Find a nontrivial solution y_h to the separable equation

 (22) $$\frac{dy}{dx} + \frac{3}{x}y = 0, \qquad x > 0.$$

 (c) Assuming (21) has a solution of the form $y_p(x) = v(x)y_h(x)$, substitute this into equation (21), and simplify to obtain $v'(x) = x^2/y_h(x)$.

(d) Now integrate to get $v(x)$.

(e) Verify that $y(x) = Cy_h(x) + v(x)y_h(x)$ is a general solution to (21).

37. Secretion of Hormones. The secretion of hormones into the blood is often a periodic activity. If a hormone is secreted on a 24-h cycle, then the rate of change of the level of the hormone in the blood may be represented by the initial value problem

$$\frac{dx}{dt} = \alpha - \beta \cos\frac{\pi t}{12} - kx, \qquad x(0) = x_0,$$

where $x(t)$ is the amount of the hormone in the blood at time t, α is the average secretion rate, β is the amount of daily variation in the secretion, and k is a positive constant reflecting the rate at which the body removes the hormone from the blood. If $\alpha = \beta = 1$, $k = 2$, and $x_0 = 10$, solve for $x(t)$.

38. Use the separation of variables technique to derive the solution (7) to the differential equation (6).

39. The temperature T (in units of $100°$ F) of a university classroom on a cold winter day varies with time t (in hours) as

$$\frac{dT}{dt} = \begin{cases} 1 - T, & \text{if heating unit is ON.} \\ -T, & \text{if heating unit is OFF.} \end{cases}$$

Suppose $T = 0$ at 9:00 A.M., the heating unit is ON from 9–10 A.M., OFF from 10–11 A.M., ON again from 11 A.M.–noon, and so on for the rest of the day. How warm will the classroom be at noon? At 5:00 P.M.?

40. The *Nobel Prize in Physiology or Medicine* in 1963 was shared by A. L. Hodgkin and A. F. Huxley in recognition of their model for the firing of neuronal synapses. As will be discussed in Chapter 12, they proposed that the opening/closing of certain ion channels in the neuron cell was governed by a combination of probabilistic "gating variables," each satisfying a differential equation that they expressed as

$$(23) \qquad \frac{du}{dt} = \alpha(1 - u) - \beta u$$

with positive parameters α, β.

(a) Use a direction field diagram (Section 1.3) to show that the solutions of equation (23) are "probabilistic" in the sense that if their initial values lie between 0 and 1, all subsequent values also lie on [0,1].

(b) Solve (23) and show that all solutions approach the value $\alpha/(\alpha + \beta)$ exponentially.

2.4 Exact Equations

Suppose the mathematical function $F(x, y)$ represents some physical quantity, such as temperature, in a region of the xy-plane. Then the level curves of F, where $F(x, y) = constant$, could be interpreted as isotherms on a weather map, as depicted in Figure 2.8.

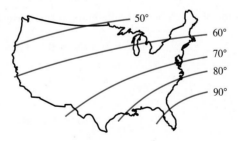

Figure 2.8 Level curves of $F(x, y)$

How does one calculate the slope of the tangent to a level curve? It is accomplished by implicit differentiation: One takes the derivative, with respect to x, of both sides of the equation $F(x, y) = C$, taking into account that y depends on x along the curve:

$$\frac{d}{dx}F(x,y) = \frac{d}{dx}(C) \qquad \text{or}$$

$$(1) \qquad \frac{\partial F}{\partial x} + \frac{\partial F}{\partial y}\frac{dy}{dx} = 0,$$

and solves for the slope:

$$(2) \qquad \frac{dy}{dx} = f(x, y) = -\frac{\partial F/\partial x}{\partial F/\partial y}.$$

The expression obtained by formally multiplying the left-hand member of (1) by dx is known as the *total differential* of F, written dF:

$$dF := \frac{\partial F}{\partial x}dx + \frac{\partial F}{\partial y}dy,$$

and our procedure for obtaining the equation for the slope $f(x, y)$ of the level curve $F(x, y) = C$ can be expressed as setting the total differential $dF = 0$ and solving.

Because equation (2) has the form of a differential equation, we should be able to reverse this logic and come up with a very easy technique for solving some differential equations. After all, any first-order differential equation $dy/dx = f(x, y)$ can be rewritten in the (differential) form

$$(3) \qquad M(x, y)\, dx + N(x, y)\, dy = 0$$

(in a variety of ways). Now, if the left-hand side of equation (3) can be identified as a total differential,

$$M(x, y)\, dx + N(x, y)\, dy = \frac{\partial F}{\partial x}dx + \frac{\partial F}{\partial y}dy = dF(x, y),$$

then its solutions are given (implicitly) by the level curves

$$F(x, y) = C$$

for an arbitrary constant C.

Example 1 Solve the differential equation

$$\frac{dy}{dx} = -\frac{2xy^2 + 1}{2x^2 y}.$$

Solution Some of the choices of differential forms corresponding to this equation are

$$(2xy^2 + 1)\, dx + 2x^2 y\, dy = 0,$$

$$\frac{2xy^2 + 1}{2x^2 y}\, dx + dy = 0,$$

$$dx + \frac{2x^2 y}{2xy^2 + 1}\, dy = 0, \quad \text{etc.}$$

However, the first form is best for our purposes because it is a total differential of the function $F(x, y) = x^2 y^2 + x$:

$$(2xy^2 + 1)\, dx + 2x^2 y\, dy = d[x^2 y^2 + x]$$

$$= \frac{\partial}{\partial x}(x^2 y^2 + x)\, dx + \frac{\partial}{\partial y}(x^2 y^2 + x)\, dy.$$

Thus, the solutions are given implicitly by the formula $x^2 y^2 + x = C$. See Figure 2.9 on page 59. ◆

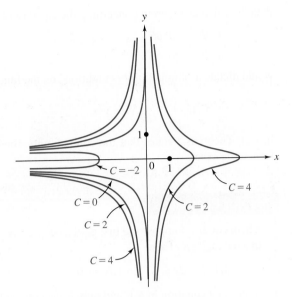

Figure 2.9 Solutions of Example 1

Next we introduce some terminology.

Exact Differential Form

Definition 2. The differential form $M(x, y)\, dx + N(x, y)\, dy$ is said to be **exact** in a rectangle R if there is a function $F(x, y)$ such that

(4) $\dfrac{\partial F}{\partial x}(x, y) = M(x, y)$ and $\dfrac{\partial F}{\partial y}(x, y) = N(x, y)$

for all (x, y) in R. That is, the total differential of $F(x, y)$ satisfies

$$dF(x, y) = M(x, y)\, dx + N(x, y)\, dy\,.$$

If $M(x, y)\, dx + N(x, y)\, dy$ is an exact differential form, then the equation

$$M(x, y)\, dx + N(x, y)\, dy = 0$$

is called an **exact equation.**

As you might suspect, in applications a differential equation is rarely given to us in exact differential form. However, the solution procedure is so quick and simple for such equations that we devote this section to it. From Example 1, we see that what is needed is (i) a test to determine if a differential form $M(x, y)\, dx + N(x, y)\, dy$ is exact and, if so, (ii) a procedure for finding the function $F(x, y)$ itself.

The test for exactness arises from the following observation. If

$$M(x, y)\, dx + N(x, y)\, dy = \frac{\partial F}{\partial x}dx + \frac{\partial F}{\partial y}dy\,,$$

then the calculus theorem concerning the equality of continuous mixed partial derivatives

$$\frac{\partial}{\partial y}\frac{\partial F}{\partial x} = \frac{\partial}{\partial x}\frac{\partial F}{\partial y}$$

would dictate a "compatibility condition" on the functions M and N:

$$\frac{\partial}{\partial y}M(x, y) = \frac{\partial}{\partial x}N(x, y) .$$

In fact, Theorem 2 states that the compatibility condition is also *sufficient* for the differential form to be exact.

Test for Exactness

Theorem 2. Suppose the first partial derivatives of $M(x, y)$ and $N(x, y)$ are continuous in a rectangle R. Then

$$M(x, y)\,dx + N(x, y)\,dy = 0$$

is an exact equation in R if and only if the compatibility condition

$$(5) \qquad \frac{\partial M}{\partial y}(x, y) = \frac{\partial N}{\partial x}(x, y)$$

holds for all (x, y) in R.[†]

Before we address the proof of Theorem 2, note that in Example 1 the differential form that led to the total differential was

$$(2xy^2 + 1)\,dx + (2x^2 y)\,dy = 0 .$$

The compatibility conditions are easily confirmed:

$$\frac{\partial M}{\partial y} = \frac{\partial}{\partial y}(2xy^2 + 1) = 4xy ,$$

$$\frac{\partial N}{\partial x} = \frac{\partial}{\partial x}(2x^2 y) = 4xy .$$

Also clear is the fact that the other differential forms considered,

$$\frac{2xy^2 + 1}{2x^2 y}\,dx + dy = 0 , \qquad dx + \frac{2x^2 y}{2xy^2 + 1}\,dy = 0 ,$$

do *not* meet the compatibility conditions.

Proof of Theorem 2. There are two parts to the theorem: Exactness implies compatibility, and compatibility implies exactness. First, we have seen that if the differential equation is exact, then the two members of equation (5) are simply the mixed second partials of a function $F(x, y)$. As such, their equality is ensured by the theorem of calculus that states that mixed second partials are equal if they are continuous. Because the hypothesis of Theorem 2 guarantees the latter condition, equation (5) is validated.

[†]*Historical Footnote:* This theorem was proven by Leonhard Euler in 1734.

Rather than proceed directly with the proof of the second part of the theorem, let's derive a formula for a function $F(x, y)$ that satisfies $\partial F/\partial x = M$ and $\partial F/\partial y = N$. Integrating the first equation with respect to x yields

$$(6) \qquad F(x, y) = \int M(x, y)\, dx + g(y) \,.$$

Notice that instead of using C to represent the constant of integration, we have written $g(y)$. This is because y is held fixed while integrating with respect to x, and so our "constant" may well depend on y. To determine $g(y)$, we differentiate both sides of (6) with respect to y to obtain

$$(7) \qquad \frac{\partial F}{\partial y}(x, y) = \frac{\partial}{\partial y} \int M(x, y)\, dx + \frac{\partial}{\partial y} g(y) \,.$$

As g is a function of y alone, we can write $\partial g/\partial y = g'(y)$, and solving (7) for $g'(y)$ gives

$$g'(y) = \frac{\partial F}{\partial y}(x, y) - \frac{\partial}{\partial y} \int M(x, y)\, dx \,.$$

Since $\partial F/\partial y = N$, this last equation becomes

$$(8) \qquad g'(y) = N(x, y) - \frac{\partial}{\partial y} \int M(x, y)\, dx \,.$$

Notice that although the right-hand side of (8) indicates a possible dependence on x, *the appearances of this variable must cancel* because the left-hand side, $g'(y)$, depends only on y. By integrating (8), we can determine $g(y)$ up to a numerical constant, and therefore we can determine the function $F(x, y)$ up to a numerical constant from the functions $M(x, y)$ and $N(x, y)$.

To finish the proof of Theorem 2, we need to show that the condition (5) implies that $M\, dx + N\, dy = 0$ is an exact equation. This we do by actually exhibiting a function $F(x, y)$ that satisfies $\partial F/\partial x = M$ and $\partial F/\partial y = N$. Fortunately, we needn't look too far for such a function. The discussion in the first part of the proof suggests (6) as a candidate, where $g'(y)$ is given by (8). Namely, we *define* $F(x, y)$ by

$$(9) \qquad F(x, y) := \int_{x_0}^{x} M(t, y)\, dt + g(y) \,,$$

where (x_0, y_0) is a fixed point in the rectangle R and $g(y)$ is determined, up to a numerical constant, by the equation

$$(10) \qquad g'(y) := N(x, y) - \frac{\partial}{\partial y} \int_{x_0}^{x} M(t, y)\, dt \,.$$

Before proceeding we must address an extremely important question concerning the definition of $F(x, y)$. That is, how can we be sure (in this portion of the proof) that $g'(y)$, as given in equation (10), is really a function of just y alone? To show that the right-hand side of (10) is independent of x (that is, that the appearances of the variable x cancel), all we need to do is show that its partial derivative with respect to x is zero. This is where condition (5) is utilized. We leave to the reader this computation and the verification that $F(x, y)$ satisfies conditions (4) (see Problems 35 and 36). ◆

The construction in the proof of Theorem 2 actually provides an explicit procedure for solving exact equations. Let's recap and look at some examples.

Method for Solving Exact Equations

(a) If $M\,dx + N\,dy = 0$ is exact, then $\partial F/\partial x = M$. Integrate this last equation with respect to x to get

$$(11) \qquad F(x, y) = \int M(x, y)\,dx + g(y)\,.$$

(b) To determine $g(y)$, take the partial derivative with respect to y of both sides of equation (11) and substitute N for $\partial F/\partial y$. We can now solve for $g'(y)$.

(c) Integrate $g'(y)$ to obtain $g(y)$ up to a numerical constant. Substituting $g(y)$ into equation (11) gives $F(x, y)$.

(d) The solution to $M\,dx + N\,dy = 0$ is given implicitly by

$$F(x, y) = C\,.$$

(Alternatively, starting with $\partial F/\partial y = N$, the implicit solution can be found by first integrating with respect to y; see Example 3.)

Example 2 Solve

$$(12) \qquad (2xy - \sec^2 x)\,dx + (x^2 + 2y)\,dy = 0\,.$$

Solution Here $M(x, y) = 2xy - \sec^2 x$ and $N(x, y) = x^2 + 2y$. Because

$$\frac{\partial M}{\partial y} = 2x = \frac{\partial N}{\partial x}\,,$$

equation (12) is exact. To find $F(x, y)$, we begin by integrating M with respect to x:

$$(13) \qquad F(x, y) = \int (2xy - \sec^2 x)\,dx + g(y)$$

$$= x^2 y - \tan x + g(y)\,.$$

Next we take the partial derivative of (13) with respect to y and substitute $x^2 + 2y$ for N:

$$\frac{\partial F}{\partial y}(x, y) = N(x, y)\,,$$

$$x^2 + g'(y) = x^2 + 2y\,.$$

Thus, $g'(y) = 2y$, and since the choice of the constant of integration is not important, we can take $g(y) = y^2$. Hence, from (13), we have $F(x, y) = x^2 y - \tan x + y^2$, and the solution to equation (12) is given implicitly by $x^2 y - \tan x + y^2 = C$. ◆

Remark. The procedure for solving exact equations requires several steps. As a check on our work, we observe that when we solve for $g'(y)$, we must obtain a function that is independent of x. If this is not the case, then we have erred either in our computation of $F(x, y)$ or in computing $\partial M/\partial y$ or $\partial N/\partial x$.

In the construction of $F(x, y)$, we can first integrate $N(x, y)$ with respect to y to get

$$(14) \qquad F(x, y) = \int N(x, y)\,dy + h(x)$$

and then proceed to find $h(x)$. We illustrate this alternative method in the next example.

Example 3 Solve

(15) $(1 + e^x y + x e^x y)\, dx + (x e^x + 2)\, dy = 0$.

Solution Here $M = 1 + e^x y + x e^x y$ and $N = x e^x + 2$. Because

$$\frac{\partial M}{\partial y} = e^x + x e^x = \frac{\partial N}{\partial x},$$

equation (15) is exact. If we now integrate $N(x, y)$ with respect to y, we obtain

$$F(x, y) = \int (x e^x + 2)\, dy + h(x) = x e^x y + 2y + h(x) .$$

When we take the partial derivative with respect to x and substitute for M, we get

$$\frac{\partial F}{\partial x}(x, y) = M(x, y)$$

$$e^x y + x e^x y + h'(x) = 1 + e^x y + x e^x y .$$

Thus, $h'(x) = 1$, so we take $h(x) = x$. Hence, $F(x, y) = x e^x y + 2y + x$, and the solution to equation (15) is given implicitly by $x e^x y + 2y + x = C$. In this case we can solve explicitly for y to obtain $y = (C - x)/(2 + x e^x)$. ◆

Remark. Since we can use either procedure for finding $F(x, y)$, it may be worthwhile to consider each of the integrals $\int M(x, y)\, dx$ and $\int N(x, y)\, dy$. If one is easier to evaluate than the other, this would be sufficient reason for us to use one method over the other. [The skeptical reader should try solving equation (15) by first integrating $M(x, y)$.]

Example 4 Show that

(16) $(x + 3x^3 \sin y)\, dx + (x^4 \cos y)\, dy = 0$

is *not* exact but that multiplying this equation by the factor x^{-1} yields an exact equation. Use this fact to solve (16).

Solution In equation (16), $M = x + 3x^3 \sin y$ and $N = x^4 \cos y$. Because

$$\frac{\partial M}{\partial y} = 3x^3 \cos y \neq 4x^3 \cos y = \frac{\partial N}{\partial x},$$

equation (16) is not exact. When we multiply (16) by the factor x^{-1}, we obtain

(17) $(1 + 3x^2 \sin y)\, dx + (x^3 \cos y)\, dy = 0$.

For this new equation, $M = 1 + 3x^2 \sin y$ and $N = x^3 \cos y$. If we test for exactness, we now find that

$$\frac{\partial M}{\partial y} = 3x^2 \cos y = \frac{\partial N}{\partial x},$$

and hence (17) is exact. Upon solving (17), we find that the solution is given implicitly by $x + x^3 \sin y = C$. Since equations (16) and (17) differ only by a factor of x, then any solution to one will be a solution for the other whenever $x \neq 0$. Hence the solution to equation (16) is given implicitly by $x + x^3 \sin y = C$. ◆

In Section 2.5 we discuss methods for finding factors that, like x^{-1} in Example 4, change inexact equations into exact equations.

2.4 EXERCISES

In Problems 1–8, classify the equation as separable, linear, exact, or none of these. Notice that some equations may have more than one classification.

1. $(x^2y + x^4\cos x)\,dx - x^3\,dy = 0$

2. $(x^{10/3} - 2y)\,dx + x\,dy = 0$

3. $\sqrt{-2y - y^2}\,dx + (3 + 2x - x^2)\,dy = 0$

4. $(ye^{xy} + 2x)\,dx + (xe^{xy} - 2y)\,dy = 0$

5. $xy\,dx + dy = 0$

6. $y^2\,dx + (2xy + \cos y)\,dy = 0$

7. $[2x + y\cos(xy)]\,dx + [x\cos(xy) - 2y]dy = 0$

8. $\theta\,dr + (3r - \theta - 1)\,d\theta = 0$

In Problems 9–20, determine whether the equation is exact. If it is, then solve it.

9. $(2xy + 3)\,dx + (x^2 - 1)\,dy = 0$

10. $(2x + y)\,dx + (x - 2y)\,dy = 0$

11. $(e^x\sin y - 3x^2)\,dx + (e^x\cos y + y^{-2/3}/3)\,dy = 0$

12. $(\cos x \cos y + 2x)\,dx - (\sin x \sin y + 2y)\,dy = 0$

13. $e^t(y - t)\,dt + (1 + e^t)\,dy = 0$

14. $(t/y)\,dy + (1 + \ln y)\,dt = 0$

15. $\cos\theta\,dr - (r\sin\theta - e^\theta)\,d\theta = 0$

16. $(ye^{xy} - 1/y)\,dx + (xe^{xy} + x/y^2)\,dy = 0$

17. $(1/y)\,dx - (3y - x/y^2)\,dy = 0$

18. $[2x + y^2 - \cos(x + y)]\,dx$
 $\quad + [2xy - \cos(x + y) - e^y]\,dy = 0$

19. $\left(2x + \dfrac{y}{1 + x^2y^2}\right)dx + \left(\dfrac{x}{1 + x^2y^2} - 2y\right)dy = 0$

20. $\left[\dfrac{2}{\sqrt{1 - x^2}} + y\cos(xy)\right]dx$
 $\quad + [x\cos(xy) - y^{-1/3}]dy = 0$

In Problems 21–26, solve the initial value problem.

21. $(1/x + 2y^2x)\,dx + (2yx^2 - \cos y)\,dy = 0$,
 $\qquad y(1) = \pi$

22. $(ye^{xy} - 1/y)\,dx + (xe^{xy} + x/y^2)\,dy = 0$,
 $\qquad y(1) = 1$

23. $(e^ty + te^ty)\,dt + (te^t + 2)\,dy = 0$, $y(0) = -1$

24. $(e^tx + 1)\,dt + (e^t - 1)\,dx = 0$, $x(1) = 1$

25. $(y^2\sin x)\,dx + (1/x - y/x)\,dy = 0$, $y(\pi) = 1$

26. $(\tan y - 2)\,dx + (x\sec^2 y + 1/y)\,dy = 0$,
 $\qquad y(0) = 1$

27. For each of the following equations, find the most general function $M(x, y)$ so that the equation is exact.
 (a) $M(x, y)\,dx + (\sec^2 y - x/y)\,dy = 0$
 (b) $M(x, y)\,dx + (\sin x \cos y - xy - e^{-y})\,dy = 0$

28. For each of the following equations, find the most general function $N(x, y)$ so that the equation is exact.
 (a) $[y\cos(xy) + e^x]\,dx + N(x, y)\,dy = 0$
 (b) $(ye^{xy} - 4x^3y + 2)\,dx + N(x, y)\,dy = 0$

29. Consider the equation
 $$(y^2 + 2xy)\,dx - x^2\,dy = 0 .$$
 (a) Show that this equation is not exact.
 (b) Show that multiplying both sides of the equation by y^{-2} yields a new equation that is exact.
 (c) Use the solution of the resulting exact equation to solve the original equation.
 (d) Were any solutions lost in the process?

30. Consider the equation
 $$(5x^2y + 6x^3y^2 + 4xy^2)\,dx$$
 $$\quad + (2x^3 + 3x^4y + 3x^2y)\,dy = 0 .$$
 (a) Show that the equation is not exact.
 (b) Multiply the equation by x^ny^m and determine values for n and m that make the resulting equation exact.
 (c) Use the solution of the resulting exact equation to solve the original equation.

31. Argue that in the proof of Theorem 2 the function g can be taken as
 $$g(y) = \int_{y_0}^{y} N(x, t)\,dt - \int_{y_0}^{y}\left[\frac{\partial}{\partial t}\int_{x_0}^{x} M(s, t)\,ds\right]dt ,$$
 which can be expressed as
 $$g(y) = \int_{y_0}^{y} N(x, t)\,dt - \int_{x_0}^{x} M(s, y)\,ds$$
 $$\quad + \int_{x_0}^{x} M(s, y_0)\,ds .$$
 This leads ultimately to the representation
 $$(18)\qquad F(x, y) = \int_{y_0}^{y} N(x, t)\,dt + \int_{x_0}^{x} M(s, y_0)\,ds .$$
 Evaluate this formula directly with $x_0 = 0$, $y_0 = 0$ to rework
 (a) Example 1.
 (b) Example 2.
 (c) Example 3.

32. Orthogonal Trajectories. A geometric problem occurring often in engineering is that of finding a family of curves (orthogonal trajectories) that intersects a given family of curves orthogonally at each point. For example, we may be given the lines of force of an electric field and want to find the equation for the equipotential curves. Consider the family of curves described by $F(x, y) = k$, where k is a parameter. Recall from the discussion of equation (2) that for each curve in the family, the slope is given by

$$\frac{dy}{dx} = -\frac{\partial F}{\partial x} \Big/ \frac{\partial F}{\partial y}.$$

(a) Recall that the slope of a curve that is orthogonal (perpendicular) to a given curve is just the negative reciprocal of the slope of the given curve. Using this fact, show that the curves orthogonal to the family $F(x, y) = k$ satisfy the differential equation

$$\frac{\partial F}{\partial y}(x, y)\, dx - \frac{\partial F}{\partial x}(x, y)\, dy = 0.$$

(b) Using the preceding differential equation, show that the orthogonal trajectories to the family of circles $x^2 + y^2 = k$ are just straight lines through the origin (see Figure 2.10).

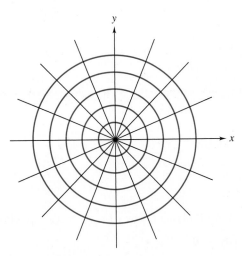

Figure 2.10 Orthogonal trajectories for concentric circles are lines through the center

(c) Show that the orthogonal trajectories to the family of hyperbolas $xy = k$ are the hyperbolas $x^2 - y^2 = k$ (see Figure 2.11).

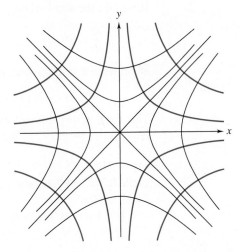

Figure 2.11 Families of orthogonal hyperbolas

33. Use the method in Problem 32 to find the orthogonal trajectories for each of the given families of curves, where k is a parameter.

(a) $2x^2 + y^2 = k$ **(b)** $y = kx^4$
(c) $y = e^{kx}$ **(d)** $y^2 = kx$

[*Hint:* First express the family in the form $F(x, y) = k$.]

34. Use the method described in Problem 32 to show that the orthogonal trajectories to the family of curves $x^2 + y^2 = kx$, k a parameter, satisfy

$$(2yx^{-1})\, dx + (y^2 x^{-2} - 1)\, dy = 0.$$

Find the orthogonal trajectories by solving the above equation. Sketch the family of curves, along with their orthogonal trajectories. [*Hint:* Try multiplying the equation by $x^m y^n$ as in Problem 30.]

35. Using condition (5), show that the right-hand side of (10) is independent of x by showing that its partial derivative with respect to x is zero. [*Hint:* Since the partial derivatives of M are continuous, Leibniz's theorem allows you to interchange the operations of integration and differentiation.]

36. Verify that $F(x, y)$ as defined by (9) and (10) satisfies conditions (4).

2.5 Special Integrating Factors

If we take the standard form for the linear differential equation of Section 2.3,

$$\frac{dy}{dx} + P(x)y = Q(x) \,,$$

and rewrite it in differential form by multiplying through by dx, we obtain

$$[P(x)y - Q(x)]\,dx + dy = 0 \,.$$

This form is certainly not exact, but it becomes exact upon multiplication by the integrating factor $\mu(x) = e^{\int P(x)\,dx}$. We have

$$[\mu(x)P(x)y - \mu(x)Q(x)]\,dx + \mu(x)\,dy = 0$$

as the form, and the compatibility condition is precisely the identity $\mu(x)P(x) = \mu'(x)$ (see Problem 20).

This leads us to generalize the notion of an integrating factor.

Integrating Factor

Definition 3. If the equation

(1) $M(x, y)\,dx + N(x, y)\,dy = 0$

is not exact, but the equation

(2) $\mu(x, y)M(x, y)\,dx + \mu(x, y)N(x, y)\,dy = 0 \,,$

which results from multiplying equation (1) by the function $\mu(x, y)$, *is* exact, then $\mu(x, y)$ is called an **integrating factor**[†] of the equation (1).

Example 1 Show that $\mu(x, y) = xy^2$ is an integrating factor for

(3) $(2y - 6x)\,dx + (3x - 4x^2y^{-1})\,dy = 0 \,.$

Use this integrating factor to solve the equation.

Solution We leave it to you to show that (3) is not exact. Multiplying (3) by $\mu(x, y) = xy^2$, we obtain

(4) $(2xy^3 - 6x^2y^2)\,dx + (3x^2y^2 - 4x^3y)\,dy = 0 \,.$

For this equation we have $M = 2xy^3 - 6x^2y^2$ and $N = 3x^2y^2 - 4x^3y$. Because

$$\frac{\partial M}{\partial y}(x, y) = 6xy^2 - 12x^2y = \frac{\partial N}{\partial x}(x, y) \,,$$

equation (4) is exact. Hence, $\mu(x, y) = xy^2$ is indeed an integrating factor of equation (3).

[†]*Historical Footnote:* A general theory of integrating factors was developed by Alexis Clairaut in 1739. Leonhard Euler also studied classes of equations that could be solved using a specific integrating factor.

Let's now solve equation (4) using the procedure of Section 2.4. To find $F(x, y)$, we begin by integrating M with respect to x:

$$F(x, y) = \int (2xy^3 - 6x^2y^2)\, dx + g(y) = x^2y^3 - 2x^3y^2 + g(y)\, .$$

When we take the partial derivative with respect to y and substitute for N, we find

$$\frac{\partial F}{\partial y}(x, y) = N(x, y)$$

$$3x^2y^2 - 4x^3y + g'(y) = 3x^2y^2 - 4x^3y\, .$$

Thus, $g'(y) = 0$, so we can take $g(y) \equiv 0$. Hence, $F(x, y) = x^2y^3 - 2x^3y^2$, and the solution to equation (4) is given implicitly by

$$x^2y^3 - 2x^3y^2 = C\, .$$

Although equations (3) and (4) have essentially the same solutions, *it is possible to lose or gain solutions when multiplying by* $\mu(x, y)$. In this case $y \equiv 0$ is a solution of equation (4) but not of equation (3). The extraneous solution arises because, when we multiply (3) by $\mu = xy^2$ to obtain (4), we are actually multiplying both sides of (3) by zero if $y \equiv 0$. This gives us $y \equiv 0$ as a solution to (4), but it is not a solution to (3). ◆

Generally speaking, when using integrating factors, you should check whether any solutions to $\mu(x, y) = 0$ are in fact solutions to the original differential equation.

How do we find an integrating factor? If $\mu(x, y)$ is an integrating factor of (1) with continuous first partial derivatives, then testing (2) for exactness, we must have

$$\frac{\partial}{\partial y}[\mu(x, y)M(x, y)] = \frac{\partial}{\partial x}[\mu(x, y)N(x, y)]\, .$$

By use of the product rule, this reduces to the equation

$$(5) \qquad M\frac{\partial \mu}{\partial y} - N\frac{\partial \mu}{\partial x} = \left(\frac{\partial N}{\partial x} - \frac{\partial M}{\partial y} \right)\mu\, .$$

But solving the partial differential equation (5) for μ is usually more difficult than solving the original equation (1). There are, however, two important exceptions.

Let's assume that equation (1) has an integrating factor that depends only on x; that is, $\mu = \mu(x)$. In this case equation (5) reduces to the separable equation

$$(6) \qquad \frac{d\mu}{dx} = \left(\frac{\partial M/\partial y - \partial N/\partial x}{N} \right)\mu\, ,$$

where $(\partial M/\partial y - \partial N/\partial x)/N$ is (presumably) just a function of x. In a similar fashion, if equation (1) has an integrating factor that depends only on y, then equation (5) reduces to the separable equation

$$(7) \qquad \frac{d\mu}{dy} = \left(\frac{\partial N/\partial x - \partial M/\partial y}{M} \right)\mu\, ,$$

where $(\partial N/\partial x - \partial M/\partial y)/M$ is just a function of y.

We can reverse the above argument. In particular, if $(\partial M/\partial y - \partial N/\partial x)/N$ is a function that depends only on x, then we can solve the separable equation (6) to obtain the integrating factor

$$\mu(x) = \exp\left[\int\left(\frac{\partial M/\partial y - \partial N/\partial x}{N}\right)dx\right]$$

for equation (1). We summarize these observations in the following theorem.

Special Integrating Factors

Theorem 3. If $(\partial M/\partial y - \partial N/\partial x)/N$ is continuous and depends only on x, then

$$(8) \qquad \mu(x) = \exp\left[\int\left(\frac{\partial M/\partial y - \partial N/\partial x}{N}\right)dx\right]$$

is an integrating factor for equation (1).
 If $(\partial N/\partial x - \partial M/\partial y)/M$ is continuous and depends only on y, then

$$(9) \qquad \mu(y) = \exp\left[\int\left(\frac{\partial N/\partial x - \partial M/\partial y}{M}\right)dy\right]$$

is an integrating factor for equation (1).

Theorem 3 suggests the following procedure.

Method for Finding Special Integrating Factors

If $M\,dx + N\,dy = 0$ is neither separable nor linear, compute $\partial M/\partial y$ and $\partial N/\partial x$. If $\partial M/\partial y = \partial N/\partial x$, then the equation is exact. If it is not exact, consider

$$(10) \qquad \frac{\partial M/\partial y - \partial N/\partial x}{N}.$$

If (10) is a function of just x, then an integrating factor is given by formula (8). If not, consider

$$(11) \qquad \frac{\partial N/\partial x - \partial M/\partial y}{M}.$$

If (11) is a function of just y, then an integrating factor is given by formula (9).

Example 2 Solve

$$(12) \qquad (2x^2 + y)\,dx + (x^2y - x)\,dy = 0.$$

Solution A quick inspection shows that equation (12) is neither separable nor linear. We also note that

$$\frac{\partial M}{\partial y} = 1 \neq (2xy - 1) = \frac{\partial N}{\partial x}.$$

Because (12) is not exact, we compute

$$\frac{\partial M/\partial y - \partial N/\partial x}{N} = \frac{1 - (2xy - 1)}{x^2 y - x} = \frac{2(1 - xy)}{-x(1 - xy)} = \frac{-2}{x}.$$

We obtain a function of only x, so an integrating factor for (12) is given by formula (8). That is,

$$\mu(x) = \exp\left(\int \frac{-2}{x} dx \right) = x^{-2}.$$

When we multiply (12) by $\mu = x^{-2}$, we get the exact equation

$$(2 + yx^{-2}) dx + (y - x^{-1}) dy = 0.$$

Solving this equation, we ultimately derive the implicit solution

(13) $2x - yx^{-1} + \dfrac{y^2}{2} = C.$

Notice that the solution $x \equiv 0$ was lost in multiplying by $\mu = x^{-2}$. Hence, (13) and $x \equiv 0$ are solutions to equation (12). ◆

There are many differential equations that are not covered by Theorem 3 but for which an integrating factor nevertheless exists. The major difficulty, however, is in finding an explicit formula for these integrating factors, which in general will depend on both x and y.

2.5 EXERCISES

In Problems 1–6, identify the equation as separable, linear, exact, or having an integrating factor that is a function of either x alone or y alone.

1. $(2x + yx^{-1}) dx + (xy - 1) dy = 0$
2. $(2y^3 + 2y^2) dx + (3y^2 x + 2xy) dy = 0$
3. $(2x + y) dx + (x - 2y) dy = 0$
4. $(y^2 + 2xy) dx - x^2 dy = 0$
5. $(x^2 \sin x + 4y) dx + x dy = 0$
6. $(2y^2 x - y) dx + x dy = 0$

In Problems 7–12, solve the equation.

7. $(2xy) dx + (y^2 - 3x^2) dy = 0$
8. $(3x^2 + y) dx + (x^2 y - x) dy = 0$
9. $(x^4 - x + y) dx - x dy = 0$
10. $(2y^2 + 2y + 4x^2) dx + (2xy + x) dy = 0$
11. $(y^2 + 2xy) dx - x^2 dy = 0$
12. $(2xy^3 + 1) dx + (3x^2 y^2 - y^{-1}) dy = 0$

In Problems 13 and 14, find an integrating factor of the form $x^n y^m$ and solve the equation.

13. $(2y^2 - 6xy) dx + (3xy - 4x^2) dy = 0$
14. $(12 + 5xy) dx + (6xy^{-1} + 3x^2) dy = 0$

15. **(a)** Show that if $(\partial N/\partial x - \partial M/\partial y)/(xM - yN)$ depends only on the product xy, that is,

$$\frac{\partial N/\partial x - \partial M/\partial y}{xM - yN} = H(xy),$$

then the equation $M(x, y) dx + N(x, y) dy = 0$ has an integrating factor of the form $\mu(xy)$. Give the general formula for $\mu(xy)$.

(b) Use your answer to part (a) to find an implicit solution to

$$(3y + 2xy^2) dx + (x + 2x^2 y) dy = 0,$$

satisfying the initial condition $y(1) = 1$.

16. **(a)** Prove that $Mdx + Ndy = 0$ has an integrating factor that depends only on the sum $x + y$ if and only if the expression

$$\frac{\partial N/\partial x - \partial M/\partial y}{M - N}$$

depends only on $x + y$.

(b) Use part (a) to solve the equation
$$(3 + y + xy) dx + (3 + x + xy) dy = 0.$$

17. **(a)** Find a condition on M and N that is necessary and sufficient for $M\,dx + N\,dy = 0$ to have an integrating factor that depends only on the product x^2y.

(b) Use part (a) to solve the equation

$$(2x + 2y + 2x^3y + 4x^2y^2)\,dx$$
$$+ (2x + x^4 + 2x^3y)\,dy = 0 \ .$$

18. If $xM(x, y) + yN(x, y) \equiv 0$, find the solution to the equation $M(x, y)\,dx + N(x, y)\,dy = 0$.

19. Fluid Flow. The streamlines associated with a certain fluid flow are represented by the family of curves $y = x - 1 + ke^{-x}$. The velocity potentials of the flow are just the orthogonal trajectories of this family.

(a) Use the method described in Problem 32 of Exercises 2.4 to show that the velocity potentials satisfy $dx + (x - y)\,dy = 0$.
[*Hint:* First express the family $y = x - 1 + ke^{-x}$ in the form $F(x, y) = k$.]

(b) Find the velocity potentials by solving the equation obtained in part (a).

20. Verify that when the linear differential equation $[P(x)y - Q(x)]\,dx + dy = 0$ is multiplied by $\mu(x) = e^{\int P(x)\,dx}$, the result is exact.

2.6 Substitutions and Transformations

When the equation

$$M(x, y)\,dx + N(x, y)\,dy = 0$$

is not a separable, exact, or linear equation, it may still be possible to transform it into one that we know how to solve. This was in fact our approach in Section 2.5, where we used an integrating factor to transform our original equation into an exact equation.

In this section we study four types of equations that can be transformed into either a separable or linear equation by means of a suitable substitution or transformation.

Substitution Procedure

(a) Identify the type of equation and determine the appropriate substitution or transformation.

(b) Rewrite the original equation in terms of new variables.

(c) Solve the transformed equation.

(d) Express the solution in terms of the original variables.

Homogeneous Equations

Homogeneous Equation

Definition 4. If the right-hand side of the equation

(1) $\dfrac{dy}{dx} = f(x, y)$

can be expressed as a function of the ratio y/x alone, then we say the equation is **homogeneous.**

For example, the equation

(2) $(x - y)\,dx + x\,dy = 0$

can be written in the form

$$\frac{dy}{dx} = \frac{y - x}{x} = \frac{y}{x} - 1 \,.$$

Since we have expressed $(y - x)/x$ as a function of the ratio y/x [that is, $(y - x)/x = G(y/x)$, where $G(v) := v - 1$], then equation (2) is homogeneous.

The equation

(3) $(x - 2y + 1)\,dx + (x - y)\,dy = 0$

can be written in the form

$$\frac{dy}{dx} = \frac{x - 2y + 1}{y - x} = \frac{1 - 2(y/x) + (1/x)}{(y/x) - 1} \,.$$

Here the right-hand side cannot be expressed as a function of y/x alone because of the term $1/x$ in the numerator. Hence, equation (3) is not homogeneous.

One test for the homogeneity of equation (1) is to replace x by tx and y by ty. Then (1) is homogeneous if and only if

$$f(tx, ty) = f(x, y)$$

for all $t \neq 0$ [see Problem 43(a)].

To solve a homogeneous equation, we make a rather obvious substitution. Let

$$v = \frac{y}{x} \,.$$

Our homogeneous equation now has the form

(4) $\dfrac{dy}{dx} = G(v)\,,$

and all we need is to express dy/dx in terms of x and v. Since $v = y/x$, then $y = vx$. Keeping in mind that both v and y are functions of x, we use the product rule for differentiation to deduce from $y = vx$ that

$$\frac{dy}{dx} = v + x\frac{dv}{dx} \,.$$

We then substitute the above expression for dy/dx into equation (4) to obtain

(5) $v + x\dfrac{dv}{dx} = G(v)\,.$

The new equation (5) is separable, and we can obtain its implicit solution from

$$\int \frac{1}{G(v) - v}\,dv = \int \frac{1}{x}\,dx \,.$$

All that remains to do is to express the solution in terms of the original variables x and y.

Example 1 Solve

$$(6) \qquad (xy + y^2 + x^2)\, dx - x^2\, dy = 0 \,.$$

Solution A check will show that equation (6) is not separable, exact, or linear. If we express (6) in the derivative form

$$(7) \qquad \frac{dy}{dx} = \frac{xy + y^2 + x^2}{x^2} = \frac{y}{x} + \left(\frac{y}{x}\right)^2 + 1\,,$$

then we see that the right-hand side of (7) is a function of just y/x. Thus, equation (6) is homogeneous.

Now let $v = y/x$ and recall that $dy/dx = v + x(dv/dx)$. With these substitutions, equation (7) becomes

$$v + x\frac{dv}{dx} = v + v^2 + 1 \,.$$

The above equation is separable, and, on separating the variables and integrating, we obtain

$$\int \frac{1}{v^2 + 1}\, dv = \int \frac{1}{x}\, dx \,,$$

$$\arctan v = \ln|x| + C \,.$$

Hence,

$$v = \tan(\ln|x| + C) \,.$$

Finally, we substitute y/x for v and solve for y to get

$$y = x \tan(\ln|x| + C)$$

as an explicit solution to equation (6). Also note that $x \equiv 0$ is a solution. ◆

Equations of the Form $\dfrac{dy}{dx} = G(ax + by)$

When the right-hand side of the equation $dy/dx = f(x, y)$ can be expressed as a function of the combination $ax + by$, where a and b are constants, that is,

$$\frac{dy}{dx} = G(ax + by) \,,$$

then the substitution

$$z = ax + by$$

transforms the equation into a separable one. The method is illustrated in the next example.

Example 2 Solve

$$(8) \qquad \frac{dy}{dx} = y - x - 1 + (x - y + 2)^{-1} \,.$$

Solution The right-hand side can be expressed as a function of $x - y$, that is,

$$y - x - 1 + (x - y + 2)^{-1} = -(x - y) - 1 + [(x - y) + 2]^{-1} \,,$$

so let $z = x - y$. To solve for dy/dx, we differentiate $z = x - y$ with respect to x to obtain $dz/dx = 1 - dy/dx$, and so $dy/dx = 1 - dz/dx$. Substituting into (8) yields

$$1 - \frac{dz}{dx} = -z - 1 + (z + 2)^{-1},$$

or

$$\frac{dz}{dx} = (z + 2) - (z + 2)^{-1}.$$

Solving this separable equation, we obtain

$$\int \frac{z + 2}{(z + 2)^2 - 1} dz = \int dx,$$

$$\frac{1}{2} \ln \left| (z + 2)^2 - 1 \right| = x + C_1,$$

from which it follows that

$$(z + 2)^2 = Ce^{2x} + 1.$$

Finally, replacing z by $x - y$ yields

$$(x - y + 2)^2 = Ce^{2x} + 1$$

as an implicit solution to equation (8). ◆

Bernoulli Equations

Bernoulli Equation

Definition 5. A first-order equation that can be written in the form

$$(9) \qquad \frac{dy}{dx} + P(x)y = Q(x)y^n,$$

where $P(x)$ and $Q(x)$ are continuous on an interval (a, b) and n is a real number, is called a **Bernoulli equation.**[†]

Notice that when $n = 0$ or 1, equation (9) is also a linear equation and can be solved by the method discussed in Section 2.3. For other values of n, the substitution

$$v = y^{1-n}$$

transforms the Bernoulli equation into a linear equation, as we now show.

[†]*Historical Footnote:* This equation was proposed for solution by James Bernoulli in 1695. It was solved by his brother John Bernoulli. (James and John were two of eight mathematicians in the Bernoulli family.) In 1696, Gottfried Leibniz showed that the Bernoulli equation can be reduced to a linear equation by making the substitution $v = y^{1-n}$.

Dividing equation (9) by y^n yields

$$\textbf{(10)} \qquad y^{-n}\frac{dy}{dx} + P(x)y^{1-n} = Q(x) .$$

Taking $v = y^{1-n}$, we find via the chain rule that

$$\frac{dv}{dx} = (1-n)y^{-n}\frac{dy}{dx} ,$$

and so equation (10) becomes

$$\frac{1}{1-n}\frac{dv}{dx} + P(x)v = Q(x) .$$

Because $1/(1-n)$ is just a constant, the last equation is indeed linear.

Example 3 Solve

$$\textbf{(11)} \qquad \frac{dy}{dx} - 5y = -\frac{5}{2}xy^3 .$$

Solution This is a Bernoulli equation with $n = 3$, $P(x) = -5$, and $Q(x) = -5x/2$. To transform (11) into a linear equation, we first divide by y^3 to obtain

$$y^{-3}\frac{dy}{dx} - 5y^{-2} = -\frac{5}{2}x .$$

Next we make the substitution $v = y^{-2}$. Since $dv/dx = -2y^{-3}\,dy/dx$, the transformed equation is

$$-\frac{1}{2}\frac{dv}{dx} - 5v = -\frac{5}{2}x ,$$

$$\textbf{(12)} \qquad \frac{dv}{dx} + 10v = 5x .$$

Equation (12) is linear, so we can solve it for v using the method discussed in Section 2.3. When we do this, it turns out that

$$v = \frac{x}{2} - \frac{1}{20} + Ce^{-10x} .$$

Substituting $v = y^{-2}$ gives the solution

$$y^{-2} = \frac{x}{2} - \frac{1}{20} + Ce^{-10x} .$$

Not included in the last equation is the solution $y \equiv 0$ that was lost in the process of dividing (11) by y^3. ◆

A general formula for the solution to the Bernoulli equation (9) is given in Problem 48.

Equations with Linear Coefficients

We have used various substitutions for y to transform the original equation into a new equation that we could solve. In some cases we must transform *both* x and y into new variables, say, u and v. This is the situation for **equations with linear coefficients**—that is, equations of the form

$$\textbf{(13)} \qquad (a_1x + b_1y + c_1)\,dx + (a_2x + b_2y + c_2)\,dy = 0 ,$$

where the a_i's, b_i's, and c_i's are constants. We leave it as an exercise to show that when $a_1 b_2 = a_2 b_1$, equation (13) can be put in the form $dy/dx = G(ax + by)$, which we solved via the substitution $z = ax + by$. Moreover, if $b_1 = a_2$, then (13) is exact and can be solved using the method of Section 2.4.

Before considering the general case when $b_1 \neq a_2$ and $a_1 b_2 \neq a_2 b_1$, let's first look at the special situation when $c_1 = c_2 = 0$. Equation (13) then becomes

$$(a_1 x + b_1 y)\, dx + (a_2 x + b_2 y)\, dy = 0\,,$$

which can be rewritten in the form

$$\frac{dy}{dx} = -\frac{a_1 x + b_1 y}{a_2 x + b_2 y} = -\frac{a_1 + b_1(y/x)}{a_2 + b_2(y/x)}\,.$$

This equation is homogeneous, so we can solve it using the method discussed earlier in this section.

The above discussion suggests the following procedure for solving (13). If $b_1 \neq a_2$ and $a_1 b_2 \neq a_2 b_1$, then we seek a translation of axes of the form

$$x = u + h \qquad \text{and} \qquad y = v + k\,,$$

where h and k are constants, that will change $a_1 x + b_1 y + c_1$ into $a_1 u + b_1 v$ and change $a_2 x + b_2 y + c_2$ into $a_2 u + b_2 v$. Some elementary algebra shows that such a transformation exists if the system of equations

(14)
$$\begin{aligned}
a_1 h + b_1 k + c_1 &= 0\,, \\
a_2 h + b_2 k + c_2 &= 0
\end{aligned}$$

has a solution. This is ensured by the assumption $a_1 b_2 \neq a_2 b_1$, which is geometrically equivalent to assuming that the two lines described by the system (14) intersect. Now if (h, k) satisfies (14), then the substitutions $x = u + h$ and $y = v + k$ transform equation (13) into the homogeneous equation

(15)
$$\frac{dv}{du} = -\frac{a_1 u + b_1 v}{a_2 u + b_2 v} = -\frac{a_1 + b_1(v/u)}{a_2 + b_2(v/u)}\,,$$

which we know how to solve.

Example 4 Solve

(16)
$$(2x - 2y - 6)\, dx + (x - 3y - 5)\, dy = 0\,.$$

Solution Since $b_1 = -2 \neq 1 = a_2$ and $a_1 b_2 = (2)(-3) \neq (1)(-2) = a_2 b_1$, we will use the translation of axes $x = u + h$, $y = v + k$, where h and k satisfy the system

$$\begin{aligned}
2h - 2k &= 6 \\
h - 3k &= 5\,.
\end{aligned}$$

Solving this system gives $h = 2$, $k = -1$. Next we substitute $x = u + 2$, $y = v - 1$ into (16), observing that $dx = du$ and $dy = dv$, and we get

$$(2u - 2v)\, du + (u - 3v)\, dv = 0$$

$$\frac{dv}{du} = \frac{2(v/u) - 2}{1 - 3(v/u)}\,.$$

The last equation is homogeneous, so we let $z = v/u$. Then $dv/du = z + u(dz/du)$, and substituting for v/u yields

$$z + u\frac{dz}{du} = \frac{2z - 2}{1 - 3z}.$$

Separating variables gives

$$\int \frac{1 - 3z}{3z^2 + z - 2}\,dz = \int \frac{1}{u}\,du\,,$$

from which, after utilizing a partial fraction expansion of the first integrand, we find

$$-\frac{4}{5}\ln|z + 1| - \frac{1}{5}\ln|3z - 2| = \ln|u| + C_1.$$

It follows after exponentiating that

$$|z + 1|^4|3z - 2| = C|u|^{-5},$$

and when we substitute back in for z, u and v, we obtain

$$\left|\frac{v}{u} + 1\right|^4\left|3\frac{v}{u} - 2\right| = C|u|^{-5},$$

$$|u + v|^4|2u - 3v| = C,$$

$$(x + y - 1)^4(2x - 3y - 7) = C.$$

The last equation gives an implicit solution to (16), with C any real constant. ◆

2.6 EXERCISES

In Problems 1–8, identify (do not solve) the equation as homogeneous, Bernoulli, linear coefficients, or of the form $y' = G(ax + by)$.

1. $2tx\,dx + (t^2 - x^2)\,dt = 0$
2. $(y - 4x - 1)^2\,dx - dy = 0$
3. $dy/dx + y/x = x^3y^2$
4. $(t + x + 2)\,dx + (3t - x - 6)\,dt = 0$
5. $\theta\,dy - y\,d\theta = \sqrt{\theta y}\,d\theta$
6. $(ye^{-2x} + y^3)\,dx - e^{-2x}\,dy = 0$
7. $\cos(x + y)\,dy = \sin(x + y)\,dx$
8. $(y^3 - \theta y^2)\,d\theta + 2\theta^2y\,dy = 0$

Use the method discussed under "Homogeneous Equations" to solve Problems 9–16.

9. $(xy + y^2)\,dx - x^2\,dy = 0$
10. $(3x^2 - y^2)\,dx + (xy - x^3y^{-1})\,dy = 0$
11. $(y^2 - xy)\,dx + x^2\,dy = 0$
12. $(x^2 + y^2)\,dx + 2xy\,dy = 0$
13. $\dfrac{dx}{dt} = \dfrac{x^2 + t\sqrt{t^2 + x^2}}{tx}$
14. $\dfrac{dy}{d\theta} = \dfrac{\theta\sec(y/\theta) + y}{\theta}$

15. $\dfrac{dy}{dx} = \dfrac{x^2 - y^2}{3xy}$
16. $\dfrac{dy}{dx} = \dfrac{y(\ln y - \ln x + 1)}{x}$

Use the method discussed under "Equations of the Form $dy/dx = G(ax + by)$" to solve Problems 17–20.

17. $dy/dx = \sqrt{x + y} - 1$
18. $dy/dx = (x + y + 2)^2$
19. $dy/dx = (x - y + 5)^2$
20. $dy/dx = \sin(x - y)$

Use the method discussed under "Bernoulli Equations" to solve Problems 21–28.

21. $\dfrac{dy}{dx} + \dfrac{y}{x} = x^2y^2$
22. $\dfrac{dy}{dx} - y = e^{2x}y^3$
23. $\dfrac{dy}{dx} = \dfrac{2y}{x} - x^2y^2$
24. $\dfrac{dy}{dx} + \dfrac{y}{x - 2} = 5(x - 2)y^{1/2}$
25. $\dfrac{dx}{dt} + tx^3 + \dfrac{x}{t} = 0$
26. $\dfrac{dy}{dx} + y = e^xy^{-2}$

27. $\dfrac{dr}{d\theta} = \dfrac{r^2 + 2r\theta}{\theta^2}$

28. $\dfrac{dy}{dx} + y^3x + y = 0$

Use the method discussed under "Equations with Linear Coefficients" to solve Problems 29–32.

29. $(x + y - 1)\,dx + (y - x - 5)\,dy = 0$

30. $(-4x - y - 1)\,dx + (x + y + 3)\,dy = 0$

31. $(2x - y)\,dx + (4x + y - 3)\,dy = 0$

32. $(2x - y + 4)\,dx + (x - 2y - 2)\,dy = 0$

In Problems 33–40, solve the equation given in:

33. Problem 1. **34.** Problem 2.

35. Problem 3. **36.** Problem 4.

37. Problem 5. **38.** Problem 6.

39. Problem 7. **40.** Problem 8.

41. Use the substitution $v = x - y + 2$ to solve equation (8).

42. Use the substitution $y = vx^2$ to solve

$$\frac{dy}{dx} = \frac{2y}{x} + \cos(y/x^2)\,.$$

43. (a) Show that the equation $dy/dx = f(x, y)$ is homogeneous if and only if $f(tx, ty) = f(x, y)$. [*Hint:* Let $t = 1/x$.]

 (b) A function $H(x, y)$ is called **homogeneous of order n** if $H(tx, ty) = t^n H(x, y)$. Show that the equation

$$M(x, y)\,dx + N(x, y)\,dy = 0$$

 is homogeneous if $M(x, y)$ and $N(x, y)$ are both homogeneous of the same order.

44. Show that equation (13) reduces to an equation of the form

$$\frac{dy}{dx} = G(ax + by)\,,$$

 when $a_1b_2 = a_2b_1$. [*Hint:* If $a_1b_2 = a_2b_1$, then $a_2/a_1 = b_2/b_1 = k$, so that $a_2 = ka_1$ and $b_2 = kb_1$.]

45. Coupled Equations. In analyzing coupled equations of the form

$$\frac{dy}{dt} = ax + by\,,$$

$$\frac{dx}{dt} = \alpha x + \beta y\,,$$

 where a, b, α, and β are constants, we may wish to determine the relationship between x and y rather than the

individual solutions $x(t), y(t)$. For this purpose, divide the first equation by the second to obtain

(17) $\quad \dfrac{dy}{dx} = \dfrac{ax + by}{\alpha x + \beta y}\,.$

This new equation is homogeneous, so we can solve it via the substitution $v = y/x$. We refer to the solutions of (17) as **integral curves**. Determine the integral curves for the system

$$\frac{dy}{dt} = -4x - y\,,$$

$$\frac{dx}{dt} = 2x - y\,.$$

46. Magnetic Field Lines. As described in Problem 20 of Exercises 1.3, the magnetic field lines of a dipole satisfy

$$\frac{dy}{dx} = \frac{3xy}{2x^2 - y^2}\,.$$

Solve this equation and sketch several of these lines.

47. Riccati Equation. An equation of the form

(18) $\quad \dfrac{dy}{dx} = P(x)y^2 + Q(x)y + R(x)$

is called a generalized Riccati equation.[†]

 (a) If one solution—say, $u(x)$—of (18) is known, show that the substitution $y = u + 1/v$ reduces (18) to a linear equation in v.

 (b) Given that $u(x) = x$ is a solution to

$$\frac{dy}{dx} = x^3(y - x)^2 + \frac{y}{x}\,,$$

 use the result of part (a) to find all the other solutions to this equation. (The particular solution $u(x) = x$ can be found by inspection or by using a Taylor series method; see Section 8.1.)

48. Derive the following general formula for the solution to the Bernoulli equation (9):

$$y = \begin{cases} \left[\dfrac{(1-n)\int e^{(1-n)\int P(x)dx}\,Q(x)dx + C_1}{e^{(1-n)\int P(x)dx}}\right]^{1/(1-n)} & \text{for } n \neq 1 \\[4mm] C_2\, e^{\int [Q(x) - P(x)]dx} & \text{for } n = 1. \end{cases}$$

[†]*Historical Footnote:* Count Jacopo Riccati studied a particular case of this equation in 1724 during his investigation of curves whose radii of curvature depend only on the variable y and not the variable x.

Chapter 2 Summary

In this chapter we have discussed various types of first-order differential equations. The most important were the separable, linear, and exact equations. Their principal features and method of solution are outlined below.

Separable Equations: $dy/dx = g(x)\,p(y)$. Separate the variables and integrate.

Linear Equations: $dy/dx + P(x)y = Q(x)$. The integrating factor $\mu = \exp[\,\int P(x)\,dx\,]$ reduces the equation to $d(\mu y)/dx = \mu Q$, so that $\mu y = \int \mu Q\,dx + C$.

Exact Equations: $dF(x, y) = 0$. Solutions are given implicitly by $F(x, y) = C$. If $\partial M/\partial y = \partial N/\partial x$, then $M\,dx + N\,dy = 0$ is exact and F is given by

$$F = \int M\,dx + g(y)\,, \quad \text{where} \quad g'(y) = N - \frac{\partial}{\partial y}\int M\,dx$$

or

$$F = \int N\,dy + h(x)\,, \quad \text{where} \quad h'(x) = M - \frac{\partial}{\partial x}\int N\,dy\,.$$

When an equation is not separable, linear, or exact, it may be possible to find an integrating factor or perform a substitution that will enable us to solve the equation.

Special Integrating Factors: $\mu M\,dx + \mu N\,dy = 0$ is exact. If $(\partial M/\partial y - \partial N/\partial x)/N$ depends only on x, then

$$\mu(x) = \exp\left[\int \left(\frac{\partial M/\partial y - \partial N/\partial x}{N}\right)dx\right]$$

is an integrating factor. If $(\partial N/\partial x - \partial M/\partial y)/M$ depends only on y, then

$$\mu(y) = \exp\left[\int \left(\frac{\partial N/\partial x - \partial M/\partial y}{M}\right)dy\right]$$

is an integrating factor.

Homogeneous Equations: $dy/dx = G(y/x)$. Let $v = y/x$. Then $dy/dx = v + x(dv/dx)$, and the transformed equation in the variables v and x is separable.

Equations of the Form: $dy/dx = G(ax + by)$. Let $z = ax + by$. Then $dz/dx = a + b(dy/dx)$, and the transformed equation in the variables z and x is separable.

Bernoulli Equations: $dy/dx + P(x)y = Q(x)y^n$. For $n \neq 0$ or 1, let $v = y^{1-n}$. Then $dv/dx = (1 - n)y^{-n}(dy/dx)$, and the transformed equation in the variables v and x is linear.

Linear Coefficients: $(a_1x + b_1y + c_1)dx + (a_2x + b_2y + c_2)dy = 0$. For $a_1b_2 \neq a_2b_1$, let $x = u + h$ and $y = v + k$, where h and k satisfy

$$a_1h + b_1k + c_1 = 0\,,$$
$$a_2h + b_2k + c_2 = 0\,.$$

Then the transformed equation in the variables u and v is homogeneous.

REVIEW PROBLEMS FOR CHAPTER 2

In Problems 1–30, solve the equation.

1. $\dfrac{dy}{dx} = \dfrac{e^{x+y}}{y-1}$

2. $\dfrac{dy}{dx} - 4y = 32x^2$

3. $(x^2 - 2y^{-3})\,dy + (2xy - 3x^2)\,dx = 0$

4. $\dfrac{dy}{dx} + \dfrac{3y}{x} = x^2 - 4x + 3$

5. $[\sin(xy) + xy\cos(xy)]\,dx + [1 + x^2\cos(xy)]\,dy = 0$

6. $2xy^3\,dx - (1 - x^2)\,dy = 0$

7. $t^3 y^2\,dt + t^4 y^{-6}\,dy = 0$

8. $\dfrac{dy}{dx} + \dfrac{2y}{x} = 2x^2 y^2$

9. $(x^2 + y^2)\,dx + 3xy\,dy = 0$

10. $[1 + (1 + x^2 + 2xy + y^2)^{-1}]\,dx$
 $+ [y^{-1/2} + (1 + x^2 + 2xy + y^2)^{-1}]\,dy = 0$

11. $\dfrac{dx}{dt} = 1 + \cos^2(t - x)$

12. $(y^3 + 4e^x y)\,dx + (2e^x + 3y^2)\,dy = 0$

13. $\dfrac{dy}{dx} - \dfrac{y}{x} = x^2 \sin 2x$

14. $\dfrac{dx}{dt} - \dfrac{x}{t-1} = t^2 + 2$

15. $\dfrac{dy}{dx} = 2 - \sqrt{2x - y + 3}$

16. $\dfrac{dy}{dx} + y\tan x + \sin x = 0$

17. $\dfrac{dy}{d\theta} + 2y = y^2$

18. $\dfrac{dy}{dx} = (2x + y - 1)^2$

19. $(x^2 - 3y^2)\,dx + 2xy\,dy = 0$

20. $\dfrac{dy}{d\theta} + \dfrac{y}{\theta} = -4\theta y^{-2}$

21. $(y - 2x - 1)\,dx + (x + y - 4)\,dy = 0$

22. $(2x - 2y - 8)\,dx + (x - 3y - 6)\,dy = 0$

23. $(y - x)\,dx + (x + y)\,dy = 0$

24. $\left(\sqrt{y/x} + \cos x\right)dx + \left(\sqrt{x/y} + \sin y\right)dy = 0$

25. $y(x - y - 2)\,dx + x(y - x + 4)\,dy = 0$

26. $\dfrac{dy}{dx} + xy = 0$

27. $(3x - y - 5)\,dx + (x - y + 1)\,dy = 0$

28. $\dfrac{dy}{dx} = \dfrac{x - y - 1}{x + y + 5}$

29. $(4xy^3 - 9y^2 + 4xy^2)\,dx + (3x^2 y^2 - 6xy + 2x^2 y)\,dy = 0$

30. $\dfrac{dy}{dx} = (x + y + 1)^2 - (x + y - 1)^2$

In Problems 31–40, solve the initial value problem.

31. $(x^3 - y)\,dx + x\,dy = 0$, $\quad y(1) = 3$

32. $\dfrac{dy}{dx} = \left(\dfrac{x}{y} + \dfrac{y}{x}\right)$, $\quad y(1) = -4$

33. $(t + x + 3)\,dt + dx = 0$, $\quad x(0) = 1$

34. $\dfrac{dy}{dx} - \dfrac{2y}{x} = x^2 \cos x$, $\quad y(\pi) = 2$

35. $(2y^2 + 4x^2)\,dx - xy\,dy = 0$, $\quad y(1) = -2$

36. $[2\cos(2x + y) - x^2]\,dx$
 $+ [\cos(2x + y) + e^y]\,dy = 0$, $\quad y(1) = 0$

37. $(2x - y)\,dx + (x + y - 3)\,dy = 0$, $\quad y(0) = 2$

38. $\sqrt{y}\,dx + (x^2 + 4)\,dy = 0$, $\quad y(0) = 4$

39. $\dfrac{dy}{dx} - \dfrac{2y}{x} = x^{-1} y^{-1}$, $\quad y(1) = 3$

40. $\dfrac{dy}{dx} - 4y = 2xy^2$, $\quad y(0) = -4$

41. Express the solution to the following initial value problem using a definite integral:

$$\frac{dy}{dt} = \frac{1}{1 + t^2} - y, \quad y(2) = 3.$$

 Then use your expression and numerical integration to estimate $y(3)$ to four decimal places.

TECHNICAL WRITING EXERCISES FOR CHAPTER 2

1. An instructor at Ivey U. asserted: "All you need to know about first-order differential equations is how to solve those that are exact." Give arguments that support and arguments that refute the instructor's claim.

2. What properties do solutions to linear equations have that are not shared by solutions to either separable or exact equations? Give some specific examples to support your conclusions.

3. Consider the differential equation

$$\frac{dy}{dx} = ay + be^{-x}, \quad y(0) = c,$$

where a, b, and c are constants. Describe what happens to the asymptotic behavior as $x \to +\infty$ of the solution when the constants a, b, and c are varied. Illustrate with figures and/or graphs.

Projects for Chapter 2

A Oil Spill in a Canal

In 1973 an oil barge collided with a bridge in the Mississippi River, leaking oil into the water at a rate estimated at 50 gallons per minute. In 1989 the Exxon Valdez spilled an estimated 11,000,000 gallons of oil into Prudhoe Bay in 6 hours[†], and in 2010 the Deepwater Horizon well leaked into the Gulf of Mexico at a rate estimated to be 15,000 barrels per day[‡] (1 barrel = 42 gallons). In this project you are going to use differential equations to analyze a simplified model of the dissipation of heavy crude oil spilled at a rate of S ft^3/sec into a flowing body of water. The flow region is a canal, namely a straight channel of rectangular cross section, w feet wide by d feet deep, having a constant flow rate of v ft/sec; the oil is presumed to float in a thin layer of thickness s (feet) on top of the water, without mixing.

In Figure 2.12, the oil that passes through the cross-section window in a short time Δt occupies a box of dimensions s by w by $v\Delta t$. To make the analysis easier, presume that the canal is conceptually partitioned into cells of length L ft. each, and that *within each particular cell* the oil instantaneously disperses and forms a *uniform* layer of thickness $s_i(t)$ in cell i (cell 1 starts at the point of the spill). So, at time t, the ith cell contains $s_i(t)wL$ ft^3 of oil. Oil flows out of cell i at a rate equal to $s_i(t)wv$ ft^3/sec, and it flows into cell i at the rate $s_{i-1}(t)wv$; it flows into the first cell at S ft^3/sec.

Figure 2.12 Oil leak in a canal.

[†]Cutler J. Cleveland (Lead Author); C Michael Hogan and Peter Saundry (Topic Editor). 2010. "Deepwater Horizon oil spill." In: *Encyclopedia of Earth*, ed. Cutler J. Cleveland (Washington, D.C.: Environmental Information Coalition, National Council for Science and the Environment).

[‡]Cutler J. Cleveland (Contributing Author); National Oceanic and Atmospheric Administration (Content source); Peter Saundry (Topic Editor). 2010. "Exxon Valdez oil spill." In: *Encyclopedia of Earth*, ed. Cutler J. Cleveland (Washington, D.C.: Environmental Information Coalition, National Council for Science and the Environment).

(a) Formulate a system of differential equations and initial conditions for the oil thickness in the first three cells. Take $S = 50$ gallons/min, which was roughly the spillage rate for the Mississippi River incident, and take $w = 200$ ft, $d = 25$ ft, and $v = 1$ mi/h (which are reasonable estimates for the Mississippi River[†]). Take $L = 1000$ ft.

(b) Solve for $s_1(t)$. [*Caution*: Make sure your units are consistent.]

(c) If the spillage lasts for T seconds, what is the maximum oil layer thickness in cell 1?

(d) Solve for $s_2(t)$. What is the maximum oil layer thickness in cell 2?

(e) Probably the least tenable simplification in this analysis lies in regarding the layer thickness as uniform over distances of length L. Reevaluate your answer to part (c) with L reduced to 500 ft. By what fraction does the answer change?

B Differential Equations in Clinical Medicine

Courtesy of Philip Crooke, Vanderbilt University

In medicine, **mechanical ventilation** is a procedure that assists or replaces spontaneous breathing for critically ill patients, using a medical device called a *ventilator*. Some people attribute the first mechanical ventilation to Andreas Vesalius in 1555. Negative pressure ventilators (iron lungs) came into use in the 1940s–1950s in response to *poliomyelitis* (polio) epidemics. Philip Drinker and Louis Shaw are credited with its invention. Modern ventilators use positive pressure to inflate the lungs of the patient. In the ICU (intensive care unit), common indications for the initiation of mechanical ventilation are acute respiratory failure, acute exacerbation of chronic obstructive pulmonary disease, coma, and neuromuscular disorders. The goals of mechanical ventilation are to provide oxygen to the lungs and to remove carbon dioxide.

In this project, we model the mechanical process performed by the ventilator. We make the following assumptions about this process of filling the lungs with air and then letting them deflate to some rest volume (see Figure 2.13).

(i) The length (in seconds) of each breath is fixed (t_{tot}) and is set by the clinician, with each breath being identical to the previous breath.

(ii) Each breath is divided into two parts: *inspiration* (air flowing into the patient) and *expiration* (air flowing out of the patient). We assume that inspiration takes place over the interval $[0, t_i]$ and expiration over the time interval $[t_i, t_{tot}]$. The time t_i is called the *inspiratory time*.

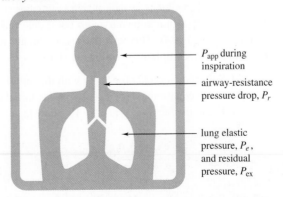

P_{app} during inspiration

airway-resistance pressure drop, P_r

lung elastic pressure, P_e, and residual pressure, P_{ex}

Figure 2.13 Lung ventilation pressures

[†]http://www.nps.gov/miss/riverfacts.htm

(iii) During inspiration the ventilator applies a constant pressure P_{app} to the patient's airway, and during expiration this pressure is zero, relative to atmospheric pressure. This is called *pressure-controlled ventilation.*

(iv) We assume that the pulmonary system (lung) is modeled by a single compartment. Hence, the action of the ventilator is similar to inflating a balloon and then releasing the pressure.

(v) At the airway there is a pressure balance:

(1) $$P_r + P_e + P_{ex} = P_{aw} \, ,$$

where P_r denotes pressure losses due to resistance to flow into and out of the lung, P_e is the elastic pressure due to changes in volume of the lung, P_{ex} is a residual pressure that remains in the lung at the completion of a breath, and P_{aw} denotes the pressure applied to the airway. ($P_{aw} = P_{app}$ during inspiration and $P_{aw} = 0$ during expiration.) The residual pressure P_{ex} is called the *end-expiratory pressure.*

(vi) Let $V(t)$ denote the volume of the lung at time t, with $V_i(t)$, $0 \le t \le t_i$, denoting its volume during inspiration and $V_e(t)$, $t_i \le t \le t_{tot}$, its volume during expiration. We assume that $V_i(0) = V_e(t_{tot}) = 0$. The number $V_i(t_i) = V_T$ is called the *tidal volume* of the breath.

(vii) We assume that the resistive pressure P_r is proportional to the flows into and out of the lung such that $P_r = R(dV/dt)$, and we assume that the proportionality constant R is the same for inspiration and expiration.

(viii) Furthermore, we assume that the elastic pressure is proportional to the instantaneous volume of the lung. That is, $P_e = (1/C)V$, where the constant C is called the *compliance* of the lung.

Using the pressure equation in (1) together with the above assumptions, a mathematical model for the instantaneous volume in the single compartment lung is given by the following pair of first-order linear differential equations:

(2) $$R\left(\frac{dV_i}{dt}\right) + \left(\frac{1}{C}\right)V_i + P_{ex} = P_{app} \, , \qquad 0 \le t \le t_i \, ,$$

(3) $$R\left(\frac{dV_e}{dt}\right) + \left(\frac{1}{C}\right)V_e + P_{ex} = 0 \, , \qquad t_i \le t \le t_{tot} \, .$$

The initial conditions, as indicated in assumption *(vi)*, are $V_i(0) = 0$ and $V_e(t_i) = V_i(t_i) = V_T$. The constant P_{ex} is not known *a priori* but is determined from the end condition on the expiratory volume: $V_e(t_{tot}) = 0$. This will make each breath identical to the previous breath. To obtain a formula for P_{ex}, complete the following steps.

(a) Solve equation (2) for $V_i(t)$ with the initial condition $V_i(0) = 0$.
(b) Solve equation (3) for $V_e(t)$ with the initial condition $V_e(t_i) = V_T$.
(c) Using the fact that $V_i(t_i) = V_T$, show that

$$P_{ex} = \frac{(e^{t_i/RC} - 1)\, P_{app}}{e^{t_{tot}/RC} - 1} \, .$$

(d) For $R = 10$ cm $(H_2O)/L/\text{sec}$, $C = 0.02$ L/cm (H_2O), $P_{app} = 20$ cm (H_2O), $t_i = 1$ sec and $t_{tot} = 3$ sec, plot the graphs of $V_i(t)$ and $V_e(t)$ over the interval $[0, t_{tot}]$. Compute P_{ex} for these parameters.

(e) The *mean alveolar pressure* is the average pressure in the lung during inspiration and is given by the formula

$$P_m = \frac{1}{t_i} \int_0^{t_i} \left(\frac{V_i(t)}{C} \right) dt + P_{\text{ex}} \, .$$

Compute this quantity using your expression for $V_i(t)$ in part (a).

C Torricelli's Law of Fluid Flow

Courtesy of Randall K. Campbell-Wright

How long does it take for water to drain through a hole in the bottom of a tank? Consider the tank pictured in Figure 2.14, which drains through a small, round hole. **Torricelli's law**[†] states that when the surface of the water is at a height h, the water drains with the velocity it would have if it fell freely from a height h (ignoring various forms of friction).

(a) Show that the standard gravity differential equation

$$\frac{d^2h}{dt^2} = -g$$

leads to the conclusion that an object that falls from a height $h(0)$ will land with a velocity of $-\sqrt{2gh(0)}$.

(b) Let $A(h)$ be the cross-sectional area of the water in the tank at height h and a the area of the drain hole. The rate at which water is flowing out of the tank at time t can be expressed as the cross-sectional area at height h times the rate at which the height of the water is changing. Alternatively, the rate at which water flows out of the hole can be expressed as the area of the hole times the velocity of the draining water. Set these two equal to each other and insert Torricelli's law to derive the differential equation

(4) $$A(h) \frac{dh}{dt} = -a\sqrt{2gh} \, .$$

(c) The conical tank of Figure 2.14 has a radius of 30 cm when it is filled to an initial depth of 50 cm. A small round hole at the bottom has a diameter of 1 cm. Determine $A(h)$ and a and then solve the differential equation in (4), thus deriving a formula relating time and the height of the water in this tank.

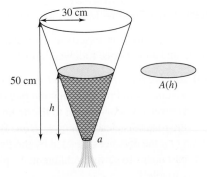

Figure 2.14 Conical tank

[†]*Historical Footnote:* Evangelista Torricelli (1608–1647) invented the barometer and worked on computing the value of the acceleration of gravity as well as observing this principle of fluid flow.

(d) Use your solution to (c) to predict how long it will take for the tank to drain entirely.

(e) Which would drain faster, the tank pictured or an upside-down conical tank of the same dimensions draining through a hole of the same size (1-cm diameter)? How long would it take to drain the upside-down tank?

(f) Find a water tank and time how long it takes to drain. (You may be able to borrow a "separatory funnel" from your chemistry department or use a large water cooler.) The tank should be large enough to take several minutes to drain, and the drain hole should be large enough to allow water to flow freely. The top of the tank should be open (so that the water will not "glug"). Repeat steps (c) and (d) for your tank and compare the prediction of Torricelli's law to your experimental results.

 The Snowplow Problem

To apply the techniques discussed in this chapter to real-world problems, it is necessary to translate these problems into questions that can be answered mathematically. The process of reformulating a real-world problem as a mathematical one often requires making certain simplifying assumptions. To illustrate this, consider the following snowplow problem:

One morning it began to snow very hard and continued snowing steadily throughout the day. A snowplow set out at 9:00 A.M. to clear a road, clearing 2 mi by 11:00 A.M. and an additional mile by 1:00 P.M. *At what time did it start snowing?*

To solve this problem, you can make two physical assumptions concerning the rate at which it is snowing and the rate at which the snowplow can clear the road. Because it is snowing steadily, it is reasonable to assume it is snowing at a constant rate. From the data given (and from our experience), the deeper the snow, the slower the snowplow moves. With this in mind, assume that the rate (in mph) at which a snowplow can clear a road is inversely proportional to the depth of the snow.

 Two Snowplows

Courtesy of Alar Toomre, Massachusetts Institute of Technology

One day it began to snow exactly at noon at a heavy and *steady* rate. A snowplow left its garage at 1:00 P.M., and another one followed in its tracks at 2:00 P.M. (see Figure 2.15 on page 85).

(a) *At what time did the second snowplow crash into the first?* To answer this question, assume as in Project D that the rate (in mph) at which a snowplow can clear the road is inversely proportional to the depth of the snow (and hence to the time elapsed since the road was clear of snow). [*Hint:* Begin by writing differential equations for $x(t)$ and $y(t)$, the distances traveled by the first and second snowplows, respectively, at t hours past noon. To solve the differential equation involving y, let t rather than y be the dependent variable!]

(b) *Could the crash have been avoided by dispatching the second snowplow at 3:00 P.M. instead?*

Figure 2.15 Method of successive snowplows

F Clairaut Equations and Singular Solutions

An equation of the form

(5) $$y = x\frac{dy}{dx} + f(dy/dx) \,,$$

where the continuously differentiable function $f(t)$ is evaluated at $t = dy/dx$, is called a **Clairaut equation.**[†] Interest in these equations is due to the fact that (5) has a one-parameter family of solutions that consist of *straight lines*. Further, the **envelope** of this family—that is, the curve whose tangent lines are given by the family—is also a solution to (5) and is called the **singular solution.**
 To solve a Clairaut equation:

(a) Differentiate equation (5) with respect to x and simplify to show that

(6) $$[x + f'(dy/dx)]\frac{d^2y}{dx^2} = 0 \,, \qquad \text{where} \qquad f'(t) = \frac{d}{dt}f(t) \,.$$

(b) From (6), conclude that $dy/dx = c$ or $f'(dy/dx) = -x$. Assume that $dy/dx = c$ and substitute back into (5) to obtain the family of *straight-line solutions*

$$y = cx + f(c) \,.$$

(c) Show that another solution to (5) is given parametrically by
$$x = -f'(p) \,,$$
$$y = f(p) - pf'(p) \,,$$

where the parameter $p = dy/dx$. This solution is the *singular solution.*

(d) Use the above method to find the family of straight-line solutions and the singular solution to the equation

$$y = x\left(\frac{dy}{dx}\right) + 2\left(\frac{dy}{dx}\right)^2 .$$

Here $f(t) = 2t^2$. Sketch several of the straight-line solutions along with the singular solution on the same coordinate system. Observe that the straight-line solutions are all tangent to the singular solution.

(e) Repeat part (d) for the equation

$$x(dy/dx)^3 - y(dy/dx)^2 + 2 = 0 \,.$$

[†]*Historical Footnote:* These equations were studied by Alexis Clairaut in 1734.

G Multiple Solutions of a First-Order Initial Value Problem

Courtesy of Bruce W. Atkinson, Samford University

The initial value problem (IVP),

$$(7) \qquad \frac{dy}{dx} = 3y^{2/3}, \qquad y(2) = 0,$$

which was discussed in Example 9 and Problem 29 of Section 1.2, is an example of an IVP that has more than one solution. The goal of this project is to find *all* the solutions to (7) on $(-\infty, +\infty)$. It turns out that there are infinitely many! These solutions can be obtained by concatenating the three functions $(x-a)^3$ for $x < a$, the constant 0 for $a \le x \le b$, and $(x-b)^3$ for $x > b$, where $a \le 2 \le b$, as can be seen by completing the following steps:

(a) Show that if $y = f(x)$ is a solution to the differential equation $dy/dx = 3y^{2/3}$ that is *not zero* on an open interval I, then on this interval $f(x)$ must be of the form $f(x) = (x-c)^3$ for some constant c not in I.

(b) Prove that if $y = f(x)$ is a solution to the differential equation $dy/dx = 3y^{2/3}$ on $(-\infty, +\infty)$ and $f(a) = f(b) = 0$, where $a < b$, then $f(x) = 0$ for $a \le x \le b$. [*Hint:* Consider the sign of f'.]

(c) Now let $y = g(x)$ be a solution to the IVP (7) on $(-\infty, +\infty)$. Of course $g(2) = 0$. If g vanishes at some point $x > 2$, then let b be the largest of such points; otherwise, set $b = 2$. Similarly, if g vanishes at some point $x < 2$, then let a be the smallest (furthest to the left) of such points; otherwise, set $a = 2$. Here we allow $b = +\infty$ and $a = -\infty$. (Because g is a continuous function, it can be proved that there always exist such largest and smallest points.) Using the results of parts (a) and (b) prove that if both a and b are finite, then g has the following form:

$$g(x) = \begin{cases} (x-a)^3 & \text{if } x \le a \\ 0 & \text{if } a < x \le b \\ (x-b)^3 & \text{if } x > b \end{cases}$$

What is the form of g if $b = +\infty$? If $a = -\infty$? If both $b = +\infty$ and $a = -\infty$?

(d) Verify directly that the above concatenated function g is indeed a solution to the IVP (7) for all choices of a and b with $a \le 2 \le b$. Also sketch the graph of several of the solution function g in part (c) for various values of a and b, including infinite values.

We have analyzed here a first-order IVP that not only fails to have a unique solution but has a solution set consisting of a doubly infinite family of functions (with a and b as the two parameters).

H Utility Functions and Risk Aversion

Courtesy of James E. Foster, George Washington University

Would you rather have $5 with certainty or a gamble involving a 50% chance of receiving $1 *and* a 50% chance of receiving $11? The gamble has a higher expected value ($6); however, it also has a greater level of risk. Economists model the behavior of consumers or other agents facing

risky decisions with the help of a (von Neumann–Morgenstern) utility function u and the criterion of *expected utility*.

Rather than using expected values of the dollar payoffs, the payoffs are first transformed into utility levels and then weighted by probabilities to obtain expected utility. Following the suggestion of Daniel Bernoulli, we might set $u(x) = \ln x$ and then compare $\ln 5 \approx 0.8047$ to $[0.5 \ln 1 + 0.5 \ln 11] \approx 0.1969$, which would result in the sure thing being chosen in this case rather than the gamble. This utility function is strictly concave, which corresponds to the agent being risk averse, or wanting to avoid gambles (unless of course the extra risk is sufficiently compensated by a high enough increase in the mean or expected payoff).

Alternatively, the utility function might be $u(x) = x^2$, which is strictly convex and corresponds to the agent being *risk loving*. This agent would surely select the above gamble. The case of $u(x) = x$ occurs when the agent is *risk neutral* and would select according to the expected value of the payoff. It is normally assumed that $u'(x) > 0$ at all payoff levels, x; in other words, higher payoffs are desirable.

In addition to knowing if an agent is risk averse or risk loving, economists are often interested in knowing *how* risk averse (or risk loving) an agent is. Clearly this has something to do with the second derivative of the utility function. The measure of *relative risk aversion* of an agent with utility function $u(x)$ and payoff x is defined as $r(x) = -u''(x)x/u'(x)$. Normally, $r(x)$ is a function of the payoff level. However, economists often find it convenient to restrict consideration to utility functions for which $r(x)$ is constant, say, $r(x) = s$ for all x. It is easily shown that each of $u(x) = \ln x$, $u(x) = x^2$, and $u(x) = x$ exhibits constant relative risk aversion (with levels $s = 1$, $s = -1$, and $s = 0$, respectively). A question naturally arises: What is the set of all utility functions that have constant relative risk aversion?

(a) State the second-order differential equation defined by the above question.

(b) Convert this into a separable first-order differential equation for $u'(x)$, solve, and use the solution to determine the possible forms that $u'(x)$ can take.

(c) Integrate to obtain the set of all constant relative risk-aversion utility functions. This class is used extensively throughout economics.

(d) An alternative measure of risk aversion is $a(x) = u''(x)/u'(x)$, the measure of *absolute risk aversion*. Find the set of all utility functions exhibiting constant absolute risk aversion.

(e) Which functions $u(x)$ are both *constant absolute* and *constant relative* risk-aversion utility functions?

For further reading, see, for example, the economics text, *Microeconomic Theory*, by A. Mas-Colell, M. Whinston, and J. Green (Oxford University Press, Oxford, 1995).

Designing a Solar Collector

You want to design a solar collector that will concentrate the Sun's rays at a point. By symmetry this surface will have a shape that is a surface of revolution obtained by revolving a curve about an axis. Without loss of generality, you can assume that this axis is the x-axis and the rays parallel to this axis are focused at the origin (see Figure 2.16). To derive the equation for the curve, proceed as follows:

(a) The law of reflection says that the angles γ and δ are equal. Use this and results from geometry to show that $\beta = 2\alpha$.

(b) From calculus recall that $dy/dx = \tan \alpha$. Use this, the fact that $y/x = \tan \beta$, and the double angle formula to show that

$$\frac{y}{x} = \frac{2\,dy/dx}{1 - (dy/dx)^2}\,.$$

(c) Now show that the curve satisfies the differential equation

(8) $$\frac{dy}{dx} = \frac{-x + \sqrt{x^2 + y^2}}{y}\,.$$

(d) Solve equation (8). [*Hint:* See Section 2.6.]

(e) Describe the solutions and identify the type of collector obtained.

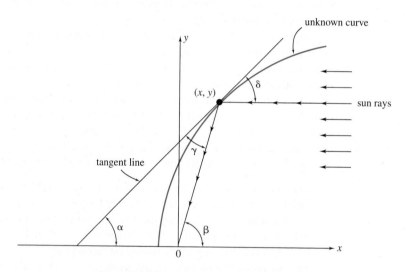

Figure 2.16 Curve that generates a solar collector

J Asymptotic Behavior of Solutions to Linear Equations

To illustrate how the asymptotic behavior of the forcing term $Q(x)$ affects the solution to a linear equation, consider the equation

(9) $$\frac{dy}{dx} + ay = Q(x)\,,$$

where the constant a is positive and $Q(x)$ is continuous on $[0, \infty]$.

(a) Show that the general solution to equation (9) can be written in the form

$$y(x) = y(x_0)e^{-a(x-x_0)} + e^{-ax}\int_{x_0}^{x} e^{at}Q(t)\,dt\,,$$

where x_0 is a nonnegative constant.

(b) If $|Q(x)| \leq k$ for $x \geq x_0$, where k and x_0 are nonnegative constants, show that

$$|y(x)| \leq |y(x_0)| e^{-a(x-x_0)} + \frac{k}{a}\left[1 - e^{-a(x-x_0)}\right] \qquad \text{for } x \geq x_0 .$$

(c) Let $z(x)$ satisfy the same equation as (9) but with forcing function $\widetilde{Q}(x)$. That is,

$$\frac{dz}{dx} + az = \widetilde{Q}(x) ,$$

where $\widetilde{Q}(x)$ is continuous on $[0, \infty]$. Show that if

$$|\widetilde{Q}(x) - Q(x)| \leq K \qquad \text{for } x \geq x_0 ,$$

then

$$|z(x) - y(x)| \leq |z(x_0) - y(x_0)| e^{-a(x-x_0)} + \frac{K}{a}\left[1 - e^{-a(x-x_0)}\right] \qquad \text{for } x \geq x_0 .$$

(d) Now show that if $Q(x) \to \beta$ as $x \to \infty$, then any solution $y(x)$ of equation (9) satisfies $y(x) \to \beta/a$ as $x \to \infty$. [*Hint:* Take $\widetilde{Q}(x) \equiv \beta$ and $z(x) \equiv \beta/a$ in part (c).]

(e) As an application of part (d), suppose a brine solution containing $q(t)$ kg of salt per liter at time t runs into a tank of water at a fixed rate and that the mixture, kept uniform by stirring, flows out at the same rate. Given that $q(t) \to \beta$ as $t \to \infty$, use the result of part (d) to determine the limiting concentration of the salt in the tank as $t \to \infty$ (see Exercises 2.3, Problem 35).

Mathematical Models and Numerical Methods Involving First-Order Equations

3.1 Mathematical Modeling

Adopting the Babylonian practices of careful measurement and detailed observations, the ancient Greeks sought to comprehend nature by logical analysis. Aristotle's convincing arguments that the world was not flat, but spherical, led the intellectuals of that day to ponder the question: What is the circumference of Earth? And it was astonishing that Eratosthenes managed to obtain a fairly accurate answer to this problem without having to set foot beyond the ancient city of Alexandria. His method involved certain assumptions and simplifications: Earth is a perfect sphere, the sun's rays travel parallel paths, the city of Syene was 5000 stadia due south of Alexandria, and so on. With these idealizations Eratosthenes created a mathematical context in which the principles of geometry could be applied.[†]

Today, as scientists seek to further our understanding of nature and as engineers seek, on a more pragmatic level, to find answers to technical problems, the technique of representing our "real world" in mathematical terms has become an invaluable tool. This process of mimicking reality by using the language of mathematics is known as **mathematical modeling.**

Formulating problems in mathematical terms has several benefits. First, it requires that we clearly state our premises. Real-world problems are often complex, involving several different and possibly interrelated processes. Before mathematical treatment can proceed, one must determine which variables are significant and which can be ignored. Often, for the relevant variables, relationships are postulated in the form of laws, formulas, theories, and the like. These assumptions constitute the **idealizations** of the model.

Mathematics contains a wealth of theorems and techniques for making logical deductions and manipulating equations. Hence, it provides a context in which analysis can take place free of any preconceived notions of the outcome. It is also of great practical importance that mathematics provides a format for obtaining numerical answers via a computer.

The process of building an effective mathematical model takes skill, imagination, and objective evaluation. Certainly an exposure to several existing models that illustrate various aspects of modeling can lead to a better feel for the process. Several excellent books and articles are devoted exclusively to the subject.[††] In this chapter we concentrate on examples of models that involve first-order differential equations. In studying these and in building your own models, the following broad outline of the process may be helpful.

[†]For further reading, see, for example, *The Mapmakers*, by John Noble Wilford (Vintage Books, New York, 2001), Chapter 2.

[††]See, for example, *A First Course in Mathematical Modeling*, 5th ed., by F. T. Giordano, W. P. Fox, S. B. Horton, and M. D. Weir (Cengage Learning, Independence, KY, 2014) or *Concepts of Mathematical Modeling*, by W. J. Meyer (Dover Publications, Mineola, New York, 2004).

Formulate the Problem

Here you must pose the problem in such a way that it can be "answered" mathematically. This requires an understanding of the problem area as well as the mathematics. At this stage you may need to confer with experts in that area and read the relevant literature.

Develop the Model

There are two things to be done here. First, you must decide which variables are important and which are not. The former are then classified as independent variables or dependent variables. The unimportant variables are those that have very little or no effect on the process. (For example, in studying the motion of a falling body, its color is usually of little interest.) The independent variables are those whose effect is significant and that will serve as input for the model.[†] For the falling body, its shape, mass, initial position, initial velocity, and time from release are possible independent variables. The dependent variables are those that are affected by the independent variables and that are important to solving the problem. Again, for a falling body, its velocity, location, and time of impact are all possible dependent variables.

Second, you must determine or specify the relationships (for example, a differential equation) that exist among the relevant variables. This requires a good background in the area and insight into the problem. You may begin with a crude model and then, based upon testing, refine the model as needed. For example, you might begin by ignoring any friction acting on the falling body. Then, if it is necessary to obtain a more acceptable answer, try to take into account any frictional forces that may affect the motion.

Test the Model

Before attempting to "verify" a model by comparing its output with experimental data, the following questions should be considered:

Are the assumptions reasonable?

Are the equations dimensionally consistent? (For example, we don't want to add units of force to units of velocity.)

Is the model internally consistent in the sense that equations do not contradict one another?

Do the relevant equations have solutions?

Are the solutions unique?

How difficult is it to obtain the solutions?

Do the solutions provide an answer for the problem being studied?

When possible, try to validate the model by comparing its predictions with any experimental data. Begin with rather simple predictions that involve little computation or analysis. Then, as the model is refined, check to see that the accuracy of the model's predictions is acceptable to you. In some cases validation is impossible or socially, politically, economically, or morally unreasonable. For example, how does one validate a model that predicts when our sun will die out?

[†]In the mathematical formulation of the model, some of the independent variables may be called **parameters**.

Each time the model is used to predict the outcome of a process and hence solve a problem, it provides a test of the model that may lead to further refinements or simplifications. In many cases a model is simplified to give a quicker or less expensive answer—provided, of course, that sufficient accuracy is maintained.

One should always keep in mind that a model is *not* reality but only a representation of reality. The more refined models *may* provide an understanding of the underlying processes of nature. For this reason applied mathematicians strive for better, more refined models. Still, the real test of a model is its ability to find an acceptable answer for the posed problem.

In this chapter we discuss various models that involve differential equations. Section 3.2, Compartmental Analysis, studies mixing problems and population models. Sections 3.3 through 3.5 are physics-based and examine heating and cooling, Newtonian mechanics, and electrical circuits. Finally Sections 3.6 and 3.7 introduce some numerical methods for solving first-order initial value problems. This will enable us to consider more realistic models that cannot be solved using the methods of Chapter 2.

3.2 Compartmental Analysis

Many complicated processes can be broken down into distinct stages and the entire system modeled by describing the interactions between the various stages. Such systems are called **compartmental** and are graphically depicted by **block diagrams**. In this section we study the basic unit of these systems, a single compartment, and analyze some simple processes that can be handled by such a model.

The basic one-compartment system consists of a function $x(t)$ that represents the amount of a substance in the compartment at time t, an input rate at which the substance enters the compartment, and an output rate at which the substance leaves the compartment (see Figure 3.1).

Because the derivative of x with respect to t can be interpreted as the rate of change in the amount of the substance in the compartment with respect to time, the one-compartment system suggests

(1) $$\frac{dx}{dt} = \textbf{input rate} - \textbf{output rate}$$

as a mathematical model for the process.

Mixing Problems

A problem for which the one-compartment system provides a useful representation is the mixing of fluids in a tank. Let $x(t)$ represent the amount of a substance in a tank (compartment) at time t. To use the compartmental analysis model, we must be able to determine the rates at which this substance enters and leaves the tank. In mixing problems one is often given the rate at which a fluid containing the substance flows into the tank, along with the concentration of the substance in that fluid. Hence, multiplying the flow rate (volume/time) by the concentration (amount/volume) yields the input rate (amount/time).

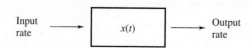

Figure 3.1 Schematic representation of a one-compartment system

The output rate of the substance is usually more difficult to determine. If we are given the exit rate of the mixture of fluids in the tank, then how do we determine the concentration of the substance in the mixture? One simplifying assumption that we might make is that the concentration is kept uniform in the mixture. Then we can compute the concentration of the substance in the mixture by dividing the amount $x(t)$ by the volume of the mixture in the tank at time t. Multiplying this concentration by the exit rate of the mixture then gives the desired output rate of the substance. This model is used in Examples 1 and 2.

Example 1 Consider a large tank holding 1000 L of pure water into which a brine solution of salt begins to flow at a constant rate of 6 L/min. The solution inside the tank is kept well stirred and is flowing out of the tank at a rate of 6 L/min. If the concentration of salt in the brine entering the tank is 0.1 kg/L, determine when the concentration of salt in the tank will reach 0.05 kg/L (see Figure 3.2).

Solution We can view the tank as a compartment containing salt. If we let $x(t)$ denote the mass of salt in the tank at time t, we can determine the concentration of salt in the tank by dividing $x(t)$ by the volume of fluid in the tank at time t. We use the mathematical model described by equation (1) to solve for $x(t)$.

First we must determine the rate at which salt enters the tank. We are given that brine flows into the tank at a rate of 6 L/min. Since the concentration is 0.1 kg/L, we conclude that the input rate of salt into the tank is

(2) $(6 \text{ L/min})(0.1 \text{ kg/L}) = 0.6 \text{ kg/min}$.

We must now determine the output rate of salt from the tank. The brine solution in the tank is kept well stirred, so let's assume that the concentration of salt in the tank is uniform. That is, the concentration of salt in any part of the tank at time t is just $x(t)$ divided by the volume of fluid in the tank. Because the tank initially contains 1000 L and the rate of flow into the tank is the same as the rate of flow out, the volume is a constant 1000 L. Hence, the output rate of salt is

(3) $(6 \text{ L/min}) \left[\dfrac{x(t)}{1000} \text{ kg/L} \right] = \dfrac{3x(t)}{500} \text{ kg/min}$.

The tank initially contained pure water, so we set $x(0) = 0$. Substituting the rates in (2) and (3) into equation (1) then gives the initial value problem

(4) $\dfrac{dx}{dt} = 0.6 - \dfrac{3x}{500}$, $x(0) = 0$,

as a mathematical model for the mixing problem.

6 L/min
0.1 kg/L

$x(t)$

1000 L

$x(0) = 0$ kg

6 L/min

Figure 3.2 Mixing problem with equal flow rates

The equation in (4) is separable (and linear) and easy to solve. Using the initial condition $x(0) = 0$ to evaluate the arbitrary constant, we obtain

(5) $x(t) = 100(1 - e^{-3t/500})$.

Thus, the concentration of salt in the tank at time t is

$$\frac{x(t)}{1000} = 0.1(1 - e^{-3t/500}) \text{ kg/L} .$$

To determine when the concentration of salt is 0.05 kg/L, we set the right-hand side equal to 0.05 and solve for t. This gives

$$0.1(1 - e^{-3t/500}) = 0.05 \quad \text{or} \quad e^{-3t/500} = 0.5 ,$$

and hence

$$t = \frac{500 \ln 2}{3} \approx 115.52 \text{ min} .$$

Consequently the concentration of salt in the tank will be 0.05 kg/L after 115.52 min. ◆

From equation (5), we observe that the mass of salt in the tank steadily increases and has the limiting value

$$\lim_{t \to \infty} x(t) = \lim_{t \to \infty} 100(1 - e^{-3t/500}) = 100 \text{ kg} .$$

Thus, the limiting concentration of salt in the tank is 0.1 kg/L, which is the same as the concentration of salt in the brine flowing into the tank. This certainly agrees with our expectations!
It might be interesting to see what would happen to the concentration if the flow rate into the tank is greater than the flow rate out.

Example 2 For the mixing problem described in Example 1, assume now that the brine leaves the tank at a rate of 5 L/min instead of 6 L/min, with all else being the same (see Figure 3.3). Determine the concentration of salt in the tank as a function of time.

Solution The difference between the rate of flow into the tank and the rate of flow out is $6 - 5 = 1$ L/min, so the volume of fluid in the tank after t minutes is $(1000 + t)$ L. Hence, the rate at which salt leaves the tank is

$$(5 \text{ L/min})\left[\frac{x(t)}{1000 + t} \text{ kg/L}\right] = \frac{5x(t)}{1000 + t} \text{ kg/min} .$$

Figure 3.3 Mixing problem with unequal flow rates

Using this in place of (3) for the output rate gives the initial value problem

(6) $\dfrac{dx}{dt} = 0.6 - \dfrac{5x}{1000 + t}$, $x(0) = 0$,

as a mathematical model for the mixing problem.

The differential equation in (6) is linear, so we can use the procedure outlined on page 50 to solve for $x(t)$. The integrating factor is $\mu(t) = (1000 + t)^5$. Thus,

$$\frac{d}{dt}\big[(1000 + t)^5 x\big] = 0.6(1000 + t)^5$$
$$(1000 + t)^5 x = 0.1(1000 + t)^6 + c$$
$$x(t) = 0.1(1000 + t) + c(1000 + t)^{-5}.$$

Using the initial condition $x(0) = 0$, we find $c = -0.1(1000)^6$, and thus the solution to (6) is

$$x(t) = 0.1\big[(1000 + t) - (1000)^6(1000 + t)^{-5}\big].$$

Hence, the concentration of salt in the tank at time t is

(7) $\dfrac{x(t)}{1000 + t} = 0.1\big[1 - (1000)^6(1000 + t)^{-6}\big]$ kg/L . ◆

As in Example 1, the concentration given by (7) approaches 0.1 kg/L as $t \rightarrow \infty$. However, in Example 2 the volume of fluid in the tank becomes unbounded, and when the tank begins to overflow, the model in (6) is no longer appropriate.

Population Models

How does one predict the growth of a population? If we are interested in a single population, we can think of the species as being contained in a compartment (a petri dish, an island, a country, etc.) and study the growth process as a one-compartment system.

Let $p(t)$ be the population at time t. While the population is always an integer, it is usually large enough so that very little error is introduced in assuming that $p(t)$ is a continuous function. We now need to determine the growth (input) rate and the death (output) rate for the population.

Let's consider a population of bacteria that reproduce by simple cell division. In our model, we assume that the growth rate is proportional to the population present. This assumption is consistent with observations of bacterial growth. As long as there are sufficient space and ample food supply for the bacteria, we can also assume that the death rate is zero. (Remember that in cell division, the parent cell does not die, but becomes two new cells.) Hence, a mathematical model for a population of bacteria is

(8) $\dfrac{dp}{dt} = k_1 p$, $p(0) = p_0$,

where $k_1 > 0$ is the proportionality constant for the growth rate and p_0 is the population at time $t = 0$. For human populations the assumption that the death rate is zero is certainly wrong! However, if we assume that people die only of natural causes, we might expect the death rate also to be proportional to the size of the population. That is, we revise (8) to read

(9) $\dfrac{dp}{dt} = k_1 p - k_2 p = (k_1 - k_2)p = kp$,

where $k := k_1 - k_2$ and k_2 is the proportionality constant for the death rate. Let's assume that $k_1 > k_2$ so that $k > 0$. This gives the mathematical model

(10) $\dfrac{dp}{dt} = kp$, $p(0) = p_0$,

which is called the **Malthusian,**[†] or **exponential, law** of population growth. This equation is separable, and solving the initial value problem for $p(t)$ gives

(11) $p(t) = p_0 e^{kt}$.

To test the Malthusian model, let's apply it to the demographic history of the United States.

Example 3 In 1790 the population of the United States was 3.93 million, and in 1890 it was 62.98 million. Using the Malthusian model, estimate the U.S. population as a function of time.

Solution If we set $t = 0$ to be the year 1790, then by formula (11) we have

(12) $p(t) = (3.93)e^{kt}$,

where $p(t)$ is the population in millions. One way to obtain a value for k would be to make the model fit the data for some specific year, such as 1890 ($t = 100$ years).[††] We have

$$p(100) = 62.98 = (3.93)e^{100k}.$$

Solving for k yields

$$k = \frac{\ln(62.98) - \ln(3.93)}{100} \approx 0.027742.$$

Substituting this value in equation (12), we find

(13) $p(t) = (3.93)e^{(0.027742)t}$. ◆

In Table 3.1 on page 97 we list the U.S. population as given by the U.S. Bureau of the Census and the population predicted by the Malthusian model using equation (13). From Table 3.1 we see that the predictions based on the Malthusian model are in reasonable agreement with the census data until about 1900. After 1900 the predicted population is too large, and the Malthusian model is unacceptable.

We remark that a Malthusian model can be generated using the census data for any two different years. We selected 1790 and 1890 for purposes of comparison with the logistic model that we now describe.

The Malthusian model considered only death by natural causes. What about premature deaths due to malnutrition, inadequate medical supplies, communicable diseases, violent crimes, etc.? These factors involve a competition within the population, so we might

[†]*Historical Footnote:* Thomas R. Malthus (1766–1834) was a British economist who studied population models.
[††]The choice of the year 1890 is purely arbitrary, of course; a more democratic (and better) way of extracting parameters from data is described after Example 4.

		Malthusian	Logistic	$\dfrac{1}{p}\dfrac{dp}{dt}$	Logistic
Year	U.S. Census	(Example 3)	(Example 4)		(Least Squares)
1790	3.93	3.93	3.93		4.11
1800	5.31	5.19	5.30	0.0312	5.42
1810	7.24	6.84	7.13	0.0299	7.14
1820	9.64	9.03	9.58	0.0292	9.39
1830	12.87	11.92	12.82	0.0289	12.33
1840	17.07	15.73	17.07	0.0302	16.14
1850	23.19	20.76	22.60	0.0310	21.05
1860	31.44	27.40	29.70	0.0265	27.33
1870	39.82	36.16	38.66	0.0235	35.28
1880	50.19	47.72	49.71	0.0231	45.21
1890	62.98	62.98	62.98	0.0207	57.41
1900	76.21	83.12	78.42	0.0192	72.11
1910	92.23	109.69	95.73	0.0162	89.37
1920	106.02	144.76	114.34	0.0146	109.10
1930	123.20	191.05	133.48	0.0106	130.92
1940	132.16	252.13	152.26	0.0106	154.20
1950	151.33	333.74	169.90	0.0156	178.12
1960	179.32	439.12	185.76	0.0145	201.75
1970	203.30	579.52	199.50	0.0116	224.21
1980	226.54	764.80	211.00	0.0100	244.79
1990	248.71	1009.33	220.38	0.0110	263.01
2000	281.42	1332.03	227.84	0.0107	278.68
2010	308.75	1757.91	233.68		291.80
2020	?	2319.95	238.17		302.56

TABLE 3.1 A Comparison of the Malthusian and Logistic Models with U.S. Census Data (Population is given in Millions)

assume that another component of the death rate is proportional to the number of two-party interactions. There are $p(p-1)/2$ such possible interactions for a population of size p. Thus, if we combine the birth rate (8) with the death rate and rearrange constants, we get the **logistic model**

$$\frac{dp}{dt} = k_1 p - k_3 \frac{p(p-1)}{2}$$

or

$$(14) \qquad \frac{dp}{dt} = -Ap(p - p_1), \qquad p(0) = p_0,$$

where $A = k_3/2$ and $p_1 = (2k_1/k_3) + 1$.

Equation (14) has two equilibrium (constant) solutions: $p(t) \equiv p_1$ and $p(t) \equiv 0$. The nonequilibrium solutions can be found by separating variables and using the integral table at the back of the book:

$$\int \frac{dp}{p(p - p_1)} = -A \int dt \quad \text{or} \quad \frac{1}{p_1} \ln \left| \frac{p - p_1}{p} \right| = -At + c_1 \quad \text{or} \quad \left| 1 - \frac{p_1}{p} \right| = c_2 e^{-Ap_1 t}.$$

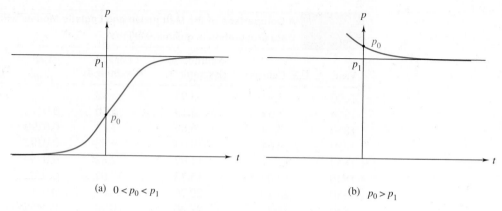

(a) $0 < p_0 < p_1$ (b) $p_0 > p_1$

Figure 3.4 The logistic curves

If $p(t) = p_0$ at $t = 0$, and $c_3 = 1 - p_1/p_0$, then solving for $p(t)$, we find

$$(15) \qquad p(t) = \frac{p_1}{1 - c_3 e^{-Ap_1 t}} = \frac{p_0 p_1}{p_0 + (p_1 - p_0)e^{-Ap_1 t}} .$$

The function $p(t)$ given in (15) is called the **logistic function,** and graphs of logistic curves are displayed in Figure 3.4.[†] Note that for $A > 0$ and $p_0 > 0$, the limit population as $t \to \infty$, is p_1.

Let's test the logistic model on the population growth of the United States.

Example 4 Taking the 1790 population of 3.93 million as the initial population and given the 1840 and 1890 populations of 17.07 and 62.98 million, respectively, use the logistic model to estimate the population at time t.

Solution With $t = 0$ corresponding to the year 1790, we know that $p_0 = 3.93$. We must now determine the parameters A, p_1 in equation (15). For this purpose, we use the given facts that $p(50) = 17.07$ and $p(100) = 62.98$; that is,

$$(16) \qquad 17.07 = \frac{3.93\, p_1}{3.93 + (p_1 - 3.93)e^{-50Ap_1}} ,$$

$$(17) \qquad 62.98 = \frac{3.93\, p_1}{3.93 + (p_1 - 3.93)e^{-100Ap_1}} .$$

Equations (16) and (17) are two nonlinear equations in the two unknowns A, p_1. To solve such a system, we would usually resort to a numerical approximation scheme such as Newton's method. However, for the case at hand, it is possible to find the solutions directly because the data are given at times t_a and t_b with $t_b = 2t_a$ (see Problem 12). Carrying out the algebra described in Problem 12, we ultimately find that

$$(18) \qquad p_1 \approx 251.7812 \quad \text{and} \quad A \approx 0.0001210 .$$

Thus, the logistic model for the given data is

$$(19) \qquad p(t) = \frac{989.50}{3.93 + (247.85)e^{-(0.030463)t}} . \quad \blacklozenge$$

[†]*Historical Footnote:* The logistic model for population growth was first developed by P. F. Verhulst around 1840.

The population data predicted by (19) are displayed in column 4 of Table 3.1 on page 97. As you can see, these predictions are in better agreement with the census data than the Malthusian model is. And, of course, the agreement is perfect in the years 1790, 1840, and 1890. However, the choice of these particular years for estimating the parameters p_0, A, and p_1 is quite arbitrary, and we would expect that a more robust model would use *all* of the data, in some way, for the estimation. One way to implement this idea is the following.

Observe from equation (14) that the logistic model predicts a linear relationship between $(dp/dt)/p$ and p:

$$\frac{1}{p}\frac{dp}{dt} = Ap_1 - Ap \,,$$

with Ap_1 as the intercept and $-A$ as the slope. In column five of Table 3.1, we list values of $(dp/dt)/p$, which are estimated from centered differences according to

$$(20) \qquad \frac{1}{p(t)}\frac{dp}{dt}(t) \approx \frac{1}{p(t)}\frac{p(t+10)-p(t-10)}{20}$$

(see Problem 16). In Figure 3.5 these estimated values of $(dp/dt)/p$ are plotted against p in what is called a *scatter diagram*. The linear relationship predicted by the logistic model suggests that we approximate the plot by a straight line. A standard technique for doing this is the so-called *least-squares linear fit,* which is discussed in Appendix E. This yields the straight line

$$\frac{1}{p}\frac{dp}{dt} \approx 0.0280960 - 0.00008231p \,,$$

which is also depicted in Figure 3.5. Now with $A = 0.00008231$ and $p_1 = (0.0280960/A)$ ≈ 341.4, we can solve equation (15) for p_0:

$$(21) \qquad p_0 = \frac{p(t)p_1 e^{-Ap_1 t}}{p_1 - p(t)\left[1 - e^{-Ap_1 t}\right]} \,.$$

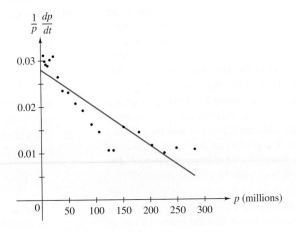

Figure 3.5 Scatter data and straight line fit

By averaging the right-hand side of (21) over all the data, we obtain the estimate $p_0 \approx 4.107$. Finally, the insertion of these estimates for the parameters in equation (15) leads to the predictions listed in column six of Table 3.1.

Note that this model yields $p_1 \approx 341.4$ million as the limit on the future population of the United States.

3.2 EXERCISES

1. A brine solution of salt flows at a constant rate of 8 L/min into a large tank that initially held 100 L of brine solution in which was dissolved 0.5 kg of salt. The solution inside the tank is kept well stirred and flows out of the tank at the same rate. If the concentration of salt in the brine entering the tank is 0.05 kg/L, determine the mass of salt in the tank after t min. When will the concentration of salt in the tank reach 0.02 kg/L?

2. A brine solution of salt flows at a constant rate of 6 L/min into a large tank that initially held 50 L of brine solution in which was dissolved 0.5 kg of salt. The solution inside the tank is kept well stirred and flows out of the tank at the same rate. If the concentration of salt in the brine entering the tank is 0.05 kg/L, determine the mass of salt in the tank after t min. When will the concentration of salt in the tank reach 0.03 kg/L?

3. A nitric acid solution flows at a constant rate of 6 L/min into a large tank that initially held 200 L of a 0.5% nitric acid solution. The solution inside the tank is kept well stirred and flows out of the tank at a rate of 8 L/min. If the solution entering the tank is 20% nitric acid, determine the volume of nitric acid in the tank after t min. When will the percentage of nitric acid in the tank reach 10%?

4. A brine solution of salt flows at a constant rate of 4 L/min into a large tank that initially held 100 L of pure water. The solution inside the tank is kept well stirred and flows out of the tank at a rate of 3 L/min. If the concentration of salt in the brine entering the tank is 0.2 kg/L, determine the mass of salt in the tank after t min. When will the concentration of salt in the tank reach 0.1 kg/L?

5. A swimming pool whose volume is 10,000 gal contains water that is 0.01% chlorine. Starting at $t = 0$, city water containing 0.001% chlorine is pumped into the pool at a rate of 5 gal/min. The pool water flows out at the same rate. What is the percentage of chlorine in the pool after 1 h? When will the pool water be 0.002% chlorine?

6. The air in a small room 12 ft by 8 ft by 8 ft is 3% carbon monoxide. Starting at $t = 0$, fresh air containing no carbon monoxide is blown into the room at a rate of 100 ft^3/min. If air in the room flows out through a vent at the same rate, when will the air in the room be 0.01% carbon monoxide?

7. Beginning at time $t = 0$, fresh water is pumped at the rate of 3 gal/min into a 60-gal tank initially filled with brine. The resulting less-and-less salty mixture overflows at the same rate into a second 60-gal tank that initially contained only pure water, and from there it eventually spills onto the ground. Assuming perfect mixing in both tanks, when will the water in the second tank taste saltiest? And exactly how salty will it then be, compared with the original brine?

8. A tank initially contains s_0 lb of salt dissolved in 200 gal of water, where s_0 is some positive number. Starting at time $t = 0$, water containing 0.5 lb of salt per gallon enters the tank at a rate of 4 gal/min, and the well-stirred solution leaves the tank at the same rate. Letting $c(t)$ be the concentration of salt in the tank at time t, show that the limiting concentration—that is, $\lim_{t \to \infty} c(t)$—is 0.5 lb/gal.

9. In 1990 the Department of Natural Resources released 1000 splake (a crossbreed of fish) into a lake. In 1997 the population of splake in the lake was estimated to be 3000. Using the Malthusian law for population growth, estimate the population of splake in the lake in the year 2020.

10. Use a sketch of the phase line (see Project B, Chapter 1, page 33) to argue that any solution to the mixing problem model

$$\frac{dx}{dt} = a - bx; \qquad a, b > 0,$$

approaches the equilibrium solution $x(t) \equiv a/b$ as t approaches $+\infty$; that is, a/b is a sink.

11. Use a sketch of the phase line (see Project B, Chapter 1) to argue that any solution to the logistic model

$$\frac{dp}{dt} = (a - bp)p; \qquad p(t_0) = p_0,$$

where a, b, and p_0 are positive constants, approaches the equilibrium solution $p(t) \equiv a/b$ as t approaches $+\infty$.

12. For the logistic curve (15), assume $p_a := p(t_a)$ and $p_b := p(t_b)$ are given with $t_b = 2t_a$ $(t_a > 0)$. Show that

$$p_1 = \left[\frac{p_a p_b - 2p_0 p_b + p_0 p_a}{p_a^2 - p_0 p_b} \right] p_a,$$

$$A = \frac{1}{p_1 t_a} \ln \left[\frac{p_b(p_a - p_0)}{p_0(p_b - p_a)} \right].$$

[*Hint:* Equate the expressions (21) for p_0 at times t_a and t_b. Set $\chi = \exp(-Ap_1 t_a)$ and $\chi^2 = \exp(-Ap_1 t_b)$ and solve for χ. Insert into one of the earlier expressions and solve for p_1.]

13. In Problem 9, suppose we have the additional information that the population of splake in 2004 was estimated to be 5000. Use a logistic model to estimate the population of splake in the year 2020. What is the predicted limiting population? [*Hint:* Use the formulas in Problem 12.]

14. In 1980 the population of alligators on the Kennedy Space Center grounds was estimated to be 1500. In 2006 the population had grown to an estimated 6000. Using the Malthusian law for population growth, estimate the alligator population on the Kennedy Space Center grounds in the year 2020.

15. In Problem 14, suppose we have the additional information that the population of alligators on the grounds of the Kennedy Space Center in 1993 was estimated to be 4100. Use a logistic model to estimate the population of alligators in the year 2020. What is the predicted limiting population? [*Hint:* Use the formulas in Problem 12.]

16. Show that for a differentiable function $p(t)$, we have

$$\lim_{h \to 0} \frac{p(t+h) - p(t-h)}{2h} = p'(t),$$

which is the basis of the centered difference approximation used in (20).

17. **(a)** For the U.S. census data, use the forward difference approximation to the derivative, that is,

$$\frac{1}{p(t)} \frac{dp}{dt}(t) \approx \frac{1}{p(t)} \frac{p(t+10) - p(t)}{10},$$

to recompute column 5 of Table 3.1 on page 97.

(b) Using the data from part (a), determine the constants A, p_1 in the least-squares fit

$$\frac{1}{p} \frac{dp}{dt} \approx Ap_1 - Ap$$

(see Appendix E).

(c) With the values for A and p_1 found in part (b), determine p_0 by averaging formula (21) over the data.

(d) Substitute A, p_1, and p_0 as determined above into the logistic formula (15) and calculate the populations predicted for each of the years listed in Table 3.1.

(e) Compare this model with that of the centered difference-based model in column 6 of Table 3.1.

18. Using the U.S. census data in Table 3.1 for 1900, 1920, and 1940 to determine parameters in the logistic equation model, what populations does the model predict for 2000 and 2010? Compare your answers with the census data for those years.

19. The initial mass of a certain species of fish is 7 million tons. The mass of fish, if left alone, would increase at a rate proportional to the mass, with a proportionality constant of 2/yr. However, commercial fishing removes fish mass at a constant rate of 15 million tons per year. When will all the fish be gone? If the fishing rate is changed so that the mass of fish remains constant, what should that rate be?

20. From theoretical considerations, it is known that light from a certain star should reach Earth with intensity I_0. However, the path taken by the light from the star to Earth passes through a dust cloud, with absorption coefficient 0.1/light-year. The light reaching Earth has intensity $1/2 \, I_0$. How thick is the dust cloud? (The rate of change of light intensity with respect to thickness is proportional to the intensity. One light-year is the distance traveled by light during 1 yr.)

21. A snowball melts in such a way that the rate of change in its volume is proportional to its surface area. If the snowball was initially 4 in. in diameter and after 30 min its diameter is 3 in., when will its diameter be 2 in.? Mathematically speaking, when will the snowball disappear?

22. Suppose the snowball in Problem 21 melts so that the rate of change in its *diameter* is proportional to its surface area. Using the same given data, determine when its diameter will be 2 in. Mathematically speaking, when will the snowball disappear?

In Problems 23–27, assume that the rate of decay of a radioactive substance is proportional to the amount of the substance present. The half-life of a radioactive substance is the time it takes for one-half of the substance to disintegrate.

23. If initially there are 50 g of a radioactive substance and after 3 days there are only 10 g remaining, what percentage of the original amount remains after 4 days?

24. If initially there are 300 g of a radioactive substance and after 5 yr there are 200 g remaining, how much time must elapse before only 10 g remain?

25. Carbon dating is often used to determine the age of a fossil. For example, a humanoid skull was found in a cave in South Africa along with the remains of a campfire. Archaeologists believe the age of the skull to be the same age as the campfire. It is determined that only 2% of the original amount of carbon-14 remains in the burnt wood of the campfire. Estimate the age of the skull if the half-life of carbon-14 is about 5600 years.

26. To see how sensitive the technique of carbon dating of Problem 25 is,

 (a) Redo Problem 25 assuming the half-life of carbon-14 is 5550 yr.

 (b) Redo Problem 25 assuming 3% of the original mass remains.

27. The only undiscovered isotopes of the two unknown elements hohum and inertium (symbols Hh and It) are radioactive. Hohum decays into inertium with a decay constant of 2/yr, and inertium decays into the nonradioactive isotope of bunkum (symbol Bu) with a decay constant of 1/yr. An initial mass of 1 kg of hohum is put into a nonradioactive container, with no other source of hohum, inertium, or bunkum. How much of each of the three elements is in the container after t yr? (The **decay constant** is the constant of proportionality in the statement that the rate of loss of mass of the element at any time is proportional to the mass of the element at that time.)

3.3 Heating and Cooling of Buildings

Our goal is to formulate a mathematical model that describes the 24-hr temperature profile inside a building as a function of the outside temperature, the heat generated inside the building, and the furnace heating or air conditioner cooling. From this model we would like to answer the following three questions:

 (a) How long does it take to change the building temperature substantially?

 (b) How does the building temperature vary during spring and fall when there is no furnace heating or air conditioning?

 (c) How does the building temperature vary in summer when there is air conditioning or in the winter when there is furnace heating?

A natural approach to modeling the temperature inside a building is to use compartmental analysis. Let $T(t)$ represent the temperature inside the building at time t and view the building as a single compartment. Then the rate of change in the temperature is determined by all the factors that generate or dissipate heat.

We will consider three main factors that affect the temperature inside the building. First is the heat produced by people, lights, and machines inside the building. This causes a rate of increase in temperature that we will denote by $H(t)$. Second is the heating (or cooling) supplied by the furnace (or air conditioner). This rate of increase (or decrease) in temperature will be represented by $U(t)$. In general, the additional heating rate $H(t)$ and the furnace (or air conditioner) rate $U(t)$ are described in terms of energy per unit time (such as British thermal units per hour). However, by multiplying by the heat capacity of the building (in units of degrees temperature change per unit heat energy), we can express the two quantities $H(t)$ and $U(t)$ in terms of temperature per unit time.

The third factor is the effect of the outside temperature $M(t)$ on the temperature inside the building. Experimental evidence has shown that this factor can be modeled using **Newton's law of cooling**. This law states that the rate of change in the temperature $T(t)$ is proportional to the difference between the outside temperature $M(t)$ and the inside temperature $T(t)$. That is, the rate of change in the building temperature due to $M(t)$ is

$$K\left[M(t) - T(t)\right].$$

The positive constant K depends on the physical properties of the building, such as the number of doors and windows and the type of insulation, but K does not depend on M, T, or t. Hence, when the outside temperature is greater than the inside temperature, $M(t) - T(t) > 0$ and there is an increase in the building temperature due to $M(t)$. On the other hand, when the

outside temperature is less than the inside temperature, then $M(t) - T(t) < 0$ and the building temperature decreases.

Summarizing, we find

(1) $\dfrac{dT}{dt} = K\big[M(t) - T(t)\big] + H(t) + U(t)\,,$

where the additional heating rate $H(t)$ is always nonnegative and $U(t)$ is positive for furnace heating and negative for air conditioner cooling. A more detailed model of the temperature dynamics of the building could involve more variables to represent different temperatures in different rooms or zones. Such an approach would use compartmental analysis, with the rooms as different compartments (see Problems 35–37, Exercises 5.2).

Because equation (1) is linear, it can be solved using the method discussed in Section 2.3. Rewriting (1) in the standard form

(2) $\dfrac{dT}{dt}(t) + P(t)T(t) = Q(t)\,,$

where

$P(t) := K\,,$

(3) $Q(t) := KM(t) + H(t) + U(t)\,,$

we find that the integrating factor is

$$\mu(t) = \exp\left(\int K\,dt\right) = e^{Kt}\,.$$

To solve (2), multiply each side by e^{Kt} and integrate:

$$e^{Kt}\dfrac{dT}{dt}(t) + Ke^{Kt}T(t) = e^{Kt}Q(t)\,,$$

$$e^{Kt}T(t) = \int e^{Kt}Q(t)\,dt + C\,.$$

Solving for $T(t)$ gives

(4) $T(t) = e^{-Kt}\int e^{Kt}Q(t)\,dt + Ce^{-Kt}$

$$= e^{-Kt}\left\{\int e^{Kt}\big[KM(t) + H(t) + U(t)\big]\,dt + C\right\}.$$

Example 1 Suppose at the end of the day (at time t_0), when people leave the building, the outside temperature stays constant at M_0, the additional heating rate H inside the building is zero, and the furnace/air conditioner rate U is zero. Determine $T(t)$, given the initial condition $T(t_0) = T_0$.

Solution With $M = M_0, H = 0$, and $U = 0$, equation (4) becomes

$$T(t) = e^{-Kt}\left[\int e^{Kt}KM_0\,dt + C\right] = e^{-Kt}\big[M_0e^{Kt} + C\big]$$

$$= M_0 + Ce^{-Kt}\,.$$

Setting $t = t_0$ and using the initial value T_0 of the temperature, we find that the constant C is $(T_0 - M_0)\exp(Kt_0)$. Hence,

(5) $T(t) = M_0 + (T_0 - M_0)e^{-K(t-t_0)}\,.$ ◆

When $M_0 < T_0$, the solution in (5) decreases exponentially from the initial temperature T_0 to the final temperature M_0. To determine a measure of the time it takes for the temperature to change "substantially," consider the simple linear equation $dA/dt = -\alpha A$, whose solutions have the form $A(t) = A(0)e^{-\alpha t}$. Now, as $t \to +\infty$, the function $A(t)$ either decays exponentially $(\alpha > 0)$ or grows exponentially $(\alpha < 0)$. In either case the time it takes for $A(t)$ to change from $A(0)$ to $A(0)/e$ ($\approx 0.368\,A(0)$) is just $1/\alpha$ because

$$A\left(\frac{1}{\alpha}\right) = A(0)e^{-\alpha(1/\alpha)} = \frac{A(0)}{e} .$$

The quantity $1/|\alpha|$, which is independent of $A(0)$, is called the **time constant** for the equation. For linear equations of the more general form $dA/dt = -\alpha A + g(t)$, we again refer to $1/|\alpha|$ as the time constant.

Returning to Example 1, we see that the temperature $T(t)$ satisfies the equations

$$\frac{dT}{dt}(t) = -KT(t) + KM_0 , \qquad \frac{d(T - M_0)}{dt}(t) = -K\left[T(t) - M_0\right] ,$$

for M_0 a constant. In either case, the time constant is just $1/K$, which represents the time it takes for the temperature difference $T - M_0$ to change from $T_0 - M_0$ to $(T_0 - M_0)/e$. We also call $1/K$ the **time constant for the building** (without heating or air conditioning). A typical value for the time constant of a building is 2 to 4 hr, but the time constant can be much shorter if windows are open or if there is a fan circulating air. Or it can be much longer if the building is well insulated.

In the context of Example 1, we can use the notion of time constant to answer our initial question (a): The building temperature changes exponentially with a time constant of $1/K$. An answer to question (b) about the temperature inside the building during spring and fall is given in the next example.

Example 2 Find the building temperature $T(t)$ if the additional heating rate $H(t)$ is equal to the constant H_0, there is no heating or cooling $\left(U(t) \equiv 0\right)$, and the outside temperature M varies as a sine wave over a 24-hr period, with its minimum at $t = 0$ (midnight) and its maximum at $t = 12$ (noon). That is,

$$M(t) = M_0 - B \cos \omega t ,$$

where B is a positive constant, M_0 is the average outside temperature, and $\omega = 2\pi/24 = \pi/12$ radians/hr. (This could be the situation during the spring or fall when there is neither furnace heating nor air conditioning.)

Solution The function $Q(t)$ in (3) is now

$$Q(t) = K(M_0 - B \cos \omega t) + H_0 .$$

Setting $B_0 := M_0 + H_0/K$, we can rewrite Q as

(6) $$Q(t) = K(B_0 - B \cos \omega t) ,$$

where KB_0 represents the daily average value of $Q(t)$; that is,

$$KB_0 = \frac{1}{24}\int_0^{24} Q(t)\,dt .$$

When the forcing function $Q(t)$ in (6) is substituted into the expression for the temperature in equation (4), the result (after using integration by parts) is

$$(7) \qquad T(t) = e^{-Kt} \left[\int e^{Kt}(KB_0 - KB \cos \omega t)\, dt + C \right]$$

$$= B_0 - BF(t) + Ce^{-Kt},$$

where

$$F(t) := \frac{\cos \omega t + (\omega/K) \sin \omega t}{1 + (\omega/K)^2}.$$

The constant C is chosen so that at midnight $(t = 0)$, the value of the temperature T is equal to some initial temperature T_0. Thus,

$$C = T_0 - B_0 + BF(0) = T_0 - B_0 + \frac{B}{1 + (\omega/K)^2}. \quad \blacklozenge$$

Notice that the third term in solution (7) involving the constant C tends to zero exponentially. The constant term B_0 in (7) is equal to $M_0 + H_0/K$ and represents the daily average temperature inside the building (neglecting the exponential term). When there is no additional heating rate inside the building $(H_0 = 0)$, this average temperature is equal to the average outside temperature M_0. The term $BF(t)$ in (7) represents the sinusoidal variation of temperature inside the building responding to the outside temperature variation. Since $F(t)$ can be written in the form

$$(8) \qquad F(t) = \left[1 + (\omega/K)^2 \right]^{-1/2} \cos(\omega t - \phi),$$

where $\tan \phi = \omega/K$ (see Problem 16), the sinusoidal variation inside the building lags behind the outside variation by ϕ/ω hours. Further, the magnitude of the variation inside the building is slightly less, by a factor of $\left[1 + (\omega/K)^2 \right]^{-1/2}$, than the outside variation. The angular frequency of variation ω is $2\pi/24$ radians/hr (which is about $1/4$). Typical values for the dimensionless ratio ω/K lie between $1/2$ and 1. For this range, the lag between inside and outside temperature is approximately 1.8 to 3 hr and the magnitude of the inside variation is between 89% and 71% of the variation outside. Figure 3.6 shows the 24-hr sinusoidal variation of the

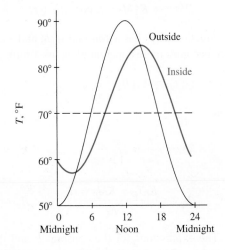

Figure 3.6 Temperature variation inside and outside an unheated building

outside temperature for a typical moderate day as well as the temperature variations inside the building for a dimensionless ratio ω/K of unity, which corresponds to a time constant $1/K$ of approximately 4 hr. In sketching the latter curve, we have assumed that the exponential term has died out.

Example 3 Suppose, in the building of Example 2, a simple thermostat is installed that is used to compare the actual temperature inside the building with a desired temperature T_D. If the actual temperature is below the desired temperature, the furnace supplies heating; otherwise, it is turned off. If the actual temperature is above the desired temperature, the air conditioner supplies cooling; otherwise, it is off. (In practice, there is some dead zone around the desired temperature in which the temperature difference is not sufficient to activate the thermostat, but that is to be ignored here.) Assuming that the amount of heating or cooling supplied is proportional to the difference in temperature—that is,

$$U(t) = K_U \left[T_D - T(t) \right],$$

where K_U is the (positive) proportionality constant—find $T(t)$.

Solution If the proportional control $U(t)$ is substituted directly into the differential equation (1) for the building temperature, we get

(9) $$\frac{dT(t)}{dt} = K \left[M(t) - T(t) \right] + H(t) + K_U \left[T_D - T(t) \right].$$

A comparison of equation (9) with the first-order linear differential equation (2) shows that for this example the quantity P is equal to $K + K_U$ and the quantity $Q(t)$ representing the forcing function includes the desired temperature T_D. That is,

$$P = K + K_U,$$
$$Q(t) = KM(t) + H(t) + K_U T_D.$$

When the additional heating rate is a constant H_0 and the outside temperature M varies as a sine wave over a 24-hr period in the same way as it did in Example 2, the forcing function is

$$Q(t) = K(M_0 - B \cos \omega t) + H_0 + K_U T_D.$$

The function $Q(t)$ has a constant term and a cosine term just as in equation (6). This equivalence becomes more apparent after substituting

(10) $$Q(t) = K_1 (B_2 - B_1 \cos \omega t),$$

where

$$\omega := \frac{2\pi}{24} = \frac{\pi}{12}, \qquad\qquad K_1 := K + K_U,$$

(11) $$B_2 := \frac{K_U T_D + KM_0 + H_0}{K_1}, \qquad B_1 := \frac{BK}{K_1}.$$

The expressions for the constant P and the forcing function $Q(t)$ of equation (10) are the same as the expressions in Example 2, except that the constants K, B_0, and B are replaced,

respectively, by the constants K_1, B_2, and B_1. Hence, the solution to the differential equation (9) will be the same as the temperature solution in Example 2, except that the three constant terms are changed. Thus,

$$(12) \qquad T(t) = B_2 - B_1 F_1(t) + C \exp(-K_1 t) \, ,$$

where

$$F_1(t) := \frac{\cos \omega t + (\omega/K_1)\sin \omega t}{1 + (\omega/K_1)^2} \, .$$

The constant C is chosen so that at time $t = 0$ the value of the temperature T equals T_0. Thus,

$$C = T_0 - B_2 + B_1 F_1(0) \, . \quad \blacklozenge$$

In the above example, the time constant for equation (9) is $1/P = 1/K_1$, where $K_1 = K + K_U$. Here $1/K_1$ is referred to as the **time constant for the building with heating and air-conditioning**. For a typical heating and cooling system, K_U is somewhat less than 2; for a typical building, the constant K is between $1/2$ and $1/4$. Hence, the sum gives a value for K_1 of about 2, and the time constant for the building with heating and air-conditioning is about $1/2$ hr.

When the heating or cooling is turned on, it takes about 30 min for the exponential term in (12) to die off. If we neglect this exponential term, the average temperature inside the building is B_2. Since K_1 is much larger than K and H_0 is small, it follows from (11) that B_2 is roughly T_D, the desired temperature. In other words, after a certain period of time, the temperature inside the building is roughly T_D, with a small sinusoidal variation. (The outside average M_0 and inside heating rate H_0 have only a small effect.) Thus, to save energy, the heating or cooling system may be left off during the night. When it is turned on in the morning, it will take roughly 30 min for the inside of the building to attain the desired temperature. These observations provide an answer to question (c), regarding the temperature inside the building during summer and winter, that was posed at the beginning of this section.

The assumption made in Example 3 that the amount of heating or cooling is $U(t) = K_U\left[T_D - T(t) \right]$ may not always be suitable. We have used it here and in the exercises to illustrate the use of the time constant. More adventuresome readers may want to experiment with other models for $U(t)$, especially if they have available the numerical techniques discussed in Sections 3.6 and 3.7.

3.3 EXERCISES

1. A cup of hot coffee initially at 95°C cools to 80°C in 5 min while sitting in a room of temperature 21°C. Using just Newton's law of cooling, determine when the temperature of the coffee will be a nice 50°C.

2. A cold beer initially at 35°F warms up to 40°F in 3 min while sitting in a room of temperature 70°F. How warm will the beer be if left out for 20 min?

3. A white wine at room temperature 70°F is chilled in ice (32°F). If it takes 15 min for the wine to chill to 60°F, how long will it take for the wine to reach 56°F?

4. A red wine is brought up from the wine cellar, which is a cool 10°C, and left to breathe in a room of temperature 23°C. If it takes 10 min for the wine to reach 15°C, when will the temperature of the wine reach 18°C?

5. It was noon on a cold December day in Tampa: 16°C. Detective Taylor arrived at the crime scene to find the sergeant leaning over the body. The sergeant said there were several suspects. If they knew the exact time of death, then they could narrow the list. Detective Taylor took out a thermometer and measured the temperature of the body: 34.5°C. He then left for lunch. Upon returning at 1:00 P.M., he found the body temperature to be 33.7°C. When did the murder occur? [*Hint:* Normal body temperature is 37°C.]

6. On a mild Saturday morning while people are working inside, the furnace keeps the temperature inside the building at 21°C. At noon the furnace is turned off, and the people go home. The temperature outside is a constant 12°C for the rest of the afternoon. If the time constant for the building is 3 hr, when will the temperature inside the building reach 16°C? If some windows are left open and the time constant drops to 2 hr, when will the temperature inside reach 16°C?

7. On a hot Saturday morning while people are working inside, the air conditioner keeps the temperature inside the building at 24°C. At noon the air conditioner is turned off, and the people go home. The temperature outside is a constant 35°C for the rest of the afternoon. If the time constant for the building is 4 hr, what will be the temperature inside the building at 2:00 P.M.? At 6:00 P.M.? When will the temperature inside the building reach 27°C?

8. A garage with no heating or cooling has a time constant of 2 hr. If the outside temperature varies as a sine wave with a minimum of 50°F at 2:00 A.M. and a maximum of 80°F at 2:00 P.M., determine the times at which the building reaches its lowest temperature and its highest temperature, assuming the exponential term has died off.

9. A warehouse is being built that will have neither heating nor cooling. Depending on the amount of insulation, the time constant for the building may range from 1 to 5 hr. To illustrate the effect insulation will have on the temperature inside the warehouse, assume the outside temperature varies as a sine wave, with a minimum of 16°C at 2:00 A.M. and a maximum of 32°C at 2:00 P.M. Assuming the exponential term (which involves the initial temperature T_0) has died off, what is the lowest temperature inside the building if the time constant is 1 hr? If it is 5 hr? What is the highest temperature inside the building if the time constant is 1 hr? If it is 5 hr?

10. Early Monday morning, the temperature in the lecture hall has fallen to 40°F, the same as the temperature outside. At 7:00 A.M., the janitor turns on the furnace with the thermostat set at 70°F. The time constant for

the building is $1/K = 2$ hr and that for the building along with its heating system is $1/K_1 = 1/2$ hr. Assuming that the outside temperature remains constant, what will be the temperature inside the lecture hall at 8:00 A.M.? When will the temperature inside the hall reach 65°F?

11. During the summer the temperature inside a van reaches 55°C, while that outside is a constant 35°C. When the driver gets into the van, she turns on the air conditioner with the thermostat set at 16°C. If the time constant for the van is $1/K = 2$ hr and that for the van with its air conditioning system is $1/K_1 = 1/3$ hr, when will the temperature inside the van reach 27°C?

12. Two friends sit down to talk and enjoy a cup of coffee. When the coffee is served, the impatient friend immediately adds a teaspoon of cream to his coffee. The relaxed friend waits 5 min before adding a teaspoon of cream (which has been kept at a constant temperature). The two now begin to drink their coffee. *Who has the hotter coffee?* Assume that the cream is cooler than the air and has the same heat capacity per unit volume as the coffee, and that Newton's law of cooling governs the heat transfer.

13. A solar hot-water-heating system consists of a hot-water tank and a solar panel. The tank is well insulated and has a time constant of 64 hr. The solar panel generates 2000 Btu/hr during the day, and the tank has a heat capacity of 2°F per thousand Btu. If the water in the tank is initially 110°F and the room temperature outside the tank is 80°F, what will be the temperature in the tank after 12 hr of sunlight?

14. In Problem 13, if a larger tank with a heat capacity of 1°F per thousand Btu and a time constant of 72 hr is used instead (with all other factors being the same), what will be the temperature in the tank after 12 hr?

15. **Stefan's law** of radiation states that the rate of change of temperature of a body at T degrees Kelvin in a medium at M degrees Kelvin is proportional to $M^4 - T^4$. That is,

$$\frac{dT}{dt} = k(M^4 - T^4),$$

where k is a positive constant. Solve this equation using separation of variables. Explain why Newton's law and Stefan's law are nearly the same when T is close to M and M is constant. $\left[\textit{Hint:} \text{ Factor } M^4 - T^4.\right]$

16. Show that $C_1 \cos \omega t + C_2 \sin \omega t$ can be written in the form $A \cos(\omega t - \phi)$, where $A = \sqrt{C_1^2 + C_2^2}$ and $\tan \phi = C_2/C_1$. [*Hint:* Use a standard trigonometric identity with $C_1 = A \cos \phi$, $C_2 = A \sin \phi$.] Use this fact to verify the alternate representation (8) of $F(t)$ discussed in Example 2 on page 104.

3.4 Newtonian Mechanics

Mechanics is the study of the motion of objects and the effect of forces acting on those objects. It is the foundation of several branches of physics and engineering. **Newtonian,** or **classical, mechanics** deals with the motion of **ordinary** objects—that is, objects that are large compared to an atom and slow moving compared with the speed of light. A model for Newtonian mechanics can be based on **Newton's laws of motion:**[†]

1. When a body is subject to no resultant external force, it moves with a constant velocity.
2. When a body is subject to one or more external forces, the time rate of change of the body's momentum is equal to the vector sum of the external forces acting on it.
3. When one body interacts with a second body, the force of the first body on the second is equal in magnitude, but opposite in direction, to the force of the second body on the first.

Experimental results for more than two centuries verify that these laws are extremely useful for studying the motion of ordinary objects in an **inertial reference frame**—that is, a reference frame in which an undisturbed body moves with a constant velocity. It is Newton's second law, which applies only to inertial reference frames, that enables us to formulate the equations of motion for a moving body. We can express Newton's second law by

$$\frac{dp}{dt} = F(t, x, v) \, ,$$

where $F(t, x, v)$ is the resultant force on the body at time t, location x, and velocity v, and $p(t)$ is the momentum of the body at time t. The momentum is the product of the mass of the body and its velocity—that is,

$$p(t) = mv(t)$$

—so we can express Newton's second law as

$$(1) \qquad m\frac{dv}{dt} = ma = F(t, x, v) \, ,$$

where $a = dv/dt$ is the acceleration of the body at time t.

Typically one substitutes $v = dx/dt$ for the velocity in (1) and obtains a second-order differential equation in the dependent variable x. However, in the present section, we will focus on situations where the force F does not depend on x. This enables us to regard (1) as a *first*-order equation

$$(2) \qquad m\frac{dv}{dt} = F(t, v)$$

in $v(t)$.

[†]For a discussion of Newton's laws of motion, see *Sears and Zemansky's University Physics,* 13th ed., by H. D. Young, R. A. Freedman, J. R. Sandin, and A. L. Ford (Pearson Addison Wesley, San Francisco, 2011).

To apply Newton's laws of motion to a problem in mechanics, the following general procedure may be useful.

> ### Procedure for Newtonian Models
>
> **(a)** Determine *all* relevant forces acting on the object being studied. It is helpful to draw a simple diagram of the object that depicts these forces.
> **(b)** Choose an appropriate axis or coordinate system in which to represent the motion of the object and the forces acting on it. Keep in mind that this coordinate system must be an inertial reference frame.
> **(c)** Apply Newton's second law as expressed in equation (2) to determine the equations of motion for the object.

In this section we express Newton's second law in either of two systems of units: the U.S. Customary System or the meter-kilogram-second (MKS) system. The various units in these systems are summarized in Table 3.2, along with approximate values for the gravitational acceleration (on the surface of Earth).

TABLE 3.2	Mechanical Units in the U.S. Customary and MKS Systems	
Unit	U.S. Customary System	MKS System
Distance	foot (ft)	meter (m)
Mass	slug	kilogram (kg)
Time	second (sec)	second (sec)
Force	pound (lb)	newton (N)
g (Earth)	32 ft/sec^2	9.81 m/sec^2

Example 1 An object of mass m is given an initial downward velocity v_0 and allowed to fall under the influence of gravity. Assuming the gravitational force is constant and the force due to air resistance is proportional to the velocity of the object, determine the equation of motion for this object.

Solution Two forces are acting on the object: a constant force due to the downward pull of gravity and a force due to air resistance that is proportional to the velocity of the object and acts in opposition to the motion of the object. Hence, the motion of the object will take place along a vertical axis. On this axis we choose the origin to be the point where the object was initially dropped and let $x(t)$ denote the distance the object has fallen in time t (see Figure 3.7).

Figure 3.7 Forces on a falling object

The forces acting on the object along this axis can be expressed as follows. The force due to gravity is

$$F_1 = mg ,$$

where g is the acceleration due to gravity at Earth's surface (see Table 3.2). The force due to air resistance is

$$F_2 = -bv(t) ,$$

where $b(>0)$ is the proportionality constant[†] and the negative sign is present because air resistance acts in opposition to the motion of the object. Hence, the net force acting on the object (see Figure 3.7) is

(3) $F = F_1 + F_2 = mg - bv(t)$.

We now apply Newton's second law by substituting (3) into (2) to obtain

$$m\frac{dv}{dt} = mg - bv .$$

Since the initial velocity of the object is v_0, a model for the velocity of the falling body is expressed by the initial value problem

(4) $m\dfrac{dv}{dt} = mg - bv , \qquad v(0) = v_0 ,$

where g and b are positive constants.

The model (4) is the same as the one we obtained in Section 2.1. Using separation of variables or the method of Section 2.3 for linear equations, we get

(5) $v(t) = \dfrac{mg}{b} + \left(v_0 - \dfrac{mg}{b} \right) e^{-bt/m}$.

Since we have taken $x = 0$ when $t = 0$, we can determine the equation of motion of the object by integrating $v = dx/dt$ with respect to t. Thus, from (5) we obtain

$$x(t) = \int v(t)\, dt = \frac{mg}{b} t - \frac{m}{b}\left(v_0 - \frac{mg}{b} \right) e^{-bt/m} + c ,$$

and setting $x = 0$ when $t = 0$, we find

$$0 = -\frac{m}{b}\left(v_0 - \frac{mg}{b} \right) + c ,$$

$$c = \frac{m}{b}\left(v_0 - \frac{mg}{b} \right) .$$

Hence, the equation of motion is

(6) $x(t) = \dfrac{mg}{b} t + \dfrac{m}{b}\left(v_0 - \dfrac{mg}{b} \right)\left(1 - e^{-bt/m} \right)$. ◆

[†]The units of b are lb-sec/ft in the U.S. system, and N-sec/m in the MKS system.

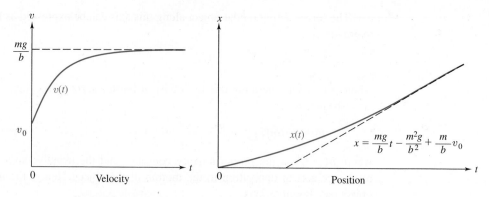

Figure 3.8 Graphs of the position and velocity of a falling object when $v_0 < mg/b$

In Figure 3.8, we have sketched the graphs of the velocity and the position as functions of t. Observe that the velocity $v(t)$ approaches the horizontal asymptote $v = mg/b$ as $t \to +\infty$ and that the position $x(t)$ asymptotically approaches the line

$$x = \frac{mg}{b}t - \frac{m^2g}{b^2} + \frac{m}{b}v_0$$

as $t \to +\infty$. The value mg/b of the horizontal asymptote for $v(t)$ is called the **limiting,** or **terminal, velocity** of the object, and, in fact, $v = mg/b = constant$ is a solution of (4). Since the two forces are in balance, this is called an "equilibrium" solution.

Notice that the terminal velocity depends on the mass but not the initial velocity of the object; the velocity of every free-falling body approaches the limiting value mg/b. Heavier objects *do,* in the presence of friction, ultimately fall faster than lighter ones, but for a given object the terminal velocity is the same whether it initially is tossed upward or downward, or simply dropped from rest.

Now that we have obtained the equation of motion for a falling object with air resistance proportional to v, we can answer a variety of questions.

Example 2 An object of mass 3 kg is released from rest 500 m above the ground and allowed to fall under the influence of gravity. Assume the gravitational force is constant, with $g = 9.81$ m/sec^2, and the force due to air resistance is proportional to the velocity of the object[†] with proportionality constant $b = 3$ N-sec/m. Determine when the object will strike the ground.

Solution We can use the model discussed in Example 1 with $v_0 = 0$, $m = 3$, $b = 3$, and $g = 9.81$. From (6), the equation of motion in this case is

$$x(t) = \frac{(3)(9.81)}{3}t - \frac{(3)^2(9.81)}{(3)^2}(1 - e^{-3t/3}) = (9.81)t - (9.81)(1 - e^{-t}).$$

Because the object is released 500 m above the ground, we can determine when the object strikes the ground by setting $x(t) = 500$ and solving for t. Thus, we put

$$500 = (9.81)t - 9.81 + (9.81)e^{-t}$$

[†]The effects of more sophisticated air resistance models (such as a quadratic drag law) are analyzed numerically in Exercises 3.6, Problem 20.

or

$$t + e^{-t} = \frac{509.81}{9.81} = 51.97,$$

where we have rounded the computations to two decimal places. Unfortunately, this equation cannot be solved explicitly for t. We might try to approximate t using Newton's approximation method (see Appendix B), but in this case, it is not necessary. Since e^{-t} will be very small for t near 51.97 ($e^{-51.97} \approx 10^{-22}$), we simply ignore the term e^{-t} and obtain as our approximation $t = 51.97$ sec. ◆

Example 3 A parachutist whose mass is 75 kg drops from a helicopter hovering 4000 m above the ground and falls toward the earth under the influence of gravity. Assume the gravitational force is constant. Assume also that the force due to air resistance is proportional to the velocity of the parachutist, with the proportionality constant $b_1 = 15$ N-sec/m when the chute is closed and with constant $b_2 = 105$ N-sec/m when the chute is open. If the chute does not open until 1 min after the parachutist leaves the helicopter, after how many seconds will she hit the ground?

Solution We are interested only in when the parachutist will hit the ground, not where. Thus, we consider only the vertical component of her descent. For this, we need to use two equations: one to describe the motion before the chute opens and the other to apply after it opens. Before the chute opens, the model is the same as in Example 1 with $v_0 = 0$, $m = 75$ kg, $b = b_1 = 15$ N-sec/m, and $g = 9.81$ m/sec². If we let $x_1(t)$ be the distance the parachutist has fallen in t sec and let $v_1 = dx_1/dt$, then substituting into equations (5) and (6), we have

$$v_1(t) = \frac{(75)(9.81)}{15}\left(1 - e^{-(15/75)t}\right)$$

$$= (49.05)\left(1 - e^{-0.2t}\right),$$

and

$$x_1(t) = \frac{(75)(9.81)}{15}t - \frac{(75)^2(9.81)}{(15)^2}\left(1 - e^{-(15/75)t}\right)$$

$$= 49.05t - 245.25\left(1 - e^{-0.2t}\right).$$

Hence, after 1 min, when $t = 60$, the parachutist is falling at the rate

$$v_1(60) = (49.05)\left(1 - e^{-0.2(60)}\right) = 49.05 \text{ m/sec},$$

and has fallen

$$x_1(60) = (49.05)(60) - (245.25)\left(1 - e^{-0.2(60)}\right) = 2697.75 \text{ m}.$$

(In these and other computations for this problem, we round our answers to two decimal places.)

Now when the chute opens, the parachutist is $4000 - 2697.75$ or 1302.25 m above the ground and traveling at a velocity of 49.05 m/sec. To determine the equation of motion after the chute opens, let $x_2(T)$ denote the position of the parachutist T sec after the chute opens (so that $T = t - 60$), taking $x_2(0) = 0$ at $x_1(60)$ (see Figure 3.9, page 114). Further assume that the initial velocity of the parachutist after the chute opens is the same as the final velocity

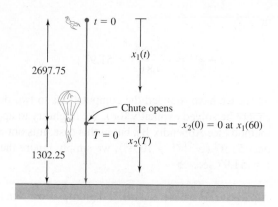

Figure 3.9 The fall of the parachutist

before it opens—that is, $x'_2(0) = x'_1(60) = 49.05$ m/sec. Because the forces acting on the parachutist are the same as those acting on the object in Example 1, we can again use equations (5) and (6). With $v_0 = 49.05$, $m = 75$, $b = b_2 = 105$, and $g = 9.81$, we find from (6) that

$$x_2(T) = \frac{(75)(9.81)}{105}T + \frac{75}{105}\left[49.05 - \frac{(75)(9.81)}{105}\right]\left(1 - e^{-(105/75)T}\right)$$

$$= 7.01T + 30.03\left(1 - e^{-1.4T}\right).$$

To determine when the parachutist will hit the ground, we set $x_2(T) = 1302.25$, the height the parachutist was above the ground when her parachute opened. This gives

$$7.01T + 30.03 - 30.03e^{-1.4T} = 1302.25$$

(7) $$T - 4.28e^{-1.4T} - 181.49 = 0.$$

Again we cannot solve (7) explicitly for T. However, observe that $e^{-1.4T}$ is very small for T near 181.49, so we ignore the exponential term and obtain $T = 181.49$. Hence, the parachutist will strike the ground 181.49 sec after the parachute opens, or 241.49 sec after dropping from the helicopter. ◆

In the computation for T in equation (7), we found that the exponential $e^{-1.4T}$ was negligible. Consequently, ignoring the corresponding exponential term in equation (5), we see that the parachutist's velocity at impact is

$$\frac{mg}{b_2} = \frac{(75)(9.81)}{105} = 7.01 \text{ m/sec},$$

which is the limiting velocity for her fall with the chute open.

Example 4 In some situations the resistive drag force on an object is proportional to a power of $|v|$ other than 1. Then when the velocity is positive, Newton's second law for a falling object generalizes to

(8) $$m\frac{dv}{dt} = mg - bv^r,$$

where m and g have their usual interpretation and b (>0) and the exponent r are experimental constants. [More generally, the drag force would be written as $-b|v|^r \text{sign}(v)$.] Express the solution to (8) for the case $r = 2$.

Solution The (positive) equilibrium solution, with the forces in balance, is $v = v_0 = \sqrt{mg/b}$. Otherwise, we write (8) as $v' = b(v_0^2 - v^2)/m$, a separable equation that we can solve using partial fractions or the integral tables at the back of the book:

$$\int \frac{dv}{v_0^2 - v^2} = \frac{1}{2v_0} \ln\left|\frac{v_0 + v}{v_0 - v}\right| = \frac{b}{m}t + c_1,$$

$$\left|\frac{v_0 + v}{v_0 - v}\right| = c_2 e^{2v_0 bt/m},$$

and after some algebra,

$$v = v_0 \frac{c_3 - e^{-2v_0 bt/m}}{c_3 + e^{-2v_0 bt/m}}.$$

Here $c_3 = c_2 \operatorname{sign}\left[(v_0 + v)/(v_0 - v)\right]$ is determined by the initial conditions. Again we see that v approaches the terminal velocity v_0 as $t \to \infty$. ◆

3.4 EXERCISES

Unless otherwise stated, in the following problems we assume that the gravitational force is constant with $g = 9.81$ m/sec^2 in the MKS system and $g = 32$ ft/sec^2 in the U.S. Customary System.

1. An object of mass 5 kg is released from rest 1000 m above the ground and allowed to fall under the influence of gravity. Assuming the force due to air resistance is proportional to the velocity of the object with proportionality constant $b = 50$ N-sec/m, determine the equation of motion of the object. When will the object strike the ground?

2. A 400-lb object is released from rest 500 ft above the ground and allowed to fall under the influence of gravity. Assuming that the force in pounds due to air resistance is $-10v$, where v is the velocity of the object in ft/sec, determine the equation of motion of the object. When will the object hit the ground?

3. If the object in Problem 1 has a mass of 500 kg instead of 5 kg, when will it strike the ground? [*Hint:* Here the exponential term is too large to ignore. Use Newton's method to approximate the time t when the object strikes the ground (see Appendix B).]

4. If the object in Problem 2 is released from rest 30 ft above the ground instead of 500 ft, when will it strike the ground? [*Hint:* Use Newton's method to solve for t.]

5. An object of mass 5 kg is given an initial downward velocity of 50 m/sec and then allowed to fall under the influence of gravity. Assume that the force in newtons due to air resistance is $-10v$, where v is the velocity of the object in m/sec. Determine the equation of motion of the object. If the object is initially 500 m above the ground, determine when the object will strike the ground.

6. An object of mass 8 kg is given an upward initial velocity of 20 m/sec and then allowed to fall under the influence of gravity. Assume that the force in newtons due to air resistance is $-16v$, where v is the velocity of the object in m/sec. Determine the equation of motion of the object. If the object is initially 100 m above the ground, determine when the object will strike the ground.

7. A parachutist whose mass is 75 kg drops from a helicopter hovering 2000 m above the ground and falls toward the ground under the influence of gravity. Assume that the force due to air resistance is proportional to the velocity of the parachutist, with the proportionality constant $b_1 = 30$ N-sec/m when the chute is closed and $b_2 = 90$ N-sec/m when the chute is open. If the chute does not open until the velocity of the parachutist reaches 20 m/sec, after how many seconds will she reach the ground?

8. A parachutist whose mass is 100 kg drops from a helicopter hovering 3000 m above the ground and falls under the influence of gravity. Assume that the force due to air resistance is proportional to the velocity of the parachutist, with the proportionality constant $b_3 = 20$ N-sec/m when the chute is closed and $b_4 = 100$ N-sec/m when the chute is open. If the chute does not open until 30 sec after the parachutist leaves the helicopter, after how many seconds will he hit the ground? If the chute does not open until 1 min after he leaves the helicopter, after how many seconds will he hit the ground?

9. An object of mass 100 kg is released from rest from a boat into the water and allowed to sink. While gravity is pulling the object down, a buoyancy force of $1/40$ times the weight of the object is pushing the object up (weight $= mg$). If we assume that water resistance exerts a force on the object that is proportional to the velocity of the object, with proportionality constant 10 N-sec/m, find the equation of motion of the object. After how many seconds will the velocity of the object be 70 m/sec?

10. An object of mass 2 kg is released from rest from a platform 30 m above the water and allowed to fall under the influence of gravity. After the object strikes the water, it begins to sink with gravity pulling down and a buoyancy force pushing up. Assume that the force of gravity is constant, no change in momentum occurs on impact with the water, the buoyancy force is $1/2$ the weight (weight $= mg$), and the force due to air resistance or water resistance is proportional to the velocity, with proportionality constant $b_1 = 10$ N-sec/m in the air and $b_2 = 100$ N-sec/m in the water. Find the equation of motion of the object. What is the velocity of the object 1 min after it is released?

11. In Example 1, we solved for the velocity of the object as a function of time (equation (5)). In some cases, it is useful to have an expression, independent of t, that relates v and x. Find this relation for the motion in Example 1. [*Hint:* Letting $v(t) = V(x(t))$, then $dv/dt = (dV/dx)V$.]

12. A shell of mass 2 kg is shot upward with an initial velocity of 200 m/sec. The magnitude of the force on the shell due to air resistance is $|v|/20$. When will the shell reach its maximum height above the ground? What is the maximum height?

13. When the velocity v of an object is very large, the magnitude of the force due to air resistance is proportional to v^2 with the force acting in opposition to the motion of the object. A shell of mass 3 kg is shot upward from the ground with an initial velocity of 500 m/sec. If the magnitude of the force due to air resistance is $(0.1)v^2$, when will the shell reach its maximum height above the ground? What is the maximum height?

14. An object of mass m is released from rest and falls under the influence of gravity. If the magnitude of the force due to air resistance is bv^n, where b and n are positive constants, find the limiting velocity of the object (assuming this limit exists). [*Hint:* Argue that the existence of a (finite) limiting velocity implies that $dv/dt \to 0$ as $t \to +\infty$.]

15. A rotating flywheel is being turned by a motor that exerts a constant torque T (see Figure 3.10). A retarding torque due to friction is proportional to the angular velocity ω. If the moment of inertia of the flywheel is I and its initial angular velocity is ω_0, find the equation for the angular velocity ω as a function of time. [*Hint:* Use Newton's second law for rotational motion,

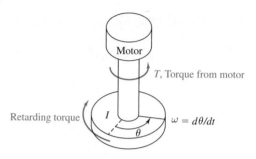

Figure 3.10 Motor-driven flywheel

that is, moment of inertia $\times$ angular acceleration $=$ sum of the torques.]

16. Find the equation for the angular velocity ω in Problem 15, assuming that the retarding torque is proportional to $\sqrt{\omega}$.

17. In Problem 16, let $I = 50$ kg-m^2 and the retarding torque be $5\sqrt{\omega}$ N-m. If the motor is turned off with the angular velocity at 225 rad/sec, determine how long it will take for the flywheel to come to rest.

18. When an object slides on a surface, it encounters a resistance force called **friction**. This force has a magnitude of μN, where μ is the **coefficient of kinetic friction** and N is the magnitude of the normal force that the surface applies to the object. Suppose an object of mass 30 kg is released from the top of an inclined plane that is inclined 30° to the horizontal (see Figure 3.11). Assume the gravitational force is constant, air resistance is negligible, and the coefficient of kinetic friction $\mu = 0.2$. Determine the equation of motion for the object as it slides down the plane. If the top surface of the plane is 5 m long, what is the velocity of the object when it reaches the bottom?

19. An object of mass 60 kg starts from rest at the top of a 45° inclined plane. Assume that the coefficient of kinetic friction is 0.05 (see Problem 18). If the force due to air resistance is proportional to the velocity of the object, say, $-3v$, find the equation of motion of the object. How long will it take the object to reach the bottom of the inclined plane if the incline is 10 m long?

Figure 3.11 Forces on an object on an inclined plane

20. An object at rest on an inclined plane will not slide until the component of the gravitational force down the incline is sufficient to overcome the force due to **static friction.** Static friction is governed by an experimental law somewhat like that of kinetic friction (Problem 18); it has a magnitude of *at most* μN, where μ is the coefficient of static friction and N is, again, the magnitude of the normal force exerted by the surface on the object. If the plane is inclined at an angle α, determine the critical value α_0 for which the object will slide if $\alpha > \alpha_0$ but will not move for $\alpha < \alpha_0$.

21. A sailboat has been running (on a straight course) under a light wind at 1 m/sec. Suddenly the wind picks up, blowing hard enough to apply a constant force of 600 N to the sailboat. The only other force acting on the boat is water resistance that is proportional to the velocity of the boat. If the proportionality constant for water resistance is $b = 100$ N-sec/m and the mass of the sailboat is 50 kg, find the equation of motion of the sailboat. What is the limiting velocity of the sailboat under this wind?

22. In Problem 21 it is observed that when the velocity of the sailboat reaches 5 m/sec, the boat begins to rise out of the water and "plane." When this happens, the proportionality constant for the water resistance drops to $b_0 = 60$ N-sec/m. Now find the equation of motion of the sailboat. What is the limiting velocity of the sailboat under this wind as it is planing?

23. Sailboats A and B each have a mass of 60 kg and cross the starting line at the same time on the first leg of a race. Each has an initial velocity of 2 m/sec. The wind applies a constant force of 650 N to each boat, and the force due to water resistance is proportional to the velocity of the boat. For sailboat A the proportionality constants are $b_1 = 80$ N-sec/m before planing when the velocity is less than 5 m/sec and $b_2 = 60$ N-sec/m when the velocity is above 5 m/sec. For sailboat B the proportionality constants are $b_3 = 100$ N-sec/m before planing when the velocity is less than 6 m/sec and $b_4 = 50$ N-sec/m when the velocity is above 6 m/sec. If the first leg of the race is 500 m long, which sailboat will be leading at the end of the first leg?

24. Rocket Flight. A model rocket having initial mass m_0 kg is launched vertically from the ground. The rocket expels gas at a constant rate of α kg/sec and at a constant velocity of β m/sec relative to the rocket. Assume that the magnitude of the gravitational force is proportional to the mass with proportionality constant g. Because the mass is not constant, Newton's second law leads to the equation

$$(m_0 - \alpha t)\frac{dv}{dt} - \alpha\beta = -g(m_0 - \alpha t),$$

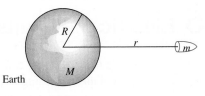

Figure 3.12 Projectile escaping from Earth

where $v = dx/dt$ is the velocity of the rocket, x is its height above the ground, and $m_0 - \alpha t$ is the mass of the rocket at t sec after launch. If the initial velocity is zero, solve the above equation to determine the velocity of the rocket and its height above ground for $0 \leq t < m_0/\alpha$.

25. Escape Velocity. According to Newton's law of gravitation, the attractive force between two objects varies inversely as the square of the distances between them. That is, $F_g = GM_1M_2/r^2$, where M_1 and M_2 are the masses of the objects, r is the distance between them (center to center), F_g is the attractive force, and G is the constant of proportionality. Consider a projectile of constant mass m being fired vertically from Earth (see Figure 3.12). Let t represent time and v the velocity of the projectile.

(a) Show that the motion of the projectile, under Earth's gravitational force, is governed by the equation

$$\frac{dv}{dt} = -\frac{gR^2}{r^2},$$

where r is the distance between the projectile and the center of Earth, R is the radius of Earth, M is the mass of Earth, and $g = GM/R^2$.

(b) Use the fact that $dr/dt = v$ to obtain

$$v\frac{dv}{dr} = -\frac{gR^2}{r^2}.$$

(c) If the projectile leaves Earth's surface with velocity v_0, show that

$$v^2 = \frac{2gR^2}{r} + v_0^2 - 2gR.$$

(d) Use the result of part (c) to show that the velocity of the projectile remains positive if and only if $v_0^2 - 2gR > 0$. The velocity $v_e = \sqrt{2gR}$ is called the **escape velocity** of Earth.

(e) If $g = 9.81$ m/sec^2 and $R = 6370$ km for Earth, what is Earth's escape velocity?

(f) If the acceleration due to gravity for the moon is $g_m = g/6$ and the radius of the moon is $R_m = 1738$ km, what is the escape velocity of the moon?

3.5 Electrical Circuits

In this section we consider the application of first-order differential equations to simple electrical circuits consisting of a voltage source (e.g., a battery or a generator), a resistor, and either an inductor or a capacitor. These so-called *RL* and *RC* circuits are shown in Figure 3.13. More general circuits will be discussed in Section 5.7.

The physical principles governing electrical circuits were formulated by G. R. Kirchhoff[†] in 1859. They are the following:

1. **Kirchhoff's current law** The algebraic sum of the currents flowing into any junction point must be zero.
2. **Kirchhoff's voltage law** The algebraic sum of the instantaneous changes in potential (voltage drops) around any closed loop must be zero.

Kirchhoff's current law implies that the same current passes through all elements in each circuit of Figure 3.13.

To apply Kirchhoff's voltage law, we need to know the voltage drop across each element of the circuit. These voltage formulas are stated below (you can consult an introductory physics text for further details).

(a) According to Ohm's law, the voltage drop E_R across a resistor is proportional to the current I passing through the resistor:

$$E_R = RI\,.$$

The proportionality constant R is called the **resistance.**

(b) It can be shown using Faraday's law and Lenz's law that the voltage drop E_L across an inductor is proportional to the instantaneous rate of change of the current I:

$$E_L = L\frac{dI}{dt}\,.$$

The proportionality constant L is called the **inductance.**

Figure 3.13 (a) *RL* circuit and (b) *RC* circuit

[†]*Historical Footnote:* Gustav Robert Kirchhoff (1824–1887) was a German physicist noted for his research in spectrum analysis, optics, and electricity.

(c) The voltage drop E_C across a capacitor is proportional to the electrical charge q on the capacitor:

$$E_C = \frac{1}{C}q \,.$$

The constant C is called the **capacitance.**

TABLE 3.3	Common Units and Symbols Used With Electrical Circuits		
Quantity	**Letter Representation**	**Units**	**Symbol Representation**
Voltage source	E	volt (V)	–○– Generator ⊣⊢ Battery
Resistance	R	ohm (Ω)	–⋀⋀–
Inductance	L	henry (H)	–⬯⬯⬯–
Capacitance	C	farad (F)	⊣⊢
Charge	q	coulomb (C)	
Current	I	ampere (A)	

The common units and symbols used for electrical circuits are listed in Table 3.3.

A voltage source is assumed to *add* voltage or potential energy to the circuit. If we let $E(t)$ denote the voltage supplied to the circuit at time t, then applying Kirchhoff's voltage law to the *RL* circuit in Figure 3.13(a) on page 118 gives

(1) $E_L + E_R = E(t) \,.$

Substituting into (1) the expressions for E_L and E_R gives

(2) $L\dfrac{dI}{dt} + RI = E(t) \,.$

Note that this equation is linear (compare Section 2.3), and upon writing it in standard form we obtain the integrating factor

$$\mu(t) = e^{\int (R/L)\,dt} = e^{Rt/L} \,,$$

which leads to the following general solution [see equation (8), Section 2.3, page 50]

(3) $I(t) = e^{-Rt/L}\left[\displaystyle\int e^{Rt/L}\dfrac{E(t)}{L}\,dt + K \right] \,.$

For the *RL* circuit, one is usually given the initial current $I(0)$ as an initial condition.

Example 1 An *RL* circuit with a 1-Ω resistor and a 0.01-H inductor is driven by a voltage $E(t) = \sin 100t$ V. If the initial inductor current is zero, determine the subsequent resistor and inductor voltages and the current.

Solution From equation (3) and the integral tables, we find that the general solution to the linear equation (2) is given by

$$I(t) = e^{-100t}\left(\int e^{100t} \frac{\sin 100t}{0.01} dt + K \right)$$

$$= e^{-100t}\left[100 \frac{e^{100t}(100 \sin 100t - 100 \cos 100t)}{10{,}000 + 10{,}000} + K \right]$$

$$= \frac{\sin 100t - \cos 100t}{2} + Ke^{-100t} .$$

For $I(0) = 0$, we obtain $-1/2 + K = 0$, so $K = 1/2$ and the current is

$$I(t) = 0.5\left(\sin 100t - \cos 100t + e^{-100t} \right) .$$

The resistor and inductor voltages are then given by

$$E_R(t) = RI(t) \equiv I(t) ,$$

$$E_L(t) = L\frac{dI}{dt} = (0.5)(\cos 100t + \sin 100t - e^{-100t}) . \quad \blacklozenge$$

Now we turn to the *RC* circuit in Figure 3.13(b), page 118. Applying Kirchhoff's voltage law yields

$$RI + q/C = E(t) .$$

The capacitor current, however, is the rate of change of its charge: $I = dq/dt$. So

(4) $$R\frac{dq}{dt} + \frac{q}{C} = E$$

is the governing differential equation for the *RC* circuit. The initial condition for a capacitor is its charge q at $t = 0$.

Example 2 Suppose a capacitor of C farads holds an initial charge of Q coulombs. To alter the charge, a constant voltage source of V volts is applied through a resistance of R ohms. Describe the capacitor charge for $t > 0$.

Solution Since $E(t) = V$ is constant in equation (4), the latter is both separable and linear, and its general solution is easily derived:

$$q(t) = CV + Ke^{-t/RC} .$$

The solution meeting the prescribed initial condition is

$$q(t) = CV + (Q - CV)e^{-t/RC} .$$

The capacitor charge changes exponentially from Q to CV as time increases. $\quad \blacklozenge$

If we take $V = 0$ in Example 2, we see that the *time constant*—that is, the time required for the capacitor charge to drop to $1/e$ times its initial value—is RC. Thus, a capacitor is a leaky energy-storage device; even the very high resistivity of the surrounding air can dissipate its charge, particularly on a humid day. Capacitors are used in cellular phones to store electrical energy from the battery while the phone is in a (more-or-less) idle receiving mode and then assist the battery in delivering energy during the transmitting mode.

The time constant for inductor current in the *RL* circuit can be gleaned from Example 1 to be L/R. An application of the *RL* circuit is the spark plug of a combustion engine. If a voltage source establishes a nonzero current in an inductor and the source is suddenly disconnected, the rapid

change of current produces a high dI/dt and, in accordance with the formula $E_L = L\,dI/dt$, the inductor generates a voltage surge sufficient to cause a spark across the terminals—thus igniting the gasoline.

If an inductor and a capacitor *both* appear in a circuit, the governing differential equation will be second order. We'll return to *RLC* circuits in Section 5.7.

3.5 EXERCISES

1. An *RL* circuit with a 5-Ω resistor and a 0.05-H inductor carries a current of 1 A at $t = 0$, at which time a voltage source $E(t) = 5 \cos 120t$ V is added. Determine the subsequent inductor current and voltage.

2. An *RC* circuit with a 1-Ω resistor and a 0.000001-F capacitor is driven by a voltage $E(t) = \sin 100t$ V. If the initial capacitor voltage is zero, determine the subsequent resistor and capacitor voltages and the current.

3. The pathway for a binary electrical signal between gates in an integrated circuit can be modeled as an *RC* circuit, as in Figure 3.13(b); the voltage source models the transmitting gate, and the capacitor models the receiving gate. Typically, the resistance is 100 Ω, and the capacitance is very small, say, 10^{-12} F (1 picofarad, pF). If the capacitor is initially uncharged and the transmitting gate changes instantaneously from 0 to 5 V, how long will it take for the voltage at the receiving gate to reach (say) 3 V? (This is the time it takes to transmit a logical "1.")

4. If the resistance in the *RL* circuit of Figure 3.13(a) is zero, show that the current $I(t)$ is directly proportional to the integral of the applied voltage $E(t)$. Similarly, show that if the resistance in the *RC* circuit of Figure 3.13(b) is zero, the current is directly proportional to the derivative of the applied voltage.

5. The *power* generated or dissipated by a circuit element equals the voltage across the element times the current through the element. Show that the power dissipated by a resistor equals I^2R, the power associated with an inductor equals the derivative of $(1/2)LI^2$, and the power associated with a capacitor equals the derivative of $(1/2)CE_C^2$.

6. Derive a power balance equation for the *RL* and *RC* circuits. (See Problem 5.) Discuss the significance of the signs of the three power terms.

7. An industrial electromagnet can be modeled as an *RL* circuit, while it is being energized with a voltage source. If the inductance is 10 H and the wire windings contain 3 Ω of resistance, how long does it take a constant applied voltage to energize the electromagnet to within 90% of its final value (that is, the current equals 90% of its asymptotic value)?

8. A 10^{-8}-F capacitor (10 nanofarads) is charged to 50 V and then disconnected. One can model the charge leakage of the capacitor with a *RC* circuit with no voltage source and the resistance of the air between the capacitor plates. On a cold dry day, the resistance of the air gap is 5×10^{13} Ω; on a humid day, the resistance is 7×10^6 Ω. How long will it take the capacitor voltage to dissipate to half its original value on each day?

3.6 Numerical Methods: A Closer Look At Euler's Algorithm

Although the analytical techniques presented in Chapter 2 were useful for the variety of mathematical models presented earlier in this chapter, *the majority of the differential equations encountered in applications cannot be solved either implicitly or explicitly.* This is especially true of higher-order equations and systems of equations, which we study in later chapters. In this section and the next, we discuss methods for obtaining a numerical *approximation* of the solution to an initial value problem for a first-order differential equation. Our goal is to develop algorithms that you can use with a calculator or computer.[†] These algorithms also extend naturally

[†]Appendix G describes various websites and commercial software that sketch direction fields and automate most of the differential equation algorithms discussed in this book.

to higher-order equations (see Section 5.3). We describe the rationale behind each method but leave the more detailed discussion to texts on numerical analysis.[†]

Consider the initial value problem

$$(1) \qquad y' = f(x, y), \qquad y(x_0) = y_0.$$

To guarantee that (1) has a unique solution, we assume that f and $\partial f / \partial y$ are continuous in a rectangle $R := \{(x, y): a < x < b, c < y < d\}$ containing (x_0, y_0). It follows from Theorem 1 in Chapter 1 (page 11) that the initial value problem (1) has a unique solution $\phi(x)$ in some interval $x_0 - \delta < x < x_0 + \delta$, where δ is a positive number. Because δ is not known *a priori*, there is no assurance that the solution will exist at a particular point x ($\neq x_0$), even if x is in the interval (a, b). However, if $\partial f / \partial y$ is continuous and *bounded*[††] on the vertical strip

$$S := \{(x, y): a < x < b, -\infty < y < \infty\},$$

then it turns out that (1) has a unique solution on the whole interval (a, b). In describing numerical methods, we assume that this last condition is satisfied and that f possesses as many continuous partial derivatives as needed.

In Section 1.4 we used the concept of direction fields to motivate a scheme for approximating the solution to the initial value problem (1). This scheme, called **Euler's method,** is one of the most basic, so it is worthwhile to discuss its advantages, disadvantages, and possible improvements. We begin with a derivation of Euler's method that is somewhat different from that presented in Section 1.4.

Let $h > 0$ be fixed (h is called the **step size**) and consider the equally spaced points

$$(2) \qquad x_n := x_0 + nh, \qquad n = 0, 1, 2, \ldots.$$

Our goal is to obtain an approximation to the solution $\phi(x)$ of the initial value problem (1) at those points x_n that lie in the interval (a, b). Namely, we will describe a method that generates values $y_0, y_1, y_2, \ldots$ that approximate $\phi(x)$ at the respective points $x_0, x_1, x_2, \ldots$; that is,

$$y_n \approx \phi(x_n), \qquad n = 0, 1, 2, \ldots.$$

Of course, the first "approximant" y_0 is exact, since $y_0 = \phi(x_0)$ is given. Thus, we must describe how to compute $y_1, y_2, \ldots$.

For Euler's method we begin by integrating both sides of equation (1) from x_n to x_{n+1} to obtain

$$\phi(x_{n+1}) - \phi(x_n) = \int_{x_n}^{x_{n+1}} \phi'(t)\, dt = \int_{x_n}^{x_{n+1}} f(t, \phi(t))\, dt,$$

where we have substituted $\phi(x)$ for y. Solving for $\phi(x_{n+1})$, we have

$$(3) \qquad \phi(x_{n+1}) = \phi(x_n) + \int_{x_n}^{x_{n+1}} f(t, \phi(t))\, dt.$$

Without knowing $\phi(t)$, we cannot integrate $f(t, \phi(t))$. Hence, we must approximate the integral in (3). Assuming we have already found $y_n \approx \phi(x_n)$, the simplest approach is to

[†]See, for example, *A First Course in the Numerical Analysis of Differential Equations*, 2nd ed., by A. Iserles (Cambridge University Press, 2008), or *Numerical Analysis*, 9th ed., by R. L. Burden and J. D. Faires (Cengage Learning, Independence, KY, 2011).

[††]A function $g(x, y)$ is bounded on S if there exists a number M such that $|g(x, y)| \leq M$ for all (x, y) in S.

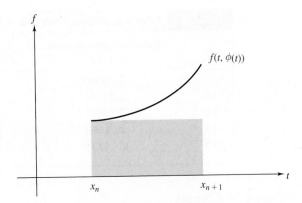

Figure 3.14 Approximation by a rectangle

approximate the area under the function $f(t, \phi(t))$ by the rectangle with base $\left[x_n, x_{n+1} \right]$ and height $f(x_n, \phi(x_n))$ (see Figure 3.14). This gives

$$\phi(x_{n+1}) \approx \phi(x_n) + (x_{n+1} - x_n)f(x_n, \phi(x_n)) \,.$$

Substituting h for $x_{n+1} - x_n$ and the approximation y_n for $\phi(x_n)$, we arrive at the numerical scheme

$$(4) \qquad y_{n+1} = y_n + hf(x_n, y_n) \,, \qquad n = 0, 1, 2, \ldots,$$

which is Euler's method.

Starting with the given value y_0, we use (4) to compute $y_1 = y_0 + hf(x_0, y_0)$ and then use y_1 to compute $y_2 = y_1 + hf(x_1, y_1)$, and so on. Several examples of Euler's method can be found in Section 1.4. (Compare page 24.)

As discussed in Section 1.4, if we wish to use Euler's method to approximate the solution to the initial value problem (1) at a particular value of x, say, $x = c$, then we must first determine a suitable step size h so that $x_0 + Nh = c$ for some integer N. For example, we can take $N = 1$ and $h = c - x_0$ in order to arrive at the approximation after just one step:

$$\phi(c) = \phi(x_0 + h) \approx y_1 \,.$$

If, instead, we wish to take 10 steps in Euler's method, we choose $h = (c - x_0)/10$ and ultimately obtain

$$\phi(c) = \phi(x_0 + 10h) = \phi(x_{10}) \approx y_{10} \,.$$

In general, depending on the size of h, we will get different approximations to $\phi(c)$. It is reasonable to expect that as h gets smaller (or, equivalently, as N gets larger), the Euler approximations approach the exact value $\phi(c)$. On the other hand, as h gets smaller, the number (and cost) of computations increases and hence so do machine errors that arise from round-off. Thus, it is important to analyze how the error in the approximation scheme varies with h.

If Euler's method is used to approximate the solution $\phi(x) = e^x$ to the problem

$$(5) \qquad y' = y \,, \qquad y(0) = 1 \,,$$

at $x = 1$, then we obtain approximations to the constant $e = \phi(1)$. It turns out that these approximations take a particularly simple form that enables us to compare the error in the approximation with the step size h. Indeed, setting $f(x, y) = y$ in (4) yields

$$y_{n+1} = y_n + hy_n = (1 + h)y_n \,, \qquad n = 0, 1, 2, \ldots.$$

	Euler's Approximation $(1+h)^{1/h}$	Error $e-(1+h)^{1/h}$	Error$/h$
TABLE 3.4	**Euler's Approximations to $e = 2.71828\ldots$**		
h			
1	2.00000	0.71828	0.71828
10^{-1}	2.59374	0.12454	1.24539
10^{-2}	2.70481	0.01347	1.34680
10^{-3}	2.71692	0.00136	1.35790
10^{-4}	2.71815	0.00014	1.35902

Since $y_0 = 1$, we get

$$y_1 = (1+h)y_0 = 1+h,$$
$$y_2 = (1+h)y_1 = (1+h)(1+h) = (1+h)^2,$$
$$y_3 = (1+h)y_2 = (1+h)(1+h)^2 = (1+h)^3,$$

and, in general,

(6) $\qquad y_n = (1+h)^n, \qquad n = 0, 1, 2, \ldots.$

For the problem in (5) we have $x_0 = 0$, so to obtain approximations at $x = 1$, we must set $nh = 1$. That is, h must be the reciprocal of an integer ($h = 1/n$). Replacing n by $1/h$ in (6), we see that Euler's method gives the (familiar) approximation $(1+h)^{1/h}$ to the constant e. In Table 3.4, we list this approximation for $h = 1, 10^{-1}, 10^{-2}, 10^{-3}$, and 10^{-4}, along with the corresponding errors

$$e - (1+h)^{1/h}.$$

From the second and third columns in Table 3.4, we see that the approximation gains roughly one decimal place in accuracy as h decreases by a factor of 10; that is, the error is roughly proportional to h. This observation is further confirmed by the entries in the last column of Table 3.4. In fact, using methods of calculus (see Exercises 1.4, Problem 13), it can be shown that

(7) $\qquad \lim_{h\to 0} \dfrac{\text{error}}{h} = \lim_{h\to 0} \dfrac{e-(1+h)^{1/h}}{h} = \dfrac{e}{2} \approx 1.35914.$

The general situation is similar: When Euler's method is used to approximate the solution to the initial value problem (1), the error in the approximation is at worst a constant times the step size h. Moreover, in view of (7), this is the best one can say.

Numerical analysts have a convenient notation for describing the convergence behavior of a numerical scheme. For fixed x we denote by $y(x; h)$ the approximation to the solution $\phi(x)$ of (1) obtained via the scheme when using a step size of h. We say that the numerical scheme **converges** at x if

$$\lim_{h\to 0} y(x; h) = \phi(x).$$

In other words, as the step size h decreases to zero, the approximations for a convergent scheme approach the exact value $\phi(x)$. The rate at which $y(x; h)$ tends to $\phi(x)$ is often expressed in terms of a suitable power of h. If the error $\phi(x) - y(x; h)$ tends to zero like a constant times h^p, we write

$$\phi(x) - y(x; h) = O(h^p)$$

and say that the **method is of order p.** Of course, the higher the power p, the faster is the rate of convergence as $h \to 0$.

As seen from our earlier discussion, the rate of convergence of Euler's method is $O(h)$; that is, *Euler's method is of order $p = 1$*. In fact, the limit in (7) shows that for equation (5), the error is roughly $1.36h$ for small h. This means that to have an error less than 0.01 requires $h < 0.01/1.36$, or $n = 1/h > 136$ computation steps. Thus Euler's method converges too slowly to be of practical use.

How can we improve Euler's method? To answer this, let's return to the derivation expressed in formulas (3) and (4) and analyze the "errors" that were introduced to get the approximation. The crucial step in the process was to approximate the integral

$$\int_{x_n}^{x_{n+1}} f\big(t, \phi(t)\big)\, dt$$

by using a rectangle (recall Figure 3.14 on page 123). This step gives rise to what is called the *local truncation error* in the method. From calculus we know that a better (more accurate) approach to approximating the integral is to use a trapezoid—that is, to apply the trapezoidal rule (see Figure 3.15). This gives

$$\int_{x_n}^{x_{n+1}} f\big(t, \phi(t)\big)\, dt \approx \frac{h}{2}\Big[f\big(x_n, \phi(x_n)\big) + f\big(x_{n+1}, \phi(x_{n+1})\big)\Big],$$

which leads to the numerical scheme

$$(8) \qquad y_{n+1} = y_n + \frac{h}{2}\Big[f(x_n, y_n) + f(x_{n+1}, y_{n+1})\Big], \qquad n = 0, 1, 2, \ldots.$$

We call equation (8) the **trapezoid scheme**. It is an example of an **implicit method;** that is, unlike Euler's method, equation (8) gives only an implicit formula for y_{n+1}, since y_{n+1} appears as an argument of f. Assuming we have already computed y_n, some root-finding technique such as Newton's method (see Appendix B) might be needed to compute y_{n+1}. Despite the inconvenience of working with an implicit method, the trapezoid scheme has two advantages over Euler's method. First, it is a method of order $p = 2$; that is, it converges at a rate that is proportional to h^2 and hence is faster than Euler's method. Second, as described in Project F, page 148, the trapezoid scheme has the desirable feature of being **stable.**

Can we somehow modify the trapezoid scheme in order to obtain an explicit method? One idea is first to get an estimate, say, y_{n+1}^*, of the value y_{n+1} using Euler's method and then use formula (8) with y_{n+1} replaced by y_{n+1}^* on the right-hand side. This two-step process is an example of a **predictor–corrector method.** That is, we predict y_{n+1} using Euler's

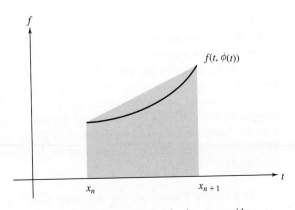

Figure 3.15 Approximation by a trapezoid

method and then use that value in (8) to obtain a "more correct" approximation. Setting $y_{n+1} = y_n + hf(x_n, y_n)$ in the right-hand side of (8), we obtain

$$(9) \qquad y_{n+1} = y_n + \frac{h}{2}\left[f(x_n, y_n) + f\left(x_n + h, y_n + hf(x_n, y_n)\right)\right], \qquad n = 0, 1, \ldots,$$

where $x_{n+1} = x_n + h$. This explicit scheme is known as the **improved Euler's method.**

Example 1 Compute the improved Euler's method approximation to the solution $\phi(x) = e^x$ of

$$y' = y, \qquad y(0) = 1$$

at $x = 1$ using step sizes of $h = 1, 10^{-1}, 10^{-2}, 10^{-3}$, and 10^{-4}.

Solution The starting values are $x_0 = 0$ and $y_0 = 1$. Since $f(x, y) = y$, formula (9) becomes

$$y_{n+1} = y_n + \frac{h}{2}\left[y_n + (y_n + hy_n)\right] = y_n + hy_n + \frac{h^2}{2}y_n;$$

that is,

$$(10) \qquad y_{n+1} = \left(1 + h + \frac{h^2}{2}\right)y_n.$$

Since $y_0 = 1$, we see inductively that

$$y_n = \left(1 + h + \frac{h^2}{2}\right)^n, \qquad n = 0, 1, 2, \ldots.$$

To obtain approximations at $x = 1$, we must have $1 = x_0 + nh = nh$, and so $n = 1/h$. Hence, the improved Euler's approximations to $e = \phi(1)$ are just

$$\left(1 + h + \frac{h^2}{2}\right)^{1/h}.$$

In Table 3.5 on page 127 we have computed this approximation for the specified values of h, along with the corresponding errors

$$e - \left(1 + h + \frac{h^2}{2}\right)^{1/h}.$$

Comparing the entries of this table with those of Table 3.4 on page 124, we observe that the improved Euler's method converges much more rapidly than the original Euler's method. In fact, from the first few entries in the second and third columns of Table 3.5, it appears that the approximation gains two decimal places in accuracy each time h is decreased by a factor of 10. In other words, the error is roughly proportional to h^2 (see the last column of the table and also Problem 4). The entries in the last row of the table must be regarded with caution. Indeed, when $h = 10^{-3}$, the true error is so small that our calculator rounded it to zero, to five decimal places. The entry in color in the last column may be inaccurate due to the loss of significant figures in the calculator arithmetic. ◆

As Example 1 suggests, the improved Euler's method converges at the rate $O(h^2)$, and indeed it can be proved that in general *this method is of order $p = 2$.*

TABLE 3.5	Improved Euler's Approximation to $e = 2.71828\ldots$		
h	Approximation $\left(1 + h + \dfrac{h^2}{2}\right)^{1/h}$	Error	Error$/h^2$
1	2.50000	0.21828	0.21828
10^{-1}	2.71408	0.00420	0.42010
10^{-2}	2.71824	0.00004	0.44966
10^{-3}	2.71828	0.00000	0.45271

A step-by-step outline for a subroutine that implements the improved Euler's method over a given interval $\left[x_0, c\right]$ is described below. For programming purposes it is usually more convenient to input the number of steps N in the interval rather than the step size h itself. For an interval starting at $x = x_0$ and ending at $x = c$, the relation between h and N is

(11) $Nh = c - x_0$.

(Note that the subroutine includes an option for printing x and y.) Of course, the implementation of this algorithm with N steps on the interval $\left[x_0, c\right]$ only produces approximations to the actual solution at $N + 1$ equally spaced points. If we wish to use these points to help graph an approximate solution over the whole interval $\left[x_0, c\right]$, then we must somehow "fill in" the gaps between these points. A crude method is to simply join the points by straight-line segments producing a polygonal line approximation to $\phi(x)$. More sophisticated techniques for prescribing the intermediate points are used in professional codes.

Improved Euler's Method Subroutine

Purpose To approximate the solution $\phi(x)$ to the initial value problem

$$y' = f(x, y), \qquad y(x_0) = y_0,$$

for $x_0 \leq x \leq c$.

INPUT x_0, y_0, c, N (number of steps), PRNTR ($= 1$ to print a table)

Step 1 Set step size $h = (c - x_0)/N$, $x = x_0$, $y = y_0$

Step 2 For $i = 1$ to N, do Steps 3–5

Step 3 Set

$$F = f(x, y)$$
$$G = f(x + h, y + hF)$$

Step 4 Set

$$x = x + h$$
$$y = y + h(F + G)/2$$

Step 5 If PRNTR $= 1$, print x, y

Now we want to devise a program that will compute $\phi(c)$ *to a desired accuracy*. As we have seen, the accuracy of the approximation depends on the step size h. Our strategy, then, will be to estimate $\phi(c)$ for a given step size and then halve the step size and recompute the estimate, halve again, and so on. When two consecutive estimates of $\phi(c)$ differ by less than some prescribed tolerance ε, we take the final estimate as our approximation to $\phi(c)$. Admittedly, this does not guarantee that $\phi(c)$ is known to within ϵ, but it provides a reasonable stopping procedure in practice.[†] The following procedure also contains a safeguard to stop if the desired tolerance is not reached after M halvings of h.

Improved Euler's Method With Tolerance

Purpose To approximate the solution to the initial value problem

$$y' = f(x, y), \qquad y(x_0) = y_0,$$

at $x = c$, with tolerance ε

INPUT $x_0, y_0, c, \varepsilon,$
 M (maximum number of halvings of step size)

Step 1 Set $z = y_0$, PRNTR $= 0$

Step 2 For $m = 0$ to M, do Steps 3–7[††]

Step 3 Set $N = 2^m$

Step 4 Call IMPROVED EULER'S METHOD SUBROUTINE

Step 5 Print h, y

Step 6 If $|y - z| < \varepsilon$, go to Step 10

Step 7 Set $z = y$

Step 8 Print "$\phi(c)$ is approximately"; y; "but may not be within the tolerance"; ε

Step 9 Go to Step 11

Step 10 Print "$\phi(c)$ is approximately"; y; "with tolerance"; ε

Step 11 STOP

OUTPUT Approximations of the solution to the initial value problem at $x = c$ using 2^m steps

If one desires a stopping procedure that simulates the **relative error**

$$\left| \frac{\text{approximation} - \text{true value}}{\text{true value}} \right| ,$$

then replace Step 6 by

Step 6′. If $\left| \dfrac{z - y}{y} \right| < \varepsilon$, go to Step 10 .

Example 2 Use the improved Euler's method with tolerance to approximate the solution to the initial value problem

(12) $y' = x + 2y, \qquad y(0) = 0.25,$

at $x = 2$. For a tolerance of $\epsilon = 0.001$, use a stopping procedure based on the absolute error.

[†]Professional codes monitor accuracy much more carefully and vary step size in an adaptive fashion for this purpose.

[††]To save time, one can start with $m = K < M$ rather than with $m = 0$.

Solution The starting values are $x_0 = 0$, $y_0 = 0.25$. Because we are computing the approximations for $c = 2$, the initial value for h is

$$h = (2-0)2^{-0} = 2.$$

For equation (12), we have $f(x, y) = x + 2y$, so the numbers F and G in the subroutine are

$$F = x + 2y,$$
$$G = (x + h) + 2(y + hF) = x + 2y + h(1 + 2x + 4y),$$

and we find

$$x = x + h,$$
$$y = y + \frac{h}{2}(F + G) = y + \frac{h}{2}(2x + 4y) + \frac{h^2}{2}(1 + 2x + 4y).$$

Thus, with $x_0 = 0$, $y_0 = 0.25$, and $h = 2$, we get for the first approximation

$$y = 0.25 + (0 + 1) + 2(1 + 1) = 5.25.$$

To describe the further outputs of the algorithm, we use the notation $y(2; h)$ for the approximation obtained with step size h. Thus, $y(2; 2) = 5.25$, and we find from the algorithm

$$
\begin{aligned}
y(2; 1) &= 11.25000 & y(2; 2^{-5}) &= 25.98132 \\
y(2; 2^{-1}) &= 18.28125 & y(2; 2^{-6}) &= 26.03172 \\
y(2; 2^{-2}) &= 23.06067 & y(2; 2^{-7}) &= 26.04468 \\
y(2; 2^{-3}) &= 25.12012 & y(2; 2^{-8}) &= 26.04797 \\
y(2; 2^{-4}) &= 25.79127 & y(2; 2^{-9}) &= 26.04880.
\end{aligned}
$$

Since $|y(2; 2^{-9}) - y(2; 2^{-8})| = 0.00083$, which is less than $\varepsilon = 0.001$, we stop.

The exact solution of (12) is $\phi(x) = \frac{1}{2}(e^{2x} - x - \frac{1}{2})$, so we have determined that

$$\phi(2) = \frac{1}{2}\left(e^4 - \frac{5}{2}\right) \approx 26.04880. \quad \blacklozenge$$

In the next section, we discuss methods with higher rates of convergence than either Euler's or the improved Euler's methods.

3.6 EXERCISES

In many of the following problems, it will be essential to have a calculator or computer available. You may use a software package[†] or write a program for solving initial value problems using the improved Euler's method algorithms on pages 127 and 128. (Remember, all trigonometric calculations are done in radians.)

1. Show that when Euler's method is used to approximate the solution of the initial value problem

 $$y' = 5y, \qquad y(0) = 1,$$

 at $x = 1$, then the approximation with step size h is $(1 + 5h)^{1/h}$.

2. Show that when Euler's method is used to approximate the solution of the initial value problem

 $$y' = -\frac{1}{2}y, \qquad y(0) = 3,$$

 at $x = 2$, then the approximation with step size h is

 $$3\left(1 - \frac{h}{2}\right)^{2/h}.$$

[†]Appendix G describes various websites and commercial software that sketch direction fields and automate most of the differential equation algorithms discussed in this book.

3. Show that when the trapezoid scheme given in formula (8) is used to approximate the solution $\phi(x) = e^x$ of

$$y' = y, \qquad y(0) = 1,$$

at $x = 1$, then we get

$$y_{n+1} = \left(\frac{1 + h/2}{1 - h/2}\right) y_n, \qquad n = 0, 1, 2, \dots,$$

which leads to the approximation

$$\left(\frac{1 + h/2}{1 - h/2}\right)^{1/h}$$

for the constant e. Compute this approximation for $h = 1$, 10^{-1}, 10^{-2}, 10^{-3}, and 10^{-4} and compare your results with those in Tables 3.4 and 3.5.

4. In Example 1, page 126, the improved Euler's method approximation to e with step size h was shown to be

$$\left(1 + h + \frac{h^2}{2}\right)^{1/h}.$$

First prove that the error $:= e - (1 + h + h^2/2)^{1/h}$ approaches zero as $h \to 0$. Then use L'Hôpital's rule to show that

$$\lim_{h \to 0} \frac{\text{error}}{h^2} = \frac{e}{6} \approx 0.45305.$$

Compare this constant with the entries in the last column of Table 3.5.

5. Show that when the improved Euler's method is used to approximate the solution of the initial value problem

$$y' = 4y, \qquad y(0) = \frac{1}{3},$$

at $x = 1/2$, then the approximation with step size h is

$$\frac{1}{3}(1 + 4h + 8h^2)^{1/(2h)}.$$

6. Since the integral $y(x) := \int_0^x f(t)\,dt$ with variable upper limit satisfies (for continuous f) the initial value problem

$$y' = f(x), \qquad y(0) = 0,$$

any numerical scheme that is used to approximate the solution at $x = 1$ will give an approximation to the definite integral

$$\int_0^1 f(t)\,dt.$$

Derive a formula for this approximation of the integral using
(a) Euler's method.
(b) the trapezoid scheme.
(c) the improved Euler's method.

7. Use the improved Euler's method subroutine with step size $h = 0.1$ to approximate the solution to the initial value problem

$$y' = x - y^2, \qquad y(1) = 0,$$

at the points $x = 1.1$, 1.2, 1.3, 1.4, and 1.5. (Thus, input $N = 5$.) Compare these approximations with those obtained using Euler's method (see Exercises 1.4, Problem 5, page 28).

8. Use the improved Euler's method subroutine with step size $h = 0.2$ to approximate the solution to the initial value problem

$$y' = \frac{1}{x}(y^2 + y), \qquad y(1) = 1,$$

at the points $x = 1.2$, 1.4, 1.6, and 1.8. (Thus, input $N = 4$.) Compare these approximations with those obtained using Euler's method (see Exercises 1.4, Problem 6, page 28).

9. Use the improved Euler's method subroutine with step size $h = 0.2$ to approximate the solution to

$$y' = x + 3\cos(xy), \qquad y(0) = 0,$$

at the points $x = 0, 0.2, 0.4, \dots, 2.0$. Use your answers to make a rough sketch of the solution on $[0, 2]$.

10. Use the improved Euler's method subroutine with step size $h = 0.1$ to approximate the solution to

$$y' = 4\cos(x + y), \qquad y(0) = 1,$$

at the points $x = 0, 0.1, 0.2, \dots, 1.0$. Use your answers to make a rough sketch of the solution on $[0, 1]$.

11. Use the improved Euler's method with tolerance to approximate the solution to

$$\frac{dx}{dt} = 1 + t\sin(tx), \qquad x(0) = 0,$$

at $t = 1$. For a tolerance of $\varepsilon = 0.01$, use a stopping procedure based on the absolute error.

12. Use the improved Euler's method with tolerance to approximate the solution to

$$y' = 1 - \sin y, \qquad y(0) = 0,$$

at $x = \pi$. For a tolerance of $\varepsilon = 0.01$, use a stopping procedure based on the absolute error.

13. Use the improved Euler's method with tolerance to approximate the solution to

$$y' = 1 - y + y^3, \qquad y(0) = 0,$$

at $x = 1$. For a tolerance of $\varepsilon = 0.003$, use a stopping procedure based on the absolute error.

14. By experimenting with the improved Euler's method subroutine, find the maximum value over the interval $[0, 2]$ of the solution to the initial value problem

$$y' = \sin(x + y), \qquad y(0) = 2.$$

Where does this maximum value occur? Give answers to two decimal places.

15. The solution to the initial value problem

$$\frac{dy}{dx} = (x + y + 2)^2, \qquad y(0) = -2$$

crosses the x-axis at a point in the interval $[0, 1.4]$. By experimenting with the improved Euler's method subroutine, determine this point to two decimal places.

16. The solution to the initial value problem

$$\frac{dy}{dx} + \frac{y}{x} = x^3 y^2, \qquad y(1) = 3$$

has a vertical asymptote ("blows up") at some point in the interval $[1, 2]$. By experimenting with the improved Euler's method subroutine, determine this point to two decimal places.

17. Use Euler's method (4) with $h = 0.1$ to approximate the solution to the initial value problem

$$y' = -20y, \qquad y(0) = 1,$$

on the interval $0 \leq x \leq 1$ (that is, at $x = 0, 0.1, \ldots, 1.0$). Compare your answers with the actual solution $y = e^{-20x}$. What went wrong? Next, try the step size $h = 0.025$ and also $h = 0.2$. What conclusions can you draw concerning the choice of step size?

18. **Local versus Global Error.** In deriving formula (4) for Euler's method, a rectangle was used to approximate the area under a curve (see Figure 3.14). With $g(t) := f(t, \phi(t))$, this approximation can be written as

$$\int_{x_n}^{x_{n+1}} g(t)\, dt \approx hg(x_n), \quad \text{where} \quad h = x_{n+1} - x_n.$$

(a) Show that if g has a continuous derivative that is bounded in absolute value by B, then the rectangle approximation has error $O(h^2)$; that is, for some constant M,

$$\left| \int_{x_n}^{x_{n+1}} g(t)\, dt - hg(x_n) \right| \leq Mh^2.$$

This is called the *local truncation error* of the scheme. [*Hint:* Write

$$\int_{x_n}^{x_{n+1}} g(t)\, dt - hg(x_n) = \int_{x_n}^{x_{n+1}} \left[g(t) - g(x_n) \right] dt.$$

Next, using the mean value theorem, show that $|g(t) - g(x_n)| \leq B|t - x_n|$. Then integrate to obtain the error bound $(B/2)h^2$.]

(b) In applying Euler's method, local truncation errors occur in each step of the process and are propagated throughout the further computations. Show that the *sum* of the local truncation errors in part (a) that arise after n steps is $O(h)$. This is the *global error*, which is the same as the convergence rate of Euler's method.

19. **Building Temperature.** In Section 3.3 we modeled the temperature inside a building by the initial value problem

$$(13) \qquad \frac{dT}{dt} = K[M(t) - T(t)] + H(t) + U(t)$$

$$T(t_0) = T_0,$$

where M is the temperature outside the building, T is the temperature inside the building, H is the additional heating rate, U is the furnace heating or air conditioner cooling rate, K is a positive constant, and T_0 is the initial temperature at time t_0. In a typical model, $t_0 = 0$ (midnight), $T_0 = 65°F$, $H(t) = 0.1$, $U(t) = 1.5[70 - T(t)]$, and

$$M(t) = 75 - 20\cos(\pi t/12).$$

The constant K is usually between $1/4$ and $1/2$, depending on such things as insulation. To study the effect of insulating this building, consider the typical building described above and use the improved Euler's method subroutine with $h = 2/3$ to approximate the solution to (13) on the interval $0 \leq t \leq 24$ (1 day) for $K = 0.2, 0.4$, and 0.6.

20. **Falling Body.** In Example 1 of Section 3.4, page 110, we modeled the velocity of a falling body by the initial value problem

$$m\frac{dv}{dt} = mg - bv, \qquad v(0) = v_0,$$

under the assumption that the force due to air resistance is $-bv$. However, in certain cases the force due to air resistance behaves more like $-bv^r$, where $r(>1)$ is some constant. This leads to the model

$$(14) \qquad m\frac{dv}{dt} = mg - bv^r, \qquad v(0) = v_0.$$

To study the effect of changing the parameter r in (14), take $m = 1$, $g = 9.81$, $b = 2$, and $v_0 = 0$. Then use the improved Euler's method subroutine with $h = 0.2$ to approximate the solution to (14) on the interval $0 \leq t \leq 5$ for $r = 1.0, 1.5$, and 2.0. What is the relationship between these solutions and the constant solution $v(t) \equiv (9.81/2)^{1/r}$?

3.7 Higher-Order Numerical Methods: Taylor and Runge–Kutta

In Sections 1.4 and 3.6, we discussed a simple numerical procedure, Euler's method, for obtaining a numerical approximation of the solution $\phi(x)$ to the initial value problem

$$(1) \qquad y' = f(x, y) , \qquad y(x_0) = y_0 .$$

Euler's method is easy to implement because it involves only linear approximations to the solution $\phi(x)$. But it suffers from slow convergence, being a method of order 1; that is, the error is $O(h)$. Even the improved Euler's method discussed in Section 3.6 has order of only 2. In this section we present numerical methods that have faster rates of convergence. These include **Taylor methods,** which are natural extensions of the Euler procedure, and **Runge–Kutta methods,** which are the more popular schemes for solving initial value problems because they have fast rates of convergence and are easy to program.

As in the previous section, we assume that f and $\partial f / \partial y$ are continuous and bounded on the vertical strip $\{(x, y): a < x < b, -\infty < y < \infty\}$ and that f possesses as many continuous partial derivatives as needed.

To derive the Taylor methods, let $\phi_n(x)$ be the *exact* solution of the related initial value problem

$$(2) \qquad \phi_n' = f(x, \phi_n) , \qquad \phi_n(x_n) = y_n .$$

The Taylor series for $\phi_n(x)$ about the point x_n is

$$\phi_n(x) = \phi_n(x_n) + h\phi_n'(x_n) + \frac{h^2}{2!}\phi_n''(x_n) + \cdots ,$$

where $h = x - x_n$. Since ϕ_n satisfies (2), we can write this series in the form

$$(3) \qquad \phi_n(x) = y_n + hf(x_n, y_n) + \frac{h^2}{2!}\phi_n''(x_n) + \cdots .$$

Observe that the recursive formula for y_{n+1} in Euler's method is obtained by truncating the Taylor series after the linear term. For a better approximation, we will use more terms in the Taylor series. This requires that we express the higher-order derivatives of the solution in terms of the function $f(x, y)$.

If y satisfies $y' = f(x, y)$, we can compute y'' by using the chain rule:

$$(4) \qquad y'' = \frac{\partial f}{\partial x}(x, y) + \frac{\partial f}{\partial y}(x, y) \, y'$$

$$= \frac{\partial f}{\partial x}(x, y) + \frac{\partial f}{\partial y}(x, y) f(x, y)$$

$$=: f_2(x, y) .$$

In a similar fashion, define $f_3, f_4, \ldots$, that correspond to the expressions $y'''(x), y^{(4)}(x)$, etc. If we truncate the expansion in (3) after the h^p term, then, with the above notation, the recursive formulas for the **Taylor method of order p** are

$$(5) \qquad x_{n+1} = x_n + h ,$$

$$(6) \qquad y_{n+1} = y_n + hf(x_n, y_n) + \frac{h^2}{2!}f_2(x_n, y_n) + \cdots + \frac{h^p}{p!}f_p(x_n, y_n) .$$

As before, $y_n \approx \phi(x_n)$, where $\phi(x)$ is the solution to the initial value problem (1). It can be shown[†] that **the Taylor method of order p has the rate of convergence $O(h^p)$.**

Example 1 Determine the recursive formulas for the Taylor method of order 2 for the initial value problem

(7) $y' = \sin(xy)$, $y(0) = \pi$.

Solution We must compute $f_2(x, y)$ as defined in (4). Since $f(x, y) = \sin(xy)$,

$$\frac{\partial f}{\partial x}(x, y) = y \cos(xy) , \qquad \frac{\partial f}{\partial y}(x, y) = x \cos(xy) .$$

Substituting into (4), we have

$$f_2(x, y) = \frac{\partial f}{\partial x}(x, y) + \frac{\partial f}{\partial y}(x, y) f(x, y)$$

$$= y \cos(xy) + x \cos(xy) \sin(xy)$$

$$= y \cos(xy) + \frac{x}{2} \sin(2xy) ,$$

and the recursive formulas (5) and (6) become

$$x_{n+1} = x_n + h ,$$

$$y_{n+1} = y_n + h \sin(x_n y_n) + \frac{h^2}{2}\left[y_n \cos(x_n y_n) + \frac{x_n}{2} \sin(2x_n y_n) \right] ,$$

where $x_0 = 0$, $y_0 = \pi$ are the starting values. ◆

The convergence rate, $O(h^p)$, of the pth-order Taylor method raises an interesting question: If we could somehow let p go to infinity, would we obtain **exact** solutions for the interval $[x_0, x_0 + h]$? This possibility is explored in depth in Chapter 8. Of course, a practical difficulty in employing high-order Taylor methods is the tedious computation of the partial derivatives needed to determine f_p (typically these computations grow exponentially with p). One way to circumvent this difficulty is to use one of the **Runge–Kutta methods.**[††]

Observe that the general Taylor method has the form

(8) $y_{n+1} = y_n + hF(x_n, y_n; h)$,

where the choice of F depends on p. In particular [compare (6)], for

$$p = 1 , \qquad F = T_1(x, y; h) := f(x, y) ,$$

(9) $$p = 2 , \qquad F = T_2(x, y; h) := f(x, y) + \frac{h}{2}\left[\frac{\partial f}{\partial x}(x, y) + \frac{\partial f}{\partial y}(x, y) f(x, y) \right] .$$

The idea behind the Runge–Kutta method of order 2 is to choose F in (8) of the form

(10) $$F = K_2(x, y; h) := f\big(x + \alpha h, y + \beta h f(x, y)\big) ,$$

where the constants α, β are to be selected so that (8) has the rate of convergence $O(h^2)$. The advantage here is that K_2 is computed by two evaluations of the original function $f(x, y)$ and does not involve the derivatives of $f(x, y)$.

[†]See *Introduction to Numerical Analysis* by J. Stoer and R. Bulirsch (Springer-Verlag, New York, 2002).

[††]*Historical Footnote:* These methods were developed by C. Runge in 1895 and W. Kutta in 1901.

To ensure $O(h^2)$ convergence, we compare this new scheme with the Taylor method of order 2 and require

$$T_2(x, y; h) - K_2(x, y; h) = O(h^2), \quad \text{as} \quad h \to 0.$$

That is, we choose α, β so that the Taylor expansions for T_2 and K_2 agree through terms of order h. For (x, y) fixed, when we expand $K_2 = K_2(h)$ as given in (10) about $h = 0$, we find

(11) $K_2(h) = K_2(0) + \dfrac{dK_2}{dh}(0)h + O(h^2)$

$$= f(x, y) + \left[\alpha \frac{\partial f}{\partial x}(x, y) + \beta \frac{\partial f}{\partial y}(x, y) f(x, y) \right] h + O(h^2),$$

where the expression in brackets for dK_2/dh, evaluated at $h = 0$, follows from the chain rule. Comparing (11) with (9), we see that for T_2 and K_2 to agree through terms of order h, we must have $\alpha = \beta = 1/2$. Thus,

$$K_2(x, y; h) = f\left(x + \frac{h}{2}, \; y + \frac{h}{2}f(x, y) \right).$$

The Runge–Kutta method we have derived is called the **midpoint method** and it has the recursive formulas

(12) $x_{n+1} = x_n + h,$

(13) $y_{n+1} = y_n + hf\left(x_n + \dfrac{h}{2}, \; y_n + \dfrac{h}{2}f(x_n, y_n) \right).$

By construction, the midpoint method has the same rate of convergence as the Taylor method of order 2; that is, $O(h^2)$. This is the same rate as the improved Euler's method.

In a similar fashion, one can work with the Taylor method of order 4 and, after some elaborate calculations, obtain the **classical fourth-order Runge–Kutta method.** The recursive formulas for this method are

$$x_{n+1} = x_n + h,$$

(14)

$$y_{n+1} = y_n + \frac{1}{6}(k_1 + 2k_2 + 2k_3 + k_4),$$

where

$$k_1 = hf(x_n, y_n),$$

$$k_2 = hf\left(x_n + \frac{h}{2}, \; y_n + \frac{k_1}{2} \right),$$

$$k_3 = hf\left(x_n + \frac{h}{2}, \; y_n + \frac{k_2}{2} \right),$$

$$k_4 = hf(x_n + h, \; y_n + k_3).$$

The classical fourth-order Runge–Kutta method is one of the more popular methods because its rate of convergence is $O(h^4)$ and it is easy to program. Typically, it produces very accurate approximations even when the number of iterations is reasonably small. However, as the number of iterations becomes large, other types of errors may creep in.

Program outlines for the fourth-order Runge–Kutta method are given below. Just as with the algorithms for the improved Euler's method, the first program (the Runge–Kutta subroutine) is useful for approximating the solution over an interval $[x_0, c]$ and takes the number of

steps in the interval as input. As in Section 3.6, the number of steps N is related to the step size h and the interval $[x_0, c]$ by

$$Nh = c - x_0 .$$

The subroutine has the option to print out a table of values of x and y. The second algorithm (Runge–Kutta with tolerance) on page 136 is used to approximate, for a given tolerance, the solution at an inputted value $x = c$. This algorithm[†] automatically halves the step sizes successively until the two approximations $y(c; h)$ and $y(c; h/2)$ differ by less than the prescribed tolerance ε. For a stopping procedure based on the relative error, Step 6 of the algorithm should be replaced by

Step 6′ If $\left| \dfrac{y - z}{y} \right| < \varepsilon$, go to Step 10 .

Classical Fourth-Order Runge–Kutta Subroutine

Purpose To approximate the solution to the initial value problem

$$y' = f(x, y) , \qquad y(x_0) = y_0$$

for $x_0 \le x \le c$

INPUT x_0, y_0, c, N (number of steps), PRNTR ($= 1$ to print a table)

Step 1 Set step size $h = (c - x_0)/N$, $x = x_0$, $y = y_0$

Step 2 For $i = 1$ to N, do Steps 3–5

Step 3 Set

$$k_1 = hf(x, y)$$

$$k_2 = hf\left(x + \frac{h}{2}, y + \frac{k_1}{2} \right)$$

$$k_3 = hf\left(x + \frac{h}{2}, y + \frac{k_2}{2} \right)$$

$$k_4 = hf(x + h, y + k_3)$$

Step 4 Set

$$x = x + h$$

$$y = y + \frac{1}{6}(k_1 + 2k_2 + 2k_3 + k_4)$$

Step 5 If PRNTR $= 1$, print x, y

[†]Note that the form of the algorithm on page 136 is the same as that for the improved Euler's method on page 128 except for Step 4, where the Runge–Kutta subroutine is called. More sophisticated stopping procedures are used in production-grade codes.

> ### Classical Fourth-Order Runge–Kutta Algorithm with Tolerance
>
> **Purpose** To approximate the solution to the initial value problem
>
> $$y' = f(x, y), \qquad y(x_0) = y_0$$
>
> at $x = c$, with tolerance ε
>
> INPUT $x_0, y_0, c, \varepsilon, M$ (maximum number of iterations)
> Step 1 Set $z = y_0$, PRNTR $= 0$
> Step 2 For $m = 0$ to M, do Steps 3–7 (or, to save time, start with $m > 0$)
> Step 3 Set $N = 2^m$
> Step 4 Call FOURTH-ORDER RUNGE–KUTTA SUBROUTINE
> Step 5 Print h, y
> Step 6 If $|z - y| < \varepsilon$, go to Step 10
> Step 7 Set $z = y$
> Step 8 Print "$\phi(c)$ is approximately"; y; "but may not be within the tolerance"; ε
> Step 9 Go to Step 11
> Step 10 Print "$\phi(c)$ is approximately"; y; "with tolerance"; ε
> Step 11 STOP
> OUTPUT Approximations of the solution to the initial value problem at $x = c$, using 2^m steps.

Example 2 Use the classical fourth-order Runge–Kutta algorithm to approximate the solution $\phi(x)$ of the initial value problem

$$y' = y, \qquad y(0) = 1,$$

at $x = 1$ with a tolerance of 0.001.

Solution The inputs are $x_0 = 0$, $y_0 = 1$, $c = 1$, $\varepsilon = 0.001$, and $M = 25$ (say). Since $f(x, y) = y$, the formulas in Step 3 of the subroutine become

$$k = hy, \qquad k_2 = h\left(y + \frac{k_1}{2}\right), \qquad k_3 = h\left(y + \frac{k_2}{2}\right), \qquad k_4 = h(y + k_3).$$

The initial value for N in this algorithm is $N = 1$, so

$$h = (1 - 0)/1 = 1.$$

Thus, in Step 3 of the subroutine, we compute

$$k_1 = (1)(1) = 1, \qquad\qquad k_2 = (1)(1 + 0.5) = 1.5,$$
$$k_3 = (1)(1 + 0.75) = 1.75, \qquad k_4 = (1)(1 + 1.75) = 2.75,$$

and, in Step 4 of the subroutine, we get for the first approximation

$$y = y_0 + \frac{1}{6}(k_1 + 2k_2 + 2k_3 + k_4)$$

$$= 1 + \frac{1}{6}\left[1 + 2(1.5) + 2(1.75) + 2.75\right]$$

$$= 2.70833,$$

where we have rounded to five decimal places. Because

$$|z - y| = |y_0 - y| = |1 - 2.70833| = 1.70833 > \varepsilon ,$$

we start over and reset $N = 2, h = 0.5$.

Doing Steps 3 and 4 for $i = 1$ and 2, we ultimately obtain (for $i = 2$) the approximation

$$y = 2.71735 .$$

Since $|z - y| = |2.70833 - 2.71735| = 0.00902 > \varepsilon$, we again start over and reset $N = 4, h = 0.25$. This leads to the approximation

$$y = 2.71821 ,$$

so that

$$|z - y| = |2.71735 - 2.71821| = 0.00086 ,$$

which is less than $\varepsilon = 0.001$. Hence $\phi(1) = e \approx 2.71821$. ◆

In Example 2 we were able to obtain a better approximation for $\phi(1) = e$ with $h = 0.25$ than we obtained in Section 3.6 using Euler's method with $h = 0.001$ (see Table 3.4, page 124) and roughly the same accuracy as we obtained in Section 3.6 using the improved Euler's method with $h = 0.01$ (see Table 3.5, page 127).

Example 3 Use the fourth-order Runge–Kutta subroutine to approximate the solution $\phi(x)$ of the initial value problem

$$(15) \quad y' = y^2 , \quad y(0) = 1 ,$$

on the interval $0 \le x \le 2$ using $N = 8$ steps (i.e., $h = 0.25$).

Solution Here the starting values are $x_0 = 0$ and $y_0 = 1$. Since $f(x, y) = y^2$, the formulas in Step 3 of the subroutine are

$$k_1 = hy^2 , \qquad k_2 = h\left(y + \frac{k_1}{2} \right)^2 ,$$

$$k_3 = h\left(y + \frac{k_2}{2} \right)^2 , \qquad k_4 = h(y + k_3)^2 .$$

From the output, we find

$x = 0.25$	$y = 1.33322 ,$
$x = 0.50$	$y = 1.99884 ,$
$x = 0.75$	$y = 3.97238 ,$
$x = 1.00$	$y = 32.82820 ,$
$x = 1.25$	$y = 4.09664 * 10^{11} ,$
$x = 1.50$	$y = \text{overflow} .$

What happened? Fortunately, the equation in (15) is separable, and, solving for $\phi(x)$, we obtain $\phi(x) = (1 - x)^{-1}$. It is now obvious where the problem lies: The true solution $\phi(x)$ is not defined at $x = 1$. If we had been more cautious, we would have realized that $\partial f/\partial y = 2y$ is *not* bounded for all y. Hence, the existence of a unique solution is not guaranteed for all x between 0 and 2, and in this case, the method does *not* give meaningful approximations for x near (or greater than) 1. ◆

Example 4 Use the fourth-order Runge–Kutta algorithm to approximate the solution $\phi(x)$ of the initial value problem

$$y' = x - y^2, \qquad y(0) = 1,$$

at $x = 2$ with a tolerance of 0.0001.

Solution This time we check to see whether $\partial f/\partial y$ is bounded. Here $\partial f/\partial y = -2y$, which is certainly unbounded in any vertical strip. However, let's consider the qualitative behavior of the solution $\phi(x)$. The solution curve starts at (0, 1), where $\phi'(0) = 0 - 1 < 0$, so $\phi(x)$ begins decreasing and continues to decrease until it crosses the curve $y = \sqrt{x}$. After crossing this curve, $\phi(x)$ begins to increase, since $\phi'(x) = x - \phi^2(x) > 0$. As $\phi(x)$ increases, it remains below the curve $y = \sqrt{x}$. This is because if the solution were to get "close" to the curve $y = \sqrt{x}$, then the derivative of $\phi(x)$ would approach zero, so that overtaking the function $\sqrt{x}$ is impossible.

Therefore, although the existence-uniqueness theorem does not guarantee a solution, we are inclined to try the algorithm anyway. The above argument shows that $\phi(x)$ probably exists for $x > 0$, so we feel reasonably sure the fourth-order Runge–Kutta method will give a good approximation of the true solution $\phi(x)$. Proceeding with the algorithm, we use the starting values $x_0 = 0$ and $y_0 = 1$. Since $f(x, y) = x - y^2$, the formulas in Step 3 of the subroutine become

$$k_1 = h(x - y^2), \qquad\qquad k_2 = h\left[\left(x + \frac{h}{2}\right) - \left(y + \frac{k_1}{2}\right)^2\right],$$

$$k_3 = h\left[\left(x + \frac{h}{2}\right) - \left(y + \frac{k_2}{2}\right)^2\right], \qquad k_4 = h\left[(x + h) - (y + k_3)^2\right].$$

In Table 3.6, we give the approximations $y(2; 2^{-m+1})$ for $\phi(2)$ for $m = 0, 1, 2, 3$, and 4. The algorithm stops at $m = 4$, since

$$\left|y(2; 0.125) - y(2; 0.25)\right| = 0.00000.$$

Hence, $\phi(2) \approx 1.25132$ with a tolerance of 0.0001. ◆

TABLE 3.6	Classical Fourth-Order Runge–Kutta Approximation for $\phi(2)$		
m	h	Approximation for $\phi(2)$	$\left\|y(2; h) - y(2; 2h)\right\|$
0	2.0	-8.33333	
1	1.0	1.27504	9.60837
2	0.5	1.25170	0.02334
3	0.25	1.25132	0.00038
4	0.125	1.25132	0.00000

3.7 Exercises

As in Exercises 3.6, for some problems you will find it essential to have a calculator or computer available.[†] For Problems 1–17, note whether or not $\partial f/\partial y$ is bounded.

1. Determine the recursive formulas for the Taylor method of order 2 for the initial value problem

$$y' = \cos(x+y), \qquad y(0) = \pi.$$

2. Determine the recursive formulas for the Taylor method of order 2 for the initial value problem

$$y' = xy - y^2, \qquad y(0) = -1.$$

3. Determine the recursive formulas for the Taylor method of order 4 for the initial value problem

$$y' = x - y, \qquad y(0) = 0.$$

4. Determine the recursive formulas for the Taylor method of order 4 for the initial value problem

$$y' = x^2 + y, \qquad y(0) = 0.$$

5. Use the Taylor methods of orders 2 and 4 with $h = 0.25$ to approximate the solution to the initial value problem

$$y' = x + 1 - y, \qquad y(0) = 1,$$

at $x = 1$. Compare these approximations to the actual solution $y = x + e^{-x}$ evaluated at $x = 1$.

6. Use the Taylor methods of orders 2 and 4 with $h = 0.25$ to approximate the solution to the initial value problem

$$y' = 1 - y, \qquad y(0) = 0,$$

at $x = 1$. Compare these approximations to the actual solution $y = 1 - e^{-x}$ evaluated at $x = 1$.

7. Use the fourth-order Runge–Kutta subroutine with $h = 0.25$ to approximate the solution to the initial value problem

$$y' = 2y - 6, \qquad y(0) = 1,$$

at $x = 1$. (Thus, input $N = 4$.) Compare this approximation to the actual solution $y = 3 - 2e^{2x}$ evaluated at $x = 1$.

8. Use the fourth-order Runge–Kutta subroutine with $h = 0.25$ to approximate the solution to the initial value problem

$$y' = 1 - y, \qquad y(0) = 0,$$

at $x = 1$. Compare this approximation with the one obtained in Problem 6 using the Taylor method of order 4.

9. Use the fourth-order Runge–Kutta subroutine with $h = 0.25$ to approximate the solution to the initial value problem

$$y' = x + 1 - y, \qquad y(0) = 1,$$

at $x = 1$. Compare this approximation with the one obtained in Problem 5 using the Taylor method of order 4.

10. Use the fourth-order Runge–Kutta algorithm to approximate the solution to the initial value problem

$$y' = 1 - xy, \qquad y(1) = 1,$$

at $x = 2$. For a tolerance of $\varepsilon = 0.001$, use a stopping procedure based on the absolute error.

11. The solution to the initial value problem

$$y' = \frac{2}{x^4} - y^2, \qquad y(1) = -0.414$$

crosses the x-axis at a point in the interval $[1, 2]$. By experimenting with the fourth-order Runge–Kutta subroutine, determine this point to two decimal places.

12. By experimenting with the fourth-order Runge–Kutta subroutine, find the maximum value over the interval $[1, 2]$ of the solution to the initial value problem

$$y' = \frac{1.8}{x^4} - y^2, \qquad y(1) = -1.$$

Where does this maximum occur? Give your answers to two decimal places.

13. The solution to the initial value problem

$$\frac{dy}{dx} = y^2 - 2e^x y + e^{2x} + e^x, \qquad y(0) = 3$$

has a vertical asymptote ("blows up") at some point in the interval $[0, 2]$. By experimenting with the fourth-order Runge–Kutta subroutine, determine this point to two decimal places.

14. Use the fourth-order Runge–Kutta algorithm to approximate the solution to the initial value problem

$$y' = y \cos x, \qquad y(0) = 1,$$

at $x = \pi$. For a tolerance of $\varepsilon = 0.01$, use a stopping procedure based on the absolute error.

[†]Appendix G describes various websites and commercial software that sketch direction fields and automate most of the differential equation algorithms discussed in this book.

15. Use the fourth-order Runge–Kutta subroutine with $h = 0.1$ to approximate the solution to

$$y' = \cos(5y) - x, \qquad y(0) = 0,$$

at the points $x = 0, 0.1, 0.2, \ldots, 3.0$. Use your answers to make a rough sketch of the solution on $[0, 3]$.

16. Use the fourth-order Runge–Kutta subroutine with $h = 0.1$ to approximate the solution to

$$y' = 3\cos(y - 5x), \qquad y(0) = 0,$$

at the points $x = 0, 0.1, 0.2, \ldots, 4.0$. Use your answers to make a rough sketch of the solution on $[0, 4]$.

17. The Taylor method of order 2 can be used to approximate the solution to the initial value problem

$$y' = y, \qquad y(0) = 1,$$

at $x = 1$. Show that the approximation y_n obtained by using the Taylor method of order 2 with the step size $1/n$ is given by the formula

$$y_n = \left(1 + \frac{1}{n} + \frac{1}{2n^2}\right)^n, \qquad n = 1, 2, \ldots.$$

The solution to the initial value problem is $y = e^x$, so y_n is an approximation to the constant e.

18. If the Taylor method of order p is used in Problem 17, show that

$$y_n = \left(1 + \frac{1}{n} + \frac{1}{2n^2} + \frac{1}{6n^3} + \cdots + \frac{1}{p!n^p}\right)^n,$$

$$n = 1, 2, \ldots.$$

19. Fluid Flow. In the study of the nonisothermal flow of a Newtonian fluid between parallel plates, the equation

$$\frac{d^2y}{dx^2} + x^2 e^y = 0, \qquad x > 0,$$

was encountered. By a series of substitutions, this equation can be transformed into the first-order equation

$$\frac{dv}{du} = u\left(\frac{u}{2} + 1\right)v^3 + \left(u + \frac{5}{2}\right)v^2.$$

Use the fourth-order Runge–Kutta algorithm to approximate $v(3)$ if $v(u)$ satisfies $v(2) = 0.1$. For a tolerance of $\varepsilon = 0.0001$, use a stopping procedure based on the relative error.

20. Chemical Reactions. The reaction between nitrous oxide and oxygen to form nitrogen dioxide is given by the balanced chemical equation $2NO + O_2 = 2NO_2$. At high temperatures the dependence of the rate of this reaction on the concentrations of NO, O_2, and NO_2 is complicated. However, at 25°C the rate at which NO_2 is formed obeys the law of mass action and is given by the rate equation

$$\frac{dx}{dt} = k(\alpha - x)^2\left(\beta - \frac{x}{2}\right),$$

where $x(t)$ denotes the concentration of NO_2 at time t, k is the rate constant, α is the initial concentration of NO, and β is the initial concentration of O_2. At 25°C, the constant k is 7.13×10^3 (liter)2/(mole)2(second). Let $\alpha = 0.0010$ mole/L, $\beta = 0.0041$ mole/L, and $x(0) = 0$ mole/L. Use the fourth-order Runge–Kutta algorithm to approximate $x(10)$. For a tolerance of $\varepsilon = 0.000001$, use a stopping procedure based on the relative error.

21. Transmission Lines. In the study of the electric field that is induced by two nearby transmission lines, an equation of the form

$$\frac{dz}{dx} + g(x)z^2 = f(x)$$

arises. Let $f(x) = 5x + 2$ and $g(x) = x^2$. If $z(0) = 1$, use the fourth-order Runge–Kutta algorithm to approximate $z(1)$. For a tolerance of $\varepsilon = 0.0001$, use a stopping procedure based on the absolute error.

A Dynamics of HIV Infection

Courtesy of Glenn Webb, Vanderbilt University

The dynamics of HIV (human immunodeficiency virus) infection within a human host involve the interaction of the HIV virions and CD4+ T lymphocytes. CD4+ T lymphocytes are long-lived white blood cells that play a major role in the defense of the human body against microbial invaders. HIV targets these very cells. When HIV first appeared as a new and major health threat, it was recognized that the disease typically exhibited a lengthy gradual progression lasting 10 or more years. It was widely believed that the dynamics of HIV destruction of the CD4+ T-cell population involved a very low rate of infection and a very slow turnover of virus and infected cells. In 1995 differential equation models of HIV-CD4+ T-cell interaction revealed that the turnover rate for the infected CD4+ T cells was very much faster than this (about 2 days)—a scientific breakthrough reported simultaneously in the papers of D. D. Ho et al., "Rapid Turnover of Plasma Virions and CD4 Lymphocytes in HIV-1 Infection," *Nature*, 1995; and of G. M. Shaw et al., "Viral Dynamics in Human Immunodeficiency Virus Type I Infection," *Nature*, 1995.

Underlying the models in these papers is the knowledge that within a person infected with HIV, the virus spends part of its existence free and part inside an infected CD4+ T cell. The time spent free was known to be very short—on the order of 30 minutes. The time spent inside an invaded CD4+ T cell was *believed* to be very long—on the order of years. When a cell was invaded, a virion (a complete viral particle, consisting of RNA surrounded by a protein shell) took over the cell's DNA and used it to replicate its own RNA, thereby creating new virions; then it budded, or burst the cell, to release multiple virus particles.

The differential equations of the model are similar to those for compartmental analysis discussed in Section 3.2. They involve compartments and parameters. The compartments of the model are

$T(t)$ = the population of uninfected CD4+ T cells at time t.
$I(t)$ = the population of infected CD4+ T cells at time t.
$V(t)$ = the population of virus at time t.

The parameters (followed by their units) of the model are

$\lambda(\text{cells} \cdot \text{day}^{-1})$ = constant input source of uninfected cells per day (the human body produces these cells daily in the thymus).
$\delta(\text{day}^{-1})$ = normal loss rate constant of uninfected cells ($1/\delta$ = the average lifespan of an uninfected cell in days).
$\beta(\text{virions}^{-1} \cdot \text{day}^{-1})$ = infection rate constant of uninfected cells per infected cell (the rate is of mass action form, i.e., $\beta V(t) T(t)$).
$\mu(\text{day}^{-1})$ = loss rate constant of infected cells ($1/\mu$ = the average lifespan of an infected cell in days).
$\gamma(\text{day}^{-1})$ = loss rate constant of free virus ($1/\gamma$ = the average lifespan of a free virion in days).
$N(\text{virions} \cdot \text{cell}^{-1})$ = number of virions produced per day per infected cell (the burst number of an infected cell).

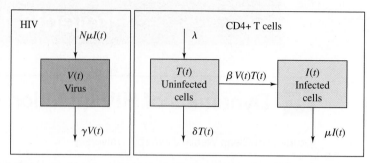

Figure 3.16 Compartmental views of virus, uninfected T cells, and infected T cells

The independent variable of the model is time t in days and the dependent variables of the model are $T(t), I(t)$, and $V(t)$. The equations are as follows (see Figure 3.16).

$$\underbrace{\frac{d}{dt}T(t)}_{\text{Time change}} = \underbrace{\lambda}_{\text{Source}} - \underbrace{\delta T(t)}_{\text{Normal loss}} - \underbrace{\beta V(t)T(t)}_{\text{Infection rate}}$$

$$\underbrace{\frac{d}{dt}I(t)}_{\text{Time change}} = \underbrace{\beta V(t)T(t)}_{\text{Gain from infection}} - \underbrace{\mu I(t)}_{\text{Loss rate}}$$

$$\underbrace{\frac{d}{dt}V(t)}_{\text{Time change}} = \underbrace{N\mu I(t)}_{\text{Viral production}} - \underbrace{\gamma V(t)}_{\text{Decay rate}}$$

As mentioned above, the average lifespan of a free virion, $1/\gamma$, is approximately 30 minutes, which means $\gamma \approx 48 \text{ day}^{-1}$. On the other hand, it was thought that $1/\mu$, the average length of time an infected CD4+ T cell lasts before bursting to produce new virions, should be several years, implying that μ must be quite small (on the order of 10^{-3} day^{-1}). However, when drugs to treat HIV infection first became available in the mid-1990s, researchers were able to deduce a surprisingly different value from patient data and the differential equation models. By completing the following steps, you will be able to determine a better approximation to $1/\mu$ in the manner utilized by Ho et al.

To incorporate the effect of treatment in the differential equations model, set $\beta = 0$; that is, assume the action of the drug completely inhibits the infection process. This is a reasonable approximation and it simplifies the analysis.

(a) With the assumption of treatment, what are the reduced forms of the differential equations for $T(t), I(t)$, and $V(t)$?

(b) Solve these reduced equations for $T(t), I(t)$, and $V(t)$, with the initial conditions $T(0) = T_0, I(0) = I_0$, and $V(0) = V_0$.

(c) Argue from your formula for $V(t)$ that the graph of $V(t)$ on a log scale (i.e., the graph of $\log V$) over an extended period of time (say, several weeks) will tend toward a graph of a straight line whose slope is either $-\gamma$ (the negative reciprocal of the average lifespan of a free virus) or $-\mu$ (the negative reciprocal of the average lifespan of an infected CD4+ T cell), according to whether γ or μ is smaller.

(d) For patients who undergo therapy, it is possible to measure the viral load V and to determine the rate at which their viral load declines. In Ho et al., the data in Figure 3.17 were presented for the decrease (on a logarithmic scale) in viral loads of three patients. Using

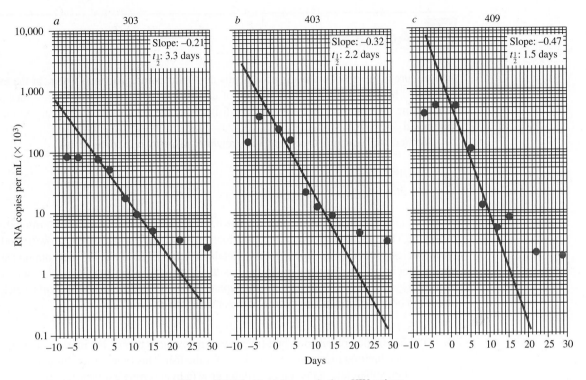

Figure 3.17 Viral load decrease in three HIV patients

these data and part (c), explain why it follows that $-\mu$ must be the approximate slope of the nearly linear curve for log V and thereby deduce the revised estimate for the average time an infected cell lasts between being invaded and bursting.

(e) Check out the recent literature of mathematical models of HIV dynamics in infected hosts (try a Google Scholar search) and find out how the estimate of the lifespan of infected CD4+ T cells has been improved using more refined ordinary differential equation models. (Other models have been used to estimate the lifespan of the free virus. Also, models have been developed to track the long-term effects of patients undergoing therapy, optimal ways to schedule treatment, and the problems that arise when drug resistance develops.)

The dynamics of HIV-1 replication in patients receiving anti-retroviral therapy is a subject of continuing investigation and mathematical modeling. It is well known that therapy does not eliminate the virus in patients, and it is necessary to continue treatment indefinitely. The reasons are complex and connected to the presence of viral reservoirs, which allow the virus population to restore if treatment is discontinued. Further investigations of this subject may be found in the following references:

Quantifying residual HIV-1 replication in patients receiving combination anti-retroviral therapy. Zhang LQ, Ramratnam B, Tenner-Racz K, He YX, Vesanen M, Lewin S, Talal A, Racz P, Perelson AS, Korber BT, Markowitz M, and Ho DD. *New England Journal of Medicine* 340:1605–1613, 1999.

The decay of the latent reservoir of replication-competent HIV-1 is inversely correlated with the extent of residual viral replication during prolonged anti-retroviral therapy. Ramratnam B, Mittler JE, Zhang LQ, Boden D, Hurley A, Fang F, Macken CA, Perelson AS, Markowitz M, and Ho DD. *Nature Medicine* 6:82–85, 2000.

B Aquaculture

Aquaculture is the art of cultivating the plants and animals indigenous to water. In the example considered here, it is assumed that a batch of catfish are raised in a pond. We are interested in determining the best time for harvesting the fish so that the cost per pound for raising the fish is minimized.

A differential equation describing the growth of fish may be expressed as

$$(1) \qquad \frac{dW}{dt} = KW^\alpha,$$

where $W(t)$ is the weight of the fish at time t and K and α are empirically determined growth constants. The functional form of this relationship is similar to that of the growth models for other species. Modeling the growth rate or metabolic rate by a term like W^α is a common assumption. Biologists often refer to equation (1) as the **allometric equation.** It can be supported by plausibility arguments such as growth rate depending on the surface area of the gut (which varies like $W^{2/3}$) or depending on the volume of the animal (which varies like W).

(a) Solve equation (1) when $\alpha \neq 1$.

(b) The solution obtained in part (a) grows large without bound, but in practice there is some limiting maximum weight W_{max} for the fish. This limiting weight may be included in the differential equation describing growth by inserting a dimensionless variable S that can range between 0 and 1 and involves an empirically determined parameter μ. Namely, we now assume that

$$(2) \qquad \frac{dW}{dt} = KW^\alpha S,$$

where $S := 1 - (W/W_{max})^\mu$. When $\mu = 1 - \alpha$, equation (2) has a closed form solution. Solve equation (2) when $K = 10$, $\alpha = 3/4$, $\mu = 1/4$, $W_{max} = 81$ (ounces), and $W(0) = 1$ (ounce). The constants are given for t measured in months.

(c) The differential equation describing the total cost in dollars $C(t)$ of raising a fish for t months has one constant term K_1 that specifies the cost per month (due to costs such as interest, depreciation, and labor) and a second constant K_2 that multiplies the growth rate (because the amount of food consumed by the fish is approximately proportional to the growth rate). That is,

$$(3) \qquad \frac{dC}{dt} = K_1 + K_2 \frac{dW}{dt}.$$

Solve equation (3) when $K_1 = 0.4$, $K_2 = 0.1$, $C(0) = 1.1$ (dollars), and $W(t)$ is as determined in part (b).

(d) Sketch the curve obtained in part (b) that represents the weight of the fish as a function of time. Next, sketch the curve obtained in part (c) that represents the total cost of raising the fish as a function of time.

(e) To determine the optimal time for harvesting the fish, sketch the ratio $C(t)/W(t)$. This ratio represents the total cost per ounce as a function of time. When this ratio reaches its minimum—that is, when the total cost per ounce is at its lowest—it is the optimal time to harvest the fish. Determine this optimal time to the nearest month.

C Curve of Pursuit

An interesting geometric model arises when one tries to determine the path of a pursuer chasing its prey. This path is called a **curve of pursuit.** These problems were analyzed using methods of calculus circa 1730 (more than two centuries after Leonardo da Vinci had considered them). The simplest problem is to find the curve along which a vessel moves in pursuing another vessel that flees along a straight line, assuming the speeds of the two vessels are constant.

Let's assume that vessel A, traveling at a speed α, is pursuing vessel B, which is traveling at a speed β. In addition, assume that vessel A begins (at time $t = 0$) at the origin and pursues vessel B, which begins at the point $(1, 0)$ and travels up the line $x = 1$. After t hours, vessel A is located at the point $P = (x, y)$, and vessel B is located at the point $Q = (1, \beta t)$ (see Figure 3.18). The goal is to describe the locus of points P; that is, to find y as a function of x.

(a) Vessel A is pursuing vessel B, so at time t, vessel A must be heading right at vessel B. That is, the tangent line to the curve of pursuit at P must pass through the point Q (see Figure 3.18). For this to be true, show that

(4) $$\frac{dy}{dx} = \frac{y - \beta t}{x - 1}.$$

(b) We know the speed at which vessel A is traveling, so we know that the distance it travels in time t is αt. This distance is also the length of the pursuit curve from $(0, 0)$ to (x, y). Using the arc length formula from calculus, show that

(5) $$\alpha t = \int_0^x \sqrt{1 + [y'(u)]^2}\, du.$$

Solving for t in equations (4) and (5), conclude that

(6) $$\frac{y - (x - 1)(dy/dx)}{\beta} = \frac{1}{\alpha} \int_0^x \sqrt{1 + [y'(u)]^2}\, du.$$

(c) Differentiating both sides of (6) with respect to x, derive the first-order equation

$$(x - 1)\frac{dw}{dx} = -\frac{\beta}{\alpha}\sqrt{1 + w^2},$$

where $w := dy/dx$.

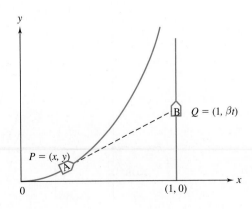

Figure 3.18 The path of vessel A as it pursues vessel B

(d) Using separation of variables and the initial conditions $x = 0$ and $w = dy/dx = 0$ when $t = 0$, show that

(7) $$\frac{dy}{dx} = w = \frac{1}{2}\left[(1-x)^{-\beta/\alpha} - (1-x)^{\beta/\alpha}\right].$$

(e) For $\alpha > \beta$—that is, the pursuing vessel A travels faster than the pursued vessel B—use equation (7) and the initial conditions $x = 0$ and $y = 0$ when $t = 0$, to derive the curve of pursuit

$$y = \frac{1}{2}\left[\frac{(1-x)^{1+\beta/\alpha}}{1+\beta/\alpha} - \frac{(1-x)^{1-\beta/\alpha}}{1-\beta/\alpha}\right] + \frac{\alpha\beta}{\alpha^2 - \beta^2}.$$

(f) Find the location where vessel A intercepts vessel B if $\alpha > \beta$.

(g) Show that if $\alpha = \beta$, then the curve of pursuit is given by

$$y = \frac{1}{2}\left\{\frac{1}{2}\left[(1-x)^2 - 1\right] - \ln(1-x)\right\}.$$

Will vessel A ever reach vessel B?

D Aircraft Guidance in a Crosswind

Courtesy of T. L. Pearson, Professor of Mathematics, Acadia University (Retired), Nova Scotia, Canada

An aircraft flying under the guidance of a nondirectional beacon (a fixed radio transmitter, abbreviated NDB) moves so that its longitudinal axis always points toward the beacon (see Figure 3.19). A pilot sets out toward an NDB from a point at which the wind is at right angles to the initial direction of the aircraft; the wind maintains this direction. Assume that the wind speed and the speed of the aircraft through the air (its "airspeed") remain constant. (Keep in mind that the latter is different from the aircraft's speed with respect to the ground.)

(a) Locate the flight in the xy-plane, placing the start of the trip at $(2, 0)$ and the destination at $(0, 0)$. Set up the differential equation describing the aircraft's path over the ground. $\left[\textit{Hint: } dy/dx = (dy/dt)/(dx/dt).\right]$

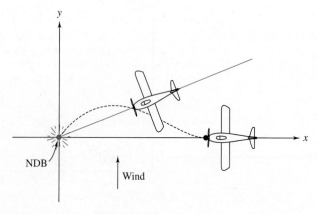

Figure 3.19 Guided aircraft

(b) Make an appropriate substitution and solve this equation. [*Hint:* See Section 2.6.]
(c) Use the fact that $x = 2$ and $y = 0$ at $t = 0$ to determine the appropriate value of the arbitrary constant in the solution set.
(d) Solve to get y explicitly in terms of x. Write your solution in terms of a hyperbolic function.
(e) Let γ be the ratio of windspeed to airspeed. Using a software package, graph the solutions for the cases $\gamma = 0.1, 0.3, 0.5$, and 0.7 all on the same set of axes. Interpret these graphs.
(f) Discuss the (terrifying!) cases $\gamma = 1$ and $\gamma > 1$.

E Market Equilibrium: Stability and Time Paths

Courtesy of James E. Foster, George Washington University

A *perfectly competitive market* is made up of many buyers and sellers of an economic product, each of whom has no control over the market price. In this model, the overall quantity demanded by the buyers of the product is taken to be a function of the price of the product (among other things) called the *demand function*. Similarly, the overall quantity supplied by the sellers of the product is a function of the price of the product (among other things) called the *supply function*. A market is in *equilibrium* at a price where the quantity demanded is just equal to the quantity supplied.

The *linear* model assumes that the demand and supply functions have the form $q_d = d_0 - d_1 p$ and $q_s = -s_0 + s_1 p$, respectively, where p is the market price of the product, q_d is the associated quantity demanded, q_s is the associated quantity supplied, and d_0, d_1, s_0, and s_1 are all positive constants. The functional forms ensure that the "laws" of downward sloping demand and upward sloping supply are being satisfied. It is easy to show that the equilibrium price is $p^* = (d_0 + s_0)/(d_1 + s_1)$.

Economists typically assume that markets are in equilibrium and justify this assumption with the help of stability arguments. For example, consider the simple *price adjustment equation*

$$\frac{dp}{dt} = \lambda(q_d - q_s),$$

where $\lambda > 0$ is a constant indicating the *speed of adjustments*. This follows the intuitive requirement that price rises when demand exceeds supply and falls when supply exceeds demand. The market equilibrium is said to be *globally stable* if, for every initial price level $p(0)$, the price adjustment path $p(t)$ satisfies $p(t) \to p^*$ as $t \to \infty$.

(a) Find the price adjustment path: Substitute the expressions for q_d and q_s into the price adjustment equation and show that the solution to the resulting differential equation is $p(t) = [p(0) - p^*]e^{ct} + p^*$, where $c = -\lambda(d_1 + s_1)$.
(b) Is the market equilibrium globally stable?

Now consider a model that takes into account the *expectations* of agents. Let the market demand and supply functions over time $t \geq 0$ be given by

$$q_d(t) = d_0 - d_1 p(t) + d_2 p'(t) \quad \text{and} \quad q_s(t) = -s_0 + s_1 p(t) - s_2 p'(t),$$

respectively, where $p(t)$ is the market price of the product, $q_d(t)$ is the associated quantity demanded, $q_s(t)$ is the associated quantity supplied, and d_0, d_1, d_2, s_0, s_1, and s_2 are all positive constants. The functional forms ensure that, when faced with an increasing price, demanders

will tend to purchase more (before prices rise further) while suppliers will tend to offer less (to take advantage of the higher prices in the future). Now given the above stability argument, we restrict consideration to *market clearing time paths* $p(t)$ satisfying $q_d(t) = q_s(t)$, for all $t \geq 0$, and explore the evolution of price over time. We say that the market is in *dynamic equilibrium* if $p'(t) = 0$ for all t. It is easy to show that the dynamic equilibrium in this model is given by $p(t) = p^*$ for all t, where p^* is the market equilibrium price defined above. However, many other market clearing time paths are possible.

(c) Find a market clearing time path: Equate $q_d(t)$ and $q_s(t)$ and solve the resulting differential equation $p(t)$ in terms of its initial value $p_0 = p(0)$.

(d) Is it true that for any market clearing time path we must have $p(t) \to p^*$ as $t \to \infty$?

(e) If the price $p(t)$ of a product is \$5 at $t = 0$ months and demand and supply functions are modeled as $q_d(t) = 30 - 2p(t) + 4p'(t)$ and $q_s(t) = -20 + p(t) - 6p'(t)$, what will be the price after 10 months? As t becomes very large? What is happening to $p'(t)$ and how are the expectations of demanders and suppliers evolving?

For further reading, see, for example, *Mathematical Economics*, 2nd ed. by Akira Takayama (Cambridge University Press, Cambridge, 1985), Chapter 3; and *Fundamental Methods of Mathematical Economics*, 4th ed. by Alpha Chiang, and Kevin Wainwright (McGraw-Hill/Irvin, Boston, 2005), Chapter 14.

F Stability of Numerical Methods

 Numerical methods are often tested on simple initial value problems of the form

(8) $y' + \lambda y = 0$, $y(0) = 1$, $(\lambda = \text{constant})$,

which has the solution $\phi(x) = e^{-\lambda x}$. Notice that for each $\lambda > 0$ the solution $\phi(x)$ tends to zero as $x \to +\infty$. Thus, a desirable property for any numerical scheme that generates approximations $y_0, y_1, y_2, y_3, \ldots$ to $\phi(x)$ at the points $0, h, 2h, 3h, \ldots$ is that, for $\lambda > 0$,

(9) $y_n \to 0$ as $n \to \infty$.

For single-step linear methods, property (9) is called **absolute stability.**

(a) Show that for $x_n = nh$, Euler's method, when applied to the initial value problem (8), yields the approximations

$$y_n = (1 - \lambda h)^n , \qquad n = 0, 1, 2, \ldots ,$$

and deduce that this method is absolutely stable only when $0 < \lambda h < 2$. (This means that for a given $\lambda > 0$, we must choose the step size h sufficiently small in order for property (9) to hold.) Further show that for $h > 2/\lambda$, the error $y_n - \phi(x_n)$ grows large exponentially!

(b) Show that for $x_n = nh$ the trapezoid scheme of Section 3.6, applied to problem (8), yields the approximations

$$y_n = \left(\frac{1 - \lambda h/2}{1 + \lambda h/2} \right)^n , \qquad n = 0, 1, 2, \ldots ,$$

and deduce that this scheme is absolutely stable for all $\lambda > 0, h > 0$.

(c) Show that the improved Euler's method applied to problem (8) is absolutely stable for $0 < \lambda h < 2$.

Multistep Methods. When multistep numerical methods are used, instability problems may arise that cannot be circumvented by simply choosing a sufficiently small step size h. This is because multistep methods yield "extraneous solutions," which may dominate the calculations. To see what can happen, consider the two-step method

(10) $y_{n+1} = y_{n-1} + 2hf(x_n, y_n)$, $n = 1, 2, \ldots,$

for the equation $y' = f(x, y)$.

 (d) Show that for the initial value problem

(11) $y' + 2y = 0$, $y(0) = 1$,

the recurrence formula (10), with $x_n = nh$, becomes

(12) $y_{n+1} + 4hy_n - y_{n-1} = 0$.

Equation (12), which is called a **difference equation,** can be solved by using the following approach. We postulate a solution of the form $y_n = r^n$, where r is a constant to be determined.

 (e) Show that substituting $y_n = r^n$ in (12) leads to the "characteristic equation"

$$r^2 + 4hr - 1 = 0 ,$$

which has roots

$$r_1 = -2h + \sqrt{1 + 4h^2} \quad \text{and} \quad r_2 = -2h - \sqrt{1 + 4h^2} .$$

By analogy with the theory for second-order differential equations, it can be shown that a general solution of (12) is

$$y_n = c_1 r_1^n + c_2 r_2^n ,$$

where c_1 and c_2 are arbitrary constants. Thus, the general solution to the difference equation (12) has two independent constants, whereas the differential equation in (11) has only one, namely, $\phi(x) = ce^{-2x}$.

 (f) Show that for each $h > 0$,

$$\lim_{n \to \infty} r_1^n = 0 \quad \text{but} \quad \lim_{n \to \infty} |r_2^n| = \infty .$$

Hence, the term r_1^n behaves like the solution $\phi(x_n) = e^{-2x_n}$ as $n \to \infty$. However, the extraneous solution r_2^n grows large without bound.

 (g) Applying the scheme of (10) to the initial value problem (11) requires two starting values y_0, y_1. The exact values are $y_0 = 1$, $y_1 = e^{-2h}$. However, regardless of the choice of starting values and the size of h, the term $c_2 r_2^n$ will eventually dominate the full solution to the recurrence equation as x_n increases. Illustrate this instability taking $y_0 = 1$, $y_1 = e^{-2h}$, and using a calculator or computer to compute $y_2, y_3, \ldots, y_{100}$ from the recurrence formula (12) for $h = 0.5$ and $h = 0.05$. (*Note:* Even if initial conditions are chosen so that $c_2 = 0$, round-off error will inevitably "excite" the extraneous dominant solution.)

G Period Doubling and Chaos

 In the study of dynamical systems, the phenomena of **period doubling** and **chaos** are observed. These phenomena can be seen when one uses a numerical scheme to approximate the solution to an initial value problem for a nonlinear differential equation such as the following logistic model for population growth:

(13) $$\frac{dp}{dt} = 10p(1-p), \qquad p(0) = 0.1 .$$

(See Section 3.2.)

(a) Solve the initial value problem (13) and show that $p(t)$ approaches 1 as $t \to +\infty$.

(b) Show that using Euler's method (see Sections 1.4 and 3.6) with step size h to approximate the solution to (13) gives

(14) $$p_{n+1} = (1 + 10h)p_n - (10h)p_n^2, \qquad p_0 = 0.1 .$$

(c) For $h = 0.18, 0.23, 0.25,$ and 0.3, show that the first 40 iterations of (14) appear to (i) converge to 1 when $h = 0.18$, (ii) jump between 1.18 and 0.69 when $h = 0.23$, (iii) jump between 1.23, 0.54, 1.16, and 0.70 when $h = 0.25$, and (iv) display no discernible pattern when $h = 0.3$.

The transitions from convergence to jumping between two numbers, then four numbers, and so on, are called **period doubling.** The phenomenon displayed when $h = 0.3$ is referred to as **chaos.** This transition from period doubling to chaos as h increases is frequently observed in dynamical systems.

The transition to chaos is nicely illustrated in the bifurcation diagram (see Figure 3.20). This diagram is generated for equation (14) as follows. Beginning at $h = 0.18$, compute the sequence

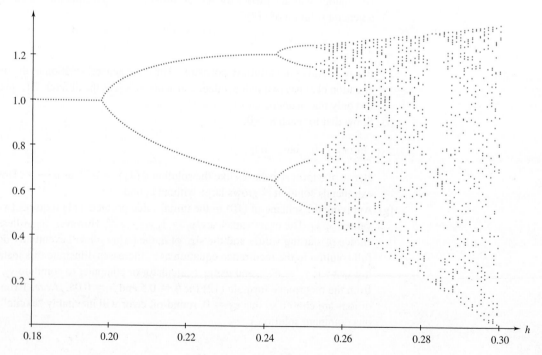

Figure 3.20 Period doubling to chaos

$\{p_n\}$ using (14) and, starting at $n = 201$, plot the next 30 values—that is, $p_{201}, p_{202}, \ldots, p_{230}$. Next, increment h by 0.001 to 0.181 and repeat. Continue this process until $h = 0.30$. Notice how the figure splits from one branch to two, then four, and then finally gives way to chaos.

Our concern is with the instabilities of the numerical procedure when h is not chosen small enough. Fortunately, the instability observed for Euler's method—the period doubling and chaos—was immediately recognized because we know that this type of behavior is not expected of a solution to the logistic equation. Consequently, if we had tried Euler's method with $h = 0.23$, 0.25, or 0.3 to solve (13) numerically, we would have realized that h was not chosen small enough.

The situation for the classical fourth-order Runge–Kutta method (see Section 3.7) is more troublesome. It may happen that for a certain choice of h period doubling occurs, but it is also possible that for other choices of h the numerical solution actually converges to a limiting value that is *not* the limiting value for any solution to the logistic equation in (13).

(d) Approximate the solution to (13) by computing the first 60 iterations of the classical fourth-order Runge–Kutta method using the step size $h = 0.3$. (Thus, for the subroutine on page 135, input $N = 60$ and $c = 60(0.3) = 18$.) Repeat with $h = 0.325$ and $h = 0.35$. Which values of h (if any) do you feel are giving the "correct" approximation to the solution? Why?

A further discussion of chaos appears in Section 5.8.

Linear Second-Order Equations

4.1 Introduction: The Mass–Spring Oscillator

> A damped mass–spring oscillator consists of a mass m attached to a spring fixed at one end, as shown in Figure 4.1. Devise a differential equation that governs the motion of this oscillator, taking into account the forces acting on it due to the spring elasticity, damping friction, and possible external influences.

Figure 4.1 Damped mass–spring oscillator

Newton's second law—force equals mass times acceleration $(F = ma)$—is without a doubt the most commonly encountered differential equation in practice. It is an ordinary differential equation of the *second order* since acceleration is the second derivative of position (y) with respect to time $\left(a = d^2y/dt^2 \right)$.

When the second law is applied to a mass–spring oscillator, the resulting motions are common experiences of everyday life, and we can exploit our familiarity with these vibrations to obtain a qualitative description of the solutions of more general second-order equations.

We begin by referring to Figure 4.1, which depicts the mass–spring oscillator. When the spring is unstretched and the inertial mass m is still, the system is at equilibrium; we measure the coordinate y of the mass by its displacement from the equilibrium position. When the mass m is displaced from equilibrium, the spring is stretched or compressed and it exerts a force that resists the displacement. For most springs this force is directly proportional to the displacement y and is thus given by

$$(1) \qquad F_{\text{spring}} = -ky \,,$$

where the positive constant k is known as the *stiffness* and the negative sign reflects the opposing nature of the force. **Hooke's law,** as equation (1) is commonly known, is only valid for sufficiently small displacements; if the spring is compressed so strongly that the coils press against each other, the opposing force obviously becomes much stronger.

Figure 4.2 (a) Sinusoidal oscillation, (b) stiffer spring, and (c) heavier mass

Practically all mechanical systems also experience friction, and for vibrational motion this force is usually modeled accurately by a term proportional to velocity:

$$(2) \qquad F_{\text{friction}} = -b\frac{dy}{dt} = -by' ,$$

where b (≥ 0) is the *damping coefficient* and the negative sign has the same significance as in equation (1).

The other forces on the oscillator are usually regarded as *external* to the system. Although they may be gravitational, electrical, or magnetic, commonly the most important external forces are transmitted to the mass by shaking the supports holding the system. For the moment we lump all the external forces into a single, *known* function $F_{\text{ext}}(t)$. Newton's law then provides the differential equation for the mass–spring oscillator:

$$my'' = -ky - by' + F_{\text{ext}}(t)$$

or

$$(3) \qquad my'' + by' + ky = F_{\text{ext}}(t) .$$

What do mass–spring motions look like? From our everyday experience with weak auto suspensions, musical gongs, and bowls of jelly, we expect that when there is no friction ($b = 0$) or external force, the (idealized) motions would be perpetual vibrations like the ones depicted in Figure 4.2. These vibrations resemble sinusoidal functions, with their amplitude depending on the initial displacement and velocity. The frequency of the oscillations increases for stiffer springs but decreases for heavier masses.

In Section 4.3 we will show how to find these solutions. Example 1 demonstrates a quick calculation that confirms our intuitive predictions.

Example 1 Verify that if $b = 0$ and $F_{\text{ext}}(t) = 0$, equation (3) has a solution of the form $y(t) = \cos \omega t$ and that the angular frequency ω increases with k and decreases with m.

Solution Under the conditions stated, equation (3) simplifies to

$$(4) \qquad my'' + ky = 0 .$$

The second derivative of $y(t)$ is $-\omega^2 \cos \omega t$, and if we insert it into (4), we find

$$my'' + ky = -m\omega^2 \cos \omega t + k \cos \omega t ,$$

which is indeed zero if $\omega = \sqrt{k/m}$. This ω increases with k and decreases with m, as predicted. ◆

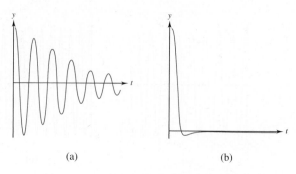

Figure 4.3 (a) Low damping and (b) high damping

When damping is present, the oscillations die out, and the motions resemble Figure 4.3. In Figure 4.3(a) the graph displays a damped oscillation; damping has slowed the frequency, and the amplitude appears to diminish exponentially with time. In Figure 4.3(b) the damping is so dominant that it has prevented the system from oscillating at all. Devices that are *supposed* to vibrate, like tuning forks or crystal oscillators, behave like Figure 4.3(a), and the damping effect is usually regarded as an undesirable loss mechanism. Good automotive suspension systems, on the other hand, behave like Figure 4.3(b); they exploit damping to *suppress* the oscillations.

The procedures for solving (unforced) mass–spring systems with damping are also described in Section 4.3, but as Examples 2 and 3 below show, the calculations are more complex. Example 2 has a relatively low damping coefficient $(b = 6)$ and illustrates the solutions for the "underdamped" case in Figure 4.3(a). In Example 3 the damping is more severe $(b = 10)$, and the solution is "overdamped" as in Figure 4.3(b).

Example 2 Verify that the exponentially damped sinusoid given by $y(t) = e^{-3t}\cos 4t$ is a solution to equation (3) if $F_{\text{ext}} = 0, m = 1, k = 25$, and $b = 6$.

Solution The derivatives of y are

$$y'(t) = -3e^{-3t}\cos 4t - 4e^{-3t}\sin 4t\,,$$
$$y''(t) = 9e^{-3t}\cos 4t + 12e^{-3t}\sin 4t + 12e^{-3t}\sin 4t - 16e^{-3t}\cos 4t$$
$$= -7e^{-3t}\cos 4t + 24e^{-3t}\sin 4t\,,$$

and insertion into (3) gives

$$my'' + by' + ky = (1)y'' + 6y' + 25y$$
$$= -7e^{-3t}\cos 4t + 24e^{-3t}\sin 4t + 6(-3e^{-3t}\cos 4t - 4e^{-3t}\sin t)$$
$$+ 25e^{-3t}\cos 4t$$
$$= 0\,. \blacklozenge$$

Example 3 Verify that the simple exponential function $y(t) = e^{-5t}$ is a solution to equation (3) if $F_{\text{ext}} = 0, m = 1, k = 25$, and $b = 10$.

Solution The derivatives of y are $y'(t) = -5e^{-5t}, y''(t) = 25e^{-5t}$ and insertion into (3) produces

$$my'' + by' + ky = (1)y'' + 10y' + 25y = 25e^{-5t} + 10(-5e^{-5t}) + 25e^{-5t} = 0\,. \blacklozenge$$

Now if a mass–spring system is driven by an external force that is sinusoidal at the angular frequency ω, our experiences indicate that although the initial response of the system may be

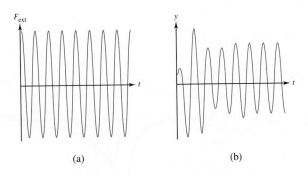

Figure 4.4 (a) Driving force and (b) response

somewhat erratic, eventually it will respond in "sync" with the driver and oscillate at the same frequency, as illustrated in Figure 4.4.

Common examples of systems vibrating in synchronization with their drivers are sound system speakers, cyclists bicycling over railroad tracks, electronic amplifier circuits, and ocean tides (driven by the periodic pull of the moon). However, there is more to the story than is revealed above. Systems can be enormously sensitive to the particular frequency ω at which they are driven. Thus, accurately tuned musical notes can shatter fine crystal, wind-induced vibrations at the right (wrong?) frequency can bring down a bridge, and a dripping faucet can cause inordinate headaches. These "resonance" responses (for which the responses have maximum amplitudes) may be quite destructive, and structural engineers have to be very careful to ensure that their products will not resonate with any of the vibrations likely to occur in the operating environment. Radio engineers, on the other hand, *do* want their receivers to resonate selectively to the desired broadcasting channel.

The calculation of these forced solutions is the subject of Sections 4.4 and 4.5. The next example illustrates some of the features of synchronous response and resonance.

Example 4 Find the synchronous response of the mass–spring oscillator with $m = 1, b = 1, k = 25$ to the force $\sin \Omega t$.

Solution We seek solutions of the differential equation

(5) $y'' + y' + 25y = \sin \Omega t$

that are sinusoids in sync with $\sin \Omega t$; so let's try the form $y(t) = A \cos \Omega t + B \sin \Omega t$. Since

$$y' = -\Omega A \sin \Omega t + \Omega B \cos \Omega t \,,$$
$$y'' = -\Omega^2 A \cos \Omega t - \Omega^2 B \sin \Omega t \,,$$

we can simply insert these forms into equation (5), collect terms, and match coefficients to obtain a solution:

$$\sin \Omega t = y'' + y' + 25y$$
$$= -\Omega^2 A \cos \Omega t - \Omega^2 B \sin \Omega t + \left[-\Omega A \sin \Omega t + \Omega B \cos \Omega t \right]$$
$$+ 25 \left[A \cos \Omega t + B \sin \Omega t \right]$$
$$= \left[-\Omega^2 B - \Omega A + 25B \right] \sin \Omega t + \left[-\Omega^2 A + \Omega B + 25A \right] \cos \Omega t \,,$$

so

$$-\Omega A + (-\Omega^2 + 25)B = 1$$
$$(-\Omega^2 + 25)A + \Omega B = 0 \,.$$

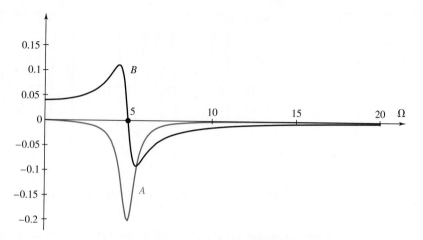

Figure 4.5 Vibration amplitudes around resonance

We find

$$A = \frac{-\Omega}{\Omega^2 + (\Omega^2 - 25)^2}, \qquad B = \frac{-\Omega^2 + 25}{\Omega^2 + (\Omega^2 - 25)^2}.$$

Figure 4.5 displays A and B as functions of the driving frequency Ω. A resonance clearly occurs around $\Omega \approx 5$. ◆

In most of this chapter, we are going to restrict our attention to differential equations of the form

(6) $ay'' + by' + cy = f(t)$,

where $y(t)$ [or $y(x)$, or $x(t)$, etc.] is the unknown function that we seek; a, b, and c are constants; and $f(t)$ [or $f(x)$] is a *known* function. The proper nomenclature for (6) is the *linear, second-order ordinary differential equation with constant coefficients*. In Sections 4.7 and 4.8, we will generalize our focus to equations with nonconstant coefficients, as well as to nonlinear equations. However, (6) is an excellent starting point because we are able to obtain explicit solutions and observe, in concrete form, the theoretical properties that are predicted for more general equations. For motivation of the mathematical procedures and theory for solving (6), we will consistently compare it with the mass–spring paradigm:

$$[\mathbf{inertia}] \times y'' + [\mathbf{damping}] \times y' + [\mathbf{stiffness}] \times y = F_{ext}.$$

4.1 EXERCISES

1. Verify that for $b = 0$ and $F_{ext}(t) = 0$, equation (3) has a solution of the form

 $$y(t) = \cos \omega t, \text{ where } \omega = \sqrt{k/m}.$$

2. If $F_{ext}(t) = 0$, equation (3) becomes

 $$my'' + by' + ky = 0.$$

 For this equation, verify the following:

 (a) If $y(t)$ is a solution, so is $cy(t)$, for any constant c.
 (b) If $y_1(t)$ and $y_2(t)$ are solutions, so is their sum $y_1(t) + y_2(t)$.

3. Show that if $F_{ext}(t) = 0, m = 1, k = 9$, and $b = 6$, then equation (3) has the "critically damped" solutions $y_1(t) = e^{-3t}$ and $y_2(t) = te^{-3t}$. What is the limit of these solutions as $t \rightarrow \infty$?

4. Verify that $y = \sin 3t + 2 \cos 3t$ is a solution to the initial value problem

$$2y'' + 18y = 0 ; \qquad y(0) = 2, \qquad y'(0) = 3 .$$

Find the maximum of $|y(t)|$ for $-\infty < t < \infty$.

5. Verify that the exponentially damped sinusoid $y(t) = e^{-3t} \sin(\sqrt{3}\, t)$ is a solution to equation (3) if $F_{\text{ext}}(t) = 0, m = 1, b = 6$, and $k = 12$. What is the limit of this solution as $t \to \infty$?

6. An external force $F(t) = 2 \cos 2t$ is applied to a mass–spring system with $m = 1, b = 0$, and $k = 4$, which is initially at rest; i.e., $y(0) = 0, y'(0) = 0$. Verify that $y(t) = \frac{1}{2} t \sin 2t$ gives the motion of this spring. What will eventually (as t increases) happen to the spring?

In Problems 7–9, find a synchronous solution of the form $A \cos \Omega t + B \sin \Omega t$ to the given forced oscillator equation using the method of Example 4 to solve for A and B.

7. $y'' + 2y' + 4y = 5 \sin 3t, \quad \Omega = 3$

8. $y'' + 2y' + 5y = -50 \sin 5t, \quad \Omega = 5$

9. $y'' + 2y' + 4y = 6 \cos 2t + 8 \sin 2t, \quad \Omega = 2$

10. Undamped oscillators that are driven at resonance have unusual (and nonphysical) solutions.

(a) To investigate this, find the synchronous solution $A \cos \Omega t + B \sin \Omega t$ to the generic forced oscillator equation

$$(7) \qquad my'' + by' + ky = \cos \Omega t .$$

(b) Sketch graphs of the coefficients A and B, as functions of Ω, for $m = 1, b = 0.1$, and $k = 25$.

(c) Now set $b = 0$ in your formulas for A and B and resketch the graphs in part (b), with $m = 1$, and $k = 25$. What happens at $\Omega = 5$? Notice that the amplitudes of the synchronous solutions grow without bound as Ω approaches 5.

(d) Show directly, by substituting the form $A \cos \Omega t + B \sin \Omega t$ into equation (7), that when $b = 0$ there are *no* synchronous solutions if $\Omega = \sqrt{k/m}$.

(e) Verify that $(2m\Omega)^{-1} t \sin \Omega t$ solves equation (7) when $b = 0$ and $\Omega = \sqrt{k/m}$. Notice that this *non*synchronous solution grows in time, without bound.

Clearly one cannot neglect damping in analyzing an oscillator forced at resonance, because otherwise the solutions, as shown in part (e), are nonphysical. This behavior will be studied later in this chapter.

4.2 Homogeneous Linear Equations: The General Solution

We begin our study of the linear second-order constant-coefficient differential equation

$$(1) \qquad ay'' + by' + cy = f(t) \quad (a \neq 0)$$

with the special case where the function $f(t)$ is zero:

$$(2) \qquad ay'' + by' + cy = 0 .$$

This case arises when we consider mass–spring oscillators vibrating freely—that is, without external forces applied. Equation (2) is called the *homogeneous form* of equation (1); $f(t)$ is the "nonhomogeneity" in (1). (This nomenclature is not related to the way we used the term for first-order equations in Section 2.6.)

A look at equation (2) tells us that a solution of (2) must have the property that its second derivative is expressible as a linear combination of its first and zeroth derivatives.[†] This suggests that we try to find a solution of the form $y = e^{rt}$, since derivatives of e^{rt} are just constants times e^{rt}. If we substitute $y = e^{rt}$ into (2), we obtain

$$ar^2 e^{rt} + br e^{rt} + c e^{rt} = 0 ,$$
$$e^{rt}(ar^2 + br + c) = 0 .$$

[†]The zeroth derivative of a function is the function itself.

Because e^{rt} is never zero, we can divide by it to obtain

(3) $ar^2 + br + c = 0$.

Consequently, $y = e^{rt}$ is a solution to (2) if and only if r satisfies equation (3). Equation (3) is called the **auxiliary equation** (also known as the **characteristic equation**) associated with the homogeneous equation (2).

Now the auxiliary equation is just a quadratic, and its roots are

$$r_1 = \frac{-b + \sqrt{b^2 - 4ac}}{2a} \quad \text{and} \quad r_2 = \frac{-b - \sqrt{b^2 - 4ac}}{2a}.$$

When the discriminant, $b^2 - 4ac$, is positive, the roots r_1 and r_2 are real and distinct. If $b^2 - 4ac = 0$, the roots are real and equal. And when $b^2 - 4ac < 0$, the roots are complex conjugate numbers. We consider the first two cases in this section; the complex case is deferred to Section 4.3.

Example 1 Find a pair of solutions to

(4) $y'' + 5y' - 6y = 0$.

Solution The auxiliary equation associated with (4) is

$$r^2 + 5r - 6 = (r - 1)(r + 6) = 0,$$

which has the roots $r_1 = 1, r_2 = -6$. Thus, e^t and e^{-6t} are solutions. ◆

Notice that the identically zero function, $y(t) \equiv 0$, is always a solution to (2). Furthermore, when we have a pair of solutions $y_1(t)$ and $y_2(t)$ to this equation, as in Example 1, we can construct an infinite number of other solutions by forming linear combinations:

(5) $y(t) = c_1 y_1(t) + c_2 y_2(t)$

for any choice of the constants c_1 and c_2. The fact that (5) is a solution to (2) can be seen by direct substitution and rearrangement:

$$\begin{aligned}
ay'' + by' + cy &= a(c_1 y_1 + c_2 y_2)'' + b(c_1 y_1 + c_2 y_2)' + c(c_1 y_1 + c_2 y_2) \\
&= a(c_1 y_1'' + c_2 y_2'') + b(c_1 y_1' + c_2 y_2') + c(c_1 y_1 + c_2 y_2) \\
&= c_1(ay_1'' + by_1' + cy_1) + c_2(ay_2'' + by_2' + cy_2) \\
&= 0 + 0.
\end{aligned}$$

The two "degrees of freedom" c_1 and c_2 in the combination (5) suggest that solutions to the differential equation (2) can be found meeting additional conditions, such as the initial conditions for the first-order equations in Chapter 1. But the presence of c_1 *and* c_2 leads one to anticipate that *two* such conditions, rather than just one, can be imposed. This is consistent with the mass–spring interpretation of equation (2), since predicting the motion of a mechanical system requires knowledge not only of the forces but also of the initial position $y(0)$ and velocity $y'(0)$ of the mass. A typical *initial value problem* for these second-order equations is given in the following example.

Example 2 Solve the initial value problem

(6) $y'' + 2y' - y = 0$; $y(0) = 0$, $y'(0) = -1$.

Solution We will first find a pair of solutions as in the previous example. Then we will adjust the constants c_1 and c_2 in (5) to obtain a solution that matches the initial conditions on $y(0)$ and $y'(0)$. The auxiliary equation is

$$r^2 + 2r - 1 = 0 .$$

Using the quadratic formula, we find that the roots of this equation are

$$r_1 = -1 + \sqrt{2} \quad \text{and} \quad r_2 = -1 - \sqrt{2} .$$

Consequently, the given differential equation has solutions of the form

(7) $y(t) = c_1 e^{(-1+\sqrt{2})t} + c_2 e^{(-1-\sqrt{2})t} .$

To find the specific solution that satisfies the initial conditions given in (6), we first differentiate y as given in (7), then plug y and y' into the initial conditions of (6). This gives

$$y(0) = c_1 e^0 + c_2 e^0 ,$$
$$y'(0) = (-1 + \sqrt{2})c_1 e^0 + (-1 - \sqrt{2})c_2 e^0 ,$$

or

$$0 = c_1 + c_2 ,$$
$$-1 = (-1 + \sqrt{2})c_1 + (-1 - \sqrt{2})c_2 .$$

Solving this system yields $c_1 = -\sqrt{2}/4$ and $c_2 = \sqrt{2}/4$. Thus,

$$y(t) = -\frac{\sqrt{2}}{4} e^{(-1+\sqrt{2})t} + \frac{\sqrt{2}}{4} e^{(-1-\sqrt{2})t}$$

is the desired solution. ◆

To gain more insight into the significance of the two-parameter solution form (5), we need to look at some of the properties of the second-order equation (2). First of all, there is an existence-and-uniqueness theorem for solutions to (2); it is somewhat like the corresponding Theorem 1 in Section 1.2 for first-order equations but updated to reflect the fact that *two* initial conditions are appropriate for *second*-order equations. As motivation for the theorem, suppose the differential equation (2) were *really easy*, with $b = 0$ and $c = 0$. Then $y'' = 0$ would merely say that the graph of $y(t)$ is simply a *straight line*, so it is uniquely determined by specifying a point on the line,

(8) $y(t_0) = Y_0 ,$

and the slope of the line,

(9) $y'(t_0) = Y_1 .$

Theorem 1 states that conditions (8) and (9) suffice to determine the solution uniquely for the more general equation (2).

Existence and Uniqueness: Homogeneous Case

Theorem 1. For any real numbers a ($\neq 0$), b, c, t_0, Y_0, and Y_1, there exists a unique solution to the initial value problem

(10) $ay'' + by' + cy = 0 ; \quad y(t_0) = Y_0 , \quad y'(t_0) = Y_1 .$

The solution is valid for all t in $(-\infty, +\infty)$.

Note in particular that if a solution $y(t)$ and its derivative vanish simultaneously at a point t_0 (i.e., $Y_0 = Y_1 = 0$), then $y(t)$ must be the identically zero solution.

In this section and the next, we will construct explicit solutions to (10), so the question of *existence* of a solution is not really an issue. It is extremely valuable to know, however, that the solution is *unique*. The proof of uniqueness is rather different from anything else in this chapter, so we defer it to Chapter 13.[†]

Now we want to use this theorem to show that, given two solutions $y_1(t)$ and $y_2(t)$ to equation (2), we can always find values of c_1 and c_2 so that $c_1 y_1(t) + c_2 y_2(t)$ meets specified initial conditions in (10) and therefore is the (unique) solution to the initial value problem. But we need to be a little more precise; if, for example, $y_2(t)$ is simply the identically zero solution, then $c_1 y_1(t) + c_2 y_2(t) = c_1 y_1(t)$ actually has only *one* constant and cannot be expected to satisfy *two* conditions. Furthermore, if $y_2(t)$ is simply a constant multiple of $y_1(t)$—say, $y_2(t) = \kappa y_1(t)$—then again $c_1 y_1(t) + c_2 y_2(t) = (c_1 + \kappa c_2) y_1(t) = C y_1(t)$ actually has only one constant. The condition we need is *linear independence*.

Linear Independence of Two Functions

Definition 1. A pair of functions $y_1(t)$ and $y_2(t)$ is said to be **linearly independent on the interval I** if and only if neither of them is a constant multiple of the other on all of I.[††] We say that y_1 and y_2 are **linearly dependent on I** if one of them *is* a constant multiple of the other on all of I.

Representation of Solutions to Initial Value Problem

Theorem 2. If $y_1(t)$ and $y_2(t)$ are any two solutions to the differential equation (2) that are linearly independent on $(-\infty, \infty)$, then unique constants c_1 and c_2 can always be found so that $c_1 y_1(t) + c_2 y_2(t)$ satisfies the initial value problem (10) on $(-\infty, \infty)$.

The proof of Theorem 2 will be easy once we establish the following technical lemma.

A Condition for Linear Dependence of Solutions

Lemma 1. For any real numbers $a\ (\neq 0)$, b, and c, if $y_1(t)$ and $y_2(t)$ are any two solutions to the differential equation (2) on $(-\infty, \infty)$ and if the equality

$$(11) \qquad y_1(\tau) y_2'(\tau) - y_1'(\tau) y_2(\tau) = 0$$

holds at *any* point τ, then y_1 and y_2 are linearly dependent on $(-\infty, \infty)$. (The expression on the left-hand side of (11) is called the *Wronskian* of y_1 and y_2 at the point τ; see Problem 34 on page 164.)

[†]All references to Chapters 11–13 refer to the expanded text, *Fundamentals of Differential Equations and Boundary Value Problems*, 7th ed.

[††]This definition will be generalized to three or more functions in Problem 35 and Chapter 6.

Proof of Lemma 1. *Case 1.* If $y_1(\tau) \neq 0$, then let κ equal $y_2(\tau)/y_1(\tau)$ and consider the solution to (2) given by $y(t) = \kappa y_1(t)$. It satisfies the same "initial conditions" at $t = \tau$ as does $y_2(t)$:

$$y(\tau) = \frac{y_2(\tau)}{y_1(\tau)} y_1(\tau) = y_2(\tau) ; \qquad y'(\tau) = \frac{y_2(\tau)}{y_1(\tau)} y_1'(\tau) = y_2'(\tau) ,$$

where the last equality follows from (11). By uniqueness, $y_2(t)$ must be the same function as $\kappa y_1(t)$ on I.

Case 2. If $y_1(\tau) = 0$ but $y_1'(\tau) \neq 0$, then (11) implies $y_2(\tau) = 0$. Let $\kappa = y_2'(\tau)/y_1'(\tau)$. Then the solution to (2) given by $y(t) = \kappa y_1(t)$ (again) satisfies the same "initial conditions" at $t = \tau$ as does $y_2(t)$:

$$y(\tau) = \frac{y_2'(\tau)}{y_1'(\tau)} y_1(\tau) = 0 = y_2(\tau) ; \qquad y'(\tau) = \frac{y_2'(\tau)}{y_1'(\tau)} y_1'(\tau) = y_2'(\tau) .$$

By uniqueness, then, $y_2(t) = \kappa y_1(t)$ on I.

Case 3. If $y_1(\tau) = y_1'(\tau) = 0$, then $y_1(t)$ is a solution to the differential equation (2) satisfying the initial conditions $y_1(\tau) = y_1'(\tau) = 0$; but $y(t) \equiv 0$ is the *unique* solution to this initial value problem. Thus, $y_1(t) \equiv 0$ [and is a constant multiple of $y_2(t)$]. ◆

Proof of Theorem 2. We already know that $y(t) = c_1 y_1(t) + c_2 y_2(t)$ is a solution to (2); we must show that c_1 and c_2 can be chosen so that

$$y(t_0) = c_1 y_1(t_0) + c_2 y_2(t_0) = Y_0$$

and

$$y'(t_0) = c_1 y_1'(t_0) + c_2 y_2'(t_0) = Y_1 .$$

But simple algebra shows these equations have the solution[†]

$$c_1 = \frac{Y_0 y_2'(t_0) - Y_1 y_2(t_0)}{y_1(t_0) y_2'(t_0) - y_1'(t_0) y_2(t_0)} \quad \text{and} \quad c_2 = \frac{Y_1 y_1(t_0) - Y_0 y_1'(t_0)}{y_1(t_0) y_2'(t_0) - y_1'(t_0) y_2(t_0)}$$

as long as the denominator is nonzero, and the technical lemma assures us that this condition is met. ◆

Now we can honestly say that if y_1 and y_2 are linearly independent solutions to (2) on $(-\infty, +\infty)$, then (5) is a **general solution,** since *any* solution $y_g(t)$ of (2) can be expressed in this form; simply pick c_1 and c_2 so that $c_1 y_1 + c_2 y_2$ matches the value and the derivative of y_g *at any point.* By uniqueness, $c_1 y_1 + c_2 y_2$ and y_g have to be the same function. See Figure 4.6 on page 162.

How do we *find* a general solution for the differential equation (2)? We already know the answer if the roots of the auxiliary equation (3) are real and distinct because clearly $y_1(t) = e^{r_1 t}$ is not a constant multiple of $y_2(t) = e^{r_2 t}$ if $r_1 \neq r_2$.

Distinct Real Roots

If the auxiliary equation (3) has distinct real roots r_1 and r_2, then both $y_1(t) = e^{r_1 t}$ and $y_2(t) = e^{r_2 t}$ are solutions to (2) and $y(t) = c_1 e^{r_1 t} + c_2 e^{r_2 t}$ is a general solution.

[†]To solve for c_1, for example, multiply the first equation by $y_2'(t_0)$ and the second by $y_2(t_0)$ and subtract.

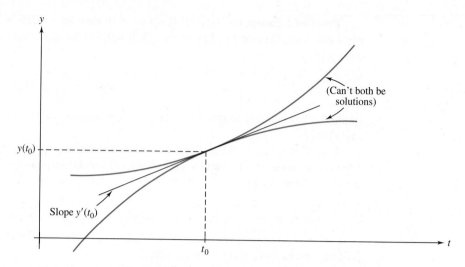

Figure 4.6 $y(t_0), y'(t_0)$ determine a *unique* solution.

When the roots of the auxiliary equation are equal, we only get one nontrivial solution, $y_1 = e^{rt}$. To satisfy *two* initial conditions, $y(t_0)$ and $y'(t_0)$, then we will need a second, linearly independent solution. The following rule is the key to finding a second solution.

Repeated Root

If the auxiliary equation (3) has a repeated root r, then both $y_1(t) = e^{rt}$ and $y_2(t) = te^{rt}$ are solutions to (2), and $y(t) = c_1 e^{rt} + c_2 te^{rt}$ is a general solution.

We illustrate this result before giving its proof.

Example 3 Find a solution to the initial value problem

(12) $y'' + 4y' + 4y = 0$; $y(0) = 1$, $y'(0) = 3$.

Solution The auxiliary equation for (12) is

$$r^2 + 4r + 4 = (r+2)^2 = 0 .$$

Because $r = -2$ is a double root, the rule says that (12) has solutions $y_1 = e^{-2t}$ and $y_2 = te^{-2t}$. Let's confirm that $y_2(t)$ is a solution:

$$y_2(t) = te^{-2t} ,$$
$$y_2'(t) = e^{-2t} - 2te^{-2t} ,$$
$$y_2''(t) = -2e^{-2t} - 2e^{-2t} + 4te^{-2t} = -4e^{-2t} + 4te^{-2t} ,$$
$$y_2'' + 4y_2' + 4y_2 = -4e^{-2t} + 4te^{-2t} + 4(e^{-2t} - 2te^{-2t}) + 4te^{-2t} = 0 .$$

Further observe that e^{-2t} and te^{-2t} are linearly independent since neither is a constant multiple of the other on $(-\infty, \infty)$. Finally, we insert the general solution $y(t) = c_1 e^{-2t} + c_2 te^{-2t}$ into the initial conditions,

$$y(0) = c_1 e^0 + c_2(0)e^0 = 1 ,$$
$$y'(0) = -2c_1 e^0 + c_2 e^0 - 2c_2(0)e^0 = 3 ,$$

and solve to find $c_1 = 1, c_2 = 5$. Thus $y = e^{-2t} + 5te^{-2t}$ is the desired solution. ◆

Why is it that $y_2(t) = te^{rt}$ is a solution to the differential equation (2) when r is a double root (and not otherwise)? In later chapters we will see a theoretical justification of this rule in very general circumstances; for present purposes, though, simply note what happens if we substitute y_2 into the differential equation (2):

$$y_2(t) = te^{rt},$$
$$y_2'(t) = e^{rt} + rte^{rt},$$
$$y''_2(t) = re^{rt} + re^{rt} + r^2te^{rt} = 2re^{rt} + r^2te^{rt},$$
$$ay_2'' + by_2' + cy_2 = [2ar + b]e^{rt} + [ar^2 + br + c]te^{rt}.$$

Now if r is a root of the auxiliary equation (3), the expression in the second brackets is zero. However, *if r is a double root, the expression in the first brackets is zero also*:

$$(13) \qquad r = \frac{-b \pm \sqrt{b^2 - 4ac}}{2a} = \frac{-b \pm (0)}{2a};$$

hence, $2ar + b = 0$ for a double root. In such a case, then, y_2 is a solution.

The method we have described for solving homogeneous linear second-order equations with constant coefficients applies to any order (even first-order) homogeneous linear equations with constant coefficients. We give a detailed treatment of such higher-order equations in Chapter 6. For now, we will be content to illustrate the method by means of an example. We remark briefly that a homogeneous linear nth-order equation has a general solution of the form

$$y(t) = c_1y_1(t) + c_2y_2(t) + \cdots + c_ny_n(t),$$

where the individual solutions $y_i(t)$ are "linearly independent." By this we mean that no y_i is expressible as a linear combination of the others; see Problem 35 on page 164.

Example 4 Find a general solution to

$$(14) \qquad y''' + 3y'' - y' - 3y = 0.$$

Solution If we try to find solutions of the form $y = e^{rt}$, then, as with second-order equations, we are led to finding roots of the auxiliary equation

$$(15) \qquad r^3 + 3r^2 - r - 3 = 0.$$

We observe that $r = 1$ is a root of the above equation, and dividing the polynomial on the left-hand side of (15) by $r - 1$ leads to the factorization

$$(r - 1)(r^2 + 4r + 3) = (r - 1)(r + 1)(r + 3) = 0.$$

Hence, the roots of the auxiliary equation are 1, -1, and -3, and so three solutions of (14) are e^t, e^{-t}, and e^{-3t}. The linear independence of these three exponential functions is proved in Problem 36. A general solution to (14) is then

$$(16) \qquad y(t) = c_1e^t + c_2e^{-t} + c_3e^{-3t}. \; \blacklozenge$$

So far we have seen only exponential solutions to the linear second-order constant coefficient equation. You may wonder where the vibratory solutions that govern mass–spring oscillators are. In the next section, it will be seen that they arise when the solutions to the auxiliary equation are complex.

4.2 EXERCISES

In Problems 1–12, find a general solution to the given differential equation.

1. $2y'' + 7y' - 4y = 0$

2. $y'' + 6y' + 9y = 0$

3. $y'' + 5y' + 6y = 0$

4. $y'' - y' - 2y = 0$

5. $y'' + 8y' + 16y = 0$

6. $y'' - 5y' + 6y = 0$

7. $6y'' + y' - 2y = 0$

8. $z'' + z' - z = 0$

9. $4y'' - 4y' + y = 0$

10. $y'' - y' - 11y = 0$

11. $4w'' + 20w' + 25w = 0$

12. $3y'' + 11y' - 7y = 0$

In Problems 13–20, solve the given initial value problem.

13. $y'' + 2y' - 8y = 0$; $y(0) = 3$, $y'(0) = -12$

14. $y'' + y' = 0$; $y(0) = 2$, $y'(0) = 1$

15. $y'' - 4y' + 3y = 0$; $y(0) = 1$, $y'(0) = 1/3$

16. $y'' - 4y' - 5y = 0$; $y(-1) = 3$, $y'(-1) = 9$

17. $y'' - 6y' + 9y = 0$; $y(0) = 2$, $y'(0) = 25/3$

18. $z'' - 2z' - 2z = 0$; $z(0) = 0$, $z'(0) = 3$

19. $y'' + 2y' + y = 0$; $y(0) = 1$, $y'(0) = -3$

20. $y'' - 4y' + 4y = 0$; $y(1) = 1$, $y'(1) = 1$

21. First-Order Constant-Coefficient Equations.

(a) Substituting $y = e^{rt}$, find the auxiliary equation for the first-order linear equation

$$ay' + by = 0,$$

where a and b are constants with $a \neq 0$.

(b) Use the result of part (a) to find the general solution.

In Problems 22–25, use the method described in Problem 21 to find a general solution to the given equation.

22. $3y' - 7y = 0$

23. $5y' + 4y = 0$

24. $3z' + 11z = 0$

25. $6w' - 13w = 0$

26. Boundary Value Problems. When the values of a solution to a differential equation are specified at two different points, these conditions are called **boundary conditions.** (In contrast, initial conditions specify the values of a function and its derivative at the same point.) The purpose of this exercise is to show that for boundary value problems there is no existence–uniqueness theorem that is analogous to Theorem 1. Given that every solution to

(17) $y'' + y = 0$

is of the form

$$y(t) = c_1 \cos t + c_2 \sin t,$$

where c_1 and c_2 are arbitrary constants, show that

(a) There is a unique solution to (17) that satisfies the boundary conditions $y(0) = 2$ and $y(\pi/2) = 0$.

(b) There is no solution to (17) that satisfies $y(0) = 2$ and $y(\pi) = 0$.

(c) There are infinitely many solutions to (17) that satisfy $y(0) = 2$ and $y(\pi) = -2$.

In Problems 27–32, use Definition 1 to determine whether the functions y_1 and y_2 are linearly dependent on the interval $(0, 1)$.

27. $y_1(t) = \cos t \sin t$, $y_2(t) = \sin 2t$

28. $y_1(t) = e^{3t}$, $y_2(t) = e^{-4t}$

29. $y_1(t) = te^{2t}$, $y_2(t) = e^{2t}$

30. $y_1(t) = t^2 \cos(\ln t)$, $y_2(t) = t^2 \sin(\ln t)$

31. $y_1(t) = \tan^2 t - \sec^2 t$, $y_2(t) \equiv 3$

32. $y_1(t) \equiv 0$, $y_2(t) = e^t$

33. Explain why two functions are linearly dependent on an interval I if and only if there exist constants c_1 and c_2, *not both zero*, such that

$$c_1 y_1(t) + c_2 y_2(t) = 0 \quad \text{for all } t \text{ in } I .$$

34. Wronskian. For any two differentiable functions y_1 and y_2, the function

(18) $$W[y_1, y_2](t) = y_1(t)y'_2(t) - y'_1(t)y_2(t)$$

is called the *Wronskian*[†] of y_1 and y_2. This function plays a crucial role in the proof of Theorem 2.

(a) Show that $W[y_1, y_2]$ can be conveniently expressed as the 2×2 determinant

$$W[y_1, y_2](t) = \begin{vmatrix} y_1(t) & y_2(t) \\ y'_1(t) & y'_2(t) \end{vmatrix}.$$

(b) Let $y_1(t), y_2(t)$ be a pair of solutions to the homogeneous equation $ay'' + by' + cy = 0$ (with $a \neq 0$) on an open interval I. Prove that $y_1(t)$ and $y_2(t)$ are linearly independent on I if and only if their Wronskian is never zero on I. [*Hint:* This is just a reformulation of Lemma 1.]

(c) Show that if $y_1(t)$ and $y_2(t)$ are any two differentiable functions that are linearly dependent on I, then their Wronskian is identically zero on I.

35. Linear Dependence of Three Functions. Three functions $y_1(t)$, $y_2(t)$, and $y_3(t)$ are said to be linearly dependent on an interval I if, on I, at least one of these functions is a linear combination of the remaining two [e.g., if $y_1(t) = c_1 y_2(t) + c_2 y_3(t)$]. Equivalently (compare Problem 33), y_1, y_2, and y_3 are linearly dependent on I if there exist constants C_1, C_2, and C_3, *not all zero*, such that

$$C_1 y_1(t) + C_2 y_2(t) + C_3 y_3(t) = 0 \quad \text{for all } t \text{ in } I.$$

Otherwise, we say that these functions are linearly independent on I.

†*Historical Footnote:* The Wronskian was named after the Polish mathematician H. Wronski (1778–1863).

For each of the following, determine whether the given three functions are linearly dependent or linearly independent on $(-\infty, \infty)$:

(a) $y_1(t) = 1$, $y_2(t) = t$, $y_3(t) = t^2$.

(b) $y_1(t) = -3$, $y_2(t) = 5\sin^2 t$, $y_3(t) = \cos^2 t$.

(c) $y_1(t) = e^t$, $y_2(t) = te^t$, $y_3(t) = t^2 e^t$.

(d) $y_1(t) = e^t$, $y_2(t) = e^{-t}$, $y_3(t) = \cosh t$.

36. Using the definition in Problem 35, prove that if $r_1, r_2,$ and r_3 are distinct real numbers, then the functions $e^{r_1 t}, e^{r_2 t},$ and $e^{r_3 t}$ are linearly independent on $(-\infty, \infty)$. [*Hint:* Assume to the contrary that, say, $e^{r_1 t} = c_1 e^{r_2 t} + c_2 e^{r_3 t}$ for all t. Divide by $e^{r_2 t}$ to get $e^{(r_1-r_2)t} = c_1 + c_2 e^{(r_3-r_2)t}$ and then differentiate to deduce that $e^{(r_1-r_2)t}$ and $e^{(r_3-r_2)t}$ are linearly dependent, which is a contradiction. (Why?)]

In Problems 37–41, find three linearly independent solutions (see Problem 35) of the given third-order differential equation and write a general solution as an arbitrary linear combination of these.

37. $y''' + y'' - 6y' + 4y = 0$

38. $y''' - 6y'' - y' + 6y = 0$

39. $z''' + 2z'' - 4z' - 8z = 0$

40. $y''' - 7y'' + 7y' + 15y = 0$

41. $y''' + 3y'' - 4y' - 12y = 0$

42. (True or False): If f_1, f_2, f_3 are three functions defined on $(-\infty, \infty)$ that are *pairwise* linearly independent on $(-\infty, \infty)$, then f_1, f_2, f_3 form a linearly independent set on $(-\infty, \infty)$. Justify your answer.

43. Solve the initial value problem:
$$y''' - y' = 0; \quad y(0) = 2,$$
$$y'(0) = 3, \quad y''(0) = -1.$$

44. Solve the initial value problem:
$$y''' - 2y'' - y' + 2y = 0;$$
$$y(0) = 2, \quad y'(0) = 3, \quad y''(0) = 5.$$

45. By using Newton's method or some other numerical procedure to approximate the roots of the auxiliary equation, find general solutions to the following equations:

(a) $3y''' + 18y'' + 13y' - 19y = 0$.

(b) $y^{iv} - 5y'' + 5y = 0$.

(c) $y^v - 3y^{iv} - 5y''' + 15y'' + 4y' - 12y = 0$.

46. One way to define hyperbolic functions is by means of differential equations. Consider the equation $y'' - y = 0$. The *hyperbolic cosine*, cosh t, is defined as the solution of this equation subject to the initial values: $y(0) = 1$ and $y'(0) = 0$. The *hyperbolic sine*, sinh t, is defined as the solution of this equation subject to the initial values: $y(0) = 0$ and $y'(0) = 1$.

(a) Solve these initial value problems to derive explicit formulas for cosh t and sinh t. Also show that
$$\frac{d}{dt} \cosh t = \sinh t \text{ and } \frac{d}{dt} \sinh t = \cosh t.$$

(b) Prove that a general solution of the equation $y'' - y = 0$ is given by $y = c_1 \cosh t + c_2 \sinh t$.

(c) Suppose a, b, and c are given constants for which $ar^2 + br + c = 0$ has two distinct real roots. If the two roots are expressed in the form $\alpha - \beta$ and $\alpha + \beta$, show that a general solution of the equation $ay'' + by' + cy = 0$ is $y = c_1 e^{\alpha t} \cosh(\beta t) + c_2 e^{\alpha t} \sinh(\beta t)$.

(d) Use the result of part (c) to solve the initial value problem: $y'' + y' - 6y = 0, y(0) = 2, y'(0) = -17/2$.

4.3 Auxiliary Equations with Complex Roots

The *simple harmonic equation* $y'' + y = 0$, so called because of its relation to the fundamental vibration of a musical tone, has as solutions $y_1(t) = \cos t$ and $y_2(t) = \sin t$. Notice, however, that the auxiliary equation associated with the harmonic equation is $r^2 + 1 = 0$, which has imaginary roots $r = \pm i$, where i denotes $\sqrt{-1}$.[†] In the previous section, we expressed the solutions to a linear second-order equation with constant coefficients in terms of exponential functions. It would appear, then, that one might be able to attribute a meaning to the forms e^{it} and e^{-it} and that these "functions" should be related to cos t and sin t. This matchup is accomplished by Euler's formula, which is discussed in this section.

When $b^2 - 4ac < 0$, the roots of the auxiliary equation

(1) $\qquad ar^2 + br + c = 0$

[†] Electrical engineers frequently use the symbol j to denote $\sqrt{-1}$.

associated with the homogeneous equation

(2) $ay'' + by' + cy = 0$

are the complex conjugate numbers

$$r_1 = \alpha + i\beta \quad \text{and} \quad r_2 = \alpha - i\beta \quad (i = \sqrt{-1}) ,$$

where α, β are the real numbers

(3) . $\alpha = -\dfrac{b}{2a} \quad \text{and} \quad \beta = \dfrac{\sqrt{4ac - b^2}}{2a} .$

As in the previous section, we would like to assert that the functions $e^{r_1 t}$ and $e^{r_2 t}$ are solutions to the equation (2). This is in fact the case, but before we can proceed, we need to address some fundamental questions. For example, if $r_1 = \alpha + i\beta$ is a complex number, what do we mean by the expression $e^{(\alpha + i\beta)t}$? If we assume that the law of exponents applies to complex numbers, then

(4) $e^{(\alpha + i\beta)t} = e^{\alpha t + i\beta t} = e^{\alpha t} e^{i\beta t} .$

We now need only clarify the meaning of $e^{i\beta t}$.

For this purpose, let's assume that the Maclaurin series for e^z is the same for complex numbers z as it is for real numbers. Observing that $i^2 = -1$, then for θ real we have

$$e^{i\theta} = 1 + (i\theta) + \frac{(i\theta)^2}{2!} + \cdots + \frac{(i\theta)^n}{n!} + \cdots$$

$$= 1 + i\theta - \frac{\theta^2}{2!} - \frac{i\theta^3}{3!} + \frac{\theta^4}{4!} + \frac{i\theta^5}{5!} + \cdots$$

$$= \left(1 - \frac{\theta^2}{2!} + \frac{\theta^4}{4!} + \cdots\right) + i\left(\theta - \frac{\theta^3}{3!} + \frac{\theta^5}{5!} + \cdots\right) .$$

Now recall the Maclaurin series for $\cos\theta$ and $\sin\theta$:

$$\cos\theta = 1 - \frac{\theta^2}{2!} + \frac{\theta^4}{4!} + \cdots ,$$

$$\sin\theta = \theta - \frac{\theta^3}{3!} + \frac{\theta^5}{5!} + \cdots .$$

Recognizing these expansions in the proposed series for $e^{i\theta}$, we make the identification

(5) $e^{i\theta} = \cos\theta + i\sin\theta ,$

which is known as **Euler's formula.**[†]

When Euler's formula (with $\theta = \beta t$) is used in equation (4), we find

(6) $e^{(\alpha + i\beta)t} = e^{\alpha t}(\cos\beta t + i\sin\beta t) ,$

which expresses the complex function $e^{(\alpha + i\beta)t}$ in terms of familiar real functions. Having made sense out of $e^{(\alpha + i\beta)t}$, we can now show (see Problem 30 on page 172) that

(7) $\dfrac{d}{dt} e^{(\alpha + i\beta)t} = (\alpha + i\beta) e^{(\alpha + i\beta)t} ,$

[†]*Historical Footnote:* This formula first appeared in Leonhard Euler's monumental two-volume *Introductio in Analysin Infinitorum* (1748).

and, with the choices of α and β as given in (3), the complex function $e^{(\alpha+i\beta)t}$ is indeed a solution to equation (2), as is $e^{(\alpha-i\beta)t}$, and a general solution is given by

$$(8) \qquad y(t) = c_1 e^{(\alpha+i\beta)t} + c_2 e^{(\alpha-i\beta)t}$$

$$= c_1 e^{\alpha t}(\cos\beta t + i\sin\beta t) + c_2 e^{\alpha t}(\cos\beta t - i\sin\beta t) .$$

Example 1 shows that in general the constants c_1 and c_2 that go into (8), for a specific initial value problem, are complex.

Example 1 Use the general solution (8) to solve the initial value problem

$$y'' + 2y' + 2y = 0 ; \qquad y(0) = 0, \qquad y'(0) = 2 .$$

Solution The auxiliary equation is $r^2 + 2r + 2 = 0$, which has roots

$$r = \frac{-2\sqrt{4-8}}{2} = -1 \pm i .$$

Hence, with $\alpha = -1$, $\beta = 1$, a general solution is given by

$$y(t) = c_1 e^{-t}(\cos t + i\sin t) + c_2 e^{-t}(\cos t - i\sin t) .$$

For initial conditions we have

$$y(0) = c_1 e^0(\cos 0 + i\sin 0) + c_2 e^0(\cos 0 - i\sin 0) = c_1 + c_2 = 0 ,$$

$$\begin{aligned} y'(0) &= -c_1 e^0(\cos 0 + i\sin 0) + c_1 e^0(-\sin 0 + i\cos 0) \\ &\quad -c_2 e^0(\cos 0 - i\sin 0) + c_2 e^0(-\sin 0 - i\cos 0) \\ &= (-1+i)c_1 + (-1-i)c_2 \\ &= 2 . \end{aligned}$$

As a result, $c_1 = -i$, $c_2 = i$, and $y(t) = -ie^{-t}(\cos t + i\sin t) + ie^{-t}(\cos t - i\sin t)$, or simply $2e^{-t}\sin t$. ◆

The final form of the answer to Example 1 suggests that we should seek an alternative pair of solutions to the differential equation (2) that don't require complex arithmetic, and we now turn to that task.

In general, if $z(t)$ is a complex-valued function of the real variable t, we can write $z(t) = u(t) + iv(t)$, where $u(t)$ and $v(t)$ are real-valued functions. The derivatives of $z(t)$ are then given by

$$\frac{dz}{dt} = \frac{du}{dt} + i\frac{dv}{dt}, \qquad \frac{d^2z}{dt^2} = \frac{d^2u}{dt^2} + i\frac{d^2v}{dt^2} .$$

With the following lemma, we show that the complex-valued solution $e^{(\alpha+i\beta)t}$ gives rise to two linearly independent *real-valued* solutions.

Real Solutions Derived from Complex Solutions

Lemma 2. Let $z(t) = u(t) + iv(t)$ be a solution to equation (2), where a, b, and c are real numbers. Then, the real part $u(t)$ and the imaginary part $v(t)$ are real-valued solutions of (2).[†]

[†] It will be clear from the proof that this property holds for any linear homogeneous differential equation having real-valued coefficients.

Proof. By assumption, $az'' + bz' + cz = 0$, and hence

$$a(u'' + iv'') + b(u' + iv') + c(u + iv) = 0,$$
$$(au'' + bu' + cu) + i(av'' + bv' + cv) = 0.$$

But a complex number is zero if and only if its real and imaginary parts are both zero. Thus, we must have

$$au'' + bu' + cu = 0 \quad \text{and} \quad av'' + bv' + cv = 0,$$

which means that both $u(t)$ and $v(t)$ are real-valued solutions of (2). ◆

When we apply Lemma 2 to the solution

$$e^{(\alpha + i\beta)t} = e^{\alpha t}\cos\beta t + ie^{\alpha t}\sin\beta t,$$

we obtain the following.

Complex Conjugate Roots

If the auxiliary equation has complex conjugate roots $\alpha \pm i\beta$, then two linearly independent solutions to (2) are

$$e^{\alpha t}\cos\beta t \quad \textbf{and} \quad e^{\alpha t}\sin\beta t,$$

and a general solution is

(9) $$y(t) = c_1 e^{\alpha t}\cos\beta t + c_2 e^{\alpha t}\sin\beta t,$$

where c_1 and c_2 are arbitrary constants.

In the preceding discussion, we glossed over some important details concerning complex numbers and complex-valued functions. In particular, further analysis is required to justify the use of the law of exponents, Euler's formula, and even the fact that the derivative of e^{rt} is re^{rt} when r is a complex constant.[†] If you feel uneasy about our conclusions, we encourage you to substitute the expression in (9) into equation (2) to verify that it is, indeed, a solution.

You may also be wondering what would have happened if we had worked with the function $e^{(\alpha - i\beta)t}$ instead of $e^{(\alpha + i\beta)t}$. We leave it as an exercise to verify that $e^{(\alpha - i\beta)t}$ gives rise to the same general solution (9). Indeed, the sum of these two complex solutions, divided by two, gives the first real-valued solution, while their difference, divided by $2i$, gives the second.

Example 2 Find a general solution to

(10) $$y'' + 2y' + 4y = 0.$$

Solution The auxiliary equation is

$$r^2 + 2r + 4 = 0,$$

which has roots

$$r = \frac{-2 \pm \sqrt{4 - 16}}{2} = \frac{-2 \pm \sqrt{-12}}{2} = -1 \pm i\sqrt{3}.$$

[†]For a detailed treatment of these topics see, for example, *Fundamentals of Complex Analysis,* 3rd ed., by E. B. Saff and A. D. Snider (Prentice Hall, Upper Saddle River, New Jersey, 2003).

Hence, with $\alpha = -1, \beta = \sqrt{3}$, a general solution for (10) is

$$y(t) = c_1 e^{-t} \cos(\sqrt{3}\, t) + c_2 e^{-t} \sin(\sqrt{3}\, t) \, . \quad \blacklozenge$$

When the auxiliary equation has complex conjugate roots, the (real) solutions oscillate between positive and negative values. This type of behavior is observed in vibrating springs.

Example 3 In Section 4.1 we discussed the mechanics of the mass–spring oscillator (Figure 4.1, page 152), and we saw how Newton's second law implies that the position $y(t)$ of the mass m is governed by the second-order differential equation

(11) $my''(t) + by'(t) + ky(t) = 0 \, ,$

where the terms are physically identified as

$$[\,\text{inertia}\,]y'' + [\,\text{damping}\,]y' + [\,\text{stiffness}\,]y = 0 \, .$$

Determine the equation of motion for a spring system when $m = 36$ kg, $b = 12$ kg/sec (which is equivalent to 12 N-sec/m), $k = 37$ kg/sec^2, $y(0) = 0.7$ m, and $y'(0) = 0.1$ m/sec. After how many seconds will the mass first cross the equilibrium point?

Solution The equation of motion is given by $y(t)$, the solution of the initial value problem for the specified values of m, b, k, $y(0)$, and $y'(0)$. That is, we seek the solution to

(12) $36y'' + 12y' + 37y = 0 \, ; \quad y(0) = 0.7 \, , \; y'(0) = 0.1 \, .$

The auxiliary equation for (12) is

$$36r^2 + 12r + 37 = 0 \, ,$$

which has roots

$$r = \frac{-12 \pm \sqrt{144 - 4(36)(37)}}{72} = \frac{-12 \pm 12\sqrt{1 - 37}}{72} = -\frac{1}{6} \pm i \, .$$

Hence, with $\alpha = -1/6, \quad \beta = 1$, the displacement $y(t)$ can be expressed in the form

(13) $y(t) = c_1 e^{-t/6} \cos t + c_2 e^{-t/6} \sin t \, .$

We can find c_1 and c_2 by substituting $y(t)$ and $y'(t)$ into the initial conditions given in (12). Differentiating (13), we get a formula for $y'(t)$:

$$y'(t) = \left(-\frac{c_1}{6} + c_2\right) e^{-t/6} \cos t + \left(-c_1 - \frac{c_2}{6}\right) e^{-t/6} \sin t \, .$$

Substituting into the initial conditions now results in the system

$$c_1 = 0.7 \, ,$$

$$-\frac{c_1}{6} + c_2 = 0.1 \, .$$

Upon solving, we find $c_1 = 0.7$ and $c_2 = 1.3/6$. With these values, the equation of motion becomes

$$y(t) = 0.7 e^{-t/6} \cos t + \frac{1.3}{6} e^{-t/6} \sin t \, .$$

To determine the times when the mass will cross the equilibrium point, we set $y(t) = 0$ and solve for t:

$$0 = y(t) = 0.7e^{-t/6}\cos t + \frac{1.3}{6}e^{-t/6}\sin t = (\cos t)\left(0.7e^{-t/6} + \frac{1.3}{6}e^{-t/6}\tan t\right).$$

But $y(t)$ is not zero for $\cos t = 0$, so only the zeros of the second factor are pertinent; that is,

$$\tan t = -\frac{4.2}{1.3}.$$

A quick glance at the graph of $\tan t$ reveals that the first positive t for which this is true lies between $\pi/2$ and π. Since the arctangent function takes only values between $-\pi/2$ and $\pi/2$, the appropriate adjustment is

$$t = \arctan\left(\frac{-4.2}{1.3}\right) + \pi \approx 1.87 \text{ seconds}. \quad \blacklozenge$$

From Example 3 we see that *any* second-order constant-coefficient differential equation $ay'' + by' + cy = 0$ can be interpreted as describing a mass–spring system with mass a, damping coefficient b, spring stiffness c, and displacement y, if these constants make sense physically; that is, if a is positive and b and c are nonnegative. From the discussion in Section 4.1, then, we expect on physical grounds to see damped oscillatory solutions in such a case. This is consistent with the display in equation (9). With $a = m$ and $c = k$, the exponential decay rate α equals $-b/(2m)$, and the angular frequency β equals $\sqrt{4mk - b^2}/(2m)$, by equation (3).

It is a little surprising, then, that the solutions to the equation $y'' + 4y' + 4y = 0$ do *not* oscillate; the general solution was shown in Example 3 of Section 4.2 (page 162) to be $c_1 e^{-2t} + c_2 t e^{-2t}$. The physical significance of this is simply that when the damping coefficient b is too high, the resulting friction prevents the mass from oscillating. Rather than overshoot the spring's equilibrium point, it merely settles in lazily. This could happen if a light mass on a weak spring were submerged in a viscous fluid.

From the above formula for the oscillation frequency β, we can see that the oscillations will not occur for $b > \sqrt{4mk}$. This *overdamping* phenomenon is discussed in more detail in Section 4.9.

It is extremely enlightening to contemplate the predictions of the mass–spring analogy when the coefficients b and c in the equation $ay'' + by' + cy = 0$ are *negative*.

Example 4 Interpret the equation

(14) $36y'' - 12y' + 37y = 0$

in terms of the mass–spring system.

Solution Equation (14) is a minor alteration of equation (12) in Example 3; the auxiliary equation $36r^2 - 12r + 37$ has roots $r = (+)\frac{1}{6} \pm i$. Thus, its general solution becomes

(15) $y(t) = c_1 e^{+t/6}\cos t + c_2 e^{+t/6}\sin t$.

Comparing equation (14) with the mass–spring model

(16) $[\text{inertia}]y'' + [\text{damping}]y' + [\text{stiffness}]y = 0$,

we have to envision a *negative* damping coefficient $b = -12$, giving rise to a friction force $F_{\text{friction}} = -by'$ that *imparts* energy to the system instead of draining it. The increase in energy over time must then reveal itself in oscillations of ever-greater amplitude–precisely in accordance with formula (15), for which a typical graph is drawn in Figure 4.7. $\quad \blacklozenge$

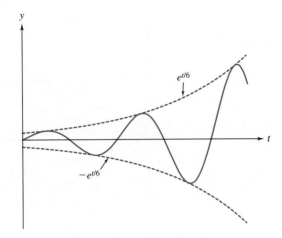

Figure 4.7 Solution graph for Example 4

Example 5 Interpret the equation

(17) $y'' + 5y' - 6y = 0$

in terms of the mass–spring system.

Solution Comparing the given equation with (16), we have to envision a spring with a *negative* stiffness $k = -6$. What does this mean? As the mass is moved away from the spring's equilibrium point, the spring *repels* the mass farther with a force $F_{\text{spring}} = -ky$ that intensifies as the displacement increases. Clearly the spring must "exile" the mass to (plus or minus) infinity, and we expect all solutions $y(t)$ to approach $\pm\infty$ as t increases (except for the equilibrium solution $y(t) \equiv 0$).

In fact, in Example 1 of Section 4.2, we showed the general solution to equation (17) to be

(18) $c_1 e^t + c_2 e^{-6t}$.

Indeed, if we examine the solutions $y(t)$ that start with a unit displacement $y(0) = 1$ and velocity $y'(0) = v_0$, we find

(19) $y(t) = \dfrac{6 + v_0}{7} e^t + \dfrac{1 - v_0}{7} e^{-6t}$,

and the plots in Figure 4.8 on page 172 confirm our prediction that all (nonequilibrium) solutions diverge—except for the one with $v_0 = -6$.

What is the physical significance of this isolated bounded solution? Evidently, if the mass is given an initial inwardly directed velocity of -6, it has barely enough energy to overcome the effect of the spring banishing it to $+\infty$ but not enough energy to cross the equilibrium point (and get pushed to $-\infty$). So it asymptotically approaches the (extremely delicate) equilibrium position $y = 0$. ◆

In Section 4.8, we will see that taking further liberties with the mass–spring interpretation enables us to predict qualitative features of more complicated equations.

Throughout this section we have assumed that the coefficients a, b, and c in the differential equation were real numbers. If we now allow them to be *complex* constants, then the roots

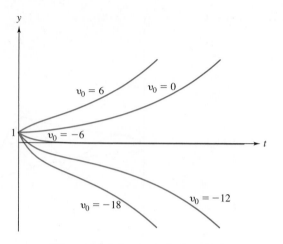

Figure 4.8 Solution graphs for Example 5

r_1, r_2 of the auxiliary equation (1) are, in general, also complex but not necessarily conjugates of each other. When $r_1 \neq r_2$, a general solution to equation (2) still has the form

$$y(t) = c_1 e^{r_1 t} + c_2 e^{r_2 t} ,$$

but c_1 and c_2 are now arbitrary complex-valued constants, and we have to resort to the clumsy calculations of Example 1.

We also remark that a complex differential equation can be regarded as a system of two real differential equations since we can always work separately with its real and imaginary parts. Systems are discussed in Chapters 5 and 9.

4.3 EXERCISES

In Problems 1–8, the auxiliary equation for the given differential equation has complex roots. Find a general solution.

1. $y'' + 9y = 0$
2. $y'' + y = 0$
3. $z'' - 6z' + 10z = 0$
4. $y'' - 10y' + 26y = 0$
5. $w'' + 4w' + 6w = 0$
6. $y'' - 4y' + 7y = 0$
7. $4y'' + 4y' + 6y = 0$
8. $4y'' - 4y' + 26y = 0$

In Problems 9–20, find a general solution.

9. $y'' - 8y' + 7y = 0$
10. $y'' + 4y' + 8y = 0$
11. $z'' + 10z' + 25z = 0$
12. $u'' + 7u = 0$
13. $y'' - 2y' + 26y = 0$
14. $y'' + 2y' + 5y = 0$
15. $y'' - 3y' - 11y = 0$
16. $y'' + 10y' + 41y = 0$
17. $y'' - y' + 7y = 0$
18. $2y'' + 13y' - 7y = 0$
19. $y''' + y'' + 3y' - 5y = 0$
20. $y''' - y'' + 2y = 0$

In Problems 21–27, solve the given initial value problem.

21. $y'' + 2y' + 2y = 0$; $y(0) = 2$, $y'(0) = 1$
22. $y'' + 2y' + 17y = 0$; $y(0) = 1$, $y'(0) = -1$
23. $w'' - 4w' + 2w = 0$; $w(0) = 0$, $w'(0) = 1$

24. $y'' + 9y = 0$; $y(0) = 1$, $y'(0) = 1$
25. $y'' - 2y' + 2y = 0$; $y(\pi) = e^{\pi}$, $y'(\pi) = 0$
26. $y'' - 2y' + y = 0$; $y(0) = 1$, $y'(0) = -2$
27. $y''' - 4y'' + 7y' - 6y = 0$; $y(0) = 1$, $y'(0) = 0$, $y''(0) = 0$

28. To see the effect of changing the parameter b in the initial value problem

$$y'' + by' + 4y = 0; \quad y(0) = 1, \quad y'(0) = 0,$$

solve the problem for $b = 5$, 4, and 2 and sketch the solutions.

29. Find a general solution to the following higher-order equations.
 (a) $y''' - y'' + y' + 3y = 0$
 (b) $y''' + 2y'' + 5y' - 26y = 0$
 (c) $y^{iv} + 13y'' + 36y = 0$

30. Using the representation for $e^{(\alpha+i\beta)t}$ in (6), verify the differentiation formula (7).

31. Using the mass–spring analogy, predict the behavior as $t \to +\infty$ of the solution to the given initial value problem. Then confirm your prediction by actually solving the problem.

(a) $y'' + 16y = 0$; $y(0) = 2$, $y'(0) = 0$
(b) $y'' + 100y' + y = 0$; $y(0) = 1$, $y'(0) = 0$
(c) $y'' - 6y' + 8y = 0$; $y(0) = 1$, $y'(0) = 0$
(d) $y'' + 2y' - 3y = 0$; $y(0) = -2$, $y'(0) = 0$
(e) $y'' - y' - 6y = 0$; $y(0) = 1$, $y'(0) = 1$

32. Vibrating Spring without Damping. A vibrating spring without damping can be modeled by the initial value problem (11) in Example 3 by taking $b = 0$.

(a) If $m = 10$ kg, $k = 250$ kg/sec^2, $y(0) = 0.3$ m, and $y'(0) = -0.1$ m/sec, find the equation of motion for this undamped vibrating spring.

(b) After how many seconds will the mass in part (a) first cross the equilibrium point?

(c) When the equation of motion is of the form displayed in (9), the motion is said to be **oscillatory** with **frequency** $\beta/2\pi$. Find the frequency of oscillation for the spring system of part (a).

33. Vibrating Spring with Damping. Using the model for a vibrating spring with damping discussed in Example 3:

(a) Find the equation of motion for the vibrating spring with damping if $m = 10$ kg, $b = 60$ kg/sec, $k = 250$ kg/sec^2, $y(0) = 0.3$ m, and $y'(0) = -0.1$ m/sec.

(b) After how many seconds will the mass in part (a) first cross the equilibrium point?

(c) Find the frequency of oscillation for the spring system of part (a). [*Hint:* See the definition of frequency given in Problem 32(c).]

(d) Compare the results of Problems 32 and 33 and determine what effect the damping has on the frequency of oscillation. What other effects does it have on the solution?

34. *RLC* Series Circuit. In the study of an electrical circuit consisting of a resistor, capacitor, inductor, and an electromotive force (see Figure 4.9), we are led to an initial value problem of the form

(20) $\quad L\dfrac{dI}{dt} + RI + \dfrac{q}{C} = E(t)$;

$\qquad q(0) = q_0$,

$\qquad I(0) = I_0$,

where L is the inductance in henrys, R is the resistance in ohms, C is the capacitance in farads, $E(t)$ is the electromotive force in volts, $q(t)$ is the charge in coulombs on the capacitor at time t, and $I = dq/dt$ is the current in amperes. Find the current at time t if the charge on the capacitor is initially zero, the initial current is zero, $L = 10$ H, $R = 20$ Ω, $C = (6260)^{-1}$ F, and $E(t) = 100$ V. [*Hint:* Differentiate both sides of the

differential equation in (20) to obtain a homogeneous linear second-order equation for $I(t)$. Then use (20) to determine dI/dt at $t = 0$.]

Figure 4.9 *RLC* series circuit

35. Swinging Door. The motion of a swinging door with an adjustment screw that controls the amount of friction on the hinges is governed by the initial value problem

$$I\theta'' + b\theta' + k\theta = 0; \quad \theta(0) = \theta_0, \quad \theta'(0) = v_0,$$

where θ is the angle that the door is open, I is the moment of inertia of the door about its hinges, $b > 0$ is a damping constant that varies with the amount of friction on the door, $k > 0$ is the spring constant associated with the swinging door, θ_0 is the initial angle that the door is opened, and v_0 is the initial angular velocity imparted to the door (see Figure 4.10). If I and k are fixed, determine for which values of b the door will *not* continually swing back and forth when closing.

Figure 4.10 Top view of swinging door

36. Although the real general solution form (9) is convenient, it is also possible to use the form

(21) $\quad d_1 e^{(\alpha+i\beta)t} + d_2 e^{(\alpha-i\beta)t}$

to solve initial value problems, as illustrated in Example 1. The coefficients d_1 and d_2 are complex constants.

(a) Use the form (21) to solve Problem 21. Verify that your form is equivalent to the one derived using (9).

(b) Show that, in general, d_1 and d_2 in (21) must be complex conjugates in order that the solution be real.

37. The auxiliary equations for the following differential equations have repeated complex roots. Adapt the "repeated root" procedure of Section 4.2 to find their general solutions:

(a) $y^{iv} + 2y'' + y = 0$.

(b) $y^{iv} + 4y''' + 12y'' + 16y' + 16y = 0$. [*Hint:* The auxiliary equation is $(r^2 + 2r + 4)^2 = 0$.]

38. Prove the sum of angles formula for the sine function by following these steps. Fix x.

(a) Let $f(t) := \sin(x+t)$. Show that $f''(t) + f(t) = 0$, $f(0) = \sin x$, and $f'(0) = \cos x$.

(b) Use the auxiliary equation technique to solve the initial value problem $y'' + y = 0$, $y(0) = \sin x$, and $y'(0) = \cos x$.

(c) By uniqueness, the solution in part (b) is the same as $f(t)$ from part (a). Write this equality; this should be the standard sum of angles formula for $\sin(x+t)$.

4.4 Nonhomogeneous Equations: the Method of Undetermined Coefficients

In this section we employ "judicious guessing" to derive a simple procedure for finding a solution to a *non*homogeneous linear equation with constant coefficients

$$(1) \qquad ay'' + by' + cy = f(t),$$

when the nonhomogeneity $f(t)$ is a single term of a special type. Our experience in Section 4.3 indicates that (1) will have an infinite number of solutions. For the moment we are content to find one particular solution. To motivate the procedure, let's first look at a few instructive examples.

Example 1 Find a particular solution to

$$(2) \qquad y'' + 3y' + 2y = 3t.$$

Solution We need to find a function $y(t)$ such that the combination $y'' + 3y' + 2y$ is a linear function of t—namely, $3t$. Now what kind of function y "ends up" as a linear function after having its zeroth, first, and second derivatives combined? One immediate answer is: *another linear function.* So we might try $y_1(t) = At$ and attempt to match up $y_1'' + 3y_1' + 2y_1$ with $3t$.

Perhaps you can see that this won't work: $y_1 = At$, $y_1' = A$ and $y_1'' = 0$ gives us

$$y_1'' + 3y_1' + 2y_1 = 3A + 2At,$$

and for this to equal $3t$, we require both that $A = 0$ and $A = 3/2$. We'll have better luck if we append a constant term to the trial function: $y_2(t) = At + B$. Then $y_2' = A$, $y_2'' = 0$, and

$$y_2'' + 3y_2' + 2y_2 = 3A + 2(At + B) = 2At + (3A + 2B),$$

which successfully matches up with $3t$ if $2A = 3$ and $3A + 2B = 0$. Solving this system gives $A = 3/2$ and $B = -9/4$. Thus, the function

$$y_2(t) = \frac{3}{2}t - \frac{9}{4}$$

is a solution to (2). ◆

Example 1 suggests the following method for finding a particular solution to the equation

$$ay'' + by' + cy = Ct^m, \quad m = 0, 1, 2, \ldots;$$

namely, we guess a solution of the form

$$y_p(t) = A_m t^m + \cdots + A_1 t + A_0,$$

with undetermined coefficients A_j, and match the corresponding powers of t in $ay'' + by' + cy$ with Ct^m.[†] This procedure involves solving $m+1$ linear equations in the $m+1$ unknowns

[†]In this case the coefficient of t^k in $ay'' + by' + cy$ will be zero for $k \ne m$ and C for $k = m$.

$A_0, A_1, \ldots, A_m$, and hopefully they have a solution. The technique is called the **method of undetermined coefficients.** Note that, as Example 1 demonstrates, we must retain *all* the powers $t^m, t^{m-1}, \ldots, t^1, t^0$ in the trial solution even though they are not present in the nonhomogeneity $f(t)$.

Example 2 Find a particular solution to

(3) $y'' + 3y' + 2y = 10e^{3t}$.

Solution We guess $y_p(t) = Ae^{3t}$ because then y'_p and y''_p will retain the same exponential form:

$$y''_p + 3y'_p + 2y_p = 9Ae^{3t} + 3(3Ae^{3t}) + 2(Ae^{3t}) = 20Ae^{3t}.$$

Setting $20Ae^{3t} = 10e^{3t}$ and solving for A gives $A = 1/2$; hence,

$$y_p(t) = \frac{e^{3t}}{2}$$

is a solution to (3). ◆

Example 3 Find a particular solution to

(4) $y'' + 3y' + 2y = \sin t$.

Solution Our initial action might be to guess $y_1(t) = A \sin t$, but this will fail because the derivatives introduce cosine terms:

$$y''_1 + 3y'_1 + 2y_1 = -A \sin t + 3A \cos t + 2A \sin t = A \sin t + 3A \cos t,$$

and matching this with $\sin t$ would require that A equal both 1 and 0. So we include the cosine term in the trial solution:

$$y_p(t) = A \sin t + B \cos t,$$
$$y'_p(t) = A \cos t - B \sin t,$$
$$y''_p(t) = -A \sin t - B \cos t,$$

and (4) becomes

$$y''_p(t) + 3y'_p(t) + 2y_p(t) = -A \sin t - B \cos t + 3A \cos t - 3B \sin t$$
$$+ 2A \sin t + 2B \cos t$$
$$= (A - 3B) \sin t + (B + 3A) \cos t$$
$$= \sin t.$$

The equations $A - 3B = 1$, $B + 3A = 0$ have the solution $A = 0.1$, $B = -0.3$. Thus, the function

$$y_p(t) = 0.1 \sin t - 0.3 \cos t$$

is a particular solution to (4). ◆

More generally, for an equation of the form

(5) $ay'' + by' + cy = C \sin \beta t$ (or $C \cos \beta t$),

the method of undetermined coefficients suggests that we guess

(6) $y_p(t) = A \cos \beta t + B \sin \beta t$

and solve (5) for the unknowns A and B.

If we compare equation (5) with the mass–spring system equation

(7) $[\text{inertia}] \times y'' + [\text{damping}] \times y' + [\text{stiffness}] \times y = F_{\text{ext}}$,

we can interpret (5) as describing a damped oscillator, shaken with a sinusoidal force. According to our discussion in Section 4.1, then, we would expect the mass ultimately to respond by moving in synchronization with the forcing sinusoid. In other words, the form (6) is suggested by physical, as well as mathematical, experience. A complete description of forced oscillators will be given in Section 4.10.

Example 4 Find a particular solution to

(8) $y'' + 4y = 5t^2 e^t$.

Solution Our experience with Example 1 suggests that we take a trial solution of the form $y_p(t) = (At^2 + Bt + C)e^t$, to match the nonhomogeneity in (8). We find

$$y_p = (At^2 + Bt + C)e^t,$$

$$y_p' = (2At + B)e^t + (At^2 + Bt + C)e^t,$$

$$y_p'' = 2Ae^t + 2(2At + B)e^t + (At^2 + Bt + C)e^t,$$

$$y_p'' + 4y_p = e^t(2A + 2B + C + 4C) + te^t(4A + B + 4B) + t^2 e^t(A + 4A)$$

$$= 5t^2 e^t.$$

Matching like terms yields $A = 1$, $B = -4/5$, and $C = -2/25$. A solution is given by

$$y_p(t) = \left(t^2 - \frac{4t}{5} - \frac{2}{25}\right)e^t.\quad \blacklozenge$$

As our examples illustrate, when the nonhomogeneous term $f(t)$ is an exponential, a sine, a cosine function, or a nonnegative integer power of t times any of these, the function $f(t)$ itself suggests the form of a particular solution. However, certain situations thwart the straightforward application of the method of undetermined coefficients. Consider, for example, the equation

(9) $y'' + y' = 5$.

Example 1 suggests that we guess $y_1(t) = A$, a zero-degree polynomial. But substitution into (9) proves futile:

$$(A)'' + (A)' = 0 \neq 5.$$

The problem arises because any constant function, such as $y_1(t) = A$, is a solution to the corresponding homogeneous equation $y'' + y' = 0$, and the undetermined coefficient A gets lost upon substitution into the equation. We would encounter the same situation if we tried to find a solution to

(10) $y'' - 6y' + 9y = e^{3t}$

of the form $y_1 = Ae^{3t}$, because e^{3t} solves the associated homogeneous equation and

$$\left[Ae^{3t}\right]'' - 6\left[Ae^{3t}\right]' + 9\left[Ae^{3t}\right] = 0 \neq e^{3t}.$$

The "trick" for refining the method of undetermined coefficients in these situations smacks of the same logic as in Section 4.2, when a method was prescribed for finding second solutions to homogeneous equations with double roots. Basically, we append an extra factor of t to the trial

solution suggested by the basic procedure. In other words, to solve (9) we try $y_p(t) = At$ instead of A:

(9') $y_p = At, \quad y'_p = A, \quad y''_p = 0,$
$$y''_p + y'_p = 0 + A = 5,$$
$$A = 5, \qquad y_p(t) = 5t.$$

Similarly, to solve (10) we try $y_p = Ate^{3t}$ instead of Ae^{3t}. The trick won't work this time, because the characteristic equation of (10) has a double root and, consequently, Ate^{3t} also solves the homogeneous equation:

$$\left[Ate^{3t}\right]'' - 6\left[Ate^{3t}\right]' + 9\left[Ate^{3t}\right] = 0 \neq e^{3t}.$$

But if we append *another* factor of t, $y_p = At^2e^{3t}$, we succeed in finding a particular solution:[†]

$$y_p = At^2e^{3t}, \quad y'_p = 2Ate^{3t} + 3At^2e^{3t}, \quad y''_p = 2Ae^{3t} + 12Ate^{3t} + 9At^2e^{3t},$$
$$y''_p - 6y'_p + 9y_p = (2Ae^{3t} + 12Ate^{3t} + 9At^2e^{3t}) - 6(2Ate^{3t} + 3At^2e^{3t}) + 9(At^2e^{3t})$$
$$= 2Ae^{3t} = e^{3t},$$

so $A = 1/2$ and $y_p(t) = t^2e^{3t}/2$.

To see why this strategy resolves the problem and to generalize it, recall the form of the original differential equation (1), $ay'' + by' + cy = f(t)$. Its associated auxiliary equation is

(11) $ar^2 + br + c = 0,$

and if r is a double root, then

(12) $2ar + b = 0$

holds also [equation (13), Section 4.2, page 163].

Now suppose the nonhomogeneity $f(t)$ has the form Ct^me^{rt}, and we seek to match this $f(t)$ by substituting $y_p(t) = (A_nt^n + A_{n-1}t^{n-1} + \cdots + A_1t + A_0)e^{rt}$ into (1), *with the power n to be determined*. For simplicity we merely list the leading terms in y_p, y'_p, and y''_p:

$$y_p = A_nt^ne^{rt} + A_{n-1}t^{n-1}e^{rt} + A_{n-2}t^{n-2}e^{rt} + \text{ (lower-order terms)}$$
$$y'_p = A_nrt^ne^{rt} + A_nnt^{n-1}e^{rt} + A_{n-1}rt^{n-1}e^{rt} + A_{n-1}(n-1)t^{n-2}e^{rt}$$
$$\quad + A_{n-2}rt^{n-2}e^{rt} + (l.o.t.),$$
$$y''_p = A_nr^2t^ne^{rt} + 2A_nnrt^{n-1}e^{rt} + A_nn(n-1)t^{n-2}e^{rt}$$
$$\quad + A_{n-1}r^2t^{n-1}e^{rt} + 2A_{n-1}r(n-1)t^{n-2}e^{rt} + A_{n-2}r^2t^{n-2}e^{rt} + (l.o.t).$$

Then the left-hand member of (1) becomes

(13) $ay''_p + by'_p + cy_p$

$$= A_n(ar^2 + br + c)t^ne^{rt} + [A_nn(2ar + b) + A_{n-1}(ar^2 + br + c)]t^{n-1}e^{rt}$$
$$+ [A_nn(n-1)a + A_{n-1}(n-1)(2ar + b) + A_{n-2}(ar^2 + br + c)]t^{n-2}e^{rt}$$
$$+ (l.o.t.),$$

[†]Indeed, the solution t^2 to the equation $y'' = 2$, computed by simple integration, can also be derived by appending two factors of t to the solution $y \equiv 1$ of the associated homogeneous equation.

and we observe the following:

Case 1. If r is *not* a root of the auxiliary equation, the leading term in (13) is $A_n(ar^2 + br + c)t^n e^{rt}$, and to match $f(t) = Ct^m e^{rt}$ we must take $n = m$:

$$y_p(t) = (A_m t^m + \cdots + A_1 t + A_0)e^{rt}.$$

Case 2. If r is a *simple* root of the auxiliary equation, (11) holds and the leading term in (13) is $A_n n(2ar + b)t^{n-1}e^{rt}$, and to match $f(t) = Ct^m e^{rt}$ we must take $n = m + 1$:

$$y_p(t) = (A_{m+1}t^{m+1} + A_m t^m + \cdots + A_1 t + A_0)e^{rt}.$$

However, now the final term $A_0 e^{rt}$ can be dropped, since it solves the associated homogeneous equation, so we can factor out t and for simplicity renumber the coefficients to write

$$y_p(t) = t(A_m t^m + \cdots + A_1 t + A_0)e^{rt}.$$

Case 3. If r is a *double* root of the auxiliary equation, (11) and (12) hold and the leading term in (13) is $A_n n(n-1)at^{n-2}e^{rt}$, and to match $f(t) = Ct^m e^{rt}$ we must take $n = m + 2$:

$$y_p(t) = (A_{m+2}t^{m+2} + A_{m+1}t^{m+1} + \cdots + A_2 t^2 + A_1 t + A_0)e^{rt},$$

but again we drop the solutions to the associated homogeneous equation and renumber to write

$$y_p(t) = t^2(A_m t^m + \cdots + A_1 t + A_0)e^{rt}.$$

We summarize with the following rule.

Method of Undetermined Coefficients

To find a particular solution to the differential equation

$$ay'' + by' + cy = Ct^m e^{rt},$$

where m is a nonnegative integer, use the form

$$(14) \qquad y_p(t) = t^s(A_m t^m + \cdots + A_1 t + A_0)e^{rt},$$

with

(i) $s = 0$ if r is not a root of the associated auxiliary equation;
(ii) $s = 1$ if r is a simple root of the associated auxiliary equation; and
(iii) $s = 2$ if r is a double root of the associated auxiliary equation.

To find a particular solution to the differential equation

$$ay'' + by' + cy = \begin{cases} Ct^m e^{\alpha t}\cos\beta t \\ \text{or} \\ Ct^m e^{\alpha t}\sin\beta t \end{cases}$$

for $\beta \neq 0$, use the form

$$(15) \qquad y_p(t) = t^s(A_m t^m + \cdots + A_1 t + A_0)e^{\alpha t}\cos\beta t$$
$$+ t^s(B_m t^m + \cdots + B_1 t + B_0)e^{\alpha t}\sin\beta t,$$

with

(iv) $s = 0$ if $\alpha + i\beta$ is not a root of the associated auxiliary equation; and
(v) $s = 1$ if $\alpha + i\beta$ is a root of the associated auxiliary equation.

[The (cos, sin) formulation (15) is easily derived from the exponential formulation (14) by putting $r = \alpha + i\beta$ and employing Euler's formula, as in Section 4.3.]

Remark 1. The nonhomogeneity Ct^m corresponds to the case when $r = 0$.

Remark 2. The rigorous justification of the method of undetermined coefficients [including the analysis of the terms we dropped in (13)] will be presented in a more general context in Chapter 6.

Example 5 Find the form for a particular solution to

(16) $y'' + 2y' - 3y = f(t)$,

where $f(t)$ equals

(a) $7\cos 3t$ **(b)** $2te^t \sin t$ **(c)** $t^2 \cos \pi t$ **(d)** $5e^{-3t}$ **(e)** $3te^t$ **(f)** $t^2 e^t$

Solution The auxiliary equation for the homogeneous equation corresponding to (16), $r^2 + 2r - 3 = 0$, has roots $r_1 = 1$ and $r_2 = -3$. Notice that the functions in (a), (b), and (c) are associated with *complex* roots (because of the trigonometric factors). These are clearly different from r_1 and r_2, so the solution forms correspond to (15) with $s = 0$:

(a) $y_p(t) = A\cos 3t + B\sin 3t$

(b) $y_p(t) = (A_1 t + A_0)e^t \cos t + (B_1 t + B_0)e^t \sin t$

(c) $y_p(t) = (A_2 t^2 + A_1 t + A_0)\cos \pi t + (B_2 t^2 + B_1 t + B_0)\sin \pi t$

For the nonhomogeneity in (d) we appeal to (ii) and take $y_p(t) = Ate^{-3t}$ since -3 is a simple root of the auxiliary equation. Similarly, for (e) we take $y_p(t) = t(A_1 t + A_0)e^t$ and for (f) we take $y_p(t) = t(A_2 t^2 + A_1 t + A_0)e^t$. ◆

Example 6 Find the form of a particular solution to

$$y'' - 2y' + y = f(t),$$

for the same set of nonhomogeneities $f(t)$ as in Example 5.

Solution Now the auxiliary equation for the corresponding homogeneous equation is $r^2 - 2r + 1 = (r-1)^2 = 0$, with the double root $r = 1$. This root is not linked with any of the nonhomogeneities (a) through (d), so the same trial forms should be used for (a), (b), and (c) as in the previous example, and $y(t) = Ae^{-3t}$ will work for (d).

Since $r = 1$ is a double root, we have $s = 2$ in (14) and the trial forms for (e) and (f) have to be changed to

(e) $y_p(t) = t^2(A_1 t + A_0)e^t$

(f) $y_p(t) = t^2(A_2 t^2 + A_1 t + A_0)e^t$

respectively, in accordance with (iii). ◆

Example 7 Find the form of a particular solution to

$$y'' - 2y' + 2y = 5te^t \cos t .$$

Solution Now the auxiliary equation for the corresponding homogeneous equation is $r^2 - 2r + 2 = 0$, and it has complex roots $r_1 = 1 + i, r_2 = 1 - i$. Since the nonhomogeneity involves $e^{\alpha t}\cos \beta t$ with $\alpha = \beta = 1$; that is, $\alpha + i\beta = 1 + i = r_1$, the solution takes the form

$$y_p(t) = t(A_1 t + A_0)e^t \cos t + t(B_1 t + B_0)e^t \sin t .$$ ◆

The nonhomogeneity $\tan t$ in an equation like $y'' + y' + y = \tan t$ is not one of the forms for which the method of undetermined coefficients can be used; the derivatives of the "trial solution" $y(t) = A \tan t$, for example, get complicated, and it is not clear what additional terms need to be added to obtain a true solution. In Section 4.6 we discuss a different procedure that can handle such nonhomogeneous terms. **Keep in mind that the method of undetermined coefficients applies only to nonhomogeneities that are polynomials, exponentials, sines or cosines, or products of these functions.** The superposition principle in Section 4.5 shows how the method can be extended to the **sums** of such nonhomogeneities. Also, it provides the key to assembling a general solution to (1) that can accommodate initial value problems, which we have avoided so far in our examples.

4.4 EXERCISES

In Problems 1–8, decide whether or not the method of undetermined coefficients can be applied to find a particular solution of the given equation.

1. $y'' + 2y' - y = t^{-1}e^t$
2. $5y'' - 3y' + 2y = t^3 \cos 4t$
3. $2y''(x) - 6y'(x) + y(x) = (\sin x)/e^{4x}$
4. $x'' + 5x' - 3x = 3^t$
5. $y''(\theta) + 3y'(\theta) - y(\theta) = \sec \theta$
6. $2\omega''(x) - 3\omega(x) = 4x \sin^2 x + 4x \cos^2 x$
7. $8z'(x) - 2z(x) = 3x^{100}e^{4x} \cos 25x$
8. $ty'' - y' + 2y = \sin 3t$

In Problems 9–26, find a particular solution to the differential equation.

9. $y'' + 3y = -9$
10. $y'' + 2y' - y = 10$
11. $y''(x) + y(x) = 2^x$
12. $2x' + x = 3t^2$
13. $y'' - y' + 9y = 3 \sin 3t$
14. $2z'' + z = 9e^{2t}$
15. $\dfrac{d^2y}{dx^2} - 5\dfrac{dy}{dx} + 6y = xe^x$
16. $\theta''(t) - \theta(t) = t \sin t$
17. $y'' + 4y = 8 \sin 2t$
18. $y'' - 2y' + y = 8e^t$
19. $4y'' + 11y' - 3y = -2te^{-3t}$
20. $y'' + 4y = 16t \sin 2t$
21. $x''(t) - 4x'(t) + 4x(t) = te^{2t}$
22. $x''(t) - 2x'(t) + x(t) = 24t^2 e^t$
23. $y''(\theta) - 7y'(\theta) = \theta^2$
24. $y''(x) + y(x) = 4x \cos x$
25. $y'' + 2y' + 4y = 111e^{2t} \cos 3t$
26. $y'' + 2y' + 2y = 4te^{-t} \cos t$

In Problems 27–32, determine the form of a particular solution for the differential equation. (Do not evaluate coefficients.)

27. $y'' + 9y = 4t^3 \sin 3t$
28. $y'' - 6y' + 9y = 5t^6 e^{3t}$
29. $y'' + 3y' - 7y = t^4 e^t$
30. $y'' - 2y' + y = 7e^t \cos t$
31. $y'' + 2y' + 2y = 8t^3 e^{-t} \sin t$
32. $y'' - y' - 12y = 2t^6 e^{-3t}$

In Problems 33–36, use the method of undetermined coefficients to find a particular solution to the given higher-order equation.

33. $y''' - y'' + y = \sin t$
34. $2y''' + 3y'' + y' - 4y = e^{-t}$
35. $y''' + y'' - 2y = te^t$
36. $y^{(4)} - 3y'' - 8y = \sin t$

4.5 The Superposition Principle and Undetermined Coefficients Revisited

The next theorem describes the superposition principle, a very simple observation which nonetheless endows the solution set for our equations with a powerful structure. It extends the applicability of the method of undetermined coefficients and enables us to solve initial value problems for nonhomogeneous differential equations.

Superposition Principle

Theorem 3. If y_1 is a solution to the differential equation

$$ay'' + by' + cy = f_1(t) \, ,$$

and y_2 is a solution to

$$ay'' + by' + cy = f_2(t) \, ,$$

then for any constants k_1 and k_2, the function $k_1 y_1 + k_2 y_2$ is a solution to the differential equation

$$ay'' + by' + cy = k_1 f_1(t) + k_2 f_2(t) \, .$$

Proof. This is straightforward; by substituting and rearranging we find

$$a(k_1 y_1 + k_2 y_2)'' + b(k_1 y_1 + k_2 y_2)' + c(k_1 y_1 + k_2 y_2)$$
$$= k_1(ay_1'' + by_1' + cy_1) + k_2(ay_2'' + by_2' + cy_2)$$
$$= k_1 f_1(t) + k_2 f_2(t) \, . \quad \blacklozenge$$

Example 1 Find a particular solution to

(1) $y'' + 3y' + 2y = 3t + 10e^{3t}$ and

(2) $y'' + 3y' + 2y = -9t + 20e^{3t} \, .$

Solution In Example 1, Section 4.4, we found that $y_1(t) = 3t/2 - 9/4$ was a solution to $y'' + 3y' + 2y = 3t$, and in Example 2 we found that $y_2(t) = e^{3t}/2$ solved $y'' + 3y' + 2y = 10e^{3t}$. By superposition, then, $y_1 + y_2 = 3t/2 - 9/4 + e^{3t}/2$ solves equation (1).

The right-hand member of (2) equals minus three times $(3t)$ plus two times $(10e^{3t})$. Therefore, this same combination of y_1 and y_2 will solve (2):

$$y(t) = -3y_1 + 2y_2 = -3(3t/2 - 9/4) + 2(e^{3t}/2) = -9t/2 + 27/4 + e^{3t} \, . \quad \blacklozenge$$

If we take a particular solution y_p to a nonhomogeneous equation like

(3) $ay'' + by' + cy = f(t)$

and add it to a general solution $c_1 y_1 + c_2 y_2$ of the homogeneous equation associated with (3),

(4) $ay'' + by' + cy = 0 \, ,$

the sum

(5) $y(t) = y_p(t) + c_1 y_1(t) + c_2 y_2(t)$

is again, according to the superposition principle, a solution to (3):

$$a(y_p + c_1 y_1 + c_2 y_2)'' + b(y_p + c_1 y_1 + c_2 y_2)' + c(y_p + c_1 y_1 + c_2 y_2)$$
$$= f(t) + 0 + 0 = f(t) \, .$$

Since (5) contains two parameters, one would suspect that c_1 and c_2 can be chosen to make it satisfy arbitrary initial conditions. It is easy to verify that this is indeed the case.

Existence and Uniqueness: Nonhomogeneous Case

Theorem 4. For any real numbers $a\,(\neq 0)$, b, c, t_0, Y_0, and Y_1, suppose $y_p(t)$ is a particular solution to (3) in an interval I containing t_0 and that $y_1(t)$ and $y_2(t)$ are linearly independent solutions to the associated homogeneous equation (4) in I. Then there exists a unique solution in I to the initial value problem

(6) $ay'' + by' + cy = f(t)$, $y(t_0) = Y_0$, $y'(t_0) = Y_1$,

and it is given by (5), for the appropriate choice of the constants c_1, c_2.

Proof. We have already seen that the superposition principle implies that (5) solves the differential equation. To satisfy the initial conditions in (6) we need to choose the constants so that

(7) $$\begin{cases} y_p(t_0) + c_1 y_1(t_0) + c_2 y_2(t_0) = Y_0, \\ y_p'(t_0) + c_1 y_1'(t_0) + c_2 y_2'(t_0) = Y_1. \end{cases}$$

But as in the proof of Theorem 2 in Section 4.2, simple algebra shows that the choice

$$c_1 = \frac{\left[Y_0 - y_p(t_0)\right] y_2'(t_0) - \left[Y_1 - y_p'(t_0)\right] y_2(t_0)}{y_1(t_0) y_2'(t_0) - y_1'(t_0) y_2(t_0)}$$

and

$$c_2 = \frac{\left[Y_1 - y_p'(t_0)\right] y_1(t_0) - \left[Y_0 - y_p(t_0)\right] y_1'(t_0)}{y_1(t_0) y_2'(t_0) - y_1'(t_0) y_2(t_0)}$$

solves (7) unless the denominator is zero; Lemma 1, Section 4.2, assures us that it is not.

Why is the solution unique? If $y_{\mathrm{I}}(t)$ were another solution to (6), then the difference $y_{\mathrm{II}}(t) := y_p(t) + c_1 y_1(t) + c_2 y_2(t) - y_{\mathrm{I}}(t)$ would satisfy

(8) $$\begin{cases} ay_{\mathrm{II}}'' + by_{\mathrm{II}}' + cy_{\mathrm{II}} = f(t) - f(t) = 0, \\ y_{\mathrm{II}}(t_0) = Y_0 - Y_0 = 0, \quad y_{\mathrm{II}}'(t_0) = Y_1 - Y_1 = 0. \end{cases}$$

But the initial value problem (8) admits the identically zero solution, and Theorem 1 in Section 4.2 applies since the differential equation in (8) is homogeneous. Consequently, (8) has *only* the identically zero solution. Thus, $y_{\mathrm{II}} \equiv 0$ and $y_{\mathrm{I}} = y_p + c_1 y_1 + c_2 y_2$. ◆

These deliberations entitle us to say that $y = y_p + c_1 y_1 + c_2 y_2$ is a **general solution** to the nonhomogeneous equation (3), since *any* solution $y_g(t)$ can be expressed in this form. (**Proof:** As in Section 4.2, we simply pick c_1 and c_2 so that $y_p + c_1 y_1 + c_2 y_2$ matches the value and the derivative of y_g at *any single point*; by uniqueness, $y_p + c_1 y_1 + c_2 y_2$ and y_g have to be the same function.)

Example 2 Given that $y_p(t) = t^2$ is a particular solution to

$$y'' - y = 2 - t^2,$$

find a general solution and a solution satisfying $y(0) = 1$, $y'(0) = 0$.

Solution The corresponding homogeneous equation,

$$y'' - y = 0 ,$$

has the associated auxiliary equation $r^2 - 1 = 0$. Because $r = \pm 1$ are the roots of this equation, a general solution to the homogeneous equation is $c_1 e^t + c_2 e^{-t}$. Combining this with the particular solution $y_p(t) = t^2$ of the nonhomogeneous equation, we find that a general solution is

$$y(t) = t^2 + c_1 e^t + c_2 e^{-t} .$$

To meet the initial conditions, set

$$y(0) = 0^2 + c_1 e^0 + c_2 e^{-0} = 1 ,$$
$$y'(0) = 2 \times 0 + c_1 e^0 - c_2 e^{-0} = 0 ,$$

which yields $c_1 = c_2 = \frac{1}{2}$. The answer is

$$y(t) = t^2 + \frac{1}{2}(e^t + e^{-t}) = t^2 + \cosh t . \quad \blacklozenge$$

Example 3 A mass–spring system is driven by a sinusoidal external force $(5 \sin t + 5 \cos t)$. The mass equals 1, the spring constant equals 2, and the damping coefficient equals 2 (in appropriate units), so the deliberations of Section 4.1 imply that the motion is governed by the differential equation

$$(9) \qquad y'' + 2y' + 2y = 5 \sin t + 5 \cos t .$$

If the mass is initially located at $y(0) = 1$, with a velocity $y'(0) = 2$, find its equation of motion.

Solution The associated homogeneous equation $y'' + 2y' + 2y = 0$ was studied in Example 1, Section 4.3; the roots of the auxiliary equation were found to be $-1 \pm i$, leading to a general solution $c_1 e^{-t} \cos t + c_2 e^{-t} \sin t$.

The method of undetermined coefficients dictates that we try to find a particular solution of the form $A \sin t + B \cos t$ for the first nonhomogeneity $5 \sin t$:

$$(10) \qquad y_p = A \sin t + B \cos t , \qquad y_p' = A \cos t - B \sin t , \qquad y_p'' = -A \sin t - B \cos t ;$$
$$y_p'' + 2y_p' + 2y_p = (-A - 2B + 2A) \sin t + (-B + 2A + 2B) \cos t = 5 \sin t .$$

Matching coefficients requires $A = 1, B = -2$ and so $y_p = \sin t - 2 \cos t$.

The second nonhomogeneity $5 \cos t$ calls for the identical form for y_p and leads to $(-A - 2B + 2A) \sin t + (-B + 2A + 2B) \cos t = 5 \cos t$, or $A = 2, B = 1$. Hence $y_p = 2 \sin t + \cos t$.

By the superposition principle, a general solution to (9) is given by the sum

$$y = c_1 e^{-t} \cos t + c_2 e^{-t} \sin t + \sin t - 2 \cos t + 2 \sin t + \cos t$$
$$= c_1 e^{-t} \cos t + c_2 e^{-t} \sin t + 3 \sin t - \cos t .$$

The initial conditions are

$$y(0) = 1 = c_1 e^{-0} \cos 0 + c_2 e^{-0} \sin 0 + 3 \sin 0 - \cos 0 = c_1 - 1 ,$$
$$y'(0) = 2 = c_1 \left[-e^{-t} \cos t - e^{-t} \sin t \right]_{t=0} + c_2 \left[-e^{-t} \sin t + e^{-t} \cos t \right]_{t=0}$$
$$+ 3 \cos 0 + \sin 0$$
$$= -c_1 + c_2 + 3 ,$$

requiring $c_1 = 2, c_2 = 1$, and thus

$$(11) \qquad y(t) = 2e^{-t} \cos t + e^{-t} \sin t + 3 \sin t - \cos t . \quad \blacklozenge$$

The solution (11) exemplifies the features of forced, damped oscillations that we antici-pated in Section 4.1. There is a sinusoidal component $(3\sin t - \cos t)$ that is synchronous with the driving force $(5\sin t + 5\cos t)$, and a component $(2e^{-t}\cos t + e^{-t}\sin t)$ that dies out. When the system is "pumped" sinusoidally, the response is a synchronous sinusoidal oscilla-tion, after an initial transient that depends on the initial conditions; the synchronous response is the particular solution supplied by the method of undetermined coefficients, and the transient is the solution to the associated homogeneous equation. This interpretation will be discussed in detail in Sections 4.9 and 4.10.

You may have observed that, since the two undetermined-coefficient forms in the last example were identical and were destined to be added together, we could have used the form (10) to match both nonhomogeneities at the same time, deriving the condition

$$y_p'' + 2y_p' + 2y_p = (-A - 2B + 2A)\sin t + (-B + 2A + 2B)\cos t = 5\sin t + 5\cos t,$$

with solution $y_p = 3\sin t - \cos t$. The next example illustrates this "streamlined" procedure.

Example 4 Find a particular solution to

(12) $y'' - y = 8te^t + 2e^t.$

Solution A general solution to the associated homogeneous equation is easily seen to be $c_1 e^t + c_2 e^{-t}$. Thus, a particular solution for matching the nonhomogeneity $8te^t$ has the form $t(A_1 t + A_0)e^t$, whereas matching $2e^t$ requires the form $A_0 te^t$. Therefore, we can match both with the first form:

$$y_p = t(A_1 t + A_0)e^t = (A_1 t^2 + A_0 t)e^t,$$
$$y_p' = (A_1 t^2 + A_0 t)e^t + (2A_1 t + A_0)e^t = [A_1 t^2 + (2A_1 + A_0)t + A_0]e^t,$$
$$y_p'' = [2A_1 t + (2A_1 + A_0)]e^t + [A_1 t^2 + (2A_1 + A_0)t + A_0]e^t$$
$$= [A_1 t^2 + (4A_1 + A_0)t + (2A_1 + 2A_0)]e^t.$$

Thus

$$y_p'' - y_p = [4A_1 t + (2A_1 + 2A_0)]e^t$$
$$= 8te^t + 2e^t,$$

which yields $A_1 = 2$, $A_0 = -1$, and so $y_p = (2t^2 - t)e^t.$ ◆

We generalize this procedure by modifying the method of undetermined coefficients as follows.

Method of Undetermined Coefficients (Revisited)

To find a particular solution to the differential equation

$$ay'' + by' + cy = P_m(t)e^{rt},$$

where $P_m(t)$ is a polynomial of degree m, use the form

(13) $y_p(t) = t^s(A_m t^m + \cdots + A_1 t + A_0)e^{rt};$

if r is not a root of the associated auxiliary equation, take $s = 0$; if r is a simple root of the associated auxiliary equation, take $s = 1$; and if r is a double root of the associated auxiliary equation, take $s = 2$.

To find a particular solution to the differential equation

$$ay'' + by' + cy = P_m(t)e^{\alpha t}\cos\beta t + Q_n(t)e^{\alpha t}\sin\beta t\,, \quad \beta \neq 0\,,$$

where $P_m(t)$ is a polynomial of degree m and $Q_n(t)$ is a polynomial of degree n, use the form

$$(14) \qquad y_p(t) = t^s(A_k t^k + \cdots + A_1 t + A_0)e^{\alpha t}\cos\beta t$$
$$+ t^s(B_k t^k + \cdots + B_1 t + B_0)e^{\alpha t}\sin\beta t\,,$$

where k is the larger of m and n. If $\alpha + i\beta$ is not a root of the associated auxiliary equation, take $s = 0$; if $\alpha + i\beta$ is a root of the associated auxiliary equation, take $s = 1$.

Example 5 Write down the form of a particular solution to the equation

$$y'' + 2y' + 2y = 5e^{-t}\sin t + 5t^3 e^{-t}\cos t\,.$$

Solution The roots of the associated homogeneous equation $y'' + 2y' + 2y = 0$ were identified in Example 3 as $-1 \pm i$. Application of (14) dictates the form

$$y_p(t) = t(A_3 t^3 + A_2 t^2 + A_1 t + A_0)e^{-t}\cos t + t(B_3 t^3 + B_2 t^2 + B_1 t + B_0)e^{-t}\sin t\,. \quad \blacklozenge$$

The method of undetermined coefficients applies to higher-order linear differential equations with constant coefficients. Details will be provided in Chapter 6, but the following example should be clear.

Example 6 Write down the form of a particular solution to the equation

$$y''' + 2y'' + y' = 5e^{-t}\sin t + 3 + 7te^{-t}\,.$$

Solution The auxiliary equation for the associated homogeneous is $r^3 + 2r^2 + r = r(r+1)^2 = 0$, with a double root $r = -1$ and a single root $r = 0$. Term by term, the nonhomogeneities call for the forms

$$A_0 e^{-t}\cos t + B_0 e^{-t}\sin t \qquad (\text{for } 5e^{-t}\sin t)\,,$$
$$t A_0 \qquad (\text{for } 3)\,,$$
$$t^2(A_1 t + A_0)e^{-t} \qquad (\text{for } 7te^{-t})\,.$$

(If -1 were a *triple* root, we would need $t^3(A_1 t + A_0)e^{-t}$ for $7te^{-t}$.) Of course, we have to rename the coefficients, so the general form is

$$y_p(t) = Ae^{-t}\cos t + Be^{-t}\sin t + tC + t^2(Dt + E)e^{-t}\,. \quad \blacklozenge$$

4.5 EXERCISES

1. Given that $y_1(t) = \cos t$ is a solution to

$$y'' - y' + y = \sin t$$

and $y_2(t) = e^{2t}/3$ is a solution to

$$y'' - y' + y = e^{2t},$$

use the superposition principle to find solutions to the following differential equations:

(a) $y'' - y' + y = 5\sin t$.

(b) $y'' - y' + y = \sin t - 3e^{2t}$.

(c) $y'' - y' + y = 4\sin t + 18e^{2t}$.

2. Given that $y_1(t) = (1/4)\sin 2t$ is a solution to $y'' + 2y' + 4y = \cos 2t$ and that $y_2(t) = t/4 - 1/8$ is a solution to $y'' + 2y' + 4y = t$, use the superposition principle to find solutions to the following:

(a) $y'' + 2y' + 4y = t + \cos 2t$.

(b) $y'' + 2y' + 4y = 2t - 3\cos 2t$.

(c) $y'' + 2y' + 4y = 11t - 12\cos 2t$.

In Problems 3–8, a nonhomogeneous equation and a particular solution are given. Find a general solution for the equation.

3. $y'' - y = t$, $y_p(t) = -t$

4. $y'' + y' = 1$, $y_p(t) = t$

5. $\theta'' - \theta' - 2\theta = 1 - 2t$, $\theta_p(t) = t - 1$

6. $y'' + 5y' + 6y = 6x^2 + 10x + 2 + 12e^x$,
$y_p(x) = e^x + x^2$

7. $y'' = 2y + 2\tan^3 x$, $y_p(x) = \tan x$

8. $y'' = 2y' - y + 2e^x$, $y_p(x) = x^2 e^x$

In Problems 9–16 decide whether the method of undetermined coefficients together with superposition can be applied to find a particular solution of the given equation. Do not solve the equation.

9. $3y'' + 2y' + 8y = t^2 + 4t - t^2 e^t \sin t$

10. $y'' - y' + y = (e^t + t)^2$

11. $y'' - 6y' - 4y = 4\sin 3t - e^{3t}t^2 + 1/t$

12. $y'' + y' + ty = e^t + 7$

13. $y'' - 2y' + 3y = \cosh t + \sin^3 t$

14. $2y'' + 3y' - 4y = 2t + \sin^2 t + 3$

15. $y'' + e^t y' + y = 7 + 3t$

16. $2y'' - y' + 6y = t^2 e^{-t}\sin t - 8t\cos 3t + 10^t$

In Problems 17–22, find a general solution to the differential equation.

17. $y'' - 2y' - 3y = 3t^2 - 5$

18. $y'' - y = -11t + 1$

19. $y''(x) - 3y'(x) + 2y(x) = e^x \sin x$

20. $y''(\theta) + 4y(\theta) = \sin\theta - \cos\theta$

21. $y''(\theta) + 2y'(\theta) + 2y(\theta) = e^{-\theta}\cos\theta$

22. $y''(x) + 6y'(x) + 10y(x)$
$= 10x^4 + 24x^3 + 2x^2 - 12x + 18$

In Problems 23–30, find the solution to the initial value problem.

23. $y' - y = 1$, $y(0) = 0$

24. $y'' = 6t$; $y(0) = 3$, $y'(0) = -1$

25. $z''(x) + z(x) = 2e^{-x}$; $z(0) = 0$, $z'(0) = 0$

26. $y'' + 9y = 27$; $y(0) = 4$, $y'(0) = 6$

27. $y''(x) - y'(x) - 2y(x) = \cos x - \sin 2x$;
$y(0) = -7/20$, $y'(0) = 1/5$

28. $y'' + y' - 12y = e^t + e^{2t} - 1$; $y(0) = 1$, $y'(0) = 3$

29. $y''(\theta) - y(\theta) = \sin\theta - e^{2\theta}$;
$y(0) = 1$, $y'(0) = -1$

30. $y'' + 2y' + y = t^2 + 1 - e^t$; $y(0) = 0$, $y'(0) = 2$

In Problems 31–36, determine the form of a particular solution for the differential equation. Do not solve.

31. $y'' + y = \sin t + t\cos t + 10^t$

32. $y'' - y = e^{2t} + te^{2t} + t^2 e^{2t}$

33. $x'' - x' - 2x = e^t \cos t - t^2 + \cos^3 t$

34. $y'' + 5y' + 6y = \sin t - \cos 2t$

35. $y'' - 4y' + 5y = e^{5t} + t\sin 3t - \cos 3t$

36. $y'' - 4y' + 4y = t^2 e^{2t} - e^{2t}$

In Problems 37–40, find a particular solution to the given higher-order equation.

37. $y''' - 2y'' - y' + 2y = 2t^2 + 4t - 9$

38. $y^{(4)} - 5y'' + 4y = 10\cos t - 20\sin t$

39. $y''' + y'' - 2y = te^t + 1$

40. $y^{(4)} - 3y''' + 3y'' - y' = 6t - 20$

41. Discontinuous Forcing Term. In certain physical models, the nonhomogeneous term, or **forcing term,** $g(t)$ in the equation

$$ay'' + by' + cy = g(t)$$

may not be continuous but may have a jump discontinuity. If this occurs, we can still obtain a reasonable solution using the following procedure. Consider the initial value problem

$$y'' + 2y' + 5y = g(t) ; y(0) = 0 , y'(0) = 0,$$

where

$$g(t) = \begin{cases} 10 & \text{if } 0 \le t \le 3\pi/2 \\ 0 & \text{if } t > 3\pi/2 \end{cases} .$$

(a) Find a solution to the initial value problem for $0 \le t \le 3\pi/2$.

(b) Find a general solution for $t > 3\pi/2$.

(c) Now choose the constants in the general solution from part (b) so that the solution from part (a) and the solution from part (b) agree, together with their first derivatives, at $t = 3\pi/2$. This gives us a continuously differentiable function that satisfies the differential equation except at $t = 3\pi/2$.

42. Forced Vibrations. As discussed in Section 4.1, a vibrating spring with damping that is under external force can be modeled by

(15) $my'' + by' + ky = g(t)$,

where $m > 0$ is the mass of the spring system, $b > 0$ is the damping constant, $k > 0$ is the spring constant, $g(t)$ is the force on the system at time t, and $y(t)$ is the displacement from the equilibrium of the spring system at time t. Assume $b^2 < 4mk$.

(a) Determine the form of the equation of motion for the spring system when $g(t) = \sin\beta t$ by finding a general solution to equation (15).

(b) Discuss the long-term behavior of this system. [*Hint:* Consider what happens to the general solution obtained in part (a) as $t \to +\infty$.]

43. A mass–spring system is driven by a sinusoidal external force $g(t) = 5\sin t$. The mass equals 1, the spring constant equals 3, and the damping coefficient equals 4. If the mass is initially located at $y(0) = 1/2$ and at rest, i.e., $y'(0) = 0$, find its equation of motion.

44. A mass–spring system is driven by the external force $g(t) = 2\sin 3t + 10\cos 3t$. The mass equals 1, the spring constant equals 5, and the damping coefficient equals 2. If the mass is initially located at $y(0) = -1$, with initial velocity $y'(0) = 5$, find its equation of motion.

Figure 4.11 Speed bump

45. Speed Bumps. Often bumps like the one depicted in Figure 4.11 are built into roads to discourage speeding. The figure suggests that a crude model of the vertical motion $y(t)$ of a car encountering the speed bump with the speed V is given by

$$y(t) = 0 \qquad \text{for } t \le -L/(2V)\,,$$

$$my'' + ky = \begin{cases} F_0 \cos\left(\pi Vt/L\right) \text{ for } |t| < L/(2V) \\ 0 \qquad\qquad \text{for } t \ge L/(2V). \end{cases}$$

(The absence of a damping term indicates that the car's shock absorbers are not functioning.)

(a) Taking $m = k = 1, L = \pi$, and $F_0 = 1$ in appropriate units, solve this initial value problem. Thereby show that the formula for the oscillatory motion *after the car has traversed the speed bump* is $y(t) = A\sin t$, where the constant A depends on the speed V.

 (b) Plot the amplitude $|A|$ of the solution $y(t)$ found in part (a) versus the car's speed V. From the graph, estimate the speed that produces the most violent shaking of the vehicle.

46. Show that the *boundary value problem*

$$y'' + \lambda^2 y = \sin t\,; \quad y(0) = 0, \quad y(\pi) = 1\,,$$

has a solution if and only if $\lambda \ne \pm 1, \pm 2, \pm 3, \dots$.

47. Find the solution(s) to

$$y'' + 9y = 27\cos 6t$$

(if it exists) satisfying the *boundary conditions*

(a) $y(0) = -1, \quad y(\pi/6) = 3$.

(b) $y(0) = -1, \quad y(\pi/3) = 5$.

(c) $y(0) = -1, \quad y(\pi/3) = -1$.

48. All that is known concerning a mysterious second-order constant-coefficient differential equation $y'' + py' + qy = g(t)$ is that $t^2 + 1 + e^t\cos t$, $t^2 + 1 + e^t\sin t$, and $t^2 + 1 + e^t\cos t + e^t\sin t$ are solutions.

(a) Determine two linearly independent solutions to the corresponding homogeneous equation.

(b) Find a suitable choice of p, q, and $g(t)$ that enables these solutions.

4.6 Variation of Parameters

We have seen that the method of undetermined coefficients is a simple procedure for determining a particular solution when the equation has constant coefficients and the nonhomogeneous term is of a special type. Here we present a more general method, called **variation of parameters,**[†] for finding a particular solution.

Consider the nonhomogeneous linear second-order equation

(1) $$ay'' + by' + cy = f(t)$$

and let $\{y_1(t), y_2(t)\}$ be two linearly independent solutions for the corresponding homogeneous equation

$$ay'' + by' + cy = 0\,.$$

[†]*Historical Footnote:* The method of variation of parameters was invented by Joseph Lagrange in 1774.

Then we know that a general solution to this homogeneous equation is given by

(2) $y_h(t) = c_1 y_1(t) + c_2 y_2(t) \, ,$

where c_1 and c_2 are constants. To find a particular solution to the nonhomogeneous equation, the strategy of variation of parameters is to replace the constants in (2) by functions of t. That is, we seek a solution of (1) of the form[†]

(3) $y_p(t) = v_1(t) y_1(t) + v_2(t) y_2(t) \, .$

Because we have introduced two unknown functions, $v_1(t)$ and $v_2(t)$, it is reasonable to expect that we can impose two equations (requirements) on these functions. Naturally, one of these equations should come from (1). Let's therefore plug $y_p(t)$ given by (3) into (1). To accomplish this, we must first compute $y_p'(t)$ and $y_p''(t)$. From (3) we obtain

$$y_p' = (v_1' y_1 + v_2' y_2) + (v_1 y_1' + v_2 y_2') \, .$$

To simplify the computation and to avoid second-order derivatives for the unknowns v_1, v_2 in the expression for y_p'', we impose the requirement

(4) $v_1' y_1 + v_2' y_2 = 0 \, .$

Thus, the formula for y_p' becomes

(5) $y_p' = v_1 y_1' + v_2 y_2' \, ,$

and so

(6) $y_p'' = v_1' y_1' + v_1 y_1'' + v_2' y_2' + v_2 y_2'' \, .$

Now, substituting y_p, y_p', and y_p'', as given in (3), (5), and (6), into (1), we find

(7) $f = a y_p'' + b y_p' + c y_p$

$\quad = a(v_1' y_1' + v_1 y_1'' + v_2' y_2' + v_2 y_2'') + b(v_1 y_1' + v_2 y_2') + c(v_1 y_1 + v_2 y_2)$

$\quad = a(v_1' y_1' + v_2' y_2') + v_1(a y_1'' + b y_1' + c y_1) + v_2(a y_2'' + b y_2' + c y_2)$

$\quad = a(v_1' y_1' + v_2' y_2') + 0 + 0$

since y_1 and y_2 are solutions to the homogeneous equation. Thus, (7) reduces to

(8) $v_1' y_1' + v_2' y_2' = \dfrac{f}{a} \, .$

To summarize, if we can find v_1 and v_2 that satisfy both (4) and (8), that is,

(9) $\begin{aligned} y_1 v_1' + y_2 v_2' &= 0 \, , \\ y_1' v_1' + y_2' v_2' &= \dfrac{f}{a} \, , \end{aligned}$

then y_p given by (3) will be a particular solution to (1). To determine v_1 and v_2, we first solve the linear system (9) for v_1' and v_2'. Algebraic manipulation or Cramer's rule (see Appendix D) immediately gives

$$v_1'(t) = \frac{-f(t) y_2(t)}{a[y_1(t) y_2'(t) - y_1'(t) y_2(t)]} \quad \text{and} \quad v_2'(t) = \frac{f(t) y_1(t)}{a[y_1(t) y_2'(t) - y_1'(t) y_2(t)]} \, ,$$

[†]In Exercises 2.3, Problem 36, we developed this approach for first-order linear equations. Because of the similarity of equations (2) and (3), this technique is sometimes known as "variation of constants."

where the bracketed expression in the denominator (the Wronskian) is never zero because of Lemma 1, Section 4.2. Upon integrating these equations, we finally obtain

(10) $v_1(t) = \int \dfrac{-f(t)y_2(t)}{a[y_1(t)y_2'(t) - y_1'(t)y_2(t)]}\, dt$ and $v_2(t) = \int \dfrac{f(t)y_1(t)}{a[y_1(t)y_2'(t) - y_1'(t)y_2(t)]}\, dt$.

Let's review this procedure.

Method of Variation of Parameters

To determine a particular solution to $ay'' + by' + cy = f$:

(a) Find two linearly independent solutions $\{y_1(t), y_2(t)\}$ to the corresponding homogeneous equation and take

$$y_p(t) = v_1(t)y_1(t) + v_2(t)y_2(t)\, .$$

(b) Determine $v_1(t)$ and $v_2(t)$ by solving the system in (9) for $v_1'(t)$ and $v_2'(t)$ and integrating.

(c) Substitute $v_1(t)$ and $v_2(t)$ into the expression for $y_p(t)$ to obtain a particular solution.

Of course, in step (b) one could use the formulas in (10), but $v_1(t)$ and $v_2(t)$ are so easy to derive that you are advised not to memorize them.

Example 1 Find a general solution on $(-\pi/2, \pi/2)$ to

(11) $\dfrac{d^2y}{dt^2} + y = \tan t$.

Solution Observe that two independent solutions to the homogeneous equation $y'' + y = 0$ are $\cos t$ and $\sin t$. We now set

(12) $y_p(t) = v_1(t)\cos t + v_2(t)\sin t$

and, referring to (9), solve the system

$$(\cos t)v_1'(t) + (\sin t)v_2'(t) = 0\, ,$$
$$(-\sin t)v_1'(t) + (\cos t)v_2'(t) = \tan t\, ,$$

for $v_1'(t)$ and $v_2'(t)$. This gives

$$v_1'(t) = -\tan t\,\sin t\, ,$$
$$v_2'(t) = \tan t\,\cos t = \sin t\, .$$

Integrating, we obtain

(13) $v_1(t) = -\displaystyle\int \tan t\,\sin t\, dt = -\int \dfrac{\sin^2 t}{\cos t}\, dt$

$= -\displaystyle\int \dfrac{1 - \cos^2 t}{\cos t}\, dt = \int (\cos t - \sec t)\, dt$

$= \sin t - \ln|\sec t + \tan t| + C_1\, ,$

(14) $v_2(t) = \displaystyle\int \sin t\, dt = -\cos t + C_2$.

We need only one particular solution, so we take both C_1 and C_2 to be zero for simplicity. Then, substituting $v_1(t)$ and $v_2(t)$ in (12), we obtain

$$y_p(t) = (\sin t - \ln|\sec t + \tan t|) \cos t - \cos t \sin t,$$

which simplifies to

$$y_p(t) = -(\cos t) \ln|\sec t + \tan t|.$$

We may drop the absolute value symbols because $\sec t + \tan t = (1 + \sin t)/\cos t > 0$ for $-\pi/2 < t < \pi/2$.

Recall that a general solution to a nonhomogeneous equation is given by the sum of a general solution to the homogeneous equation and a particular solution. Consequently, a general solution to equation (11) on the interval $(-\pi/2, \pi/2)$ is

(15) $y(t) = c_1 \cos t + c_2 \sin t - (\cos t) \ln(\sec t + \tan t).$ ◆

Note that in the above example the constants C_1 and C_2 appearing in (13) and (14) were chosen to be zero. If we had retained these arbitrary constants, the ultimate effect would be just to add $C_1 \cos t + C_2 \sin t$ to (15), which is clearly redundant.

Example 2 Find a particular solution on $(-\pi/2, \pi/2)$ to

(16) $$\frac{d^2y}{dt^2} + y = \tan t + 3t - 1.$$

Solution With $f(t) = \tan t + 3t - 1$, the variation of parameters procedure will lead to a solution. But it is simpler in this case to consider separately the equations

(17) $$\frac{d^2y}{dt^2} + y = \tan t,$$

(18) $$\frac{d^2y}{dt^2} + y = 3t - 1$$

and then use the superposition principle (Theorem 3, page 181).

In Example 1 we found that

$$y_q(t) = -(\cos t) \ln(\sec t + \tan t)$$

is a particular solution for equation (17). For equation (18) the method of undetermined coefficients can be applied. On seeking a solution to (18) of the form $y_r(t) = At + B$, we quickly obtain

$$y_r(t) = 3t - 1.$$

Finally, we apply the superposition principle to get

$$\begin{aligned} y_p(t) &= y_q(t) + y_r(t) \\ &= -(\cos t) \ln(\sec t + \tan t) + 3t - 1 \end{aligned}$$

as a particular solution for equation (16). ◆

Note that we could not have solved Example 1 by the method of undetermined coefficients; the nonhomogeneity $\tan t$ is unsuitable. Another important advantage of the method of variation of parameters is its applicability to linear equations whose coefficients a, b, c are functions of t. Indeed, on reviewing the derivation of the system (9) and the formulas (10), one

can check that we did not make any use of the constant coefficient property; i.e., the method works provided we know a pair of linearly independent solutions to the corresponding homogeneous equation. We illustrate the method in the next example.

Example 3 Find a particular solution of the variable coefficient linear equation

(19) $t^2 y'' - 4ty' + 6y = 4t^3$, $t > 0$,

given that $y_1(t) = t^2$ and $y_2(t) = t^3$ are solutions to the corresponding homogeneous equation.

Solution The functions t^2 and t^3 are linearly independent solutions to the corresponding homogeneous equation on $(0, \infty)$ (verify this!) and so (19) has a particular solution of the form

$$y_p(t) = v_1(t)t^2 + v_2(t)t^3 .$$

To determine the unknown functions v_1 and v_2, we solve the system (9) with $f(t) = 4t^3$ and $a = a(t) = t^2$:

$$t^2 v_1'(t) + t^3 v_2'(t) = 0$$

$$2t v_1'(t) + 3t^2 v_2'(t) = f/a = 4t .$$

The solutions are readily found to be $v_1'(t) = -4$ and $v_2'(t) = 4/t$, which gives $v_1(t) = -4t$ and $v_2(t) = 4 \ln t$. Consequently,

$$y_p(t) = (-4t)t^2 + (4 \ln t)t^3 = 4t^3(-1 + \ln t)$$

is a solution to (19). ◆

Variable coefficient linear equations will be discussed in more detail in the next section.

4.6 EXERCISES

In Problems 1–8, find a general solution to the differential equation using the method of variation of parameters.

1. $y'' + 4y = \tan 2t$

2. $y'' + y = \sec t$

3. $y'' - 2y' + y = t^{-1}e^t$

4. $y'' + 2y' + y = e^{-t}$

5. $y''(\theta) + 16y(\theta) = \sec 4\theta$

6. $y'' + 9y = \sec^2(3t)$

7. $y'' + 4y' + 4y = e^{-2t} \ln t$

8. $y'' + 4y = \csc^2(2t)$

In Problems 9 and 10, find a particular solution first by undetermined coefficients, and then by variation of parameters. Which method was quicker?

9. $y'' - y = 2t + 4$

10. $2x''(t) - 2x'(t) - 4x(t) = 2e^{2t}$

In Problems 11–18, find a general solution to the differential equation.

11. $y'' + y = \tan t + e^{3t} - 1$

12. $y'' + y = \tan^2 t$

13. $v'' + 4v = \sec^4(2t)$

14. $y''(\theta) + y(\theta) = \sec^3 \theta$

15. $y'' + y = 3 \sec t - t^2 + 1$

16. $y'' + 5y' + 6y = 18t^2$

17. $\frac{1}{2}y'' + 2y = \tan 2t - \frac{1}{2}e^t$

18. $y'' - 6y' + 9y = t^{-3}e^{3t}$

 19. Express the solution to the initial value problem

$$y'' - y = \frac{1}{t}, \quad y(1) = 0, \quad y'(1) = -2,$$

using definite integrals. Using numerical integration (Appendix C) to approximate the integrals, find an approximation for $y(2)$ to two decimal places.

20. Use the method of variation of parameters to show that

$$y(t) = c_1 \cos t + c_2 \sin t + \int_0^t f(s) \sin(t-s) \, ds$$

is a general solution to the differential equation

$$y'' + y = f(t) ,$$

where $f(t)$ is a continuous function on $(-\infty, \infty)$. [*Hint:* Use the trigonometric identity $\sin(t-s) = \sin t \cos s - \sin s \cos t$.]

21. Suppose y satisfies the equation $y'' + 10y' + 25y = e^{t^3}$ subject to $y(0) = 1$ and $y'(0) = -5$. Estimate $y(0.2)$ to within ± 0.0001 by numerically approximating the integrals in the variation of parameters formula.

In Problems 22 through 25, use variation of parameters to find a general solution to the differential equation given that the functions y_1 and y_2 are linearly independent solutions to the corresponding homogeneous equation for $t > 0$.

22. $t^2 y'' - 4ty' + 6y = t^3 + 1$;
$\quad y_1 = t^2$, $\quad y_2 = t^3$

23. $ty'' - (t+1)y' + y = t^2$;
$\quad y_1 = e^t$, $\quad y_2 = t + 1$

24. $ty'' + (1 - 2t)y' + (t - 1)y = te^t$;
$\quad y_1 = e^t$, $\quad y_2 = e^t \ln t$

25. $ty'' + (5t - 1)y' - 5y = t^2 e^{-5t}$;
$\quad y_1 = 5t - 1$, $\quad y_2 = e^{-5t}$

4.7 Variable-Coefficient Equations

The techniques of Sections 4.2 and 4.3 have explicitly demonstrated that solutions to a linear homogeneous constant-coefficient differential equation,

$$(1) \qquad ay'' + by' + cy = 0,$$

are defined and satisfy the equation over the whole interval $(-\infty, +\infty)$. After all, such solutions are combinations of exponentials, sinusoids, and polynomials.

The variation of parameters formula of Section 4.6 extended this to nonhomogeneous constant-coefficient problems,

$$(2) \qquad ay'' + by' + cy = f(t),$$

yielding solutions valid *over all intervals where $f(t)$ is continuous* (ensuring that the integrals in (10) of Section 4.6 containing $f(t)$ exist and are differentiable). We could hardly hope for more; indeed, it is debatable what *meaning* the differential equation (2) would have at a point where $f(t)$ is undefined, or discontinuous.

Therefore, when we move to the realm of equations with *variable* coefficients of the form

$$(3) \qquad a_2(t)y'' + a_1(t)y' + a_0(t)y = f(t),$$

the most we can expect is that there are solutions that are valid over intervals where all four "governing" functions—$a_2(t)$, $a_1(t)$, $a_0(t)$, and $f(t)$—are continuous. Fortunately, this expectation is fulfilled except for an important technical requirement—namely, that the coefficient function $a_2(t)$ must be nonzero over the interval.[†]

Typically, one divides by the nonzero coefficient $a_2(t)$ and expresses the theorem for the equation in **standard form** [see (4), below] as follows.

Existence and Uniqueness of Solutions

Theorem 5. If $p(t)$, $q(t)$, and $g(t)$ are continuous on an interval (a, b) that contains the point t_0, then for any choice of the initial values Y_0 and Y_1, there exists a unique solution $y(t)$ on the same interval (a, b) to the initial value problem

$$(4) \qquad y''(t) + p(t)y'(t) + q(t)y(t) = g(t); \quad y(t_0) = Y_0, \quad y'(t_0) = Y_1.$$

[†]Indeed, the whole nature of the equation—reduction from *second*-order to *first*-order—changes at points where $a_2(t)$ is zero.

Example 1 Determine the largest interval for which Theorem 5 ensures the existence and uniqueness of a solution to the initial value problem

$$(5) \qquad (t-3)\frac{d^2y}{dt^2} + \frac{dy}{dt} + \sqrt{t}\,y = \ln t; \quad y(1) = 3, \quad y'(1) = -5.$$

Solution The data $p(t)$, $q(t)$, and $g(t)$ in the standard form of the equation,

$$y'' + py' + qy = \frac{d^2y}{dt^2} + \frac{1}{(t-3)}\frac{dy}{dt} + \frac{\sqrt{t}}{(t-3)}y = \frac{\ln t}{(t-3)} = g,$$

are simultaneously continuous in the intervals $0 < t < 3$ and $3 < t < \infty$. The former contains the point $t_0 = 1$, where the initial conditions are specified, so Theorem 5 guarantees (5) has a unique solution in $0 < t < 3$. ◆

Theorem 5, embracing existence and uniqueness for the variable-coefficient case, is difficult to prove because we can't construct explicit solutions in the general case. So the proof is deferred to Chapter 13.[†] However, it is instructive to examine a special case that we can solve explicitly.

Cauchy–Euler, or Equidimensional, Equations

Definition 2. A linear second-order equation that can be expressed in the form

$$(6) \qquad at^2y''(t) + bty'(t) + cy = f(t),$$

where a, b, and c are constants, is called a **Cauchy–Euler, or equidimensional, equation.**

For example, the differential equation

$$3t^2y'' + 11ty' - 3y = \sin t$$

is a Cauchy–Euler equation, whereas

$$2y'' - 3ty' + 11y = 3t - 1$$

is *not* because the coefficient of y'' is 2, which is not a constant times t^2.

The nomenclature *equidimensional* comes about because if y has the dimensions of, say, meters and t has dimensions of time, then each term t^2y'', ty', and y has the same dimensions (meters). The coefficient of $y''(t)$ in (6) is at^2, and it is zero at $t = 0$; equivalently, the standard form

$$y'' + \frac{b}{at}y' + \frac{c}{at^2}y = \frac{f(t)}{at^2}$$

has discontinuous coefficients at $t = 0$. Therefore, we can expect the solutions to be valid only for $t > 0$ or $t < 0$. Discontinuities in f, of course, will impose further restrictions.

[†]All references to Chapters 11–13 refer to the expanded text, *Fundamentals of Differential Equations and Boundary Value Problems*, 7th ed.

To solve a *homogeneous* Cauchy–Euler equation, for $t > 0$, we exploit the equidimensional feature by looking for solutions of the form $y = t^r$, because then t^2y'', ty', and y each have the form *(constant)* $\times t^r$:

$$y = t^r, \quad ty' = trt^{r-1} = rt^r, \quad t^2y'' = t^2r(r-1)t^{r-2} = r(r-1)t^r,$$

and substitution into the homogeneous form of (6) (that is, with $g = 0$) yields a simple quadratic equation for r:

$$ar(r-1)t^r + brt^r + ct^r = [ar^2 + (b-a)r + c]t^r = 0, \quad \text{or}$$

(7) $$ar^2 + (b-a)r + c = 0,$$

which we call the associated *characteristic equation*.

Example 2 Find two linearly independent solutions to the equation

$$3t^2y'' + 11ty' - 3y = 0, \quad t > 0.$$

Solution Inserting $y = t^r$ yields, according to (7),

$$3r^2 + (11 - 3)r - 3 = 3r^2 + 8r - 3 = 0,$$

whose roots $r = 1/3$ and $r = -3$ produce the independent solutions

$$y_1(t) = t^{1/3}, \quad y_2(t) = t^{-3} \quad (\text{for } t > 0). \quad \blacklozenge$$

Clearly, the substitution $y = t^r$ into a homogeneous *equidimensional* equation has the same simplifying effect as the insertion of $y = e^{rt}$ into the homogeneous *constant-coefficient* equation in Section 4.2. That means we will have to deal with the same encumbrances:

1. What to do when the roots of (7) are complex
2. What to do when the roots of (7) are equal

If r is complex, $r = \alpha + i\beta$, we can interpret $t^{\alpha+i\beta}$ by using the identity $t = e^{\ln t}$ and invoking Euler's formula [equation (5), Section 4.3]:

$$t^{\alpha+i\beta} = t^\alpha t^{i\beta} = t^\alpha e^{i\beta \ln t} = t^\alpha [\cos(\beta \ln t) + i \sin(\beta \ln t)].$$

Then we simplify as in Section 4.3 by taking the real and imaginary parts to form independent solutions:

(8) $$y_1 = t^\alpha \cos(\beta \ln t), \quad y_2 = t^\alpha \sin(\beta \ln t).$$

If r is a double root of the characteristic equation (7), then independent solutions of the Cauchy–Euler equation on $(0, \infty)$ are given by

(9) $$y_1 = t^r, \quad y_2 = t^r \ln t.$$

This can be verified by direct substitution into the differential equation. Alternatively, the second, linearly independent, solution can be obtained by *reduction of order*, a procedure to be discussed shortly in Theorem 8. Furthermore, Problem 23 demonstrates that the substitution $t = e^x$ changes the homogeneous Cauchy–Euler equation into a homogeneous constant-coefficient equation, and the formats (8) and (9) then follow from our earlier deliberations.

We remark that if a homogeneous Cauchy–Euler equation is to be solved for $t < 0$, then one simply introduces the change of variable $t = -\tau$, where $\tau > 0$. The reader should verify via the chain rule that the identical characteristic equation (7) arises when $\tau^r = (-t)^r$ is substituted in the equation. Thus the solutions take the same form as (8), (9), but with t replaced

by $-t$; for example, if r is a double root of (7), we get $(-t)^r$ and $(-t)^r \ln(-t)$ as two linearly independent solutions on $(-\infty, 0)$.

Example 3 Find a pair of linearly independent solutions to the following Cauchy–Euler equations for $t > 0$.

$\quad$ **(a)** $t^2 y'' + 5ty' + 5y = 0$ $\qquad$ **(b)** $t^2 y'' + ty' = 0$

Solution For part (a), the characteristic equation becomes $r^2 + 4r + 5 = 0$, with the roots $r = -2 \pm i$, and (8) produces the real solutions $t^{-2} \cos(\ln t)$ and $t^{-2} \sin(\ln t)$.

$\quad$ For part (b), the characteristic equation becomes simply $r^2 = 0$ with the double root $r = 0$, and (9) yields the solutions $t^0 = 1$ and $\ln t$. $\blacklozenge$

$\quad$ In Chapter 8 we will see how one can obtain power series expansions for solutions to variable-coefficient equations when the coefficients are *analytic* functions. But, as we said, there is no procedure for explicitly solving the general case. Nonetheless, thanks to the existence/uniqueness result of Theorem 5, most of the other theorems and concepts of the preceding sections are easily extended to the variable-coefficient case, with the proviso that they apply only over intervals in which the governing functions $p(t), q(t), g(t)$ are continuous. Thus we have the following analog of Lemma 1, page 160.

A Condition for Linear Dependence of Solutions

Lemma 3. $\quad$ If $y_1(t)$ and $y_2(t)$ are any two solutions to the homogeneous differential equation

$(10) \qquad y''(t) + p(t)y'(t) + q(t)y(t) = 0$

on an interval I where the functions $p(t)$ and $q(t)$ are continuous and if the Wronskian[†]

$$W[y_1, y_2](t) := y_1(t)y_2'(t) - y_1'(t)y_2(t) = \begin{vmatrix} y_1(t) & y_2(t) \\ y_1'(t) & y_2'(t) \end{vmatrix}$$

is zero at *any* point t of I, then y_1 and y_2 are linearly dependent **on I**.

$\quad$ As in the constant-coefficient case, the Wronskian of two solutions is either identically zero or never zero on I, with the latter implying linear independence on I.

$\quad$ Precisely as in the proof for the constant-coefficient case, it can be verified that any linear combination $c_1 y_1 + c_2 y_2$ of solutions y_1 and y_2 to (10) is also a solution. In fact, these are the only solutions to (10) as stated in the following result.

Representation of Solutions to Initial Value Problems

Theorem 6. $\quad$ If $y_1(t)$ and $y_2(t)$ are any two solutions to the homogeneous differential equation (10) that are linearly independent on an interval I, then every solution to (10) on I is expressible as a linear combination of y_1 and y_2. Moreover, the initial value problem consisting of equation (10) and the initial conditions $y(t_0) = Y_0, y'(t_0) = Y_1$ has a unique solution on I for any point t_0 in I and any constants Y_0, Y_1.

[†]The determinant representation of the Wronskian was introduced in Problem 34, Section 4.2.

As in the constant-coefficient case, the linear combination $y_h = c_1 y_1 + c_2 y_2$ is called a **general solution** to (10) on I if y_1, y_2 are linearly independent solutions on I.

For the nonhomogeneous equation

$$(11) \quad y''(t) + p(t)y'(t) + q(t)y(t) = g(t) \,,$$

a general solution on I is given by $y = y_p + y_h$, **where** $y_h = c_1 y_1 + c_2 y_2$ **is a general solution to the corresponding homogeneous equation (10) on I and** y_p **is a particular solution to (11) on I.** In other words, the solution to the initial value problem stated in Theorem 5 must be of this form for a suitable choice of the constants c_1, c_2. This follows, just as before, from a straightforward extension of the superposition principle for variable-coefficient equations described in Problem 30.

As illustrated at the end of the Section 4.6, if linearly independent solutions to the homogeneous equation (10) are known, then y_p can be determined for (11) by the variation of parameters method.

Variation of Parameters

Theorem 7. If y_1 and y_2 are two linearly independent solutions to the homogeneous equation (10) on an interval I where $p(t)$, $q(t)$, and $g(t)$ are continuous, then a particular solution to (11) is given by $y_p = v_1 y_1 + v_2 y_2$, where v_1 and v_2 are determined up to a constant by the pair of equations

$$y_1 v_1' + y_2 v_2' = 0 \,,$$
$$y_1' v_1' + y_2' v_2' = g \,,$$

which have the solution

$$(12) \quad v_1(t) = \int \frac{-g(t)\, y_2(t)}{W[y_1, y_2](t)}\, dt \,, \quad v_2(t) = \int \frac{g(t) y_1(t)}{W[y_1, y_2](t)}\, dt \,.$$

Note the formulation (12) presumes that the differential equation has been put into standard form [that is, divided by $a_2(t)$].

The proofs of the constant-coefficient versions of these theorems in Sections 4.2 and 4.5 did not make use of the constant-coefficient property, so one can prove them in the general case by literally copying those proofs but interpreting the coefficients as variables. Unfortunately, however, there is no construction analogous to the method of undetermined coefficients for the variable-coefficient case.

What does all this mean? The only stumbling block for our completely solving nonhomogeneous initial value problems for equations with variable coefficients,

$$y'' + p(t)y' + q(t)y = g(t) \,; \quad y(t_0) = Y_0 \,, \quad y'(t_0) = Y_1 \,,$$

is the lack of an explicit procedure for constructing independent solutions to the associated homogeneous equation (10). If we *had* y_1 and y_2 as described in the variation of parameters formula, we could implement (12) to find y_p, formulate the general solution of (11) as $y_p + c_1 y_1 + c_2 y_2$, and (with the assurance that the Wronskian is nonzero) fit the constants to the initial conditions. But with the exception of the Cauchy–Euler equation and the ponderous power series machinery of Chapter 8, we are stymied at the outset; there is no general procedure for finding y_1 and y_2.

Ironically, we only need *one* nontrivial solution to the associated homogeneous equation, thanks to a procedure known as *reduction of order* that constructs a second, linearly independent solution y_2 from a known one y_1. So one might well feel that the following theorem rubs salt into the wound.

Reduction of Order

Theorem 8. If $y_1(t)$ is a solution, not identically zero, to the homogeneous differential equation (10) in an interval I (see page 195), then

$$(13) \qquad y_2(t) = y_1(t) \int \frac{e^{-\int p(t)dt}}{y_1(t)^2} \, dt$$

is a second, linearly independent solution.

This remarkable formula can be confirmed directly, but the following derivation shows how the procedure got its name.

Proof of Theorem 8. Our strategy is similar to that used in the derivation of the variation of parameters formula, Section 4.6. Bearing in mind that cy_1 is a solution of (10) for any constant c, we replace c by a *function* $v(t)$ and propose the trial solution $y_2(t) = v(t)y_1(t)$, spawning the formulas

$$y_2' = vy_1' + v'y_1 \,, \quad y_2'' = vy_1'' + 2v'y_1' + v''y_1 \,.$$

Substituting these expressions into the differential equation (10) yields

$$(vy_1'' + 2v'y_1' + v''y_1) + p(vy_1' + v'y_1) + qvy_1 = 0 \,,$$

or, on regrouping,

$$(14) \qquad (y_1'' + py_1' + qy_1)v + y_1v'' + (2y_1' + py_1)v' = 0 \,.$$

The group in front of the undifferentiated $v(t)$ is simply a copy of the left-hand member of the original differential equation (10), so it is zero.[†] Thus (14) reduces to

$$(15) \qquad y_1v'' + (2y_1' + py_1)v' = 0 \,,$$

which is actually a *first*-order equation in the variable $w \equiv v'$:

$$(16) \qquad y_1w' + (2y_1' + py_1)w = 0 \,.$$

Indeed, (16) is separable and can be solved immediately using the procedure of Section 2.2. Problem 48 on page 201 requests the reader to carry out the details of this procedure to complete the derivation of (13). ◆

Example 4 Given that $y_1(t) = t$ is a solution to

$$(17) \qquad y'' - \frac{1}{t}y' + \frac{1}{t^2}y = 0 \,,$$

use the reduction of order procedure to determine a second linearly independent solution for $t > 0$.

[†]This is hardly a surprise; if v were constant, vy would be a solution with $v' = v'' = 0$ in (14).

Solution Rather than implementing the formula (13), let's apply the strategy used to derive it. We set $y_2(t) = v(t)y_1(t) = v(t)t$ and substitute $y_2' = v't + v$, $y_2'' = v''t + 2v'$ into (17) to find

(18) $$v''t + 2v' - \frac{1}{t}(v't + v) + \frac{1}{t^2}vt = v''t + (2v' - v') = v''t + v' = 0 .$$

As promised, (18) is a *separable first*-order equation in v', simplifying to $(v')'/(v') = -1/t$ with a solution $v' = 1/t$, or $v = \ln t$ (taking integration constants to be zero). Therefore, a second solution to (17) is $y_2 = vt = t \ln t$.

Of course (17) is a Cauchy–Euler equation for which (7) has equal roots:

$$ar^2 + (b - a)r + c = r^2 - 2r + 1 = (r - 1)^2 = 0 ,$$

and y_2 is precisely the form for the independent solution predicted by (9). ◆

Example 5 The following equation arises in the mathematical modeling of reverse osmosis.[†]

(19) $$(\sin t)y'' - 2(\cos t)y' - (\sin t)y = 0 , \quad 0 < t < \pi .$$

Find a general solution.

Solution As we indicated above, the tricky part is to find a single nontrivial solution. Inspection of (19) suggests that $y = \sin t$ or $y = \cos t$, combined with a little luck with trigonometric identities, might be solutions. In fact, trial and error shows that the cosine function works:

$$y_1 = \cos t , \quad y_1' = -\sin t , \quad y_1'' = -\cos t ,$$
$$(\sin t)y_1'' - 2(\cos t)y_1' - (\sin t)y_1 = (\sin t)(-\cos t) - 2(\cos t)(-\sin t) - (\sin t)(\cos t) = 0 .$$

Unfortunately, the sine function fails (try it).

So we use reduction of order to construct a second, independent solution. Setting $y_2(t) = v(t)y_1(t) = v(t)\cos t$ and computing $y_2' = v'\cos t - v\sin t$, $y_2'' = v''\cos t - 2v'\sin t - v\cos t$, we substitute into (19) to derive

$$(\sin t)[v''\cos t - 2v'\sin t - v\cos t] - 2(\cos t)[v'\cos t - v\sin t] - (\sin t)[v\cos t]$$
$$= v''(\sin t)(\cos t) - 2v'(\sin^2 t + \cos^2 t) = 0,$$

which is equivalent to the separated first-order equation

$$\frac{(v')'}{(v')} = \frac{2}{(\sin t)(\cos t)} = 2\frac{\sec^2 t}{\tan t} .$$

Taking integration constants to be zero yields $\ln v' = 2\ln(\tan t)$ or $v' = \tan^2 t$, and $v = \tan t - t$. Therefore, a second solution to (19) is $y_2 = (\tan t - t)\cos t = \sin t - t\cos t$. We conclude that a general solution is $c_1 \cos t + c_2 (\sin t - t\cos t)$. ◆

In this section we have seen that the *theory* for variable-coefficient equations differs only slightly from the constant-coefficient case (in that solution domains are restricted to intervals), but explicit solutions can be hard to come by. In the next section, we will supplement our exposition by describing some nonrigorous procedures that sometimes can be used to predict qualitative features of the solutions.

[†]Reverse osmosis is a process used to fortify the alcoholic content of wine, among other applications.

4.7 EXERCISES

In Problems 1 through 4, use Theorem 5 to discuss the existence and uniqueness of a solution to the differential equation that satisfies the initial conditions $y(1) = Y_0$, $y'(1) = Y_1$, *where* Y_0 *and* Y_1 *are real constants.*

1. $(1 + t^2)y'' + ty' - y = \tan t$

2. $t(t-3)y'' + 2ty' - y = t^2$

3. $t^2 y'' + y = \cos t$

4. $e^t y'' - \dfrac{y'}{t-3} + y = \ln t$

In Problems 5 through 8, determine whether Theorem 5 applies. If it does, then discuss what conclusions can be drawn. If it does not, explain why.

5. $t^2 z'' + tz' + z = \cos t$; $z(0) = 1$, $z'(0) = 0$

6. $y'' + yy' = t^2 - 1$; $y(0) = 1$, $y'(0) = -1$

7. $y'' + ty' - t^2 y = 0$; $y(0) = 0$, $y(1) = 0$

8. $(1-t)y'' + ty' - 2y = \sin t$; $y(0) = 1$, $y'(0) = 1$

In Problems 9 through 14, find a general solution to the given Cauchy–Euler equation for $t > 0$.

9. $t^2 y''(t) + 7ty'(t) - 7y(t) = 0$

10. $t^2 \dfrac{d^2 y}{dt^2} + 2t \dfrac{dy}{dt} - 6y = 0$

11. $t^2 \dfrac{d^2 z}{dt^2} + 5t \dfrac{dz}{dt} + 4z = 0$

12. $\dfrac{d^2 w}{dt^2} + \dfrac{6}{t} \dfrac{dw}{dt} + \dfrac{4}{t^2} w = 0$

13. $9t^2 y''(t) + 15ty'(t) + y(t) = 0$

14. $t^2 y''(t) - 3ty'(t) + 4y(t) = 0$

In Problems 15 through 18, find a general solution for $t < 0$.

15. $y''(t) - \dfrac{1}{t} y'(t) + \dfrac{5}{t^2} y(t) = 0$

16. $t^2 y''(t) - 3ty'(t) + 6y(t) = 0$

17. $t^2 y''(t) + 9ty'(t) + 17y(t) = 0$

18. $t^2 y''(t) + 3ty'(t) + 5y(t) = 0$

In Problems 19 and 20, solve the given initial value problem for the Cauchy–Euler equation.

19. $t^2 y''(t) - 4ty'(t) + 4y(t) = 0$;
 $y(1) = -2$, $y'(1) = -11$

20. $t^2 y''(t) + 7ty'(t) + 5y(t) = 0$;
 $y(1) = -1$, $y'(1) = 13$

In Problems 21 and 22, devise a modification of the method for Cauchy–Euler equations to find a general solution to the given equation.

21. $(t-2)^2 y''(t) - 7(t-2)y'(t) + 7y(t) = 0$, $t > 2$

22. $(t+1)^2 y''(t) + 10(t+1)y'(t) + 14y(t) = 0$,
 $t > -1$

23. To justify the solution formulas (8) and (9), perform the following analysis.
 (a) Show that if the substitution $t = e^x$ is made in the function $y(t)$ and x is regarded as the new independent variable in $Y(x) := y(e^x)$, the chain rule implies the following relationships:
$$t \frac{dy}{dt} = \frac{dY}{dx}, \quad t^2 \frac{d^2 y}{dt^2} = \frac{d^2 Y}{dx^2} - \frac{dY}{dx}.$$
 (b) Using part (a), show that if the substitution $t = e^x$ is made in the Cauchy–Euler differential equation (6), the result is a constant-coefficient equation for $Y(x) = y(e^x)$, namely,
$$(20) \qquad a \frac{d^2 Y}{dx^2} + (b-a) \frac{dY}{dx} + cY = f(e^x).$$
 (c) Observe that the auxiliary equation (recall Section 4.2) for the homogeneous form of (20) is the same as (7) in this section. If the roots of the former are complex, linearly independent solutions of (20) have the form $e^{\alpha x} \cos \beta x$ and $e^{\alpha x} \sin \beta x$; if they are equal, linearly independent solutions of (20) have the form e^{rx} and xe^{rx}. Express x in terms of t to derive the corresponding solution forms (8) and (9).

24. Solve the following Cauchy–Euler equations by using the substitution described in Problem 23 to change them to constant coefficient equations, finding their general solutions by the methods of the preceding sections, and restoring the original independent variable t.
 (a) $t^2 y'' + ty' - 9y = 0$
 (b) $t^2 y'' + 3ty' + 10y = 0$
 (c) $t^2 y'' + 3ty' + y = t + t^{-1}$
 (d) $t^2 y'' + ty' + 9y = -\tan(3 \ln t)$

25. Let y_1 and y_2 be two functions defined on $(-\infty, \infty)$.
 (a) True or False: If y_1 and y_2 are linearly dependent on the interval $[a, b]$, then y_1 and y_2 are linearly dependent on the smaller interval $[c, d] \subset [a, b]$.
 (b) True or False: If y_1 and y_2 are linearly dependent on the interval $[a, b]$, then y_1 and y_2 are linearly dependent on the larger interval $[C, D] \supset [a, b]$.

26. Let $y_1(t) = t^3$ and $y_2(t) = |t^3|$. Are y_1 and y_2 linearly independent on the following intervals?
 (a) $[0, \infty)$ **(b)** $(-\infty, 0]$ **(c)** $(-\infty, \infty)$
 (d) Compute the Wronskian $W[y_1, y_2](t)$ on the interval $(-\infty, \infty)$.

27. Consider the linear equation
$$(21) \qquad t^2 y'' - 3ty' + 3y = 0,$$
 for $-\infty < t < \infty$.

(a) Verify that $y_1(t) := t$ and $y_2(t) := t^3$ are two solutions to (21) on $(-\infty, \infty)$. Furthermore, show that $y_1(t_0)y_2'(t_0) - y_1'(t_0)y_2(t_0) \neq 0$ for $t_0 = 1$.

(b) Prove that $y_1(t)$ and $y_2(t)$ are linearly independent on $(-\infty, \infty)$.

(c) Verify that the function $y_3(t) := |t|^3$ is also a solution to (21) on $(-\infty, \infty)$.

(d) Prove that there is *no* choice of constants c_1, c_2 such that $y_3(t) = c_1 y_1(t) + c_2 y_2(t)$ for *all* t in $(-\infty, \infty)$. [*Hint:* Argue that the contrary assumption leads to a contradiction.]

(e) From parts (c) and (d), we see that there is at least one solution to (21) on $(-\infty, \infty)$ that is not expressible as a linear combination of the solutions $y_1(t)$, $y_2(t)$. Does this provide a counterexample to the theory in this section? Explain.

28. Let $y_1(t) = t^2$ and $y_2(t) = 2t|t|$. Are y_1 and y_2 linearly independent on the interval:

(a) $[0, \infty)$? **(b)** $(-\infty, 0]$? **(c)** $(-\infty, \infty)$?

(d) Compute the Wronskian $W[y_1, y_2](t)$ on the interval $(-\infty, \infty)$.

29. Prove that if y_1 and y_2 are linearly independent solutions of $y'' + py' + qy = 0$ on (a, b), then they cannot both be zero at the same point t_0 in (a, b).

30. Superposition Principle. Let y_1 be a solution to

$$y''(t) + p(t)y'(t) + q(t)y(t) = g_1(t)$$

on the interval I and let y_2 be a solution to

$$y''(t) + p(t)y'(t) + q(t)y(t) = g_2(t)$$

on the same interval. Show that for any constants k_1 and k_2, the function $k_1 y_1 + k_2 y_2$ is a solution on I to

$$y''(t) + p(t)y'(t) + q(t)y(t) = k_1 g_1(t) + k_2 g_2(t) .$$

31. Determine whether the following functions can be Wronskians on $-1 < t < 1$ for a pair of solutions to some equation $y'' + py' + qy = 0$ (with p and q continuous).

(a) $w(t) = 6e^{4t}$ **(b)** $w(t) = t^3$

(c) $w(t) = (t+1)^{-1}$ **(d)** $w(t) \equiv 0$

32. By completing the following steps, prove that the Wronskian of any two solutions y_1, y_2 to the equation $y'' + py' + qy = 0$ on (a, b) is given by **Abel's formula**[†]

$$W[y_1, y_2](t) = C \exp\left\{-\int_{t_0}^{t} p(\tau)\, d\tau\right\} ,$$

t_0 and t in (a, b),

where the constant C depends on y_1 and y_2.

(a) Show that the Wronskian W satisfies the equation $W' + pW = 0$.

(b) Solve the separable equation in part (a).

(c) How does Abel's formula clarify the fact that the Wronskian is either identically zero or never zero on (a, b)?

33. Use Abel's formula (Problem 32) to determine (up to a constant multiple) the Wronskian of two solutions on $(0, \infty)$ to

$$ty'' + (t-1)y' + 3y = 0 .$$

34. All that is known concerning a mysterious differential equation $y'' + p(t)y' + q(t)y = g(t)$ is that the functions t, t^2, and t^3 are solutions.

(a) Determine two linearly independent solutions to the corresponding homogeneous differential equation.

(b) Find the solution to the original equation satisfying the initial conditions $y(2) = 2$, $y'(2) = 5$.

(c) What is $p(t)$? [*Hint:* Use Abel's formula for the Wronskian, Problem 32.]

35. Given that $1 + t$, $1 + 2t$, and $1 + 3t^2$ are solutions to the differential equation $y'' + p(t)y' + q(t)y = g(t)$, find the solution to this equation that satisfies $y(1) = 2$, $y'(1) = 0$.

36. Verify that the given functions y_1 and y_2 are linearly independent solutions of the following differential equation and find the solution that satisfies the given initial conditions.

$$ty'' - (t+2)y' + 2y = 0 ;$$
$$y_1(t) = e^t, \qquad y_2(t) = t^2 + 2t + 2 ;$$
$$y(1) = 0, \qquad y'(1) = 1$$

In Problems 37 through 39, find general solutions to the nonhomogeneous Cauchy–Euler equations using variation of parameters.

37. $t^2 z'' + tz' + 9z = -\tan(3 \ln t)$

38. $t^2 y'' + 3ty' + y = t^{-1}$

39. $t^2 z'' - tz' + z = t\left(1 + \dfrac{3}{\ln t}\right)$

40. The **Bessel equation** of order one-half

$$t^2 y'' + ty' + \left(t^2 - \dfrac{1}{4}\right)y = 0, \quad t > 0$$

has two linearly independent solutions,

$$y_1(t) = t^{-1/2}\cos t, \quad y_2(t) = t^{-1/2}\sin t .$$

Find a general solution to the nonhomogeneous equation

$$t^2 y'' + ty' + \left(t^2 - \dfrac{1}{4}\right)y = t^{5/2}, \quad t > 0 .$$

In Problems 41 through 44, a differential equation and a nontrivial solution f are given. Find a second linearly independent solution using reduction of order.

41. $t^2 y'' - 2ty' - 4y = 0$, $t > 0$; $f(t) = t^{-1}$

42. $t^2 y'' + 6ty' + 6y = 0$, $t > 0$; $f(t) = t^{-2}$

43. $tx'' - (t+1)x' + x = 0$, $t > 0$; $f(t) = e^t$

44. $ty'' + (1 - 2t)y' + (t - 1)y = 0$, $t > 0$; $f(t) = e^t$

[†]*Historical Footnote:* Niels Abel derived this identity in 1827.

45. Find a particular solution to the nonhomogeneous equation

$$ty'' - (t+1)y' + y = t^2 e^{2t},$$

given that $f(t) = e^t$ is a solution to the corresponding homogeneous equation.

46. Find a particular solution to the nonhomogeneous equation

$$(1-t)y'' + ty' - y = (1-t)^2,$$

given that $f(t) = t$ is a solution to the corresponding homogeneous equation.

47. In quantum mechanics, the study of the Schrödinger equation for the case of a harmonic oscillator leads to a consideration of **Hermite's equation,**

$$y'' - 2ty' + \lambda y = 0,$$

where λ is a parameter. Use the reduction of order formula to obtain an integral representation of a second linearly independent solution to Hermite's equation for the given value of λ and corresponding solution $f(t)$.
(a) $\lambda = 4$, $f(t) = 1 - 2t^2$
(b) $\lambda = 6$, $f(t) = 3t - 2t^3$

48. Complete the proof of Theorem 8 by solving equation (16).

49. The reduction of order procedure can be used more generally to reduce a homogeneous linear nth-order equation to a homogeneous linear $(n-1)$th-order equation. For the equation

$$ty''' - ty'' + y' - y = 0,$$

which has $f(t) = e^t$ as a solution, use the substitution $y(t) = v(t)f(t)$ to reduce this third-order equation to a homogeneous linear second-order equation in the variable $w = v'$.

50. The equation

$$ty''' + (1-t)y'' + ty' - y = 0$$

has $f(t) = t$ as a solution. Use the substitution $y(t) = v(t)f(t)$ to reduce this third-order equation to a homogeneous linear second-order equation in the variable $w = v'$.

51. Isolated Zeros. Let $\phi(t)$ be a solution to $y'' + py' + qy = 0$ on (a, b), where p, q are continuous on (a, b). By completing the following steps, prove that if ϕ is not identically zero, then its zeros in (a, b) are *isolated*, i.e., if $\phi(t_0) = 0$, then there exists a $\delta > 0$ such that $\phi(t) \neq 0$ for $0 < |t - t_0| < \delta$.
(a) Suppose $\phi(t_0) = 0$ and assume to the contrary that for each $n = 1, 2, \ldots$, the function ϕ has a zero at t_n, where $0 < |t_0 - t_n| < 1/n$. Show that this implies $\phi'(t_0) = 0$. [*Hint:* Consider the difference quotient for ϕ at t_0.]
(b) With the assumptions of part (a), we have $\phi(t_0) = \phi'(t_0) = 0$. Conclude from this that ϕ must be identically zero, which is a contradiction. Hence, there is some integer n_0 such that $\phi(t)$ is not zero for $0 < |t - t_0| < 1/n_0$.

52. The reduction of order formula (13) can also be derived from Abels' identity (Problem 32). Let $f(t)$ be a nontrivial solution to (10) and $y(t)$ a second linearly independent solution. Show that

$$\left(\frac{y}{f} \right)' = \frac{W[f, y]}{f^2}$$

and then use Abel's identity for the Wronskian $W[f, y]$ to obtain the reduction of order formula.

4.8 Qualitative Considerations for Variable-Coefficient and Nonlinear Equations

There are no techniques for obtaining explicit, closed-form solutions to second-order linear differential equations with variable coefficients (with certain exceptions) or for nonlinear equations. In general, we will have to settle for numerical solutions or power series expansions. So it would be helpful to be able to derive, with simple calculations, some nonrigorous, qualitative conclusions about the behavior of the solutions before we launch the heavy computational machinery. In this section we first display a few examples that illustrate the profound differences that can occur when the equations have variable coefficients or are nonlinear. Then we show how the mass–spring analogy, discussed in Section 4.1, can be exploited to predict some of the attributes of solutions of these more complicated equations.

To begin our discussion we display a linear constant-coefficient, a linear variable-coefficient, and two nonlinear equations.

(a) The equation

(1) $\qquad 3y'' + 2y' + 4y = 0$

is linear, homogeneous with constant coefficients. We know everything about such equations; the solutions are, at worst, polynomials times exponentials times sinusoids in t, and unique solutions can be found to match any prescribed data $y(a), y'(a)$ at any instant $t = a$. It has the superposition property: If $y_1(t)$ and $y_2(t)$ are solutions, so is $y(t) = c_1 y_1(t) + c_2 y_2(t)$.

(b) The equation

(2) $\qquad (1 - t^2)y'' - 2ty' + 2y = 0$

also has the superposition property (Problem 30, Exercises 4.7). It is a linear variable-coefficient equation and is a special case of **Legendre's equation** $(1 - t^2)y'' - 2ty' + \lambda y = 0$, which arises in the analysis of wave and diffusion phenomena in spherical coordinates.

(c) The equations

(3) $\qquad y'' - 6y^2 = 0$,

(4) $\qquad y'' - 24y^{1/3} = 0$

do not share the superposition property because of the square and the cube root of y terms (e.g., the quadratic term y^2 does not reduce to $y_1^2 + y_2^2$). They are *nonlinear*[†] equations.

The Legendre equation (2) has one simple solution, $y_1(t) = t$, as can easily be verified by mental calculation. A second, linearly independent, solution for $-1 < t < 1$ can be derived by the reduction of order procedure of Section 4.7. Traditionally, the second solution is taken to be

(5) $\qquad y_2(t) = \dfrac{t}{2} \ln\left(\dfrac{1 + t}{1 - t}\right) - 1$.

Notice in particular the behavior near $t = \pm 1$; none of the solutions of our *constant*-coefficient equations ever diverged at a finite point!

We would have anticipated troublesome behavior for (2) at $t = \pm 1$ if we had rewritten it in standard form as

(6) $\qquad y'' - \dfrac{2t}{1 - t^2} y' + \dfrac{2}{1 - t^2} y = 0$,

since Theorem 5, page 192, only promises existence and uniqueness *between* these points.

As we have noted, there are no general solution procedures for solving nonlinear equations. However, the following lemma is very useful in some situations such as equations (3), (4). It has an extremely significant physical interpretation, which we will explore in Project D of Chapter 5; for now we will merely tantalize the reader by giving it a suggestive name.

[†]Although the quadratic y^2 renders equation (3) nonlinear, the occurrence of t^2 in (2) does not spoil its linearity (in y).

The Energy Integral Lemma

Lemma 4. Let $y(t)$ be a solution to the differential equation

(7) $y'' = f(y)$,

where $f(y)$ is a continuous function that does not depend on y' or the independent variable t. Let $F(y)$ be an indefinite integral of $f(y)$, that is,

$$f(y) = \frac{d}{dy} F(y) .$$

Then the quantity

(8) $E(t) := \frac{1}{2} y'(t)^2 - F\big(y(t)\big)$

is constant; i.e.,

(9) $\dfrac{d}{dt} E(t) = 0$.

Proof. This is immediate; we insert (8), differentiate, and apply the differential equation (7):

$$\begin{aligned}
\frac{d}{dt} E(t) &= \frac{d}{dt}\left[\frac{1}{2} y'(t)^2 - F\big(y(t)\big)\right] \\
&= \frac{1}{2} 2 y' y'' - \frac{dF}{dy} y' \\
&= y' [y'' - f(y)] \\
&= 0 . \;\blacklozenge
\end{aligned}$$

As a result, an equation of the form (7) can be reduced to

(10) $\dfrac{1}{2}(y')^2 - F(y) = K$,

for some constant K, which is equivalent to the *separable* first-order equation

$$\frac{dy}{dt} = \pm \sqrt{2[F(y) + K]}$$

having the implicit two-parameter solution (Section 2.2)

(11) $t = \pm \displaystyle\int \frac{dy}{\sqrt{2[F(y) + K]}} + c$.

We will use formula (11) to illustrate some startling features of nonlinear equations.

Example 1 Apply the energy integral lemma to explore the solutions of the nonlinear equation $y'' = 6y^2$, given in (3).

Solution Since $6y^2 = \dfrac{d}{dy}(2y^3)$, the solution form (11) becomes

$$t = \pm \int \frac{dy}{\sqrt{2[y^3 + K]}} + c \,.$$

For simplicity we take the plus sign and focus attention on solutions with $K = 0$. Then we find $t = \int \frac{1}{2}y^{-3/2}\, dy = -y^{-1/2} + c$, or

(12) $y(t) = (c - t)^{-2} \,,$

for any value of the constant c.

Clearly, this equation is an enigma; solutions can blow up at $t = 1, t = 2, t = \pi$, or anywhere—and there is no clue in equation (3) as to why this should happen! Moreover, we have found an *infinite* family of pairwise linearly independent solutions (rather than the expected *two*). Yet we still cannot assemble, out of these, a solution matching (say) $y(0) = 1, y'(0) = 3$; all our solutions (12) have $y'(0) = 2y(0)^{3/2}$, and the absence of a superposition principle voids the use of linear combinations $c_1 y_1(t) + c_2 y_2(t)$. ◆

Example 2 Apply the energy integral lemma to explore the solutions of the nonlinear equation $y'' = 24y^{1/3}$, given in (4).

Solution Since $24y^{1/3} = \dfrac{d}{dy}(18y^{4/3})$, formula (11) gives

$$t = \pm \int \frac{dy}{\sqrt{2[18y^{4/3} + K]}} + c \,.$$

Again we take the plus sign and focus attention on solutions with $K = 0$. Then we find $t = y^{1/3}/2 + c$. In particular, $y_1(t) = 8t^3$ is a solution, and it satisfies the initial conditions $y(0) = 0, y'(0) = 0$. But note that $y_2(t) \equiv 0$ is also a solution to (4), *and it satisfies the same initial conditions*! Hence, the *uniqueness* feature, guaranteed for linear equations by Theorem 5 on page 192 of Section 4.7, can fail in the nonlinear case. ◆

So these examples have demonstrated violations of the existence and uniqueness properties (as well as "finiteness") that we have come to expect from the constant-coefficient case. It should not be surprising, then, that the solution techniques for variable-coefficient and nonlinear second-order equations are more complicated—when, indeed, they exist. Recall, however, that in Section 4.3 we saw that our familiarity with the mass–spring oscillator equation

(13) $F_{\text{ext}} = [\,\text{inertia}\,]y'' + [\,\text{damping}\,]y' + [\,\text{stiffness}\,]y$
 $= my'' + by' + ky$

was helpful in picturing the qualitative features of the solutions of other constant-coefficient equations. (See Figure 4.1, page 152.) By pushing these analogies further, we can also anticipate some of the features of the solutions in the variable-coefficient and nonlinear cases. One of the simplest linear second-order differential equations with variable coefficients is

(14) $y'' + ty = 0 \,.$

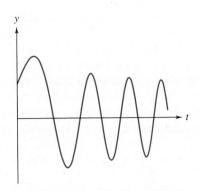

Figure 4.12 Solution to equation (14)

Example 3 Using the mass–spring analogy, predict the nature of the solutions to equation (14) for $t > 0$.

Solution Comparing (13) with (14), we see that the latter equation describes a mass–spring oscillator where the *spring stiffness* "k" varies in time—in fact, it stiffens as time passes ["k" $= t$ in equation (14)]. Physically, then, we would expect to see oscillations whose frequency increases with time, while the amplitude of the oscillations diminishes (because the spring gets harder to stretch). The numerically computed solution in Figure 4.12 displays precisely this behavior. ◆

Remark Schemes for numerically computing solutions to second-order equations will be discussed in Section 5.3. If such schemes are available, then why is there a need for the qualitative analysis discussed in this section? The answer is that numerical methods provide only *approximations* to solutions of initial value problems, and their accuracy is sometimes difficult to predict (especially for nonlinear equations). For example, numerical methods are often ineffective near points of discontinuity or over the long time intervals needed to study the asymptotic behavior. And this is precisely when qualitative arguments can lend insight into the reasonableness of the computed solution.

It is easy to verify (Problem 1) that if $y(t)$ is a solution of the **Airy equation**

(15) $y'' - ty = 0$,

then $y(-t)$ solves $y'' + ty = 0$, so "Airy functions" exhibit the behavior shown in Figure 4.12 for *negative time*. For positive t the Airy equation has a *negative* stiffness "k" $= -t$, with magnitude increasing in time. As we observed in Example 5 of Section 4.3, negative stiffness tends to reinforce, rather than oppose, displacements, and the solutions $y(t)$ grow rapidly with (positive) time. The solution known as the Airy function of the second kind $Bi(t)$, depicted in Figure 4.13, behaves exactly as expected.

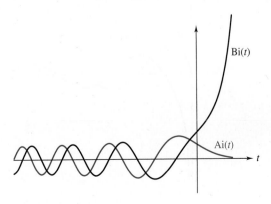

Figure 4.13 Airy functions

In Section 4.3 we also pointed out that mass–spring systems with negative spring stiffness can have isolated *bounded* solutions if the initial displacement $y(0)$ and velocity $y'(0)$ are precisely selected to counteract the repulsive spring force. The Airy function of the first kind $\text{Ai}(t)$, also depicted in Figure 4.13, is such a solution for the "Airy spring"; the initial inwardly directed velocity is just adequate to overcome the outward push of the stiffening spring, and the mass approaches a delicate equilibrium state $y(t) \equiv 0$.

Now let's look at **Bessel's equation.** It arises in the analysis of wave or diffusion phenomena in cylindrical coordinates. The *Bessel equation of order n* is written

$$(16) \qquad y'' + \frac{1}{t}y' + \left(1 - \frac{n^2}{t^2}\right)y = 0 \,.$$

Clearly, there are irregularities at $t = 0$, analogous to those at $t = \pm 1$ for the Legendre equation (2); we will explore these in depth in Chapter 8.

Example 4 Apply the mass–spring analogy to predict qualitative features of solutions to Bessel's equation for $t > 0$.

Solution Comparing (16) with the paradigm (13), we observe that

- the inertia "m" $= 1$ is fixed at unity;
- there is positive damping ("b" $= 1/t$), although it weakens with time; and
- the stiffness ("k" $= 1 - n^2/t^2$) is positive when $t > n$ and tends to 1 as $t \to +\infty$.

Solutions, then, should be expected to oscillate with amplitudes that diminish slowly (due to the damping), and the frequency of the oscillations should settle at a constant value (given, according to the procedures of Section 4.3, by $\sqrt{k/m} = 1$ radian per unit time). The graphs of the Bessel functions $J_n(t)$ and $Y_n(t)$ of the first and second kind of order $n = \frac{1}{2}$ exemplify these qualitative predictions; see Figure 4.14. The effect of the singularities in the coefficients at $t = 0$ is manifested in the graph of $Y_{1/2}(t)$. ◆

Although most Bessel functions have to be computed by power series methods, if the *order n* is a half-integer, then $J_n(t)$ and $Y_n(t)$ have closed-form expressions. In fact, $J_{1/2}(t) = \sqrt{2/(\pi t)}\sin t$ and $Y_{1/2}(t) = -\sqrt{2/(\pi t)}\cos t$. You can verify directly that these functions solve equation (16).

Figure 4.14 Bessel functions

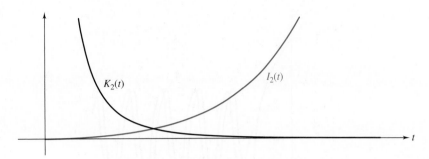

Figure 4.15 Modified Bessel functions

Example 5 Give a qualitative analysis of the **modified Bessel equation** of order n:

$$(17) \qquad y'' + \frac{1}{t}y' - \left(1 - \frac{n^2}{t^2}\right)y = 0 \ .$$

Solution This equation also exhibits unit mass and positive, diminishing, damping. However, the stiffness now converges to *negative* 1. Accordingly, we expect typical solutions to diverge as $t \to +\infty$. The modified Bessel function of the first kind, $I_n(t)$ of order $n = 2$ in Figure 4.15 follows this prediction, whereas the modified Bessel function of the second kind, $K_n(t)$ of order $n = 2$ in Figure 4.15, exhibits the same sort of balance of initial position and velocity as we saw for the Airy function $\mathrm{Ai}(t)$. Again, the effect of the singularity at $t = 0$ is evident. ◆

Example 6 Use the mass–spring model to predict qualitative features of the solutions to the nonlinear **Duffing equation**

$$(18) \qquad y'' + y + y^3 = y'' + (1 + y^2)y = 0 \ .$$

Solution Although equation (18) is nonlinear, it can be matched with the paradigm (13) if we envision unit mass, no damping, and a (positive) stiffness "k" $= 1 + y^2$, which increases as the *displacement y* gets larger. (This increasing-stiffness effect is built into some popular mattresses for therapeutic reasons.)[†] Such a spring grows stiffer as the mass moves farther away, but it restores to its original value when the mass returns. Thus, high-amplitude excursions should oscillate faster than low-amplitude ones, and the sinusoidal shapes in the graphs of $y(t)$ should be "pinched in" somewhat at their peaks. These qualitative predictions are demonstrated by the numerically computed solutions plotted in Figure 4.16. ◆

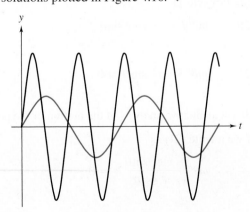

Figure 4.16 Solution graphs for the Duffing equation

[†]Graphic depictions of oscillations on a therapeutic mattress are best left to one's imagination, the editors say.

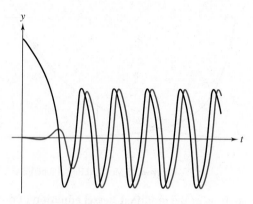

Figure 4.17 Solutions to the van der Pol equation

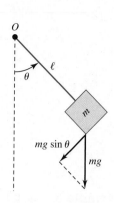

Figure 4.18 A pendulum

The fascinating **van der Pol equation**

(19) $y'' - (1 - y^2)y' + y = 0$

originated in the study of the electrical oscillations observed in vacuum tubes.

Example 7 Predict the behavior of the solutions to equation (19) using the mass–spring model.

Solution By comparison to the paradigm (13), we observe unit mass and stiffness, positive damping $["b" = -(1 - y^2)]$ when $|y(t)| > 1$, and negative damping when $|y(t)| < 1$. *Friction thus dampens large-amplitude motions but energizes small oscillations.* The result, then, is that all (nonzero) solutions tend to a *limit cycle* whose friction penalty incurred while $|y(t)| > 1$ is balanced by the negative-friction boost received while $|y(t)| < 1$. The computer-generated Figure 4.17 illustrates the convergence to the limit cycle for some solutions to the van der Pol equation[†]. ◆

Finally, we consider the motion of the pendulum depicted in Figure 4.18. This motion is measured by the angle $\theta(t)$ that the pendulum makes with the vertical line through O at time t. As the diagram shows, the component of gravity, which exerts a torque on the pendulum and thus accelerates the angular velocity $d\theta/dt$, is given by $-mg \sin\theta$. Consequently, the rotational analog of Newton's second law, *torque equals rate of change of angular momentum,* dictates (see Problem 7)

(20) $m\ell^2\theta'' = -\ell mg \sin\theta$,

or

(21) $m\theta'' + m\dfrac{g}{\ell}\sin\theta = 0$.

Example 8 Give a qualitative analysis of the motion of the pendulum.

Solution If we rewrite (21) as

$$m\theta'' + \frac{mg}{\ell}\ \frac{\sin\theta}{\theta}\ \theta = 0$$

[†]The vdP equation also provides a model for the world's most accurate pendulum clock, a detail of which appears on the cover.

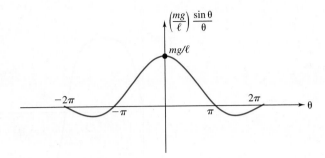

Figure 4.19 Pendulum "stiffness".

and compare with the paradigm (13), we see fixed mass m, no damping, and a stiffness given by

$$"k" = \frac{mg}{\ell}\frac{\sin\theta}{\theta}.$$

This stiffness is plotted in Figure 4.19, where we see that small-amplitude motions are driven by a nearly constant spring stiffness of value mg/ℓ, and the considerations of Section 4.3 dictate the familiar formula

$$\omega = \sqrt{\frac{k}{m}} = \sqrt{\frac{g}{\ell}}$$

for the angular frequency of the nearly sinusoidal oscillations. See Figure 4.20, which compares a computer-generated solution to equation (21) to the solution of the constant-stiffness equation with the same initial conditions.[†]

For larger motions, however, the diminishing stiffness distorts the sinusoidal nature of the graph of $\theta(t)$, and lowers the frequency. This is evident in Figure 4.21 on page 210.

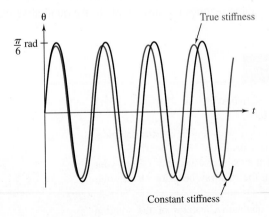

Figure 4.20 Small-amplitude pendulum motion

[†]The latter is identified as the *linearized equation* in Project D on page 236.

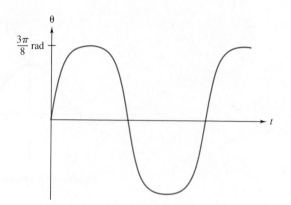

Figure 4.21 Large-amplitude pendulum motion

Finally, if the motion is so energetic that θ reaches the value π, the stiffness changes sign and abets the displacement; the pendulum passes the apex and gains speed as it falls, and this spinning motion repeats continuously. See Figure 4.22. ◆

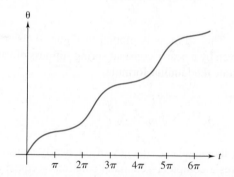

Figure 4.22 Very-large-amplitude pendulum motion

The computation of the solutions of the Legendre, Bessel, and Airy equations and the analysis of the nonlinear equations of Duffing, van der Pol, and the pendulum have challenged many of the great mathematicians of the past. It is gratifying, then, to note that so many of their salient features are susceptible to the qualitative reasoning we have used herein.

4.8 EXERCISES

1. Show that if $y(t)$ satisfies $y'' - ty = 0$, then $y(-t)$ satisfies $y'' + ty = 0$.

2. Using the paradigm (13), what are the inertia, damping, and stiffness for the equation $y'' - 6y^2 = 0$? If $y > 0$, what is the sign of the "stiffness constant"? Does your answer help explain the runaway behavior of the solutions $y(t) = 1/(c-t)^2$?

3. Try to predict the qualitative features of the solution to $y'' - 6y^2 = 0$ that satisfies the initial conditions $y(0) = -1, y'(0) = -1$. Compare with the computer-generated Figure 4.23. [*Hint:* Consider the sign of the spring stiffness.]

Figure 4.23 Solution for Problem 3

4. Show that the *three* solutions $1/(1-t)^2$, $1/(2-t)^2$, and $1/(3-t)^2$ to $y'' - 6y^2 = 0$ are linearly independent on $(-1, 1)$. (See Problem 35, Exercises 4.2, page 164.)

5. **(a)** Use the energy integral lemma to derive the family of solutions $y(t) = 1/(t-c)$ to the equation $y'' = 2y^3$.

 (b) For $c \neq 0$ show that these solutions are pairwise linearly independent for different values of c in an appropriate interval around $t = 0$.

 (c) Show that none of these solutions satisfies the initial conditions $y(0) = 1, y'(0) = 2$.

6. Use the energy integral lemma to show that motions of the free undamped mass–spring oscillator $my'' + ky = 0$ obey

$$m(y')^2 + ky^2 = \text{constant}.$$

7. **Pendulum Equation.** To derive the pendulum equation (21), complete the following steps.

 (a) The *angular momentum* of the pendulum mass m measured about the support O in Figure 4.18 on page 208 is given by the product of the "lever arm" length ℓ and the component of the vector momentum mv perpendicular to the lever arm. Show that this gives

$$\text{angular momentum} = m\ell^2 \frac{d\theta}{dt}.$$

 (b) The *torque* produced by gravity equals the product of the lever arm length ℓ and the component of gravitational (vector) force $m\mathbf{g}$ perpendicular to the lever arm. Show that this gives

$$\text{torque} = -\ell mg \sin\theta.$$

 (c) Now use Newton's law of rotational motion to deduce the pendulum equation (20).

8. Use the energy integral lemma to show that pendulum motions obey

$$\frac{(\theta')^2}{2} - \frac{g}{\ell} \cos\theta = \text{constant}.$$

9. Use the result of Problem 8 to find the value of $\theta'(0)$, the initial velocity, that must be imparted to a pendulum at rest to make it approach (but not cross over) the apex of its motion. Take $\ell = g$ for simplicity.

10. Use the result of Problem 8 to prove that if the pendulum in Figure 4.18 on page 208 is released *from rest* at the angle $\alpha, 0 < \alpha < \pi$, then $|\theta(t)| \leq \alpha$ for all t. [*Hint:* The initial conditions are $\theta(0) = \alpha, \theta'(0) = 0$; argue that the *constant* in Problem 8 equals $-(g/\ell)\cos\alpha$.]

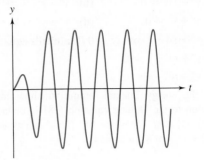

Figure 4.24 Solution to the Rayleigh equation

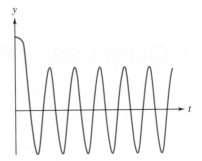

Figure 4.25 Solution to the Rayleigh equation

11. Use the mass–spring analogy to explain the qualitative nature of the solutions to the **Rayleigh equation**

(22) $$y'' - [1 - (y')^2]y' + y = 0$$

depicted in Figures 4.24 and 4.25.

12. Use reduction of order to derive the solution $y_2(t)$ in equation (5) for Legendre's equation.

13. Figure 4.26 contains graphs of solutions to the Duffing, Airy, and van der Pol equations. Try to match the solution to the equation.

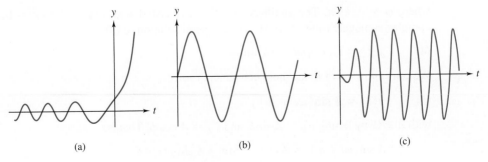

(a) (b) (c)

Figure 4.26 Solution graphs for Problem 13

14. Verify that the formulas for the Bessel functions $J_{1/2}(t)$, $Y_{1/2}(t)$ do indeed solve equation (16).

15. Use the mass–spring oscillator analogy to decide whether all solutions to each of the following differential equations are bounded as $t \to +\infty$.

(a) $y'' + t^2 y = 0$ (b) $y'' - t^2 y = 0$

(c) $y'' + y^5 = 0$ (d) $y'' + y^6 = 0$

(e) $y'' + (4 + 2\cos t)y = 0$ (**Mathieu's equation**)

(f) $y'' + ty' + y = 0$

(g) $y'' - ty' - y = 0$

16. Use the energy integral lemma to show that every solution to the Duffing equation (18) is bounded; that is, $|y(t)| \le M$ for some M. [*Hint:* First argue that $y^2/2 + y^4/4 \le K$ for some K.]

17. Armageddon. Earth revolves around the sun in an approximately circular orbit with radius $r = a$, completing a revolution in the time $T = 2\pi(a^3/GM)^{1/2}$, which is one Earth year; here M is the mass of the sun and G is the universal gravitational constant. The gravitational force of the sun on Earth is given by GMm/r^2, where m is the mass of Earth. Therefore, if Earth "stood still," losing its orbital velocity, it would fall on a straight line into the sun in accordance with Newton's second law:

$$m\frac{d^2r}{dt^2} = -\frac{GMm}{r^2}.$$

If this calamity occurred, what fraction of the normal year T would it take for Earth to splash into the sun (i.e., achieve $r = 0$)? [*Hint:* Use the energy integral lemma and the initial conditions $r(0) = a$, $r'(0) = 0$.]

4.9 A Closer Look at Free Mechanical Vibrations

In this section we return to the mass–spring system depicted in Figure 4.1 (page 152) and analyze its motion in more detail. The governing equation is

(1) $F_{\text{ext}} = [\,\text{inertia}\,]\dfrac{d^2y}{dt^2} + [\,\text{damping}\,]\dfrac{dy}{dt} + [\,\text{stiffness}\,]y$

$= my'' + by' + ky\,.$

Let's focus on the simple case in which $b = 0$ and $F_{\text{ext}} = 0$, the so-called **undamped, free** case. Then equation (1) reduces to

(2) $m\dfrac{d^2y}{dt^2} + ky = 0$

and, when divided by m, becomes

(3) $\dfrac{d^2y}{dt^2} + \omega^2 y = 0\,,$

where $\omega = \sqrt{k/m}$. The auxiliary equation associated with (3) is $r^2 + \omega^2 = 0$, which has complex conjugate roots $\pm \omega i$. Hence, a general solution to (3) is

(4) $y(t) = C_1 \cos \omega t + C_2 \sin \omega t\,.$

We can express $y(t)$ in the more convenient form

(5) $y(t) = A \sin(\omega t + \phi)\,,$

with $A \ge 0$, by letting $C_1 = A \sin\phi$ and $C_2 = A \cos\phi$. That is,

$A \sin(\omega t + \phi) = A \cos \omega t \sin\phi + A \sin \omega t \cos\phi$
$= C_1 \cos \omega t + C_2 \sin \omega t\,.$

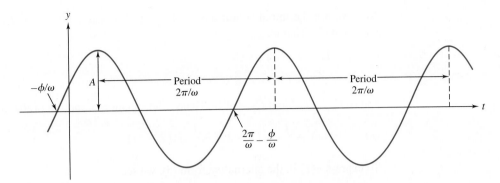

Figure 4.27 Simple harmonic motion of undamped, free vibrations

Solving for A and ϕ in terms of C_1 and C_2, we find

(6) $A = \sqrt{C_1^2 + C_2^2}$ and $\tan \phi = \dfrac{C_1}{C_2}$,

where the quadrant in which ϕ lies is determined by the signs of C_1 and C_2. This is because $\sin \phi$ has the same sign as C_1 ($\sin \phi = C_1/A$) and $\cos \phi$ has the same sign as C_2 ($\cos \phi = C_2/A$). For example, if $C_1 > 0$ and $C_2 < 0$, then ϕ is in Quadrant II. (Note, in particular, that ϕ is not simply the arctangent of C_1/C_2, which would lie in Quadrant IV.)

It is evident from (5) that, as we predicted in Section 4.1, the motion of a mass in an *undamped, free* system is simply a sine wave, or what is called **simple harmonic motion.** (See Figure 4.27.) The constant A is the amplitude of the motion and ϕ is the phase angle. The motion is periodic with **period** $2\pi/\omega$ and **natural frequency** $\omega/2\pi$, where $\omega = \sqrt{k/m}$. The period is measured in units of time, and the natural frequency has the dimensions of periods (or cycles) per unit time. The constant ω is the **angular frequency** for the sine function in (5) and has dimensions of radians per unit time. To summarize:

$$\text{angular frequency} = \omega = \sqrt{k/m} \quad (\text{rad/sec}) \,,$$
$$\text{natural frequency} = \omega/2\pi \quad (\text{cycles/sec}) \,,$$
$$\text{period} = 2\pi/\omega \quad (\text{sec}) \,.$$

Observe that the amplitude and phase angle depend on the constants C_1 and C_2, which, in turn, are determined by the initial position and initial velocity of the mass. However, the period and frequency depend only on k and m and not on the initial conditions.

Example 1 A 1/8-kg mass is attached to a spring with stiffness $k = 16$ N/m, as depicted in Figure 4.1. The mass is displaced 1/2 m to the right of the equilibrium point and given an outward velocity (to the right) of $\sqrt{2}$ m/sec. Neglecting any damping or external forces that may be present, determine the equation of motion of the mass along with its amplitude, period, and natural frequency. How long after release does the mass pass through the equilibrium position?

Solution Because we have a case of undamped, free vibration, the equation governing the motion is (3). Thus, we find the angular frequency to be

$$\omega = \sqrt{\frac{k}{m}} = \sqrt{\frac{16}{1/8}} = 8\sqrt{2} \text{ rad/sec} \,.$$

Substituting this value for ω into (4) gives

(7) $y(t) = C_1 \cos(8\sqrt{2}\, t) + C_2 \sin(8\sqrt{2}\, t) \,.$

Now we use the initial conditions, $y(0) = 1/2$ m and $y'(0) = \sqrt{2}$ m/sec, to solve for C_1 and C_2 in (7). That is,

$$1/2 = y(0) = C_1 \,,$$
$$\sqrt{2} = y'(0) = 8\sqrt{2}C_2 \,,$$

and so $C_1 = 1/2$ and $C_2 = 1/8$. Hence, the equation of motion of the mass is

(8) $\qquad y(t) = \dfrac{1}{2}\cos(8\sqrt{2}t) + \dfrac{1}{8}\sin(8\sqrt{2}t) \,.$

To express $y(t)$ in the alternative form (5), we set

$$A = \sqrt{C_1^2 + C_2^2} = \sqrt{(1/2)^2 + (1/8)^2} = \frac{\sqrt{17}}{8} \,,$$

$$\tan\phi = \frac{C_1}{C_2} = \frac{1/2}{1/8} = 4 \,.$$

Since both C_1 and C_2 are positive, ϕ is in Quadrant I, so $\phi = \arctan 4 \approx 1.326$. Hence,

(9) $\qquad y(t) = \dfrac{\sqrt{17}}{8}\sin(8\sqrt{2}t + \phi) \,.$

Thus, the amplitude A is $\sqrt{17}/8$ m, and the phase angle ϕ is approximately 1.326 rad. The period is $P = 2\pi/\omega = 2\pi/(8\sqrt{2}) = \sqrt{2}\pi/8$ sec, and the natural frequency is $1/P = 8/(\sqrt{2}\pi)$ cycles per sec.

Finally, to determine when the mass will pass through the equilibrium position, $y = 0$, we must solve the trigonometric equation

(10) $\qquad y(t) = \dfrac{\sqrt{17}}{8}\sin(8\sqrt{2}t + \phi) = 0$

for t. Equation (10) will be satisfied whenever

(11) $\qquad 8\sqrt{2}t + \phi = n\pi \qquad \text{or} \qquad t = \dfrac{n\pi - \phi}{8\sqrt{2}} \approx \dfrac{n\pi - 1.326}{8\sqrt{2}} \,,$

n an integer. Putting $n = 1$ in (11) determines the first time t when the mass crosses its equilibrium position:

$$t = \frac{\pi - \phi}{8\sqrt{2}} \approx 0.16 \text{ sec} \,. \quad \blacklozenge$$

In most applications of vibrational analysis, of course, there is some type of frictional or damping force affecting the vibrations. This force may be due to a component in the system, such as a shock absorber in a car, or to the medium that surrounds the system, such as air or some liquid. So we turn to a study of the effects of damping on free vibrations, and equation (2) generalizes to

(12) $\qquad m\dfrac{d^2y}{dt^2} + b\dfrac{dy}{dt} + ky = 0 \,.$

The auxiliary equation associated with (12) is

(13) $\qquad mr^2 + br + k = 0 \,,$

and its roots are

$$(14) \qquad \frac{-b \pm \sqrt{b^2 - 4mk}}{2m} = -\frac{b}{2m} \pm \frac{1}{2m} \sqrt{b^2 - 4mk} \,.$$

As we found in Sections 4.2 and 4.3, the form of the solution to (12) depends on the nature of these roots and, in particular, on the discriminant $b^2 - 4mk$.

Underdamped or Oscillatory Motion ($b^2 < 4mk$)

When $b^2 < 4mk$, the discriminant $b^2 - 4mk$ is negative, and there are two complex conjugate roots to the auxiliary equation (13). These roots are $\alpha \pm i\beta$, where

$$(15) \qquad \alpha := -\frac{b}{2m}, \quad \beta := \frac{1}{2m} \sqrt{4mk - b^2} \,.$$

Hence, a general solution to (12) is

$$(16) \qquad y(t) = e^{\alpha t}(C_1 \cos \beta t + C_2 \sin \beta t) \,.$$

As we did with simple harmonic motion, we can express $y(t)$ in the alternate form

$$(17) \qquad y(t) = A e^{\alpha t} \sin(\beta t + \phi) \,,$$

where $A = \sqrt{C_1^2 + C_2^2}$ and $\tan \phi = C_1/C_2$. It is now evident that $y(t)$ is the product of an exponential **damping factor,**

$$A e^{\alpha t} = A e^{-(b/2m)t} \,,$$

and a sine factor, $\sin(\beta t + \phi)$, which accounts for the oscillatory motion. Because the sine factor varies between –1 and 1 with period $2\pi/\beta$, the solution $y(t)$ varies between $-A e^{\alpha t}$ and $A e^{\alpha t}$ with **quasiperiod** $P = 2\pi/\beta = 4m\pi/\sqrt{4mk - b^2}$ and **quasifrequency** $1/P$. Moreover, since b and m are positive, $\alpha = -b/2m$ is negative, and thus the exponential factor tends to zero as $t \to +\infty$. A graph of a typical solution $y(t)$ is given in Figure 4.28. The system is called **underdamped** because there is not enough damping present (b is too small) to prevent the system from oscillating.

It is easily seen that as $b \to 0$ the damping factor approaches the constant A and the quasifrequency approaches the natural frequency of the corresponding undamped harmonic motion. Figure 4.28 demonstrates that the values of t where the graph of $y(t)$ touches the exponential curves $\pm A e^{\alpha t}$ are close to (but not exactly) the same values of t at which $y(t)$ attains its relative maximum and minimum values (see Problem 13 on page 221).

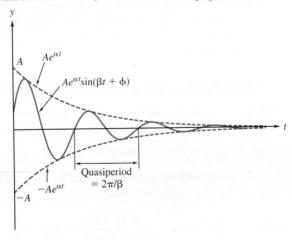

Figure 4.28 Damped oscillatory motion

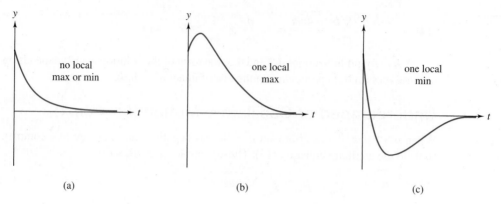

Figure 4.29 Overdamped vibrations

Overdamped Motion ($b^2 > 4mk$)

When $b^2 > 4mk$, the discriminant $b^2 - 4mk$ is positive, and there are two distinct real roots to the auxiliary equation (13):

$$(18) \qquad r_1 = -\frac{b}{2m} + \frac{1}{2m}\sqrt{b^2 - 4mk}\,, \quad r_2 = -\frac{b}{2m} - \frac{1}{2m}\sqrt{b^2 - 4mk}\,.$$

Hence, a general solution to (12) in this case is

$$(19) \qquad y(t) = C_1 e^{r_1 t} + C_2 e^{r_2 t}\,.$$

Obviously, r_2 is negative. And since $b^2 > b^2 - 4mk$ (that is, $b > \sqrt{b^2 - 4mk}$), it follows that r_1 is also negative. Therefore, as $t \to +\infty$, both of the exponentials in (19) decay and $y(t) \to 0$. Moreover, since

$$y'(t) = C_1 r_1 e^{r_1 t} + C_2 r_2 e^{r_2 t} = e^{r_1 t}(C_1 r_1 + C_2 r_2 e^{(r_2 - r_1)t})\,,$$

we see that the derivative is either identically zero (when $C_1 = C_2 = 0$) or vanishes for at most one value of t (when the factor in parentheses is zero). If the trivial solution $y(t) \equiv 0$ is ignored, it follows that $y(t)$ has at most one local maximum or minimum for $t > 0$. Therefore, $y(t)$ *does not oscillate*. This leaves, qualitatively, only three possibilities for the motion of $y(t)$, depending on the initial conditions. These are illustrated in Figure 4.29. This case where $b^2 > 4mk$ is called **overdamped** motion.

Critically Damped Motion ($b^2 = 4mk$)

When $b^2 = 4mk$, the discriminant $b^2 - 4mk$ is zero, and the auxiliary equation has the repeated root $-b/2m$. Hence, a general solution to (12) is now

$$(20) \qquad y(t) = (C_1 + C_2 t)e^{-(b/2m)t}\,.$$

To understand the motion described by $y(t)$ in (20), we first consider the behavior of $y(t)$ as $t \to +\infty$. By L'Hôpital's rule,

$$(21) \qquad \lim_{t \to +\infty} y(t) = \lim_{t \to +\infty} \frac{C_1 + C_2 t}{e^{(b/2m)t}} = \lim_{t \to +\infty} \frac{C_2}{(b/2m)e^{(b/2m)t}} = 0$$

(recall that $b/2m > 0$). Hence, $y(t)$ dies off to zero as $t \to +\infty$. Next, since

$$y'(t) = \left(C_2 - \frac{b}{2m}C_1 - \frac{b}{2m}C_2 t\right)e^{-(b/2m)t},$$

we see again that a nontrivial solution can have at most one local maximum or minimum for $t > 0$, so motion is *non*oscillatory. If b were any smaller, oscillation would occur. Thus, the special case where $b^2 = 4mk$ is called **critically damped** motion. Qualitatively, critically damped motions are similar to overdamped motions (see Figure 4.29 again).

Example 2 Assume that the motion of a mass–spring system with damping is governed by

(22) $$\frac{d^2y}{dt^2} + b\frac{dy}{dt} + 25y = 0; \quad y(0) = 1, \quad y'(0) = 0.$$

Find the equation of motion and sketch its graph for the three cases where $b = 6$, 10, and 12.

Solution The auxiliary equation for (22) is

(23) $$r^2 + br + 25 = 0,$$

whose roots are

(24) $$r = -\frac{b}{2} \pm \frac{1}{2}\sqrt{b^2 - 100}.$$

 Case 1. When $b = 6$, the roots (24) are $-3 \pm 4i$. This is thus a case of underdamping, and the equation of motion has the form

(25) $$y(t) = C_1 e^{-3t}\cos 4t + C_2 e^{-3t}\sin 4t.$$

Setting $y(0) = 1$ and $y'(0) = 0$ gives the system

$$C_1 = 1, \quad -3C_1 + 4C_2 = 0,$$

whose solution is $C_1 = 1$, $C_2 = 3/4$. To express $y(t)$ as the product of a damping factor and a sine factor [recall equation (17)], we set

$$A = \sqrt{C_1^2 + C_2^2} = \frac{5}{4}, \quad \tan\phi = \frac{C_1}{C_2} = \frac{4}{3},$$

where ϕ is a Quadrant I angle, since C_1 and C_2 are both positive. Then

(26) $$y(t) = \frac{5}{4}e^{-3t}\sin(4t + \phi),$$

where $\phi = \arctan(4/3) \approx 0.9273$. The underdamped spring motion is shown in Figure 4.30(a) on page 218.

 Case 2. When $b = 10$, there is only one (repeated) root to the auxiliary equation (23), namely, $r = -5$. This is a case of critical damping, and the equation of motion has the form

(27) $$y(t) = (C_1 + C_2 t)e^{-5t}.$$

Setting $y(0) = 1$ and $y'(0) = 0$ now gives

$$C_1 = 1, \quad C_2 - 5C_1 = 0,$$

and so $C_1 = 1$, $C_2 = 5$. Thus,

(28) $$y(t) = (1 + 5t)e^{-5t}.$$

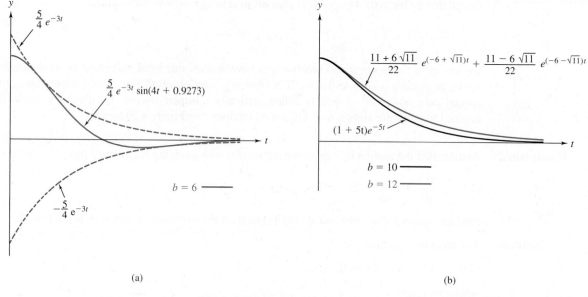

Figure 4.30 Solutions for various values of b

The graph of $y(t)$ given in (28) is represented by the lower curve in Figure 4.30(b). Notice that $y(t)$ is zero only for $t = -1/5$ and hence does not cross the t-axis for $t > 0$.

Case 3. When $b = 12$, the roots to the auxiliary equation are $-6 \pm \sqrt{11}$. This is a case of overdamping, and the equation of motion has the form

(29) $y(t) = C_1 e^{(-6+\sqrt{11})t} + C_2 e^{(-6-\sqrt{11})t}$.

Setting $y(0) = 1$ and $y'(0) = 0$ gives

$$C_1 + C_2 = 1, \quad (-6 + \sqrt{11})C_1 + (-6 - \sqrt{11})C_2 = 0,$$

from which we find $C_1 = (11 + 6\sqrt{11})/22$ and $C_2 = (11 - 6\sqrt{11})/22$. Hence,

(30) $\displaystyle y(t) = \frac{11 + 6\sqrt{11}}{22}e^{(-6+\sqrt{11})t} + \frac{11 - 6\sqrt{11}}{22}e^{(-6-\sqrt{11})t}$

$\displaystyle = \frac{e^{(-6+\sqrt{11})t}}{22}\left\{ 11 + 6\sqrt{11} + (11 - 6\sqrt{11})e^{-2\sqrt{11}t} \right\}$.

The graph of this overdamped motion is represented by the upper curve in Figure 4.30(b). ◆

It is interesting to observe in Example 2 that when the system is underdamped ($b = 6$), the solution goes to zero like e^{-3t}; when the system is critically damped ($b = 10$), the solution tends to zero roughly like e^{-5t}; and when the system is overdamped ($b = 12$), the solution goes to zero like $e^{(-6+\sqrt{11})t} \approx e^{-2.68t}$. This means that if the system is underdamped, it not only oscillates but also dies off slower than if it were critically damped. Moreover, if the system is overdamped, it again dies off more slowly than if it were critically damped (in agreement with our physical intuition that the damping forces hinder the return to equilibrium).

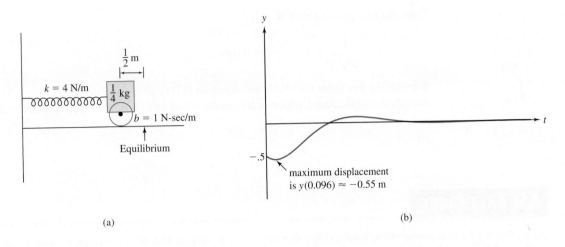

Figure 4.31 Mass–spring system and graph of motion for Example 3

Example 3 A 1/4-kg mass is attached to a spring with a stiffness 4 N/m as shown in Figure 4.31(a). The damping constant b for the system is 1 N-sec/m. If the mass is displaced 1/2 m to the left and given an initial velocity of 1 m/sec to the left, find the equation of motion. What is the maximum displacement that the mass will attain?

Solution Substituting the values for m, b, and k into equation (12) and enforcing the initial conditions, we obtain the initial value problem

(31) $$\frac{1}{4}\frac{d^2y}{dt^2} + \frac{dy}{dt} + 4y = 0 \, ; \quad y(0) = -\frac{1}{2}, \quad y'(0) = -1 \, .$$

The negative signs for the initial conditions reflect the facts that the initial displacement and push are to the left.

It can readily be verified that the solution to (31) is

(32) $$y(t) = -\frac{1}{2}e^{-2t}\cos\left(2\sqrt{3}t\right) - \frac{1}{\sqrt{3}}e^{-2t}\sin\left(2\sqrt{3}t\right) \, ,$$

or

(33) $$y(t) = \sqrt{\frac{7}{12}}e^{-2t}\sin\left(2\sqrt{3}t + \phi\right) \, ,$$

where $\tan\phi = \sqrt{3}/2$ and ϕ lies in Quadrant III because $C_1 = -1/2$ and $C_2 = -1/\sqrt{3}$ are both negative. [See Figure 4.31(b) for a sketch of $y(t)$.]

To determine the maximum displacement from equilibrium, we must determine the maximum value of $|y(t)|$ on the graph in Figure 4.31(b). Because $y(t)$ dies off exponentially, this will occur at the first critical point of $y(t)$. Computing $y'(t)$ from (32), setting it equal to zero, and solving gives

$$y'(t) = e^{-2t}\left\{\frac{5}{\sqrt{3}}\sin\left(2\sqrt{3}t\right) - \cos\left(2\sqrt{3}t\right)\right\} = 0 \, ,$$

$$\frac{5}{\sqrt{3}}\sin(2\sqrt{3}t) = \cos(2\sqrt{3}t) \, ,$$

$$\tan(2\sqrt{3}t) = \frac{\sqrt{3}}{5} \, .$$

Thus, the first positive root is

$$t = \frac{1}{2\sqrt{3}} \arctan \frac{\sqrt{3}}{5} \approx 0.096 \,.$$

Substituting this value for t back into equation (32) or (33) gives $y(0.096) \approx -0.55$. Hence, the maximum displacement, which occurs to the left of equilibrium, is approximately 0.55 m. ◆

4.9 EXERCISES

All problems refer to the mass–spring configuration depicted in Figure 4.1, page 152.

1. A 2-kg mass is attached to a spring with stiffness $k = 50$ N/m. The mass is displaced $1/4$ m to the left of the equilibrium point and given a velocity of 1 m/sec to the left. Neglecting damping, find the equation of motion of the mass along with the amplitude, period, and frequency. How long after release does the mass pass through the equilibrium position?

2. A 3-kg mass is attached to a spring with stiffness $k = 48$ N/m. The mass is displaced $1/2$ m to the left of the equilibrium point and given a velocity of 2 m/sec to the right. The damping force is negligible. Find the equation of motion of the mass along with the amplitude, period, and frequency. How long after release does the mass pass through the equilibrium position?

3. The motion of a mass–spring system with damping is governed by

$$y''(t) + by'(t) + 16y(t) = 0 \,;$$
$$y(0) = 1 \,, \qquad y'(0) = 0 \,.$$

Find the equation of motion and sketch its graph for $b = 0, 6, 8,$ and 10.

4. The motion of a mass–spring system with damping is governed by

$$y''(t) + by'(t) + 64y(t) = 0 \,;$$
$$y(0) = 1 \,, \qquad y'(0) = 0 \,.$$

Find the equation of motion and sketch its graph for $b = 0, 10, 16,$ and 20.

5. The motion of a mass–spring system with damping is governed by

$$y''(t) + 10y'(t) + ky(t) = 0 \,;$$
$$y(0) = 1 \,, \qquad y'(0) = 0 \,.$$

Find the equation of motion and sketch its graph for $k = 20, 25,$ and 30.

6. The motion of a mass–spring system with damping is governed by

$$y''(t) + 4y'(t) + ky(t) = 0 \,;$$
$$y(0) = 1 \,, \qquad y'(0) = 0 \,.$$

Find the equation of motion and sketch its graph for $k = 2, 4,$ and 6.

7. A $1/8$-kg mass is attached to a spring with stiffness 16 N/m. The damping constant for the system is 2 N-sec/m. If the mass is moved $3/4$ m to the left of equilibrium and given an initial leftward velocity of 2 m/sec, determine the equation of motion of the mass and give its damping factor, quasiperiod, and quasifrequency.

8. A 20-kg mass is attached to a spring with stiffness 200 N/m. The damping constant for the system is 140 N-sec/m. If the mass is pulled 25 cm to the right of equilibrium and given an initial leftward velocity of 1 m/sec, when will it first return to its equilibrium position?

9. A 2-kg mass is attached to a spring with stiffness 40 N/m. The damping constant for the system is $8\sqrt{5}$ N-sec/m. If the mass is pulled 10 cm to the right of equilibrium and given an initial rightward velocity of 2 m/sec, what is the maximum displacement from equilibrium that it will attain?

10. A $1/4$-kg mass is attached to a spring with stiffness 8 N/m. The damping constant for the system is $1/4$ N-sec/m. If the mass is moved 1 m to the left of equilibrium and released, what is the maximum displacement to the right that it will attain?

11. A 1-kg mass is attached to a spring with stiffness 100 N/m. The damping constant for the system is 0.2 N-sec/m. If the mass is pushed rightward from the equilibrium position with a velocity of 1 m/sec, when will it attain its maximum displacement to the right?

12. A $1/4$-kg mass is attached to a spring with stiffness 8 N/m. The damping constant for the system is 2 N-sec/m. If the mass is pushed 50 cm to the left of equilibrium and given a leftward velocity of 2 m/sec, when will the mass attain its maximum displacement to the left?

13. Show that for the underdamped system of Example 3, the times when the solution curve $y(t)$ in (33) touches the exponential curves $\pm \sqrt{7/12}e^{-2t}$ are not the same values of t for which the function $y(t)$ attains its relative extrema.

14. For an underdamped system, verify that as $b \to 0$ the damping factor approaches the constant A and the quasifrequency approaches the natural frequency $\sqrt{k/m}/(2\pi)$.

15. How can one deduce the value of the damping constant b by observing the motion of an underdamped system? Assume that the mass m is known.

16. A mass attached to a spring oscillates with a period of 3 sec. After 2 kg are added, the period becomes 4 sec. Assuming that we can neglect any damping or external forces, determine how much mass was originally attached to the spring.

17. Consider the equation for free mechanical vibration, $my'' + by' + ky = 0$, and assume the motion is critically damped. Let $y(0) = y_0$, $y'(0) = v_0$ and assume $y_0 \neq 0$.

(a) Prove that the mass will pass through its equilibrium at exactly one positive time if and only if
$$\frac{-2my_0}{2mv_0 + by_0} > 0.$$

(b) Use computer software to illustrate part (a) for a specific choice of m, b, k, y_0, and v_0. Be sure to include an appropriate graph in your illustration.

18. Consider the equation for free mechanical vibration, $my'' + by' + ky = 0$, and assume the motion is overdamped. Suppose $y(0) > 0$ and $y'(0) > 0$. Prove that the mass will never pass through its equilibrium at any positive time.

4.10 A Closer Look at Forced Mechanical Vibrations

We now consider the vibrations of a mass–spring system when an external force is applied. Of particular interest is the response of the system to a *sinusoidal* forcing term. As a paradigm, let's investigate the effect of a cosine forcing function on the system governed by the differential equation

$$(1) \qquad m\frac{d^2y}{dt^2} + b\frac{dy}{dt} + ky = F_0 \cos \gamma t ,$$

where F_0 and γ are nonnegative constants and $0 < b^2 < 4mk$ (so the system is *underdamped*).

A solution to (1) has the form $y = y_h + y_p$, where y_p is a particular solution and y_h is a general solution to the corresponding homogeneous equation. We found in equation (17) of Section 4.9 that

$$(2) \qquad y_h(t) = Ae^{-(b/2m)t} \sin\left(\frac{\sqrt{4mk - b^2}}{2m}t + \phi\right),$$

where A and ϕ are constants.

To determine y_p, we can use the method of undetermined coefficients (Section 4.4). From the form of the nonhomogeneous term, we know that

$$(3) \qquad y_p(t) = A_1 \cos \gamma t + A_2 \sin \gamma t ,$$

where A_1 and A_2 are constants to be determined. Substituting this expression into equation (1) and simplifying gives

$$(4) \qquad \left[\left(k - m\gamma^2\right) A_1 + b\gamma A_2\right]\cos \gamma t + \left[\left(k - m\gamma^2\right)A_2 - b\gamma A_1\right]\sin \gamma t = F_0 \cos \gamma t .$$

Setting the corresponding coefficients on both sides equal, we have

$$(k - m\gamma^2)A_1 + b\gamma A_2 = F_0 ,$$
$$-b\gamma A_1 + (k - m\gamma^2)A_2 = 0 .$$

OK

.

Let me write.

Solving, we obtain

$$(5) \qquad A_1 = \frac{F_0(k - m\gamma^2)}{(k - m\gamma^2)^2 + b^2\gamma^2}, \qquad A_2 = \frac{F_0 b\gamma}{(k - m\gamma^2)^2 + b^2\gamma^2}.$$

Hence, a particular solution to (1) is

$$(6) \qquad y_p(t) = \frac{F_0}{(k - m\gamma^2)^2 + b^2\gamma^2}\left[(k - m\gamma^2)\cos\gamma t + b\gamma\sin\gamma t\right].$$

The expression in brackets can also be written as

$$\sqrt{(k - m\gamma^2)^2 + b^2\gamma^2}\,\sin(\gamma t + \theta),$$

so we can express y_p in the alternative form

$$(7) \qquad y_p(t) = \frac{F_0}{\sqrt{(k - m\gamma^2)^2 + b^2\gamma^2}}\sin(\gamma t + \theta),$$

where $\tan\theta = A_1/A_2 = (k - m\gamma^2)/(b\gamma)$ and the quadrant in which θ lies is determined by the signs of A_1 and A_2.

Combining equations (2) and (7), we have the following representation of a general solution to (1) in the case $0 < b^2 < 4mk$:

$$(8) \qquad y(t) = Ae^{-(b/2m)t}\sin\left(\frac{\sqrt{4mk - b^2}}{2m}t + \phi\right) + \frac{F_0}{\sqrt{(k - m\gamma^2)^2 + b^2\gamma^2}}\sin(\gamma t + \theta).$$

The solution (8) is the sum of two terms. The first term, y_h, represents damped oscillation and depends only on the parameters of the system and the initial conditions. Because of the damping factor $Ae^{-(b/2m)t}$, this term tends to zero as $t \to +\infty$. Consequently, it is referred to as the **transient** part of the solution. The second term, y_p, in (8) is the offspring of the external forcing function $f(t) = F_0\cos\gamma t$. Like the forcing function, y_p is a sinusoid with angular frequency γ. It is the synchronous solution that we anticipated in Section 4.1. However, y_p is out of phase with $f(t)$ (by the angle $\theta - \pi/2$), and its magnitude is different by the factor

$$(9) \qquad \frac{1}{\sqrt{(k - m\gamma^2)^2 + b^2\gamma^2}}.$$

As the transient term dies off, the motion of the mass–spring system becomes essentially that of the second term y_p (see Figure 4.32, page 223). Hence, this term is called the **steady-state** solution. The factor appearing in (9) is referred to as the **frequency gain,** or **gain factor,** since it represents the ratio of the magnitude of the sinusoidal response to that of the input force. Note that this factor depends on the frequency γ and has units of length/force.

Example 1 A 10-kg mass is attached to a spring with stiffness $k = 49$ N/m. At time $t = 0$, an external force $f(t) = 20\cos 4t$ N is applied to the system. The damping constant for the system is 3 N-sec/m. Determine the steady-state solution for the system.

Solution Substituting the given parameters into equation (1), we obtain

$$(10) \qquad 10\frac{d^2y}{dt^2} + 3\frac{dy}{dt} + 49y = 20\cos 4t,$$

where $y(t)$ is the displacement (from equilibrium) of the mass at time t.

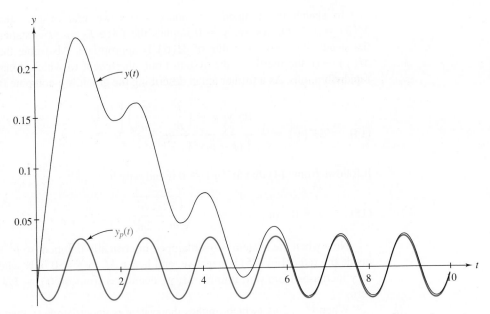

Figure 4.32 Convergence of $y(t)$ to the steady-state solution $y_p(t)$ when $m = 4, b = 6, k = 3, F_0 = 2, \gamma = 4$

To find the steady-state response, we must produce a particular solution to (10) that is a sinusoid. We can do this using the method of undetermined coefficients, guessing a solution of the form $A_1 \cos 4t + A_2 \sin 4t$. But this is precisely how we derived equation (7). Thus, we substitute directly into (7) and find

$$(11) \qquad y_p(t) = \frac{20}{\sqrt{(49 - 160)^2 + (9)(16)}} \sin(4t + \theta) \approx (0.18)\sin(4t + \theta),$$

where $\tan \theta = (49 - 160)/12 \approx -9.25$. Since the numerator, $(49 - 160)$, is negative and the denominator, 12, is positive, θ is a Quadrant IV angle. Thus,

$$\theta \approx \arctan(-9.25) \approx -1.46,$$

and the steady-state solution is given (approximately) by

$$(12) \qquad y_p(t) = (0.18)\sin(4t - 1.46). \quad \blacklozenge$$

The above example illustrates an important point made earlier: The steady-state response (12) to the sinusoidal forcing function $20 \cos 4t$ is a sinusoid of the same frequency but different amplitude. The gain factor [see (9)] in this case is $(0.18)/20 = 0.009$ m/N.

In general, the amplitude of the steady-state solution to equation (1) depends on the angular frequency γ of the forcing function and is given by $A(\gamma) = F_0 M(\gamma)$, where

$$(13) \qquad M(\gamma) := \frac{1}{\sqrt{(k - m\gamma^2)^2 + b^2\gamma^2}}$$

is the frequency gain [see (9)]. This formula is valid even when $b^2 \geq 4mk$. For a given system (m, b, and k fixed), it is often of interest to know how this system reacts to sinusoidal inputs of various frequencies (γ is a variable). For this purpose the graph of the gain $M(\gamma)$, called the **frequency response curve**, or **resonance curve**, for the system, is enlightening.

To sketch the frequency response curve, we first observe that for $\gamma = 0$ we find $M(0) = 1/k$. Of course, $\gamma = 0$ implies the force $F_0 \cos \gamma t$ is static; there is no motion in the steady state, so this value of $M(0)$ is appropriate. Also note that as $\gamma \to \infty$ the gain $M(\gamma) \to 0$; the inertia of the system limits the extent to which it can respond to extremely rapid vibrations. As a further aid in describing the graph, we compute from (13)

$$(14) \qquad M'(\gamma) = -\frac{2m^2\gamma\left[\gamma^2 - \left(\dfrac{k}{m} - \dfrac{b^2}{2m^2}\right)\right]}{\left[(k - m\gamma^2)^2 + b^2\gamma^2\right]^{3/2}} .$$

It follows from (14) that $M'(\gamma) = 0$ if and only if

$$(15) \qquad \gamma = 0 \quad \text{or} \quad \gamma = \gamma_r := \sqrt{\frac{k}{m} - \frac{b^2}{2m^2}} .$$

Now when the system is overdamped or critically damped, so $b^2 \geq 4mk > 2mk$, the term inside the radical in (15) is negative, and hence $M'(\gamma) = 0$ only when $\gamma = 0$. In this case, as γ increases from 0 to infinity, $M(\gamma)$ decreases from $M(0) = 1/k$ to a limiting value of zero.

When $b^2 < 2mk$ (which implies the system is underdamped), then γ_r is real and positive, and it is easy to verify that $M(\gamma)$ has a *maximum* at γ_r. Substituting γ_r into (13) gives

$$(16) \qquad M(\gamma_r) = \frac{1/b}{\sqrt{\dfrac{k}{m} - \dfrac{b^2}{4m^2}}} .$$

The value $\gamma_r/2\pi$ is called the **resonance frequency** for the system. When the system is stimulated by an external force at this frequency, it is said to be **at resonance.**

To illustrate the effect of the damping constant b on the resonance curve, we consider a system in which $m = k = 1$. In this case the frequency response curves are given by

$$(17) \qquad M(\gamma) = \frac{1}{\sqrt{(1 - \gamma^2)^2 + b^2\gamma^2}} ,$$

and, for $b < \sqrt{2}$, the resonance frequency is $\gamma_r/2\pi = (1/2\pi)\sqrt{1 - b^2/2}$. Figure 4.33 displays the graphs of these frequency response curves for $b = 1/4$, $1/2$, 1, $3/2$, and 2. Observe that as $b \to 0$ the maximum magnitude of the frequency gain increases and the resonance frequency $\gamma_r/2\pi$ for the damped system approaches $\sqrt{k/m}/2\pi = 1/2\pi$, the natural frequency for the undamped system.

To understand what is occurring, consider the undamped system $(b = 0)$ with forcing term $F_0 \cos \gamma t$. This system is governed by

$$(18) \qquad m\frac{d^2y}{dt^2} + ky = F_0 \cos \gamma t .$$

A general solution to (18) is the sum of a particular solution and a general solution to the homogeneous equation. In Section 4.9 we showed that the latter describes simple harmonic motion:

$$(19) \qquad y_h(t) = A \sin(\omega t + \phi) , \qquad \omega := \sqrt{k/m} .$$

The formula for the particular solution given in (7) is valid for $b = 0$, provided $\gamma \neq \omega = \sqrt{k/m}$. However, when $b = 0$ and $\gamma = \omega$, then the form we used with undetermined coefficients

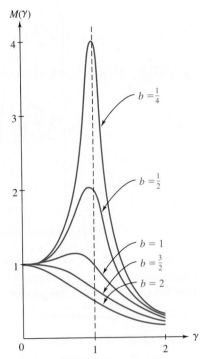

Figure 4.33 Frequency response curves for various values of b

to derive (7) does not work because $\cos \omega t$ and $\sin \omega t$ are solutions to the corresponding homogeneous equation. The correct form is

$$(20) \qquad y_p(t) = A_1 t \cos \omega t + A_2 t \sin \omega t \, ,$$

which leads to the solution

$$(21) \qquad y_p(t) = \frac{F_0}{2m\omega} t \sin \omega t \, .$$

[The verification of (21) is straightforward.] Hence, in the *undamped resonant case* (when $\gamma = \omega$), a general solution to (18) is

$$(22) \qquad y(t) = A \sin(\omega t + \phi) + \frac{F_0}{2m\omega} t \sin \omega t \, .$$

Returning to the question of resonance, observe that the particular solution in (21) oscillates between $\pm (F_0 t)/(2m\omega)$. Hence, as $t \to +\infty$ the maximum magnitude of (21) approaches ∞ (see Figure 4.34 on page 226).

It is obvious from the above discussion that **if the damping constant b is very small, the system is subject to large oscillations when the forcing function has a frequency near the resonance frequency for the system.** It is these large vibrations at resonance that concern engineers. Indeed, resonance vibrations have been known to cause airplane wings to snap, bridges to collapse,[†] and (less catastrophically) wine glasses to shatter.

[†]An interesting discussion of one such disaster, involving the Tacoma Narrows bridge in Washington State, can be found in *Differential Equations and Their Applications*, 4th ed., by M. Braun (Springer-Verlag, New York, 1993). See also the articles, "Large-Amplitude Periodic Oscillations in Suspension Bridges: Some New Connections with Nonlinear Analysis," by A. C. Lazer and P. J. McKenna, *SIAM Review*, Vol. 32 (1990): 537–578; or "Still Twisting," by Henry Petroski, *American Scientist*, Vol. 19 (1991): 398–401.

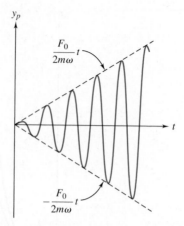

Figure 4.34 Undamped oscillation of the particular solution in (21)

When the mass–spring system is hung vertically as in Figure 4.35, the force of gravity must be taken into account. This is accomplished very easily. With y measured downward from the unstretched spring position, the governing equation is

$$my'' + by' + ky = mg \,,$$

and if the right-hand side is recognized as a sinusoidal forcing term with frequency zero $(mg \cos 0t)$, then the synchronous steady-state response is a constant, which is easily seen to be $y_p(t) = mg/k$. Now if we redefine $y(t)$ to be measured *from this (true) equilibrium level*, as indicated in Figure 4.35(c),

$$y_{\text{new}}(t) := y(t) - mg/k \,,$$

then the governing equation

$$my'' + by' + ky = mg + F_{\text{ext}}(t)$$

simplifies; we find

$$
\begin{aligned}
my''_{\text{new}} + by'_{\text{new}} + ky_{\text{new}} &= m(y - mg/k)'' + b(y - mg/k)' + k(y - mg/k) \\
&= my'' + by' + ky - mg \\
&= mg + F_{\text{ext}}(t) - mg \,,
\end{aligned}
$$

(a)

$y_p = mg/k$

(b)

mg/k

y_{new}

(c)

Figure 4.35 Spring (a) in natural position, (b) in equilibrium, and (c) in motion

or

(23) $my''_{\text{new}} + by'_{\text{new}} + ky_{\text{new}} = F_{\text{ext}}(t)$.

Thus, the gravitational force can be ignored if $y(t)$ is measured from the equilibrium position. Adopting this convention, we drop the "new" subscript hereafter.

Example 2 Suppose the mass–spring system in Example 1 is hung vertically. Find the steady-state solution.

Solution This is trivial; the steady-state solution is identical to what we derived before,

$$y_p(t) = \frac{20}{\sqrt{(49 - 160)^2 + (9)(16)}} \sin(4t + \theta)$$

[equation (11)], but now y_p is measured from the equilibrium position, which is $mg/k \approx 10 \times 9.8/49 = 2$ m below the unstretched spring position. ◆

Example 3 A 64-lb weight is attached to a vertical spring, causing it to stretch 3 in. upon coming to rest at equilibrium. The damping constant for the system is 3 lb-sec/ft. An external force $F(t) = 3 \cos 12t$ lb is applied to the weight. Find the steady-state solution for the system.

Solution If a weight of 64 lb stretches a spring by 3 in. (0.25 ft), then the spring stiffness must be $64/0.25 = 256$ lb/ft. Thus, if we measured the displacement from the (true) equilibrium level, equation (23) becomes

(24) $my'' + by' + ky = 3 \cos 12t$,

with $b = 3$ and $k = 256$. But recall that the unit of mass in the U.S. Customary System is the *slug*, which equals the weight divided by the gravitational acceleration constant $g \approx 32$ ft/sec^2 (Table 3.2, page 110). Therefore, m in equation (24) is $64/32 = 2$ slugs, and we have

$$2y'' + 3y' + 256y = 3 \cos 12t .$$

The steady-state solution is given by equation (6) with $F_0 = 3$ and $\gamma = 12$:

$$y_p(t) = \frac{3}{(256 - 2 \cdot 12^2)^2 + 3^2 \cdot 12^2} \left[(256 - 2 \cdot 12^2) \cos 12t + 3 \cdot 12 \sin 12t \right]$$

$$= \frac{3}{580} (-8 \cos 12t + 9 \sin 12t) . ◆$$

4.10 EXERCISES

In the following problems, take $g = 32$ ft/sec^2 for the U.S. Customary System and $g = 9.8$ m/sec^2 for the MKS system.

1. Sketch the frequency response curve (13) for the system in which $m = 4$, $k = 1$, $b = 2$.

2. Sketch the frequency response curve (13) for the system in which $m = 2$, $k = 3$, $b = 3$.

3. Determine the equation of motion for an undamped system at resonance governed by

$$\frac{d^2y}{dt^2} + 9y = 2 \cos 3t \ ;$$

$$y(0) = 1 , \quad y'(0) = 0 .$$

Sketch the solution.

4. Determine the equation of motion for an undamped system at resonance governed by

$$\frac{d^2y}{dt^2} + y = 5\cos t \; ;$$

$$y(0) = 0, \quad y'(0) = 1 .$$

Sketch the solution.

5. An undamped system is governed by

$$m\frac{d^2y}{dt^2} + ky = F_0\cos\gamma t \; ;$$

$$y(0) = y'(0) = 0,$$

where $\gamma \neq \omega := \sqrt{k/m}$.

(a) Find the equation of motion of the system.

(b) Use trigonometric identities to show that the solution can be written in the form

$$y(t) = \frac{2F_0}{m(\omega^2 - \gamma^2)}\sin\left(\frac{\omega + \gamma}{2}t\right)\sin\left(\frac{\omega - \gamma}{2}t\right).$$

(c) When γ is near ω, then $\omega - \gamma$ is small, while $\omega + \gamma$ is relatively large compared with $\omega - \gamma$. Hence, $y(t)$ can be viewed as the product of a slowly varying sine function, $\sin[(\omega - \gamma)t/2]$, and a rapidly varying sine function, $\sin[(\omega + \gamma)t/2]$. The net effect is a sine function $y(t)$ with frequency $(\omega - \gamma)/4\pi$, which serves as the time-varying amplitude of a sine function with frequency $(\omega + \gamma)/4\pi$. This vibration phenomenon is referred to as **beats** and is used in tuning stringed instruments. This same phenomenon in electronics is called **amplitude modulation.** To illustrate this phenomenon, sketch the curve $y(t)$ for $F_0 = 32, m = 2, \omega = 9$, and $\gamma = 7$.

6. Derive the formula for $y_p(t)$ given in (21).

7. Shock absorbers in automobiles and aircraft can be described as forced *overdamped* mass–spring systems. Derive an expression analogous to equation (8) for the general solution to the differential equation (1) when $b^2 > 4mk$.

8. The response of an overdamped system to a constant force is governed by equation (1) with $m = 2, b = 8, k = 6$, $F_0 = 18$, and $\gamma = 0$. If the system starts from rest $[y(0) = y'(0) = 0]$, compute and sketch the displacement $y(t)$. What is the limiting value of $y(t)$ as $t \to +\infty$? Interpret this physically.

9. An 8-kg mass is attached to a spring hanging from the ceiling, thereby causing the spring to stretch 1.96 m upon coming to rest at equilibrium. At time $t = 0$, an external force $F(t) = \cos 2t$ N is applied to the system. The damping constant for the system is 3 N-sec/m. Determine the steady-state solution for the system.

10. Show that the period of the simple harmonic motion of a mass hanging from a spring is $2\pi\sqrt{l/g}$, where l denotes

the amount (beyond its natural length) that the spring is stretched when the mass is at equilibrium.

11. A mass weighing 8 lb is attached to a spring hanging from the ceiling and comes to rest at its equilibrium position. At $t = 0$, an external force $F(t) = 2\cos 2t$ lb is applied to the system. If the spring constant is 10 lb/ft and the damping constant is 1 lb-sec/ft, find the equation of motion of the mass. What is the resonance frequency for the system?

12. A 2-kg mass is attached to a spring hanging from the ceiling, thereby causing the spring to stretch 20 cm upon coming to rest at equilibrium. At time $t = 0$, the mass is displaced 5 cm below the equilibrium position and released. At this same instant, an external force $F(t) = 0.3\cos t$ N is applied to the system. If the damping constant for the system is 5 N-sec/m, determine the equation of motion for the mass. What is the resonance frequency for the system?

13. A mass weighing 32 lb is attached to a spring hanging from the ceiling and comes to rest at its equilibrium position. At time $t = 0$, an external force $F(t) = 3\cos 4t$ lb is applied to the system. If the spring constant is 5 lb/ft and the damping constant is 2 lb-sec/ft, find the steady-state solution for the system.

14. An 8-kg mass is attached to a spring hanging from the ceiling and allowed to come to rest. Assume that the spring constant is 40 N/m and the damping constant is 3 N-sec/m. At time $t = 0$, an external force of $2\sin(2t + \pi/4)$ N is applied to the system. Determine the amplitude and frequency of the steady-state solution.

15. An 8-kg mass is attached to a spring hanging from the ceiling and allowed to come to rest. Assume that the spring constant is 40 N/m and the damping constant is 3 N-sec/m. At time $t = 0$, an external force of $2\sin 2t\cos 2t$ N is applied to the system. Determine the amplitude and frequency of the steady-state solution.

16. A helium-filled balloon on a cord, hanging y km above a level surface, is subjected to three forces:

(i) the (constant) buoyant force B exerted by the external air pressure;

(ii) the weight of the balloon mg, where m is the mass of the balloon;

(iii) the weight of the part of the cord that has been lifted off the surface, dgy, where d is the linear density (kg/m) of the cord. (We assume that the cord is longer than the elevation of the balloon.)

Express Newton's second law for the balloon and show that it is exactly analogous to the governing equation for a mass vertically hung from a spring (page 226). Find the equation of motion. What is the frequency of the balloon's oscillations?

Chapter **4** Summary

In this chapter we discussed the theory of second-order linear differential equations and presented explicit solution methods for equations with constant coefficients. Much of the methodology can also be applied to the more general case of variable coefficients. We also studied the mathematical description of vibrating mechanical systems, and we saw how the mass–spring analogy could be used to predict qualitative features of solutions to some variable-coefficient and nonlinear equations.

The important features and solution techniques for the constant-coefficient case are listed below.

Homogeneous Linear Equations (Constant Coefficients)

$$ay'' + by' + cy = 0, \quad a(\neq 0), b, c \text{ constants.}$$

Linearly Independent Solutions: y_1, y_2. Two solutions y_1 and y_2 to the homogeneous equation on the interval I are said to be linearly independent on I if neither function is a constant times the other on I. This will be true provided their Wronskian,

$$W[y_1, y_2](t) := y_1(t)y_2'(t) - y_1'(t)y_2(t) ,$$

is different from zero for some (and hence all) t in I.

General Solution to Homogeneous Equation: $c_1 y_1 + c_2 y_2$. If y_1 and y_2 are linearly independent solutions to the homogeneous equation, then a general solution is

$$y(t) = c_1 y_1(t) + c_2 y_2(t) ,$$

where c_1 and c_2 are arbitrary constants.

Form of General Solution. The form of a general solution for a homogeneous equation with constant coefficients depends on the roots

$$r_1 = \frac{-b + \sqrt{b^2 - 4ac}}{2a} , \qquad r_2 = \frac{-b - \sqrt{b^2 - 4ac}}{2a}$$

of the auxiliary equation

$$ar^2 + br + c = 0 , \qquad a \neq 0 .$$

(a) When $b^2 - 4ac > 0$, the auxiliary equation has two distinct real roots r_1 and r_2 and a general solution is

$$y(t) = c_1 e^{r_1 t} + c_2 e^{r_2 t} .$$

(b) When $b^2 - 4ac = 0$, the auxiliary equation has a repeated real root $r = r_1 = r_2$ and a general solution is

$$y(t) = c_1 e^{rt} + c_2 t e^{rt} .$$

(c) When $b^2 - 4ac < 0$, the auxiliary equation has complex conjugate roots $r = \alpha \pm i\beta$ and a general solution is

$$y(t) = c_1 e^{\alpha t} \cos \beta t + c_2 e^{\alpha t} \sin \beta t .$$

Nonhomogeneous Linear Equations (Constant Coefficients)

$$ay'' + by' + cy = f(t)$$

General Solution to Nonhomogeneous Equation: $y_p + c_1y_1 + c_2y_2$. If y_p is any particular solution to the nonhomogeneous equation and y_1 and y_2 are linearly independent solutions to the corresponding homogeneous equation, then a general solution is

$$y(t) = y_p(t) + c_1y_1(t) + c_2y_2(t),$$

where c_1 and c_2 are arbitrary constants.

Two methods for finding a particular solution y_p are those of undetermined coefficients and variation of parameters.

Undetermined Coefficients: $f(t) = p_n(t)e^{\alpha t}\begin{Bmatrix} \cos \beta t \\ \sin \beta t \end{Bmatrix}$. If the right-hand side $f(t)$ of a nonhomogeneous equation with constant coefficients is a polynomial $p_n(t)$, an exponential of the form $e^{\alpha t}$, a trigonometric function of the form $\cos \beta t$ or $\sin \beta t$, or any product of these special types of functions, then a particular solution of an appropriate form can be found. The form of the particular solution involves unknown coefficients and depends on whether $\alpha + i\beta$ is a root of the corresponding auxiliary equation. See the summary box on page 184. The unknown coefficients are found by substituting the form into the differential equation and equating coefficients of like terms.

Variation of Parameters: $y(t) = v_1(t)y_1(t) + v_2(t)y_2(t)$. If y_1 and y_2 are two linearly independent solutions to the corresponding homogeneous equation, then a particular solution to the nonhomogeneous equation is

$$y(t) = v_1(t)y_1(t) + v_2(t)y_2(t),$$

where v_1' and v_2' are determined by the equations

$$v_1'y_1 + v_2'y_2 = 0$$
$$v_1'y_1' + v_2'y_2' = f(t)/a .$$

Superposition Principle. If y_1 and y_2 are solutions to the equations

$$ay'' + by' + cy = f_1 \quad \text{and} \quad ay'' + by' + cy = f_2,$$

respectively, then $k_1y_1 + k_2y_2$ is a solution to the equation

$$ay'' + by' + cy = k_1 f_1 + k_2 f_2 .$$

The superposition principle facilitates finding a particular solution when the nonhomogeneous term is the sum of nonhomogeneities for which particular solutions can be determined.

Cauchy–Euler (Equidimensional) Equations

$$at^2y'' + bty' + cy = f(t)$$

Substituting $y = t^r$ yields the associated characteristic equation

$$ar^2 + (b - a)r + c = 0$$

for the corresponding *homogeneous* Cauchy–Euler equation. A general solution to the homogeneous equation for $t > 0$ is given by

(i) $c_1t^{r_1} + c_2t^{r_2}$, if r_1 and r_2 are distinct real roots;

(ii) $c_1t^r + c_2t^r \ln t$, if r is a repeated root;

(iii) $c_1t^{\alpha} \cos(\beta \ln t) + c_2t^{\alpha} \sin(\beta \ln t)$, if $\alpha + i\beta$ is a complex root.

A general solution to the nonhomogeneous equation is $y = y_p + y_h$, where y_p is a particular solution and y_h is a general solution to the corresponding homogeneous equation. The method of variation of parameters (but not the method of undetermined coefficients) can be used to find a particular solution.

REVIEW PROBLEMS FOR CHAPTER 4

In Problems 1–28, find a general solution to the given differential equation.

1. $y'' + 8y' - 9y = 0$
2. $49y'' + 14y' + y = 0$
3. $4y'' - 4y' + 10y = 0$
4. $9y'' - 30y' + 25y = 0$
5. $6y'' - 11y' + 3y = 0$
6. $y'' + 8y' - 14y = 0$
7. $36y'' + 24y' + 5y = 0$
8. $25y'' + 20y' + 4y = 0$
9. $16z'' - 56z' + 49z = 0$
10. $u'' + 11u = 0$
11. $t^2 x''(t) + 5x(t) = 0$, $\quad t > 0$
12. $2y''' - 3y'' - 12y' + 20y = 0$
13. $y'' + 16y = te^t$
14. $v'' - 4v' + 7v = 0$
15. $3y''' + 10y'' + 9y' + 2y = 0$
16. $y''' + 3y'' + 5y' + 3y = 0$
17. $y''' + 10y' - 11y = 0$
18. $y^{(4)} = 120t$
19. $4y''' + 8y'' - 11y' + 3y = 0$
20. $2y'' - y = t \sin t$
21. $y'' - 3y' + 7y = 7t^2 - e^t$
22. $y'' - 8y' - 33y = 546 \sin t$
23. $y''(\theta) + 16y(\theta) = \tan 4\theta$
24. $10y'' + y' - 3y = t - e^{t/2}$
25. $4y'' - 12y' + 9y = e^{5t} + e^{3t}$
26. $y'' + 6y' + 15y = e^{2t} + 75$
27. $x^2 y'' + 2xy' - 2y = 6x^{-2} + 3x$, $\quad x > 0$
28. $y'' = 5x^{-1}y' - 13x^{-2}y$, $\quad x > 0$

In Problems 29–36, find the solution to the given initial value problem.

29. $y'' + 4y' + 7y = 0$;
$y(0) = 1$, $\quad y'(0) = -2$
30. $y''(\theta) + 2y'(\theta) + y(\theta) = 2 \cos \theta$;
$y(0) = 3$, $\quad y'(0) = 0$
31. $y'' - 2y' + 10y = 6 \cos 3t - \sin 3t$;
$y(0) = 2$, $\quad y'(0) = -8$

32. $4y'' - 4y' + 5y = 0$;
$y(0) = 1$, $\quad y'(0) = -11/2$
33. $y''' - 12y'' + 27y' + 40y = 0$;
$y(0) = -3$, $\quad y'(0) = -6$, $\quad y''(0) = -12$
34. $y'' + 5y' - 14y = 0$;
$y(0) = 5$, $\quad y'(0) = 1$
35. $y''(\theta) + y(\theta) = \sec \theta$; $\quad y(0) = 1$, $\quad y'(0) = 2$
36. $9y'' + 12y' + 4y = 0$;
$y(0) = -3$, $\quad y'(0) = 3$

37. Use the mass–spring oscillator analogy to decide whether all solutions to each of the following differential equations are bounded as $t \to +\infty$.
 (a) $y'' + t^4 y = 0$
 (b) $y'' - t^4 y = 0$
 (c) $y'' + y^7 = 0$
 (d) $y'' + y^8 = 0$
 (e) $y'' + (3 + \sin t)y = 0$
 (f) $y'' + t^2 y' + y = 0$
 (g) $y'' - t^2 y' - y = 0$

38. A 3-kg mass is attached to a spring with stiffness $k = 75$ N/m, as in Figure 4.1, page 152. The mass is displaced $1/4$ m to the left and given a velocity of 1 m/sec to the right. The damping force is negligible. Find the equation of motion of the mass along with the amplitude, period, and frequency. How long after release does the mass pass through the equilibrium position?

39. A 32-lb weight is attached to a vertical spring, causing it to stretch 6 in. upon coming to rest at equilibrium. The damping constant for the system is 2 lb-sec/ft. An external force $F(t) = 4 \cos 8t$ lb is applied to the weight. Find the steady-state solution for the system. What is its resonant frequency?

TECHNICAL WRITING EXERCISES FOR CHAPTER 4

1. Compare the two methods—undetermined coefficients and variation of parameters—for determining a particular solution to a nonhomogeneous equation. What are the advantages and disadvantages of each?

2. Consider the differential equation

$$\frac{d^2y}{dx^2} + 2b\frac{dy}{dx} + y = 0,$$

where b is a constant. Describe how the behavior of solutions to this equation changes as b varies.

3. Consider the differential equation

$$\frac{d^2y}{dx^2} + cy = 0,$$

where c is a constant. Describe how the behavior of solutions to this equation changes as c varies.

4. For students with a background in linear algebra: Compare the theory for linear second-order equations with that for systems of n linear equations in n unknowns whose coefficient matrix has rank $n - 2$. Use the terminology from linear algebra; for example, subspace, basis, dimension, linear transformation, and kernel. Discuss both homogeneous and nonhomogeneous equations.

A Nonlinear Equations Solvable by First-Order Techniques

Certain *nonlinear* second-order equations—namely, those with dependent or independent variables missing—can be solved by reducing them to a pair of *first-order equations*. This is accomplished by making the substitution $w = dy/dx$, where x is the independent variable.

(a) To solve an equation of the form $y'' = F(x, y')$ in which the dependent variable y is missing, setting $w = y'$ (so that $w' = y''$) yields the pair of equations

$$w' = F(x, w) \,,$$

$$y' = w \,.$$

Because $w' = F(x, w)$ is a first-order equation, we have available the techniques of Chapter 2 to solve it for $w(x)$. Once $w(x)$ is determined, we integrate it to obtain $y(x)$.

Using this method, solve

$$2xy'' - y' + \frac{1}{y'} = 0 \,, \qquad x > 0 \,.$$

(b) To solve an equation of the form $y'' = F(y, y')$ in which the independent variable x is missing, setting $w = dy/dx$ yields, via the chain rule,

$$\frac{d^2y}{dx^2} = \frac{dw}{dx} = \frac{dw}{dy}\frac{dy}{dx} = w\frac{dw}{dy} \,.$$

Thus, $y'' = F(y, y')$ is equivalent to the pair of equations

(1) $\qquad w\dfrac{dw}{dy} = F(y, w) \,,$

(2) $\qquad \dfrac{dy}{dx} = w \,.$

In equation (1) notice that y plays the role of the *independent* variable; hence, solving it yields $w(y)$. Then substituting $w(y)$ into (2), we obtain a separable equation that determines $y(x)$.

Using this method, solve the following equations:

(i) $\quad 2y\dfrac{d^2y}{dx^2} = 1 + \left(\dfrac{dy}{dx}\right)^2 .$ **(ii)** $\dfrac{d^2y}{dx^2} + y\dfrac{dy}{dx} = 0 \,.$

(c) Suspended Cable. In the study of a cable suspended between two fixed points (see Figure 4.36 on page 234), one encounters the initial value problem

$$\frac{d^2y}{dx^2} = \frac{1}{a}\sqrt{1 + \left(\frac{dy}{dx}\right)^2} \,; \qquad y(0) = a \,, \qquad y'(0) = 0 \,,$$

where a ($\neq 0$) is a constant. Solve this initial value problem for y. The resulting curve is called a **catenary**.

Figure 4.36 Suspended cable

B *Apollo* Reentry

Courtesy of Alar Toomre, Massachusetts Institute of Technology

Each time the Apollo astronauts returned from the moon circa 1970, they took great care to reenter Earth's atmosphere along a path that was only a small angle α from the horizontal. (See Figure 4.37.) This was necessary in order to avoid intolerably large "g" forces during their reentry.

To appreciate their grounds for concern, consider the idealized problem

$$\frac{d^2s}{dt^2} = -Ke^{s/H}\left(\frac{ds}{dt}\right)^2 ,$$

where K and H are constants and distance s is measured downrange from some reference point on the trajectory, as shown in the figure. This approximate equation pretends that the only force on the capsule during reentry is air drag. For a bluff body such as the *Apollo,* drag is proportional to the square of the speed and to the local atmospheric density, which falls off exponentially with height. Intuitively, one might expect that the deceleration predicted by this model would depend heavily on the constant K (which takes into account the vehicle's mass, area, etc.); but, remarkably, for capsules entering the atmosphere (at "$s = -\infty$") with a common speed V_0, the *maximum* deceleration turns out to be independent of K.

(a) Verify this last assertion by demonstrating that this maximum deceleration is just $V_0^2/(2eH)$. [*Hint:* The independent variable t does not appear in the differential equation, so it is helpful to make the substitution $v = ds/dt$; see Project A, part (b).]

(b) Also verify that any such spacecraft at the instant when it is decelerating most fiercely will be traveling exactly with speed $V_0/\sqrt{e}$, having by then lost almost 40% of its original velocity.

(c) Using the plausible data $V_0 = 11$ km/sec and $H = 10/(\sin\alpha)$ km, estimate how small α had to be chosen so as to inconvenience the returning travelers with no more than 10 g's.

Figure 4.37 Reentry path

C Simple Pendulum

In Section 4.8, we discussed the simple pendulum consisting of a mass m suspended by a rod of length ℓ having negligible mass and derived the nonlinear initial value problem

(3)
$$\frac{d^2\theta}{dt^2} + \frac{g}{\ell}\sin\theta = 0; \quad \theta(0) = \alpha, \quad \theta'(0) = 0,$$

where g is the acceleration due to gravity and $\theta(t)$ is the angle the rod makes with the vertical at time t (see Figure 4.18, page 208). Here it is assumed that the mass is released with zero velocity at an initial angle α, $0 < \alpha < \pi$. We would like to determine the equation of motion for the pendulum and its period of oscillation.

(a) Use equation (3) and the energy integral lemma discussed in Section 4.8 to show that

$$\left(\frac{d\theta}{dt}\right)^2 = \frac{2g}{\ell}(\cos\theta - \cos\alpha)$$

and hence

$$dt = -\sqrt{\frac{\ell}{2g}}\,\frac{d\theta}{\sqrt{\cos\theta - \cos\alpha}}.$$

(b) Use the trigonometric identity $\cos x = 1 - 2\sin^2(x/2)$ to express dt by

$$dt = -\frac{1}{2}\sqrt{\frac{\ell}{g}}\,\frac{d\theta}{\sqrt{\sin^2(\alpha/2) - \sin^2(\theta/2)}}.$$

(c) Make the change of variables $\sin(\theta/2) = \sin(\alpha/2)\sin\phi$ and show that the elapsed time, T, for the pendulum to fall from the angle $\theta = \alpha$ (corresponding to $\phi = \pi/2$) to the angle $\theta = \beta$ (corresponding to $\phi = \Phi$), when $\alpha \geq \beta \geq 0$, is given by

(4)
$$T = \int_0^T dt = -\int_\alpha^\beta \frac{1}{2}\sqrt{\frac{\ell}{g}}\,\frac{d\theta}{\sqrt{\sin^2(\alpha/2) - \sin^2(\theta/2)}} = -\sqrt{\frac{\ell}{g}}\int_{\pi/2}^\Phi \frac{d\phi}{\sqrt{1 - k^2\sin^2\phi}},$$

where $k := \sin(\alpha/2)$.

(d) The period P of the pendulum is defined to be the time required for it to swing from one extreme to the other and back—that is, from α to $-\alpha$ and back to α. Show that the period is given by

(5)
$$P = 4\sqrt{\frac{\ell}{g}}\int_0^{\pi/2} \frac{d\phi}{\sqrt{1 - k^2\sin^2\phi}}.$$

The integral in (5) is called an **elliptic integral of the first kind** and is denoted by $F(k, \pi/2)$. As you might expect, the period of the simple pendulum depends on the length ℓ of the rod and the initial displacement α. In fact, a check of an elliptic integral table will show that the period nearly doubles as the initial displacement increases from $\pi/8$ to $15\pi/16$ (for fixed ℓ). What happens as α approaches π?

(e) From equation (5) show that

(6)
$$-T + P/4 = \sqrt{\frac{\ell}{g}}\,F(k, \Phi), \text{ where } F(k, \Phi) := \int_{\phi=0}^\Phi \frac{d\phi}{\sqrt{1 - k^2\sin^2\phi}}.$$

For fixed k, $F(k, \Phi)$ has an "inverse," denoted by $\text{sn}(k, u)$, that satisfies $u = F(k, \Phi)$ if and only if $\text{sn}(k, u) = \sin \Phi$. The function $\text{sn}(k, u)$ is called a **Jacobi elliptic function** and has many properties that resemble those of the sine function. Using the Jacobi elliptic function $\text{sn}(k, u)$, express the equation of motion for the pendulum in the form

$$(7) \qquad \beta = 2 \arcsin \left\{ k \, \text{sn}\left[k, \sqrt{\frac{g}{\ell}}(-T + P/4) \right] \right\}, \quad 0 \le T \le P/4 \,.$$

(f) Take $\ell = 1$ m, $g = 9.8$ m/sec, $\alpha = \pi/4$ radians. Use Runge–Kutta algorithms or tabulated values of the Jacobi elliptic function to determine the period of the pendulum.

D Linearization of Nonlinear Problems

A useful approach to analyzing a nonlinear equation is to study its **linearized equation,** which is obtained by replacing the nonlinear terms by linear approximations. For example, the nonlinear equation

$$(8) \qquad \frac{d^2\theta}{dt^2} + \sin \theta = 0 \,,$$

which governs the motion of a simple pendulum, has

$$(9) \qquad \frac{d^2\theta}{dt^2} + \theta = 0$$

as a linearization for small θ. (The nonlinear term $\sin \theta$ has been replaced by the linear approximation θ.)

A general solution to equation (8) involves Jacobi elliptic functions (see Project C), which are rather complicated, so let's try to approximate the solutions. For this purpose we consider two methods: Taylor series and linearization.

(a) Derive the first six terms of the Taylor series about $t = 0$ of the solution to equation (8) with initial conditions $\theta(0) = \pi/12$, $\theta'(0) = 0$. (The Taylor series method is discussed in Project D of Chapter 1 and Section 8.1.)

(b) Solve equation (9) subject to the same initial conditions $\theta(0) = \pi/12$, $\theta'(0) = 0$.

(c) On the same coordinate axes, graph the two approximations found in parts (a) and (b).

(d) Discuss the advantages and disadvantages of the Taylor series method and the linearization method.

(e) Give a linearization for the initial value problem.

$$x''(t) + 0.1[1 - x^2(t)]x'(t) + x(t) = 0 \quad x(0) = 0.4 \,, \quad x'(0) = 0 \,,$$

for x small. Solve this linearized problem to obtain an approximation for the nonlinear problem.

E Convolution Method

The **convolution** of two functions g and f is the function $g * f$ defined by

$$(g * f)(t) := \int_0^t g(t - v)f(v)\,dv.$$

The aim of this project is to show how convolutions can be used to obtain a particular solution to a nonhomogeneous equation of the form

(10) $ay'' + by' + cy = f(t)$, where a, b, and c are constants, $a \neq 0$.

(a) Use Leibniz's rule,

$$\frac{d}{dt}\int_a^t h(t, v)\,dv = \int_a^t \frac{\partial h}{\partial t}(t, v)\,dv + h(t, t),$$

to show the following:

$$(y * f)'(t) = (y' * f)(t) + y(0)f(t)$$
$$(y * f)''(t) = (y'' * f)(t) + y'(0)f(t) + y(0)f'(t),$$

assuming y and f are sufficiently differentiable.

(b) Let $y_s(t)$ be the solution to the homogeneous equation $ay'' + by' + cy = 0$ that satisfies $y_s(0) = 0$, $y_s'(0) = 1/a$. Show that $y_s * f$ is the particular solution to equation (10) satisfying $y(0) = y'(0) = 0$.

(c) Let $y_k(t)$ be the solution to the homogeneous equation $ay'' + by' + cy = 0$ that satisfies $y(0) = Y_0$, $y'(0) = Y_1$, and let y_s be as defined in part (b). Show that

$$(y_s * f)(t) + y_k(t)$$

is the unique solution to the initial value problem

(11) $ay'' + by' + cy = f(t)$; $y(0) = Y_0$, $y'(0) = Y_1$.

(d) Use the result of part (c) to determine the solution to each of the following initial value problems. Carry out all integrations and express your answers in terms of elementary functions.

(i) $y'' + y = \tan t$; $y(0) = 0$, $y'(0) = -1$

(ii) $2y'' + y' - y = e^{-t}\sin t$; $y(0) = 1$, $y'(0) = 1$

(iii) $y'' - 2y' + y = \sqrt{t}\,e^t$; $y(0) = 2$, $y'(0) = 0$

F Undetermined Coefficients Using Complex Arithmetic

The technique of undetermined coefficients described in Section 4.5 can be streamlined with the aid of complex arithmetic and the properties of the complex exponential function. The essential formulas are

$$e^{(\alpha+i\beta)t} = e^{\alpha t}(\cos\beta t + i\sin\beta t),\qquad \frac{d}{dt}e^{(\alpha+i\beta)t} = (\alpha + i\beta)e^{(\alpha+i\beta)t},$$

$$\mathrm{Re}\,e^{(\alpha+i\beta)t} = e^{\alpha t}\cos\beta t,\qquad \mathrm{Im}\,e^{(\alpha+i\beta)t} = e^{\alpha t}\sin\beta t.$$

(a) From the preceding formulas derive the equations

(12) $\qquad \text{Re}\left[(a + ib)e^{(\alpha+i\beta)t} \right] = e^{\alpha t}(a \cos\beta t - b \sin\beta t)\,,$

(13) $\qquad \text{Im}\left[(a + ib)e^{(\alpha+i\beta)t} \right] = e^{\alpha t}(b \cos\beta t + a \sin\beta t)\,.$

Now consider a second-order equation of the form

(14) $\qquad L[y] := ay'' + by' + cy = g\,,$

where a, b, and c are real numbers and g is of the special form

(15) $\qquad g(t) = e^{\alpha t}\left[(a_n t^n + \cdots + a_1 t + a_0) \cos\beta t + (b_n t^n + \cdots + b_1 t + b_0) \sin\beta t \right]\,,$

with the a_j's, b_j's, α, and β real numbers. Such a function can always be expressed as the real or imaginary part of a function involving the complex exponential. For example, using equation (12), one can quickly check that

(16) $\qquad g(t) = \text{Re}\left[G(t) \right]\,,$

where

(17) $\qquad G(t) = e^{(\alpha - i\beta)t}\left[(a_n + ib_n)t^n + \cdots + (a_1 + ib_1)t + (a_0 + ib_0) \right]\,.$

Now suppose for the moment that we can find a complex-valued solution Y to the equation

(18) $\qquad L[Y] = aY'' + bY' + cY = G\,.$

Then, since a, b, and c are real numbers, we get a real-valued solution y to (14) by simply taking the real part of Y; that is, $y = \text{Re}\,Y$ solves (14). (Recall that in Lemma 2, page 167, we proved this fact for homogeneous equations.) Thus, we need focus only on finding a solution to (18).

The method of undetermined coefficients implies that any differential equation of the form

(19) $\qquad L[Y] = e^{(\alpha \pm i\beta)t}\left[(a_n + ib_n)t^n + \cdots + (a_1 + ib_1)t + (a_0 + ib_0) \right]$

has a solution of the form

(20) $\qquad Y_p(t) = t^s e^{(\alpha \pm i\beta)t}\left[A_n t^n + \cdots + A_1 t + A_0 \right]\,,$

where $A_n, \ldots, A_0$ are complex constants and s is the multiplicity of $\alpha + i\beta$ as a root of the auxiliary equation for the corresponding homogeneous equation $L[Y] = 0$. We can solve for the unknown constants A_j by substituting (20) into (19) and equating coefficients of like terms. With these facts in mind, we can (for the small price of using complex arithmetic) dispense with the methods of Section 4.5 and avoid the unpleasant task of computing derivatives of a function like $e^{3t}(2 + 3t + t^2)\sin(2t)$, which involves both exponential and trigonometric factors.

Carry out this procedure to determine particular solutions to the following equations:

(b) $y'' - y' - 2y = \cos t - \sin 2t$.
(c) $y'' + y = e^{-t}(\cos 2t - 3 \sin 2t)$.
(d) $y'' - 2y' + 10y = te^t \sin 3t$.

The use of complex arithmetic not only streamlines the computations but also proves very useful in analyzing the response of a linear system to a sinusoidal input. Electrical engineers make good use of this in their study of *RLC* circuits by introducing the concept of **impedance.**

G Asymptotic Behavior of Solutions

In the application of linear systems theory to mechanical problems, we have encountered the equation

(21) $y'' + py' + qy = f(t)$,

where p and q are positive constants with $p^2 < 4q$ and $f(t)$ is a forcing function for the system. In many cases it is important for the design engineer to know that a bounded forcing function gives rise only to bounded solutions. More specifically, how does the behavior of $f(t)$ for large values of t affect the asymptotic behavior of the solution? To answer this question, do the following:

(a) Show that the homogeneous equation associated with equation (21) has two linearly independent solutions given by

$$e^{\alpha t} \cos \beta t , \qquad e^{\alpha t} \sin \beta t ,$$

where $\alpha = -p/2 < 0$ and $\beta = \frac{1}{2}\sqrt{4q - p^2}$.

(b) Let $f(t)$ be a continuous function defined on the interval $[0, \infty)$. Use the variation of parameters formula to show that any solution to (21) on $[0, \infty)$ can be expressed in the form

$$(22) \qquad y(t) = c_1 e^{\alpha t} \cos \beta t + c_2 e^{\alpha t} \sin \beta t$$

$$-\frac{1}{\beta} e^{\alpha t} \cos \beta t \int_0^t f(v) e^{-\alpha v} \sin \beta v \, dv$$

$$+\frac{1}{\beta} e^{\alpha t} \sin \beta t \int_0^t f(v) e^{-\alpha v} \cos \beta v \, dv .$$

(c) Assuming that f is bounded on $[0, \infty)$ (that is, there exists a constant K such that $|f(v)| \le K$ for all $v \ge 0$), use the triangle inequality and other properties of the absolute value to show that $y(t)$ given in (22) satisfies

$$|y(t)| \le (|c_1| + |c_2|) e^{\alpha t} + \frac{2K}{|\alpha|\beta} (1 - e^{\alpha t})$$

for all $t > 0$.

(d) In a similar fashion, show that if $f_1(t)$ and $f_2(t)$ are two bounded continuous functions on $[0, \infty)$ such that $|f_1(t) - f_2(t)| \le \varepsilon$ for all $t > t_0$, and if ϕ_1 is a solution to (21) with $f = f_1$ and ϕ_2 is a solution to (21) with $f = f_2$, then

$$|\phi_1(t) - \phi_2(t)| \le Me^{\alpha t} + \frac{2\varepsilon}{|\alpha|\beta} (1 - e^{\alpha(t-t_0)})$$

for all $t > t_0$, where M is a constant that depends on ϕ_1 and ϕ_2 but not on t.

(e) Now assume $f(t) \to F_0$ as $t \to +\infty$, where F_0 is a constant. Use the result of part (d) to prove that any solution ϕ to (21) must satisfy $\phi(t) \to F_0/q$ as $t \to +\infty$. [*Hint:* Choose $f_1 = f$, $f_2 \equiv F_0$, $\phi_1 = \phi$, $\phi_2 \equiv F_0/q$.]

H GRAVITY TRAIN[†]

Legend has it that Robert Hooke proposed to Isaac Newton that a straight tunnel be drilled from Moscow to St. Petersburg to accommodate a "gravity train" serving the two cities. Upon departure gravity would accelerate the train toward the center of the tunnel, then decelerate it as it continued toward the terminal. It is amusing to investigate the mechanics of such a conveyance, under idealized conditions.

As depicted in Figure 4.38, x denotes the distance of the train from the center of the tunnel, R denotes its distance from the center of the earth, and mg denotes the pull of gravity.

(a) Show that Newton's law yields the differential equation

$$(23) \qquad mx'' = -mgx/R$$

if friction is neglected.

(b) Assume that the tunnel lies so close to the surface of the earth that it is reasonable to regard g and R as constants—in which case (23) is analogous to the undamped mass–spring equation (2) on page 212 of Section 4.9, with mg/R as the effective "spring constant" k. Take $g = 9.8$ km/s^2 and $R = 6400$ km. The overland (surface) distance from Moscow to St. Petersburg is about 650 km. Find the equation of motion for the train departing at rest from Moscow at time $t = 0$.

(c) How long will it take the train to get to St. Petersburg?

(d) The American mathematician/entertainer Tom Lehrer composed a humorous song[‡] about Nicolai Ivanovich Lobachevsky, in which a plagiarized dissertation travels from Dnepropetrovsk to Petropavlovsk (16,000 km) to Iliysk (8,500 km) to Novorossiysk (17,000 km) to Alexandrovsk (15,000 km) to Akmolinsk (1,000 km) to Tomsk (1,300 km) to Omsk (900 km) to Pinsk (21,000 km) to Minsk (3,000 km) to Moscow (5,000 km). If gravity trains were to connect all these cities, what would be unusual about their timetables?

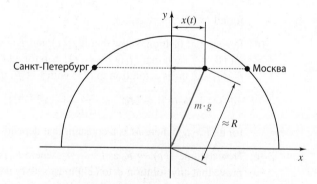

Figure 4.38 Force diagram of gravity train

[†]*Suggested by Valery Ochkov, National Research University Moscow; and Katarina Pisacic, Sveucilisni Centar Varazdin.*

[‡]As we go to press the song is archived at https://en.wikipedia.org/wiki/Lobachevsky_%28song%29, and Lehrer's performance can be viewed at https://www.youtube.com/watch?v=UQHaGhC7C2E.

Introduction to Systems and Phase Plane Analysis

5.1 Interconnected Fluid Tanks

Two large tanks, each holding 24 liters of a brine solution, are interconnected by pipes as shown in Figure 5.1. Fresh water flows into tank A at a rate of 6 L/min, and fluid is drained out of tank B at the same rate; also 8 L/min of fluid are pumped from tank A to tank B, and 2 L/min from tank B to tank A. The liquids inside each tank are kept well stirred so that each mixture is homogeneous. If, initially, the brine solution in tank A contains x_0 kg of salt and that in tank B initially contains y_0 kg of salt, determine the mass of salt in each tank at time $t > 0$.[†]

Figure 5.1 Interconnected fluid tanks

Note that the *volume* of liquid in each tank remains constant at 24 L because of the balance between the inflow and outflow volume rates. Hence, we have two unknown functions of t: the mass of salt $x(t)$ in tank A and the mass of salt $y(t)$ in tank B. By focusing attention on one tank at a time, we can derive two equations relating these unknowns. Since the system is being flushed with freshwater, we expect that the salt content of each tank will diminish to zero as $t \to +\infty$.

To formulate the equations for this system, we equate the rate of change of salt in each tank with the *net* rate at which salt is transferred to that tank. The salt *concentration* in tank A is $x(t)/24$ kg/L, so the upper interconnecting pipe carries salt out of tank A at a rate of $8x/24$ kg/min; similarly, the lower interconnecting pipe brings salt into tank A at the rate $2y/24$ kg/min (the concentration of salt in tank B is $y/24$ kg/L). The freshwater inlet, of course, transfers no salt (it simply maintains the volume in tank A at 24 L). From our premise,

$$\frac{dx}{dt} = \text{input rate} - \text{output rate},$$

[†]For this application we simplify the analysis by assuming the lengths and volumes of the pipes are sufficiently small that we can ignore the diffusive and advective dynamics taking place therein.

so the rate of change of the mass of salt in tank A is

$$\frac{dx}{dt} = \frac{2}{24}y - \frac{8}{24}x = \frac{1}{12}y - \frac{1}{3}x.$$

The rate of change of salt in tank B is determined by the same interconnecting pipes *and* by the drain pipe, carrying away $6y/24$ kg/min:

$$\frac{dy}{dt} = \frac{8}{24}x - \frac{2}{24}y - \frac{6}{24}y = \frac{1}{3}x - \frac{1}{3}y.$$

The interconnected tanks are thus governed by a *system* of differential equations:

$$
\begin{aligned}
x' &= -\frac{1}{3}x + \frac{1}{12}y, \\
y' &= \frac{1}{3}x - \frac{1}{3}y.
\end{aligned}
\tag{1}
$$

Although both unknowns $x(t)$ and $y(t)$ appear in each of equations (1) (they are "coupled"), the structure is so transparent that we can obtain an equation for y alone by solving the second equation for x,

$$x = 3y' + y,\tag{2}$$

and substituting (2) in the first equation to eliminate x:

$$(3y' + y)' = -\frac{1}{3}(3y' + y) + \frac{1}{12}y,$$

$$3y'' + y' = -y' - \frac{1}{3}y + \frac{1}{12}y,$$

or

$$3y'' + 2y' + \frac{1}{4}y = 0.$$

This last equation, which is linear with constant coefficients, is readily solved by the methods of Section 4.2. Since the auxiliary equation

$$3r^2 + 2r + \frac{1}{4} = 0$$

has roots $-1/2$, $-1/6$, a general solution is given by

$$y(t) = c_1 e^{-t/2} + c_2 e^{-t/6}.\tag{3}$$

Having determined y, we use equation (2) to deduce a formula for x:

$$x(t) = 3\left(-\frac{c_1}{2}e^{-t/2} - \frac{c_2}{6}e^{-t/6}\right) + c_1 e^{-t/2} + c_2 e^{-t/6} = -\frac{1}{2}c_1 e^{-t/2} + \frac{1}{2}c_2 e^{-t/6}.\tag{4}$$

Formulas (3) and (4) contain two undetermined parameters, c_1 and c_2, which can be adjusted to meet the specified initial conditions:

$$x(0) = -\frac{1}{2}c_1 + \frac{1}{2}c_2 = x_0, \qquad\qquad y(0) = c_1 + c_2 = y_0,$$

or

$$c_1 = \frac{y_0 - 2x_0}{2}, \quad c_2 = \frac{y_0 + 2x_0}{2}.$$

Thus, the mass of salt in tanks A and B at time t are, respectively,

$$x(t) = -\left(\frac{y_0 - 2x_0}{4}\right)e^{-t/2} + \left(\frac{y_0 + 2x_0}{4}\right)e^{-t/6},$$

(5)

$$y(t) = \left(\frac{y_0 - 2x_0}{2}\right)e^{-t/2} + \left(\frac{y_0 + 2x_0}{2}\right)e^{-t/6}.$$

The ad hoc elimination procedure that we used to solve this example will be generalized and formalized in the next section, to find solutions of all *linear systems with constant coefficients*. Furthermore, in later sections we will show how to extend our numerical algorithms for first-order equations to *general* systems and will consider applications to coupled oscillators and electrical systems.

It is interesting to note from (5) that all solutions of the interconnected-tanks problem tend to the constant solution $x(t) \equiv 0$, $y(t) \equiv 0$ as $t \to +\infty$. (This is of course consistent with our physical expectations.) This constant solution will be identified as a *stable equilibrium solution* in Section 5.4, in which we introduce phase plane analysis. It turns out that, for a general class of systems, equilibria can be identified and classified so as to give qualitative information about the other solutions even when we cannot solve the system explicitly.

5.2 Differential Operators and the Elimination Method* for Systems

The notation $y'(t) = \dfrac{dy}{dt} = \dfrac{d}{dt}y$ was devised to suggest that the derivative of a function y is the result of *operating* on the function y with the differentiation operator $\dfrac{d}{dt}$. Indeed, second derivatives are formed by iterating the operation: $y''(t) = \dfrac{d^2y}{dt^2} = \dfrac{d}{dt}\dfrac{d}{dt}y$. Commonly, the symbol D is used instead of $\dfrac{d}{dt}$, and the second-order differential equation

$$y'' + 4y' + 3y = 0$$

is represented[†] by

$$D^2y + 4Dy + 3y = (D^2 + 4D + 3)[y] = 0.$$

So, we have implicitly adopted the convention that the operator "product," D times D, is interpreted as the *composition* of D with itself when it operates on functions: D^2y means $D(D[y])$; i.e., the second derivative. Similarly, the product $(D + 3)(D + 1)$ operates on a function via

$$
\begin{aligned}
(D + 3)(D + 1)[y] &= (D + 3)[(D + 1)[y]] = (D + 3)[y' + y] \\
&= D[y' + y] + 3[y' + y] \\
&= (y'' + y') + (3y' + 3y) = y'' + 4y' + 3y = (D^2 + 4D + 3)[y].
\end{aligned}
$$

*An alternative procedure to the methodology of this section will be described in Chapter 9. Although it involves the machinery of matrix analysis, it is preferable for large systems.

[†]Some authors utilize the identity operator I, defined by $I[y] = y$, and write more formally $D^2 + 4D + 3I$ instead of $D^2 + 4D + 3$.

Thus, $(D+3)(D+1)$ is the same operator as $D^2 + 4D + 3$; when they are applied to twice-differentiable functions, the results are identical.

Example 1 Show that the operator $(D+1)(D+3)$ is also the same as $D^2 + 4D + 3$.

Solution For any twice-differentiable function $y(t)$, we have

$$(D+1)(D+3)[y] = (D+1)[(D+3)[y]] = (D+1)[y' + 3y]$$
$$= D[y' + 3y] + 1[y' + 3y] = (y'' + 3y') + (y' + 3y)$$
$$= y'' + 4y' + 3y = (D^2 + 4D + 3)[y].$$

Hence, $(D+1)(D+3) = D^2 + 4D + 3$. ◆

Since $(D+1)(D+3) = (D+3)(D+1) = D^2 + 4D + 3$, it is tempting to generalize and propose that one can treat expressions like $aD^2 + bD + c$ as if they were ordinary polynomials in D. This is true, as long as we restrict the coefficients a, b, c to be *constants*. The following example, which has *variable* coefficients, is instructive.

Example 2 Show that $(D + 3t)D$ is *not* the same as $D(D + 3t)$.

Solution With $y(t)$ as before,

$$(D + 3t)D[y] = (D + 3t)[y'] = y'' + 3ty';$$
$$D(D + 3t)[y] = D[y' + 3ty] = y'' + 3y + 3ty'.$$

They are not the same! ◆

Because the coefficient $3t$ is not a constant, it "interrupts" the interaction of the differentiation operator D with the function $y(t)$. As long as we only deal with expressions like $aD^2 + bD + c$ with *constant* coefficients a, b, and c, the "algebra" of differential operators follows the same rules as the algebra of polynomials. (See Problem 39 for elaboration on this point.)

This means that the familiar elimination method, used for solving *algebraic* systems like

$$3x - 2y + z = 4,$$
$$x + y - z = 0,$$
$$2x - y + 3z = 6,$$

can be adapted to solve any system of *linear differential equations with constant coefficients*. In fact, we used this approach in solving the system that arose in the interconnected tanks problem of Section 5.1. Our goal in this section is to formalize this **elimination method** so that we can tackle more general linear constant coefficient systems.

We first demonstrate how the method applies to a linear system of two first-order differential equations of the form

$$a_1 x'(t) + a_2 x(t) + a_3 y'(t) + a_4 y(t) = f_1(t),$$
$$a_5 x'(t) + a_6 x(t) + a_7 y'(t) + a_8 y(t) = f_2(t),$$

where $a_1, a_2, \ldots, a_8$ are constants and $x(t)$, $y(t)$ is the function pair to be determined. In operator notation this becomes

$$(a_1 D + a_2)[x] + (a_3 D + a_4)[y] = f_1,$$
$$(a_5 D + a_6)[x] + (a_7 D + a_8)[y] = f_2.$$

Example 3 Solve the system

(1)
$$x'(t) = 3x(t) - 4y(t) + 1 ,$$
$$y'(t) = 4x(t) - 7y(t) + 10t .$$

Solution The alert reader may observe that since y' is absent from the first equation, we could use the latter to express y in terms of x and x' and substitute into the second equation to derive an "uncoupled" equation containing only x and its derivatives. However, this simple trick will not work on more general systems (Problem 18 is an example).

To utilize the elimination method, we first write the system using the operator notation:

(2)
$$(D-3)[x] + 4y = 1 ,$$
$$-4x + (D+7)[y] = 10t .$$

Imitating the elimination procedure for algebraic systems, we can eliminate x from this system by adding 4 times the first equation to $(D-3)$ applied to the second equation. This gives

$$\left(16 + (D-3)(D+7) \right)[y] = 4 \cdot 1 + (D-3)[10t] = 4 + 10 - 30t ,$$

which simplifies to

(3)
$$(D^2 + 4D - 5)[y] = 14 - 30t .$$

Now equation (3) is just a second-order linear equation in y with constant coefficients that has the general solution

(4)
$$y(t) = C_1 e^{-5t} + C_2 e^t + 6t + 2 ,$$

which can be found using undetermined coefficients.

To find $x(t)$, we have two options.

Method 1. We return to system (2) and eliminate y. This is accomplished by "multiplying" the first equation in (2) by $(D+7)$ and the second equation by -4 and then adding to obtain

$$(D^2 + 4D - 5)[x] = 7 - 40t .$$

This equation can likewise be solved using undetermined coefficients to yield

(5)
$$x(t) = K_1 e^{-5t} + K_2 e^t + 8t + 5 ,$$

where we have taken K_1 and K_2 to be the arbitrary constants, which are not necessarily the same as C_1 and C_2 used in formula (4).

It is reasonable to expect that system (1) will involve only *two* arbitrary constants, since it consists of two first-order equations. Thus, the four constants C_1, C_2, K_1, and K_2 are not independent. To determine the relationships, we substitute the expressions for $x(t)$ and $y(t)$ given in (4) and (5) into one of the equations in (1)—say, the first one. This yields

$$-5K_1 e^{-5t} + K_2 e^t + 8 =$$
$$3K_1 e^{-5t} + 3K_2 e^t + 24t + 15 - 4C_1 e^{-5t} - 4C_2 e^t - 24t - 8 + 1 ,$$

which simplifies to

$$(4C_1 - 8K_1)e^{-5t} + (4C_2 - 2K_2)e^t = 0 .$$

Because e^t and e^{-5t} are linearly independent functions on any interval, this last equation holds for all t only if

$$4C_1 - 8K_1 = 0 \quad \text{and} \quad 4C_2 - 2K_2 = 0 \,.$$

Therefore, $K_1 = C_1/2$ and $K_2 = 2C_2$.

A solution to system (1) is then given by the pair

(6) $x(t) = \dfrac{1}{2}C_1 e^{-5t} + 2C_2 e^t + 8t + 5 \,, \quad y(t) = C_1 e^{-5t} + C_2 e^t + 6t + 2 \,.$

As you might expect, this pair is a **general solution** to (1) in the sense that *any* solution to (1) can be expressed in this fashion.

Method 2. A simpler method for determining $x(t)$ once $y(t)$ is known is to use the system to obtain an equation for $x(t)$ in terms of $y(t)$ and $y'(t)$. In this example we can directly solve the second equation in (1) for $x(t)$:

$$x(t) = \frac{1}{4}y'(t) + \frac{7}{4}y(t) - \frac{5}{2}t \,.$$

Substituting $y(t)$ as given in (4) yields

$$x(t) = \frac{1}{4}\left[-5C_1 e^{-5t} + C_2 e^t + 6\right] + \frac{7}{4}\left[C_1 e^{-5t} + C_2 e^t + 6t + 2\right] - \frac{5}{2}t$$

$$= \frac{1}{2}C_1 e^{-5t} + 2C_2 e^t + 8t + 5 \,,$$

which agrees with (6). ◆

The above procedure works, more generally, for any linear system of two equations and two unknowns with *constant coefficients* regardless of the order of the equations. For example, if we let L_1, L_2, L_3, and L_4 denote linear differential operators with constant coefficients (i.e., polynomials in D), then the method can be applied to the linear system

$$L_1[x] + L_2[y] = f_1 \,,$$
$$L_3[x] + L_4[y] = f_2 \,.$$

Because the system has constant coefficients, the operators commute (e.g., $L_2 L_4 = L_4 L_2$) and we can eliminate variables in the usual algebraic fashion. Eliminating the variable y gives

(7) $(L_1 L_4 - L_2 L_3)[x] = g_1 \,,$

where $g_1 := L_4[f_1] - L_2[f_2]$. Similarly, eliminating the variable x yields

(8) $(L_1 L_4 - L_2 L_3)[y] = g_2 \,,$

where $g_2 := L_1[f_2] - L_3[f_1]$. Now if $L_1 L_4 - L_2 L_3$ is a differential operator of order n, then a general solution for (7) contains n arbitrary constants, and a general solution for (8) also contains n arbitrary constants. Thus, a total of $2n$ constants arise. However, as we saw in Example 3, there are only n of these that are independent for the system; the remaining constants can be expressed in terms of these.[†] The pair of general solutions to (7) and (8) written in terms of the n independent constants is called a **general solution for the system.**

[†]For a proof of this fact, see *Ordinary Differential Equations*, by M. Tenenbaum and H. Pollard (Dover, New York, 1985), Chapter 7.

If it turns out that $L_1L_4 - L_2L_3$ is the zero operator, the system is said to be **degenerate.** As with the anomalous problem of solving for the points of intersection of two parallel or coincident lines, a degenerate system may have no solutions, or if it does possess solutions, they may involve any number of arbitrary constants (see Problems 23 and 24).

Elimination Procedure for 2 × 2 Systems

To find a general solution for the system

$$L_1[x] + L_2[y] = f_1,$$
$$L_3[x] + L_4[y] = f_2,$$

where $L_1, L_2, L_3,$ and L_4 are polynomials in $D = d/dt$:

(a) Make sure that the system is written in operator form.

(b) Eliminate one of the variables, say, y, and solve the resulting equation for $x(t)$. If the system is degenerate, stop! A separate analysis is required to determine whether or not there are solutions.

(c) (*Shortcut*) If possible, use the system to derive an equation that involves $y(t)$ but not its derivatives. [Otherwise, go to step (d).] Substitute the found expression for $x(t)$ into this equation to get a formula for $y(t)$. The expressions for $x(t), y(t)$ give the desired general solution. ◆

(d) Eliminate x from the system and solve for $y(t)$. [Solving for $y(t)$ gives more constants—in fact, twice as many as needed.]

(e) Remove the extra constants by substituting the expressions for $x(t)$ and $y(t)$ into one or both of the equations in the system. Write the expressions for $x(t)$ and $y(t)$ in terms of the remaining constants. ◆

Example 4 Find a general solution for

$$(9) \qquad \begin{aligned} x''(t) + y'(t) - x(t) + y(t) &= -1, \\ x'(t) + y'(t) - x(t) &= t^2. \end{aligned}$$

Solution We begin by expressing the system in operator notation:

$$(10) \qquad \begin{aligned} (D^2 - 1)[x] + (D + 1)[y] &= -1, \\ (D - 1)[x] + D[y] &= t^2. \end{aligned}$$

Here $L_1 := D^2 - 1$, $L_2 := D + 1$, $L_3 := D - 1$, and $L_4 := D$.

Eliminating y gives [see (7)]:

$$\big((D^2 - 1)D - (D + 1)(D - 1)\big)[x] = D[-1] - (D + 1)[t^2],$$

which reduces to

$$(11) \qquad \begin{aligned} (D^2 - 1)(D - 1)[x] &= -2t - t^2, \quad \text{or} \\ (D - 1)^2(D + 1)[x] &= -2t - t^2. \end{aligned}$$

Since $(D - 1)^2(D + 1)$ is third order, we should expect three arbitrary constants in a general solution to system (9).

Although the methods of Chapter 4 focused on solving second-order equations, we have seen several examples of how they extend in a natural way to higher-order

equations.[†] Applying this strategy to the third-order equation (11), we observe that the corresponding homogeneous equation has the auxiliary equation $(r-1)^2(r+1) = 0$ with roots $r = 1, 1, -1$. Hence, a general solution for the homogeneous equation is

$$x_h(t) = C_1 e^t + C_2 t e^t + C_3 e^{-t}.$$

To find a particular solution to (11), we use the method of undetermined coefficients with $x_p(t) = At^2 + Bt + C$. Substituting into (11) and solving for A, B, and C yields (after a little algebra)

$$x_p(t) = -t^2 - 4t - 6.$$

Thus, a general solution to equation (11) is

(12) $x(t) = x_h(t) + x_p(t) = C_1 e^t + C_2 t e^t + C_3 e^{-t} - t^2 - 4t - 6.$

To find $y(t)$, we take the shortcut described in step (c) of the elimination procedure box. Subtracting the second equation in (10) from the first, we find

$$(D^2 - D)[x] + y = -1 - t^2,$$

so that

$$y = (D - D^2)[x] - 1 - t^2.$$

Inserting the expression for $x(t)$, given in (12), we obtain

$$y(t) = C_1 e^t + C_2(te^t + e^t) - C_3 e^{-t} - 2t - 4$$
$$- [C_1 e^t + C_2(te^t + 2e^t) + C_3 e^{-t} - 2] - 1 - t^2, \quad \text{or}$$

(13) $y(t) = -C_2 e^t - 2C_3 e^{-t} - t^2 - 2t - 3.$

The formulas for $x(t)$ in (12) and $y(t)$ in (13) give the desired general solution to (9). ◆

The elimination method also applies to linear systems with three or more equations and unknowns; however, the process becomes more cumbersome as the number of equations and unknowns increases. The matrix methods presented in Chapter 9 are better suited for handling larger systems. Here we illustrate the elimination technique for a 3×3 system.

Example 5 Find a general solution to

(14)
$$x'(t) = x(t) + 2y(t) - z(t),$$
$$y'(t) = x(t) + z(t),$$
$$z'(t) = 4x(t) - 4y(t) + 5z(t).$$

Solution We begin by expressing the system in operator notation:

(15)
$$(D-1)[x] - 2y + z = 0,$$
$$-x + D[y] - z = 0,$$
$$-4x + 4y + (D-5)[z] = 0.$$

Eliminating z from the first two equations (by adding them) and then from the last two equations yields (after some algebra, which we omit)

(16)
$$(D-2)[x] + (D-2)[y] = 0,$$
$$-(D-1)[x] + (D-1)(D-4)[y] = 0.$$

[†]More detailed treatment of higher-order equations is given in Chapter 6.

On eliminating x from this 2×2 system, we eventually obtain

$$(D-1)(D-2)(D-3)[y] = 0 ,$$

which has the general solution

(17) $y(t) = C_1 e^t + C_2 e^{2t} + C_3 e^{3t} .$

Taking the shortcut approach, we add the two equations in (16) to get an expression for x in terms of y and its derivatives, which simplifies to

$$x = (D^2 - 4D + 2)[y] = y'' - 4y' + 2y .$$

When we substitute the expression (17) for $y(t)$ into this equation, we find

(18) $x(t) = -C_1 e^t - 2C_2 e^{2t} - C_3 e^{3t} .$

Finally, using the second equation in (14) to solve for $z(t)$, we get

$$z(t) = y'(t) - x(t) ,$$

and substituting in for $y(t)$ and $x(t)$ yields

(19) $z(t) = 2C_1 e^t + 4C_2 e^{2t} + 4C_3 e^{3t} .$

The expressions for $x(t)$ in (18), $y(t)$ in (17), and $z(t)$ in (19) give a general solution with C_1, C_2, and C_3 as arbitrary constants. ◆

5.2 EXERCISES

1. Let $A = D - 1$, $B = D + 2$, $C = D^2 + D - 2$, where $D = d/dt$. For $y = t^3 - 8$, compute
 (a) $A[y]$ **(b)** $B[A[y]]$ **(c)** $B[y]$
 (d) $A[B[y]]$ **(e)** $C[y]$

2. Show that the operator $(D-1)(D+2)$ is the same as the operator $D^2 + D - 2$.

In Problems 3–18, use the elimination method to find a general solution for the given linear system, where differentiation is with respect to t.

3. $x' + 2y = 0 ,$
 $x' - y' = 0$

4. $x' = x - y ,$
 $y' = y - 4x$

5. $x' + y' - x = 5 ,$
 $x' + y' + y = 1$

6. $x' = 3x - 2y + \sin t ,$
 $y' = 4x - y - \cos t$

7. $(D+1)[u] - (D+1)[v] = e^t ,$
 $(D-1)[u] + (2D+1)[v] = 5$

8. $(D-3)[x] + (D-1)[y] = t ,$
 $(D+1)[x] + (D+4)[y] = 1$

9. $x' + y' + 2x = 0 ,$
 $x' + y' - x - y = \sin t$

10. $2x' + y' - x - y = e^{-t} ,$
 $x' + y' + 2x + y = e^t$

11. $(D^2 - 1)[u] + 5v = e^t ,$
 $2u + (D^2 + 2)[v] = 0$

12. $D^2[u] + D[v] = 2 ,$
 $4u + D[v] = 6$

13. $\dfrac{dx}{dt} = x - 4y ,$

 $\dfrac{dy}{dt} = x + y$

14. $\dfrac{dx}{dt} + y = t^2 ,$

 $-x + \dfrac{dy}{dt} = 1$

15. $\dfrac{dw}{dt} = 5w + 2z + 5t ,$

 $\dfrac{dz}{dt} = 3w + 4z + 17t$

16. $\dfrac{dx}{dt} + x + \dfrac{dy}{dt} = e^{4t} ,$

 $2x + \dfrac{d^2y}{dt^2} = 0$

17. $x'' + 5x - 4y = 0 ,$
 $-x + y'' + 2y = 0$

18. $x'' + y'' - x' = 2t ,$
 $x'' + y' - x + y = -1$

In Problems 19–21, solve the given initial value problem.

19. $\dfrac{dx}{dt} = 4x + y ; \quad x(0) = 1 ,$

 $\dfrac{dy}{dt} = -2x + y ; \quad y(0) = 0$

20. $\dfrac{dx}{dt} = 2x + y - e^{2t} ; \quad x(0) = 1 ,$

 $\dfrac{dy}{dt} = x + 2y ; \quad y(0) = -1$

21. $\dfrac{d^2x}{dt^2} = y$; $x(0) = 3$, $x'(0) = 1$,

$\dfrac{d^2y}{dt^2} = x$; $y(0) = 1$, $y'(0) = -1$

22. Verify that the solution to the initial value problem

$$x' = 5x - 3y - 2 \; ; \quad x(0) = 2 \; ,$$
$$y' = 4x - 3y - 1 \; ; \quad y(0) = 0$$

satisfies $|x(t)| + |y(t)| \to +\infty$ as $t \to +\infty$.

In Problems 23 and 24, show that the given linear system is degenerate. In attempting to solve the system, determine whether it has no solutions or infinitely many solutions.

23. $(D - 1)[x] + (D - 1)[y] = -3e^{-2t}$,
$\quad (D + 2)[x] + (D + 2)[y] = 3e^t$

24. $D[x] + (D + 1)[y] = e^t$,
$\quad D^2[x] + (D^2 + D)[y] = 0$

In Problems 25–28, use the elimination method to find a general solution for the given system of three equations in the three unknown functions $x(t)$, $y(t)$, $z(t)$.

25. $x' = x + 2y - z$,
 $y' = x + z$,
 $z' = 4x - 4y + 5z$

26. $x' = 3x + y - z$,
 $y' = x + 2y - z$,
 $z' = 3x + 3y - z$

27. $x' = 4x - 4z$,
 $y' = 4y - 2z$,
 $z' = -2x - 4y + 4z$

28. $x' = x + 2y + z$,
 $y' = 6x - y$,
 $z' = -x - 2y - z$

*In Problems 29 and 30, determine the range of values (if any) of the parameter λ that will ensure **all** solutions $x(t)$, $y(t)$ of the given system remain bounded as $t \to +\infty$.*

29. $\dfrac{dx}{dt} = \lambda x - y$,

$\dfrac{dy}{dt} = 3x + y$

30. $\dfrac{dx}{dt} = -x + \lambda y$,

$\dfrac{dy}{dt} = x - y$

31. Two large tanks, each holding 100 L of liquid, are interconnected by pipes, with the liquid flowing from tank A into tank B at a rate of 3 L/min and from B into A at a rate of 1 L/min (see Figure 5.2). The liquid inside each tank is kept well stirred. A brine solution with a concentration of 0.2 kg/L of salt flows into tank A at a rate of 6 L/min. The (diluted) solution flows out of the system from tank A at 4 L/min and from tank B at 2 L/min. If, initially, tank A contains pure water and tank B contains 20 kg of salt, determine the mass of salt in each tank at time $t \geq 0$.

32. In Problem 31, 3 L/min of liquid flowed from tank A into tank B and 1 L/min from B into A. Determine the mass of salt in each tank at time $t \geq 0$ if, instead, 5 L/min flows from A into B and 3 L/min flows from B into A, with all other data the same.

33. In Problem 31, assume that no solution flows out of the system from tank B, only 1 L/min flows from A into B, and only 4 L/min of brine flows into the system at tank A, other data being the same. Determine the mass of salt in each tank at time $t \geq 0$.

34. Feedback System with Pooling Delay. Many physical and biological systems involve time delays. A pure time delay has its output the same as its input but shifted in time. A more common type of delay is *pooling delay*. An example of such a feedback system is shown in Figure 5.3 on page 251. Here the level of fluid in tank B determines the rate at which fluid enters tank A. Suppose this rate is given by $R_1(t) = \alpha[V - V_2(t)]$, where α and V are positive constants and $V_2(t)$ is the volume of fluid in tank B at time t.

(a) If the outflow rate R_3 from tank B is constant and the flow rate R_2 from tank A into B is $R_2(t) = KV_1(t)$, where K is a positive constant and $V_1(t)$ is the volume of fluid in tank A at time t, then show that this feedback system is governed by the system

$$\dfrac{dV_1}{dt} = \alpha\big(V - V_2(t)\big) - KV_1(t) \; ,$$

$$\dfrac{dV_2}{dt} = KV_1(t) - R_3 \; .$$

6 L/min
0.2 kg/L

A

3 L/min

B

$x(t)$

100 L

$y(t)$

100 L

4 L/min

$x(0) = 0$ kg

$y(0) = 20$ kg

2 L/min

1 L/min

Figure 5.2 Mixing problem for interconnected tanks

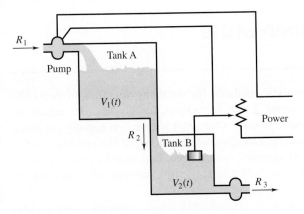

Figure 5.3 Feedback system with pooling delay

(b) Find a general solution for the system in part (a) when $\alpha = 5\,(\min)^{-1}$, $V = 20\,\text{L}$, $K = 2\,(\min)^{-1}$, and $R_3 = 10\,\text{L/min}$.

(c) Using the general solution obtained in part (b), what can be said about the volume of fluid in each of the tanks as $t \to +\infty$?

35. A house, for cooling purposes, consists of two zones: the attic area zone A and the living area zone B (see Figure 5.4). The living area is cooled by a 2-ton air conditioning unit that removes 24,000 Btu/hr. The heat capacity of zone B is $1/2°F$ per thousand Btu. The time constant for heat transfer between zone A and the outside is 2 hr, between zone B and the outside is 4 hr, and between the two zones is 4 hr. If the outside temperature stays at 100°F, how warm does it eventually get in the attic zone A? (Heating and cooling of buildings was treated in Section 3.3 on page 102.)

36. A building consists of two zones A and B (see Figure 5.5). Only zone A is heated by a furnace, which generates 80,000 Btu/hr. The heat capacity of zone A is $1/4°F$ per thousand Btu. The time constant for heat transfer between

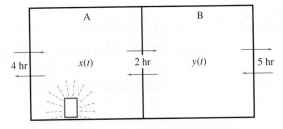

Figure 5.5 Two-zone building with one zone heated

zone A and the outside is 4 hr, between the unheated zone B and the outside is 5 hr, and between the two zones is 2 hr. If the outside temperature stays at 0°F, how cold does it eventually get in the unheated zone B?

37. In Problem 36, if a small furnace that generates 1000 Btu/hr is placed in zone B, determine the coldest it would eventually get in zone B if zone B has a heat capacity of 2°F per thousand Btu.

38. Arms Race. A simplified mathematical model for an arms race between two countries whose expenditures for defense are expressed by the variables $x(t)$ and $y(t)$ is given by the linear system

$$\frac{dx}{dt} = 2y - x + a\,; \qquad x(0) = 1\,,$$
$$\frac{dy}{dt} = 4x - 3y + b\,; \qquad y(0) = 4\,,$$

where a and b are constants that measure the trust (or distrust) each country has for the other. Determine whether there is going to be disarmament (x and y approach 0 as t increases), a stabilized arms race (x and y approach a constant as $t \to +\infty$), or a runaway arms race (x and y approach $+\infty$ as $t \to +\infty$).

39. Let A, B, and C represent three linear differential operators with constant coefficients; for example,

$$A := a_2 D^2 + a_1 D + a_0,\ B := b_2 D^2 + b_1 D + b_0\,,$$
$$C := c_2 D^2 + c_1 D + c_0\,,$$

where the a's, b's, and c's are constants. Verify the following properties:[†]

(a) Commutative laws:
$$A + B = B + A\,,$$
$$AB = BA\,.$$

(b) Associative laws:
$$(A + B) + C = A + (B + C)\,,$$
$$(AB)C = A(BC)\,.$$

(c) Distributive law: $A(B + C) = AB + AC\,.$

Figure 5.4 Air-conditioned house with attic

[†]We say that two operators A and B are *equal* if $A[y] = B[y]$ for all functions y with the necessary derivatives.

5.3 Solving Systems and Higher-Order Equations Numerically

Although we studied a half-dozen analytic methods for obtaining solutions to first-order ordinary differential equations in Chapter 2, the techniques for higher-order equations, or systems of equations, are much more limited. Chapter 4 focused on solving the linear constant-coefficient second-order equation. The elimination method of the previous section is also restricted to constant-coefficient systems. And, indeed, higher-order linear constant-coefficient equations and systems can be solved analytically by extensions of these methods, as we will see in Chapters 6, 7, and 9.

However, if the equations—even a single second-order linear equation—have variable coefficients, the solution process is much less satisfactory. As will be seen in Chapter 8, the solutions are expressed as infinite series, and their computation can be very laborious (with the notable exception of the Cauchy–Euler, or equidimensional, equation). And we know virtually nothing about how to obtain exact solutions to nonlinear second-order equations.

Fortunately, all the cases that arise (constant or variable coefficients, nonlinear, higher-order equations or systems) can be addressed by a single formulation that lends itself to a multitude of *numerical* approaches. In this section we'll see how to express differential equations as a *system in normal form* and then show how the basic Euler method for computer solution can be easily "vectorized" to apply to such systems. Although subsequent chapters will return to analytic solution methods, the vectorized version of the Euler technique or the more efficient Runge–Kutta technique will hereafter be available as fallback methods for numerical exploration of intractable problems.

Normal Form

A system of m differential equations in the m unknown functions $x_1(t), x_2(t), \ldots, x_m(t)$ expressed as

$$
\begin{aligned}
(1) \qquad x_1'(t) &= f_1(t, x_1, x_2, \ldots, x_m) \ , \\
x_2'(t) &= f_2(t, x_1, x_2, \ldots, x_m) \ , \\
&\ \ \vdots \\
x_m'(t) &= f_m(t, x_1, x_2, \ldots, x_m)
\end{aligned}
$$

is said to be in **normal form.** Notice that (1) consists of m first-order equations that collectively look like a *vectorized* version of the single generic first-order equation

$$
(2) \qquad x' = f(t, x) \ ,
$$

and that the system expressed in equation (1) of Section 5.1 takes this form, as do equations (1) and (14) in Section 5.2. An initial value problem for (1) entails finding a solution to this system that satisfies the initial conditions

$$
x_1(t_0) = a_1, \quad x_2(t_0) = a_2, \quad \ldots, \quad x_m(t_0) = a_m
$$

for prescribed values $t_0, a_1, a_2, \ldots, a_m$.

The importance of the normal form is underscored by the fact that most professional codes for initial value problems presume that the system is written in this form. Furthermore, for a *linear* system in normal form, the powerful machinery of linear algebra can be readily applied. [Indeed, in Chapter 9 we will show how the solutions $x(t) = ce^{at}$ of the simple equation $x' = ax$ can be generalized to constant-coefficient systems in normal form.]

For these reasons it is gratifying to note that a (single) higher-order equation can always be converted to an equivalent system of first-order equations.

To convert an mth-order differential equation

(3) $$y^{(m)}(t) = f\left(t, y, y', \ldots, y^{(m-1)}\right)$$

into a first-order system, we introduce, as additional unknowns, the sequence of derivatives of y:

$$x_1(t) := y(t), \quad x_2(t) := y'(t), \quad \ldots, \quad x_m(t) := y^{(m-1)}(t) .$$

With this scheme, we obtain $m - 1$ first-order equations quite trivially:

(4)
$$\begin{aligned} x_1'(t) &= y'(t) = x_2(t) , \\ x_2'(t) &= y''(t) = x_3(t) , \\ &\vdots \\ x_{m-1}'(t) &= y^{(m-1)}(t) = x_m(t) . \end{aligned}$$

The mth and final equation then constitutes a restatement of the original equation (3) in terms of the new unknowns:

(5) $$x_m'(t) = y^{(m)}(t) = f(t, x_1, x_2, \ldots, x_m) .$$

If equation (3) has initial conditions $y(t_0) = a_1, y'(t_0) = a_2, \ldots, y^{(m-1)}(t_0) = a_m$, then the system (4)–(5) has initial conditions $x_1(t_0) = a_1, x_2(t_0) = a_2, \ldots, x_m(t_0) = a_m$.

Example 1 Convert the initial value problem

(6) $$y''(t) + 3ty'(t) + y(t)^2 = \sin t ; \quad y(0) = 1 , \quad y'(0) = 5$$

into an initial value problem for a system in normal form.

Solution We first express the differential equation in (6) as

$$y''(t) = -3ty'(t) - y(t)^2 + \sin t .$$

Setting $x_1(t) := y(t)$ and $x_2(t) := y'(t)$, we obtain

$$\begin{aligned} x_1'(t) &= x_2(t) , \\ x_2'(t) &= -3tx_2(t) - x_1(t)^2 + \sin t . \end{aligned}$$

The initial conditions transform to $x_1(0) = 1, x_2(0) = 5$. ◆

Euler's Method for Systems in Normal Form

Recall from Section 1.4 that Euler's method for solving a single first-order equation (2) is based on estimating the solution x at time $(t_0 + h)$ using the approximation

(7) $$x(t_0 + h) \approx x(t_0) + hx'(t_0) = x(t_0) + hf\left(t_0, x(t_0)\right) ,$$

and that as a consequence the algorithm can be summarized by the recursive formulas

(8) $$t_{n+1} = t_n + h ,$$

(9) $$x_{n+1} = x_n + hf(t_n, x_n), \quad n = 0, 1, 2, \ldots$$

[compare equations (2) and (3), Section 1.4]. Now we can apply the approximation (7) to each of the equations in the system (1):

(10) $$x_k(t_0 + h) \approx x_k(t_0) + hx_k'(t_0) = x_k(t_0) + hf_k\left(t_0, x_1(t_0), x_2(t_0), \ldots, x_m(t_0)\right) ,$$

and for $k = 1, 2, \ldots m$, we are led to the recursive formulas

(11) $t_{n+1} = t_n + h$,

(12)
$$x_{1;n+1} = x_{1;n} + hf_1(t_n, x_{1;n}, x_{2;n}, \ldots, x_{m;n}),$$
$$x_{2;n+1} = x_{2;n} + hf_2(t_n, x_{1;n}, x_{2;n}, \ldots, x_{m;n}),$$
$$\vdots$$
$$x_{m;n+1} = x_{m;n} + hf_m(t_n, x_{1;n}, x_{2;n}, \ldots, x_{m;n}) \qquad (n = 0, 1, 2, \ldots).$$

Here we are burdened with the ungainly notation $x_{p;n}$ for the approximation to the value of the pth-function x_p at time $t = t_0 + nh$; i.e., $x_{p;n} \approx x_p(t_0 + nh)$. However, if we treat the unknowns and right-hand members of (1) as components of vectors

$$\mathbf{x}(t) := [x_1(t), x_2(t), \ldots, x_m(t)],$$
$$\mathbf{f}(t, \mathbf{x}) := [f_1(t, x_1, x_2, \ldots, x_m), f_2(t, x_1, x_2, \ldots, x_m), \ldots, f_m(t, x_1, x_2, \ldots, x_m)],$$

then (12) can be expressed in the much neater form

(13) $\mathbf{x}_{n+1} = \mathbf{x}_n + h\mathbf{f}(t_n, \mathbf{x}_n)$.

Example 2 Use the vectorized Euler method with step size $h = 0.1$ to find an approximation for the solution to the initial value problem

(14) $y''(t) + 4y'(t) + 3y(t) = 0;$ $y(0) = 1.5$, $y'(0) = -2.5$,

on the interval $[0, 1]$.

Solution For the given step size, the method will yield approximations for $y(0.1), y(0.2), \ldots, y(1.0)$. To apply the vectorized Euler method to (14), we first convert it to normal form. Setting $x_1 = y$ and $x_2 = y'$, we obtain the system

(15)
$$x_1' = x_2; \qquad x_1(0) = 1.5,$$
$$x_2' = -4x_2 - 3x_1; \qquad x_2(0) = -2.5.$$

Comparing (15) with (1) we see that $f_1(t, x_1, x_2) = x_2$ and $f_2(t, x_1, x_2) = -4x_2 - 3x_1$. With the starting values of $t_0 = 0$, $x_{1;0} = 1.5$, and $x_{2;0} = -2.5$, we compute

$$\begin{cases} x_1(0.1) \approx x_{1;1} = x_{1;0} + hx_{2;0} = 1.5 + 0.1(-2.5) = 1.25, \\ x_2(0.1) \approx x_{2;1} = x_{2;0} + h(-4x_{2;0} - 3x_{1;0}) = -2.5 + 0.1[-4(-2.5) - 3 \cdot 1.5] = -1.95; \end{cases}$$

$$\begin{cases} x_1(0.2) \approx x_{1;2} = x_{1;1} + hx_{2;1} = 1.25 + 0.1(-1.95) = 1.055, \\ x_2(0.2) \approx x_{2;2} = x_{2;1} + h(-4x_{2;1} - 3x_{1;1}) = -1.95 + 0.1[-4(-1.95) - 3 \cdot 1.25] = -1.545. \end{cases}$$

Continuing the algorithm we compute the remaining values. These are listed in Table 5.1 on page 255, along with the exact values calculated via the methods of Chapter 4. Note that the $x_{2;n}$ column gives approximations to $y'(t)$, since $x_2(t) \equiv y'(t)$. ◆

TABLE 5.1	Approximations of the Solution to (14) in Example 2			
$t = n(0.1)$	$x_{1;n}$	y **Exact**	$x_{2;n}$	y' **Exact**
0	1.5	1.5	-2.5	-2.5
0.1	1.25	1.275246528	-1.95	-2.016064749
0.2	1.055	1.093136571	-1.545	-1.641948207
0.3	0.9005	0.944103051	-1.2435	-1.35067271
0.4	0.77615	0.820917152	-1.01625	-1.122111364
0.5	0.674525	0.71809574	-0.842595	-0.9412259
0.6	0.5902655	0.63146108	-0.7079145	-0.796759968
0.7	0.51947405	0.557813518	-0.60182835	-0.680269946
0.8	0.459291215	0.494687941	-0.516939225	-0.585405894
0.9	0.407597293	0.440172416	-0.4479509	-0.507377929
1	0.362802203	0.392772975	-0.391049727	-0.442560044

Euler's method is modestly accurate for this problem with a step size of $h = 0.1$. The next example demonstrates the effects of using a sequence of smaller values of h to improve the accuracy.

Example 3 For the initial value problem of Example 2, use Euler's method to estimate $y(1)$ for successively halved step sizes $h = 0.1, 0.05, 0.025, 0.0125, 0.00625$.

Solution Using the same scheme as in Example 2, we find the following approximations, denoted by $y(1;h)$ (obtained with step size h):

h	0.1	0.05	0.025	0.0125	0.00625
$y(1;h)$	0.36280	0.37787	0.38535	0.38907	0.39092

[Recall that the exact value, rounded to 5 decimal places, is $y(1) = 0.39277$.] ◆

The Runge–Kutta scheme described in Section 3.7 is easy to vectorize also; details are given on the following page. As would be expected, its performance is considerably more accurate, yielding five-decimal agreement with the exact solution for a step size of 0.05:

h	0.1	0.05	0.025	0.0125	0.00625
$y(1;h)$	0.39278	0.39277	0.39277	0.39277	0.39277

As in Section 3.7, both algorithms can be coded so as to repeat the calculation of $y(1)$ with a sequence of smaller step sizes until two consecutive estimates agree to within some pre-specified tolerance ε. Here one should interpret "two estimates agree to within ε" to mean that *each component* of the successive vector approximants [i.e., approximants to $y(1)$ and $y'(1)$] should agree to within ε.

An Application to Population Dynamics

A mathematical model for the population dynamics of competing species, one a predator with population $x_2(t)$ and the other its prey with population $x_1(t)$, was developed independently in

the early 1900s by A. J. Lotka and V. Volterra. It assumes that there is plenty of food available for the prey to eat, so the birthrate of the prey should follow the Malthusian or exponential law (see Section 3.2); that is, the birthrate of the prey is Ax_1, where A is a positive constant. The death rate of the prey depends on the number of interactions between the predators and the prey. This is modeled by the expression Bx_1x_2, where B is a positive constant. Therefore, the rate of change in the population of the prey per unit time is $dx_1/dt = Ax_1 - Bx_1x_2$. Assuming that the predators depend entirely on the prey for their food, it is argued that the birthrate of the predators depends on the number of interactions with the prey; that is, the birthrate of predators is Dx_1x_2, where D is a positive constant. The death rate of the predators is assumed to be Cx_2 because without food the population would die off at a rate proportional to the population present. Hence, the rate of change in the population of predators per unit time is $dx_2/dt = -Cx_2 + Dx_1x_2$. Combining these two equations, we obtain the Volterra–Lotka system for the population dynamics of two competing species:

$$(16) \qquad \begin{aligned} x_1' &= Ax_1 - Bx_1x_2 \,, \\ x_2' &= -Cx_2 + Dx_1x_2 \,. \end{aligned}$$

Such systems are in general not explicitly solvable. In the following example, we obtain an approximate solution for such a system by utilizing the vectorized form of the Runge–Kutta algorithm.

For the system of two equations

$$\begin{aligned} x_1' &= f_1(t, x_1, x_2) \,, \\ x_2' &= f_2(t, x_1, x_2) \,, \end{aligned}$$

with initial conditions $x_1(t_0) = x_{1;0}, x_2(t_0) = x_{2;0}$, the vectorized form of the Runge–Kutta recursive equations (cf. (14), page 254) becomes

$$(17) \qquad \begin{cases} t_{n+1} := t_n + h \qquad (n = 0, 1, 2, \dots) \,, \\[2mm] x_{1;n+1} := x_{1;n} + \dfrac{1}{6}(k_{1,1} + 2k_{1,2} + 2k_{1,3} + k_{1,4}) \,, \\[2mm] x_{2;n+1} := x_{2;n} + \dfrac{1}{6}(k_{2,1} + 2k_{2,2} + 2k_{2,3} + k_{2,4}) \,, \end{cases}$$

where h is the step size and, for $i = 1$ and 2,

$$(18) \qquad \begin{cases} k_{i,1} := hf_i(t_n, x_{1;n}, x_{2;n}) \,, \\[2mm] k_{i,2} := hf_i\left(t_n + \dfrac{h}{2},\ x_{1;n} + \dfrac{1}{2}k_{1,1},\ x_{2;n} + \dfrac{1}{2}k_{2,1}\right) \,, \\[2mm] k_{i,3} := hf_i\left(t_n + \dfrac{h}{2},\ x_{1;n} + \dfrac{1}{2}k_{1,2},\ x_{2;n} + \dfrac{1}{2}k_{2,2}\right) \,, \\[2mm] k_{i,4} := hf_i\left(t_n + h,\ x_{1;n} + k_{1,3},\ x_{2;n} + k_{2,3}\right) \,. \end{cases}$$

It is important to note that both $k_{1,1}$ and $k_{2,1}$ must be computed before either $k_{1,2}$ or $k_{2,2}$. Similarly, both $k_{1,2}$ and $k_{2,2}$ are needed to compute $k_{1,3}$ and $k_{2,3}$, etc. In Appendix F, program outlines are given for applying the method to graph approximate solutions over a specified interval $[t_0, t_1]$ or to obtain approximations of the solutions at a specified point to within a desired tolerance.

Example 4 Use the classical fourth-order Runge–Kutta algorithm for systems to approximate the solution of the initial value problem

(19)
$$x_1' = 2x_1 - 2x_1x_2\,; \qquad x_1(0) = 1\,,$$
$$x_2' = x_1x_2 - x_2\,; \qquad x_2(0) = 3$$

at $t = 1$. Starting with $h = 1$, continue halving the step size until two successive approximations of $x_1(1)$ and of $x_2(1)$ differ by at most 0.0001.

Solution Here $f_1(t, x_1, x_2) = 2x_1 - 2x_1x_2$ and $f_2(t, x_1, x_2) = x_1x_2 - x_2$. With the inputs $t_0 = 0, x_{1;0} = 1$, $x_{2;0} = 3$, we proceed with the algorithm to compute $x_1(1; 1)$ and $x_2(1; 1)$, the approximations to $x_1(1), x_2(1)$ using $h = 1$. We find from the formulas in (18) that

$$k_{1,1} = h(2x_{1;0} - 2x_{1;0}x_{2;0}) = 2(1) - 2(1)(3) = -4\,,$$

$$k_{2,1} = h(x_{1;0}x_{2;0} - x_{2;0}) = (1)(3) - 3 = 0\,,$$

$$k_{1,2} = h\left[2\left(x_{1;0} + \tfrac{1}{2}k_{1,1}\right) - 2\left(x_{1;0} + \tfrac{1}{2}k_{1,1}\right)\left(x_{2;0} + \tfrac{1}{2}k_{2,1}\right)\right]$$

$$= 2\left[1 + \tfrac{1}{2}(-4)\right] - 2\left[1 + \tfrac{1}{2}(-4)\right]\left[3 + \tfrac{1}{2}(0)\right]$$

$$= -2 + 2(3) = 4\,,$$

$$k_{2,2} = h\left[\left(x_{1;0} + \tfrac{1}{2}k_{1,1}\right)\left(x_{2;0} + \tfrac{1}{2}k_{2,1}\right) - \left(x_{2;0} + \tfrac{1}{2}k_{2,1}\right)\right]$$

$$= \left[1 + \tfrac{1}{2}(-4)\right]\left[3 + \tfrac{1}{2}(0)\right] - \left[3 + \tfrac{1}{2}(0)\right]$$

$$= (-1)(3) - 3 = -6\,,$$

and similarly we compute

$$k_{1,3} = h\left[2\left(x_{1;0} + \tfrac{1}{2}k_{1,2}\right) - 2\left(x_{1;0} + \tfrac{1}{2}k_{1,2}\right)\left(x_{2;0} + \tfrac{1}{2}k_{2,2}\right)\right] = 6\,,$$

$$k_{2,3} = h\left[\left(x_{1;0} + \tfrac{1}{2}k_{1,2}\right)\left(x_{2;0} + \tfrac{1}{2}k_{2,2}\right) - \left(x_{2;0} + \tfrac{1}{2}k_{2,2}\right)\right] = 0\,,$$

$$k_{1,4} = h\left[2(x_{1;0} + k_{1,3}) - 2(x_{1;0} + k_{1,3})(x_{2;0} + k_{2,3})\right] = -28\,,$$

$$k_{2,4} = h\left[(x_{1;0} + k_{1,3})(x_{2;0} + k_{2,3}) - (x_{2;0} + k_{2,3})\right] = -18\,.$$

Inserting these values into formula (17), we get

$$x_{1;1} = x_{1;0} + \frac{1}{6}(k_{1,1} + 2k_{1,2} + 2k_{1,3} + k_{1,4})$$

$$= 1 + \frac{1}{6}(-4 + 8 + 12 - 28) = -1\,,$$

$$x_{2;1} = x_{2;0} + \frac{1}{6}(k_{2,1} + 2k_{2,2} + 2k_{2,3} + k_{2,4})\,,$$

$$= 3 + \frac{1}{6}(0 - 12 + 0 + 18) = 4\,,$$

as the respective approximations to $x_1(1)$ and $x_2(1)$.

Repeating the algorithm with $h = 1/2$ ($N = 2$) we obtain the approximations $x_1(1; 2^{-1})$ and $x_2(1; 2^{-1})$ for $x_1(1)$ and $x_2(1)$. In Table 5.2, we list the approximations $x_1(1; 2^{-m})$ and $x_2(1; 2^{-m})$ for $x_1(1)$ and $x_2(1)$ using step size $h = 2^{-m}$ for $m = 0, 1, 2, 3$, and 4. We stopped at $m = 4$, since both

$$|x_1(1; 2^{-3}) - x_1(1; 2^{-4})| = 0.00006 < 0.0001$$

and

$$|x_2(1; 2^{-3}) - x_2(1; 2^{-4})| = 0.00001 < 0.0001 .$$

Hence, $x_1(1) \approx 0.07735$ and $x_2(1) \approx 1.46445$, with tolerance 0.0001. ◆

TABLE 5.2	Approximations of the Solution to System (19) in Example 4		
m	h	$x_1(1; h)$	$x_2(1; h)$
0	1.0	-1.0	4.0
1	0.5	0.14662	1.47356
2	0.25	0.07885	1.46469
3	0.125	0.07741	1.46446
4	0.0625	0.07735	1.46445

To get a better feel for the solution to system (19), we have graphed in Figure 5.6 an approximation of the solution for $0 \leq t \leq 12$, using linear interpolation to connect the vectorized Runge–Kutta approximants for the points $t = 0, 0.125, 0.25, \ldots, 12.0$ (i.e., with $h = 0.125$). From the graph it appears that the components x_1 and x_2 are periodic in the variable t. Phase plane analysis is used in Section 5.5 to show that, indeed, Volterra–Lotka equations have periodic solutions.

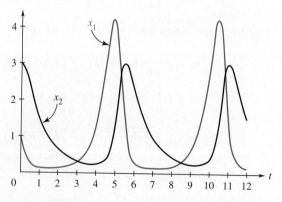

Figure 5.6 Graphs of the components of an approximate solution to the Volterra–Lotka system (17)

5.3 EXERCISES

In Problems 1–7, convert the given initial value problem into an initial value problem for a system in normal form.

1. $y''(t) + ty'(t) - 3y(t) = t^2$;
 $y(0) = 3$, $y'(0) = -6$

2. $y''(t) = \cos(t - y) + y^2(t)$;
 $y(0) = 1$, $y'(0) = 0$

3. $y^{(4)}(t) - y^{(3)}(t) + 7y(t) = \cos t$;
 $y(0) = y'(0) = 1$, $y''(0) = 0$, $y^{(3)}(0) = 2$

4. $y^{(6)}(t) = [y'(t)]^3 - \sin(y(t)) + e^{2t}$;
 $y(0) = y'(0) = \cdots = y^{(5)}(0) = 0$

5. $x'' + y - x' = 2t$; $x(3) = 5$, $x'(3) = 2$,
 $y'' - x + y = -1$; $y(3) = 1$, $y'(3) = -1$
 [*Hint:* Set $x_1 = x$, $x_2 = x'$, $x_3 = y$, $x_4 = y'$.]

6. $3x'' + 5x - 2y = 0$; $x(0) = -1$, $x'(0) = 0$,
 $4y'' + 2y - 6x = 0$; $y(0) = 1$, $y'(0) = 2$

7. $x''' - y = t$; $x(0) = x'(0) = x''(0) = 4$,
 $2x'' + 5y'' - 2y = 1$; $y(0) = y'(0) = 1$

8. Sturm–Liouville Form. A second-order equation is said to be in **Sturm–Liouville form** if it is expressed as

$$[p(t)y'(t)]' + q(t)y(t) = 0.$$

Show that the substitutions $x_1 = y, x_2 = py'$ result in the normal form

$$x_1' = x_2/p,$$
$$x_2' = -qx_1.$$

If $y(0) = a$ and $y'(0) = b$ are the initial values for the Sturm–Liouville problem, what are $x_1(0)$ and $x_2(0)$?

9. In Section 3.6, we discussed the improved Euler's method for approximating the solution to a first-order equation. Extend this method to normal systems and give the recursive formulas for solving the initial value problem.

In Problems 10–13, use the vectorized Euler method with $h = 0.25$ to find an approximation for the solution to the given initial value problem on the specified interval.

10. $y'' + ty' + y = 0$;
 $y(0) = 1$, $y'(0) = 0$ on $[0, 1]$

11. $(1 + t^2)y'' + y' - y = 0$;
 $y(0) = 1$, $y'(0) = -1$ on $[0, 1]$

12. $t^2y'' + y = t + 2$;
 $y(1) = 1$, $y'(1) = -1$ on $[1, 2]$

13. $y'' = t^2 - y^2$;
 $y(0) = 0$, $y'(0) = 1$ on $[0, 1]$
 (Can you guess the solution?)

In Problems 14–24, you will need a computer and a programmed version of the vectorized classical fourth-order Runge–Kutta algorithm. (At the instructor's discretion, other algorithms may be used.)[†]

14. Using the vectorized Runge–Kutta algorithm with $h = 0.5$, approximate the solution to the initial value problem

$$3t^2y'' - 5ty' + 5y = 0;$$
$$y(1) = 0, y'(1) = \frac{2}{3}$$

at $t = 8$. Compare this approximation to the actual solution $y(t) = t^{5/3} - t$.

15. Using the vectorized Runge–Kutta algorithm, approximate the solution to the initial value problem

$$y'' = t^2 + y^2; y(0) = 1, y'(0) = 0$$

at $t = 1$. Starting with $h = 1$, continue halving the step size until two successive approximations [of both $y(1)$ and $y'(1)$] differ by at most 0.01.

16. Using the vectorized Runge–Kutta algorithm for systems with $h = 0.125$, approximate the solution to the initial value problem

$$x' = 2x - y; x(0) = 0,$$
$$y' = 3x + 6y; y(0) = -2$$

at $t = 1$. Compare this approximation to the actual solution

$$x(t) = e^{5t} - e^{3t}, y(t) = e^{3t} - 3e^{5t}.$$

17. Using the vectorized Runge–Kutta algorithm, approximate the solution to the initial value problem

$$\frac{du}{dx} = 3u - 4v; u(0) = 1,$$
$$\frac{dv}{dx} = 2u - 3v; v(0) = 1$$

at $x = 1$. Starting with $h = 1$, continue halving the step size until two successive approximations of $u(1)$ and $v(1)$ differ by at most 0.001.

[†]Appendix G describes various websites and commercial software that sketch direction fields and automate most of the differential equation algorithms discussed in this book.

18. Combat Model. A simplified mathematical model for conventional versus guerrilla combat is given by the system

$$x_1' = -(0.1)x_1x_2 ; \qquad x_1(0) = 10 ,$$
$$x_2' = -x_1 ; \qquad\qquad x_2(0) = 15 ,$$

where x_1 and x_2 are the strengths of guerrilla and conventional troops, respectively, and 0.1 and 1 are the *combat effectiveness coefficients*. Who will win the conflict: the conventional troops or the guerrillas? [*Hint:* Use the vectorized Runge–Kutta algorithm for systems with $h = 0.1$ to approximate the solutions.]

19. Predator–Prey Model. The Volterra–Lotka predator–prey model predicts some rather interesting behavior that is evident in certain biological systems. For example, suppose you fix the initial population of prey but increase the initial population of predators. Then the population cycle for the prey becomes more severe in the sense that there is a long period of time with a reduced population of prey followed by a short period when the population of prey is very large. To demonstrate this behavior, use the vectorized Runge–Kutta algorithm for systems with $h = 0.5$ to approximate the populations of prey x and of predators y over the period $[0, 5]$ that satisfy the Volterra–Lotka system

$$x' = x(3 - y) ,$$
$$y' = y(x - 3)$$

under each of the following initial conditions:
(a) $x(0) = 2 ,$ $\qquad y(0) = 4 .$
(b) $x(0) = 2 ,$ $\qquad y(0) = 5 .$
(c) $x(0) = 2 ,$ $\qquad y(0) = 7 .$

20. In Project C of Chapter 4, it was shown that the simple pendulum equation

$$\theta''(t) + \sin \theta(t) = 0$$

has periodic solutions when the initial displacement and velocity are small. Show that the period of the solution may depend on the initial conditions by using the vectorized Runge–Kutta algorithm with $h = 0.02$ to approximate the solutions to the simple pendulum problem on $[0, 4]$ for the initial conditions:
(a) $\theta(0) = 0.1 ,$ $\qquad \theta'(0) = 0 .$
(b) $\theta(0) = 0.5 ,$ $\qquad \theta'(0) = 0 .$
(c) $\theta(0) = 1.0 ,$ $\qquad \theta'(0) = 0 .$

[*Hint:* Approximate the length of time it takes to reach $-\theta(0)$.]

21. Fluid Ejection. In the design of a sewage treatment plant, the following equation arises:[†]

$$60 - H = (77.7)H'' + (19.42)(H')^2 ;$$
$$H(0) = H'(0) = 0 ,$$

where H is the level of the fluid in an ejection chamber and t is the time in seconds. Use the vectorized Runge–Kutta algorithm with $h = 0.5$ to approximate $H(t)$ over the interval $[0, 5]$.

22. Oscillations and Nonlinear Equations. For the initial value problem

$$x'' + (0.1)(1 - x^2)x' + x = 0 ;$$
$$x(0) = x_0 , \qquad x'(0) = 0 ,$$

use the vectorized Runge–Kutta algorithm with $h = 0.02$ to illustrate that as t increases from 0 to 20, the solution x exhibits damped oscillations when $x_0 = 1$, whereas x exhibits expanding oscillations when $x_0 = 2.1$.

23. Nonlinear Spring. The **Duffing equation**

$$y'' + y + ry^3 = 0 ,$$

where r is a constant, is a model for the vibrations of a mass attached to a *nonlinear* spring. For this model, does the period of vibration vary as the parameter r is varied? Does the period vary as the initial conditions are varied? [*Hint:* Use the vectorized Runge–Kutta algorithm with $h = 0.1$ to approximate the solutions for $r = 1$ and 2, with initial conditions $y(0) = a$, $y'(0) = 0$ for $a = 1$, 2, and 3.]

24. Pendulum with Varying Length. A pendulum is formed by a mass m attached to the end of a wire that is attached to the ceiling. Assume that the length $l(t)$ of the wire varies with time in some predetermined fashion. If $\theta(t)$ is the angle in radians between the pendulum and the vertical, then the motion of the pendulum is governed for small angles by the initial value problem

$$l^2(t)\theta''(t) + 2l(t)l'(t)\theta'(t) + gl(t)\sin(\theta(t)) = 0 ;$$
$$\theta(0) = \theta_0 , \qquad \theta'(0) = \theta_1 ,$$

where g is the acceleration due to gravity. Assume that

$$l(t) = l_0 + l_1 \cos(\omega t - \phi) ,$$

where l_1 is much smaller than l_0. (This might be a model for a person on a swing, where the *pumping* action changes the distance from the center of mass of the swing to the point where the swing is attached.) To simplify the computations, take $g = 1$. Using the Runge–Kutta algorithm with $h = 0.1$, study the motion of the pendulum when $\theta_0 = 0.05$, $\theta_1 = 0$, $l_0 = 1$, $l_1 = 0.1$, $\omega = 1$, and $\phi = 0.02$. In particular, does the pendulum ever attain an angle greater in absolute value than the initial angle θ_0?

[†]See *Numerical Solution of Differential Equations*, by William Milne (Dover, New York, 1970), p. 82.

In Problems 25–30, use a software package or the SUB-ROUTINE in Appendix F.

25. Using the Runge–Kutta algorithm for systems with $h = 0.05$, approximate the solution to the initial value problem

$$y''' + y'' + y^2 = t \,;$$
$$y(0) = 1 \,, \qquad y'(0) = 0 \,, \qquad y''(0) = 1$$

at $t = 1$.

26. Use the Runge–Kutta algorithm for systems with $h = 0.1$ to approximate the solution to the initial value problem

$$\begin{aligned} x' &= yz \,; & x(0) &= 0 \,, \\ y' &= -xz \,; & y(0) &= 1 \,, \\ z' &= -xy/2 \,; & z(0) &= 1 \,, \end{aligned}$$

at $t = 1$.

27. Generalized Blasius Equation. H. Blasius, in his study of laminar flow of a fluid, encountered an equation of the form

$$y''' + yy'' = (y')^2 - 1 \,.$$

Use the Runge–Kutta algorithm for systems with $h = 0.1$ to approximate the solution that satisfies the initial conditions $y(0) = 0, y'(0) = 0$, and $y''(0) = 1.32824$. Sketch this solution on the interval $[0, 2]$.

28. Lunar Orbit. The motion of a moon moving in a planar orbit about a planet is governed by the equations

$$\frac{d^2 x}{dt^2} = -G \frac{mx}{r^3} \,, \qquad \frac{d^2 y}{dt^2} = -G \frac{my}{r^3} \,,$$

where $r := (x^2 + y^2)^{1/2}$, G is the gravitational constant, and m is the mass of the planet. Assume $Gm = 1$. When $x(0) = 1, x'(0) = y(0) = 0$, and $y'(0) = 1$, the motion is a circular orbit of radius 1 and period 2π.
 (a) Setting $x_1 = x$, $x_2 = x'$, $x_3 = y$, $x_4 = y'$, express the governing equations as a first-order system in normal form.
 (b) Using $h = 2\pi/100 \approx 0.0628318$, compute one orbit of this moon (i.e., do $N = 100$ steps.). Do your approximations agree with the fact that the orbit is a circle of radius 1?

29. Competing Species. Let $p_i(t)$ denote, respectively, the populations of three competing species S_i, $i = 1, 2, 3$.

Suppose these species have the same growth rates, and the maximum population that the habitat can support is the same for each species. (We assume it to be one unit.) Also suppose the competitive advantage that S_1 has over S_2 is the same as that of S_2 over S_3 and S_3 over S_1. This situation is modeled by the system

$$\begin{aligned} p_1' &= p_1(1 - p_1 - ap_2 - bp_3) \,, \\ p_2' &= p_2(1 - bp_1 - p_2 - ap_3) \,, \\ p_3' &= p_3(1 - ap_1 - bp_2 - p_3) \,, \end{aligned}$$

where a and b are positive constants. To demonstrate the population dynamics of this system when $a = b = 0.5$, use the Runge–Kutta algorithm for systems with $h = 0.1$ to approximate the populations p_i over the time interval $[0, 10]$ under each of the following initial conditions:

 (a) $p_1(0) = 1.0 \,, \quad p_2(0) = 0.1 \,, \quad p_3(0) = 0.1 \,.$
 (b) $p_1(0) = 0.1 \,, \quad p_2(0) = 1.0 \,, \quad p_3(0) = 0.1 \,.$
 (c) $p_1(0) = 0.1 \,, \quad p_2(0) = 0.1 \,, \quad p_3(0) = 1.0 \,.$

On the basis of the results of parts (a)–(c), decide what you think will happen to these populations as $t \to +\infty$.

30. Spring Pendulum. Let a mass be attached to one end of a spring with spring constant k and the other end attached to the ceiling. Let l_0 be the natural length of the spring and let $l(t)$ be its length at time t. If $\theta(t)$ is the angle between the pendulum and the vertical, then the motion of the spring pendulum is governed by the system

$$l''(t) - l(t)\theta'(t) - g\cos\theta(t) + \frac{k}{m}(l - l_0) = 0 \,,$$
$$l^2(t)\theta''(t) + 2l(t)l'(t)\theta'(t) + gl(t)\sin\theta(t) = 0 \,.$$

Assume $g = 1$, $k = m = 1$, and $l_0 = 4$. When the system is at rest, $l = l_0 + mg/k = 5$.
 (a) Describe the motion of the pendulum when $l(0) = 5.5, l'(0) = 0, \theta(0) = 0$, and $\theta'(0) = 0$.
 (b) When the pendulum is both stretched and given an angular displacement, the motion of the pendulum is more complicated. Using the Runge–Kutta algorithm for systems with $h = 0.1$ to approximate the solution, sketch the graphs of the length l and the angular displacement θ on the interval $[0, 10]$ if $l(0) = 5.5$, $l'(0) = 0, \theta(0) = 0.5$, and $\theta'(0) = 0$.

5.4 Introduction to the Phase Plane

In this section, we study systems of two first-order equations of the form

$$\frac{dx}{dt} = f(x,y) ,$$

(1)

$$\frac{dy}{dt} = g(x,y) .$$

Note that the independent variable t does not appear in the right-hand terms $f(x, y)$ and $g(x, y)$; such systems are called **autonomous.** For example, the system that modeled the interconnected tanks problem in Section 5.1,

$$x' = -\frac{1}{3}x + \frac{1}{12}y ,$$
$$y' = \frac{1}{3}x - \frac{1}{3}y ,$$

is autonomous. So is the Volterra–Lotka system,

$$x' = Ax - Bxy ,$$
$$y' = -Cy + Dxy ,$$

(with A, B, C, D constants), which was discussed in Example 4 of Section 5.3 as a model for population dynamics.

For future reference, we note that the solutions to autonomous systems have a "time-shift immunity," in the sense that if the pair $x(t), y(t)$ solves (1), so does the **time-shifted** pair $x(t+c), y(t+c)$ for any constant c. Specifically, if we let $X(t) := x(t+c)$ and $Y(t) := y(t+c)$, then by the chain rule

$$\frac{dX}{dt}(t) = \frac{dx}{dt}(t+c) = f\big(x(t+c), y(t+c)\big) = f\big(X(t), Y(t)\big) ,$$

$$\frac{dY}{dt}(t) = \frac{dy}{dt}(t+c) = g\big(x(t+c), y(t+c)\big) = g\big(X(t), Y(t)\big) ,$$

proving that $X(t), Y(t)$ is also a solution to (1).

Since a solution to the autonomous system is a pair of functions $x(t), y(t)$ that satisfies (1) for all t in some interval I, it can be visualized as a pair of graphs, as shown in Figure 5.7. For purposes of comparing one of the solution pair with its mate, we can compress these two graphs into one by charting the path in the plane traced out by the ordered pair $(x(t), y(t))$ as t increases.

Trajectories and the Phase Plane

Definition 1. If $x(t), y(t)$ is a solution pair to (1) for t in the interval I, then a plot in the xy-plane of the parametrized curve $x = x(t), y = y(t)$ for t in I, together with arrows indicating its direction with increasing t, is said to be a **trajectory** of the system (1). In such a context we call the xy-plane the **phase plane** and refer to a representative set of trajectories in this plane as a **phase portrait** of the system.

A trajectory provides a visualization of the motion of a particle in the plane that starts at some point (x_0, y_0) and follows a path as time increases that is determined by the solution to the autonomous system with initial conditions $x(0) = x_0, y(0) = y_0$. The trajectory for the

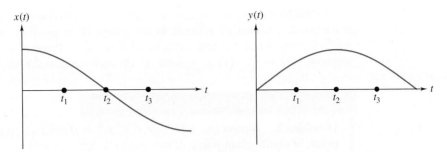

Figure 5.7 Solution pair for system (1)

solution pair of Figure 5.7 is shown in Figure 5.8. Note, however, that the trajectory contains less information than the individual graphs for $x(t)$ and $y(t)$, because the t-dependence has been suppressed. (We can't tell if the particle's speed is fast or slow.) In principle we can construct, point by point, the trajectory from the solution graphs, but we cannot reconstruct the solution graphs from the phase plane trajectory alone (because we would not know what value of t to assign to each point). Nonetheless, trajectories provide valuable information about the qualitative behavior of solutions to the autonomous system.

Can we find trajectories without knowing a solution pair to the system? Yes; the key is to observe that since t does not appear explicitly on the right-hand side of system (1), if we divide the two equations and invoke the chain rule

$$\frac{dy}{dx} = \frac{dy/dt}{dx/dt},$$

we get the "related phase plane differential equation" (*without t*)

(2) $$\frac{dy}{dx} = \frac{g(x,y)}{f(x,y)}.$$

We mastered equations like (2) in Chapter 2; their solution curves provide highways along which the trajectories travel. What is missing from these highways is the orientation of a trajectory, which can be gleaned from the autonomous system (1) by determining the sign of f or g at a point and recalling that $(dx/dt, dy/dt) = (f, g)$ is the velocity of the moving particle. Furthermore, as illustrated in Example 1 below, a single curve arising from the related phase plane differential equation might well accommodate several different trajectories.

As we shall see in the examples to follow as well as in Figure 5.12, trajectories are typically of the following types: (i) a single point, indicating a stationary particle; (ii) a non-closed arc, without self-intersections; or (iii) a closed curve (cycle), indicating periodic motion. Moreover, thanks to the Existence-Uniqueness Theorem for first-order equations in Chapter 1 (page 11), as long as the functions f and g in (1) are continuously differentiable in the plane, the trajectories of a phase portrait will not intersect each other (unless they coincide).

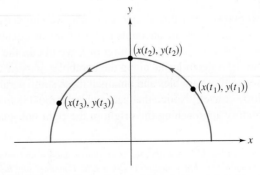

Figure 5.8 Phase plane trajectory of the solution pair for system (1)

Except for the very special case of linear systems with constant coefficients that was discussed in Section 5.2, finding all solutions to the system (1) is generally an impossible task. But it is relatively easy to find *constant solutions;* if $f(x_0, y_0) = 0$ and $g(x_0, y_0) = 0$, then the constant functions $x(t) \equiv x_0$, $y(t) \equiv y_0$ solve (1). For such solutions the following terminology is used.

Critical Points and Equilibrium Solutions

Definition 2. A point (x_0, y_0) where $f(x_0, y_0) = 0$ and $g(x_0, y_0) = 0$ is called a **critical point,** or **equilibrium point,** of the system $dx/dt = f(x, y)$, $dy/dt = g(x, y)$, and the corresponding constant solution $x(t) \equiv x_0, y(t) \equiv y_0$ is called an **equilibrium solution.**

Notice that trajectories of equilibrium solutions consist of just single points (the equili-brium points). But what can be said about the other trajectories? Can we predict any of their features from closer examination of the equilibrium points? To explore this we focus on the related phase plane differential equation (2) and exploit its direction field (recall Section 1.3, page 15). However, we'll augment the direction field plot by attaching arrowheads to the line segments, indicating the direction of the "flow" of solutions as t increases. This is easy: When dx/dt is positive, $x(t)$ increases so the trajectory flows to the right. Therefore, according to (1), all direction field segments drawn in a region where $f(x, y)$ is positive should point to the right [and, of course, they point to the left if $f(x, y)$ is negative]. If $f(x, y)$ is zero, we can use $g(x, y)$ to decide if the flow is upward [$y(t)$ increases] or downward [$y(t)$ decreases]. [*What if both $f(x, y)$ and $g(x, y)$ are zero?*]

In the examples that follow, one can use computers or calculators for generating these direction fields.

Example 1 Sketch the direction field in the phase plane for the system

$$(3) \quad \begin{aligned} \frac{dx}{dt} &= -x, \\[1mm] \frac{dy}{dt} &= -2y \end{aligned}$$

and identify its critical point.

Solution Here $f(x, y) = -x$ and $g(x, y) = -2y$ are both zero when $x = y = 0$, so $(0, 0)$ is the critical point. The direction field for the related phase plane differential equation

$$(4) \quad \frac{dy}{dx} = \frac{-2y}{-x} = \frac{2y}{x}$$

is given in Figure 5.9 on page 265. Since $dx/dt = -x$ in (3), trajectories in the right half-plane (where $x > 0$) flow to the left, and vice versa. From the figure we can see that all solutions "flow into" the critical point $(0, 0)$. Such a critical point is called **asymptotically stable.**[†] ◆

Remark. For this simple example, we can actually solve the system (3) explicitly; indeed, (3) constitutes an uncoupled pair of linear equations whose solutions are $x(t) = c_1 e^{-t}$ and $y(t) = c_2 e^{-2t}$. By elimination of t, we obtain the equation $y = c_2 e^{-2t} = c_2[x(t)/c_1]^2 = cx^2$. So the trajectories lie along the parabolas $y = cx^2$. [Alternatively, we could have solved (4) by separating variables and obtained these same parabolas as the "highways" on which the trajectories travel.] Notice that each such parabola is made up of three trajectories: an incoming trajectory approaching the origin in the right half-plane; its mirror-image trajectory approaching

[†]See Section 12.3 for a rigorous exposition of stability and critical points. All references to Chapters 11–13 refer to the expanded text, *Fundamentals of Differential Equations and Boundary Value Problems,* 7th ed.

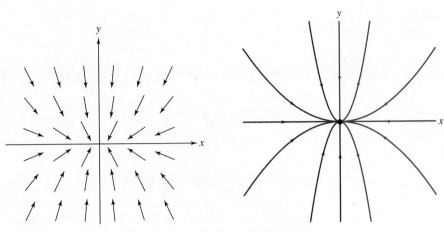

Figure 5.9 Direction field for Example 1 **Figure 5.10** Trajectories for Example 1

the origin in the left half-plane; and the origin itself, an equilibrium point. Sample trajectories are indicated in Figure 5.10.

Example 2 Sketch the direction field in the phase plane for the system

$$(5) \qquad \begin{aligned} \frac{dx}{dt} &= x\,, \\ \frac{dy}{dt} &= 2y \end{aligned}$$

and describe the behavior of solutions near the critical point $(0, 0)$.

Solution This example is almost identical to the previous one; in fact, one could say we have merely "reversed time" in (3). The direction field segments for

$$(6) \qquad \frac{dy}{dx} = \frac{2y}{x}$$

are the same as those of (4), but the direction arrows are reversed. Now all solutions flow *away* from the critical point $(0, 0)$; the equilibrium is **unstable.** ◆

Example 3 For the system (7) below, find the critical points, sketch the direction field in the phase plane, and predict the asymptotic nature (i.e., behavior as $t \to +\infty$) of the solution starting at $x = 2$, $y = 0$ when $t = 0$.

$$(7) \qquad \begin{aligned} \frac{dx}{dt} &= 5x - 3y - 2\,, \\ \frac{dy}{dt} &= 4x - 3y - 1\,. \end{aligned}$$

Solution The only critical point is the solution of the simultaneous equations $f(x, y) = g(x, y) = 0$:

$$(8) \qquad \begin{aligned} 5x_0 - 3y_0 - 2 &= 0\,, \\ 4x_0 - 3y_0 - 1 &= 0\,, \end{aligned}$$

from which we find $x_0 = y_0 = 1$. The direction field for the related phase plane differential equation

Figure 5.11 Direction field, trajectories, and a typical solution pair for a trajectory (Example 3)

(9) $$\dfrac{dy}{dx} = \dfrac{4x - 3y - 1}{5x - 3y - 2}$$

is shown in Figure 5.11, along with some trajectories and a typical solution pair.[†] Note that solutions flow to the right for $5x - 3y - 2 > 0$, i.e., for all points *below* the line $5x - 3y - 2 = 0$, and that they slow down near the critical point.

The solution curve to (9) passing through $(2, 0)$ in Figure 5.11 apparently is a "highway" extending to infinity. Does this imply the corresponding *system* solution $x(t), y(t)$ also approaches infinity in the sense that $|x(t)| + |y(t)| \to +\infty$ as $t \to +\infty$, or could its trajectory "stall" at some point along the highway, or possibly even "backtrack?" *It cannot backtrack,* because the direction arrows along the trajectory point, unambiguously, to the right. And if $\big(x(t), y(t)\big)$ stalls at some point (x_1, y_1), then intuitively we would conclude that (x_1, y_1) was an equilibrium point (since the "speeds" dx/dt and dy/dt would approach zero there). But we have already found the only critical point. So we conclude, with a high degree of confidence,[‡] that the system solution does indeed go to infinity.

The critical point $(1, 1)$ is **unstable** because, although many solutions get arbitrarily close to $(1, 1)$, most of them eventually flow away. Solutions that lie on the line $y = 2x - 1$, however, do converge to $(1, 1)$. Such an equilibrium is an example of a **saddle point.** ◆

In the preceding example, we informally argued that if a trajectory "stalls"—that is, if it has an endpoint—then this endpoint would have to be a critical point. This is more carefully stated in the following theorem, whose proof is outlined in Problem 30.

Endpoints Are Critical Points

Theorem 1. Let the pair $x(t), y(t)$ be a solution on $[0, +\infty)$ to the autonomous system $dx/dt = f(x, y), dy/dt = g(x, y)$, where f and g are continuous in the plane. If the limits

$$x^* := \lim_{t \to +\infty} x(t) \qquad \text{and} \qquad y^* := \lim_{t \to +\infty} y(t)$$

exist and are finite, then the point (x^*, y^*) is a critical point for the system.

[†]Solutions to equation (9) can be obtained analytically using the methods of Section 2.6.

[‡]These informal arguments are made more rigorous in Chapter 12. All references to Chapters 11–13 refer to the expanded text, *Fundamentals of Differential Equations and Boundary Value Problems,* 7th edition.

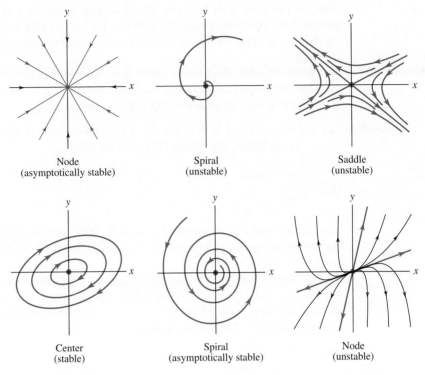

Figure 5.12 Examples of different trajectory behaviors near critical point at origin

Some typical trajectory configurations near critical points are displayed and classified in Figure 5.12. These phase plane portraits were generated by software packages having trajectory-sketching options[†]. A more complete discussion of the nature of various types of equilibrium solutions and their stability is deferred to Chapter 12.[‡] For the moment, however, notice that unstable critical points are distinguished by "runaway" trajectories emanating from arbitrarily nearby points, while stable equilibria "trap" all neighboring trajectories. The asymptotically stable critical points *attract* their neighboring trajectories as $t \rightarrow +\infty$.

Historically, the phase plane was introduced to facilitate the analysis of mechanical systems governed by Newton's second law, force equals mass times acceleration. An autonomous mechanical system arises when this force is independent of time and can be modeled by a second-order equation of the form

(10) $y'' = f(y, y')$.

As we have seen in Section 5.3, this equation can be converted to a normal first-order system by introducing the velocity $v = dy/dt$ and writing

(11)
$$\frac{dy}{dt} = v ,$$
$$\frac{dv}{dt} = f(y, v) .$$

[†]Appendix G describes various websites and commercial software that sketch direction fields and automate most of the differential equation algorithms discussed in this book.

[‡]All references to Chapters 11–13 refer to the expanded text, *Fundamentals of Differential Equations and Boundary Value Problems,* 7th edition.

Thus, we can analyze the behavior of an autonomous mechanical system by studying its phase portrait in the yv-plane. Notice that with v as the vertical axis, trajectories $(y(t), v(t))$ flow to the right in the upper half-plane (where $v > 0$), and to the left in the lower half-plane.

Example 4 Sketch the direction field in the phase plane for the first-order system corresponding to the unforced, undamped mass–spring oscillator described in Section 4.1 (Figure 4.1, page 152). Sketch several trajectories and interpret them physically.

Solution The equation derived in Section 4.1 for this oscillator is $my'' + ky = 0$ or, equivalently, $y'' = -ky/m$. Hence, the system (11) takes the form

$$
\begin{aligned}
y' &= v, \\
v' &= -\frac{ky}{m}.
\end{aligned}
\tag{12}
$$

The critical point is at the origin $y = v = 0$. The direction field in Figure 5.13 indicates that the trajectories appear to be either closed curves (ellipses?) or spirals that encircle the critical point.

We saw in Section 4.9 that the undamped oscillator motions are periodic; they cycle repeatedly through the same sets of points, with the same velocities. Their trajectories in the phase plane, then, must be closed curves.[†] Let's confirm this mathematically by solving the related phase plane differential equation

$$
\frac{dv}{dy} = -\frac{ky}{mv}.
\tag{13}
$$

Equation (13) is separable, and we find

$$
v\, dv = -\frac{ky}{m} dy \qquad \text{or} \qquad d\left(\frac{v^2}{2}\right) = -\frac{k}{m} d\left(\frac{y^2}{2}\right),
$$

Figure 5.13 Direction field for Example 4

[†]By the same reasoning, *underdamped* oscillations would correspond to spiral trajectories asymptotically approaching the origin as $t \to +\infty$.

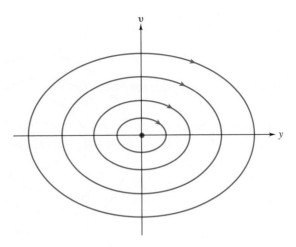

Figure 5.14 Trajectories for Example 4

so its solutions are the ellipses $v^2/2 + ky^2/2m = C$ as shown in Figure 5.14. The trajectories of (12) are confined to these ellipses, and hence neither come arbitrarily close to nor wander arbitrarily far from the equilibrium solution. The critical point is thus identified as a **center** in Figure 5.12 on page 267.

Furthermore, the system solutions must continually circulate around the ellipses, since there are no critical points to stop them. This confirms that all solutions are periodic. ◆

Remark. More generally, we argue that if a solution to an autonomous system like (1) passes through a point in the phase plane *twice* and if it is sufficiently well behaved to satisfy a *uniqueness theorem,* then the second "tour" satisfies the same initial conditions as the first tour and so must replicate it. In other words, *closed trajectories containing no critical points correspond to periodic solutions.*

Through these examples we have seen how, by studying the phase plane, one can often anticipate some of the features (boundedness, periodicity, etc.) of solutions of autonomous systems without solving them explicitly. Much of this information can be predicted simply from the critical points and the direction field (oriented by arrowheads), which are obtainable through standard software packages. The final example ties together several of these ideas.

Example 5 Find the critical points and solve the related phase plane equation for

(14)
$$\frac{dx}{dt} = -y(y-2) \,,$$
$$\frac{dy}{dt} = (x-2)(y-2) \,.$$

What is the asymptotic behavior of the solutions starting from $(3,0)$, $(5,0)$, and $(2,3)$?

Solution To find the critical points, we solve the system

$$-y(y-2) = 0, \qquad (x-2)(y-2) = 0 \,.$$

One family of solutions to this system is given by $y = 2$ with x arbitrary; that is, the line $y = 2$. If $y \neq 2$, then the system simplifies to $-y = 0$, and $x - 2 = 0$, which has the solution $x = 2$, $y = 0$. Hence, the critical point set consists of the isolated point $(2,0)$ and the horizontal line $y = 2$. The corresponding equilibrium solutions are $x(t) \equiv 2, y(t) \equiv 0$, and the family $x(t) \equiv c, y(t) \equiv 2$, where c is an arbitrary constant.

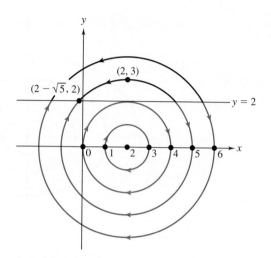

Figure 5.15 Phase portrait for Example 5

The related phase plane differential equation for the system (14) is

$$\textbf{(15)} \qquad \frac{dy}{dx} = \frac{(x-2)(y-2)}{-y(y-2)} = -\frac{x-2}{y}.$$

Solving this equation by separating variables,

$$y\,dy = -(x-2)\,dx \qquad \text{or} \qquad y^2 + (x-2)^2 = C,$$

demonstrates that the trajectories lie on concentric circles centered at $(2,0)$. See Figure 5.15.

Next we analyze the flow along each trajectory. From the equation $dx/dt = -y(y-2)$, we see that x is decreasing when $y > 2$. This means the flow is from right to left along the arc of a circle that lies above the line $y = 2$. For $0 < y < 2$, we have $dx/dt > 0$, so in this region the flow is from left to right. Furthermore, for $y < 0$, we have $dx/dt < 0$, and again the flow is from right to left.

We now observe in Figure 5.15 that there are four types of trajectories associated with system (14): (a) those that begin above the line $y = 2$ and follow the arc of a circle counterclockwise back to that line; (b) those that begin below the line $y = 2$ and follow the arc of a circle clockwise back to that line; (c) those that continually move clockwise around a circle centered at $(2,0)$ with radius less than 2 (i.e., they do not intersect the line $y = 2$); and finally, (d) the critical points $(2,0)$ and $y = 2$, x arbitrary.

The solution starting at $(3,0)$ lies on a circle with no critical points; therefore, it is a *periodic* solution, and the critical point $(2,0)$ is a center. But the circle containing the solutions starting at $(5,0)$ and at $(2,3)$ has critical points at $(2-\sqrt{5},2)$ and $(2+\sqrt{5},2)$. The direction arrows indicate that both solutions approach $(2-\sqrt{5},2)$ asymptotically (as $t \to +\infty$). They lie on the same solution curve for the related phase plane differential equation (a circle), but they are quite different trajectories. ◆

Note that for the system (14) the critical points on the line $y = 2$ are not isolated, so they do not fit into any of the categories depicted in Figure 5.12 on page 267. Observe also that all solutions of this system are bounded, since they are confined to circles.

5.4 EXERCISES

In Problems 1 and 2, verify that the pair $x(t)$, $y(t)$ is a solution to the given system. Sketch the trajectory of the given solution in the phase plane.

1. $\dfrac{dx}{dt} = 3y^3$, $\quad \dfrac{dy}{dt} = y$;

$x(t) = e^{3t}$, $\quad y(t) = e^t$

2. $\dfrac{dx}{dt} = 1$, $\quad \dfrac{dy}{dt} = 3x^2$;

$x(t) = t + 1$, $\quad y(t) = t^3 + 3t^2 + 3t$

In Problems 3–6, find the critical point set for the given system.

3. $\dfrac{dx}{dt} = x - y$,

$\dfrac{dy}{dt} = x^2 + y^2 - 1$

4. $\dfrac{dx}{dt} = y - 1$,

$\dfrac{dy}{dt} = x + y + 5$

5. $\dfrac{dx}{dt} = x^2 - 2xy$,

$\dfrac{dy}{dt} = 3xy - y^2$

6. $\dfrac{dx}{dt} = y^2 - 3y + 2$,

$\dfrac{dy}{dt} = (x - 1)(y - 2)$

In Problems 7–9, solve the related phase plane differential equation (2), page 263, for the given system.

7. $\dfrac{dx}{dt} = y - 1$,

$\dfrac{dy}{dt} = e^{x+y}$

8. $\dfrac{dx}{dt} = x^2 - 2y^{-3}$,

$\dfrac{dy}{dt} = 3x^2 - 2xy$

9. $\dfrac{dx}{dt} = 2y - x$,

$\dfrac{dy}{dt} = e^x + y$

10. Find all the critical points of the system

$\dfrac{dx}{dt} = x^2 - 1$,

$\dfrac{dy}{dt} = xy$,

and the solution curves for the related phase plane differential equation. Thereby prove that there are two trajectories that lie on semicircles. What are the endpoints of the semicircles?

In Problems 11–14, solve the related phase plane differential equation for the given system. Then sketch by hand several representative trajectories (with their flow arrows).

11. $\dfrac{dx}{dt} = 2y$,

$\dfrac{dy}{dt} = 2x$

12. $\dfrac{dx}{dt} = -8y$,

$\dfrac{dy}{dt} = 18x$

13. $\dfrac{dx}{dt} = (y - x)(y - 1)$,

$\dfrac{dy}{dt} = (x - y)(x - 1)$

14. $\dfrac{dx}{dt} = \dfrac{3}{y}$,

$\dfrac{dy}{dt} = \dfrac{2}{x}$

 In Problems 15–18, find all critical points for the given system. Then use a software package to sketch the direction field in the phase plane and from this describe the stability of the critical points (i.e., compare with Figure 5.12).

15. $\dfrac{dx}{dt} = 2x + y + 3$,

$\dfrac{dy}{dt} = -3x - 2y - 4$

16. $\dfrac{dx}{dt} = -5x + 2y$,

$\dfrac{dy}{dt} = x - 4y$

17. $\dfrac{dx}{dt} = 2x + 13y$,

$\dfrac{dy}{dt} = -x - 2y$

18. $\dfrac{dx}{dt} = x(7 - x - 2y)$,

$\dfrac{dy}{dt} = y(5 - x - y)$

 In Problems 19–24, convert the given second-order equation into a first-order system by setting $v = y'$. Then find all the critical points in the yv-plane. Finally, sketch (by hand or software) the direction fields, and describe the stability of the critical points (i.e., compare with Figure 5.12).

19. $\dfrac{d^2y}{dt^2} - y = 0$

20. $\dfrac{d^2y}{dt^2} + y = 0$

21. $\dfrac{d^2y}{dt^2} + y + y^5 = 0$

22. $\dfrac{d^2y}{dt^2} + y^3 = 0$

23. $y''(t) + y(t) - y(t)^4 = 0$

24. $y''(t) + y(t) - y(t)^3 = 0$

25. Using software, sketch the direction field in the phase plane for the system

$dx/dt = y$,

$dy/dt = -x + x^3$.

From the sketch, conjecture whether the solution passing through each given point is periodic:

(a) $(0.25, 0.25)$ **(b)** $(2, 2)$ **(c)** $(1, 0)$

26. Using software, sketch the direction field in the phase plane for the system

$dx/dt = y$,

$dy/dt = -x - x^3$.

From the sketch, conjecture whether all solutions of this system are bounded. Solve the related phase plane differential equation and confirm your conjecture.

27. Using software, sketch the direction field in the phase plane for the system

$$dx/dt = -2x + y,$$
$$dy/dt = -5x - 4y.$$

From the sketch, predict the asymptotic limit (as $t \to +\infty$) of the solution starting at $(1, 1)$.

28. Figure 5.16 displays some trajectories for the system

$$dx/dt = y,$$
$$dy/dt = -x + x^2.$$

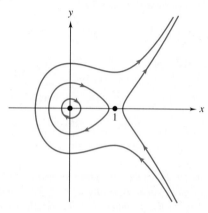

Figure 5.16 Phase plane for Problem 28

What types of critical points (compare Figure 5.12 on page 267) occur at $(0, 0)$ and $(1, 0)$?

29. Find the critical points and solve the related phase plane differential equation for the system

$$\frac{dx}{dt} = (x - 1)(y - 1)$$
$$\frac{dy}{dt} = y(y - 1).$$

Describe (without using computer software) the asymptotic behavior of trajectories (as $t \to \infty$) that start at **(a)** $(3, 2)$, **(b)** $(2, 1/2)$, **(c)** $(-2, 1/2)$, **(d)** $(3, -2)$.

30. A proof of Theorem 1, page 266, is outlined below. The goal is to show that $f(x^*, y^*) = g(x^*, y^*) = 0$. Justify each step.
 (a) From the given hypotheses, deduce that $\lim_{t \to +\infty} x'(t) = f(x^*, y^*)$ and $\lim_{t \to +\infty} y'(t) = g(x^*, y^*)$.
 (b) Suppose $f(x^*, y^*) > 0$. Then, by continuity, $x'(t) > f(x^*, y^*)/2$ for all large t (say, for $t \geq T$). Deduce from this that $x(t) > tf(x^*, y^*)/2 + C$ for $t > T$, where C is some constant.
 (c) Conclude from part (b) that $\lim_{t \to +\infty} x(t) = +\infty$, contradicting the fact that this limit is the finite number x^*. Thus, $f(x^*, y^*)$ cannot be positive.

(d) Argue similarly that the supposition that $f(x^*, y^*) < 0$ also leads to a contradiction; hence, $f(x^*, y^*)$ must be zero.
 (e) In the same manner, argue that $g(x^*, y^*)$ must be zero.

Therefore, $f(x^*, y^*) = g(x^*, y^*) = 0$, and (x^*, y^*) is a critical point.

31. Phase plane analysis provides a quick derivation of the energy integral lemma of Section 4.8 (page 201). By completing the following steps, prove that solutions of equations of the special form $y'' = f(y)$ satisfy

$$\frac{1}{2}(y')^2 - F(y) = \text{constant},$$

where $F(y)$ is an antiderivative of $f(y)$.
 (a) Set $v = y'$ and write $y'' = f(y)$ as an equivalent first-order system.
 (b) Show that the solutions to the yv-phase plane equation for the system in part (a) satisfy $v^2/2 = F(y) + K$. Replacing v by y' then completes the proof.

32. Use the result of Problem 31 to prove that all solutions to the equation

$$y'' + y^3 = 0$$

remain bounded. [*Hint:* Argue that $y^4/4$ is bounded above by the constant appearing in Problem 31.]

33. A Problem of Current Interest. The motion of an iron bar attracted by the magnetic field produced by a parallel current wire and restrained by springs (see Figure 5.17) is governed by the equation

$$\frac{d^2x}{dt^2} = -x + \frac{1}{\lambda - x}, \quad \text{for} \quad -x_0 < x < \lambda,$$

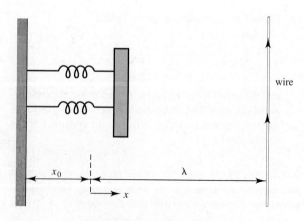

Figure 5.17 Bar restrained by springs and attracted by a parallel current

where the constants x_0 and λ are, respectively, the distances from the bar to the wall and to the wire *when the bar is at equilibrium (rest) with the current off.*

(a) Setting $v = dx/dt$, convert the second-order equation to an equivalent first-order system.

(b) Solve the related phase plane differential equation for the system in part (a) and thereby show that its solutions are given by

$$v = \pm \sqrt{C - x^2 - 2\ln(\lambda - x)}\,,$$

where C is a constant.

(c) Show that if $\lambda < 2$ there are no critical points in the xv-phase plane, whereas if $\lambda > 2$ there are two critical points. For the latter case, determine these critical points.

(d) Physically, the case $\lambda < 2$ corresponds to a current so high that the magnetic attraction completely overpowers the spring. To gain insight into this, use software to plot the phase plane diagrams for the system when $\lambda = 1$ and when $\lambda = 3$.

(e) From your phase plane diagrams in part (d), describe the possible motions of the bar when $\lambda = 1$ and when $\lambda = 3$, under various initial conditions.

34. Falling Object. The motion of an object moving vertically through the air is governed by the equation

$$\frac{d^2y}{dt^2} = -g - \frac{g}{V^2}\frac{dy}{dt}\left|\frac{dy}{dt}\right|,$$

where y is the upward vertical displacement and V is a constant called the terminal speed. Take $g = 32$ ft/sec^2 and $V = 50$ ft/sec. Sketch trajectories in the yv-phase plane for $-100 \le y \le 100$, $-100 \le v \le 100$, starting from $y = 0$ and $v = -75, -50, -25, 0, 25, 50,$ and 75 ft/sec. Interpret the trajectories physically; why is V called the terminal speed?

35. Sticky Friction. An alternative for the damping friction model $F = -by'$ discussed in Section 4.1 is the "sticky friction" model. For a mass sliding on a surface as depicted in Figure 5.18, the contact friction is more complicated than simply $-by'$. We observe, for example, that even if the mass is displaced slightly off the equilibrium location $y = 0$, it may nonetheless remain stationary due to the fact that the spring force $-ky$ is insufficient to break the static friction's grip. If the maximum force that the friction can exert is denoted by μ, then a feasible model is given by

$$F_{\text{friction}} = \begin{cases} ky\,, & \text{if } |ky| < \mu \\ & \text{and } y' = 0\,, \\ \mu\,\text{sign}(y)\,, & \text{if } |ky| \ge \mu \\ & \text{and } y' = 0\,, \\ -\mu\,\text{sign}(y')\,, & \text{if } y' \ne 0\,. \end{cases}$$

(The function sign (s) is $+1$ when $s > 0$, -1 when $s < 0$, and 0 when $s = 0$.) The motion is governed by the equation

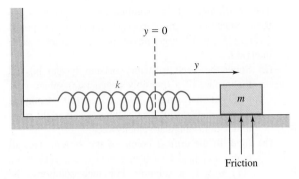

Figure 5.18 Mass–spring system with friction

(16) $$m\frac{d^2y}{dt^2} = -ky + F_{\text{friction}}\,.$$

Thus, if the mass is at rest, friction *balances* the spring force if $|y| < \mu/k$ but simply *opposes* it with intensity μ if $|y| \ge \mu/k$. If the mass is moving, friction opposes the velocity with the same intensity μ.

(a) Taking $m = \mu = k = 1$, convert (16) into the first-order system

$$y' = v\,,$$

(17) $$v' = \begin{cases} 0\,, & \text{if } |y| < 1 \\ & \text{and } v = 0\,, \\ -y + \text{sign}(y)\,, & \text{if } |y| \ge 1 \\ & \text{and } v = 0\,, \\ -y - \text{sign}(v)\,, & \text{if } v \ne 0\,. \end{cases}$$

(b) Form the phase plane equation for (17) when $v \ne 0$ and solve it to derive the solutions

$$v^2 + (y \pm 1)^2 = c\,,$$

where the plus sign prevails for $v > 0$ and the minus sign for $v < 0$.

(c) Identify the trajectories in the phase plane as two families of concentric semicircles. What is the center of the semicircles in the upper half-plane? The lower half-plane?

(d) What are the critical points for (17)?

(e) Sketch the trajectory in the phase plane of the mass released from rest at $y = 7.5$. At what value for y does the mass come to rest?

36. Rigid Body Nutation. Euler's equations describe the motion of the principal-axis components of the angular velocity of a freely rotating rigid body (such as a space station), as seen by an observer rotating with the body (the astronauts, for example). This motion is called *nutation.* If the angular velocity components are denoted by x, y, and z, then an example of Euler's equations is the three-dimensional autonomous system

$$dx/dt = yz\,,$$
$$dy/dt = -2xz\,,$$
$$dz/dt = xy\,.$$

The *trajectory* of a solution $x(t), y(t), z(t)$ to these equations is the curve generated by the points $\big(x(t), y(t), z(t)\big)$ in xyz-phase space as t varies over an interval I.

(a) Show that each trajectory of this system lies on the surface of a (possibly degenerate) sphere centered at the origin $(0, 0, 0)$. [*Hint:* Compute $\frac{d}{dt}(x^2 + y^2 + z^2)$.] What does this say about the magnitude of the angular velocity vector?

(b) Find all the critical points of the system, i.e., all points (x_0, y_0, z_0) such that $x(t) \equiv x_0$, $y(t) \equiv y_0$, $z(t) \equiv z_0$ is a solution. For such solutions, the angular velocity vector remains constant in the body system.

(c) Show that the trajectories of the system lie along the intersection of a sphere and an elliptic cylinder of the form $y^2 + 2x^2 = C$, for some constant C. [*Hint:* Consider the expression for dy/dx implied by Euler's equations.]

(d) Using the results of parts (b) and (c), argue that the trajectories of this system are *closed* curves. What does this say about the corresponding solutions?

(e) Figure 5.19 displays some typical trajectories for this system. Discuss the stability of the three critical points indicated on the positive axes.

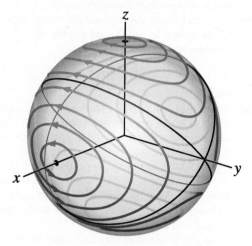

Figure 5.19 Trajectories for Euler's system

5.5 Applications to Biomathematics: Epidemic and Tumor Growth Models

In this section we are going to survey some issues in biological systems that have been successfully modeled by differential equations. We begin by reviewing the population models described in Sections 3.2 and 5.3.

In the *Malthusian model*, the rate of growth of a population $p(t)$ is proportional to the size of the existing population:

$$(1) \qquad \frac{dp}{dt} = kp \quad (k > 0) .$$

Cells that reproduce by splitting, such as amoebae and bacteria, are obvious biological examples of this type of growth. Equation (1) implies that a Malthusian population grows exponentially; there is no mechanism for constraining the growth. In Section 3.2 we saw that certain populations exhibit Malthusian growth over limited periods of time (as does compound interest).[†]

Inserting a negative growth rate,

$$(2) \qquad \frac{dp}{dt} = -kp ,$$

results in solutions that *decay* exponentially. Their average lifetime is $1/k$, and their half-life is $(\ln 2)/k$ (Problems 6 and 8). In animals, certain organs such as the kidney serve to cleanse

[†]Gordon E. Moore (1929–) has observed that the number of transistors on new integrated circuits produced by the electronics industry doubles every 24 months. "Moore's law" is commonly cited by industrialists.

the bloodstream of unwanted components (creatinine clearance, renal clearance), and their concentrations diminish exponentially. As a general rule, the body tends to dissipate ingested drugs in such a manner. (Of course, the most familiar physical instance of Malthusian disintegration is radioactive decay.) Note that if there are both growth and extinction processes, $dp/dt = k_+ p - k_- p$ and the equation in (1) still holds with $k = k_+ - k_-$.

When there are two-party interactions occurring in the population that decrease the growth rate, such as competition for resources or violent crime, the *logistic model* might be applicable; it assumes that the extinction rate is proportional to the number of possible pairs in the population, $p(p-1)/2$:

$$(3) \qquad \frac{dp}{dt} = k_1 p - k_2 \frac{p(p-1)}{2} \quad \text{or, equivalently,} \quad \frac{dp}{dt} = -Ap(p - p_1).^\dagger$$

Rodent, bird, and plant populations exhibit logistic growth rates due to social structure, territoriality, and competition for light and space, respectively. The logistic function

$$p(t) = \frac{p_0 p_1}{p_0 + (p_1 - p_0)e^{-Ap_1 t}}, \qquad p_0 := p(0)$$

was shown in Section 3.2 to be the solution of (3), and typical graphs of $p(t)$ were displayed there.

In Section 5.3 we observed that the *Volterra–Lotka model* for *two different* populations, a *predator* $x_2(t)$ and a *prey* $x_1(t)$, postulates a Malthusian growth rate for the prey and an extinction rate governed by $x_1 x_2$, the number of possible pairings of one from each population,

$$(4) \qquad \frac{dx_1}{dt} = Ax_1 - Bx_1 x_2,$$

while predators follow a Malthusian extinction rate and pairwise growth rate

$$(5) \qquad \frac{dx_2}{dt} = -Cx_2 + Dx_1 x_2.$$

Volterra–Lotka dynamics have been observed in blood vessel growth (predator = new capillary tips; prey = chemoattractant), fish populations, and several animal–plant interactions.

Systems like (4)–(5) were studied in Section 5.3 with the aid of the Runge–Kutta algorithm. Now, armed with the insights of Section 5.4, we can further explore this model theoretically.

First, we perform a "reality check" by proving that the populations $x_1(t), x_2(t)$ in the Volterra–Lotka model never change sign. Separating (4) leads to

$$\frac{1}{x_1} \frac{dx_1}{dt} = \frac{d \ln x_1}{dt} = A - Bx_2,$$

while integrating from 0 to t results in

$$(6) \qquad x_1(t) = x_1(0)e^{\int_0^t \{A - Bx_2(\tau)\} d\tau},$$

and the exponential factor is always positive. Thus $x_1(t)$ [and similarly $x_2(t)$] retains its initial sign (*negative* populations never arise).

†One might propose that the *growth* rate in animal populations is due to *two*–party interactions as well. (Wink, wink.) However, in monogamous societies, the number of pairs participating in procreation is proportional to $p/2$, leading to (1). A growth rate determined by *all* possible pairings $p(p-1)/2$ would indicate a highly hedonistic social order.

Example 1 Find and interpret the critical points for the Volterra–Lotka model (4)–(5).

Solution The system

$$
(7)\quad
\begin{aligned}
\frac{dx_1}{dt} &= Ax_1 - Bx_1x_2 = -Bx_1\!\left(x_2 - \frac{A}{B}\right) = 0, \\
\frac{dx_2}{dt} &= -Cx_2 + Dx_1x_2 = Dx_2\!\left(x_1 - \frac{C}{D}\right) = 0
\end{aligned}
$$

has the trivial solution $x_1(t) \equiv x_2(t) \equiv 0$, with an obvious interpretation in terms of populations. If all four coefficients A, B, C, and D are positive, there is also the more interesting solution

$$(8)\qquad x_2(t) \equiv \frac{A}{B}, \quad x_1(t) \equiv \frac{C}{D}.$$

At these population levels, the growth and extinction rates for each species cancel. The direction field diagram in Figure 5.20 for the phase plane equation

$$(9)\qquad \frac{dx_2}{dx_1} = \frac{-Cx_2 + Dx_1x_2}{Ax_1 - Bx_1x_2} = \frac{x_2}{x_1}\frac{-C + Dx_1}{A - Bx_2}$$

suggests that this equilibrium is a *center* (compare Figure 5.12 on page 267) with closed (periodic) neighboring trajectories, in accordance with the simulations in Section 5.3. However it is conceivable that some *spiral* trajectories might snake through the field pattern and approach the critical point asymptotically. A rather tricky argument in Problem 4 demonstrates that this is *not* the case. ◆

The SIR Epidemic Model. The SIR[†] model for an epidemic addresses the spread of diseases that are only contracted by contact with an infected individual; its victims, once recovered, are immune to further infection and are themselves noninfectious. So the members of a population of size N fall into three classes:

$S(t) =$ the number of susceptible individuals—that is, those who have not been infected; $s := S/N$ is the fraction of susceptibles.

$I(t) =$ the number of individuals who are currently infected, comprising a fraction $i := I/N$ of the population.

$R(t) =$ the number of individuals who have recovered from infection, comprising the fraction $r := R/N$.

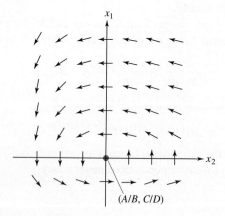

Figure 5.20 Typical direction field diagram for the Volterra–Lotka system

[†]Introduced by W. O. Kermack and A. G. McKendrick in "A Contribution to the Mathematical Theory of Epidemics," *Proc. Royal Soc. London*, Vol. A115 (1927): 700–721.

The classic SIR epidemic model assumes that on the average, an infectious individual encounters a people per unit time (usually per week). Thus, a total of aI people per week are contacted by infectees, but only a fraction $s = S/N$ of them are susceptible. So the susceptible population diminishes at a rate

(10) $$\frac{dS}{dt} = -saI \text{ or (dividing by } N)$$

(11) $$\frac{ds}{dt} = -asi \, .$$

The parameter a is crucial in disease control. Crowded conditions, or high a, make it difficult to combat the spread of infection. Ideally, we would quarantine the infectees (low a) to fight the epidemic.

The infected population is (obviously) increased whenever a susceptible individual is infected. Additionally, infectees recover in a Malthusian-disintegrative manner over an average time of, say, $1/k$ weeks [recall (1)], so the infected population changes at a rate

(12) $$\frac{dI}{dt} = saI - kI = a\left(s - \frac{k}{a}\right)I \text{ or}$$

(13) $$\frac{di}{dt} = a\left(s - \frac{k}{a}\right)i \, .$$

And, of course, the population of recovered individuals increases whenever an infectee is healed:

(14) $$\frac{dR}{dt} = kI \text{ or } \frac{dr}{dt} = ki \, .$$

With the SIR model, the total population count remains unchanged:

$$\frac{d(S + I + R)}{dt} = -saI + saI - kI + kI = 0 \, .$$

Thus, any fatalities are tallied in the "recovered/noninfectious" population R.

Interestingly, equations (11) and (13) do not contain R or r, so they are suitable for phase plane analysis. In fact they constitute a Volterra–Lotka system with $A = 0$, $B = D = a$, and $C = k$. Because the coefficient A is zero, the critical point structure is different from that discussed in Example 1. Specifically, if $-asi$ in (11) is zero, then only $-ki$ remains on the right in (13), so $I(t) = i(t) \equiv 0$ is necessary and sufficient for a critical point, with S unrestricted. (Physically, this means the populations remain stable only if there are no carriers of the infection.)

Our earlier argument has shown that if $s(t)$ and $i(t)$ are initially positive, they remain so. As a result we conclude from (11) immediately that $s(t)$ decreases monotonically; as such, it has a limiting value $s(\infty)$ as $t \to \infty$. Does $i(t)$ have a limiting value also? If so, $\{s(\infty), i(\infty)\}$ would be a critical point by Theorem 1 of Section 5.4, page 266, and thus $i(\infty) = 0$.

To analyze $i(t)$ consider the phase plane equation for (11) and (13):

(15) $$\frac{di}{ds} = \frac{asi - ki}{-asi} = -1 + \frac{k}{as} \, ,$$

which has solutions

(16) $$i = -s + \frac{k}{a}\ln s + K \, .$$

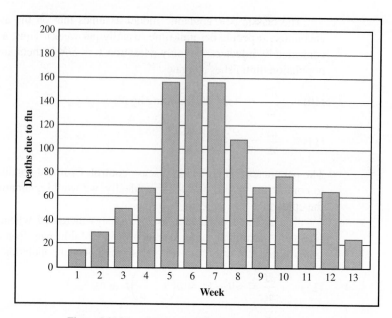

Figure 5.21 Mortality data for Hong Kong flu, New York City

From (16) we see that $s(\infty)$ cannot be zero; otherwise the right-hand side would eventually be negative, contradicting $i(t) > 0$. Therefore, (16) demonstrates that $i(t)$ *does* have a limiting value $i(\infty) = -s(\infty) + (k/a)\ln s(\infty) + K$. As noted, $i(\infty)$ must then be zero.

From (13) we further conclude that if $s(0)$ exceeds the "threshold value" k/a, the infected fraction $i(t)$ will initially increase $(di/dt > 0$ at $t = 0)$ before eventually dying out. The peak value of $i(t)$ occurs when $di/dt = 0 = a[s - (k/a)]i$, i.e., when $s(t)$ passes through the value k/a. In the jargon of epidemiology, this phenomenon defines an "epidemic." You will be directed in Problem 10 to show that if $s(0) \leq k/a$, the infected population diminishes monotonically, and no epidemic develops.

Example 2[†] According to data issued by the Centers for Disease Control and Prevention (CDC) in Atlanta, Georgia, the Hong Kong flu epidemic during the winter of 1968–1969 was responsible for 1035 deaths in New York City (population 7,900,000), according to the time chart in Figure 5.21. Analyze this data with the SIR model.

Solution Of course, we need to make some assumptions about the parameters. First of all, only a small percentage of people who contract Hong Kong flu perish, so let's assume that the chart reflects a scaled version of the infected population fraction $i(t)$. It is known that the recovery period for this flu is around 5 days, or 5/7 week, so we try $k = 7/5 = 1.4$. And since the infectees spend much of their convalescence in bed, the average contact rate a is probably less than 1 person per day or 7 per week. The CDC estimated that the initial infected population $I(0)$ was about 10, so the initial data for (11), (13), and (14) are

$$s(0) = \frac{7{,}900{,}000 - 10}{7{,}900{,}000} \approx 0.9999987, \quad i(0) = \frac{10}{7{,}900{,}000} \approx 1.2658 \times 10^{-6}, \quad r(0) = 0.$$

[†]We borrow liberally from "The SIR Model for Spread of Disease" by Duke University's David Smith and Lang Moore, *Journal of Online Mathematics and Its Applications*, The MAA Mathematical Sciences Digital Library, http://mathdl.maa.org/mathDL/4/?pa=content&sa=viewDocument&nodeId=479&bodyId=612, copyright 2000, CCP and the authors, published December, 2001. The article contains much interesting information about this epidemic.

Figure 5.22 SIR simulations (a) $k = 1.4$, $a = 2.0$; (b) $k = 1.4$, $a = 3.5$; (c) $k = 1.4$, $a = 6.0$

Numerical simulations of this system are displayed in Figure 5.22.[†] The contact rate $a = 3.5$ per week generates an infection fraction curve that closely matches the mortality data's characteristics: time of peak and duration of epidemic. ◆

A Tumor Growth Model.[‡] The observed growth of certain tumors can be explained by a model that is mathematically similar to the epidemic model. The total number of cells N in the tumor subdivides into a population P that proliferates by splitting (Malthusian growth) and a population Q that remains quiescent. However, the proliferating cells also can make a transition to the quiescent state, and this occurrence is modeled as a Malthusian-like decay with a "rate" $r(N)$ *that increases with the overall size of the tumor:*

$$(17) \qquad \frac{dP}{dt} = cP - r(N)P \,,$$

$$(18) \qquad \frac{dQ}{dt} = r(N)P \,.$$

Thus the total population N increases only when the proliferating cells split, as can be seen by adding the equations (17) and (18):

$$(19) \qquad \frac{dN}{dt} = \frac{d(P + Q)}{dt} = cP \,.$$

We take (17) and (19) as the system for our analysis. The phase plane equation

$$(20) \qquad \frac{dP}{dN} = \frac{cP - r(N)P}{cP} = 1 - \frac{r(N)}{c}$$

can be integrated, leading to a formula for P in terms of N

$$(21) \qquad P = N - \frac{1}{c} \int r(N) \, dN + K \,.$$

[†]Appendix G describes various websites and commercial software that sketch direction fields and automate most of the differential equation algorithms discussed in this book.

[‡]The authors wish to thank Dr. Glenn Webb of Vanderbilt University for this application. See M. Gyllenberg and G. F. Webb, "Quiescence as an Explanation of Gompertz Tumor Growth," *Growth, Development, and Aging*, Vol. 53 (1989): 25–55.

Suppose the initial conditions are $P(0) = 1$, $Q(0) = 0$, and $N(0) = 1$ (a single prolif-erating cell). Then we can eliminate the nuisance constant K by taking the indefinite integral in (21) to run from 1 to N and evaluating at $t = 0$:

$$1 = 1 - \frac{1}{c}\int_1^1 r(N)\,dN + K \implies K = 0.$$

Insertion of (21) with $K = 0$ into (19) produces a differential equation for N alone:

$$(22) \qquad \frac{dN}{dt} = cN - \int_1^N r(u)\,du.$$

Example 3 The *Gompertz law*

$$(23) \qquad N(t) = e^{c(1-e^{-bt})/b}$$

has been observed experimentally for the growth of some tumors. Show that a transition rate $r(N)$ of the form $b(1 + \ln N)$ predicts Gompertzian growth.

Solution If the indicated integral of the rate $r(N)$ is carried out, (22) becomes

$$(24) \qquad \frac{dN}{dt} = cN - b(N - 1) - b(N \ln N - N + 1) = (c - b \ln N)\,N.$$

Dividing by N we obtain a linear differential equation for the function $\ln N$

$$\frac{d \ln N}{dt} = -b \ln N + c$$

whose solution, for the initial condition $N(0) = 1$, is found by the methods of Section 2.3 to be

$$\ln N(t) = \frac{c}{b}(1 - e^{-bt}),$$

confirming (23). ◆

Problem 9 invites the reader to show that if the growth rate is modeled as $r(N) = s(2N - 1)$, then the solution of (22) describes logistic growth. Typical curves for the Gompertz and logistic models are displayed in Figure 5.23. See also Figure 3.4 on page 98.

Other applications of differential equations to biomathematics appear in the discussions of artificial respiration (Project B, page 81) in Chapter 2, HIV infection (Project A, page 141) and aquaculture (Project B, page 144) in Chapter 3, and spread of staph infections (Project B) and the Ebola epidemic (Project F, page 314) in this chapter.

(a)　　　　　　　　　　　　　　(b)

Figure 5.23 (a) Gompertz and (b) logistic curves

5.5 EXERCISES

1. **Logistic Model.** In Section 3.2 we discussed the logistic equation

$$\frac{dp}{dt} = A p_1 p - A p^2 , \qquad p(0) = p_0 ,$$

and its use in modeling population growth. A more general model might involve the equation

(25) $\qquad \dfrac{dp}{dt} = A p_1 p - A p^r , \qquad p(0) = p_0 ,$

where $r > 1$. To see the effect of changing the parameter r in (25), take $p_1 = 3$, $A = 1$, and $p_0 = 1$. Then use a numerical scheme such as Runge–Kutta with $h = 0.25$ to approximate the solution to (25) on the interval $0 \le t \le 5$ for $r = 1.5$, 2, and 3. What is the limiting population in each case? For $r > 1$, determine a general formula for the limiting population.

2. **Radioisotopes and Cancer Detection.** A radioisotope commonly used in the detection of breast cancer is technetium-99m. This radionuclide is attached to a chemical that upon injection into a patient accumulates at cancer sites. The isotope's radiation is then detected and the site located, using gamma cameras or other tomographic devices.

 Technetium-99m decays radioactively in accordance with the equation $dy/dt = -ky$, with $k = 0.1155/\text{h}$. The short half-life of technetium-99m has the advantage that its radioactivity does not endanger the patient. A disadvantage is that the isotope must be manufactured in a cyclotron. Since hospitals are not equipped with cyclotrons, doses of technetium-99m have to be ordered in advance from medical suppliers.

 Suppose a dosage of 5 millicuries (mCi) of technetium-99m is to be administered to a patient. Estimate the delivery time from production at the manufacturer to arrival at the hospital treatment room to be 24 hours and calculate the amount of the radionuclide that the hospital must order, to be able to administer the proper dosage.

3. **Secretion of Hormones.** The secretion of hormones into the blood is often a periodic activity. If a hormone is secreted on a 24-h cycle, then the rate of change of the level of the hormone in the blood may be represented by the initial value problem

$$\frac{dx}{dt} = \alpha - \beta \cos \frac{\pi t}{12} - kx , \quad x(0) = x_0 ,$$

where $x(t)$ is the amount of the hormone in the blood at time t, α is the average secretion rate, β is the amount of daily variation in the secretion, and k is a positive constant reflecting the rate at which the body removes the hormone from the blood. If $\alpha = \beta = 1, k = 2$, and $x_0 = 10$, solve for $x(t)$.

4. Prove that the critical point (8) of the Volterra–Lotka system is a center; that is, the neighboring trajectories are periodic. [*Hint:* Observe that (9) is separable and show that its solutions can be expressed as

(26) $\qquad [x_2^A \, e^{-Bx_2}] \cdot [x_1^C \, e^{-Dx_1}] = K .]$

 Prove that the maximum of the function $x^p e^{-qx}$ is $(p/qe)^p$, occurring at the unique value $x = p/q$ (see Figure 5.24), so the critical values (8) maximize the factors on the left in (26). Argue that if K takes the corresponding maximum value $(A/Be)^A (C/De)^C$, the critical point (8) is the (unique) solution of (26), and it cannot be an endpoint of any trajectory for (26) with a lower value of K.[†]

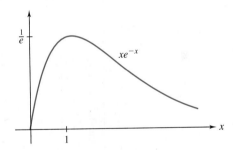

Figure 5.24 Graph of xe^{-x}

5. Suppose for a certain disease described by the SIR model it is determined that $a = 0.003$ and $b = 0.5$.
 (a) In the *SI*-phase plane, sketch the trajectory corresponding to the initial condition that one person is infected and 700 persons are susceptible.

[†]In fact, the periodic fluctuations predicted by the Volterra–Lotka model were observed in fish populations by Lotka's son-in-law, Humberto D'Ancona.

(b) From your graph in part (a), estimate the peak number of infected persons. Compare this with the theoretical prediction $S = k/a \approx 167$ persons when the epidemic is at its peak.

6. Show that the half-life of solutions to (2)—that is, the time required for the solution to decay to one-half of its value—equals $(\ln 2)/k$.

7. Complete the solution of the tumor growth model for Example 3 on page 280 by finding $P(t)$ and $Q(t)$.

8. If $p(t)$ is a Malthusian population that diminishes according to (2), then $p(t_2) - p(t_1)$ is the number of individuals in the population whose lifetime lies between t_1 and t_2. Argue that the *average* lifetime of the population is given by the formula

$$\frac{\displaystyle\int_0^\infty t \left|\frac{dp(t)}{dt}\right| dt}{p(0)}$$

and show that this equals $1/k$.

9. Show that with the transition rate formula $r(N) = s(2N - 1)$, equation (22) takes the form of the equation for the logistic model (Section 3.2, equation (14)). Solve (22) for this case.

10. Prove that the infected population $I(t)$ in the SIR model does not increase if $S(0)$ is less than or equal to k/a.

11. An epidemic reported by the British Communicable Disease Surveillance Center in the *British Medical Journal* (March 4, 1978, p. 587) took place in a boarding school with 763 residents.[†] The statistics for the infected population are shown in the graph in Figure 5.25.

 Assuming that the average duration of the infection is 2 days, use a numerical differential equation solver (see Appendix G) to try to reproduce the data. Take $S(0) = 762, I(0) = 1, R(0) = 0$ as initial conditions. Experiment with reasonable estimates for the average number of contacts per day by the infected students, who were confined to bed after the infection was detected. What value of this parameter seems to fit the curve best?

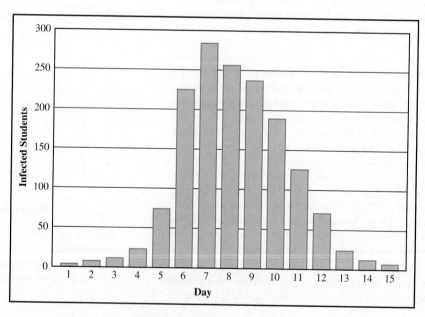

Figure 5.25 Flu data for Problem 11

[†]See also the discussion of this epidemic in *Mathematical Biology I, An Introduction*, by J. D. Murray (Springer-Verlag, New York, 2002), 325–326.

5.6 Coupled Mass–Spring Systems

In this section we extend the mass–spring model of Chapter 4 to include situations in which coupled springs connect two masses, both of which are free to move. The resulting motions can be very intriguing. For simplicity we'll neglect the effects of friction, gravity, and external forces. Let's analyze the following experiment.

Example 1 On a smooth horizontal surface, a mass $m_1 = 2$ kg is attached to a fixed wall by a spring with spring constant $k_1 = 4$ N/m. Another mass $m_2 = 1$ kg is attached to the first object by a spring with spring constant $k_2 = 2$ N/m. The objects are aligned horizontally so that the springs are their natural lengths (Figure 5.26). If both objects are displaced 3 m to the right of their equilibrium positions (Figure 5.27) and then released, what are the equations of motion for the two objects?

Figure 5.26 Coupled system at equilibrium **Figure 5.27** Coupled system at initial displacement

Solution From our assumptions, the only forces we need to consider are those due to the springs themselves. Recall that Hooke's law asserts that the force acting on an object due to a spring has magnitude proportional to the displacement of the spring from its natural length and has direction opposite to its displacement. That is, if the spring is either stretched or compressed, then it tries to return to its natural length.

Because each mass is free to move, we apply Newton's second law to each object. Let $x(t)$ denote the displacement (to the right) of the 2-kg mass from its equilibrium position and similarly, let $y(t)$ denote the corresponding displacement for the 1-kg mass. The 2-kg mass has a force F_1 acting on its left side due to one spring and a force F_2 acting on its right side due to the second spring. Referring to Figure 5.27 and applying Hooke's law, we see that

$$F_1 = -k_1 x, \qquad F_2 = +k_2(y - x),$$

since $(y - x)$ is the net displacement of the second spring from its natural length. There is only one force acting on the 1-kg mass: the force due to the second spring, which is

$$F_3 = -k_2(y - x).$$

Applying Newton's second law to these objects, we obtain the system

(1)
$$m_1 \frac{d^2 x}{dt^2} = F_1 + F_2 = -k_1 x + k_2(y - x),$$
$$m_2 \frac{d^2 y}{dt^2} = F_3 = -k_2(y - x),$$

or

(2)
$$m_1 \frac{d^2 x}{dt^2} + (k_1 + k_2)x - k_2 y = 0,$$
$$m_2 \frac{d^2 y}{dt^2} + k_2 y - k_2 x = 0.$$

In this problem, we know that $m_1 = 2$, $m_2 = 1$, $k_1 = 4$, and $k_2 = 2$. Substituting these values into system (2) yields

(3) $$2\frac{d^2x}{dt^2} + 6x - 2y = 0 \,,$$

(4) $$\frac{d^2y}{dt^2} + 2y - 2x = 0 \,.$$

We'll use the elimination method of Section 5.2 to solve (3)–(4). With $D := d/dt$ we rewrite the system as

(5) $$(2D^2 + 6)[x] - 2y = 0 \,,$$

(6) $$-2x + (D^2 + 2)[y] = 0 \,.$$

Adding $(D^2 + 2)$ applied to equation (5) to 2 times equation (6) eliminates y:

$$[(D^2 + 2)(2D^2 + 6) - 4][x] = 0 \,,$$

which simplifies to

(7) $$2\frac{d^4x}{dt^4} + 10\frac{d^2x}{dt^2} + 8x = 0 \,.$$

Notice that equation (7) is linear with constant coefficients. To solve it let's proceed as we did with linear second-order equations and try to find solutions of the form $x = e^{rt}$. Substituting e^{rt} in equation (7) gives

$$2(r^4 + 5r^2 + 4)e^{rt} = 0 \,.$$

Thus, we get a solution to (7) when r satisfies the auxiliary equation

$$r^4 + 5r^2 + 4 = 0 \,.$$

From the factorization $r^4 + 5r^2 + 4 = (r^2 + 1)(r^2 + 4)$, we see that the roots of the auxiliary equation are complex numbers i, $-i$, $2i$, $-2i$. Using Euler's formula, it follows that

$$z_1(t) = e^{it} = \cos t + i \sin t \qquad \text{and} \qquad z_2(t) = e^{2it} = \cos 2t + i \sin 2t$$

are complex-valued solutions to equation (7). To obtain real-valued solutions, we take the real and imaginary parts of $z_1(t)$ and $z_2(t)$. Thus, four real-valued solutions are

$$x_1(t) = \cos t \,, \qquad x_2(t) = \sin t \,, \qquad x_3(t) = \cos 2t \,, \qquad x_4(t) = \sin 2t \,,$$

and a general solution is

(8) $$x(t) = a_1 \cos t + a_2 \sin t + a_3 \cos 2t + a_4 \sin 2t \,,$$

where a_1, a_2, a_3, and a_4 are arbitrary constants.[†]
 To obtain a formula for $y(t)$, we use equation (3) to express y in terms of x:

$$y(t) = \frac{d^2x}{dt^2} + 3x$$

$$= -a_1 \cos t - a_2 \sin t - 4a_3 \cos 2t - 4a_4 \sin 2t$$
$$+ 3a_1 \cos t + 3a_2 \sin t + 3a_3 \cos 2t + 3a_4 \sin 2t \,,$$

[†]A more detailed discussion of general solutions is given in Chapter 6.

and so

(9) $y(t) = 2a_1 \cos t + 2a_2 \sin t - a_3 \cos 2t - a_4 \sin 2t$.

To determine the constants a_1, a_2, a_3, and a_4, let's return to the original problem. We were told that the objects were originally displaced 3 m to the right and then released. Hence,

(10) $x(0) = 3$, $\dfrac{dx}{dt}(0) = 0$; $y(0) = 3$, $\dfrac{dy}{dt}(0) = 0$.

On differentiating equations (8) and (9), we find

$$\frac{dx}{dt} = -a_1 \sin t + a_2 \cos t - 2a_3 \sin 2t + 2a_4 \cos 2t,$$

$$\frac{dy}{dt} = -2a_1 \sin t + 2a_2 \cos t + 2a_3 \sin 2t - 2a_4 \cos 2t.$$

Now, if we put $t = 0$ in the formulas for $x, dx/dt, y$, and dy/dt, the initial conditions (10) give the four equations

$$x(0) = a_1 + a_3 = 3, \qquad \frac{dx}{dt}(0) = a_2 + 2a_4 = 0,$$

$$y(0) = 2a_1 - a_3 = 3, \qquad \frac{dy}{dt}(0) = 2a_2 - 2a_4 = 0.$$

From this system, we find $a_1 = 2, a_2 = 0, a_3 = 1$, and $a_4 = 0$. Hence, the equations of motion for the two objects are

$$x(t) = 2\cos t + \cos 2t,$$
$$y(t) = 4\cos t - \cos 2t,$$

which are depicted in Figure 5.28. ◆

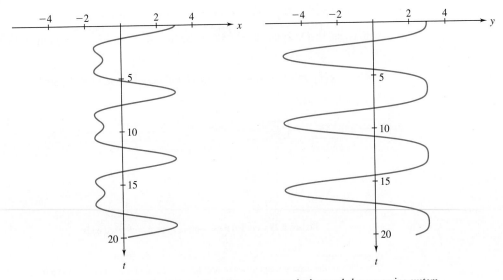

Figure 5.28 Graphs of the motion of the two masses in the coupled mass–spring system

The general solution pair (8), (9) that we have obtained is a combination of sinusoids oscillating at two different angular frequencies: 1 rad/sec and 2 rad/sec. These frequencies extend the notion of the *natural frequency* of the single (free, undamped) mass–spring oscillator (page 212, Section 4.9) and are called the **natural (or normal) angular frequencies**[†] of the system. A complex system consisting of more masses and springs would have many normal frequencies.

Notice that if the initial conditions were altered so that the constants a_3 and a_4 in (8) and (9) were zero, the motion would be a pure sinusoid oscillating at the single frequency 1 rad/sec. Similarly, if a_1 and a_2 were zero, only the 2 rad/sec oscillation would be "excited." Such solutions, wherein the entire motion is described by a single sinusoid, are called the **normal modes** of the system.[‡] The normal modes in the following example are particularly easy to visualize because we will take all the masses and all the spring constants to be equal.

Example 2 Three identical springs with spring constant k and two identical masses m are attached in a straight line with the ends of the outside springs fixed (see Figure 5.29). Determine and interpret the normal modes of the system.

Solution We define the displacements from equilibrium, x and y, as in Example 1 on page 283. The equations expressing Newton's second law for the masses are quite analogous to (1), except for the effect of the third spring on the second mass:

(11) $mx'' = -kx + k(y - x)$,

(12) $my'' = -k(y - x) - ky$,

or

$$(mD^2 + 2k)[x] - ky = 0,$$
$$-kx + (mD^2 + 2k)[y] = 0.$$

Eliminating y in the usual manner results in

(13) $\left[(mD^2 + 2k)^2 - k^2\right][x] = 0$.

This has the auxiliary equation

$$(mr^2 + 2k)^2 - k^2 = (mr^2 + k)(mr^2 + 3k) = 0,$$

with roots $\pm i\sqrt{k/m}$, $\pm i\sqrt{3k/m}$. Setting $\omega := \sqrt{k/m}$, we get the following general solution to (13):

(14) $x(t) = C_1 \cos \omega t + C_2 \sin \omega t + C_3 \cos\left(\sqrt{3}\omega t\right) + C_4 \sin\left(\sqrt{3}\omega t\right)$.

Figure 5.29 Coupled mass–spring system with fixed ends

[†]The study of the natural frequencies of oscillations of complex systems is known in engineering as **modal analysis.**
[‡]The normal modes are more naturally characterized in terms of *eigenvalues* (see Section 9.5).

(a) (b)

Figure 5.30 Normal modes for Example 2

To obtain $y(t)$, we solve for $y(t)$ in (11) and substitute $x(t)$ as given in (14). Upon simplifying, we get

(15) $y(t) = C_1 \cos \omega t + C_2 \sin \omega t - C_3 \cos\left(\sqrt{3}\omega t\right) - C_4 \sin\left(\sqrt{3}\omega t\right).$

From the formulas (14) and (15), we see that the normal angular frequencies are ω and $\sqrt{3}\omega$. Indeed, if $C_3 = C_4 = 0$, we have a solution where $y(t) \equiv x(t)$, oscillating at the angular frequency $\omega = \sqrt{k/m}$ rad/sec (equivalent to a frequency $\sqrt{k/m}/2\pi$ periods/sec). Now if $x(t) \equiv y(t)$ in Figure 5.29, the two masses are moving as if they were a single rigid body of mass $2m$, forced by a "double spring" with a spring constant given by $2k$. And indeed, according to equation (4) of Section 4.9 (page 212), we would expect such a system to oscillate at the angular frequency $\sqrt{2k/2m} = \sqrt{k/m}$(!). This motion is depicted in Figure 5.30(a).

Similarly, if $C_1 = C_2 = 0$, we find the second normal mode where $y(t) = -x(t)$, so that in Figure 5.29 there are two mirror-image systems, each with mass m and a "spring and a half" with spring constant $k + 2k = 3k$. (The half-spring will be twice as stiff.) Section 4.9's equation (4) then predicts an angular oscillation frequency for each system of $\sqrt{3k/m} = \sqrt{3}\omega$, which again is consistent with (14) and (15). This motion is shown in Figure 5.30(b). ◆

5.6 EXERCISES

1. Two springs and two masses are attached in a straight line on a horizontal frictionless surface as illustrated in Figure 5.31. The system is set in motion by holding the mass m_2 at its equilibrium position and pulling the mass m_1 to the left of its equilibrium position a distance 1 m and then releasing both masses. Express Newton's law for the system and determine the equations of motion for the two masses if $m_1 = 1$ kg, $m_2 = 2$ kg, $k_1 = 4$ N/m, and $k_2 = 10/3$ N/m.

Figure 5.31 Coupled mass–spring system with one end free

2. Determine the equations of motion for the two masses described in Problem 1 if $m_1 = 1$ kg, $m_2 = 1$ kg, $k_1 = 3$ N/m, and $k_2 = 2$ N/m.

3. Four springs with the same spring constant and three equal masses are attached in a straight line on a horizontal frictionless surface as illustrated in Figure 5.32. Determine the normal frequencies for the system and describe the three normal modes of vibration.

Figure 5.32 Coupled mass–spring system with three degrees of freedom

4. Two springs, two masses, and a dashpot are attached in a straight line on a horizontal frictionless surface as shown in Figure 5.33. The dashpot provides a damping force on mass m_2, given by $F = -by'$. Derive the system of differential equations for the displacements x and y.

Figure 5.33 Coupled mass–spring system with one end damped

5. Two springs, two masses, and a dashpot are attached in a straight line on a horizontal frictionless surface as shown in Figure 5.34. The system is set in motion by holding the mass m_2 at its equilibrium position and pushing the mass m_1 to the left of its equilibrium position a distance 2 m and then releasing both masses. Determine the equations of motion for the two masses if $m_1 = m_2 = 1$ kg, $k_1 = k_2 = 1$ N/m, and $b = 1$ N-sec/m. [*Hint:* The dashpot damps both m_1 and m_2 with a force whose magnitude equals $b|y' - x'|$.]

Figure 5.34 Coupled mass–spring system with damping between the masses

6. Referring to the coupled mass–spring system discussed in Example 1, suppose an external force $E(t) = 37 \cos 3t$

is applied to the second object of mass 1 kg. The displacement functions $x(t)$ and $y(t)$ now satisfy the system

(16) $\qquad 2x''(t) + 6x(t) - 2y(t) = 0 ,$

(17) $\qquad y''(t) + 2y(t) - 2x(t) = 37 \cos 3t .$

(a) Show that $x(t)$ satisfies the equation

(18) $\qquad x^{(4)}(t) + 5x''(t) + 4x(t) = 37 \cos 3t .$

(b) Find a general solution $x(t)$ to equation (18). [*Hint:* Use undetermined coefficients with $x_p = A \cos 3t + B \sin 3t$.]

(c) Substitute $x(t)$ back into (16) to obtain a formula for $y(t)$.

(d) If both masses are displaced 2 m to the right of their equilibrium positions and then released, find the displacement functions $x(t)$ and $y(t)$.

7. Suppose the displacement functions $x(t)$ and $y(t)$ for a coupled mass–spring system (similar to the one discussed in Problem 6) satisfy the initial value problem

$$x''(t) + 5x(t) - 2y(t) = 0 ,$$
$$y''(t) + 2y(t) - 2x(t) = 3 \sin 2t ;$$
$$x(0) = x'(0) = 0 ,$$
$$y(0) = 1 , \qquad y'(0) = 0 .$$

Solve for $x(t)$ and $y(t)$.

8. A double pendulum swinging in a vertical plane under the influence of gravity (see Figure 5.35) satisfies the system

$$(m_1 + m_2)l_1^2\theta_1'' + m_2 l_1 l_2 \theta_2'' + (m_1 + m_2)l_1 g\theta_1 = 0 ,$$
$$m_2 l_2^2 \theta_2'' + m_2 l_1 l_2 \theta_1'' + m_2 l_2 g\theta_2 = 0 ,$$

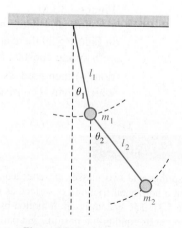

Figure 5.35 Double pendulum

when θ_1 and θ_2 are small angles. Solve the system when $m_1 = 3$ kg, $m_2 = 2$ kg, $l_1 = l_2 = 5$ m, $\theta_1(0) = \pi/6$, $\theta_2(0) = \theta_1'(0) = \theta_2'(0) = 0$.

9. The motion of a pair of identical pendulums coupled by a spring is modeled by the system

$$mx_1'' = -\frac{mg}{l}x_1 - k(x_1 - x_2),$$

$$mx_2'' = -\frac{mg}{l}x_2 + k(x_1 - x_2)$$

for small displacements (see Figure 5.36). Determine the two normal frequencies for the system.

10. Suppose the coupled mass–spring system of Problem 1 (Figure 5.31) is hung *vertically* from a support (with mass m_2 above m_1), as in Section 4.10, page 226.
 (a) Argue that at equilibrium, the lower spring is stretched a distance l_1 from its natural length L_1, given by $l_1 = m_1g/k_1$.
 (b) Argue that at equilibrium, the upper spring is stretched a distance $l_2 = (m_1 + m_2)g/k_2$.
 (c) Show that if x_1 and x_2 are redefined to be displacements from the equilibrium positions of the masses m_1 and m_2, then the equations of motion are identical with those derived in Problem 1.

Figure 5.36 Coupled pendulums

5.7 Electrical Systems

The equations governing the voltage–current relations for the resistor, inductor, and capacitor were given in Section 3.5, together with Kirchhoff's laws that constrain how these quantities behave when the elements are electrically connected into a circuit. Now that we have the tools for solving linear equations of higher-order systems, we are in a position to analyze more complex electrical circuits.

Example 1 The series *RLC* circuit in Figure 5.37 on page 290 has a voltage source given by $E(t) = \sin 100t$ volts (V), a resistor of 0.02 ohms (Ω), an inductor of 0.001 henrys (H), and a capacitor of 2 farads (F). (These values are selected for numerical convenience; typical capacitance values are much smaller.) If the initial current and the initial charge on the capacitor are both zero, determine the current in the circuit for $t > 0$.

Solution Using the notation of Section 3.5, we have $L = 0.001$ H, $R = 0.02\ \Omega$, $C = 2$ F, and $E(t) = \sin 100t$. According to Kirchhoff's current law, the same current I passes through each circuit element. The current through the capacitor equals the instantaneous rate of change of its charge q:

(1) $I = dq/dt$.

From the physics equations in Section 3.5, we observe that the voltage drops across the capacitor (E_C), the resistor (E_R), and the inductor (E_L) are expressed as

(2) $E_C = \dfrac{q}{C},\qquad E_R = RI,\qquad E_L = L\dfrac{dI}{dt}.$

Therefore, Kirchhoff's voltage law, which implies $E_L + E_R + E_C = E$, can be expressed as

(3) $L\dfrac{dI}{dt} + RI + \dfrac{1}{C}q = E(t).$

Figure 5.37 Schematic representation of an *RLC* series circuit

In most applications we will be interested in determining the current $I(t)$. If we differentiate (3) with respect to t and substitute I for dq/dt, we obtain

(4) $$L\frac{d^2I}{dt^2} + R\frac{dI}{dt} + \frac{1}{C}I = \frac{dE}{dt}.$$

After substitution of the given values this becomes

$$(0.001)\frac{d^2I}{dt^2} + (0.02)\frac{dI}{dt} + (0.5)I = 100\cos 100t,$$

or, equivalently,

(5) $$\frac{d^2I}{dt^2} + 20\frac{dI}{dt} + 500I = 100{,}000\cos 100t.$$

The homogeneous equation associated with (5) has the auxiliary equation

$$r^2 + 20r + 500 = (r+10)^2 + (20)^2 = 0,$$

whose roots are $-10 \pm 20i$. Hence, the solution to the homogeneous equation is

(6) $$I_h(t) = C_1 e^{-10t}\cos 20t + C_2 e^{-10t}\sin 20t.$$

To find a particular solution for (5), we can use the method of undetermined coefficients. Setting

$$I_p(t) = A\cos 100t + B\sin 100t$$

and carrying out the procedure in Section 4.5, we ultimately find, to three decimals,

$$A = -10.080, \qquad B = 2.122.$$

Hence, a particular solution to (5) is given by

(7) $$I_p(t) = -10.080\cos 100t + 2.122\sin 100t.$$

Since $I = I_h + I_p$, we find from (6) and (7) that

(8) $$I(t) = e^{-10t}(C_1\cos 20t + C_2\sin 20t) - 10.080\cos 100t + 2.122\sin 100t.$$

To determine the constants C_1 and C_2, we need the values $I(0)$ and $I'(0)$. We were given $I(0) = q(0) = 0$. To find $I'(0)$, we substitute the values for L, R, and C into equation (3) and equate the two sides at $t = 0$. This gives

$$(0.001)I'(0) + (0.02)I(0) + (0.5)q(0) = \sin 0.$$

Because $I(0) = q(0) = 0$, we find $I'(0) = 0$. Finally, using $I(t)$ in (8) and the initial conditions $I(0) = I'(0) = 0$, we obtain the system

$$I(0) = C_1 - 10.080 = 0 ,$$

$$I'(0) = -10C_1 + 20C_2 + 212.2 = 0 .$$

Solving this system yields $C_1 = 10.080$ and $C_2 = -5.570$. Hence, the current in the *RLC* series circuit is

(9) $\qquad I(t) = e^{-10t}(10.080 \cos 20t - 5.570 \sin 20t) - 10.080 \cos 100t + 2.122 \sin 100t .$ ◆

Observe that, as was the case with forced mechanical vibrations, the current in (9) is made up of two components. The first, I_h, is a **transient current** that tends to zero as $t \to +\infty$. The second,

$$I_p(t) = -10.080 \cos 100t + 2.122 \sin 100t ,$$

is a sinusoidal **steady-state current** that remains.

It is straightforward to verify that the steady-state solution $I_p(t)$ that arises from the more general voltage source $E(t) = E_0 \sin \gamma t$ is

(10) $\qquad I_p(t) = \dfrac{E_0 \sin(\gamma t + \theta)}{\sqrt{R^2 + [\gamma L - 1/(\gamma C)]^2}} ,$

where $\tan \theta = (1/C - L\gamma^2)/(\gamma R)$ (compare Section 4.10, page 221).

Example 2 At time $t = 0$, the charge on the capacitor in the electrical network shown in Figure 5.38 is 2 coulombs (C), while the current through the capacitor is zero. Determine the charge on the capacitor and the currents in the various branches of the network at any time $t > 0$.

Solution To determine the charge and currents in the electrical network, we begin by observing that the network consists of three closed circuits: loop 1 through the battery, resistor, and inductor; loop 2 through the battery, resistor, and capacitor; and loop 3 containing the capacitor and inductor. Taking advantage of Kirchhoff's current law, we denote the current passing through the battery and the resistor by I_1, the current through the inductor by I_2, and the current through the capacitor by I_3. For consistency of notation, we denote the charge on the capacitor by q_3. Hence, $I_3 = dq_3/dt$.

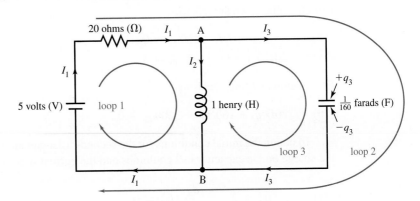

Figure 5.38 Schematic of an electrical network

As discussed at the beginning of this section, the voltage drop at a resistor is RI, at an inductor $L\,dI/dt$, and at a capacitor q/C. So, applying Kirchhoff's voltage law to the electrical network in Figure 5.38, we find for loop 1,

(11) $\dfrac{dI_2}{dt} + 20I_1 = 5 \; ;$
 (inductor) (resistor) (battery)

for loop 2,

(12) $20I_1 + 160q_3 = 5 \; ;$
 (resistor) (capacitor) (battery)

and for loop 3,

(13) $-\dfrac{dI_2}{dt} + 160q_3 = 0 \, .$
 (inductor) (capacitor)

[The minus sign in (13) arises from taking a clockwise path around loop 3 so that the current passing through the inductor is $-I_2$.] Notice that these three equations are not independent: We can obtain equation (13) by subtracting (11) from (12). Hence, we have only two equations from which to determine the three unknowns I_1, I_2, and q_3. If we now apply Kirchhoff's current law to the two junction points in the network, we find at point A that $I_1 - I_2 - I_3 = 0$ and at point B that $I_2 + I_3 - I_1 = 0$. In both cases, we get

(14) $I_1 - I_2 - \dfrac{dq_3}{dt} = 0 \, ,$

since $I_3 = dq_3/dt$. Assembling equations (11), (12), and (14) into a system, we have (with $D = d/dt$)

(15) $DI_2 + 20I_1 = 5 \, ,$
(16) $20I_1 + 160q_3 = 5 \, ,$
(17) $-I_2 + I_1 - Dq_3 = 0 \, .$

We solve these by the elimination method of Section 5.2. Using equation (16) to eliminate I_1 from the others, we are left with

(18) $DI_2 - 160q_3 = 0 \, ,$
(19) $20I_2 + (20D + 160)q_3 = 5 \, .$

Elimination of I_2 then leads to

(20) $20D^2 q_3 + 160Dq_3 + 3200q_3 = 0 \, .$

To obtain the initial conditions for the second-order equation (20), recall that at time $t = 0$, the charge on the capacitor is 2 coulombs and the current is zero. Hence,

(21) $q_3(0) = 2 \, , \qquad \dfrac{dq_3}{dt}(0) = 0 \, .$

We can now solve the initial value problem (20)–(21) using the techniques of Chapter 4. Ultimately, we find

$$q_3(t) = 2e^{-4t}\cos 12t + \frac{2}{3}e^{-4t}\sin 12t \,,$$

$$I_3(t) = \frac{dq_3}{dt}(t) = -\frac{80}{3}e^{-4t}\sin 12t \,.$$

Next, to determine I_2, we substitute these expressions into (19) and obtain

$$I_2(t) = \frac{1}{4} - \frac{dq_3}{dt}(t) - 8q_3(t)$$

$$= \frac{1}{4} - 16e^{-4t}\cos 12t + \frac{64}{3}e^{-4t}\sin 12t \,.$$

Finally, from $I_1 = I_2 + I_3$, we get

$$I_1(t) = \frac{1}{4} - 16e^{-4t}\cos 12t - \frac{16}{3}e^{-4t}\sin 12t \,. \quad \blacklozenge$$

Note that the differential equations that describe mechanical vibrations and *RLC* series circuits are essentially the same. And, in fact, there is a natural identification of the parameters m, b, and k for a mass–spring system with the parameters L, R, and C that describe circuits. This is illustrated in Table 5.3. Moreover, the terms *transient, steady-state, overdamped, critically damped, underdamped,* and *resonant frequency* described in Sections 4.9 and 4.10 apply to electrical circuits as well.

This analogy between a mechanical system and an electrical circuit extends to large-scale systems and circuits. An interesting consequence of this is the use of **analog simulation** and, in particular, analog computers to analyze mechanical systems. Large-scale mechanical systems are modeled by building a corresponding electrical system and then measuring the charges $q(t)$ and currents $I(t)$.

Although such analog simulations are important, *both* large-scale mechanical and electrical systems are currently modeled using digital computer simulation. This involves the numerical solution of the initial value problem governing the system. Still, the analogy between mechanical and electrical systems means that basically the same computer software can be used to analyze both systems.

TABLE 5.3	Analogy Between Mechanical and Electrical Systems		
Mechanical Mass–Spring System with Damping		**Electrical *RLC* Series Circuit**	
$mx'' + bx' + kx = f(t)$		$Lq'' + Rq' + (1/C)q = E(t)$	
Displacement	x	Charge	q
Velocity	x'	Current	$q' = I$
Mass	m	Inductance	L
Damping constant	b	Resistance	R
Spring constant	k	(Capacitance)$^{-1}$	$1/C$
External force	$f(t)$	Voltage source	$E(t)$

5.7 EXERCISES

1. An *RLC* series circuit has a voltage source given by $E(t) = 20$ V, a resistor of 100 Ω, an inductor of 4 H, and a capacitor of 0.01 F. If the initial current is zero and the initial charge on the capacitor is 4 C, determine the current in the circuit for $t > 0$.

2. An *RLC* series circuit has a voltage source given by $E(t) = 40 \cos 2t$ V, a resistor of 2 Ω, an inductor of 1/4 H, and a capacitor of 1/13 F. If the initial current is zero and the initial charge on the capacitor is 3.5 C, determine the charge on the capacitor for $t > 0$.

3. An *RLC* series circuit has a voltage source given by $E(t) = 10 \cos 20t$ V, a resistor of 120 Ω, an inductor of 4 H, and a capacitor of $(2200)^{-1}$ F. Find the steady-state current (solution) for this circuit. What is the resonance frequency of the circuit?

4. An *LC* series circuit has a voltage source given by $E(t) = 30 \sin 50t$ V, an inductor of 2 H, and a capacitor of 0.02 F (but no resistor). What is the current in this circuit for $t > 0$ if at $t = 0, I(0) = q(0) = 0$?

5. An *RLC* series circuit has a voltage source of the form $E(t) = E_0 \cos \gamma t$ V, a resistor of 10 Ω, an inductor of 4 H, and a capacitor of 0.01 F. Sketch the frequency response curve for this circuit.

6. Show that when the voltage source in (4) is of the form $E(t) = E_0 \sin \gamma t$, then the steady-state solution I_p is as given in equation (10).

7. A mass–spring system with damping consists of a 7-kg mass, a spring with spring constant 3 N/m, a frictional component with damping constant 2 N-sec/m, and an external force given by $f(t) = 10 \cos 10t$ N. Using a 10-Ω resistor, construct an *RLC* series circuit that is the analog of this mechanical system in the sense that the two systems are governed by the same differential equation.

8. A mass–spring system with damping consists of a 16-lb weight, a spring with spring constant 64 lb/ft, a frictional component with damping constant 10 lb-sec/ft, and an external force given by $f(t) = 20 \cos 8t$ lb. Using an inductor of 0.01 H, construct an *RLC* series circuit that is the analog of this mechanical system.

9. Because of Euler's formula, $e^{i\theta} = \cos\theta + i\sin\theta$, it is often convenient to treat the voltage sources $E_0 \cos \gamma t$ and $E_0 \sin \gamma t$ simultaneously, using $E(t) = E_0 e^{i\gamma t}$. In this case, equation (3) becomes

(22) $$L\frac{d^2q}{dt^2} + R\frac{dq}{dt} + \frac{1}{C}q = E_0 e^{i\gamma t},$$

where q is now complex (recall $I = q', I' = q''$).

(a) Use the method of undetermined coefficients to show that the steady-state solution to (22) is
$$q_p(t) = \frac{E_0}{1/C - \gamma^2 L + i\gamma R}e^{i\gamma t}.$$
The technique is discussed in detail in Project F of Chapter 4, page 237.

(b) Now show that the steady-state current is
$$I_p(t) = \frac{E_0}{R + i[\gamma L - 1/(\gamma C)]}e^{i\gamma t}.$$

(c) Use the relation $\alpha + i\beta = \sqrt{\alpha^2 + \beta^2}e^{i\theta}$, where $\tan\theta = \beta/\alpha$, to show that I_p can be expressed in the form
$$I_p(t) = \frac{E_0}{\sqrt{R^2 + [\gamma L - 1/(\gamma C)]^2}}e^{i(\gamma t + \theta)},$$
where $\tan\theta = (1/C - L\gamma^2)/(\gamma R)$.

(d) The imaginary part of $e^{i\gamma t}$ is $\sin \gamma t$, so the imaginary part of the solution to (22) must be the solution to equation (3) for $E(t) = E_0 \sin \gamma t$. Verify that this is also the case for the current by showing that the imaginary part of I_p in part (c) is the same as that given in equation (10).

In Problems 10–13, find a system of differential equations and initial conditions for the currents in the networks given in the schematic diagrams (Figures 5.39–5.42 on pages 294–295). Assume that all initial currents are zero. Solve for the currents in each branch of the network.

10.

Figure 5.39 *RL* network for Problem 10

11.

Figure 5.40 *RLC* network for Problem 11

12.

Figure 5.41 *RL* network for Problem 12

13.

Figure 5.42 *RLC* network for Problem 13

5.8 Dynamical Systems, Poincaré Maps, and Chaos

In this section we take an excursion through an area of mathematics that has received a lot of attention both for the interesting mathematical phenomena being observed and for its application to fields such as meteorology, heat conduction, fluid mechanics, lasers, chemical reactions, and nonlinear circuits, among others. The area is that of nonlinear dynamical systems.[†]

A **dynamical system** is any system that allows one to determine (at least theoretically) the future states of the system given its present or past state. For example, the recursive formula (difference equation)

$$x_{n+1} = (1.05)x_n, \qquad n = 0, 1, 2, \dots$$

is a dynamical system, since we can determine the next state, x_{n+1}, given the previous state, x_n. If we know x_0, then we can compute any future state $[$ indeed, $x_{n+1} = x_0(1.05)^{n+1}]$.

Another example of a dynamical system is provided by the differential equation

$$\frac{dx}{dt} = -2x,$$

where the solution $x(t)$ specifies the state of the system at "time" t. If we know $x(t_0) = x_0$, then we can determine the state of the system at any future time $t > t_0$ by solving the initial value problem

$$\frac{dx}{dt} = -2x, \qquad x(t_0) = x_0.$$

Indeed, a simple calculation yields $x(t) = x_0 e^{-2(t-t_0)}$ for $t \geq t_0$.

For a dynamical system defined by a differential equation, it is often helpful to work with a related dynamical system defined by a difference equation. For example, when we cannot express the solution to a differential equation using elementary functions, we can use a numerical technique such as the improved Euler's method or Runge–Kutta to approximate the solution to an initial value problem. This numerical scheme defines a new (but related) dynamical system that is often easier to study.

In Section 5.4, we used phase plane diagrams to study autonomous systems in the plane. Many important features of the system can be detected just by looking at these diagrams. For example, a closed trajectory corresponds to a periodic solution. The trajectories for *non*autonomous systems in

[†]For a more detailed study of dynamical systems, see *An Introduction to Chaotic Dynamical Systems*, by R. L. Devaney (Westview Press, 2003) and *Nonlinear Oscillations, Dynamical Systems and Bifurcations of Vector Fields*, by J. Guckenheimer and P. J. Holmes (Springer-Verlag, New York, 1983).

the phase plane are much more complicated to decipher. One technique that is helpful in this regard is the so-called **Poincaré map.** As we will see, these maps replace the study of a nonautonomous system with the study of a dynamical system defined by the location in the xv-plane $(v = dx/dt)$ of the solution at regularly spaced moments in time such as $t = 2\pi n$, where $n = 0, 1, 2, \ldots$. The advantage in using the Poincaré map will become clear when the method is applied to a nonlinear problem for which no explicit solution is known. In such a case, the trajectories are computed using a numerical scheme such as Runge–Kutta. Several software packages have options that will construct Poincaré maps for a given system.

To illustrate the Poincaré map, consider the equation

$$(1) \qquad x''(t) + \omega^2 x(t) = F \cos t ,$$

where F and ω are positive constants. We studied similar equations in Section 4.10 and found that a general solution for $\omega \neq 1$ is given by

$$(2) \qquad x(t) = A \sin(\omega t + \phi) + \frac{F}{\omega^2 - 1} \cos t ,$$

where the amplitude A and the phase angle ϕ are arbitrary constants. Since $v = x'$,

$$v(t) = \omega A \cos(\omega t + \phi) - \frac{F}{\omega^2 - 1} \sin t .$$

Because the forcing function $F \cos t$ is 2π-periodic, it is natural to seek 2π-periodic solutions to (1). For this purpose, we define the Poincaré map as

$$(3) \qquad \begin{aligned} x_n &:= x(2\pi n) = A \sin(2\pi\omega n + \phi) + F/(\omega^2 - 1) , \\ v_n &:= v(2\pi n) = \omega A \cos(2\pi\omega n + \phi) , \end{aligned}$$

for $n = 0, 1, 2, \ldots$. In Figure 5.43 on page 297, we plotted the first 100 values of (x_n, v_n) in the xv-plane for different choices of ω. For simplicity, we have taken $A = F = 1$ and $\phi = 0$. These graphs are called **Poincaré sections.** We will interpret them shortly.

Now let's play the following game. We agree to ignore the fact that we already know the formula for $x(t)$ for all $t \geq 0$. We want to see what information about the solution we can glean just from the Poincaré section and the form of the differential equation.

Notice that the first two Poincaré sections in Figure 5.43, corresponding to $\omega = 2$ and 3, consist of a single point. This tells us that, starting with $t = 0$, every increment 2π of t returns us to the same point in the phase plane. This in turn implies that equation (1) has a 2π-periodic solution, which can be proved as follows: For $\omega = 2$, let $x(t)$ be the solution to (1) with $\big(x(0), v(0)\big) = (1/3, 2)$ and let $X(t) := x(t + 2\pi)$. Since the Poincaré section is just the point $(1/3, 2)$, we have $X(0) = x(2\pi) = 1/3$ and $X'(0) = x'(2\pi) = 2$. Thus, $x(t)$ and $X(t)$ have the same initial values at $t = 0$. Further, because $\cos t$ is 2π-periodic, we also have

$$X''(t) + \omega^2 X(t) = x''(t + 2\pi) + \omega^2 x(t + 2\pi) = \cos(t + 2\pi) = \cos t .$$

Consequently, $x(t)$ and $X(t)$ satisfy the same initial value problem. By the uniqueness theorems of Sections 4.2 and 4.5, these functions must agree on the interval $[0, \infty)$. Hence, $x(t) = X(t) = x(t + 2\pi)$ for all $t \geq 0$; that is, $x(t)$ is 2π-periodic. (The same reasoning works for $\omega = 3$.) With a similar argument, it follows from the Poincaré section for $\omega = 1/2$ that there is a solution of period 4π that alternates between the two points displayed in Figure 5.43(c) as t is incremented by 2π. For the case $\omega = 1/3$, we deduce that there is a solution of period 6π rotating among three points, and for $\omega = 1/4$, there is an 8π-periodic solution rotating among four points. We call these last three solutions **subharmonics.**

The case $\omega = \sqrt{2}$ is different. So far, in Figure 5.43(f), none of the points has repeated. Did we stop too soon? Will the points ever repeat? Here, the fact that $\sqrt{2}$ is irrational plays

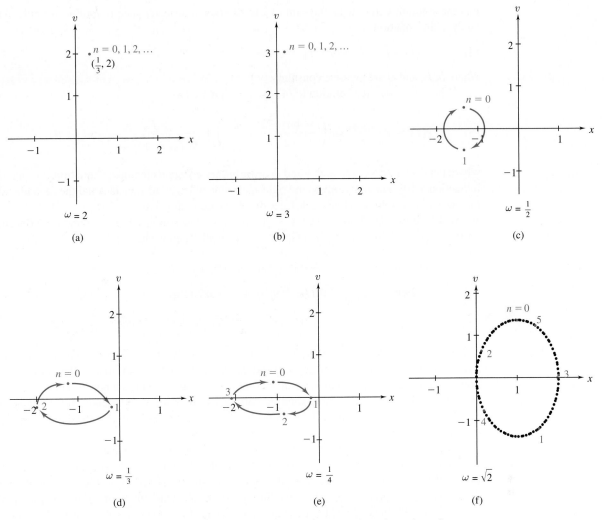

Figure 5.43 Poincaré sections for equation (1) for various values of ω

a crucial role. It turns out that every integer n yields a distinct point in the Poincaré section (see Problem 8). However, there is a pattern developing. The points all appear to lie on a simple curve, possibly an ellipse. To see that this is indeed the case, notice that when $\omega = \sqrt{2}$, $A = F = 1$, and $\phi = 0$, we have

$$ x_n = \sin\left(2\sqrt{2}\,\pi n\right) + 1\,, \qquad v_n = \sqrt{2}\cos\left(2\sqrt{2}\,\pi n\right)\,, \qquad n = 0, 1, 2, \ldots\,. $$

It is then an easy computation to show that each (x_n, v_n) lies on the ellipse

$$ (x-1)^2 + \frac{v^2}{2} = 1\,. $$

In our investigation of equation (1), we concentrated on 2π-periodic solutions because the forcing term $F\cos t$ has period 2π. [We observed subharmonics when $\omega = 1/2, 1/3$, and $1/4$—that is, solutions with periods $2(2\pi), 3(2\pi)$, and $4(2\pi)$.] When a damping term is introduced into the differential equation, the Poincaré map displays a different behavior. Recall

that the solution will now be the sum of a transient and a steady-state term. For example, let's analyze the equation

(4) $x''(t) + bx'(t) + \omega^2 x(t) = F \cos t$,

where b, F, and ω are positive constants.

When $b^2 < 4\omega^2$, the solution to (4) can be expressed as

(5) $x(t) = Ae^{-(b/2)t} \sin\left(\dfrac{\sqrt{4\omega^2 - b^2}}{2} t + \phi\right) + \dfrac{F}{\sqrt{(\omega^2 - 1)^2 + b^2}} \sin(t + \theta)$,

where $\tan\theta = (\omega^2 - 1)/b$ and A and ϕ are arbitrary constants [compare equations (7) and (8) in Section 4.10]. The first term on the right-hand side of (5) is the transient and the second term, the steady-state solution. Let's construct the Poincaré map using $t = 2\pi n$, $n = 0, 1, 2, \ldots$. We will take $b = 0.22$, $\omega = A = F = 1$, and $\phi = 0$ to simplify the computations. Because $\tan\theta = (\omega^2 - 1)/b = 0$, we will take $\theta = 0$ as well. Then we have

$$x(2\pi n) = x_n = e^{-0.22\pi n}\sin\left(\sqrt{0.9879}\, 2\pi n\right),$$
$$x'(2\pi n) = v_n = -0.11 e^{-0.22\pi n}\sin\left(\sqrt{0.9879}\, 2\pi n\right)$$
$$+ \sqrt{0.9879}\, e^{-0.22\pi n}\cos\left(\sqrt{0.9879}\, 2\pi n\right) + \dfrac{1}{(0.22)}.$$

The Poincaré section for $n = 0, 1, 2, \ldots, 10$ is shown in Figure 5.44 (black points). After just a few iterations, we observe that $x_n \approx 0$ and $v_n \approx 1/(0.22) \approx 4.545$; that is, the points of the Poincaré section are approaching a single point in the xv-plane (colored point). Thus, we might expect that there is a 2π-periodic solution corresponding to a particular choice of A and ϕ. [In this example, where we can explicitly represent the solution, we see that indeed a 2π-periodic solution arises when we take $A = 0$ in (5).]

There is an important difference between the Poincaré sections for equation (1) and those for equation (4). In Figure 5.43, the location of all of the points in (a)–(e) depends on the initial value selected (here $A = 1$ and $\phi = 0$). (See Problem 10.) However, in Figure 5.44, the first few points (black points) depend on the initial conditions, while the limit point (colored point) does *not* (see Problem 6). The latter behavior is typical for equations that have a "damping"

Figure 5.44 Poincaré section for equation (4) with $F = 1$, $b = 0.22$, and $\omega = 1$

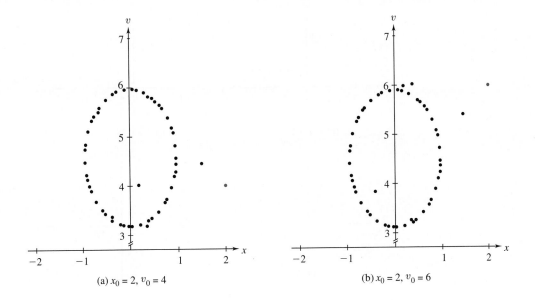

(a) $x_0 = 2$, $v_0 = 4$ (b) $x_0 = 2$, $v_0 = 6$

Figure 5.45 Poincaré section for equation (6) with initial values x_0, v_0

term (i.e., $b > 0$); namely, the Poincaré section has a limit set[†] that is essentially independent of the initial conditions.

For equations with damping, the limit set may be more complicated than just a point. For example, the Poincaré map for the equation

$$(6) \qquad x''(t) + (0.22)x'(t) + x(t) = \cos t + \cos\left(\sqrt{2}t\right)$$

has a limit set consisting of an ellipse (see Problem 11). This is illustrated in Figure 5.45 for the initial values $x_0 = 2$, $v_0 = 4$ and $x_0 = 2$, $v_0 = 6$.

So far we have seen limit sets for the Poincaré map that were either a single point or an ellipse—independent of the initial values. These particular limit sets are **attractors.** In general, an attractor is a set A with the property that there exists an open set[‡] B containing A such that whenever the Poincaré map enters B, its points remain in B *and* the limit set of the Poincaré map is a subset of A. Further, we assume A has the *invariance property:* Whenever the Poincaré map starts at a point in A, it remains in A.

In the previous examples, the attractors of the dynamical system (Poincaré map) were easy to describe. In recent years, however, many investigators, working on a variety of applications, have encountered dynamical systems that do *not* behave orderly—their attractor sets are very complicated (not just isolated points or familiar geometric objects such as ellipses). The behavior of such systems is called **chaotic,** and the corresponding limit sets are referred to as **strange attractors.**

[†]The **limit set** for a map (x_n, v_n), $n = 1, 2, 3, \ldots$, is the set of points (p, q) such that $\lim_{k \to \infty}(x_{n_k}, v_{n_k}) = (p, q)$, where $n_1 < n_2 < n_3 < \ldots$ is some subsequence of the positive integers.

[‡]A set $B \subset \mathbf{R}^2$ is an **open set** if for each point $p \in B$ there is an open disk V containing p such that $V \subset B$.

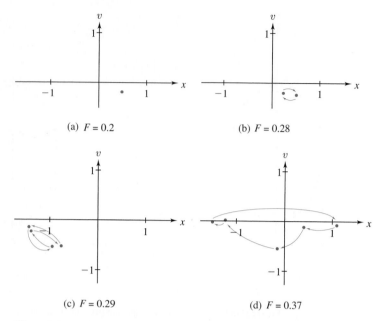

(a) $F = 0.2$ (b) $F = 0.28$

(c) $F = 0.29$ (d) $F = 0.37$

Figure 5.46 Poincaré sections for the Duffing equation (7) with $b = 0.3$ and $\gamma = 1.2$

To illustrate chaotic behavior and what is meant by a strange attractor, we discuss two non-linear differential equations and a simple difference equation. First, let's consider the **forced Duffing equation**

$$(7) \qquad x''(t) + bx'(t) - x(t) + x^3(t) = F \sin \gamma t .$$

We cannot express the solution to (7) in any explicit form, so we must obtain the Poincaré map by numerically approximating the solution to (7) for fixed initial values and then plot the approximations for $x(2\pi n/\gamma)$ and $v(2\pi n/\gamma) = x'(2\pi n/\gamma)$. (Because the forcing term $F \sin \gamma t$ has period $2\pi/\gamma$, we seek $2\pi/\gamma$-periodic solutions and subharmonics.) In Figure 5.46, we display the limit sets (attractors) when $b = 0.3$ and $\gamma = 1.2$ in the cases (a) $F = 0.2$, (b) $F = 0.28$, (c) $F = 0.29$, and (d) $F = 0.37$.

Notice that as the constant F increases, the Poincaré map changes character. When $F = 0.2$, the Poincaré section tells us that there is a $2\pi/\gamma$-periodic solution. For $F = 0.28$, there is a subharmonic of period $4\pi/\gamma$, and for $F = 0.29$ and 0.37, there are subharmonics with periods $8\pi/\gamma$ and $10\pi/\gamma$, respectively.

Things are dramatically different when $F = 0.5$: The solution is neither $2\pi/\gamma$-periodic nor subharmonic. The Poincaré section for $F = 0.5$ is illustrated in Figure 5.47 on page 301. This section was generated by numerically approximating the solution to (7) when $\gamma = 1.2, b = 0.3$, and $F = 0.5$, for fixed initial values.[†] Not all of the approximations $x(2\pi n/\gamma)$ and $v(2\pi n/\gamma)$ that were calculated are graphed; because of the presence of a transient solution, the first few points were omitted. It turns out that the plotted set is essentially independent of the initial values and has the property that once a point is in the set, all subsequent points lie in the set.

[†]*Historical Footnote:* When researchers first encountered these strange-looking Poincaré sections, they would check their computations using different computers and different numerical schemes [see Hénon and Heiles, "The Applicability of the Third Integral of Motion: Some Numerical Experiments," *Astronomical Journal*, Vol. 69 (1964): 75]. For special types of dynamical systems, such as the Hénon map, it can be shown that there exists a true trajectory that *shadows* the numerical trajectory [see M. Hammel, J. A. Yorke, and C. Grebogi, "Numerical Orbits of Chaotic Processes Represent True Orbits," *Bulletin American Mathematical Society*, Vol. 19 (1988): 466–469].

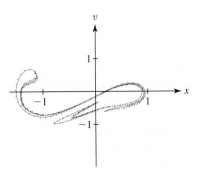

Figure 5.47 Poincaré section for the Duffing equation (7) with $b = 0.3$, $\gamma = 1.2$, and $F = 0.5$

Because of the complicated shape of the set, it is indeed a strange attractor. While the shape of the strange attractor does not depend on the initial values, the picture does change if we consider different sections; for example, $t = (2\pi n + \pi/2)/\gamma$, $n = 0, 1, 2, \ldots$ yields a different configuration.

Another example of a strange attractor occurs when we consider the **forced pendulum equation**

$$(8) \qquad x''(t) + bx'(t) + \sin\big(x(t)\big) = F\cos t,$$

where the $x(t)$ term in (4) has been replaced by $\sin\big(x(t)\big)$. Here $x(t)$ is the angle between the pendulum and the vertical rest position, b is related to damping, and F represents the strength of the forcing function (see Figure 5.48). For $F = 2.7$ and $b = 0.22$, we have graphed in Figure 5.49 approximately 90,000 points in the Poincaré map. Since we cannot express the solution to (8) in any explicit form, the Poincaré map was obtained by numerically approximating the solution to (8) for fixed initial values and plotting the approximations for $x(2\pi n)$ and $v(2\pi n) = x'(2\pi n)$.

The Poincaré maps for the forced Duffing equation and for the forced pendulum equation not only illustrate the idea of a strange attractor; they also exhibit another peculiar behavior called **chaos.** Chaos occurs when small changes in the initial conditions lead to major changes in the behavior of the solution. Henri Poincaré described the situation as follows:

> It may happen that small differences in the initial conditions will produce very large ones in the final phenomena. A small error in the former produces an enormous error in the latter. Prediction becomes impossible

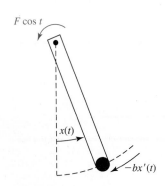

Figure 5.48 Forced damped pendulum

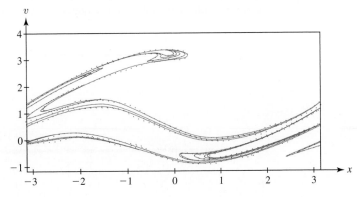

Figure 5.49 Poincaré section for the forced pendulum equation (8) with $b = 0.22$ and $F = 2.7$

In a physical experiment, we can never *exactly* (with infinite accuracy) reproduce the same initial conditions. Consequently, if the behavior is chaotic, even a slight difference in the initial conditions may lead to quite different values for the corresponding Poincaré map when n is large. Such behavior does not occur for solutions to either equation (4) or equation (1) (see Problems 6 and 7). However, two solutions to the Duffing equation (7) with $F = 0.5$ that correspond to two different but close initial values have Poincaré maps that do *not* remain close together. Although they both are attracted to the same set, their locations with respect to this set may be relatively far apart.

The phenomenon of chaos can also be illustrated by the following simple map. Let x_0 lie in $[0, 1)$ and define

(9) $x_{n+1} = 2x_n \,(\text{mod } 1)\,,$

where by (mod 1) we mean the decimal part of the number if it is greater than or equal to 1; that is

$$x_{n+1} = \begin{cases} 2x_n, & \text{for } 0 \le x_n < 1/2\,, \\ 2x_n - 1, & \text{for } 1/2 \le x_n < 1\,. \end{cases}$$

When $x_0 = 1/3$, we find

$$x_1 = 2 \cdot (1/3)\,(\text{mod } 1) = 2/3\,,$$
$$x_2 = 2 \cdot (2/3)\,(\text{mod } 1) = 4/3\,(\text{mod } 1) = 1/3\,,$$
$$x_3 = 2 \cdot (1/3)\,(\text{mod } 1) = 2/3\,,$$
$$x_4 = 2 \cdot (2/3)\,(\text{mod } 1) = 1/3\,, \quad \text{etc.}$$

Written as a sequence, we get $\{1/3, 2/3, \overline{1/3, 2/3}, \ldots\}$, where the overbar denotes the repeated pattern.

What happens when we pick a starting value x_0 near $1/3$? Does the sequence cluster about $1/3$ and $2/3$ as does the mapping when $x_0 = 1/3$? For example, when $x_0 = 0.3$, we get the sequence

$$\{0.3, 0.6, 0.2, 0.4, 0.8, \overline{0.6, 0.2, 0.4, 0.8}, \ldots\}\,.$$

In Figure 5.50, we have plotted the values of x_n for $x_0 = 0.3$, 0.33, and 0.333. We have not plotted the first few terms, but only those that repeat. (This omission of the first few terms parallels the situation depicted in Figure 5.47 on page 301, where transient solutions arise.)

It is clear from Figure 5.50 that while the values for x_0 are getting closer to $1/3$, the corresponding maps are spreading out over the whole interval $[0, 1]$ and *not* clustering near $1/3$ and $2/3$. This behavior is chaotic, since the Poincaré maps for initial values near $1/3$ behave quite differently from the map for $x_0 = 1/3$. If we had selected x_0 to be irrational (which we can't do with a calculator), the sequence would *not* repeat and would be dense in $[0, 1]$.

Figure 5.50 Plots of the map $x_{n+1} = 2x_n\,(\text{mod } 1)$ for $x_0 = 0.3$, 0.33, and 0.333

Systems that exhibit chaotic behavior arise in many applications. The challenge to engineers is to design systems that avoid this chaos and, instead, enjoy the property of **stability.** The topic of stable systems is discussed at length in Chapter 12.[†]

5.8 EXERCISES

A software package that supports the construction of Poincaré maps is required for the problems in this section.

1. Compute and graph the points of the Poincaré map with $t = 2\pi n$, $n = 0, 1, \ldots, 20$ for equation (1), taking $A = F = 1$, $\phi = 0$, and $\omega = 3/2$. Repeat, taking $\omega = 3/5$. Do you think the equation has a 2π-periodic solution for either choice of ω? A subharmonic solution?

2. Compute and graph the points of the Poincaré map with $t = 2\pi n$, $n = 0, 1, \ldots, 20$ for equation (1), taking $A = F = 1$, $\phi = 0$, and $\omega = 1/\sqrt{3}$. Describe the limit set for this system.

3. Compute and graph the points of the Poincaré map with $t = 2\pi n$, $n = 0, 1, \ldots, 20$ for equation (4), taking $A = F = 1$, $\phi = 0$, $\omega = 1$, and $b = -0.1$. What is happening to these points as $n \to \infty$?

4. Compute and graph the Poincaré map with $t = 2\pi n$, $n = 0, 1, \ldots, 20$ for equation (4), taking $A = F = 1$, $\phi = 0$, $\omega = 1$, and $b = 0.1$. Describe the attractor for this system.

5. Compute and graph the Poincaré map with $t = 2\pi n$, $n = 0, 1, \ldots, 20$ for equation (4), taking $A = F = 1$, $\phi = 0$, $\omega = 1/3$, and $b = 0.22$. Describe the attractor for this system.

6. Show that for $b > 0$, the Poincaré map for equation (4) is not chaotic by showing that as t gets large

$$x_n = x(2\pi n) \approx \frac{F}{\sqrt{(\omega^2 - 1)^2 + b^2}} \sin(2\pi n + \theta),$$

$$v_n = x'(2\pi n) \approx \frac{F}{\sqrt{(\omega^2 - 1)^2 + b^2}} \cos(2\pi n + \theta)$$

independent of the initial values $x_0 = x(0)$ and $v_0 = x'(0)$.

7. Show that the Poincaré map for equation (1) is not chaotic by showing that if (x_0, v_0) and (x_0^*, v_0^*) are two initial values that define the Poincaré maps $\{(x_n, v_n)\}$ and $\{(x_n^*, v_n^*)\}$, respectively, using the recursive formulas in (3), then one can make the distance between (x_n, v_n) and (x_n^*, v_n^*) small by making the distance between (x_0, v_0) and (x_0^*, v_0^*) small. [*Hint:* Let (A, ϕ) and (A^*, ϕ^*) be the polar coordinates of two points in the plane. From the law of cosines, it follows that the distance d between them is given by $d^2 = (A - A^*)^2 + 2AA^*[1 - \cos(\phi - \phi^*)]$.]

8. Consider the Poincaré maps defined in (3) with $\omega = \sqrt{2}$, $A = F = 1$, and $\phi = 0$. If this map were ever to repeat, then for two distinct positive integers n and m, $\sin(2\sqrt{2}\pi n) = \sin(2\sqrt{2}\pi m)$. Using basic properties of the sine function, show that this would imply that $\sqrt{2}$ is rational. It follows from this contradiction that the points of the Poincaré map do not repeat.

9. The doubling modulo 1 map defined by equation (9) exhibits some fascinating behavior. Compute the sequence obtained when
 (a) $x_0 = k/7$ for $k = 1, 2, \ldots, 6$.
 (b) $x_0 = k/15$ for $k = 1, 2, \ldots, 14$.
 (c) $x_0 = k/2^j$, where j is a positive integer and $k = 1, 2, \ldots, 2^j - 1$.

 Numbers of the form $k/2^j$ are called **dyadic numbers** and are dense in $[0, 1]$. That is, there is a dyadic number arbitrarily close to any real number (rational or irrational).

10. To show that the limit set of the Poincaré map given in (3) depends on the initial values, do the following:
 (a) Show that when $\omega = 2$ or 3, the Poincaré map consists of the single point

 $$(x, v) = \left(A \sin\phi + \frac{F}{\omega^2 - 1}, \ \omega A \cos\phi \right).$$

 (b) Show that when $\omega = 1/2$, the Poincaré map alternates between the two points

 $$\left(\frac{F}{\omega^2 - 1} \pm A \sin\phi, \ \pm \omega A \cos\phi \right).$$

 (c) Use the results of parts (a) and (b) to show that when $\omega = 2, 3$, or $1/2$, the Poincaré map (3) depends on the initial values (x_0, v_0).

[†]All references to Chapters 11–13 refer to the expanded text, *Fundamentals of Differential Equations and Boundary Value Problems*, 7th ed.

11. To show that the limit set for the Poincaré map
$x_n := x(2\pi n)$, $v_n := x'(2\pi n)$, where $x(t)$ is a solution
to equation (6), is an ellipse and that this ellipse is the
same for any initial values x_0, v_0, do the following:
 (a) Argue that since the initial values affect only the
 transient solution to (6), the limit set for the Poin-
 caré map is independent of the initial values.
 (b) Now show that for n large,

$$x_n \approx a \sin(2\sqrt{2}\pi n + \psi),$$

$$v_n \approx c + \sqrt{2}a \cos(2\sqrt{2}\pi n + \psi),$$

 where $a = \left(1 + 2(0.22)^2\right)^{-1/2}$, $c = (0.22)^{-1}$,
 and $\psi = \arctan\left\{-\left[(0.22)\sqrt{2}\right]^{-1}\right\}$.
 (c) Use the result of part (b) to conclude that the ellipse

$$x^2 + \frac{(v-c)^2}{2} = a^2$$

 contains the limit set of the Poincaré map.

12. Using a numerical scheme such as Runge–Kutta or a
software package, calculate the Poincaré map for equa-
tion (7) when $b = 0.3$, $\gamma = 1.2$, and $F = 0.2$. (Notice
that the closer you start to the limiting point, the sooner
the transient part will die out.) Compare your map with
Figure 5.46(a) on page 300. Redo for $F = 0.28$.

13. Redo Problem 12 with $F = 0.31$. What kind of behavior
does the solution exhibit?

14. Redo Problem 12 with $F = 0.65$. What kind of behavior
does the solution exhibit?

15. **Chaos Machine.** Chaos can be illustrated using a long
ruler, a short ruler, a pin, and a tie tack (pivot). Construct
the double pendulum as shown in Figure 5.51(a). The
pendulum is set in motion by releasing it from a posi-
tion such as the one shown in Figure 5.51(b). Repeatedly

(a) double pendulum

(b) release position

Figure 5.51 Double pendulum as a chaos machine

set the pendulum in motion, each time trying to release
it from the same position. Record the number of times
the short ruler flips over and the direction in which it
was moving. If the pendulum was released in *exactly* the
same position each time, then the motion would be the
same. However, from your experiments you will observe
that even beginning close to the same position leads to
very different motions. This double pendulum exhibits
chaotic behavior.

Chapter 5 Summary

Systems of differential equations arise in the study of biological systems, coupled mass–spring
oscillators, electrical circuits, ecological models, and many other areas.

Linear systems with constant coefficients can be solved explicitly using a variant of the
Gauss elimination process. For this purpose we begin by writing the system with opera-
tor notation, using $D := d/dt$, $D^2 := d^2/dt^2$, and so on. A system of two equations in two
unknown functions then takes the form

$$L_1[x] + L_2[y] = f_1, \qquad L_3[x] + L_4[y] = f_2,$$

where L_1, L_2, L_3, and L_4 are polynomial expressions in D. Applying L_4 to the first equation, L_2
to the second, and subtracting, we get a single (typically higher-order) equation in $x(t)$, namely,

$$(L_4 L_1 - L_2 L_3)[x] = L_4[f_1] - L_2[f_2].$$

We then solve this constant coefficient equation for $x(t)$. Similarly, we can eliminate x from the system to obtain a single equation for $y(t)$, which we can also solve. This procedure introduces some extraneous constants, but by substituting the expressions for x and y back into one of the original equations, we can determine the relationships among these constants.

A preliminary step for the application of numerical algorithms for solving systems or single equations of higher order is to rewrite them as an equivalent system of first-order equations in **normal form:**

$$\begin{aligned} x_1'(t) &= f_1(t, x_1, x_2, \ldots, x_m)\,, \\ x_2'(t) &= f_2(t, x_1, x_2, \ldots, x_m)\,, \\ &\;\;\vdots \\ x_m'(t) &= f_m(t, x_1, x_2, \ldots, x_m)\,. \end{aligned}$$

(1)

For example, by setting $v = y'$, we can rewrite the second-order equation $y'' = f(t, y, y')$ as the normal system

$$\begin{aligned} y' &= v\,, \\ v' &= f(t, y, v)\,. \end{aligned}$$

The normal system (1) has the outward appearance of a vectorized version of a single first-order equation, and as such it suggests how to generalize numerical algorithms such as those of Euler and Runge–Kutta.

A technique for studying the qualitative behavior of solutions to the **autonomous system**

$$(2) \qquad \frac{dx}{dt} = f(x, y)\,, \qquad \frac{dy}{dt} = g(x, y)$$

is phase plane analysis. We begin by finding the **critical points** of (2)—namely, points (x_0, y_0) where

$$f(x_0, y_0) = 0 \quad \text{and} \quad g(x_0, y_0) = 0\,.$$

The corresponding constant solution pairs $x(t) \equiv x_0$, $y(t) \equiv y_0$ are called **equilibrium solutions** to (2). We then sketch the direction field for the **related phase plane differential equation**

$$(3) \qquad \frac{dy}{dx} = \frac{g(x, y)}{f(x, y)}$$

with appropriate direction arrows (oriented by the sign of dx/dt or dy/dt). From this we can usually conjecture qualitative features of the solutions and of the critical points, such as stability and asymptotic behavior. Software is typically used to visualize the solution curves to (3), which provide the highways along which the **trajectories** of the system (2) travel.

Nonautonomous systems can be studied by considering a **Poincaré map** for the system. A Poincaré map can be used to detect periodic and subharmonic solutions and to study systems whose solutions exhibit chaotic behavior.

REVIEW PROBLEMS FOR CHAPTER 5

In Problems 1–4, find a general solution $x(t)$, $y(t)$ for the given system.

1. $x' + y'' + y = 0$,
 $x'' + y' = 0$

2. $x' = x + 2y$,
 $y' = -4x - 3y$

3. $2x' - y' = y + 3x + e^t$,
 $3y' - 4x' = y - 15x + e^{-t}$

4. $x'' + x - y'' = 2e^{-t}$,
 $x'' - x + y'' = 0$

In Problems 5 and 6, solve the given initial value problem.

5. $x' = z - y$; $x(0) = 0$,
 $y' = z$; $y(0) = 0$,
 $z' = z - x$; $z(0) = 2$

6. $x' = y + z$; $x(0) = 2$,
 $y' = x + z$; $y(0) = 2$,
 $z' = x + y$; $z(0) = -1$

7. For the interconnected tanks problem of Section 5.1, page 241, suppose that instead of pure water being fed into tank *A*, a brine solution with concentration 0.2 kg/L is used; all other data remain the same. Determine the mass of salt in each tank at time *t* if the initial masses are $x_0 = 0.1$ kg and $y_0 = 0.3$ kg.

In Problems 8–11, write the given higher-order equation or system in an equivalent normal form (compare Section 5.3).

8. $2y'' - ty' + 8y = \sin t$

9. $3y''' + 2y' - e^t y = 5$

10. $x'' - x + y = 0$,
 $x' - y + y'' = 0$

11. $x''' + y' + y'' = t$,
 $x'' - x' + y''' = 0$

In Problems 12 and 13, solve the related phase plane equation for the given system. Then sketch by hand several representative trajectories (with their flow arrows) and describe the stability of the critical points (i.e., compare with Figure 5.12, page 267).

12. $x' = y - 2$,
 $y' = 2 - x$

13. $x' = 4 - 4y$,
 $y' = -4x$

14. Find all the critical points and determine the phase plane solution curves for the system
$$\frac{dx}{dt} = \sin x \cos y,$$
$$\frac{dy}{dt} = \cos x \sin y.$$

In Problems 15 and 16, sketch some typical trajectories for the given system, and by comparing with Figure 5.12, page 267, identify the type of critical point at the origin.

15. $x' = -2x - y$,
 $y' = 3x - y$

16. $x' = -x + 2y$,
 $y' = x + y$

17. In the electrical circuit of Figure 5.52, take $R_1 = R_2 = 1\ \Omega$, $C = 1$ F, and $L = 1$ H. Derive three equations for the unknown currents I_1, I_2, and I_3 by writing Kirchhoff's voltage law for loops 1 and 2, and Kirchhoff's current law for the top juncture. Find the general solution.

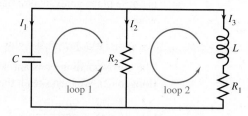

Figure 5.52 Electrical circuit for Problem 17

18. In the coupled mass–spring system depicted in Figure 5.26, page 283, take each mass to be 1 kg and let $k_1 = 8$ N/m, while $k_2 = 3$ N/m. What are the natural angular frequencies of the system? What is the general solution?

A Designing a Landing System for Interplanetary Travel

Courtesy of Alfred Clark, Jr., Professor Emeritus, University of Rochester, Rochester, NY

You are a second-year Starfleet Academy Cadet aboard the U.S.S. *Enterprise* on a continuing study of the star system Glia. The object of study on the present expedition is the large airless planet Glia-4. A class 1 sensor probe of mass m is to be sent to the planet's surface to collect data. The probe has a modifiable landing system so that it can be used on planets of different gravity. The system consists of a linear spring (force $= -kx$, where x is displacement), a nonlinear spring (force $= -ax^3$), and a shock-damper (force $= -b\dot{x}$),[†] all in parallel. Figure 5.53 shows a schematic of the system. During the landing process, the probe's thrusters are used to create a constant rate of descent. The velocity at impact varies; the symbol V_L is used to denote the largest velocity likely to happen in practice. At the instant of impact, (1) the thrust is turned off, and (2) the suspension springs are at their unstretched natural length.

(a) Let the displacement x be measured from the unstretched length of the springs and be taken negative downward (i.e., compression gives a negative x). Show that the equation governing the oscillations after impact is

$$m\ddot{x} + b\dot{x} + kx + ax^3 = -mg \ .$$

(b) The probe has a mass $m = 1220$ kg. The linear spring is permanently installed and has a stiffness $k = 35{,}600$ N/m. The gravity on the surface of Glia-4 is $g = 17.5$ m/sec^2. The nonlinear spring is removable; an appropriate spring must be chosen for each mission. These nonlinear springs are made of coralidium, a rare and difficult-to-fabricate alloy. Therefore, the *Enterprise* stocks only four different strengths: $a = 150{,}000,\ 300{,}000,\ 450{,}000,$ and $600{,}000$ N/m^3. Determine which springs give a compression as close as

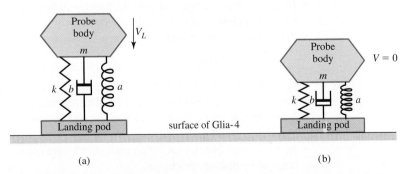

(a) surface of Glia-4 (b)

Figure 5.53 Schematic of the probe landing system. (a) The system at the instant of impact. The springs are not stretched or compressed, the thrusters have been turned off, and the velocity is V_L downward. (b) The probe has reached a state of rest on the surface, and the springs are compressed enough to support the weight. Between states (a) and (b), the probe oscillates relative to the landing pod.

[†]The symbol $\dot{x}$ denotes dx/dt.

possible to 0.3 m without exceeding 0.3 m, when the ship is resting on the surface of Glia-4. (The limit of 0.3 m is imposed by unloading clearance requirements.)

 (c) The other adjustable component on the landing system is the linear shock-damper, which may be adjusted in increments of $\Delta b = 500$ N-sec/m, from a low value of 1000 N-sec/m to a high value of 10,000 N-sec/m. It is desirable to make b as small as possible because a large b produces large forces at impact. However, if b is too small, there is some danger that the probe will rebound after impact. To minimize the chance of this, find the smallest value of b such that the springs are always in compression during the oscillations after impact. Use a minimum impact velocity $V_L = 5$ m/sec downward. To find this value of b, you will need to use a software package to integrate the differential equation.

B Spread of Staph Infections in Hospitals—Part I

Courtesy of Joanna Wares, University of Richmond, and Glenn Webb, Vanderbilt University

Methicillin-resistant *Staphylococcus aureus* (MRSA), commonly referred to as staph, is a bacterium that causes serious infections in humans and is resistant to treatment with the widely used antibiotic methicillin. MRSA has traditionally been a problem inside hospitals, where elderly patients or patients with compromised immune systems could more easily contract the bacteria and develop bloodstream infections. MRSA is implicated in a large percentage of hospital fatalities, causing more deaths per year than AIDS. Recently, a genetically different strain of MRSA has been found in the community at large. The new strain (CA-MRSA) is able to infect healthy and young people, which the traditional strain (HA-MRSA) rarely does. As CA-MRSA appears in the community, it is inevitably being spread into hospitals. Some studies suggest that CA-MRSA will overtake HA-MRSA in the hospital, which would increase the severity of the problem and likely cause more deaths per year.

To predict whether or not CA-MRSA will overtake HA-MRSA, a compartmental model has been developed by mathematicians in collaboration with medical professionals (see references [1], [2] on page 310). This model classifies all patients in the hospital into three groups:

- $H(t)$ = patients colonized with the traditional hospital strain, HA-MRSA.
- $C(t)$ = patients colonized with the community strain, CA-MRSA.
- $S(t)$ = susceptible patients, those not colonized with either strain.

The *parameters* of the model are

- β_C = the rate (per day) at which CA-MRSA is transmitted between patients.
- β_H = the rate (per day) at which HA-MRSA is transmitted between patients.
- δ_C = the rate (per day) at which patients who are colonized with CA-MRSA exit the hospital by death or discharge.
- δ_H = the rate (per day) at which patients who are colonized with HA-MRSA exit the hospital by death or discharge.
- δ_S = the rate (per day) at which susceptible patients exit the hospital by death or discharge.
- α_C = the rate (per day) at which patients who are colonized with CA-MRSA successfully undergo decolonization measures.
- α_H = the rate (per day) at which patients who are colonized with HA-MRSA successfully undergo decolonization measures.
- N = the total number of patients in the hospital.
- Λ = the rate (per day) at which patients enter the hospital.

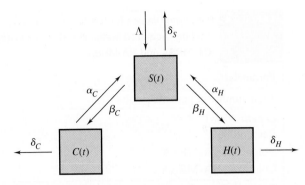

Figure 5.54 A diagram of how patients transit between the compartments

Patients move between compartments as they become colonized or decolonized (see Figure 5.54). This type of model is typically known as a SIS (susceptible-infected-susceptible) model, in which patients who become colonized can become susceptible again and colonized again (there is no immunity).

The transition between states is described by the following system of differential equations:

$$\frac{dS}{dt} = \underbrace{\Lambda}_{\text{entrance rate}} - \underbrace{\frac{\beta_H S(t) H(t)}{N}}_{\text{acquire HA-MRSA}} - \underbrace{\frac{\beta_C S(t) C(t)}{N}}_{\text{acquire CA-MRSA}}$$

$$+ \underbrace{\alpha_H H(t)}_{\text{HA-MRSA decolonized}} + \underbrace{\alpha_C C(t)}_{\text{CA-MRSA decolonized}} - \underbrace{\delta_S S(t)}_{\text{exit hospital}}$$

$$\frac{dH}{dt} = \underbrace{\frac{\beta_H S(t) H(t)}{N}}_{\text{from S}} - \underbrace{\alpha_H H(t)}_{\text{decolonized}} - \underbrace{\delta_H H(t)}_{\text{exit hospital}}$$

$$\frac{dC}{dt} = \underbrace{\frac{\beta_C S(t) C(t)}{N}}_{\text{from S}} - \underbrace{\alpha_C C(t)}_{\text{decolonized}} - \underbrace{\delta_C C(t)}_{\text{exit hospital}}.$$

If we assume that the hospital is always full, we can conserve the system by letting $\Lambda = \delta_S S(t) + \delta_H H(t) + \delta_C C(t)$. In this case $S(t) + C(t) + H(t) = N$ for all t (assuming you start with a population of size N).

(a) Show that this assumption simplifies the above system of equations to

(1)
$$\frac{dH}{dt} = (\beta_H/N)(N - C - H)H - (\delta_H + \alpha_H)H ,$$
$$\frac{dC}{dt} = (\beta_C/N)(N - C - H)C - (\delta_C + \alpha_C)C .$$

S is then determined by the equation $S(t) = N - H(t) - C(t)$.

Parameter values obtained from the Beth Israel Deaconess Medical Center are given in Table 5.4 on page 310. Plug these values into the model and then complete the following problems.

(b) Find the three equilibria (critical points) of the system (1).
(c) Using a computer, sketch the direction field for the system (1).
(d) Which trajectory configuration exists near each critical point (node, spiral, saddle, or center)? What do they represent in terms of how many patients are susceptible, colonized with HA-MRSA, and colonized with CA-MRSA over time?
(e) Examining the direction field, do you think CA-MRSA will overtake HA-MRSA in the hospital?

Further discussion of this model appears in Project E of Chapter 12.[†]

[†]All references to Chapters 11–13 refer to the expanded text, *Fundamentals of Differential Equations and Boundary Value Problems*, 7th ed.

TABLE 5.4	Parameter Values for the Transmission Dynamics of Community-Acquired and Hospital-Acquired Methicillin-Resistant *Staphylococcus aureus* Colonization (CA-MRSA and HA-MRSA)	
Parameter	**Symbol**	**Baseline Value**
Total number of patients	N	**400**
Length of stay		
Susceptible	$1/\delta_S$	5 days
Colonized CA-MRSA	$1/\delta_C$	7 days
Colonized HA-MRSA	$1/\delta_H$	5 days
Transmission rate per susceptible patient to		
Colonized CA-MRSA per colonized CA-MRSA	β_C	0.45 per day
Colonized HA-MRSA per colonized HA-MRSA	β_H	0.4 per day
Decolonization rate per colonized patient per day per length of stay		
CA-MRSA	α_C	0.1 per day
HA-MRSA	α_H	0.1 per day

References

1. D'Agata, E. M. C., Webb, G. F., Pressley, J. 2010. "Rapid emergence of co-colonization with community-acquired and hospital-acquired methicillin-resistant *Staphylococcus aureus* strains in the hospital setting". *Mathematical Modelling of Natural Phenomena* 5(3): 76–93.
2. Pressley, J., D'Agata, E. M. C., Webb, G. F. 2010. "The effect of co-colonization with community-acquired and hospital-acquired methicillin-resistant *Staphylococcus aureus* strains on competitive exclusion". *Journal of Theoretical Biology* 265(3): 645–656.

C Things That Bob

Courtesy of Richard Bernatz, Department of Mathematics, Luther College

The motion of various-shaped objects that bob in a pool of water can be modeled by a second-order differential equation derived from Newton's second law of motion, $F = ma$. The forces acting on the object include the force due to gravity, a frictional force due to the motion of the object in the water, and a buoyant force based on **Archimedes' principle:** An object that is completely or partially submerged in a fluid is acted on by an upward (buoyant) force equal to the weight of the water it displaces.

(a) The first step is to write down the governing differential equation. The dependent variable is the depth z of the object's lowest point in the water. Take z to be negative downward so that $z = -1$ means 1 ft of the object has submerged. Let $V(z)$ be the submerged volume of the object, m be the mass of the object, ρ be the density of water (in pounds per cubic foot), g be the acceleration due to gravity, and γ_w be the coefficient of friction for water. Assuming that the frictional force is proportional to the vertical velocity of the object, write down the governing second-order ODE.

(b) For the time being, neglect the effect of friction and assume the object is a cube measuring L feet on a side. Write down the governing differential equation for this case. Next, designate $z = l$ (a negative number) to be the depth of submersion such that the buoyant force is equal and opposite the gravitational force. Introduce a new variable, ζ, that gives

the displacement of the object from its equilibrium position l (that is, $z = \zeta + l$). You can now write the ODE in a more familiar form. [*Hint:* Recall the mass–spring system and the equilibrium case.] Now you should recognize the type of solution for this problem. What is the natural frequency?

(c) In this task you consider the effect of friction. The bobbing object is a cube, 1 ft on a side, that weighs 32 lb. Let $\gamma_w = 3$ lb-sec/ft, $\rho = 62.57$ lb/ft^3, and suppose the object is initially placed on the surface of the water. Solve the governing ODE by hand to find the general solution. Next, find the particular solution for the case in which the cube is initially placed on the surface of the water and is given no initial velocity. Provide a plot of the position of the object as a function of time t.

(d) In this step of the project, you develop a numerical solution to the same problem presented in part (c). The numerical solution will be useful (indeed necessary) for subsequent parts of the project. This case provides a trial to verify that your numerical solution is correct. Go back to the initial ODE you developed in part (a). Using parameter values given in part (c), solve the initial value problem for the cube starting on the surface with no initial velocity. To solve this problem numerically, you will have to write the second-order ODE as a system of two first-order ODEs, one for vertical position z and one for vertical velocity w. Plot your results for vertical position as a function of time t for the first 3 or 4 sec and compare with the analytical solution you found in part (c). Are they in close agreement? What might you have to do in order to compare these solutions? Provide a plot of both your analytical and numerical solutions on the same graph.

(e) Suppose a sphere of radius R is allowed to bob in the water. Derive the governing second-order equation for the sphere using Archimedes' principle and allowing for friction due to its motion in the water. Suppose a sphere weighs 32 lb, has a radius of $1/2$ ft, and $\gamma_w = 3.0$ lb-sec/ft. Determine the limiting value of the position of the sphere without solving the ODE. Next, solve numerically the governing ODE for the velocity and position of the sphere as a function of time for a sphere placed on the surface of the water. You will need to write the governing second-order ODE as a system of two ODEs, one for velocity and one for position. What is the limiting position of the sphere for your solution? Does it agree with the equilibrium solution you found above? How does it compare with the equilibrium position of the cube? If it is different, explain why.

(f) Suppose the sphere in part (d) is a volleyball. Calculate the position of the sphere as a function of time t for the first 3 sec if the ball is submerged so that its lowest point is 5 ft under water. Will the ball leave the water? How high will it go? Next, calculate the ball's trajectory for initial depths lower than 5 ft and higher than 5 ft. Provide plots of velocity and position for each case and comment on what you see. Specifically, comment on the relationship between the initial depth of the ball and the maximum height the ball eventually attains.

 You might consider taking a volleyball into a swimming pool to gather real data in order to verify and improve on your model. If you do so, report the data you found and explain how you used it for verification and improvement of your model.

D Hamiltonian Systems

The problems in this project explore the **Hamiltonian**[†] **formulation** of the laws of motion of a system and its phase plane implications. This formulation replaces Newton's second law $F = ma = my''$ and is based on three mathematical manipulations:

[†]*Historical Footnote:* Sir William Rowan Hamilton (1805–1865) was an Irish mathematical physicist. Besides his work in mechanics, he invented quaternions and discovered the anticommutative law for vector products.

(i) It is presumed that the force $F(t, y, y')$ depends only on y and has an antiderivative $-V(y)$, that is, $F = F(y) = -dV(y)/dy$.

(ii) The velocity variable y' is replaced throughout by the momentum $p = my'$ (so $y' = p/m$).

(iii) The **Hamiltonian** of the system is defined as

$$H = H(y, p) = \frac{p^2}{2m} + V(y) \, .$$

(a) Express Newton's law $F = my''$ as an equivalent first-order system in the manner prescribed in Section 5.3.

(b) Show that this system is equivalent to **Hamilton's equations**

$$(2) \quad \frac{dy}{dt} = \frac{\partial H}{\partial p} \quad \left(= \frac{p}{m} \right),$$

$$(3) \quad \frac{dp}{dt} = -\frac{\partial H}{\partial y} \quad \left(= -\frac{dV}{dy} \right).$$

(c) Using Hamilton's equations and the chain rule, show that the Hamiltonian remains constant along the solution curves:

$$\frac{d}{dt} H(y, p) = 0 \, .$$

In the formula for the Hamiltonian function $H(y, p)$, the first term, $p^2/(2m) = m(y')^2/2$, is the **kinetic energy** of the mass. By analogy, then, the second term $V(y)$ is known as the **potential energy** of the mass, and the Hamiltonian is the total energy. The total (mechanical)[†] energy is constant—hence "conserved"—when the forces $F(y)$ do not depend on time t or velocity y'; such forces are called **conservative.** The energy integral lemma of Section 4.8 (page 203) is simply an alternate statement of the conservation of energy.

Hamilton's formulation for mechanical systems and the conservation of energy principle imply that the phase plane trajectories of conservative systems lie on the curves where the Hamiltonian $H(y, p)$ is constant, and plotting these curves may be considerably easier than solving for the trajectories directly (which, in turn, is easier than solving the original system!).

(d) For the mass–spring oscillator of Section 4.1, the spring force is given by $F = -ky$ (where k is the spring constant). Find the Hamiltonian, express Hamilton's equations, and show that the phase plane trajectories $H(y, p) = $ constant for this system are the ellipses given by $p^2/(2m) + ky^2/2 = $ constant. See Figure 5.14, page 269.

The **damping** force $-by'$ considered in Section 4.1 is not conservative, of course. Physically speaking, we know that damping drains the energy from a system until it grinds to a halt at an equilibrium point. In the phase plane, we can qualitatively describe the trajectory as continuously migrating to successively lower constant-energy orbits; stable centers become asymptotically stable spiral points when damping is taken into consideration.

(e) The second Hamiltonian equation (3), which effectively states $p' = my'' = F$, has to be changed to

$$p' = -\frac{\partial H}{\partial y} - by' = -\frac{\partial H}{\partial y} - \frac{bp}{m}$$

when damping is present. Show that the Hamiltonian decreases along trajectories in this case (for $b > 0$):

$$\frac{d}{dt} H(y, p) = -b \left(\frac{p}{m} \right)^2 = -b(y')^2 \, .$$

[†]Physics states that when all forms of energy, such as heat and radiation, are taken into account, energy is conserved even when the forces are not conservative.

 (f) The force on a mass–spring system suspended vertically in a gravitational field was shown in Section 4.10 (page 226) to be $F = -ky + mg$. Derive the Hamiltonian and sketch the phase plane trajectories. Sketch the trajectories when damping is present.

 (g) As indicated in Section 4.8 (page 207), the Duffing spring force is modeled by $F = -y - y^3$. Derive the Hamiltonian and sketch the phase plane trajectories. Sketch the trajectories when damping is present.

 (h) For the pendulum system studied in Section 4.8, Example 8, the force is given by (cf. Figure 4.18, page 208)

$$F = -\ell mg \sin\theta = -\frac{\partial}{\partial\theta}(-\ell mg \cos\theta) = -\frac{\partial}{\partial\theta}V(\theta)$$

(where ℓ is the length of the pendulum). For angular variables, the Hamiltonian formulation dictates expressing the *angular velocity variable* θ' *in terms of the angular momentum* $p = m\ell^2\theta'$; the kinetic energy, mass × velocity²/2, is expressed as $m(\ell\theta')^2/2 = p^2/(2m\ell^2)$. Derive the Hamiltonian for the pendulum and sketch the phase plane trajectories. Sketch the trajectories when damping is present.

 (i) The Coulomb force field is a force that varies as the reciprocal square of the distance from the origin: $F = k/y^2$. The force is *attractive* if $k < 0$ and *repulsive* if $k > 0$. Sketch the phase plane trajectories for this motion. Sketch the trajectories when damping is present.

(j) For an attractive Coulomb force field, what is the *escape velocity* for a particle situated at a position y? That is, what is the minimal (outward-directed) velocity required for the trajectory to reach $y = \infty$?

E Cleaning Up the Great Lakes

A simple mathematical model that can be used to determine the time it would take to clean up the Great Lakes can be developed using a multiple compartmental analysis approach.[†] In particular, we can view each lake as a tank that contains a liquid in which is dissolved a particular pollutant (DDT, phosphorus, mercury). Schematically, we view the lakes as consisting of five tanks connected as indicated in Figure 5.55.

Figure 5.55 Compartmental model of the Great Lakes with flow rates (mi³/yr) and volumes (mi³)

[†]For a detailed discussion of this model, see *An Introduction to Mathematical Modeling* by Edward A. Bender (Dover Publications, New York, 2000), Chapter 8.

For our model, we make the following assumptions:

1. The volume of each lake remains constant.
2. The flow rates are constant throughout the year.
3. When a liquid enters the lake, perfect mixing occurs and the pollutants are uniformly distributed.
4. Pollutants are dissolved in the water and enter or leave by inflow or outflow of solution.

Before using this model to obtain estimates on the cleanup times for the lakes, we consider some simpler models:

(a) Use the outflow rates given in Figure 5.55 to determine the time it would take to "drain" each lake. This gives a lower bound on how long it would take to remove all the pollutants.
(b) A better estimate is obtained by assuming that each lake is a separate tank with *only* clean water flowing in. Use this approach to determine how long it would take the pollution level in each lake to be reduced to 50% of its original level. How long would it take to reduce the pollution to 5% of its original level?
(c) Finally, to take into account the fact that pollution from one lake flows into the next lake in the chain, use the entire multiple compartment model given in Figure 5.55 to determine when the pollution level in each lake has been reduced to 50% of its original level, assuming pollution has ceased (that is, inflows not from a lake are clean water). Assume that all the lakes initially have the same pollution concentration p. How long would it take for the pollution to be reduced to 5% of its original level?

F The 2014-2015 Ebola Epidemic

Courtesy of Glenn Webb, Vanderbilt University

We develop a model for the 2014–2015 Ebola epidemic in West Africa, which began in the spring of 2014 and spread though the countries of Sierra Leone, Liberia, and Guinea. Initially there was great concern that the epidemic might develop with great severity, and even spread globally.[†] Instead, by early 2015 the epidemic had subsided in these West African countries, and in January 2016, the World Health Organization (WHO) declared that the epidemic was contained in all three countries. The reasons that Ebola subsided are complex, but a key role was increased identification and isolation of infected cases, and contact tracing of these cases to identify additional cases.

Our model incorporates the principal features of contact tracing, namely, the number of contacts per identified infectious case, the likelihood that a traced contact is infectious, and the efficiency of the isolation of contact-traced individuals. The model consists of a system of ordinary differential equations for the compartments of the epidemic population, and is based on an earlier version of the model.[‡] The model incorporates the unique features of the Ebola outbreaks in this region. These features include the rates of transmission to susceptibles from both infectious cases and improperly handled deceased cases, the rates of reporting cases, and the rates of recovery and mortality of unreported cases.

[†]H. Nishiura and G. Chowell, "Early transmission dynamics of Ebola virus disease (EVD), West Africa, March to August 2014." *Eurosurveil.*, Vol. 19 no. 36 (Sept. 11, 2014).

[‡]C. Browne, X. Huo, P. Magal. M. Seydi, O. Seydi, G. Webb, "A model of the 2014 Ebola epidemic in West Africa with contact tracing," *PLOS Currents Outbreaks*, published online, January 30, 2015.

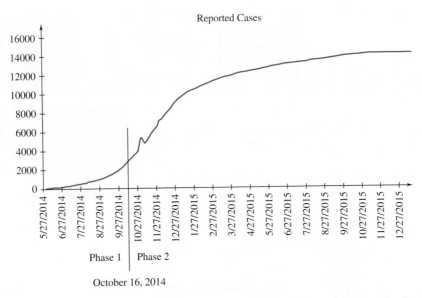

Figure 5.56 WHO Ebola Situation Data: reported cases. (Center for Disease Control and Prevention, Previous Case Counts: http://www.cdc.gov/vhf/ebola/outbreaks/2014-west-africa/cumulative-cases-graphs.html)

The model extends the SIR paradigm of Section 5.5, by refining the population classes into susceptibles $S(t)$ (those who have not been affected), exposed $E(t)$ (incubating cases), $I(t)$ (infectious cases), and contaminated deceased $C(t)$ (improperly handled corpses of cases). The successful containment of the epidemic was due to measures isolating infected individuals and contaminated corpses, implemented around October 16, 2014. The WHO data in Figure 5.56 traces the chronology of the number of *reported* cases $R(t)$; the inflection point clearly highlights the initiation of isolation and contact tracing measures.

Figure 5.57 on page 316 schematically depicts the transitions between the classifications, as quantified by the subsequent system of differential equations:

TABLE 5.5	Ebola Model Parameters in Sierra Leone
Parameter	**Description**
$N = 6 \times 10^6$	Population of Sierra Leone (assumed to be constant)
$\beta = 0.321$	Transmission rate exclusive of improper handling of deceased cases
$\varepsilon = 0.002$	Transmission rate due to improper handling of deceased cases
$\kappa = 0$ (1st phase)	Average number of contacts traced per reported and hospitalized
$\kappa = 8$ (2nd phase)	infectious case
$\alpha = 0.08$ (1st phase)	Rate of reporting and hospitalization of infectious cases not
$\alpha = 0.145$ (2nd phase)	resulting from contact tracing
$\pi = 0.5$	Probability a contact-traced infected individual is isolated without causing a new case
$\omega = 0.1$	Probability a contact-traced individual is infected
$\gamma = 0.0333$	Rate of recovery of unreported infectious cases
$V = 0.125$	Rate of mortality of unreported infectious cases
$\sigma = 1/9$	Average incubation period (denominator)
$\psi = 1/5$ (1st phase)	Average number of days before proper handling of deceased
$\psi = 1/2$ (2nd phase)	unreported cases (denominator)

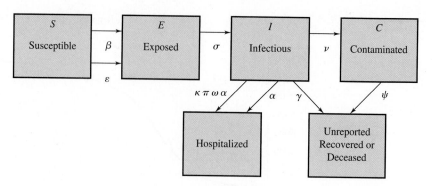

Figure 5.57 Schematic of Ebola transitions

$$\frac{dS(t)}{dt} = -\beta S(t)\frac{I(t)}{N} \quad -\varepsilon S(t)\frac{C(t)}{N}$$

$$\uparrow \qquad\qquad \uparrow$$

susceptible- susceptible-
infected contaminated
interaction interaction

$$\downarrow \qquad\qquad \downarrow$$

$$\frac{dE(t)}{dt} = \quad \beta S(t)\frac{I(t)}{N} \quad + \varepsilon S(t)\frac{C(t)}{N} \quad -\sigma E(t)$$

$$\uparrow$$

incubation

$$\downarrow$$

$$\frac{dI(t)}{dt} = \qquad\qquad\qquad\qquad\qquad \sigma E(t) \quad -\nu I(t) \quad -(\alpha + \gamma + \alpha\kappa\pi\omega)I(t)$$

$$\uparrow \qquad\qquad \uparrow \qquad\qquad \text{proper}$$

mortality, hospitaliza- handling of
unreported tion, recovery, unreported
infected or isolation by corpses
 contact tracing

$$\downarrow \qquad\qquad\qquad\qquad\qquad\qquad \downarrow$$

$$\frac{dC(t)}{dt} = \qquad\qquad\qquad\qquad\qquad\qquad\qquad \nu I(t) \qquad\qquad -\psi C(t)$$

$$\frac{dR(t)}{dt} = \qquad\qquad\qquad\qquad\qquad\qquad\qquad \alpha I(t) \qquad\qquad +\psi C(t)$$

By trial and error and numerical experimentation, the following parameter values were found to give a good fit, to the WHO data, of the solution of this system. The initial values were taken to be $S(0) = N, E(0) = 20, I(0) = 20, C(0) = 10$. Note the change in the values of κ, α, and ψ on the 142nd day, reflecting the initiation of contact tracing.

 (a) Use a software package to solve the system for the time interval indicated in Figure 5.56 on page 315. Plot the number of reported cases $R(t)$. Note the high quality of the fit of the simulations to the WHO data.

(b) Plot the number of persons affected by the epidemic, $6 \times 10^6 - S(t)$. You should see that about 42% of all cases were *unreported* on October 16, 2014, but this fell to 10% by January, 2016.

G Phase-Locked Loops

Today's high-speed communications technology requires circuitry that will measure or match incoming radio-frequency oscillations in the gigahertz range. Mathematically, that means one must determine the frequency ω in a sinusoidal signal $A \sin(\omega t + \alpha)$ when ω is in the neighborhood of $2\pi \times 10^9$ rad/s. In theory this can be accomplished by taking three measurements of the sinusoid within a quarter-period. However, obtaining three accurate measurements of a signal in an interval of 10^{-10} seconds is impractical. The *phase-locked loop* accomplishes the task in hardware, using a voltage-controlled oscillator (VCO) to synthesize another sinusoid $B \cos\theta(t)$ whose *phase*, $\theta(t)$, matches that of the incoming signal $(\omega t + \alpha)$. The frequency $d\theta(t)/dt$ of the VCO output is linearly related to the voltage applied to the VCO (say, $d\theta(t)/dt = E + Fv(t)$), and it matches ω once synchronization is achieved. Thus one can determine the incoming frequency by reading off the VCO voltage.

A diagram of the phase-locked loop circuit is given in Figure 5.58. In brief: the mixer combines the signals $A \sin(\omega t + \alpha)$ and $B \cos\theta(t)$ to create a *control signal* $\dfrac{AB}{2} \sin[\omega t + \alpha - \theta(t)]$ whose phase equals the *difference* between the phase of the incoming signal and the phase of the VCO output.[†] This control signal provides a measure of the lack of synchronization; it is zero when $\theta(t) = \omega t + \alpha (\pm 2\pi n)$. Thus to help $\theta(t)$ catch up to $\omega t + \alpha$, the circuit adjusts the VCO voltage $v(t)$ so as to increase the output frequency, $d\theta(t)/dt$, in proportion to a weighted combination of the control signal and its derivative:

$$(4) \quad \frac{d}{dt}\frac{d\theta(t)}{dt} = \frac{d^2\theta}{dt^2} = \frac{d(E + Fv)}{dt}$$

$$= K_1 \frac{AB}{2} \sin[\omega t + \alpha - \theta(t)] + K_2 \frac{d}{dt}\left\{\frac{AB}{2} \sin[\omega t + \alpha - \theta(t)]\right\}.$$

We are going to employ the phase plane to analyze how the phase-locked loop approaches synchronicity when it is controlled by equation (4).

(a) Convert (4) into an autonomous equation by introducing the *error* $e(t) := \omega t + \alpha - \theta(t)$ and deriving

$$(5) \quad \frac{d^2 e}{dt^2} = -\frac{d^2\theta}{dt^2} = -K_1 \frac{AB}{2} \sin e - K_2 \frac{AB}{2}(\cos e)\frac{de}{dt}.$$

(b) Put equation (5) into normal form (two first-order differential equations in $x_1 := e(t)$ and $x_2 = de/dt$) and find all the critical points.

Figure 5.58 Phase-locked loop

[†]A classic trigonometric identity shows that multiplication of the incoming and VCO signals yields the control signal plus a high-frequency oscillation, which is easily filtered out.

(c) Use software to sketch the direction field in this system's phase plane, with $K_1 \dfrac{AB}{2} = 4$, $K_2 \dfrac{AB}{2} = 1$. (The duplicitous nomenclature is unintended: *phase plane* does not refer to the *phase* of the sinusoid.) Trace a few trajectories to get a feel for how the phase-locked loop performs. Which critical points are stable, and which are unstable? Are all the stable critical points suitable for synchronization?

(d) Apply the mass–spring analogy (Section 4.8) to equation (5). The mass m is one, the spring force is $-K_1 \dfrac{AB}{2} \sin e$ instead of $-ke$, and the damping force is $-K_2 \dfrac{AB}{2} (\cos e) \dfrac{de}{dt}$ instead of $-b \dfrac{de}{dt}$. If e is small, the mass–spring parameters and the phase-locked loop parameters are approximately the same (recall Project D, Chapter 4). Sketch the direction field for the (normal form of the) damped mass–spring system

$$\frac{d^2 e}{dt^2} = -b \frac{de}{dt} - ke$$

with $b = 1$, $k = 4$, and compare with that of the phase-locked loop. Explain why the number of equilibria is different, and why the nature of the phase-locked loop's "forces" renders certain of its equilibria to be unstable.

CHAPTER 6

Theory of Higher-Order Linear Differential Equations

In this chapter we discuss the basic theory of linear higher-order differential equations. The material is a generalization of the results we obtained in Chapter 4 for second-order constant-coefficient equations. In the statements and proofs of these results, we use concepts usually covered in an elementary linear algebra course—namely, linear dependence, determinants, and methods for solving systems of linear equations. These concepts also arise in the matrix approach for solving systems of differential equations and are discussed in Chapter 9, which includes a brief review of linear algebraic equations and determinants.

Since this chapter is more mathematically oriented—that is, not tied to any particular physical application—we revert to the customary practice of calling the independent variable "x" and the dependent variable "y."

6.1 Basic Theory of Linear Differential Equations

A *linear* differential equation of order n is an equation that can be written in the form

(1) $\qquad a_n(x)y^{(n)}(x) + a_{n-1}(x)y^{(n-1)}(x) + \cdots + a_0(x)y(x) = b(x)$,

where $a_0(x), a_1(x), \ldots, a_n(x)$ and $b(x)$ depend only on x, not y. When $a_0, a_1, \ldots, a_n$ are all constants, we say equation (1) has **constant coefficients;** otherwise it has **variable coefficients.** If $b(x) \equiv 0$, equation (1) is called **homogeneous;** otherwise it is **nonhomogeneous.**

In developing a basic theory, we assume that $a_0(x), a_1(x), \ldots, a_n(x)$ and $b(x)$ are all continuous on an interval I and $a_n(x) \neq 0$ on I. Then, on dividing by $a_n(x)$, we can rewrite (1) in the **standard form**

(2) $\qquad y^{(n)}(x) + p_1(x)y^{(n-1)}(x) + \cdots + p_n(x)y(x) = g(x)$,

where the functions $p_1(x), \ldots, p_n(x)$, and $g(x)$ are continuous on I.

For a linear higher-order differential equation, the initial value problem always has a unique solution.

Existence and Uniqueness

Theorem 1. Suppose $p_1(x), \ldots, p_n(x)$ and $g(x)$ are each continuous on an interval (a, b) that contains the point x_0. Then, for any choice of the initial values $\gamma_0, \gamma_1, \ldots, \gamma_{n-1}$, there exists a unique solution $y(x)$ on the whole interval (a, b) to the initial value problem

(3) $\qquad y^{(n)}(x) + p_1(x)y^{(n-1)}(x) + \cdots + p_n(x)y(x) = g(x)$,

(4) $\qquad y(x_0) = \gamma_0, y'(x_0) = \gamma_1, \ldots, y^{(n-1)}(x_0) = \gamma_{n-1}$.

The proof of Theorem 1 can be found in Chapter 13.[†]

Example 1 For the initial value problem

(5) $x(x-1)y''' - 3xy'' + 6x^2y' - (\cos x)y = \sqrt{x+5}\,;$

(6) $y(x_0) = 1\,,\quad y'(x_0) = 0\,,\quad y''(x_0) = 7\,,$

determine the values of x_0 and the intervals (a, b) containing x_0 for which Theorem 1 guarantees the existence of a unique solution on (a, b).

Solution Putting equation (5) in standard form, we find that $p_1(x) = -3/(x-1)$, $p_2(x) = 6x/(x-1)$, $p_3(x) = -(\cos x)/[x(x-1)]$, and $g(x) = \sqrt{x+5}/[x(x-1)]$. Now $p_1(x)$ and $p_2(x)$ are continuous on every interval not containing $x = 1$, while $p_3(x)$ is continuous on every interval not containing $x = 0$ or $x = 1$. The function $g(x)$ is not defined for $x < -5$, $x = 0$, and $x = 1$, but is continuous on $(-5, 0)$, $(0, 1)$, and $(1, \infty)$. Hence, the functions p_1, p_2, p_3, and g are *simultaneously* continuous on the intervals $(-5, 0)$, $(0, 1)$, and $(1, \infty)$. From Theorem 1 it follows that if we choose $x_0 \in (-5, 0)$, then there exists a unique solution to the initial value problem (5)–(6) on the whole interval $(-5, 0)$. Similarly, for $x_0 \in (0, 1)$, there is a unique solution on $(0, 1)$ and, for $x_0 \in (1, \infty)$, a unique solution on $(1, \infty)$. ◆

If we let the left-hand side of equation (3) define the *differential operator L*,

(7) $L[y] := \dfrac{d^n y}{dx^n} + p_1 \dfrac{d^{n-1}y}{dx^{n-1}} + \cdots + p_n y = (D^n + p_1 D^{n-1} + \cdots + p_n)[y]\,,$

then we can express equation (3) in the operator form

(8) $L[y](x) = g(x)\,.$

It is essential to keep in mind that L is a *linear* operator—that is, it satisfies

(9) $L[y_1 + y_2 + \cdots + y_m] = L[y_1] + L[y_2] + \cdots + L[y_m]\,,$

(10) $L[cy] = cL[y]$ (c any constant) .

These are familiar properties for the differentiation operator D, from which (9) and (10) follow (see Problem 25).

As a consequence of this linearity, if $y_1, \ldots, y_m$ are solutions to the *homogeneous* equation

(11) $L[y](x) = 0\,,$

then any linear combination of these functions, $C_1 y_1 + \cdots + C_m y_m$, is also a solution, because

$$L[C_1 y_1 + C_2 y_2 + \cdots + C_m y_m] = C_1 \cdot 0 + C_2 \cdot 0 + \cdots + C_m \cdot 0 = 0\,.$$

Imagine now that we have found n solutions $y_1, \ldots, y_n$ to the nth-order linear equation (11). Is it true that *every* solution to (11) can be represented by

(12) $C_1 y_1 + C_2 y_2 + \cdots + C_n y_n$

[†]All references to Chapters 11–13 refer to the expanded text, *Fundamentals of Differential Equations and Boundary Value Problems*, 7th ed.

for appropriate choices of the constants $C_1, \ldots, C_n$? The answer is yes, provided the solutions $y_1, \ldots, y_n$ satisfy a certain property that we now derive.

Let $\phi(x)$ be a solution to (11) on the interval (a, b) and let x_0 be a fixed number in (a, b). If it is possible to choose the constants $C_1, \ldots, C_n$ so that

$$
\begin{aligned}
C_1 y_1(x_0) \quad &+ \cdots + C_n y_n(x_0) \quad = \phi(x_0)\,, \\
C_1 y_1'(x_0) \quad &+ \cdots + C_n y_n'(x_0) \quad = \phi'(x_0)\,, \\
&\vdots \\
C_1 y_1^{(n-1)}(x_0) &+ \cdots + C_n y_n^{(n-1)}(x_0) = \phi^{(n-1)}(x_0)\,,
\end{aligned}
\tag{13}
$$

then, since $\phi(x)$ and $C_1 y_1(x) + \cdots + C_n y_n(x)$ are two solutions satisfying the same initial conditions at x_0, the uniqueness conclusion of Theorem 1 gives

$$
\phi(x) = C_1 y_1(x) + \cdots + C_n y_n(x)
\tag{14}
$$

for all x in (a, b).

The system (13) consists of n linear equations in the n unknowns $C_1, \ldots, C_n$. It has a unique solution for all possible values of $\phi(x_0), \phi'(x_0), \ldots, \phi^{(n-1)}(x_0)$ if and only if the determinant[†] of the coefficients is different from zero; that is, if and only if

$$
\begin{vmatrix}
y_1(x_0) & y_2(x_0) & \cdots & y_n(x_0) \\
y_1'(x_0) & y_2'(x_0) & \cdots & y_n'(x_0) \\
\vdots & \vdots & & \vdots \\
y_1^{(n-1)}(x_0) & y_2^{(n-1)}(x_0) & \cdots & y_n^{(n-1)}(x_0)
\end{vmatrix} \neq 0\,.
\tag{15}
$$

Hence, if $y_1, \ldots, y_n$ are solutions to equation (11) and there is some point x_0 in (a, b) such that (15) holds, then every solution $\phi(x)$ to (11) is a linear combination of $y_1, \ldots, y_n$. Before formulating this fact as a theorem, it is convenient to identify the determinant by name.

Wronskian

Definition 1. Let $f_1, \ldots, f_n$ be any n functions that are $(n-1)$ times differentiable. The function

$$
W[f_1, \ldots, f_n](x) := \begin{vmatrix}
f_1(x) & f_2(x) & \cdots & f_n(x) \\
f_1'(x) & f_2'(x) & \cdots & f_n'(x) \\
\vdots & \vdots & & \vdots \\
f_1^{(n-1)}(x) & f_2^{(n-1)}(x) & \cdots & f_n^{(n-1)}(x)
\end{vmatrix}
\tag{16}
$$

is called the **Wronskian** of $f_1, \ldots, f_n$.

We now state the representation theorem that we proved above for solutions to homogeneous linear differential equations.

[†]Readers unfamiliar with determinants and cofactor expansions can find these topics discussed in any linear algebra book (such as *Fundamentals of Matrix Analysis with Applications*, by Edward Barry Saff and Arthur David Snider, John Wiley & Sons, Hoboken, New Jersey, 2016.)

Representation of Solutions (Homogeneous Case)

Theorem 2. Let $y_1, \ldots, y_n$ be n solutions on (a, b) of

(17) $y^{(n)}(x) + p_1(x)y^{(n-1)}(x) + \cdots + p_n(x)y(x) = 0 \,,$

where $p_1, \ldots, p_n$ are continuous on (a, b). If at some point x_0 in (a, b) these solutions satisfy

(18) $W[y_1, \ldots, y_n](x_0) \neq 0 \,,$

then every solution of (17) on (a, b) can be expressed in the form

(19) $y(x) = C_1 y_1(x) + \cdots + C_n y_n(x) \,,$

where $C_1, \ldots, C_n$ are constants.

The linear combination of $y_1, \ldots, y_n$ in (19), written with arbitrary constants $C_1, \ldots, C_n$, is referred to as a **general solution** to (17).

In linear algebra a set of m column vectors $\{\mathbf{v}_1, \mathbf{v}_2, \ldots, \mathbf{v}_m\}$, each having m components, is said to be linearly dependent if and only if at least one of them can be expressed as a linear combination of the others.[†] A basic theorem then states that if a determinant is zero, its column vectors are linearly dependent, and conversely. So if a Wronskian of solutions to (17) is zero at a point x_0, one of its columns (the final column, say; we can always renumber!) equals a linear combination of the others:

(20) $$\begin{bmatrix} y_n(x_0) \\ y_n'(x_0) \\ \vdots \\ y_n^{(n-1)}(x_0) \end{bmatrix} = d_1 \begin{bmatrix} y_1(x_0) \\ y_1'(x_0) \\ \vdots \\ y_1^{(n-1)}(x_0) \end{bmatrix} + d_2 \begin{bmatrix} y_2(x_0) \\ y_2'(x_0) \\ \vdots \\ y_2^{(n-1)}(x_0) \end{bmatrix} + \cdots + d_{n-1} \begin{bmatrix} y_{n-1}(x_0) \\ y_{n-1}'(x_0) \\ \vdots \\ y_{n-1}^{(n-1)}(x_0) \end{bmatrix} .$$

Now consider the two functions $y_n(x)$ and $[d_1 y_1(x) + d_2 y_2(x) + \cdots + d_{n-1} y_{n-1}(x)]$. They are both solutions to (17), and we can interpret (20) as stating that they satisfy the same initial conditions at $x = x_0$. By the uniqueness theorem, then, they are one and the same function:

(21) $y_n(x) = d_1 y_1(x) + d_2 y_2(x) + \cdots + d_{n-1} y_{n-1}(x)$

for all x in the interval I. Consequently, their derivatives are the same also, and so

(22) $$\begin{bmatrix} y_n(x) \\ y_n'(x) \\ \vdots \\ y_n^{(n-1)}(x) \end{bmatrix} = d_1 \begin{bmatrix} y_1(x) \\ y_1'(x) \\ \vdots \\ y_1^{(n-1)}(x) \end{bmatrix} + d_2 \begin{bmatrix} y_2(x) \\ y_2'(x) \\ \vdots \\ y_2^{(n-1)}(x) \end{bmatrix} + \cdots + d_{n-1} \begin{bmatrix} y_{n-1}(x) \\ y_{n-1}'(x) \\ \vdots \\ y_{n-1}^{(n-1)}(x) \end{bmatrix}$$

for all x in I. Hence, the final column of the Wronskian $W[y_1, y_2, \ldots, y_n]$ is *always* a linear combination of the other columns, and consequently the Wronskian is always zero.

In summary, the Wronskian of n solutions to the homogeneous equation (17) is either identically zero, or never zero, on the interval (a, b). We have also shown that, in the former

[†] This is equivalent to saying there exist constants $c_1, c_2, \ldots, c_m$ **not all zero**, such that $c_1 \mathbf{v}_1 + c_2 \mathbf{v}_2 + \cdots + c_m \mathbf{v}_m$ equals the zero vector.

case, (21) holds throughout (a, b). Such a relationship among functions is an extension of the notion of linear dependence introduced in Section 4.2. We employ the same nomenclature for the general case.

Linear Dependence of Functions

Definition 2. The m functions $f_1, f_2, \ldots, f_m$ are said to be **linearly dependent on an interval I** if at least one of them can be expressed as a linear combination of the others on I; equivalently, they are linearly dependent if there exist constants $c_1, c_2, \ldots, c_m$, *not all zero*, such that

$$(23) \qquad c_1 f_1(x) + c_2 f_2(x) + \cdots + c_m f_m(x) = 0$$

for all x in I. Otherwise, they are said to be **linearly independent on I**.

Example 2 Show that the functions $f_1(x) = e^x, f_2(x) = e^{-2x}$, and $f_3(x) = 3e^x - 2e^{-2x}$ are linearly dependent on $(-\infty, \infty)$.

Solution Obviously, f_3 is a linear combination of f_1 and f_2:

$$f_3(x) = 3e^x - 2e^{-2x} = 3f_1(x) - 2f_2(x) .$$

Note further that the corresponding identity $3f_1(x) - 2f_2(x) - f_3(x) = 0$ matches the pattern (23). Moreover, observe that f_1, f_2, and f_3 are *pairwise* linearly independent on $(-\infty, \infty)$, but this does not suffice to make the triplet independent. ◆

To prove that functions $f_1, f_2, \ldots, f_m$ are linearly *independent* on the interval (a, b), a convenient approach is the following: *Assume* that equation (23) holds on (a, b) and show that this forces $c_1 = c_2 = \cdots = c_m = 0$.

Example 3 Show that the functions $f_1(x) = x, f_2(x) = x^2$, and $f_3(x) = 1 - 2x^2$ are linearly independent on $(-\infty, \infty)$.

Solution Assume c_1, c_2, and c_3 are constants for which

$$(24) \qquad c_1 x + c_2 x^2 + c_3(1 - 2x^2) = 0$$

holds at every x. If we can prove that (24) implies $c_1 = c_2 = c_3 = 0$, then linear independence follows. Let's set $x = 0, 1$, and -1 in equation (24). These x values are, essentially, "picked out of a hat," but will get the job done. Substituting in (24) gives

$$
\begin{aligned}
c_3 &= 0 \quad (x = 0), \\
(25) \qquad c_1 + c_2 - c_3 &= 0 \quad (x = 1), \\
-c_1 + c_2 - c_3 &= 0 \quad (x = -1).
\end{aligned}
$$

When we solve this system (or compute the determinant of the coefficients), we find that the only possible solution is $c_1 = c_2 = c_3 = 0$. Consequently, the functions f_1, f_2, and f_3 are linearly independent on $(-\infty, \infty)$.

A neater solution is to note that if (24) holds for all x, so do its first and second derivatives. At $x = 0$ these conditions are $c_3 = 0$, $c_1 = 0$, and $2c_2 - 4c_3 = 0$. Obviously, each coefficient must be zero. ◆

Linear dependence of functions is, *prima facie*, different from linear dependence of vectors in the Euclidean space $\mathbf{R}^n$, because (23) is a functional equation that imposes a condition at every point of an interval. However, we have seen in (21) that when the functions are all solutions to the same homogeneous differential equation, linear dependence of the column vectors of the Wronskian (at *any* point x_0) implies linear dependence of the functions. The converse is also true, as demonstrated by (21) and (22). Theorem 3 summarizes our deliberations.

Linear Dependence and the Wronskian

Theorem 3. If $y_1, y_2, \ldots, y_n$ are n solutions to $y^{(n)} + p_1 y^{(n-1)} + \cdots + p_n y = 0$ on the interval (a, b), with $p_1, p_2, \ldots, p_n$ continuous on (a, b), then the following statements are equivalent:

(i) $y_1, y_2, \ldots, y_n$ are linearly dependent on (a, b).
(ii) The Wronskian $W[y_1, y_2, \ldots, y_n](x_0)$ is zero at some point x_0 in (a, b).
(iii) The Wronskian $W[y_1, y_2, \ldots, y_n](x)$ is identically zero on (a, b).

The contrapositives of these statements are also equivalent:

(iv) $y_1, y_2, \ldots, y_n$ are linearly independent on (a, b).
(v) The Wronskian $W[y_1, y_2, \ldots, y_n](x_0)$ is nonzero at some point x_0 in (a, b).
(vi) The Wronskian $W[y_1, y_2, \ldots, y_n](x)$ is never zero on (a, b).

Whenever (iv), (v), or (vi) is met, $\{y_1, y_2, \ldots, y_n\}$ is called a **fundamental solution set** for (17) on (a, b).

The Wronskian is a curious function. If we take $W[f_1, f_2, \ldots, f_n](x)$ for n *arbitrary* functions, we simply get a function of x with no particularly interesting properties. But if the n functions are all solutions to the same homogeneous differential equation, then either it is identically zero or never zero. In fact, one can prove **Abel's identity** when the functions are all solutions to (17):

$$(26) \qquad W[y_1, y_2, \ldots, y_n](x) = W[y_1, y_2, \ldots, y_n](x_0) \exp\left(\int_{x_0}^{x} p_1(t)\,dt \right),$$

which clearly exhibits this property. Problem 30 on page 327 outlines a proof of (26) for $n = 3$.[†]
It is useful to keep in mind that the following sets consist of functions that are linearly independent on every open interval (a, b):

$$\{1, x, x^2, \ldots, x^n\}$$
$$\{1, \cos x, \sin x, \cos 2x, \sin 2x, \ldots, \cos nx, \sin nx\},$$
$$\{e^{\alpha_1 x}, e^{\alpha_2 x}, \ldots, e^{\alpha_n x}\} \quad (\alpha_i\text{'s distinct constants}).$$

[See Problems 27 and 28 on page 326, and Section 6.2 (page 327).]
If we combine the linearity (superposition) properties (9) and (10) with the representation theorem for solutions of the homogeneous equation, we obtain the following representation theorem for nonhomogeneous equations.

[†]See Problem 32, Exercises 4.7, for the case $n = 2$.

Representation of Solutions (Nonhomogeneous Case)

Theorem 4. Let $y_p(x)$ be a particular solution to the nonhomogeneous equation

(27) $y^{(n)}(x) + p_1(x)y^{(n-1)}(x) + \cdots + p_n(x)y(x) = g(x)$

on the interval (a, b) with $p_1, p_2, \ldots, p_n$ continuous on (a, b), and let $\{y_1, \ldots, y_n\}$ be a fundamental solution set for the corresponding homogeneous equation

(28) $y^{(n)}(x) + p_1(x)y^{(n-1)}(x) + \cdots + p_n(x)y(x) = 0$.

Then every solution of (27) on the interval (a, b) can be expressed in the form

(29) $y(x) = y_p(x) + C_1y_1(x) + \cdots + C_ny_n(x)$.

Proof. Let $\phi(x)$ be any solution to (27). Because both $\phi(x)$ and $y_p(x)$ are solutions to (27), by linearity the difference $\phi(x) - y_p(x)$ is a solution to the homogeneous equation (28). It then follows from Theorem 2 that

$$\phi(x) - y_p(x) = C_1y_1(x) + \cdots + C_ny_n(x)$$

for suitable constants $C_1, \ldots, C_n$. The last equation is equivalent to (29) [with $\phi(x)$ in place of $y(x)$], so the theorem is proved. ◆

The linear combination of $y_p, y_1, \ldots, y_n$ in (29) written with arbitrary constants $C_1, \ldots, C_n$ is, for obvious reasons, referred to as a **general solution** to (27). Theorem 4 can be easily generalized. For example, if L denotes the operator appearing as the left-hand side in equation (27) and if $L[y_{p1}] = g_1$ and $L[y_{p2}] = g_2$, then any solution of $L[y] = c_1g_1 + c_2g_2$ can be expressed as

$$y(x) = c_1y_{p1}(x) + c_2y_{p2}(x) + C_1y_1(x) + C_2y_2(x) + \cdots + C_ny_n(x),$$

for a suitable choice of the constants $C_1, C_2, \ldots, C_n$.

Example 4 Find a general solution on the interval $(-\infty, \infty)$ to

(30) $L[y] := y''' - 2y'' - y' + 2y = 2x^2 - 2x - 4 - 24e^{-2x}$,

given that $y_{p1}(x) = x^2$ is a particular solution to $L[y] = 2x^2 - 2x - 4$, that $y_{p2}(x) = e^{-2x}$ is a particular solution to $L[y] = -12e^{-2x}$, and that $y_1(x) = e^{-x}$, $y_2(x) = e^x$, and $y_3(x) = e^{2x}$ are solutions to the corresponding homogeneous equation.

Solution We previously remarked that the functions e^{-x}, e^x, e^{2x} are linearly independent because the exponents $-1, 1,$ and 2 are distinct. Since each of these functions is a solution to the corresponding homogeneous equation, then $\{e^{-x}, e^x, e^{2x}\}$ is a fundamental solution set. It now follows from the remarks above for nonhomogeneous equations that a general solution to (30) is

(31) $y(x) = y_{p1} + 2y_{p2} + C_1y_1 + C_2y_2 + C_3y_3$
 $= x^2 + 2e^{-2x} + C_1e^{-x} + C_2e^x + C_3e^{2x}$. ◆

6.1 EXERCISES

In Problems 1–6, determine the largest interval (a, b) for which Theorem 1 guarantees the existence of a unique solution on (a, b) to the given initial value problem.

1. $xy''' - 3y' + e^x y = x^2 - 1$;
 $y(-2) = 1$, $\quad y'(-2) = 0$, $\quad y''(-2) = 2$

2. $y''' - \sqrt{x}\, y = \sin x$;
 $y(\pi) = 0$, $\quad y'(\pi) = 11$, $\quad y''(\pi) = 3$

3. $y''' - y'' + \sqrt{x-1}\, y = \tan x$;
 $y(5) = y'(5) = y''(5) = 1$

4. $x(x+1)y''' - 3xy' + y = 0$;
 $y(-1/2) = 1$, $\quad y'(-1/2) = y''(-1/2) = 0$

5. $x\sqrt{x+1}\, y'' - y' + xy = 0$;
 $y(1/2) = y'(1/2) = -1$, $\quad y''(1/2) = 1$

6. $(x^2 - 1)y''' + e^x y = \ln x$;
 $y(3/4) = 1$, $\quad y'(3/4) = y''(3/4) = 0$

In Problems 7–14, determine whether the given functions are linearly dependent or linearly independent on the specified interval. Justify your decisions.

7. $\{e^{3x}, e^{5x}, e^{-x}\}$ on $(-\infty, \infty)$

8. $\{x^2, x^2 - 1, 5\}$ on $(-\infty, \infty)$

9. $\{\sin^2 x, \cos^2 x, 1\}$ on $(-\infty, \infty)$

10. $\{\sin x, \cos x, \tan x\}$ on $(-\pi/2, \pi/2)$

11. $\{x^{-1}, x^{1/2}, x\}$ on $(0, \infty)$

12. $\{\cos 2x, \cos^2 x, \sin^2 x\}$ on $(-\infty, \infty)$

13. $\{x, x^2, x^3, x^4\}$ on $(-\infty, \infty)$

14. $\{x, xe^x, 1\}$ on $(-\infty, \infty)$

Using the Wronskian in Problems 15–18, verify that the given functions form a fundamental solution set for the given differential equation and find a general solution.

15. $y''' + 2y'' - 11y' - 12y = 0$;
 $\{e^{3x}, e^{-x}, e^{-4x}\}$

16. $y''' - y'' + 4y' - 4y = 0$;
 $\{e^x, \cos 2x, \sin 2x\}$

17. $x^3 y''' - 3x^2 y'' + 6xy' - 6y = 0$, $\quad x > 0$;
 $\{x, x^2, x^3\}$

18. $y^{(4)} - y = 0$; $\quad \{e^x, e^{-x}, \cos x, \sin x\}$

In Problems 19–22, a particular solution and a fundamental solution set are given for a nonhomogeneous equation and its corresponding homogeneous equation. (a) Find a general solution to the nonhomogeneous equation. (b) Find the solution that satisfies the specified initial conditions.

19. $y''' + y'' + 3y' - 5y = 2 + 6x - 5x^2$;
 $y(0) = -1$, $\quad y'(0) = 1$, $\quad y''(0) = -3$;
 $y_p = x^2$; $\quad \{e^x, e^{-x}\cos 2x, e^{-x}\sin 2x\}$

20. $xy''' - y'' = -2$; $\quad y(1) = 2$, $\quad y'(1) = -1$,
 $y''(1) = -4$; $\quad y_p = x^2$; $\quad \{1, x, x^3\}$

21. $x^3 y''' + xy' - y = 3 - \ln x$, $\quad x > 0$;
 $y(1) = 3$, $\quad y'(1) = 3$, $\quad y''(1) = 0$;
 $y_p = \ln x$; $\quad \{x, x\ln x, x(\ln x)^2\}$

22. $y^{(4)} + 4y = 5\cos x$;
 $y(0) = 2$, $\quad y'(0) = 1$, $\quad y''(0) = -1$,
 $y'''(0) = -2$; $\quad y_p = \cos x$;
 $\{e^x\cos x, e^x\sin x, e^{-x}\cos x, e^{-x}\sin x\}$

23. Let $\quad L[y] := y''' + y' + xy$, $\quad y_1(x) := \sin x$, and $y_2(x) := x$. Verify that $L[y_1](x) = x\sin x$ and $L[y_2](x) = x^2 + 1$. Then use the superposition principle (linearity) to find a solution to the differential equation:
 (a) $L[y] = 2x\sin x - x^2 - 1$.
 (b) $L[y] = 4x^2 + 4 - 6x\sin x$.

24. Let $L[y] := y''' - xy'' + 4y' - 3xy$, $y_1(x) = \cos 2x$, and $y_2(x) := -1/3$. Verify that $L[y_1](x) = x\cos 2x$ and $L[y_2](x) = x$. Then use the superposition principle (linearity) to find a solution to the differential equation:
 (a) $L[y] = 7x\cos 2x - 3x$.
 (b) $L[y] = -6x\cos 2x + 11x$.

25. Prove that L defined in (7) is a linear operator by verifying that properties (9) and (10) hold for any n-times differentiable functions $y, y_1, \ldots, y_m$ on (a, b).

26. **Existence of Fundamental Solution Sets.** By Theorem 1, for each $j = 1, 2, \ldots, n$ there is a unique solution $y_j(x)$ to equation (17) satisfying the initial conditions
 $$y_j^{(k)}(x_0) = \begin{cases} 1, & \text{for } k = j - 1, \\ 0, & \text{for } k \neq j - 1, \quad 0 \leq k \leq n - 1. \end{cases}$$
 (a) Show that $\{y_1, y_2, \ldots, y_n\}$ is a fundamental solution set for (17). [*Hint*: Write out the Wronskian at x_0.]
 (b) For given initial values $\gamma_0, \gamma_1, \ldots, \gamma_{n-1}$, express the solution $y(x)$ to (17) satisfying $y^{(k)}(x_0) = \gamma_k$, $k = 0, \ldots, n - 1$, [as in equations (4)] in terms of this fundamental solution set.

27. Show that the set of functions $\{1, x, x^2, \ldots, x^n\}$, where n is a positive integer, is linearly independent on every open interval (a, b). [*Hint*: Use the fact that a polynomial of degree at most n has no more than n zeros unless it is identically zero.]

28. The set of functions
 $$\{1, \cos x, \sin x, \ldots, \cos nx, \sin nx\},$$
 where n is a positive integer, is linearly independent on every interval (a, b). Prove this in the special case $n = 2$ and $(a, b) = (-\infty, \infty)$.

29. (a) Show that if $f_1, \ldots, f_m$ are linearly independent on $(-1, 1)$, then they are linearly independent on $(-\infty, \infty)$.

(b) Give an example to show that if $f_1, \ldots, f_m$ are linearly independent on $(-\infty, \infty)$, then they need not be linearly independent on $(-1, 1)$.

30. To prove Abel's identity (26) for $n = 3$, proceed as follows:

(a) Let $W(x) := W[y_1, y_2, y_3](x)$. Use the product rule for differentiation to show

$$W'(x) = \begin{vmatrix} y_1' & y_2' & y_3' \\ y_1' & y_2' & y_3' \\ y_1'' & y_2'' & y_3'' \end{vmatrix} + \begin{vmatrix} y_1 & y_2 & y_3 \\ y_1'' & y_2'' & y_3'' \\ y_1'' & y_2'' & y_3'' \end{vmatrix}$$

$$+ \begin{vmatrix} y_1 & y_2 & y_3 \\ y_1' & y_2' & y_3' \\ y_1''' & y_2''' & y_3''' \end{vmatrix}.$$

(b) Show that the above expression reduces to

$$(32) \qquad W'(x) = \begin{vmatrix} y_1 & y_2 & y_3 \\ y_1' & y_2' & y_3' \\ y_1''' & y_2''' & y_3''' \end{vmatrix}.$$

(c) Since each y_i satisfies (17), show that

$$(33) \qquad y_i^{(3)}(x) = -\sum_{k=1}^{3} p_k(x) y_i^{(3-k)}(x)$$
$$(i = 1, 2, 3).$$

(d) Substituting the expressions in (33) into (32), show that

$$(34) \qquad W'(x) = -p_1(x) W(x).$$

(e) Deduce Abel's identity by solving the first-order differential equation (34).

31. Reduction of Order. If a nontrivial solution $f(x)$ is known for the *homogeneous* equation

$$y^{(n)} + p_1(x) y^{(n-1)} + \cdots + p_n(x) y = 0,$$

the substitution $y(x) = v(x) f(x)$ can be used to reduce the order of the equation, as was shown in Section 4.7

for second-order equations. By completing the following steps, demonstrate the method for the third-order equation

$$(35) \qquad y''' - 2y'' - 5y' + 6y = 0,$$

given that $f(x) = e^x$ is a solution.

(a) Set $y(x) = v(x) e^x$ and compute y', y'', and y'''.

(b) Substitute your expressions from (a) into (35) to obtain a *second-order* equation in $w := v'$.

(c) Solve the second-order equation in part (b) for w and integrate to find v. Determine two linearly independent choices for v, say, v_1 and v_2.

(d) By part (c), the functions $y_1(x) = v_1(x) e^x$ and $y_2(x) = v_2(x) e^x$ are two solutions to (35). Verify that the three solutions e^x, $y_1(x)$, and $y_2(x)$ are linearly independent on $(-\infty, \infty)$.

32. Given that the function $f(x) = x$ is a solution to $y''' - x^2 y' + xy = 0$, show that the substitution $y(x) = v(x) f(x) = v(x) x$ reduces this equation to $xw'' + 3w' - x^3 w = 0$, where $w = v'$.

33. Use the reduction of order method described in Problem 31 to find three linearly independent solutions to $y''' - 2y'' + y' - 2y = 0$, given that $f(x) = e^{2x}$ is a solution.

34. Constructing Differential Equations. Given three functions $f_1(x), f_2(x), f_3(x)$ that are each three times differentiable and whose Wronskian is never zero on (a, b), show that the equation

$$\begin{vmatrix} f_1(x) & f_2(x) & f_3(x) & y \\ f_1'(x) & f_2'(x) & f_3'(x) & y' \\ f_1''(x) & f_2''(x) & f_3''(x) & y'' \\ f_1'''(x) & f_2'''(x) & f_3'''(x) & y''' \end{vmatrix} = 0$$

is a third-order linear differential equation for which $\{f_1, f_2, f_3\}$ is a fundamental solution set. What is the coefficient of y''' in this equation?

35. Use the result of Problem 34 to construct a third-order differential equation for which $\{x, \sin x, \cos x\}$ is a fundamental solution set.

6.2 Homogeneous Linear Equations with Constant Coefficients

Our goal in this section is to obtain a general solution to an nth-order linear differential equation with constant coefficients. Based on the experience gained with second-order equations in Section 4.2, you should have little trouble guessing the form of such a solution. However, our interest here is to help you understand *why* these techniques work. This is done using an operator approach—a technique that is useful in tackling many other problems in analysis such as solving partial differential equations.

Let's consider the homogeneous linear nth-order differential equation

$$(1) \qquad a_n y^{(n)}(x) + a_{n-1} y^{(n-1)}(x) + \cdots + a_1 y'(x) + a_0 y(x) = 0,$$

where $a_n(\neq 0), a_{n-1}, \ldots, a_0$ are real constants.[†] Since constant functions are everywhere continuous, equation (1) has solutions defined for all x in $(-\infty, \infty)$ (recall Theorem 1 in Section 6.1). If we can find n linearly independent solutions to (1) on $(-\infty, \infty)$, say, $y_1, \ldots, y_n$, then we can express a general solution to (1) in the form

$$(2) \qquad y(x) = C_1 y_1(x) + \cdots + C_n y_n(x),$$

with $C_1, \ldots, C_n$ as arbitrary constants.

To find these n linearly independent solutions, we capitalize on our previous success with second-order equations. Namely, experience suggests that we begin by trying a function of the form $y = e^{rx}$.

If we let L be the differential operator defined by the left-hand side of (1), that is,

$$(3) \qquad L[y] := a_n y^{(n)} + a_{n-1} y^{(n-1)} + \cdots + a_1 y' + a_0 y,$$

then we can write (1) in the operator form

$$(4) \qquad L[y](x) = 0.$$

For $y = e^{rx}$, we find

$$(5) \qquad L[e^{rx}](x) = a_n r^n e^{rx} + a_{n-1} r^{n-1} e^{rx} + \cdots + a_0 e^{rx}$$
$$= e^{rx}(a_n r^n + a_{n-1} r^{n-1} + \cdots + a_0) = e^{rx} P(r),$$

where $P(r)$ is the polynomial $a_n r^n + a_{n-1} r^{n-1} + \cdots + a_0$. Thus, e^{rx} is a solution to equation (4), provided r is a root of the **auxiliary** (or **characteristic**) **equation**

$$(6) \qquad P(r) = a_n r^n + a_{n-1} r^{n-1} + \cdots + a_0 = 0.$$

According to the fundamental theorem of algebra, the auxiliary equation has n roots (counting multiplicities), which may be either real or complex. However, there are no formulas for determining the zeros of an arbitrary polynomial of degree greater than four, although if we can determine one zero r_1, then we can divide out the factor $r - r_1$ and be left with a polynomial of lower degree. (For convenience, we have chosen most of our examples and exercises so that $0, \pm 1$, or ± 2 are zeros of any polynomial of degree greater than two that we must factor.) When a zero cannot be exactly determined, numerical algorithms such as Newton's method or the quotient-difference algorithm can be used to compute approximate roots of the polynomial equation.[‡] Some pocket calculators even have these algorithms built in.

We proceed to discuss the various possibilities.

Distinct Real Roots

If the roots $r_1, \ldots, r_n$ of the auxiliary equation (6) are real and distinct, then n solutions to equation (1) are

$$(7) \qquad y_1(x) = e^{r_1 x}, y_2(x) = e^{r_2 x}, \ldots, y_n(x) = e^{r_n x}.$$

[†]*Historical Footnote:* In a letter to John Bernoulli dated September 15, 1739, Leonhard Euler claimed to have solved the general case of the homogeneous linear nth-order equation with constant coefficients.

[‡]See, for example, *Applied and Computational Complex Analysis*, by P. Henrici (Wiley-Interscience, New York, 1993), Volume 1, or *Numerical Analysis*, 9th ed., by R. L. Burden and J. D. Faires (Brooks/Cole Cengage Learning, 2011).

As stated in the previous section, these functions are linearly independent on $(-\infty, \infty)$, a fact that we now officially verify. Let's assume that $c_1, \ldots, c_n$ are constants such that

(8) $c_1 e^{r_1 x} + \cdots + c_n e^{r_n x} = 0$

for all x in $(-\infty, \infty)$. Our goal is to prove that $c_1 = c_2 = \cdots = c_n = 0$.

One way to show this is to construct a linear operator L_k that annihilates (maps to zero) everything on the left-hand side of (8) except the kth term. For this purpose, we note that since $r_1, \ldots, r_n$ are the zeros of the auxiliary polynomial $P(r)$, then $P(r)$ can be factored as

(9) $P(r) = a_n(r - r_1) \cdots (r - r_n)$.

Consequently, the operator $L[y] = a_n y^{(n)} + a_{n-1} y^{(n-1)} + \cdots + a_0 y$ can be expressed in terms of the differentiation operator D as the following composition:[†]

(10) $L = P(D) = a_n(D - r_1) \cdots (D - r_n)$.

We now construct the polynomial $P_k(r)$ by deleting the factor $(r - r_k)$ from $P(r)$. Then we set $L_k := P_k(D)$; that is,

(11) $L_k := P_k(D) = a_n(D - r_1) \cdots (D - r_{k-1})(D - r_{k+1}) \cdots (D - r_n)$.

Applying L_k to both sides of (8), we get, via linearity,

(12) $c_1 L_k[e^{r_1 x}] + \cdots + c_n L_k[e^{r_n x}] = 0$.

Also, since $L_k = P_k(D)$, we find [just as in equation (5)] that $L_k[e^{rx}](x) = e^{rx} P_k(r)$ for all r. Thus (12) can be written as

$$c_1 e^{r_1 x} P_k(r_1) + \cdots + c_n e^{r_n x} P_k(r_n) = 0,$$

which simplifies to

(13) $c_k e^{r_k x} P_k(r_k) = 0$,

because $P_k(r_i) = 0$ for $i \neq k$. Since r_k is not a root of $P_k(r)$, then $P_k(r_k) \neq 0$. It now follows from (13) that $c_k = 0$. But as k is arbitrary, all the constants $c_1, \ldots, c_n$ must be zero. Thus, $y_1(x), \ldots, y_n(x)$ as given in (7) are linearly independent. (See Problem 26 for an alternative proof.)

We have proved that, in the case of n distinct real roots, a general solution to (1) is

(14) $y(x) = C_1 e^{r_1 x} + \cdots + C_n e^{r_n x}$,

where $C_1, \ldots, C_n$ are arbitrary constants.

Example 1 Find a general solution to

(15) $y''' - 2y'' - 5y' + 6y = 0$.

Solution The auxiliary equation is

(16) $r^3 - 2r^2 - 5r + 6 = 0$.

By inspection we find that $r = 1$ is a root. Then, using polynomial division, we get

$$r^3 - 2r^2 - 5r + 6 = (r - 1)(r^2 - r - 6),$$

which further factors into $(r - 1)(r + 2)(r - 3)$. Hence the roots of equation (16) are $r_1 = 1$, $r_2 = -2, r_3 = 3$. Since these roots are real and distinct, a general solution to (15) is

$$y(x) = C_1 e^x + C_2 e^{-2x} + C_3 e^{3x}. \quad \blacklozenge$$

[†]*Historical Footnote:* The symbolic notation $P(D)$ was introduced by Augustin Cauchy in 1827.

Complex Roots

If $\alpha + i\beta$ (α, β real) is a complex root of the auxiliary equation (6), then so is its complex conjugate $\alpha - i\beta$, since the coefficients of $P(r)$ are real-valued (see Problem 24). If we accept complex-valued functions as solutions, then both $e^{(\alpha+i\beta)x}$ and $e^{(\alpha-i\beta)x}$ are solutions to (1). Moreover, if there are no repeated roots, then a general solution to (1) is again given by (14). To find two real-valued solutions corresponding to the roots $\alpha \pm i\beta$, we can just take the real and imaginary parts of $e^{(\alpha+i\beta)x}$. That is, since

$$(17) \qquad e^{(\alpha+i\beta)x} = e^{\alpha x}\cos\beta x + ie^{\alpha x}\sin\beta x,$$

then two linearly independent solutions to (1) are

$$(18) \qquad e^{\alpha x}\cos\beta x, \qquad e^{\alpha x}\sin\beta x.$$

In fact, using these solutions in place of $e^{(\alpha+i\beta)x}$ and $e^{(\alpha-i\beta)x}$ in (14) preserves the linear independence of the set of n solutions. Thus, treating each of the conjugate pairs of roots in this manner, we obtain a real-valued general solution to (1).

Example 2 Find a general solution to

$$(19) \qquad y''' + y'' + 3y' - 5y = 0.$$

Solution The auxiliary equation is

$$(20) \qquad r^3 + r^2 + 3r - 5 = (r-1)(r^2 + 2r + 5) = 0,$$

which has distinct roots $r_1 = 1$, $r_2 = -1 + 2i$, $r_3 = -1 - 2i$. Thus, a general solution is

$$(21) \qquad y(x) = C_1 e^x + C_2 e^{-x}\cos 2x + C_3 e^{-x}\sin 2x. \quad \blacklozenge$$

Repeated Roots

If r_1 is a root of multiplicity m, then the n solutions given in (7) are not even distinct, let alone linearly independent. Recall that for a second-order equation, when we had a repeated root r_1 to the auxiliary equation, we obtained two linearly independent solutions by taking $e^{r_1 x}$ and $xe^{r_1 x}$. So if r_1 is a root of (6) of multiplicity m, we might expect that m linearly independent solutions are

$$(22) \qquad e^{r_1 x}, \quad xe^{r_1 x}, \quad x^2 e^{r_1 x}, \quad \ldots, \quad x^{m-1}e^{r_1 x}.$$

To see that this is the case, observe that if r_1 is a root of multiplicity m, then the auxiliary equation can be written in the form

$$(23) \qquad a_n(r-r_1)^m(r-r_{m+1})\cdots(r-r_n) = (r-r_1)^m\tilde{P}(r) = 0,$$

where $\tilde{P}(r) := a_n(r-r_{m+1})\cdots(r-r_n)$ and $\tilde{P}(r_1) \neq 0$. With this notation, we have the identity

$$(24) \qquad L[e^{rx}](x) = e^{rx}(r-r_1)^m\tilde{P}(r)$$

[see (5) on page 328]. Setting $r = r_1$ in (24), we again see that $e^{r_1 x}$ is a solution to $L[y] = 0$.

To find other solutions, we take the kth partial derivative with respect to r of both sides of (24):

$$(25) \qquad \frac{\partial^k}{\partial r^k}L[e^{rx}](x) = \frac{\partial^k}{\partial r^k}\left[e^{rx}(r-r_1)^m\tilde{P}(r)\right].$$

Carrying out the differentiation on the right-hand side of (25), we find that the resulting expression will still have $(r-r_1)$ as a factor, provided $k \leq m-1$. Thus, setting $r = r_1$ in (25) gives

(26) $\left. \dfrac{\partial^k}{\partial r^k} L[e^{rx}](x) \right|_{r=r_1} = 0 \quad \text{if} \quad k \le m-1$.

Now notice that the function e^{rx} has continuous partial derivatives of all orders with respect to r and x. Hence, for mixed partial derivatives of e^{rx}, it makes no difference whether the differentiation is done first with respect to x, then with respect to r, or vice versa. Since L involves derivatives with respect to x, this means we can interchange the order of differentiation in (26) to obtain

$$L\left[\left. \dfrac{\partial^k}{\partial r^k}(e^{rx}) \right|_{r=r_1} \right](x) = 0 .$$

Thus,

(27) $\left. \dfrac{\partial^k}{\partial r^k}(e^{rx}) \right|_{r=r_1} = x^k e^{r_1 x}$

will be a solution to (1) for $k = 0, 1, \dots, m-1$. So m distinct solutions to (1), due to the root $r = r_1$ of multiplicity m, are indeed given by (22). We leave it as an exercise to show that the m functions in (22) are linearly independent on $(-\infty, \infty)$ (see Problem 25).

If $\alpha + i\beta$ is a repeated complex root of multiplicity m, then we can replace the $2m$ complex-valued functions

$$e^{(\alpha+i\beta)x}, \quad xe^{(\alpha+i\beta)x}, \quad \dots, \quad x^{m-1}e^{(\alpha+i\beta)x},$$
$$e^{(\alpha-i\beta)x}, \quad xe^{(\alpha-i\beta)x}, \quad \dots, \quad x^{m-1}e^{(\alpha-i\beta)x}$$

by the $2m$ linearly independent real-valued functions

(28) $\begin{aligned} &e^{\alpha x}\cos\beta x, \quad xe^{\alpha x}\cos\beta x, \quad \dots, \quad x^{m-1}e^{\alpha x}\cos\beta x, \\ &e^{\alpha x}\sin\beta x, \quad xe^{\alpha x}\sin\beta x, \quad \dots, \quad x^{m-1}e^{\alpha x}\sin\beta x. \end{aligned}$

Using the results of the three cases discussed above, we can obtain a set of n linearly independent solutions that yield a real-valued general solution for (1).

Example 3 Find a general solution to

(29) $y^{(4)} - y^{(3)} - 3y'' + 5y' - 2y = 0$.

Solution The auxiliary equation is

$$r^4 - r^3 - 3r^2 + 5r - 2 = (r-1)^3(r+2) = 0 ,$$

which has roots $r_1 = 1, r_2 = 1, r_3 = 1, r_4 = -2$. Because the root at 1 has multiplicity 3, a general solution is

(30) $y(x) = C_1 e^x + C_2 x e^x + C_3 x^2 e^x + C_4 e^{-2x}$. ◆

Example 4 Find a general solution to

(31) $y^{(4)} - 8y^{(3)} + 26y'' - 40y' + 25y = 0$,

whose auxiliary equation can be factored as

(32) $r^4 - 8r^3 + 26r^2 - 40r + 25 = (r^2 - 4r + 5)^2 = 0$.

Solution The auxiliary equation (32) has repeated complex roots: $r_1 = 2+i, r_2 = 2+i, r_3 = 2-i$, and $r_4 = 2-i$. Hence, a general solution is

$$y(x) = C_1 e^{2x}\cos x + C_2 x e^{2x}\cos x + C_3 e^{2x}\sin x + C_4 x e^{2x}\sin x .$$ ◆

6.2 EXERCISES

In Problems 1–14, find a general solution for the differential equation with x as the independent variable.

1. $y''' + 2y'' - 8y' = 0$

2. $y''' - 3y'' - y' + 3y = 0$

3. $6z''' + 7z'' - z' - 2z = 0$

4. $y''' + 2y'' - 19y' - 20y = 0$

5. $y''' + 3y'' + 28y' + 26y = 0$

6. $y''' - y'' + 2y = 0$

7. $2y''' - y'' - 10y' - 7y = 0$

8. $y''' + 5y'' - 13y' + 7y = 0$

9. $u''' - 9u'' + 27u' - 27u = 0$

10. $y''' + 3y'' - 4y' - 6y = 0$

11. $y^{(4)} + 4y''' + 6y'' + 4y' + y = 0$

12. $y''' + 5y'' + 3y' - 9y = 0$

13. $y^{(4)} + 4y'' + 4y = 0$

14. $y^{(4)} + 2y''' + 10y'' + 18y' + 9y = 0$
 [*Hint:* $y(x) = \sin 3x$ is a solution.]

In Problems 15–18, find a general solution to the given homogeneous equation.

15. $(D-1)^2(D+3)(D^2+2D+5)^2[y] = 0$

16. $(D+1)^2(D-6)^3(D+5)(D^2+1)(D^2+4)[y] = 0$

17. $(D+4)(D-3)(D+2)^3(D^2+4D+5)^2D^5[y] = 0$

18. $(D-1)^3(D-2)(D^2+D+1)$
$\qquad \cdot (D^2+6D+10)^3[y] = 0$

In Problems 19–21, solve the given initial value problem.

19. $y''' - y'' - 4y' + 4y = 0$;
$\quad y(0) = -4, \quad y'(0) = -1, \quad y''(0) = -19$

20. $y''' + 7y'' + 14y' + 8y = 0$;
$\quad y(0) = 1, \quad y'(0) = -3, \quad y''(0) = 13$

21. $y''' - 4y'' + 7y' - 6y = 0$;
$\quad y(0) = 1, \quad y'(0) = 0, \quad y''(0) = 0$

In Problems 22 and 23, find a general solution for the given linear system using the elimination method of Section 5.2.

22. $d^2x/dt^2 - x + 5y = 0$,
$\quad 2x + d^2y/dt^2 + 2y = 0$

23. $d^3x/dt^3 - x + dy/dt + y = 0$,
$\quad dx/dt - x + y = 0$

24. Let $P(r) = a_n r^n + \cdots + a_1 r + a_0$ be a polynomial with real coefficients $a_n, \ldots, a_0$. Prove that if r_1 is a zero of $P(r)$, then so is its complex conjugate $\bar{r}_1$. [*Hint:* Show that $\overline{P(r)} = P(\bar{r})$, where the bar denotes complex conjugation.]

25. Show that the m functions $e^{rx}, xe^{rx}, \ldots, x^{m-1}e^{rx}$ are linearly independent on $(-\infty, \infty)$. [*Hint:* Show that these functions are linearly independent if and only if $1, x, \ldots, x^{m-1}$ are linearly independent.]

26. As an alternative proof that the functions $e^{r_1 x}, e^{r_2 x}, \ldots, e^{r_n x}$ are linearly independent on $(-\infty, \infty)$ when $r_1, r_2, \ldots, r_n$ are distinct, assume

(33) $\qquad C_1 e^{r_1 x} + C_2 e^{r_2 x} + \cdots + C_n e^{r_n x} = 0$

holds for all x in $(-\infty, \infty)$ and proceed as follows:

(a) Because the r_i's are distinct we can (if necessary) relabel them so that

$$r_1 > r_2 > \cdots > r_n.$$

Divide equation (33) by $e^{r_1 x}$ to obtain

$$C_1 + C_2 \frac{e^{r_2 x}}{e^{r_1 x}} + \cdots + C_n \frac{e^{r_n x}}{e^{r_1 x}} = 0.$$

Now let $x \to +\infty$ on the left-hand side to obtain $C_1 = 0$.

(b) Since $C_1 = 0$, equation (33) becomes

$$C_2 e^{r_2 x} + C_3 e^{r_3 x} + \cdots + C_n e^{r_n x} = 0$$

for all x in $(-\infty, \infty)$. Divide this equation by $e^{r_2 x}$ and let $x \to +\infty$ to conclude that $C_2 = 0$.

(c) Continuing in the manner of (b), argue that all the coefficients, $C_1, C_2, \ldots, C_n$ are zero and hence $e^{r_1 x}, e^{r_2 x}, \ldots, e^{r_n x}$ are linearly independent on $(-\infty, \infty)$.

 27. Find a general solution to

$$y^{(4)} + 2y''' - 3y'' - y' + \frac{1}{2}y = 0$$

by using Newton's method (Appendix B) or some other numerical procedure to approximate the roots of the auxiliary equation.

 28. Find a general solution to $y''' - 3y' - y = 0$ by using Newton's method or some other numerical procedure to approximate the roots of the auxiliary equation.

 29. Find a general solution to

$$y^{(4)} + 2y^{(3)} + 4y'' + 3y' + 2y = 0$$

by using Newton's method to approximate numerically the roots of the auxiliary equation. [*Hint:* To find complex roots, use the Newton recursion formula $z_{n+1} = z_n - f(z_n)/f'(z_n)$ and start with a *complex* initial guess z_0.]

30. (a) Derive the form

$$y(x) = A_1 e^x + A_2 e^{-x} + A_3 \cos x + A_4 \sin x$$

for the general solution to the equation $y^{(4)} = y$, from the observation that the fourth roots of unity are $1, -1, i,$ and $-i$.

(b) Derive the form

$$y(x) = A_1 e^x + A_2 e^{-x/2} \cos\left(\sqrt{3}x/2\right)$$
$$+ A_3 e^{-x/2} \sin\left(\sqrt{3}x/2\right)$$

for the general solution to the equation $y^{(3)} = y$, from the observation that the cube roots of unity are $1, e^{i2\pi/3},$ and $e^{-i2\pi/3}$.

31. Higher-Order Cauchy–Euler Equations. A differential equation that can be expressed in the form

$$a_n x^n y^{(n)}(x) + a_{n-1} x^{n-1} y^{(n-1)}(x) + \cdots + a_0 y(x) = 0,$$

where $a_n, a_{n-1}, \ldots, a_0$ are constants, is called a homogeneous **Cauchy–Euler** equation. (The second-order case is discussed in Section 4.7.) Use the substitution $y = x^r$ to help determine a fundamental solution set for the following Cauchy–Euler equations:

(a) $x^3 y''' + x^2 y'' - 2xy' + 2y = 0$, $x > 0$.

(b) $x^4 y^{(4)} + 6x^3 y''' + 2x^2 y'' - 4xy' + 4y = 0$, $x > 0$

(c) $x^3 y''' - 2x^2 y'' + 13xy' - 13y = 0$, $x > 0$

[*Hint:* $x^{\alpha + i\beta} = e^{(\alpha + i\beta)\ln x}$
$= x^\alpha \{\cos(\beta \ln x) + i \sin(\beta \ln x)\}$.]

32. Let $y(x) = Ce^{rx}$, where C $(\neq 0)$ and r are real numbers, be a solution to a differential equation. Suppose we cannot determine r exactly but can only approximate it by $\tilde{r}$. Let $\tilde{y}(x) := Ce^{\tilde{r}x}$ and consider the error $|y(x) - \tilde{y}(x)|$.

(a) If r and $\tilde{r}$ are positive, $r \neq \tilde{r}$, show that the error grows exponentially large as x approaches $+\infty$.

(b) If r and $\tilde{r}$ are negative, $r \neq \tilde{r}$, show that the error goes to zero exponentially as x approaches $+\infty$.

33. On a smooth horizontal surface, a mass of m_1 kg is attached to a fixed wall by a spring with spring constant k_1 N/m. Another mass of m_2 kg is attached to the first object by a spring with spring constant k_2 N/m. The objects are aligned horizontally so that the springs are their natural lengths. As we showed in Section 5.6, this coupled mass–spring system is governed by the system of differential equations

$$(34) \qquad m_1 \frac{d^2 x}{dt^2} + (k_1 + k_2)x - k_2 y = 0,$$

$$(35) \qquad m_2 \frac{d^2 y}{dt^2} - k_2 x + k_2 y = 0.$$

Let's assume that $m_1 = m_2 = 1, k_1 = 3,$ and $k_2 = 2$. If both objects are displaced 1 m to the right of their equilibrium positions (compare Figure 5.26, page 283) and then released, determine the equations of motion for the objects as follows:

(a) Show that $x(t)$ satisfies the equation

$$(36) \qquad x^{(4)}(t) + 7x''(t) + 6x(t) = 0.$$

(b) Find a general solution $x(t)$ to (36).

(c) Substitute $x(t)$ back into (34) to obtain a general solution for $y(t)$.

(d) Use the initial conditions to determine the solutions, $x(t)$ and $y(t)$, which are the equations of motion.

34. Suppose the two springs in the coupled mass–spring system discussed in Problem 33 are switched, giving the new data $m_1 = m_2 = 1, k_1 = 2,$ and $k_2 = 3$. If both objects are now displaced 1 m to the right of their equilibrium positions and then released, determine the equations of motion of the two objects.

35. Vibrating Beam. In studying the transverse vibrations of a beam, one encounters the homogeneous equation

$$EI \frac{d^4 y}{dx^4} - ky = 0,$$

where $y(x)$ is related to the displacement of the beam at position x, the constant E is Young's modulus, I is the area moment of inertia, and k is a parameter. Assuming $E, I,$ and k are positive constants, find a general solution in terms of sines, cosines, hyperbolic sines, and hyperbolic cosines.

6.3 Undetermined Coefficients and the Annihilator Method

In Sections 4.4 and 4.5 we mastered an easy method for obtaining a particular solution to a nonhomogeneous linear second-order constant-coefficient equation,

(1) $L[y] = (aD^2 + bD + c)[y] = f(x)$,

when the nonhomogeneity $f(x)$ had a particular form (namely, a product of a polynomial, an exponential, and a sinusoid). Roughly speaking, we were motivated by the observation that if a function f, of this type, resulted from operating on y with an operator L of the form $(aD^2 + bD + c)$, then we must have started with a y of the same type. So we solved (1) by postulating a solution form y_p that resembled f, but with *undetermined coefficients,* and we inserted this form into the equation to fix the values of these coefficients. Eventually, we realized that we had to make certain accommodations when f was a solution to the homogeneous equation $L[y] = 0$.

In this section we are going to reexamine the method of undetermined coefficients from another, more rigorous, point of view—partly with the objective of tying up the loose ends in our previous exposition and more importantly with the goal of extending the method to higher-order equations (with constant coefficients). At the outset we'll describe the new point of view that will be adopted for the analysis. Then we illustrate its implications and ultimately derive a simplified set of rules for its implementation: rules that justify and extend the procedures of Section 4.4. The rigorous approach is known as the **annihilator method.**

The first premise of the annihilator method is the observation, gleaned from the analysis of the previous section, that all of the "suitable types" of nonhomogeneities $f(x)$ (products of polynomials times exponentials times sinusoids) are themselves *solutions to **homogeneous** differential equations with constant coefficients.* Observe the following:

(i) Any nonhomogeneous term of the form $f(x) = e^{rx}$ satisfies $(D - r)[f] = 0$.

(ii) Any nonhomogeneous term of the form $f(x) = x^k e^{rx}$ satisfies $(D - r)^m[f] = 0$ for $k = 0, 1, \ldots, m - 1$.

(iii) Any nonhomogeneous term of the form $f(x) = \cos \beta x$ or $\sin \beta x$ satisfies $(D^2 + \beta^2)[f] = 0$.

(iv) Any nonhomogeneous term of the form $f(x) = x^k e^{\alpha x} \cos \beta x$ or $x^k e^{\alpha x} \sin \beta x$ satisfies $[(D - \alpha)^2 + \beta^2]^m[f] = 0$ for $k = 0, 1, \ldots, m - 1$.

In other words, each of these nonhomogeneities is *annihilated* by a differential operator with constant coefficients.

Annihilator

Definition 3. A linear differential operator A is said to **annihilate** a function f if

(2) $A[f](x) = 0$,

for all x. That is, A annihilates f if f is a solution to the homogeneous linear differential equation (2) on $(-\infty, \infty)$.

Example 1 Find a differential operator that annihilates

(3) $6xe^{-4x} + 5e^x \sin 2x$.

Solution Consider the two functions whose sum appears in (3). Observe that $(D+4)^2$ annihilates the function $f_1(x) := 6xe^{-4x}$. Further, $f_2(x) := 5e^x \sin 2x$ is annihilated by the operator $(D-1)^2 + 4$. Hence, the composite operator

$$A := (D+4)^2 [(D-1)^2 + 4] ,$$

which is the same as the operator

$$[(D-1)^2 + 4](D+4)^2 ,$$

annihilates both f_1 and f_2. But then, by linearity, A also annihilates the sum $f_1 + f_2$. ◆

We now show how annihilators can be used to determine particular solutions to certain *non-homogeneous* equations. Consider the nth-order differential equation with constant coefficients

(4) $a_n y^{(n)}(x) + a_{n-1} y^{(n-1)}(x) + \cdots + a_0 y(x) = f(x)$,

which can be written in the operator form

(5) $L[y](x) = f(x)$,

where

$$L = a_n D^n + a_{n-1} D^{n-1} + \cdots + a_0 .$$

Assume that A is a linear differential operator with constant coefficients that annihilates $f(x)$. Then

$$A\big[L[y]\big](x) = A[f](x) = 0 ,$$

so any solution to (5) is also a solution to the *homogeneous* equation

(6) $AL[y](x) = 0$,

involving the composition of the operators A and L. But we are experts on *homogeneous* differential equations (with constant coefficients)! In particular, we can use the methods of Section 6.2 to write down a general solution of (6). From this we can deduce the *form* of a particular solution to (5). Let's look at some examples and then summarize our findings. The differential equation in the next example is second order, so we will be able to see exactly how the annihilator method is related to the techniques of Sections 4.4 and 4.5.

Example 2 Find a general solution to

(7) $y'' - y = xe^x + \sin x$.

Solution First let's solve this by the methods of Sections 4.4 and 4.5, to get a perspective for the annihilator method. The homogeneous equation corresponding to (7) is $y'' - y = 0$, with the general solution $C_1 e^{-x} + C_2 e^x$. Since e^x is a solution of the homogeneous equation, the nonhomogeneity xe^x demands a solution form $x(C_3 + C_4 x)e^x$. To accommodate the nonhomogeneity $\sin x$, we need an undetermined coefficient form $C_5 \sin x + C_6 \cos x$. Values for C_3 through C_6 in the *particular solution* are determined by substitution:

$$y_p'' - y_p = \left[C_3 xe^x + C_4 x^2 e^x + C_5 \sin x + C_6 \cos x \right]''$$
$$- \left[C_3 xe^x + C_4 x^2 e^x + C_5 \sin x + C_6 \cos x \right] = \sin x + xe^x ,$$

eventually leading to the conclusion $C_3 = -1/4$, $C_4 = 1/4$, $C_5 = -1/2$, and $C_6 = 0$. Thus (for future reference), a general solution to (7) is

$$(8) \qquad y(x) = C_1 e^{-x} + C_2 e^x + x\left(-\frac{1}{4} + \frac{1}{4}x\right)e^x - \frac{1}{2}\sin x.$$

For the annihilator method, observe that $(D^2 + 1)$ annihilates $\sin x$ and $(D-1)^2$ annihilates xe^x. Therefore, any solution to (7), expressed for convenience in operator form as $(D^2 - 1)[y](x) = xe^x + \sin x$, is annihilated by the composition $(D^2 + 1)(D-1)^2(D^2 - 1)$; that is, it satisfies the constant-coefficient homogeneous equation

$$(9) \qquad (D^2 + 1)(D-1)^2(D^2 - 1)[y] = (D+1)(D-1)^3(D^2 + 1)[y] = 0.$$

From Section 6.2 we deduce that the general solution to (9) is given by

$$(10) \qquad y = C_1 e^{-x} + C_2 e^x + C_3 xe^x + C_4 x^2 e^x + C_5 \sin x + C_6 \cos x.$$

This is precisely the solution form generated by the methods of Chapter 4; the first two terms are the general solution to the associated homogeneous equation, and the remaining four terms express the particular solution to the nonhomogeneous equation with undetermined coefficients. Substitution of (10) into (7) will lead to the quoted values for C_3 through C_6, and indeterminant values for C_1 and C_2; the latter are available to fit initial conditions.

Note how the annihilator method automatically accounts for the fact that the nonhomogeneity xe^x requires the form $C_3 xe^x + C_4 x^2 e^x$ in the particular solution, by counting the total number of factors of $(D-1)$ in the annihilator and the original differential operator. ◆

Example 3 Find a general solution, using the annihilator method, to

$$(11) \qquad y''' - 3y'' + 4y = xe^{2x}.$$

Solution The associated homogeneous equation takes the operator form

$$(12) \qquad (D^3 - 3D^2 + 4)[y] = (D+1)(D-2)^2[y] = 0.$$

The nonhomogeneity xe^{2x} is annihilated by $(D-2)^2$. Therefore, every solution of (11) also satisfies

$$(13) \qquad (D-2)^2(D^3 - 3D^2 + 4)[y] = (D+1)(D-2)^4[y] = 0.$$

A general solution to (13) is

$$(14) \qquad y(x) = C_1 e^{-x} + C_2 e^{2x} + C_3 xe^{2x} + C_4 x^2 e^{2x} + C_5 x^3 e^{2x}.$$

Comparison with (12) shows that the first three terms of (14) give a general solution to the associated homogeneous equation and the last two terms constitute a particular solution form with undetermined coefficients. Direct substitution reveals $C_4 = -1/18$ and $C_5 = 1/18$ and so a general solution to (11) is

$$y(x) = C_1 e^{-x} + C_2 e^{2x} + C_3 xe^{2x} - \frac{1}{18}x^2 e^{2x} + \frac{1}{18}x^3 e^{2x}. \quad ◆$$

The annihilator method, then, rigorously justifies the method of undetermined coefficients of Section 4.4. It also tells us how to upgrade that procedure for higher-order equations with constant coefficients. Note that we don't have to implement the annihilator method directly; we simply need to introduce the following modifications to the method of undetermined coefficients for second-order equations that was described in the procedural box on page 178.

Method of Undetermined Coefficients

To find a particular solution to the constant-coefficient differential equation $L[y] = Cx^m e^{rx}$, where m is a nonnegative integer, use the form

(15) $y_p(x) = x^s [A_m x^m + \cdots + A_1 x + A_0] e^{rx}$,

with $s = 0$ if r is not a root of the associated auxiliary equation; otherwise, take s equal to the multiplicity of this root.

 To find a particular solution to the constant-coefficient differential equation $L[y] = Cx^m e^{\alpha x} \cos \beta x$ or $L[y] = Cx^m e^{\alpha x} \sin \beta x$, where $\beta \neq 0$, use the form

(16) $y_p(x) = x^s [A_m x^m + \cdots + A_1 x + A_0] e^{\alpha x} \cos \beta x$
 $+ x^s [B_m x^m + \cdots + B_1 x + B_0] e^{\alpha x} \sin \beta x$,

with $s = 0$ if $\alpha + i\beta$ is not a root of the associated auxiliary equation; otherwise, take s equal to the multiplicity of this root.

As in Chapter 4, the superposition principle can be used to streamline the method for sums of nonhomogeneous terms of the above types.

6.3 EXERCISES

In Problems 1–4, use the method of undetermined coefficients to determine the form of a particular solution for the given equation.

1. $y''' - 2y'' - 5y' + 6y = e^x + x^2$
2. $y''' + y'' - 5y' + 3y = e^{-x} + \sin x$
3. $y''' + 3y'' - 4y = e^{-2x}$
4. $y''' + y'' - 2y = xe^x + 1$

In Problems 5–10, find a general solution to the given equation.

5. $y''' - 2y'' - 5y' + 6y = e^x + x^2$
6. $y''' + y'' - 5y' + 3y = e^{-x} + \sin x$
7. $y''' + 3y'' - 4y = e^{-2x}$
8. $y''' + y'' - 2y = xe^x + 1$
9. $y''' - 3y'' + 3y' - y = e^x$
10. $y''' + 4y'' + y' - 26y = e^{-3x} \sin 2x + x$

In Problems 11–20, find a differential operator that annihilates the given function.

11. $x^4 - x^2 + 11$
12. $3x^2 - 6x + 1$
13. e^{-7x}
14. e^{5x}
15. $e^{2x} - 6e^x$
16. $x^2 - e^x$
17. $x^2 e^{-x} \sin 2x$
18. $xe^{3x} \cos 5x$
19. $xe^{-2x} + xe^{-5x} \sin 3x$
20. $x^2 e^x - x \sin 4x + x^3$

In Problems 21–30, use the annihilator method to determine the form of a particular solution for the given equation.

21. $u'' - 5u' + 6u = \cos 2x + 1$
22. $y'' + 6y' + 8y = e^{3x} - \sin x$
23. $y'' - 5y' + 6y = e^{3x} - x^2$
24. $\theta'' - \theta = xe^x$
25. $y'' - 6y' + 9y = \sin 2x + x$
26. $y'' + 2y' + y = x^2 - x + 1$
27. $y'' + 2y' + 2y = e^{-x} \cos x + x^2$
28. $y'' - 6y' + 10y = e^{3x} - x$
29. $z''' - 2z'' + z' = x - e^x$
30. $y''' + 2y'' - y' - 2y = e^x - 1$

In Problems 31–33, solve the given initial value problem.

31. $y''' + 2y'' - 9y' - 18y = -18x^2 - 18x + 22$;
 $y(0) = -2$, $y'(0) = -8$, $y''(0) = -12$
32. $y''' - 2y'' + 5y' = -24e^{3x}$;
 $y(0) = 4$, $y'(0) = -1$, $y''(0) = -5$
33. $y''' - 2y'' - 3y' + 10y$
 $= 34xe^{-2x} - 16e^{-2x} - 10x^2 + 6x + 34$;
 $y(0) = 3$, $y'(0) = 0$, $y''(0) = 0$

34. Use the annihilator method to show that if $a_0 \neq 0$ in equation (4) and $f(x)$ has the form

 (17) $f(x) = b_m x^m + b_{m-1} x^{m-1} + \cdots + b_1 x + b_0$,

 then

 $y_p(x) = B_m x^m + B_{m-1} x^{m-1} + \cdots + B_1 x + B_0$

 is the form of a particular solution to equation (4).

35. Use the annihilator method to show that if $a_0 = 0$ and $a_1 \neq 0$ in (4) and $f(x)$ has the form given in (17), then equation (4) has a particular solution of the form

 $y_p(x) = x \{ B_m x^m + B_{m-1} x^{m-1} + \cdots + B_1 x + B_0 \}$.

36. Use the annihilator method to show that if $f(x)$ in (4) has the form $f(x) = Be^{\alpha x}$, then equation (4) has a particular solution of the form $y_p(x) = x^s Be^{\alpha x}$, where s is chosen to be the smallest nonnegative integer such that $x^s e^{\alpha x}$ is not a solution to the corresponding homogeneous equation.

37. Use the annihilator method to show that if $f(x)$ in (4) has the form

$$f(x) = a \cos \beta x + b \sin \beta x \,,$$

then equation (4) has a particular solution of the form

(18) $y_p(x) = x^s\{A \cos \beta x + B \sin \beta x\} \,,$

where s is chosen to be the smallest nonnegative integer such that $x^s \cos \beta x$ and $x^s \sin \beta x$ are not solutions to the corresponding homogeneous equation.

In Problems 38 and 39, use the elimination method of Section 5.2 to find a general solution to the given system.

38. $x - d^2y/dt^2 = t + 1 \,,$
 $dx/dt + dy/dt - 2y = e^t$

39. $d^2x/dt^2 - x + y = 0 \,,$
 $x + d^2y/dt^2 - y = e^{3t}$

40. The currents in the electrical network in Figure 6.1 satisfy the system

$$\frac{1}{9}I_1 + 64I_2'' = -2 \sin \frac{t}{24} \,,$$

$$\frac{1}{64}I_3 + 9I_3'' - 64I_2'' = 0 \,,$$

$$I_1 = I_2 + I_3 \,,$$

where I_1, I_2, and I_3 are the currents through the different branches of the network. Using the elimination method of Section 5.2, determine the currents if initially $I_1(0) = I_2(0) = I_3(0) = 0$, $I_1'(0) = 73/12$, $I_2'(0) = 3/4$, and $I_3'(0) = 16/3$.

Figure 6.1 An electrical network

6.4 Method Of Variation of Parameters

In the previous section, we discussed the method of undetermined coefficients and the annihilator method. These methods work only for linear equations with constant coefficients *and* when the nonhomogeneous term is a solution to some homogeneous linear equation with constant coefficients. In this section we show how the method of **variation of parameters** discussed in Sections 4.6 and 4.7 generalizes to higher-order linear equations with variable coefficients.

Our goal is to find a particular solution to the standard form equation

(1) $L[y](x) = g(x) \,,$

where $L[y] := y^{(n)} + p_1 y^{(n-1)} + \cdots + p_n y$ and the coefficient functions $p_1, \ldots, p_n$, as well as g, are continuous on (a, b). The method to be described requires that we already know a fundamental solution set $\{y_1, \ldots, y_n\}$ for the corresponding homogeneous equation

(2) $L[y](x) = 0 \,.$

A general solution to (2) is then

(3) $y_h(x) = C_1 y_1(x) + \cdots + C_n y_n(x) \,,$

where $C_1, \ldots, C_n$ are arbitrary constants. In the method of variation of parameters, we assume there exists a particular solution to (1) of the form

(4) $y_p(x) = v_1(x) y_1(x) + \cdots + v_n(x) y_n(x)$

and try to determine the functions $v_1, \ldots, v_n$.

There are n unknown functions, so we will need n conditions (equations) to determine them. These conditions are obtained as follows. Differentiating y_p in (4) gives

(5) $y_p' = (v_1 y_1' + \cdots + v_n y_n') + (v_1' y_1 + \cdots + v_n' y_n) \,.$

To prevent second derivatives of the unknowns $v_1, \ldots, v_n$ from entering the formula for y_p'', we impose the condition

$$v_1' y_1 + \cdots + v_n' y_n = 0 \,.$$

In a like manner, as we compute $y_p'', y_p''', \ldots, y_p^{(n-1)}$, we impose $(n-2)$ additional conditions involving $v_1', \ldots, v_n'$; namely,

$$v_1' y_1' + \cdots + v_n' y_n' = 0, \ldots, v_1' y_1^{(n-2)} + \cdots + v_n' y_n^{(n-2)} = 0 \,.$$

Finally, the nth condition that we impose is that y_p satisfy the given equation (1). Using the previous conditions and the fact that $y_1, \ldots, y_n$ are solutions to the homogeneous equation, then $L[y_p] = g$ reduces to

(6) $\qquad v_1' y_1^{(n-1)} + \cdots + v_n' y_n^{(n-1)} = g$

(see Problem 12, page 341). We therefore seek n functions $v_1', \ldots, v_n'$ that satisfy the system

(7)
$$
\begin{aligned}
y_1 v_1' + &\cdots + y_n v_n' = 0 \,, \\
&\vdots \\
y_1^{(n-2)} v_1' + &\cdots + y_n^{(n-2)} v_n' = 0 \,, \\
y_1^{(n-1)} v_1' + &\cdots + y_n^{(n-1)} v_n' = g \,.
\end{aligned}
$$

Caution. This system was derived under the assumption that the coefficient of the highest derivative $y^{(n)}$ in (1) is one. If, instead, the coefficient of this term is the constant a, then in the last equation in (7) the right-hand side becomes g/a.

A sufficient condition for the existence of a solution to system (7) for x in (a, b) is that the determinant of the matrix made up of the coefficients of $v_1', \ldots, v_n'$ be different from zero for all x in (a, b). But this determinant is just the Wronskian:

(8)
$$
\begin{vmatrix}
y_1 & \cdots & y_n \\
\vdots & & \vdots \\
y_1^{(n-2)} & \cdots & y_n^{(n-2)} \\
y_1^{(n-1)} & \cdots & y_n^{(n-1)}
\end{vmatrix} = W[y_1, \ldots, y_n](x) \,,
$$

which is never zero on (a, b) because $\{y_1, \ldots, y_n\}$ is a fundamental solution set. Solving (7) via Cramer's rule (Appendix D), we find

(9) $\qquad v_k'(x) = \dfrac{g(x) W_k(x)}{W[y_1, \ldots, y_n](x)} \,, \qquad k = 1, \ldots, n \,,$

where $W_k(x)$ is the determinant of the matrix obtained from the Wronskian $W[y_1, \ldots y_n](x)$ by replacing the kth column by $\mathrm{col}[0, \ldots, 0, 1]$. Using a cofactor expansion about this column, we can express $W_k(x)$ in terms of an $(n-1)$th-order Wronskian:

(10) $\qquad W_k(x) = (-1)^{n-k} W[y_1, \ldots, y_{k-1}, y_{k+1}, \ldots, y_n](x) \,, \qquad k = 1, \ldots, n \,.$

Integrating $v_k'(x)$ in (9) gives

(11) $\qquad v_k(x) = \displaystyle\int \dfrac{g(x) W_k(x)}{W[y_1, \ldots, y_n](x)} dx \,, \qquad k = 1, \ldots, n \,.$

Finally, substituting the v_k's back into (4), we obtain a particular solution to equation (1):

(12) $\qquad y_p(x) = \displaystyle\sum_{k=1}^{n} y_k(x) \int \dfrac{g(x) W_k(x)}{W[y_1, \ldots, y_n](x)} dx \,.$

Note that in equation (1) we presumed that the coefficient of the leading term, $y^{(n)}$, was unity. If, instead, it is $p_0(x)$, we must replace $g(x)$ by $g(x)/p_0(x)$ in (12).

Although equation (12) gives a neat formula for a particular solution to (1), its implementation requires one to evaluate $n + 1$ determinants and then perform n integrations. This may entail several tedious computations. However, the method does work in cases when the technique of undetermined coefficients does not apply (provided, of course, we know a fundamental solution set).

Example 1 Find a general solution to the Cauchy–Euler equation

(13) $x^3 y''' + x^2 y'' - 2xy' + 2y = x^3 \sin x$, $x > 0$,

given that $\{x, x^{-1}, x^2\}$ is a fundamental solution set to the corresponding homogeneous equation.

Solution An important first step is to divide (13) by x^3 to obtain the standard form

(14) $y''' + \dfrac{1}{x} y'' - \dfrac{2}{x^2} y' + \dfrac{2}{x^3} y = \sin x$, $x > 0$,

from which we see that $g(x) = \sin x$. Since $\{x, x^{-1}, x^2\}$ is a fundamental solution set, we can obtain a particular solution of the form

(15) $y_p(x) = v_1(x) x + v_2(x) x^{-1} + v_3(x) x^2$.

To use formula (12), we must first evaluate the four determinants:

$$W[x, x^{-1}, x^2](x) = \begin{vmatrix} x & x^{-1} & x^2 \\ 1 & -x^{-2} & 2x \\ 0 & 2x^{-3} & 2 \end{vmatrix} = -6x^{-1},$$

$$W_1(x) = (-1)^{(3-1)} W[x^{-1}, x^2](x) = (-1)^2 \begin{vmatrix} x^{-1} & x^2 \\ -x^{-2} & 2x \end{vmatrix} = 3,$$

$$W_2(x) = (-1)^{(3-2)} \begin{vmatrix} x & x^2 \\ 1 & 2x \end{vmatrix} = -x^2,$$

$$W_3(x) = (-1)^{(3-3)} \begin{vmatrix} x & x^{-1} \\ 1 & -x^{-2} \end{vmatrix} = -2x^{-1}.$$

Substituting the above expressions into (12), we find

$$y_p(x) = x \int \frac{(\sin x) 3}{-6x^{-1}} dx + x^{-1} \int \frac{(\sin x)(-x^2)}{-6x^{-1}} dx + x^2 \int \frac{(\sin x)(-2x^{-1})}{-6x^{-1}} dx$$

$$= x \int \left(-\frac{1}{2} x \sin x \right) dx + x^{-1} \int \frac{1}{6} x^3 \sin x \, dx + x^2 \int \frac{1}{3} \sin x \, dx,$$

which after some labor simplifies to

(16) $y_p(x) = \cos x - x^{-1} \sin x + C_1 x + C_2 x^{-1} + C_3 x^2$,

where C_1, C_2, and C_3 denote the constants of integration. Since $\{x, x^{-1}, x^2\}$ is a fundamental solution set for the homogeneous equation, we can take C_1, C_2, and C_3 to be arbitrary constants. The right-hand side of (16) then gives the desired general solution. ◆

In the preceding example, the fundamental solution set $\{x, x^{-1}, x^2\}$ can be derived by substituting $y = x^r$ into the homogeneous equation corresponding to (13) (see Problem 31, Exercises 6.2). However, in dealing with other equations that have variable coefficients, the determination of a fundamental set may be extremely difficult. In Chapter 8 we tackle this problem using power series methods.

6.4 EXERCISES

In Problems 1–6, use the method of variation of parameters to determine a particular solution to the given equation.

1. $y''' - 3y'' + 4y = e^{2x}$

2. $y''' - 2y'' + y' = x$

3. $z''' + 3z'' - 4z = e^{2x}$

4. $y''' - 3y'' + 3y' - y = e^x$

5. $y''' + y' = \tan x$, $0 < x < \pi/2$

6. $y''' + y' = \sec\theta\tan\theta$, $0 < \theta < \pi/2$

7. Find a general solution to the Cauchy–Euler equation

$$x^3 y''' - 3x^2 y'' + 6xy' - 6y = x^{-1}, \qquad x > 0,$$

given that $\{x, x^2, x^3\}$ is a fundamental solution set for the corresponding homogeneous equation.

8. Find a general solution to the Cauchy–Euler equation

$$x^3 y''' - 2x^2 y'' + 3xy' - 3y = x^2, \qquad x > 0,$$

given that $\{x, x\ln x, x^3\}$ is a fundamental solution set for the corresponding homogeneous equation.

9. Given that $\{e^x, e^{-x}, e^{2x}\}$ is a fundamental solution set for the homogeneous equation corresponding to the equation

$$y''' - 2y'' - y' + 2y = g(x),$$

determine a formula involving integrals for a particular solution.

10. Given that $\{x, x^{-1}, x^4\}$ is a fundamental solution set for the homogeneous equation corresponding to the equation

$$x^3 y''' - x^2 y'' - 4xy' + 4y = g(x), \qquad x > 0,$$

determine a formula involving integrals for a particular solution.

11. Find a general solution to the Cauchy–Euler equation

$$x^3 y''' - 3xy' + 3y = x^4 \cos x, \qquad x > 0$$

12. Derive the system (7) in the special case when $n = 3$. [*Hint:* To determine the last equation, require that $L[y_p] = g$ and use the fact that y_1, y_2, and y_3 satisfy the corresponding homogeneous equation.]

13. Show that

$$W_k(x) = (-1)^{(n-k)} W[y_1, \ldots, y_{k-1}, y_{k+1}, \ldots, y_n](x).$$

14. **Deflection of a Beam Under Axial Force.** A uniform beam under a load and subject to a constant axial force is governed by the differential equation

$$y^{(4)}(x) - k^2 y''(x) = q(x), \qquad 0 < x < L,$$

where $y(x)$ is the deflection of the beam, L is the length of the beam, k^2 is proportional to the axial force, and $q(x)$ is proportional to the load (see Figure 6.2).

(a) Show that a general solution can be written in the form

$$y(x) = C_1 + C_2 x + C_3 e^{kx} + C_4 e^{-kx}$$
$$+ \frac{1}{k^2}\int q(x)\, x\, dx - \frac{x}{k^2}\int q(x)\, dx$$
$$+ \frac{e^{kx}}{2k^3}\int q(x) e^{-kx}\, dx - \frac{e^{-kx}}{2k^3}\int q(x) e^{kx}\, dx\,.$$

(b) Show that the general solution in part (a) can be rewritten in the form

$$y(x) = c_1 + c_2 x + c_3 e^{kx} + c_4 e^{-kx}$$
$$+ \int_0^x q(s) G(s, x)\, ds\,,$$

where

$$G(s, x) := \frac{s-x}{k^2} - \frac{\sinh[k(s-x)]}{k^3}\,.$$

(c) Let $q(x) \equiv 1$. First compute the general solution using the formula in part (a) and then using the formula in part (b). Compare these two general solutions with the general solution

$$y(x) = B_1 + B_2 x + B_3 e^{kx} + B_4 e^{-kx} - \frac{1}{2k^2}x^2\,,$$

which one would obtain using the method of undetermined coefficients.

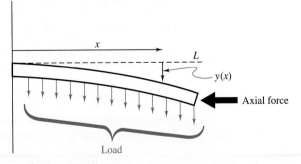

Figure 6.2 Deformation of a beam under axial force and load

Chapter 6 Summary

The theory and techniques for solving an nth-order linear differential equation

$$(1) \qquad y^{(n)} + p_1(x)y^{(n-1)} + \cdots + p_n(x)y = g(x)$$

are natural extensions of the development for linear second-order equations given in Chapter 4. Assuming that $p_1, \ldots, p_n$ and g are continuous functions on an open interval I, there is a unique solution to (1) on I that satisfies the n initial conditions: $y(x_0) = \gamma_0$, $y'(x_0) = \gamma_1, \ldots, y^{(n-1)}(x_0) = \gamma_{n-1}$, where $x_0 \in I$.

For the corresponding homogeneous equation

$$(2) \qquad y^{(n)} + p_1(x)y^{(n-1)} + \cdots + p_n(x)y = 0,$$

there exists a set of n **linearly independent** solutions $\{y_1, \ldots, y_n\}$ on I. Such functions are said to form a **fundamental solution set,** and every solution to (2) can be written as a linear combination of these functions:

$$y(x) = C_1 y_1(x) + C_2 y_2(x) + \cdots + C_n y_n(x).$$

The linear independence of solutions to (2) is equivalent to the nonvanishing on I of the **Wronskian**

$$W[y_1, \ldots, y_n](x) := \det \begin{bmatrix} y_1(x) & \cdots & y_n(x) \\ y_1'(x) & \cdots & y_n'(x) \\ \vdots & & \vdots \\ y_1^{(n-1)}(x) & \cdots & y_n^{(n-1)}(x) \end{bmatrix}.$$

When equation (2) has (real) constant coefficients so that it is of the form

$$(3) \qquad a_n y^{(n)} + a_{n-1} y^{(n-1)} + \cdots + a_0 y = 0, \qquad a_n \neq 0,$$

then the problem of determining a fundamental solution set can be reduced to the algebraic problem of solving the **auxiliary** or **characteristic equation**

$$(4) \qquad a_n r^n + a_{n-1} r^{n-1} + \cdots + a_0 = 0.$$

If the n roots of (4)—say, $r_1, r_2, \ldots, r_n$—are all distinct, then

$$(5) \qquad \{e^{r_1 x}, e^{r_2 x}, \ldots, e^{r_n x}\}$$

is a fundamental solution set for (3). If some real root—say, r_1—occurs with multiplicity m (e.g., $r_1 = r_2 = \cdots = r_m$), then m of the functions in (5) are replaced by

$$e^{r_1 x}, xe^{r_1 x}, \ldots, x^{m-1} e^{r_1 x}.$$

When a complex root $\alpha + i\beta$ to (4) occurs with multiplicity m, then so does its conjugate and $2m$ members of the set (5) are replaced by the real-valued functions

$$e^{\alpha x} \sin \beta x, xe^{\alpha x} \sin \beta x, \ldots, x^{m-1} e^{\alpha x} \sin \beta x,$$
$$e^{\alpha x} \cos \beta x, xe^{\alpha x} \cos \beta x, \ldots, x^{m-1} e^{\alpha x} \cos \beta x.$$

A general solution to the nonhomogeneous equation (1) can be written as

$$y(x) = y_p(x) + y_h(x),$$

where y_p is some particular solution to (1) and y_h is a general solution to the corresponding homogeneous equation. Two useful techniques for finding particular solutions are the **annihilator method** (undetermined coefficients) and the method of **variation of parameters**.

The annihilator method applies to equations of the form

(6) $L[y] = g(x)$,

where L is a linear differential operator with constant coefficients and the forcing term $g(x)$ is a polynomial, exponential, sine, or cosine, or a linear combination of products of these. Such a function $g(x)$ is annihilated (mapped to zero) by a linear differential operator A that also has constant coefficients. Every solution to the nonhomogeneous equation (6) is then a solution to the homogeneous equation $AL[y] = 0$, and, by comparing the solutions of the latter equation with a general solution to $L[y] = 0$, we can obtain the *form* of a particular solution to (6). These forms have previously been studied in Section 4.4 for the method of undetermined coefficients.

The method of variation of parameters is more general in that it applies to arbitrary equations of the form (1). The idea is, starting with a fundamental solution set $\{y_1, \ldots, y_n\}$ for (2), to determine functions $v_1, \ldots, v_n$ such that

(7) $y_p(x) = v_1(x)y_1(x) + \cdots + v_n(x)y_n(x)$

satisfies (1). This method leads to the formula

(8) $y_p(x) = \sum_{k=1}^{n} y_k(x) \int \dfrac{g(x)W_k(x)}{W[y_1, \ldots, y_n](x)} dx$,

where

$$W_k(x) = (-1)^{n-k}W[y_1, \ldots, y_{k-1}, y_{k+1}, \ldots, y_n](x), \qquad k = 1, \ldots, n.$$

REVIEW PROBLEMS FOR CHAPTER 6

1. Determine the intervals for which Theorem 1 on page 319 guarantees the existence of a solution in that interval.
 (a) $y^{(4)} - (\ln x)y'' + xy' + 2y = \cos 3x$
 (b) $(x^2 - 1)y''' + (\sin x)y'' + \sqrt{x+4}\, y' + e^x y = x^2 + 3$

2. Determine whether the given functions are linearly dependent or linearly independent on the interval $(0, \infty)$.
 (a) $\{e^{2x}, x^2 e^{2x}, e^{-x}\}$
 (b) $\{e^x \sin 2x, xe^x \sin 2x, e^x, xe^x\}$
 (c) $\{2e^{2x} - e^x, e^{2x} + 1, e^{2x} - 3, e^x + 1\}$

3. Show that the set of functions $\{\sin x, x\sin x, x^2 \sin x, x^3 \sin x\}$ is linearly independent on $(-\infty, \infty)$.

4. Find a general solution for the given differential equation.
 (a) $y^{(4)} + 2y''' - 4y'' - 2y' + 3y = 0$
 (b) $y''' + 3y'' - 5y' + y = 0$

(c) $y^{(5)} - y^{(4)} + 2y''' - 2y'' + y' - y = 0$
(d) $y''' - 2y'' - y' + 2y = e^x + x$

5. Find a general solution for the homogeneous linear differential equation with constant coefficients whose auxiliary equation is
 (a) $(r+5)^2(r-2)^3(r^2+1)^2 = 0$.
 (b) $r^4(r-1)^2(r^2+2r+4)^2 = 0$.

6. Given that $y_p = \sin(x^2)$ is a particular solution to
 $$y^{(4)} + y = (16x^4 - 11)\sin(x^2) - 48x^2 \cos(x^2)$$
 on $(0, \infty)$, find a general solution.

7. Find a differential operator that annihilates the given function.
 (a) $x^2 - 2x + 5$
 (b) $e^{3x} + x - 1$
 (c) $x \sin 2x$
 (d) $x^2 e^{-2x} \cos 3x$
 (e) $x^2 - 2x + xe^{-x} + \sin 2x - \cos 3x$

8. Use the annihilator method to determine the form of a particular solution for the given equation.

 (a) $y'' + 6y' + 5y = e^{-x} + x^2 - 1$
 (b) $y''' + 2y'' - 19y' - 20y = xe^{-x}$
 (c) $y^{(4)} + 6y'' + 9y = x^2 - \sin 3x$
 (d) $y''' - y'' + 2y = x \sin x$

9. Find a general solution to the Cauchy–Euler equation

$$x^3 y''' - 2x^2 y'' - 5xy' + 5y = x^{-2}, \\ x > 0,$$

given that $\{x, x^5, x^{-1}\}$ is a fundamental solution set to the corresponding homogeneous equation.

10. Find a general solution to the given Cauchy–Euler equation.

 (a) $4x^3 y''' + 8x^2 y'' - xy' + y = 0, \quad x > 0$
 (b) $x^3 y''' + 2x^2 y'' + 2xy' + 4y = 0, \quad x > 0$

TECHNICAL WRITING EXERCISES FOR CHAPTER 6

1. Describe the differences and similarities between second-order and higher-order linear differential equations. Include in your comparisons both theoretical results and the methods of solution. For example, what complications arise in solving higher-order equations that are not present for the second-order case?

2. Explain the relationship between the method of undetermined coefficients and the annihilator method. What difficulties would you encounter in applying the annihilator method if the linear equation did not have constant coefficients?

3. For students with a background in linear algebra: Compare the theory for kth-order linear differential equations with that for systems of n linear equations in n unknowns whose coefficient matrix has rank $n - k$. Use the terminology from linear algebra; for example, subspaces, basis, dimension, linear transformation, and kernel. Discuss both homogeneous and nonhomogeneous equations.

Projects for Chapter 6

A | Computer Algebra Systems and Exponential Shift

Courtesy of Bruce W. Atkinson, Samford University

This project shows that the solution of constant-coefficient equations can be obtained by performing successive integrations, each of which, at worst, involves integration by parts, which is nicely automated with a computer algebra system (CAS). This technique could be programmed on a computer and thus sheds some light on the inner workings of the CAS.

(a) Use mathematical induction to prove the following **exponential shift property**. Given a nonnegative integer n, an n-times differentiable function $y(x)$, and a real number r, then

$$D^n[e^{-rx}y] = e^{-rx}(D - r)^n[y].$$

(b) Let n be a positive integer and r a real number. Use the exponential shift property to prove that the solution of the homogeneous equation

$$(D - r)^n[y] = 0$$

is given by

$$y(x) = (C_0 + C_1 x + \cdots + C_{n-1} x^{n-1}) e^{rx},$$

where $C_0, C_1, \ldots, C_{n-1}$ are real constants.

We can apply the exponential shift property to nonhomogeneous equations.

(c) Use the exponential shift property to find the general solution to the third-order nonhomogeneous equation:

$$y'''(x) + 9y''(x) + 27y'(x) + 27y(x) = (24 + 47x)e^x \cos x$$
$$+ (45 + 52x)e^x \sin x - 6xe^{-3x} \cos x$$
$$+ (-6 + x^2)e^{-3x} \sin x.$$

[*Hint:* Rewrite the equation in operator form; then for the factor of the form $(D - r)^n$, multiply by e^{-rx}, apply the exponential shift property, and perform n integrals using the CAS. These integrals would involve integration by parts if done by hand.]

Note that using the superposition principle together with the method of undetermined coefficients, in part (c) we would have to set up and solve two linear systems, one 4×4 and the other 6×6. For the equation in part (c), this new method is much faster.

We can use a "peeling off" method to solve equations for which the corresponding auxiliary equation has more than one root.

(d) Find a general solution to the equation

$$y^{(4)}(x) - 2y'''(x) - 3y''(x) + 4y'(x) + 4y(x) = 6e^{-x}[-6 + 9x + 2e^{3x}(2 + 12x + 9x^2)].$$

[*Hint:* Rewrite the equation in operator form; then for the first factor of the form $(D - r)^n$, multiply by e^{-rx}, apply the exponential shift property, and perform n integrals using the CAS. Repeat the procedure for each new factor.]

The exponential shift property also holds with r replaced by a complex number $\alpha + i\beta$. In this case we have to use Euler's formula on page 166. Now, suppose the factored form of an operator has a power of $(D - \alpha)^2 + \beta^2$. Then we could write $(D - \alpha)^2 + \beta^2 = (D - z)(D - \bar{z})$, where $z = \alpha + i\beta$.

(e) Find a general solution to the equation

$$[(D - 1)^2 + 4]^2[y] = \sin^2 x \,.$$

[*Hint:* So as to only involve integrating polynomials times complex exponentials, write $\sin x = (e^{ix} - e^{-ix})/(2i)$. But make sure your answer is given as a real-valued function.]

B Justifying the Method of Undetermined Coefficients

The annihilator method discussed in Section 6.3 can be used to derive the rules in Sections 4.4 and 4.5 for the method of undetermined coefficients. To show this, it suffices to work with functions of the form

(1) $g(x) = p_n(x)e^{\alpha x}\cos\beta x + q_m(x)e^{\alpha x}\sin\beta x \,,$

where p_n and q_m are polynomials of degrees n and m, respectively—since the other types in Section 4.5 are just special cases of (1).

Consider the nonhomogeneous equation

(2) $L[y](x) = g(x) \,,$

where L is the linear operator

(3) $L[y] := a_n y^{(n)} + a_{n-1} y^{(n-1)} + \cdots + a_0 y \,,$

with $a_n, a_{n-1}, \ldots, a_0$ constants and $g(x)$ as given in equation (1). Let $N := \max(n, m)$.

(a) Show that

$$A := [(D - \alpha)^2 + \beta^2]^{N+1}$$

is an annihilator for g.

(b) Show that the auxiliary equation associated with $AL[y] = 0$ is of the form

(4) $a_n[(r - \alpha)^2 + \beta^2]^{s+N+1}(r - r_{2s+1}) \cdots (r - r_n) = 0 \,,$

where $s(\geq 0)$ is the multiplicity of $\alpha \pm i\beta$ as roots of the auxiliary equation associated with $L[y] = 0$, and $r_{2s+1}, \ldots, r_n$ are the remaining roots of this equation.

(c) Find a general solution for $AL[y] = 0$ and compare it with a general solution for $L[y] = 0$ to verify that equation (2) has a particular solution of the form

$$y_p(x) = x^s e^{\alpha x}\{P_N(x)\cos\beta x + Q_N(x)\sin\beta x\} \,,$$

where P_N and Q_N are polynomials of degree N.

C Transverse Vibrations of a Beam

In applying elasticity theory to study the transverse vibrations of a beam, one encounters the equation

$$EIy^{(4)}(x) - \gamma\lambda y(x) = 0,$$

where $y(x)$ is related to the displacement of the beam at position x; the constant E is Young's modulus; I is the area moment of inertia, which we assume is constant; γ is the constant mass per unit length of the beam; and λ is a positive parameter to be determined. We can simplify the equation by letting $r^4 := \gamma\lambda/EI$; that is, we consider

(5) $y^{(4)}(x) - r^4 y(x) = 0.$

When the beam is clamped at each end, we seek a solution to (5) that satisfies the boundary conditions

(6) $y(0) = y'(0) = 0$ and $y(L) = y'(L) = 0,$

where L is the length of the beam. The problem is to determine those nonnegative values of r for which equation (5) has a nontrivial solution $(y(x) \not\equiv 0)$ that satisfies (6). To do this, proceed as follows:

(a) Show that there are no nontrivial solutions to the boundary value problem (5)–(6) when $r = 0$.

(b) Represent the general solution to (5) for $r > 0$ in terms of sines, cosines, hyperbolic sines, and hyperbolic cosines.

(c) Substitute the general solution obtained in part (b) into the equations (6) to obtain four linear algebraic equations for the four coefficients appearing in the general solution.

(d) Show that the system of equations in part (c) has nontrivial solutions only for those values of r satisfying

(7) $\cosh(rL) = \sec(rL).$

(e) On the same coordinate system, sketch the graphs of $\cosh(rL)$ and $\sec(rL)$ versus r for $L = 1$ and argue that equation (7) has an infinite number of positive solutions.

 (f) For $L = 1$, determine the first two positive solutions to (7) numerically, and plot the corresponding solutions to the boundary value problem (5)–(6). [*Hint:* You may want to use Newton's method in Appendix B.]

D Higher-Order Difference Equations

Difference equations occur in mathematical models of physical processes and as tools in numerical analysis. The theory of linear difference equations parallels the theory of linear differential equations. Using the results of this chapter as models, both for the statements of theorems and for their proofs, we can develop a theory for linear difference equations.

A Kth order linear difference equation is an equation of the form

(8) $a_K(n)y_{n+K} + a_{K-1}(n)y_{n+K-1} + \cdots + a_1(n)y_{n+1} + a_0(n)y_n = g_n, \qquad n \geq 0,$

where $a_K(n), \ldots, a_0(n)$, and g_n are defined for all nonnegative integers n. By a solution to (8) we mean a sequence of real numbers $\{y_n\}_{n=0}^{\infty}$ that satisfies (8) for all integers $n \geq 0$.

(a) Prove the following existence and uniqueness result.

Theorem Let $c_0, c_1, \ldots, c_{K-1}$ be given constants and assume that $a_K(n) \neq 0$ for all integers $n \geq 0$. Then there exists a unique solution $\{y_n\}$ to (8) that satisfies the initial conditions $y_0 = c_0, y_1 = c_1, \ldots, y_{K-1} = c_{K-1}$. [*Hint:* Begin by solving for y_K in terms of $y_{K-1}, \ldots, y_0$.]

(b) The homogeneous equation associated with (8) is

$$(9) \qquad a_K(n)y_{n+K} + a_{K-1}(n)y_{n+K-1} + \cdots + a_0(n)y_n = 0, \qquad n \geq 0.$$

Prove that any linear combination of solutions to (9) is also a solution to (9). (By a linear combination of sequences $\{y_n\}_{n=0}^{\infty}$ and $\{z_n\}_{n=0}^{\infty}$ we mean a sequence of the form $\{ay_n + bz_n\}_{n=0}^{\infty}$, where a, b are constants.)

(c) An analog of the Wronskian for difference equations is the **Casoratian** of K sequences $\{y_n^{(1)}\}, \{y_n^{(2)}\}, \ldots, \{y_n^{(K)}\}$:

$$(10) \qquad \mathscr{C}_n[\{y_n^{(1)}\}, \ldots, \{y_n^{(K)}\}] := \begin{vmatrix} y_n^{(1)} & y_n^{(2)} & \cdots & y_n^{(K)} \\ y_{n+1}^{(1)} & y_{n+1}^{(2)} & \cdots & y_{n+1}^{(K)} \\ \vdots & \vdots & & \vdots \\ y_{n+K-1}^{(1)} & y_{n+K-1}^{(2)} & \cdots & y_{n+K-1}^{(K)} \end{vmatrix}$$

Prove the following representation theorem.

Theorem Assume $a_K(n) \neq 0$ for all integers $n \geq 0$, and let $\{y_n^{(1)}\}, \ldots, \{y_n^{(K)}\}$ be any K solutions to (9) satisfying

$$(11) \qquad \mathscr{C}_n[\{y_n^{(1)}\}, \ldots, \{y_n^{(K)}\}] \neq 0$$

for all integers $n \geq 0$. Then every solution to (9) can be expressed in the form

$$y_n = C_1 y_n^{(1)} + C_2 y_n^{(2)} + \cdots + C_K y_n^{(K)},$$

where $C_1, \ldots C_K$ are constants.

(d) When the $a_j(n)$'s are independent of n—that is, when $a_K, a_{K-1}, \ldots, a_0$ are constants—then equation (8) (or (9)) is said to have **constant coefficients**. In this case, equation (9) looks like

$$(12) \qquad a_K y_{n+K} + a_{K-1}y_{n+K-1} + \cdots + a_0 y_n = 0 \, .$$

Show that equation (12) has a solution of the form $y_n = r^n$, where $r \neq 0$ is a fixed number (real or complex), if and only if r satisfies the **auxiliary equation**

$$(13) \qquad a_K r^K + a_{K-1}r^{K-1} + \cdots + a_1 r + a_0 = 0.$$

In parts (e) through (h), use the results of parts (c) and (d) and your experience with linear differential equations with constant coefficients to find a "general solution" to the given homogeneous difference equation. When necessary, use Euler's formula to obtain real-valued solutions.

(e) $y_{n+3} + y_{n+2} - 4y_{n+1} - 4y_n = 0$.

(f) $y_{n+3} - 4y_{n+2} + 5y_{n+1} - 2y_n = 0$.

(g) $y_{n+3} - 3y_{n+2} + 4y_{n+1} - 2y_n = 0$.

(h) $y_{n+3} + 3y_{n+2} + 3y_{n+1} + y_n = 0$.

In 1202 the mathematician Fibonacci (Leonardo da Pisa) modeled the growth of a rabbit population by assuming that each mating pair produces one new pair (one male, one female) every month. A newly born pair of rabbits, one male, one female, are put in a field, and they begin mating at the age of one month. (And, in this Utopian setting, they are immortal.)

(i) Show that the numbers of *pairs* of rabbits y_n after n months satisfies the difference equation system

(14) $$y_{n+2} = y_{n+1} + y_n, \qquad y_0 = 0, \qquad y_1 = 1.$$

(j) Equations (14) generate the *Fibonacci sequence* 0, 1, 1, 2, 3, 5, 8, 13, Show that

$$y_n = \frac{\phi^n - (-\phi)^{-n}}{\sqrt{5}}, \text{ where } \phi = \frac{1 + \sqrt{5}}{2} \approx 1.618 .$$

The number ϕ is known as the *golden ratio*. If you rough-sketch a rectangle, without thinking about it, its sides will probably be in the proportion 1.618; it's just an aesthetically appealing dimension. (The Parthenon's dimensions conform to the golden ratio. Next time you visit Athens, you will understand.)

Many biological phenomena can be modeled by Fibonacci's rule (14), and you will see the sequence appearing in tree branching, arrangement of leaves on a stem, pineapple and artichoke fruitlets, ferns, pine cones, honeybee populations, and nautilus spirals. For further discussion visit the website http://www.maths.surrey.ac.uk/hosted-sites/R.Knott/Fibonacci/fibInArt.html

Laplace Transforms

7.1 Introduction: A Mixing Problem

Figure 7.1 depicts a mixing problem with valved input feeders. At time $t = 0$, valve A is opened, delivering 6 L/min of a brine solution containing 0.04 kg of salt per liter. At $t = 10$ min, valve A is closed and valve B is opened, delivering 6 L/min of brine at a concentration of 0.02 kg/L. Initially, 30 kg of salt are dissolved in 1000 L of water in the tank. The exit valve C, which empties the tank at 6 L/min, maintains the contents of the tank at constant volume. Assuming the solution is kept well stirred, determine the amount of salt in the tank at all times $t > 0$.

6 L/min,
0.04 kg/L

A

$x(t)$

B

1000 L

6 L/min,
0.02 kg/L

C

$x(0) = 30$ kg

6 L/min

Figure 7.1 Mixing tank with valve A open

We analyzed a simpler version of this problem in Example 1 of Section 3.2. Let $x(t)$ be the amount of salt (in kilograms) in the tank at time t. Then of course $x(t)/1000$ is the concentration, in kilograms per liter. The salt content is depleted at the rate $(6\ \text{L/min}) \times (x(t)/1000\ \text{kg/L}) = 3x(t)/500$ kg/min through the exit valve. Simultaneously, it is enriched through valves A and B at the rate $g(t)$, given by

(1) $g(t) = \begin{cases} 0.04\ \text{kg/L} \times 6\ \text{L/min} = 0.24\ \text{kg/min}, & 0 < t < 10\ (\text{valve } A), \\ 0.02\ \text{kg/L} \times 6\ \text{L/min} = 0.12\ \text{kg/min}, & t > 10\ (\text{valve } B). \end{cases}$

Thus, $x(t)$ changes at a rate

$$\frac{d}{dt}x(t) = g(t) - \frac{3x(t)}{500},$$

or

(2) $$\frac{dx}{dt} + \frac{3}{500}x = g(t),$$

with initial condition

(3) $x(0) = 30$.

To solve the initial value problem (2)–(3) using the techniques of Chapter 4, we would have to break up the time interval $(0, \infty)$ into two subintervals $(0, 10)$ and $(10, \infty)$. On these subintervals, the nonhomogeneous term $g(t)$ is constant, and the method of undetermined coefficients could be applied to equation (2) to determine general solutions for each subinterval, each containing one arbitrary constant (in the associated homogeneous solutions). The initial condition (3) would fix this constant for $0 < t < 10$, but then we would need to evaluate $x(10)$ and use it to reset the constant in the general solution for $t > 10$.

Our purpose here is to illustrate a new approach using Laplace transforms. As we will see, this method offers several advantages over the previous techniques. For one thing, it is much more convenient in solving initial value problems for linear constant coefficient equations when the forcing term contains jump discontinuities.

The **Laplace transform** of a function $f(t)$, defined on $[0, \infty)$, is given by[†]

(4) $F(s) := \displaystyle\int_0^{\infty} e^{-st} f(t)\, dt$.

Thus we multiply $f(t)$ by e^{-st} and integrate with respect to t from 0 to ∞. This takes a function of t and produces a function of s.

In this chapter we'll scrutinize many of the details on this "exchange of functions," but for now let's simply state the main advantage of executing the transform. *The Laplace transform replaces linear constant coefficient differential equations in the t-domain by (simpler) algebraic equations in the s-domain!* In particular, if $X(s)$ is the Laplace transform of $x(t)$, then the transform of $x'(t)$ is simply $sX(s) - x(0)$. Therefore, the information in the differential equation (2) and initial condition (3) is transformed from the *t*-domain to the *s*-domain simply as

	t-Domain		*s*-Domain

(5) $x'(t) + \dfrac{3}{500} x(t) = g(t) ,\quad x(0) = 30 ;\qquad sX(s) - 30 + \dfrac{3}{500} X(s) = G(s) ,$

where $G(s)$ is the Laplace transform of $g(t)$. (Notice that we have taken certain linearity properties for granted, such as the fact that the transform preserves sums and multiplications by constants.) We can find $X(s)$ in the *s*-domain without solving any differential equations: the solution is simply

(6) $X(s) = \dfrac{30}{s + 3/500} + \dfrac{G(s)}{s + 3/500}$.

For this procedure to be useful, there has to be an easy way to convert from the *t*-domain to the *s*-domain and vice versa. There are, in fact, tables and theorems that facilitate this conversion in many useful circumstances. We'll see, for instance, that the transform of $g(t)$, despite its unpleasant piecewise specification in equation (1), is given by the single formula

$G(s) = \dfrac{0.24}{s} - \dfrac{0.12}{s} e^{-10s}$,

and as a consequence the transform of $x(t)$ equals

$X(s) = \dfrac{30}{s + 3/500} + \dfrac{0.24}{s(s + 3/500)} - \dfrac{0.12 e^{-10s}}{s(s + 3/500)}$.

[†]*Historical Footnote:* The Laplace transform was first introduced by Pierre Laplace in 1779 in his research on probability. G. Doetsch helped develop the use of Laplace transforms to solve differential equations. His work in the 1930s served to justify the operational calculus procedures earlier used by Oliver Heaviside.

Again by table lookup (and a little theory), we can deduce that

(7) $$x(t) = 200 - 170e^{-3t/500} - 100 \cdot \begin{cases} 0, & t \le 10, \\ [1 - e^{-3(t-10)/500}], & t \ge 10. \end{cases}$$

See Figure 7.2.

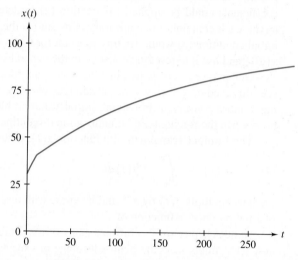

Figure 7.2 Solution to mixing tank example

Note that to arrive at (7) we did not have to take derivatives of trial solutions, break up intervals, or evaluate constants through initial data. The Laplace transform machinery replaces all of these operations by basic algebra: addition, subtraction, multiplication, division—and, of course, the judicious use of the table. Figure 7.3 depicts the advantages of the transform method.

Unfortunately, the Laplace transform method is less helpful with equations containing variable coefficients or nonlinear equations (and sometimes determining inverse transforms can be a Herculean task!). But it is ideally suited for many problems arising in applications. Thus, we devote the present chapter to this important topic.

Figure 7.3 Comparison of solution methods

7.2 Definition of the Laplace Transform

In earlier chapters we studied differential operators. These operators took a function and mapped or transformed it (via differentiation) into another function. The Laplace transform, which is an integral operator, is another such transformation.

Laplace Transform

Definition 1. Let $f(t)$ be a function on $[0, \infty)$. The **Laplace transform** of f is the function F defined by the integral

$$(1) \qquad F(s) := \int_0^\infty e^{-st} f(t)\, dt .$$

The domain of $F(s)$ is all the values of s for which the integral in (1) exists.[†] The Laplace transform of f is denoted by both F and $\mathscr{L}\{f\}$.

Notice that the integral in (1) is an **improper** integral. More precisely,

$$\int_0^\infty e^{-st} f(t)\, dt := \lim_{N \to \infty} \int_0^N e^{-st} f(t)\, dt$$

whenever the limit exists.

Example 1 Determine the Laplace transform of the constant function $f(t) = 1, t \ge 0$.

Solution Using the definition of the transform, we compute

$$F(s) = \int_0^\infty e^{-st} \cdot 1\, dt = \lim_{N \to \infty} \int_0^N e^{-st}\, dt$$

$$= \lim_{N \to \infty} \frac{-e^{-st}}{s} \bigg|_{t=0}^{t=N} = \lim_{N \to \infty} \left[\frac{1}{s} - \frac{e^{-sN}}{s} \right].$$

Since $e^{-sN} \to 0$ when $s > 0$ is fixed and $N \to \infty$, we get

$$F(s) = \frac{1}{s} \qquad \text{for} \qquad s > 0 .$$

When $s \le 0$, the integral $\int_0^\infty e^{-st}\, dt$ diverges. (Why?) Hence $F(s) = 1/s$, with the domain of $F(s)$ being all $s > 0$. ◆

[†]We treat s as real-valued, but in certain applications s may be a complex variable. For a detailed treatment of complex-valued Laplace transforms, see *Complex Variables and the Laplace Transform for Engineers*, by Wilbur R. LePage (Dover Publications, New York, 2010), or *Fundamentals of Complex Analysis with Applications to Engineering and Science* (3rd ed.), by E. B. Saff and A. D. Snider (Pearson Education, Boston, MA, 2003).

Example 2 Determine the Laplace transform of $f(t) = e^{at}$, where a is a constant.

Solution Using the definition of the transform,

$$F(s) = \int_0^\infty e^{-st} e^{at} dt = \int_0^\infty e^{-(s-a)t} dt$$

$$= \lim_{N \to \infty} \int_0^N e^{-(s-a)t} dt = \lim_{N \to \infty} \left. \frac{-e^{-(s-a)t}}{s-a} \right|_0^N$$

$$= \lim_{N \to \infty} \left[\frac{1}{s-a} - \frac{e^{-(s-a)N}}{s-a} \right]$$

$$= \frac{1}{s-a} \quad \text{for} \quad s > a.$$

Again, if $s \le a$ the integral diverges, and hence the domain of $F(s)$ is all $s > a$. ◆

It is comforting to note from Example 2 that the transform of the constant function $f(t) = 1 = e^{0t}$ is $1/(s-0) = 1/s$, which agrees with the solution in Example 1.

Example 3 Find $\mathscr{L}\{\sin bt\}$, where b is a nonzero constant.

Solution We need to compute

$$\mathscr{L}\{\sin bt\}(s) = \int_0^\infty e^{-st} \sin bt \, dt = \lim_{N \to \infty} \int_0^N e^{-st} \sin bt \, dt.$$

Referring to the table of integrals at the back of the book, we see that

$$\mathscr{L}\{\sin bt\}(s) = \lim_{N \to \infty} \left[\frac{e^{-st}}{s^2 + b^2} (-s \sin bt - b \cos bt) \Big|_0^N \right]$$

$$= \lim_{N \to \infty} \left[\frac{b}{s^2 + b^2} - \frac{e^{-sN}}{s^2 + b^2} (s \sin bN + b \cos bN) \right]$$

$$= \frac{b}{s^2 + b^2} \quad \text{for} \quad s > 0$$

(since for such s we have $\lim_{N \to \infty} e^{-sN}(s \sin bN + b \cos bN) = 0$; see Problem 32). ◆

Example 4 Determine the Laplace transform of

$$f(t) = \begin{cases} 2, & 0 < t < 5, \\ 0, & 5 < t < 10, \\ e^{4t}, & 10 < t. \end{cases}$$

Solution Since $f(t)$ is defined by a different formula on different intervals, we begin by breaking up the integral in (1) into three separate parts.[†] Thus,

$$F(s) = \int_0^\infty e^{-st} f(t)\,dt$$

$$= \int_0^5 e^{-st} \cdot 2\,dt + \int_5^{10} e^{-st} \cdot 0\,dt + \int_{10}^\infty e^{-st} e^{4t}\,dt$$

$$= 2\int_0^5 e^{-st}\,dt + \lim_{N\to\infty}\int_{10}^N e^{-(s-4)t}\,dt$$

$$= \frac{2}{s} - \frac{2e^{-5s}}{s} + \lim_{N\to\infty}\left[\frac{e^{-10(s-4)}}{s-4} - \frac{e^{-(s-4)N}}{s-4}\right]$$

$$= \frac{2}{s} - \frac{2e^{-5s}}{s} + \frac{e^{-10(s-4)}}{s-4} \qquad \text{for} \quad s > 4. \;\blacklozenge$$

Notice that the function $f(t)$ of Example 4 has jump discontinuities at $t = 5$ and $t = 10$. These values are reflected in the exponential terms e^{-5s} and e^{-10s} that appear in the formula for $F(s)$. We'll make this connection more precise when we discuss the unit step function in Section 7.6.

An important property of the Laplace transform is its **linearity.** That is, the Laplace transform $\mathcal{L}$ is a linear operator.

Linearity of the Transform

Theorem 1. Let f, f_1, and f_2 be functions whose Laplace transforms exist for $s > \alpha$ and let c be a constant. Then, for $s > \alpha$,

(2) $\mathcal{L}\{f_1 + f_2\} = \mathcal{L}\{f_1\} + \mathcal{L}\{f_2\}\,,$

(3) $\mathcal{L}\{cf\} = c\mathcal{L}\{f\}\,.$

Proof. Using the linearity properties of integration, we have for $s > \alpha$

$$\mathcal{L}\{f_1 + f_2\}(s) = \int_0^\infty e^{-st}[f_1(t) + f_2(t)]\,dt$$

$$= \int_0^\infty e^{-st}f_1(t)\,dt + \int_0^\infty e^{-st}f_2(t)\,dt$$

$$= \mathcal{L}\{f_1\}(s) + \mathcal{L}\{f_2\}(s)\,.$$

Hence, equation (2) is satisfied. In a similar fashion, we see that

$$\mathcal{L}\{cf\}(s) = \int_0^\infty e^{-st}[cf(t)]\,dt = c\int_0^\infty e^{-st}f(t)\,dt$$

$$= c\mathcal{L}\{f\}(s)\,. \;\blacklozenge$$

[†]Notice that $f(t)$ is not defined at the points $t = 0, 5$, and 10. Nevertheless, the integral in (1) is still meaningful and unaffected by the function's values at finitely many points.

Example 5 Determine $\mathscr{L}\{11 + 5e^{4t} - 6\sin 2t\}$.

Solution From the linearity property, we know that the Laplace transform of the sum of any finite number of functions is the sum of their Laplace transforms. Thus,

$$\mathscr{L}\{11 + 5e^{4t} - 6\sin 2t\} = \mathscr{L}\{11\} + \mathscr{L}\{5e^{4t}\} + \mathscr{L}\{-6\sin 2t\}$$

$$= 11\mathscr{L}\{1\} + 5\mathscr{L}\{e^{4t}\} - 6\mathscr{L}\{\sin 2t\} .$$

In Examples 1, 2, and 3, we determined that

$$\mathscr{L}\{1\}(s) = \frac{1}{s}, \quad \mathscr{L}\{e^{4t}\}(s) = \frac{1}{s-4}, \quad \mathscr{L}\{\sin 2t\}(s) = \frac{2}{s^2 + 2^2}.$$

Using these results, we find

$$\mathscr{L}\{11 + 5e^{4t} - 6\sin 2t\}(s) = 11\left(\frac{1}{s}\right) + 5\left(\frac{1}{s-4}\right) - 6\left(\frac{2}{s^2+4}\right)$$

$$= \frac{11}{s} + \frac{5}{s-4} - \frac{12}{s^2+4}.$$

Since $\mathscr{L}\{1\}, \mathscr{L}\{e^{4t}\}$, and $\mathscr{L}\{\sin 2t\}$ are all defined for $s > 4$, so is the transform $\mathscr{L}\{11 + 5e^{4t} - 6\sin 2t\}$. ◆

Table 7.1 lists the Laplace transforms of some of the elementary functions. You should become familiar with these, since they are frequently encountered in solving linear differential equations with constant coefficients. The entries in the table can be derived from the definition of the Laplace transform. A more elaborate table of transforms is given on the inside back cover of this book.

TABLE 7.1 Brief Table of Laplace Transforms

$f(t)$	$F(s) = \mathscr{L}\{f\}(s)$
1	$\dfrac{1}{s}, \quad s>0$
e^{at}	$\dfrac{1}{s-a}, \quad s>a$
$t^n, \quad n=1,2,\ldots$	$\dfrac{n!}{s^{n+1}}, \quad s>0$
$\sin bt$	$\dfrac{b}{s^2+b^2}, \quad s>0$
$\cos bt$	$\dfrac{s}{s^2+b^2}, \quad s>0$
$e^{at}t^n, \quad n=1,2,\ldots$	$\dfrac{n!}{(s-a)^{n+1}}, \quad s>a$
$e^{at}\sin bt$	$\dfrac{b}{(s-a)^2+b^2}, \quad s>a$
$e^{at}\cos bt$	$\dfrac{s-a}{(s-a)^2+b^2}, \quad s>a$

Example 6 Use Table 7.1 to determine $\mathcal{L}\{5t^2 e^{-3t} - e^{12t}\cos 8t\}$.

Solution From the table,

$$\mathcal{L}\{t^2 e^{-3t}\} = \frac{2!}{[s-(-3)]^{2+1}} = \frac{2}{(s+3)^3} \quad \text{for } s > -3,$$

and

$$\mathcal{L}\{e^{12t}\cos 8t\} = \frac{s-12}{(s-12)^2 + 8^2} \quad \text{for } s > 12 .$$

Therefore, by linearity,

$$\mathcal{L}\{5t^2 e^{-3t} - e^{12t}\cos 8t\} = \frac{10}{(s+3)^3} - \frac{s-12}{(s-12)^2 + 64} \quad \text{for } s > 12 . \quad \blacklozenge$$

Existence of the Transform

There are functions for which the improper integral in (1) fails to converge for any value of s. For example, this is the case for the function $f(t) = 1/t$, which grows too fast near zero. Likewise, no Laplace transform exists for the function $f(t) = e^{t^2}$, which increases too rapidly as $t \to \infty$. Fortunately, the set of functions for which the Laplace transform is defined includes many of the functions that arise in applications involving linear differential equations. We now discuss some properties that will (collectively) ensure the existence of the Laplace transform.

A function $f(t)$ on $[a, b]$ is said to have a **jump discontinuity** at $t_0 \in (a, b)$ if $f(t)$ is discontinuous at t_0, but the one-sided limits

$$\lim_{t \to t_0^-} f(t) \quad \text{and} \quad \lim_{t \to t_0^+} f(t)$$

exist as finite numbers. We have encountered jump discontinuities in Example 4 (page 354) and in the input to the mixing tank in Section 7.1 (page 350). If the discontinuity occurs at an endpoint, $t_0 = a$ (or b), a jump discontinuity occurs if the one-sided limit of $f(t)$ as $t \to a^+$ ($t \to b^-$) exists as a finite number. We can now define piecewise continuity.

Piecewise Continuity

Definition 2. A function $f(t)$ is said to be **piecewise continuous on a finite interval** $[a, b]$ if $f(t)$ is continuous at every point in $[a, b]$, except possibly for a finite number of points at which $f(t)$ has a jump discontinuity.

A function $f(t)$ is said to be **piecewise continuous on $[0, \infty)$** if $f(t)$ is piecewise continuous on $[0, N]$ for all $N > 0$.

Example 7 Show that

$$f(t) = \begin{cases} t, & 0 < t < 1, \\ 2, & 1 < t < 2, \\ (t-2)^2, & 2 \le t \le 3, \end{cases}$$

whose graph is sketched in Figure 7.4 (on page 358), is piecewise continuous on $[0, 3]$.

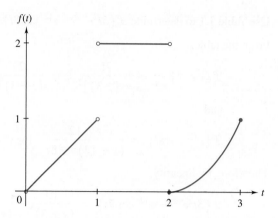

Figure 7.4 Graph of $f(t)$ in Example 7

Solution From the graph of $f(t)$ we see that $f(t)$ is continuous on the intervals $(0, 1)$, $(1, 2)$, and $(2, 3]$. Moreover, at the points of discontinuity, $t = 0, 1$, and 2, the function has jump discontinuities, since the one-sided limits exist as finite numbers. In particular, at $t = 1$, the left-hand limit is 1 and the right-hand limit is 2. Therefore $f(t)$ is piecewise continuous on $[0, 3]$. ◆

Observe that the function $f(t)$ of Example 4 on page 354 is piecewise continuous on $[0, \infty)$ because it is piecewise continuous on every finite interval of the form $[0, N]$, with $N > 0$. In contrast, the function $f(t) = 1/t$ is not piecewise continuous on any interval containing the origin, since it has an "infinite jump" at the origin (see Figure 7.5).

A function that is piecewise continuous on a *finite* interval is necessarily integrable over that interval. However, piecewise continuity on $[0, \infty)$ is not enough to guarantee the existence (as a finite number) of the improper integral over $[0, \infty)$; we also need to consider the

Figure 7.5 Infinite jump at origin

growth of the integrand for large t. Roughly speaking, we'll show that the Laplace transform of a piecewise continuous function exists, provided the function does not grow "faster than an exponential."

Exponential Order α

Definition 3. A function $f(t)$ is said to be of **exponential order** α if there exist positive constants T and M such that

(4) $|f(t)| \leq Me^{\alpha t}$, for all $t \geq T$.

For example, $f(t) = e^{5t} \sin 2t$ is of exponential order $\alpha = 5$ since

$$|e^{5t} \sin 2t| \leq e^{5t},$$

and hence (4) holds with $M = 1$ and T any positive constant.

We use the phrase $f(t)$ *is of exponential order* to mean that for *some* value of α, the function $f(t)$ satisfies the conditions of Definition 3; that is, $f(t)$ grows no faster than a function of the form $Me^{\alpha t}$. The function e^{t^2} is *not* of exponential order. To see this, observe that

$$\lim_{t \to \infty} \frac{e^{t^2}}{e^{\alpha t}} = \lim_{t \to \infty} e^{t(t - \alpha)} = +\infty$$

for any α. Consequently, e^{t^2} grows faster than $e^{\alpha t}$ for every choice of α.

The functions usually encountered in solving linear differential equations with constant coefficients (e.g., polynomials, exponentials, sines, and cosines) are both piecewise continuous and of exponential order. As we now show, the Laplace transforms of such functions exist for large enough values of s.

Conditions for Existence of the Transform

Theorem 2. If $f(t)$ is piecewise continuous on $[0, \infty)$ and of exponential order α, then $\mathcal{L}\{f\}(s)$ exists for $s > \alpha$.

Proof. We need to show that the integral

$$\int_0^\infty e^{-st} f(t)\, dt$$

converges for $s > \alpha$. We begin by breaking up this integral into two separate integrals:

(5) $\displaystyle \int_0^T e^{-st} f(t)\, dt + \int_T^\infty e^{-st} f(t)\, dt$,

where T is chosen so that inequality (4) holds. The first integral in (5) exists because $f(t)$ and hence $e^{-st} f(t)$ are piecewise continuous on the interval $[0, T]$ for any fixed s. To see that the second integral in (5) converges, we use the **comparison test for improper integrals.**

Since $f(t)$ is of exponential order α, we have for $t \geq T$

$$|f(t)| \leq Me^{\alpha t},$$

and hence

$$\left|e^{-st}f(t)\right| = e^{-st}\left|f(t)\right| \le Me^{-(s-\alpha)t},$$

for all $t \ge T$. Now for $s > \alpha$.

$$\int_T^\infty Me^{-(s-\alpha)t}\,dt = M\int_T^\infty e^{-(s-\alpha)t}\,dt = \frac{Me^{-(s-\alpha)T}}{s-\alpha} < \infty.$$

Since $\left|e^{-st}f(t)\right| \le Me^{-(s-\alpha)t}$ for $t \ge T$ and the improper integral of the larger function converges for $s > \alpha$, then, by the comparison test, the integral

$$\int_T^\infty e^{-st}f(t)\,dt$$

converges for $s > \alpha$. Finally, because the two integrals in (5) exist, the Laplace transform $\mathcal{L}\{f\}(s)$ exists for $s > \alpha$. ◆

7.2 EXERCISES

In Problems 1–12, use Definition 1 to determine the Laplace transform of the given function.

1. t

2. t^2

3. e^{6t}

4. te^{3t}

5. $\cos 2t$

6. $\cos bt$, b a constant

7. $e^{2t}\cos 3t$

8. $e^{-t}\sin 2t$

9. $f(t) = \begin{cases} 0, & 0 < t < 2, \\ t, & 2 < t \end{cases}$

10. $f(t) = \begin{cases} 1-t, & 0 < t < 1, \\ 0, & 1 < t \end{cases}$

11. $f(t) = \begin{cases} \sin t, & 0 < t < \pi, \\ 0, & \pi < t \end{cases}$

12. $f(t) = \begin{cases} e^{2t}, & 0 < t < 3, \\ 1, & 3 < t \end{cases}$

In Problems 13–20, use the Laplace transform table and the linearity of the Laplace transform to determine the following transforms.

13. $\mathcal{L}\{6e^{-3t} - t^2 + 2t - 8\}$

14. $\mathcal{L}\{5 - e^{2t} + 6t^2\}$

15. $\mathcal{L}\{t^3 - te^t + e^{4t}\cos t\}$

16. $\mathcal{L}\{t^2 - 3t - 2e^{-t}\sin 3t\}$

17. $\mathcal{L}\{e^{3t}\sin 6t - t^3 + e^t\}$

18. $\mathcal{L}\{t^4 - t^2 - t + \sin\sqrt{2}t\}$

19. $\mathcal{L}\{t^4 e^{5t} - e^t \cos\sqrt{7}t\}$

20. $\mathcal{L}\{e^{-2t}\cos\sqrt{3}t - t^2e^{-2t}\}$

In Problems 21–28, determine whether $f(t)$ is continuous, piecewise continuous, or neither on $[0, 10]$ and sketch the graph of $f(t)$.

21. $f(t) = \begin{cases} 1, & 0 \le t \le 1, \\ (t-2)^2, & 1 < t \le 10 \end{cases}$

22. $f(t) = \begin{cases} 0, & 0 \le t < 2, \\ t, & 2 \le t \le 10 \end{cases}$

23. $f(t) = \begin{cases} 1, & 0 \le t < 1, \\ t-1, & 1 < t < 3, \\ t^2 - 4, & 3 < t \le 10 \end{cases}$

24. $f(t) = \dfrac{t^2 - 3t + 2}{t^2 - 4}$

25. $f(t) = \dfrac{t^2 - t - 20}{t^2 + 7t + 10}$

26. $f(t) = \dfrac{t}{t^2 - 1}$

27. $f(t) = \begin{cases} 1/t, & 0 < t < 1, \\ 1, & 1 \le t \le 2, \\ 1-t, & 2 < t \le 10 \end{cases}$

28. $f(t) = \begin{cases} \dfrac{\sin t}{t}, & t \ne 0, \\ 1, & t = 0 \end{cases}$

29. Which of the following functions are of exponential order?
 (a) $t^3 \sin t$ (b) $100e^{49t}$ (c) e^{t^3}
 (d) $t \ln t$ (e) $\cosh(t^2)$ (f) $\dfrac{1}{t^2+1}$
 (g) $\sin(t^2) + t^4 e^{6t}$ (h) $3 - e^{t^2} + \cos 4t$
 (i) $\exp\{t^2/(t+1)\}$ (j) $\sin(e^{t^2}) + e^{\sin t}$

30. For the transforms $F(s)$ in Table 7.1, what can be said about $\lim_{s\to\infty} F(s)$?

31. Thanks to Euler's formula (page 166) and the algebraic properties of complex numbers, several of the entries of Table 7.1 can be derived from a single formula; namely,

(6) $\mathcal{L}\{e^{(a+ib)t}\}(s) = \dfrac{s - a + ib}{(s - a)^2 + b^2}$, $s > a$.

(a) By computing the integral in the definition of the Laplace transform on page 353 with $f(t) = e^{(a+ib)t}$, show that

$\mathcal{L}\{e^{(a+ib)t}\}(s) = \dfrac{1}{s - (a + ib)}$, $s > a$.

(b) Deduce (6) from part (a) by showing that

$\dfrac{1}{s - (a + ib)} = \dfrac{s - a + ib}{(s - a)^2 + b^2}$.

(c) By equating the real and imaginary parts in formula (6), deduce the last two entries in Table 7.1.

32. Prove that for fixed $s > 0$, we have

$$\lim_{N \to \infty} e^{-sN}(s \sin bN + b \cos bN) = 0.$$

33. Prove that if f is piecewise continuous on $[a, b]$ and g is continuous on $[a, b]$, then the product fg is piecewise continuous on $[a, b]$.

7.3 Properties of the Laplace Transform

In the previous section, we defined the Laplace transform of a function $f(t)$ as

$$\mathcal{L}\{f\}(s) := \int_0^\infty e^{-st} f(t)\, dt.$$

Using this definition to get an explicit expression for $\mathcal{L}\{f\}$ requires the evaluation of the improper integral—frequently a tedious task! We have already seen how the linearity property of the transform can help relieve this burden. In this section we discuss some further properties of the Laplace transform that simplify its computation. These new properties will also enable us to use the Laplace transform to solve initial value problems.

Translation in s

Theorem 3. If the Laplace transform $\mathcal{L}\{f\}(s) = F(s)$ exists for $s > \alpha$, then

(1) $\mathcal{L}\{e^{at}f(t)\}(s) = F(s - a)$

for $s > \alpha + a$.

Proof. We simply compute

$$\mathcal{L}\{e^{at}f(t)\}(s) = \int_0^\infty e^{-st} e^{at} f(t)\, dt$$

$$= \int_0^\infty e^{-(s-a)t} f(t)\, dt$$

$$= F(s - a). \quad \blacklozenge$$

Theorem 3 illustrates the effect on the Laplace transform of multiplication of a function $f(t)$ by e^{at}.

Example 1 Determine the Laplace transform of $e^{at}\sin bt$.

Solution In Example 3 in Section 7.2, page 354, we found that

$$\mathscr{L}\{\sin bt\}(s) = F(s) = \frac{b}{s^2 + b^2}.$$

Thus, by the translation property of $F(s)$, we have

$$\mathscr{L}\{e^{at}\sin bt\}(s) = F(s-a) = \frac{b}{(s-a)^2 + b^2}. \quad \blacklozenge$$

Laplace Transform of the Derivative

Theorem 4. Let $f(t)$ be continuous on $[0, \infty)$ and $f'(t)$ be piecewise continuous on $[0, \infty)$, with both of exponential order α. Then, for $s > \alpha$,

(2) $\mathscr{L}\{f'\}(s) = s\mathscr{L}\{f\}(s) - f(0)$.

Proof. Since $\mathscr{L}\{f'\}$ exists, we can use integration by parts $\big[$with $u = e^{-st}$ and $dv = f'(t)dt\big]$ to obtain

(3) $$\mathscr{L}\{f'\}(s) = \int_0^\infty e^{-st}f'(t)\,dt = \lim_{N\to\infty}\int_0^N e^{-st}f'(t)\,dt$$

$$= \lim_{N\to\infty}\left[e^{-st}f(t)\Big|_0^N + s\int_0^N e^{-st}f(t)\,dt\right]$$

$$= \lim_{N\to\infty} e^{-sN}f(N) - f(0) + s\lim_{N\to\infty}\int_0^N e^{-st}f(t)\,dt$$

$$= \lim_{N\to\infty} e^{-sN}f(N) - f(0) + s\mathscr{L}\{f\}(s).$$

To evaluate $\lim_{N\to\infty} e^{-sN}f(N)$, we observe that since $f(t)$ is of exponential order α, there exists a constant M such that for N large,

$$\left|e^{-sN}f(N)\right| \le e^{-sN}Me^{\alpha N} = Me^{-(s-\alpha)N}.$$

Hence, for $s > \alpha$,

$$0 \le \lim_{N\to\infty}\left|e^{-sN}f(N)\right| \le \lim_{N\to\infty} Me^{-(s-\alpha)N} = 0,$$

so

$$\lim_{N\to\infty} e^{-sN}f(N) = 0$$

for $s > \alpha$. Equation (3) now reduces to

$$\mathscr{L}\{f'\}(s) = s\mathscr{L}\{f\}(s) - f(0). \quad \blacklozenge$$

Using induction, we can extend the last theorem to higher-order derivatives of $f(t)$. For example,

$$\mathscr{L}\{f''\}(s) = s\mathscr{L}\{f'\}(s) - f'(0)$$
$$= s[s\mathscr{L}\{f\}(s) - f(0)] - f'(0),$$

which simplifies to

$$\mathcal{L}\{f''\}(s) = s^2\mathcal{L}\{f\}(s) - sf(0) - f'(0).$$

In general, we obtain the following result.

Laplace Transform of Higher-Order Derivatives

Theorem 5. Let $f(t), f'(t), \ldots, f^{(n-1)}(t)$ be continuous on $[0, \infty)$ and let $f^{(n)}(t)$ be piecewise continuous on $[0, \infty)$, with all these functions of exponential order α. Then, for $s > \alpha$,

(4) $$\mathcal{L}\{f^{(n)}\}(s) = s^n\mathcal{L}\{f\}(s) - s^{n-1}f(0) - s^{n-2}f'(0) - \cdots - f^{(n-1)}(0).$$

The last two theorems shed light on the reason why the Laplace transform is such a useful tool in solving initial value problems. Roughly speaking, they tell us that by using the Laplace transform we can replace "differentiation with respect to t" with "multiplication by s," thereby converting a differential equation into an algebraic one. This idea is explored in Section 7.5. For now, we show how Theorem 4 can be helpful in computing a Laplace transform.

Example 2 Using Theorem 4 and the fact that

$$\mathcal{L}\{\sin bt\}(s) = \frac{b}{s^2 + b^2},$$

determine $\mathcal{L}\{\cos bt\}$.

Solution Let $f(t) := \sin bt$. Then $f(0) = 0$ and $f'(t) = b\cos bt$. Substituting into equation (2), we have

$$\mathcal{L}\{f'\}(s) = s\mathcal{L}\{f\}(s) - f(0),$$
$$\mathcal{L}\{b\cos bt\}(s) = s\mathcal{L}\{\sin bt\}(s) - 0,$$
$$b\mathcal{L}\{\cos bt\}(s) = \frac{sb}{s^2 + b^2}.$$

Dividing by b gives

$$\mathcal{L}\{\cos bt\}(s) = \frac{s}{s^2 + b^2}. \quad \blacklozenge$$

Example 3 Prove the following identity for continuous functions $f(t)$ (assuming the transforms exist):

(5) $$\mathcal{L}\left\{\int_0^t f(\tau)d\tau\right\}(s) = \frac{1}{s}\mathcal{L}\{f(t)\}(s).$$

Use it to verify the solution to Example 2.

Solution Define the function $g(t)$ by the integral

$$g(t) := \int_0^t f(\tau)\,d\tau.$$

364 Chapter 7 Laplace Transforms

Observe that $g(0) = 0$ and $g'(t) = f(t)$. Thus, if we apply Theorem 4 to $g(t)$ [instead of $f(t)$], equation (2) on page 362 reads

$$\mathcal{L}\{f(t)\}(s) = s\mathcal{L}\left\{\int_0^t f(\tau)\,d\tau\right\}(s) - 0,$$

which is equivalent to equation (5).

Now since

$$\sin bt = \int_0^t b\cos b\tau\, d\tau,$$

equation (5) predicts

$$\mathcal{L}\{\sin bt\}(s) = \frac{1}{s}\mathcal{L}\{b\cos bt\}(s) = \frac{b}{s}\mathcal{L}\{\cos bt\}(s).$$

This identity is indeed valid for the transforms in Example 2. ◆

Another question arises concerning the Laplace transform. If $F(s)$ is the Laplace transform of $f(t)$, is $F'(s)$ also a Laplace transform of some function of t? The answer is yes:

$$F'(s) = \mathcal{L}\{-tf(t)\}(s).$$

In fact, the following more general assertion holds.

Derivatives of the Laplace Transform

Theorem 6. Let $F(s) = \mathcal{L}\{f\}(s)$ and assume $f(t)$ is piecewise continuous on $[0, \infty)$ and of exponential order α. Then, for $s > \alpha$,

(6) $$\mathcal{L}\{t^n f(t)\}(s) = (-1)^n \frac{d^n F}{ds^n}(s).$$

Proof. Consider the identity

$$\frac{dF}{ds}(s) = \frac{d}{ds}\int_0^\infty e^{-st}f(t)\,dt.$$

Because of the assumptions on $f(t)$, we can apply a theorem from advanced calculus (sometimes called **Leibniz's rule**) to interchange the order of integration and differentiation:

$$\frac{dF}{ds}(s) = \int_0^\infty \frac{d}{ds}(e^{-st})f(t)\,dt$$

$$= -\int_0^\infty e^{-st}tf(t)\,dt = -\mathcal{L}\{tf(t)\}(s).$$

Thus,

$$\mathcal{L}\{tf(t)\}(s) = (-1)\frac{dF}{ds}(s).$$

The general result (6) now follows by induction on n. ◆

A consequence of the above theorem is that if $f(t)$ is piecewise continuous and of exponential order, then its transform $F(s)$ has derivatives of all orders.

Example 4 Determine $\mathcal{L}\{t\sin bt\}$.

Solution We already know that

$$\mathcal{L}\{\sin bt\}(s) = F(s) = \frac{b}{s^2 + b^2}.$$

Differentiating $F(s)$, we obtain

$$\frac{dF}{ds}(s) = \frac{-2bs}{(s^2 + b^2)^2}.$$

Hence, using formula (6), we have

$$\mathcal{L}\{t\sin bt\}(s) = -\frac{dF}{ds}(s) = \frac{2bs}{(s^2 + b^2)^2}. \quad \blacklozenge$$

For easy reference, Table 7.2 lists some of the basic properties of the Laplace transform derived so far.

TABLE 7.2	Properties of Laplace Transforms

$$\mathcal{L}\{f+g\} = \mathcal{L}\{f\} + \mathcal{L}\{g\}.$$

$$\mathcal{L}\{cf\} = c\mathcal{L}\{f\} \qquad \text{for any constant } c.$$

$$\mathcal{L}\{e^{at}f(t)\}(s) = \mathcal{L}\{f\}(s-a).$$

$$\mathcal{L}\{f'\}(s) = s\mathcal{L}\{f\}(s) - f(0).$$

$$\mathcal{L}\{f''\}(s) = s^2\mathcal{L}\{f\}(s) - sf(0) - f'(0).$$

$$\mathcal{L}\{f^{(n)}\}(s) = s^n\mathcal{L}\{f\}(s) - s^{n-1}f(0) - s^{n-2}f'(0) - \cdots - f^{(n-1)}(0).$$

$$\mathcal{L}\{t^nf(t)\}(s) = (-1)^n\frac{d^n}{ds^n}\left(\mathcal{L}\{f\}(s)\right).$$

7.3 EXERCISES

In Problems 1–20, determine the Laplace transform of the given function using Table 7.1 on page 356 and the properties of the transform given in Table 7.2. [Hint: In Problems 12–20, use an appropriate trigonometric identity.]

1. $t^2 + e^t\sin 2t$
2. $3t^2 - e^{2t}$
3. $e^{-t}\cos 3t + e^{6t} - 1$
4. $3t^4 - 2t^2 + 1$
5. $2t^2e^{-t} - t + \cos 4t$
6. $e^{-2t}\sin 2t + e^{3t}t^2$
7. $(t-1)^4$
8. $(1 + e^{-t})^2$
9. $e^{-t}t\sin 2t$
10. $te^{2t}\cos 5t$
11. $\cosh bt$
12. $\sin 3t\cos 3t$
13. $\sin^2 t$
14. $e^{7t}\sin^2 t$
15. $\cos^3 t$
16. $t\sin^2 t$
17. $\sin 2t\sin 5t$
18. $\cos nt\cos mt$, $m \neq n$
19. $\cos nt\sin mt$, $m \neq n$
20. $t\sin 2t\sin 5t$

21. Given that $\mathcal{L}\{\cos bt\}(s) = s/(s^2 + b^2)$, use the translation property to compute $\mathcal{L}\{e^{at}\cos bt\}$.

22. Starting with the transform $\mathcal{L}\{1\}(s) = 1/s$, use formula (6) for the derivatives of the Laplace transform to show that $\mathcal{L}\{t\}(s) = 1/s^2$, $\mathcal{L}\{t^2\}(s) = 2!/s^3$, and, by using induction, that $\mathcal{L}\{t^n\}(s) = n!/s^{n+1}$, $n = 1, 2, \ldots$.

23. Use Theorem 4 on page 362 to show how entry 32 follows from entry 31 in the Laplace transform table on the inside back cover of the text.

24. Show that $\mathcal{L}\{e^{at}t^n\}(s) = n!/(s-a)^{n+1}$ in two ways:
 (a) Use the translation property for $F(s)$.
 (b) Use formula (6) for the derivatives of the Laplace transform.

25. Use formula (6) to help determine

 (a) $\mathcal{L}\{t\cos bt\}$. **(b)** $\mathcal{L}\{t^2\cos bt\}$.

26. Let $f(t)$ be piecewise continuous on $[0,\infty)$ and of exponential order.

 (a) Show that there exist constants K and α such that

$$|f(t)| \le Ke^{\alpha t} \qquad \text{for all } t \ge 0 .$$

 (b) By using the definition of the transform and estimating the integral with the help of part (a), prove that

$$\lim_{s\to\infty}\mathcal{L}\{f\}(s) = 0 .$$

27. Let $f(t)$ be piecewise continuous on $[0,\infty)$ and of exponential order α and assume $\lim_{t\to 0^+}[f(t)/t]$ exists. Show that

$$\mathcal{L}\left\{\frac{f(t)}{t}\right\}(s) = \int_s^\infty F(u)\,du ,$$

 where $F(s) = \mathcal{L}\{f\}(s)$. [*Hint:* First show that $\frac{d}{ds}\mathcal{L}\{f(t)/t\}(s) = -F(s)$ and then use the result of Problem 26.]

28. Verify the identity in Problem 27 for the following functions. (Use the table of Laplace transforms on the inside back cover.)

 (a) $f(t) = t^5$ **(b)** $f(t) = t^{3/2}$

29. The **transfer function** of a linear system is defined as the ratio of the Laplace transform of the output function $y(t)$ to the Laplace transform of the input function $g(t)$, when all initial conditions are zero. If a linear system is governed by the differential equation

$$y''(t) + 6y'(t) + 10y(t) = g(t) , \qquad t > 0 ,$$

use the linearity property of the Laplace transform and Theorem 5 on page 363 on the Laplace transform of higher-order derivatives to determine the transfer function $H(s) = Y(s)/G(s)$ for this system.

30. Find the transfer function, as defined in Problem 29, for the linear system governed by

$$y''(t) + 5y'(t) + 6y(t) = g(t) , \qquad t > 0 .$$

31. **Translation in t.** Show that for $c > 0$, the translated function

$$g(t) = \begin{cases} 0 , & 0 < t < c , \\ f(t-c) , & c < t \end{cases}$$

has Laplace transform

$$\mathcal{L}\{g\}(s) = e^{-cs}\mathcal{L}\{f\}(s) .$$

In Problems 32–35, let $g(t)$ be the given function $f(t)$ translated to the right by c units. Sketch $f(t)$ and $g(t)$ and find $\mathcal{L}\{g(t)\}(s)$. (See Problem 31.)

32. $f(t) \equiv 1$, $c = 2$

33. $f(t) = t$, $c = 1$

34. $f(t) = \sin t$, $c = \pi$

35. $f(t) = \sin t$, $c = \pi/2$

36. Use equation (5) to provide another derivation of the formula $\mathcal{L}\{t^n\}(s) = n!/s^{n+1}$. [*Hint:* Start with $\mathcal{L}\{1\}(s) = 1/s$ and use induction.]

37. **Initial Value Theorem.** Apply the relation

$$(7)\quad \mathcal{L}\{f'\}(s) = \int_0^\infty e^{-st}f'(t)\,dt = s\mathcal{L}\{f\}(s) - f(0)$$

to argue that for any function $f(t)$ whose derivative is piecewise continuous and of exponential order on $[0,\infty)$,

$$f(0) = \lim_{s\to\infty} s\mathcal{L}\{f\}(s) .$$

38. Verify the initial value theorem (Problem 37) for the following functions. (Use the table of Laplace transforms on the inside back cover.)

 (a) 1 **(b)** e^t **(c)** e^{-t} **(d)** $\cos t$
 (e) $\sin t$ **(f)** t^2 **(g)** $t\cos t$

7.4 Inverse Laplace Transform

In Section 7.2 we defined the Laplace transform as an integral operator that maps a function $f(t)$ into a function $F(s)$. In this section we consider the problem of finding the function $f(t)$ when we are given the transform $F(s)$. That is, we seek an **inverse mapping** for the Laplace transform.

 To see the usefulness of such an inverse, let's consider the simple initial value problem

$$(1)\qquad y'' - y = -t ; \qquad y(0) = 0 , \qquad y'(0) = 1 .$$

 If we take the transform of both sides of equation (1) and use the linearity property of the transform, we find

$$\mathcal{L}\{y''\}(s) - Y(s) = -\frac{1}{s^2} ,$$

where $Y(s) := \mathcal{L}\{y\}(s)$. We know the initial values of the solution $y(t)$, so we can use Theorem 5, page 363, on the Laplace transform of higher-order derivatives to express

$$\mathcal{L}\{y''\}(s) = s^2 Y(s) - sy(0) - y'(0) = s^2 Y(s) - 1 .$$

Substituting for $\mathcal{L}\{y''\}(s)$ yields

$$s^2 Y(s) - 1 - Y(s) = -\frac{1}{s^2} .$$

Solving this algebraic equation for $Y(s)$ gives

$$Y(s) = \frac{1 - \left(\dfrac{1}{s^2}\right)}{s^2 - 1} = \frac{s^2 - 1}{s^2(s^2 - 1)} = \frac{1}{s^2} .$$

We now recall that $\mathcal{L}\{t\}(s) = 1/s^2$, and since $Y(s) = \mathcal{L}\{y\}(s)$, we have

$$\mathcal{L}\{y\}(s) = 1/s^2 = \mathcal{L}\{t\}(s) .$$

It therefore seems reasonable to conclude that $y(t) = t$ is the solution to the initial value problem (1). A quick check confirms this!

Notice that in the above procedure, a crucial step is to determine $y(t)$ from its Laplace transform $Y(s) = 1/s^2$. As we noted, $y(t) = t$ is such a function, but it is *not* the only function whose Laplace function is $1/s^2$. For example, the transform of

$$g(t) := \begin{cases} t , & t \neq 6 , \\ 0 , & t = 6 \end{cases}$$

is also $1/s^2$. This is because the transform is an integral, and integrals are not affected by changing a function's values at isolated points. The significant difference between $y(t)$ and $g(t)$ as far as we are concerned is that $y(t)$ is continuous on $[0, \infty)$, whereas $g(t)$ is not. Naturally, we prefer to work with continuous functions, since solutions to differential equations are continuous. Fortunately, it can be shown that if two different functions have the same Laplace transform, at most one of them can be continuous.[†] With this in mind we give the following definition.

Inverse Laplace Transform

Definition 4. Given a function $F(s)$, if there is a function $f(t)$ that is continuous on $[0, \infty)$ and satisfies

(2) $\mathcal{L}\{f\} = F ,$

then we say that $f(t)$ is the **inverse Laplace transform** of $F(s)$ and employ the notation $f = \mathcal{L}^{-1}\{F\}$.

In case every function $f(t)$ satisfying (2) is discontinuous (and hence not a solution of a differential equation), one could choose any one of them to be the inverse transform; the distinction among them has no physical significance. [Indeed, two *piecewise* continuous functions satisfying (2) can only differ at their points of discontinuity.]

[†]For this result and further properties of the Laplace transform and its inverse, we refer you to *Operational Mathematics*, 3rd ed., by R. V. Churchill (McGraw-Hill, New York, 1971).

Naturally the Laplace transform tables will be a great help in determining the inverse Laplace transform of a given function $F(s)$.

Example 1 Determine $\mathcal{L}^{-1}\{F\}$, where

(a) $F(s) = \dfrac{2}{s^3}$. (b) $F(s) = \dfrac{3}{s^2+9}$. (c) $F(s) = \dfrac{s-1}{s^2-2s+5}$.

Solution To compute $\mathcal{L}^{-1}\{F\}$, we refer to the Laplace transform table on page 356.

(a) $\mathcal{L}^{-1}\left\{\dfrac{2}{s^3}\right\}(t) = \mathcal{L}^{-1}\left\{\dfrac{2!}{s^3}\right\}(t) = t^2$

(b) $\mathcal{L}^{-1}\left\{\dfrac{3}{s^2+9}\right\}(t) = \mathcal{L}^{-1}\left\{\dfrac{3}{s^2+3^2}\right\}(t) = \sin 3t$

(c) $\mathcal{L}^{-1}\left\{\dfrac{s-1}{s^2-2s+5}\right\}(t) = \mathcal{L}^{-1}\left\{\dfrac{s-1}{(s-1)^2+2^2}\right\}(t) = e^t\cos 2t$

In part (c) we used the technique of completing the square to rewrite the denominator in a form that we could find in the table. ◆

In practice, we do not always encounter a transform $F(s)$ that exactly corresponds to an entry in the second column of the Laplace transform table. To handle more complicated functions $F(s)$, we use properties of $\mathcal{L}^{-1}$, just as we used properties of $\mathcal{L}$. One such tool is the linearity of the inverse Laplace transform, a property that is inherited from the linearity of the operator $\mathcal{L}$.

Linearity of the Inverse Transform

Theorem 7. Assume that $\mathcal{L}^{-1}\{F\}$, $\mathcal{L}^{-1}\{F_1\}$, and $\mathcal{L}^{-1}\{F_2\}$ exist and are continuous on $[0,\infty)$ and let c be any constant. Then

(3) $\mathcal{L}^{-1}\{F_1+F_2\} = \mathcal{L}^{-1}\{F_1\} + \mathcal{L}^{-1}\{F_2\}$,

(4) $\mathcal{L}^{-1}\{cF\} = c\mathcal{L}^{-1}\{F\}$.

The proof of Theorem 7 is outlined in Problem 37. We illustrate the usefulness of this theorem in the next example.

Example 2 Determine $\mathcal{L}^{-1}\left\{\dfrac{5}{s-6} - \dfrac{6s}{s^2+9} + \dfrac{3}{2s^2+8s+10}\right\}$.

Solution We begin by using the linearity property. Thus,

$$\mathcal{L}^{-1}\left\{\dfrac{5}{s-6} - \dfrac{6s}{s^2+9} + \dfrac{3}{2(s^2+4s+5)}\right\}$$

$$= 5\mathcal{L}^{-1}\left\{\dfrac{1}{s-6}\right\} - 6\mathcal{L}^{-1}\left\{\dfrac{s}{s^2+9}\right\} + \dfrac{3}{2}\mathcal{L}^{-1}\left\{\dfrac{1}{s^2+4s+5}\right\}.$$

Referring to the Laplace transform tables, we see that

$$\mathscr{L}^{-1}\left\{\frac{1}{s-6}\right\}(t) = e^{6t} \qquad \text{and} \qquad \mathscr{L}^{-1}\left\{\frac{s}{s^2+3^2}\right\}(t) = \cos 3t \,.$$

This gives us the first two terms. To determine $\mathscr{L}^{-1}\{1/(s^2+4s+5)\}$, we complete the square of the denominator to obtain $s^2 + 4s + 5 = (s+2)^2 + 1$. We now recognize from the tables that

$$\mathscr{L}^{-1}\left\{\frac{1}{(s+2)^2+1^2}\right\}(t) = e^{-2t}\sin t \,.$$

Hence,

$$\mathscr{L}^{-1}\left\{\frac{5}{s-6} - \frac{6s}{s^2+9} + \frac{3}{2s^2+8s+10}\right\}(t) = 5e^{6t} - 6\cos 3t + \frac{3e^{-2t}}{2}\sin t \,. \; \blacklozenge$$

Example 3 Determine $\mathscr{L}^{-1}\left\{\dfrac{5}{(s+2)^4}\right\}$.

Solution The $(s+2)^4$ in the denominator suggests that we work with the formula

$$\mathscr{L}^{-1}\left\{\frac{n!}{(s-a)^{n+1}}\right\}(t) = e^{at}t^n \,.$$

Here we have $a = -2$ and $n = 3$, so $\mathscr{L}^{-1}\{6/(s+2)^4\}(t) = e^{-2t}t^3$. Using the linearity property, we find

$$\mathscr{L}^{-1}\left\{\frac{5}{(s+2)^4}\right\}(t) = \frac{5}{6}\mathscr{L}^{-1}\left\{\frac{3!}{(s+2)^4}\right\}(t) = \frac{5}{6}e^{-2t}t^3 \,. \; \blacklozenge$$

Example 4 Determine $\mathscr{L}^{-1}\left\{\dfrac{3s+2}{s^2+2s+10}\right\}$.

Solution By completing the square, the quadratic in the denominator can be written as

$$s^2 + 2s + 10 = s^2 + 2s + 1 + 9 = (s+1)^2 + 3^2 \,.$$

The form of $F(s)$ now suggests that we use one or both of the formulas

$$\mathscr{L}^{-1}\left\{\frac{s-a}{(s-a)^2+b^2}\right\}(t) = e^{at}\cos bt \,,$$

$$\mathscr{L}^{-1}\left\{\frac{b}{(s-a)^2+b^2}\right\}(t) = e^{at}\sin bt \,.$$

In this case, $a = -1$ and $b = 3$. The next step is to express

$$(5) \qquad \frac{3s+2}{s^2+2s+10} = A\frac{s+1}{(s+1)^2+3^2} + B\frac{3}{(s+1)^2+3^2} \,,$$

where A, B are constants to be determined. Multiplying both sides of (5) by $s^2 + 2s + 10$ leaves

$$3s + 2 = A(s+1) + 3B = As + (A+3B) \,,$$

which is an identity between two polynomials in s. Equating the coefficients of like terms gives

$$A = 3 \,, \qquad A + 3B = 2 \,,$$

so $A = 3$ and $B = -1/3$. Finally, from (5) and the linearity property, we find

$$\mathcal{L}^{-1}\left\{\frac{3s+2}{s^2+2s+10}\right\}(t) = 3\mathcal{L}^{-1}\left\{\frac{s+1}{(s+1)^2+3^2}\right\}(t) - \frac{1}{3}\mathcal{L}^{-1}\left\{\frac{3}{(s+1)^2+3^2}\right\}(t)$$

$$= 3e^{-t}\cos 3t - \frac{1}{3}e^{-t}\sin 3t. \quad \blacklozenge$$

Given the choice of finding the inverse Laplace transform of

$$F_1(s) = \frac{7s^2+10s-1}{s^3+3s^2-s-3}$$

or of

$$F_2(s) = \frac{2}{s-1} + \frac{1}{s+1} + \frac{4}{s+3},$$

which would you select? No doubt $F_2(s)$ is the easier one. Actually, the two functions $F_1(s)$ and $F_2(s)$ are identical. This can be checked by combining the simple fractions that form $F_2(s)$. Thus, if we are faced with the problem of computing $\mathcal{L}^{-1}$ of a rational function such as $F_1(s)$, we will first express it, as we did $F_2(s)$, as a sum of simple rational functions. This is accomplished by the **method of partial fractions.**

We briefly review this method. Recall from calculus that a rational function of the form $P(s)/Q(s)$, where $P(s)$ and $Q(s)$ are polynomials with the degree of P less than the degree of Q, has a partial fraction expansion whose form is based on the linear and quadratic factors of $Q(s)$. (We assume the coefficients of the polynomials to be real numbers.) There are three cases to consider:

1. Nonrepeated linear factors.
2. Repeated linear factors.
3. Quadratic factors.

1. Nonrepeated Linear Factors

If $Q(s)$ can be factored into a product of distinct linear factors,

$$Q(s) = (s-r_1)(s-r_2)\cdots(s-r_n),$$

where the r_i's are all distinct real numbers, then the partial fraction expansion has the form

$$\frac{P(s)}{Q(s)} = \frac{A_1}{s-r_1} + \frac{A_2}{s-r_2} + \cdots + \frac{A_n}{s-r_n},$$

where the A_i's are real numbers. There are various ways of determining the constants $A_1, \ldots, A_n$. In the next example, we demonstrate two such methods.

Example 5 Determine $\mathcal{L}^{-1}\{F\}$, where

$$F(s) = \frac{7s-1}{(s+1)(s+2)(s-3)}.$$

Solution We begin by finding the partial fraction expansion for $F(s)$. The denominator consists of three distinct linear factors, so the expansion has the form

(6) $$\frac{7s-1}{(s+1)(s+2)(s-3)} = \frac{A}{s+1} + \frac{B}{s+2} + \frac{C}{s-3},$$

where A, B, and C are real numbers to be determined.

One procedure that works for all partial fraction expansions is first to multiply the expansion equation by the denominator of the given rational function. This leaves us with two identical polynomials. Equating the coefficients of s^k leads to a system of linear equations that we can solve to determine the unknown constants. In this example, we multiply (6) by $(s+1)(s+2)(s-3)$ and find

(7) $$7s - 1 = A(s+2)(s-3) + B(s+1)(s-3) + C(s+1)(s+2),^\dagger$$

which reduces to

$$7s - 1 = (A + B + C)s^2 + (-A - 2B + 3C)s + (-6A - 3B + 2C).$$

Equating the coefficients of s^2, s, and 1 gives the system of linear equations

$$A + B + C = 0,$$
$$-A - 2B + 3C = 7,$$
$$-6A - 3B + 2C = -1.$$

Solving this system yields $A = 2$, $B = -3$, and $C = 1$. Hence,

(8) $$\frac{7s - 1}{(s+1)(s+2)(s-3)} = \frac{2}{s+1} - \frac{3}{s+2} + \frac{1}{s-3}.$$

An alternative method for finding the constants A, B, and C from (7) is to choose three values for s and substitute them into (7) to obtain three linear equations in the three unknowns. If we are careful in our choice of the values for s, the system is easy to solve. In this case, equation (7) obviously simplifies if $s = -1, -2$, or 3. Putting $s = -1$ gives

$$-7 - 1 = A(1)(-4) + B(0) + C(0),$$
$$-8 = -4A.$$

Hence $A = 2$. Next, setting $s = -2$ gives

$$-14 - 1 = A(0) + B(-1)(-5) + C(0),$$
$$-15 = 5B,$$

and so $B = -3$. Finally, letting $s = 3$, we similarly find that $C = 1$. In the case of nonrepeated linear factors, the alternative method is easier to use.

Now that we have obtained the partial fraction expansion (8), we use linearity to compute

$$\mathscr{L}^{-1}\left\{\frac{7s - 1}{(s+1)(s+2)(s-3)}\right\}(t) = \mathscr{L}^{-1}\left\{\frac{2}{s+1} - \frac{3}{s+2} + \frac{1}{s-3}\right\}(t)$$

$$= 2\mathscr{L}^{-1}\left\{\frac{1}{s+1}\right\}(t) - 3\mathscr{L}^{-1}\left\{\frac{1}{s+2}\right\}(t)$$

$$+ \mathscr{L}^{-1}\left\{\frac{1}{s-3}\right\}(t)$$

$$= 2e^{-t} - 3e^{-2t} + e^{3t}. \quad \blacklozenge$$

†Rigorously speaking, equation (7) was derived for s different from $-1, -2$, and 3, but by continuity it holds for these values as well.

2. Repeated Linear Factors

If $s - r$ is a factor of $Q(s)$ and $(s - r)^m$ is the highest power of $s - r$ that divides $Q(s)$, then the portion of the partial fraction expansion of $P(s)/Q(s)$ that corresponds to the term $(s - r)^m$ is

$$\frac{A_1}{s - r} + \frac{A_2}{(s - r)^2} + \cdots + \frac{A_m}{(s - r)^m},$$

where the A_i's are real numbers.

Example 6 Determine $\mathscr{L}^{-1}\left\{ \dfrac{s^2 + 9s + 2}{(s - 1)^2(s + 3)} \right\}$.

Solution Since $s - 1$ is a repeated linear factor with multiplicity two and $s + 3$ is a nonrepeated linear factor, the partial fraction expansion has the form

$$\frac{s^2 + 9s + 2}{(s - 1)^2(s + 3)} = \frac{A}{s - 1} + \frac{B}{(s - 1)^2} + \frac{C}{s + 3}.$$

We begin by multiplying both sides by $(s - 1)^2(s + 3)$ to obtain

(9) $s^2 + 9s + 2 = A(s - 1)(s + 3) + B(s + 3) + C(s - 1)^2.$

Now observe that when we set $s = 1$ (or $s = -3$), two terms on the right-hand side of (9) vanish, leaving a linear equation that we can solve for B (or C). Setting $s = 1$ in (9) gives

$$1 + 9 + 2 = A(0) + 4B + C(0),$$
$$12 = 4B,$$

and, hence, $B = 3$. Similarly, setting $s = -3$ in (9) gives

$$9 - 27 + 2 = A(0) + B(0) + 16C$$
$$-16 = 16C.$$

Thus, $C = -1$. Finally, to find A, we pick a different value for s, say $s = 0$. Then, since $B = 3$ and $C = -1$, plugging $s = 0$ into (9) yields

$$2 = -3A + 3B + C = -3A + 9 - 1$$

so that $A = 2$. Hence,

(10) $$\frac{s^2 + 9s + 2}{(s - 1)^2(s + 3)} = \frac{2}{s - 1} + \frac{3}{(s - 1)^2} - \frac{1}{s + 3}.$$

We could also have determined the constants A, B, and C by first rewriting equation (9) in the form

$$s^2 + 9s + 2 = (A + C)s^2 + (2A + B - 2C)s + (-3A + 3B + C).$$

Then, equating the corresponding coefficients of s^2, s, and 1 and solving the resulting system, we again find $A = 2$, $B = 3$, and $C = -1$.

Now that we have derived the partial fraction expansion (10) for the given rational function, we can determine its inverse Laplace transform:

$$\mathscr{L}^{-1}\left\{\frac{s^2 + 9s + 2}{(s-1)^2(s+3)}\right\}(t) = \mathscr{L}^{-1}\left\{\frac{2}{s-1} + \frac{3}{(s-1)^2} - \frac{1}{s+3}\right\}(t)$$

$$= 2\mathscr{L}^{-1}\left\{\frac{1}{s-1}\right\}(t) + 3\mathscr{L}^{-1}\left\{\frac{1}{(s-1)^2}\right\}(t)$$

$$- \mathscr{L}^{-1}\left\{\frac{1}{s+3}\right\}(t)$$

$$= 2e^t + 3te^t - e^{-3t}. \quad \blacklozenge$$

3. Quadratic Factors

If $(s-\alpha)^2 + \beta^2$ is a quadratic factor of $Q(s)$ that cannot be reduced to linear factors with real coefficients and m is the highest power of $(s-\alpha)^2 + \beta^2$ that divides $Q(s)$, then the portion of the partial fraction expansion that corresponds to $(s-\alpha)^2 + \beta^2$ is

$$\frac{C_1 s + D_1}{(s-\alpha)^2 + \beta^2} + \frac{C_2 s + D_2}{[(s-\alpha)^2 + \beta^2]^2} + \cdots + \frac{C_m s + D_m}{[(s-\alpha)^2 + \beta^2]^m}.$$

As we saw in Example 4, page 369, it is more convenient to express $C_i s + D_i$ in the form $A_i(s-\alpha) + \beta B_i$ when we look up the Laplace transforms. So let's agree to write this portion of the partial fraction expansion in the equivalent form

$$\frac{A_1(s-\alpha) + \beta B_1}{(s-\alpha)^2 + \beta^2} + \frac{A_2(s-\alpha) + \beta B_2}{[(s-\alpha)^2 + \beta^2]^2} + \cdots + \frac{A_m(s-\alpha) + \beta B_m}{[(s-\alpha)^2 + \beta^2]^m}.$$

Example 7 Determine $\mathscr{L}^{-1}\left\{\dfrac{2s^2 + 10s}{(s^2 - 2s + 5)(s+1)}\right\}$.

Solution We first observe that the quadratic factor $s^2 - 2s + 5$ is irreducible (check the sign of the discriminant in the quadratic formula). Next we write the quadratic in the form $(s-\alpha)^2 + \beta^2$ by completing the square:

$$s^2 - 2s + 5 = (s-1)^2 + 2^2.$$

Since $s^2 - 2s + 5$ and $s + 1$ are nonrepeated factors, the partial fraction expansion has the form

$$\frac{2s^2 + 10s}{(s^2 - 2s + 5)(s+1)} = \frac{A(s-1) + 2B}{(s-1)^2 + 2^2} + \frac{C}{s+1}.$$

When we multiply both sides by the common denominator, we obtain

(11) $2s^2 + 10s = [A(s-1) + 2B](s+1) + C(s^2 - 2s + 5).$

In equation (11), let's put $s = -1$, 1, and 0. With $s = -1$, we find

$$2 - 10 = [A(-2) + 2B](0) + C(8),$$
$$-8 = 8C,$$

and, hence, $C = -1$. With $s = 1$ in (11), we obtain

$$2 + 10 = [A(0) + 2B](2) + C(4),$$

and since $C = -1$, the last equation becomes $12 = 4B - 4$. Thus $B = 4$. Finally, setting $s = 0$ in (11) and using $C = -1$ and $B = 4$ gives

$$0 = [A(-1) + 2B](1) + C(5),$$
$$0 = -A + 8 - 5,$$
$$A = 3.$$

Hence, $A = 3$, $B = 4$, and $C = -1$ so that

$$\frac{2s^2 + 10s}{(s^2 - 2s + 5)(s + 1)} = \frac{3(s - 1) + 2(4)}{(s - 1)^2 + 2^2} - \frac{1}{s + 1}.$$

With this partial fraction expansion in hand, we can immediately determine the inverse Laplace transform:

$$\mathcal{L}^{-1}\left\{\frac{2s^2 + 10s}{(s^2 - 2s + 5)(s + 1)}\right\}(t) = \mathcal{L}^{-1}\left\{\frac{3(s - 1) + 2(4)}{(s - 1)^2 + 2^2} - \frac{1}{s + 1}\right\}(t)$$

$$= 3\mathcal{L}^{-1}\left\{\frac{s - 1}{(s - 1)^2 + 2^2}\right\}(t)$$

$$+ 4\mathcal{L}^{-1}\left\{\frac{2}{(s - 1)^2 + 2^2}\right\}(t) - \mathcal{L}^{-1}\left\{\frac{1}{s + 1}\right\}(t)$$

$$= 3e^t \cos 2t + 4e^t \sin 2t - e^{-t}. \quad \blacklozenge$$

In Section 7.8, we discuss a different method (involving convolutions) for computing inverse transforms that does not require partial fraction decompositions. Moreover, the convolution method is convenient in the case of a rational function with a repeated quadratic factor in the denominator. Other helpful tools are described in Problems 33–36 and 38–43.

7.4 EXERCISES

In Problems 1–10, determine the inverse Laplace transform of the given function.

1. $\dfrac{6}{(s - 1)^4}$

2. $\dfrac{2}{s^2 + 4}$

3. $\dfrac{s + 1}{s^2 + 2s + 10}$

4. $\dfrac{4}{s^2 + 9}$

5. $\dfrac{1}{s^2 + 4s + 8}$

6. $\dfrac{3}{(2s + 5)^3}$

7. $\dfrac{2s + 16}{s^2 + 4s + 13}$

8. $\dfrac{1}{s^5}$

9. $\dfrac{3s - 15}{2s^2 - 4s + 10}$

10. $\dfrac{s - 1}{2s^2 + s + 6}$

In Problems 11–20, determine the partial fraction expansion for the given rational function.

11. $\dfrac{s^2 - 26s - 47}{(s - 1)(s + 2)(s + 5)}$

12. $\dfrac{-s - 7}{(s + 1)(s - 2)}$

13. $\dfrac{-2s^2 - 3s - 2}{s(s + 1)^2}$

14. $\dfrac{-8s^2 - 5s + 9}{(s + 1)(s^2 - 3s + 2)}$

15. $\dfrac{8s - 2s^2 - 14}{(s + 1)(s^2 - 2s + 5)}$

16. $\dfrac{-5s - 36}{(s + 2)(s^2 + 9)}$

17. $\dfrac{3s + 5}{s(s^2 + s - 6)}$

18. $\dfrac{3s^2 + 5s + 3}{s^4 + s^3}$

19. $\dfrac{1}{(s-3)(s^2+2s+2)}$ **20.** $\dfrac{s}{(s-1)(s^2-1)}$

In Problems 21–30, determine $\mathcal{L}^{-1}\{F\}$.

21. $F(s) = \dfrac{6s^2-13s+2}{s(s-1)(s-6)}$

22. $F(s) = \dfrac{s+11}{(s-1)(s+3)}$

23. $F(s) = \dfrac{5s^2+34s+53}{(s+3)^2(s+1)}$

24. $F(s) = \dfrac{7s^2-41s+84}{(s-1)(s^2-4s+13)}$

25. $F(s) = \dfrac{7s^2+23s+30}{(s-2)(s^2+2s+5)}$

26. $F(s) = \dfrac{7s^3-2s^2-3s+6}{s^3(s-2)}$

27. $s^2F(s) - 4F(s) = \dfrac{5}{s+1}$

28. $s^2F(s) + sF(s) - 6F(s) = \dfrac{s^2+4}{s^2+s}$

29. $sF(s) + 2F(s) = \dfrac{10s^2+12s+14}{s^2-2s+2}$

30. $sF(s) - F(s) = \dfrac{2s+5}{s^2+2s+1}$

31. Determine the Laplace transform of each of the following functions:

(a) $f_1(t) = \begin{cases} 0, & t=2, \\ t, & t\neq 2. \end{cases}$

(b) $f_2(t) = \begin{cases} 5, & t=1, \\ 2, & t=6, \\ t, & t\neq 1,6. \end{cases}$

(c) $f_3(t) = t$.

Which of the preceding functions is the inverse Laplace transform of $1/s^2$?

32. Determine the Laplace transform of each of the following functions:

(a) $f_1(t) = \begin{cases} t, & t=1,2,3,\ldots, \\ e^t, & t\neq 1,2,3,\ldots. \end{cases}$

(b) $f_2(t) = \begin{cases} e^t, & t\neq 5,8, \\ 6, & t=5, \\ 0, & t=8. \end{cases}$

(c) $f_3(t) = e^t$.

Which of the preceding functions is the inverse Laplace transform of $1/(s-1)$?

Theorem 6 in Section 7.3 on page 364 can be expressed in terms of the inverse Laplace transform as

$$\mathcal{L}^{-1}\left\{\frac{d^n F}{ds^n}\right\}(t) = (-t)^n f(t),$$

where $f = \mathcal{L}^{-1}\{F\}$. Use this equation in Problems 33–36 to compute $\mathcal{L}^{-1}\{F\}$.

33. $F(s) = \ln\left(\dfrac{s+2}{s-5}\right)$ **34.** $F(s) = \ln\left(\dfrac{s-4}{s-3}\right)$

35. $F(s) = \ln\left(\dfrac{s^2+9}{s^2+1}\right)$ **36.** $F(s) = \arctan(1/s)$

37. Prove Theorem 7, page 368, on the linearity of the inverse transform. [*Hint:* Show that the right-hand side of equation (3) is a continuous function on $[0,\infty)$ whose Laplace transform is $F_1(s) + F_2(s)$.]

38. Residue Computation. Let $P(s)/Q(s)$ be a rational function with deg $P <$ deg Q and suppose $s-r$ is a nonrepeated linear factor of $Q(s)$. Prove that the portion of the partial fraction expansion of $P(s)/Q(s)$ corresponding to $s-r$ is

$$\frac{A}{s-r},$$

where A (called the **residue**) is given by the formula

$$A = \lim_{s\to r}\frac{(s-r)P(s)}{Q(s)} = \frac{P(r)}{Q'(r)}.$$

39. Use the residue computation formula derived in Problem 38 to determine quickly the partial fraction expansion for

$$F(s) = \frac{2s+1}{s(s-1)(s+2)}.$$

40. Heaviside's Expansion Formula.[†] Let $P(s)$ and $Q(s)$ be polynomials with the degree of $P(s)$ less than the degree of $Q(s)$. Let

$$Q(s) = (s-r_1)(s-r_2)\cdots(s-r_n),$$

where the r_i's are distinct real numbers. Show that

$$\mathcal{L}^{-1}\left\{\frac{P}{Q}\right\}(t) = \sum_{i=1}^{n}\frac{P(r_i)}{Q'(r_i)}e^{r_i t}.$$

(c) $f_3(t) = e^t$.

Which of the preceding functions is the inverse Laplace transform of $1/(s-1)$?

[†]*Historical Footnote:* This formula played an important role in the "operational solution" to ordinary differential equations developed by Oliver Heaviside in the 1890s.

41. Use Heaviside's expansion formula derived in Problem 40 to determine the inverse Laplace transform of

$$F(s) = \frac{3s^2 - 16s + 5}{(s+1)(s-3)(s-2)}.$$

42. Complex Residues. Let $P(s)/Q(s)$ be a rational function with deg $P <$ deg Q and suppose $(s - \alpha)^2 + \beta^2$ is a nonrepeated quadratic factor of Q. (That is, $\alpha \pm i\beta$ are complex conjugate zeros of Q.) Prove that the portion of the partial fraction expansion of $P(s)/Q(s)$ corresponding to $(s - \alpha)^2 + \beta^2$ is

$$\frac{A(s - \alpha) + \beta B}{(s - \alpha)^2 + \beta^2},$$

where the **complex residue** $\beta B + i\beta A$ is given by the formula

$$\beta B + i\beta A = \lim_{s \to \alpha + i\beta} \frac{[(s - \alpha)^2 + \beta^2]P(s)}{Q(s)}.$$

(Thus we can determine B and A by taking the real and imaginary parts of the limit and dividing them by β.)

43. Use the residue formulas derived in Problems 38 and 42 to determine the partial fraction expansion for

$$F(s) = \frac{6s^2 + 28}{(s^2 - 2s + 5)(s + 2)}.$$

7.5 Solving Initial Value Problems

Our goal is to show how Laplace transforms can be used to solve initial value problems for linear differential equations. Recall that we have already studied ways of solving such initial value problems in Chapter 4. These previous methods required that we first find a *general solution* of the differential equation and then use the initial conditions to determine the desired solution. As we will see, the method of Laplace transforms leads to the solution of the initial value problem *without* first finding a general solution.

Other advantages to the transform method are worth noting. For example, the technique can easily handle equations involving forcing functions having jump discontinuities, as illustrated in Section 7.1. Further, the method can be used for certain linear differential equations with variable coefficients, a special class of integral equations, systems of differential equations, and partial differential equations.

> ### Method of Laplace Transforms
>
> To solve an initial value problem:
>
> **(a)** Take the Laplace transform of both sides of the equation.
> **(b)** Use the properties of the Laplace transform and the initial conditions to obtain an equation for the Laplace transform of the solution and then solve this equation for the transform.
> **(c)** Determine the inverse Laplace transform of the solution by looking it up in a table or by using a suitable method (such as partial fractions) in combination with the table.

In step (a) we are tacitly assuming the solution is piecewise continuous on $[0, \infty)$ and of exponential order. Once we have obtained the inverse Laplace transform in step (c), we can verify that these tacit assumptions are satisfied.

Example 1 Solve the initial value problem

(1) $y'' - 2y' + 5y = -8e^{-t}$; $y(0) = 2$, $y'(0) = 12$.

Solution The differential equation in (1) is an identity between two functions of t. Hence equality holds for the Laplace transforms of these functions:

$$\mathcal{L}\{y'' - 2y' + 5y\} = \mathcal{L}\{-8e^{-t}\} \, .$$

Using the linearity property of $\mathcal{L}$ and the previously computed transform of the exponential function, we can write

(2) $$\mathcal{L}\{y''\}(s) - 2\mathcal{L}\{y'\}(s) + 5\mathcal{L}\{y\}(s) = \frac{-8}{s+1} \, .$$

Now let $Y(s) := \mathcal{L}\{y\}(s)$. From the formulas for the Laplace transform of higher-order derivatives (see Section 7.3) and the initial conditions in (1), we find

$$\mathcal{L}\{y'\}(s) = sY(s) - y(0) = sY(s) - 2 \, ,$$
$$\mathcal{L}\{y''\}(s) = s^2Y(s) - sy(0) - y'(0) = s^2Y(s) - 2s - 12 \, .$$

Substituting these expressions into (2) and solving for $Y(s)$ yields

$$\left[s^2Y(s) - 2s - 12 \right] - 2\left[sY(s) - 2 \right] + 5Y(s) = \frac{-8}{s+1}$$

$$(s^2 - 2s + 5)Y(s) = 2s + 8 - \frac{8}{s+1}$$

$$(s^2 - 2s + 5)Y(s) = \frac{2s^2 + 10s}{s+1}$$

$$Y(s) = \frac{2s^2 + 10s}{(s^2 - 2s + 5)(s+1)} \, .$$

Our remaining task is to compute the inverse transform of the rational function $Y(s)$. This was done in Example 7 of Section 7.4, page 373, where, using a partial fraction expansion, we found

(3) $$y(t) = 3e^t \cos 2t + 4e^t \sin 2t - e^{-t} \, ,$$

which is the solution to the initial value problem (1). ◆

As a quick check on the accuracy of our computations, the reader is advised to verify that the computed solution satisfies the given initial conditions.

The reader is probably questioning the wisdom of using the Laplace transform method to solve an initial value problem that can be easily handled by the methods discussed in Chapter 4. The objective of the first few examples in this section is simply to make the reader familiar with the Laplace transform procedure. We will see in Example 4 and in later sections that the method is applicable to problems that cannot be readily handled by the techniques discussed in the previous chapters.

Example 2 Solve the initial value problem

(4) $$y'' + 4y' - 5y = te^t \, ; \qquad y(0) = 1 \, , \qquad y'(0) = 0 \, .$$

Solution Let $Y(s) := \mathcal{L}\{y\}(s)$. Taking the Laplace transform of both sides of the differential equation in (4) gives

(5) $$\mathcal{L}\{y''\}(s) + 4\mathcal{L}\{y'\}(s) - 5Y(s) = \frac{1}{(s-1)^2} \, .$$

Using the initial conditions, we can express $\mathscr{L}\{y'\}(s)$ and $\mathscr{L}\{y''\}(s)$ in terms of $Y(s)$. That is,

$$\mathscr{L}\{y'\}(s) = sY(s) - y(0) = sY(s) - 1,$$
$$\mathscr{L}\{y''\}(s) = s^2Y(s) - sy(0) - y'(0) = s^2Y(s) - s.$$

Substituting back into (5) and solving for $Y(s)$ gives

$$[s^2Y(s) - s] + 4[sY(s) - 1] - 5Y(s) = \frac{1}{(s-1)^2}$$

$$(s^2 + 4s - 5)Y(s) = s + 4 + \frac{1}{(s-1)^2}$$

$$(s+5)(s-1)Y(s) = \frac{s^3 + 2s^2 - 7s + 5}{(s-1)^2}$$

$$Y(s) = \frac{s^3 + 2s^2 - 7s + 5}{(s+5)(s-1)^3}.$$

The partial fraction expansion for $Y(s)$ has the form

(6) $$\frac{s^3 + 2s^2 - 7s + 5}{(s+5)(s-1)^3} = \frac{A}{s+5} + \frac{B}{s-1} + \frac{C}{(s-1)^2} + \frac{D}{(s-1)^3}.$$

Solving for the numerators, we ultimately obtain $A = 35/216$, $B = 181/216$, $C = -1/36$, and $D = 1/6$. Substituting these values into (6) gives

$$Y(s) = \frac{35}{216}\left(\frac{1}{s+5}\right) + \frac{181}{216}\left(\frac{1}{s-1}\right) - \frac{1}{36}\left(\frac{1}{(s-1)^2}\right) + \frac{1}{12}\left(\frac{2}{(s-1)^3}\right),$$

where we have written $D = 1/6 = (1/12)2$ to facilitate the final step of taking the inverse transform. From the tables, we now obtain

(7) $$y(t) = \frac{35}{216}e^{-5t} + \frac{181}{216}e^t - \frac{1}{36}te^t + \frac{1}{12}t^2e^t$$

as the solution to the initial value problem (4). ◆

Example 3 Solve the initial value problem

(8) $$w''(t) - 2w'(t) + 5w(t) = -8e^{\pi - t}; \quad w(\pi) = 2, \quad w'(\pi) = 12.$$

Solution To use the method of Laplace transforms, we first move the initial conditions to $t = 0$. This can be done by setting $y(t) := w(t + \pi)$. Then

$$y'(t) = w'(t + \pi), \qquad y''(t) = w''(t + \pi).$$

Replacing t by $t + \pi$ in the differential equation in (8), we have

(9) $$w''(t + \pi) - 2w'(t + \pi) + 5w(t + \pi) = -8e^{\pi - (t + \pi)} = -8e^{-t}.$$

Substituting $y(t) = w(t + \pi)$ in (9), the initial value problem in (8) becomes

$$y''(t) - 2y'(t) + 5y(t) = -8e^{-t}; \quad y(0) = 2, \quad y'(0) = 12.$$

Because the initial conditions are now given at the origin, the Laplace transform method is applicable. In fact, we carried out the procedure in Example 1, page 376, where we found

(10) $$y(t) = 3e^t \cos 2t + 4e^t \sin 2t - e^{-t}.$$

Since $w(t + \pi) = y(t)$, then $w(t) = y(t - \pi)$. Hence, replacing t by $t - \pi$ in (10) gives

$$w(t) = y(t - \pi) = 3e^{t-\pi}\cos[2(t-\pi)] + 4e^{t-\pi}\sin[2(t-\pi)] - e^{-(t-\pi)}$$
$$= 3e^{t-\pi}\cos 2t + 4e^{t-\pi}\sin 2t - e^{\pi-t}. \quad \blacklozenge$$

Thus far we have applied the Laplace transform method only to linear equations with constant coefficients. Yet several important equations in mathematical physics involve linear equations whose coefficients are polynomials in t. To solve such equations using Laplace transforms, we apply Theorem 6, page 364, where we proved that

(11) $$\mathcal{L}\{t^n f(t)\}(s) = (-1)^n \frac{d^n \mathcal{L}\{f\}}{ds^n}(s).$$

If we let $n = 1$ and $f(t) = y'(t)$, we find

$$\mathcal{L}\{ty'(t)\}(s) = -\frac{d}{ds}\mathcal{L}\{y'\}(s)$$

$$= -\frac{d}{ds}[sY(s) - y(0)] = -sY'(s) - Y(s).$$

Similarly, with $n = 1$ and $f(t) = y''(t)$, we obtain from (11)

$$\mathcal{L}\{ty''(t)\}(s) = -\frac{d}{ds}\mathcal{L}\{y''\}(s)$$

$$= -\frac{d}{ds}[s^2Y(s) - sy(0) - y'(0)]$$

$$= -s^2Y'(s) - 2sY(s) + y(0).$$

Thus, we see that for a linear differential equation in $y(t)$ whose coefficients are polynomials in t, the method of Laplace transforms will convert the given equation into a linear differential equation in $Y(s)$ whose coefficients are polynomials in s. Moreover, if the coefficients of the given equation are polynomials of degree 1 in t, then (regardless of the order of the given equation) the differential equation for $Y(s)$ is just a linear *first-order* equation. Since we know how to solve this first-order equation, the only serious obstacle we may encounter is obtaining the inverse Laplace transform of $Y(s)$. [This problem may be insurmountable, since the solution $y(t)$ may *not* have a Laplace transform.]

In illustrating the technique, we make use of the following fact. *If $f(t)$ is piecewise continuous on $[0, \infty)$ and of exponential order, then*

(12) $$\lim_{s \to \infty} \mathcal{L}\{f\}(s) = 0.$$

(You may have already guessed this from the entries in Table 7.1, page 356.) An outline of the proof of (12) is given in Exercises 7.3, page 366, Problem 26.

Example 4 Solve the initial value problem

(13) $$y'' + 2ty' - 4y = 1, \quad y(0) = y'(0) = 0.$$

Solution Let $Y(s) = \mathcal{L}\{y\}(s)$ and take the Laplace transform of both sides of the equation in (13):

(14) $$\mathcal{L}\{y''\}(s) + 2\mathcal{L}\{ty'(t)\}(s) - 4Y(s) = \frac{1}{s}.$$

Using the initial conditions, we find

$$\mathcal{L}\{y''\}(s) = s^2Y(s) - sy(0) - y'(0) = s^2Y(s)$$

and

$$\mathcal{L}\{ty'(t)\}(s) = -\frac{d}{ds}\mathcal{L}\{y'\}(s)$$

$$= -\frac{d}{ds}[sY(s) - y(0)] = -sY'(s) - Y(s) .$$

Substituting these expressions into (14) gives

$$s^2Y(s) + 2[-sY'(s) - Y(s)] - 4Y(s) = \frac{1}{s}$$

$$-2sY'(s) + (s^2 - 6)Y(s) = \frac{1}{s}$$

(15) $$Y'(s) + \left(\frac{3}{s} - \frac{s}{2}\right)Y(s) = \frac{-1}{2s^2} .$$

Equation (15) is a linear first-order equation and has the integrating factor

$$\mu(s) = e^{\int (3/s - s/2)ds} = e^{\ln s^3 - s^2/4} = s^3 e^{-s^2/4}$$

(see Section 2.3). Multiplying (15) by $\mu(s)$, we obtain

$$\frac{d}{ds}\{\mu(s)Y(s)\} = \frac{d}{ds}\{s^3 e^{-s^2/4}Y(s)\} = -\frac{s}{2}e^{-s^2/4} .$$

Integrating and solving for $Y(s)$ yields

$$s^3 e^{-s^2/4}Y(s) = -\int \frac{s}{2}e^{-s^2/4} ds = e^{-s^2/4} + C$$

(16) $$Y(s) = \frac{1}{s^3} + C\frac{e^{s^2/4}}{s^3} .$$

Now if $Y(s)$ is the Laplace transform of a piecewise continuous function of exponential order, then it follows from equation (12) that

$$\lim_{s \to \infty} Y(s) = 0 .$$

For this to occur, the constant C in equation (16) must be zero. Hence, $Y(s) = 1/s^3$, and taking the inverse transform gives $y(t) = t^2/2$. We can easily verify that $y(t) = t^2/2$ is the solution to the given initial value problem by substituting it into (13). ◆

We end this section with an application from **control theory.** Let's consider a servomechanism that models an automatic pilot. Such a mechanism applies a torque to the steering control shaft so that a plane or boat will follow a prescribed course. If we let $y(t)$ be the true direction (angle) of the craft at time t and $g(t)$ be the desired direction at time t, then

$$e(t) := y(t) - g(t)$$

denotes the **error** or **deviation** between the desired direction and the true direction.

Let's assume that the servomechanism can measure the error $e(t)$ and feed back to the steering shaft a component of torque that is proportional to $e(t)$ but opposite in sign (see Figure 7.6 on page 381). Newton's second law, expressed in terms of torques, states that

(**moment of inertia**) × (**angular acceleration**) = total torque.

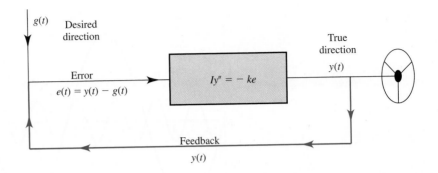

Figure 7.6 Servomechanism with feedback

For the servomechanism described, this becomes

(17) $Iy''(t) = -ke(t)$,

where I is the moment of inertia of the steering shaft and k is a positive proportionality constant.

Example 5 Determine the error $e(t)$ for the automatic pilot if the steering shaft is initially at rest in the zero direction and the desired direction is given by $g(t) = at$, where a is a constant.

Solution Based on the discussion leading to equation (17), a model for the mechanism is given by the initial value problem

(18) $Iy''(t) = -ke(t)$; $y(0) = 0$, $y'(0) = 0$,

where $e(t) = y(t) - g(t) = y(t) - at$. We begin by taking the Laplace transform of both sides of (18):

$$I\mathscr{L}\{y''\}(s) = -k\mathscr{L}\{e\}(s)$$
$$I[s^2 Y(s) - sy(0) - y'(0)] = -kE(s)$$
(19) $$s^2 IY(s) = -kE(s) ,$$

where $Y(s) = \mathscr{L}\{y\}(s)$ and $E(s) = \mathscr{L}\{e\}(s)$. Since

$$E(s) = \mathscr{L}\{y(t) - at\}(s) = Y(s) - \mathscr{L}\{at\}(s) = Y(s) - as^{-2} ,$$

we find from (19) that

$$s^2 IE(s) + aI = -kE(s) .$$

Solving this equation for $E(s)$ gives

$$E(s) = -\frac{aI}{s^2 I + k} = \frac{-a}{\sqrt{k/I}} \frac{\sqrt{k/I}}{s^2 + k/I} .$$

Hence, on taking the inverse Laplace transform, we obtain the error

(20) $$e(t) = -\frac{a}{\sqrt{k/I}} \sin\left(\sqrt{k/I}\, t\right) . \quad \blacklozenge$$

As we can see from equation (20), the automatic pilot will oscillate back and forth about the desired course, always "oversteering" by the factor $a/\sqrt{k/I}$. Clearly, we can make the

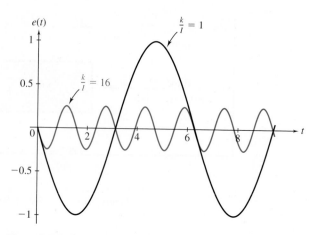

Figure 7.7 Error for automatic pilot when $k/I = 1$ and when $k/I = 16$

error small by making k large relative to I, but then the term $\sqrt{k/I}$ becomes large, causing the error to oscillate more rapidly. (See Figure 7.7.) As with vibrations, the oscillations or over-steering can be controlled by introducing a damping torque proportional to $e'(t)$ but opposite in sign (see Problem 40).

7.5 EXERCISES

In Problems 1–14, solve the given initial value problem using the method of Laplace transforms.

1. $y'' - 2y' + 5y = 0$; $y(0) = 2$, $y'(0) = 4$

2. $y'' - y' - 2y = 0$; $y(0) = -2$, $y'(0) = 5$

3. $y'' + 6y' + 9y = 0$; $y(0) = -1$, $y'(0) = 6$

4. $y'' + 6y' + 5y = 12e^t$; $y(0) = -1$, $y'(0) = 7$

5. $w'' + w = t^2 + 2$; $w(0) = 1$, $w'(0) = -1$

6. $y'' - 4y' + 5y = 4e^{3t}$; $y(0) = 2$, $y'(0) = 7$

7. $y'' - 7y' + 10y = 9\cos t + 7\sin t$;
$y(0) = 5$, $y'(0) = -4$

8. $y'' + 4y = 4t^2 - 4t + 10$;
$y(0) = 0$, $y'(0) = 3$

9. $z'' + 5z' - 6z = 21e^{t-1}$;
$z(1) = -1$, $z'(1) = 9$

10. $y'' - 4y = 4t - 8e^{-2t}$; $y(0) = 0$, $y'(0) = 5$

11. $y'' - y = t - 2$; $y(2) = 3$, $y'(2) = 0$

12. $w'' - 2w' + w = 6t - 2$;
$w(-1) = 3$; $w'(-1) = 7$

13. $y'' - y' - 2y = -8\cos t - 2\sin t$;
$y(\pi/2) = 1$, $y'(\pi/2) = 0$

14. $y'' + y = t$; $y(\pi) = 0$, $y'(\pi) = 0$

In Problems 15–24, solve for $Y(s)$, the Laplace transform of the solution $y(t)$ to the given initial value problem.

15. $y'' - 3y' + 2y = \cos t$; $y(0) = 0$, $y'(0) = -1$

16. $y'' + 6y = t^2 - 1$; $y(0) = 0$, $y'(0) = -1$

17. $y'' + y' - y = t^3$; $y(0) = 1$, $y'(0) = 0$

18. $y'' - 2y' - y = e^{2t} - e^t$; $y(0) = 1$, $y'(0) = 3$

19. $y'' + 5y' - y = e^t - 1$; $y(0) = 1$, $y'(0) = 1$

20. $y'' + 3y = t^3$; $y(0) = 0$, $y'(0) = 0$

21. $y'' - 2y' + y = \cos t - \sin t$; $y(0) = 1$, $y'(0) = 3$

22. $y'' - 6y' + 5y = te^t$; $y(0) = 2$, $y'(0) = -1$
23. $y'' + 4y = g(t)$; $y(0) = -1$, $y'(0) = 0$,
where

$$g(t) = \begin{cases} t, & t < 2, \\ 5, & t > 2 \end{cases}$$

24. $y'' - y = g(t)$; $y(0) = 1$, $y'(0) = 2$,
where

$$g(t) = \begin{cases} 1, & t < 3, \\ t, & t > 3 \end{cases}$$

In Problems 25–28, solve the given third-order initial value problem for $y(t)$ using the method of Laplace transforms.

25. $y''' - y'' + y' - y = 0$;

 $y(0) = 1$, $y'(0) = 1$, $y''(0) = 3$

26. $y''' + 4y'' + y' - 6y = -12$;

 $y(0) = 1$, $y'(0) = 4$, $y''(0) = -2$

27. $y''' + 3y'' + 3y' + y = 0$;

 $y(0) = -4$, $y'(0) = 4$, $y''(0) = -2$

28. $y''' + y'' + 3y' - 5y = 16e^{-t}$;

 $y(0) = 0$, $y'(0) = 2$, $y''(0) = -4$

In Problems 29–32, use the method of Laplace transforms to find a general solution to the given differential equation by assuming $y(0) = a$ and $y'(0) = b$, where a and b are arbitrary constants.

29. $y'' - 4y' + 3y = 0$ **30.** $y'' + 6y' + 5y = t$

31. $y'' + 2y' + 2y = 5$ **32.** $y'' - 5y' + 6y = -6te^{2t}$

33. Use Theorem 6 in Section 7.3, page 364, to show that

$$\mathcal{L}\{t^2 y'(t)\}(s) = sY''(s) + 2Y'(s) ,$$

where $Y(s) = \mathcal{L}\{y\}(s)$.

34. Use Theorem 6 in Section 7.3, page 364, to show that

$$\mathcal{L}\{t^2 y''(t)\}(s) = s^2 Y''(s) + 4sY'(s) + 2Y(s) ,$$

where $Y(s) = \mathcal{L}\{y\}(s)$.

In Problems 35–38, find solutions to the given initial value problem.

35. $y'' + 3ty' - 6y = 1$; $y(0) = 0$, $y'(0) = 0$
36. $ty'' - ty' + y = 2$; $y(0) = 2$, $y'(0) = -1$
37. $ty'' - 2y' + ty = 0$; $y(0) = 1$, $y'(0) = 0$
 [*Hint:* $\mathcal{L}^{-1}\{1/(s^2+1)^2\}(t) = (\sin t - t\cos t)/2$.]
38. $y'' + ty' - y = 0$;

 $y(0) = 0$, $y'(0) = 3$

39. Determine the error $e(t)$ for the automatic pilot in Example 5, page 381, if the shaft is initially at rest in the zero direction and the desired direction is $g(t) = a$, where a is a constant.

40. In Example 5 assume that in order to control oscillations, a component of torque proportional to $e'(t)$, but opposite in sign, is also fed back to the steering shaft. Show that equation (17) is now replaced by

$$Iy''(t) = -ke(t) - \mu e'(t) ,$$

where μ is a positive constant. Determine the error $e(t)$ for the automatic pilot with mild damping (i.e., $\mu < 2\sqrt{Ik}$) if the steering shaft is initially at rest in the zero direction and the desired direction is given by $g(t) = a$, where a is a constant.

41. In Problem 40 determine the error $e(t)$ when the desired direction is given by $g(t) = at$, where a is a constant.

7.6 Transforms of Discontinuous Functions

In this section we study special functions that often arise when the method of Laplace transforms is applied to physical problems. Of particular interest are methods for handling functions with jump discontinuities. As we saw in the mixing problem of Section 7.1, jump discontinuities occur naturally in any physical situation that involves switching. Finding the Laplace transforms of such functions is straightforward; however, we need some theory for inverting these transforms. To facilitate this, Oliver Heaviside introduced the following step function.

Unit Step Function

Definition 5. The **unit step function** $u(t)$ is defined by

$$(1) \qquad u(t) := \begin{cases} 0, & t < 0, \\ 1, & 0 < t. \end{cases}$$

(Any Riemann integral, like the Laplace transform, of a function is unaffected if the integrand's value at a single point is changed by a finite amount. Therefore, we do not specify a value for $u(t)$ at $t = 0$.)

By shifting the argument of $u(t)$, the jump can be moved to a different location. That is,

$$(2) \qquad u(t - a) = \begin{cases} 0, & t - a < 0, \\ 1, & 0 < t - a \end{cases} = \begin{cases} 0, & t < a \\ 1, & a < t \end{cases}$$

has its jump at $t = a$. By multiplying by a constant M, the height of the jump can also be modified:

$$Mu(t - a) = \begin{cases} 0, & t < a, \\ M, & a < t. \end{cases}$$

See Figure 7.8.

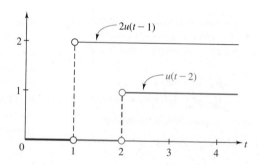

Figure 7.8 Two-step functions expressed using the unit step function

To simplify the formulas for piecewise continuous functions, we employ the rectangular window, which turns the step function on and then turns it back off.

Rectangular Window Function

Definition 6. The **rectangular window function** $\Pi_{a,b}(t)$ is defined by[†]

$$(3) \qquad \Pi_{a,b}(t) := u(t - a) - u(t - b) = \begin{cases} 0, & t < a, \\ 1, & a < t < b, \\ 0, & b < t. \end{cases}$$

[†] Also known as the **square pulse**, or the **boxcar function**.

The function $\Pi_{a,b}(t)$ is displayed in Figure 7.9, and Figure 7.10, illustrating multiplication of a function by $\Pi_{a,b}(t)$, justifies its name.

Figure 7.9 The rectangular window

Figure 7.10 The windowing effect of $\Pi_{a,b}(t)$

Any piecewise continuous function can be expressed in terms of window and step functions.

Example 1 Write the function

$$(4) \qquad f(t) = \begin{cases} 3, & t < 2, \\ 1, & 2 < t < 5, \\ t, & 5 < t < 8, \\ t^2/10, & 8 < t \end{cases}$$

(see Figure 7.11 on page 386) in terms of window and step functions.

Solution Clearly, from the figure we want to window the function in the intervals $(0, 2)$, $(2, 5)$, and $(5, 8)$, and to introduce a step for $t > 8$. From (5) we read off the desired representation as

$$(5) \qquad f(t) = 3\Pi_{0,2}(t) + 1\Pi_{2,5}(t) + t\Pi_{5,8}(t) + (t^2/10)u(t - 8). \quad \blacklozenge$$

The Laplace transform of $u(t - a)$ with $a \geq 0$ is

$$(6) \qquad \mathscr{L}\{u(t - a)\}(s) = \frac{e^{-as}}{s},$$

since, for $s > 0$,

$$\mathscr{L}\{u(t - a)\}(s) = \int_0^\infty e^{-st}u(t - a)\,dt = \int_a^\infty e^{-st}\,dt$$

$$= \lim_{N \to \infty} \left. \frac{-e^{-st}}{s} \right|_a^N = \frac{e^{-as}}{s}.$$

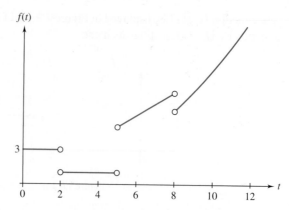

Figure 7.11 Graph of $f(t)$ in equation (4)

Conversely, for $a > 0$, we say that the piecewise continuous function $u(t - a)$ is an inverse Laplace transform for e^{-as}/s and we write

$$\mathscr{L}^{-1}\left\{\frac{e^{-as}}{s}\right\}(t) = u(t - a) \, .$$

For the rectangular window function, we deduce from (6) that

(7) $\mathscr{L}\{\Pi_{a,b}(t)\}(s) = \mathscr{L}\{u(t - a) - u(t - b)\}(s) = [e^{-sa} - e^{-sb}]/s \, , \; 0 < a < b.$

The translation property of $F(s)$ discussed in Section 7.3 described the effect on the Laplace transform of multiplying a function by e^{at}. The next theorem illustrates an analogous effect of multiplying the Laplace transform of a function by e^{-as}.

Translation in t

Theorem 8. Let $F(s) = \mathscr{L}\{f\}(s)$ exist for $s > \alpha \geq 0$. If a is a positive constant, then

(8) $\mathscr{L}\{f(t - a)u(t - a)\}(s) = e^{-as}F(s) \, ,$

and, conversely, an inverse Laplace transform[†] of $e^{-as}F(s)$ is given by

(9) $\mathscr{L}^{-1}\{e^{-as}F(s)\}(t) = f(t - a)u(t - a) \, .$

Proof. By the definition of the Laplace transform,

(10) $\mathscr{L}\{f(t - a)u(t - a)\}(s) = \displaystyle\int_0^\infty e^{-st}f(t - a)u(t - a)\,dt$

$$= \int_a^\infty e^{-st}f(t - a)\,dt \, ,$$

[†]This inverse transform is in fact a *continuous* function of t if $f(0) = 0$ and $f(t)$ is continuous for $t \geq 0$; the values of $f(t)$ for $t < 0$ are of no consequence, since the factor $u(t - a)$ is zero there.

where, in the last equation, we used the fact that $u(t-a)$ is zero for $t<a$ and equals 1 for $t>a$. Now let $v=t-a$. Then we have $dv=dt$, and equation (10) becomes

$$\mathcal{L}\{f(t-a)u(t-a)\}(s) = \int_0^\infty e^{-as}e^{-sv}f(v)\,dv$$

$$= e^{-as}\int_0^\infty e^{-sv}f(v)\,dv = e^{-as}F(s). \;\blacklozenge$$

Notice that formula (8) includes as a special case the formula for $\mathcal{L}\{u(t-a)\}$; indeed, if we take $f(t)\equiv 1$, then $F(s)=1/s$ and (8) becomes $\mathcal{L}\{u(t-a)\}(s)=e^{-as}/s$.

In practice it is more common to be faced with the problem of computing the transform of a function expressed as $g(t)u(t-a)$ rather than $f(t-a)u(t-a)$. To compute $\mathcal{L}\{g(t)u(t-a)\}$, we simply identify $g(t)$ with $f(t-a)$ so that $f(t)=g(t+a)$. Equation (8) then gives

(11) $$\mathcal{L}\{g(t)u(t-a)\}(s) = e^{-as}\mathcal{L}\{g(t+a)\}(s).$$

Example 2 Determine the Laplace transform of $t^2u(t-1)$.

Solution To apply equation (11), we take $g(t)=t^2$ and $a=1$. Then

$$g(t+a) = g(t+1) = (t+1)^2 = t^2+2t+1.$$

Now the Laplace transform of $g(t+a)$ is

$$\mathcal{L}\{g(t+a)\}(s) = \mathcal{L}\{t^2+2t+1\}(s) = \frac{2}{s^3}+\frac{2}{s^2}+\frac{1}{s}.$$

So, by formula (11), we have

$$\mathcal{L}\{t^2u(t-1)\}(s) = e^{-s}\left\{\frac{2}{s^3}+\frac{2}{s^2}+\frac{1}{s}\right\}. \;\blacklozenge$$

Example 3 Determine $\mathcal{L}\{(\cos t)u(t-\pi)\}$.

Solution Here $g(t)=\cos t$ and $a=\pi$. Hence,

$$g(t+a) = g(t+\pi) = \cos(t+\pi) = -\cos t,$$

and so the Laplace transform of $g(t+a)$ is

$$\mathcal{L}\{g(t+a)\}(s) = -\mathcal{L}\{\cos t\}(s) = -\frac{s}{s^2+1}.$$

Thus, from formula (11), we get

$$\mathcal{L}\{(\cos t)u(t-\pi)\}(s) = -e^{-\pi s}\frac{s}{s^2+1}. \;\blacklozenge$$

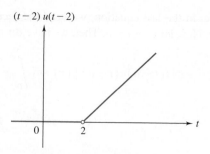

$(t-2)\,u(t-2)$

Figure 7.12 Graph of solution to Example 4

In Examples 2 and 3, we could also have computed the Laplace transform directly from the definition. In dealing with inverse transforms, however, we do not have a simple alternative formula[†] upon which to rely, and so formula (9) is especially useful whenever the transform has e^{-as} as a factor.

Example 4 Determine $\mathscr{L}^{-1}\left\{\dfrac{e^{-2s}}{s^2}\right\}$ and sketch its graph.

Solution To use the translation property (9), we first express e^{-2s}/s^2 as the product $e^{-as}F(s)$. For this purpose, we put $e^{-as} = e^{-2s}$ and $F(s) = 1/s^2$. Thus, $a = 2$ and

$$f(t) = \mathscr{L}^{-1}\left\{\frac{1}{s^2}\right\}(t) = t\,.$$

It now follows from the translation property that

$$\mathscr{L}^{-1}\left\{\frac{e^{-2s}}{s^2}\right\}(t) = f(t-2)u(t-2) = (t-2)u(t-2)\,.$$

See Figure 7.12. ◆

As we anticipated in the beginning of this section, step functions arise in the modeling of on/off switches, changes in polarity, etc.

Example 5 The current I in an LC series circuit is governed by the initial value problem

$$(12)\qquad I''(t) + 4I(t) = g(t)\,;\qquad I(0) = 0\,,\qquad I'(0) = 0\,,$$

where

$$g(t) := \begin{cases} 1\,, & 0 < t < 1\,, \\ -1\,, & 1 < t < 2\,, \\ 0\,, & 2 < t\,. \end{cases}$$

Determine the current as a function of time t.

Solution Let $J(s) := \mathscr{L}\{I\}(s)$. Then we have $\mathscr{L}\{I''\}(s) = s^2 J(s)$.

[†]Under certain conditions, the inverse transform is given by the contour integral

$$\mathscr{L}^{-1}\{F\}(t) = \frac{1}{2\pi i}\int_{a-i\infty}^{a+i\infty} e^{st}F(s)\,ds\,.$$

See, for example, *Complex Variables and the Laplace Transform for Engineers,* by Wilbur R. LePage (Dover Publications, New York, 2010), or *Fundamentals of Complex Analysis with Applications to Engineering and Science,* 3rd ed., by E. B. Saff and A. D. Snider (Pearson Education, Boston. MA, 2003).

Writing $g(t)$ in terms of the rectangular window function $\Pi_{a,b}(t) = u(t-a) - u(t-b)$, we get

$$g(t) = \Pi_{0,1}(t) + (-1)\Pi_{1,2}(t) = u(t) - u(t-1) - [u(t-1) - u(t-2)]$$
$$= 1 - 2u(t-1) + u(t-2),$$

and so

$$\mathcal{L}\{g\}(s) = \frac{1}{s} - \frac{2e^{-s}}{s} + \frac{e^{-2s}}{s}.$$

Thus, when we take the Laplace transform of both sides of (12), we obtain

$$\mathcal{L}\{I''\}(s) + 4\mathcal{L}\{I\}(s) = \mathcal{L}\{g\}(s)$$
$$s^2 J(s) + 4J(s) = \frac{1}{s} - \frac{2e^{-s}}{s} + \frac{e^{-2s}}{s}$$
$$J(s) = \frac{1}{s(s^2+4)} - \frac{2e^{-s}}{s(s^2+4)} + \frac{e^{-2s}}{s(s^2+4)}.$$

To find $I = \mathcal{L}^{-1}\{J\}$, we first observe that

$$J(s) = F(s) - 2e^{-s}F(s) + e^{-2s}F(s),$$

where

$$F(s) := \frac{1}{s(s^2+4)} = \frac{1}{4}\left(\frac{1}{s}\right) - \frac{1}{4}\left(\frac{s}{s^2+4}\right).$$

Computing the inverse transform of $F(s)$ gives

$$f(t) := \mathcal{L}^{-1}\{F\}(t) = \frac{1}{4} - \frac{1}{4}\cos 2t.$$

Hence, via the translation property (9), we find

$$I(t) = \mathcal{L}^{-1}\{F(s) - 2e^{-s}F(s) + e^{-2s}F(s)\}(t)$$
$$= f(t) - 2f(t-1)u(t-1) + f(t-2)u(t-2)$$
$$= \left(\frac{1}{4} - \frac{1}{4}\cos 2t\right) - \left[\frac{1}{2} - \frac{1}{2}\cos 2(t-1)\right]u(t-1)$$
$$+ \left[\frac{1}{4} - \frac{1}{4}\cos 2(t-2)\right]u(t-2).$$

The current is graphed in Figure 7.13. Note that $I(t)$ is smoother than $g(t)$; the former has discontinuities in its second derivative at the points where the latter has jumps. ◆

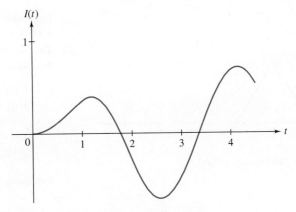

Figure 7.13 Solution to Example 5

In Problems 1–4, sketch the graph of the given function and determine its Laplace transform.

1. $(t-1)^2 u(t-1)$ **2.** $u(t-1) - u(t-4)$

3. $t^2 u(t-2)$ **4.** $tu(t-1)$

In Problems 5–10, express the given function using window and step functions and compute its Laplace transform.

5. $g(t) = \begin{cases} 0, & 0 < t < 1, \\ 2, & 1 < t < 2, \\ 1, & 2 < t < 3, \\ 3, & 3 < t \end{cases}$

6. $g(t) = \begin{cases} 0, & 0 < t < 2, \\ t+1, & 2 < t \end{cases}$

7.

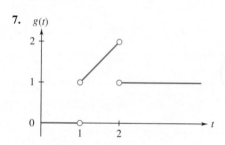

Figure 7.14 Function in Problem 7

8.

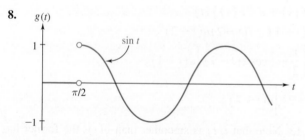

Figure 7.15 Function in Problem 8

9.

Figure 7.16 Function in Problem 9

10.

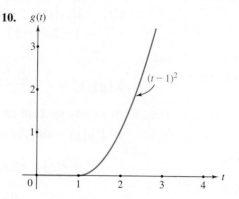

Figure 7.17 Function in Problem 10

In Problems 11–18, determine an inverse Laplace transform of the given function.

11. $\dfrac{e^{-2s}}{s-1}$ **12.** $\dfrac{e^{-3s}}{s^2}$

13. $\dfrac{e^{-2s} - 3e^{-4s}}{s+2}$ **14.** $\dfrac{e^{-3s}}{s^2+9}$

15. $\dfrac{se^{-3s}}{s^2+4s+5}$ **16.** $\dfrac{e^{-s}}{s^2+4}$

17. $\dfrac{e^{-3s}(s-5)}{(s+1)(s+2)}$ **18.** $\dfrac{e^{-s}(3s^2-s+2)}{(s-1)(s^2+1)}$

19. The current $I(t)$ in an *RLC* series circuit is governed by the initial value problem

$$I''(t) + 2I'(t) + 2I(t) = g(t) ;$$
$$I(0) = 10, \qquad I'(0) = 0,$$

where

$$g(t) := \begin{cases} 20, & 0 < t < 3\pi, \\ 0, & 3\pi < t < 4\pi, \\ 20, & 4\pi < t. \end{cases}$$

Determine the current as a function of time t. Sketch $I(t)$ for $0 < t < 8\pi$.

20. The current $I(t)$ in an *LC* series circuit is governed by the initial value problem

$$I''(t) + 4I(t) = g(t) ;$$
$$I(0) = 1, \qquad I'(0) = 3,$$

where

$$g(t) := \begin{cases} 3\sin t, & 0 \le t \le 2\pi, \\ 0, & 2\pi < t. \end{cases}$$

Determine the current as a function of time t.

In Problems 21–24, solve the given initial value problem using the method of Laplace transforms. Sketch the graph of the solution.

21. $y'' + y = u(t-3)$;
 $y(0) = 0, \quad y'(0) = 1$

22. $w'' + w = u(t-2) - u(t-4)$;
 $w(0) = 1, \quad w'(0) = 0$

23. $y'' + y = t - (t-4)u(t-2)$;
 $y(0) = 0, \quad y'(0) = 1$

24. $y'' + y = 3 \sin 2t - 3(\sin 2t)u(t - 2\pi)$;
 $y(0) = 1, \quad y'(0) = -2$

In Problems 25–32, solve the given initial value problem using the method of Laplace transforms.

25. $y'' + 2y' + 2y = u(t - 2\pi) - u(t - 4\pi)$;
 $y(0) = 1, \quad y'(0) = 1$

26. $y'' + 4y' + 4y = u(t - \pi) - u(t - 2\pi)$;
 $y(0) = 0, \quad y'(0) = 0$

27. $z'' + 3z' + 2z = e^{-3t}u(t-2)$;
 $z(0) = 2, \quad z'(0) = -3$

28. $y'' + 5y' + 6y = tu(t-2)$;
 $y(0) = 0, \quad y'(0) = 1$

29. $y'' + 4y = g(t)$; $y(0) = 1, \quad y'(0) = 3$,
 where $g(t) = \begin{cases} \sin t, & 0 \le t \le 2\pi, \\ 0, & 2\pi < t \end{cases}$

30. $y'' + 2y' + 10y = g(t)$;
 $y(0) = -1, \quad y'(0) = 0$,
 where $g(t) = \begin{cases} 10, & 0 \le t \le 10, \\ 20, & 10 < t < 20, \\ 0, & 20 < t \end{cases}$

31. $y'' + 5y' + 6y = g(t)$;
 $y(0) = 0, \qquad y'(0) = 2$,
 where $g(t) = \begin{cases} 0, & 0 \le t < 1, \\ t, & 1 < t < 5, \\ 1, & 5 < t \end{cases}$

32. $y'' + 3y' + 2y = g(t)$;
 $y(0) = 2, \quad y'(0) = -1$,
 where $g(t) = \begin{cases} e^{-t}, & 0 \le t < 3, \\ 1, & 3 < t \end{cases}$

33. The mixing tank in Figure 7.18 initially holds 500 L of a brine solution with a salt concentration of 0.02 kg/L. For the first 10 min of operation, valve A is open, adding 12 L/min of brine containing a 0.04 kg/L salt concentration. After 10 min, valve B is switched in, adding a 0.06 kg/L concentration at 12 L/min. The exit valve C removes 12 L/min, thereby keeping the volume constant. Find the concentration of salt in the tank as a function of time.

Figure 7.18 Mixing tank

34. Suppose in Problem 33 valve B is initially opened for 10 min and then valve A is switched in for 10 min. Finally, valve B is switched back in. Find the concentration of salt in the tank as a function of time.

35. Suppose valve C removes only 6 L/min in Problem 33. Can Laplace transforms be used to solve the problem? Discuss.

36. The **unit triangular pulse** $\Lambda(t)$ is defined by

$$\Lambda(t) := \begin{cases} 0, & t < 0, \\ 2t, & 0 < t < 1/2, \\ 2 - 2t, & 1/2 < t < 1, \\ 0, & t > 1. \end{cases}$$

 (a) Sketch the graph of $\Lambda(t)$. Why is it so named? Why is its symbol appropriate?

 (b) Show that $\Lambda(t) = \displaystyle\int_{-\infty}^{t} 2\{\Pi_{0,1/2}(\tau) - \Pi_{1/2,1}(\tau)\} \, d\tau$.

 (c) Find the Laplace transform of $\Lambda(t)$.

7.7 Transforms of Periodic and Power Functions

Periodic functions arise frequently in physical situations such as sinusoidal vibrations in structures, and in electromagnetic oscillations in AC machinery and microwave transmission. *Power* functions (t^n) occur in more specialized applications: the square-cube law of biomechanics[*], the cube rule of electoral politics,[†] Coulomb's inverse-square force, and, most significantly, the Taylor series of Section 3.7 and Chapter 8. The manipulation of these functions' transforms (when they exist) is facilitated by the techniques described in this section.

Periodic Function

Definition 7. A function $f(t)$ is said to be **periodic of period T** $(\neq 0)$ if

$$f(t + T) = f(t)$$

for all t in the domain of f.

As we know, the sine and cosine functions are periodic with period 2π and the tangent function is periodic with period π.[‡] To specify a periodic function, it is sufficient to give its values over one period. For example, the square wave function in Figure 7.19 can be expressed as

$$(1) \qquad f(t) := \begin{cases} 1, & 0 < t < 1, \\ -1, & 1 < t < 2, \end{cases} \qquad \text{and } f(t) \text{ has period 2.}$$

Figure 7.19 Graph of square wave function $f(t)$

Figure 7.20 Windowed version of periodic function

[*]The volume of a body increases as the cube of its length; its surface area increases as the square of the length. First formulated by Galileo in 1638 (*Discourses and Mathematical Demonstrations Relating to Two New Sciences*), this principle is useful in explaining the limitations on animal growth.

[†]In a two-party system, the ratio of the seats won equals the cube of the ratio of the votes cast. (G. Upton, "Blocks of voters and the cube law," *British Journal of Political Science*. Vol. 15, Issue 03 (1985): 388–398.)

[‡]A function that has period T will also have period $2T$, $3T$, etc. For example, the sine function has periods 2π, 4π, 6π, etc. Some authors refer to the smallest period as the **fundamental period** or just the period of the function.

It is convenient to introduce a notation for the "windowed" version of a periodic function $f(t)$, using a rectangular window whose width is the period:

$$(2) \qquad f_T(t) := f(t)\Pi_{0,T}(t) = f(t)[u(t) - u(t-T)] = \begin{cases} f(t), & 0 < t < T, \\ 0, & \text{otherwise}. \end{cases}$$

(See Figure 7.20 on page 392.) The Laplace transform of $f_T(t)$ is given by

$$F_T(s) = \int_0^\infty e^{-st}f_T(t)\,dt = \int_0^T e^{-st}f(t)\,dt.$$

It is related to the Laplace transform of $f(t)$ as follows.

Transform of Periodic Function

Theorem 9. If f has period T and is piecewise continuous on $[0, T]$, then the Laplace transforms

$$F(s) = \int_0^\infty e^{-st}f(t)\,dt \quad \text{and} \quad F_T(s) = \int_0^\infty e^{-st}f_T(t)\,dt = \int_0^T e^{-st}f(t)\,dt$$

are related by

$$(3) \qquad F_T(s) = F(s)\left[1 - e^{-sT}\right] \text{ or } F(s) = \frac{F_T(s)}{1 - e^{-sT}}.$$

Proof. From (2) and the periodicity of f, we have

$$(4) \qquad f_T(t) = f(t)u(t) - f(t)u(t-T) = f(t)u(t) - f(t-T)u(t-T),$$

so taking transforms and applying the translation-in-t property (Theorem 8, page 386) yields $F_T(s) = F(s) - e^{-sT}F(s)$, which is equivalent to (3). ◆

Example 1 Determine $\mathcal{L}\{f\}$, where f is the periodic square wave function in Figure 7.19.

Solution Here $T = 2$. Windowing the function results in $f_T(t) = \Pi_{0,1}(t) - \Pi_{1,2}(t)$, so from the formula for the transform of the window function (equation (7) in Section 7.6, page 386) we get $F_T(s) = (1 - e^{-s})/s - (e^{-s} - e^{-2s})/s = (1 - e^{-s})^2/s$. Therefore (3) implies

$$\mathcal{L}\{f\}(s) = \frac{(1 - e^{-s})^2/s}{1 - e^{-2s}} = \frac{1 - e^{-s}}{(1 + e^{-s})s}. \quad ◆$$

We next turn to the problem of finding transforms of functions given by a power series. Our approach is simply to apply the formula $\mathcal{L}\{t^n\}(s) = n!/s^{n+1}, n = 0, 1, 2, \ldots,$ to the terms of the series.

Example 2 Determine $\mathcal{L}\{f\}$, where

$$f(t) := \begin{cases} \dfrac{\sin t}{t}, & t \neq 0, \\ 1, & t = 0. \end{cases}$$

Solution We begin by expressing $f(t)$ in a Taylor series[†] about $t = 0$. Since

$$\sin t = t - \frac{t^3}{3!} + \frac{t^5}{5!} - \frac{t^7}{7!} + \cdots,$$

then dividing by t, we obtain

$$f(t) = \frac{\sin t}{t} = 1 - \frac{t^2}{3!} + \frac{t^4}{5!} - \frac{t^6}{7!} + \cdots$$

for $t > 0$. This representation also holds at $t = 0$ since

$$\lim_{t \to 0} f(t) = \lim_{t \to 0} \frac{\sin t}{t} = 1.$$

Observe that $f(t)$ is continuous on $[0, \infty)$ and of exponential order. Hence, its Laplace transform exists for all s sufficiently large. Because of the linearity of the Laplace transform, we would expect that

$$\mathscr{L}\{f\}(s) = \mathscr{L}\{1\}(s) - \frac{1}{3!}\mathscr{L}\{t^2\}(s) + \frac{1}{5!}\mathscr{L}\{t^4\}(s) + \cdots$$

$$= \frac{1}{s} - \frac{2!}{3!s^3} + \frac{4!}{5!s^5} - \frac{6!}{7!s^7} + \cdots$$

$$= \frac{1}{s} - \frac{1}{3s^3} + \frac{1}{5s^5} - \frac{1}{7s^7} + \cdots.$$

Indeed, using tools from analysis, it can be verified that this series representation is valid for all $s > 1$. Moreover, one can show that the series converges to the function $\arctan(1/s)$ (see Problem 22). Thus,

$$(5) \qquad \mathscr{L}\left\{\frac{\sin t}{t}\right\}(s) = \arctan\frac{1}{s}. \qquad \blacklozenge$$

A similar procedure involving the series expansion for $F(s)$ in powers of $1/s$ can be used to compute $f(t) = \mathscr{L}^{-1}\{F\}(t)$ (see Problems 23–25).

We have previously shown, for every nonnegative integer n, that $\mathscr{L}\{t^n\}(s) = n!/s^{n+1}$. But what if the power of t is *not* an integer? Is this formula still valid? To answer this question, we need to extend the idea of "factorial." This is accomplished by the gamma function.[‡]

Gamma Function

Definition 8. The **gamma function** $\Gamma(r)$ is defined by

$$(6) \qquad \Gamma(r) := \int_0^\infty e^{-u} u^{r-1} du, \qquad r > 0.$$

It can be shown that the integral in (6) converges for $r > 0$. A useful property of the gamma function is the recursive relation

$$(7) \qquad \Gamma(r + 1) = r\Gamma(r).$$

[†]For a discussion of Taylor series, see Sections 8.1 and 8.2.

[‡]*Historical Footnote:* The gamma function was introduced by Leonhard Euler.

This identity follows from the definition (6) after performing an integration by parts:

$$\Gamma(r+1) = \int_0^\infty e^{-u}u^r\,du = \lim_{N\to\infty}\int_0^N e^{-u}u^r\,du$$

$$= \lim_{N\to\infty}\left\{ e^{-u}u^r\Big|_0^N + \int_0^N re^{-u}u^{r-1}\,du \right\}$$

$$= \lim_{N\to\infty}\left(e^{-N}N^r\right) + r\lim_{N\to\infty}\int_0^N e^{-u}u^{r-1}\,du$$

$$= 0 + r\Gamma(r) = r\Gamma(r)\,.$$

When r is a positive integer, say $r = n$, then the recursive relation (7) can be repeatedly applied to obtain

$$\Gamma(n+1) = n\Gamma(n) = n(n-1)\Gamma(n-1) = \cdots$$
$$= n(n-1)(n-2)\cdots 2\Gamma(1)\,.$$

It follows from the definition (6) that $\Gamma(1) = 1$, so we find

$$\Gamma(n+1) = n!\,.$$

Thus, the gamma function extends the notion of factorial.

As an application of the gamma function, let's return to the problem of determining the Laplace transform of an arbitrary power of t. We will verify that the formula

(8) $$\mathscr{L}\{t^r\}(s) = \frac{\Gamma(r+1)}{s^{r+1}}$$

holds for every constant $r > -1$.

By definition,

$$\mathscr{L}\{t^r\}(s) = \int_0^\infty e^{-st}t^r\,dt\,.$$

Let's make the substitution $u = st$. Then $du = s\,dt$, and we find

$$\mathscr{L}\{t^r\}(s) = \int_0^\infty e^{-u}\left(\frac{u}{s}\right)^r\left(\frac{1}{s}\right)du$$

$$= \frac{1}{s^{r+1}}\int_0^\infty e^{-u}u^r\,du = \frac{\Gamma(r+1)}{s^{r+1}}\,.$$

Notice that when $r = n$ is a nonnegative integer, then $\Gamma(n+1) = n!$, and so formula (8) reduces to the familiar formula for $\mathscr{L}\{t^n\}$.

Example 3 Given that $\Gamma(1/2) = \sqrt{\pi}$ (see Problem 26), find the Laplace transform of $f(t) = t^{3/2}e^{2t}$.

Solution We'll apply the translation-in-s property (Theorem 3, page 361) to the transform for $t^{3/2}$, which from (8) is given by $\Gamma\left(\frac{3}{2}+1\right)/s^{\frac{3}{2}+1}$. Thanks to the basic gamma function property (7), we can write

$$\Gamma\left(\frac{3}{2}+1\right) = \frac{3}{2}\Gamma\left(\frac{3}{2}\right) = \frac{3}{2}\Gamma\left(\frac{1}{2}+1\right) = \left(\frac{3}{2}\right)\left(\frac{1}{2}\right)\Gamma\left(\frac{1}{2}\right) = \frac{3}{4}\sqrt{\pi}\,.$$

Hence $\mathscr{L}\{t^{3/2}\}(s) = \frac{3\sqrt{\pi}}{4s^{5/2}}$, and so

$$\mathscr{L}\{t^{3/2}e^{2t}\}(s) = \frac{3\sqrt{\pi}}{4(s-2)^{5/2}}\,.\quad\blacklozenge$$

7.7 EXERCISES

In Problems 1–4, determine $\mathscr{L}\{f\}$, where $f(t)$ is periodic with the given period. Also graph $f(t)$.

1. $f(t) = t$, $0 < t < 2$, and $f(t)$ has period 2.

2. $f(t) = e^t$, $0 < t < 1$, and $f(t)$ has period 1.

3. $f(t) = \begin{cases} e^{-t}, & 0 < t < 1, \\ 1, & 1 < t < 2, \end{cases}$ and $f(t)$ has period 2.

4. $f(t) = \begin{cases} t, & 0 < t < 1, \\ 1 - t, & 1 < t < 2, \end{cases}$ and $f(t)$ has period 2.

In Problems 5–8, determine $\mathscr{L}\{f\}$, where the periodic function is described by its graph.

5. $f(t)$

Figure 7.21 Square wave

6. $f(t)$

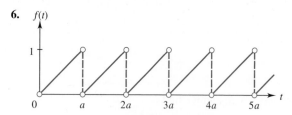

Figure 7.22 Sawtooth wave

7. $f(t)$

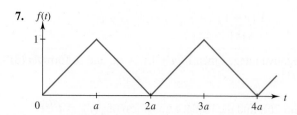

Figure 7.23 Triangular wave

8. $f(t)$

Figure 7.24 Half-rectified sine wave

9. Show that if $\mathscr{L}\{g\}(s) = [(s + \alpha)(1 - e^{-Ts})]^{-1}$, where $T > 0$ is fixed, then

 (9) $g(t) = e^{-\alpha t} + e^{-\alpha(t - T)}u(t - T)$
 $+ e^{-\alpha(t - 2T)}u(t - 2T)$
 $+ e^{-\alpha(t - 3T)}u(t - 3T) + \cdots .$

 [*Hint:* Use the fact that $1 + x + x^2 + \cdots = 1/(1 - x)$.]

10. The function $g(t)$ in (9) can be expressed in a more convenient form as follows:

 (a) Show that for each $n = 0, 1, 2, \ldots$,

 $$g(t) = e^{-\alpha t}\left[\frac{e^{(n+1)\alpha T} - 1}{e^{\alpha T} - 1}\right] \text{ for } nT < t < (n+1)T.$$

 [*Hint:* Use the fact that $1 + x + x^2 + \cdots + x^n = (x^{n+1} - 1)/(x - 1)$.]

 (b) Let $v = t - (n + 1)T$. Show that when $nT < t < (n+1)T$, then $-T < v < 0$ and

 $$\textbf{(10)} \quad g(t) = \frac{e^{-\alpha v}}{e^{\alpha T} - 1} - \frac{e^{-\alpha t}}{e^{\alpha T} - 1}.$$

 (c) Use the facts that the first term in (10) is periodic with period T and the second term is independent of n to sketch the graph of $g(t)$ in (10) for $\alpha = 1$ and $T = 2$.

11. Show that if $\mathscr{L}\{g\}(s) = \beta[(s^2 + \beta^2)(1 - e^{-Ts})]^{-1}$, then

 $g(t) = \sin \beta t + [\sin \beta(t - T)]u(t - T)$
 $+ [\sin \beta(t - 2T)]u(t - 2T)$
 $+ [\sin \beta(t - 3T)]u(t - 3T) + \cdots .$

12. Use the result of Problem 11 to show that

 $$\mathscr{L}^{-1}\left\{\frac{1}{(s^2 + 1)(1 - e^{-\pi s})}\right\}(t) = g(t),$$

 where $g(t)$ is periodic with period 2π and

 $$g(t) := \begin{cases} \sin t, & 0 \le t \le \pi, \\ 0, & \pi \le t \le 2\pi . \end{cases}$$

In Problems 13 and 14, use the method of Laplace transforms and the results of Problems 9 and 10 to solve the initial value problem.

$$y'' + 3y' + 2y = f(t);$$
$$y(0) = 0, \quad y'(0) = 0,$$

where $f(t)$ is the periodic function defined in the stated problem.

13. Problem 2

14. Problem 5 with $a = 1$

In Problems 15–18, find a Taylor series for f(t) about t = 0. Assuming the Laplace transform of f(t) can be computed term by term, find an expansion for $\mathcal{L}\{f\}(s)$ in powers of 1/s. If possible, sum the series.

15. $f(t) = e^t$

16. $f(t) = \sin t$

17. $f(t) = \dfrac{1 - \cos t}{t}$

18. $f(t) = e^{-t^2}$

19. Using the recursive relation (7) and the fact that $\Gamma(1/2) = \sqrt{\pi}$, determine
 (a) $\mathcal{L}\{t^{-1/2}\}$.
 (b) $\mathcal{L}\{t^{7/2}\}$.

20. Use the recursive relation (7) and the fact that $\Gamma(1/2) = \sqrt{\pi}$ to show that

$$\mathcal{L}^{-1}\{s^{-(n+1/2)}\}(t) = \frac{2^n t^{n-1/2}}{1 \cdot 3 \cdot 5 \cdots (2n-1)\sqrt{\pi}},$$

 where n is a positive integer.

21. Verify (3) in Theorem 9 for the function $f(t) = \sin t$, taking the period as 2π. Repeat, taking the period as 4π.

22. By replacing s by 1/s in the Maclaurin series expansion for arctan s, show that

$$\arctan \frac{1}{s} = \frac{1}{s} - \frac{1}{3s^3} + \frac{1}{5s^5} - \frac{1}{7s^7} + \cdots .$$

23. Find an expansion for $e^{-1/s}$ in powers of 1/s. Use the expansion for $e^{-1/s}$ to obtain an expansion for $s^{-1/2}e^{-1/s}$

in terms of $1/s^{n+1/2}$. Assuming the inverse Laplace transform can be computed term by term, show that

$$\mathcal{L}^{-1}\{s^{-1/2}e^{-1/s}\}(t) = \frac{1}{\sqrt{\pi t}} \cos 2\sqrt{t} .$$

 [*Hint:* Use the result of Problem 20.]

24. Use the procedure discussed in Problem 23 to show that

$$\mathcal{L}^{-1}\{s^{-3/2}e^{-1/s}\}(t) = \frac{1}{\sqrt{\pi}} \sin 2\sqrt{t} .$$

25. Find an expansion for $\ln[1 + (1/s^2)]$ in powers of 1/s. Assuming the inverse Laplace transform can be computed term by term, show that

$$\mathcal{L}^{-1}\left\{\ln\left(1 + \frac{1}{s^2}\right)\right\}(t) = \frac{2}{t}(1 - \cos t) .$$

26. Evaluate $\Gamma(1/2)$ by setting $r = x^2$ in (6) and relating it to the Gaussian integral $\int_{-\infty}^{\infty} e^{-x^2} dx = \sqrt{\pi}$. (The latter formula is proved by using polar coordinates to evaluate its square; type "Gaussian integral" into your web browser.)

27. Which of these periodic functions coincides with the square wave in Figure 7.19?
 (a) $f(t) = -1$, $-1 < t < 0$; $f(t) = 1$, $0 < t < 1$; and f has period 2.
 (b) $f(t) = 1$, $2 < t < 3$; $f(t) = -1$, $3 < t < 4$; and f has period 2.
 (c) $f(t) = 1$, $3 < t < 4$; $f(t) = -1$, $4 < t < 5$; and f has period 2.

7.8 Convolution

Consider the initial value problem

$$(1) \qquad y'' + y = g(t) ; \qquad y(0) = 0 , \qquad y'(0) = 0 .$$

If we let $Y(s) = \mathcal{L}\{y\}(s)$ and $G(s) = \mathcal{L}\{g\}(s)$, then taking the Laplace transform of both sides of (1) yields

$$s^2 Y(s) + Y(s) = G(s) ,$$

and hence

$$(2) \qquad Y(s) = \left(\frac{1}{s^2 + 1}\right) G(s) .$$

That is, the Laplace transform of the solution to (1) is the product of the Laplace transform of $\sin t$ and the Laplace transform of the forcing term $g(t)$. What we would now like to have is a simple formula for $y(t)$ in terms of $\sin t$ and $g(t)$. Just as the integral of a product is not the product of the integrals, $y(t)$ is not the product of $\sin t$ and $g(t)$. However, we can express $y(t)$ as the "convolution" of $\sin t$ and $g(t)$.

Convolution

Definition 9. Let $f(t)$ and $g(t)$ be piecewise continuous on $[0, \infty)$. The **convolution** of $f(t)$ and $g(t)$, denoted $f * g$, is defined by

$$(3) \qquad (f * g)(t) := \int_0^t f(t - v)g(v)\, dv \,.$$

For example, the convolution of t and t^2 is

$$t * t^2 = \int_0^t (t - v)v^2\, dv = \int_0^t (tv^2 - v^3)\, dv$$

$$= \left(\frac{tv^3}{3} - \frac{v^4}{4} \right)\Bigg|_0^t = \frac{t^4}{3} - \frac{t^4}{4} = \frac{t^4}{12}\,.$$

Convolution is certainly different from ordinary multiplication. For example, $1 * 1 = t \neq 1$ and in general $1 * f \neq f$. However, convolution does satisfy some of the same properties as multiplication.

Properties of Convolution

Theorem 10. Let $f(t)$, $g(t)$, and $h(t)$ be piecewise continuous on $[0, \infty)$. Then

$$(4) \qquad f * g = g * f \,,$$
$$(5) \qquad f * (g + h) = (f * g) + (f * h) \,,$$
$$(6) \qquad (f * g) * h = f * (g * h) \,,$$
$$(7) \qquad f * 0 = 0 \,.$$

Proof. To prove equation (4), we begin with the definition

$$(f * g)(t) := \int_0^t f(t - v)g(v)\, dv \,.$$

Using the change of variables $w = t - v$, we have

$$(f * g)(t) = \int_t^0 f(w)g(t - w)(-dw) = \int_0^t g(t - w)f(w)\, dw = (g * f)(t) \,,$$

which proves (4). The proofs of equations (5) and (6) are left to the exercises (see Problems 33 and 34). Equation (7) is obvious, since $f(t - v) \cdot 0 \equiv 0$. ◆

Returning to our original goal, we now prove that if $Y(s)$ is the product of the Laplace transforms $F(s)$ and $G(s)$, then $y(t)$ is the convolution $(f * g)(t)$.

Convolution Theorem

Theorem 11. Let $f(t)$ and $g(t)$ be piecewise continuous on $[0, \infty)$ and of exponential order α and set $F(s) = \mathscr{L}\{f\}(s)$ and $G(s) = \mathscr{L}\{g\}(s)$. Then

(8) $\mathscr{L}\{f * g\}(s) = F(s)G(s)$,

or, equivalently,

(9) $\mathscr{L}^{-1}\{F(s)G(s)\}(t) = (f * g)(t)$.

Proof. Starting with the left-hand side of (8), we use the definition of convolution to write for $s > \alpha$

$$\mathscr{L}\{f * g\}(s) = \int_0^\infty e^{-st}\left[\int_0^t f(t-v)g(v)\,dv\right]dt\ .$$

To simplify the evaluation of this iterated integral, we introduce the unit step function $u(t-v)$ and write

$$\mathscr{L}\{f * g\}(s) = \int_0^\infty e^{-st}\left[\int_0^\infty u(t-v)f(t-v)g(v)\,dv\right]dt\ ,$$

where we have used the fact that $u(t-v) = 0$ if $v > t$. Reversing the order of integration[†] gives

(10) $$\mathscr{L}\{f * g\}(s) = \int_0^\infty g(v)\left[\int_0^\infty e^{-st}u(t-v)f(t-v)\,dt\right]dv\ .$$

Recall from the translation property in Section 7.6 that the integral in brackets in equation (10) equals $e^{-sv}F(s)$. Hence,

$$\mathscr{L}\{f * g\}(s) = \int_0^\infty g(v)e^{-sv}F(s)\,dv = F(s)\int_0^\infty e^{-sv}g(v)\,dv = F(s)G(s)\ .$$

This proves formula (8). ◆

For the initial value problem (1), recall that we found

$$Y(s) = \left(\frac{1}{s^2+1}\right)G(s) = \mathscr{L}\{\sin t\}(s)\,\mathscr{L}\{g\}(s)\ .$$

It now follows from the convolution theorem that

$$y(t) = \sin t * g(t) = \int_0^t \sin(t-v)g(v)\,dv\ .$$

Thus we have obtained an integral representation for the solution to the initial value problem (1) for any forcing function $g(t)$ that is piecewise continuous on $[0, \infty)$ and of exponential order.

Example 1 Use the convolution theorem to solve the initial value problem

(11) $y'' - y = g(t)$; $y(0) = 1$, $y'(0) = 1$,

where $g(t)$ is piecewise continuous on $[0, \infty)$ and of exponential order.

[†]This is permitted since, for each $s > \alpha$, the absolute value of the integrand is integrable on $(0, \infty) \times (0, \infty)$.

Solution Let $Y(s) = \mathcal{L}\{y\}(s)$ and $G(s) = \mathcal{L}\{g\}(s)$. Taking the Laplace transform of both sides of the differential equation in (11) and using the initial conditions gives

$$s^2Y(s) - s - 1 - Y(s) = G(s) \,.$$

Solving for $Y(s)$, we have

$$Y(s) = \frac{s+1}{s^2-1} + \left(\frac{1}{s^2-1}\right)G(s) = \frac{1}{s-1} + \left(\frac{1}{s^2-1}\right)G(s) \,.$$

Hence,

$$y(t) = \mathcal{L}^{-1}\left\{\frac{1}{s-1}\right\}(t) + \mathcal{L}^{-1}\left\{\frac{1}{s^2-1}\,G(s)\right\}(t)$$

$$= e^t + \mathcal{L}^{-1}\left\{\frac{1}{s^2-1}\,G(s)\right\}(t) \,.$$

Referring to the table of Laplace transforms on the inside back cover, we find

$$\mathcal{L}\{\sinh t\}(s) = \frac{1}{s^2-1} \,,$$

so we can now express

$$\mathcal{L}^{-1}\left\{\frac{1}{s^2-1}\,G(s)\right\}(t) = \sinh t * g(t) \,.$$

Thus,

$$y(t) = e^t + \int_0^t \sinh(t-v)g(v)\,dv$$

is the solution to the initial value problem (11). ◆

Example 2 Use the convolution theorem to find $\mathcal{L}^{-1}\{1/(s^2+1)^2\}$.

Solution Write

$$\frac{1}{(s^2+1)^2} = \left(\frac{1}{s^2+1}\right)\left(\frac{1}{s^2+1}\right) \,.$$

Since $\mathcal{L}\{\sin t\}(s) = 1/(s^2+1)$, it follows from the convolution theorem that

$$\mathcal{L}^{-1}\left\{\frac{1}{(s^2+1)^2}\right\}(t) = \sin t * \sin t = \int_0^t \sin(t-v)\sin v\,dv$$

$$= \frac{1}{2}\int_0^t [\cos(2v-t) - \cos t]\,dv^\dagger$$

$$= \frac{1}{2}\left[\frac{\sin(2v-t)}{2}\right]_0^t - \frac{1}{2}t\cos t$$

$$= \frac{1}{2}\left[\frac{\sin t}{2} - \frac{\sin(-t)}{2}\right] - \frac{1}{2}t\cos t$$

$$= \frac{\sin t - t\cos t}{2} \,. \quad ◆$$

†Here we used the identity $\sin\alpha\sin\beta = \frac{1}{2}[\cos(\beta-\alpha) - \cos(\beta+\alpha)]$.

As the preceding example attests, the convolution theorem is useful in determining the inverse transforms of rational functions of s. In fact, it provides an alternative to the method of partial fractions. For example,

$$\mathcal{L}^{-1}\left\{\frac{1}{(s-a)(s-b)}\right\}(t) = \mathcal{L}^{-1}\left\{\left(\frac{1}{s-a}\right)\left(\frac{1}{s-b}\right)\right\}(t) = e^{at} * e^{bt},$$

and all that remains in finding the inverse is to compute the convolution $e^{at} * e^{bt}$.

In the early 1900s, V. Volterra introduced **integro-differential** equations in his study of population growth. These equations enabled him to take into account "hereditary influences." In certain cases, these equations involved a convolution. As the next example shows, the convolution theorem helps to solve such integro-differential equations.

Example 3 Solve the integro-differential equation

(12) $\quad y'(t) = 1 - \displaystyle\int_0^t y(t-v)e^{-2v}\,dv\,, \qquad y(0) = 1\,.$

Solution Equation (12) can be written as

(13) $\quad y'(t) = 1 - y(t) * e^{-2t}\,.$

Let $Y(s) = \mathcal{L}\{y\}(s)$. Taking the Laplace transform of (13) (with the help of the convolution theorem) and solving for $Y(s)$, we obtain

$$sY(s) - 1 = \frac{1}{s} - Y(s)\left(\frac{1}{s+2}\right)$$

$$sY(s) + \left(\frac{1}{s+2}\right)Y(s) = 1 + \frac{1}{s}$$

$$\left(\frac{s^2 + 2s + 1}{s+2}\right)Y(s) = \frac{s+1}{s}$$

$$Y(s) = \frac{(s+1)(s+2)}{s(s+1)^2} = \frac{s+2}{s(s+1)}$$

$$Y(s) = \frac{2}{s} - \frac{1}{s+1}\,.$$

Hence, $y(t) = 2 - e^{-t}$. ◆

The **transfer function** $H(s)$ of a linear system is defined as the ratio of the Laplace transform of the output function $y(t)$ to the Laplace transform of the input function $g(t)$, under the assumption that all initial conditions are zero. That is, $H(s) = Y(s)/G(s)$. If the linear system is governed by the differential equation

(14) $\quad ay'' + by' + cy = g(t)\,, \qquad t > 0\,,$

where a, b, and c are constants, we can compute the transfer function as follows. Take the Laplace transform of both sides of (14) to get

$$as^2Y(s) - asy(0) - ay'(0) + bsY(s) - by(0) + cY(s) = G(s)\,.$$

Because the initial conditions are assumed to be zero, the equation reduces to

$$(as^2 + bs + c)Y(s) = G(s)\,.$$

Thus the transfer function for equation (14) is

(15) $$H(s) = \frac{Y(s)}{G(s)} = \frac{1}{as^2 + bs + c} \,.$$

You may note the similarity of these calculations to those for finding the auxiliary equation for the homogeneous equation associated with (14) (recall Section 4.2, page 157). Indeed, the first step in inverting $Y(s) = G(s)/(as^2 + bs + c)$ would be to find the roots of the denominator $as^2 + bs + c$, which is identical to solving the characteristic equation for (14).

The function $h(t) := \mathcal{L}^{-1}\{H\}(t)$ is called the **impulse response function** for the system because, physically speaking, it describes the solution when a mass–spring system is struck by a hammer (see Section 7.9). We can also characterize $h(t)$ as the unique solution to the homogeneous problem

(16) $ah'' + bh' + ch = 0 \,; \qquad h(0) = 0 \,, \qquad h'(0) = 1/a \,.$

Indeed, observe that taking the Laplace transform of the equation in (16) gives

(17) $a[s^2H(s) - sh(0) - h'(0)] + b[sH(s) - h(0)] + cH(s) = 0 \,.$

Substituting in $h(0) = 0$ and $h'(0) = 1/a$ and solving for $H(s)$ yields

$$H(s) = \frac{1}{as^2 + bs + c} \,,$$

which is the same as the formula for the transfer function given in equation (15).

One nice feature of the impulse response function h is that it can help us describe the solution to the *general* initial value problem

(18) $ay'' + by' + cy = g(t) \,; \qquad y(0) = y_0 \,, \qquad y'(0) = y_1 \,.$

From the discussion of equation (14), we can see that the convolution $h * g$ is the solution to (18) in the special case when the initial conditions are zero (i.e., $y_0 = y_1 = 0$). To deal with nonzero initial conditions, let y_k denote the solution to the corresponding *homogeneous* initial value problem; that is, y_k solves

(19) $ay'' + by' + cy = 0 \,; \qquad y(0) = y_0 \,, \qquad y'(0) = y_1 \,.$

Then, the desired solution to the general initial value problem (18) must be $h * g + y_k$. Indeed, it follows from the superposition principle (see Theorem 3 in Section 4.5) that since $h * g$ is a solution to equation (14) and y_k is a solution to the corresponding homogeneous equation, then $h * g + y_k$ is a solution to equation (14). Moreover, since $h * g$ has initial conditions zero,

$$(h * g)(0) + y_k(0) = 0 + y_0 = y_0 \,,$$
$$(h * g)'(0) + y_k'(0) = 0 + y_1 = y_1 \,.$$

We summarize these observations in the following theorem.

Solution Using Impulse Response Function

Theorem 12. Let I be an interval containing the origin. The unique solution to the initial value problem

$$ay'' + by' + cy = g \; ; \qquad y(0) = y_0 \, , \qquad y'(0) = y_1 \, ,$$

where a, b, and c are constants and g is continuous on I, is given by

$$(20) \qquad y(t) = (h * g)(t) + y_k(t) = \int_0^t h(t - v)g(v)dv + y_k(t) \, ,$$

where h is the impulse response function for the system and y_k is the unique solution to (19).

Equation (20) is instructive in that it highlights how the value of y at time t depends on the initial conditions (through $y_k(t)$) and on the nonhomogeneity $g(t)$ (through the convolution integral). It even displays the *causal* nature of the dependence, in that the value of $g(v)$ cannot influence $y(t)$ until $t \geq v$.

A proof of Theorem 12 that does not involve Laplace transforms is outlined in Project E in Chapter 4.

In the next example, we use Theorem 12 to find a formula for the solution to an initial value problem.

Example 4 A linear system is governed by the differential equation

$$(21) \qquad y'' + 2y' + 5y = g(t) \; ; \qquad y(0) = 2 \, , \qquad y'(0) = -2 \, .$$

Find the transfer function for the system, the impulse response function, and a formula for the solution.

Solution According to formula (15), the transfer function for (21) is

$$H(s) = \frac{1}{as^2 + bs + c} = \frac{1}{s^2 + 2s + 5} = \frac{1}{(s + 1)^2 + 2^2} \, .$$

The inverse Laplace transform of $H(s)$ is the impulse response function

$$h(t) = \mathscr{L}^{-1}\{H\}(t) = \frac{1}{2}\mathscr{L}^{-1}\left\{ \frac{2}{(s + 1)^2 + 2^2} \right\}(t)$$

$$= \frac{1}{2}e^{-t}\sin 2t \, .$$

To solve the initial value problem, we need the solution to the corresponding homogeneous problem. The auxiliary equation for the homogeneous equation is $r^2 + 2r + 5 = 0$, which has roots $r = -1 \pm 2i$. Thus a general solution is $C_1 e^{-t}\cos 2t + C_2 e^{-t}\sin 2t$. Choosing C_1 and C_2 so that the initial conditions in (21) are satisfied, we obtain $y_k(t) = 2e^{-t}\cos 2t$.

Hence, a formula for the solution to the initial value problem (21) is

$$(h * g)(t) + y_k(t) = \frac{1}{2}\int_0^t e^{-(t-v)}\sin[2(t - v)]g(v)\,dv + 2e^{-t}\cos 2t \, . \; \blacklozenge$$

7.8 EXERCISES

In Problems 1–4, use the convolution theorem to obtain a formula for the solution to the given initial value problem, where $g(t)$ is piecewise continuous on $[0, \infty)$ and of exponential order.

1. $y'' - 2y' + y = g(t)$; $y(0) = -1$, $y'(0) = 1$

2. $y'' + 9y = g(t)$; $y(0) = 1$, $y'(0) = 0$

3. $y'' + 4y' + 5y = g(t)$; $y(0) = 1$, $y'(0) = 1$

4. $y'' + y = g(t)$; $y(0) = 0$, $y'(0) = 1$

In Problems 5–12, use the convolution theorem to find the inverse Laplace transform of the given function.

5. $\dfrac{1}{s(s^2 + 1)}$

6. $\dfrac{1}{(s + 1)(s + 2)}$

7. $\dfrac{14}{(s + 2)(s - 5)}$

8. $\dfrac{1}{(s^2 + 4)^2}$

9. $\dfrac{s}{(s^2 + 1)^2}$

10. $\dfrac{1}{s^3(s^2 + 1)}$

11. $\dfrac{s}{(s - 1)(s + 2)}$ $\left[\text{Hint: } \dfrac{s}{s - 1} = 1 + \dfrac{1}{s - 1}. \right]$

12. $\dfrac{s + 1}{(s^2 + 1)^2}$

13. Find the Laplace transform of
$$f(t) := \int_0^t (t - v)e^{3v}\,dv.$$

14. Find the Laplace transform of
$$f(t) := \int_0^t e^v \sin(t - v)\,dv.$$

In Problems 15–22, solve the given integral equation or integro-differential equation for $y(t)$.

15. $y(t) + 3 \displaystyle\int_0^t y(v)\sin(t - v)\,dv = t$

16. $y(t) + \displaystyle\int_0^t e^{t-v}y(v)\,dv = \sin t$

17. $y(t) + \displaystyle\int_0^t (t - v)y(v)\,dv = 1$

18. $y(t) + \displaystyle\int_0^t (t - v)y(v)\,dv = t^2$

19. $y(t) + \displaystyle\int_0^t (t - v)^2 y(v)\,dv = t^3 + 3$

20. $y'(t) + \displaystyle\int_0^t (t - v)y(v)\,dv = t$, $y(0) = 0$

21. $y'(t) + y(t) - \displaystyle\int_0^t y(v)\sin(t - v)\,dv = -\sin t$, $y(0) = 1$

22. $y'(t) - 2 \displaystyle\int_0^t e^{t-v}y(v)\,dv = t$, $y(0) = 2$

In Problems 23–28, a linear system is governed by the given initial value problem. Find the transfer function $H(s)$ for the system and the impulse response function $h(t)$ and give a formula for the solution to the initial value problem.

23. $y'' + 9y = g(t)$; $y(0) = 2$, $y'(0) = -3$

24. $y'' - 9y = g(t)$; $y(0) = 2$, $y'(0) = 0$

25. $y'' - y' - 6y = g(t)$; $y(0) = 1$, $y'(0) = 8$

26. $y'' + 2y' - 15y = g(t)$; $y(0) = 0$, $y'(0) = 8$

27. $y'' - 2y' + 5y = g(t)$; $y(0) = 0$, $y'(0) = 2$

28. $y'' - 4y' + 5y = g(t)$; $y(0) = 0$, $y'(0) = 1$

In Problems 29 and 30, the current $I(t)$ in an RLC circuit with voltage source $E(t)$ is governed by the initial value problem
$$LI''(t) + RI'(t) + \frac{1}{C}I(t) = e(t),$$
$$I(0) = a, \qquad I'(0) = b,$$

where $e(t) = E'(t)$ (see Figure 7.25). For the given constants R, L, C, a, and b, find a formula for the solution $I(t)$ in terms of $e(t)$.

Resistance R

Voltage source E

Inductance L

Capacitance C

Figure 7.25 Schematic representation of an *RLC* series circuit

29. $R = 20$ Ω, $L = 5$ H, $C = 0.005$ F, $a = -1$ A, $b = 8$ A/sec.

30. $R = 80$ Ω, $L = 10$ H, $C = 1/410$ F, $a = 2$ A, $b = -8$ A/sec.

31. Use the convolution theorem and Laplace transforms to compute $1 * 1 * 1$.

32. Use the convolution theorem and Laplace transforms to compute $1 * t * t^2$.

33. Prove property (5) in Theorem 10.

34. Prove property (6) in Theorem 10.

35. Use the convolution theorem to show that

$$\mathcal{L}^{-1}\left\{\frac{F(s)}{s}\right\}(t) = \int_0^t f(v)\, dv\,,$$

where $F(s) = \mathcal{L}\{f\}(s)$.

36. Using Theorem 5 in Section 7.3 and the convolution theorem, show that

$$\int_0^t \int_0^v f(z)\, dz\, dv = \mathcal{L}^{-1}\left\{\frac{F(s)}{s^2}\right\}(t)$$

$$= t \int_0^t f(v)\, dv - \int_0^t v f(v)\, dv\,,$$

where $F(s) = \mathcal{L}\{f\}(s)$.

37. Prove directly that if $h(t)$ is the impulse response function characterized by equation (16), then for any continuous $g(t)$, we have $(h * g)(0) = (h * g)'(0) = 0$. [*Hint:* Use Leibniz's rule, described in Project E of Chapter 4.]

7.9 Impulses and the Dirac Delta Function

In mechanical systems, electrical circuits, bending of beams, and other applications, one encounters functions that have a very large value over a very short interval. For example, the strike of a hammer exerts a relatively large force over a relatively short time, and a heavy weight concentrated at a spot on a suspended beam exerts a large force over a very small section of the beam. To deal with violent forces of short duration, physicists and engineers use the delta function introduced by Paul A. M. Dirac. Relaxing our standards of rigor for the moment, we present the following somewhat informal definition.

Dirac Delta Function

Definition 10. The **Dirac delta function** $\delta(t)$ is characterized by the following two properties:

$$(1) \qquad \delta(t) = \begin{cases} 0\,, & t \neq 0\,, \\ \text{``infinite,''} & t = 0\,, \end{cases}$$

and

$$(2) \qquad \int_{-\infty}^{\infty} f(t)\delta(t)\, dt = f(0)$$

for any function $f(t)$ that is continuous on an open interval containing $t = 0$.

By shifting the argument of $\delta(t)$, we have $\delta(t - a) = 0$, $t \neq a$, and

$$(3) \qquad \int_{-\infty}^{\infty} f(t)\delta(t - a)\, dt = f(a)$$

for any function $f(t)$ that is continuous on an interval containing $t = a$.

It is obvious that $\delta(t - a)$ is not a function in the usual sense; instead it is an example of what is called a **generalized function** or a **distribution**. Despite this shortcoming, the Dirac

delta function was successfully used for several years to solve various physics and engineering problems before Laurent Schwartz mathematically justified its use!

A heuristic argument for the existence of the Dirac delta function can be made by considering the impulse of a force over a short interval. If a force $\mathscr{F}(t)$ is applied from time t_0 to time t_1, then the **impulse** due to $\mathscr{F}$ is the integral

$$\text{Impulse} := \int_{t_0}^{t_1} \mathscr{F}(t)\,dt\,.$$

By Newton's second law, we see that

$$(4) \qquad \int_{t_0}^{t_1} \mathscr{F}(t)\,dt = \int_{t_0}^{t_1} m\frac{dv}{dt}\,dt = mv(t_1) - mv(t_0)\,,$$

where m denotes mass and v denotes velocity. Since mv represents the momentum, we can interpret equation (4) as saying: **The impulse equals the change in momentum.**

When a hammer strikes an object, it transfers momentum to the object. This change in momentum takes place over a very short period of time, say, $[t_0, t_1]$. If we let $\mathscr{F}_1(t)$ represent the force due to the hammer, then the *area* under the curve $\mathscr{F}_1(t)$ is the impulse or change in momentum (see Figure 7.26). If, as is illustrated in Figure 7.27, the same change in momentum takes place over shorter and shorter time intervals—say, $[t_0, t_2]$ or $[t_0, t_3]$—then the average force must get greater and greater in order for the impulses (the areas under the curves $\mathscr{F}_n$) to remain the same. In fact, if the forces $\mathscr{F}_n$ having the same impulse act, respectively, over the intervals $[t_0, t_n]$, where $t_n \to t_0$ as $n \to \infty$, then $\mathscr{F}_n$ approaches a function that is zero for $t \neq t_0$ but has an infinite value for $t = t_0$. Moreover, the areas under the $\mathscr{F}_n$'s have a common value. Normalizing this value to be 1 gives

$$\int_{-\infty}^{\infty} \mathscr{F}_n(t)\,dt = 1 \qquad \text{for all } n\,.$$

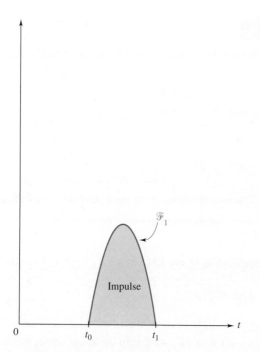

Figure 7.26 Force due to a blow from a hammer

Figure 7.27 Forces with the same impulse

When $t_0 = 0$, we derive from the limiting properties of the $\mathscr{F}_n$'s a "function" δ that satisfies property (1) and the integral condition

$$(5) \qquad \int_{-\infty}^{\infty} \delta(t)\,dt = 1 \,.$$

Notice that (5) is a special case of property (2) that is obtained by taking $f(t) \equiv 1$. It is interesting to note that (1) and (5) actually imply the general property (2) (see Problem 33).

The Laplace transform of the Dirac delta function can be quickly derived from property (3). Since $\delta(t - a) = 0$ for $t \neq a$, then setting $f(t) = e^{-st}$ in (3), we find for $a \geq 0$

$$\int_0^{\infty} e^{-st}\delta(t-a)\,dt = \int_{-\infty}^{\infty} e^{-st}\delta(t-a)\,dt = e^{-as} \,.$$

Thus, for $a \geq 0$,

$$(6) \qquad \mathscr{L}\{\delta(t-a)\}(s) = e^{-as} \,.$$

An interesting connection exists between the unit step function and the Dirac delta function. Observe that as a consequence of equation (5) and the fact that $\delta(x - a)$ is zero for $x < a$ and for $x > a$, we have

$$(7) \qquad \int_{-\infty}^{t} \delta(x-a)\,dx = \begin{cases} 0, & t < a \,, \\ 1, & t > a \end{cases}$$

$$= u(t-a) \,.$$

If we formally differentiate both sides of (7) with respect to t (in the spirit of the fundamental theorem of calculus), we find

$$\delta(t-a) = u'(t-a) \,.$$

Thus it appears that the Dirac delta function is the derivative of the unit step function. That is, in fact, the case if we consider "differentiation" in a more general sense.[†]

The Dirac delta function is used in modeling mechanical vibration problems involving an impulse. For example, a mass–spring system at rest that is struck by a hammer exerting an impulse on the mass might be governed by the *symbolic* initial value problem

$$(8) \qquad x'' + x = \delta(t) \,; \qquad x(0) = 0 \,, \qquad x'(0) = 0 \,,$$

where $x(t)$ denotes the displacement from equilibrium at time t. We refer to this as a symbolic problem because while the left-hand side of equation (8) represents an ordinary function, the right-hand side does *not*. Consequently, it is not clear what we mean by a solution to problem (8). Because $\delta(t)$ is zero everywhere except at $t = 0$, one might be tempted to treat (8) as a homogeneous equation with zero initial conditions. But the solution to the latter is zero everywhere, which certainly does not describe the motion of the spring after the mass is struck by the hammer.

To define what is meant by a solution to (8), recall that $\delta(t)$ is depicted as the limit of forces $\mathscr{F}_n(t)$ having unit impulse and acting over shorter and shorter intervals. If we let $y_n(t)$ be the solution to the initial value problem

$$(9) \qquad y_n'' + y_n = \mathscr{F}_n(t) \,; \qquad y_n(0) = 0 \,, \qquad y_n'(0) = 0 \,,$$

where δ is replaced by $\mathscr{F}_n$, then we can think of the solution $x(t)$ to (8) as the limit (as $n \to \infty$) of the solutions $y_n(t)$.

[†]See *Distributions, Complex Variables, and Fourier Transforms*, by H. J. Bremermann (Addison-Wesley, Reading, MA, 1965).

For example, let

$$\mathcal{F}_n(t) := n - nu(t - 1/n) = \begin{cases} n, & 0 < t < 1/n, \\ 0, & \text{otherwise}. \end{cases}$$

Taking the Laplace transform of equation (9), we find

$$(s^2 + 1)Y_n(s) = \frac{n}{s}(1 - e^{-s/n}),$$

and so

$$Y_n(s) = \frac{n}{s(s^2 + 1)} - e^{-s/n}\frac{n}{s(s^2 + 1)}.$$

Now

$$\frac{n}{s(s^2 + 1)} = \frac{n}{s} - \frac{ns}{s^2 + 1} = \mathcal{L}\{n - n\cos t\}(s).$$

Hence,

(10)
$$\begin{aligned}
y_n(t) &= n - n\cos t - [n - n\cos(t - 1/n)]\,u(t - 1/n) \\
&= \begin{cases} n - n\cos t, & 0 < t < 1/n, \\ n\cos(t - 1/n) - n\cos t, & 1/n < t. \end{cases}
\end{aligned}$$

Fix $t > 0$. Then for n large enough, we have $1/n < t$. Thus,

$$\begin{aligned}
\lim_{n\to\infty} y_n(t) &= \lim_{n\to\infty}[n\cos(t - 1/n) - n\cos t] \\
&= -\lim_{n\to\infty}\frac{\cos(t - 1/n) - \cos t}{-1/n} \\
&= -\lim_{h\to\infty}\frac{\cos(t + h) - \cos t}{h} \qquad \text{(where } h = -1/n), \\
&= -\frac{d}{dt}(\cos t) = \sin t.
\end{aligned}$$

Also, for $t = 0$, we have $\lim_{n\to\infty} y_n(0) = 0 = \sin 0$. Therefore,

$$\lim_{n\to\infty} y_n(t) = \sin t.$$

Hence, the solution to the symbolic initial value problem (8) is $x(t) = \sin t$.

Fortunately, we do not have to go through the tedious process of solving for each y_n in order to find the solution x of the symbolic problem. It turns out that the Laplace transform method when applied directly to (8) yields the derived solution $x(t)$. Indeed, simply taking the Laplace transform of both sides of (8), we obtain from (6) (with $a = 0$)

$$(s^2 + 1)X(s) = 1,$$

$$X(s) = \frac{1}{s^2 + 1},$$

which gives

$$x(t) = \mathcal{L}^{-1}\left\{\frac{1}{s^2 + 1}\right\}(t) = \sin t.$$

A peculiarity of using the Dirac delta function is that the solution $x(t) = \sin t$ of the symbolic initial value problem (8) does not satisfy both initial conditions; that is, $x'(0) = 1 \neq 0$. This reflects the fact that the impulse $\delta(t)$ is applied at $t = 0$. Thus, the momentum x' (observe that in equation (8) the mass equals one) jumps abruptly from $x'(0) = 0$ to $x'(0^+) = 1$.

In the next example, the Dirac delta function is used in modeling a mechanical vibration problem.

Example 1 A mass attached to a spring is released from rest 1 m below the equilibrium position for the mass–spring system and begins to vibrate. After π seconds, the mass is struck by a hammer exerting an impulse on the mass. The system is governed by the symbolic initial value problem

(11) $$\frac{d^2x}{dt^2} + 9x = 3\delta(t - \pi)\,; \qquad x(0) = 1\,, \qquad \frac{dx}{dt}(0) = 0\,,$$

where $x(t)$ denotes the displacement from equilibrium at time t. Determine $x(t)$.

Solution Let $X(s) = \mathscr{L}\{x\}(s)$. Since

$$\mathscr{L}\{x''\}(s) = s^2X(s) - s \qquad \text{and} \qquad \mathscr{L}\{\delta(t-\pi)\}(s) = e^{-\pi s}\,,$$

taking the Laplace transform of both sides of (11) and solving for $X(s)$ yields

$$s^2X(s) - s + 9X(s) = 3e^{-\pi s}$$

$$X(s) = \frac{s}{s^2+9} + e^{-\pi s}\frac{3}{s^2+9}$$

$$= \mathscr{L}\{\cos 3t\}(s) + e^{-\pi s}\mathscr{L}\{\sin 3t\}(s)\,.$$

Using the translation property (cf. page 386) to determine the inverse Laplace transform of $X(s)$, we find

$$x(t) = \cos 3t + [\sin 3(t-\pi)]u(t-\pi)$$

$$= \begin{cases} \cos 3t\,, & t < \pi\,, \\ \cos 3t - \sin 3t\,, & \pi < t \end{cases}$$

$$= \begin{cases} \cos 3t\,, & t < \pi\,, \\ \sqrt{2}\cos\left(3t + \dfrac{\pi}{4}\right), & \pi < t\,. \end{cases}$$

The graph of $x(t)$ is given in color in Figure 7.28. For comparison, the dashed curve depicts the displacement of an undisturbed vibrating spring. Note that the impulse effectively adds 3 units to the momentum at time $t = \pi$. ◆

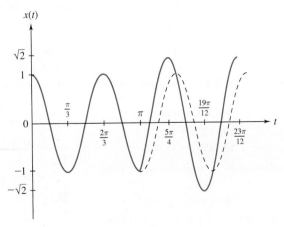

Figure 7.28 Displacement of a vibrating spring that is struck by a hammer at $t = \pi$

In Section 7.8 we defined the **impulse response function** for

(12) $ay'' + by' + cy = g(t)$

as the function $h(t) := \mathcal{L}^{-1}\{H\}(t)$, where $H(s)$ is the **transfer function.** Recall that $H(s)$ is the ratio

$$H(s) := \frac{Y(s)}{G(s)},$$

where $Y(s)$ is the Laplace transform of the solution to (12) with zero initial conditions and $G(s)$ is the Laplace transform of $g(t)$. It is important to note that $H(s)$, and hence $h(t)$, does not depend on the choice of the function $g(t)$ in (12) [see equation (15) in Section 7.8, page 402]. However, it is useful to think of the impulse response function as the solution of the symbolic initial value problem

(13) $ay'' + by' + cy = \delta(t) ;$ $y(0) = 0 ,$ $y'(0) = 0 .$

Indeed, with $g(t) = \delta(t)$, we have $G(s) = 1$, and hence $H(s) = Y(s)$. Consequently, $h(t) = y(t)$. So we see that the function $h(t)$ is the response to the impulse $\delta(t)$ for a mechanical system governed by the symbolic initial value problem (13).

7.9 EXERCISES

In Problems 1–6, evaluate the given integral.

1. $\displaystyle\int_{-\infty}^{\infty} (t^2 - 1)\delta(t)\,dt$

2. $\displaystyle\int_{-\infty}^{\infty} e^{3t}\delta(t)\,dt$

3. $\displaystyle\int_{-\infty}^{\infty} (\sin 3t)\delta\left(t - \frac{\pi}{2}\right)dt$

4. $\displaystyle\int_{-\infty}^{\infty} e^{-2t}\delta(t + 1)\,dt$

5. $\displaystyle\int_{0}^{\infty} e^{-2t}\delta(t - 1)\,dt$

6. $\displaystyle\int_{-1}^{1} (\cos 2t)\delta(t)\,dt$

In Problems 7–12, determine the Laplace transform of the given generalized function.

7. $\delta(t - 1) - \delta(t - 3)$ 8. $3\delta(t - 1)$

9. $t\delta(t - 1)$ 10. $t^3\delta(t - 3)$

11. $\delta(t - \pi)\sin t$ 12. $e^t\delta(t - 3)$

In Problems 13–20, solve the given symbolic initial value problem.

13. $w'' + w = \delta(t - \pi) ;$ $w(0) = 0 ,$ $w'(0) = 0$

14. $y'' + 2y' + 2y = \delta(t - \pi) ;$ $y(0) = 1 ,$ $y'(0) = 1$

15. $y'' + 2y' - 3y = \delta(t - 1) - \delta(t - 2) ;$
 $y(0) = 2 ,$ $y'(0) = -2$

16. $y'' - 2y' - 3y = 2\delta(t - 1) - \delta(t - 3) ;$
 $y(0) = 2 ,$ $y'(0) = 2$

17. $y'' - y = 4\delta(t - 2) + t^2 ;$
 $y(0) = 0 ,$ $y'(0) = 2$

18. $y'' - y' - 2y = 3\delta(t - 1) + e^t ;$
 $y(0) = 0 ,$ $y'(0) = 3$

19. $w'' + 6w' + 5w = e^t\delta(t - 1) ;$
 $w(0) = 0 ,$ $w'(0) = 4$

20. $y'' + 5y' + 6y = e^{-t}\delta(t - 2) ;$
 $y(0) = 2 ,$ $y'(0) = -5$

In Problems 21–24, solve the given symbolic initial value problem and sketch a graph of the solution.

21. $y'' + y = \delta(t - 2\pi) ;$
 $y(0) = 0 ,$ $y'(0) = 1$

22. $y'' + y = \delta(t - \pi/2) ;$
 $y(0) = 0 ,$ $y'(0) = 1$

23. $y'' + y = -\delta(t - \pi) + \delta(t - 2\pi) ;$
 $y(0) = 0 ,$ $y'(0) = 1$

24. $y'' + y = \delta(t - \pi) - \delta(t - 2\pi) ;$
 $y(0) = 0 ,$ $y'(0) = 1$

In Problems 25–28, find the impulse response function $h(t)$ by using the fact that $h(t)$ is the solution to the symbolic initial value problem with $g(t) = \delta(t)$ and zero initial conditions.

25. $y'' + 4y' + 8y = g(t)$ **26.** $y'' - 6y' + 13y = g(t)$

27. $y'' - 2y' + 5y = g(t)$ **28.** $y'' - y = g(t)$

29. A mass attached to a spring is released from rest 1 m below the equilibrium position for the mass–spring system and begins to vibrate. After $\pi/2$ sec, the mass is struck by a hammer exerting an impulse on the mass. The system is governed by the symbolic initial value problem

$$\frac{d^2x}{dt^2} + 9x = -3\delta\left(t - \frac{\pi}{2}\right);$$

$$x(0) = 1, \quad \frac{dx}{dt}(0) = 0,$$

where $x(t)$ denotes the displacement from equilibrium at time t. What happens to the mass after it is struck?

30. You have probably heard that soldiers are told not to march in cadence when crossing a bridge. By solving the symbolic initial value problem

$$y'' + y = \sum_{k=1}^{\infty} \delta(t - 2k\pi); \quad y(0) = 0, \quad y'(0) = 0,$$

explain why soldiers are so instructed. [*Hint:* See Section 4.10.]

31. A linear system is said to be **stable** if its impulse response function $h(t)$ remains bounded as $t \to +\infty$. If the linear system is governed by

$$ay'' + by' + cy = g(t),$$

where b and c are not both zero, show that the system is stable if and only if the real parts of the roots to

$$ar^2 + br + c = 0$$

are less than or equal to zero.

32. A linear system is said to be **asymptotically stable** if its impulse response function satisfies $h(t) \to 0$ as $t \to +\infty$. If the linear system is governed by

$$ay'' + by' + cy = g(t),$$

show that the system is asymptotically stable if and only if the real parts of the roots to

$$ar^2 + br + c = 0$$

are strictly less than zero.

33. The Dirac delta function may also be characterized by the properties

$$\delta(t) = \begin{cases} 0, & t \neq 0, \\ \text{``infinite,''} & t = 0, \end{cases}$$

and $$\int_{-\infty}^{\infty} \delta(t)\, dt = 1.$$

Formally using the mean value theorem for definite integrals, verify that if $f(t)$ is continuous, then the above properties imply

$$\int_{-\infty}^{\infty} f(t)\delta(t)\, dt = f(0).$$

34. Formally using integration by parts, show that

$$\int_{-\infty}^{\infty} f(t)\delta'(t)\, dt = -f'(0).$$

Also show that, in general,

$$\int_{-\infty}^{\infty} f(t)\delta^{(n)}(t)\, dt = (-1)^n f^{(n)}(0).$$

35. Figure 7.29 shows a beam of length 2λ that is imbedded in a support on the left side and free on the right. The vertical deflection of the beam a distance x from the support is denoted by $y(x)$. If the beam has a concentrated load L acting on it in the center of the beam, then the deflection must satisfy the symbolic boundary value problem

$$EIy^{(4)}(x) = L\delta(x - \lambda);$$

$$y(0) = y'(0) = y''(2\lambda) = y'''(2\lambda) = 0,$$

where E, the modulus of elasticity, and I, the moment of inertia, are constants. Find a formula for the displacement $y(x)$ in terms of the constants $\lambda, L, E,$ and I. [*Hint:* Let $y''(0) = A$ and $y'''(0) = B$. First solve the fourth-order symbolic initial value problem and then use the conditions $y''(2\lambda) = y'''(2\lambda) = 0$ to determine A and B.]

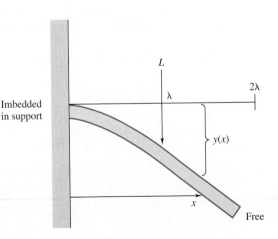

Figure 7.29 Beam imbedded in a support under a concentrated load at $x = \lambda$

7.10 Solving Linear Systems with Laplace Transforms

We can use the Laplace transform to reduce certain *systems* of linear differential equations with initial conditions to a system of linear algebraic equations, where again the unknowns are the transforms of the functions that make up the solution. Solving for these unknowns and taking their inverse Laplace transforms, we can then obtain the solution to the initial value problem for the system.

Example 1 Solve the initial value problem

(1)
$$x'(t) - 2y(t) = 4t \, ; \qquad x(0) = 4 \, ,$$
$$y'(t) + 2y(t) - 4x(t) = -4t - 2 \, ; \qquad y(0) = -5 \, .$$

Solution Taking the Laplace transform of both sides of the differential equations gives

(2)
$$\mathscr{L}\{x'\}(s) - 2\mathscr{L}\{y\}(s) = \frac{4}{s^2} \, ,$$

$$\mathscr{L}\{y'\}(s) + 2\mathscr{L}\{y\}(s) - 4\mathscr{L}\{x\}(s) = -\frac{4}{s^2} - \frac{2}{s} \, .$$

Let $X(s) := \mathscr{L}\{x\}(s)$ and $Y(s) := \mathscr{L}\{y\}(s)$. Then, by Theorem 4 on page 362,

$$\mathscr{L}\{x'\}(s) = sX(s) - x(0) = sX(s) - 4 \, ,$$
$$\mathscr{L}\{y'\}(s) = sY(s) - y(0) = sY(s) + 5 \, .$$

Substituting these expressions into system (2) and simplifying, we find

(3)
$$sX(s) - 2Y(s) = \frac{4s^2 + 4}{s^2} \, ,$$

$$-4X(s) + (s+2)Y(s) = -\frac{5s^2 + 2s + 4}{s^2} \, .$$

To eliminate $Y(s)$ from the system, we multiply the first equation by $(s+2)$ and the second by 2 and then add to obtain

$$[s(s+2) - 8]X(s) = \frac{(s+2)(4s^2+4)}{s^2} - \frac{10s^2 + 4s + 8}{s^2} \, .$$

This simplifies to

$$X(s) = \frac{4s - 2}{(s+4)(s-2)} \, .$$

To compute the inverse transform, we first write $X(s)$ in the partial fraction form

$$X(s) = \frac{3}{s+4} + \frac{1}{s-2} \, .$$

Hence, from the Laplace transform table on the inside back cover, we find that

(4) $x(t) = 3e^{-4t} + e^{2t} \, .$

To determine $y(t)$, we could solve system (3) for $Y(s)$ and then compute its inverse Laplace transform. However, it is easier just to solve the first equation in system (1) for $y(t)$ in terms of $x(t)$. Thus,

$$y(t) = \frac{1}{2}x'(t) - 2t\,.$$

Substituting $x(t)$ from equation (4), we find that

(5) $y(t) = -6e^{-4t} + e^{2t} - 2t\,.$

The solution to the initial value problem (1) consists of the pair of functions $x(t)$, $y(t)$ given by equations (4) and (5). ◆

7.10 EXERCISES

In Problems 1–19, use the method of Laplace transforms to solve the given initial value problem. Here x', y', etc., denotes differentiation with respect to t; so does the symbol D.

1. $x' = 3x - 2y\,;$ $x(0) = 1\,,$
 $y' = 3y - 2x\,;$ $y(0) = 1$

2. $x' = x - y\,;$ $x(0) = -1\,,$
 $y' = 2x + 4y\,;$ $y(0) = 0$

3. $z' + w' = z - w\,;$ $z(0) = 1\,,$
 $z' - w' = z - w\,;$ $w(0) = 0$

4. $x' - 3x + 2y = \sin t\,;$ $x(0) = 0\,,$
 $4x - y' - y = \cos t\,;$ $y(0) = 0$

5. $x' = y + \sin t\,;$ $x(0) = 2\,,$
 $y' = x + 2\cos t\,;$ $y(0) = 0$

6. $x' - x - y = 1\,;$ $x(0) = 0\,,$
 $-x + y' - y = 0\,;$ $y(0) = -5/2$

7. $(D - 4)[x] + 6y = 9e^{-3t}\,;$ $x(0) = -9\,,$
 $x - (D - 1)[y] = 5e^{-3t}\,;$ $y(0) = 4$

8. $D[x] + y = 0\,;$ $x(0) = 7/4\,,$
 $4x + D[y] = 3\,;$ $y(0) = 4$

9. $x'' + 2y' = -x\,;$ $x(0) = 2\,,$ $x'(0) = -7\,,$
 $-3x'' + 2y'' = 3x - 4y\,;$ $y(0) = 4\,,$ $y'(0) = -9$

10. $x'' + y = 1\,;$ $x(0) = 1\,,$ $x'(0) = 1\,,$
 $x + y'' = -1\,;$ $y(0) = 1\,,$ $y'(0) = -1$

11. $x' + y = 1 - u(t - 2)\,;$ $x(0) = 0\,,$
 $x + y' = 0\,;$ $y(0) = 0$

12. $x' + y = x\,;$ $x(0) = 0\,,$ $y(0) = 1\,,$
 $2x' + y'' = u(t - 3)\,;$ $y'(0) = -1$

13. $x' - y' = (\sin t)u(t - \pi)\,;$ $x(0) = 1\,,$
 $x + y' = 0\,;$ $y(0) = 1$

14. $x'' = y + u(t - 1)\,;$ $x(0) = 1\,,$ $x'(0) = 0\,,$
 $y'' = x + 1 - u(t - 1)\,;$ $y(0) = 0\,,$ $y'(0) = 0$

15. $x' - 2y = 2\,;$ $x(1) = 1\,,$
 $x + x - y' = t^2 + 2t - 1\,;$ $y(1) = 0$

16. $x' - 2x + y' = -(\cos t + 4\sin t)\,;$ $x(\pi) = 0\,,$
 $2x + y' + y = \sin t + 3\cos t\,;$ $y(\pi) = 3$

17. $x' + x - y' = 2(t - 2)e^{t-2}\,;$ $x(2) = 0\,,$
 $x'' - x' - 2y = -e^{t-2}\,;$ $x'(2) = 1\,,$ $y(2) = 1$

18. $x' - 2y = 0\,;$ $x(0) = 0\,,$
 $x' - z' = 0\,;$ $y(0) = 0\,,$
 $x + y' - z = 3\,;$ $z(0) = -2$

19. $x' = 3x + y - 2z\,;$ $x(0) = -6\,,$
 $y' = -x + 2y + z\,;$ $y(0) = 2\,,$
 $z' = 4x + y - 3z\,;$ $z(0) = -12$

20. Use the method of Laplace transforms to solve
 $x'' + y' = 2\,;$ $x(0) = 3\,,$ $x'(0) = 0\,,$
 $4x + y' = 6\,;$ $y(1) = 4\,.$
 [*Hint:* Let $y(0) = c$ and then solve for c.]

21. For the interconnected tanks problem of Section 5.1, page 241, suppose that the input to tank A is now controlled by a valve which for the first 5 min delivers 6 L/min of pure water, but thereafter delivers 6 L/min of brine at a concentration of 0.02 kg/L. Assuming that all other data remain the same (see Figure 5.1, page 241), determine the mass of salt in each tank for $t > 0$ if $x_0 = 0$ and $y_0 = 0.04$.

22. Recompute the coupled mass–spring oscillator motion in Problem 1, Exercises 5.6 (page 287), using Laplace transforms.

In Problems 23 and 24, find a system of differential equations and initial conditions for the currents in the networks given by the schematic diagrams; the initial currents are all assumed to be zero. Solve for the currents in each branch of the network. (See Section 5.7 for a discussion of electrical networks.)

23.

Figure 7.30 *RL* network for Problem 23

24.

Figure 7.31 *RL* network for Problem 24

Chapter **7** Summary

The use of the Laplace transform helps to simplify the process of solving initial value problems for certain differential and integral equations, especially when a forcing function with jump discontinuities is involved. The Laplace transform $\mathcal{L}\{f\}$ of a function $f(t)$ is defined by

$$\mathcal{L}\{f\}(s) := \int_0^\infty e^{-st} f(t)\, dt$$

for all values of s for which the improper integral exists. If $f(t)$ is piecewise continuous on $[0, \infty)$ and of exponential order α (that is, $|f(t)|$ grows no faster than a constant times $e^{\alpha t}$ as $t \to \infty$), then $\mathcal{L}\{f\}(s)$ exists for all $s > \alpha$.

The Laplace transform can be interpreted as an integral operator that maps a function $f(t)$ to a function $F(s)$. The transforms of commonly occurring functions appear in Table 7.1, page 356, and on the inside back cover of this book. The use of these tables is enhanced by several important properties of the operator $\mathcal{L}$.

Linearity: $\mathcal{L}\{af + bg\} = a\mathcal{L}\{f\} + b\mathcal{L}\{g\}$.

Translation in s: $\mathcal{L}\{e^{at}f(t)\}(s) = F(s - a)$, where $F = \mathcal{L}\{f\}$.

Translation in t: $\mathcal{L}\{g(t)u(t - a)\}(s) = e^{-as}\mathcal{L}\{g(t + a)\}(s)$, where $u(t - a)$ is the step function that equals 1 for $t > a$ and 0 for $t < a$. If $f(t)$ is continuous and $f(0) = 0$, then

$$\mathcal{L}^{-1}\{e^{-as}F(s)\}(t) = f(t - a)u(t - a),$$

where $f = \mathcal{L}^{-1}\{F\}$.

Convolution Property: $\mathcal{L}\{f * g\} = \mathcal{L}\{f\}\mathcal{L}\{g\}$, where $f * g$ denotes the convolution function

$$(f * g)(t) := \int_0^t f(t - v)g(v)\, dv.$$

One reason for the usefulness of the Laplace transform lies in the simple formula for the transform of the derivative f':

(1) $\qquad \mathscr{L}\{f'\}(s) = sF(s) - f(0)\,, \qquad$ where $\qquad F = \mathscr{L}\{f\}\,.$

This formula shows that by using the Laplace transform, "differentiation with respect to t" can essentially be replaced by the simple operation of "multiplication by s." The extension of (1) to higher-order derivatives is

(2) $\qquad \mathscr{L}\{f^{(n)}\}(s) = s^n F(s) - s^{n-1}f(0) - s^{n-2}f'(0) - \cdots - f^{(n-1)}(0)\,.$

To solve an initial value problem of the form

(3) $\qquad ay'' + by' + cy = f(t)\,; \qquad y(0) = \alpha\,, \qquad y'(0) = \beta$

via the Laplace transform method, one takes the transform of both sides of the differential equation in (3). Using the linearity of $\mathscr{L}$ and formula (2) leads to an equation involving the Laplace transform $Y(s)$ of the (unknown) solution $y(t)$. The next step is to solve this simpler equation for $Y(s)$. Finally, one computes the *inverse* Laplace transform of $Y(s)$ to obtain the desired solution. This last step of finding $\mathscr{L}^{-1}\{Y\}$ is often the most difficult. Sometimes it requires a partial fractions decomposition, a judicious use of the properties of the transform, an excursion through the Laplace transform tables, or the evaluation of a contour integral in the complex plane.

For the special problem in (3), where a, b, and c are constants, the differential equation is transformed to a simple *algebraic* equation for $Y(s)$. Another nice feature of this latter equation is that it incorporates the initial conditions. When the coefficients of the equation in (3) depend on t, the following formula may be helpful in taking the transform:

$$\mathscr{L}\{t^n f(t)\}(s) = (-1)^n \frac{d^n F}{ds^n}(s)\,, \qquad \text{where} \qquad F = \mathscr{L}\{f\}\,.$$

If the forcing function $f(t)$ in equation (3) has jump discontinuities, it is often convenient to write $f(t)$ in terms of unit step functions $u(t-a)$ before proceeding with the Laplace transform method. For this purpose, the rectangular window function $\Pi_{a,b}(t) = u(t-a) - u(t-b)$ is useful. The transform of a periodic function $f(t)$ with period T is given by

$$\mathscr{L}\{f\}(s) = \frac{\displaystyle\int_0^T e^{-st}f(t)\,dt}{1 - e^{-sT}}\,.$$

The Dirac delta function $\delta(t)$ is useful in modeling a system that is excited by a large force applied over a short time interval. It is not a function in the usual sense but can be roughly interpreted as the derivative of a unit step function. The transform of $\delta(t-a)$ is

$$\mathscr{L}\{\delta(t-a)\}(s) = e^{-as}\,, \qquad a \geq 0\,.$$

REVIEW PROBLEMS FOR CHAPTER 7

In Problems 1 and 2, use the definition of the Laplace transform to determine $\mathscr{L}\{f\}$.

1. $f(t) = \begin{cases} 3\,, & 0 \leq t \leq 2\,, \\ 6 - t\,, & 2 < t \end{cases}$

2. $f(t) = \begin{cases} e^{-t}\,, & 0 \leq t \leq 5\,, \\ -1\,, & 5 < t \end{cases}$

In Problems 3–10, determine the Laplace transform of the given function.

3. $t^2 e^{-9t}$

4. $e^{3t} \sin 4t$

5. $e^{2t} - t^3 + t^2 - \sin 5t$

6. $7e^{2t} \cos 3t - 2e^{7t} \sin 5t$

7. $t \cos 6t$

8. $(t+3)^2 - (e^t + 3)^2$

9. $t^2 u(t-4)$

10. $f(t) = \cos t$, $-\pi/2 \leq t \leq \pi/2$ and $f(t)$ has period π.

In Problems 11–17, determine the inverse Laplace transform of the given function.

11. $\dfrac{7}{(s+3)^3}$

12. $\dfrac{2s-1}{s^2-4s+6}$

13. $\dfrac{4s^2+13s+19}{(s-1)(s^2+4s+13)}$

14. $\dfrac{s^2+16s+9}{(s+1)(s+3)(s-2)}$

15. $\dfrac{2s^2+3s-1}{(s+1)^2(s+2)}$

16. $\dfrac{1}{(s^2+9)^2}$

17. $\dfrac{e^{-2s}(4s+2)}{(s-1)(s+2)}$

18. Find the Taylor series for $f(t)=e^{-t^2}$ about $t=0$. Then, assuming that the Laplace transform of $f(t)$ can be computed term by term, find an expansion for $\mathcal{L}\{f\}(s)$ in powers of $1/s$.

In Problems 19–24, solve the given initial value problem for $y(t)$ using the method of Laplace transforms.

19. $y''-7y'+10y=0$;
$\quad y(0)=0$, $y'(0)=-3$

20. $y''+6y'+9y=0$;
$\quad y(0)=-3$, $y'(0)=10$

21. $y''+2y'+2y=t^2+4t$;
$\quad y(0)=0$, $y'(0)=-1$

22. $y''+9y=10e^{2t}$;
$\quad y(0)=-1$, $y'(0)=5$

23. $y''+3y'+4y=u(t-1)$;
$\quad y(0)=0$, $y'(0)=1$

24. $y''-4y'+4y=t^2e^t$;
$\quad y(0)=0$, $y'(0)=0$

In Problems 25 and 26, find solutions to the given initial value problem.

25. $ty''+2(t-1)y'-2y=0$;
$\quad y(0)=0$, $y'(0)=0$

26. $ty''+2(t-1)y'+(t-2)y=0$;
$\quad y(0)=1$, $y'(0)=-1$

In Problems 27 and 28, solve the given equation for $y(t)$.

27. $y(t)+\displaystyle\int_0^t (t-v)y(v)\,dv=e^{-3t}$

28. $y'(t)-2\displaystyle\int_0^t y(v)\sin(t-v)\,dv=1$;
$\quad y(0)=-1$

29. A linear system is governed by

$$y''-5y'+6y=g(t)\,.$$

Find the transfer function and the impulse response function.

30. Solve the symbolic initial value problem

$$y''+4y=\delta\!\left(t-\frac{\pi}{2}\right);$$

$$y(0)=0\,,\qquad y'(0)=1\,.$$

In Problems 31 and 32, use Laplace transforms to solve the given system.

31. $x'+y=0$; $x(0)=0$,
$\quad x+y'=1-u(t-2)$; $y(0)=0$

32. $x''+2y'=u(t-3)$; $x(0)=1$, $x'(0)=-1$,
$\quad x+y'=y$; $y(0)=0$

TECHNICAL WRITING EXERCISES FOR CHAPTER 7

1. Compare the use of Laplace transforms in solving linear differential equations with constant coefficients with the use of logarithms in solving algebraic equations of the form $x^r=a$.

2. Explain why the method of Laplace transforms works so well for linear differential equations with constant coefficients *and* integro-differential equations involving a convolution.

3. Discuss several examples of initial value problems in which the method of Laplace transforms cannot be applied.

4. A linear system is said to be **asymptotically stable** if its impulse response function $h(t)\to 0$ as $t\to +\infty$. Assume $H(s)$, the Laplace transform of $h(t)$, is a rational function in reduced form with the degree of its numerator less than the degree of its denominator. Explain in detail how the asymptotic stability of the linear system can be characterized in terms of the zeros of the denominator of $H(s)$. Give examples.

5. Compare and contrast the solution of initial value problems by Laplace transforms versus the methods of Chapter 4.

A Duhamel's Formulas

For a linear system governed by the equation

$$(1) \qquad ay'' + by' + cy = g(t) ,$$

where a, b, c are real constants, the function

$$(2) \qquad H(s) := \frac{\mathscr{L}\{y\}(s)}{\mathscr{L}\{g\}(s)} = \frac{\mathscr{L}\{\text{output}\}}{\mathscr{L}\{\text{input}\}} ,$$

where all initial conditions are taken to be zero, is called the **transfer function** for the system. (As mentioned in Section 7.8, the transfer function $H(s)$ depends only on the constants a, b, c of the system; it is not affected by the choice of g.) If the output function $g(t)$ is the unit step function $u(t)$, then equation (2) yields

$$\mathscr{L}\{y\}(s) = \mathscr{L}\{u\}(s)H(s) = \frac{H(s)}{s} .$$

The solution (output function) in this special case is called the **indicial admittance** and is denoted by $A(t)$. Hence, $\mathscr{L}\{A\}(s) = H(s)/s$.

It is possible to express the response $y(t)$ of the system to a general input function $g(t)$ in terms of $g(t)$ and the indicial admittance $A(t)$. To derive these relations, proceed as follows:

(a) Show that

$$(3) \qquad \mathscr{L}\{y\}(s) = s\mathscr{L}\{A\}(s)\,\mathscr{L}\{g\}(s) .$$

(b) Now apply the convolution theorem to (3) and show that

$$(4) \qquad y(t) = \frac{d}{dt}\left[\int_0^t A(t - v)g(v)\,dv \right] = \frac{d}{dt}\left[\int_0^t A(v)g(t - v)\,dv \right] .$$

(c) To perform the differentiation indicated in (4), one can use **Leibniz's rule:**

$$\frac{d}{dt}\left[\int_{a(t)}^{b(t)} f(v, t)\,dv \right] = \int_{a(t)}^{b(t)} \frac{\partial f}{\partial t}(v, t)\,dv + f(b(t), t)\frac{db}{dt}(t) - f(a(t), t)\frac{da}{dt}(t) ,$$

where f and $\partial f/\partial t$ are assumed continuous in v and t, and $a(t)$, $b(t)$ are differentiable functions of t. Applying this rule to (4), derive the formulas

$$(5) \qquad y(t) = \int_0^t A'(t - v)g(v)\,dv ,$$

$$(6) \qquad y(t) = \int_0^t A(v)g'(t - v)\,dv + A(t)g(0) .$$

[*Hint:* Recall that the initial conditions on A are $A(0) = A'(0) = 0$.]

(d) In equations (5) and (6), make the change of variables $w = t - v$ and show that

$$(7) \qquad y(t) = \int_0^t A'(w)g(t-w)\,dw\,,$$

$$(8) \qquad y(t) = \int_0^t A(t-w)g'(w)\,dw + A(t)g(0)\,.$$

Equations (5)–(8) are referred to as **Duhamel's formulas,** in honor of the French mathematician J. M. C. Duhamel. These formulas are helpful in determining the response of the system to a general input $g(t)$, since the indicial admittance of the system can be determined experimentally by measuring the response of the system to a unit step function.

(e) The impulse response function $h(t)$ is defined as $h(t) := \mathcal{L}^{-1}\{H\}(t)$, where $H(s)$ is the transfer function. Show that $h(t) = A'(t)$, so that equations (5) and (7) can be written in the form

$$(9) \qquad y(t) = \int_0^t h(t-v)g(v)\,dv = \int_0^t h(v)g(t-v)\,dv\,.$$

We remark that the indicial admittance is the response of the system to a unit step function, and the impulse response function is the response to the unit impulse or delta function (see Section 7.9). But the delta function is the derivative (in a generalized sense) of the unit step function. Therefore, the fact that $h(t) = A'(t)$ is not really surprising.

B Frequency Response Modeling

Frequency response modeling of a linear system is based on the premise that the dynamics of a linear system can be recovered from a knowledge of how the system responds to sinusoidal inputs. (This will be made mathematically precise in Theorem 13.) In other words, to determine (or identify) a linear system, all one has to do is observe how the system reacts to sinusoidal inputs.

Let's assume that we have a linear system governed by

$$(10) \qquad y'' + py' + qy = g(t)\,,$$

where p and q are real constants. The function $g(t)$ is called the **forcing function** or **input function.** When $g(t)$ is a sinusoid, the particular solution to (10) obtained by the method of undetermined coefficients is the **steady-state solution** or **output function** $y_{ss}(t)$ corresponding to $g(t)$. We can think of a linear system as a compartment or block into which goes an input function g and out of which comes the output function y_{ss} (see Figure 7.32). To **identify** a linear system means to determine the coefficients p and q in equation (10).

Figure 7.32 Block diagram depicting a linear system

It will be convenient for us to work with complex variables. A complex number z is usually expressed in the form $z = \alpha + i\beta$, with α, β real numbers and i denoting $\sqrt{-1}$. We can also express z in polar form, $z = re^{i\theta}$, where $r^2 = \alpha^2 + \beta^2$ and $\tan\theta = \beta/\alpha$. Here $r \ (\geq 0)$ is called the **magnitude** and θ the **phase angle** of z.

The following theorem gives the relationship between the linear system and its response to sinusoidal inputs in terms of the **transfer function** $H(s)$ [see Project A, equation (2), page 417].

Steady-State Solutions to Sinusoidal Inputs

Theorem 13. Let $H(s)$ be the transfer function for equation (10). If $H(s)$ is finite at $s = i\omega$, with ω real, then the steady-state solution to (10) for $g(t) = e^{i\omega t}$ is

(11) $y_{ss}(t) = H(i\omega)e^{i\omega t} = H(i\omega)\{\cos\omega t + i\sin\omega t\}$.

(a) Prove Theorem 13. [*Hint:* Guess $y_{ss}(t) = Ae^{i\omega t}$ and show that $A = H(i\omega)$.]
(b) Use Theorem 13 to show that if $g(t) = \sin\omega t$, then the steady-state solution to (10) is
$M(\omega)\sin[\omega t + N(\omega)]$, where $H(i\omega) = M(\omega)e^{iN(\omega)}$ is the polar form for $H(i\omega)$.
(c) Solve for $M(\omega)$ and $N(\omega)$ in terms of p and q.
(d) Experimental results for modeling done by frequency response methods are usually presented in **frequency response**[†] or **Bode plots.** There are two types of Bode plots. The first is the log of the magnitude $M(\omega)$ of $H(i\omega)$ versus the angular frequency ω using a log scale for ω. The second is a plot of the phase angle or argument $N(\omega)$ of $H(i\omega)$ versus the angular frequency using a log scale for ω. The Bode plots for the transfer function $H(s) = (1 + 0.2s + s^2)^{-1}$ are given in Figure 7.33.

 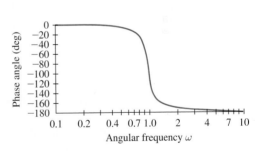

Figure 7.33 Bode plots for $H(i\omega) = [1 + 0.2(i\omega) + (i\omega)^2]^{-1}$

Sketch the Bode plots of the linear system governed by equation (10) with $p = 0.4$ and $q = 1.0$. Use $\omega = 0.3, 0.6, 0.9, 1.2,$ and 1.5 for the plot of $M(\omega)$ and $\omega = 0.5, 0.8, 1, 2,$ and 5 for the plot of $N(\omega)$.
(e) Assume we know that $q = 1$. When we input a sine wave with $\omega = 2$, the system settles into a steady-state sinusoidal output with magnitude $M(2) = 0.333$. Find p and thus identify the linear system.
(f) Suppose a sine wave input with $\omega = 2$ produces a steady-state sinusoidal output with magnitude $M(2) = 0.5$ and that when $\omega = 4$, then $M(4) = 0.1$. Find p and q and thus identify the system.

[†]Frequency response curves are also discussed in Section 4.10.

In most applications there are some inaccuracies in the measurement of the magnitudes and frequencies. To compensate for these errors, sinusoids with several different frequencies are used as input. A least-squares approximation for p and q is then found. For a discussion of frequency response modeling as a mathematical modeling tool, see the chapter by W. F. Powers, "Modeling Linear Systems by Frequency Response Methods," in *Differential Equations Models*, by M. Braun, C. Coleman, and D. Drew (eds.) (Springer-Verlag, New York, 1983), Chapter 9. Additional examples may be found in *Schaum's Outline on Feedback and Control Systems*, by J. J. DiStefano, A. R. Stubberud, and I. J. Williams (McGraw-Hill, New York, 1995, 3rd edition), Chapter 15.

C Determining System Parameters

In mechanical design one sometimes must determine system parameters even though information on the system forces is incomplete. For example, the differential equation governing the motion $x(t)$ of an externally forced damped mass–spring oscillator was shown in Section 4.1 (page 152) to be

$$m\frac{d^2x}{dt^2} + b\frac{dx}{dt} + kx = f(t) ,$$

where $f(t)$ is the external force; the other parameters are defined in Section 4.1. Assume the system is underdamped ($b^2 < 4mk$; see Section 4.9) and starts from rest $[x(0) = 0, x'(0) = 0]$ and that the force is bounded: $|f(t)| \leq A$ for all t.

(a) Show that the transforms $X(s)$ and $F(s)$ of $x(t)$ and $f(t)$ are related by

$$X(s) = \frac{1}{ms^2 + bs + k}F(s) .$$

(b) Use the convolution theorem to derive the formula

$$x(t) = \frac{1}{\beta m} \int_0^t f(t - v)e^{-bv/2m} \sin\beta v \, dv ,$$

where

$$\beta := \frac{1}{2m}\sqrt{4mk - b^2} .$$

(c) Show that the motion $x(t)$ is bounded under these circumstances by

$$|x(t)| \leq \frac{At}{\beta m}$$

and also by

$$|x(t)| \leq \frac{2A}{\beta b} .$$

(d) Suppose the mass $m = 5$ kg, the spring constant $k = 3000$ N/m, and the force is bounded by $A = 10$ N. What range of values for the damping constant b will ensure a displacement of 1 cm (0.01 m) or less?

CHAPTER 8

Series Solutions of Differential Equations

8.1 Introduction: The Taylor Polynomial Approximation

Probably the best tool for numerically approximating a function $f(x)$ near a particular point x_0 is the *Taylor polynomial*. The formula for the Taylor polynomial of degree n centered at x_0, approximating a function $f(x)$ possessing n derivatives at x_0, is given by

$$
(1) \qquad p_n(x) = f(x_0) + f'(x_0)(x - x_0) + \frac{f''(x_0)}{2!}(x - x_0)^2
$$

$$
+ \frac{f'''(x_0)}{3!}(x - x_0)^3 + \cdots + \frac{f^{(n)}(x_0)}{n!}(x - x_0)^n
$$

$$
= \sum_{j=0}^{n} \frac{f^{(j)}(x_0)}{j!}(x - x_0)^j \,.
$$

This polynomial matches the value of f and the values of its derivatives, up to the order of the polynomial, at the point x_0:

$$
p_n(x_0) = f(x_0) \,,
$$
$$
p_n'(x_0) = f'(x_0) \,,
$$
$$
p_n''(x_0) = f''(x_0) \,,
$$
$$
\vdots
$$
$$
p_n^{(n)}(x_0) = f^{(n)}(x_0) \,.
$$

For example, the first four Taylor polynomials for e^x, expanded around $x_0 = 0$, are

$$
(2) \qquad
\begin{aligned}
p_0(x) &= 1 \,, \\
p_1(x) &= 1 + x \,, \\
p_2(x) &= 1 + x + \frac{x^2}{2} \,, \\
p_3(x) &= 1 + x + \frac{x^2}{2} + \frac{x^3}{6} \,.
\end{aligned}
$$

Their efficacy in approximating the exponential function is demonstrated in Figure 8.1, page 422.

The Taylor polynomial of degree n differs from the polynomial of the next lower degree only in the addition of a single term:

$$
p_n(x) = p_{n-1}(x) + \frac{f^{(n)}(x_0)}{n!}(x - x_0)^n \,,
$$

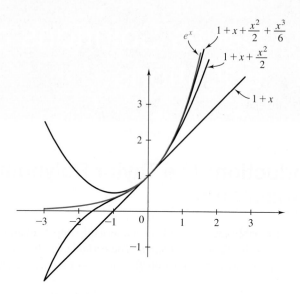

Figure 8.1 Graphs of Taylor polynomials for e^x

so a listing like (2) is clearly redundant—one can read off $p_0(x), p_1(x)$, and $p_2(x)$ from the formula for $p_3(x)$. In fact, if f is infinitely differentiable, $p_n(x)$ is just the $(n+1)$st partial sum of the **Taylor series**[†]

(3)
$$\sum_{j=0}^{\infty} \frac{f^{(j)}(x_0)}{j!} (x - x_0)^j .$$

Example 1 Determine the fourth-degree Taylor polynomials matching the functions e^x, $\cos x$, and $\sin x$ at $x_0 = 2$.

Solution For $f(x) = e^x$ we have $f^{(j)}(2) = e^2$ for each $j = 0, 1, \ldots$, so from (1) we obtain

$$e^x \approx e^2 + e^2(x-2) + \frac{e^2}{2!}(x-2)^2 + \frac{e^2}{3!}(x-2)^3 + \frac{e^2}{4!}(x-2)^4 .$$

For $f(x) = \cos x$, we have $f'(x) = -\sin x, f''(x) = -\cos x, f'''(x) = \sin x, f^{(4)}(x) = \cos x$, so that

$$\cos x \approx \cos 2 - (\sin 2)(x-2) - \frac{\cos 2}{2!}(x-2)^2 + \frac{\sin 2}{3!}(x-2)^3 + \frac{\cos 2}{4!}(x-2)^4 .$$

In a similar fashion we find

$$\sin x \approx \sin 2 + (\cos 2)(x-2) - \frac{\sin 2}{2!}(x-2)^2 - \frac{\cos 2}{3!}(x-2)^3$$
$$+ \frac{\sin 2}{4!}(x-2)^4 . \quad \blacklozenge$$

[†]Truncated Taylor series were introduced in Section 3.7 (page 132) as a tool for constructing recursive formulas for approximate solutions of differential equations.

To relate this approximation scheme to our theme (the solution of differential equations), we alter our point of view; we regard a differential equation *not* as a "condition to be satisfied," but as a prescription for constructing the Taylor polynomials for its solutions. Besides providing a very general method for computing accurate approximate solutions to the equation near any particular "starting" point, this interpretation also provides insight into the role of the initial conditions. The following example illustrates the method.

Example 2 Find the first few Taylor polynomials approximating the solution around $x_0 = 0$ of the initial value problem

$$y'' = 3y' + x^2 y; \qquad y(0) = 10, \qquad y'(0) = 5.$$

Solution To construct

$$p_n(x) = y(0) + y'(0)x + \frac{y''(0)}{2!}x^2 + \frac{y'''(0)}{3!}x^3 + \cdots + \frac{y^{(n)}(0)}{n!}x^n,$$

we need the values of $y(0), y'(0), y''(0), y'''(0)$, etc. The first two are provided by the given initial conditions. The value of $y''(0)$ can be deduced from the differential equation itself and the values of the lower derivatives:

$$y''(0) = 3y'(0) + 0^2 y(0) = 3 \cdot 5 + 0 \cdot 10 = 15.$$

Now since $y'' = 3y' + x^2 y$ holds for some interval around $x_0 = 0$, we can differentiate both sides to derive

$$y''' = 3y'' + 2xy + x^2 y',$$
$$y^{(4)} = 3y''' + 2y + 2xy' + 2xy' + x^2 y'' = 3y''' + 2y + 4xy' + x^2 y'',$$
$$y^{(5)} = 3y^{(4)} + 2y' + 4y' + 4xy'' + 2xy'' + x^2 y''' = 3y^{(4)} + 6y' + 6xy'' + x^2 y'''.$$

Thus on substituting $x = 0$ we deduce, in turn, that

$$y'''(0) = 3 \cdot 15 + 2 \cdot 0 \cdot 10 + 0^2 \cdot 5 = 45,$$
$$y^{(4)}(0) = 3 \cdot 45 + 2 \cdot 10 + 4 \cdot 0 \cdot 5 + 0^2 \cdot 15 = 155,$$
$$y^{(5)}(0) = 3 \cdot 155 + 6 \cdot 5 + 6 \cdot 0 \cdot 15 + 0^2 \cdot 45 = 495.$$

Consequently the Taylor polynomial of degree 5 for the solution is given by

$$p_5(x) = 10 + 5x + \frac{15}{2!}x^2 + \frac{45}{3!}x^3 + \frac{155}{4!}x^4 + \frac{495}{5!}x^5$$

$$= 10 + 5x + \frac{15}{2}x^2 + \frac{15}{2}x^3 + \frac{155}{24}x^4 + \frac{33}{8}x^5. \quad \blacklozenge$$

It is of interest to note that if the original equation in Example 2 were replaced by $y'' = 3y' + x^{1/3}y$, the third derivative would look like $y''' = 3y'' + y/(3x^{2/3}) + x^{1/3}y'$, and $y'''(0)$ would not exist. Only Taylor polynomials of degree 0 through 2 can be constructed for the solution to this problem.

The next example demonstrates the application of the Taylor polynomial method to a *nonlinear* equation.

Example 3 Determine the Taylor polynomial of degree 3 for the solution to the initial value problem

$$(4) \qquad y' = \frac{1}{x + y + 1}, \qquad y(0) = 0.$$

Solution Using $y(0) = 0$, we substitute $x = 0$ and $y = 0$ into equation (4) and find that $y'(0) = 1$. To determine $y''(0)$, we differentiate both sides of the equation in (4) with respect to x, thereby getting an expression for $y''(x)$ in terms of x, $y(x)$, and $y'(x)$. That is,

(5) $$y''(x) = (-1)[x + y(x) + 1]^{-2}[1 + y'(x)] .$$

Substituting $x = 0$, $y(0) = 0$, and $y'(0) = 1$ in (5), we obtain

$$y''(0) = (-1)(1)^{-2}(1 + 1) = -2 .$$

Similarly, differentiating (5) and substituting, we obtain

$$y'''(x) = 2[x + y(x) + 1]^{-3}[1 + y'(x)]^2 - [x + y(x) + 1]^{-2}y''(x) ,$$
$$y'''(0) = 2(1)^{-3}(1 + 1)^2 - (1)^{-2}(-2) = 10 .$$

Thus, the Taylor polynomial of degree 3 is

$$p_3(x) = 0 + x - x^2 + \frac{10}{6}x^3 = x - x^2 + \frac{5}{3}x^3 . \quad \blacklozenge$$

In a theoretical sense we can estimate the accuracy to which a Taylor polynomial $p_n(x)$ approximates its target function $f(x)$ for x near x_0. Indeed, if we let $\varepsilon_n(x)$ measure the accuracy of the approximation,

$$\varepsilon_n(x) := f(x) - p_n(x) ,$$

then calculus provides us with several formulas for estimating ε_n. The most transparent is due to Lagrange: *if the $(n + 1)$st derivative of f exists and is continuous on an interval containing x_0 and x, then*

(6) $$\varepsilon_n(x) = \frac{f^{(n+1)}(\xi)}{(n + 1)!}(x - x_0)^{n+1} ,$$

where ξ, although unknown, is guaranteed to lie between x_0 and x.[†]

Figure 8.1 on page 422 and equation (6) suggest that one might control the error in the Taylor polynomial approximation by increasing the degree n of the polynomial (i.e., taking more terms), thereby increasing the factor $(n + 1)!$ in the denominator. This possibility is limited, of course, by the number of times f can be differentiated. In Example 2, for instance, the solution did not have a fifth derivative at $x_0 = 0$ ($f^{(5)}(0)$ is "infinite"). Thus, we could not construct $p_5(x)$, nor could we conclude anything about the accuracy of $p_4(x)$ from the Lagrange formula.

However, for Example 3 we could, in theory, compute *every* derivative of the solution $y(x)$ at $x_0 = 0$, and speculate on the *convergence* of the Taylor series

$$\sum_{j=0}^{\infty} \frac{y^{(j)}(x_0)}{j!}(x - x_0)^j = \lim_{n \to \infty} \sum_{j=0}^{n} \frac{y^{(j)}(x_0)}{j!}(x - x_0)^j$$

to the solution $y(x)$. Now for nonlinear equations such as (4), the factor $f^{(n+1)}(\xi)$ in the Lagrange error formula may grow too rapidly with n, and the convergence can be thwarted. But if the differential equation is linear and its coefficients and nonhomogeneous term enjoy a feature known as *analyticity,* our wish is granted; the error does indeed diminish to zero as the degree n goes to infinity, and the sequence of Taylor polynomials can be guaranteed to converge to the actual solution on a certain (known) interval. For instance, the exponential,

[†]Equation (6) is proved by invoking the mean value theorem; see, e.g., *Principles of Mathematical Analysis,* 3rd ed., by Walter Rudin (McGraw-Hill, New York, 1976).

sine, and cosine functions in Example 1 all satisfy linear differential equations with constant coefficients, and their Taylor series converge to the corresponding function values. (Indeed, *all* their derivatives are bounded on any interval of finite length, so the Lagrange formula for their approximation errors approaches zero as n increases, for each value of x.) This topic is the theme for the early sections of this chapter.

In Sections 8.5 and 8.6 we'll see that solutions to second-order linear equations can exhibit very wild behavior near points x_0 where the coefficient of y'' is zero; so wild, in fact, that no Euler or Runge–Kutta algorithm could hope to keep up with them. But a clever modification of the Taylor polynomial method, due to Frobenius, provides very accurate approximations to the solutions in such regions. It is this latter feature, perhaps, that underscores the value of the Taylor methodology in the current practice of applied mathematics.

8.1 EXERCISES

In Problems 1–8, determine the first three nonzero terms in the Taylor polynomial approximations for the given initial value problem.

1. $y' = x^2 + y^2$; $y(0) = 1$

2. $y' = y^2$; $y(0) = 2$

3. $y' = \sin y + e^x$; $y(0) = 0$

4. $y' = \sin(x + y)$; $y(0) = 0$

5. $x'' + tx = 0$; $x(0) = 1$, $x'(0) = 0$

6. $y'' + y = 0$; $y(0) = 0$, $y'(0) = 1$

7. $y''(\theta) + y(\theta)^3 = \sin\theta$;
 $y(0) = 0$, $y'(0) = 0$

8. $y'' + \sin y = 0$; $y(0) = 1$, $y'(0) = 0$

9. **(a)** Construct the Taylor polynomial $p_3(x)$ of degree 3 for the function $f(x) = \ln x$ around $x = 1$.
 (b) Using the error formula (6), show that
 $$|\ln(1.5) - p_3(1.5)| \le \frac{(0.5)^4}{4} = 0.015625.$$
 (c) Compare the estimate in part (b) with the actual error by calculating $|\ln(1.5) - p_3(1.5)|$.
 (d) Sketch the graphs of $\ln x$ and $p_3(x)$ (on the same axes) for $0 < x < 2$.

10. **(a)** Construct the Taylor polynomial $p_3(x)$ of degree 3 for the function $f(x) = 1/(2-x)$ around $x = 0$.
 (b) Using the error formula (6), show that
 $$\left| f\left(\frac{1}{2}\right) - p_3\left(\frac{1}{2}\right) \right| = \left| \frac{2}{3} - p_3\left(\frac{1}{2}\right) \right| \le \frac{2}{3^5}.$$
 (c) Compare the estimate in part (b) with the actual error
 $$\left| \frac{2}{3} - p_3\left(\frac{1}{2}\right) \right|.$$

(d) Sketch the graphs of $1/(2-x)$ and $p_3(x)$ (on the same axes) for $-2 < x < 2$.

11. Argue that if $y = \phi(x)$ is a solution to the differential equation $y'' + p(x)y' + q(x)y = g(x)$ on the interval (a, b), where p, q, and g are each twice-differentiable, then the fourth derivative of $\phi(x)$ exists on (a, b).

12. Argue that if $y = \phi(x)$ is a solution to the differential equation $y'' + p(x)y' + q(x)y = g(x)$ on the interval (a, b), where p, q, and g possess derivatives of all orders, then ϕ has derivatives of all orders on (a, b).

13. **Duffing's Equation.** In the study of a nonlinear spring with periodic forcing, the following equation arises:

$$y'' + ky + ry^3 = A\cos\omega t.$$

 Let $k = r = A = 1$ and $\omega = 10$. Find the first three nonzero terms in the Taylor polynomial approximations to the solution with initial values $y(0) = 0$, $y'(0) = 1$.

14. **Soft versus Hard Springs.** For Duffing's equation given in Problem 13, the behavior of the solutions changes as r changes sign. When $r > 0$, the restoring force $ky + ry^3$ becomes stronger than for the linear spring ($r = 0$). Such a spring is called **hard.** When $r < 0$, the restoring force becomes weaker than the linear spring and the spring is called **soft.** Pendulums act like soft springs.

 (a) Redo Problem 13 with $r = -1$. Notice that for the initial conditions $y(0) = 0, y'(0) = 1$, the soft and hard springs appear to respond in the same way for t small.
 (b) Keeping $k = A = 1$ and $\omega = 10$, change the initial conditions to $y(0) = 1$ and $y'(0) = 0$. Now redo Problem 13 with $r = \pm 1$.
 (c) Based on the results of part (b), is there a difference between the behavior of soft and hard springs for t small? Describe.

15. The solution to the initial value problem

$$xy''(x) + 2y'(x) + xy(x) = 0 \ ;$$
$$y(0) = 1 \ , \qquad y'(0) = 0$$

has derivatives of all orders at $x = 0$ (although this is far from obvious). Use L'Hôpital's rule to compute the Taylor polynomial of degree 2 approximating this solution.

16. van der Pol Equation. In the study of the vacuum tube, the following equation is encountered:

$$y'' + (0.1)(y^2 - 1)y' + y = 0 \ .$$

Find the Taylor polynomial of degree 4 approximating the solution with the initial values $y(0) = 1$, $y'(0) = 0$.

8.2 Power Series and Analytic Functions

The differential equations studied in earlier sections often possessed solutions $y(x)$ that could be written in terms of elementary functions such as polynomials, exponentials, sines, and cosines. However, many important equations arise whose solutions cannot be so expressed. In the previous chapters, when we encountered such an equation we either settled for expressing the solution as an integral (see Exercises 2.2, Problem 27, page 46) or as a numerical approximation (Sections 3.6, 3.7, and 5.3). However, the Taylor polynomial approximation scheme of the preceding section suggests another possibility. Suppose the differential equation (and initial conditions) permit the computation of *every* derivative $y^{(n)}$ at the expansion point x_0. Are there any conditions that would guarantee that the sequence of Taylor polynomials would *converge* to the solution $y(x)$ as the degree of the polynomials tends to infinity:

$$\lim_{n \to \infty} \sum_{j=0}^{n} \frac{y^{(j)}(x_0)}{j!}(x - x_0)^j = \sum_{j=0}^{\infty} \frac{y^{(j)}(x_0)}{j!}(x - x_0)^j = y(x) \ ?$$

In other words, when can we be sure that a solution to a differential equation is represented by its *Taylor series*? As we'll see, the answer is quite favorable, and it enables a powerful new technique for solving equations.

The most efficient way to begin an exploration of this issue is by investigating the algebraic and convergence properties of generic expressions that include Taylor series—"long polynomials" so to speak, or more conventionally, *power series*.

Power Series

A **power series** about the point x_0 is an expression of the form

$$(1) \qquad \sum_{n=0}^{\infty} a_n(x - x_0)^n = a_0 + a_1(x - x_0) + a_2(x - x_0)^2 + \cdots \ ,$$

where x is a variable and the a_n's are constants. We say that (1) **converges** at the point $x = c$ if the infinite series (of real numbers) $\sum_{n=0}^{\infty} a_n(c - x_0)^n$ converges; that is, the limit of the partial sums,

$$\lim_{N \to \infty} \sum_{n=0}^{N} a_n(c - x_0)^n \ ,$$

exists (as a finite number). If this limit does not exist, the power series is said to **diverge** at $x = c$. Observe that (1) converges at $x = x_0$, since

$$\sum_{n=0}^{\infty} a_n(x_0 - x_0)^n = a_0 + 0 + 0 + \cdots = a_0 \ .$$

But what about convergence for other values of x? As stated in the following Theorem 1, a power series of the form (1) converges for all values of x in some "interval" centered at x_0 and diverges for x outside this interval. Moreover, at the interior points of this interval, the power series **converges absolutely** in the sense that $\sum_{n=0}^{\infty}|a_n(x-x_0)^n|$ converges. [Recall that absolute convergence of a series implies (ordinary) convergence of the series.]

Radius of Convergence

Theorem 1. For each power series of the form (1), there is a number ρ $(0 \le \rho \le \infty)$, called the **radius of convergence** of the power series, such that (1) converges absolutely for $|x-x_0| < \rho$ and diverges for $|x-x_0| > \rho$. (See Figure 8.2.)

 If the series (1) converges for all values of x, then $\rho = \infty$. When the series (1) converges only at x_0, then $\rho = 0$.

Figure 8.2 Interval of convergence

Notice that Theorem 1 settles the question of convergence except at the endpoints $x_0 \pm \rho$. Thus, these two points require separate analysis. To determine the radius of convergence ρ, one method that is often easy to apply is the ratio test.

Ratio Test for Power Series

Theorem 2. If, for n large, the coefficients a_n are nonzero and satisfy

$$\lim_{n\to\infty}\left|\frac{a_n}{a_{n+1}}\right| = L \qquad (0 \le L \le \infty),$$

then the radius of convergence of the power series $\sum_{n=0}^{\infty} a_n(x-x_0)^n$ is $\rho = L$.

Remark. We caution that if the ratio $|a_n/a_{n+1}|$ does not have a limit, then methods other than the ratio test (e.g., root test) must be used to determine ρ. In particular, if infinitely many of the a_n's are zero, then the ratio test cannot be directly applied. (However, Problem 7 demonstrates how to apply the result for series containing only "even-order" or "odd-order" terms.)

Example 1 Determine the convergence set of

(2) $$\sum_{n=0}^{\infty} \frac{(-2)^n}{n+1}(x-3)^n.$$

Solution Since $a_n = (-2)^n/(n+1)$, we have

$$\lim_{n\to\infty}\left|\frac{a_n}{a_{n+1}}\right| = \lim_{n\to\infty}\left|\frac{(-2)^n(n+2)}{(-2)^{n+1}(n+1)}\right|$$

$$= \lim_{n\to\infty}\frac{n+2}{2(n+1)} = \frac{1}{2} = L.$$

By the ratio test, the radius of convergence is $\rho = 1/2$. Hence, the series (2) converges absolutely for $|x-3| < 1/2$ and diverges when $|x-3| > 1/2$. It remains only to determine what happens when $|x-3| = 1/2$, that is, when $x = 5/2$ and $x = 7/2$.

Set $x = 5/2$, and the series (2) becomes the harmonic series $\sum_{n=0}^{\infty}(n+1)^{-1}$, which is known to diverge. When $x = 7/2$, the series (2) becomes an **alternating** harmonic series, which is known to converge. Thus, the power series converges for each x in the half-open interval $(5/2, 7/2]$; outside this interval it diverges. ◆

For each value of x for which the power series $\sum_{n=0}^{\infty} a_n(x-x_0)^n$ converges, we get a number that is the sum of the series. It is appropriate to denote this sum by $f(x)$, since its value depends on the choice of x. Thus, we write

$$f(x) = \sum_{n=0}^{\infty} a_n(x-x_0)^n$$

for all numbers x in the convergence interval. For example, the **geometric series** $\sum_{n=0}^{\infty} x^n$ has the radius of convergence $\rho = 1$ and the sum function $f(x) = 1/(1-x)$; that is,

(3) $$\frac{1}{1-x} = 1 + x + x^2 + \cdots = \sum_{n=0}^{\infty} x^n \quad \text{for} \quad -1 < x < 1.$$

In this chapter we'll frequently appeal to the following basic property of power series.

Power Series Vanishing on an Interval

Theorem 3. If $\sum_{n=0}^{\infty} a_n(x-x_0)^n = 0$ for all x in some open interval, then each coefficient a_n equals zero.

Given two power series

(4) $$f(x) = \sum_{n=0}^{\infty} a_n(x-x_0)^n, \quad g(x) = \sum_{n=0}^{\infty} b_n(x-x_0)^n,$$

with nonzero radii of convergence, we want to find power series representations for the sum, product, and quotient of the functions $f(x)$ and $g(x)$. The sum is simply obtained by termwise addition:

$$f(x) + g(x) = \sum_{n=0}^{\infty} (a_n + b_n)(x-x_0)^n$$

for all x in the common interval of convergence of the power series in (4). The power series representation for the product $f(x)g(x)$ is a bit more complicated. To provide motivation for

the formula, we treat the power series for $f(x)$ and $g(x)$ as "long polynomials," apply the distributive law, and group the terms in powers of $(x - x_0)$:

$$[a_0 + a_1(x - x_0) + a_2(x - x_0)^2 + \cdots] \cdot [b_0 + b_1(x - x_0) + b_2(x - x_0)^2 + \cdots]$$
$$= a_0b_0 + (a_0b_1 + a_1b_0)(x - x_0) + (a_0b_2 + a_1b_1 + a_2b_0)(x - x_0)^2 + \cdots.$$

The general formula for the product is

(5) $$f(x)g(x) = \sum_{n=0}^{\infty} c_n(x - x_0)^n,$$

where

(6) $$c_n := \sum_{k=0}^{n} a_k b_{n-k}.$$

The power series in (5) is called the **Cauchy product,** and it will converge for all x in the common *open* interval of convergence for the power series of f and g.[†]

The quotient $f(x)/g(x)$ will also have a power series expansion about x_0, provided $g(x_0) \neq 0$. However, the radius of convergence for this quotient series may be smaller than that for $f(x)$ or $g(x)$. Unfortunately, there is no nice formula for obtaining the coefficients in the power series for $f(x)/g(x)$. However, we can use the Cauchy product to divide power series indirectly (see Problem 15 on page 434). The quotient series can also be obtained by formally carrying out polynomial long division (see Problem 16).

The next theorem explains, in part, why power series are so useful.

Differentiation and Integration of Power Series

Theorem 4. If the series $f(x) = \sum_{n=0}^{\infty} a_n(x - x_0)^n$ has a positive radius of convergence ρ, then f is differentiable in the interval $|x - x_0| < \rho$ and termwise differentiation gives the power series for the derivative:

$$f'(x) = \sum_{n=1}^{\infty} na_n(x - x_0)^{n-1} \qquad \text{for} \qquad |x - x_0| < \rho.$$

Furthermore, termwise integration gives the power series for the integral of f:

$$\int f(x)\,dx = \sum_{n=0}^{\infty} \frac{a_n}{n+1}(x - x_0)^{n+1} + C \quad \text{for} \quad |x - x_0| < \rho.$$

Example 2 Starting with the geometric series (3) for $1/(1-x)$, find a power series for each of the following functions:

(a) $\dfrac{1}{1 + x^2}$. (b) $\dfrac{1}{(1-x)^2}$. (c) $\arctan x$.

Solution (a) Replacing x by $-x^2$ in (3) immediately gives

(7) $$\frac{1}{1 + x^2} = 1 - x^2 + x^4 - x^6 + \cdots + (-1)^n x^{2n} + \cdots.$$

[†]Actually, it may happen that the radius of convergence of the power series for $f(x)g(x)$ or $f(x) + g(x)$ is larger than that for the power series of f or g.

(b) Notice that $1/(1-x)^2$ is the derivative of the function $f(x) = 1/(1-x)$. Hence, on differentiating (3) term by term, we obtain

(8) $\qquad f'(x) = \dfrac{1}{(1-x)^2} = 1 + 2x + 3x^2 + 4x^3 + \cdots + nx^{n-1} + \cdots$.

(c) Since

$$\arctan x = \int_0^x \frac{1}{1+t^2}\,dt ,$$

we can integrate the series in (7) termwise to obtain the series for $\arctan x$. Thus,

$$\int_0^x \frac{1}{1+t^2}\,dt = \int_0^x \{1 - t^2 + t^4 - t^6 + \cdots + (-1)^n t^{2n} + \cdots\}\,dt$$

(9) $\qquad \arctan x = x - \dfrac{1}{3}x^3 + \dfrac{1}{5}x^5 - \dfrac{1}{7}x^7 + \cdots + \dfrac{(-1)^n x^{2n+1}}{2n+1} + \cdots$. ◆

It is important to keep in mind that since the geometric series (3) has the (open) interval of convergence $(-1, 1)$, the representations (7), (8), and (9) are at least valid in this interval. [Actually, the series (9) for $\arctan x$ converges for all $|x| \le 1$.]

Shifting the Summation Index

The index of summation in a power series is a dummy index just like the variable of integration in a definite integral. Hence,

$$\sum_{n=0}^{\infty} a_n(x - x_0)^n = \sum_{k=0}^{\infty} a_k(x - x_0)^k = \sum_{i=0}^{\infty} a_i(x - x_0)^i .$$

Just as there are times when we want to change the variable of integration, there are situations (and we will encounter many in this chapter) when it is desirable to change or shift the index of summation. This is particularly important when one has to combine two different power series.

Example 3 Express the series

$$\sum_{n=2}^{\infty} n(n-1)a_n x^{n-2}$$

as a series where the generic term is x^k instead of x^{n-2}.

Solution Writing out a few terms of the series, we have

$$\sum_{n=2}^{\infty} n(n-1)a_n x^{n-2}$$

$$= 2 \cdot 1 a_2 x^{2-2} + 3 \cdot 2 a_3 x^{3-2} + 4 \cdot 3 a_4 x^{4-2} + 5 \cdot 4 a_5 x^{5-2} + 6 \cdot 5 a_6 x^{6-2} + \cdots$$

$$= 2 \cdot 1 a_2 x^0 \quad + 3 \cdot 2 a_3 x^1 \quad + 4 \cdot 3 a_4 x^2 \quad + 5 \cdot 4 a_5 x^3 \quad + 6 \cdot 5 a_6 x^4 \quad + \cdots .$$

Relabeling the terms by the exponents of x, we get

$$\sum_{k=0}^{\infty} (k+2)(k+1)a_{k+2} x^k$$

In effect we have implemented the change of variables $k = n - 2$, or $n = k + 2$; thus when $n = 2$, then $k = 0$. ◆

We employ this technique for changing indices in the next two examples.

Example 4 Show that

$$x^3 \sum_{n=0}^{\infty} n^2(n-2)a_n x^n = \sum_{n=3}^{\infty} (n-3)^2(n-5)a_{n-3} x^n .$$

Solution We start by taking the x^3 inside the summation on the left-hand side:

$$x^3 \sum_{n=0}^{\infty} n^2(n-2)a_n x^n = \sum_{n=0}^{\infty} n^2(n-2)a_n x^{n+3} .$$

To rewrite this with generic term x^k, we set $k = n + 3$. Thus $n = k - 3$, and $n = 0$ corresponds to $k = 3$. Straightforward substitution thus yields

$$\sum_{n=0}^{\infty} n^2(n-2)a_n x^{n+3} = \sum_{k=3}^{\infty} (k-3)^2(k-5)a_{k-3} x^k .$$

By replacing k by n, we obtain the desired form. ◆

Example 5 Show that the identity

$$\sum_{n=1}^{\infty} n a_{n-1} x^{n-1} + \sum_{n=2}^{\infty} b_n x^{n+1} = 0$$

implies that $a_0 = a_1 = a_2 = 0$ and $a_n = -b_{n-1}/(n+1)$ for $n \geq 3$.

Solution First, we rewrite both series in terms of x^k. For the first series, we set $k = n - 1$, and hence $n = k + 1$, to write

$$\sum_{n=1}^{\infty} n a_{n-1} x^{n-1} = \sum_{k=0}^{\infty} (k+1)a_k x^k .$$

Then with $k = n + 1$, $n = k - 1$, the second series becomes

$$\sum_{n=2}^{\infty} b_n x^{n+1} = \sum_{k=3}^{\infty} b_{k-1} x^k .$$

The identity thus states

$$\sum_{k=0}^{\infty} (k+1)a_k x^k + \sum_{k=3}^{\infty} b_{k-1} x^k = 0 ,$$

and so we have a power series that sums to zero; consequently, by Theorem 3, each of its coefficients equals zero. For $k = 3, 4, \ldots$, both series contribute to the coefficient of x^k, and thus we confirm that

$$(k+1)a_k + b_{k-1} = 0$$

or $a_k = -b_{k-1}/(k+1)$ for $k \geq 3$. For $k = 0, 1,$ or 2, only the first series contributes, and we find, in turn,

$$(0+1)a_0 = 0 ,$$
$$(1+1)a_1 = 0 ,$$
$$(2+1)a_2 = 0 .$$

Hence, $a_0 = a_1 = a_2 = 0$. ◆

Analytic Functions

Not all functions are expressible as power series. Those distinguished functions that can be so represented are called **analytic.**

Analytic Function

Definition 1. A function f is said to be **analytic at x_0** if, in an open interval about x_0, this function is the sum of a power series $\sum_{n=0}^{\infty} a_n(x - x_0)^n$ that has a positive radius of convergence.

For example, a polynomial function $b_0 + b_1 x + \cdots + b_n x^n$ is analytic at every x_0, since we can always rewrite it in the form $a_0 + a_1(x - x_0) + \cdots + a_n(x - x_0)^n$. A rational function $P(x)/Q(x)$, where $P(x)$ and $Q(x)$ are polynomials without a common factor, is an analytic function except at those x_0 for which $Q(x_0) = 0$. As you may recall from calculus, the elementary functions e^x, $\sin x$, and $\cos x$ are analytic for all x, while $\ln x$ is analytic for $x > 0$. Some familiar representations are

$$(10) \qquad e^x = 1 + x + \frac{x^2}{2!} + \frac{x^3}{3!} + \cdots \qquad = \sum_{n=0}^{\infty} \frac{x^n}{n!},$$

$$(11) \qquad \sin x = x - \frac{x^3}{3!} + \frac{x^5}{5!} - \cdots \qquad = \sum_{n=0}^{\infty} \frac{(-1)^n}{(2n+1)!} x^{2n+1},$$

$$(12) \qquad \cos x = 1 - \frac{x^2}{2!} + \frac{x^4}{4!} - \cdots \qquad = \sum_{n=0}^{\infty} \frac{(-1)^n}{(2n)!} x^{2n},$$

$$(13) \qquad \ln x = (x - 1) - \frac{1}{2}(x - 1)^2 + \frac{1}{3}(x - 1)^3 - \cdots = \sum_{n=1}^{\infty} \frac{(-1)^{n-1}}{n}(x - 1)^n,$$

where (10), (11), and (12) are valid for all x, whereas (13) is valid for x in the half-open interval $(0, 2]$.

From Theorem 4 on the differentiation of power series, we see that a function f analytic at x_0 is differentiable in a neighborhood of x_0. Moreover, because f' has a power series representation in this neighborhood, it too is analytic at x_0. Repeating this argument, we see that f'', $f^{(3)}$, etc., exist and are analytic at x_0. Consequently, if a function does not have derivatives of all orders at x_0, then it cannot be analytic at x_0. The function $f(x) = |x - 1|$ is not analytic at $x_0 = 1$ because $f'(1)$ does not exist; and $f(x) = x^{7/3}$ is not analytic at $x_0 = 0$ because $f'''(0)$ does not exist.

Now we can deduce something very specific about the possible power series that can represent an analytic function. If $f(x)$ is analytic at x_0, then (by definition) it is the sum of *some* power series that converges in a neighborhood of x_0:

$$f(x) = \sum_{n=0}^{\infty} a_n(x - x_0)^n.$$

By the reasoning in the previous paragraph, the derivatives of f have convergent power series representations

$$f'(x) = \sum_{n=0}^{\infty} na_n(x-x_0)^{n-1} = 0 + \sum_{n=1}^{\infty} na_n(x-x_0)^{n-1},$$

$$f''(x) = \sum_{n=0}^{\infty} n(n-1)a_n(x-x_0)^{n-2} = 0 + 0 + \sum_{n=2}^{\infty} n(n-1)a_n(x-x_0)^{n-2},$$

$$\vdots$$

$$f^{(j)}(x) = \sum_{n=0}^{\infty} n(n-1)\cdots\left(n-[j-1]\right)a_n(x-x_0)^{n-j},$$

$$= \sum_{n=j}^{\infty} n(n-1)\cdots\left(n-[j-1]\right)a_n(x-x_0)^{n-j},$$

$$\vdots$$

But if we evaluate these series at $x = x_0$, we learn that

$$f(x_0) = a_0,$$
$$f'(x_0) = 1\cdot a_1,$$
$$f''(x_0) = 2\cdot 1\cdot a_2,$$
$$\vdots$$
$$f^{(j)}(x_0) = j!\cdot a_j,$$
$$\vdots$$

that is, $a_j = f^{(j)}(x_0)/j!$ and the power series *must coincide* with the Taylor series[†]

$$\sum_{j=0}^{\infty} \frac{f^{(j)}(x_0)}{j!}(x-x_0)^j$$

about x_0. *Any power series—regardless of how it is derived—that converges in some neighborhood of x_0 to a function has to be the Taylor series of that function.* For example, the expansion for arctan x given in (9) of Example 2 must be its Taylor expansion.

With these facts in mind we are ready to turn to the study of the effectiveness of power series techniques for solving differential equations. In the next sections, you will find it helpful to keep in mind that if f and g are analytic at x_0, then so are $f + g$, cf, fg, and f/g if $g(x_0) \neq 0$. These facts follow from the algebraic properties of power series discussed earlier.

8.2 EXERCISES

In Problems 1–6, determine the convergence set of the given power series.

1. $\displaystyle\sum_{n=0}^{\infty} \frac{2^{-n}}{n+1}(x-1)^n$

2. $\displaystyle\sum_{n=0}^{\infty} \frac{3^n}{n!}x^n$

3. $\displaystyle\sum_{n=0}^{\infty} \frac{n^2}{2^n}(x+2)^n$

4. $\displaystyle\sum_{n=1}^{\infty} \frac{4}{n^2+2n}(x-3)^n$

5. $\displaystyle\sum_{n=1}^{\infty} \frac{3}{n^3}(x-2)^n$

6. $\displaystyle\sum_{n=0}^{\infty} \frac{(n+2)!}{n!}(x+2)^n$

[†]When the expansion point x_0 is zero, the Taylor series is also known as the **Maclaurin series.**

7. Sometimes the ratio test (Theorem 2) can be applied to a power series containing an infinite number of zero coefficients, provided the zero pattern is regular. Use Theorem 2 to show, for example, that the series

$$a_0 + a_2 x^2 + a_4 x^4 + a_6 x^6 + \cdots = \sum_{k=0}^{\infty} a_{2k} x^{2k}$$

has a radius of convergence $\rho = \sqrt{L}$, if

$$\lim_{n \to \infty} \left| \frac{a_{2k}}{a_{2k+2}} \right| = L ,$$

and that

$$a_1 x + a_3 x^3 + a_5 x^5 + a_7 x^7 + \cdots$$
$$= \sum_{k=0}^{\infty} a_{2k+1} x^{2k+1}$$

has a radius of convergence $\rho = \sqrt{M}$, if

$$\lim_{k \to \infty} \left| \frac{a_{2k+1}}{a_{2k+3}} \right| = M .$$

[*Hint:* Let $z = x^2$.]

8. Determine the convergence set of the given power series.

(a) $\displaystyle\sum_{k=0}^{\infty} 2^{2k} x^{2k}$ (b) $\displaystyle\sum_{k=0}^{\infty} 2^{2k+1} x^{2k+1}$

(c) $\sin x$ [equation (11)] (d) $\cos x$ [equation (12)]

(e) $(\sin x)/x = \displaystyle\sum_{n=0}^{\infty} (-1)^n x^{2n}/(2n+1)!$

(f) $\displaystyle\sum_{k=0}^{\infty} 2^{2k} x^{4k}$

In Problems 9 and 10, find the power series expansion $\sum_{n=0}^{\infty} a_n x^n$ for $f(x) + g(x)$, given the expansions for $f(x)$ and $g(x)$.

9. $f(x) = \displaystyle\sum_{n=0}^{\infty} \frac{1}{n+1} x^n$, $g(x) = \displaystyle\sum_{n=1}^{\infty} 2^{-n} x^{n-1}$

10. $f(x) = \displaystyle\sum_{n=3}^{\infty} \frac{2^n}{n!} (x-1)^{n-2}$, $g(x) = \displaystyle\sum_{n=1}^{\infty} \frac{n^2}{2^n} (x-1)^{n-1}$

In Problems 11–14, find the first three nonzero terms in the power series expansion for the product $f(x)g(x)$.

11. $f(x) = e^x = \displaystyle\sum_{n=0}^{\infty} \frac{1}{n!} x^n$,

$g(x) = \sin x = \displaystyle\sum_{k=0}^{\infty} \frac{(-1)^k}{(2k+1)!} x^{2k+1}$

12. $f(x) = \sin x = \displaystyle\sum_{k=0}^{\infty} \frac{(-1)^k}{(2k+1)!} x^{2k+1}$,

$g(x) = \cos x = \displaystyle\sum_{k=0}^{\infty} \frac{(-1)^k}{(2k)!} x^{2k}$

13. $f(x) = e^{-x} = \displaystyle\sum_{n=0}^{\infty} \frac{(-1)^n}{n!} x^n$,

$g(x) = (1+x)^{-1} = \displaystyle\sum_{n=0}^{\infty} (-1)^n x^n$

14. $f(x) = e^x = \displaystyle\sum_{n=0}^{\infty} \frac{1}{n!} x^n$,

$g(x) = e^{-x} = \displaystyle\sum_{n=0}^{\infty} \frac{(-1)^n}{n!} x^n$

15. Find the first few terms of the power series for the quotient

$$q(x) = \left(\sum_{n=0}^{\infty} \frac{1}{2^n} x^n \right) \bigg/ \left(\sum_{n=0}^{\infty} \frac{1}{n!} x^n \right)$$

by completing the following:

(a) Let $q(x) = \sum_{n=0}^{\infty} a_n x^n$, where the coefficients a_n are to be determined. Argue that $\sum_{n=0}^{\infty} x^n/2^n$ is the Cauchy product of $q(x)$ and $\sum_{n=0}^{\infty} x^n/n!$.

(b) Use formula (6) of the Cauchy product (page 429) to deduce the equations

$$\frac{1}{2^0} = a_0 , \quad \frac{1}{2} = a_0 + a_1 , \quad \frac{1}{2^2} = \frac{a_0}{2} + a_1 + a_2 ,$$

$$\frac{1}{2^3} = \frac{a_0}{6} + \frac{a_1}{2} + a_2 + a_3 , \ \dots .$$

(c) Solve the equations in part (b) to determine the constants a_0, a_1, a_2, a_3.

16. To find the first few terms in the power series for the quotient $q(x)$ in Problem 15, treat the power series in the numerator and denominator as "long polynomials" and carry out long division. That is, perform

$$1 + x + \frac{1}{2} x^2 + \cdots \ \overline{\left)\ 1 + \frac{1}{2} x + \frac{1}{4} x^2 + \cdots \right.} .$$

In Problems 17–20, find a power series expansion for $f'(x)$, given the expansion for $f(x)$.

17. $f(x) = (1+x)^{-1} = \displaystyle\sum_{n=0}^{\infty} (-1)^n x^n$

18. $f(x) = \sin x = \displaystyle\sum_{k=0}^{\infty} \frac{(-1)^k}{(2k+1)!} x^{2k+1}$

19. $f(x) = \displaystyle\sum_{k=0}^{\infty} a_k x^{2k}$ 20. $f(x) = \displaystyle\sum_{n=1}^{\infty} n a_n x^{n-1}$

In Problems 21 and 22, find a power series expansion for $g(x) := \int_0^x f(t) \, dt$, given the expansion for $f(x)$.

21. $f(x) = (1+x)^{-1} = \displaystyle\sum_{n=0}^{\infty} (-1)^n x^n$

22. $f(x) = \displaystyle\frac{\sin x}{x} = \sum_{k=0}^{\infty} \frac{(-1)^k}{(2k+1)!} x^{2k}$

In Problems 23–26, express the given power series as a series with generic term x^k.

23. $\displaystyle\sum_{n=1}^{\infty} n a_n x^{n-1}$

24. $\displaystyle\sum_{n=2}^{\infty} n(n-1) a_n x^{n+2}$

25. $\displaystyle\sum_{n=0}^{\infty} a_n x^{n+1}$

26. $\displaystyle\sum_{n=1}^{\infty} \frac{a_n}{n+3} x^{n+3}$

27. Show that

$$x^2 \sum_{n=0}^{\infty} n(n+1) a_n x^n = \sum_{n=2}^{\infty} (n-2)(n-1) a_{n-2} x^n .$$

28. Show that

$$2 \sum_{n=0}^{\infty} a_n x^{n+1} + \sum_{n=1}^{\infty} n b_n x^{n-1}$$

$$= b_1 + \sum_{n=1}^{\infty} [2 a_{n-1} + (n+1) b_{n+1}] x^n.$$

In Problems 29–34, determine the Taylor series about the point x_0 for the given functions and values of x_0.

29. $f(x) = \cos x$, $x_0 = \pi$

30. $f(x) = x^{-1}$, $x_0 = 1$

31. $f(x) = \dfrac{1+x}{1-x}$, $x_0 = 0$

32. $f(x) = \ln(1+x)$, $x_0 = 0$

33. $f(x) = x^3 + 3x - 4$, $x_0 = 1$

34. $f(x) = \sqrt{x}$, $x_0 = 1$

35. The Taylor series for $f(x) = \ln x$ about $x_0 = 1$ given in equation (13) can also be obtained as follows:

(a) Starting with the expansion $1/(1-s) = \sum_{n=0}^{\infty} s^n$ and observing that

$$\frac{1}{x} = \frac{1}{1+(x-1)},$$

obtain the Taylor series for $1/x$ about $x_0 = 1$.

(b) Since $\ln x = \int_1^x 1/t \, dt$, use the result of part (a) and termwise integration to obtain the Taylor series for $f(x) = \ln x$ about $x_0 = 1$.

36. Let $f(x)$ and $g(x)$ be analytic at x_0. Determine whether the following statements are always true or sometimes false:

(a) $3f(x) + g(x)$ is analytic at x_0

(b) $f(x)/g(x)$ is analytic at x_0

(c) $f'(x)$ is analytic at x_0

(d) $[f(x)]^3 - \int_{x_0}^x g(t) \, dt$ is analytic at x_0

37. Let

$$f(x) = \begin{cases} e^{-1/x^2}, & x \neq 0, \\ 0, & x = 0. \end{cases}$$

Show that $f^{(n)}(0) = 0$ for $n = 0, 1, 2, \ldots$ and hence that the Maclaurin series for $f(x)$ is $0 + 0 + 0 + \cdots$, which converges for all x but is equal to $f(x)$ only when $x = 0$. This is an example of a function possessing derivatives of all orders (at $x_0 = 0$), whose Taylor series converges, but the Taylor series (about $x_0 = 0$) does not converge to the original function! Consequently, this function is not analytic at $x = 0$.

38. Compute the Taylor series for $f(x) = \ln(1+x^2)$ about $x_0 = 0$. [*Hint:* Multiply the series for $(1+x^2)^{-1}$ by $2x$ and integrate.]

8.3 Power Series Solutions to Linear Differential Equations

In this section we demonstrate a method for obtaining a power series solution to a linear differential equation with polynomial coefficients. This method is easier to use than the Taylor series method discussed in Section 8.1 and sometimes gives a nice expression for the general term in the power series expansion. Knowing the form of the general term also allows us to test for the radius of convergence of the power series.

We begin by writing the linear differential equation

(1) $\qquad a_2(x) y'' + a_1(x) y' + a_0(x) y = 0$

in the standard form

(2) $\qquad y'' + p(x) y' + q(x) y = 0 ,$

where $p(x) := a_1(x)/a_2(x)$ and $q(x) := a_0(x)/a_2(x)$.

Ordinary and Singular Points

> **Definition 2.** A point x_0 is called an **ordinary point** of equation (2) if both p and q are analytic at x_0. If x_0 is not an ordinary point, it is called a **singular point** of the equation.

Example 1 Determine all the singular points of

$$xy'' + x(1-x)^{-1}y' + (\sin x)y = 0 .$$

Solution Dividing the equation by x, we find that

$$p(x) = \frac{x}{x(1-x)} , \qquad q(x) = \frac{\sin x}{x} .$$

The singular points are those points where $p(x)$ *or* $q(x)$ fails to be analytic. Observe that $p(x)$ and $q(x)$ are the ratios of functions that are everywhere analytic. Hence, $p(x)$ and $q(x)$ are analytic except, *perhaps,* when their denominators are zero. For $p(x)$ this occurs at $x = 0$ and $x = 1$. But since we can cancel an x in the numerator and denominator of $p(x)$, that is,

$$p(x) = \frac{x}{x(1-x)} = \frac{1}{1-x} ,$$

we see that $p(x)$ is actually analytic at $x = 0$.[†] Therefore, $p(x)$ is analytic except at $x = 1$. For $q(x)$, the denominator is zero at $x = 0$. Just as with $p(x)$, this zero is removable since $q(x)$ has the power series expansion

$$q(x) = \frac{\sin x}{x} = \frac{x - \dfrac{x^3}{3!} + \dfrac{x^5}{5!} - \cdots}{x} = 1 - \frac{x^2}{3!} + \frac{x^4}{5!} - \cdots .$$

Thus, $q(x)$ is everywhere analytic. Consequently, the only singular point of the given equation is $x = 1$. ◆

At an ordinary point x_0 of equation (1) (or (2)), the coefficient functions $p(x)$ and $q(x)$ are analytic. Hence, we might expect that the solutions to these equations inherit this property. From the discussion in Section 6.1 on linear equations, the continuity of p and q in a neighborhood of x_0 is sufficient to imply that equation (2) has two linearly independent solutions defined in that neighborhood. But analytic functions are not merely continuous—they possess derivatives of all orders in a neighborhood of x_0. Thus we can differentiate equation (2) to show that $y^{(3)}$ exists and, by a "bootstrap" argument, prove that solutions to (2) must likewise possess derivatives of all orders. Although we cannot conclude by this reasoning that the solutions enjoy the stronger property of analyticity, this is nonetheless the case (see Theorem 5 in Section 8.4, page 445). Hence, in a neighborhood of an ordinary point x_0, the solutions to (1) (or (2)) can be expressed as a power series about x_0.

To illustrate the power series method about an ordinary point, let's look at a simple *first-order* linear differential equation.

[†]Such points are called **removable singularities.** In this chapter we assume in such cases that the function has been defined (or redefined) so that it is analytic at the point.

Example 2 Find a power series solution about $x = 0$ to

(3) $y' + 2xy = 0$.

Solution The coefficient of y is the polynomial $2x$, which is analytic everywhere, so $x = 0$ is an ordinary point[†] of equation (3). Thus, we expect to find a power series solution of the form

(4) $y(x) = a_0 + a_1 x + a_2 x^2 + \cdots = \displaystyle\sum_{n=0}^{\infty} a_n x^n$.

Our task is to determine the coefficients a_n.

For this purpose we need the expansion for $y'(x)$ that is given by termwise differentiation of (4):

$$y'(x) = 0 + a_1 + 2a_2 x + 3a_3 x^2 + \cdots = \sum_{n=0}^{\infty} n a_n x^{n-1} .$$

We now substitute the series expansions for y and y' into (3) and obtain

$$\sum_{n=0}^{\infty} n a_n x^{n-1} + 2x \sum_{n=0}^{\infty} a_n x^n = 0 ,$$

which simplifies to

(5) $\displaystyle\sum_{n=0}^{\infty} n a_n x^{n-1} + \sum_{n=0}^{\infty} 2 a_n x^{n+1} = 0$.

To add the two power series in (5), we add the coefficients of like powers of x. If we write out the first few terms of these summations and add, we get

$$(a_1 + 2a_2 x + 3a_3 x^2 + 4a_4 x^3 + \cdots) + (2a_0 x + 2a_1 x^2 + 2a_2 x^3 + \cdots) = 0 ,$$

(6) $a_1 + (2a_2 + 2a_0) x + (3a_3 + 2a_1) x^2 + (4a_4 + 2a_2) x^3 + \cdots = 0$.

For the power series on the left-hand side of equation (6) to be identically zero, we must have all the coefficients equal to zero. Thus,

$$a_1 = 0 , \qquad 2a_2 + 2a_0 = 0 ,$$
$$3a_3 + 2a_1 = 0 , \qquad 4a_4 + 2a_2 = 0 , \qquad \text{etc.}$$

Solving the preceding system, we find

$$a_1 = 0 , \qquad a_2 = -a_0 , \qquad a_3 = -\frac{2}{3} a_1 = 0 ,$$

$$a_4 = -\frac{1}{2} a_2 = -\frac{1}{2}(-a_0) = \frac{1}{2} a_0 .$$

Hence, the power series for the solution takes the form

(7) $y(x) = a_0 - a_0 x^2 + \dfrac{1}{2} a_0 x^4 + \cdots$.

Although the first few terms displayed in (7) are useful, it is sometimes advantageous to have a formula for the *general term* in the power series expansion for the solution. To achieve this goal, let's return to equation (5). This time, instead of just writing out a few terms, let's

[†] By an ordinary point of a first-order equation $y' + q(x)y = 0$, we mean a point where $q(x)$ is analytic.

shift the indices in the two power series so that they sum over the same powers of x, say, x^k. To do this we shift the index in the first summation in (5) by letting $k = n - 1$. Then $n = k + 1$ and $k = 0$ when $n = 1$. Hence, the first summation in (5) becomes

$$(8) \qquad \sum_{n=0}^{\infty} n a_n x^{n-1} = \sum_{k=0}^{\infty} (k+1) a_{k+1} x^k \,.$$

In the second summation of (5), we put $k = n + 1$ so that $n = k - 1$ and $k = 1$ when $n = 0$. This gives

$$(9) \qquad \sum_{n=0}^{\infty} 2 a_n x^{n+1} = \sum_{k=1}^{\infty} 2 a_{k-1} x^k \,.$$

Substituting (8) and (9) into (5) yields

$$(10) \qquad \sum_{k=0}^{\infty} (k+1) a_{k+1} x^k + \sum_{k=1}^{\infty} 2 a_{k-1} x^k = 0 \,.$$

Since the first summation in (10) begins at $k = 0$ and the second at $k = 1$, we break up the first into

$$\sum_{k=0}^{\infty} (k+1) a_{k+1} x^k = a_1 + \sum_{k=0}^{\infty} (k+1) a_{k+1} x^k \,.$$

Then (10) becomes

$$(11) \qquad a_1 + \sum_{k=1}^{\infty} [(k+1) a_{k+1} + 2 a_{k-1}] x^k = 0 \,.$$

When we set all the coefficients in (11) equal to zero, we find

$$a_1 = 0 \,,$$

and, for all $k \geq 1$,

$$(12) \qquad (k+1) a_{k+1} + 2 a_{k-1} = 0 \,.$$

Equation (12) is a **recurrence relation** that we can use to determine the coefficient a_{k+1} in terms of a_{k-1}; that is,

$$a_{k+1} = -\frac{2}{k+1} a_{k-1} \,.$$

Setting $k = 1, 2, \ldots, 8$ and using the fact that $a_1 = 0$, we find

$$a_2 = -\frac{2}{2} a_0 = -a_0 \qquad (k = 1) \,, \qquad a_3 = -\frac{2}{3} a_1 = 0 \qquad (k = 2) \,,$$

$$a_4 = -\frac{2}{4} a_2 = \frac{1}{2} a_0 \qquad (k = 3) \,, \qquad a_5 = -\frac{2}{5} a_3 = 0 \qquad (k = 4) \,,$$

$$a_6 = -\frac{2}{6} a_4 = -\frac{1}{3!} a_0 \qquad (k = 5) \,, \qquad a_7 = -\frac{2}{7} a_5 = 0 \qquad (k = 6) \,,$$

$$a_8 = -\frac{2}{8} a_6 = \frac{1}{4!} a_0 \qquad (k = 7) \,, \qquad a_9 = -\frac{2}{9} a_7 = 0 \qquad (k = 8) \,.$$

After a moment's reflection, we realize that

$$a_{2n} = \frac{(-1)^n}{n!} a_0 \,, \qquad n = 1, 2, \ldots \,,$$

$$a_{2n+1} = 0 \,, \qquad n = 0, 1, 2, \ldots \,.$$

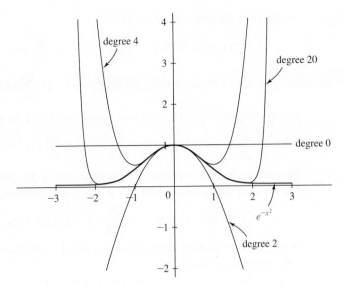

Figure 8.3 Partial sum approximations to e^{-x^2}

Substituting back into the expression (4), we obtain the power series solution

$$(13) \qquad y(x) = a_0 - a_0 x^2 + \frac{1}{2!} a_0 x^4 + \cdots = a_0 \sum_{n=0}^{\infty} \frac{(-1)^n}{n!} x^{2n}.$$

Since a_0 is left undetermined, it serves as an arbitrary constant, and hence (13) gives a general solution to equation (3). ◆

Applying the ratio test as described in Problem 7, Exercises 8.2 (page 434), we can verify that the power series in (13) has radius of convergence $\rho = \infty$. Moreover, (13) is reminiscent of the expansion for the exponential function; you can check that it converges to

$$y(x) = a_0 e^{-x^2}.$$

This general solution to the simple equation (3) can also be obtained by the method of separation of variables.

The convergence of the partial sums of (13), with $a_0 = 1$, to the actual solution e^{-x^2} is depicted in Figure 8.3. Notice that taking more terms results in better approximations around $x = 0$. Note also, however, that every partial sum is a *polynomial* and hence must diverge at $x = \pm\infty$, while e^{-x^2} converges to zero, of course. Thus, each partial sum approximation eventually deteriorates for large $|x|$. This is a typical feature of power series approximations.

In the next example we use the power series method to obtain a general solution to a linear second-order differential equation.

Example 3 Find a general solution to

$$(14) \qquad 2y'' + xy' + y = 0$$

in the form of a power series about the ordinary point $x = 0$.

Solution Writing

$$(15) \qquad y(x) = a_0 + a_1 x + a_2 x^2 + \cdots = \sum_{n=0}^{\infty} a_n x^n,$$

we differentiate termwise to obtain

$$y'(x) = a_1 + 2a_2x + 3a_3x^2 + \cdots = \sum_{n=1}^{\infty} na_n x^{n-1},$$

$$y''(x) = 2a_2 + 6a_3x + 12a_4x^2 + \cdots = \sum_{n=2}^{\infty} n(n-1)a_n x^{n-2}.$$

Substituting these power series into equation (14), we find

(16) $$\sum_{n=2}^{\infty} 2n(n-1)a_n x^{n-2} + \sum_{n=1}^{\infty} na_n x^n + \sum_{n=0}^{\infty} a_n x^n = 0.$$

To simplify the addition of the three summations in (16), let's shift the indices so that the general term in each is a constant times x^k. For the first summation, we substitute $k = n - 2$ and get

$$\sum_{n=2}^{\infty} 2n(n-1)a_n x^{n-2} = \sum_{k=0}^{\infty} 2(k+2)(k+1)a_{k+2} x^k.$$

In the second and third summations, we simply substitute k for n. With these changes of indices, equation (16) becomes

$$\sum_{k=0}^{\infty} 2(k+2)(k+1)a_{k+2} x^k + \sum_{k=1}^{\infty} ka_k x^k + \sum_{k=0}^{\infty} a_k x^k = 0.$$

Next, we separate the x^0 terms from the others and then combine the like powers of x in the three summations to get

$$4a_2 + a_0 + \sum_{k=1}^{\infty} [2(k+2)(k+1)a_{k+2} + ka_k + a_k]x^k = 0.$$

Setting the coefficients of this power series equal to zero yields

(17) $$4a_2 + a_0 = 0$$

and the recurrence relation

(18) $$2(k+2)(k+1)a_{k+2} + (k+1)a_k = 0, \qquad k \geq 1.$$

We can now use (17) and (18) to determine all the coefficients a_k of the solution in terms of a_0 and a_1. Solving (18) for a_{k+2} gives

(19) $$a_{k+2} = \frac{-1}{2(k+2)}a_k, \qquad k \geq 1.$$

Thus,

$$a_2 = \frac{-1}{2^2}a_0,$$

$$a_3 = \frac{-1}{2 \cdot 3}a_1 \qquad\qquad\qquad\qquad (k = 1),$$

$$a_4 = \frac{-1}{2 \cdot 4}a_2 = \frac{1}{2^2 \cdot 2 \cdot 4}a_0 \qquad\qquad (k = 2),$$

$$a_5 = \frac{-1}{2 \cdot 5}a_3 = \frac{1}{2^2 \cdot 3 \cdot 5}a_1 \qquad\qquad (k = 3),$$

$$a_6 = \frac{-1}{2 \cdot 6}a_4 = \frac{-1}{2^3 \cdot 2 \cdot 4 \cdot 6}a_0 = \frac{-1}{2^6 \cdot 3!}a_0 \qquad (k = 4),$$

$$a_7 = \frac{-1}{2 \cdot 7}a_5 = \frac{-1}{2^3 \cdot 3 \cdot 5 \cdot 7}a_1 \qquad\qquad (k = 5),$$

$$a_8 = \frac{-1}{2 \cdot 8}a_6 = \frac{1}{2^4 \cdot 2 \cdot 4 \cdot 6 \cdot 8}a_0 = \frac{1}{2^8 \cdot 4!}a_0 \qquad (k = 6).$$

The pattern for the coefficients is now apparent. Since a_0 and a_1 are not restricted, we find

$$a_{2n} = \frac{(-1)^n}{2^{2n}n!}a_0, \qquad n \geq 1,$$

and

$$a_{2n+1} = \frac{(-1)^n}{2^n[1 \cdot 3 \cdot 5 \cdots (2n+1)]}a_1, \qquad n \geq 1.$$

From this, two linearly independent solutions emerge; namely,

$$(20) \qquad y_1(x) = \sum_{n=0}^{\infty} \frac{(-1)^n}{2^{2n}n!}x^{2n} \qquad (\text{take } a_0 = 1, a_1 = 0),$$

$$(21) \qquad y_2(x) = \sum_{n=0}^{\infty} \frac{(-1)^n}{2^n[1 \cdot 3 \cdot 5 \cdots (2n+1)]}x^{2n+1} \qquad (\text{take } a_0 = 0, a_1 = 1).$$

Hence, a general solution to (14) is $a_0 y_1(x) + a_1 y_2(x)$. Approximations to the solutions $y_1(x), y_2(x)$ are depicted in Figure 8.4. ◆

The method illustrated in Example 3 can also be used to solve initial value problems. Suppose we are given the values of $y(0)$ and $y'(0)$; then, from equation (15), we see that $a_0 = y(0)$ and $a_1 = y'(0)$. Knowing these two coefficients leads to a unique power series solution for the initial value problem.

The recurrence relation (18) in Example 3 involved just two of the coefficients, a_{k+2} and a_k, and we were fortunate in being able to deduce from this relation the general form for the coefficient a_n. However, many cases arise that lead to more complicated two-term or even to many-term recurrence relations. When this occurs, it may be impossible to determine the general form for the coefficients a_n. In the next example, we consider an equation that gives rise to a three-term recurrence relation.

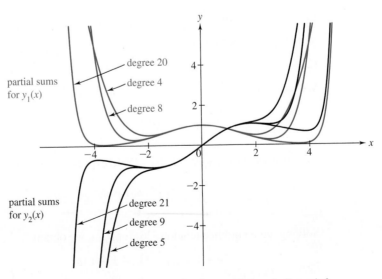

Figure 8.4 Partial sum approximations to solutions for Example 3

Example 4 Find the first few terms in a power series expansion about $x = 0$ for a general solution to

(22) $(1 + x^2)y'' - y' + y = 0$.

Solution Since $p(x) = -(1 + x^2)^{-1}$ and $q(x) = (1 + x^2)^{-1}$ are analytic at $x = 0$, then $x = 0$ is an ordinary point for equation (22). Hence, we can express its general solution in the form

$$y(x) = \sum_{n=0}^{\infty} a_n x^n .$$

Substituting this expansion into (22) yields

$$(1 + x^2) \sum_{n=2}^{\infty} n(n-1)a_n x^{n-2} - \sum_{n=1}^{\infty} n a_n x^{n-1} + \sum_{n=0}^{\infty} a_n x^n = 0 ,$$

(23) $$\sum_{n=2}^{\infty} n(n-1)a_n x^{n-2} + \sum_{n=2}^{\infty} n(n-1)a_n x^n - \sum_{n=1}^{\infty} n a_n x^{n-1} + \sum_{n=0}^{\infty} a_n x^n = 0 .$$

To sum over like powers x^k, we put $k = n - 2$ in the first summation of (23), $k = n - 1$ in the third, and $k = n$ in the second and fourth. This gives

$$\sum_{k=0}^{\infty} (k+2)(k+1)a_{k+2} x^k + \sum_{k=2}^{\infty} k(k-1)a_k x^k - \sum_{k=0}^{\infty} (k+1)a_{k+1} x^k + \sum_{k=0}^{\infty} a_k x^k = 0 .$$

Separating the terms corresponding to $k = 0$ and $k = 1$ and combining the rest under one summation, we have

$$(2a_2 - a_1 + a_0) + (6a_3 - 2a_2 + a_1)x$$
$$+ \sum_{k=2}^{\infty} [(k+2)(k+1)a_{k+2} - (k+1)a_{k+1} + (k(k-1) + 1)a_k]x^k = 0 .$$

Setting the coefficients equal to zero gives

(24) $2a_2 - a_1 + a_0 = 0$,

(25) $6a_3 - 2a_2 + a_1 = 0$,

and the recurrence relation

(26) $(k+2)(k+1)a_{k+2} - (k+1)a_{k+1} + (k^2 - k + 1)a_k = 0$, $k \geq 2$.

We can solve (24) for a_2 in terms of a_0 and a_1:

$$a_2 = \frac{a_1 - a_0}{2} .$$

Now that we have a_2, we can use (25) to express a_3 in terms of a_0 and a_1:

$$a_3 = \frac{2a_2 - a_1}{6} = \frac{(a_1 - a_0) - a_1}{6} = \frac{-a_0}{6} .$$

Solving the recurrence relation (26) for a_{k+2}, we obtain

(27) $$a_{k+2} = \frac{(k+1)a_{k+1} - (k^2 - k + 1)a_k}{(k+2)(k+1)} , k \geq 2 .$$

For $k = 2, 3$, and 4, this gives

$$a_4 = \frac{3a_3 - 3a_2}{4 \cdot 3} = \frac{a_3 - a_2}{4}$$

$$= \frac{-\dfrac{a_0}{6} - \left(\dfrac{a_1 - a_0}{2}\right)}{4} = \frac{2a_0 - 3a_1}{24} \qquad (k = 2) \ ,$$

$$a_5 = \frac{4a_4 - 7a_3}{5 \cdot 4} = \frac{3a_0 - a_1}{40} \qquad (k = 3) \ ,$$

$$a_6 = \frac{5a_5 - 13a_4}{6 \cdot 5} = \frac{36a_1 - 17a_0}{720} \qquad (k = 4) \ .$$

We can now express a general solution in terms up to order 6, using a_0 and a_1 as the arbitrary constants. Thus,

$$(28) \qquad y(x) = a_0 + a_1 x + \left(\frac{a_1 - a_0}{2}\right)x^2 - \frac{a_0}{6}x^3$$

$$+ \left(\frac{2a_0 - 3a_1}{24}\right)x^4 + \left(\frac{3a_0 - a_1}{40}\right)x^5 + \left(\frac{36a_1 - 17a_0}{720}\right)x^6 + \cdots$$

$$= a_0\left(1 - \frac{1}{2}x^2 - \frac{1}{6}x^3 + \frac{1}{12}x^4 + \frac{3}{40}x^5 - \frac{17}{720}x^6 + \cdots\right)$$

$$+ a_1\left(x + \frac{1}{2}x^2 - \frac{1}{8}x^4 - \frac{1}{40}x^5 + \frac{1}{20}x^6 + \cdots\right). \quad \blacklozenge$$

Given specific values for a_0 and a_1, will the partial sums of the power series representation (28) yield useful approximations to the solution when $x = 0.5$? What about when $x = 2.3$ or $x = 7.8$? The answers to these questions certainly depend on the radius of convergence of the power series in (28). But since we were not able to determine a general form for the coefficients a_n in this example, we cannot use the ratio test (or other methods such as the root test, integral test, or comparison test) to compute the radius ρ. In the next section we remedy this situation by giving a simple procedure that determines a lower bound for the radius of convergence of power series solutions.

8.3 EXERCISES

In Problems 1–10, determine all the singular points of the given differential equation.

1. $(x + 1)y'' - x^2 y' + 3y = 0$

2. $x^2 y'' + 3y' - xy = 0$

3. $(\theta^2 - 2)y'' + 2y' + (\sin\theta)y = 0$

4. $(x^2 + x)y'' + 3y' - 6xy = 0$

5. $(t^2 - t - 2)x'' + (t + 1)x' - (t - 2)x = 0$

6. $(x^2 - 1)y'' + (1 - x)y' + (x^2 - 2x + 1)y = 0$

7. $(\sin x)y'' + (\cos x)y = 0$

8. $e^x y'' - (x^2 - 1)y' + 2xy = 0$

9. $(\sin\theta)y'' - (\ln\theta)y = 0$

10. $[\ln(x - 1)]y'' + (\sin 2x)y' - e^x y = 0$

In Problems 11–18, find at least the first four nonzero terms in a power series expansion about $x = 0$ for a general solution to the given differential equation.

11. $y' + (x + 2)y = 0$

12. $y' - y = 0$

13. $z'' - x^2 z = 0$

14. $(x^2 + 1)y'' + y = 0$

15. $y'' + (x - 1)y' + y = 0$

16. $y'' - 2y' + y = 0$

17. $w'' - x^2 w' + w = 0$

18. $(2x - 3)y'' - xy' + y = 0$

In Problems 19–24, find a power series expansion about $x = 0$ for a general solution to the given differential equation. Your answer should include a general formula for the coefficients.

19. $y' - 2xy = 0$

20. $y'' + y = 0$

21. $y'' - xy' + 4y = 0$

22. $y'' - xy = 0$

23. $z'' - x^2z' - xz = 0$

24. $(x^2 + 1)y'' - xy' + y = 0$

In Problems 25–28, find at least the first four nonzero terms in a power series expansion about $x = 0$ for the solution to the given initial value problem.

25. $w'' + 3xw' - w = 0$;

$\quad w(0) = 2$, $\quad w'(0) = 0$

26. $(x^2 - x + 1)y'' - y' - y = 0$;

$\quad y(0) = 0$, $\quad y'(0) = 1$

27. $(x + 1)y'' - y = 0$;

$\quad y(0) = 0$, $\quad y'(0) = 1$

28. $y'' + (x - 2)y' - y = 0$;

$\quad y(0) = -1$, $\quad y'(0) = 0$

In Problems 29–31, use the first few terms of the power series expansion to find a cubic polynomial approximation for the solution to the given initial value problem. Graph the linear, quadratic, and cubic polynomial approximations for $-5 \le x \le 5$.

29. $y'' + y' - xy = 0$;

$\quad y(0) = 1$, $\quad y'(0) = -2$

30. $y'' - 4xy' + 5y = 0$;

$\quad y(0) = -1$, $\quad y'(0) = 1$

31. $(x^2 + 2)y'' + 2xy' + 3y = 0$;

$\quad y(0) = 1$, $\quad y'(0) = 2$

32. Consider the initial value problem

$$y'' - 2xy' - 2y = 0;$$
$$y(0) = a_0, \quad y'(0) = a_1,$$

where a_0 and a_1 are constants.

(a) Show that if $a_0 = 0$, then the solution will be an odd function [that is, $y(-x) = -y(x)$ for all x]. What happens when $a_1 = 0$?

(b) Show that if a_0 and a_1 are positive, then the solution is increasing on $(0, \infty)$.

(c) Show that if a_0 is negative and a_1 is positive, then the solution is increasing on $(-\infty, 0)$.

(d) What conditions on a_0 and a_1 would guarantee that the solution is increasing on $(-\infty, \infty)$?

33. Use the ratio test to show that the radius of convergence of the series in equation (13) is infinite. [*Hint:* See Problem 7, Exercises 8.2, page 434.]

34. Emden's Equation. A classical nonlinear equation that occurs in the study of the thermal behavior of a spherical cloud is **Emden's equation**

$$y'' + \frac{2}{x}y' + y^n = 0,$$

with initial conditions $y(0) = 1, y'(0) = 0$. Even though $x = 0$ is *not* an ordinary point for this equation (which is nonlinear for $n \ne 1$), it turns out that there does exist a solution analytic at $x = 0$. Assuming that n is a positive integer, show that the first few terms in a power series solution are

$$y = 1 - \frac{x^2}{3!} + n\frac{x^4}{5!} + \cdots .$$

[*Hint:* Substitute $y = 1 + c_2x^2 + c_3x^3 + c_4x^4 + c_5x^5 + \cdots$ into the equation and carefully compute the first few terms in the expansion for y^n.]

35. Variable Resistor. In Section 5.7, we showed that the charge q on the capacitor in a simple *RLC* circuit is governed by the equation

$$Lq''(t) + Rq'(t) + \frac{1}{C}q(t) = E(t),$$

where L is the inductance, R the resistance, C the capacitance, and E the voltage source. Since the resistance of a resistor increases with temperature, let's assume that the resistor is heated so that the resistance at time t is $R(t) = 1 + t/10\ \Omega$ (see Figure 8.5). If $L = 0.1$ H, $C = 2$ F, $E(t) \equiv 0, q(0) = 10$ C, and $q'(0) = 0$ A, find at least the first four nonzero terms in a power series expansion about $t = 0$ for the charge on the capacitor.

Figure 8.5 An *RLC* circuit whose resistor is being heated

36. Variable Spring Constant. As a spring is heated, its spring "constant" decreases. Suppose the spring is heated so that the spring "constant" at time t is $k(t) = 6 - t$ N/m (see Figure 8.6). If the unforced mass–spring system has mass $m = 2$ kg and a damping constant $b = 1$ N-sec/m with initial conditions $x(0) = 3$ m and $x'(0) = 0$ m/sec, then the displacement $x(t)$ is governed by the initial value problem

$$2x''(t) + x'(t) + (6 - t)x(t) = 0 \,;$$
$$x(0) = 3 \,, \qquad x'(0) = 0 \,.$$

Find at least the first four nonzero terms in a power series expansion about $t = 0$ for the displacement.

Figure 8.6 A mass–spring system whose spring is being heated

8.4 Equations with Analytic Coefficients

In Section 8.3 we introduced a method for obtaining a power series solution about an ordinary point. In this section we continue the discussion of this procedure. We begin by stating a basic existence theorem for the equation

(1) $y''(x) + p(x)y'(x) + q(x)y(x) = 0 \,,$

which justifies the power series method.

Existence of Analytic Solutions

Theorem 5. Suppose x_0 is an ordinary point for equation (1). Then (1) has two linearly independent analytic solutions of the form

(2) $y(x) = \displaystyle\sum_{n=0}^{\infty} a_n (x - x_0)^n \,.$

Moreover, the radius of convergence of any power series solution of the form given by (2) is at least as large as the distance from x_0 to the nearest singular point (real or complex-valued) of equation (1).

The key element in the proof of Theorem 5 is the construction of a convergent geometric series that dominates the series expansion (2) of a solution to equation (1). The convergence of the series in (2) then follows by the comparison test. The details of the proof can be found in more advanced books on differential equations.[†]

As we saw in Section 8.3, the power series method gives us a general solution in the same form as (2), with a_0 and a_1 as arbitrary constants. The two linearly independent solutions referred to in Theorem 5 can be obtained by taking $a_0 = 1$, $a_1 = 0$ for the first and $a_0 = 0$, $a_1 = 1$ for the second. Thus, we can extend Theorem 5 by saying that *equation (1) has a general solution of the form (2) with a_0 and a_1 as the arbitrary constants.*

[†]See, for example, *Ordinary Differential Equations*, 4th ed., by G. Birkhoff and G.-C. Rota (Wiley, New York, 1989).

The second part of Theorem 5 gives a simple way of determining a minimum value for the radius of convergence of the power series. We need only find the singular points of equation (1) and then determine the distance between the ordinary point x_0 and the nearest singular point.

Example 1 Find a minimum value for the radius of convergence of a power series solution about $x = 0$ to

$$(3) \qquad 2y'' + xy' + y = 0 .$$

Solution For this equation, $p(x) = x/2$ and $q(x) = 1/2$. Both of these functions are analytic for all real or complex values of x. Since equation (3) has no singular points, the distance between the ordinary point $x = 0$ and the nearest singular point is infinite. Hence, the radius of convergence is infinite. ◆

The next example helps to answer the questions posed at the end of the last section.

Example 2 Find a minimum value for the radius of convergence of a power series solution about $x = 0$ to

$$(4) \qquad (1 + x^2)y'' - y' + y = 0 .$$

Solution Here $p(x) = -1/(1 + x^2)$, $q(x) = 1/(1 + x^2)$, and so the singular points of equation (4) occur when $1 + x^2 = 0$; that is, when $x = \pm\sqrt{-1} = \pm i$. Since the only singular points of equation (4) are the complex numbers $\pm i$, we see that $x = 0$ is an ordinary point. Moreover, the distance[†] from 0 to either $\pm i$ is 1. Thus, the radius of convergence of a power series solution about $x = 0$ is at least 1. ◆

In equation (28) of Section 8.3, page 443, we found the first few terms of a power series solution to equation (4). Because we now know that this series has radius of convergence at least 1, the partial sums of this series will converge to the solution for $|x| < 1$. However, when $|x| \geq 1$, we have no basis on which to decide whether we can use the series to approximate the solution.

Power series expansions about $x_0 = 0$ are somewhat easier to manipulate than expansions about nonzero points. As the next example shows, a simple shift in variable enables us always to expand about the origin.

Example 3 Find the first few terms in a power series expansion about $x = 1$ for a general solution to

$$(5) \qquad 2y'' + xy' + y = 0 .$$

Also determine the radius of convergence of the series.

Solution As seen in Example 1, there are no singular points for equation (5). Thus, $x = 1$ is an ordinary point, and, as a consequence of Theorem 5, equation (5) has a general solution of the form

$$(6) \qquad y(x) = \sum_{n=0}^{\infty} a_n(x - 1)^n .$$

Moreover, the radius of convergence of the series in (6) must be infinite.

[†]Recall that the distance between the two complex numbers $z = a + bi$ and $w = c + di$ is given by $\sqrt{(a - c)^2 + (b - d)^2}$.

We can simplify the computation of the coefficients a_n by shifting the center of the expansion (6) from $x_0 = 1$ to $t_0 = 0$. This is accomplished by the substitution $x = t + 1$. Setting $Y(t) = y(t + 1)$, we find via the chain rule

$$\frac{dy}{dx} = \frac{dY}{dt} , \qquad \frac{d^2y}{dx^2} = \frac{d^2Y}{dt^2} ,$$

and, hence, equation (5) is changed into

(7) $$2\frac{d^2Y}{dt^2} + (t + 1) \frac{dY}{dt} + Y = 0 .$$

We now seek a general solution of the form

(8) $$Y(t) = \sum_{n=0}^{\infty} a_n t^n ,$$

where the a_n's in equations (6) and (8) are the same. Proceeding as usual, we substitute the power series for $Y(t)$ into (7), derive a recurrence relation for the coefficients, and ultimately find that

$$Y(t) = a_0 \left\{ 1 - \frac{1}{4}t^2 + \frac{1}{24}t^3 + \cdots \right\} + a_1 \left\{ t - \frac{1}{4}t^2 - \frac{1}{8}t^3 + \cdots \right\}$$

(the details are left as an exercise). Thus, restoring $t = x - 1$ we have

(9) $$y(x) = a_0 \left\{ 1 - \frac{1}{4}(x - 1)^2 + \frac{1}{24}(x - 1)^3 + \cdots \right\}$$

$$+ a_1 \left\{ (x - 1) - \frac{1}{4}(x - 1)^2 - \frac{1}{8}(x - 1)^3 + \cdots \right\} . \; \blacklozenge$$

When the coefficients of a linear equation are not polynomials in x, but are analytic functions, we can still find analytic solutions by essentially the same method.

Example 4 Find a power series expansion for the solution to

(10) $$y''(x) + e^x y'(x) + (1 + x^2)y(x) = 0 ; \qquad y(0) = 1 , \qquad y'(0) = 0 .$$

Solution Here $p(x) = e^x$ and $q(x) = 1 + x^2$, and both are analytic for all x. Thus, by Theorem 5, the initial value problem (10) has a power series solution

(11) $$y(x) = \sum_{n=0}^{\infty} a_n x^n$$

that converges for all x ($\rho = \infty$). To find the first few terms of this series, we first expand $p(x) = e^x$ in its Maclaurin series:

$$e^x = 1 + x + \frac{x^2}{2!} + \frac{x^3}{3!} + \cdots .$$

Substituting the expansions for $y(x), y'(x), y''(x)$, and e^x into (10) gives

(12) $$\sum_{n=2}^{\infty} n(n - 1)a_n x^{n-2} + \left(1 + x + \frac{x^2}{2} + \frac{x^3}{6} + \frac{x^4}{24} + \cdots \right) \sum_{n=1}^{\infty} n a_n x^{n-1}$$

$$+ (1 + x^2) \sum_{n=0}^{\infty} a_n x^n = 0 .$$

Because of the computational difficulties due to the appearance of the product of the power series for e^x and $y'(x)$, we concern ourselves with just those terms up to order 4. Writing out (12) and keeping track of all such terms, we find

(13) $(2a_2 + 6a_3x + 12a_4x^2 + 20a_5x^3 + 30a_6x^4 + \cdots)$

$\qquad + (a_1 + 2a_2x + 3a_3x^2 + 4a_4x^3 + 5a_5x^4 + \cdots)$ ⟵ $\qquad 1 \cdot \Sigma\, na_n x^{n-1}$

$\qquad\quad + (a_1x + 2a_2x^2 + 3a_3x^3 + 4a_4x^4 + \cdots)$ ⟵ $\qquad x \cdot \Sigma\, na_n x^{n-1}$

$\qquad\qquad + \left(\dfrac{1}{2}a_1x^2 + a_2x^3 + \dfrac{3}{2}a_3x^4 + \cdots\right)$ ⟵ $\qquad \dfrac{x^2}{2} \cdot \Sigma\, na_n x^{n-1}$

$\qquad\qquad\quad + \left(\dfrac{1}{6}a_1x^3 + \dfrac{1}{3}a_2x^4 + \cdots\right)$

$\qquad\qquad\qquad + \left(\dfrac{1}{24}a_1x^4 + \cdots\right)$ $\qquad\qquad\qquad\qquad\qquad \vdots$

$\qquad + (a_0 + a_1x + a_2x^2 + a_3x^3 + a_4x^4 + \cdots)$ ⟵ $\qquad 1 \cdot \Sigma\, a_n x^n$

$\qquad\quad + (a_0x^2 + a_1x^3 + a_2x^4 + \cdots) = 0 .$ ⟵ $\qquad x^2 \cdot \Sigma\, a_n x^n$

Grouping the like powers of x in equation (13) (for example, the x^2 terms are shown in color) and then setting the coefficients equal to zero yields the system of equations

$$2a_2 + a_1 + a_0 = 0 \qquad (x^0 \text{ term}) ,$$

$$6a_3 + 2a_2 + 2a_1 = 0 \qquad (x^1 \text{ term}) ,$$

$$12a_4 + 3a_3 + 3a_2 + \frac{1}{2}a_1 + a_0 = 0 \qquad (x^2 \text{ term}) ,$$

$$20a_5 + 4a_4 + 4a_3 + a_2 + \frac{7}{6}a_1 = 0 \qquad (x^3 \text{ term}) ,$$

$$30a_6 + 5a_5 + 5a_4 + \frac{3}{2}a_3 + \frac{4}{3}a_2 + \frac{1}{24}a_1 = 0 \qquad (x^4 \text{ term}) .$$

The initial conditions in (10) imply that $y(0) = a_0 = 1$ and $y'(0) = a_1 = 0$. Using these values for a_0 and a_1, we can solve the above system first for a_2, then a_3, and so on:

$$2a_2 + 0 + 1 = 0 \Rightarrow a_2 = -\frac{1}{2} ,$$

$$6a_3 - 1 + 0 = 0 \Rightarrow a_3 = \frac{1}{6} ,$$

$$12a_4 + \frac{1}{2} - \frac{3}{2} + 0 + 1 = 0 \Rightarrow a_4 = 0 ,$$

$$20a_5 + 0 + \frac{2}{3} - \frac{1}{2} + 0 = 0 \Rightarrow a_5 = -\frac{1}{120} ,$$

$$30a_6 - \frac{1}{24} + 0 + \frac{1}{4} - \frac{2}{3} + 0 = 0 \Rightarrow a_6 = \frac{11}{720} .$$

Thus, the solution to the initial value problem in (10) is

(14) $y(x) = 1 - \dfrac{1}{2}x^2 + \dfrac{1}{6}x^3 - \dfrac{1}{120}x^5 + \dfrac{11}{720}x^6 + \cdots .$ ◆

Thus far we have used the power series method only for homogeneous equations. But the same method applies, with obvious modifications, to nonhomogeneous equations of the form

(15) $y''(x) + p(x)y'(x) + q(x)y(x) = g(x)$,

provided the forcing term $g(x)$ and the coefficient functions are analytic at x_0. For example, to find a power series about $x = 0$ for a general solution to

(16) $y''(x) - xy'(x) - y(x) = \sin x$,

we use the substitution $y(x) = \Sigma a_n x^n$ to obtain a power series expansion for the left-hand side of (16). We then equate the coefficients of this series with the corresponding coefficients of the Maclaurin expansion for sin x:

$$\sin x = \sum_{n=0}^{\infty} \frac{(-1)^n}{(2n+1)!} x^{2n+1} .$$

Carrying out the details (see Problem 20 on page 450), we ultimately find that an expansion for a general solution to (16) is

(17) $y(x) = a_0 y_1(x) + a_1 y_2(x) + y_p(x)$,

where

(18) $y_1(x) = 1 + \dfrac{1}{2}x^2 + \dfrac{1}{8}x^4 + \dfrac{1}{48}x^6 + \cdots$,

(19) $y_2(x) = x + \dfrac{1}{3}x^3 + \dfrac{1}{15}x^5 + \dfrac{1}{105}x^7 + \cdots$

are the solutions to the homogeneous equation associated with equation (16) and

(20) $y_p(x) = \dfrac{1}{6}x^3 + \dfrac{1}{40}x^5 + \dfrac{19}{5040}x^7 + \cdots$

is a particular solution to equation (16).

8.4 EXERCISES

In Problems 1–6, find a minimum value for the radius of convergence of a power series solution about x_0.

1. $(x+1)y'' - 3xy' + 2y = 0$; $x_0 = 1$
2. $y'' - xy' - 3y = 0$; $x_0 = 2$
3. $(1 + x + x^2)y'' - 3y = 0$; $x_0 = 1$
4. $(x^2 - 5x + 6)y'' - 3xy' - y = 0$; $x_0 = 0$
5. $y'' - (\tan x)y' + y = 0$; $x_0 = 0$
6. $(1 + x^3)y'' - xy' + 3x^2 y = 0$; $x_0 = 1$

In Problems 7–12, find at least the first four nonzero terms in a power series expansion about x_0 for a general solution to the given differential equation with the given value for x_0.

7. $y' + 2(x-1)y = 0$; $x_0 = 1$
8. $y' - 2xy = 0$; $x_0 = -1$

9. $(x^2 - 2x)y'' + 2y = 0$; $x_0 = 1$
10. $x^2 y'' - xy' + 2y = 0$; $x_0 = 2$
11. $x^2 y'' - y' + y = 0$; $x_0 = 2$
12. $y'' + (3x - 1)y' - y = 0$; $x_0 = -1$

In Problems 13–19, find at least the first four nonzero terms in a power series expansion of the solution to the given initial value problem.

13. $x' + (\sin t)x = 0$; $x(0) = 1$
14. $y' - e^x y = 0$; $y(0) = 1$
15. $(x^2 + 1)y'' - e^x y' + y = 0$; $y(0) = 1$, $y'(0) = 1$
16. $y'' + ty' + e^t y = 0$; $y(0) = 0$, $y'(0) = -1$
17. $y'' - (\sin x)y = 0$; $y(\pi) = 1$, $y'(\pi) = 0$

18. $y'' - (\cos x)y' - y = 0$;

$y(\pi/2) = 1$, $\qquad y'(\pi/2) = 1$

19. $y'' - e^{2x}y' + (\cos x)y = 0$;

$y(0) = -1$, $\qquad y'(0) = 1$

20. To derive the general solution given by equations (17)–(20) for the nonhomogeneous equation (16), complete the following steps:

(a) Substitute $y(x) = \sum_{n=0}^{\infty} a_n x^n$ and the Maclaurin series for $\sin x$ into equation (16) to obtain

$$(2a_2 - a_0) + \sum_{k=1}^{\infty} [(k+2)(k+1)a_{k+2} - (k+1)a_k]x^k$$

$$= \sum_{n=0}^{\infty} \frac{(-1)^n}{(2n+1)!}x^{2n+1}.$$

(b) Equate the coefficients of like powers of x on both sides of the equation in part (a) and thereby deduce the equations

$$a_2 = \frac{a_0}{2}, \qquad a_3 = \frac{1}{6} + \frac{a_1}{3}, \qquad a_4 = \frac{a_0}{8},$$

$$a_5 = \frac{1}{40} + \frac{a_1}{15}, \qquad a_6 = \frac{a_0}{48},$$

$$a_7 = \frac{19}{5040} + \frac{a_1}{105}.$$

(c) Show that the relations in part (b) yield the general solution to (16) given in equations (17)–(20).

In Problems 21–28, use the procedure illustrated in Problem 20 to find at least the first four nonzero terms in a power series expansion about $x = 0$ of a general solution to the given differential equation.

21. $y' - xy = \sin x$

22. $w' + xw = e^x$

23. $z'' + xz' + z = x^2 + 2x + 1$

24. $y'' - 2xy' + 3y = x^2$

25. $(1 + x^2)y'' - xy' + y = e^{-x}$

26. $y'' - xy' + 2y = \cos x$

27. $(1 - x^2)y'' - y' + y = \tan x$

28. $y'' - (\sin x)y = \cos x$

29. The equation

$$(1 - x^2)y'' - 2xy' + n(n+1)y = 0,$$

where n is an unspecified parameter, is called **Legendre's equation.** This equation occurs in applications of differential equations to engineering systems in spherical coordinates.

(a) Find a power series expansion about $x = 0$ for a solution to Legendre's equation.

(b) Show that for n a nonnegative integer, there exists an nth-degree polynomial that is a solution to Legendre's equation. These polynomials, up to a constant multiple, are called **Legendre polynomials.**

(c) Determine the first three Legendre polynomials (up to a constant multiple).

30. Aging Spring. As a spring ages, its spring "constant" decreases in value. One such model for a mass–spring system with an aging spring is

$$mx''(t) + bx'(t) + ke^{-\eta t}x(t) = 0,$$

where m is the mass, b the damping constant, k and η positive constants, and $x(t)$ the displacement of the spring from its equilibrium position. Let $m = 1$ kg, $b = 2$ N-sec/m, $k = 1$ N/m, and $\eta = 1(\sec)^{-1}$. The system is set in motion by displacing the mass 1 m from its equilibrium position and then releasing it $(x(0) = 1, x'(0) = 0)$. Find at least the first four nonzero terms in a power series expansion about $t = 0$ for the displacement.

31. Aging Spring without Damping. In the mass–spring system for an aging spring discussed in Problem 30, assume that there is no damping (i.e., $b = 0$), $m = 1$, and $k = 1$. To see the effect of aging, consider η as a positive parameter.

(a) Redo Problem 30 with $b = 0$ and η arbitrary but fixed.

(b) Set $\eta = 0$ in the expansion obtained in part (a). Does this expansion agree with the expansion for the solution to the problem with $\eta = 0$? [*Hint:* When $\eta = 0$, the solution is $x(t) = \cos t$.]

8.5 Cauchy–Euler (Equidimensional) Equations

In the previous sections we considered methods for obtaining power series solutions about an ordinary point for a linear second-order equation. However, in certain cases we may want a series expansion about a *singular point* of the equation. To motivate a procedure for finding such expansions, we consider the class of **Cauchy–Euler equations.** In Section 4.7 we studied this topic briefly. Here we will use the operator approach to rederive and extend our conclusions.

A second-order homogeneous Cauchy–Euler equation has the form

(1) $ax^2y''(x) + bxy'(x) + cy(x) = 0$, $x > 0$,

where a ($\neq 0$), b, and c are (real) constants. Since here $p(x) = b/(ax)$ and $q(x) = c/(ax^2)$, it follows that $x = 0$ is a singular point for (1).

Equation (1) has solutions of the form $y = x^r$. To determine the values for r, we can proceed as follows. Let L be the differential operator defined by the left-hand side of equation (1), that is,

(2) $L[y](x) := ax^2y''(x) + bxy'(x) + cy(x),$

and set

(3) $w(r, x) := x^r.$

When we substitute $w(r, x)$ for $y(x)$ in (2), we find

$$L[w](x) = ax^2r(r-1)x^{r-2} + bxrx^{r-1} + cx^r$$
$$= \{ar^2 + (b-a)r + c\}x^r.$$

From this, we see that $w = x^r$ is a solution to (1) if and only if r satisfies

(4) $ar^2 + (b-a)r + c = 0.$

This equation is referred to as the **characteristic**, or **indicial, equation** for (1).

When the indicial equation has two distinct roots, we have

$$L[w](x) = a(r - r_1)(r - r_2)x^r,$$

from which it follows that equation (1) has the two solutions

(5) $y_1(x) = w(r_1, x) = x^{r_1},$ $x > 0,$

(6) $y_2(x) = w(r_2, x) = x^{r_2},$ $x > 0,$

which are linearly independent.

When r_1 and r_2 are complex conjugates, $\alpha \pm i\beta$, we can use Euler's formula to express

$$x^{\alpha+i\beta} = e^{(\alpha+i\beta)\ln x} = e^{\alpha\ln x}\cos(\beta \ln x) + ie^{\alpha\ln x}\sin(\beta \ln x)$$
$$= x^\alpha \cos(\beta \ln x) + ix^\alpha \sin(\beta \ln x).$$

Since the real and imaginary parts of $x^{\alpha+i\beta}$ must also be solutions to (1), we can replace (5) and (6) by the two linearly independent real-valued solutions

$$y_1(x) = x^\alpha \cos(\beta \ln x),\qquad y_2(x) = x^\alpha \sin(\beta \ln x).$$

If the indicial equation (4) has a repeated real root r_0, then it turns out that x^{r_0} and $x^{r_0} \ln x$ are two linearly independent solutions. This can be deduced using the reduction of order approach of Section 4.7. However, it is more instructive for later applications to see how these two linearly independent solutions can be obtained via the operator approach. If r_0 is a repeated root, then

(7) $L[w](x) = a(r - r_0)^2 x^r.$

Setting $r = r_0$ gives the solution

(8) $y_1(x) = w(r_0, x) = x^{r_0},$ $x > 0.$

To find a second linearly independent solution, we make the following observation. The right-hand side of (7) has the factor $(r - r_0)^2$, so taking the partial derivative of (7) with respect to r and setting $r = r_0$, we get *zero*. That is,

(9) $\dfrac{\partial}{\partial r}\{L[w](x)\}\bigg|_{r=r_0} = \{a(r-r_0)^2 x^r \ln x + 2a(r-r_0)x^r\}\bigg|_{r=r_0} = 0.$

While it may not appear that we have made any progress toward finding a second solution, a closer look at the expression on the left in (9) will soon vindicate our efforts.

First note that $w(r, x) = x^r$ has continuous partial derivatives of all orders with respect to both r and x. Hence, the mixed partial derivatives are equal:

$$\frac{\partial^3 w}{\partial r \partial x^2} = \frac{\partial^3 w}{\partial x^2 \partial r}, \qquad \frac{\partial^2 w}{\partial r \partial x} = \frac{\partial^2 w}{\partial x \partial r}.$$

Consequently, for the differential operator L, we have

$$\frac{\partial}{\partial r} L[w] = \frac{\partial}{\partial r} \left\{ ax^2 \frac{\partial^2 w}{\partial x^2} + bx \frac{\partial w}{\partial x} + cw \right\}$$

$$= ax^2 \frac{\partial^3 w}{\partial r \partial x^2} + bx \frac{\partial^2 w}{\partial r \partial x} + c \frac{\partial w}{\partial r}$$

$$= ax^2 \frac{\partial^3 w}{\partial x^2 \partial r} + bx \frac{\partial^2 w}{\partial x \partial r} + c \frac{\partial w}{\partial r}$$

$$= L\left[\frac{\partial w}{\partial r} \right];$$

that is, the operators $\partial/\partial r$ and L commute. With this fact, (9) can be written as

$$L\left[\frac{\partial w}{\partial r} \right]\bigg|_{r=r_0} = 0.$$

Thus for the case of a repeated root r_0, a second linearly independent solution to (1) is

(10) $\qquad y_2(x) = \frac{\partial w}{\partial r}(r_0, x) = \frac{\partial}{\partial r}(x^r)\bigg|_{r=r_0} = x^{r_0} \ln x, \qquad x > 0.$

Example 1 Find a general solution to

(11) $\qquad 4x^2 y''(x) + y(x) = 0, \qquad x > 0.$

Solution Let $w(r, x) = x^r$ and let L denote the left-hand side of (11). A short calculation gives

$$L[w](x) = (4r^2 - 4r + 1)x^r.$$

Solving the indicial equation

$$4r^2 - 4r + 1 = (2r - 1)^2 = 4\left(r - \frac{1}{2} \right)^2 = 0$$

yields the repeated root $r_0 = 1/2$. Thus, a general solution to (11) is obtained from equations (8) and (10) by setting $r_0 = 1/2$. That is,

$$y(x) = C_1 \sqrt{x} + C_2 \sqrt{x} \ln x, \qquad x > 0. \quad \blacklozenge$$

In closing we note that the solutions to Cauchy–Euler equations can exhibit some extraordinary features near their singular points—features unlike anything we have encountered in the case of constant coefficient equations. Observe that:

1. x^r and $x^r \ln x$, for $0 < r < 1$, have infinite slope at the origin;
2. x^r, for $r < 0$, grows without bound near the origin; and
3. $x^\alpha \cos(\beta \ln x)$ and $x^\alpha \sin(\beta \ln x)$ oscillate "infinitely rapidly" near the origin. (See Figure 8.7 on page 453.)

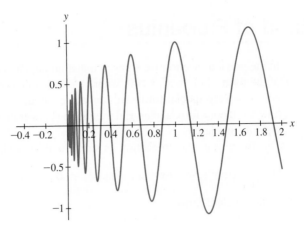

Figure 8.7 Graph of $x^{0.3}\cos(12 \ln x)$

Numerical solvers based on Euler or Runge–Kutta procedures would be completely frustrated in trying to chart such behavior; analytic methods are essential for characterizing solutions near singular points.

In Section 8.7 we discuss the problem of finding a second linearly independent *series* solution to certain differential equations. As we will see, operator methods similar to those described in this section will lead to the desired second solution.

8.5 EXERCISES

In Problems 1–10, use the substitution $y = x^r$ to find a general solution to the given equation for $x > 0$.

1. $x^2 y''(x) + 6xy'(x) + 6y(x) = 0$

2. $2x^2 y''(x) + 13xy'(x) + 15y(x) = 0$

3. $x^2 y''(x) - xy'(x) + 17y(x) = 0$

4. $x^2 y''(x) + 2xy'(x) - 3y(x) = 0$

5. $\dfrac{d^2 y}{dx^2} = \dfrac{5}{x}\dfrac{dy}{dx} - \dfrac{13}{x^2} y$ 6. $\dfrac{d^2 y}{dx^2} = \dfrac{1}{x}\dfrac{dy}{dx} - \dfrac{4}{x^2} y$

7. $x^3 y'''(x) + 4x^2 y''(x) + 10xy'(x) - 10y(x) = 0$

8. $x^3 y'''(x) + 4x^2 y''(x) + xy'(x) = 0$

9. $x^3 y'''(x) + 3x^2 y''(x) + 5xy'(x) - 5y(x) = 0$

10. $x^3 y'''(x) + 9x^2 y''(x) + 19xy'(x) + 8y(x) = 0$

In Problems 11 and 12, use a substitution of the form $y = (x - c)^r$ to find a general solution to the given equation for $x > c$.

11. $2(x - 3)^2 y''(x) + 5(x - 3)y'(x) - 2y(x) = 0$

12. $4(x + 2)^2 y''(x) + 5y(x) = 0$

In Problems 13 and 14, use variation of parameters to find a general solution to the given equation for $x > 0$.

13. $x^2 y''(x) - 2xy'(x) + 2y(x) = x^{-1/2}$

14. $x^2 y''(x) + 2xy'(x) - 2y(x) = 6x^{-2} + 3x$

In Problems 15–17, solve the given initial value problem.

15. $t^2 x''(t) - 12x(t) = 0$;

 $x(1) = 3, \quad x'(1) = 5$

16. $x^2 y''(x) + 5xy'(x) + 4y(x) = 0$;

 $y(1) = 3, \quad y'(1) = 7$

17. $x^3 y'''(x) + 6x^2 y''(x) + 29xy'(x) - 29y(x) = 0$;

 $y(1) = 2, \quad y'(1) = -3, \quad y''(1) = 19$

18. Suppose r_0 is a repeated root of the *auxiliary* equation $ar^2 + br + c = 0$. Then, as we well know, $y_1(t) = e^{r_0 t}$ is a solution to the equation $ay'' + by' + cy = 0$, where a, b, and c are constants. Use a derivation similar to the one given in this section for the case when the *indicial* equation has a repeated root to show that a second linearly independent solution is $y_2(t) = te^{r_0 t}$.

19. Let $L[y](x) := x^3 y'''(x) + xy'(x) - y(x)$.
 (a) Show that $L[x^r](x) = (r - 1)^3 x^r$.
 (b) Using an extension of the argument given in this section for the case when the indicial equation has a double root, show that $L[y] = 0$ has the general solution

 $$y(x) = C_1 x + C_2 x \ln x + C_3 x (\ln x)^2.$$

8.6 Method of Frobenius

In the previous section we showed that a homogeneous Cauchy–Euler equation has a solution of the form $y(x) = x^r, x > 0$, where r is a certain constant. Cauchy–Euler equations have, of course, a very special form with only one singular point (at $x = 0$). In this section we show how the theory for Cauchy–Euler equations generalizes to other equations that have a special type of singularity.

To motivate the procedure, let's rewrite the Cauchy–Euler equation,

$$(1) \qquad ax^2y''(x) + bxy'(x) + cy(x) = 0, \qquad x > 0,$$

in the standard form

$$(2) \qquad y''(x) + p(x)y'(x) + q(x)y(x) = 0, \qquad x > 0,$$

where

$$p(x) = \frac{p_0}{x}, \qquad q(x) = \frac{q_0}{x^2},$$

and p_0, q_0 are the constants b/a and c/a, respectively. When we substitute $w(r, x) = x^r$ for y into equation (2), we get

$$[r(r-1) + p_0 r + q_0]x^{r-2} = 0,$$

which yields the indicial equation

$$(3) \qquad r(r-1) + p_0 r + q_0 = 0.$$

Thus, if r_1 is a root of (3), then $w(r_1, x) = x^{r_1}$ is a solution to equations (1) and (2).

Let's now assume, more generally, that (2) is an equation for which $xp(x)$ and $x^2q(x)$, instead of being constants, are *analytic functions*. That is, in some open interval about $x = 0$,

$$(4) \qquad xp(x) = p_0 + p_1 x + p_2 x^2 + \cdots = \sum_{n=0}^{\infty} p_n x^n,$$

$$(5) \qquad x^2q(x) = q_0 + q_1 x + q_2 x^2 + \cdots = \sum_{n=0}^{\infty} q_n x^n.$$

It follows from (4) and (5) that

$$(6) \qquad \lim_{x \to 0} xp(x) = p_0 \qquad \text{and} \qquad \lim_{x \to 0} x^2q(x) = q_0,$$

and hence, for x near 0 we have $xp(x) \approx p_0$ and $x^2q(x) \approx q_0$. Therefore, it is reasonable to expect that the solutions to (2) will behave (for x near 0) like the solutions to the Cauchy–Euler equation

$$x^2y'' + p_0 xy' + q_0 y = 0.$$

When $p(x)$ and $q(x)$ satisfy (4) and (5), we say that the singular point at $x = 0$ is regular. More generally, we state the following.

Regular Singular Point

Definition 3. A singular point x_0 of

(7) $y''(x) + p(x)y'(x) + q(x)y(x) = 0$

is said to be a **regular singular point** if both $(x - x_0)p(x)$ and $(x - x_0)^2q(x)$ are analytic at x_0.[†] Otherwise x_0 is called an **irregular singular point.**

Example 1 Classify the singular points of the equation

(8) $(x^2 - 1)^2 y''(x) + (x + 1)y'(x) - y(x) = 0$.

Solution Here

$$p(x) = \frac{x + 1}{(x^2 - 1)^2} = \frac{1}{(x + 1)(x - 1)^2},$$

$$q(x) = \frac{-1}{(x^2 - 1)^2} = \frac{-1}{(x + 1)^2(x - 1)^2},$$

from which we see that ± 1 are the singular points of (8). For the singularity at 1, we have

$$(x - 1)p(x) = \frac{1}{(x + 1)(x - 1)},$$

which is not analytic at $x = 1$. Therefore, $x = 1$ is an irregular singular point.

For the singularity at -1, we have

$$(x + 1)p(x) = \frac{1}{(x - 1)^2}, \qquad (x + 1)^2 q(x) = \frac{-1}{(x - 1)^2},$$

both of which are analytic at $x = -1$. Hence, $x = -1$ is a regular singular point. ◆

Let's assume that $x = 0$ is a regular singular point for equation (7) so that $p(x)$ and $q(x)$ satisfy (4) and (5); that is,

(9) $p(x) = \displaystyle\sum_{n=0}^{\infty} p_n x^{n-1}$, $q(x) = \displaystyle\sum_{n=0}^{\infty} q_n x^{n-2}$.

The idea of the mathematician Frobenius was that since Cauchy–Euler equations have solutions of the form x^r, then for the regular singular point $x = 0$, there should be solutions to (7) of the form x^r *times an analytic function*.[‡] Hence we seek solutions to (7) of the form

(10) $w(r, x) = x^r \displaystyle\sum_{n=0}^{\infty} a_n x^n = \sum_{n=0}^{\infty} a_n x^{n+r}$, $x > 0$.

In writing (10), we have assumed a_0 is the first nonzero coefficient, so we are left with determining r and the coefficients a_n, $n \geq 1$. Differentiating $w(r, x)$ with respect to x, we have

(11) $w'(r, x) = \displaystyle\sum_{n=0}^{\infty} (n + r)a_n x^{n+r-1}$,

(12) $w''(r, x) = \displaystyle\sum_{n=0}^{\infty} (n + r)(n + r - 1)a_n x^{n+r-2}$.

[†]In the terminology of complex variables, p has a pole of order at most 1, and q has a pole of order at most 2, at x_0.

[‡]*Historical Footnote:* George Frobenius (1848–1917) developed this method in 1873. He is also known for his research on group theory.

If we substitute the above expansions for $w(r, x)$, $w'(r, x)$, $w''(r, x)$, $p(x)$, and $q(x)$ into (7), we obtain

(13)
$$\sum_{n=0}^{\infty} (n+r)(n+r-1)a_n x^{n+r-2} + \left(\sum_{n=0}^{\infty} p_n x^{n-1} \right) \left(\sum_{n=0}^{\infty} (n+r)a_n x^{n+r-1} \right)$$
$$+ \left(\sum_{n=0}^{\infty} q_n x^{n-2} \right) \left(\sum_{n=0}^{\infty} a_n x^{n+r} \right) = 0.$$

Now we use the Cauchy product to perform the series multiplications and then group like powers of x, starting with the lowest power, x^{r-2}. This gives

(14)
$$[r(r-1) + p_0 r + q_0]a_0 x^{r-2}$$
$$+ [(r+1)ra_1 + (r+1)p_0 a_1 + p_1 ra_0 + q_0 a_1 + q_1 a_0]x^{r-1} + \cdots = 0.$$

For the expansion on the left-hand side of equation (14) to sum to zero, each coefficient must be zero. Considering the first term, x^{r-2}, we find

(15)
$$[r(r-1) + p_0 r + q_0]a_0 = 0.$$

We have assumed that $a_0 \neq 0$, so the quantity in brackets must be zero. This gives the indicial equation; it is the same as the one we derived for Cauchy–Euler equations.

Indicial Equation

Definition 4. If x_0 is a regular singular point of $y'' + py' + qy = 0$, then the **indicial equation** for this point is

(16)
$$r(r-1) + p_0 r + q_0 = 0,$$

where

$$p_0 := \lim_{x \to x_0} (x - x_0)p(x), \qquad q_0 := \lim_{x \to x_0} (x - x_0)^2 q(x).$$

The roots of the indicial equation are called the **exponents (indices)** of the singularity x_0.

Example 2 Find the indicial equation and the exponents at the singularity $x = -1$ of

(17)
$$(x^2 - 1)^2 y''(x) + (x+1)y'(x) - y(x) = 0.$$

Solution In Example 1 we showed that $x = -1$ is a regular singular point. Since $p(x) = (x+1)^{-1}(x-1)^{-2}$ and $q(x) = -(x+1)^{-2}(x-1)^{-2}$, we find

$$p_0 = \lim_{x \to -1} (x+1)p(x) = \lim_{x \to -1} (x-1)^{-2} = \frac{1}{4},$$

$$q_0 = \lim_{x \to -1} (x+1)^2 q(x) = \lim_{x \to -1} [-(x-1)^{-2}] = -\frac{1}{4}.$$

Substituting these values for p_0 and q_0 into (16), we obtain the indicial equation

(18)
$$r(r-1) + \frac{1}{4}r - \frac{1}{4} = 0.$$

Multiplying by 4 and factoring gives $(4r+1)(r-1) = 0$. Hence, $r = 1, -1/4$ are the exponents. ◆

As we have seen, we can use the indicial equation to determine those values of r for which the coefficient of x^{r-2} in (14) is zero. If we set the coefficient of x^{r-1} in (14) equal to zero, we have

(19) $[(r+1)r + (r+1)p_0 + q_0]a_1 + (p_1 r + q_1)a_0 = 0$.

Since a_0 is arbitrary and we know the p_i's, q_i's, and r, we can solve equation (19) for a_1, provided the coefficient of a_1 in (19) is not zero. This will be the case if we take r to be the *larger* of the two roots of the indicial equation (see Problem 43, page 464).[†] Similarly, when we set the coefficient of x^r equal to zero, we can solve for a_2 in terms of the p_i's, q_i's, r, a_0, and a_1. Continuing in this manner, we can recursively solve for the a_n's. The procedure is illustrated in the following example.

Example 3 Find a series expansion about the regular singular point $x = 0$ for a solution to

(20) $(x+2)x^2 y''(x) - xy'(x) + (1+x)y(x) = 0$, $x > 0$.

Solution Here $p(x) = -x^{-1}(x+2)^{-1}$ and $q(x) = x^{-2}(x+2)^{-1}(1+x)$, so

$$p_0 = \lim_{x \to 0} xp(x) = \lim_{x \to 0} [-(x+2)^{-1}] = -\frac{1}{2} ,$$

$$q_0 = \lim_{x \to 0} x^2 q(x) = \lim_{x \to 0} (x+2)^{-1}(1+x) = \frac{1}{2} .$$

Since $x = 0$ is a regular singular point, we seek a solution to (20) of the form

(21) $w(r, x) = x^r \sum_{n=0}^{\infty} a_n x^n = \sum_{n=0}^{\infty} a_n x^{n+r}$.

By the previous discussion, r must satisfy the indicial equation (16). Substituting for p_0 and q_0 in (16), we obtain

$$r(r-1) - \frac{1}{2}r + \frac{1}{2} = 0 ,$$

which simplifies to $2r^2 - 3r + 1 = (2r-1)(r-1) = 0$. Thus, $r = 1$ and $r = 1/2$ are the roots of the indicial equation associated with $x = 0$.

Let's use the larger root $r = 1$ and solve for a_1, a_2, etc., to obtain the solution $w(1, x)$. We can simplify the computations by substituting $w(r, x)$ directly into equation (20), where the coefficients are polynomials in x, rather than dividing by $(x+2)x^2$ and having to work with the rational functions $p(x)$ and $q(x)$. Inserting $w(r, x)$ in (20) and recalling the formulas for $w'(r, x)$ and $w''(r, x)$ in (11) and (12) gives (with $r = 1$)

(22) $(x+2)x^2 \sum_{n=0}^{\infty} (n+1)na_n x^{n-1} - x \sum_{n=0}^{\infty} (n+1)a_n x^n$

$\qquad + (1+x) \sum_{n=0}^{\infty} a_n x^{n+1} = 0$,

which we can write as

(23) $\sum_{n=0}^{\infty} (n+1)na_n x^{n+2} + \sum_{n=0}^{\infty} 2(n+1)na_n x^{n+1} - \sum_{n=0}^{\infty} (n+1)a_n x^{n+1}$

$\qquad + \sum_{n=0}^{\infty} a_n x^{n+1} + \sum_{n=0}^{\infty} a_n x^{n+2} = 0$.

[†]"Larger" in the sense of Problem 43.

Next we shift the indices so that each summation in (23) is in powers x^k. With $k = n + 2$ in the first and last summations and $k = n + 1$ in the rest, (23) becomes

(24) $\displaystyle\sum_{k=2}^{\infty} [(k-1)(k-2) + 1]a_{k-2}x^k + \sum_{k=1}^{\infty} [2k(k-1) - k + 1]a_{k-1}x^k = 0$.

Separating off the $k = 1$ term and combining the rest under one summation yields

(25) $\displaystyle [2(1)(0) - 1 + 1]a_0 x + \sum_{k=2}^{\infty} [(k^2 - 3k + 3)a_{k-2} + (2k-1)(k-1)a_{k-1}]x^k = 0$.

Notice that the coefficient of x in (25) is zero. This is because $r = 1$ is a root of the indicial equation, which is the equation we obtained by setting the coefficient of the lowest power of x equal to zero.

We can now determine the a_k's in terms of a_0 by setting the coefficients of x^k in equation (25) equal to zero for $k = 2, 3$, etc. This gives the recurrence relation

(26) $(k^2 - 3k + 3)a_{k-2} + (2k - 1)(k - 1)a_{k-1} = 0$,

or, equivalently,

(27) $\displaystyle a_{k-1} = -\frac{k^2 - 3k + 3}{(2k-1)(k-1)}a_{k-2}$, $\qquad k \geq 2$.

Setting $k = 2, 3$, and 4 in (27), we find

$$a_1 = -\frac{1}{3}a_0 \qquad\qquad (k = 2) ,$$

$$a_2 = -\frac{3}{10}a_1 = \frac{1}{10}a_0 \qquad (k = 3) ,$$

$$a_3 = -\frac{1}{3}a_2 = -\frac{1}{30}a_0 \qquad (k = 4) .$$

Substituting these values for r, a_1, a_2, and a_3 into (21) gives

(28) $\displaystyle w(1, x) = a_0 x^1 \left(1 - \frac{1}{3}x + \frac{1}{10}x^2 - \frac{1}{30}x^3 + \cdots \right)$,

where a_0 is arbitrary. In particular, for $a_0 = 1$, we get the solution

$$y_1(x) = x - \frac{1}{3}x^2 + \frac{1}{10}x^3 - \frac{1}{30}x^4 + \cdots \qquad (x > 0) .$$

See Figure 8.8 on page 459. ◆

To find a second linearly independent solution to equation (20), we could try setting $r = 1/2$ and solving for $a_1, a_2, \ldots$ to obtain a solution $w(1/2, x)$ (see Problem 44, page 464). In this particular case, the approach would work. However, if we encounter an indicial equation that has a repeated root, then the method of Frobenius would yield just one solution (apart from constant multiples). To find the desired second solution, we must use another technique, such as the reduction of order procedure discussed in Section 4.7 or Exercises 6.1, Problem 31, page 327. We tackle the problem of finding a second linearly independent solution in the next section.

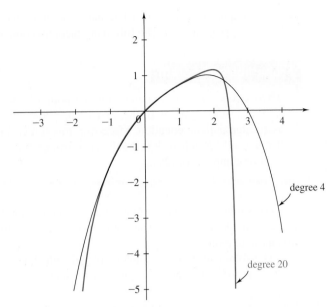

Figure 8.8 Partial sums approximating the solution $y_1(x)$ of Example 3

The method of Frobenius can be summarized as follows.

Method of Frobenius

To derive a series solution about the singular point x_0 of

$$(29) \qquad a_2(x)y''(x) + a_1(x)y'(x) + a_0(x)y(x) = 0, \qquad x > x_0:$$

(a) Set $p(x) := a_1(x)/a_2(x)$, $q(x) := a_0(x)/a_2(x)$. If both $(x - x_0)p(x)$ and $(x - x_0)^2 q(x)$ are analytic at x_0, then x_0 is a regular singular point and the remaining steps apply.

(b) Let

$$(30) \qquad w(r, x) = (x - x_0)^r \sum_{n=0}^{\infty} a_n(x - x_0)^n = \sum_{n=0}^{\infty} a_n(x - x_0)^{n+r},$$

and, using termwise differentiation, substitute $w(r, x)$ into equation (29) to obtain an equation of the form

$$A_0(x - x_0)^{r+J} + A_1(x - x_0)^{r+J+1} + \cdots = 0.$$

(c) Set the coefficients $A_0, A_1, A_2, \ldots$ equal to zero. [Notice that the equation $A_0 = 0$ is just a constant multiple of the indicial equation $r(r - 1) + p_0 r + q_0 = 0$.]

(d) Use the system of equations

$$A_0 = 0, \qquad A_1 = 0, \qquad \ldots, \qquad A_k = 0$$

to find a recurrence relation involving a_k and $a_0, a_1, \ldots, a_{k-1}$.

(e) Take $r = r_1$, the larger root of the indicial equation, and use the relation obtained in step (d) to determine $a_1, a_2, \ldots$ recursively in terms of a_0 and r_1.

(f) A series expansion of a solution to (29) is

$$(31) \qquad w(r_1, x) = (x - x_0)^{r_1} \sum_{n=0}^{\infty} a_n(x - x_0)^n, \qquad x > x_0,$$

where a_0 is arbitrary and the a_n's are defined in terms of a_0 and r_1.

One important question that remains concerns the radius of convergence of the power series that appears in (31). The following theorem contains an answer.[†]

Frobenius's Theorem

Theorem 6. If x_0 is a regular singular point of equation (29), then there exists at least one series solution of the form (30), where $r = r_1$ is the larger root of the associated indicial equation. Moreover, this series converges for all x such that $0 < x - x_0 < R$, where R is the distance from x_0 to the nearest other singular point (real or complex) of (29).

For simplicity, in the examples that follow we consider only series expansions about the regular singular point $x = 0$ and only those equations for which the associated indicial equation has real roots.

The following three examples not only illustrate the method of Frobenius but also are important models to which we refer in later sections.

Example 4 Find a series solution about the regular singular point $x = 0$ of

$$(32) \qquad x^2 y''(x) - xy'(x) + (1 - x)y(x) = 0, \qquad x > 0.$$

Solution Here $p(x) = -x^{-1}$ and $q(x) = (1 - x)x^{-2}$. It is easy to check that $x = 0$ is a regular singular point of (32), so we compute

$$p_0 = \lim_{x \to 0} xp(x) = \lim_{x \to 0}(-1) = -1,$$
$$q_0 = \lim_{x \to 0} x^2 q(x) = \lim_{x \to 0}(1 - x) = 1.$$

Then the indicial equation is

$$r(r - 1) - r + 1 = r^2 - 2r + 1 = (r - 1)^2 = 0,$$

which has the roots $r_1 = r_2 = 1$.

Next we substitute

$$(33) \qquad w(r, x) = x^r \sum_{n=0}^{\infty} a_n x^n = \sum_{n=0}^{\infty} a_n x^{n+r}$$

into (32) and obtain

$$(34) \qquad x^2 \sum_{n=0}^{\infty} (n + r)(n + r - 1)a_n x^{n+r-2} - x \sum_{n=0}^{\infty} (n + r)a_n x^{n+r-1}$$
$$+ (1 - x) \sum_{n=0}^{\infty} a_n x^{n+r} = 0,$$

which we write as

$$(35) \qquad \sum_{n=0}^{\infty} (n + r)(n + r - 1)a_n x^{n+r} - \sum_{n=0}^{\infty} (n + r)a_n x^{n+r}$$
$$+ \sum_{n=0}^{\infty} a_n x^{n+r} - \sum_{n=0}^{\infty} a_n x^{n+r+1} = 0.$$

[†]For a proof of this theorem, see *Ordinary Differential Equations*, by E. L. Ince (Dover Publications, New York, 1956), Chapter 16.

Shifting the indices so that each summation in (35) is in powers x^{k+r}, we take $k = n + 1$ in the last summation and $k = n$ in the rest. This gives

(36) $\displaystyle\sum_{k=0}^{\infty} [(k+r)(k+r-1) - (k+r) + 1]a_k x^{k+r} - \sum_{k=1}^{\infty} a_{k-1} x^{k+r} = 0$.

Singling out the term corresponding to $k = 0$ and combining the rest under one summation yields

(37) $[r(r-1) - r + 1]a_0 x^r$

$\qquad + \displaystyle\sum_{k=1}^{\infty} \{[(k+r)(k+r-1) - (k+r) + 1]a_k - a_{k-1}\} x^{k+r} = 0$.

When we set the coefficients equal to zero, we recover the indicial equation

(38) $[r(r-1) - r + 1]a_0 = 0$,

and obtain, for $k \geq 1$, the recurrence relation

(39) $[(k+r)^2 - 2(k+r) + 1]a_k - a_{k-1} = 0$,

which reduces to

(40) $(k+r-1)^2 a_k - a_{k-1} = 0$.

Relation (40) can be used to solve for a_k in terms of a_{k-1}:

(41) $a_k = \dfrac{1}{(k+r-1)^2} a_{k-1}$, $k \geq 1$.

Setting $r = r_1 = 1$ in (38) gives (as expected) $0 \cdot a_0 = 0$, and in (41) gives

(42) $a_k = \dfrac{1}{k^2} a_{k-1}$, $k \geq 1$.

For $k = 1, 2,$ and 3, we now find

$\qquad a_1 = \dfrac{1}{1^2} a_0 = a_0$ $\qquad\qquad\qquad\qquad (k = 1)$,

$\qquad a_2 = \dfrac{1}{2^2} a_1 = \dfrac{1}{(2 \cdot 1)^2} a_0 = \dfrac{1}{4} a_0$ $\qquad (k = 2)$,

$\qquad a_3 = \dfrac{1}{3^2} a_2 = \dfrac{1}{(3 \cdot 2 \cdot 1)^2} a_0 = \dfrac{1}{36} a_0$ $\qquad (k = 3)$.

In general, we have

(43) $a_k = \dfrac{1}{(k!)^2} a_0$.

Hence, equation (32) has a series solution given by

(44) $w(1, x) = a_0 x \left\{ 1 + x + \dfrac{1}{4} x^2 + \dfrac{1}{36} x^3 + \cdots \right\}$

$\qquad\qquad = a_0 x \displaystyle\sum_{k=0}^{\infty} \dfrac{1}{(k!)^2} x^k$, $x > 0$. ◆

Since $x = 0$ is the only singular point for equation (32), it follows from Frobenius's theorem or directly by the ratio test that the series solution (44) converges for all $x > 0$.

In the next two examples, we only outline the method; we leave it to you to furnish the intermediate steps.

Example 5 Find a series solution about the regular singular point $x = 0$ of

(45) $xy''(x) + 4y'(x) - xy(x) = 0$, $x > 0$.

Solution Since $p(x) = 4/x$ and $q(x) = -1$, we see that $x = 0$ is indeed a regular singular point and

$$p_0 = \lim_{x \to 0} xp(x) = 4, \qquad q_0 = \lim_{x \to 0} x^2 q(x) = 0.$$

The indicial equation is

$$r(r-1) + 4r = r^2 + 3r = r(r+3) = 0,$$

with roots $r_1 = 0$ and $r_2 = -3$.

Now we substitute

(46) $w(r, x) = x^r \sum_{n=0}^{\infty} a_n x^n = \sum_{n=0}^{\infty} a_n x^{n+r}$

into (45). After a little algebra and a shift in indices, we get

(47) $[r(r-1) + 4r]a_0 x^{r-1} + [(r+1)r + 4(r+1)]a_1 x^r$

$$+ \sum_{k=1}^{\infty} [(k+r+1)(k+r+4)a_{k+1} - a_{k-1}]x^{k+r} = 0.$$

Next we set the coefficients equal to zero and find

(48) $[r(r-1) + 4r]a_0 = 0$,

(49) $[(r+1)r + 4(r+1)]a_1 = 0$,

and, for $k \geq 1$, the recurrence relation

(50) $(k+r+1)(k+r+4)a_{k+1} - a_{k-1} = 0$.

For $r = r_1 = 0$, equation (48) becomes $0 \cdot a_0 = 0$ and (49) becomes $4 \cdot a_1 = 0$. Hence, although a_0 is arbitrary, a_1 must be zero. Setting $r = r_1 = 0$ in (50), we find

(51) $a_{k+1} = \dfrac{1}{(k+1)(k+4)} a_{k-1}$, $k \geq 1$,

from which it follows (after a few experimental computations) that $a_{2k+1} = 0$ for $k = 0, 1, \ldots,$ and

(52) $a_{2k} = \dfrac{1}{[2 \cdot 4 \cdots (2k)][5 \cdot 7 \cdots (2k+3)]} a_0$

$$= \dfrac{1}{2^k k![5 \cdot 7 \cdots (2k+3)]} a_0, \quad k \geq 1.$$

Hence equation (45) has the power series solution

(53) $w(0, x) = a_0 \left\{ 1 + \sum_{k=1}^{\infty} \dfrac{1}{2^k k![5 \cdot 7 \cdots (2k+3)]} x^{2k} \right\}$, $x > 0$. ◆

If in Example 5 we had worked with the root $r = r_2 = -3$, then we would actually have obtained *two* linearly independent solutions (see Problem 45).

Example 6 Find a series solution about the regular singular point of $x = 0$ of

$$(54) \qquad xy''(x) + 3y'(x) - xy(x) = 0, \qquad x > 0.$$

Solution Since $p(x) = 3/x$ and $q(x) = -1$, we see that $x = 0$ is a regular singular point. Moreover,

$$p_0 = \lim_{x \to 0} xp(x) = 3, \qquad q_0 = \lim_{x \to 0} x^2 q(x) = 0.$$

So the indicial equation is

$$(55) \qquad r(r-1) + 3r = r^2 + 2r = r(r+2) = 0,$$

with roots $r_1 = 0$ and $r_2 = -2$.
 Substituting

$$(56) \qquad w(r, x) = x^r \sum_{n=0}^{\infty} a_n x^n = \sum_{n=0}^{\infty} a_n x^{n+r}$$

into (54) ultimately gives

$$(57) \qquad [r(r-1) + 3r]a_0 x^{r-1} + [(r+1)r + 3(r+1)]a_1 x^r$$
$$+ \sum_{k=1}^{\infty} [(k+r+1)(k+r+3)a_{k+1} - a_{k-1}]x^{k+r} = 0.$$

Setting the coefficients equal to zero, we have

$$(58) \qquad [r(r-1) + 3r]a_0 = 0,$$

$$(59) \qquad [(r+1)r + 3(r+1)]a_1 = 0,$$

and, for $k \geq 1$, the recurrence relation

$$(60) \qquad (k+r+1)(k+r+3)a_{k+1} - a_{k-1} = 0.$$

With $r = r_1 = 0$, these equations lead to the following formulas: $a_{2k+1} = 0, \ k = 0, 1, \ldots,$ and

$$(61) \qquad a_{2k} = \frac{1}{[2 \cdot 4 \cdots (2k)][4 \cdot 6 \cdots (2k+2)]}a_0 = \frac{1}{2^{2k}k!(k+1)!}a_0, \qquad k \geq 0.$$

Hence equation (54) has the power series solution

$$(62) \qquad w(0, x) = a_0 \sum_{k=0}^{\infty} \frac{1}{2^{2k}k!(k+1)!}x^{2k}, \qquad x > 0. \ \blacklozenge$$

Unlike in Example 5, if we work with the second root $r = r_2 = -2$ in Example 6, then we do *not* obtain a second linearly independent solution (see Problem 46).
 In the preceding examples we were able to use the method of Frobenius to find a series solution valid to the right ($x > 0$) of the regular singular point $x = 0$. For $x < 0$, we can use the change of variables $x = -t$ and then solve the resulting equation for $t > 0$.
 The method of Frobenius also applies to higher-order linear equations (see Problems 35–38).

8.6 EXERCISES

In Problems 1–10, classify each singular point (real or complex) of the given equation as regular or irregular.

1. $(x^2 - 1)y'' + xy' + 3y = 0$

2. $x^2y'' + 8xy' - 3xy = 0$

3. $(x^2 + 1)z'' + 7x^2z' - 3xz = 0$

4. $x^2y'' - 5xy' + 7y = 0$

5. $(x^2 - 1)^2y'' - (x - 1)y' + 3y = 0$

6. $(x^2 - 4)y'' + (x + 2)y' + 3y = 0$

7. $(t^2 - t - 2)^2x'' + (t^2 - 4)x' - tx = 0$

8. $(x^2 - x)y'' + xy' + 7y = 0$

9. $(x^2 + 2x - 8)^2y'' + (3x + 12)y' - x^2y = 0$

10. $x^3(x - 1)y'' + (x^2 - 3x)(\sin x)y' - xy = 0$

In Problems 11–18, find the indicial equation and the exponents for the specified singularity of the given differential equations.

11. $x^2y'' - 2xy' - 10y = 0$, at $x = 0$

12. $x^2y'' + 4xy' + 2y = 0$, at $x = 0$

13. $(x^2 - x - 2)^2z'' + (x^2 - 4)z' - 6xz = 0$, at $x = 2$

14. $(x^2 - 4)y'' + (x + 2)y' + 3y = 0$, at $x = -2$

15. $\theta^3y'' + \theta(\sin\theta)y' - (\tan\theta)y = 0$, at $\theta = 0$

16. $(x^2 - 1)y'' - (x - 1)y' - 3y = 0$, at $x = 1$

17. $(x - 1)^2y'' + (x^2 - 1)y' - 12y = 0$, at $x = 1$

18. $4x(\sin x)y'' - 3y = 0$, at $x = 0$

In Problems 19–24, use the method of Frobenius to find at least the first four nonzero terms in the series expansion about $x = 0$ for a solution to the given equation for $x > 0$.

19. $9x^2y'' + 9x^2y' + 2y = 0$

20. $2x(x - 1)y'' + 3(x - 1)y' - y = 0$

21. $x^2y'' + xy' + x^2y = 0$

22. $xy'' + y' - 4y = 0$

23. $x^2z'' + (x^2 + x)z' - z = 0$

24. $3xy'' + (2 - x)y' - y = 0$

In Problems 25–30, use the method of Frobenius to find a general formula for the coefficient a_n in a series expansion about $x = 0$ for a solution to the given equation for $x > 0$.

25. $4x^2y'' + 2x^2y' - (x + 3)y = 0$

26. $x^2y'' + (x^2 - x)y' + y = 0$

27. $xw'' - w' - xw = 0$

28. $3x^2y'' + 8xy' + (x - 2)y = 0$

29. $xy'' + (x - 1)y' - 2y = 0$

30. $x(x + 1)y'' + (x + 5)y' - 4y = 0$

In Problems 31–34, first determine a recurrence formula for the coefficients in the (Frobenius) series expansion of the solution about $x = 0$. Use this recurrence formula to determine if there exists a solution to the differential equation that is decreasing for $x > 0$.

31. $xy'' + (1 - x)y' - y = 0$

32. $x^2y'' - x(1 + x)y' + y = 0$

33. $3xy'' + 2(1 - x)y' - 4y = 0$

34. $xy'' + (x + 2)y' - y = 0$

In Problems 35–38, use the method of Frobenius to find at least the first four nonzero terms in the series expansion about $x = 0$ for a solution to the given linear third-order equation for $x > 0$.

35. $6x^3y''' + 13x^2y'' + (x + x^2)y' + xy = 0$

36. $6x^3y''' + 11x^2y'' - 2xy' - (x - 2)y = 0$

37. $6x^3y''' + 13x^2y'' - (x^2 + 3x)y' - xy = 0$

38. $6x^3y''' + (13x^2 - x^3)y'' + xy' - xy = 0$

In Problems 39 and 40, try to use the method of Frobenius to find a series expansion about the irregular singular point $x = 0$ for a solution to the given differential equation. If the method works, give at least the first four nonzero terms in the expansion. If the method does not work, explain what went wrong.

39. $x^2y'' + (3x - 1)y' + y = 0$

40. $x^2y'' + y' - 2y = 0$

*In certain applications, it is desirable to have **an expansion about the point at infinity**. To obtain such an expansion, we use the change of variables $z = 1/x$ and expand about $z = 0$. In Problems 41 and 42, show that infinity is a regular singular point of the given differential equation by showing that $z = 0$ is a regular singular point for the transformed equation in z. Also find at least the first four nonzero terms in the series expansion about infinity of a solution to the original equation in x.*

41. $x^3y'' - x^2y' - y = 0$

42. $18(x - 4)^2(x - 6)y'' + 9x(x - 4)y' - 32y = 0$

43. Show that if r_1 and r_2 are roots of the indicial equation (16) on page 456, with r_1 the larger root (Re $r_1 \geq$ Re r_2), then the coefficient of a_1 in equation (19) on page 457 is not zero when $r = r_1$.

44. To obtain a second linearly independent solution to equation (20):

 (a) Substitute $w(r, x)$ given in (21) into (20) and conclude that the coefficients a_k, $k \geq 1$, must satisfy the recurrence relation

$$(k + r - 1)(2k + 2r - 1)a_k$$
$$+ [(k + r - 1)(k + r - 2) + 1]a_{k-1} = 0.$$

(b) Use the recurrence relation with $r = 1/2$ to derive the second series solution

$$w\left(\frac{1}{2}, x\right) =$$

$$a_0\left(x^{1/2} - \frac{3}{4}x^{3/2} + \frac{7}{32}x^{5/2} - \frac{133}{1920}x^{7/2} + \cdots\right).$$

(c) Use the recurrence relation with $r = 1$ to obtain $w(1, x)$ in (28) on page 458.

45. In Example 5, show that if we choose $r = r_2 = -3$, then we obtain *two* linearly independent solutions to equation (45). [*Hint:* a_0 and a_3 are arbitrary constants.]

46. In Example 6, page 463, show that if we choose $r = r_2 = -2$, then we obtain a solution that is a constant multiple of the solution given in (62). [*Hint:* Show that a_0 and a_1 must be zero while a_2 is arbitrary.]

47. In applying the method of Frobenius, the following recurrence relation arose: $a_{k+1} = 15^7 a_k / (k+1)^9$, $k = 0, 1, 2, \ldots$.

(a) Show that the coefficients are given by the formula $a_k = 15^{7k} a_0 / (k!)^9$, $k = 0, 1, 2, \ldots$.

(b) Use the formula obtained in part (a) with $a_0 = 1$ to compute $a_5, a_{10}, a_{15}, a_{20}$, and a_{25} on your computer or calculator. What goes wrong?

(c) Now use the recurrence relation to compute a_k for $k = 1, 2, 3, \ldots, 25$, assuming $a_0 = 1$.

(d) What advantage does the recurrence relation have over the formula?

8.7 Finding a Second Linearly Independent Solution

In the previous section we showed that if $x = 0$ is a regular singular point of

$$\text{(1)} \qquad y''(x) + p(x)y'(x) + q(x)y(x) = 0, \qquad x > 0,$$

then the method of Frobenius can be used to find a series solution valid for x near zero. The first step in the method is to find the roots r_1 and r_2 ($\operatorname{Re} r_1 \geq \operatorname{Re} r_2$) of the associated indicial equation

$$\text{(2)} \qquad r(r - 1) + p_0 r + q_0 = 0.$$

Then, utilizing the larger root r_1, equation (1) has a series solution of the form

$$\text{(3)} \qquad w(r_1, x) = x^{r_1} \sum_{n=0}^{\infty} a_n x^n = \sum_{n=0}^{\infty} a_n x^{n+r_1},$$

where $a_0 \neq 0$. To find a second linearly independent solution, our first inclination is to set $r = r_2$ and seek a solution of the form

$$\text{(4)} \qquad w(r_2, x) = x^{r_2} \sum_{n=0}^{\infty} a_n x^n = \sum_{n=0}^{\infty} a_n x^{n+r_2}.$$

We'll see that this procedure works, *provided* $r_1 - r_2$ *is not an integer.* However, when $r_1 - r_2$ is an integer, the Frobenius method with $r = r_2$ may just lead to the same solution that we obtained using the root r_1. (This is obviously true when $r_1 = r_2$.)

Example 1 Find the first few terms in the series expansion about the regular singular point $x = 0$ for a general solution to

$$\text{(5)} \qquad (x + 2)x^2 y''(x) - xy'(x) + (1 + x)y(x) = 0, \qquad x > 0.$$

Solution In Example 3 of Section 8.6, we used the method of Frobenius to find a series solution for (5). In the process we determined that $p_0 = -1/2$, $q_0 = 1/2$ and that the indicial equation

has roots $r_1 = 1, r_2 = 1/2$. Since these roots do not differ by an integer $(r_1 - r_2 = 1/2)$, the method of Frobenius will give two linearly independent solutions of the form

(6) $$w(r, x) := x^r \sum_{n=0}^{\infty} a_n x^n = \sum_{n=0}^{\infty} a_n x^{n+r} .$$

In Problem 44 of Exercises 8.6, page 464, you were requested to show that substituting $w(r, x)$ into (5) leads to the recurrence relation

(7) $$(k + r - 1)(2k + 2r - 1)a_k + [(k + r - 1)(k + r - 2) + 1]a_{k-1} = 0 , \quad k \geq 1 .$$

With $r = r_1 = 1$ and $a_0 = 1$, we find

(8) $$y_1(x) = x - \frac{1}{3}x^2 + \frac{1}{10}x^3 - \frac{1}{30}x^4 + \cdots$$

as obtained in the previous section. Moreover, taking $r = r_2 = 1/2$ and $a_0 = 1$ in (7) leads to the second solution,

(9) $$y_2(x) = x^{1/2} - \frac{3}{4}x^{3/2} + \frac{7}{32}x^{5/2} - \frac{133}{1920}x^{7/2} + \cdots .$$

Consequently, a general solution to equation (5) is

(10) $$y(x) = c_1 y_1(x) + c_2 y_2(x) , \quad x > 0 ,$$

where $y_1(x)$ and $y_2(x)$ are the series solutions given in equations (8) and (9). See Figure 8.9. ◆

When the indicial equation has repeated roots, $r_1 = r_2$, obviously substituting $r = r_2$ just gives us back the first solution and gets us nowhere. One possibility is to use the reduction of order method described in Section 4.7 or Exercises 6.1, Problem 31. However, this method has drawbacks in that it requires manipulations of series that often make it difficult to determine the general term in the series expansion for the second linearly independent solution. A more direct approach is to use the next theorem, which provides the *form* of this second solution. You will not be surprised to see that, analogous to the situation for a Cauchy–Euler equation whose indicial equation has repeated roots, a second linearly independent solution will involve the first solution multiplied by a logarithmic function.

In the following theorem, we give the *general form* of two linearly independent solutions for the three cases where the roots of the indicial equation (a) do not differ by an integer, (b) are equal, or (c) differ by a nonzero integer.

Figure 8.9 Partial sums approximating solutions to Example 1

Form of Second Linearly Independent Solution

Theorem 7.[†] Let x_0 be a regular singular point for $y'' + py' + qy = 0$ and let r_1 and r_2 be the roots of the associated indicial equation, where $\text{Re } r_1 \geq \text{Re } r_2$.

(a) If $r_1 - r_2$ is not an integer, then there exist two linearly independent solutions of the form

$$(11) \qquad y_1(x) = \sum_{n=0}^{\infty} a_n(x - x_0)^{n+r_1}\,, \qquad a_0 \neq 0\,,$$

$$(12) \qquad y_2(x) = \sum_{n=0}^{\infty} b_n(x - x_0)^{n+r_2}\,, \qquad b_0 \neq 0\,.$$

(b) If $r_1 = r_2$, then there exist two linearly independent solutions of the form

$$(13) \qquad y_1(x) = \sum_{n=0}^{\infty} a_n(x - x_0)^{n+r_1}\,, \qquad a_0 \neq 0\,,$$

$$(14) \qquad y_2(x) = y_1(x)\ln(x - x_0) + \sum_{n=1}^{\infty} b_n(x - x_0)^{n+r_1}\,.$$

(c) If $r_1 - r_2$ is a positive integer, then there exist two linearly independent solutions of the form

$$(15) \qquad y_1(x) = \sum_{n=0}^{\infty} a_n(x - x_0)^{n+r_1}\,, \qquad a_0 \neq 0\,,$$

$$(16) \qquad y_2(x) = Cy_1(x)\ln(x - x_0) + \sum_{n=0}^{\infty} b_n(x - x_0)^{n+r_2}\,, \qquad b_0 \neq 0\,,$$

where C is a constant that could be zero.

In each case of the theorem, $y_1(x)$ is just the series solution obtained by the method of Frobenius, with $r = r_1$. When $r_1 - r_2$ is not an integer, the method of Frobenius yields a second linearly independent solution by taking $r = r_2$. Let's see how knowing the form of the second solution enables us to obtain it. Again, for simplicity, we consider only indicial equations having real roots.

Example 2 Find the first few terms in the series expansion about the regular singular point $x = 0$ for two linearly independent solutions to

$$(17) \qquad x^2 y''(x) - xy'(x) + (1 - x)y(x) = 0\,, \qquad x > 0\,.$$

Solution In Example 4 of Section 8.6, page 460, we used the method of Frobenius to obtain a series solution to equation (17). In the process, we found the indicial equation to be $r^2 - 2r + 1 = 0$, which has roots $r_1 = r_2 = 1$. Working with $r_1 = 1$, we derived the series solution

$$(18) \qquad y_1(x) = x + x^2 + \frac{1}{4}x^3 + \frac{1}{36}x^4 + \frac{1}{576}x^5 + \cdots = \sum_{k=0}^{\infty} \frac{1}{(k!)^2} x^{k+1}$$

[see equation (44) in Section 8.6 with $a_0 = 1$].

[†]For a proof of Theorem 7, see the text *Ordinary Differential Equations*, 4th ed., by G. Birkhoff and G.-C. Rota (Wiley, New York, 1989).

To find a second linearly independent solution $y_2(x)$, we appeal to Theorem 7, part (b), which asserts that (with $x_0 = 0$) $y_2(x)$ has the form

(19) $y_2(x) = y_1(x)\ln x + \displaystyle\sum_{n=1}^{\infty} b_n x^{n+1}$.

Our goal is to determine the coefficients b_n by substituting $y_2(x)$ directly into equation (17). We begin by differentiating $y_2(x)$ in (19) to obtain

$$y_2'(x) = y_1'(x)\ln x + x^{-1}y_1(x) + \sum_{n=1}^{\infty} (n+1)b_n x^n ,$$

$$y_2''(x) = y_1''(x)\ln x - x^{-2}y_1(x) + 2x^{-1}y_1'(x) + \sum_{n=1}^{\infty} n(n+1)b_n x^{n-1} .$$

Substituting $y_2(x)$ into (17) yields

(20) $x^2\left\{ y_1''(x)\ln x - x^{-2}y_1(x) + 2x^{-1}y_1'(x) + \displaystyle\sum_{n=1}^{\infty} n(n+1)b_n x^{n-1} \right\}$

$$- x\left\{ y_1'(x)\ln x + x^{-1}y_1(x) + \sum_{n=1}^{\infty} (n+1)b_n x^n \right\}$$

$$+ (1-x)\left\{ y_1(x)\ln x + \sum_{n=1}^{\infty} b_n x^{n+1} \right\} = 0 ,$$

which simplifies to

(21) $\{x^2 y_1''(x) - xy_1'(x) + (1-x)y_1(x)\}\ln x - 2y_1(x) + 2xy_1'(x)$

$$+ \sum_{n=1}^{\infty} n(n+1)b_n x^{n+1} - \sum_{n=1}^{\infty} (n+1)b_n x^{n+1} + \sum_{n=1}^{\infty} b_n x^{n+1} - \sum_{n=1}^{\infty} b_n x^{n+2} = 0 .$$

Notice that the factor in front of $\ln x$ is just the left-hand side of equation (17) with $y = y_1$. Since y_1 is a solution to (17), this factor is zero. With this observation and a shift in the indices of summation, equation (21) can be rewritten as

(22) $2xy_1'(x) - 2y_1(x) + b_1 x^2 + \displaystyle\sum_{k=2}^{\infty} (k^2 b_k - b_{k-1})x^{k+1} = 0 .$

To identify the coefficients in (22) so that we can set them equal to zero, we must substitute back in the series expansions for $y_1(x)$ and $y_1'(x)$. From (18), we get that $y_1'(x) = \sum_{k=0}^{\infty}(k+1)x^k/(k!)^2$. Inserting this series together with (18) into (22), we find

(23) $\displaystyle\sum_{k=0}^{\infty} \frac{2(k+1)-2}{(k!)^2}x^{k+1} + b_1 x^2 + \sum_{k=2}^{\infty} (k^2 b_k - b_{k-1})x^{k+1} = 0 .$

Listing separately the $k = 0$ and $k = 1$ terms and combining the remaining terms gives

(24) $(2 + b_1)x^2 + \displaystyle\sum_{k=2}^{\infty} \left[\frac{2k}{(k!)^2} + k^2 b_k - b_{k-1} \right] x^{k+1} = 0 .$

Next, we set the coefficients in (24) equal to zero. From the x^2 term we have $2 + b_1 = 0$, so $b_1 = -2$. From the x^{k+1} term, we obtain

$$\frac{2k}{(k!)^2} + k^2 b_k - b_{k-1} = 0$$

or

(25) $b_k = \dfrac{1}{k^2}\left[b_{k-1} - \dfrac{2k}{(k!)^2} \right]$ $(k \geq 2) .$

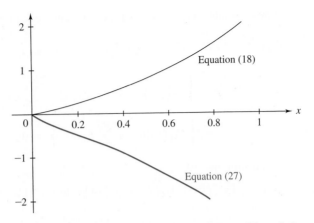

Figure 8.10 Partial sum approximations to solutions of Example 2

Taking $k = 2$ and 3, we compute

(26) $b_2 = \dfrac{1}{2^2}[b_1 - 1] = \dfrac{-3}{4}, \qquad b_3 = \dfrac{1}{9}\left[-\dfrac{3}{4} - \dfrac{6}{36}\right] = \dfrac{-11}{108}.$

Hence, a second linearly independent solution to (17) is

(27) $y_2(x) = y_1(x)\ln x - 2x^2 - \dfrac{3}{4}x^3 - \dfrac{11}{108}x^4 + \cdots.$

See Figure 8.10. ◆

In the next two examples we consider the case when the difference between the roots of the indicial equation is a positive integer. In Example 3, it will turn out that the constant C in formula (16) must be taken to be zero (i.e., no ln x term is present), while in Example 4, this constant is nonzero (i.e., a ln x term is present). Since the solutions to these examples require several intermediate computations, we do not display all the details. Rather, we encourage you to take an active part by bridging the gaps.

Example 3 Find the first few terms in the series expansion about the regular singular point $x = 0$ for a general solution to

(28) $xy''(x) + 4y'(x) - xy(x) = 0, \qquad x > 0.$

Solution In Example 5 of Section 8.6, we used the method of Frobenius to find a series expansion about $x = 0$ for a solution to equation (28). There we found the indicial equation to be $r^2 + 3r = 0$, which has roots $r_1 = 0$ and $r_2 = -3$. Working with $r_1 = 0$, we obtained the series solution

(29) $y_1(x) = 1 + \dfrac{1}{10}x^2 + \dfrac{1}{280}x^4 + \cdots$

[see equation (53) in Section 8.6, page 462, with $a_0 = 1$].

Since $r_1 - r_2 = 3$ is a positive integer, it follows from Theorem 7 that equation (28) has a second linearly independent solution of the form

(30) $y_2(x) = Cy_1(x)\ln x + \displaystyle\sum_{n=0}^{\infty} b_n x^{n-3}.$

When we substitute this expression for y_2 into equation (28), we obtain

(31)　$x\left\{ Cy_1''(x)\ln x + 2Cx^{-1}y_1'(x) - Cx^{-2}y_1(x) + \sum_{n=0}^{\infty}(n-3)(n-4)b_n x^{n-5}\right\}$

$\qquad + 4\left\{ Cy_1'(x)\ln x + Cx^{-1}y_1(x) + \sum_{n=0}^{\infty}(n-3)b_n x^{n-4}\right\}$

$\qquad - x\left\{ Cy_1(x)\ln x + \sum_{n=0}^{\infty}b_n x^{n-3}\right\} = 0,$

which simplifies to

(32)　$\{xy_1''(x) + 4y_1'(x) - xy_1(x)\}C\ln x + 3Cx^{-1}y_1(x) + 2Cy_1'(x)$

$\qquad + \sum_{n=0}^{\infty}(n-3)(n-4)b_n x^{n-4} + \sum_{n=0}^{\infty}4(n-3)b_n x^{n-4} - \sum_{n=0}^{\infty}b_n x^{n-2} = 0.$

The factor in braces is zero because $y_1(x)$ satisfies equation (28). After we combine the summations and simplify, equation (32) becomes

(33)　$3Cx^{-1}y_1(x) + 2Cy_1'(x) - 2b_1 x^{-3} + \sum_{k=2}^{\infty}\left[k(k-3)b_k - b_{k-2}\right]x^{k-4} = 0.$

Substituting in the series for $y_1(x)$ and writing out the first few terms of the summation in (33), we obtain

$\qquad -2b_1 x^{-3} + (-2b_2 - b_0)x^{-2} + (3C - b_1)x^{-1} + (4b_4 - b_2)$

$\qquad + \left(\frac{7}{10}C + 10b_5 - b_3\right)x + (18b_6 - b_4)x^2 + \left(\frac{11}{280}C + 28b_7 - b_5\right)x^3$

$\qquad + \cdots = 0.$

Next, we set the coefficients equal to zero:

$\qquad -2b_1 = 0 \Rightarrow b_1 = 0, \qquad -2b_2 - b_0 = 0 \Rightarrow b_2 = -\frac{1}{2}b_0,$

$\qquad 3C - b_1 = 0 \Rightarrow C = \frac{1}{3}b_1 = 0, \qquad 4b_4 - b_2 = 0 \Rightarrow b_4 = \frac{1}{4}b_2 = -\frac{1}{8}b_0,$

$\qquad \frac{7}{10}C + 10b_5 - b_3 = 0 \Rightarrow b_5 = \frac{b_3 - \dfrac{7}{10}C}{10} = \frac{1}{10}b_3,$

$\qquad 18b_6 - b_4 = 0 \Rightarrow b_6 = \frac{1}{18}b_4 = -\frac{1}{144}b_0,$

$\qquad \frac{11}{280}C + 28b_7 - b_5 = 0 \Rightarrow b_7 = \frac{b_5 - \dfrac{11}{280}C}{28} = \frac{1}{280}b_3.$

Note, in particular, that C must equal zero. Substituting the above values for the b_n's and $C = 0$ back into equation (30) gives

(34)　$y_2(x) = b_0\left\{ x^{-3} - \frac{1}{2}x^{-1} - \frac{1}{8}x - \frac{1}{144}x^3 + \cdots\right\}$

$\qquad + b_3\left\{ 1 + \frac{1}{10}x^2 + \frac{1}{280}x^4 + \cdots\right\},$

Figure 8.11 Partial sum approximations to solutions of Example 3

where b_0 and b_3 are arbitrary constants. Observe that the expression in braces following b_3 is just the series expansion for $y_1(x)$ given in equation (29). Hence, in order to obtain a second linearly independent solution, we must choose b_0 to be nonzero. Taking $b_0 = 1$ and $b_3 = 0$ gives

$$(35) \qquad y_2(x) = x^{-3} - \frac{1}{2}x^{-1} - \frac{1}{8}x - \frac{1}{144}x^3 + \cdots .$$

Therefore, a general solution to equation (28) is

$$(36) \qquad y(x) = c_1 y_1(x) + c_2 y_2(x) , \qquad x > 0 ,$$

where $y_1(x)$ and $y_2(x)$ are given in (29) and (35). [Notice that the right-hand side of (34) coincides with (36) if we identify b_0 as c_2 and b_3 as c_1.] See Figure 8.11. ◆

Example 4 Find the first few terms in the series expansion about the regular singular point $x = 0$ for two linearly independent solutions to

$$(37) \qquad xy''(x) + 3y'(x) - xy(x) = 0 , \qquad x > 0 .$$

Solution In Example 6 of Section 8.6, we used the method of Frobenius to find a series expansion about $x = 0$ for a solution to equation (37). The indicial equation turned out to be $r^2 + 2r = 0$, which has roots $r_1 = 0$ and $r_2 = -2$. Using $r_1 = 0$, we obtained the series solution

$$(38) \qquad y_1(x) = 1 + \frac{1}{8}x^2 + \frac{1}{192}x^4 + \frac{1}{9216}x^6 + \cdots$$

(see equation (62) in Section 8.6, page 463, and put $a_0 = 1$).

Because $r_1 - r_2 = 2$ is a positive integer, it follows from Theorem 7 that equation (37) has a second linearly independent solution of the form

$$(39) \qquad y_2(x) = C y_1(x) \ln x + \sum_{n=0}^{\infty} b_n x^{n-2} .$$

Plugging the expansion for $y_2(x)$ into equation (37) and simplifying yields

$$(40) \qquad \{ xy_1''(x) + 3y_1'(x) - xy_1(x) \} C \ln x + 2Cx^{-1} y_1(x) + 2C y_1'(x)$$

$$+ \sum_{n=0}^{\infty} (n-2)(n-3) b_n x^{n-3} + \sum_{n=0}^{\infty} 3(n-2) b_n x^{n-3} - \sum_{n=0}^{\infty} b_n x^{n-1} = 0 .$$

Again the factor in braces is zero because $y_1(x)$ is a solution to equation (37). If we combine the summations and simplify, equation (40) becomes

(41) $$2Cx^{-1}y_1(x) + 2Cy_1'(x) - b_1x^{-2} + \sum_{k=2}^{\infty} [k(k-2)b_k - b_{k-2}]x^{k-3} = 0 .$$

Substituting in the series expansions for $y_1(x)$ and $y_1'(x)$ and writing out the first few terms of the summation in (41) leads to

(42) $$-b_1x^{-2} + (2C - b_0)x^{-1} + (3b_3 - b_1) + \left(\frac{3}{4}C + 8b_4 - b_2\right)x$$
$$+ (15b_5 - b_3)x^2 + \left(\frac{5}{96}C + 24b_6 - b_4\right)x^3 + \cdots = 0 .$$

When we set the coefficients in (42) equal to zero, it turns out that we are free to choose C and b_2 as arbitrary constants:

$$-b_1 = 0 \Rightarrow b_1 = 0 ,$$
$$2C - b_0 = 0 \Rightarrow b_0 = 2C \quad (C \text{ arbitrary}),$$
$$3b_3 - b_1 = 0 \Rightarrow b_3 = \frac{1}{3}b_1 = 0 ,$$
$$8b_4 - b_2 + \frac{3}{4}C = 0 \Rightarrow b_4 = \frac{b_2 - \dfrac{3}{4}C}{8} = \frac{1}{8}b_2 - \frac{3}{32}C \quad (b_2 \text{ arbitrary}),$$
$$15b_5 - b_3 = 0 \Rightarrow b_5 = \frac{1}{15}b_3 = 0 ,$$
$$24b_6 - b_4 + \frac{5}{96}C = 0 \Rightarrow b_6 = \frac{b_4 - \dfrac{5}{96}C}{24} = \frac{1}{192}b_2 - \frac{7}{1152}C .$$

Substituting these values for the b_n's back into (39), we obtain the solution

(43) $$y_2(x) = C\left\{y_1(x)\ln x + 2x^{-2} - \frac{3}{32}x^2 - \frac{7}{1152}x^4 + \cdots\right\}$$
$$+ b_2\left\{1 + \frac{1}{8}x^2 + \frac{1}{192}x^4 + \cdots\right\} ,$$

where C and b_2 are arbitrary constants. Since the factor multiplying b_2 is the first solution $y_1(x)$, we can obtain a second linearly independent solution by choosing $C = 1$ and $b_2 = 0$:

(44) $$y_2(x) = y_1(x)\ln x + 2x^{-2} - \frac{3}{32}x^2 - \frac{7}{1152}x^4 + \cdots .$$

See Figure 8.12 on page 473. ◆

In closing we note that if the roots r_1 and r_2 of the indicial equation associated with a differential equation are complex, then they are complex conjugates. Thus, the difference $r_1 - r_2$ is imaginary and, hence, not an integer, and we are in case (a) of Theorem 7 on page 467. However, rather than employing the display (11)–(12) for the linearly independent solutions, one usually takes the real and imaginary parts of (11); Problem 26 provides an elaboration of this situation.

Figure 8.12 Partial sum approximations to the solutions of Example 4

8.7 EXERCISES

In Problems 1–14, find at least the first three nonzero terms in the series expansion about x = 0 for a general solution to the given equation for x > 0. (These are the same equations as in Problems 19–32 of Exercises 8.6.)

1. $9x^2y'' + 9x^2y' + 2y = 0$

2. $2x(x - 1)y'' + 3(x - 1)y' - y = 0$

3. $x^2y'' + xy' + x^2y = 0$

4. $xy'' + y' - 4y = 0$

5. $x^2z'' + (x^2 + x)z' - z = 0$

6. $3xy'' + (2 - x)y' - y = 0$

7. $4x^2y'' + 2x^2y' - (x + 3)y = 0$

8. $x^2y'' + (x^2 - x)y' + y = 0$

9. $xw'' - w' - xw = 0$

10. $3x^2y'' + 8xy' + (x - 2)y = 0$

11. $xy'' + (x - 1)y' - 2y = 0$

12. $x(x + 1)y'' + (x + 5)y' - 4y = 0$

13. $xy'' + (1 - x)y' - y = 0$

14. $x^2y'' - x(1 + x)y' + y = 0$

In Problems 15 and 16, determine whether the given equation has a solution that is bounded near the origin, all solutions are bounded near the origin, or none of the solutions are bounded near the origin. (These are the same equations as in Problems 33 and 34 of Exercises 8.6.) Note that you need to analyze only the indicial equation in order to answer the question.

15. $3xy'' + 2(1 - x)y' - 4y = 0$

16. $xy'' + (x + 2)y' - y = 0$

In Problems 17–20, find at least the first three nonzero terms in the series expansion about x = 0 for a general solution to the given linear third-order equation for x > 0. (These are the same equations as in Problems 35–38 in Exercises 8.6.)

17. $6x^3y''' + 13x^2y'' + (x + x^2)y' + xy = 0$

18. $6x^3y''' + 11x^2y'' - 2xy' - (x - 2)y = 0$

19. $6x^3y''' + 13x^2y'' - (x^2 + 3x)y' - xy = 0$

20. $6x^3y''' + (13x^2 - x^3)y'' + xy' - xy = 0$

21. **Buckling Columns.** In the study of the buckling of a column whose cross section varies, one encounters the equation

 (45) $x^ny''(x) + \alpha^2y(x) = 0,$ $x > 0,$

 where x is related to the height above the ground and y is the deflection away from the vertical. The positive constant α depends on the rigidity of the column, its moment of inertia at the top, and the load. The positive integer n depends on the type of column. For example, when the column is a truncated cone [see Figure 8.13(a) on page 474], we have n = 4.

 (a) Use the substitution $x = t^{-1}$ to reduce (45) with n = 4 to the form

 $$\frac{d^2y}{dt^2} + \frac{2}{t}\frac{dy}{dt} + \alpha^2y = 0, \qquad t > 0.$$

 (b) Find at least the first six nonzero terms in the series expansion about t = 0 for a general solution to the equation obtained in part (a).

 (c) Use the result of part (b) to give an expansion about $x = \infty$ for a general solution to (45).

(a) Truncated
cone

(b) Truncated pyramid with
fixed thickness T and varying width

Figure 8.13 Buckling columns

22. In Problem 21 consider a column with a rectangular cross section with two sides constant and the other two changing linearly [see Figure 8.13(b)]. In this case, $n = 1$. Find at least the first four nonzero terms in the series expansion about $x = 0$ for a general solution to equation (45) when $n = 1$.

23. Use the method of Frobenius and a reduction of order procedure (see page 197) to find at least the first three nonzero terms in the series expansion about the irregular singular point $x = 0$ for a general solution to the differential equation

$$x^2 y'' + y' - 2y = 0 .$$

24. The equation

$$xy''(x) + (1 - x)y'(x) + ny(x) = 0 ,$$

where n is a nonnegative integer, is called **Laguerre's differential equation**. Show that for each n, this equation has a polynomial solution of degree n. These polynomials are denoted by $L_n(x)$ and are called **Laguerre polynomials**. The first few Laguerre polynomials are

$$L_0(x) = 1 , \quad L_1(x) = -x + 1 .$$
$$L_2(x) = x^2 - 4x + 2 .$$

25. Use the results of Problem 24 to obtain the first few terms in a series expansion about $x = 0$ for a general solution for $x > 0$ to Laguerre's differential equation for $n = 0$ and 1.

26. To obtain two linearly independent solutions to

$$(46) \quad x^2 y'' + (x + x^2)y' + y = 0 , \quad x > 0 ,$$

complete the following steps.

(a) Verify that (46) has a regular singular point at $x = 0$ and that the associated indicial equation has complex roots $\pm i$.

(b) As discussed in Section 8.5, we can express

$$x^{\alpha + i\beta} = x^\alpha x^{i\beta}$$
$$= x^\alpha \cos(\beta \ln x) + ix^\alpha \sin(\beta \ln x) .$$

Deduce from this formula that $\dfrac{d}{dx} x^{\alpha + i\beta} = (\alpha + i\beta)x^{\alpha - 1 + i\beta}$.

(c) Set $y(x) = \sum_{n=0}^{\infty} a_n x^{n+i}$, where the coefficients now are *complex* constants, and substitute this series into equation (46) using the result of part (b).

(d) Setting the coefficients of like powers equal to zero, derive the recurrence relation

$$a_n = -\frac{n - 1 + i}{(n + i)^2 + 1} a_{n-1} , \quad \text{for } n \geq 1 .$$

(e) Taking $a_0 = 1$, compute the coefficients a_1 and a_2 and thereby obtain the first few terms of a complex solution to (46).

(f) By computing the real and imaginary parts of the solution obtained in part (e), derive the following linearly independent real solutions to (46):

$$y_1(x) = [\cos(\ln x)]\left\{1 - \frac{2}{5}x + \frac{1}{10}x^2 + \cdots\right\}$$
$$+ [\sin(\ln x)]\left\{\frac{1}{5}x - \frac{1}{20}x^2 + \cdots\right\} ,$$

$$y_2(x) = [\cos(\ln x)]\left\{-\frac{1}{5}x + \frac{1}{20}x^2 + \cdots\right\}$$
$$+ [\sin(\ln x)]\left\{1 - \frac{2}{5}x + \frac{1}{10}x^2 + \cdots\right\} .$$

8.8 Special Functions

In advanced work in applied mathematics, engineering, and physics, a few special second-order equations arise very frequently. These equations have been extensively studied, and volumes have been written on the properties of their solutions, which are often referred to as **special functions.**

For reference purposes we include a brief survey of three of these equations: the hypergeometric equation, Bessel's equation, and Legendre's equation. Bessel's equation governs the radial dependence of the solutions to the classical partial differential equations of physics in spherical coordinate systems (see Section 10.7, page 624); Legendre's equation governs their latitudinal dependence. C. F. Gauss formulated the hypergeometric equation as a generic equation whose solutions include the special functions of Legendre, Chebyshev, Gegenbauer, and Jacobi. For a more detailed study of special functions, we refer you to *Basic Hypergeometric Series*, 2nd ed., by G. Gasper and M. Rahman (Cambridge University Press, Cambridge, 2004); *Special Functions*, by E. D. Rainville (Chelsea, New York, 1971); and *Higher Transcendental Functions*, by A. Erdelyi (ed.) (McGraw-Hill, New York, 1953), 3 volumes.

Hypergeometric Equation

The linear second-order differential equation

$$(1) \qquad x(1-x)y'' + [\gamma - (\alpha + \beta + 1)x]y' - \alpha\beta y = 0,$$

where α, β, and γ are fixed parameters, is called the **hypergeometric equation**. This equation has singular points at $x = 0$ and 1, both of which are regular. Thus a series expansion about $x = 0$ for a solution to (1) obtained by the method of Frobenius will converge at least for $0 < x < 1$ (see Theorem 6, page 460). To find this expansion, observe that the indicial equation associated with $x = 0$ is

$$r(r-1) + \gamma r = r[r - (1-\gamma)] = 0,$$

which has roots 0 and $1 - \gamma$. Let us assume that γ is not an integer and use the root $r = 0$ to obtain a solution to (1) of the form

$$(2) \qquad y_1(x) = \sum_{n=0}^{\infty} a_n x^n.$$

Substituting $y_1(x)$ given in (2) into (1), shifting indices, and simplifying ultimately leads to the equation

$$(3) \qquad \sum_{n=1}^{\infty} [n(n+\gamma-1)a_n - (n+\alpha-1)(n+\beta-1)a_{n-1}]x^{n-1} = 0.$$

Setting the series coefficients equal to zero yields the recurrence relation

$$(4) \qquad n(n+\gamma-1)a_n - (n+\alpha-1)(n+\beta-1)a_{n-1} = 0, \qquad n \geq 1.$$

Since $n \geq 1$ and γ is not an integer, there is no fear of dividing by zero when we rewrite (4) as

$$(5) \qquad a_n = \frac{(n+\alpha-1)(n+\beta-1)}{n(n+\gamma-1)}a_{n-1}.$$

Solving recursively for a_n, we obtain

$$(6) \qquad a_n = \frac{\alpha(\alpha+1)\cdots(\alpha+n-1)\beta(\beta+1)\cdots(\beta+n-1)}{n!\gamma(\gamma+1)\cdots(\gamma+n-1)}a_0, \qquad n \geq 1.$$

If we employ the **factorial function** $(t)_n$, which is defined for nonnegative integers n by

$$(7) \qquad (t)_n := t(t+1)(t+2)\cdots(t+n-1), \qquad n \geq 1,$$
$$(t)_0 := 1, \qquad t \neq 0,$$

then we can express a_n more compactly as

$$(8) \qquad a_n = \frac{(\alpha)_n (\beta)_n}{n! (\gamma)_n} a_0, \qquad n \geq 1.$$

[In fact, we can write $n!$ as $(1)_n$.] If we take $a_0 = 1$ and then substitute the expression for a_n in (8) into (2), we obtain the following solution to the hypergeometric equation:

$$(9) \qquad y_1(x) = 1 + \sum_{n=1}^{\infty} \frac{(\alpha)_n (\beta)_n}{n! (\gamma)_n} x^n.$$

The solution given in (9) is called a **Gaussian hypergeometric function** and is denoted by the symbol $F(\alpha, \beta; \gamma; x)$.[†] That is,

$$(10) \qquad F(\alpha, \beta; \gamma; x) := 1 + \sum_{n=1}^{\infty} \frac{(\alpha)_n (\beta)_n}{n! (\gamma)_n} x^n.$$

Hypergeometric functions are generalizations of the geometric series. To see this, observe that for any constant β that is not zero or a negative integer,

$$F(1, \beta; \beta; x) = 1 + x + x^2 + x^3 + \cdots.$$

It is interesting to note that many other familiar functions can be expressed in terms of the hypergeometric function. For example,

$$(11) \qquad F(\alpha, \beta; \beta; x) = (1 - x)^{-\alpha},$$

$$(12) \qquad F(1, 1; 2; x) = -x^{-1} \ln(1 - x),$$

$$(13) \qquad F\left(\frac{1}{2}, 1; \frac{3}{2}; x^2\right) = \frac{1}{2} x^{-1} \ln\left(\frac{1 + x}{1 - x}\right),$$

$$(14) \qquad F\left(\frac{1}{2}, 1; \frac{3}{2}; -x^2\right) = x^{-1} \arctan x.$$

We omit the verification of these formulas.

To obtain a second linearly independent solution to (1) when γ is not an integer, we use the other root, $1 - \gamma$, of the indicial equation and seek a solution of the form

$$(15) \qquad y_2(x) = \sum_{n=0}^{\infty} b_n x^{n+1-\gamma}.$$

Substituting $y_2(x)$ into equation (1) and solving for b_n, we eventually arrive at

$$(16) \qquad y_2(x) = x^{1-\gamma} + \sum_{n=1}^{\infty} \frac{(\alpha + 1 - \gamma)_n (\beta + 1 - \gamma)_n}{n! (2 - \gamma)_n} x^{n+1-\gamma}.$$

Factoring out $x^{1-\gamma}$, we see that the second solution $y_2(x)$ can be expressed in terms of a hypergeometric function. That is,

$$(17) \qquad y_2(x) = x^{1-\gamma} F(\alpha + 1 - \gamma, \beta + 1 - \gamma; 2 - \gamma; x).$$

[†]*Historical Footnote:* A detailed study of this function was done by Carl Friedrich Gauss in 1813. In *Men of Mathematics* (Simon & Schuster, London, 1986), the mathematical historian E. T. Bell refers to Gauss as the Prince of Mathematicians.

When γ is an integer, one of the formulas given in (9) or (16) (corresponding to the larger root, 0 or $1 - \gamma$) still gives a solution. We then use the techniques of this chapter to obtain a second linearly independent solution, which may or may not involve a logarithmic term. We omit a discussion of these solutions.

In many books, the hypergeometric function is expressed in terms of the gamma function $\Gamma(x)$ instead of the factorial function. Recall that in Section 7.7, we defined

$$(18) \qquad \Gamma(x) := \int_0^\infty e^{-u} u^{x-1}\, du\,, \qquad x > 0$$

and showed that

$$(19) \qquad \Gamma(x+1) = x\,\Gamma(x)\,, \qquad x > 0\,.$$

It follows from repeated use of relation (19) that

$$(20) \qquad (t)_n = \frac{\Gamma(t+n)}{\Gamma(t)}\,,$$

for $t > 0$ and n any nonnegative integer. Using relation (20), we can express the hypergeometric function as

$$(21) \qquad F(\alpha, \beta; \gamma; x) = \frac{\Gamma(\gamma)}{\Gamma(\alpha)\Gamma(\beta)} \sum_{n=0}^\infty \frac{\Gamma(\alpha+n)\Gamma(\beta+n)}{n!\,\Gamma(\gamma+n)} x^n\,.$$

Bessel's Equation

The linear second-order differential equation

$$(22) \qquad x^2 y'' + xy' + (x^2 - \nu^2)y = 0\,,$$

where $\nu \geq 0$ is a fixed parameter, is called **Bessel's equation of order** ν. This equation has a regular singular point at $x = 0$ and no other singular points in the complex plane. Hence, a series solution for (22) obtained by the method of Frobenius will converge for $0 < x < \infty$. The indicial equation for (22) is

$$r(r-1) + r - \nu^2 = (r-\nu)(r+\nu) = 0\,,$$

which has roots $r_1 = \nu$ and $r_2 = -\nu$. If 2ν is not an integer, then the method of Frobenius yields two linearly independent solutions, given by

$$(23) \qquad y_1(x) = a_0 \sum_{n=0}^\infty \frac{(-1)^n}{2^{2n} n!\,(1+\nu)_n} x^{2n+\nu}\,,$$

$$(24) \qquad y_2(x) = b_0 \sum_{n=0}^\infty \frac{(-1)^n}{2^{2n} n!\,(1-\nu)_n} x^{2n-\nu}\,.$$

If in (23) we take

$$a_0 = \frac{1}{2^\nu \Gamma(1+\nu)}\,,$$

then it follows from relation (20) that the function

$$(25) \qquad J_\nu(x) := \sum_{n=0}^\infty \frac{(-1)^n}{n!\,\Gamma(1+\nu+n)} \left(\frac{x}{2}\right)^{2n+\nu}$$

is a solution to (22). We call $J_\nu(x)$ the **Bessel function of the first kind of order ν.**[†] Similarly, taking

$$b_0 = \frac{1}{2^{-\nu}\Gamma(1-\nu)}$$

in equation (24) gives the solution

(26) $\quad J_{-\nu}(x) := \sum_{n=0}^{\infty} \frac{(-1)^n}{n!\,\Gamma(1-\nu+n)}\left(\frac{x}{2}\right)^{2n-\nu},$

which is the Bessel function of the first kind of order $-\nu$. When $r_1 - r_2 = 2\nu$ is not an integer, we know by Theorem 7 of Section 8.7 that $J_\nu(x)$ and $J_{-\nu}(x)$ are linearly independent. Moreover, it can be shown that if ν is not an integer, even though 2ν is, then $J_\nu(x)$ and $J_{-\nu}(x)$ are still linearly independent.

What happens in the remaining case when ν is a nonnegative integer, say $\nu = m$? While $J_m(x)$ is still a solution, the function $J_{-m}(x)$ is not even properly defined because formula (26) will involve the gamma function evaluated at a nonpositive integer. From a more in-depth study of the gamma function, it turns out that $1/\Gamma(k)$ should be 0, for $k = 0, -1, -2, \ldots$. Hence, (26) becomes

(27) $\quad J_{-m}(x) = \sum_{n=0}^{\infty} \frac{(-1)^n}{n!\,\Gamma(1-m+n)}\left(\frac{x}{2}\right)^{2n-m} = \sum_{n=m}^{\infty} \frac{(-1)^n}{n!\,\Gamma(1-m+n)}\left(\frac{x}{2}\right)^{2n-m}.$

Comparing (27) with the formula (25) for $J_m(x)$, we see (after a shift in index) that

(28) $\quad J_{-m}(x) = (-1)^m J_m(x),$

which means that $J_{-m}(x)$ and $J_m(x)$ are linearly *dependent*. To resolve the problem of finding a second linearly independent solution in the case when ν is a nonnegative integer, we can use Theorem 7 in Section 8.7, page 467. There is, however, another approach to this problem, which we now describe.

For ν not an integer, we can take linear combinations of $J_\nu(x)$ and $J_{-\nu}(x)$ to obtain other solutions to (22). In particular, let

(29) $\quad Y_\nu(x) := \frac{\cos(\nu\pi)J_\nu(x) - J_{-\nu}(x)}{\sin(\nu\pi)}, \qquad \nu \text{ not an integer,}$

for $x > 0$. The function $Y_\nu(x)$ is called the **Bessel function of the second kind of order ν,** and, as can be verified, $J_\nu(x)$ and $Y_\nu(x)$ are linearly independent. Notice that when ν is an integer, the denominator in (29) is zero; but, by formula (28), so is the numerator! Hence, it is reasonable to hope that, in a limiting sense, formula (29) is still meaningful. In fact, using L'Hôpital's rule, it is possible to show that for m a nonnegative integer, the function defined by

(30) $\quad Y_m(x) := \lim_{\nu \to m} \frac{\cos(\nu\pi)J_\nu(x) - J_{-\nu}(x)}{\sin(\nu\pi)}$

for $x > 0$ is a solution to (22) with $\nu = m$. Furthermore, $J_m(x)$ and $Y_m(x)$ are linearly independent. We again call $Y_m(x)$ the Bessel function of the second kind of order m. In the literature the function $Y_m(x)$ is sometimes denoted by $N_m(x)$; variations of it have been called the **Neumann function** and the **Weber function.**

[†]*Historical Footnote:* Frederic Wilhelm Bessel (1784–1846) started his career in commercial navigation and later became an astronomer. In 1817 Bessel introduced the functions $J_\nu(x)$ in his study of planetary orbits.

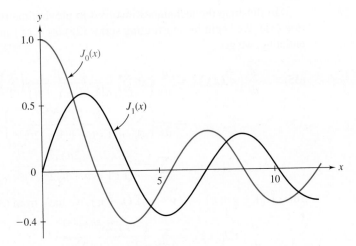

Figure 8.14 Graphs of the Bessel functions $J_0(x)$ and $J_1(x)$

Figure 8.14 shows the graphs of $J_0(x)$ and $J_1(x)$ and Figure 8.15 the graphs of $Y_0(x)$ and $Y_1(x)$. Notice that the curves for $J_0(x)$ and $J_1(x)$ behave like damped sine waves and have zeros that interlace (see Problem 28, page 485). The fact that the Bessel functions J_n have an infinite number of zeros is helpful in certain applications (see Project A in Chapter 10).

There are several useful recurrence relations involving Bessel functions. For example,

$$(31) \quad \frac{d}{dx}[x^\nu J_\nu(x)] = x^\nu J_{\nu-1}(x) \,,$$

$$(32) \quad \frac{d}{dx}[x^{-\nu} J_\nu(x)] = -x^{-\nu} J_{\nu+1}(x) \,,$$

$$(33) \quad J_{\nu+1}(x) = \frac{2\nu}{x} J_\nu(x) - J_{\nu-1}(x) \,,$$

$$(34) \quad J_{\nu+1}(x) = J_{\nu-1}(x) - 2J_\nu'(x) \,.$$

Furthermore, analogous equations hold for Bessel functions of the second kind.

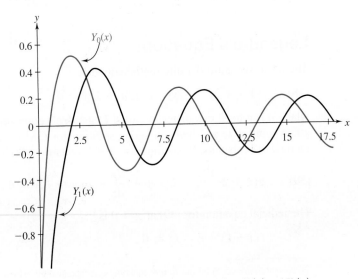

Figure 8.15 Graphs of the Bessel functions $Y_0(x)$ and $Y_1(x)$

To illustrate the techniques involved in proving the recurrence relations, let's verify relation (31). We begin by substituting series (25) for $J_\nu(x)$ into the left-hand side of (31). Differentiating, we get

$$(35) \qquad \frac{d}{dx}[x^\nu J_\nu(x)] = \frac{d}{dx}\left\{ x^\nu \sum_{n=0}^{\infty} \frac{(-1)^n}{n!\,\Gamma(1+\nu+n)} \left(\frac{x}{2}\right)^{2n+\nu} \right\}$$

$$= \frac{d}{dx}\left\{ \sum_{n=0}^{\infty} \frac{(-1)^n x^{2n+2\nu}}{n!\,\Gamma(1+\nu+n)2^{2n+\nu}} \right\}$$

$$= \sum_{n=0}^{\infty} \frac{(-1)^n(2n+2\nu)x^{2n+2\nu-1}}{n!\,\Gamma(1+\nu+n)2^{2n+\nu}}\,.$$

Since $\Gamma(1+\nu+n) = (\nu+n)\Gamma(\nu+n)$, we have from (35)

$$\frac{d}{dx}[x^\nu J_\nu(x)] = \sum_{n=0}^{\infty} \frac{(-1)^n 2 x^{2n+2\nu-1}}{n!\,\Gamma(\nu+n)2^{2n+\nu}}\,.$$

Factoring out an x^ν gives

$$\frac{d}{dx}[x^\nu J_\nu(x)] = x^\nu \sum_{n=0}^{\infty} \frac{(-1)^n}{n!\,\Gamma(1+(\nu-1)+n)} \left(\frac{x}{2}\right)^{2n+\nu-1} = x^\nu J_{\nu-1}(x)\,,$$

as claimed in equation (31). We leave the verifications of the remaining relations as exercises (see Problems 22–24 on page 485).

The following approximations are useful when analyzing the behavior of Bessel functions for large arguments:

$$(36) \qquad J_\nu(x) \approx \sqrt{\frac{2}{\pi x}} \cos\left(x - \frac{\nu\pi}{2} - \frac{\pi}{4}\right),\quad Y_\nu(x) \approx \sqrt{\frac{2}{\pi x}} \sin\left(x - \frac{\nu\pi}{2} - \frac{\pi}{4}\right),\quad x \gg 1\,.$$

This oscillatory behavior is demonstrated in Figures 8.14 and 8.15. Problem 41 provides some justification of formula (36). Of course, for small x the leading terms of the power series expansion dominate and we have

$$(37) \qquad J_\nu(x) \approx x^\nu/[2^\nu \Gamma(1+\nu)]\,,\quad Y_0(x) \approx (2\ln x)/\pi\,,\quad Y_{\nu>0}(x) \approx -\Gamma(\nu)2^\nu/(\pi x^\nu)\,,$$
$$0 < x \ll 1\,.$$

Legendre's Equation

The linear second-order differential equation

$$(38) \qquad (1 - x^2)y'' - 2xy' + n(n+1)y = 0\,,$$

where n is a fixed parameter, is called **Legendre's equation.**[†] This equation has a regular singular point at 1, and hence a series solution for (38) about $x = 1$ may be obtained by the method of Frobenius. By setting $z = x - 1$, equation (38) is transformed into

$$(39) \qquad z(z+2)\frac{d^2y}{dz^2} + 2(z+1)\frac{dy}{dz} - n(n+1)y = 0\,.$$

The indicial equation for (39) at $z = 0$ is

$$r(r-1) + r = r^2 = 0\,,$$

[†]*Historical Footnote:* Solutions to this equation were obtained by Adrien Marie Legendre (1752–1833) in 1785 and are referred to as Legendre functions.

which has roots $r_1 = r_2 = 0$. Upon substituting

$$y(z) = \sum_{k=0}^{\infty} a_k z^k$$

into (39) and proceeding as usual, we arrive at the solution

(40) $$y_1(x) = 1 + \sum_{k=1}^{\infty} \frac{(-n)_k (n+1)_k}{k!(1)_k} \left(\frac{1-x}{2}\right)^k ,$$

where we have expressed y_1 in terms of the original variable x and have taken $a_0 = 1$. We have written (40) in a "hybrid" form to facilitate the observation that

(41) $$y_1(x) = F\left(-n, n+1; 1; \frac{1-x}{2}\right) ,$$

where F is the Gaussian hypergeometric function defined in (10) on page 476.

For n a nonnegative integer, the factor

$$(-n)_k = (-n)(-n+1)(-n+2) \cdots (-n+k-1)$$

will be zero for $k \geq n+1$. Hence, the solution given in (40) and (41) is a *polynomial* of degree n. Moreover, $y_1(1) = 1$. These polynomial solutions of equation (38) are called the **Legendre polynomials** or **spherical polynomials** and are traditionally denoted by $P_n(x)$. That is,

(42) $$P_n(x) := 1 + \sum_{k=1}^{\infty} \frac{(-n)_k (n+1)_k}{k!(1)_k} \left(\frac{1-x}{2}\right)^k .$$

If we expand about $x = 0$, then $P_n(x)$ takes the form

(43) $$P_n(x) = 2^{-n} \sum_{m=0}^{[n/2]} \frac{(-1)^m (2n-2m)!}{(n-m)! m! (n-2m)!} x^{n-2m} ,$$

where $[n/2]$ is the greatest integer less than or equal to $n/2$ (see Problem 34). The first three Legendre polynomials are

$$P_0(x) = 1, \qquad P_1(x) = x, \qquad P_2(x) = \frac{3}{2}x^2 - \frac{1}{2} .$$

The Legendre polynomials satisfy the **orthogonality condition**

(44) $$\int_{-1}^{1} P_m(x) P_n(x)\, dx = 0 \quad \text{for} \quad n \neq m .$$

To see this, we first rewrite equation (38) in what is called the **selfadjoint form:**

(45) $$[(1-x^2)y']' + n(n+1)y = 0 .$$

Since $P_n(x)$ and $P_m(x)$ satisfy (45) with parameters n and m, respectively, we have

(46) $$[(1-x^2)P_n'(x)]' + n(n+1)P_n(x) = 0 ,$$

(47) $$[(1-x^2)P_m'(x)]' + m(m+1)P_m(x) = 0 .$$

Multiplying (46) by $P_m(x)$ and (47) by $P_n(x)$ and then subtracting, we find

$$P_m(x)[(1-x^2)P_n'(x)]' - P_n(x)[(1-x^2)P_m'(x)]'$$
$$+ [n(n+1) - m(m+1)]P_m(x)P_n(x) = 0 ,$$

which can be rewritten in the form

(48) $$(n^2 - m^2 + n - m)P_m(x)P_n(x) = P_n(x)[(1-x^2)P_m'(x)]' - P_m(x)[(1-x^2)P_n'(x)]' .$$

It is a straightforward calculation to show that the right-hand side of (48) is just

$$\{(1-x^2)[P_n(x)P'_m(x) - P'_n(x)P_m(x)]\}'.$$

Using this fact and the identity $n^2 - m^2 + n - m = (n-m)(n+m+1)$, we rewrite equation (48) as

(49) $(n-m)(n+m+1)P_m(x)P_n(x) = \{(1-x^2)[P_n(x)P'_m(x) - P'_n(x)P_m(x)]\}'.$

Integrating both sides of (49) from $x = -1$ to $x = 1$ yields

(50) $\displaystyle (n-m)(n+m+1)\int_{-1}^{1} P_m(x)P_n(x)\, dx$

$$= \int_{-1}^{1} \{(1-x^2)[P_n(x)P'_m(x) - P'_n(x)P_m(x)]\}'dx$$

$$= \{(1-x^2)[P_n(x)P'_m(x) - P'_n(x)P_m(x)]\}\Big|_{-1}^{1}$$

$$= 0$$

because $1 - x^2 = 0$ for $x = \pm 1$. Since n and m are nonnegative integers with $n \neq m$, then $(n-m)(n+m+1) \neq 0$, and so equation (44) follows from (50).

Legendre polynomials also satisfy the **recurrence formula**

(51) $(n+1)P_{n+1}(x) = (2n+1)xP_n(x) - nP_{n-1}(x)$

and **Rodrigues's formula**

(52) $\displaystyle P_n(x) = \frac{1}{2^n n!}\frac{d^n}{dx^n}\{(x^2-1)^n\}$

(see Problems 32 and 33).

The Legendre polynomials are generated by the function $(1 - 2xz + z^2)^{-1/2}$ in the sense that

(53) $\displaystyle (1 - 2xz + z^2)^{-1/2} = \sum_{n=0}^{\infty} P_n(x)z^n, \qquad |z| < 1, \qquad |x| < 1.$

That is, if we expand $(1 - 2xz + z^2)^{-1/2}$ in a Taylor series about $z = 0$, treating x as a fixed parameter, then the coefficients of z^n are the Legendre polynomials $P_n(x)$. The function $(1 - 2xz + z^2)^{-1/2}$ is called a **generating function** for $P_n(x)$ and can be derived from the recurrence formula (51) (see Problem 35).

The Legendre polynomials are an example of a class of special functions called **classical orthogonal polynomials.**[†] The latter includes, for example, the **Jacobi polynomials,** $P_n^{(\alpha,\beta)}(x)$; the **Gegenbauer** or **ultraspherical polynomials,** $C_n^\lambda(x)$; the **Chebyshev** (Tchebichef) **polynomials,** $T_n(x)$ and $U_n(x)$; the **Laguerre polynomials,** $L_n^\alpha(x)$; and the **Hermite polynomials,** $H_n(x)$. The properties of the classical orthogonal polynomials can be found in the books mentioned earlier in this section or in the *Handbook of Mathematical Functions with Formulas, Graphs, and Mathematical Tables,* by M. Abramowitz and I. A. Stegun (eds.) (Dover Publications, New York, 1964), Chapter 22. The latter has been updated by the National Institute of Standards and Technology and is freely available from their web site http://dlmf.nist.gov/.

We close this chapter with a table listing the forms of various well-known differential equations, the names of the researchers associated with their solutions, and the areas of application in which they arise. Many of them have been discussed in the body or exercises of our text.

[†]Here orthogonality is used in the more general sense that $\int_a^b P_n(x)P_m(x)w(x)dx = 0$ for $n \neq m$, where $w(x)$ is a weight function on the interval (a, b).

Differential Equation	Researchers	Areas of Application	Solutions
$y'' + \lambda y = 0$	(harmonic oscillator)	Vibrations, waves in Cartesian coordinates	$\cos\sqrt{\lambda}x,\ \sin\sqrt{\lambda}x,\ e^{\pm\sqrt{-\lambda}x},$ $\cosh\sqrt{-\lambda}x,\ \sinh\sqrt{-\lambda}x$
$my'' + by' + ky = 0$	(damped oscillator)	Vibrations	$e^{\alpha x}\cos\beta x,\ e^{\alpha x}\sin\beta x$
$ax^2y'' + bxy' + cy = 0$	Cauchy, Euler, Mellin	Electrostatics in polar coordinates	$x^\alpha\cos(\beta\ln x),\ x^\alpha\sin(\beta\ln x)$
$y'' - xy = 0$	Airy	Caustics	$\text{Ai}(x),\ \text{Bi}(x)$
$x^2y'' + xy' + (x^2 - \nu^2)y = 0$	Bessel, Weber, Neumann, Hankel	Waves in cylindrical coordinates	$J_\nu(x),\ Y_\nu(x),\ H_\nu^{(1)}(x),\ H_\nu^{(2)}(x)$
$x^2y'' + xy' - (x^2 + \nu^2)y = 0$	(Modified Bessel)	Electrostatics in cylindrical coordinates	$I_\nu(x),\ K_\nu(x)$
$x^2y'' + (a + 2bx^r)xy'$ $+\ [c + dx^{2s} - b(1 - a - r)x^r + b^2x^{2r}]y = 0$	(Generalized Bessel)		$x^{(1-a)/2}e^{-bx^r/r}J_p\left(\sqrt{d}x^s/s\right),$ $p = \sqrt{(1-a)^2/4 - c}/s$
$(1 - x^2)y'' - 2xy'$ $-\ [\lambda + m^2/(1 - x^2)]y = 0$	Legendre	Spherical coordinates $(x = \cos\phi)$	$P_\ell^m(x),\ Q_\ell^m(x),\ \lambda = -\ell(\ell + 1)$
$xy'' + (k + 1 - x)y' + ny = 0$	Laguerre	Hydrogen atom	$L_n^k(x)$
$y'' - 2xy' + 2ny = 0$	Hermite	Quantum mechanical harmonic oscillator	$H_n(x)$
$y'' + (2n + 1 - x^2)y = 0$	Weber	Quantum mechanical harmonic oscillator	$e^{-x^2/2}H_n(x)$
$(1 - x^2)y'' - xy' + n^2y = 0$	Chebyshev	Approximation theory, filters	$T_n(x),\ U_n(x)$

(Continued)

Differential Equation

Differential Equation	Researchers	Areas of Application	Solutions
$y'' + (a - 2q\cos 2x)y = 0$	Mathieu	Waves in elliptic coordinates	$F(a, q, x)$
$x(1-x)y'' + [\gamma - (\alpha+\beta+1)x]y' - \alpha\beta y = 0$	Gauss	Hypergeometric equation	$F(\alpha, \beta; \gamma; x)$
$xy'' + (c-x)y' - ay = 0$	Kummer, Whittaker	Confluent hypergeometric equation	$_1F_1(a; c; x)$
$(1-x^2)y'' + [b - a - (a+b+2)x]y' + n(n+a+b+1)y = 0$	Jacobi	(A form of hypergeometric equation)	$P_n^{(a,b)}(x)$
$(1-x^2)y'' - (2a+1)xy' + n(n+2a)y = 0$	Gegenbauer	Ultraspherical electrostatics (in $2a+2$ dimensions)	$C_n^{(a)}(x)$
$y'' + \left(\dfrac{1-a_1-a_2}{x-a} + \dfrac{1-b_1-b_2}{x-b} + \dfrac{1-c_1-c_2}{x-c}\right)y'$ $+ \left[\dfrac{a_1a_2(a-b)(a-c)}{x-a} + \dfrac{b_1b_2(b-c)(b-a)}{x-b}\right.$ $\left. + \dfrac{c_1c_2(c-a)(c-b)}{x-c}\right]\dfrac{y}{(x-a)(x-b)(x-c)} = 0,$ $a_1+a_2+b_1+b_2+c_1+c_2 = 1.$	Riemann	(Generalization of hypergeometric)	$P\left\{\begin{matrix} a & b & c \\ a_1 & b_1 & c_1 \\ a_2 & b_2 & c_2 \end{matrix}\; x\right\}$
$y'' - \mu(1 - y'^2)y' + y = 0$	Rayleigh	Limit cycle	
$y'' - \mu(1 - y^2)y' + y = 0$	van der Pol	Vacuum tubes	
$y'' + y + \varepsilon y^3 = 0$	Duffing	Nonlinear spring	
$(x^\rho y')' - x^\sigma y^n = 0$	Emden, Fowler	Reaction/diffusion	
$y' - p(x) - q(x)y - r(x)y^2 = 0$	Ricatti	Circuit synthesis, geometry	
$y' + p(x)y - q(x)y^n = 0$	Bernoulli	Economics, tumor growth	(Exercise 48, page 77)

8.8 EXERCISES

In Problems 1–4, express a general solution to the given equation using Gaussian hypergeometric functions.

1. $x(1-x)y'' + \left(\dfrac{1}{2} - 4x\right)y' - 2y = 0$

2. $3x(1-x)y'' + (1-27x)y' - 45y = 0$

3. $2x(1-x)y'' + (1-6x)y' - 2y = 0$

4. $2x(1-x)y'' + (3-10x)y' - 6y = 0$

In Problems 5–8, verify the following formulas by expanding each function in a power series about $x = 0$.

5. $F(1, 1; 2; x) = -x^{-1}\ln(1-x)$

6. $F(\alpha, \beta; \beta; x) = (1-x)^{-\alpha}$

7. $F\left(\dfrac{1}{2}, 1; \dfrac{3}{2}; x^2\right) = \dfrac{1}{2}x^{-1}\ln\left(\dfrac{1+x}{1-x}\right)$

8. $F\left(\dfrac{1}{2}, 1; \dfrac{3}{2}; -x^2\right) = x^{-1}\arctan x$

In Problems 9 and 10, use the method in Section 8.7 to obtain two linearly independent solutions to the given hypergeometric equation.

9. $x(1-x)y'' + (1-3x)y' - y = 0$

10. $x(1-x)y'' + (2-2x)y' - \dfrac{1}{4}y = 0$

11. Show that the **confluent hypergeometric equation**

$$xy'' + (\gamma - x)y' - \alpha y = 0,$$

where α and γ are fixed parameters and γ is not an integer, has two linearly independent solutions

$$y_1(x) = {}_1F_1(\alpha; \gamma; x) := 1 + \sum_{n=1}^{\infty} \frac{(\alpha)_n}{n!(\gamma)_n}x^n$$

and

$$y_2(x) = x^{1-\gamma}{}_1F_1(\alpha + 1 - \gamma; 2 - \gamma; x).$$

12. Use the property of the gamma function given in (19) to derive relation (20).

In Problems 13–18, express a general solution to the given equation using Bessel functions of either the first or second kind.

13. $4x^2y'' + 4xy' + (4x^2 - 1)y = 0$

14. $9x^2y'' + 9xy' + (9x^2 - 16)y = 0$

15. $x^2y'' + xy' + (x^2 - 1)y = 0$

16. $x^2y'' + xy' + x^2y = 0$

17. $9t^2x'' + 9tx' + (9t^2 - 4)x = 0$

18. $x^2z'' + xz' + (x^2 - 16)z = 0$

In Problems 19 and 20, a Bessel equation is given. For the appropriate choice of ν, the Bessel function $J_\nu(x)$ is one solution. Use the method in Section 8.7 to obtain a second linearly independent solution.

19. $x^2y'' + xy' + (x^2 - 1)y = 0$

20. $x^2y'' + xy' + (x^2 - 4)y = 0$

21. Show that $x^\nu J_\nu(x)$ satisfies the equation

$$xy'' + (1 - 2\nu)y' + xy = 0, \quad x > 0,$$

and use this result to find a solution for the equation

$$xy'' - 2y' + xy = 0, \quad x > 0.$$

In Problems 22 through 24, derive the indicated recurrence formulas.

22. Formula (32)

23. Formula (33)

24. Formula (34)

25. Show that

$$J_{1/2}(x) = (2/\pi x)^{1/2}\sin x \quad \text{and}$$
$$J_{-1/2}(x) = (2/\pi x)^{1/2}\cos x.$$

26. The Bessel functions of order $\nu = n + 1/2$, n any integer, are related to the **spherical Bessel functions.** Use relation (33) and the results of Problem 25 to show that such Bessel functions can be represented in terms of $\sin x$, $\cos x$, and powers of x. Demonstrate this by determining a closed form for $J_{-3/2}(x)$ and $J_{5/2}(x)$.

27. Use Theorem 7 in Section 8.7 to determine a second linearly independent solution to Bessel's equation of order 0 in terms of the Bessel function $J_0(x)$.

28. Show that between two consecutive positive roots (zeros) of $J_1(x)$, there is a root of $J_0(x)$. This interlacing property of the roots of Bessel functions is illustrated in Figure 8.14 on page 479. [*Hint:* Use relation (31) and Rolle's theorem from calculus.]

29. Use formula (43) to determine the first five Legendre polynomials.

30. Show that the Legendre polynomials of even degree are even functions of x, while those of odd degree are odd functions.

31. (a) Show that the orthogonality condition (44) for Legendre polynomials implies that

$$\int_{-1}^{1} P_n(x)q(x)\,dx = 0$$

for *any* polynomial $q(x)$ of degree at most $n - 1$. [*Hint:* The polynomials $P_0, P_1, \ldots, P_{n-1}$ are linearly

independent and hence span the space of all polynomials of degree at most $n-1$. Thus, $q(x) = a_0P_0(x) + \cdots + a_{n-1}P_{n-1}(x)$ for suitable constants a_k.]

(b) Prove that if $Q_n(x)$ is a polynomial of degree n such that

$$\int_{-1}^{1} Q_n(x)P_k(x)dx = 0$$

for $k = 0, 1, \ldots, n-1$, then $Q_n(x) = cP_n(x)$ for some constant c.

[*Hint:* Select c so that the coefficient of x^n for $Q_n(x) - cP_n(x)$ is zero. Then, since $P_0, \ldots, P_{n-1}$ is a basis,

$$Q_n(x) - cP_n(x) = a_0P_0(x) + \cdots + a_{n-1}P_{n-1}(x).$$

Multiply the last equation by $P_k(x)$ $(0 \le k \le n-1)$ and integrate from $x = -1$ to $x = 1$ to show that each a_k is zero.]

32. Deduce the recurrence formula (51) for Legendre polynomials by completing the following steps.

(a) Show that the function

$$Q_{n-1}(x) := (n+1)P_{n+1}(x) - (2n+1)xP_n(x)$$

is a polynomial of degree $n-1$. [*Hint:* Compute the coefficient of the x^{n+1} term using the representation (42). The coefficient of x^n is also zero because $P_{n+1}(x)$ and $xP_n(x)$ are both odd or both even functions, a consequence of Problem 30.]

(b) Using the result of Problem 31(a), show that

$$\int_{-1}^{1} Q_{n-1}(x)P_k(x)dx = 0 \text{ for } k = 0, 1, \ldots, n-2.$$

(c) From Problem 31(b), conclude that $Q_{n-1}(x) = cP_{n-1}(x)$ and, by taking $x = 1$, show that $c = -n$. [*Hint:* Recall that $P_m(1) = 1$ for all m.] From the definition of $Q_{n-1}(x)$ in part (a), the recurrence formula now follows.

33. To prove Rodrigues's formula (52) for Legendre polynomials, complete the following steps.

(a) Let $v_n := (d^n/dx^n)\{(x^2-1)^n\}$ and show that $v_n(x)$ is a polynomial of degree n with the coefficient of x^n equal to $(2n)!/n!$.

(b) Use integration by parts n times to show that, for any polynomial $q(x)$ of degree less than n,

$$\int_{-1}^{1} v_n(x)q(x)dx = 0.$$

Hint: For example, when $n = 2$,

$$\int_{-1}^{1} \frac{d^2}{dx^2}\{(x^2-1)^2\}q(x)dx$$

$$= q(x)\frac{d}{dx}\{(x^2-1)^2\}\Big|_{-1}^{1} - \{q'(x)(x^2-1)^2\}\Big|_{-1}^{1}$$

$$+ \int_{-1}^{1} q''(x)(x^2-1)^2dx.$$

Since $n = 2$, the degree of $q(x)$ is at most 1, and so $q''(x) \equiv 0$. Thus

$$\int_{-1}^{1} \frac{d^2}{dx^2}\{(x^2-1)^2\}q(x)dx = 0.$$

(c) Use the result of Problem 31(b) to conclude that $P_n(x) = cv_n(x)$ and show that $c = 1/2^n n!$ by comparing the coefficients of x^n in $P_n(x)$ and $v_n(x)$.

34. Use Rodrigues's formula (52) to obtain the representation (43) for the Legendre polynomials $P_n(x)$. *Hint:* From the binomial formula,

$$P_n(x) = \frac{1}{2^n n!}\frac{d^n}{dx^n}\{(x^2-1)^n\}$$

$$= \frac{1}{2^n n!}\frac{d^n}{dx^n}\left\{\sum_{m=0}^{n}\frac{n!(-1)^m}{(n-m)!m!}x^{2n-2m}\right\}.$$

35. The generating function in (53) for Legendre polynomials can be derived from the recurrence formula (51) as follows. Let x be fixed and set $f(z) := \sum_{n=0}^{\infty} P_n(x)z^n$. The goal is to determine an explicit formula for $f(z)$.

(a) Show that multiplying each term in the recurrence formula (51) by z^n and summing the terms from $n = 1$ to ∞ leads to the differential equation

$$\frac{df}{dz} = \frac{x-z}{1-2xz+z^2}f.$$

Hint:

$$\sum_{n=1}^{\infty}(n+1)P_{n+1}(x)z^n$$

$$= \sum_{n=0}^{\infty}(n+1)P_{n+1}(x)z^n - P_1(x)$$

$$= \frac{df}{dz} - x.$$

(b) Solve the differential equation derived in part (a) and use the initial conditions $f(0) = P_0(x) \equiv 1$ to obtain $f(z) = (1-2xz+z^2)^{-1/2}$.

36. Find a general solution about $x = 0$ for the equation

$$(1 - x^2)y'' - 2xy' + 2y = 0$$

by first finding a polynomial solution and then using the reduction of order formula given on page 197 to find a second (series) solution.

37. The **Hermite polynomials** $H_n(x)$ are polynomial solutions to Hermite's equation

$$y'' - 2xy' + 2ny = 0.$$

The Hermite polynomials are generated by

$$e^{2tx - t^2} = \sum_{n=0}^{\infty} \frac{H_n(x)}{n!} t^n.$$

Use this equation to determine the first four Hermite polynomials.

38. The **Chebyshev (Tchebichef) polynomials** $T_n(x)$ are polynomial solutions to Chebyshev's equation

$$(1 - x^2)y'' - xy' + n^2 y = 0.$$

The Chebyshev polynomials satisfy the recurrence relation

$$T_{n+1}(x) = 2xT_n(x) - T_{n-1}(x),$$

with $T_0(x) = 1$ and $T_1(x) = x$. Use this recurrence relation to determine the next three Chebyshev polynomials.

39. The **Laguerre polynomials** $L_n(x)$ are polynomial solutions to Laguerre's equation

$$xy'' + (1 - x)y' + ny = 0.$$

The Laguerre polynomials satisfy Rodrigues's formula,

$$L_n(x) = \frac{e^x}{n!} \frac{d^n}{dx^n} (x^n e^{-x}).$$

Use this formula to determine the first four Laguerre polynomials.

40. Reduction to Bessel's Equation. The class of equations of the form

$$(54) \qquad y''(x) + cx^n y(x) = 0, \qquad x > 0,$$

where c and n are positive constants, can be solved by transforming the equation into Bessel's equation.

(a) First, use the substitution $y = x^{1/2} z$ to transform (54) into an equation involving x and z.

(b) Second, use the substitution

$$s = \frac{2\sqrt{c}}{n + 2} x^{n/2 + 1}$$

to transform the equation obtained in part (a) into the Bessel equation

$$s^2 \frac{d^2 z}{ds^2} + s \frac{dz}{ds} + \left(s^2 - \frac{1}{(n + 2)^2} \right) z = 0, \quad s > 0.$$

(c) A general solution to the equation in part (b) can be given in terms of Bessel functions of the first and second kind. Substituting back in for s and z, obtain a general solution for equation (54).

41. (a) Show that the substitution $z(x) = \sqrt{x}\, y(x)$ renders Bessel's equation (22) in the form

$$(55) \qquad z'' + \left(1 + \frac{1 - 4\nu^2}{4x^2} \right) z = 0.$$

(b) For $x \gg 1$, equation (55) would apparently be approximated by the equation

$$(56) \qquad z'' + z = 0.$$

Write down the general solution to (56), reset $y(x) = z(x)/\sqrt{x}$, and argue the plausibility of formula (36).

(c) For $\nu = \pm 1/2$, equation (55) reduces to equation (56) exactly. Relate this observation to Problem 25.

Chapter **8** Summary

Initial value problems that do not fall into the "solvable" categories that have been studied (such as constant-coefficient or equidimensional equations) can often be analyzed by interpreting the differential equation as a prescription for computing the higher derivatives of the unknown function. The **Taylor polynomial method** uses the equation to construct a polynomial approximation matching the initial values of a finite number of derivatives of the unknown. If the equation permits the extrapolation of this procedure to polynomials of arbitrarily high degree, **power series** representations of the solution can be constructed.

Power Series

Every power series $\sum_{n=0}^{\infty} a_n (x - x_0)^n$ has a **radius of convergence** ρ, $0 \le \rho \le \infty$, such that the series converges absolutely for $|x - x_0| < \rho$ and diverges when $|x - x_0| > \rho$. By the ratio test,

$$\rho = \lim_{n \to \infty} \frac{|a_n|}{|a_{n+1}|} \, ,$$

provided that this limit exists as an extended real number. A function $f(x)$ that is the sum of a power series in some open interval about x_0 is said to be **analytic at x_0.** If f is analytic at x_0, its power series representation about x_0 is the **Taylor series**

$$f(x) = \sum_{n=0}^{\infty} \frac{f^{(n)}(x_0)}{n!} (x - x_0)^n \, .$$

Power Series Method for an Ordinary Point

In the case of a linear equation of the form

$$(1) \qquad y'' + p(x)y' + q(x)y = 0 \, ,$$

where p and q are analytic at x_0, the point x_0 is called an **ordinary point,** and the equation has a pair of linearly independent solutions expressible as power series about x_0. The radii of convergence of these series solutions are at least as large as the distance from x_0 to the nearest singularity (real or complex) of the equation. To find power series solutions to (1), we substitute $y(x) = \sum_{n=0}^{\infty} a_n (x - x_0)^n$ into (1), group like terms, and set the coefficients of the resulting power series equal to zero. This leads to a recurrence relation for the coefficients a_n, which, in some cases, may even yield a general formula for the a_n. The same method applies to the non-homogeneous version of (1), provided the forcing function is also analytic at x_0.

Regular Singular Points

If, in equation (1), either p or q fails to be analytic at x_0, then x_0 is a **singular point** of (1). If x_0 is a singular point for which $(x - x_0)p(x)$ and $(x - x_0)^2 q(x)$ are both analytic at x_0, then x_0 is a **regular singular point.** The Cauchy–Euler equation,

$$(2) \qquad ax^2 \frac{d^2 y}{dx^2} + bx \frac{dy}{dx} + cy = 0 \, , \qquad x > 0 \, ,$$

has a regular singular point at $x = 0$, and a general solution to (2) can be obtained by substituting $y = x^r$ and examining the roots of the resulting indicial equation $ar^2 + (b - a)r + c = 0$.

Method of Frobenius

For an equation of the form (1) with a regular singular point at x_0, a series solution can be found by the **method of Frobenius.** This is obtained by substituting

$$w(r, x) = (x - x_0)^r \sum_{n=0}^{\infty} a_n (x - x_0)^n$$

into (1), finding a recurrence relation for the coefficients, and choosing $r = r_1$, the larger root of the **indicial equation**

$$(3) \qquad r(r - 1) + p_0 r + q_0 = 0 \, ,$$

where $p_0 := \lim_{x \to x_0} (x - x_0)p(x)$, $q_0 := \lim_{x \to x_0} (x - x_0)^2 q(x)$.

Finding a Second Linearly Independent Solution

If the two roots r_1, r_2 of the indicial equation (3) do not differ by an integer, then a second linearly independent solution to (1) can be found by taking $r = r_2$ in the method of Frobenius. However, if $r_1 = r_2$ or $r_1 - r_2$ is a positive integer, then discovering a second solution requires a different approach. This may be a reduction of order procedure or the utilization of Theorem 7, page 467, which gives the *forms* of the solutions.

Special Functions

Some special functions in physics and engineering that arise from series solutions to linear second-order equations with polynomial coefficients are Gaussian hypergeometric functions, $F(\alpha, \beta; \gamma; x)$; Bessel functions $J_\nu(x)$; and orthogonal polynomials such as those of Legendre, Chebyshev, Laguerre, and Hermite.

REVIEW PROBLEMS FOR CHAPTER 8

1. Find the first four nonzero terms in the Taylor polynomial approximation for the given initial value problem.

 (a) $y' = xy - y^2$; $y(0) = 1$

 (b) $z'' - x^3 z' + xz^2 = 0$; $z(0) = -1$, $z'(0) = 1$

2. Determine all the singular points of the given equation and classify them as regular or irregular.

 (a) $(x^2 - 4)^2 y'' + (x - 4)y' + xy = 0$

 (b) $(\sin x)y'' + y = 0$

3. Find at least the first four nonzero terms in a power series expansion about $x = 0$ for a general solution to the given equation.

 (a) $y'' + x^2 y' - 2y = 0$

 (b) $y'' + e^{-x} y' - y = 0$

4. Find a general formula for the coefficient a_n in a power series expansion about $x = 0$ for a general solution to the given equation.

 (a) $(1 - x^2)y'' + xy' + 3y = 0$

 (b) $(x^2 - 2)y'' + 3y = 0$

5. Find at least the first four nonzero terms in a power series expansion about $x = 2$ for a general solution to

$$w'' + (x - 2)w' - w = 0.$$

6. Use the substitution $y = x^r$ to find a general solution to the given equation for $x > 0$.

 (a) $2x^2 y''(x) + 5xy'(x) - 12y(x) = 0$

 (b) $x^3 y'''(x) + 3x^2 y''(x) - 2xy'(x) - 2y(x) = 0$

7. Use the method of Frobenius to find at least the first four nonzero terms in the series expansion about $x = 0$ for a solution to the given equation for $x > 0$.

 (a) $x^2 y'' - 5xy' + (9 - x)y = 0$

 (b) $x^2 y'' + (x^2 + 2x)y' - 2y = 0$

8. Find the indicial equation and its roots and state (but do not compute) the form of the series expansion about $x = 0$ (as in Theorem 7 on page 467) for two linearly independent solutions of the given equation for $x > 0$.

 (a) $x^2 y'' + (\sin x)y' - 4y = 0$

 (b) $2xy'' + 5y' + xy = 0$

 (c) $(x \sin x)y'' + xy' + (\tan x)y = 0$

9. Find at least the first three nonzero terms in the series expansion about $x = 0$ for a general solution to the given equation for $x > 0$.

 (a) $x^2 y'' - x(1 + x)y' + y = 0$

 (b) $xy'' + y' - 2y = 0$

 (c) $2xy'' + 6y' + y = 0$

 (d) $x^2 y'' + (x - 2)y = 0$

10. Express a general solution to the given equation using Gaussian hypergeometric functions or Bessel functions.

 (a) $x(1 - x)y'' + \left(\dfrac{1}{2} - 6x\right)y' - 6y = 0$

 (b) $9\theta^2 y'' + 9\theta y' + (9\theta^2 - 1)y = 0$

TECHNICAL WRITING EXERCISES FOR CHAPTER 8

1. Knowing that a general solution to a nonhomogeneous linear second-order equation can be expressed as a particular solution plus a general solution to the corresponding homogeneous equation, what can you say about the form of a general power series solution to the nonhomogeneous equation about an ordinary point?

2. Discuss advantages and disadvantages of power series solutions over numerical solutions generated by the Euler or Runge–Kutta methods.

3. The factors $x^{0.3}$, $x^{0.3} \ln x$, $x^{-0.3}$, and $x^{0.3+12i}$ all can arise in solutions generated using the method of Frobenius to solve a differential equation with 0 as a regular singular point. Discuss the problems encountered in attempting to approximate such solutions using the methods of Euler and of Runge–Kutta.

4. Discuss the issues involved when attempting to use the Taylor polynomials (expanded around $x_0 = 0$) to study the *asymptotic* behavior (as $t \to \pm \infty$) of solutions to the differential equation $y'' + y = 0$.

Projects for Chapter 8

A | Alphabetization Algorithms

In 1961, C. A. R. Hoare published the algorithm Quicksort, for alphabetizing a large list of words ("Algorithm 64: Quicksort," by C. A. R. Hoare, *Comm. ACM* 4 (1961): 324). First, one of the n words is selected at random, and the rest of the list is separated into two sublists—those that precede the selected word and those that follow it. This entails $(n-1)$ comparisons. Then the same procedure is applied to each sublist, and so on.

If the selected words happen to belong in the middle of the collections—say, 50th out of 100—then the sort is very efficient. One sort involving 99 comparisons, then 2 sorts of 49 comparisons each, and so forth. But in the worst case, if the randomly selected words happen to belong at the beginning or end of the list, the number of comparisons is $99 + 98 + 97 + \cdots + 1$.

- **(a)** Show that, for a list of length n, if all the selected words happen to belong at the midpoint of their sublists, then the total number of comparisons is roughly $n \ln_2 n$.
- **(b)** Show that if all the selected words happen to belong at the beginning of their sublists, the number of comparisons is $\sum_{k=1}^{n-1} k = n(n-1)/2$.
- **(c)** Show that if e_n is the *expected number of comparisons* (in the statistical sense) to alphabetize n words, then $e_1 = 0$, $e_2 = 1$, and for all $n > 0$,

$$e_n = n - 1 + \frac{1}{n}\sum_{r=0}^{n-1}[e_r + e_{n-1-r}] = \frac{2}{n}\sum_{r=0}^{n-1}e_r + n - 1,$$

where $e_0 = 0$ by definition.

To express e_n explicitly, we are going to construct a differential equation whose solutions have Taylor coefficients satisfying the same recursion as in part (c), solve the equation explicitly, and perform the Taylor expansion to extract e_n.

Consider the nonhomogeneous linear equation $y' = 2y/(1-x) + f(x)$. It is singular at $x = 1$, of course, but it is regular at $x = 0$ if (the known function) $f(x)$ is analytic there. Assume $y(x) = \sum_{m=0}^{\infty} a_m x^m$ and $f(x) = \sum_{m=0}^{\infty} b_m x^m$.

- **(d)** Substitute these series directly into the differential equation and, recalling the geometric series $1/(1-x) = \sum_{m=0}^{\infty} x^m$, derive the recursion relation $a_n = \frac{2}{n}\sum_{r=0}^{n-1} a_r + b_{n-1}/n$.

- **(e)** Identify the value of b_n that renders the recursion relations in (c) and (d) identical, and explicitly construct $f(x) = \sum_{m=0}^{\infty} b_m x^m$. [*Hint*: Use the derivatives of the geometric series.]

- **(f)** Solve the above differential equation, with initial condition $y(0) = e_0 = 0$, explicitly to obtain $y(x) = -[2x + 2\ln(1-x)]/(1-x)^2$. [You may find it convenient to refer to Section 2.3.]

- **(g)** Multiply out the Taylor series for $\ln(1-x)$ and $(1-x)^{-2}$ in $y(x)$ and deduce the Quicksort comparisons formula

$$e_n = 2\left(\frac{n-1}{2} + \frac{n-2}{3} + \frac{n-3}{4} + \cdots + \frac{1}{n}\right) = 2\sum_{r=2}^{n}\frac{n-r+1}{r}, \quad n \geq 2.$$

Several authors have devised improvements to Quicksort to reduce the probability of worst-case scenarios. (For example, instead of choosing one word at random, one chooses three words and uses the median for comparison.) The article "An Asymptotic Theory for Cauchy-Euler Differential Equations with Applications to the Analysis of Algorithms," by H.-H. Chern, H.-K. Hwang, and T.-H. Tsai (*Journal of Algorithms*, 44 (2002): 177–225) shows how the methodology in this project can be extended to analyze many of these strategies.

B Spherically Symmetric Solutions to Schrödinger's Equation for the Hydrogen Atom

In quantum mechanics one is interested in determining the wave function and energy states of an atom. These are determined from Schrödinger's equation. In the case of the hydrogen atom, it is possible to find wave functions ψ that are functions only of r, the distance from the proton to the electron. Such functions are called **spherically symmetric** and satisfy the simpler equation

$$(1) \qquad \frac{1}{r}\frac{d^2}{dr^2}(r\psi) = \frac{-8m\pi^2}{h^2}\left(E + \frac{e_0^2}{r}\right)\psi \,,$$

where e_0^2, m, and h are constants and E, also a constant, represents the energy of the atom, which we assume here to be negative.

(a) Show that with the substitutions

$$r = \frac{h^2}{4\pi^2 m e_0^2}\rho \,, \qquad E = \frac{2\pi^2 m e_0^4}{h^2}\varepsilon \,,$$

where ε is a negative constant, equation (1) reduces to

$$\frac{d^2(\rho\psi)}{d\rho^2} = -\left(\varepsilon + \frac{2}{\rho}\right)\rho\psi \,.$$

(b) If $f := \rho\psi$, then the preceding equation becomes

$$(2) \qquad \frac{d^2 f}{d\rho^2} = -\left(\varepsilon + \frac{2}{\rho}\right)f \,.$$

Show that the substitution $f(\rho) = e^{-\alpha\rho}g(\rho)$, where α is a positive constant, transforms (2) into

$$(3) \qquad \frac{d^2 g}{d\rho^2} - 2\alpha\frac{dg}{d\rho} + \left(\frac{2}{\rho} + \varepsilon + \alpha^2\right)g = 0 \,.$$

(c) If we choose $\alpha^2 = -\varepsilon$ (ε negative), then (3) becomes

$$(4) \qquad \frac{d^2 g}{d\rho^2} - 2\alpha\frac{dg}{d\rho} + \frac{2}{\rho}g = 0 \,.$$

Show that a power series solution $g(\rho) = \sum_{k=1}^{\infty} a_k \rho^k$ (starting with $k = 1$, as per Frobenius) for (4) must have coefficients a_k that satisfy the recurrence relation

$$(5) \qquad a_{k+1} = \frac{2(\alpha k - 1)}{k(k+1)}a_k \,, \qquad k \geq 1 \,.$$

(d) Now for $a_1 = 1$ and k very large, $a_{k+1} \approx (2\alpha/k)a_k$ and so $a_{k+1} \approx (2\alpha)^k/k!$, which are the coefficients for $\rho e^{2\alpha\rho}$. Hence, g acts like $\rho e^{2\alpha\rho}$, so $f(\rho) = e^{-\alpha\rho}g(\rho)$ is like $\rho e^{\alpha\rho}$. Going back further, we then see that $\psi \approx e^{\alpha\rho}$. Therefore, when $r = h^2\rho/4\pi^2 m e_0^2$ is large, so is ψ. Roughly speaking, $\psi^2(r)$ is proportional to the probability of finding an electron a distance r from the proton. Thus, the above argument would imply that the electron in a hydrogen atom is more likely to be found at a very large distance from the proton! Since this makes no sense physically, we ask: Do there exist positive values for α for which ψ remains bounded as r becomes large?

Show that when $\alpha = 1/n$, $n = 1, 2, 3, \ldots$, then $g(\rho)$ is a polynomial of degree n and argue that ψ is therefore bounded.

(e) Let E_n and $\psi_n(\rho)$ denote, respectively, the energy state and wave function corresponding to $\alpha = 1/n$. Find E_n (in terms of the constants e_0^2, m, and h) and $\psi_n(\rho)$ for $n = 1$, 2, and 3.

C Airy's Equation

 In aerodynamics one encounters the following initial value problem for **Airy's equation:**

$$y'' + xy = 0, \qquad y(0) = 1, \qquad y'(0) = 0.$$

(a) Find the first five nonzero terms in a power series expansion about $x = 0$ for the solution and graph this polynomial for $-10 \le x \le 10$.

(b) Using the Runge–Kutta subroutine (see Section 5.3) with $h = 0.05$, approximate the solution on the interval $[0, 10]$, i.e., at the points $0.05, 0.1, 0.15$, etc.

(c) Using the Runge–Kutta subroutine with $h = 0.05$, approximate the solution on the interval $[-10, 0]$. [*Hint:* With the change of variables $z = -x$, it suffices to approximate the solution to $y'' - zy = 0$; $y(0) = 1$, $y'(0) = 0$, on the interval $[0, 10]$.]

(d) Using your knowledge of constant-coefficient equations as a basis for guessing the behavior of the solutions to Airy's equation, decide whether the power series approximation obtained in part (a) or the numerical approximation obtained in parts (b) and (c) better describes the true behavior of the solution on the interval $[-10, 10]$.

D Buckling of a Tower

A tower is constructed of four angle beams connected by diagonals (see Figure 8.16 on page 494). The deflection curve $y(x)$ for the tower is governed by the equation

$$(6) \qquad x^2\frac{d^2y}{dx^2} + \frac{Pa^2}{EI}y = 0, \qquad a < x < a + L,$$

where x is the vertical coordinate measured down from the extended top of the tower, y is the deflection from the vertical passing through the center of the unbuckled tower, L is the tower height, a is the length of the truncation, P is the load, E is the modulus of elasticity, and I is the moment of inertia. The appropriate boundary conditions for this design are

$$(7) \qquad y(a) = 0,$$

$$(8) \qquad y'(a + L) = 0.$$

(a) Unbuckled

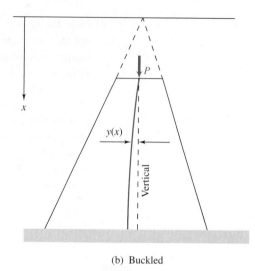

(b) Buckled

Figure 8.16 Buckling tower

Clearly, the solution $y(x) \equiv 0$ is always at hand. However, when the load P is heavy enough, the tower can *buckle* and a nontrivial solution $y(x)$ is also possible. We want to predict this phenomenon.

(a) Solve equation (6). [*Hint:* Equation (6) is a Cauchy–Euler equation.]

(b) Show that the first boundary condition (7) implies

$$y = Ax^{1/2} \sin[\beta \ln(x/a)],$$

where A is an arbitrary constant and $\beta := \sqrt{Pa^2/EI - 1/4}$.

(c) Show that the second boundary condition (8) gives

$$A\left\{ \tan\left[\beta \ln\left(\frac{a+L}{a} \right) \right] + 2\beta \right\} = 0 \,.$$

(d) Use the result of part (c) to argue that *no* buckling takes place (i.e., the only possibility is $A = 0$) if $0 < \beta < \beta_c$, where β_c is the smallest positive real number that makes the expression in braces zero.

(e) The value of the load corresponding to β_c is called the **critical load** P_c. Solve for P_c in terms of β_c, a, E, and I.

(f) Find the critical load P_c if $a = 10$, $L = 40$, and $EI = 1000$.

E Aging Spring and Bessel Functions

In Problems 30 and 31 in Exercises 8.4, page 450, we discussed a model for a mass–spring system with an aging spring. Without damping, the displacement $x(t)$ at time t is governed by the equation

$$(9) \qquad mx''(t) + ke^{-\eta t}x(t) = 0 \, ,$$

where m, k, and η are positive constants. The general solution to this equation can be expressed using Bessel functions.

(a) The coefficient of x suggests a change of variables of the form $s = \alpha e^{\beta t}$. Show that (9) transforms into

$$(10) \qquad \beta^2 s^2 \frac{d^2 x}{ds^2} + \beta^2 s \frac{dx}{ds} + \frac{k}{m}\left(\frac{s}{\alpha}\right)^{-\eta/\beta} x = 0 \, .$$

(b) Show that choosing α and β to satisfy

$$\frac{-\eta}{\beta} = 2 \qquad \text{and} \qquad \frac{k}{m\beta^2 \alpha^{-\eta/\beta}} = 1$$

transforms (10) into the Bessel equation of order 0 in s and x.

(c) Using the result of part (b), show that a general solution to (9) is given by

$$x(t) = c_1 J_0\left(\frac{2}{\eta}\sqrt{k/m}\,e^{-\eta t/2}\right) + c_2 Y_0\left(\frac{2}{\eta}\sqrt{k/m}\,e^{-\eta t/2}\right),$$

where J_0 and Y_0 are the Bessel functions of order 0 of the first and second kind, respectively.

(d) Discuss the behavior of the displacement $x(t)$ for c_2 positive, negative, and zero.

(e) Compare the behavior of the displacement $x(t)$ for η a small positive number and η a large positive number.

Matrix Methods for Linear Systems

9.1 Introduction

In this chapter we return to the analysis of systems of differential equations. When the equations in the system are *linear*, matrix algebra provides a compact notation for expressing the system. In fact, the notation itself suggests new and elegant ways of characterizing the solution properties, as well as novel, efficient techniques for explicitly obtaining solutions.

> **In Chapter 5 we analyzed physical situations wherein two fluid tanks containing brine solutions were interconnected and pumped so as ultimately to deplete the salt content in each tank. By accounting for the various influxes and outfluxes of brine, a system of differential equations for the salt contents ($x(t)$ and $y(t)$) of each tank was derived; a typical model is**
>
> (1) $\qquad \begin{aligned} dx/dt &= -4x + 2y\,, \\ dy/dt &= 4x - 4y\,. \end{aligned}$
>
> **Express this system in matrix notation as a single equation.**

The right-hand side of the first member of (1) possesses a mathematical structure that is familiar from vector calculus; namely, it is the **dot product**[†] of two vectors:

(2) $\qquad -4x + 2y = \begin{bmatrix} -4 & 2 \end{bmatrix} \cdot \begin{bmatrix} x & y \end{bmatrix}\,.$

Similarly, the second right-hand side in (1) is the dot product

$$4x - 4y = \begin{bmatrix} 4 & -4 \end{bmatrix} \cdot \begin{bmatrix} x & y \end{bmatrix}\,.$$

The frequent occurrence in mathematics of arrays of dot products, such as evidenced in the system (1), led to the development of **matrix algebra**, a mathematical discipline whose basic operation—the matrix product—is the arrangement of a set of dot products according to the following plan:

$$\begin{bmatrix} -4 & 2 \\ 4 & -4 \end{bmatrix} \begin{bmatrix} x \\ y \end{bmatrix} = \begin{bmatrix} \begin{bmatrix} -4 & 2 \end{bmatrix} \cdot \begin{bmatrix} x & y \end{bmatrix} \\ \begin{bmatrix} 4 & -4 \end{bmatrix} \cdot \begin{bmatrix} x & y \end{bmatrix} \end{bmatrix} = \begin{bmatrix} -4x + 2y \\ 4x - 4y \end{bmatrix}\,.$$

In general, the product of a **matrix**—i.e., an *m* by *n* rectangular array of numbers—and a *column vector* is defined to be the collection of dot products of the *rows* of the matrix with the

[†]Recall that the dot product of two vectors **u** and **v** equals the length of **u** times the length of **v** times the cosine of the angle between **u** and **v**. However, it is more conveniently computed from the *components* of **u** and **v** by the "inner product" indicated in equation (2).

vector, arranged as a column vector:

$$
\begin{bmatrix} \mathbf{row\ \#\ 1} \\ \mathbf{row\ \#\ 2} \\ \vdots \\ \mathbf{row\ \#\ m} \end{bmatrix} \begin{bmatrix} \\ \mathbf{v} \\ \\ \end{bmatrix} = \begin{bmatrix} [\,\mathbf{row\ \#\ 1}\,]\cdot\mathbf{v} \\ [\,\mathbf{row\ \#\ 2}\,]\cdot\mathbf{v} \\ \vdots \\ [\,\mathbf{row\ \#\ m}\,]\cdot\mathbf{v} \end{bmatrix} ,
$$

where the vector $\mathbf{v}$ has n components; the dot product of two n-dimensional vectors is computed in the obvious way:

$$
\begin{bmatrix} a_1 & a_2 & \cdots & a_n \end{bmatrix} \cdot \begin{bmatrix} x_1 & x_2 & \cdots & x_n \end{bmatrix} = a_1 x_1 + a_2 x_2 + \cdots + a_n x_n .
$$

Using the notation for the matrix product, we can write the system (1) for the interconnected tanks as

$$
\begin{bmatrix} x' \\ y' \end{bmatrix} = \begin{bmatrix} -4 & 2 \\ 4 & -4 \end{bmatrix} \begin{bmatrix} x \\ y \end{bmatrix} .
$$

The following example demonstrates a four-dimensional implementation of this notation. Note that the coefficients in the linear system need not be constants.

Example 1 Express the system

(3)
$$
\begin{aligned}
x_1' &= 2x_1 + t^2 x_2 + (4t + e^t)x_4 , \\
x_2' &= (\sin t)x_2 + (\cos t)x_3 , \\
x_3' &= x_1 + x_2 + x_3 + x_4 , \\
x_4' &= 0
\end{aligned}
$$

as a matrix equation.

Solution We express the right-hand side of the first member of (3) as the dot product

$$
2x_1 + t^2 x_2 + (4t + e^t)x_4 = \begin{bmatrix} 2 & t^2 & 0 & (4t + e^t) \end{bmatrix} \cdot \begin{bmatrix} x_1 & x_2 & x_3 & x_4 \end{bmatrix} .
$$

The other dot products are similarly identified, and the matrix form is given by

$$
\begin{bmatrix} x_1' \\ x_2' \\ x_3' \\ x_4' \end{bmatrix} = \begin{bmatrix} 2 & t^2 & 0 & (4t + e^t) \\ 0 & \sin t & \cos t & 0 \\ 1 & 1 & 1 & 1 \\ 0 & 0 & 0 & 0 \end{bmatrix} \begin{bmatrix} x_1 \\ x_2 \\ x_3 \\ x_4 \end{bmatrix} . \quad \blacklozenge
$$

In general, if a system of differential equations is expressed as

$$
\begin{aligned}
x_1' &= a_{11}(t)x_1 + a_{12}(t)x_2 + \cdots + a_{1n}(t)x_n \\
x_2' &= a_{21}(t)x_1 + a_{22}(t)x_2 + \cdots + a_{2n}(t)x_n \\
&\ \ \vdots \\
x_n' &= a_{n1}(t)x_1 + a_{n2}(t)x_2 + \cdots + a_{nn}(t)x_n ,
\end{aligned}
$$

it is said to be a linear homogeneous system in **normal form**.[†] The matrix formulation of such

[†]The normal form was defined for general systems in Section 5.3, page 256.

a system is then

$$\mathbf{x}' = \mathbf{A}\mathbf{x},$$

where $\mathbf{A}$ is the **coefficient matrix**

$$\mathbf{A} = \mathbf{A}(t) = \begin{bmatrix} a_{11}(t) & a_{12}(t) & \cdots & a_{1n}(t) \\ a_{21}(t) & a_{22}(t) & \cdots & a_{2n}(t) \\ \vdots & \vdots & & \vdots \\ a_{n1}(t) & a_{n2}(t) & \cdots & a_{nn}(t) \end{bmatrix}$$

and $\mathbf{x}$ is the solution vector

$$\mathbf{x} = \begin{bmatrix} x_1 \\ x_2 \\ \vdots \\ x_n \end{bmatrix}.$$

Note that we have used $\mathbf{x}'$ to denote the vector of derivatives

$$\mathbf{x}' = \begin{bmatrix} x_1 \\ x_2 \\ \vdots \\ x_n \end{bmatrix}' = \begin{bmatrix} x_1' \\ x_2' \\ \vdots \\ x_n' \end{bmatrix}.$$

Example 2 Express the differential equation for the undamped, unforced mass–spring oscillator (recall Section 4.1, page 152)

$$(4) \qquad my'' + ky = 0$$

as an equivalent system of first-order equations in normal form, expressed in matrix notation.

Solution We have to express the *second* derivative, y'', as a *first* derivative in order to formulate (4) as a first-order system. This is easy; the acceleration y'' is the derivative of the *velocity* $v = y'$, so (4) becomes

$$(5) \qquad mv' + ky = 0.$$

The first-order system is then assembled by identifying v with y', and appending it to (5):

$$y' = v$$
$$mv' = -ky.$$

To put this system in normal form and express it as a matrix equation, we need to divide the second equation by the mass m:

$$\begin{bmatrix} y \\ v \end{bmatrix}' = \begin{bmatrix} 0 & 1 \\ -k/m & 0 \end{bmatrix} \begin{bmatrix} y \\ v \end{bmatrix}. \quad \blacklozenge$$

In general, the customary way to write an nth-order linear homogeneous differential equation

$$a_n(t)y^{(n)} + a_{n-1}(t)y^{(n-1)} + \cdots + a_1(t)y' + a_0(t)y = 0$$

as an equivalent system in normal form is to define the first $(n-1)$ derivatives of y (including y, the zeroth derivative, itself) to be new unknowns:

$$x_1(t) = y(t),$$
$$x_2(t) = y'(t),$$
$$\vdots$$
$$x_n(t) = y^{(n-1)}(t).$$

Then the system consists of the identification of $x_j(t)$ as the derivative of $x_{j-1}(t)$, together with the original differential equation expressed in these variables (and divided by $a_n(t)$):

$$x_1' = x_2,$$
$$x_2' = x_3,$$
$$\vdots$$
$$x_{n-1}' = x_n,$$
$$x_n' = -\frac{a_0(t)}{a_n(t)}x_1 - \frac{a_1(t)}{a_n(t)}x_2 - \cdots - \frac{a_{n-1}(t)}{a_n(t)}x_n.$$

For systems of two or more higher-order differential equations, the same procedure is applied to each unknown function in turn; an example will make this clear.

Example 3 The coupled mass–spring oscillator depicted in Figure 5.26 on page 283 was shown to be governed by the system

$$\text{(6)} \qquad 2\frac{d^2x}{dt^2} + 6x - 2y = 0,$$
$$\frac{d^2y}{dt^2} + 2y - 2x = 0.$$

Write (6) in matrix notation.

Solution We introduce notation for the lower-order derivatives:

$$\text{(7)} \qquad x_1 = x, \qquad x_2 = x', \qquad x_3 = y, \qquad x_4 = y'.$$

In these variables, the system (6) states

$$\text{(8)} \qquad 2x_2' + 6x_1 - 2x_3 = 0,$$
$$x_4' + 2x_3 - 2x_1 = 0.$$

The normal form is then

$$x_1' = x_2,$$
$$x_2' = -3x_1 + x_3,$$
$$x_3' = x_4,$$
$$x_4' = 2x_1 - 2x_3$$

or in matrix notation

$$\begin{bmatrix} x_1 \\ x_2 \\ x_3 \\ x_4 \end{bmatrix}' = \begin{bmatrix} 0 & 1 & 0 & 0 \\ -3 & 0 & 1 & 0 \\ 0 & 0 & 0 & 1 \\ 2 & 0 & -2 & 0 \end{bmatrix} \begin{bmatrix} x_1 \\ x_2 \\ x_3 \\ x_4 \end{bmatrix}. \; \blacklozenge$$

9.1 EXERCISES

In Problems 1–6, express the given system of differential equations in matrix notation.

1. $x' = 7x + 2y$,
 $y' = 3x - 2y$

2. $x' = y$,
 $y' = -x$

3. $x' = x + y + z$,
 $y' = 2z - x$,
 $z' = 4y$

4. $x_1' = x_1 - x_2 + x_3 - x_4$,
 $x_2' = x_1 + x_4$,
 $x_3' = \sqrt{\pi}x_1 - x_3$,
 $x_4' = 0$

5. $x' = (\sin t)x + e^t y$,
 $y' = (\cos t)x + (a + bt^3)y$

6. $x_1' = (\cos 2t)x_1$,
 $x_2' = (\sin 2t)x_2$,
 $x_3' = x_1 - x_2$

In Problems 7–10, express the given higher-order differential equation as a matrix system in normal form.

7. The damped mass–spring oscillator equation $my'' + by' + ky = 0$

8. Legendre's equation $(1 - t^2)y'' - 2ty' + 2y = 0$

9. The Airy equation $y'' - ty = 0$

10. Bessel's equation $y'' + \dfrac{1}{t}y' + \left(1 - \dfrac{n^2}{t^2}\right)y = 0$

In Problems 11–13, express the given system of higher-order differential equations as a matrix system in normal form.

11. $x'' + 3x + 2y = 0$,
 $y'' - 2x = 0$

12. $x'' + 3x' - y' + 2y = 0$,
 $y'' + x' + 3y' + y = 0$

13. $x'' - 3x' + t^2 y - (\cos t)x = 0$,
 $y''' + y'' - tx' + y' + e^t x = 0$

9.2 Review 1: Linear Algebraic Equations

Here and in the next section we review some basic facts concerning linear algebraic systems and matrix algebra that will be useful in solving linear systems of differential equations in normal form. Readers competent in these areas may proceed to Section 9.4.

A set of equations of the form

$$a_{11}x_1 + a_{12}x_2 + \cdots + a_{1n}x_n = b_1 ,$$
$$a_{21}x_1 + a_{22}x_2 + \cdots + a_{2n}x_n = b_2 ,$$
$$\vdots$$
$$a_{n1}x_1 + a_{n2}x_2 + \cdots + a_{nn}x_n = b_n$$

(where the a_{ij}'s and b_i's are given constants) is called a *linear system of n algebraic equations* in the *n* unknowns $x_1, x_2, \ldots, x_n$. The procedure for solving the system using elimination methods is well known. Herein we describe a particularly convenient implementation of the method called the *Gauss–Jordan elimination algorithm.*[†] The basic idea of this formulation is to use the first equation to eliminate x_1 in all the other equations; then use the second equation to eliminate x_2 in all the others; and so on. If all goes well, the resulting system will be "uncoupled," and the values of the unknowns $x_1, x_2, \ldots, x_n$ will be apparent. A short example will make this clear.

[†]The Gauss–Jordan algorithm is neither the fastest nor the most accurate computer algorithm for solving a linear system of algebraic equations, but for solutions executed by hand it has many pedagogical advantages. Usually it is much faster than *Cramer's rule*, described in Appendix D.

Example 1 Solve the system

$$2x_1 + 6x_2 + 8x_3 = 16\,,$$
$$4x_1 + 15x_2 + 19x_3 = 38\,,$$
$$2x_1 \qquad\quad + 3x_3 = 6\,.$$

Solution By subtracting 2 times the first equation from the second, we eliminate x_1 from the latter. Similarly, x_1 is eliminated from the third equation by subtracting 1 times the first equation from it:

$$2x_1 + 6x_2 + 8x_3 = 16\,,$$
$$3x_2 + 3x_3 = 6\,,$$
$$-6x_2 - 5x_3 = -10\,.$$

Next we subtract multiples of the second equation from the first and third to eliminate x_2 in them; the appropriate multiples are 2 and -2, respectively:

$$2x_1 \qquad\quad + 2x_3 = 4\,,$$
$$3x_2 + 3x_3 = 6\,,$$
$$x_3 = 2\,.$$

Finally, we eliminate x_3 from the first two equations by subtracting multiples (2 and 3, respectively) of the third equation:

$$2x_1 \qquad\quad = 0\,,$$
$$3x_2 \qquad = 0\,,$$
$$x_3 = 2\,.$$

The system is now uncoupled; i.e., we can solve each equation separately:

$$x_1 = 0\,, \quad x_2 = 0\,, \quad x_3 = 2\,. \;\blacklozenge$$

Two complications can disrupt the straightforward execution of the Gauss–Jordan algorithm. The first occurs when the impending variable to be eliminated (say, x_j) does not occur in the jth equation. The solution is usually obvious; we employ one of the subsequent equations to eliminate x_j. Example 2 illustrates this maneuver.

Example 2 Solve the system

$$x_1 + 2x_2 + 4x_3 + x_4 = 0\,,$$
$$-x_1 - 2x_2 - 2x_3 \qquad = 1\,,$$
$$-2x_1 - 4x_2 - 8x_3 + 2x_4 = 4\,,$$
$$x_1 + 4x_2 + 2x_3 \qquad = -3\,.$$

Solution The first unknown x_1 is eliminated from the last three equations by subtracting multiples of the first equation:

$$x_1 + 2x_2 + 4x_3 + x_4 = 0\,,$$
$$2x_3 + x_4 = 1\,,$$
$$4x_4 = 4\,,$$
$$2x_2 - 2x_3 - x_4 = -3\,.$$

Now, we cannot use the second equation to eliminate the second unknown because x_2 is not present. The next equation that *does* contain x_2 is the fourth, so we switch the second and fourth equation:

$$x_1 + 2x_2 + 4x_3 + x_4 = 0,$$
$$2x_2 - 2x_3 - x_4 = -3,$$
$$4x_4 = 4,$$
$$2x_3 + x_4 = 1,$$

and proceed to eliminate x_2:

$$x_1 \quad + 6x_3 + 2x_4 = 3,$$
$$2x_2 - 2x_3 - \quad x_4 = -3,$$
$$4x_4 = 4,$$
$$2x_3 + x_4 = 1.$$

To eliminate x_3, we have to switch again,

$$x_1 \quad + 6x_3 + 2x_4 = 3,$$
$$2x_2 - 2x_3 - \quad x_4 = -3,$$
$$2x_3 + \quad x_4 = 1,$$
$$4x_4 = 4,$$

and eliminate, in turn, x_3 and x_4. This gives

$$
\begin{array}{ll}
x_1 \qquad\quad - x_4 = 0, & x_1 \qquad\qquad = 1, \\
2x_2 \qquad\quad = -2, \quad \text{and} & 2x_2 \qquad\qquad = -2, \\
2x_3 + x_4 = 1, & 2x_3 \qquad = 0, \\
4x_4 = 4, & 4x_4 = 4.
\end{array}
$$

The solution to the uncoupled equations is

$$x_1 = 1, \quad x_2 = -1, \quad x_3 = 0, \quad x_4 = 1. \quad \blacklozenge$$

The other complication that can disrupt the Gauss–Jordan algorithm is much more profound. What if, when we are "scheduled" to eliminate the unknown x_j, it is absent from *all* of the subsequent equations? The first thing to do is to move on to the elimination of the *next* unknown x_{j+1}, as demonstrated in Example 3.

Example 3 Apply the Gauss–Jordan algorithm to the system

$$2x_1 + 4x_2 + x_3 = 8,$$
(1)
$$2x_1 + 4x_2 \quad\;\; = 6,$$
$$-4x_1 - 8x_2 + x_3 = -10.$$

Solution Elimination of x_1 proceeds as usual:

$$2x_1 + 4x_2 + x_3 = 8,$$
$$- x_3 = -2,$$
$$3x_3 = 6.$$

Now since x_2 is absent from the second *and* third equations, we use the second equation to eliminate x_3 :

$$(2) \quad \begin{aligned} 2x_1 + 4x_2 \quad\;\; &= 6 \,, \\ -x_3 &= -2 \,, \\ 0 &= 0 \,. \end{aligned}$$

How do we interpret the system (2)? The final equation contains no information, of course, and we ignore it.[†] The second equation implies that $x_3 = 2$.

The first equation implies that $x_1 = 3 - 2x_2$, but there is no equation for x_2. Evidently, x_2 is a "free" variable, and we can assign *any* value to it—as long as we take x_1 to be $3 - 2x_2$. Thus (1) has an infinite number of solutions, and a convenient way of characterizing them is

$$x_1 = 3 - 2s \,, \quad x_2 = s \,, \quad x_3 = 2 \,; \quad -\infty < s < \infty \,.$$

We remark that an equivalent solution can be obtained by treating x_1 as the free variable, say $x_1 = s$, and taking $x_2 = (3 - s)/2, x_3 = 2$. ◆

The final example is contrived to demonstrate all the features that we have encountered.

Example 4　Find all solutions to the system

$$\begin{aligned} x_1 - x_2 + 2x_3 + 2x_4 &= 0 \,, \\ 2x_1 - 2x_2 + 4x_3 + 3x_4 &= 1 \,, \\ 3x_1 - 3x_2 + 6x_3 + 9x_4 &= -3 \,, \\ 4x_1 - 4x_2 + 8x_3 + 8x_4 &= 0 \,. \end{aligned}$$

Solution　We use the first equation to eliminate x_1:

$$\begin{aligned} x_1 - x_2 + 2x_3 + 2x_4 &= 0 \,, \\ - x_4 &= 1 \,, \\ 3x_4 &= -3 \,, \\ 0 &= 0 \,. \end{aligned}$$

Now, both x_2 *and* x_3 are absent from all subsequent equations, so we use the second equation to eliminate x_4.

$$\begin{aligned} x_1 - x_2 + 2x_3 \quad\;\; &= 2 \,, \\ -x_4 &= 1 \,, \\ 0 &= 0 \,, \\ 0 &= 0 \,. \end{aligned}$$

There are no constraints on either x_2 or x_3; thus we take them to be free variables and characterize the solutions by

$$x_1 = 2 + s - 2t \,, \quad x_2 = s \,, \quad x_3 = t \,, \quad x_4 = -1 \,, \quad -\infty < s, t < \infty \,. ◆$$

In closing, we note that if the execution of the Gauss–Jordan algorithm results in a display of the form $0 = 1$ (or $0 = k$, where $k \neq 0$), the original system has no solutions; it is *inconsistent*. This is explored in Problem 12.

[†] The occurrence of the identity $0 = 0$ in the Gauss–Jordan algorithm implies that one of the original equations was *redundant*. In this case you may observe that the final equation in (1) can be derived by subtracting 3 times the second equation from the first.

9.2 EXERCISES

In Problems 1–11, find all solutions to the system using the Gauss–Jordan elimination algorithm.

1. $x_1 + 2x_2 + 2x_3 = 6$,
 $2x_1 + x_2 + x_3 = 6$,
 $x_1 + x_2 + 3x_3 = 6$

2. $x_1 + x_2 + x_3 + x_4 = 1$,
 $x_1 \qquad\qquad + x_4 = 0$,
 $2x_1 + 2x_2 - x_3 + x_4 = 0$,
 $x_1 + 2x_2 - x_3 + x_4 = 0$

3. $x_1 + x_2 - x_3 = 0$,
 $-x_1 - x_2 + x_3 = 0$,
 $x_1 + x_2 - x_3 = 0$

4. $\qquad\qquad x_3 + x_4 = 0$,
 $x_1 + x_2 + x_3 + x_4 = 1$,
 $2x_1 - x_2 + x_3 + 2x_4 = 0$,
 $2x_1 - x_2 + x_3 + x_4 = 0$

5. $-x_1 + 2x_2 = 0$,
 $2x_1 + 3x_2 = 0$

6. $-2x_1 + 2x_2 - x_3 = 0$,
 $x_1 - 3x_2 + x_3 = 0$,
 $4x_1 - 4x_2 + 2x_3 = 0$

7. $-x_1 + 3x_2 = 0$,
 $-3x_1 + 9x_2 = 0$

8. $x_1 + 2x_2 + x_3 = -3$,
 $2x_1 + 4x_2 - x_3 = 0$,
 $x_1 + 3x_2 - 2x_3 = 3$

9. $(1 - i)x_1 + 2x_2 = 0$,
 $-x_1 - (1 + i)x_2 = 0$

10. $x_1 + x_2 + x_3 = i$,
 $2x_1 + 3x_2 - ix_3 = 0$,
 $x_1 + 2x_2 + x_3 = i$

11. $2x_1 \qquad\quad + x_3 = -1$,
 $-3x_1 + x_2 + 4x_3 = 1$,
 $-x_1 + x_2 + 5x_3 = 0$

12. Use the Gauss–Jordan elimination algorithm to show that the following systems of equations are inconsistent. That is, demonstrate that the existence of a solution would imply a mathematical contradiction.

 (a) $2x_1 - x_2 = 2$,
 $-6x_1 + 3x_2 = 4$

 (b) $2x_1 \qquad\quad + x_3 = -1$,
 $-3x_1 + x_2 + 4x_3 = 1$,
 $-x_1 + x_2 + 5x_3 = 1$

13. Use the Gauss–Jordan elimination algorithm to show that the following system of equations has a unique solution for $r = 2$, but an infinite number of solutions for $r = 1$.

$$2x_1 - 3x_2 = rx_1,$$
$$x_1 - 2x_2 = rx_2$$

14. Use the Gauss–Jordan elimination algorithm to show that the following system of equations has a unique solution for $r = -1$, but an infinite number of solutions for $r = 2$.

$$x_1 + 2x_2 - x_3 = rx_1,$$
$$x_1 \qquad\quad + x_3 = rx_2,$$
$$4x_1 - 4x_2 + 5x_3 = rx_3$$

9.3 Review 2: Matrices and Vectors

A **matrix** is a rectangular array of numbers arranged in rows and columns. An $m \times n$ matrix—that is, a matrix with m rows and n columns—is usually denoted by

$$\mathbf{A} := \begin{bmatrix} a_{11} & a_{12} & a_{13} & \cdots & a_{1n} \\ a_{21} & a_{22} & a_{23} & \cdots & a_{2n} \\ \vdots & \vdots & \vdots & \cdots & \vdots \\ a_{m1} & a_{m2} & a_{m3} & \cdots & a_{mn} \end{bmatrix},$$

where the element in the ith row and jth column is a_{ij}. The notation $[a_{ij}]$ is also used to designate $\mathbf{A}$. The matrices we will work with usually consist of real numbers, but in certain instances we allow complex-number entries.

Some matrices of special interest are **square matrices**, which have the same number of rows and columns; **diagonal matrices**, which are square matrices with only zero entries off the main diagonal (that is, $a_{ij} = 0$ if $i \neq j$); and (column) **vectors**, which are $n \times 1$ matrices. For example, if

$$\mathbf{A} = \begin{bmatrix} 3 & 4 & -1 \\ 2 & 6 & 5 \\ 0 & 1 & 4 \end{bmatrix}, \quad \mathbf{B} = \begin{bmatrix} 3 & 0 & 0 \\ 0 & 0 & 0 \\ 0 & 0 & 7 \end{bmatrix}, \quad \mathbf{x} = \begin{bmatrix} 4 \\ 2 \\ 1 \end{bmatrix},$$

then $\mathbf{A}$ is a square matrix, $\mathbf{B}$ is a diagonal matrix, and $\mathbf{x}$ is a vector. An $m \times n$ matrix whose entries are all zero is called a **zero matrix** and is denoted by $\mathbf{0}$. For consistency, we denote matrices by boldfaced capitals, such as $\mathbf{A}$, $\mathbf{B}$, $\mathbf{C}$, $\mathbf{I}$, $\mathbf{X}$, and $\mathbf{Y}$, and reserve boldfaced lower-case letters, such as $\mathbf{c}$, $\mathbf{x}$, $\mathbf{y}$, and $\mathbf{z}$, for vectors.

Algebra of Matrices

Matrix Addition and Scalar Multiplication. The operations of matrix addition and scalar multiplication are very straightforward. Addition is performed by adding corresponding elements:

$$\begin{bmatrix} 1 & 2 & 3 \\ 4 & 5 & 6 \end{bmatrix} + \begin{bmatrix} 1 & 1 & 1 \\ 1 & 1 & 1 \end{bmatrix} = \begin{bmatrix} 2 & 3 & 4 \\ 5 & 6 & 7 \end{bmatrix}.$$

Formally, the *sum* of two $m \times n$ matrices is given by

$$\mathbf{A} + \mathbf{B} = [a_{ij}] + [b_{ij}] = [a_{ij} + b_{ij}].$$

(The sole novelty here is that addition is not defined for two matrices whose dimensions m, n differ.)

To multiply a matrix by a scalar (number), we simply multiply each element in the matrix by the number:

$$3 \begin{bmatrix} 1 & 2 & 3 \\ 4 & 5 & 6 \end{bmatrix} = \begin{bmatrix} 3 & 6 & 9 \\ 12 & 15 & 18 \end{bmatrix}.$$

In other words, $r\mathbf{A} = r[a_{ij}] = [ra_{ij}]$. The notation $-\mathbf{A}$ stands for $(-1)\mathbf{A}$.

Properties of Matrix Addition and Scalar Multiplication. Matrix addition and scalar multiplication are nothing more than mere bookkeeping, and the usual algebraic properties hold. If $\mathbf{A}$, $\mathbf{B}$, and $\mathbf{C}$ are $m \times n$ matrices and r, s are scalars, then

$$\mathbf{A} + (\mathbf{B} + \mathbf{C}) = (\mathbf{A} + \mathbf{B}) + \mathbf{C}, \quad \mathbf{A} + \mathbf{B} = \mathbf{B} + \mathbf{A},$$
$$\mathbf{A} + \mathbf{0} = \mathbf{A}, \quad \mathbf{A} + (-\mathbf{A}) = \mathbf{0},$$
$$r(\mathbf{A} + \mathbf{B}) = r\mathbf{A} + r\mathbf{B}, \quad (r + s)\mathbf{A} = r\mathbf{A} + s\mathbf{A},$$
$$r(s\mathbf{A}) = (rs)\mathbf{A} = s(r\mathbf{A}).$$

Matrix Multiplication. The matrix product is what makes matrix algebra interesting and useful. We indicated in Section 9.1 that the product of a matrix $\mathbf{A}$ and a column vector $\mathbf{x}$ is the column vector composed of dot products of the rows of $\mathbf{A}$ with $\mathbf{x}$:

$$\begin{bmatrix} 1 & 2 & 3 \\ 4 & 5 & 6 \end{bmatrix} \begin{bmatrix} 1 \\ 0 \\ 2 \end{bmatrix} = \begin{bmatrix} 1 \cdot 1 + 2 \cdot 0 + 3 \cdot 2 \\ 4 \cdot 1 + 5 \cdot 0 + 6 \cdot 2 \end{bmatrix} = \begin{bmatrix} 7 \\ 16 \end{bmatrix}.$$

More generally, the product of two matrices $\mathbf{A}$ and $\mathbf{B}$ is formed by taking the array of dot products of the *rows* of the first "factor" $\mathbf{A}$ with the *columns* of the second factor $\mathbf{B}$; the dot product of the ith row of $\mathbf{A}$ with the jth column of $\mathbf{B}$ is written as the ijth entry of the product $\mathbf{AB}$:

$$\begin{bmatrix} 1 & 0 & 1 \\ 3 & -1 & 2 \end{bmatrix}\begin{bmatrix} 1 & 2 & x \\ -1 & -1 & y \\ 4 & 1 & z \end{bmatrix} = \begin{bmatrix} 1+0+4 & 2+0+1 & x+0+z \\ 3+1+8 & 6+1+2 & 3x-y+2z \end{bmatrix}$$

$$= \begin{bmatrix} 5 & 3 & x+z \\ 12 & 9 & 3x-y+2z \end{bmatrix}.$$

Note that $\mathbf{AB}$ is only defined when the number of columns of $\mathbf{A}$ matches the number of rows of $\mathbf{B}$. A useful formula for the product of an $m \times n$ matrix $\mathbf{A}$ and an $n \times p$ matrix $\mathbf{B}$ is

$$\mathbf{AB} := [c_{ij}], \qquad \text{where} \quad c_{ij} := \sum_{k=1}^{n} a_{ik}b_{kj}.$$

The dot product of the ith row of $\mathbf{A}$ and the jth column of $\mathbf{B}$ is seen in the "sum of products" expression for c_{ij}.

Since $\mathbf{AB}$ is computed in terms of the *rows* of the first factor and the *columns* of the second factor, it should not be surprising that, in general, $\mathbf{AB}$ does not equal $\mathbf{BA}$ (matrix multiplication does not *commute*):

$$\begin{bmatrix} 1 & 2 \\ 3 & 4 \end{bmatrix}\begin{bmatrix} 0 & 1 \\ 1 & 0 \end{bmatrix} = \begin{bmatrix} 2 & 1 \\ 4 & 3 \end{bmatrix}, \qquad \text{but} \qquad \begin{bmatrix} 0 & 1 \\ 1 & 0 \end{bmatrix}\begin{bmatrix} 1 & 2 \\ 3 & 4 \end{bmatrix} = \begin{bmatrix} 3 & 4 \\ 1 & 2 \end{bmatrix}.$$

In fact, the dimensions of $\mathbf{A}$ and $\mathbf{B}$ may render one or the other of these products undefined:

$$\begin{bmatrix} 1 & 2 \\ 3 & 4 \end{bmatrix}\begin{bmatrix} 0 \\ 1 \end{bmatrix} = \begin{bmatrix} 2 \\ 4 \end{bmatrix}; \qquad \begin{bmatrix} 0 \\ 1 \end{bmatrix}\begin{bmatrix} 1 & 2 \\ 3 & 4 \end{bmatrix} \text{ not defined }.$$

By the same token, one might not expect $(\mathbf{AB})\mathbf{C}$ to equal $\mathbf{A}(\mathbf{BC})$, since in $(\mathbf{AB})\mathbf{C}$ we take dot products with the *columns* of $\mathbf{B}$, whereas in $\mathbf{A}(\mathbf{BC})$ we employ the *rows* of $\mathbf{B}$. So it is a pleasant surprise that *this complication does not arise*, and the "parenthesis grouping" rules are the customary ones:

Properties of Matrix Multiplication

$(\mathbf{AB})\mathbf{C} = \mathbf{A}(\mathbf{BC})$	(Associativity)
$(\mathbf{A} + \mathbf{B})\mathbf{C} = \mathbf{AC} + \mathbf{BC}$	(Distributivity)
$\mathbf{A}(\mathbf{B} + \mathbf{C}) = \mathbf{AB} + \mathbf{AC}$	(Distributivity)
$(r\mathbf{A})\mathbf{B} = r(\mathbf{AB}) = \mathbf{A}(r\mathbf{B})$	(Associativity)

To summarize, the algebra of matrices proceeds much like the standard algebra of numbers, *except that we should never presume that we can switch the order of matrix factors.* (If you think $(\mathbf{A} + \mathbf{B})^2 = \mathbf{A}^2 + \mathbf{B}^2 + 2\mathbf{AB}$, what error have you made?)

Matrices as Linear Operators. Let $\mathbf{A}$ be an $m \times n$ matrix and let $\mathbf{x}$ and $\mathbf{y}$ be $n \times 1$ vectors. Then $\mathbf{Ax}$ is an $m \times 1$ vector, and so we can think of multiplication by $\mathbf{A}$ as defining an operator that maps $n \times 1$ vectors into $m \times 1$ vectors. A consequence of the distributivity and associativity properties is that multiplication by $\mathbf{A}$ defines a **linear operator**, since $\mathbf{A}(\mathbf{x} + \mathbf{y}) = \mathbf{Ax} + \mathbf{Ay}$ and $\mathbf{A}(r\mathbf{x}) = r\mathbf{Ax}$. Moreover, if $\mathbf{A}$ is an $m \times n$ matrix and $\mathbf{B}$ is an $n \times p$ matrix, then the $m \times p$ matrix $\mathbf{AB}$ defines a linear operator that is the composition of the linear operator defined by $\mathbf{B}$

with the linear operator defined by **A**. That is, $(\mathbf{AB})\mathbf{x} = \mathbf{A}(\mathbf{Bx})$, where **x** is a $p \times 1$ vector.

Examples of linear operations are
(i) stretching or contracting the components of a vector by constant factors;
(ii) rotating a vector through some angle about a fixed axis;
(iii) reflecting a vector in a plane mirror.

The Matrix Formulation of Linear Algebraic Systems. Matrix algebra was developed to provide a convenient tool for expressing and analyzing linear algebraic systems. Note that the set of equations

$$x_1 + 2x_2 + x_3 = 1,$$
$$x_1 + 3x_2 + 2x_3 = -1,$$
$$x_1 \qquad + x_3 = 0$$

can be written using the matrix product

(1)
$$\begin{bmatrix} 1 & 2 & 1 \\ 1 & 3 & 2 \\ 1 & 0 & 1 \end{bmatrix} \begin{bmatrix} x_1 \\ x_2 \\ x_3 \end{bmatrix} = \begin{bmatrix} 1 \\ -1 \\ 0 \end{bmatrix}.$$

In general, we express the linear system

$$a_{11}x_1 + a_{12}x_2 + \cdots + a_{1n}x_n = b_1,$$
$$a_{21}x_1 + a_{22}x_2 + \cdots + a_{2n}x_n = b_2,$$
$$\vdots$$
$$a_{n1}x_1 + a_{n2}x_2 + \cdots + a_{nn}x_n = b_n$$

in matrix notation as $\mathbf{Ax} = \mathbf{b}$, where **A** is the *coefficient matrix*, **x** is the vector of unknowns, and **b** is the vector of constants occurring on the right-hand side:

$$\mathbf{A} = \begin{bmatrix} a_{11} & a_{12} & \cdots & a_{1n} \\ a_{21} & a_{22} & \cdots & a_{2n} \\ \vdots & \vdots & & \vdots \\ a_{n1} & a_{n2} & \cdots & a_{nn} \end{bmatrix}, \quad \mathbf{x} = \begin{bmatrix} x_1 \\ x_2 \\ \vdots \\ x_n \end{bmatrix}, \quad \mathbf{b} = \begin{bmatrix} b_1 \\ b_2 \\ \vdots \\ b_n \end{bmatrix}.$$

If $\mathbf{b} = \mathbf{0}$, the system $\mathbf{Ax} = \mathbf{b}$ is said to be *homogeneous* (analogous to the nomenclature of Section 4.2).

Matrix Transpose. The matrix obtained from **A** by interchanging its rows and columns is called the **transpose** of **A** and is denoted by $\mathbf{A}^T$. For example, if

$$\mathbf{A} = \begin{bmatrix} 1 & 2 & 6 \\ -1 & 2 & -1 \end{bmatrix}, \quad \text{then}$$

$$\mathbf{A}^T = \begin{bmatrix} 1 & -1 \\ 2 & 2 \\ 6 & -1 \end{bmatrix}.$$

In general, we have $[a_{ij}]^T = [b_{ij}]$, where $b_{ij} = a_{ji}$. Properties of the transpose are explored in Problem 7.

Matrix Identity. There is a "multiplicative identity" in matrix algebra, namely, a square diagonal matrix $\mathbf{I}$ with ones down the main diagonal. Multiplying $\mathbf{I}$ on the right or left by any other matrix (with compatible dimensions) reproduces the latter matrix:

$$\begin{bmatrix} 1 & 0 & 0 \\ 0 & 1 & 0 \\ 0 & 0 & 1 \end{bmatrix}\begin{bmatrix} 1 & 2 & 1 \\ 1 & 3 & 2 \\ 1 & 0 & 1 \end{bmatrix} = \begin{bmatrix} 1 & 2 & 1 \\ 1 & 3 & 2 \\ 1 & 0 & 1 \end{bmatrix} = \begin{bmatrix} 1 & 2 & 1 \\ 1 & 3 & 2 \\ 1 & 0 & 1 \end{bmatrix}\begin{bmatrix} 1 & 0 & 0 \\ 0 & 1 & 0 \\ 0 & 0 & 1 \end{bmatrix}.$$

(The notation $\mathbf{I}_n$ is used if it is convenient to specify the dimensions, $n \times n$, of the identity matrix.)

Matrix Inverse. Some *square* matrices $\mathbf{A}$ can be paired with other (square) matrices $\mathbf{B}$ having the property that $\mathbf{BA} = \mathbf{I}$:

$$\text{(2)} \qquad \begin{bmatrix} \frac{3}{2} & -1 & \frac{1}{2} \\ \frac{1}{2} & 0 & -\frac{1}{2} \\ -\frac{3}{2} & 1 & \frac{1}{2} \end{bmatrix}\begin{bmatrix} 1 & 2 & 1 \\ 1 & 3 & 2 \\ 1 & 0 & 1 \end{bmatrix} = \begin{bmatrix} 1 & 0 & 0 \\ 0 & 1 & 0 \\ 0 & 0 & 1 \end{bmatrix}.$$

When this happens, it can be shown that

 (i) $\mathbf{B}$ is the *unique* matrix satisfying $\mathbf{BA} = \mathbf{I}$, and
 (ii) $\mathbf{B}$ also satisfies $\mathbf{AB} = \mathbf{I}$.

In such a case, we say that $\mathbf{B}$ is the **inverse** of $\mathbf{A}$ and write $\mathbf{B} = \mathbf{A}^{-1}$.

Not every matrix possesses an inverse; the zero matrix $\mathbf{0}$, for example, can never satisfy the equation $\mathbf{0B} = \mathbf{I}$. A matrix that has no inverse is said to be **singular.**

If we know an inverse for the coefficient matrix $\mathbf{A}$ in a system of linear equations $\mathbf{Ax} = \mathbf{b}$, the solution can be calculated directly by computing $\mathbf{A}^{-1}\mathbf{b}$, as the following derivation shows:

$$\mathbf{Ax} = \mathbf{b} \text{ implies } \mathbf{A}^{-1}\mathbf{Ax} = \mathbf{A}^{-1}\mathbf{b} \text{ implies } \mathbf{x} = \mathbf{A}^{-1}\mathbf{b} .$$

Using (2), for example, we can solve equation (1) quite efficiently:

$$\begin{bmatrix} x_1 \\ x_2 \\ x_3 \end{bmatrix} = \begin{bmatrix} \frac{3}{2} & -1 & \frac{1}{2} \\ \frac{1}{2} & 0 & -\frac{1}{2} \\ -\frac{3}{2} & 1 & \frac{1}{2} \end{bmatrix}\begin{bmatrix} 1 \\ -1 \\ 0 \end{bmatrix} = \begin{bmatrix} \frac{5}{2} \\ \frac{1}{2} \\ -\frac{5}{2} \end{bmatrix}.$$

On the other hand, the coefficient matrix for any *inconsistent* system has no inverse. For example, the coefficient matrices

$$\begin{bmatrix} 2 & -1 \\ -6 & 3 \end{bmatrix} \text{ and } \begin{bmatrix} 2 & 0 & 1 \\ -3 & 1 & 4 \\ -1 & 1 & 5 \end{bmatrix}$$

for the inconsistent systems of Problem 12, Exercises 9.2 (page 504), are necessarily singular.

When $\mathbf{A}^{-1}$ is known, solving $\mathbf{Ax} = \mathbf{b}$ by multiplying $\mathbf{b}$ by $\mathbf{A}^{-1}$ is certainly easier than applying the Gauss–Jordan algorithm of the previous section[†]. So it appears advantageous to be able to find matrix inverses. Some inverses can be obtained directly from the interpretation of the matrix as a linear operator. For example, the inverse of a matrix that rotates a vector is

[†](When the effort to compute the inverse is accounted for, Gauss–Jordan emerges as the winner.)

the matrix that rotates it in the opposite direction. A matrix that performs a mirror reflection is its own inverse (what do you get if you reflect twice?). But in general one must employ an algorithm to compute a matrix inverse. The underlying strategy for this algorithm is based on the observation that if $\mathbf{X}$ denotes the inverse of $\mathbf{A}$, then $\mathbf{X}$ must satisfy the equation $\mathbf{AX} = \mathbf{I}$; finding $\mathbf{X}$ amounts to solving n linear systems of equations for the columns $\{\mathbf{x}_1, \mathbf{x}_2, \ldots, \mathbf{x}_n\}$ of $\mathbf{X}$:

$$\mathbf{Ax}_1 = \begin{bmatrix} 1 \\ 0 \\ 0 \\ \vdots \\ 0 \\ 0 \end{bmatrix}, \quad \mathbf{Ax}_2 = \begin{bmatrix} 0 \\ 1 \\ 0 \\ \vdots \\ 0 \\ 0 \end{bmatrix}, \quad \cdots, \quad \mathbf{Ax}_n = \begin{bmatrix} 0 \\ 0 \\ 0 \\ \vdots \\ 0 \\ 1 \end{bmatrix}.$$

The implementation of this operation is neatly executed by the following variation of the Gauss–Jordan algorithm.

Finding the Inverse of a Matrix. By a **row operation**, we mean any one of the following:
 (a) Interchanging two rows of the matrix
 (b) Multiplying a row of the matrix by a nonzero scalar
 (c) Adding a scalar multiple of one row of the matrix to another row.

If the $n \times n$ matrix $\mathbf{A}$ has an inverse, then $\mathbf{A}^{-1}$ can be determined by performing row operations on the $n \times 2n$ matrix $[\mathbf{A} \mid \mathbf{I}]$ obtained by writing $\mathbf{A}$ and $\mathbf{I}$ side by side. In particular, we perform row operations on the matrix $[\mathbf{A} \mid \mathbf{I}]$ until the first n rows and columns form the identity matrix; that is, the new matrix is $[\mathbf{I} \mid \mathbf{B}]$. Then $\mathbf{A}^{-1} = \mathbf{B}$. We remark that if this procedure fails to produce a matrix of the form $[\mathbf{I} \mid \mathbf{B}]$, then $\mathbf{A}$ has no inverse.

Example 1 Find the inverse of $\mathbf{A} = \begin{bmatrix} 1 & 2 & 1 \\ 1 & 3 & 2 \\ 1 & 0 & 1 \end{bmatrix}$.

Solution We first form the matrix $[\mathbf{A} \mid \mathbf{I}]$ and *row-reduce* the matrix to $[\mathbf{I} \mid \mathbf{A}^{-1}]$. Computing, we find the following:

The matrix $[\mathbf{A} \mid \mathbf{I}]$
$$\left[\begin{array}{ccc:ccc} 1 & 2 & 1 & 1 & 0 & 0 \\ 1 & 3 & 2 & 0 & 1 & 0 \\ 1 & 0 & 1 & 0 & 0 & 1 \end{array}\right].$$

Subtract the first row from the second and third to obtain
$$\left[\begin{array}{ccc:ccc} 1 & 2 & 1 & 1 & 0 & 0 \\ 0 & 1 & 1 & -1 & 1 & 0 \\ 0 & -2 & 0 & -1 & 0 & 1 \end{array}\right].$$

Add 2 times the second row to the third row to obtain
$$\left[\begin{array}{ccc:ccc} 1 & 2 & 1 & 1 & 0 & 0 \\ 0 & 1 & 1 & -1 & 1 & 0 \\ 0 & 0 & 2 & -3 & 2 & 1 \end{array}\right].$$

Subtract 2 times the second row from the first to obtain
$$\left[\begin{array}{ccc:ccc} 1 & 0 & -1 & 3 & -2 & 0 \\ 0 & 1 & 1 & -1 & 1 & 0 \\ 0 & 0 & 2 & -3 & 2 & 1 \end{array}\right].$$

Multiply the third row by $1/2$ to obtain

$$\begin{bmatrix} 1 & 0 & -1 & \vdots & 3 & -2 & 0 \\ 0 & 1 & 1 & \vdots & -1 & 1 & 0 \\ 0 & 0 & 1 & \vdots & -\frac{3}{2} & 1 & \frac{1}{2} \end{bmatrix}.$$

Add the third row to the first and then subtract the third row from the second to obtain

$$\begin{bmatrix} 1 & 0 & 0 & \vdots & \frac{3}{2} & -1 & \frac{1}{2} \\ 0 & 1 & 0 & \vdots & \frac{1}{2} & 0 & -\frac{1}{2} \\ 0 & 0 & 1 & \vdots & -\frac{3}{2} & 1 & \frac{1}{2} \end{bmatrix}.$$

The matrix shown in color is $\mathbf{A}^{-1}$. [Compare equation (2).] ◆

It is convenient to have an expression for the inverse of a generic 2×2 matrix. The following formula is easily verified by mental arithmetic:

(3) $$\begin{bmatrix} a & b \\ c & d \end{bmatrix}^{-1} = \frac{1}{ad - bc} \begin{bmatrix} d & -b \\ -c & a \end{bmatrix} \quad \text{if } ad - bc \neq 0.$$

The denominator in (3), whose nonvanishing is the crucial condition for the existence of the inverse, is known as the determinant.

Determinants. The **determinant** of a 2×2 matrix $\mathbf{A}$, denoted det $\mathbf{A}$ or $|\mathbf{A}|$, is defined by

$$\det \mathbf{A} = \begin{vmatrix} a_{11} & a_{12} \\ a_{21} & a_{22} \end{vmatrix} = a_{11}a_{22} - a_{12}a_{21}.$$

The determinants of higher-order square matrices can be defined recursively in terms of lower-order determinants, using the concept of the *minor*; the **minor** of a particular entry is the determinant of the submatrix formed when that entry's row and column are deleted. Then the determinant of an $n \times n$ matrix equals the alternating-sign sum of the products of the entries of the first row with their minors. For a 3×3 matrix $\mathbf{A}$ this looks like

$$\det \mathbf{A} := \begin{vmatrix} a_{11} & a_{12} & a_{13} \\ a_{21} & a_{22} & a_{23} \\ a_{31} & a_{32} & a_{33} \end{vmatrix} = a_{11} \begin{vmatrix} a_{22} & a_{23} \\ a_{32} & a_{33} \end{vmatrix} - a_{12} \begin{vmatrix} a_{21} & a_{23} \\ a_{31} & a_{33} \end{vmatrix} + a_{13} \begin{vmatrix} a_{21} & a_{22} \\ a_{31} & a_{32} \end{vmatrix}.$$

For example,

$$\begin{vmatrix} 1 & 2 & 1 \\ 0 & 3 & 5 \\ 2 & 1 & -1 \end{vmatrix} = 1 \begin{vmatrix} 3 & 5 \\ 1 & -1 \end{vmatrix} - 2 \begin{vmatrix} 0 & 5 \\ 2 & -1 \end{vmatrix} + 1 \begin{vmatrix} 0 & 3 \\ 2 & 1 \end{vmatrix}$$

$$= 1(-3 - 5) - 2(0 - 10) + 1(0 - 6) = 6.$$

For a 4×4 matrix, we have

$$\begin{vmatrix} 4 & 3 & 2 & -6 \\ -2 & 1 & 2 & 1 \\ 3 & 0 & 3 & 5 \\ 5 & 2 & 1 & -1 \end{vmatrix} = 4 \begin{vmatrix} 1 & 2 & 1 \\ 0 & 3 & 5 \\ 2 & 1 & -1 \end{vmatrix} - 3 \begin{vmatrix} -2 & 2 & 1 \\ 3 & 3 & 5 \\ 5 & 1 & -1 \end{vmatrix}$$

$$+ 2 \begin{vmatrix} -2 & 1 & 1 \\ 3 & 0 & 5 \\ 5 & 2 & -1 \end{vmatrix} - (-6) \begin{vmatrix} -2 & 1 & 2 \\ 3 & 0 & 3 \\ 5 & 2 & 1 \end{vmatrix}.$$

As we have shown, the first minor is 6; and the others are computed the same way, resulting in

$$\begin{vmatrix} 4 & 3 & 2 & -6 \\ -2 & 1 & 2 & 1 \\ 3 & 0 & 3 & 5 \\ 5 & 2 & 1 & -1 \end{vmatrix} = 4(6) - 3(60) + 2(54) - (-6)(36) = 168 \ .$$

Although higher-order determinants can be calculated similarly, a more practical way to evaluate them involves the row-reduction of the matrix to upper triangular form. Here we will deal mainly with low-order determinants, and direct the reader to a linear algebra text for further discussion.[†]

Determinants have a geometric interpretation: $\det \mathbf{A}$ is the volume (in n-dimensional space) of the parallelepiped whose edges are given by the column vectors of $\mathbf{A}$. But their chief value lies in the role they play in the following theorem, which summarizes many of the results from linear algebra that we shall need, and in Cramer's rule, described in Appendix D.

Matrices and Systems of Equations

Theorem 1. Let $\mathbf{A}$ be an $n \times n$ matrix. The following statements are equivalent:

(a) $\mathbf{A}$ is singular (does not have an inverse).

(b) The determinant of $\mathbf{A}$ is zero.

(c) $\mathbf{Ax} = \mathbf{0}$ has nontrivial solutions ($\mathbf{x} \neq \mathbf{0}$).

(d) The columns (rows) of $\mathbf{A}$ form a linearly dependent set.

In part (d), the statement that the n columns of $\mathbf{A}$ are linearly dependent means that there exist scalars $c_1, \ldots, c_n$, *not all zero*, such that

$$c_1 \mathbf{a}_1 + c_2 \mathbf{a}_2 + \cdots + c_n \mathbf{a}_n = \mathbf{0} \ ,$$

where $\mathbf{a}_j$ is the vector forming the jth column of $\mathbf{A}$.

If $\mathbf{A}$ is a singular square matrix (so $\det \mathbf{A} = 0$), then $\mathbf{Ax} = \mathbf{0}$ has infinitely many solutions. Indeed, Theorem 1 asserts that there is a vector $\mathbf{x}_0 \neq \mathbf{0}$ such that $\mathbf{Ax}_0 = \mathbf{0}$, and we can get infinitely many other solutions by multiplying $\mathbf{x}_0$ by any scalar, i.e., taking $\mathbf{x} = c\mathbf{x}_0$. Furthermore, $\mathbf{Ax} = \mathbf{b}$ either has no solutions or it has infinitely many of them of the form

$$\mathbf{x} = \mathbf{x}_p + \mathbf{x}_h \ ,$$

where $\mathbf{x}_p$ is a *particular* solution to $\mathbf{Ax} = \mathbf{b}$ and $\mathbf{x}_h$ is any of the infinity of solutions to $\mathbf{Ax} = \mathbf{0}$ (see Problem 15). The resemblance of this situation to that of solving nonhomogeneous linear differential equations should be quite apparent.

To illustrate, in Example 3 of Section 9.2 (page 502) we saw that the system

$$\begin{bmatrix} 2 & 4 & 1 \\ 2 & 4 & 0 \\ -4 & -8 & 1 \end{bmatrix} \begin{bmatrix} x_1 \\ x_2 \\ x_3 \end{bmatrix} = \begin{bmatrix} 8 \\ 6 \\ -10 \end{bmatrix}$$

has solutions

$$x_1 = 3 - 2s \ , \qquad x_2 = s \ , \qquad x_3 = 2 \ ; \quad -\infty < s < \infty \ .$$

[†]Your authors' personal favorite is *Fundamentals of Matrix Analysis with Applications*, by Edward Barry Saff and Arthur David Snider (John Wiley & Sons, Hoboken, New Jersey, 2016).

Writing these in matrix notation, we can identify the vectors $\mathbf{x}_p$ and $\mathbf{x}_h$ mentioned above:

$$\mathbf{x} = \begin{bmatrix} 3 - 2s \\ s \\ 2 \end{bmatrix} = \begin{bmatrix} 3 \\ 0 \\ 2 \end{bmatrix} + s \begin{bmatrix} -2 \\ 1 \\ 0 \end{bmatrix} = \mathbf{x}_p + \mathbf{x}_h \, .$$

Note further that the determinant of $\mathbf{A}$ is indeed zero,

$$\det \mathbf{A} = 2 \begin{vmatrix} 4 & 0 \\ -8 & 1 \end{vmatrix} - 4 \begin{vmatrix} 2 & 0 \\ -4 & 1 \end{vmatrix} + 1 \begin{vmatrix} 2 & 4 \\ -4 & -8 \end{vmatrix} = 2 \cdot 4 - 4 \cdot 2 + 1 \cdot 0 = 0 \, ,$$

and that the linear dependence of the columns of $\mathbf{A}$ is exhibited by the identity

$$-2 \begin{bmatrix} 2 \\ 2 \\ -4 \end{bmatrix} + 1 \begin{bmatrix} 4 \\ 4 \\ -8 \end{bmatrix} + 0 \begin{bmatrix} 1 \\ 0 \\ 1 \end{bmatrix} = \begin{bmatrix} 0 \\ 0 \\ 0 \end{bmatrix} \, .$$

If $\mathbf{A}$ is a nonsingular square matrix (i.e., $\mathbf{A}$ has an inverse and $\det \mathbf{A} \neq 0$), then the homogeneous system $\mathbf{A}\mathbf{x} = \mathbf{0}$ has $\mathbf{x} = \mathbf{0}$ as its only solution. More generally, when $\det \mathbf{A} \neq 0$, the system $\mathbf{A}\mathbf{x} = \mathbf{b}$ has a unique solution (namely, $\mathbf{x} = \mathbf{A}^{-1}\mathbf{b}$).

Calculus of Matrices

If we allow the entries $a_{ij}(t)$ in a matrix $\mathbf{A}(t)$ to be functions of the variable t, then $\mathbf{A}(t)$ is a **matrix function of** t. Similarly, if the entries $x_i(t)$ of a vector $\mathbf{x}(t)$ are functions of t, then $\mathbf{x}(t)$ is a **vector function of** t.

These matrix and vector functions have a calculus much like that of real-valued functions. A matrix $\mathbf{A}(t)$ is said to be **continuous at** t_0 if each entry $a_{ij}(t)$ is continuous at t_0. Moreover, $\mathbf{A}(t)$ is **differentiable at** t_0 if each entry $a_{ij}(t)$ is differentiable at t_0, and we write

$$(4) \qquad \frac{d\mathbf{A}}{dt}(t_0) = \mathbf{A}'(t_0) := \left[a_{ij}'(t_0) \right] \, .$$

Similarly, we define

$$(5) \qquad \int_a^b \mathbf{A}(t)\, dt := \left[\int_a^b a_{ij}(t)\, dt \right] \, .$$

Example 2 Let $\mathbf{A}(t) = \begin{bmatrix} t^2 + 1 & \cos t \\ e^t & 1 \end{bmatrix}$.

Find: **(a)** $\mathbf{A}'(t)$. **(b)** $\displaystyle\int_0^1 \mathbf{A}(t)\, dt$.

Solution Using formulas (4) and (5), we compute

$$\textbf{(a)} \ \ \mathbf{A}'(t) = \begin{bmatrix} 2t & -\sin t \\ e^t & 0 \end{bmatrix} . \qquad \textbf{(b)} \ \ \int_0^1 \mathbf{A}(t)\, dt = \begin{bmatrix} \frac{4}{3} & \sin 1 \\ e - 1 & 1 \end{bmatrix} . \ \ \blacklozenge$$

Example 3 Show that $\mathbf{x}(t) = \begin{bmatrix} \cos \omega t \\ \sin \omega t \end{bmatrix}$ is a solution of the matrix differential equation $\mathbf{x}' = \mathbf{A}\mathbf{x}$, where

$$\mathbf{A} = \begin{bmatrix} 0 & -\omega \\ \omega & 0 \end{bmatrix} .$$

Solution We simply verify that $\mathbf{x}'(t)$ and $\mathbf{A}\mathbf{x}(t)$ are the same vector function:

$$\mathbf{x}'(t) = \begin{bmatrix} -\omega\sin\omega t \\ \omega\cos\omega t \end{bmatrix}; \qquad \mathbf{A}\mathbf{x} = \begin{bmatrix} 0 & -\omega \\ \omega & 0 \end{bmatrix}\begin{bmatrix} \cos\omega t \\ \sin\omega t \end{bmatrix} = \begin{bmatrix} -\omega\sin\omega t \\ \omega\cos\omega t \end{bmatrix}. \quad \blacklozenge$$

The basic properties of differentiation are valid for matrix functions.

Differentiation Formulas for Matrix Functions

$$\frac{d}{dt}(\mathbf{CA}) = \mathbf{C}\frac{d\mathbf{A}}{dt} \qquad (\mathbf{C}\ \text{a constant matrix}).$$

$$\frac{d}{dt}(\mathbf{A}+\mathbf{B}) = \frac{d\mathbf{A}}{dt} + \frac{d\mathbf{B}}{dt}.$$

$$\frac{d}{dt}(\mathbf{AB}) = \mathbf{A}\frac{d\mathbf{B}}{dt} + \frac{d\mathbf{A}}{dt}\mathbf{B}.$$

In the last formula, the order in which the matrices are written is very important because, as we have emphasized, matrix multiplication does not always commute.

9.3 EXERCISES

1. Let $\mathbf{A} := \begin{bmatrix} 2 & 1 \\ 3 & 5 \end{bmatrix}$ and $\mathbf{B} := \begin{bmatrix} -1 & 0 \\ 2 & -3 \end{bmatrix}$.
Find: (a) $\mathbf{A}+\mathbf{B}$. (b) $3\mathbf{A}-\mathbf{B}$.

2. Let $\mathbf{A} := \begin{bmatrix} 2 & 0 & 5 \\ 2 & 1 & 1 \end{bmatrix}$ and $\mathbf{B} := \begin{bmatrix} 1 & -1 & 2 \\ 0 & 3 & -2 \end{bmatrix}$.
Find: (a) $\mathbf{A}+\mathbf{B}$. (b) $7\mathbf{A}-4\mathbf{B}$.

3. Let $\mathbf{A} := \begin{bmatrix} 2 & 4 \\ 1 & 1 \end{bmatrix}$ and $\mathbf{B} := \begin{bmatrix} -1 & 3 \\ 5 & 2 \end{bmatrix}$.
Find: (a) $\mathbf{AB}$. (b) $\mathbf{A}^2=\mathbf{AA}$. (c) $\mathbf{B}^2=\mathbf{BB}$.

4. Let $\mathbf{A} := \begin{bmatrix} 2 & 1 \\ 0 & 4 \\ -1 & 3 \end{bmatrix}$ and $\mathbf{B} := \begin{bmatrix} 1 & 1 & -1 \\ 0 & 3 & 1 \end{bmatrix}$.
Find: (a) $\mathbf{AB}$. (b) $\mathbf{BA}$.

5. Let $\mathbf{A} := \begin{bmatrix} 1 & -2 \\ 2 & -3 \end{bmatrix}$, $\mathbf{B} := \begin{bmatrix} 1 & 0 \\ 1 & 1 \end{bmatrix}$, and $\mathbf{C} := \begin{bmatrix} -1 & 1 \\ 2 & 1 \end{bmatrix}$.
Find: (a) $\mathbf{AB}$. (b) $\mathbf{AC}$. (c) $\mathbf{A}(\mathbf{B}+\mathbf{C})$.

6. Let $\mathbf{A} := \begin{bmatrix} 1 & 2 \\ 1 & 1 \end{bmatrix}$, $\mathbf{B} := \begin{bmatrix} 0 & 3 \\ 1 & 2 \end{bmatrix}$, and $\mathbf{C} := \begin{bmatrix} 1 & -4 \\ 1 & 1 \end{bmatrix}$.
Find: (a) $\mathbf{AB}$. (b) $(\mathbf{AB})\mathbf{C}$. (c) $(\mathbf{A}+\mathbf{B})\mathbf{C}$.

7. (a) Show that if $\mathbf{u}$ and $\mathbf{v}$ are each $n\times 1$ column vectors, then the matrix product $\mathbf{u}^T\mathbf{v}$ is the same as the dot product $\mathbf{u}\cdot\mathbf{v}$.
(b) Let $\mathbf{v}$ be a 3×1 column vector with $\mathbf{v}^T = \begin{bmatrix} 2 & 3 & 5 \end{bmatrix}$. Show that, for $\mathbf{A}$ as given in Example 1 (page 509), $(\mathbf{A}\mathbf{v})^T = \mathbf{v}^T\mathbf{A}^T$.
(c) Does $(\mathbf{A}\mathbf{v})^T = \mathbf{v}^T\mathbf{A}^T$ hold for every $m\times n$ matrix $\mathbf{A}$ and $n\times 1$ vector $\mathbf{v}$?
(d) Does $(\mathbf{AB})^T = \mathbf{B}^T\mathbf{A}^T$ hold for every pair of matrices $\mathbf{A}$, $\mathbf{B}$ such that both matrix products are defined? Justify your answer.

8. Let $\mathbf{A} := \begin{bmatrix} 2 & -1 \\ -3 & 4 \end{bmatrix}$ and $\mathbf{B} := \begin{bmatrix} 1 & 2 \\ 3 & 2 \end{bmatrix}$.
Verify that $\mathbf{AB}\neq\mathbf{BA}$.

In Problems 9–14, use the method of Example 1 to compute the inverse of the given matrix, if it exists. For Problems 9 and 10, confirm your answer by comparison with formula (3).

9. $\begin{bmatrix} 2 & 1 \\ -1 & 4 \end{bmatrix}$ **10.** $\begin{bmatrix} 4 & 1 \\ 5 & 9 \end{bmatrix}$

11. $\begin{bmatrix} 1 & 1 & 1 \\ 1 & 2 & 1 \\ 2 & 3 & 2 \end{bmatrix}$ **12.** $\begin{bmatrix} 1 & 1 & 1 \\ 1 & 2 & 3 \\ 0 & 1 & 1 \end{bmatrix}$

13. $\begin{bmatrix} -2 & -1 & 1 \\ 2 & 1 & 0 \\ 3 & 1 & -1 \end{bmatrix}$ **14.** $\begin{bmatrix} 1 & 1 & 1 \\ 1 & -1 & 2 \\ 1 & 1 & 4 \end{bmatrix}$

15. Prove that if $\mathbf{x}_p$ satisfies $\mathbf{Ax}_p = \mathbf{b}$, then every solution to the nonhomogeneous system $\mathbf{Ax} = \mathbf{b}$ is of the form $\mathbf{x} = \mathbf{x}_p + \mathbf{x}_h$, where $\mathbf{x}_h$ is a solution to the corresponding homogeneous system $\mathbf{Ax} = \mathbf{0}$.

16. Let $\mathbf{A} = \begin{bmatrix} 2 & -1 & 1 \\ -1 & 2 & 1 \\ 1 & 1 & 2 \end{bmatrix}$.

(a) Show that $\mathbf{A}$ is singular.

(b) Show that $\mathbf{Ax} = \begin{bmatrix} 3 \\ 1 \\ 3 \end{bmatrix}$ has no solutions.

(c) Show that $\mathbf{Ax} = \begin{bmatrix} 3 \\ 0 \\ 3 \end{bmatrix}$ has infinitely many solutions.

In Problems 17–20, find the matrix function $\mathbf{X}^{-1}(t)$ *whose value at t is the inverse of the given matrix* $\mathbf{X}(t)$.

17. $\mathbf{X}(t) = \begin{bmatrix} e^t & e^{4t} \\ e^t & 4e^{4t} \end{bmatrix}$

18. $\mathbf{X}(t) = \begin{bmatrix} \sin 2t & \cos 2t \\ 2\cos 2t & -2\sin 2t \end{bmatrix}$

19. $\mathbf{X}(t) = \begin{bmatrix} e^t & e^{-t} & e^{2t} \\ e^t & -e^{-t} & 2e^{2t} \\ e^t & e^{-t} & 4e^{2t} \end{bmatrix}$

20. $\mathbf{X}(t) = \begin{bmatrix} e^{3t} & 1 & t \\ 3e^{3t} & 0 & 1 \\ 9e^{3t} & 0 & 0 \end{bmatrix}$

In Problems 21–26, evaluate the given determinant.

21. $\begin{vmatrix} 4 & 3 \\ -1 & 2 \end{vmatrix}$

22. $\begin{vmatrix} 12 & 8 \\ 3 & 2 \end{vmatrix}$

23. $\begin{vmatrix} 1 & 0 & 0 \\ 3 & 1 & 2 \\ 1 & 5 & -2 \end{vmatrix}$

24. $\begin{vmatrix} 1 & 0 & 2 \\ 0 & 3 & -1 \\ -1 & 2 & 1 \end{vmatrix}$

25. $\begin{vmatrix} 1 & 4 & 3 \\ -1 & -1 & 2 \\ 4 & 5 & 2 \end{vmatrix}$

26. $\begin{vmatrix} 1 & 4 & 4 \\ 3 & 0 & -3 \\ 1 & 6 & 2 \end{vmatrix}$

In Problems 27–29, determine the values of r for which $\det(\mathbf{A} - r\mathbf{I}) = 0$.

27. $\mathbf{A} = \begin{bmatrix} 1 & 1 \\ -2 & 4 \end{bmatrix}$

28. $\mathbf{A} = \begin{bmatrix} 3 & 3 \\ 2 & 4 \end{bmatrix}$

29. $\mathbf{A} = \begin{bmatrix} 0 & 0 & 0 \\ 0 & 1 & 0 \\ 1 & 0 & 1 \end{bmatrix}$

30. Illustrate the equivalence of the assertions (a)–(d) in Theorem 1 (page 511) for the matrix

$$\mathbf{A} = \begin{bmatrix} 4 & -2 & 2 \\ -2 & 4 & 2 \\ 2 & 2 & 4 \end{bmatrix}$$

as follows.

(a) Show that the row-reduction procedure applied to $[\mathbf{A} \mid \mathbf{I}]$ fails to produce the inverse of $\mathbf{A}$.

(b) Calculate $\det \mathbf{A}$.

(c) Determine a nontrivial solution $\mathbf{x}$ to $\mathbf{Ax} = \mathbf{0}$.

(d) Find scalars c_1, c_2, and c_3, *not all zero*, so that $c_1\mathbf{a}_1 + c_2\mathbf{a}_2 + c_3\mathbf{a}_3 = \mathbf{0}$, where $\mathbf{a}_1, \mathbf{a}_2$, and $\mathbf{a}_3$ are the columns of $\mathbf{A}$.

In Problems 31 and 32, find $d\mathbf{x}/dt$ *for the given vector functions.*

31. $\mathbf{x}(t) = \begin{bmatrix} e^{3t} \\ 2e^{3t} \\ -e^{3t} \end{bmatrix}$

32. $\mathbf{x}(t) = \begin{bmatrix} e^{-t}\sin 3t \\ 0 \\ -e^{-t}\sin 3t \end{bmatrix}$

In Problems 33 and 34, find $d\mathbf{X}/dt$ *for the given matrix functions.*

33. $\mathbf{X}(t) = \begin{bmatrix} e^{5t} & 3e^{2t} \\ -2e^{5t} & -e^{2t} \end{bmatrix}$

34. $\mathbf{X}(t) = \begin{bmatrix} \sin 2t & \cos 2t & e^{-2t} \\ -\sin 2t & 2\cos 2t & 3e^{-2t} \\ 3\sin 2t & \cos 2t & e^{-2t} \end{bmatrix}$

In Problems 35 and 36, verify that the given vector function satisfies the given system.

35. $\mathbf{x}' = \begin{bmatrix} 1 & 1 \\ -2 & 4 \end{bmatrix}\mathbf{x}$, $\mathbf{x}(t) = \begin{bmatrix} e^{3t} \\ 2e^{3t} \end{bmatrix}$

36. $\mathbf{x}' = \begin{bmatrix} 0 & 0 & 0 \\ 0 & 1 & 0 \\ 1 & 0 & 1 \end{bmatrix}\mathbf{x}$, $\mathbf{x}(t) = \begin{bmatrix} 0 \\ e^t \\ -3e^t \end{bmatrix}$

In Problems 37 and 38, verify that the given matrix function satisfies the given matrix differential equation.

37. $\mathbf{X}' = \begin{bmatrix} 1 & -1 \\ 2 & 4 \end{bmatrix}\mathbf{X}$, $\mathbf{X}(t) = \begin{bmatrix} e^{2t} & e^{3t} \\ -e^{2t} & -2e^{3t} \end{bmatrix}$

38. $\mathbf{X}' = \begin{bmatrix} 1 & 0 & 0 \\ 0 & 3 & -2 \\ 0 & -2 & 3 \end{bmatrix}\mathbf{X}$,

$$\mathbf{X}(t) = \begin{bmatrix} e^t & 0 & 0 \\ 0 & e^t & e^{5t} \\ 0 & e^t & -e^{5t} \end{bmatrix}$$

In Problems 39 and 40, the matrices $\mathbf{A}(t)$ *and* $\mathbf{B}(t)$ *are given. Find*

(a) $\displaystyle\int \mathbf{A}(t)\,dt$. (b) $\displaystyle\int_0^1 \mathbf{B}(t)\,dt$. (c) $\dfrac{d}{dt}[\mathbf{A}(t)\mathbf{B}(t)]$.

39. $\mathbf{A}(t) = \begin{bmatrix} t & e^t \\ 1 & e^t \end{bmatrix}$, $\mathbf{B}(t) = \begin{bmatrix} \cos t & -\sin t \\ \sin t & \cos t \end{bmatrix}$

40. $\mathbf{A}(t) = \begin{bmatrix} 1 & e^{-2t} \\ 3 & e^{-2t} \end{bmatrix}$, $\mathbf{B}(t) = \begin{bmatrix} e^{-t} & e^{-t} \\ -e^{-t} & 3e^{-t} \end{bmatrix}$

41. An $n \times n$ matrix $\mathbf{A}$ is called **symmetric** if $\mathbf{A}^T = \mathbf{A}$; that is, if $a_{ij} = a_{ji}$, for all $i, j = 1, \ldots, n$. Show that if $\mathbf{A}$ is an $n \times n$ matrix, then $\mathbf{A} + \mathbf{A}^T$ is a symmetric matrix.

42. Let $\mathbf{A}$ be an $m \times n$ matrix. Show that $\mathbf{A}^T\mathbf{A}$ is a symmetric $n \times n$ matrix and $\mathbf{A}\mathbf{A}^T$ is a symmetric $m \times m$ matrix (see Problem 41).

43. The **inner product** of two vectors is a generalization of the dot product, for vectors with complex entries. It is defined by

$$(\mathbf{x}, \mathbf{y}) := \sum_{i=1}^{n} x_i \bar{y}_i, \qquad \text{where}$$

$\mathbf{x} = \mathrm{col}\,(x_1, x_2, \ldots, x_n)$, $\mathbf{y} = \mathrm{col}\,(y_1, y_2, \ldots, y_n)$ are complex vectors and the overbar denotes complex conjugation.

(a) Show that $(\mathbf{x}, \mathbf{y}) = \mathbf{x}^T \bar{\mathbf{y}}$, where $\bar{\mathbf{y}} = \mathrm{col}\,(\bar{y}_1, \bar{y}_2, \ldots, \bar{y}_n)$.

(b) Prove that for any $n \times 1$ vectors $\mathbf{x}, \mathbf{y}, \mathbf{z}$ and any complex number λ, we have

$$(\mathbf{x}, \mathbf{y}) = \overline{(\mathbf{y}, \mathbf{x})},$$
$$(\mathbf{x}, \mathbf{y} + \mathbf{z}) = (\mathbf{x}, \mathbf{y}) + (\mathbf{x}, \mathbf{z}),$$
$$(\lambda\mathbf{x}, \mathbf{y}) = \lambda(\mathbf{x}, \mathbf{y}), \quad (\mathbf{x}, \lambda\mathbf{y}) = \bar{\lambda}(\mathbf{x}, \mathbf{y}).$$

9.4 Linear Systems in Normal Form

In keeping with the introduction presented in Section 9.1, we say that a system of n linear differential equations is in **normal form** if it is expressed as

$$(1) \qquad \mathbf{x}'(t) = \mathbf{A}(t)\mathbf{x}(t) + \mathbf{f}(t),$$

where $\mathbf{x}(t) = \mathrm{col}\,(x_1(t), \ldots, x_n(t))$, $\mathbf{f}(t) = \mathrm{col}\,(f_1(t), \ldots, f_n(t))$, and $\mathbf{A}(t) = [a_{ij}(t)]$ is an $n \times n$ matrix. As with a scalar linear differential equation, a system is called **homogeneous** when $\mathbf{f}(t) \equiv \mathbf{0}$; otherwise, it is called **nonhomogeneous**. When the elements of $\mathbf{A}$ are all constants, the system is said to have **constant coefficients**. Recall that an nth-order linear differential equation

$$(2) \qquad y^{(n)}(t) + p_{n-1}(t)y^{(n-1)}(t) + \cdots + p_0(t)y(t) = g(t)$$

can be rewritten as a first-order system in normal form using the substitution $x_1(t) := y(t)$, $x_2(t) := y'(t), \ldots, x_n(t) := y^{(n-1)}(t)$; indeed, equation (2) is equivalent to $\mathbf{x}'(t) = \mathbf{A}(t)\mathbf{x}(t) + \mathbf{f}(t)$, where $\mathbf{x}(t) = \mathrm{col}\,(x_1(t), \ldots, x_n(t))$, $\mathbf{f}(t) := \mathrm{col}\,(0, \ldots, 0, g(t))$, and

$$\mathbf{A}(t) := \begin{bmatrix} 0 & 1 & 0 & \cdots & 0 & 0 \\ 0 & 0 & 1 & & 0 & 0 \\ \vdots & \vdots & \vdots & & \vdots & \vdots \\ 0 & 0 & 0 & \cdots & 0 & 1 \\ -p_0(t) & -p_1(t) & -p_2(t) & \cdots & -p_{n-2}(t) & -p_{n-1}(t) \end{bmatrix}.$$

The theory for systems in normal form parallels very closely the theory of linear differential equations presented in Chapters 4 and 6. In many cases the proofs for scalar linear differential equations carry over to normal systems with appropriate modifications. Conversely, results for normal systems apply to scalar linear equations since, as we showed, any scalar linear equation can be expressed as a normal system. This is the case with the existence and uniqueness theorems for linear differential equations.

The **initial value problem** for the normal system (1) is the problem of finding a differentiable vector function $\mathbf{x}(t)$ that satisfies the system on an interval I and also satisfies the **initial condition** $\mathbf{x}(t_0) = \mathbf{x}_0$, where t_0 is a given point of I and $\mathbf{x}_0 = \mathrm{col}\,(x_{1,0}, \ldots, x_{n,0})$ is a given vector.

Existence and Uniqueness

Theorem 2. If $\mathbf{A}(t)$ and $\mathbf{f}(t)$ are continuous on an open interval I that contains the point t_0, then for any choice of the initial vector $\mathbf{x}_0$, there exists a unique solution $\mathbf{x}(t)$ on the whole interval I to the initial value problem

$$\mathbf{x}'(t) = \mathbf{A}(t)\mathbf{x}(t) + \mathbf{f}(t) \,, \qquad \mathbf{x}(t_0) = \mathbf{x}_0 \,.$$

We give a proof of this result in Chapter 13[†] and obtain as corollaries the existence and uniqueness theorems for second-order equations (Theorem 4, Section 4.5, page 182) and higher-order linear equations (Theorem 1, Section 6.1, page 319).

If we rewrite system (1) as $\mathbf{x}' - \mathbf{A}\mathbf{x} = \mathbf{f}$ and define the operator $L[\mathbf{x}] := \mathbf{x}' - \mathbf{A}\mathbf{x}$, then we can express system (1) in the operator form $L[\mathbf{x}] = \mathbf{f}$. Here the operator L maps vector functions into vector functions. Moreover, L is a *linear* operator in the sense that for any scalars a, b and differentiable vector functions $\mathbf{x}$, $\mathbf{y}$, we have

$$L[a\mathbf{x} + b\mathbf{y}] = aL[\mathbf{x}] + bL[\mathbf{y}] \,.$$

The proof of this linearity follows from the properties of matrix multiplication (see Problem 27).

As a consequence of the linearity of L, if $\mathbf{x}_1, \ldots, \mathbf{x}_n$ are solutions to the *homogeneous* system $\mathbf{x}' = \mathbf{A}\mathbf{x}$, or $L[\mathbf{x}] = \mathbf{0}$ in operator notation, then any linear combination of these vectors, $c_1\mathbf{x}_1 + \cdots + c_n\mathbf{x}_n$, is also a solution. Moreover, we will see that if the solutions $\mathbf{x}_1, \ldots, \mathbf{x}_n$ are linearly independent, then *every* solution to $L[\mathbf{x}] = \mathbf{0}$ can be expressed as $c_1\mathbf{x}_1 + \cdots + c_n\mathbf{x}_n$ for an appropriate choice of the constants $c_1, \ldots, c_n$.

Linear Dependence of Vector Functions

Definition 1. The m vector functions $\mathbf{x}_1, \ldots, \mathbf{x}_m$ are said to be **linearly dependent on an interval** I if there exist constants $c_1, \ldots, c_m$, not all zero, such that

$$(3) \qquad c_1\mathbf{x}_1(t) + \cdots + c_m\mathbf{x}_m(t) = \mathbf{0}$$

for all t in I. If the vectors are not linearly dependent, they are said to be **linearly independent on** I.

Example 1 Show that the vector functions $\mathbf{x}_1(t) = \mathrm{col}(e^t, 0, e^t)$, $\mathbf{x}_2(t) = \mathrm{col}(3e^t, 0, 3e^t)$, and $\mathbf{x}_3(t) = \mathrm{col}(t, 1, 0)$ are linearly dependent on $(-\infty, \infty)$.

Solution Notice that $\mathbf{x}_2$ is just 3 times $\mathbf{x}_1$ and therefore $3\mathbf{x}_1(t) - \mathbf{x}_2(t) + 0 \cdot \mathbf{x}_3(t) = \mathbf{0}$ for all t. Hence, $\mathbf{x}_1$, $\mathbf{x}_2$, and $\mathbf{x}_3$ are linearly dependent on $(-\infty, \infty)$. ◆

Example 2 Show that

$$\mathbf{x}_1(t) = \begin{bmatrix} t \\ |t| \end{bmatrix} \,, \qquad \mathbf{x}_2(t) = \begin{bmatrix} |t| \\ t \end{bmatrix}$$

are linearly independent on $(-\infty, \infty)$.

[†]All references to Chapters 11–13 refer to the expanded text, *Fundamentals of Differential Equations and Boundary Value Problems*, 7th ed.

Solution Note that at every instant t_0, the column vector $\mathbf{x}_1(t_0)$ is a multiple of $\mathbf{x}_2(t_0)$; indeed, $\mathbf{x}_1(t_0) = \mathbf{x}_2(t_0)$ for $t_0 \geq 0$, and $\mathbf{x}_1(t_0) = -\mathbf{x}_2(t_0)$ for $t_0 \leq 0$. Nonetheless, the vector functions are not dependent, because the c's in condition (3) are not allowed to change with t; for $t < 0$, the equation $c_1\mathbf{x}_1(t) + c_2\mathbf{x}_2(t) = \mathbf{0}$ implies $c_1 - c_2 = 0$, but for $t > 0$ it implies $c_1 + c_2 = 0$. Thus $c_1 = c_2 = 0$ and the functions are independent. ◆

Example 3 Show that the vector functions $\mathbf{x}_1(t) = \mathrm{col}(e^{2t}, 0, e^{2t})$, $\mathbf{x}_2(t) = \mathrm{col}(e^{2t}, e^{2t}, -e^{2t})$, and $\mathbf{x}_3(t) = \mathrm{col}(e^t, 2e^t, e^t)$ are linearly independent on $(-\infty, \infty)$.

Solution To prove independence, we *assume* c_1, c_2, and c_3 are constants for which

$$c_1\mathbf{x}_1(t) + c_2\mathbf{x}_2(t) + c_3\mathbf{x}_3(t) = \mathbf{0}$$

holds at every t in $(-\infty, \infty)$ and show that this forces $c_1 = c_2 = c_3 = 0$. In particular, when $t = 0$ we obtain

$$c_1\begin{bmatrix}1\\0\\1\end{bmatrix} + c_2\begin{bmatrix}1\\1\\-1\end{bmatrix} + c_3\begin{bmatrix}1\\2\\1\end{bmatrix} = \mathbf{0} \,,$$

which is equivalent to the system of linear equations

$$
\begin{aligned}
c_1 + c_2 + c_3 &= 0\,,\\
c_2 + 2c_3 &= 0\,,\\
c_1 - c_2 + c_3 &= 0\,.
\end{aligned}
$$

(4)

Either by solving (4) or by checking that the determinant of its coefficients is nonzero (recall Theorem 1 on page 511), we can verify that (4) has only the trivial solution $c_1 = c_2 = c_3 = 0$. Therefore the vector functions $\mathbf{x}_1$, $\mathbf{x}_2$, and $\mathbf{x}_3$ are linearly independent on $(-\infty, \infty)$ (in fact, on any interval containing $t = 0$). ◆

As Example 3 illustrates, if $\mathbf{x}_1(t), \mathbf{x}_2(t), \ldots, \mathbf{x}_n(t)$ are n vector functions, each having n components, we can establish their linear independence on an interval I if we can find *one* point t_0 in I where the determinant

$$\det[\mathbf{x}_1(t_0) \ldots \mathbf{x}_n(t_0)]$$

is not zero. Because of the analogy with scalar equations, we call this determinant the **Wronskian**.

Wronskian

Definition 2. The **Wronskian** of n vector functions $\mathbf{x}_1(t) = \mathrm{col}(x_{1,1}, \ldots, x_{n,1}), \ldots,$ $\mathbf{x}_n(t) = \mathrm{col}(x_{1,n}, \ldots, x_{n,n})$ is defined to be the function

$$W[\mathbf{x}_1, \ldots, \mathbf{x}_n](t) := \begin{vmatrix} x_{1,1}(t) & x_{1,2}(t) & \cdots & x_{1,n}(t) \\ x_{2,1}(t) & x_{2,2}(t) & \cdots & x_{2,n}(t) \\ \vdots & \vdots & & \vdots \\ x_{n,1}(t) & x_{n,2}(t) & \cdots & x_{n,n}(t) \end{vmatrix}.$$

We see that n vector functions are linearly independent on an interval if their Wronskian is nonzero *at any point in the interval.* But now we show that if these functions happen to be independent solutions to a homogeneous system $\mathbf{x}' = \mathbf{A}\mathbf{x}$, where $\mathbf{A}$ is an $n \times n$ matrix of continuous functions, then the Wronskian is *never* zero on I. For suppose to the contrary that $W(t_0) = 0$. Then by Theorem 1 the vanishing of the determinant implies that the column vectors $\mathbf{x}_1(t_0), \mathbf{x}_2(t_0), \ldots, \mathbf{x}_n(t_0)$ are linearly dependent. Thus there exist scalars $c_1, \ldots, c_n$ *not all zero*, such that at t_0

$$c_1 \mathbf{x}_1(t_0) + \cdots + c_n \mathbf{x}_n(t_0) = \mathbf{0}\,.$$

However, $c_1 \mathbf{x}_1(t) + \cdots + c_n \mathbf{x}_n(t)$ and the vector function $\mathbf{z}(t) \equiv \mathbf{0}$ are both solutions to $\mathbf{x}' = \mathbf{A}\mathbf{x}$ on I, and they agree at the point t_0. So these solutions must be identical on I according to the existence-uniqueness theorem (Theorem 2, page 516). That is,

$$c_1 \mathbf{x}_1(t) + \cdots + c_n \mathbf{x}_n(t) = \mathbf{0}$$

for *all* t in I. But this contradicts the given information that $\mathbf{x}_1, \ldots, \mathbf{x}_n$ are linearly independent on I. We have shown that $W(t_0) \neq 0$, and since t_0 is an arbitrary point, it follows that $W(t) \neq 0$ for all $t \in I$.

The preceding argument has two important implications that parallel the scalar case. First, the Wronskian of *solutions* to $\mathbf{x}' = \mathbf{A}\mathbf{x}$ *is either identically zero or never zero on I* (see also Problem 33). Second, a set of n solutions $\mathbf{x}_1, \ldots, \mathbf{x}_n$ to $\mathbf{x}' = \mathbf{A}\mathbf{x}$ on I *is linearly independent on I if and only if their Wronskian is never zero on I.* With these facts in hand, we can imitate the proof given for the scalar case in Section 6.1 (Theorem 2, page 322) to obtain the following representation theorem for the solutions to $\mathbf{x}' = \mathbf{A}\mathbf{x}$.

Representation of Solutions (Homogeneous Case)

Theorem 3. Let $\mathbf{x}_1, \ldots, \mathbf{x}_n$ be n linearly independent solutions to the homogeneous system

(5) $\mathbf{x}'(t) = \mathbf{A}(t)\mathbf{x}(t)$

on the interval I, where $\mathbf{A}(t)$ is an $n \times n$ matrix function continuous on I. Then every solution to (5) on I can be expressed in the form

(6) $\mathbf{x}(t) = c_1 \mathbf{x}_1(t) + \cdots + c_n \mathbf{x}_n(t)\,,$

where $c_1, \ldots, c_n$ are constants.

A set of solutions $\{\mathbf{x}_1, \ldots, \mathbf{x}_n\}$ that are linearly independent on I or, equivalently, whose Wronskian does not vanish on I, is called a **fundamental solution set** for (5) on I. The linear combination in (6), written with arbitrary constants, is referred to as a **general solution** to (5).

If we take the vectors in a fundamental solution set and let them form the columns of a matrix $\mathbf{X}(t)$, that is,

$$\mathbf{X}(t) = [\mathbf{x}_1(t)\,\mathbf{x}_2(t) \ldots \mathbf{x}_n(t)] = \begin{bmatrix} x_{1,1}(t) & x_{1,2}(t) & \cdots & x_{1,n}(t) \\ x_{2,1}(t) & x_{2,2}(t) & \cdots & x_{2,n}(t) \\ \vdots & \vdots & & \vdots \\ x_{n,1}(t) & x_{n,2}(t) & \cdots & x_{n,n}(t) \end{bmatrix},$$

then the matrix $\mathbf{X}(t)$ is called a **fundamental matrix** for (5). We can use it to express the general solution (6) as

$$\mathbf{x}(t) = \mathbf{X}(t)\mathbf{c}\,,$$

where $\mathbf{c} = \text{col}(c_1, \ldots, c_n)$ is an arbitrary constant vector. Since $\det \mathbf{X} = W[\mathbf{x}_1, \ldots, \mathbf{x}_n]$ is never zero on I, it follows from Theorem 1 on page 511 that $\mathbf{X}(t)$ is invertible for every t in I.

Example 4 Verify that the set

$$S = \left\{ \begin{bmatrix} e^{2t} \\ e^{2t} \\ e^{2t} \end{bmatrix}, \begin{bmatrix} -e^{-t} \\ 0 \\ e^{-t} \end{bmatrix}, \begin{bmatrix} 0 \\ e^{-t} \\ -e^{-t} \end{bmatrix} \right\}$$

is a fundamental solution set for the system

$$(7) \qquad \mathbf{x}'(t) = \begin{bmatrix} 0 & 1 & 1 \\ 1 & 0 & 1 \\ 1 & 1 & 0 \end{bmatrix} \mathbf{x}(t)$$

on the interval $(-\infty, \infty)$ and find a fundamental matrix for (7). Also determine a general solution for (7).

Solution Substituting the first vector in the set S into the right-hand side of (7) gives

$$\mathbf{Ax} = \begin{bmatrix} 0 & 1 & 1 \\ 1 & 0 & 1 \\ 1 & 1 & 0 \end{bmatrix} \begin{bmatrix} e^{2t} \\ e^{2t} \\ e^{2t} \end{bmatrix} = \begin{bmatrix} 2e^{2t} \\ 2e^{2t} \\ 2e^{2t} \end{bmatrix} = \mathbf{x}'(t) .$$

Hence this vector satisfies system (7) for all t. Similar computations verify that the remaining vectors in S are also solutions to (7) on $(-\infty, \infty)$. For us to show that S is a fundamental solution set, it is enough to observe that the Wronskian

$$W(t) = \begin{vmatrix} e^{2t} & -e^{-t} & 0 \\ e^{2t} & 0 & e^{-t} \\ e^{2t} & e^{-t} & -e^{-t} \end{vmatrix} = e^{2t} \begin{vmatrix} 0 & e^{-t} \\ e^{-t} & -e^{-t} \end{vmatrix} + e^{-t} \begin{vmatrix} e^{2t} & e^{-t} \\ e^{2t} & -e^{-t} \end{vmatrix} = -3$$

is never zero.

A fundamental matrix $\mathbf{X}(t)$ for (7) is just the matrix we used to compute the Wronskian; that is,

$$(8) \qquad \mathbf{X}(t) := \begin{bmatrix} e^{2t} & -e^{-t} & 0 \\ e^{2t} & 0 & e^{-t} \\ e^{2t} & e^{-t} & -e^{-t} \end{bmatrix} .$$

A general solution to (7) can now be expressed as

$$\mathbf{x}(t) = \mathbf{X}(t)\mathbf{c} = c_1 \begin{bmatrix} e^{2t} \\ e^{2t} \\ e^{2t} \end{bmatrix} + c_2 \begin{bmatrix} -e^{-t} \\ 0 \\ e^{-t} \end{bmatrix} + c_3 \begin{bmatrix} 0 \\ e^{-t} \\ -e^{-t} \end{bmatrix} . \quad \blacklozenge$$

It is easy to check that the fundamental matrix in (8) satisfies the equation

$$\mathbf{X}'(t) = \begin{bmatrix} 0 & 1 & 1 \\ 1 & 0 & 1 \\ 1 & 1 & 0 \end{bmatrix} \mathbf{X}(t) ;$$

indeed, this is equivalent to showing that $\mathbf{x}' = \mathbf{Ax}$ for each column $\mathbf{x}$ in S. In general, a fundamental matrix for a system $\mathbf{x}' = \mathbf{Ax}$ satisfies the corresponding **matrix differential equation** $\mathbf{X}' = \mathbf{AX}$.

Another consequence of the linearity of the operator L defined by $L[\mathbf{x}] := \mathbf{x}' - \mathbf{Ax}$ is the **superposition principle** for linear systems. It states that if $\mathbf{x}_1$ and $\mathbf{x}_2$ are solutions, respectively, to the nonhomogeneous systems

$$L[\mathbf{x}] = \mathbf{g}_1 \text{ and } L[\mathbf{x}] = \mathbf{g}_2 ,$$

then $c_1\mathbf{x}_1 + c_2\mathbf{x}_2$ is a solution to

$$L[\mathbf{x}] = c_1\mathbf{g}_1 + c_2\mathbf{g}_2 \, .$$

Using the superposition principle and the representation theorem for homogeneous systems, we can prove the following theorem.

Representation of Solutions (Nonhomogeneous Case)

Theorem 4. If $\mathbf{x}_p$ is a particular solution to the nonhomogeneous system

(9) $\mathbf{x}'(t) = \mathbf{A}(t)\mathbf{x}(t) + \mathbf{f}(t)$

on the interval I and $\{\mathbf{x}_1, \ldots, \mathbf{x}_n\}$ is a fundamental solution set on I for the corresponding homogeneous system $\mathbf{x}(t) = \mathbf{A}(t)\mathbf{x}(t)$, then every solution to (9) on I can be expressed in the form

(10) $\mathbf{x}(t) = \mathbf{x}_p(t) + c_1\mathbf{x}_1(t) + \cdots + c_n\mathbf{x}_n(t) \, ,$

where $c_1, \ldots, c_n$ are constants.

The proof of this theorem is almost identical to the proofs of Theorem 4 in Section 4.5 (page 182) and Theorem 4 in Section 6.1 (page 325). We leave the proof as an exercise.

The linear combination of $\mathbf{x}_p, \mathbf{x}_1, \ldots, \mathbf{x}_n$ in (10) written with arbitrary constants $c_1, \ldots, c_n$ is called a **general solution** of (9). This general solution can also be expressed as $\mathbf{x} = \mathbf{x}_p + \mathbf{X}\mathbf{c}$, where $\mathbf{X}$ is a fundamental matrix for the homogeneous system and $\mathbf{c}$ is an arbitrary constant vector.

We now summarize the results of this section as they apply to the problem of finding a general solution to a system of n linear first-order differential equations in normal form.

Approach to Solving Normal Systems

1. To determine a general solution to the $n \times n$ homogeneous system $\mathbf{x}' = \mathbf{A}\mathbf{x}$:

 (a) Find a fundamental solution set $\{\mathbf{x}_1, \ldots, \mathbf{x}_n\}$ that consists of n linearly independent solutions to the homogeneous system.
 (b) Form the linear combination

 $\mathbf{x} = \mathbf{X}\mathbf{c} = c_1\mathbf{x}_1 + \cdots + c_n\mathbf{x}_n \, ,$

 where $\mathbf{c} = \mathrm{col}(c_1, \ldots, c_n)$ is any constant vector and $\mathbf{X} = \begin{bmatrix} \mathbf{x}_1 \ldots \mathbf{x}_n \end{bmatrix}$ is the fundamental matrix, to obtain a general solution.

2. To determine a general solution to the nonhomogeneous system $\mathbf{x}' = \mathbf{A}\mathbf{x} + \mathbf{f}$:

 (a) Find a particular solution $\mathbf{x}_p$ to the nonhomogeneous system.
 (b) Form the sum of the particular solution and the general solution $\mathbf{X}\mathbf{c} = c_1\mathbf{x}_1 + \cdots + c_n\mathbf{x}_n$ to the corresponding homogeneous system in part 1,

 $\mathbf{x} = \mathbf{x}_p + \mathbf{X}\mathbf{c} = \mathbf{x}_p + c_1\mathbf{x}_1 + \cdots + c_n\mathbf{x}_n \, ,$

 to obtain a general solution to the given system.

We devote the rest of this chapter to methods for finding fundamental solution sets for homogeneous systems and particular solutions for nonhomogeneous systems.

9.4 EXERCISES

In Problems 1–4, write the given system in the matrix form
$\mathbf{x}' = \mathbf{A}\mathbf{x} + \mathbf{f}$.

1. $x'(t) = 3x(t) - y(t) + t^2$,
$\quad y'(t) = -x(t) + 2y(t) + e^t$

2. $r'(t) = 2r(t) + \sin t$,
$\quad \theta'(t) = r(t) - \theta(t) + 1$

3. $\dfrac{dx}{dt} = t^2 x - y - z + t$,

$\quad \dfrac{dy}{dt} = e^t z + 5$,

$\quad \dfrac{dz}{dt} = tx - y + 3z - e^t$

4. $\dfrac{dx}{dt} = x + y + z$,

$\quad \dfrac{dy}{dt} = 2x - y + 3z$,

$\quad \dfrac{dz}{dt} = x + 5z$

In Problems 5–8, rewrite the given scalar equation as a first-order system in normal form. Express the system in the matrix form $\mathbf{x}' = \mathbf{A}\mathbf{x} + \mathbf{f}$.

5. $y''(t) - 3y'(t) - 10y(t) = \sin t$

6. $x''(t) + x(t) = t^2$

7. $\dfrac{d^4 w}{dt^4} + w = t^2$

8. $\dfrac{d^3 y}{dt^3} - \dfrac{dy}{dt} + y = \cos t$

In Problems 9–12, write the given system as a set of scalar equations.

9. $\mathbf{x}' = \begin{bmatrix} 5 & 0 \\ -2 & 4 \end{bmatrix} \mathbf{x} + e^{-2t} \begin{bmatrix} 2 \\ -3 \end{bmatrix}$

10. $\mathbf{x}' = \begin{bmatrix} 2 & 1 \\ -1 & 3 \end{bmatrix} \mathbf{x} + e^t \begin{bmatrix} t \\ 1 \end{bmatrix}$

11. $\mathbf{x}' = \begin{bmatrix} 1 & 0 & 1 \\ -1 & 2 & 5 \\ 0 & 5 & 1 \end{bmatrix} \mathbf{x} + e^t \begin{bmatrix} 1 \\ 0 \\ 0 \end{bmatrix} + t \begin{bmatrix} 0 \\ 1 \\ 0 \end{bmatrix}$

12. $\mathbf{x}' = \begin{bmatrix} 0 & 1 & 0 \\ 0 & 0 & 1 \\ -1 & 1 & 2 \end{bmatrix} \mathbf{x} + t \begin{bmatrix} 1 \\ -1 \\ 2 \end{bmatrix} + \begin{bmatrix} 3 \\ 1 \\ 0 \end{bmatrix}$

In Problems 13–19, determine whether the given vector functions are linearly dependent (LD) or linearly independent (LI) on the interval $(-\infty, \infty)$.

13. $\begin{bmatrix} t \\ 3 \end{bmatrix}, \begin{bmatrix} 4 \\ 1 \end{bmatrix}$

14. $\begin{bmatrix} te^{-t} \\ e^{-t} \end{bmatrix}, \begin{bmatrix} e^{-t} \\ e^{-t} \end{bmatrix}$

15. $e^t \begin{bmatrix} 1 \\ 5 \end{bmatrix}, e^t \begin{bmatrix} -3 \\ -15 \end{bmatrix}$

16. $\begin{bmatrix} \sin t \\ \cos t \end{bmatrix}, \begin{bmatrix} \sin 2t \\ \cos 2t \end{bmatrix}$

17. $e^{2t} \begin{bmatrix} 1 \\ 0 \\ 5 \end{bmatrix}, e^{2t} \begin{bmatrix} 1 \\ 1 \\ -1 \end{bmatrix}, e^{3t} \begin{bmatrix} 0 \\ 1 \\ 0 \end{bmatrix}$

18. $\begin{bmatrix} \sin t \\ \cos t \end{bmatrix}, \begin{bmatrix} \sin t \\ \sin t \end{bmatrix}, \begin{bmatrix} \cos t \\ \cos t \end{bmatrix}$

19. $\begin{bmatrix} 1 \\ 0 \\ 1 \end{bmatrix}, \begin{bmatrix} t \\ 0 \\ t \end{bmatrix}, \begin{bmatrix} t^2 \\ 0 \\ t^2 \end{bmatrix}$

20. Let

$$\mathbf{x}_1 = \begin{bmatrix} \cos t \\ 0 \\ 0 \end{bmatrix}, \quad \mathbf{x}_2 = \begin{bmatrix} \sin t \\ \cos t \\ \cos t \end{bmatrix}, \quad \mathbf{x}_3 = \begin{bmatrix} \cos t \\ \sin t \\ \cos t \end{bmatrix}.$$

(a) Compute the Wronskian.

(b) Are these vector functions linearly independent on $(-\infty, \infty)$?

(c) Is there a first-order homogeneous linear system for which these functions are solutions?

In Problems 21–24, the given vector functions are solutions to a system $\mathbf{x}'(t) = \mathbf{A}\mathbf{x}(t)$. Determine whether they form a fundamental solution set. If they do, find a fundamental matrix for the system and give a general solution.

21. $\mathbf{x}_1 = e^{2t} \begin{bmatrix} 1 \\ -2 \end{bmatrix}, \quad \mathbf{x}_2 = e^{2t} \begin{bmatrix} -2 \\ 4 \end{bmatrix}$

22. $\mathbf{x}_1 = e^{-t} \begin{bmatrix} 3 \\ 2 \end{bmatrix}, \quad \mathbf{x}_2 = e^{4t} \begin{bmatrix} 1 \\ -1 \end{bmatrix}$

23. $\mathbf{x}_1 = \begin{bmatrix} e^{-t} \\ 2e^{-t} \\ e^{-t} \end{bmatrix}, \quad \mathbf{x}_2 = \begin{bmatrix} e^t \\ 0 \\ e^t \end{bmatrix}, \quad \mathbf{x}_3 = \begin{bmatrix} e^{3t} \\ -e^{3t} \\ 2e^{3t} \end{bmatrix}$

24. $\mathbf{x}_1 = \begin{bmatrix} e^t \\ e^t \\ e^t \end{bmatrix}, \quad \mathbf{x}_2 = \begin{bmatrix} \sin t \\ \cos t \\ -\sin t \end{bmatrix}, \quad \mathbf{x}_3 = \begin{bmatrix} -\cos t \\ \sin t \\ \cos t \end{bmatrix}$

25. Verify that the vector functions

$$\mathbf{x}_1 = \begin{bmatrix} e^t \\ e^t \end{bmatrix} \quad \text{and} \quad \mathbf{x}_2 = \begin{bmatrix} e^{-t} \\ 3e^{-t} \end{bmatrix}$$

are solutions to the homogeneous system

$$\mathbf{x}' = \mathbf{A}\mathbf{x} = \begin{bmatrix} 2 & -1 \\ 3 & -2 \end{bmatrix} \mathbf{x},$$

on $(-\infty, \infty)$, and that

$$\mathbf{x}_p = \frac{3}{2} \begin{bmatrix} te^t \\ te^t \end{bmatrix} - \frac{1}{4} \begin{bmatrix} e^t \\ 3e^t \end{bmatrix} + \begin{bmatrix} t \\ 2t \end{bmatrix} - \begin{bmatrix} 0 \\ 1 \end{bmatrix}$$

is a particular solution to the nonhomogeneous system $\mathbf{x}' = \mathbf{A}\mathbf{x} + \mathbf{f}(t)$, where $\mathbf{f}(t) = \operatorname{col}(e^t, t)$. Find a general solution to $\mathbf{x}' = \mathbf{A}\mathbf{x} + \mathbf{f}(t)$.

26. Verify that the vector functions

$$\mathbf{x}_1 = \begin{bmatrix} e^{3t} \\ 0 \\ e^{3t} \end{bmatrix}, \quad \mathbf{x}_2 = \begin{bmatrix} -e^{3t} \\ e^{3t} \\ 0 \end{bmatrix}, \quad \mathbf{x}_3 = \begin{bmatrix} -e^{-3t} \\ -e^{-3t} \\ e^{-3t} \end{bmatrix}$$

are solutions to the homogeneous system

$$\mathbf{x}' = \mathbf{A}\mathbf{x} = \begin{bmatrix} 1 & -2 & 2 \\ -2 & 1 & 2 \\ 2 & 2 & 1 \end{bmatrix} \mathbf{x}$$

on $(-\infty, \infty)$, and that

$$\mathbf{x}_p = \begin{bmatrix} 5t+1 \\ 2t \\ 4t+2 \end{bmatrix}$$

is a particular solution to $\mathbf{x}' = \mathbf{A}\mathbf{x} + \mathbf{f}(t)$, where $\mathbf{f}(t) = \mathrm{col}(-9t, 0, -18t)$. Find a general solution to $\mathbf{x}' = \mathbf{A}\mathbf{x} + \mathbf{f}(t)$.

27. Prove that the operator defined by $L[\mathbf{x}] := \mathbf{x}' - \mathbf{A}\mathbf{x}$, where $\mathbf{A}$ is an $n \times n$ matrix function and $\mathbf{x}$ is an $n \times 1$ differentiable vector function, is a linear operator.

28. Let $\mathbf{X}(t)$ be a fundamental matrix for the system $\mathbf{x}' = \mathbf{A}\mathbf{x}$. Show that $\mathbf{x}(t) = \mathbf{X}(t)\mathbf{X}^{-1}(t_0)\mathbf{x}_0$ is the solution to the initial value problem $\mathbf{x}' = \mathbf{A}\mathbf{x}$, $\mathbf{x}(t_0) = \mathbf{x}_0$.

In Problems 29–30, verify that $\mathbf{X}(t)$ is a fundamental matrix for the given system and compute $\mathbf{X}^{-1}(t)$. Use the result of Problem 28 to find the solution to the given initial value problem.

29. $\mathbf{x}' = \begin{bmatrix} 0 & 6 & 0 \\ 1 & 0 & 1 \\ 1 & 1 & 0 \end{bmatrix} \mathbf{x}$, $\quad \mathbf{x}(0) = \begin{bmatrix} -1 \\ 0 \\ 1 \end{bmatrix}$;

$$\mathbf{X}(t) = \begin{bmatrix} 6e^{-t} & -3e^{-2t} & 2e^{3t} \\ -e^{-t} & e^{-2t} & e^{3t} \\ -5e^{-t} & e^{-2t} & e^{3t} \end{bmatrix}$$

30. $\mathbf{x}' = \begin{bmatrix} 2 & 3 \\ 3 & 2 \end{bmatrix} \mathbf{x}$, $\quad \mathbf{x}(0) = \begin{bmatrix} 3 \\ -1 \end{bmatrix}$;

$$\mathbf{X}(t) = \begin{bmatrix} e^{-t} & e^{5t} \\ -e^{-t} & e^{5t} \end{bmatrix}$$

31. Show that

$$\begin{vmatrix} t^2 & t|t| \\ 2t & 2|t| \end{vmatrix} \equiv 0$$

on $(-\infty, \infty)$, but that the two vector functions

$$\begin{bmatrix} t^2 \\ 2t \end{bmatrix}, \quad \begin{bmatrix} t|t| \\ 2|t| \end{bmatrix}$$

are linearly independent on $(-\infty, \infty)$.

32. **Abel's Formula.** If $\mathbf{x}_1, \ldots, \mathbf{x}_n$ are any n solutions to the $n \times n$ system $\mathbf{x}'(t) = \mathbf{A}(t)\mathbf{x}(t)$, then Abel's formula gives a representation for the Wronskian $W(t) := W[\mathbf{x}_1, \ldots, \mathbf{x}_n](t)$. Namely,

$$W(t) = W(t_0)\exp\left(\int_{t_0}^{t} \{a_{11}(s) + \cdots + a_{nn}(s)\}\, ds \right),$$

where $a_{11}(s), \ldots, a_{nn}(s)$ are the main diagonal elements of $\mathbf{A}(s)$. Prove this formula in the special case when $n = 3$. [*Hint:* Follow the outline in Problem 30 of Exercises 6.1, page 327.]

33. Using Abel's formula (Problem 32), confirm that the Wronskian of n solutions to $\mathbf{x}' = \mathbf{A}\mathbf{x}$ on the interval I is either identically zero on I or never zero on I.

34. Prove that a fundamental solution set for the homogeneous system $\mathbf{x}'(t) = \mathbf{A}(t)\mathbf{x}(t)$ always exists on an interval I, provided $\mathbf{A}(t)$ is continuous on I. [*Hint:* Use the existence and uniqueness theorem (Theorem 2) and make judicious choices for $\mathbf{x}_0$.]

35. Prove Theorem 3 on the representation of solutions of the homogeneous system.

36. Prove Theorem 4 on the representation of solutions of the nonhomogeneous system.

37. To illustrate the connection between a higher-order equation and the equivalent first-order system, consider the equation

(11) $y'''(t) - 6y''(t) + 11y'(t) - 6y(t) = 0$.

(a) Show that $\{e^t, e^{2t}, e^{3t}\}$ is a fundamental solution set for (11).

(b) Using the definition in Section 6.1, compute the Wronskian of $\{e^t, e^{2t}, e^{3t}\}$.

(c) Setting $x_1 = y, x_2 = y', x_3 = y''$, show that equation (11) is equivalent to the first-order system

(12) $\mathbf{x}' = \mathbf{A}\mathbf{x}$,

where

$$\mathbf{A} := \begin{bmatrix} 0 & 1 & 0 \\ 0 & 0 & 1 \\ 6 & -11 & 6 \end{bmatrix}.$$

(d) The substitution used in part (c) suggests that

$$S := \left\{ \begin{bmatrix} e^t \\ e^t \\ e^t \end{bmatrix}, \begin{bmatrix} e^{2t} \\ 2e^{2t} \\ 4e^{2t} \end{bmatrix}, \begin{bmatrix} e^{3t} \\ 3e^{3t} \\ 9e^{3t} \end{bmatrix} \right\}$$

is a fundamental solution set for system (12). Verify that this is the case.

(e) Compute the Wronskian of S. How does it compare with the Wronskian computed in part (b)?

38. Define $\mathbf{x}_1(t)$, $\mathbf{x}_2(t)$, and $\mathbf{x}_3(t)$, for $-\infty < t < \infty$, by

$$\mathbf{x}_1(t) = \begin{bmatrix} \sin t \\ \sin t \\ 0 \end{bmatrix}, \quad \mathbf{x}_2(t) = \begin{bmatrix} \sin t \\ 0 \\ \sin t \end{bmatrix}, \quad \mathbf{x}_3(t) = \begin{bmatrix} 0 \\ \sin t \\ \sin t \end{bmatrix}.$$

(a) Show that for the three scalar functions *in each individual row* there are nontrivial linear combinations that sum to zero for all t.

(b) Show that, nonetheless, the three vector functions are linearly independent. (No single nontrivial combination works *for each row, for all t*.)

(c) Calculate the Wronskian $W[\mathbf{x}_1, \mathbf{x}_2, \mathbf{x}_3](t)$.

(d) Is there a linear third-order homogeneous differential equation system having $\mathbf{x}_1(t), \mathbf{x}_2(t)$, and $\mathbf{x}_3(t)$ as solutions?

9.5 Homogeneous Linear Systems with Constant Coefficients

In this section we discuss a procedure for obtaining a general solution for the homogeneous system

(1) $\mathbf{x}'(t) = \mathbf{A}\mathbf{x}(t)$,

where $\mathbf{A}$ is a (real) *constant* $n \times n$ matrix. The general solution we seek will be defined for all t because the elements of $\mathbf{A}$ are just constant functions, which are continuous on $(-\infty, \infty)$ (recall Theorem 2, page 516). In Section 9.4 we showed that a general solution to (1) can be constructed from a fundamental solution set consisting of n linearly independent solutions to (1). Thus our goal is to find n such vector solutions.

In Chapter 4 we were successful in solving homogeneous linear equations with constant coefficients by guessing that the equation had a solution of the form e^{rt}. Because any scalar linear equation can be expressed as a system, it is reasonable to expect system (1) to have solutions of the form

$$\mathbf{x}(t) = e^{rt}\mathbf{u},$$

where r is a constant and $\mathbf{u}$ is a constant vector, both of which must be determined. Substituting $e^{rt}\mathbf{u}$ for $\mathbf{x}(t)$ in (1) gives

$$re^{rt}\mathbf{u} = \mathbf{A}e^{rt}\mathbf{u} = e^{rt}\mathbf{A}\mathbf{u}.$$

Canceling the factor e^{rt} and rearranging terms, we find that

(2) $(\mathbf{A} - r\mathbf{I})\mathbf{u} = \mathbf{0}$,

where $r\mathbf{I}$ denotes the diagonal matrix with r's along its main diagonal.

The preceding calculation shows that $\mathbf{x}(t) = e^{rt}\mathbf{u}$ is a solution to (1) if and only if r and $\mathbf{u}$ satisfy equation (2). Since the trivial case, $\mathbf{u} = \mathbf{0}$, is of no help in finding linearly independent solutions to (1), we require that $\mathbf{u} \neq \mathbf{0}$. Such vectors are given a special name, as follows.

Eigenvalues and Eigenvectors

Definition 3. Let $\mathbf{A} = [a_{ij}]$ be an $n \times n$ constant matrix. The **eigenvalues** of $\mathbf{A}$ are those (real or complex) numbers r for which $(\mathbf{A} - r\mathbf{I})\mathbf{u} = \mathbf{0}$ has at least one nontrivial (real or complex) solution $\mathbf{u}$. The corresponding nontrivial solutions $\mathbf{u}$ are called the **eigenvectors** of $\mathbf{A}$ associated with r.

As stated in Theorem 1 of Section 9.3, a linear homogeneous system of n *algebraic* equations in n unknowns has a nontrivial solution if and only if the determinant of its coefficients is zero. Hence, a necessary and sufficient condition for (2) to have a nontrivial solution is that

(3) $|\mathbf{A} - r\mathbf{I}| = 0$.

Expanding the determinant of $\mathbf{A} - r\mathbf{I}$ in terms of its cofactors, we find that it is an nth-degree polynomial in r; that is,

(4) $|\mathbf{A} - r\mathbf{I}| = p(r)$.

Therefore, *finding the eigenvalues of a matrix* **A** *is equivalent to finding the zeros of the polynomial* $p(r)$. Equation (3) is called the **characteristic equation** of **A**, and $p(r)$ in (4) is the **characteristic polynomial** of **A**. The characteristic equation plays a role for systems similar to the role played by the auxiliary equation for scalar equations.

Many commercially available software packages can be used to compute the eigenvalues and eigenvectors for a given matrix. Three such packages are MATLAB®, available from The MathWorks, Inc.; MATHEMATICA®, available from Wolfram Research; and MAPLESOFT®, available from Waterloo Maple Inc. Although you are encouraged to make use of such packages, the examples and most exercises in this text can be easily carried out without them. Those exercises for which a computer package is desirable are flagged with the icon 🖥 .

Example 1 Find the eigenvalues and eigenvectors of the matrix

$$\mathbf{A} := \begin{bmatrix} 2 & -3 \\ 1 & -2 \end{bmatrix}.$$

Solution The characteristic equation for **A** is

$$|\mathbf{A} - r\mathbf{I}| = \begin{vmatrix} 2-r & -3 \\ 1 & -2-r \end{vmatrix} = (2-r)(-2-r) + 3 = r^2 - 1 = 0.$$

Hence the eigenvalues of **A** are $r_1 = 1$, $r_2 = -1$. To find the eigenvectors corresponding to $r_1 = 1$, we must solve $(\mathbf{A} - r_1\mathbf{I})\mathbf{u} = \mathbf{0}$. Substituting for **A** and r_1 gives

(5)
$$\begin{bmatrix} 1 & -3 \\ 1 & -3 \end{bmatrix} \begin{bmatrix} u_1 \\ u_2 \end{bmatrix} = \begin{bmatrix} 0 \\ 0 \end{bmatrix}.$$

Notice that this matrix equation is equivalent to the single scalar equation $u_1 - 3u_2 = 0$. Therefore, the solutions to (5) are obtained by assigning an arbitrary value for u_2 (say, $u_2 = s$) and setting $u_1 = 3u_2 = 3s$. Consequently, the eigenvectors associated with $r_1 = 1$ can be expressed as

(6) $$\mathbf{u}_1 = s \begin{bmatrix} 3 \\ 1 \end{bmatrix}.$$

For $r_2 = -1$, the equation $(\mathbf{A} - r_2\mathbf{I})\mathbf{u} = \mathbf{0}$ becomes

$$\begin{bmatrix} 3 & -3 \\ 1 & -1 \end{bmatrix} \begin{bmatrix} u_1 \\ u_2 \end{bmatrix} = \begin{bmatrix} 0 \\ 0 \end{bmatrix}.$$

Solving, we obtain $u_1 = s$ and $u_2 = s$, with s arbitrary. Therefore, the eigenvectors associated with the eigenvalue $r_2 = -1$ are

(7) $$\mathbf{u}_2 = s \begin{bmatrix} 1 \\ 1 \end{bmatrix}. \quad \blacklozenge$$

We remark that in the above example the collection (6) of all eigenvectors associated with $r_1 = 1$ forms a one-dimensional subspace when the zero vector is adjoined. The same is true for $r_2 = -1$. These subspaces are called **eigenspaces**.

Example 2 Find the eigenvalues and eigenvectors of the matrix

$$\mathbf{A} := \begin{bmatrix} 1 & 2 & -1 \\ 1 & 0 & 1 \\ 4 & -4 & 5 \end{bmatrix}.$$

Solution The characteristic equation for $\mathbf{A}$ is

$$|\mathbf{A} - r\mathbf{I}| = \begin{vmatrix} 1-r & 2 & -1 \\ 1 & -r & 1 \\ 4 & -4 & 5-r \end{vmatrix} = 0,$$

which simplifies to $(r-1)(r-2)(r-3) = 0$. Hence, the eigenvalues of $\mathbf{A}$ are $r_1 = 1$, $r_2 = 2$, and $r_3 = 3$. To find the eigenvectors corresponding to $r_1 = 1$, we set $r = 1$ in $(\mathbf{A} - r\mathbf{I})\mathbf{u} = \mathbf{0}$. This gives

$$(8) \qquad \begin{bmatrix} 0 & 2 & -1 \\ 1 & -1 & 1 \\ 4 & -4 & 4 \end{bmatrix} \begin{bmatrix} u_1 \\ u_2 \\ u_3 \end{bmatrix} = \begin{bmatrix} 0 \\ 0 \\ 0 \end{bmatrix}.$$

Using elementary row operations (Gaussian elimination), we see that (8) is equivalent to the two equations

$$u_1 - u_2 + u_3 = 0,$$
$$2u_2 - u_3 = 0.$$

Thus, we can obtain the solutions to (8) by assigning an arbitrary value to u_2 (say, $u_2 = s$), solving $2u_2 - u_3 = 0$ for u_3 to get $u_3 = 2s$, and then solving $u_1 - u_2 + u_3 = 0$ for u_1 to get $u_1 = -s$. Hence, the eigenvectors associated with $r_1 = 1$ are

$$(9) \qquad \mathbf{u}_1 = s \begin{bmatrix} -1 \\ 1 \\ 2 \end{bmatrix}.$$

For $r_2 = 2$, we solve

$$\begin{bmatrix} -1 & 2 & -1 \\ 1 & -2 & 1 \\ 4 & -4 & 3 \end{bmatrix} \begin{bmatrix} u_1 \\ u_2 \\ u_3 \end{bmatrix} = \begin{bmatrix} 0 \\ 0 \\ 0 \end{bmatrix}$$

in a similar fashion to obtain the eigenvectors

$$(10) \qquad \mathbf{u}_2 = s \begin{bmatrix} -2 \\ 1 \\ 4 \end{bmatrix}.$$

Finally, for $r_3 = 3$, we solve

$$\begin{bmatrix} -2 & 2 & -1 \\ 1 & -3 & 1 \\ 4 & -4 & 2 \end{bmatrix} \begin{bmatrix} u_1 \\ u_2 \\ u_3 \end{bmatrix} = \begin{bmatrix} 0 \\ 0 \\ 0 \end{bmatrix}$$

and get the eigenvectors

$$(11) \qquad \mathbf{u}_3 = s \begin{bmatrix} -1 \\ 1 \\ 4 \end{bmatrix}. \quad \blacklozenge$$

Let's return to the problem of finding a general solution to a homogeneous system of differential equations. We have already shown that $e^{rt}\mathbf{u}$ is a solution to (1) if r is an eigenvalue and $\mathbf{u}$ a corresponding eigenvector. The question is: Can we obtain n linearly independent solutions to the homogeneous system by finding all the eigenvalues and eigenvectors of $\mathbf{A}$? The answer is yes, *if* $\mathbf{A}$ *has n linearly independent eigenvectors.*

n Linearly Independent Eigenvectors

Theorem 5. Suppose the $n \times n$ constant matrix $\mathbf{A}$ has n linearly independent eigenvectors $\mathbf{u}_1, \mathbf{u}_2, \ldots, \mathbf{u}_n$. Let r_i be the eigenvalue[†] corresponding to $\mathbf{u}_i$. Then

(12) $\{e^{r_1 t}\mathbf{u}_1, e^{r_2 t}\mathbf{u}_2, \ldots, e^{r_n t}\mathbf{u}_n\}$

is a fundamental solution set (and $\mathbf{X}(t) = \begin{bmatrix} e^{r_1 t}\mathbf{u}_1 & e^{r_2 t}\mathbf{u}_2 & \cdots & e^{r_n t}\mathbf{u}_n \end{bmatrix}$ is a fundamental matrix) on $(-\infty, \infty)$ for the homogeneous system $\mathbf{x}' = \mathbf{Ax}$. Consequently, a general solution of $\mathbf{x}' = \mathbf{Ax}$ is

(13) $\mathbf{x}(t) = c_1 e^{r_1 t}\mathbf{u}_1 + c_2 e^{r_2 t}\mathbf{u}_2 + \cdots + c_n e^{r_n t}\mathbf{u}_n$,

where $c_1, \ldots, c_n$ are arbitrary constants.

Proof. As we have seen, the vector functions listed in (12) are solutions to the homogeneous system. Moreover, their Wronskian is

$$W(t) = \det[e^{r_1 t}\mathbf{u}_1, \ldots, e^{r_n t}\mathbf{u}_n] = e^{(r_1 + \cdots + r_n)t}\det[\mathbf{u}_1, \ldots, \mathbf{u}_n].$$

Since the eigenvectors are assumed to be linearly independent, it follows from Theorem 1 in Section 9.3 that $\det[\mathbf{u}_1, \ldots, \mathbf{u}_n]$ is not zero. Hence the Wronskian $W(t)$ is never zero. This shows that (12) is a fundamental solution set, and consequently a general solution is given by (13). ◆

An application of Theorem 5 is given in the next example.

Example 3 Find a general solution of

(14) $\mathbf{x}'(t) = \mathbf{Ax}(t)$, where $\mathbf{A} = \begin{bmatrix} 2 & -3 \\ 1 & -2 \end{bmatrix}$.

Solution In Example 1 we showed that the matrix $\mathbf{A}$ has eigenvalues $r_1 = 1$ and $r_2 = -1$. Taking, say, $s = 1$ in equations (6) and (7), we get the corresponding eigenvectors

$$\mathbf{u}_1 = \begin{bmatrix} 3 \\ 1 \end{bmatrix} \quad \text{and} \quad \mathbf{u}_2 = \begin{bmatrix} 1 \\ 1 \end{bmatrix}.$$

Because $\mathbf{u}_1$ and $\mathbf{u}_2$ are linearly independent, it follows from Theorem 5 that a general solution to (14) is

(15) $\mathbf{x}(t) = c_1 e^t \begin{bmatrix} 3 \\ 1 \end{bmatrix} + c_2 e^{-t} \begin{bmatrix} 1 \\ 1 \end{bmatrix}$. ◆

[†]The eigenvalues $r_1, \ldots, r_n$ may be real or complex and need not be distinct. In this section the cases we discuss have real eigenvalues. We consider complex eigenvalues in Section 9.6.

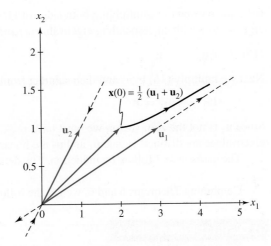

Figure 9.1 Trajectories of solutions for Example 3

If we sum the vectors on the right-hand side of equation (15) and then write out the expressions for the components of $\mathbf{x}(t) = \mathrm{col}\big(x_1(t), x_2(t)\big)$, we get

$$x_1(t) = 3c_1e^t + c_2e^{-t},$$
$$x_2(t) = c_1e^t + c_2e^{-t}.$$

This is the familiar form of a general solution for a system, as discussed in Section 5.2.

Example 3 nicely illustrates the geometric role played by the eigenvectors $\mathbf{u}_1$ and $\mathbf{u}_2$. If the initial vector $\mathbf{x}(0)$ is a scalar multiple of $\mathbf{u}_1$ (i.e., $\mathbf{x}(0) = c_1\mathbf{u}_1$), then the vector solution to the system, $\mathbf{x}(t) = c_1e^t\mathbf{u}_1$, will always have the same or opposite direction as $\mathbf{u}_1$. That is, it will lie along the straight line determined by $\mathbf{u}_1$ (see Figure 9.1). Furthermore, the trajectory of this solution, as t increases, will tend to infinity, since the corresponding eigenvalue $r_1 = 1$ is positive (observe the e^t term). A similar assertion holds if the initial vector is a scalar multiple of $\mathbf{u}_2$, except that since $r_2 = -1$ is negative, the trajectory $\mathbf{x}(t) = c_2e^{-t}\mathbf{u}_2$ will approach the origin as t increases (because of e^{-t}). For an initial vector that involves both $\mathbf{u}_1$ and $\mathbf{u}_2$, such as $\mathbf{x}(0) = \frac{1}{2}(\mathbf{u}_1 + \mathbf{u}_2)$, the resulting trajectory is a blend of the above motions, with the contribution due to the larger eigenvalue $r_1 = 1$ dominating as t increases; see Figure 9.1.

The straight-line trajectories in the x_1x_2-plane (the phase plane), then, point along the directions of the eigenvectors of the matrix $\mathbf{A}$. (See Section 5.4, Figure 5.11, page 266, for example.)

A useful property of eigenvectors that concerns their linear independence is stated in the next theorem.

Linear Independence of Eigenvectors

Theorem 6. If $r_1, \ldots, r_m$ are *distinct* eigenvalues for the matrix $\mathbf{A}$ and $\mathbf{u}_i$ is an eigenvector associated with r_i, then $\mathbf{u}_1, \ldots, \mathbf{u}_m$ are linearly independent.

Proof. Let's first treat the case $m = 2$. Suppose, to the contrary, that $\mathbf{u}_1$ and $\mathbf{u}_2$ are linearly dependent so that

$$(16) \qquad \mathbf{u}_1 = c\mathbf{u}_2$$

for some constant c. Multiplying both sides of (16) by $\mathbf{A}$ and using the fact that $\mathbf{u}_1$ and $\mathbf{u}_2$ are eigenvectors with corresponding eigenvalues r_1 and r_2, we obtain

(17) $r_1\mathbf{u}_1 = cr_2\mathbf{u}_2$.

Next we multiply (16) by r_2 and then subtract from (17) to get

$$(r_1 - r_2)\mathbf{u}_1 = \mathbf{0} .$$

Since $\mathbf{u}_1$ is not the zero vector, we must have $r_1 = r_2$. But this violates the assumption that the eigenvalues are distinct! Hence $\mathbf{u}_1$ and $\mathbf{u}_2$ are linearly independent.

The cases $m > 2$ follow by induction. The details of the proof are left as Problem 48. ◆

Combining Theorems 5 and 6, we get the following corollary.

n Distinct Eigenvalues

Corollary 1. If the $n \times n$ constant matrix $\mathbf{A}$ has n distinct eigenvalues $r_1, \ldots, r_n$ and $\mathbf{u}_i$ is an eigenvector associated with r_i, then

$$\{e^{r_1 t}\mathbf{u}_1, \ldots, e^{r_n t}\mathbf{u}_n\}$$

is a fundamental solution set for the homogeneous system $\mathbf{x}' = \mathbf{A}\mathbf{x}$.

Example 4 Solve the initial value problem

$$(18) \quad \mathbf{x}'(t) = \begin{bmatrix} 1 & 2 & -1 \\ 1 & 0 & 1 \\ 4 & -4 & 5 \end{bmatrix} \mathbf{x}(t), \quad \mathbf{x}(0) = \begin{bmatrix} -1 \\ 0 \\ 0 \end{bmatrix}.$$

Solution In Example 2 we showed that the 3×3 coefficient matrix $\mathbf{A}$ has the three distinct eigenvalues $r_1 = 1, r_2 = 2$, and $r_3 = 3$. If we set $s = 1$ in equations (9), (10), and (11), we obtain the corresponding eigenvectors

$$\mathbf{u}_1 = \begin{bmatrix} -1 \\ 1 \\ 2 \end{bmatrix}, \quad \mathbf{u}_2 = \begin{bmatrix} -2 \\ 1 \\ 4 \end{bmatrix}, \quad \mathbf{u}_3 = \begin{bmatrix} -1 \\ 1 \\ 4 \end{bmatrix},$$

whose linear independence is guaranteed by Theorem 6. Hence, a general solution to (18) is

$$(19) \quad \mathbf{x}(t) = c_1 e^t \begin{bmatrix} -1 \\ 1 \\ 2 \end{bmatrix} + c_2 e^{2t} \begin{bmatrix} -2 \\ 1 \\ 4 \end{bmatrix} + c_3 e^{3t} \begin{bmatrix} -1 \\ 1 \\ 4 \end{bmatrix}$$

$$= \begin{bmatrix} -e^t & -2e^{2t} & -e^{3t} \\ e^t & e^{2t} & e^{3t} \\ 2e^t & 4e^{2t} & 4e^{3t} \end{bmatrix} \begin{bmatrix} c_1 \\ c_2 \\ c_3 \end{bmatrix}.$$

To satisfy the initial condition in (18), we solve

$$\mathbf{x}(0) = \begin{bmatrix} -1 & -2 & -1 \\ 1 & 1 & 1 \\ 2 & 4 & 4 \end{bmatrix} \begin{bmatrix} c_1 \\ c_2 \\ c_3 \end{bmatrix} = \begin{bmatrix} -1 \\ 0 \\ 0 \end{bmatrix}$$

and find that $c_1 = 0, c_2 = 1$, and $c_3 = -1$. Inserting these values into (19) gives the desired solution. ◆

There is a special class of $n \times n$ matrices that *always* have n linearly independent eigenvectors. These are the real symmetric matrices.

Real Symmetric Matrices

Definition 4. A **real symmetric matrix A** is a matrix with real entries that satisfies $A^T = A$.

Taking the transpose of a matrix interchanges its rows and columns. Doing this is equivalent to "flipping" the matrix about its main diagonal. Consequently, $A^T = A$ if and only if A is symmetric about its main diagonal.

If A is an $n \times n$ real symmetric matrix, it is known[†] that all its eigenvalues are real and that there always exist n linearly independent eigenvectors. In such a case, Theorem 5 applies and a general solution to $x' = Ax$ is given by (13).

Example 5 Find a general solution of

$$(20) \quad x'(t) = Ax(t) \text{, where } A = \begin{bmatrix} 1 & -2 & 2 \\ -2 & 1 & 2 \\ 2 & 2 & 1 \end{bmatrix}.$$

Solution A is symmetric, so we are assured that A has three linearly independent eigenvectors. To find them, we first compute the characteristic equation for A:

$$|A - rI| = \begin{vmatrix} 1-r & -2 & 2 \\ -2 & 1-r & 2 \\ 2 & 2 & 1-r \end{vmatrix} = -(r-3)^2(r+3) = 0.$$

Thus the eigenvalues of A are $r_1 = r_2 = 3$ and $r_3 = -3$.

Notice that the eigenvalue $r = 3$ has multiplicity 2 when considered as a root of the characteristic equation. Therefore, we must find *two* linearly independent eigenvectors associated with $r = 3$. Substituting $r = 3$ in $(A - rI)u = 0$ gives

$$\begin{bmatrix} -2 & -2 & 2 \\ -2 & -2 & 2 \\ 2 & 2 & -2 \end{bmatrix} \begin{bmatrix} u_1 \\ u_2 \\ u_3 \end{bmatrix} = \begin{bmatrix} 0 \\ 0 \\ 0 \end{bmatrix}.$$

This system is equivalent to the single equation $-u_1 - u_2 + u_3 = 0$, so we can obtain its solutions by assigning an arbitrary value to u_2, say $u_2 = v$, and an arbitrary value to u_3, say $u_3 = s$. Solving for u_1, we find $u_1 = u_3 - u_2 = s - v$. Therefore, the eigenvectors associated with $r_1 = r_2 = 3$ can be expressed as

$$u = \begin{bmatrix} s - v \\ v \\ s \end{bmatrix} = s \begin{bmatrix} 1 \\ 0 \\ 1 \end{bmatrix} + v \begin{bmatrix} -1 \\ 1 \\ 0 \end{bmatrix}.$$

[†]See *Fundamentals of Matrix Analysis with Applications*, by Edward Barry Saff and Arthur David Snider (John Wiley & Sons, Hoboken, New Jersey, 2016).

By first taking $s = 1, v = 0$ and then taking $s = 0, v = 1$, we get the two linearly independent eigenvectors

(21) $\quad \mathbf{u}_1 = \begin{bmatrix} 1 \\ 0 \\ 1 \end{bmatrix}, \quad \mathbf{u}_2 = \begin{bmatrix} -1 \\ 1 \\ 0 \end{bmatrix}.$

For $r_3 = -3$, we solve

$$(\mathbf{A} + 3\mathbf{I})\mathbf{u} = \begin{bmatrix} 4 & -2 & 2 \\ -2 & 4 & 2 \\ 2 & 2 & 4 \end{bmatrix} \begin{bmatrix} u_1 \\ u_2 \\ u_3 \end{bmatrix} = \begin{bmatrix} 0 \\ 0 \\ 0 \end{bmatrix}$$

to obtain the eigenvectors $\text{col}(-s, -s, s)$. Taking $s = 1$ gives

$$\mathbf{u}_3 = \begin{bmatrix} -1 \\ -1 \\ 1 \end{bmatrix}.$$

Since the eigenvectors $\mathbf{u}_1, \mathbf{u}_2$, and $\mathbf{u}_3$ are linearly independent, a general solution to (20) is

$$\mathbf{x}(t) = c_1 e^{3t} \begin{bmatrix} 1 \\ 0 \\ 1 \end{bmatrix} + c_2 e^{3t} \begin{bmatrix} -1 \\ 1 \\ 0 \end{bmatrix} + c_3 e^{-3t} \begin{bmatrix} -1 \\ -1 \\ 1 \end{bmatrix}. \quad \blacklozenge$$

If a matrix $\mathbf{A}$ *is not symmetric, it is possible for* $\mathbf{A}$ *to have a repeated eigenvalue but not to have two linearly independent corresponding eigenvectors.* In particular, the matrix

(22) $\quad \mathbf{A} = \begin{bmatrix} 1 & -1 \\ 4 & -3 \end{bmatrix}$

has the repeated eigenvalue $r_1 = r_2 = -1$, but Problem 35 shows that all the eigenvectors associated with $r = -1$ are of the form $\mathbf{u} = s\, \text{col}(1, 2)$. Consequently, no two eigenvectors are linearly independent.

A procedure for finding a general solution in such a case is illustrated in Problems 35–40, but the underlying theory is deferred to Section 9.8, where we discuss the matrix exponential.

A final note. If an $n \times n$ matrix $\mathbf{A}$ has n linearly independent eigenvectors $\mathbf{u}_i$ with eigenvalues r_i, a little inspection reveals that property (2) is expressed columnwise by the equation

(23) $\begin{bmatrix} & \mathbf{A} & \end{bmatrix} \begin{bmatrix} \vdots & \vdots & \cdots & \vdots \\ \mathbf{u}_1 & \mathbf{u}_2 & \cdots & \mathbf{u}_n \\ \vdots & \vdots & \cdots & \vdots \end{bmatrix} = \begin{bmatrix} \vdots & \vdots & \cdots & \vdots \\ \mathbf{u}_1 & \mathbf{u}_2 & \cdots & \mathbf{u}_n \\ \vdots & \vdots & \cdots & \vdots \end{bmatrix} \begin{bmatrix} r_1 & 0 & \cdots & 0 \\ 0 & r_2 & \cdots & 0 \\ & & \ddots & \\ 0 & 0 & \cdots & r_n \end{bmatrix}$

or $\mathbf{AU} = \mathbf{UD}$, where $\mathbf{U}$ is the matrix whose column vectors are eigenvectors and $\mathbf{D}$ is a diagonal matrix whose diagonal entries are the eigenvalues. Since $\mathbf{U}$'s columns are independent, $\mathbf{U}$ is invertible and we can write

(24) $\quad \mathbf{A} = \mathbf{UDU}^{-1} \text{ or } \mathbf{D} = \mathbf{U}^{-1}\mathbf{AU},$

and we say that $\mathbf{A}$ is *diagonalizable*. [In this context equation (24) expresses a *similarity transformation*.] Because the argument that leads from (2) to (23) to (24) can be reversed, we have a new characterization: *An* $n \times n$ *matrix has n linearly independent eigenvectors if, and only if, it is diagonalizable.*

9.5 EXERCISES

In Problems 1–8, find the eigenvalues and eigenvectors of the given matrix.

1. $\begin{bmatrix} -4 & 2 \\ 2 & -1 \end{bmatrix}$

2. $\begin{bmatrix} 6 & -3 \\ 2 & 1 \end{bmatrix}$

3. $\begin{bmatrix} 1 & -1 \\ 2 & 4 \end{bmatrix}$

4. $\begin{bmatrix} 1 & 5 \\ 1 & -3 \end{bmatrix}$

5. $\begin{bmatrix} 1 & 0 & 0 \\ 0 & 0 & 2 \\ 0 & 2 & 0 \end{bmatrix}$

6. $\begin{bmatrix} 0 & 1 & 1 \\ 1 & 0 & 1 \\ 1 & 1 & 0 \end{bmatrix}$

7. $\begin{bmatrix} 1 & 0 & 0 \\ 2 & 3 & 1 \\ 0 & 2 & 4 \end{bmatrix}$

8. $\begin{bmatrix} -3 & 1 & 0 \\ 0 & -3 & 1 \\ 4 & -8 & 2 \end{bmatrix}$

In Problems 9 and 10, some of the eigenvalues of the given matrix are complex. Find all the eigenvalues and eigenvectors.

9. $\begin{bmatrix} 0 & -1 \\ 1 & 0 \end{bmatrix}$

10. $\begin{bmatrix} 1 & 2 & -1 \\ 0 & 1 & 1 \\ 0 & -1 & 1 \end{bmatrix}$

In Problems 11–16, find a general solution of the system $\mathbf{x}'(t) = \mathbf{A}\mathbf{x}(t)$ for the given matrix $\mathbf{A}$.

11. $\mathbf{A} = \begin{bmatrix} -1 & \frac{3}{4} \\ -5 & 3 \end{bmatrix}$

12. $\mathbf{A} = \begin{bmatrix} 1 & 3 \\ 12 & 1 \end{bmatrix}$

13. $\mathbf{A} = \begin{bmatrix} 1 & 2 & 2 \\ 2 & 0 & 3 \\ 2 & 3 & 0 \end{bmatrix}$

14. $\mathbf{A} = \begin{bmatrix} -1 & 1 & 0 \\ 1 & 2 & 1 \\ 0 & 3 & -1 \end{bmatrix}$

15. $\mathbf{A} = \begin{bmatrix} 1 & 2 & 3 \\ 0 & 1 & 0 \\ 2 & 1 & 2 \end{bmatrix}$

16. $\mathbf{A} = \begin{bmatrix} -7 & 0 & 6 \\ 0 & 5 & 0 \\ 6 & 0 & 2 \end{bmatrix}$

17. Consider the system $\mathbf{x}'(t) = \mathbf{A}\mathbf{x}(t), t \geq 0$, with

$$\mathbf{A} = \begin{bmatrix} 1 & \sqrt{3} \\ \sqrt{3} & -1 \end{bmatrix}.$$

 (a) Show that the matrix $\mathbf{A}$ has eigenvalues $r_1 = 2$ and $r_2 = -2$ with corresponding eigenvectors $\mathbf{u}_1 = \text{col}(\sqrt{3}, 1)$ and $\mathbf{u}_2 = \text{col}(1, -\sqrt{3})$.

 (b) Sketch the trajectory of the solution having initial vector $\mathbf{x}(0) = -\mathbf{u}_1$.

 (c) Sketch the trajectory of the solution having initial vector $\mathbf{x}(0) = \mathbf{u}_2$.

 (d) Sketch the trajectory of the solution having initial vector $\mathbf{x}(0) = \mathbf{u}_2 - \mathbf{u}_1$.

18. Consider the system $\mathbf{x}'(t) = \mathbf{A}\mathbf{x}(t), t \geq 0$, with

$$\mathbf{A} = \begin{bmatrix} -2 & 1 \\ 1 & -2 \end{bmatrix}.$$

 (a) Show that the matrix $\mathbf{A}$ has eigenvalues $r_1 = -1$ and $r_2 = -3$ with corresponding eigenvectors $\mathbf{u}_1 = \text{col}(1, 1)$ and $\mathbf{u}_2 = \text{col}(1, -1)$.

 (b) Sketch the trajectory of the solution having initial vector $\mathbf{x}(0) = \mathbf{u}_1$.

 (c) Sketch the trajectory of the solution having initial vector $\mathbf{x}(0) = -\mathbf{u}_2$.

 (d) Sketch the trajectory of the solution having initial vector $\mathbf{x}(0) = \mathbf{u}_1 - \mathbf{u}_2$.

In Problems 19–24, find a fundamental matrix for the system $\mathbf{x}'(t) = \mathbf{A}\mathbf{x}(t)$ for the given matrix $\mathbf{A}$.

19. $\mathbf{A} = \begin{bmatrix} -1 & 1 \\ 8 & 1 \end{bmatrix}$

20. $\mathbf{A} = \begin{bmatrix} 5 & 4 \\ -1 & 0 \end{bmatrix}$

21. $\mathbf{A} = \begin{bmatrix} 0 & 1 & 0 \\ 0 & 0 & 1 \\ 8 & -14 & 7 \end{bmatrix}$

22. $\mathbf{A} = \begin{bmatrix} 3 & 1 & -1 \\ 1 & 3 & -1 \\ 3 & 3 & -1 \end{bmatrix}$

23. $\mathbf{A} = \begin{bmatrix} 2 & 1 & 1 & -1 \\ 0 & -1 & 0 & 1 \\ 0 & 0 & 3 & 1 \\ 0 & 0 & 0 & 7 \end{bmatrix}$

24. $\mathbf{A} = \begin{bmatrix} 4 & -1 & 0 & 0 \\ 0 & 0 & 0 & 0 \\ 0 & 0 & 2 & -3 \\ 0 & 0 & 1 & -2 \end{bmatrix}$

25. Using matrix algebra techniques, find a general solution of the system

$$\begin{aligned} x' &= x + 2y - z \,, \\ y' &= x + z \,, \\ z' &= 4x - 4y + 5z \,. \end{aligned}$$

26. Using matrix algebra techniques, find a general solution of the system

$$\begin{aligned} x' &= 3x - 4y \,, \\ y' &= 4x - 7y \,. \end{aligned}$$

In Problems 27–30, use a linear algebra software package such as MATLAB®, MAPLESOFT®, or MATHEMATICA® to compute the required eigenvalues and eigenvectors and then give a fundamental matrix for the system $\mathbf{x}'(t) = \mathbf{A}\mathbf{x}(t)$ for the given matrix $\mathbf{A}$.

27. $\mathbf{A} = \begin{bmatrix} 0 & 1.1 & 0 \\ 0 & 0 & 1.3 \\ 0.9 & 1.1 & -6.9 \end{bmatrix}$

28. $\mathbf{A} = \begin{bmatrix} 2 & 1 & 1 \\ -1 & 1 & 0 \\ 3 & 3 & 3 \end{bmatrix}$

29. $\mathbf{A} = \begin{bmatrix} 0 & 1 & 0 & 0 \\ 0 & 0 & 1 & 0 \\ 0 & 0 & 0 & 1 \\ 2 & -6 & 3 & 3 \end{bmatrix}$

30. $\mathbf{A} = \begin{bmatrix} 0 & 1 & 0 & 0 \\ 1 & -1 & 0 & 0 \\ 0 & 0 & 0 & 1 \\ 0 & 0 & -2 & 4 \end{bmatrix}$

In Problems 31–34, solve the given initial value problem.

31. $\mathbf{x}'(t) = \begin{bmatrix} 1 & 3 \\ 3 & 1 \end{bmatrix} \mathbf{x}(t), \qquad \mathbf{x}(0) = \begin{bmatrix} 3 \\ 1 \end{bmatrix}$

32. $\mathbf{x}'(t) = \begin{bmatrix} 6 & -3 \\ 2 & 1 \end{bmatrix} \mathbf{x}(t), \qquad \mathbf{x}(0) = \begin{bmatrix} -10 \\ -6 \end{bmatrix}$

33. $\mathbf{x}'(t) = \begin{bmatrix} 1 & -2 & 2 \\ -2 & 1 & -2 \\ 2 & -2 & 1 \end{bmatrix} \mathbf{x}(t), \qquad \mathbf{x}(0) = \begin{bmatrix} -2 \\ -3 \\ 2 \end{bmatrix}$

34. $\mathbf{x}'(t) = \begin{bmatrix} 0 & 1 & 1 \\ 1 & 0 & 1 \\ 1 & 1 & 0 \end{bmatrix} \mathbf{x}(t), \qquad \mathbf{x}(0) = \begin{bmatrix} -1 \\ 4 \\ 0 \end{bmatrix}$

35. (a) Show that the matrix

$$\mathbf{A} = \begin{bmatrix} 1 & -1 \\ 4 & -3 \end{bmatrix}$$

has the repeated eigenvalue $r = -1$ and that all the eigenvectors are of the form $\mathbf{u} = s\,\text{col}(1, 2)$.

(b) Use the result of part (a) to obtain a nontrivial solution $\mathbf{x}_1(t)$ to the system $\mathbf{x}' = \mathbf{A}\mathbf{x}$.

(c) To obtain a second linearly independent solution to $\mathbf{x}' = \mathbf{A}\mathbf{x}$, try $\mathbf{x}_2(t) = te^{-t}\mathbf{u}_1 + e^{-t}\mathbf{u}_2$. [*Hint:* Substitute $\mathbf{x}_2$ into the system $\mathbf{x}' = \mathbf{A}\mathbf{x}$ and derive the relations

$$(\mathbf{A} + \mathbf{I})\mathbf{u}_1 = \mathbf{0}, \qquad (\mathbf{A} + \mathbf{I})\mathbf{u}_2 = \mathbf{u}_1 \,.$$

Since $\mathbf{u}_1$ must be an eigenvector, set $\mathbf{u}_1 = \text{col}(1, 2)$ and solve for $\mathbf{u}_2$.]

(d) What is $(\mathbf{A} + \mathbf{I})^2 \mathbf{u}_2$? (In Section 9.8, $\mathbf{u}_2$ will be identified as a *generalized eigenvector*.)

36. Use the method discussed in Problem 35 to find a general solution to the system

$$\mathbf{x}'(t) = \begin{bmatrix} 5 & -3 \\ 3 & -1 \end{bmatrix} \mathbf{x}(t) \,.$$

37. (a) Show that the matrix

$$\mathbf{A} = \begin{bmatrix} 2 & 1 & 6 \\ 0 & 2 & 5 \\ 0 & 0 & 2 \end{bmatrix}$$

has the repeated eigenvalue $r = 2$ with multiplicity 3 and that all the eigenvectors of $\mathbf{A}$ are of the form $\mathbf{u} = s\,\text{col}(1, 0, 0)$.

(b) Use the result of part (a) to obtain a solution to the system $\mathbf{x}' = \mathbf{A}\mathbf{x}$ of the form $\mathbf{x}_1(t) = e^{2t}\mathbf{u}_1$.

(c) To obtain a second linearly independent solution to $\mathbf{x}' = \mathbf{A}\mathbf{x}$, try $\mathbf{x}_2(t) = te^{2t}\mathbf{u}_1 + e^{2t}\mathbf{u}_2$. [*Hint:* Show that $\mathbf{u}_1$ and $\mathbf{u}_2$ must satisfy

$$(\mathbf{A} - 2\mathbf{I})\mathbf{u}_1 = \mathbf{0}, \qquad (\mathbf{A} - 2\mathbf{I})\mathbf{u}_2 = \mathbf{u}_1.]$$

(d) To obtain a third linearly independent solution to $\mathbf{x}' = \mathbf{A}\mathbf{x}$, try

$$\mathbf{x}_3(t) = \frac{t^2}{2}e^{2t}\mathbf{u}_1 + te^{2t}\mathbf{u}_2 + e^{2t}\mathbf{u}_3 \,.$$

[*Hint:* Show that $\mathbf{u}_1$, $\mathbf{u}_2$, and $\mathbf{u}_3$ must satisfy

$$(\mathbf{A} - 2\mathbf{I})\mathbf{u}_1 = \mathbf{0}, \qquad (\mathbf{A} - 2\mathbf{I})\mathbf{u}_2 = \mathbf{u}_1 \,,$$

$$(\mathbf{A} - 2\mathbf{I})\mathbf{u}_3 = \mathbf{u}_2.]$$

(e) Show that $(\mathbf{A} - 2\mathbf{I})^2 \mathbf{u}_2 = (\mathbf{A} - 2\mathbf{I})^3 \mathbf{u}_3 = \mathbf{0}$.

38. Use the method discussed in Problem 37 to find a general solution to the system

$$\mathbf{x}'(t) = \begin{bmatrix} 3 & -2 & 1 \\ 2 & -1 & 1 \\ -4 & 4 & 1 \end{bmatrix} \mathbf{x}(t).$$

39. (a) Show that the matrix

$$\mathbf{A} = \begin{bmatrix} 2 & 1 & 1 \\ 1 & 2 & 1 \\ -2 & -2 & -1 \end{bmatrix}$$

has the repeated eigenvalue $r = 1$ of multiplicity 3 and that all the eigenvectors of $\mathbf{A}$ are of the form $\mathbf{u} = s\,\text{col}(-1, 1, 0) + v\,\text{col}(-1, 0, 1)$.

(b) Use the result of part (a) to obtain two linearly independent solutions to the system $\mathbf{x}' = \mathbf{A}\mathbf{x}$ of the form

$$\mathbf{x}_1(t) = e^t\mathbf{u}_1 \quad \text{and} \quad \mathbf{x}_2(t) = e^t\mathbf{u}_2.$$

(c) To obtain a third linearly independent solution to $\mathbf{x}' = \mathbf{A}\mathbf{x}$, try $\mathbf{x}_3(t) = te^t\mathbf{u}_3 + e^t\mathbf{u}_4$. [*Hint:* Show that $\mathbf{u}_3$ and $\mathbf{u}_4$ must satisfy

$$(\mathbf{A} - \mathbf{I})\mathbf{u}_3 = \mathbf{0}, \qquad (\mathbf{A} - \mathbf{I})\mathbf{u}_4 = \mathbf{u}_3 \,.$$

Choose $\mathbf{u}_3$, an eigenvector of $\mathbf{A}$, so that you can solve for $\mathbf{u}_4$.]

(d) What is $(\mathbf{A} - \mathbf{I})^2 \mathbf{u}_4$?

40. Use the method discussed in Problem 39 to find a general solution to the system

$$\mathbf{x}'(t) = \begin{bmatrix} 1 & 3 & -2 \\ 0 & 7 & -4 \\ 0 & 9 & -5 \end{bmatrix} \mathbf{x}(t) \,.$$

41. Use the substitution $x_1 = y$, $x_2 = y'$ to convert the linear equation $ay'' + by' + cy = 0$, where a, b, and c are constants, into a normal system. Show that the characteristic equation for this system is the same as the auxiliary equation for the original equation.

42. (a) Show that the Cauchy–Euler equation $at^2y'' + bty' + cy = 0$ can be written as a **Cauchy–Euler system**

(25) $t\mathbf{x}' = \mathbf{A}\mathbf{x}$

with a constant coefficient matrix $\mathbf{A}$, by setting $x_1 = y/t$ and $x_2 = y'$.

(b) Show that for $t > 0$ any system of the form (25) with $\mathbf{A}$ an $n \times n$ constant matrix has nontrivial solutions of the form $\mathbf{x}(t) = t^r\mathbf{u}$ if and only if r is an eigenvalue of $\mathbf{A}$ and $\mathbf{u}$ is a corresponding eigenvector.

In Problems 43 and 44, use the result of Problem 42 to find a general solution of the given system.

43. $t\mathbf{x}'(t) = \begin{bmatrix} 1 & 3 \\ -1 & 5 \end{bmatrix} \mathbf{x}(t), \quad t > 0$

44. $t\mathbf{x}'(t) = \begin{bmatrix} -4 & 2 \\ 2 & -1 \end{bmatrix} \mathbf{x}(t), \quad t > 0$

45. Mixing Between Interconnected Tanks. Two tanks, each holding 50 L of liquid, are interconnected by pipes with liquid flowing from tank A into tank B at a rate of 4 L/min and from tank B into tank A at 1 L/min (see Figure 9.2). The liquid inside each tank is kept well stirred. Pure water flows into tank A at a rate of 3 L/min, and the solution flows out of tank B at 3 L/min. If, initially, tank A contains 2.5 kg of salt and tank B contains no salt (only water), determine the mass of salt in each tank at time $t \geq 0$. Graph on the same axes the two quantities $x_1(t)$ and $x_2(t)$, where $x_1(t)$ is the mass of salt in tank A and $x_2(t)$ is the mass of salt in tank B.

Figure 9.2 Mixing problem for interconnected tanks

46. Mixing with a Common Drain. Two tanks, each holding 1 L of liquid, are connected by a pipe through which liquid flows from tank A into tank B at a rate of $3 - \alpha$ L/min $(0 < \alpha < 3)$. The liquid inside each tank is kept well stirred. Pure water flows into tank A at a rate of 3 L/min. Solution flows out of tank A at α L/min and out of tank B at $3 - \alpha$ L/min. If, initially, tank B contains no salt (only water) and tank A contains 0.1 kg of salt, determine the mass of salt in each tank at time $t \geq 0$. How does the mass of salt in tank A depend on the choice of α? What is the maximum mass of salt in tank B? (See Figure 9.3.)

Figure 9.3 Mixing problem for a common drain, $0 < \alpha < 3$

47. To find a general solution to the system

$$\mathbf{x}' = \mathbf{A}\mathbf{x} = \begin{bmatrix} 1 & 3 & -1 \\ 3 & 0 & 1 \\ -1 & 1 & 2 \end{bmatrix} \mathbf{x},$$

proceed as follows:

(a) Use a numerical root-finding procedure to approximate the eigenvalues.

(b) If r is an eigenvalue, then let $\mathbf{u} = \text{col}(u_1, u_2, u_3)$ be an eigenvector associated with r. To solve for $\mathbf{u}$, assume $u_1 = 1$. (If not u_1, then either u_2 or u_3 may be chosen to be 1. Why?) Now solve the system

$$(\mathbf{A} - r\mathbf{I}) \begin{bmatrix} 1 \\ u_2 \\ u_3 \end{bmatrix} = \begin{bmatrix} 0 \\ 0 \\ 0 \end{bmatrix}.$$

for u_2 and u_3. Use this procedure to find approximations for three linearly independent eigenvectors for $\mathbf{A}$.

(c) Use these approximations to give a general solution to the system.

48. To complete the proof of Theorem 6, page 527, assume the induction hypothesis that $\mathbf{u}_1, \ldots, \mathbf{u}_k, 2 \leq k$, are linearly independent.

(a) Show that if

$$c_1\mathbf{u}_1 + \cdots + c_k\mathbf{u}_k + c_{k+1}\mathbf{u}_{k+1} = \mathbf{0},$$

then

$$c_1(r_1 - r_{k+1})\mathbf{u}_1 + \cdots + c_k(r_k - r_{k+1})\mathbf{u}_k = \mathbf{0}.$$

(b) Use the result of part (a) and the induction hypothesis to conclude that $\mathbf{u}_1, \ldots, \mathbf{u}_{k+1}$ are linearly independent. The theorem follows by induction.

49. Stability. A homogeneous system $\mathbf{x}' = \mathbf{A}\mathbf{x}$ with constant coefficients is **stable** if it has a fundamental matrix whose entries all remain bounded as $t \to +\infty$. (It will follow from Theorem 9 in Section 9.8 that if one fundamental matrix of the system has this property, then all fundamental matrices for the system do.) Otherwise, the system is **unstable**. A stable system is **asymptotically stable** if all solutions approach the zero solution as $t \to +\infty$. Stability is discussed in more detail in Chapter 12.[†]

[†]All references to Chapters 11–13 refer to the expanded text, *Fundamentals of Differential Equations and Boundary Value Problems*, 7th ed.

(a) Show that if **A** has all distinct real eigenvalues, then $\mathbf{x}'(t) = \mathbf{A}\mathbf{x}(t)$ is stable if and only if all eigenvalues are nonpositive.

(b) Show that if **A** has all distinct real eigenvalues, then $\mathbf{x}'(t) = \mathbf{A}\mathbf{x}(t)$ is asymptotically stable if and only if all eigenvalues are negative.

(c) Argue that in parts (a) and (b), we can replace "has distinct real eigenvalues" by "is symmetric" and the statements are still true.

50. In an ice tray, the water level in any particular ice cube cell will change at a rate proportional to the *difference* between that cell's water level and the level in the adjacent cells.

(a) Argue that a reasonable differentiable equation model for the water levels x, y, and z in the simplified three-cell tray depicted in Figure 9.4 is given by

$$x' = y - x, \quad y' = x + z - 2y, \quad z' = y - z.$$

(b) Use eigenvectors to solve this system for the initial conditions $x(0) = 3$, $y(0) = z(0) = 0$.

Figure 9.4 Ice tray

9.6 Complex Eigenvalues

In the previous section, we showed that the homogeneous system

(1) $\mathbf{x}'(t) = \mathbf{A}\mathbf{x}(t)$,

where **A** is a constant $n \times n$ matrix, has a solution of the form $\mathbf{x}(t) = e^{rt}\mathbf{u}$ if and only if r is an eigenvalue of **A** and **u** is a corresponding eigenvector. In this section we show how to obtain two real vector solutions to system (1) when **A** is real and has a pair[†] of complex conjugate eigenvalues $\alpha + i\beta$ and $\alpha - i\beta$.

Suppose $r_1 = \alpha + i\beta(\alpha$ and β real numbers) is an eigenvalue of **A** with corresponding eigenvector $\mathbf{z} = \mathbf{a} + i\mathbf{b}$, where **a** and **b** are real constant vectors. We first observe that the complex conjugate of **z**, namely $\bar{\mathbf{z}} := \mathbf{a} - i\mathbf{b}$, is an eigenvector associated with the eigenvalue $r_2 = \alpha - i\beta$. To see this, note that taking the complex conjugate of $(\mathbf{A} - r_1\mathbf{I})\mathbf{z} = \mathbf{0}$ yields $(\mathbf{A} - \bar{r}_1\mathbf{I})\bar{\mathbf{z}} = \mathbf{0}$ because the conjugate of the product is the product of the conjugates and **A** and **I** have real entries ($\bar{\mathbf{A}} = \mathbf{A}, \bar{\mathbf{I}} = \mathbf{I}$). Since $r_2 = \bar{r}_1$, we see that $\bar{\mathbf{z}}$ is an eigenvector associated with r_2. Therefore, two linearly independent complex vector solutions to (1) are

(2) $\mathbf{w}_1(t) = e^{r_1 t}\mathbf{z} = e^{(\alpha + i\beta)t}(\mathbf{a} + i\mathbf{b})$,

(3) $\mathbf{w}_2(t) = e^{r_2 t}\bar{\mathbf{z}} = e^{(\alpha - i\beta)t}(\mathbf{a} - i\mathbf{b})$.

As in Section 4.3, where we handled complex roots to the auxiliary equation, let's use one of these complex solutions and Euler's formula to obtain two real vector solutions. With the aid of Euler's formula, we rewrite $\mathbf{w}_1(t)$ as

$$\mathbf{w}_1(t) = e^{\alpha t}(\cos\beta t + i\sin\beta t)(\mathbf{a} + i\mathbf{b})$$
$$= e^{\alpha t}\{(\cos\beta t\,\mathbf{a} - \sin\beta t\,\mathbf{b}) + i(\sin\beta t\,\mathbf{a} + \cos\beta t\,\mathbf{b})\} .$$

We have thereby expressed $\mathbf{w}_1(t)$ in the form $\mathbf{w}_1(t) = \mathbf{x}_1(t) + i\mathbf{x}_2(t)$, where $\mathbf{x}_1(t)$ and $\mathbf{x}_2(t)$ are the two *real* vector functions

(4) $\mathbf{x}_1(t) := e^{\alpha t}\cos\beta t\,\mathbf{a} - e^{\alpha t}\sin\beta t\,\mathbf{b}$,

(5) $\mathbf{x}_2(t) := e^{\alpha t}\sin\beta t\,\mathbf{a} + e^{\alpha t}\cos\beta t\,\mathbf{b}$.

[†]Recall that the complex roots of a polynomial equation with real coefficients must occur in complex conjugate pairs.

Since $\mathbf{w}_1(t)$ is a solution to (1), then

$$\mathbf{w}_1'(t) = \mathbf{A}\mathbf{w}_1(t) ,$$
$$\mathbf{x}_1' + i\mathbf{x}_2' = \mathbf{A}\mathbf{x}_1 + i\mathbf{A}\mathbf{x}_2 .$$

Equating the real and imaginary parts yields

$$\mathbf{x}_1'(t) = \mathbf{A}\mathbf{x}_1(t) \quad \text{and} \quad \mathbf{x}_2'(t) = \mathbf{A}\mathbf{x}_2(t) .$$

Hence, $\mathbf{x}_1(t)$ and $\mathbf{x}_2(t)$ are real vector solutions to (1) associated with the complex conjugate eigenvalues $\alpha \pm i\beta$. Because $\mathbf{a}$ and $\mathbf{b}$ are not both the zero vector, it can be shown that $\mathbf{x}_1(t)$ and $\mathbf{x}_2(t)$ are linearly independent vector functions on $(-\infty, \infty)$ (see Problem 15).

Let's summarize our findings.

Complex Eigenvalues

If the real matrix $\mathbf{A}$ has complex conjugate eigenvalues $\alpha \pm i\beta$ with corresponding eigenvectors $\mathbf{a} \pm i\mathbf{b}$, then two linearly independent real vector solutions to $\mathbf{x}'(t) = \mathbf{A}\mathbf{x}(t)$ are

(6) $e^{\alpha t}\cos\beta t\, \mathbf{a} - e^{\alpha t}\sin\beta t\, \mathbf{b}$,

(7) $e^{\alpha t}\sin\beta t\, \mathbf{a} + e^{\alpha t}\cos\beta t\, \mathbf{b}$.

Example 1 Find a general solution of

(8) $\mathbf{x}'(t) = \begin{bmatrix} -1 & 2 \\ -1 & -3 \end{bmatrix} \mathbf{x}(t) .$

Solution The characteristic equation for $\mathbf{A}$ is

$$|\mathbf{A} - r\mathbf{I}| = \begin{vmatrix} -1 - r & 2 \\ -1 & -3 - r \end{vmatrix} = r^2 + 4r + 5 = 0 .$$

Hence, $\mathbf{A}$ has eigenvalues $r = -2 \pm i$.

To find a general solution, we need only find an eigenvector associated with the eigenvalue $r = -2 + i$. Substituting $r = -2 + i$ into $(\mathbf{A} - r\mathbf{I})\mathbf{z} = \mathbf{0}$ gives

$$\begin{bmatrix} 1 - i & 2 \\ -1 & -1 - i \end{bmatrix} \begin{bmatrix} z_1 \\ z_2 \end{bmatrix} = \begin{bmatrix} 0 \\ 0 \end{bmatrix} .$$

The solutions can be expressed as $z_1 = 2s$ and $z_2 = (-1 + i)s$, with s arbitrary. Hence, the eigenvectors associated with $r = -2 + i$ are $\mathbf{z} = s\,\text{col}(2, -1 + i)$. Taking $s = 1$ gives the eigenvector

$$\mathbf{z} = \begin{bmatrix} 2 \\ -1 + i \end{bmatrix} = \begin{bmatrix} 2 \\ -1 \end{bmatrix} + i \begin{bmatrix} 0 \\ 1 \end{bmatrix} .$$

We have found that $\alpha = -2, \beta = 1$, and $\mathbf{z} = \mathbf{a} + i\mathbf{b}$ with $\mathbf{a} = \text{col}(2, -1)$, and $\mathbf{b} = \text{col}(0, 1)$, so a general solution to (8) is

$$\mathbf{x}(t) = c_1 \left\{ e^{-2t}\cos t \begin{bmatrix} 2 \\ -1 \end{bmatrix} - e^{-2t}\sin t \begin{bmatrix} 0 \\ 1 \end{bmatrix} \right\}$$

$$+ c_2 \left\{ e^{-2t}\sin t \begin{bmatrix} 2 \\ -1 \end{bmatrix} + e^{-2t}\cos t \begin{bmatrix} 0 \\ 1 \end{bmatrix} \right\}$$

(9) $\mathbf{x}(t) = c_1 \begin{bmatrix} 2e^{-2t}\cos t \\ -e^{-2t}(\cos t + \sin t) \end{bmatrix} + c_2 \begin{bmatrix} 2e^{-2t}\sin t \\ e^{-2t}(\cos t - \sin t) \end{bmatrix} .$ ◆

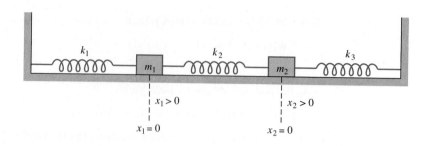

Figure 9.5 Coupled mass–spring system with fixed ends

Complex eigenvalues occur in modeling coupled mass–spring systems. For example, the motion of the mass–spring system illustrated in Figure 9.5 is governed by the second-order system

$$m_1 x_1'' = -k_1 x_1 + k_2(x_2 - x_1) ,$$
(10)
$$m_2 x_2'' = -k_2(x_2 - x_1) - k_3 x_2 ,$$

where x_1 and x_2 represent the displacements of the masses m_1 and m_2 to the right of their equilibrium positions and k_1, k_2, k_3 are the spring constants of the three springs (see the discussion in Section 5.6). If we introduce the new variables $y_1 := x_1, y_2 := x_1', y_3 := x_2, y_4 := x_2'$, then we can rewrite the system in the normal form

$$\text{(11)} \quad \mathbf{y}'(t) = \mathbf{A}\mathbf{y}(t) = \begin{bmatrix} 0 & 1 & 0 & 0 \\ -(k_1 + k_2)/m_1 & 0 & k_2/m_1 & 0 \\ 0 & 0 & 0 & 1 \\ k_2/m_2 & 0 & -(k_2 + k_3)/m_2 & 0 \end{bmatrix} \mathbf{y}(t) .$$

For such a system, it turns out that $\mathbf{A}$ has only imaginary eigenvalues and they occur in complex conjugate pairs: $\pm i\beta_1, \pm i\beta_2$. Hence, any solution will consist of sums of sine and cosine functions. The frequencies of these functions

$$f_1 := \frac{\beta_1}{2\pi} \quad \text{and} \quad f_2 := \frac{\beta_2}{2\pi}$$

are called the **normal** or **natural frequencies** of the system (β_1 and β_2 are the **angular frequencies** of the system).

In some engineering applications, the only information that is required about a particular device is a knowledge of its normal frequencies; one must ensure that they are far from the frequencies that occur naturally in the device's operating environment (so that no resonances will be excited).

Example 2 Determine the normal frequencies for the coupled mass–spring system governed by system (11) when $m_1 = m_2 = 1$ kg, $k_1 = 1$ N/m, $k_2 = 2$ N/m, and $k_3 = 3$ N/m.

Solution To find the eigenvalues of $\mathbf{A}$, we must solve the characteristic equation

$$|\mathbf{A} - r\mathbf{I}| = \begin{vmatrix} -r & 1 & 0 & 0 \\ -3 & -r & 2 & 0 \\ 0 & 0 & -r & 1 \\ 2 & 0 & -5 & -r \end{vmatrix} = r^4 + 8r^2 + 11 = 0 .$$

From the quadratic formula we find $r^2 = -4 \pm \sqrt{5}$, so the four eigenvalues of $\mathbf{A}$ are $\pm i\sqrt{4 - \sqrt{5}}$ and $\pm i\sqrt{4 + \sqrt{5}}$. Hence, the two normal frequencies for this system are

$$\frac{\sqrt{4 - \sqrt{5}}}{2\pi} \approx 0.211 \quad \text{and} \quad \frac{\sqrt{4 + \sqrt{5}}}{2\pi} \approx 0.397 \text{ cycles per second.} \quad \blacklozenge$$

9.6 EXERCISES

In Problems 1–4, find a general solution of the system $\mathbf{x}'(t) = \mathbf{A}\mathbf{x}(t)$ for the given matrix $\mathbf{A}$.

1. $\mathbf{A} = \begin{bmatrix} 2 & -4 \\ 2 & -2 \end{bmatrix}$

2. $\mathbf{A} = \begin{bmatrix} -2 & -5 \\ 1 & 2 \end{bmatrix}$

3. $\mathbf{A} = \begin{bmatrix} 1 & 2 & -1 \\ 0 & 1 & 1 \\ 0 & -1 & 1 \end{bmatrix}$

4. $\mathbf{A} = \begin{bmatrix} 5 & -5 & -5 \\ -1 & 4 & 2 \\ 3 & -5 & -3 \end{bmatrix}$

In Problems 5–9, find a fundamental matrix for the system $\mathbf{x}'(t) = \mathbf{A}\mathbf{x}(t)$ for the given matrix $\mathbf{A}$.

5. $\mathbf{A} = \begin{bmatrix} -1 & -2 \\ 8 & -1 \end{bmatrix}$

6. $\mathbf{A} = \begin{bmatrix} -2 & -2 \\ 4 & 2 \end{bmatrix}$

7. $\mathbf{A} = \begin{bmatrix} 0 & 0 & 1 \\ 0 & 0 & -1 \\ 0 & 1 & 0 \end{bmatrix}$

8. $\mathbf{A} = \begin{bmatrix} 0 & 1 & 0 & 0 \\ 1 & 0 & 0 & 0 \\ 0 & 0 & 0 & 1 \\ 0 & 0 & -13 & 4 \end{bmatrix}$

9. $\mathbf{A} = \begin{bmatrix} 0 & 1 & 0 & 0 \\ -1 & 0 & 0 & 0 \\ 0 & 0 & 0 & 1 \\ 0 & 0 & -1 & 0 \end{bmatrix}$

In Problems 10–12, use a linear algebra software package to compute the required eigenvalues and eigenvectors for the given matrix $\mathbf{A}$ and then give a fundamental matrix for the system $\mathbf{x}'(t) = \mathbf{A}\mathbf{x}(t)$.

10. $\mathbf{A} = \begin{bmatrix} 0 & 1 & 0 & 0 \\ 0 & 0 & 1 & 0 \\ 0 & 0 & 0 & 1 \\ 13 & -4 & -12 & 4 \end{bmatrix}$

11. $\mathbf{A} = \begin{bmatrix} 0 & 1 & 0 & 0 \\ 0 & 0 & 1 & 0 \\ 0 & 0 & 0 & 1 \\ -2 & 2 & -3 & 2 \end{bmatrix}$

12. $\mathbf{A} = \begin{bmatrix} 1 & 0 & 0 & 0 & 0 \\ 0 & 0 & 1 & 0 & 0 \\ 0 & 1 & 0 & 0 & 0 \\ 0 & 0 & 0 & 0 & 1 \\ 0 & 0 & 0 & -29 & -4 \end{bmatrix}$

In Problems 13 and 14, find the solution to the given system that satisfies the given initial condition.

13. $\mathbf{x}'(t) = \begin{bmatrix} -3 & -1 \\ 2 & -1 \end{bmatrix} \mathbf{x}(t)$,

 (a) $\mathbf{x}(0) = \begin{bmatrix} -1 \\ 0 \end{bmatrix}$ (b) $\mathbf{x}(\pi) = \begin{bmatrix} 1 \\ -1 \end{bmatrix}$

 (c) $\mathbf{x}(-2\pi) = \begin{bmatrix} 2 \\ 1 \end{bmatrix}$ (d) $\mathbf{x}(\pi/2) = \begin{bmatrix} 0 \\ 1 \end{bmatrix}$

14. $\mathbf{x}'(t) = \begin{bmatrix} 1 & 0 & -1 \\ 0 & 2 & 0 \\ 1 & 0 & 1 \end{bmatrix} \mathbf{x}(t)$,

 (a) $\mathbf{x}(0) = \begin{bmatrix} -2 \\ 2 \\ -1 \end{bmatrix}$ (b) $\mathbf{x}(-\pi) = \begin{bmatrix} 0 \\ 1 \\ 1 \end{bmatrix}$

15. Show that $\mathbf{x}_1(t)$ and $\mathbf{x}_2(t)$ given by equations (4) and (5) are linearly independent on $(-\infty, \infty)$, provided $\beta \neq 0$ and $\mathbf{a}$ and $\mathbf{b}$ are not both the zero vector.

16. Show that $\mathbf{x}_1(t)$ and $\mathbf{x}_2(t)$ given by equations (4) and (5) can be obtained as linear combinations of $\mathbf{w}_1(t)$ and $\mathbf{w}_2(t)$ given by equations (2) and (3). [Hint: Show that

$$\mathbf{x}_1(t) = \frac{\mathbf{w}_1(t) + \mathbf{w}_2(t)}{2}, \qquad \mathbf{x}_2(t) = \frac{\mathbf{w}_1(t) - \mathbf{w}_2(t)}{2i}.]$$

In Problems 17 and 18, use the results of Problem 42 in Exercises 9.5 to find a general solution to the given Cauchy–Euler system for $t > 0$.

17. $t\mathbf{x}'(t) = \begin{bmatrix} -1 & -1 & 0 \\ 2 & -1 & 1 \\ 0 & 1 & -1 \end{bmatrix} \mathbf{x}(t)$

18. $t\mathbf{x}'(t) = \begin{bmatrix} -1 & -1 \\ 9 & -1 \end{bmatrix} \mathbf{x}(t)$

19. For the coupled mass–spring system governed by system (10), assume $m_1 = m_2 = 1$ kg, $k_1 = k_2 = 2$ N/m, and $k_3 = 3$ N/m. Determine the normal frequencies for this coupled mass–spring system.

20. For the coupled mass–spring system governed by system (10), assume $m_1 = m_2 = 1$ kg, $k_1 = k_2 = k_3 = 1$ N/m, and assume initially that $x_1(0) = 0$ m, $x_1'(0) = 0$ m/sec, $x_2(0) = 2$ m, and $x_2'(0) = 0$ m/sec. Using matrix algebra techniques, solve this initial value problem.

21. *RLC* Network. The currents in the *RLC* network given by the schematic diagram in Figure 9.6 are governed by the following equations:

$$4I_2'(t) + 52q_1(t) = 10,$$
$$13I_3(t) + 52q_1(t) = 10,$$
$$I_1(t) = I_2(t) + I_3(t),$$

Figure 9.6 *RLC* network for Problem 21

where $q_1(t)$ is the charge on the capacitor, $I_1(t) = q_1'(t)$, and initially $q_1(0) = 0$ coulombs and $I_1(0) = 0$ amps. Solve for the currents I_1, I_2, and I_3. [*Hint:* Differentiate the first two equations, eliminate I_1, and form a normal system with $x_1 = I_2$, $x_2 = I_2'$, and $x_3 = I_3$.]

22. *RLC* Network. The currents in the *RLC* network given by the schematic diagram in Figure 9.7 are governed by the following equations:

$$50I_1'(t) + 80I_2(t) = 160,$$
$$50I_1'(t) + 800q_3(t) = 160,$$
$$I_1(t) = I_2(t) + I_3(t),$$

Figure 9.7 *RLC* network for Problem 22

where $q_3(t)$ is the charge on the capacitor, $I_3(t) = q_3'(t)$, and initially $q_3(0) = 0.5$ coulombs and $I_3(0) = 0$ amps. Solve for the currents I_1, I_2, and I_3. [*Hint:* Differentiate the first two equations, use the third equation to eliminate I_3, and form a normal system with $x_1 = I_1$, $x_2 = I_1'$, and $x_3 = I_2$.]

23. Stability. In Problem 49 of Exercises 9.5, (page 542), we discussed the notion of stability and asymptotic stability for a linear system of the form $\mathbf{x}'(t) = \mathbf{A}\mathbf{x}(t)$. Assume that $\mathbf{A}$ has all distinct eigenvalues (real or complex).

(a) Show that the system is stable if and only if all the eigenvalues of $\mathbf{A}$ have nonpositive real part.

(b) Show that the system is asymptotically stable if and only if all the eigenvalues of $\mathbf{A}$ have negative real part.

24. (a) For Example 1, page 535, verify that

$$\mathbf{x}(t) = c_1 \begin{bmatrix} -e^{-2t}\cos t + e^{-2t}\sin t \\ e^{-2t}\cos t \end{bmatrix}$$
$$+ c_2 \begin{bmatrix} -e^{-2t}\sin t - e^{-2t}\cos t \\ e^{-2t}\sin t \end{bmatrix}$$

is another general solution to equation (8).

(b) How can the general solution of part (a) be directly obtained from the general solution derived in (9) on page 535?

9.7 Nonhomogeneous Linear Systems

The techniques discussed in Chapters 4 and 6 for finding a particular solution to the nonhomogeneous equation $y'' + p(x)y' + q(x)y = g(x)$ have natural extensions to nonhomogeneous linear systems.

Undetermined Coefficients

The method of undetermined coefficients can be used to find a particular solution to the nonhomogeneous linear system

$$\mathbf{x}'(t) = \mathbf{A}\mathbf{x}(t) + \mathbf{f}(t)$$

when $\mathbf{A}$ is an $n \times n$ *constant* matrix and the entries of $\mathbf{f}(t)$ are polynomials, exponential functions, sines and cosines, or finite sums and products of these functions. We can use the procedure box in Section 4.5 (page 184) and reproduced at the back of the book as a *guide* in choosing the form of a particular solution $\mathbf{x}_p(t)$. Some exceptions are discussed in the exercises (see Problems 25–28).

Example 1 Find a general solution of

$$(1) \qquad \mathbf{x}'(t) = \mathbf{A}\mathbf{x}(t) + t\mathbf{g}, \quad \text{where} \quad \mathbf{A} = \begin{bmatrix} 1 & -2 & 2 \\ -2 & 1 & 2 \\ 2 & 2 & 1 \end{bmatrix} \quad \text{and} \quad \mathbf{g} = \begin{bmatrix} -9 \\ 0 \\ -18 \end{bmatrix}.$$

Solution In Example 5 in Section 9.5, page 529, we found that a general solution to the corresponding homogeneous system $\mathbf{x}' = \mathbf{A}\mathbf{x}$ is

$$(2) \qquad \mathbf{x}_h(t) = c_1 e^{3t} \begin{bmatrix} 1 \\ 0 \\ 1 \end{bmatrix} + c_2 e^{3t} \begin{bmatrix} -1 \\ 1 \\ 0 \end{bmatrix} + c_3 e^{-3t} \begin{bmatrix} -1 \\ -1 \\ 1 \end{bmatrix}.$$

Since the entries in $\mathbf{f}(t) := t\mathbf{g}$ are just linear functions of t, we are inclined to seek a particular solution of the form

$$\mathbf{x}_p(t) = t\mathbf{a} + \mathbf{b} = t \begin{bmatrix} a_1 \\ a_2 \\ a_3 \end{bmatrix} + \begin{bmatrix} b_1 \\ b_2 \\ b_3 \end{bmatrix},$$

where the constant vectors $\mathbf{a}$ and $\mathbf{b}$ are to be determined. Substituting this expression for $\mathbf{x}_p(t)$ into system (1) yields

$$\mathbf{a} = \mathbf{A}(t\mathbf{a} + \mathbf{b}) + t\mathbf{g},$$

which can be written as

$$t(\mathbf{A}\mathbf{a} + \mathbf{g}) + (\mathbf{A}\mathbf{b} - \mathbf{a}) = \mathbf{0}.$$

Setting the "coefficients" of this vector polynomial equal to zero yields the two systems

$$(3) \qquad \mathbf{A}\mathbf{a} = -\mathbf{g},$$

$$(4) \qquad \mathbf{A}\mathbf{b} = \mathbf{a}.$$

By Gaussian elimination or by using a linear algebra software package, we can solve (3) for $\mathbf{a}$ and we find $\mathbf{a} = \text{col}(5, 2, 4)$. Next we substitute for $\mathbf{a}$ in (4) and solve for $\mathbf{b}$ to obtain $\mathbf{b} = \text{col}(1, 0, 2)$. Hence a particular solution for (1) is

$$(5) \qquad \mathbf{x}_p(t) = t\mathbf{a} + \mathbf{b} = t \begin{bmatrix} 5 \\ 2 \\ 4 \end{bmatrix} + \begin{bmatrix} 1 \\ 0 \\ 2 \end{bmatrix} = \begin{bmatrix} 5t + 1 \\ 2t \\ 4t + 2 \end{bmatrix}.$$

A general solution for (1) is $\mathbf{x}(t) = \mathbf{x}_h(t) + \mathbf{x}_p(t)$, where $\mathbf{x}_h(t)$ is given in (2) and $\mathbf{x}_p(t)$ in (5). ◆

In the preceding example, the nonhomogeneous term $\mathbf{f}(t)$ was a vector polynomial. If, instead, $\mathbf{f}(t)$ has the form

$$\mathbf{f}(t) = \text{col}(1, t, \sin t),$$

then, using the superposition principle, we would seek a particular solution of the form

$$\mathbf{x}_p(t) = t\mathbf{a} + \mathbf{b} + (\sin t)\mathbf{c} + (\cos t)\mathbf{d}.$$

Similarly, if

$$\mathbf{f}(t) = \text{col}(t, e^t, t^2),$$

we would take

$$\mathbf{x}_p(t) = t^2\mathbf{a} + t\mathbf{b} + \mathbf{c} + e^t\mathbf{d}.$$

Of course, we must modify our guess, should one of the terms be a solution to the corresponding homogeneous system. If this is the case, the annihilator method [equations (15) and (16) of Section 6.3, page 337] would appear to suggest that for a nonhomogeneity $\mathbf{f}(t)$ of the form $e^{rt}t^m\mathbf{g}$, where r is an eigenvalue of $\mathbf{A}$, m is a nonnegative integer, and $\mathbf{g}$ is a constant vector, a particular solution of $\mathbf{x}' = \mathbf{Ax} + \mathbf{f}$ can be found in the form

$$\mathbf{x}_p(t) = e^{rt}\{t^{m+s}\mathbf{a}_{m+s} + t^{m+s-1}\mathbf{a}_{m+s-1} + \cdots + t\mathbf{a}_1 + \mathbf{a}_0\},$$

for a suitable choice of s. We omit the details.

Variation of Parameters

In Section 4.6 we discussed the method of variation of parameters for a general constant-coefficient second-order linear equation. Simply put, the idea is that if a general solution to the homogeneous equation has the form $x_h(t) = c_1x_1(t) + c_2x_2(t)$, where $x_1(t)$ and $x_2(t)$ are linearly independent solutions to the homogeneous equation, then a particular solution to the nonhomogeneous equation would have the form $x_p(t) = v_1(t)x_1(t) + v_2(t)x_2(t)$, where $v_1(t)$ and $v_2(t)$ are certain functions of t. A similar idea can be used for systems.

Let $\mathbf{X}(t)$ be a fundamental matrix for the homogeneous system

(6) $$\mathbf{x}'(t) = \mathbf{A}(t)\mathbf{x}(t),$$

where now *the entries of* $\mathbf{A}$ *may be any continuous functions of t*. Because a general solution to (6) is given by $\mathbf{X}(t)\mathbf{c}$, where $\mathbf{c}$ is a constant $n \times 1$ vector, we seek a particular solution to the nonhomogeneous system

(7) $$\mathbf{x}'(t) = \mathbf{A}(t)\mathbf{x}(t) + \mathbf{f}(t)$$

of the form

(8) $$\mathbf{x}_p(t) = \mathbf{X}(t)\boldsymbol{v}(t),$$

where $\boldsymbol{v}(t) = \text{col}(v_1(t), \ldots, v_n(t))$ is a vector function of t to be determined.

To derive a formula for $\boldsymbol{v}(t)$, we first differentiate (8) using the matrix version of the product rule to obtain

$$\mathbf{x}_p'(t) = \mathbf{X}(t)\boldsymbol{v}'(t) + \mathbf{X}'(t)\boldsymbol{v}(t).$$

Substituting the expressions for $\mathbf{x}_p(t)$ and $\mathbf{x}_p'(t)$ into (7) yields

(9) $$\mathbf{X}(t)\boldsymbol{v}'(t) + \mathbf{X}'(t)\boldsymbol{v}(t) = \mathbf{A}(t)\mathbf{X}(t)\boldsymbol{v}(t) + \mathbf{f}(t).$$

Since $\mathbf{X}(t)$ satisfies the matrix equation $\mathbf{X}'(t) = \mathbf{A}(t)\mathbf{X}(t)$, equation (9) becomes

$$\mathbf{X}\boldsymbol{v}' + \mathbf{AX}\boldsymbol{v} = \mathbf{AX}\boldsymbol{v} + \mathbf{f},$$
$$\mathbf{X}\boldsymbol{v}' = \mathbf{f}.$$

Multiplying both sides of the last equation by $\mathbf{X}^{-1}(t)$ [which exists since the columns of $\mathbf{X}(t)$ are linearly independent] gives

$$\boldsymbol{v}'(t) = \mathbf{X}^{-1}(t)\mathbf{f}(t).$$

Integrating, we obtain

$$\boldsymbol{v}(t) = \int \mathbf{X}^{-1}(t)\,\mathbf{f}(t)\,dt \;.$$

Hence, a particular solution to (7) is

(10) $\mathbf{x}_p(t) = \mathbf{X}(t)\boldsymbol{v}(t) = \mathbf{X}(t) \displaystyle\int \mathbf{X}^{-1}(t)\,\mathbf{f}(t)\,dt \;.$

Combining (10) with the solution $\mathbf{X}(t)\mathbf{c}$ to the homogeneous system yields the following general solution to (7):

(11) $\mathbf{x}(t) = \mathbf{X}(t)\mathbf{c} + \mathbf{X}(t) \displaystyle\int \mathbf{X}^{-1}(t)\,\mathbf{f}(t)\,dt \;.$

The elegance of the derivation of the variation of parameters formula (10) for systems becomes evident when one compares it with the more lengthy derivations for the scalar case in Sections 4.6 and 6.4.

Given an initial value problem of the form

(12) $\mathbf{x}'(t) = \mathbf{A}(t)\mathbf{x}(t) + \mathbf{f}(t)\,, \qquad \mathbf{x}(t_0) = \mathbf{x}_0\,,$

we can use the initial condition $\mathbf{x}(t_0) = \mathbf{x}_0$ to solve for $\mathbf{c}$ in (11). Expressing $\mathbf{x}(t)$ using a definite integral, we have

$$\mathbf{x}(t) = \mathbf{X}(t)\mathbf{c} + \mathbf{X}(t) \int_{t_0}^{t} \mathbf{X}^{-1}(s)\,\mathbf{f}(s)\,ds \;.$$

From the initial condition $\mathbf{x}(t_0) = \mathbf{x}_0$, we find

$$\mathbf{x}_0 = \mathbf{x}(t_0) = \mathbf{X}(t_0)\mathbf{c} + \mathbf{X}(t_0) \int_{t_0}^{t_0} \mathbf{X}^{-1}(s)\,\mathbf{f}(s)\,ds = \mathbf{X}(t_0)\mathbf{c} \;.$$

Solving for $\mathbf{c}$, we have $\mathbf{c} = \mathbf{X}^{-1}(t_0)\mathbf{x}_0$. Thus, the solution to (12) is given by the formula

(13) $\mathbf{x}(t) = \mathbf{X}(t)\mathbf{X}^{-1}(t_0)\mathbf{x}_0 + \mathbf{X}(t) \displaystyle\int_{t_0}^{t} \mathbf{X}^{-1}(s)\,\mathbf{f}(s)\,ds \;.$

To apply the variation of parameters formulas, we first must determine a fundamental matrix $\mathbf{X}(t)$ for the homogeneous system. In the case when the coefficient matrix $\mathbf{A}$ is constant, we have discussed methods for finding $\mathbf{X}(t)$. However, if the entries of $\mathbf{A}$ depend on t, the determination of $\mathbf{X}(t)$ may be extremely difficult (entailing, perhaps, a matrix power series!).

Example 2 Find the solution to the initial value problem

(14) $\mathbf{x}'(t) = \begin{bmatrix} 2 & -3 \\ 1 & -2 \end{bmatrix} \mathbf{x}(t) + \begin{bmatrix} e^{2t} \\ 1 \end{bmatrix}, \qquad \mathbf{x}(0) = \begin{bmatrix} -1 \\ 0 \end{bmatrix}.$

Solution In Example 3 in Section 9.5, we found two linearly independent solutions to the corresponding homogeneous system; namely,

$$\mathbf{x}_1(t) = \begin{bmatrix} 3e^t \\ e^t \end{bmatrix} \quad \text{and} \quad \mathbf{x}_2(t) = \begin{bmatrix} e^{-t} \\ e^{-t} \end{bmatrix}.$$

Hence a fundamental matrix for the homogeneous system is

$$\mathbf{X}(t) = \begin{bmatrix} 3e^t & e^{-t} \\ e^t & e^{-t} \end{bmatrix}.$$

Although the solution to (14) can be found via the method of undetermined coefficients, we shall find it directly from formula (13). For this purpose, we need $\mathbf{X}^{-1}(t)$. By formula (3) of Section 9.3 (page 510):

$$\mathbf{X}^{-1}(t) = \begin{bmatrix} \frac{1}{2}e^{-t} & -\frac{1}{2}e^{-t} \\ -\frac{1}{2}e^t & \frac{3}{2}e^t \end{bmatrix}.$$

Substituting into formula (13), we obtain the solution

$$\mathbf{x}(t) = \begin{bmatrix} 3e^t & e^{-t} \\ e^t & e^{-t} \end{bmatrix} \begin{bmatrix} \frac{1}{2} & -\frac{1}{2} \\ -\frac{1}{2} & \frac{3}{2} \end{bmatrix} \begin{bmatrix} -1 \\ 0 \end{bmatrix}$$

$$+ \begin{bmatrix} 3e^t & e^{-t} \\ e^t & e^{-t} \end{bmatrix} \int_0^t \begin{bmatrix} \frac{1}{2}e^{-s} & -\frac{1}{2}e^{-s} \\ -\frac{1}{2}e^s & \frac{3}{2}e^s \end{bmatrix} \begin{bmatrix} e^{2s} \\ 1 \end{bmatrix} ds$$

$$= \begin{bmatrix} -\frac{3}{2}e^t + \frac{1}{2}e^{-t} \\ -\frac{1}{2}e^t + \frac{1}{2}e^{-t} \end{bmatrix} + \begin{bmatrix} 3e^t & e^{-t} \\ e^t & e^{-t} \end{bmatrix} \int_0^t \begin{bmatrix} \frac{1}{2}e^s - \frac{1}{2}e^{-s} \\ -\frac{1}{2}e^{3s} + \frac{3}{2}e^s \end{bmatrix} ds$$

$$= \begin{bmatrix} -\frac{3}{2}e^t + \frac{1}{2}e^{-t} \\ -\frac{1}{2}e^t + \frac{1}{2}e^{-t} \end{bmatrix} + \begin{bmatrix} 3e^t & e^{-t} \\ e^t & e^{-t} \end{bmatrix} \begin{bmatrix} \frac{1}{2}e^t + \frac{1}{2}e^{-t} - 1 \\ \frac{3}{2}e^t - \frac{1}{6}e^{3t} - \frac{4}{3} \end{bmatrix}$$

$$= \begin{bmatrix} -\frac{9}{2}e^t - \frac{5}{6}e^{-t} + \frac{4}{3}e^{2t} + 3 \\ -\frac{3}{2}e^t - \frac{5}{6}e^{-t} + \frac{1}{3}e^{2t} + 2 \end{bmatrix}. \quad \blacklozenge$$

9.7 EXERCISES

In Problems 1–6, use the method of undetermined coefficients to find a general solution to the system $\mathbf{x}'(t) = \mathbf{A}\mathbf{x}(t) + \mathbf{f}(t)$, where $\mathbf{A}$ and $\mathbf{f}(t)$ are given.

1. $\mathbf{A} = \begin{bmatrix} 6 & 1 \\ 4 & 3 \end{bmatrix}$, $\quad \mathbf{f}(t) = \begin{bmatrix} -11 \\ -5 \end{bmatrix}$

2. $\mathbf{A} = \begin{bmatrix} 1 & 1 \\ 4 & 1 \end{bmatrix}$, $\quad \mathbf{f}(t) = \begin{bmatrix} -t-1 \\ -4t-2 \end{bmatrix}$

3. $\mathbf{A} = \begin{bmatrix} 1 & -2 & 2 \\ -2 & 1 & 2 \\ 2 & 2 & 1 \end{bmatrix}$, $\quad \mathbf{f}(t) = \begin{bmatrix} 2e^t \\ 4e^t \\ -2e^t \end{bmatrix}$

4. $\mathbf{A} = \begin{bmatrix} 2 & 2 \\ 2 & 2 \end{bmatrix}$, $\quad \mathbf{f}(t) = \begin{bmatrix} -4\cos t \\ -\sin t \end{bmatrix}$

5. $\mathbf{A} = \begin{bmatrix} 0 & -1 & 0 \\ -1 & 0 & 0 \\ 0 & 0 & 1 \end{bmatrix}$, $\mathbf{f}(t) = \begin{bmatrix} e^{2t} \\ \sin t \\ t \end{bmatrix}$

6. $\mathbf{A} = \begin{bmatrix} 1 & 1 \\ 0 & 2 \end{bmatrix}$, $\mathbf{f}(t) = e^{-2t} \begin{bmatrix} t \\ 3 \end{bmatrix}$

In Problems 7–10, use the method of undetermined coefficients to determine only the form of a particular solution for the system $\mathbf{x}'(t) = \mathbf{A}\mathbf{x}(t) + \mathbf{f}(t)$, *where* $\mathbf{A}$ *and* $\mathbf{f}(t)$ *are given.*

7. $\mathbf{A} = \begin{bmatrix} 0 & 1 \\ 2 & 0 \end{bmatrix}$, $\mathbf{f}(t) = \begin{bmatrix} \sin 3t \\ t \end{bmatrix}$

8. $\mathbf{A} = \begin{bmatrix} -1 & 0 \\ 2 & 2 \end{bmatrix}$, $\mathbf{f}(t) = \begin{bmatrix} t^2 \\ t+1 \end{bmatrix}$

9. $\mathbf{A} = \begin{bmatrix} 0 & -1 & 0 \\ -1 & 0 & 1 \\ 0 & 0 & 1 \end{bmatrix}$, $\mathbf{f}(t) = \begin{bmatrix} e^{2t} \\ \sin t \\ t \end{bmatrix}$

10. $\mathbf{A} = \begin{bmatrix} 2 & -1 \\ 1 & 5 \end{bmatrix}$, $\mathbf{f}(t) = \begin{bmatrix} te^{-t} \\ 3e^{-t} \end{bmatrix}$

In Problems 11–16, use the variation of parameters formula (11) to find a general solution of the system $\mathbf{x}'(t) = \mathbf{A}\mathbf{x}(t) + \mathbf{f}(t)$, *where* $\mathbf{A}$ *and* $\mathbf{f}(t)$ *are given.*

11. $\mathbf{A} = \begin{bmatrix} 0 & 1 \\ -1 & 0 \end{bmatrix}$, $\mathbf{f}(t) = \begin{bmatrix} 1 \\ 0 \end{bmatrix}$

12. $\mathbf{A} = \begin{bmatrix} 1 & 2 \\ 3 & 2 \end{bmatrix}$, $\mathbf{f}(t) = \begin{bmatrix} 1 \\ -1 \end{bmatrix}$

13. $\mathbf{A} = \begin{bmatrix} 8 & -4 \\ 4 & -2 \end{bmatrix}$, $\mathbf{f}(t) = \begin{bmatrix} t^{-2}/2 \\ t^{-2} \end{bmatrix}$

14. $\mathbf{A} = \begin{bmatrix} 0 & -1 \\ 1 & 0 \end{bmatrix}$, $\mathbf{f}(t) = \begin{bmatrix} t^2 \\ 1 \end{bmatrix}$

15. $\mathbf{A} = \begin{bmatrix} -4 & 2 \\ 2 & -1 \end{bmatrix}$, $\mathbf{f}(t) = \begin{bmatrix} t^{-1} \\ 4 + 2t^{-1} \end{bmatrix}$

16. $\mathbf{A} = \begin{bmatrix} 0 & 1 \\ -1 & 0 \end{bmatrix}$, $\mathbf{f}(t) = \begin{bmatrix} 8 \sin t \\ 0 \end{bmatrix}$

In Problems 17–20, use the variation of parameters formulas (11) and possibly a linear algebra software package to find a general solution of the system $\mathbf{x}'(t) = \mathbf{A}\mathbf{x}(t) + \mathbf{f}(t)$, *where* $\mathbf{A}$ *and* $\mathbf{f}(t)$ *are given.*

17. $\mathbf{A} = \begin{bmatrix} 0 & 1 & 1 \\ 1 & 0 & 1 \\ 1 & 1 & 0 \end{bmatrix}$, $\mathbf{f}(t) = \begin{bmatrix} 3e^t \\ -e^t \\ -e^t \end{bmatrix}$

18. $\mathbf{A} = \begin{bmatrix} 1 & -1 & 1 \\ 0 & 0 & 1 \\ 0 & -1 & 2 \end{bmatrix}$, $\mathbf{f}(t) = \begin{bmatrix} 0 \\ e^t \\ e^t \end{bmatrix}$

19. $\mathbf{A} = \begin{bmatrix} 0 & 1 & 0 & 0 \\ -1 & 0 & 0 & 0 \\ 0 & 0 & 0 & 1 \\ 0 & 0 & 1 & 0 \end{bmatrix}$, $\mathbf{f}(t) = \begin{bmatrix} t \\ 0 \\ e^{-t} \\ t \end{bmatrix}$

20. $\mathbf{A} = \begin{bmatrix} 0 & 1 & 0 & 0 \\ 0 & 0 & 1 & 0 \\ 0 & 0 & 0 & 1 \\ 8 & -4 & -2 & -1 \end{bmatrix}$, $\mathbf{f}(t) = \begin{bmatrix} e^t \\ 0 \\ 1 \\ 0 \end{bmatrix}$

In Problems 21 and 22, find the solution to the given system that satisfies the given initial condition.

21. $\mathbf{x}'(t) = \begin{bmatrix} 0 & 2 \\ -1 & 3 \end{bmatrix} \mathbf{x}(t) + \begin{bmatrix} e^t \\ -e^t \end{bmatrix}$,

 (a) $\mathbf{x}(0) = \begin{bmatrix} 5 \\ 4 \end{bmatrix}$ **(b)** $\mathbf{x}(1) = \begin{bmatrix} 0 \\ 1 \end{bmatrix}$

 (c) $\mathbf{x}(5) = \begin{bmatrix} 1 \\ 0 \end{bmatrix}$ **(d)** $\mathbf{x}(-1) = \begin{bmatrix} -4 \\ 5 \end{bmatrix}$

22. $\mathbf{x}'(t) = \begin{bmatrix} 0 & 2 \\ 4 & -2 \end{bmatrix} \mathbf{x}(t) + \begin{bmatrix} 4t \\ -4t - 2 \end{bmatrix}$,

 (a) $\mathbf{x}(0) = \begin{bmatrix} 4 \\ -5 \end{bmatrix}$ **(b)** $\mathbf{x}(2) = \begin{bmatrix} 1 \\ 1 \end{bmatrix}$

23. Using matrix algebra techniques and the method of undetermined coefficients, find a general solution for

$$x''(t) + y'(t) - x(t) + y(t) = -1\,,$$
$$x'(t) + y'(t) - x(t) = t^2\,.$$

Compare your solution with the solution in Example 4 in Section 5.2, page 247.

24. Using matrix algebra techniques and the method of undetermined coefficients, solve the initial value problem

$$x'(t) - 2y(t) = 4t\,, \qquad x(0) = 4\,;$$
$$y'(t) + 2y(t) - 4x(t) = -4t - 2\,, \qquad y(0) = -5\,.$$

Compare your solution with the solution in Example 1 in Section 7.10, page 412.

25. To find a general solution to the system

$$\mathbf{x}'(t) = \begin{bmatrix} 0 & 1 \\ -2 & 3 \end{bmatrix} \mathbf{x}(t) + \mathbf{f}(t)\,, \quad \text{where} \quad \mathbf{f}(t) = \begin{bmatrix} e^t \\ 0 \end{bmatrix}\,,$$

proceed as follows:

 (a) Find a fundamental solution set for the corresponding homogeneous system.

 (b) The obvious choice for a particular solution would be a vector function of the form $\mathbf{x}_p(t) = e^t \mathbf{a}$; however, the homogeneous system has a solution of this form. The next choice would be $\mathbf{x}_p(t) = te^t \mathbf{a}$. Show that this choice does *not* work.

 (c) For systems, multiplying by t is not always sufficient. The proper guess is

$$\mathbf{x}_p(t) = te^t \mathbf{a} + e^t \mathbf{b}\,.$$

 Use this guess to find a particular solution of the given system.

 (d) Use the results of parts (a) and (c) to find a general solution of the given system.

26. For the system of Problem 25, we found that a proper guess for a particular solution is $x_p(t) = te^t\mathbf{a} + e^t\mathbf{b}$. In some cases $\mathbf{a}$ or $\mathbf{b}$ may be zero.

 (a) Find a particular solution for the system of Problem 25 if $\mathbf{f}(t) = \text{col}(3e^t, 6e^t)$.

 (b) Find a particular solution for the system of Problem 25 if $\mathbf{f}(t) = \text{col}(e^t, e^t)$.

27. Find a general solution of the system

$$\mathbf{x}'(t) = \begin{bmatrix} 0 & 1 & 1 \\ 1 & 0 & 1 \\ 1 & 1 & 0 \end{bmatrix} \mathbf{x}(t) + \begin{bmatrix} -1 \\ -1 - e^{-t} \\ -2e^{-t} \end{bmatrix}.$$

 [*Hint:* Try $\mathbf{x}_p(t) = e^{-t}\mathbf{a} + te^{-t}\mathbf{b} + \mathbf{c}$.]

28. Find a particular solution for the system

$$\mathbf{x}'(t) = \begin{bmatrix} 1 & -1 \\ -1 & 1 \end{bmatrix} \mathbf{x}(t) + \begin{bmatrix} -3 \\ 1 \end{bmatrix}.$$

 [*Hint:* Try $\mathbf{x}_p(t) = t\mathbf{a} + \mathbf{b}$.]

In Problems 29 and 30, find a general solution to the given Cauchy–Euler system for $t > 0$. (See Problem 42 in Exercises 9.5, page 533.) Remember to express the system in the form $\mathbf{x}'(t) = \mathbf{A}(t)\mathbf{x}(t) + \mathbf{f}(t)$ before using the variation of parameters formula.

29. $t\mathbf{x}'(t) = \begin{bmatrix} 2 & -1 \\ 3 & -2 \end{bmatrix} \mathbf{x}(t) + \begin{bmatrix} t^{-1} \\ 1 \end{bmatrix}$

30. $t\mathbf{x}'(t) = \begin{bmatrix} 4 & -3 \\ 8 & -6 \end{bmatrix} \mathbf{x}(t) + \begin{bmatrix} t \\ 2t \end{bmatrix}$

31. Use the variation of parameters formula (10) to derive a formula for a particular solution y_p to the scalar equation $y'' + p(t)y' + q(t)y = g(t)$ in terms of two linearly independent solutions $y_1(t), y_2(t)$ of the corresponding homogeneous equation. Show that your answer agrees with the formulas derived in Section 4.6. [*Hint:* First write the scalar equation in system form.]

32. Conventional Combat Model. A simplistic model of a pair of conventional forces in combat yields the following system:

$$\mathbf{x}' = \begin{bmatrix} -a & -b \\ -c & -d \end{bmatrix} \mathbf{x} + \begin{bmatrix} p \\ q \end{bmatrix},$$

where $\mathbf{x} = \text{col}(x_1, x_2)$. The variables $x_1(t)$ and $x_2(t)$ represent the strengths of opposing forces at time t. The terms $-ax_1$ and $-dx_2$ represent the *operational loss rates*, and the terms $-bx_2$ and $-cx_1$ represent the *combat loss rates* for the troops x_1 and x_2, respectively. The constants p and q represent the respective rates of reinforcement. Let $a = 1$, $b = 4$, $c = 3$, $d = 2$, and $p = q = 5$. By solving the appropriate initial value problem, determine which forces will win if

 (a) $x_1(0) = 20$, $x_2(0) = 20$.
 (b) $x_1(0) = 21$, $x_2(0) = 20$.
 (c) $x_1(0) = 20$, $x_2(0) = 21$.

33. *RL* Network. The currents in the *RL* network given by the schematic diagram in Figure 9.8 are governed by the following equations:

$$2I_1'(t) + 90I_2(t) = 9,$$
$$I_3'(t) + 30I_4(t) - 90I_2(t) = 0,$$
$$60I_5(t) - 30I_4(t) = 0,$$
$$I_1(t) = I_2(t) + I_3(t),$$
$$I_3(t) = I_4(t) + I_5(t).$$

Figure 9.8 *RL* network for Problem 33

Assume the currents are initially zero. Solve for the five currents $I_1, \ldots, I_5$. [*Hint:* Eliminate all unknowns except I_2 and I_5, and form a normal system with $x_1 = I_2$ and $x_2 = I_5$.]

34. Mixing Problem. Two tanks A and B, each holding 50 L of liquid, are interconnected by pipes. The liquid flows from tank A into tank B at a rate of 4 L/min and from B into A at a rate of 1 L/min (see Figure 9.9). The liquid inside each tank is kept well stirred. A brine solution that has a concentration of 0.2 kg/L of salt flows into tank A at a rate of 4 L/min. A brine solution that has a concentration of 0.1 kg/L of salt flows into tank B at a rate of 1 L/min. The solutions flow out of the system from both tanks—from tank A at 1 L/min and from tank B at 4 L/min. If, initially, tank A contains pure water and tank B contains 0.5 kg of salt, determine the mass of salt in each tank at time $t \geq 0$. After several minutes have elapsed, which tank has the higher concentration of salt? What is its limiting concentration?

Figure 9.9 Mixing problem for interconnected tanks

9.8 The Matrix Exponential Function

In this chapter we have developed various ways to extend techniques for scalar differential equations to systems. In this section we take a substantial step further by showing that with the right notation, the formulas for solving normal systems with constant coefficients are identical to the formulas for solving first-order equations with constant coefficients. For example, we know that a general solution to the equation $x'(t) = ax(t)$, where a is a constant, is $x(t) = ce^{at}$. Analogously, we show that a general solution to the normal system

$$(1) \qquad \mathbf{x}'(t) = \mathbf{A}\mathbf{x}(t) \,,$$

where $\mathbf{A}$ is a constant $n \times n$ matrix, is $\mathbf{x}(t) = e^{\mathbf{A}t}\mathbf{c}$. Our first task is to define the matrix exponential $e^{\mathbf{A}t}$.

If $\mathbf{A}$ is a constant $n \times n$ matrix, we define $e^{\mathbf{A}t}$ by taking the series expansion for e^{at} and replacing a by $\mathbf{A}$; that is,

$$(2) \qquad e^{\mathbf{A}t} := \mathbf{I} + \mathbf{A}t + \mathbf{A}^2\frac{t^2}{2!} + \cdots + \mathbf{A}^n\frac{t^n}{n!} + \cdots .$$

(Note that we also replace 1 by $\mathbf{I}$.) By the right-hand side of (2), we mean the $n \times n$ matrix whose elements are power series with coefficients given by the corresponding entries in the matrices $\mathbf{I}$, $\mathbf{A}$, $\mathbf{A}^2/2!$, $\ldots$.

If $\mathbf{A}$ is a diagonal matrix, then the computation of $e^{\mathbf{A}t}$ is straightforward. For example, if

$$\mathbf{A} = \begin{bmatrix} -1 & 0 \\ 0 & 2 \end{bmatrix},$$

then

$$\mathbf{A}^2 = \mathbf{A}\mathbf{A} = \begin{bmatrix} 1 & 0 \\ 0 & 4 \end{bmatrix}, \qquad \mathbf{A}^3 = \begin{bmatrix} -1 & 0 \\ 0 & 8 \end{bmatrix}, \qquad \ldots, \qquad \mathbf{A}^n = \begin{bmatrix} (-1)^n & 0 \\ 0 & 2^n \end{bmatrix},$$

and so

$$e^{\mathbf{A}t} = \sum_{n=0}^{\infty} \mathbf{A}^n \frac{t^n}{n!} = \begin{bmatrix} \sum_{n=0}^{\infty} (-1)^n \dfrac{t^n}{n!} & 0 \\ 0 & \sum_{n=0}^{\infty} 2^n \dfrac{t^n}{n!} \end{bmatrix} = \begin{bmatrix} e^{-t} & 0 \\ 0 & e^{2t} \end{bmatrix}.$$

More generally, if $\mathbf{A}$ is an $n \times n$ *diagonal* matrix with $r_1, r_2, \ldots, r_n$ down its main diagonal, then $e^{\mathbf{A}t}$ is the diagonal matrix with $e^{r_1 t}, e^{r_2 t}, \ldots, e^{r_n t}$ down its main diagonal (see Problem 26). If $\mathbf{A}$ is not a diagonal matrix, the computation of $e^{\mathbf{A}t}$ is more involved. We deal with this important problem later in this section.

It can be shown that the series (2) converges for all t and has many of the same properties[†] as the scalar exponential e^{at}.

[†]For proofs of these and other properties of the matrix exponential function, see *Fundamentals of Matrix Analysis with Applications*, by Edward Barry Saff and Arthur David Snider (John Wiley & Sons, Hoboken, New Jersey, 2016), Chapter 7. See also the amusing articles, "Nineteen Dubious Ways to Compute the Exponential of a Matrix," by Cleve Moler and Charles van Loan, *SIAM Review*, Vol. 20, No. 4 (Oct. 1978), and "Nineteen . . . Matrix, Twenty-Five Years Later," ibid., Vol. 45, No. 1 (Jan. 2003).

Properties of the Matrix Exponential Function

Theorem 7. Let $\mathbf{A}$ and $\mathbf{B}$ be $n \times n$ constant matrices and r, s, and t be real (or complex) numbers. Then,

(a) $e^{\mathbf{A}0} = e^{0} = \mathbf{I}$.

(b) $e^{\mathbf{A}(t+s)} = e^{\mathbf{A}t}e^{\mathbf{A}s}$.

(c) $\left(e^{\mathbf{A}t}\right)^{-1} = e^{-\mathbf{A}t}$.

(d) $e^{(\mathbf{A}+\mathbf{B})t} = e^{\mathbf{A}t}e^{\mathbf{B}t}$, provided that $\mathbf{AB} = \mathbf{BA}$.

(e) $e^{r\mathbf{I}t} = e^{rt}\mathbf{I}$.

Property (c) has profound implications. First, it asserts that for any matrix $\mathbf{A}$, the matrix $e^{\mathbf{A}t}$ *has* an inverse for all t. Moreover, this inverse is obtained by simply replacing t by $-t$. (Note that (c) follows from (a) and (b) with $s = -t$.) In applying property (d) (the law of exponents), one must exercise care because of the stipulation that the matrices $\mathbf{A}$ and $\mathbf{B}$ commute (see Problem 25).

Another important property of the matrix exponential arises from the fact that we can differentiate the series in (2) term by term. This gives

$$\frac{d}{dt}\left(e^{\mathbf{A}t}\right) = \frac{d}{dt}\left(\mathbf{I} + \mathbf{A}t + \mathbf{A}^2\frac{t^2}{2} + \cdots + \mathbf{A}^n\frac{t^n}{n!} + \cdots\right)$$

$$= \mathbf{A} + \mathbf{A}^2 t + \mathbf{A}^3\frac{t^2}{2} + \cdots + \mathbf{A}^n\frac{t^{n-1}}{(n-1)!} + \cdots$$

$$= \mathbf{A}\left[\mathbf{I} + \mathbf{A}t + \mathbf{A}^2\frac{t^2}{2} + \cdots + \mathbf{A}^{n-1}\frac{t^{n-1}}{(n-1)!} + \cdots\right].$$

Hence,

$$\frac{d}{dt}\left(e^{\mathbf{A}t}\right) = \mathbf{A}e^{\mathbf{A}t},$$

and so $e^{\mathbf{A}t}$ is *a solution to the matrix differential equation* $\mathbf{X}' = \mathbf{A}\mathbf{X}$. Since $e^{\mathbf{A}t}$ is invertible [property (c)], it follows that the columns of $e^{\mathbf{A}t}$ are linearly independent solutions to system (1). Combining these facts we have the following.

e^{At} Is a Fundamental Matrix

Theorem 8. If $\mathbf{A}$ is an $n \times n$ constant matrix, then the columns of the matrix exponential $e^{\mathbf{A}t}$ form a fundamental solution set for the system $\mathbf{x}'(t) = \mathbf{A}\mathbf{x}(t)$. Therefore, $e^{\mathbf{A}t}$ is a fundamental matrix for the system, and a general solution is $\mathbf{x}(t) = e^{\mathbf{A}t}\mathbf{c}$.

If a fundamental matrix $\mathbf{X}(t)$ for the system $\mathbf{x}' = \mathbf{A}\mathbf{x}$ has somehow been determined, it is easy to compute $e^{\mathbf{A}t}$, as the next theorem describes.

Relationship Between Fundamental Matrices

Theorem 9. Let $\mathbf{X}(t)$ and $\mathbf{Y}(t)$ be two fundamental matrices for the same system $\mathbf{x}' = \mathbf{A}\mathbf{x}$. Then there exists a constant matrix $\mathbf{C}$ such that $\mathbf{Y}(t) = \mathbf{X}(t)\mathbf{C}$ for all t. In particular,

(3) $e^{\mathbf{A}t} = \mathbf{X}(t)\mathbf{X}(0)^{-1}$.

Proof. Since $\mathbf{X}(t)$ is a fundamental matrix, every column of $\mathbf{Y}(t)$ can be expressed as $\mathbf{X}(t)\mathbf{c}$ for a suitable constant vector $\mathbf{c}$, so column-by-column we have

$$
\left[\ \ \mathbf{Y}(t)\ \ \right] = \left[\ \ \mathbf{X}(t)\ \ \right]\left[\begin{array}{ccccc} \vdots & \vdots & \cdots & \vdots \\ \mathbf{c}_1 & \mathbf{c}_2 & \cdots & \mathbf{c}_n \\ \vdots & \vdots & \cdots & \vdots \end{array}\right] = \mathbf{X}(t)\mathbf{C}\,.
$$

If we choose $\mathbf{Y}(t) = e^{\mathbf{A}t} = \mathbf{X}(t)\mathbf{C}$, then (3) follows by setting $t = 0$. ◆

If the $n \times n$ matrix $\mathbf{A}$ has n linearly independent eigenvectors $\mathbf{u}_i$, then Theorem 5 of Section 9.5 (page 526) provides us with $\mathbf{X}(t)$ and (3) gives us

$$
(4) \qquad e^{\mathbf{A}t} = \left[e^{r_1 t}\mathbf{u}_1\ e^{r_2 t}\mathbf{u}_2\ \cdots\ e^{r_n t}\mathbf{u}_n\right]\left[\mathbf{u}_1\ \mathbf{u}_2\ \cdots\ \mathbf{u}_n\right]^{-1}.
$$

Are there any other ways that we can compute $e^{\mathbf{A}t}$? As we observed, if $\mathbf{A}$ is a diagonal matrix, then we simply exponentiate the diagonal elements of $\mathbf{A}t$ to obtain $e^{\mathbf{A}t}$. Also, if $\mathbf{A}$ is a **nilpotent** matrix, that is, $\mathbf{A}^k = \mathbf{0}$ for some positive integer k, then the series for $e^{\mathbf{A}t}$ has only a finite number of terms—it "truncates"—since $\mathbf{A}^k = \mathbf{A}^{k+1} = \cdots = \mathbf{0}$. In such cases, $e^{\mathbf{A}t}$ reduces to

$$
e^{\mathbf{A}t} = \mathbf{I} + \mathbf{A}t + \cdots + \mathbf{A}^{k-1}\frac{t^{k-1}}{(k-1)!} + 0 + 0 + \cdots = \mathbf{I} + \mathbf{A}t + \cdots + \mathbf{A}^{k-1}\frac{t^{k-1}}{(k-1)!}\,.
$$

Thus $e^{\mathbf{A}t}$ can be calculated in finite terms if $\mathbf{A}$ is diagonal or nilpotent. Can we take this any further? Yes; a consequence of the Cayley–Hamilton theorem[†] is that when the characteristic polynomial for $\mathbf{A}$ has the form $p(r) = (r_1 - r)^n$, then $(r_1\mathbf{I} - \mathbf{A})^n = \mathbf{0} = (-1)^n(\mathbf{A} - r_1\mathbf{I})^n$. So if $\mathbf{A}$ has only one (multiple) eigenvalue r_1, then $\mathbf{A} - r_1\mathbf{I}$ is nilpotent, and we exploit that by writing $\mathbf{A} = r_1\mathbf{I} + \mathbf{A} - r_1\mathbf{I}$:

$$
e^{\mathbf{A}t} = e^{r_1\mathbf{I}t}e^{(\mathbf{A}-r_1\mathbf{I})t} = e^{r_1 t}\left[\mathbf{I} + (\mathbf{A} - r_1\mathbf{I})t + \cdots + (\mathbf{A} - r_1\mathbf{I})^{n-1}\frac{t^{n-1}}{(n-1)!}\right].
$$

Example 1 Find the fundamental matrix $e^{\mathbf{A}t}$ for the system

$$
\mathbf{x}' = \mathbf{A}\mathbf{x}\,, \quad \text{where} \quad \mathbf{A} = \begin{bmatrix} 2 & 1 & 1 \\ 1 & 2 & 1 \\ -2 & -2 & -1 \end{bmatrix}.
$$

Solution We begin by computing the characteristic polynomial for $\mathbf{A}$:

$$
p(r) = |\mathbf{A} - r\mathbf{I}| = \begin{vmatrix} 2-r & 1 & 1 \\ 1 & 2-r & 1 \\ -2 & -2 & -1-r \end{vmatrix} = -r^3 + 3r^2 - 3r + 1 = -(r-1)^3.
$$

Thus, $r = 1$ is an eigenvalue of $\mathbf{A}$ with multiplicity 3. By the Cayley–Hamilton theorem, $(\mathbf{A} - \mathbf{I})^3 = \mathbf{0}$, and so

$$
(5) \qquad e^{\mathbf{A}t} = e^t e^{(\mathbf{A}-\mathbf{I})t} = e^t\left\{\mathbf{I} + (\mathbf{A} - \mathbf{I})t + (\mathbf{A} - \mathbf{I})^2\frac{t^2}{2}\right\}.
$$

[†]The Cayley–Hamilton theorem states that a matrix satisfies its own characteristic equation, that is, $p(\mathbf{A}) = \mathbf{0}$. For a discussion of this theorem, see *Fundamentals of Matrix Analysis with Applications*, by Edward Barry Saff and Arthur David Snider (John Wiley & Sons, Hoboken, New Jersey, 2016), Section 6.3.

Computing, we find

$$\mathbf{A} - \mathbf{I} = \begin{bmatrix} 1 & 1 & 1 \\ 1 & 1 & 1 \\ -2 & -2 & -2 \end{bmatrix} \quad \text{and} \quad (\mathbf{A} - \mathbf{I})^2 = \begin{bmatrix} 0 & 0 & 0 \\ 0 & 0 & 0 \\ 0 & 0 & 0 \end{bmatrix}.$$

Substituting into (5) yields

$$(6) \qquad e^{\mathbf{A}t} = e^t \begin{bmatrix} 1 & 0 & 0 \\ 0 & 1 & 0 \\ 0 & 0 & 1 \end{bmatrix} + te^t \begin{bmatrix} 1 & 1 & 1 \\ 1 & 1 & 1 \\ -2 & -2 & -2 \end{bmatrix} = \begin{bmatrix} e^t + te^t & te^t & te^t \\ te^t & e^t + te^t & te^t \\ -2te^t & -2te^t & e^t - 2te^t \end{bmatrix}. \quad \blacklozenge$$

We are not through with nilpotency yet. What if we have a nonzero vector $\mathbf{u}$, an exponent m, and a scalar r satisfying $(\mathbf{A} - r\mathbf{I})^m \mathbf{u} = \mathbf{0}$, so that $\mathbf{A} - r\mathbf{I}$ is "nilpotent when restricted to $\mathbf{u}$"? Such vectors are given a (predictable) name.

Generalized Eigenvectors

Definition 5. Let $\mathbf{A}$ be a square matrix. A nonzero vector $\mathbf{u}$ satisfying

$$(7) \qquad (\mathbf{A} - r\mathbf{I})^m \mathbf{u} = \mathbf{0}$$

for some scalar r and some positive integer m is called a **generalized eigenvector** associated with r.

[Note that r must be an eigenvalue of $\mathbf{A}$, since the final *nonzero* vector in the list $\mathbf{u}, (\mathbf{A} - r\mathbf{I})\mathbf{u}, (\mathbf{A} - r\mathbf{I})^2\mathbf{u}, \ldots, (\mathbf{A} - r\mathbf{I})^{m-1}\mathbf{u}$ is a "regular" eigenvector.]

A valuable feature of generalized eigenvectors $\mathbf{u}$ is that we can compute $e^{\mathbf{A}t}\mathbf{u}$ in finite terms *without knowing* $e^{\mathbf{A}t}$, because

$$e^{\mathbf{A}t}\mathbf{u} = e^{r\mathbf{I}t}e^{(\mathbf{A}-r\mathbf{I})t}\mathbf{u}$$

$$(8) \qquad = e^{rt}\left[\mathbf{I}\mathbf{u} + t(\mathbf{A} - r\mathbf{I})\mathbf{u} + \cdots + \frac{t^{m-1}}{(m-1)!}(\mathbf{A} - r\mathbf{I})^{m-1}\mathbf{u} + \frac{t^m}{m!}(\mathbf{A} - r\mathbf{I})^m\mathbf{u} + \cdots \right]$$

$$= e^{rt}\left[\mathbf{u} + t(\mathbf{A} - r\mathbf{I})\mathbf{u} + \cdots + \frac{t^{m-1}}{(m-1)!}(\mathbf{A} - r\mathbf{I})^{m-1}\mathbf{u} + \mathbf{0} + \cdots \right].$$

Moreover, $e^{\mathbf{A}t}\mathbf{u}$ is a solution to the system $\mathbf{x}' = \mathbf{A}\mathbf{x}$ (recall Theorem 8). Hence if we can find n generalized eigenvectors $\mathbf{u}_i$ for the $n \times n$ matrix $\mathbf{A}$ that are linearly independent, the corresponding solutions $\mathbf{x}_i(t) = e^{\mathbf{A}t}\mathbf{u}_i$ will form a fundamental solution set and can be assembled into a fundamental matrix $\mathbf{X}(t)$. (Since the $\mathbf{x}_i$'s are solutions that reduce to the linearly independent $\mathbf{u}_i$'s at $t = 0$, they are always linearly independent.) Finally, we get the matrix exponential by applying (3) from Theorem 9:

$$(9) \qquad e^{\mathbf{A}t} = \mathbf{X}(t)\mathbf{X}(0)^{-1} = [e^{\mathbf{A}t}\mathbf{u}_1 \; e^{\mathbf{A}t}\mathbf{u}_2 \; \cdots \; e^{\mathbf{A}t}\mathbf{u}_n][\mathbf{u}_1 \; \mathbf{u}_2 \; \cdots \; \mathbf{u}_n]^{-1},$$

computed in a finite number of steps.

Of course, any (regular) eigenvector is a generalized eigenvector (corresponding to $m = 1$), and if $\mathbf{A}$ has a full set of n linearly independent eigenvectors, then (9) is simply the representation (4). But what if $\mathbf{A}$ is *defective*, that is, possesses fewer than n linearly

independent eigenvectors? Luckily, the primary decomposition theorem in advanced linear algebra* guarantees that when the characteristic polynomial of **A** is

$$(10) \qquad p(r) = (r_1 - r)^{m_1}(r_2 - r)^{m_2} \cdots (r_k - r)^{m_k},$$

where the r_i's are the distinct eigenvalues of **A** and m_i is the multiplicity of the eigenvalue r_i, then for each i there exist m_i linearly independent generalized eigenvectors satisfying

$$(11) \qquad (\mathbf{A} - r_i\mathbf{I})^{m_i}\mathbf{u} = \mathbf{0}.$$

Furthermore, the conglomeration of these $n = m_1 + m_2 + \cdots + m_k$ generalized eigenvectors is linearly independent.[†]

We're home! The following scheme will always yield a fundamental solution set, for *any* square matrix **A**.

Solving x′ = Ax

To obtain a fundamental solution set for $\mathbf{x}' = \mathbf{Ax}$ for any constant square matrix **A**:

(a) Compute the characteristic polynomial $p(r) = |\mathbf{A} - r\mathbf{I}|$.

(b) Find the zeros of $p(r)$ and express it as $p(r) = (r_1 - r)^{m_1}(r_2 - r)^{m_2} \cdots (r_k - r)^{m_k}$, where $r_1, r_2, \ldots, r_k$ are the distinct zeros (i.e., eigenvalues) and $m_1, m_2, \ldots, m_k$ are their multiplicities.

(c) For each eigenvalue r_i find m_i linearly independent generalized eigenvectors by applying the Gauss–Jordan algorithm to the system $(\mathbf{A} - r_i\mathbf{I})^{m_i}\mathbf{u} = \mathbf{0}$.

(d) Form $n = m_1 + m_2 + \cdots + m_k$ linearly independent solutions to $\mathbf{x}' = \mathbf{Ax}$ by computing

$$\mathbf{x}(t) := e^{\mathbf{A}t}\mathbf{u} = e^{rt}\left[\mathbf{u} + t(\mathbf{A} - r\mathbf{I})\mathbf{u} + \frac{t^2}{2!}(\mathbf{A} - r\mathbf{I})^2\mathbf{u} + \cdots\right]$$

for each generalized eigenvector **u** found in part (c) and corresponding eigenvalue r. If r has multiplicity m, this series terminates after m or fewer terms.

We can then, if desired, assemble the fundamental matrix $\mathbf{X}(t)$ from the n solutions and obtain the matrix exponential $e^{\mathbf{A}t}$ using (9).[‡]

Example 2　Find the fundamental matrix $e^{\mathbf{A}t}$ for the system

$$(12) \qquad \mathbf{x}' = \mathbf{Ax}, \quad \text{where} \quad \mathbf{A} = \begin{bmatrix} 1 & 0 & 0 \\ 1 & 3 & 0 \\ 0 & 1 & 1 \end{bmatrix}.$$

Solution　We begin by finding the characteristic polynomial for **A**:

$$p(r) = |\mathbf{A} - r\mathbf{I}| = \begin{vmatrix} 1-r & 0 & 0 \\ 1 & 3-r & 0 \\ 0 & 1 & 1-r \end{vmatrix} = -(r-1)^2(r-3).$$

*See *Fundamentals of Matrix Analysis with Applications*, by Edward Barry Saff and Arthur David Snider (John Wiley & Sons, Hoboken, New Jersey, 2016), Section 7.3.

[†]Some of the generalized eigenvectors satisfy (11) with a lower exponent, but that will not affect the calculation.

[‡]With defective matrices this algorithm may prove ineffective for computer implementation; rounding effects inevitably foil the machine's ability to detect multiple eigenvalues.

Hence, the eigenvalues of $\mathbf{A}$ are $r = 1$ with multiplicity 2 and $r = 3$ with multiplicity 1.

Since $r = 1$ has multiplicity 2, we must determine two linearly independent associated generalized eigenvectors satisfying $(\mathbf{A} - \mathbf{I})^2 \mathbf{u} = \mathbf{0}$. From

$$(\mathbf{A} - \mathbf{I})^2 \mathbf{u} = \begin{bmatrix} 0 & 0 & 0 \\ 2 & 4 & 0 \\ 1 & 2 & 0 \end{bmatrix} \begin{bmatrix} u_1 \\ u_2 \\ u_3 \end{bmatrix} = \begin{bmatrix} 0 \\ 0 \\ 0 \end{bmatrix},$$

we find $u_2 = s$, $u_1 = -2u_2 = -2s$, and $u_3 = v$, where s and v are arbitrary.

Taking $s = 0$ and $v = 1$, we obtain the generalized eigenvector $\mathbf{u}_1 = \text{col}(0, 0, 1)$. The corresponding solution to (12) is

$$(13) \qquad \mathbf{x}_1(t) = e^t \{ \mathbf{u}_1 + t(\mathbf{A} - \mathbf{I})\mathbf{u}_1 \} = e^t \begin{bmatrix} 0 \\ 0 \\ 1 \end{bmatrix} + te^t \begin{bmatrix} 0 & 0 & 0 \\ 1 & 2 & 0 \\ 0 & 1 & 0 \end{bmatrix} \begin{bmatrix} 0 \\ 0 \\ 1 \end{bmatrix} = \begin{bmatrix} 0 \\ 0 \\ e^t \end{bmatrix}$$

($\mathbf{u}_1$ is, in fact, a regular eigenvector).

Next we take $s = 1$ and $v = 0$ to derive the second linearly independent generalized eigenvector $\mathbf{u}_2 = \text{col}(-2, 1, 0)$ and (linearly independent) solution

$$(14) \qquad \mathbf{x}_2(t) = e^{\mathbf{A}t}\mathbf{u}_2 = e^t \{ \mathbf{u}_2 + t(\mathbf{A} - \mathbf{I})\mathbf{u}_2 \}$$

$$= e^t \begin{bmatrix} -2 \\ 1 \\ 0 \end{bmatrix} + te^t \begin{bmatrix} 0 & 0 & 0 \\ 1 & 2 & 0 \\ 0 & 1 & 0 \end{bmatrix} \begin{bmatrix} -2 \\ 1 \\ 0 \end{bmatrix}$$

$$= e^t \begin{bmatrix} -2 \\ 1 \\ 0 \end{bmatrix} + te^t \begin{bmatrix} 0 \\ 0 \\ 1 \end{bmatrix} = \begin{bmatrix} -2e^t \\ e^t \\ te^t \end{bmatrix}.^{\dagger}$$

For the eigenvalue $r = 3$, we solve $(\mathbf{A} - 3\mathbf{I})\mathbf{u} = \mathbf{0}$, that is,

$$\begin{bmatrix} -2 & 0 & 0 \\ 1 & 0 & 0 \\ 0 & 1 & -2 \end{bmatrix} \begin{bmatrix} u_1 \\ u_2 \\ u_3 \end{bmatrix} = \begin{bmatrix} 0 \\ 0 \\ 0 \end{bmatrix},$$

to obtain the eigenvector $\mathbf{u}_3 = \text{col}(0, 2, 1)$. Hence, a third linearly independent solution to (12) is

$$(15) \qquad \mathbf{x}_3(t) = e^{3t}\mathbf{u}_3 = e^{3t} \begin{bmatrix} 0 \\ 2 \\ 1 \end{bmatrix} = \begin{bmatrix} 0 \\ 2e^{3t} \\ e^{3t} \end{bmatrix}.$$

The matrix $\mathbf{X}(t)$ whose columns are the vectors $\mathbf{x}_1(t)$, $\mathbf{x}_2(t)$, and $\mathbf{x}_3(t)$,

$$\mathbf{X}(t) = \begin{bmatrix} 0 & -2e^t & 0 \\ 0 & e^t & 2e^{3t} \\ e^t & te^t & e^{3t} \end{bmatrix},$$

†Note that $\mathbf{x}_2(t)$ has been expressed in the format discussed in Problem 35 of Exercises 9.5, page 532.

is a fundamental matrix for (12). Setting $t = 0$ and then computing $\mathbf{X}^{-1}(0)$, we find

$$\mathbf{X}(0) = \begin{bmatrix} 0 & -2 & 0 \\ 0 & 1 & 2 \\ 1 & 0 & 1 \end{bmatrix} \quad \text{and} \quad \mathbf{X}^{-1}(0) = \begin{bmatrix} -\frac{1}{4} & -\frac{1}{2} & 1 \\ -\frac{1}{2} & 0 & 0 \\ \frac{1}{4} & \frac{1}{2} & 0 \end{bmatrix}.$$

It now follows from formula (3) that

$$e^{\mathbf{A}t} = \mathbf{X}(t)\mathbf{X}^{-1}(0) = \begin{bmatrix} 0 & -2e^t & 0 \\ 0 & e^t & 2e^{3t} \\ e^t & te^t & e^{3t} \end{bmatrix} \begin{bmatrix} -\frac{1}{4} & -\frac{1}{2} & 1 \\ -\frac{1}{2} & 0 & 0 \\ \frac{1}{4} & \frac{1}{2} & 0 \end{bmatrix}$$

$$= \begin{bmatrix} e^t & 0 & 0 \\ -\frac{1}{2}e^t + \frac{1}{2}e^{3t} & e^{3t} & 0 \\ -\frac{1}{4}e^t - \frac{1}{2}te^t + \frac{1}{4}e^{3t} & -\frac{1}{2}e^t + \frac{1}{2}e^{3t} & e^t \end{bmatrix}. \quad \blacklozenge$$

Use of the fundamental matrix $e^{\mathbf{A}t}$ simplifies many computations. For example, the properties $e^{\mathbf{A}t}e^{-\mathbf{A}s} = e^{\mathbf{A}(t-s)}$ and $\left(e^{\mathbf{A}t_0}\right)^{-1} = e^{-\mathbf{A}t_0}$ enable us to rewrite the variation of parameters formula (13) in Section 9.7 (page 541) in a simpler form. Namely, the solution to the initial value problem $\mathbf{x}' = \mathbf{A}\mathbf{x} + \mathbf{f}(t)$, $\mathbf{x}(t_0) = \mathbf{x}_0$ is given by

$$(16) \qquad \mathbf{x}(t) = e^{\mathbf{A}(t-t_0)}\mathbf{x}_0 + \int_{t_0}^{t} e^{\mathbf{A}(t-s)}\mathbf{f}(s)\,ds\,,$$

which is a system version of the formula for the solution to the scalar initial value problem $x' = ax + f(t)$, $x(t_0) = x_0$.

In closing, we remark that the software packages for eigenvalue computation listed in Section 9.5 (page 524) also contain subroutines for computing the matrix exponential.

9.8 EXERCISES

In Problems 1–6, (a) show that the given matrix $\mathbf{A}$ satisfies $(\mathbf{A} - r\mathbf{I})^k = \mathbf{0}$ for some number r and some positive integer k and (b) use this fact to determine the matrix $e^{\mathbf{A}t}$. [Hint: Compute the characteristic polynomial and use the Cayley–Hamilton theorem.]

1. $\mathbf{A} = \begin{bmatrix} 3 & -2 \\ 0 & 3 \end{bmatrix}$

2. $\mathbf{A} = \begin{bmatrix} 1 & -1 \\ 1 & 3 \end{bmatrix}$

3. $\mathbf{A} = \begin{bmatrix} 2 & 1 & -1 \\ -3 & -1 & 1 \\ 9 & 3 & -4 \end{bmatrix}$

4. $\mathbf{A} = \begin{bmatrix} 2 & 1 & 3 \\ 0 & 2 & -1 \\ 0 & 0 & 2 \end{bmatrix}$

5. $\mathbf{A} = \begin{bmatrix} -2 & 0 & 0 \\ 4 & -2 & 0 \\ 1 & 0 & -2 \end{bmatrix}$

6. $\mathbf{A} = \begin{bmatrix} 0 & 1 & 0 \\ 0 & 0 & 1 \\ -1 & -3 & -3 \end{bmatrix}$

In Problems 7–10, determine e^{At} by first finding a fundamental matrix $\mathbf{X}(t)$ for $\mathbf{x}' = \mathbf{Ax}$ and then using formula (3).

7. $\mathbf{A} = \begin{bmatrix} 0 & 1 \\ -1 & 0 \end{bmatrix}$

8. $\mathbf{A} = \begin{bmatrix} 1 & 1 \\ 4 & 1 \end{bmatrix}$

9. $\mathbf{A} = \begin{bmatrix} 0 & 1 & 0 \\ 0 & 0 & 1 \\ 1 & -1 & 1 \end{bmatrix}$

10. $\mathbf{A} = \begin{bmatrix} 0 & 2 & 2 \\ 2 & 0 & 2 \\ 2 & 2 & 0 \end{bmatrix}$

In Problems 11 and 12, determine e^{At} by using generalized eigenvectors to find a fundamental matrix and then using formula (3).

11. $\mathbf{A} = \begin{bmatrix} 5 & -4 & 0 \\ 1 & 0 & 2 \\ 0 & 2 & 5 \end{bmatrix}$

12. $\mathbf{A} = \begin{bmatrix} 1 & 1 & 1 \\ 2 & 1 & -1 \\ 0 & -1 & 1 \end{bmatrix}$

In Problems 13–16 determine e^{At}, using a linear software package to find the eigenvectors.

13. $\mathbf{A} = \begin{bmatrix} 0 & 1 & 0 & 0 & 0 \\ 0 & 0 & 1 & 0 & 0 \\ 1 & -3 & 3 & 0 & 0 \\ 0 & 0 & 0 & 0 & 1 \\ 0 & 0 & 0 & -1 & 0 \end{bmatrix}$

14. $\mathbf{A} = \begin{bmatrix} 1 & 0 & 0 & 0 & 0 \\ 0 & 0 & 1 & 0 & 0 \\ 0 & -1 & -2 & 0 & 0 \\ 0 & 0 & 0 & 0 & 1 \\ 0 & 0 & 0 & -1 & 0 \end{bmatrix}$

15. $\mathbf{A} = \begin{bmatrix} 0 & 1 & 0 & 0 & 0 \\ 0 & 0 & 1 & 0 & 0 \\ -1 & -3 & -3 & 0 & 0 \\ 0 & 0 & 0 & 0 & 1 \\ 0 & 0 & 0 & -4 & -4 \end{bmatrix}$

16. $\mathbf{A} = \begin{bmatrix} -1 & 0 & 0 & 0 & 0 \\ 0 & 0 & 1 & 0 & 0 \\ 0 & -1 & -2 & 0 & 0 \\ 0 & 0 & 0 & 0 & 1 \\ 0 & 0 & 0 & -4 & -4 \end{bmatrix}$

In Problems 17–20, use the generalized eigenvectors of $\mathbf{A}$ to find a general solution to the system $\mathbf{x}'(t) = \mathbf{Ax}(t)$, where $\mathbf{A}$ is given.

17. $\mathbf{A} = \begin{bmatrix} 0 & 1 & 0 \\ 0 & 0 & 1 \\ -2 & -5 & -4 \end{bmatrix}$

18. $\mathbf{A} = \begin{bmatrix} 0 & 0 & 1 \\ 0 & 1 & 2 \\ 0 & 0 & 1 \end{bmatrix}$

19. $\mathbf{A} = \begin{bmatrix} 1 & 0 & 1 & 2 \\ 1 & 1 & 2 & 1 \\ 0 & 0 & 2 & 0 \\ 0 & 0 & 1 & 1 \end{bmatrix}$

20. $\mathbf{A} = \begin{bmatrix} -1 & -8 & 1 \\ -1 & -3 & 2 \\ -4 & -16 & 7 \end{bmatrix}$

21. Use the results of Problem 5 to find the solution to the initial value problem
$$\mathbf{x}'(t) = \begin{bmatrix} -2 & 0 & 0 \\ 4 & -2 & 0 \\ 1 & 0 & -2 \end{bmatrix} \mathbf{x}(t), \quad \mathbf{x}(0) = \begin{bmatrix} 1 \\ 1 \\ -1 \end{bmatrix}.$$

22. Use your answer to Problem 12 to find the solution to the initial value problem
$$\mathbf{x}'(t) = \begin{bmatrix} 1 & 1 & 1 \\ 2 & 1 & -1 \\ 0 & -1 & 1 \end{bmatrix} \mathbf{x}(t), \quad \mathbf{x}(0) = \begin{bmatrix} -1 \\ 0 \\ 3 \end{bmatrix}.$$

23. Use the results of Problem 3 and the variation of parameters formula (16) to find the solution to the initial value problem
$$\mathbf{x}'(t) = \begin{bmatrix} 2 & 1 & -1 \\ -3 & -1 & 1 \\ 9 & 3 & -4 \end{bmatrix} \mathbf{x}(t) + \begin{bmatrix} 0 \\ t \\ 0 \end{bmatrix},$$
$$\mathbf{x}(0) = \begin{bmatrix} 0 \\ 3 \\ 0 \end{bmatrix}.$$

24. Use your answer to Problem 9 and the variation of parameters formula (16) to find the solution to the initial value problem
$$\mathbf{x}'(t) = \begin{bmatrix} 0 & 1 & 0 \\ 0 & 0 & 1 \\ 1 & -1 & 1 \end{bmatrix} \mathbf{x}(t) + \begin{bmatrix} 0 \\ 0 \\ t \end{bmatrix},$$
$$\mathbf{x}(0) = \begin{bmatrix} 1 \\ -1 \\ 0 \end{bmatrix}.$$

25. Let
$$\mathbf{A} = \begin{bmatrix} 1 & 2 \\ -1 & 3 \end{bmatrix} \quad \text{and} \quad \mathbf{B} = \begin{bmatrix} 2 & 1 \\ 0 & 1 \end{bmatrix}.$$
(a) Show that $\mathbf{AB} \neq \mathbf{BA}$.
(b) Show that property (d) in Theorem 7 does not hold for these matrices. That is, show that $e^{(\mathbf{A}+\mathbf{B})t} \neq e^{\mathbf{A}t}e^{\mathbf{B}t}$.

26. Let $\mathbf{A}$ be a diagonal $n \times n$ matrix with entries $r_1, \ldots, r_n$ down its main diagonal. To compute $e^{\mathbf{A}t}$, proceed as follows:
(a) Show that $\mathbf{A}^k$ is the diagonal matrix with entries $r_1^k, \ldots, r_n^k$ down its main diagonal.
(b) Use the result of part (a) and the defining equation (2) to show that $e^{\mathbf{A}t}$ is the diagonal matrix with entries $e^{r_1 t}, \ldots, e^{r_n t}$ down its main diagonal.

27. In Problems 35–40 of Exercises 9.5, page 532, some ad hoc formulas were invoked to find general solutions to the system $\mathbf{x}' = \mathbf{Ax}$ when $\mathbf{A}$ had repeated eigenvalues.

Using the generalized eigenvector procedure outlined on page 548, justify the ad hoc formulas proposed in

(a) Problem 35, Exercises 9.5.
(b) Problem 37, Exercises 9.5.
(c) Problem 39, Exercises 9.5.

28. Let

$$\mathbf{A} = \begin{bmatrix} 5 & 2 & -4 \\ 0 & 3 & 0 \\ 4 & -5 & -5 \end{bmatrix}$$

(a) Find a general solution to $\mathbf{x}' = \mathbf{A}\mathbf{x}$.

(b) Determine which initial conditions $\mathbf{x}(0) = \mathbf{x}_0$ yield a solution $\mathbf{x}(t) = \mathrm{col}\big(x_1(t), x_2(t), x_3(t)\big)$ that remains bounded for all $t \geq 0$; that is, satisfies

$$||\mathbf{x}(t)|| := \sqrt{x_1^2(t) + x_2^2(t) + x_3^2(t)} \leq M$$

for some constant M and all $t \geq 0$.

29. For the matrix $\mathbf{A}$ in Problem 28, solve the initial value problem

$$\mathbf{x}'(t) = \mathbf{A}\mathbf{x}(t) + \sin(2t)\begin{bmatrix} 2 \\ 0 \\ 4 \end{bmatrix},$$

$$\mathbf{x}(0) = \begin{bmatrix} 0 \\ 1 \\ -1 \end{bmatrix}.$$

Chapter **9** Summary

In this chapter we discussed the theory of linear systems in normal form and presented methods for solving such systems. The theory and methods are natural extensions of the development for second-order and higher-order linear equations. The important properties and techniques are as follows.

Homogeneous Normal Systems

$$\mathbf{x}'(t) = \mathbf{A}(t)\mathbf{x}(t)$$

The $n \times n$ matrix function $\mathbf{A}(t)$ is assumed to be continuous on an interval I.

Fundamental Solution Set: $\{\mathbf{x}_1, \ldots, \mathbf{x}_n\}$. A set of n vector solutions $\mathbf{x}_1(t), \ldots, \mathbf{x}_n(t)$ of the homogeneous system on the interval I form a **fundamental solution set** provided they are linearly independent on I or, equivalently, their **Wronskian**

$$W[\mathbf{x}_1, \ldots, \mathbf{x}_n](t) := \det[\mathbf{x}_1, \ldots, \mathbf{x}_n] = \begin{vmatrix} x_{1,1}(t) & x_{1,2}(t) & \cdots & x_{1,n}(t) \\ x_{2,1}(t) & x_{2,2}(t) & \cdots & x_{2,n}(t) \\ \vdots & \vdots & & \vdots \\ x_{n,1}(t) & x_{n,2}(t) & \cdots & x_{n,n}(t) \end{vmatrix}$$

is never zero on I.

Fundamental Matrix: $\mathbf{X}(t)$. An $n \times n$ matrix function $\mathbf{X}(t)$ whose column vectors form a fundamental solution set for the homogeneous system is called a **fundamental matrix**. The determinant of $\mathbf{X}(t)$ is the Wronskian of the fundamental solution set. Since the Wronskian is never zero on the interval I, then $\mathbf{X}^{-1}(t)$ exists for t in I.

General Solution to Homogeneous System: $\mathbf{Xc} = c_1\mathbf{x}_1 + \cdots + c_n\mathbf{x}_n$. If $\mathbf{X}(t)$ is a fundamental matrix whose column vectors are $\mathbf{x}_1, \ldots, \mathbf{x}_n$, then a general solution to the homogeneous system is

$$\mathbf{x}(t) = \mathbf{X}(t)\mathbf{c} = c_1\mathbf{x}_1(t) + c_2\mathbf{x}_2(t) + \cdots + c_n\mathbf{x}_n(t) \, ,$$

where $\mathbf{c} = \text{col}(c_1, \ldots, c_n)$ is an arbitrary constant vector.

Homogeneous Systems with Constant Coefficients. The form of a general solution for a homogeneous system with constant coefficients depends on the eigenvalues and eigenvectors of the $n \times n$ constant matrix $\mathbf{A}$. An **eigenvalue** of $\mathbf{A}$ is a number r such that the system $\mathbf{Au} = r\mathbf{u}$ has a nontrivial solution $\mathbf{u}$, called an **eigenvector** of $\mathbf{A}$ associated with the eigenvalue r. Finding the eigenvalues of A is equivalent to finding the roots of the **characteristic equation**

$$|\mathbf{A} - r\mathbf{I}| = 0 \, .$$

The corresponding eigenvectors are found by solving the system $(\mathbf{A} - r\mathbf{I})\mathbf{u} = \mathbf{0}$.

If the matrix $\mathbf{A}$ has n linearly independent eigenvectors $\mathbf{u}_1, \ldots, \mathbf{u}_n$ and r_i is the eigenvalue corresponding to the eigenvector $\mathbf{u}_i$, then

$$\{e^{r_1 t}\mathbf{u}_1, e^{r_2 t}\mathbf{u}_2, \ldots, e^{r_n t}\mathbf{u}_n\}$$

is a fundamental solution set for the homogeneous system. A class of matrices that always has n linearly independent eigenvectors is the set of **symmetric** matrices—that is, matrices that satisfy $\mathbf{A} = \mathbf{A}^T$.

If $\mathbf{A}$ has complex conjugate eigenvalues $\alpha \pm i\beta$ and associated eigenvectors $\mathbf{z} = \mathbf{a} \pm i\mathbf{b}$, where $\mathbf{a}$ and $\mathbf{b}$ are real vectors, then two linearly independent real vector solutions to the homogeneous system are

$$e^{\alpha t}\cos\beta t\, \mathbf{a} - e^{\alpha t}\sin\beta t\, \mathbf{b} \, , \qquad e^{\alpha t}\sin\beta t\, \mathbf{a} + e^{\alpha t}\cos\beta t\, \mathbf{b} \, .$$

When $\mathbf{A}$ has a repeated eigenvalue r of multiplicity m, then it is possible that $\mathbf{A}$ does *not* have n linearly independent eigenvectors. However, associated with r are m linearly independent **generalized eigenvectors** that can be used to generate m linearly independent solutions to the homogeneous system (see page 548 under "Generalized Eigenvectors").

Nonhomogeneous Normal Systems

$$\mathbf{x}'(t) = \mathbf{A}(t)\mathbf{x}(t) + \mathbf{f}(t)$$

The $n \times n$ matrix function $\mathbf{A}(t)$ and the vector function $\mathbf{f}(t)$ are assumed continuous on an interval I.

General Solution to Nonhomogeneous System: $\mathbf{x}_p + \mathbf{Xc}$. If $\mathbf{x}_p(t)$ is any particular solution for the nonhomogeneous system and $\mathbf{X}(t)$ is a fundamental matrix for the associated homogeneous system, then a general solution for the nonhomogeneous system is

$$\mathbf{x}(t) = \mathbf{x}_p(t) + \mathbf{X}(t)\mathbf{c} = \mathbf{x}_p(t) + c_1\mathbf{x}_1(t) + \cdots + c_n\mathbf{x}_n(t) \, ,$$

where $\mathbf{x}_1(t), \ldots, \mathbf{x}_n(t)$ are the column vectors of $\mathbf{X}(t)$ and $\mathbf{c} = \text{col}(c_1, \ldots, c_n)$ is an arbitrary constant vector.

Undetermined Coefficients. If the nonhomogeneous term $\mathbf{f}(t)$ is a vector whose components are polynomials, exponential or sinusoidal functions, and $\mathbf{A}$ is a constant matrix, then one can use an extension of the method of undetermined coefficients to decide the form of a particular solution to the nonhomogeneous system.

Variation of Parameters: $X(t)v(t)$. Let $X(t)$ be a fundamental matrix for the homogeneous system. A particular solution to the nonhomogeneous system is given by the **variation of parameters** formula

$$x_p(t) = X(t)v(t) = X(t)\int X^{-1}(t)f(t)\,dt\,.$$

Matrix Exponential Function

If A is a constant $n \times n$ matrix, then the matrix exponential function

$$e^{At} := I + At + A^2\frac{t^2}{2!} + \cdots + A^n\frac{t^n}{n!} + \cdots$$

is a fundamental matrix for the homogeneous system $x'(t) = Ax(t)$. The matrix exponential has some of the same properties as the scalar exponential e^{at}. In particular,

$$e^0 = I\,, \qquad e^{A(t+s)} = e^{At}e^{As}\,, \qquad (e^{At})^{-1} = e^{-At}\,.$$

If $(A - rI)^k = 0$ for some r and k, then the series for e^{At} has only a finite number of terms:

$$e^{At} = e^{rt}\left\{I + (A - rI)t + \cdots + (A - rI)^{k-1}\frac{t^{k-1}}{(k-1)!}\right\}\,.$$

The matrix exponential function e^{At} can also be computed from any fundamental matrix $X(t)$ via the formula

$$e^{At} = X(t)X^{-1}(0)\,.$$

Generalized Eigenvectors

If an eigenvalue r_i of a constant $n \times n$ matrix A has multiplicity m_i there exist m_i linearly independent generalized eigenvectors u satisfying $(A - r_iI)^{m_i}u = 0$. Each such u determines a solution to the system $x' = Ax$ of the form

$$x(t) = e^{r_it}\left[I + (A - r_iI)t + \cdots + (A - r_iI)^{m_i-1}\frac{t^{m_i-1}}{(m_i - 1)!}\right]u$$

and the totality of all such solutions is linearly independent on $(-\infty, \infty)$ and forms a fundamental solution set for the system.

REVIEW PROBLEMS FOR CHAPTER 9

In Problems 1–4, find a general solution for the system $x'(t) = Ax(t)$, *where A is given.*

1. $A = \begin{bmatrix} 6 & -3 \\ 2 & 1 \end{bmatrix}$

2. $A = \begin{bmatrix} 3 & 2 \\ -5 & 1 \end{bmatrix}$

3. $A = \begin{bmatrix} 1 & 2 & 0 & 0 \\ 2 & 1 & 0 & 0 \\ 0 & 0 & 1 & 2 \\ 0 & 0 & 2 & 1 \end{bmatrix}$

4. $A = \begin{bmatrix} 1 & 1 & 0 \\ 0 & 1 & 0 \\ 0 & 0 & 2 \end{bmatrix}$

In Problems 5 and 6, find a fundamental matrix for the system $x'(t) = Ax(t)$, *where A is given.*

5. $A = \begin{bmatrix} 1 & -1 \\ 2 & 4 \end{bmatrix}$

6. $A = \begin{bmatrix} 5 & 0 & 0 \\ 0 & -4 & 3 \\ 0 & 3 & 4 \end{bmatrix}$

In Problems 7–10, find a general solution for the system $\mathbf{x}'(t) = \mathbf{A}\mathbf{x}(t) + \mathbf{f}(t)$, *where* $\mathbf{A}$ *and* $\mathbf{f}(t)$ *are given.*

7. $\mathbf{A} = \begin{bmatrix} 1 & 1 \\ 4 & 1 \end{bmatrix}$, $\quad \mathbf{f}(t) = \begin{bmatrix} 5 \\ 6 \end{bmatrix}$

8. $\mathbf{A} = \begin{bmatrix} -4 & 2 \\ 2 & -1 \end{bmatrix}$, $\quad \mathbf{f}(t) = \begin{bmatrix} e^{4t} \\ 3e^{4t} \end{bmatrix}$

9. $\mathbf{A} = \begin{bmatrix} 2 & 1 & -1 \\ -3 & -1 & 1 \\ 9 & 3 & -4 \end{bmatrix}$, $\quad \mathbf{f}(t) = \begin{bmatrix} t \\ 0 \\ 1 \end{bmatrix}$

10. $\mathbf{A} = \begin{bmatrix} 2 & -2 & 3 \\ 0 & 3 & 2 \\ 0 & -1 & 2 \end{bmatrix}$, $\quad \mathbf{f}(t) = \begin{bmatrix} e^{-t} \\ 2 \\ 1 \end{bmatrix}$

In Problems 11 and 12, solve the given initial value problem.

11. $\mathbf{x}'(t) = \begin{bmatrix} 0 & 1 \\ -2 & 3 \end{bmatrix}\mathbf{x}(t)$, $\quad \mathbf{x}(0) = \begin{bmatrix} 1 \\ -1 \end{bmatrix}$

12. $\mathbf{x}'(t) = \begin{bmatrix} 2 & 1 \\ -4 & 2 \end{bmatrix}\mathbf{x}(t) + \begin{bmatrix} te^{2t} \\ e^{2t} \end{bmatrix}$, $\quad \mathbf{x}(0) = \begin{bmatrix} 2 \\ 2 \end{bmatrix}$

In Problems 13 and 14, find a general solution for the Cauchy–Euler system $t\mathbf{x}'(t) = \mathbf{A}\mathbf{x}(t)$, *where* $\mathbf{A}$ *is given.*

13. $\mathbf{A} = \begin{bmatrix} 0 & 3 & 1 \\ 1 & 2 & 1 \\ 1 & 3 & 0 \end{bmatrix}$ $\qquad$ **14.** $\mathbf{A} = \begin{bmatrix} 1 & 2 & -1 \\ 2 & 1 & 1 \\ -1 & 1 & 0 \end{bmatrix}$

In Problems 15 and 16, find the fundamental matrix $e^{\mathbf{A}t}$ *for the system* $\mathbf{x}'(t) = \mathbf{A}\mathbf{x}(t)$, *where* $\mathbf{A}$ *is given.*

15. $\mathbf{A} = \begin{bmatrix} 4 & 2 & 3 \\ 2 & 1 & 2 \\ -1 & 2 & 0 \end{bmatrix}$ $\qquad$ **16.** $\mathbf{A} = \begin{bmatrix} 0 & 1 & 4 \\ 0 & 0 & 2 \\ 0 & 0 & 0 \end{bmatrix}$

17. For each of the following, determine if the statement made is *True* or *False*.

(a) Every $n \times n$ matrix of real numbers has n linearly independent eigenvectors.

(b) If $\mathbf{X}(t)$ is a fundamental matrix for $\mathbf{x}' = \mathbf{A}\mathbf{x}$, where $\mathbf{A} = \begin{bmatrix} 1 & 3 \\ 8 & 4 \end{bmatrix}$, then the product $\mathbf{X}(t)\begin{bmatrix} 2 & 3 \\ 4 & 6 \end{bmatrix}$ is also a fundamental matrix for this system.

(c) The vector functions $\begin{bmatrix} 0 \\ 1 \\ e^t \end{bmatrix}$, $\begin{bmatrix} 0 \\ t \\ te^t \end{bmatrix}$, $\begin{bmatrix} 0 \\ 1 \\ 2e^t \end{bmatrix}$ are linearly dependent on $(-\infty, \infty)$.

(d) If $\mathbf{A}$ is an $n \times n$ matrix of real numbers and $\mathbf{x}^*(t)$ satisfies the initial value problem $\mathbf{x}' = \mathbf{A}\mathbf{x}$, $\mathbf{x}(5) = \mathbf{0}$, then $\mathbf{x}^*(t) = \mathbf{0}$ for all t.

(e) If $\mathbf{A}$ is a 3×3 matrix of real numbers whose eigenvalues are 2, 2, and 1, then $\mathbf{A}$ has two linearly independent generalized eigenvectors that correspond to the eigenvalue 2.

(f) The Wronskian $W[\mathbf{x}_1, \mathbf{x}_2, \ldots, \mathbf{x}_n](t)$ of n linearly independent $n \times 1$ vector functions $\mathbf{x}_i$ on an open interval I is never zero on I.

(g) If a 3×3 matrix of real numbers has eigenvalues $2 + 3i$, $2 - 3i$, and 5, then the matrix has three linearly independent eigenvectors.

TECHNICAL WRITING EXERCISES FOR CHAPTER 9

1. Explain how the theory of homogeneous linear differential equations (as described in Section 6.1) follows from the theory of linear systems in normal form (as described in Section 9.4).

2. Discuss the similarities and differences between the method for finding solutions to a linear constant-coefficient differential equation (see Chapters 4 and 6) and the method for finding solutions to a linear system in normal form that has constant coefficients (see Sections 9.5 and 9.6).

3. Explain how the variation of parameters formulas for linear second-order equations derived in Section 4.6 follow

from the formulas derived in Section 9.7 for linear systems in normal form.

4. Explain how you would define the matrix functions $\sin \mathbf{A}t$ and $\cos \mathbf{A}t$, where $\mathbf{A}$ is a constant $n \times n$ matrix. How are these functions related to the matrix exponential and how are they connected to the solutions of the system $\mathbf{x}'' + \mathbf{A}^2\mathbf{x} = \mathbf{0}$? Your discussion should include the cases when $\mathbf{A}$ is

$$\begin{bmatrix} 0 & 0 \\ 0 & 0 \end{bmatrix}, \begin{bmatrix} 0 & 1 \\ 0 & 0 \end{bmatrix}, \begin{bmatrix} 0 & 0 \\ 1 & 0 \end{bmatrix}, \text{ and } \begin{bmatrix} 1 & -1 \\ 1 & -1 \end{bmatrix}.$$

Projects for Chapter 9

A Uncoupling Normal Systems

The easiest normal systems to solve are systems of the form

(1) $\quad \mathbf{x}'(t) = \mathbf{D}\mathbf{x}(t)$,

where $\mathbf{D}$ is an $n \times n$ diagonal matrix. Such a system actually consists of n uncoupled equations

(2) $\quad x_i'(t) = d_{ii}x_i(t)$, $\quad i = 1, \ldots, n$,

whose solution is

$$x_i(t) = c_i e^{d_{ii}t},$$

where the c_i's are arbitrary constants. This raises the following question: When can we *uncouple* a normal system?

To answer this question, we need the following result from linear algebra. An $n \times n$ matrix $\mathbf{A}$ is diagonalizable if and only if $\mathbf{A}$ has n linearly independent eigenvectors $\mathbf{p}_1, \ldots, \mathbf{p}_n$. Moreover, if $\mathbf{P}$ is the matrix whose columns are $\mathbf{p}_1, \ldots, \mathbf{p}_n$, then

(3) $\quad \mathbf{P}^{-1}\mathbf{A}\mathbf{P} = \mathbf{D}$,

where $\mathbf{D}$ is the diagonal matrix whose entry d_{ii} is the eigenvalue associated with the vector $\mathbf{p}_i$.

(a) Use the above result to show that the system

 (4) $\quad \mathbf{x}'(t) = \mathbf{A}\mathbf{x}(t)$,

 where $\mathbf{A}$ is an $n \times n$ diagonalizable matrix, is equivalent to an uncoupled system

 (5) $\quad \mathbf{y}'(t) = \mathbf{D}\mathbf{y}(t)$,

 where $\mathbf{y} = \mathbf{P}^{-1}\mathbf{x}$ and $\mathbf{D} = \mathbf{P}^{-1}\mathbf{A}\mathbf{P}$.

(b) Solve system (5).

(c) Use the results of parts (a) and (b) to show that a general solution to (4) is given by

$$\mathbf{x}(t) = c_1 e^{d_{11}t}\mathbf{p}_1 + c_2 e^{d_{22}t}\mathbf{p}_2 + \cdots + c_n e^{d_{nn}t}\mathbf{p}_n.$$

(d) Use the procedure discussed in parts (a)–(c) to obtain a general solution for the system

$$\mathbf{x}'(t) = \begin{bmatrix} 5 & -4 & 4 \\ 1 & 0 & 1 \\ -1 & 2 & 1 \end{bmatrix} \mathbf{x}(t).$$

Specify $\mathbf{P}$, $\mathbf{D}$, $\mathbf{P}^{-1}$, and $\mathbf{y}$.

B | Matrix Laplace Transform Method

The Laplace transform method for solving systems of linear differential equations with constant coefficients was discussed in Chapter 7. To apply the procedure for equations given in matrix form, we first extend the definition of the Laplace operator $\mathscr{L}$ to a column vector of functions $\mathbf{x} = \text{col}(x_1(t), \dots, x_n(t))$ by taking the transform of each component:

$$\mathscr{L}\{\mathbf{x}\}(s) := \text{col}\left(\mathscr{L}\{x_1\}(s), \dots, \mathscr{L}\{x_n\}(s)\right) .$$

With this notation, the vector analogue of the important property relating the Laplace transform of the derivative of a function (see Theorem 4, Chapter 7, page 362) becomes

(6) $$\mathscr{L}\{\mathbf{x}'\}(s) = s\mathscr{L}\{\mathbf{x}\}(s) - \mathbf{x}(0) .$$

Now suppose we are given the initial value problem

(7) $$\mathbf{x}' = \mathbf{A}\mathbf{x} + \mathbf{f}(t) , \qquad \mathbf{x}(0) = \mathbf{x}_0 ,$$

where $\mathbf{A}$ is a constant $n \times n$ matrix. Let $\hat{\mathbf{x}}(s)$ denote the Laplace transform of $\mathbf{x}(t)$ and $\hat{\mathbf{f}}(s)$ denote the transform of $\mathbf{f}(t)$. Then, taking the transform of the system and using the relation (6), we get

$$\mathscr{L}\{\mathbf{x}'\} = \mathscr{L}\{\mathbf{A}\mathbf{x} + \mathbf{f}\} ,$$
$$s\hat{\mathbf{x}} - \mathbf{x}_0 = \mathbf{A}\hat{\mathbf{x}} + \hat{\mathbf{f}} .$$

Next we collect the $\hat{\mathbf{x}}$ terms and solve for $\hat{\mathbf{x}}$ by premultiplying by $(s\mathbf{I} - \mathbf{A})^{-1}$:

$$(s\mathbf{I} - \mathbf{A})\hat{\mathbf{x}} = \hat{\mathbf{f}} + \mathbf{x}_0 ,$$
$$\hat{\mathbf{x}} = (s\mathbf{I} - \mathbf{A})^{-1}(\hat{\mathbf{f}} + x_0) .$$

Finally, we obtain the solution $\mathbf{x}(t)$ by taking the inverse Laplace transform:

(8) $$\mathbf{x} = \mathscr{L}^{-1}\{\hat{\mathbf{x}}\} = \mathscr{L}^{-1}\{(s\mathbf{I} - \mathbf{A})^{-1}(\hat{\mathbf{f}} + x_0)\} .$$

In applying the matrix Laplace transform method, it is straightforward (but possibly tedious) to compute $(s\mathbf{I} - \mathbf{A})^{-1}$, but the computation of the inverse transform may require some of the special techniques (such as partial fractions) discussed in Chapter 7.

(a) In the above procedure, we used the property that $\mathscr{L}\{\mathbf{A}\mathbf{x}\} = \mathbf{A}\mathscr{L}\{\mathbf{x}\}$ for any constant $n \times n$ matrix $\mathbf{A}$. Show that this property follows from the linearity of the transform in the scalar case.

(b) Use the matrix Laplace transform method to solve the following initial value problems:

(i) $\mathbf{x}'(t) = \begin{bmatrix} 0 & 2 \\ -1 & 3 \end{bmatrix} \mathbf{x}(t) , \qquad \mathbf{x}(0) = \begin{bmatrix} -1 \\ 3 \end{bmatrix} .$

(ii) $\mathbf{x}'(t) = \begin{bmatrix} 3 & -2 \\ 4 & -1 \end{bmatrix} \mathbf{x}(t) + \begin{bmatrix} \sin t \\ -\cos t \end{bmatrix} , \qquad \mathbf{x}(0) = \begin{bmatrix} 0 \\ 0 \end{bmatrix} .$

(c) By comparing the Laplace transform solution formula (8) with the matrix exponential solution formula given in Section 9.8 [relation (16), page 551] for the homogeneous case $\mathbf{f}(t) \equiv \mathbf{0}$ and $t_0 = 0$, derive the Laplace transform formula for the matrix exponential

(9) $$e^{\mathbf{A}t} = \mathscr{L}^{-1}\{(s\mathbf{I} - \mathbf{A})^{-1}\}(t) .$$

(d) What is $e^{\mathbf{A}t}$ for the coefficient matrices in part (b) above?

(e) Use (9) to rework Problems 1, 2, 7, and 8 in Exercises 9.8, page 551.

C Undamped Second-Order Systems

We have seen that the coupled mass–spring system depicted in Figure 9.5, page 536, is governed by equations (10) of Section 9.6, which we reproduce here:

$$m_1 x_1'' = -k_1 x_1 + k_2(x_2 - x_1),$$
$$m_2 x_2'' = -k_2(x_2 - x_1) - k_3 x_2.$$

This system was rewritten in normal form as equation (11) in Section 9.6; however, there is some advantage in expressing it as a second-order system in the form

$$\begin{bmatrix} m_1 & 0 \\ 0 & m_2 \end{bmatrix} \begin{bmatrix} x_1 \\ x_2 \end{bmatrix}'' = -\begin{bmatrix} k_1 + k_2 & -k_2 \\ -k_2 & k_2 + k_3 \end{bmatrix} \begin{bmatrix} x_1 \\ x_2 \end{bmatrix}.$$

This structure,

(10) $\mathbf{Mx}'' = -\mathbf{Kx}$,

with a diagonal *mass matrix* $\mathbf{M}$ and a symmetric *stiffness matrix* $\mathbf{K}$, is typical for most undamped vibrating systems. Our experience with other mass–spring systems (Section 5.6) suggests that we seek solutions to (10) of the form

(11) $\mathbf{x} = (\cos \omega t)\mathbf{v}$ or $\mathbf{x} = (\sin \omega t)\mathbf{v}$,

where $\mathbf{v}$ is a constant vector and ω is a positive constant.

(a) Show that the system (10) has a nontrivial solution of the form (11) if and only if ω and $\mathbf{v}$ satisfy the "generalized eigenvalue problem" $\mathbf{Kv} = \omega^2 \mathbf{Mv}$.

(b) By employing the inverse of the mass matrix, one can rewrite (10) as

$$\mathbf{x}'' = -\mathbf{M}^{-1}\mathbf{Kx} =: \mathbf{Bx}.$$

Show that $-\omega^2$ must be an eigenvalue of $\mathbf{B}$ if (11) is a nontrivial solution.

(c) If $\mathbf{B}$ is an $n \times n$ constant matrix, then $\mathbf{x}'' = \mathbf{Bx}$ can be written as a system of $2n$ first-order equations in normal form. Thus, a general solution can be formed from $2n$ linearly independent solutions. Use the observation in part (b) to find a general solution to the following second-order systems:

(i) $\mathbf{x}'' = \begin{bmatrix} 0 & 1 \\ -4 & -5 \end{bmatrix} \mathbf{x}$.

(ii) $\mathbf{x}'' = \begin{bmatrix} -2 & 2 \\ 2 & -5 \end{bmatrix} \mathbf{x}$.

(iii) $\mathbf{x}'' = \begin{bmatrix} -5 & 4 & -4 \\ -1 & 0 & -1 \\ 1 & -2 & -1 \end{bmatrix} \mathbf{x}$.

(iv) $\mathbf{x}'' = \begin{bmatrix} 2 & -1 & -1 \\ -1 & 2 & 0 \\ -1 & 0 & 2 \end{bmatrix} \mathbf{x}$.

Partial Differential Equations

10.1 Introduction: A Model for Heat Flow

> **Develop a model for the flow of heat through a thin, insulated wire whose ends are kept at a constant temperature of 0°C and whose initial temperature distribution is to be specified.**

Suppose the wire is placed along the x-axis with $x = 0$ at the left end of the wire and $x = L$ at the right end (see Figure 10.1). If we let u denote the temperature of the wire, then u depends on the time t and on the position x within the wire. (We will assume the wire is thin and hence u is constant throughout a cross section of the wire corresponding to a fixed value of x.) Because the wire is insulated, we assume no heat enters or leaves through the sides of the wire.

To develop a model for heat flow through the thin wire, let's consider the small volume element V of wire between the two cross-sectional planes A and B that are perpendicular to the x-axis, with plane A located at x and plane B located at $x + \Delta x$ (see Figure 10.1).

The temperature on plane A at time t is $u(x, t)$ and on plane B is $u(x + \Delta x, t)$. We will need the following principles of physics that describe heat flow.[†]

1. **Heat Conduction:** The rate of heat flow (the amount of heat per unit time flowing through a unit of cross-sectional area at A) is proportional to $\partial u/\partial x$, the temperature gradient at A (Fick's law). The proportionality constant k is called the **thermal conductivity** of the material. In general, the thermal conductivity can vary from point to point: $k = k(x)$.

2. **Direction of Heat Flow:** The direction of heat flow is always from points of higher temperature to points of lower temperature.

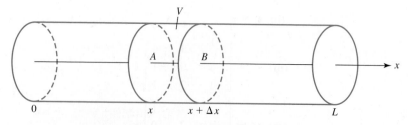

Figure 10.1 Heat flow through a thin piece of wire

[†]For a discussion of heat transfer, see *University Physics with Modern Physics*™, 13th ed., by H. D. Young and R. A. Freedman (Addison-Wesley, Reading, Mass., 2011).

3. **Specific Heat Capacity:** The amount of heat necessary to raise the temperature of an object of mass m by an amount Δu is $cm \, \Delta u$, where the constant c is the **specific heat capacity** of the material. The specific heat capacity, like the thermal conductivity, can vary with position: $c = c(x)$.

If we let H represent the amount of heat flowing from left to right through the surface A during an interval of time Δt, then the formula for heat conduction becomes

$$H(x) = -k(x) \, a \, \Delta t \frac{\partial u}{\partial x}(x, t) \,,$$

where a is the cross-sectional area of the wire. The negative sign follows from the second principle—if $\partial u / \partial x$ is positive, then heat flows from right to left (hotter to colder).

Similarly, the amount of heat flowing from left to right across plane B during an interval of time Δt is

$$H(x + \Delta x) = -k(x + \Delta x) \, a \, \Delta t \frac{\partial u}{\partial x}(x + \Delta x, t) \,.$$

The net change in the heat ΔE in volume V is the amount entering at end A minus the amount leaving at end B, plus any heat generated by sources (such as electric currents, chemical reactions, heaters, etc.). The latter is modeled by a term $Q(x, t) \Delta x a \, \Delta t$, where Q is the energy rate (power) density. Therefore,

(1) $\Delta E = H(x) - H(x + \Delta x) + Q(x, t) \Delta x a \, \Delta t$

$$= a \, \Delta t \left[k(x + \Delta x) \frac{\partial u}{\partial x}(x + \Delta x, t) - k(x) \frac{\partial u}{\partial x}(x, t) \right] + Q(x, t) \Delta x \, a \, \Delta t \,.$$

Now by the third principle, the net change is given by $\Delta E = cm \, \Delta u$, where Δu is the change in temperature and c is the specific heat capacity. If we assume that the change in temperature in the volume V is essentially equal to the change in temperature at x—that is, $\Delta u = u(x, t + \Delta t) - u(x, t)$, and that the mass of the volume V of wire is $a\rho \, \Delta x$, where $\rho = \rho(x)$ is the density of the wire, then

(2) $\Delta E = c(x) a \rho(x) \, \Delta x [u(x, t + \Delta t) - u(x, t)] \,.$

Equating the two expressions for ΔE given in equations (1) and (2) yields

$$a \, \Delta t \left[k(x + \Delta x) \frac{\partial u}{\partial x}(x + \Delta x, t) - k(x) \frac{\partial u}{\partial x}(x, t) \right] + Q(x, t) \Delta x \, a \, \Delta t$$

$$= c(x) a \rho(x) \, \Delta x [u(x, t + \Delta t) - u(x, t)] \,.$$

Now dividing both sides by $a \, \Delta x \, \Delta t$ and then taking the limits as Δx and Δt approach zero, we obtain

(3) $\dfrac{\partial}{\partial x} \left[k(x) \dfrac{\partial u}{\partial x}(x, t) \right] + Q(x, t) = c(x) \rho(x) \dfrac{\partial u}{\partial t}(x, t) \,.$

If the physical parameters k, c, and ρ are uniform along the length of the wire, then (3) reduces to the **one-dimensional heat flow equation**

(4) $\dfrac{\partial u}{\partial t}(x, t) = \beta \dfrac{\partial^2 u}{\partial x^2}(x, t) + P(x, t) \,,$

where the positive constant $\beta := k/(c\rho)$ is the **diffusivity** of the material and $P(x, t) := Q(x, t)/(c\rho)$.

Equation (4) governs the flow of heat in the wire. We have two other constraints in our original problem. First, we are keeping the ends of the wire at 0°C. Thus, we require that

(5) $u(0,t) = u(L,t) = 0$

for all t. These are called **boundary conditions.** Second, we must be given the initial temperature distribution $f(x)$. That is, we require

(6) $u(x,0) = f(x)\,, \qquad 0 < x < L\,.$

Equation (6) is referred to as the **initial condition** on u.

Combining equations (4), (5), and (6), we have the following mathematical model for the heat flow in a uniform wire without internal sources $(P = 0)$ whose ends are kept at the constant temperature 0°C:

(7) $\dfrac{\partial u}{\partial t}(x,t) = \beta \dfrac{\partial^2 u}{\partial x^2}(x,t)\,, \qquad 0 < x < L\,, \qquad t > 0\,,$

(8) $u(0,t) = u(L,t) = \mathbf{0}\,, \qquad t > 0\,,$

(9) $u(x,0) = f(x)\,, \qquad 0 < x < L\,.$

This model is an example of an **initial-boundary value problem.** Intuitively we expect that equations (7)–(9) completely and unambiguously specify the temperature in the wire. Once we have found a function $u(x,t)$ that meets all three of these conditions, we can be assured that u is the temperature. (Theorem 7 in Section 10.5, page 602, will bear this out.)

In higher dimensions, the heat flow equation (or just **heat equation**) is adjusted to account for the additional heat flow contribution along the other axes by the simple modification

$$\frac{\partial u}{\partial t} = \beta \Delta u + P(x,y,z,t)\,,$$

where Δu, known as the **Laplacian** in this context,[†] is defined in two and three dimensions, respectively, as

$$\Delta u := \frac{\partial^2 u}{\partial x^2} + \frac{\partial^2 u}{\partial y^2} \quad \text{and} \quad \Delta u := \frac{\partial^2 u}{\partial x^2} + \frac{\partial^2 u}{\partial y^2} + \frac{\partial^2 u}{\partial z^2}\,.$$

When the temperature reaches a steady state—that is, when u does not depend on time, and there are no sources—then $\partial u/\partial t = 0$ and the temperature satisfies **Laplace's equation**

$$\Delta u = 0\,.$$

One classical technique for solving the initial-boundary value problem for the heat equation (7)–(9) is the method of *separation of variables,* which effectively allows us to replace the partial differential equation by ordinary differential equations. This technique is discussed in the next section. In using separation of variables, one is often required to express a given function as a trigonometric series. Such series are called *Fourier series;* their properties are discussed in Sections 10.3 and 10.4. We devote the remaining three sections to the three basic partial differential equations that arise most commonly in applications: the heat equation, the wave equation, and Laplace's equation.

Many computer-based algorithms for solving partial differential equations have been developed. These are based on finite differences, finite elements, variational principles, and projection methods that include the method of moments and its wavelet-based implementations. Like the Runge–Kutta or Euler methods for ordinary differential equations, such techniques are more universally applicable than the analytic procedures of separation of variables, but their accuracy is often difficult to assess. Indeed, it is customary practice to gauge any newly proposed numerical procedure by comparing its predictions with those of separation of variables.

[†]Regrettably, it is the same notation as we just used for $u(x,t+\Delta t) - u(x,t)$. Some authors prefer the symbol ∇^2 for the Laplacian.

10.2 Method of Separation of Variables

The method of separation of variables is a classical technique that is effective in solving several types of partial differential equations. The idea is roughly the following. We think of a solution, say $u(x, t)$, to a partial differential equation as being a linear combination of simple component functions $u_n(x, t), n = 0, 1, 2, \ldots$, which also satisfy the equation and certain boundary conditions. (This is a reasonable assumption provided the partial differential equation and the boundary conditions are linear.) To determine a component solution, $u_n(x, t)$, we assume it can be written with its variables separated; that is, as

$$u_n(x, t) = X_n(x)T_n(t) .$$

Substituting this form for a solution into the partial differential equation and using the boundary conditions leads, in many circumstances, to two *ordinary* differential equations for the unknown functions $X_n(x)$ and $T_n(t)$. In this way we have reduced the problem of solving a partial differential equation to the more familiar problem of solving a differential equation that involves only one variable. In this section we will illustrate this technique for the heat equation and the wave equation.

In the previous section we derived the following initial-boundary value problem as a mathematical model for the sourceless heat flow in a uniform wire whose ends are kept at the constant temperature zero:

(1) $\quad \dfrac{\partial u}{\partial t}(x, t) = \beta \dfrac{\partial^2 u}{\partial x^2}(x, t) , \qquad 0 < x < L, \qquad t > 0 ,$

(2) $\quad u(0, t) = u(L, t) = 0 , \qquad t > 0 ,$

(3) $\quad u(x, 0) = f(x) , \qquad 0 < x < L .$

To solve this problem by the method of separation of variables, we begin by addressing equation (1). We propose that it has solutions of the form

$$u(x, t) = X(x)T(t) ,$$

where X is a function of x alone and T is a function of t alone. To determine X and T, we first compute the partial derivatives of u to obtain

$$\frac{\partial u}{\partial t} = X(x)T'(t) \quad \text{and} \quad \frac{\partial^2 u}{\partial x^2} = X''(x)T(t) .$$

Substituting these expressions into (1) gives

$$X(x)T'(t) = \beta X''(x)T(t) ,$$

and separating variables yields

(4) $\quad \dfrac{T'(t)}{\beta T(t)} = \dfrac{X''(x)}{X(x)} .$

We now observe that the functions on the left-hand side of (4) depend only on t, while those on the right-hand side depend only on x. If we fix t and vary x, the ratio on the right cannot change; it must be *constant*. Most authors define this constant with a minus sign, so we adhere to the convention:

$$\frac{X''(x)}{X(x)} = -\lambda \quad \text{and} \quad \frac{T'(t)}{\beta T(t)} = -\lambda ,$$

or

(5) $X''(x) = -\lambda X(x)$ and $T'(t) = -\lambda \beta T(t)$.

Consequently, for separable solutions, we have reduced the problem of solving the partial differential equation (1) to solving the two *ordinary* differential equations in (5).

Next we consider the boundary conditions in (2). Since $u(x, t) = X(x)T(t)$, these conditions are

$$X(0)T(t) = 0 \quad \text{and} \quad X(L)T(t) = 0, \qquad t > 0 .$$

Hence, either $T(t) = 0$ for all $t > 0$, which implies that $u(x, t) \equiv 0$, or

(6) $X(0) = X(L) = 0$.

Ignoring the trivial solution $u(x, t) \equiv 0$, we combine the boundary conditions in (6) with the differential equation for X in (5) and obtain the *boundary value problem*

(7) $X''(x) + \lambda X(x) = 0 , \qquad X(0) = X(L) = 0 ,$

where λ can be any constant.

Notice that the function $X(x) \equiv 0$ is a solution of (7) for every λ. Depending on the choice of λ, this may be the *only* solution to the boundary value problem (7). Thus, if we seek a nontrivial solution $u(x, t) = X(x)T(t)$ to (1)–(2), we must first determine those values of λ for which the boundary value problem (7) has nontrivial solutions. These solutions are called the **eigenfunctions** of the problem; the **eigenvalues** are the special values of λ. If we write the first equation in (7) as $-D^2[X] = \lambda X$ and compare with the matrix eigenvector equation $\mathbf{Au} = r\mathbf{u}$ (Section 9.5, page 523), the designation of λ as an eigenvalue (of $-D^2$) becomes clear.

To solve the (constant-coefficient) equation in (7), we try $X(x) = e^{rx}$, derive the auxiliary equation $r^2 + \lambda = 0$, and consider three cases.

Case 1. $\lambda < 0$. In this case, the roots of the auxiliary equation are $\pm\sqrt{-\lambda}$, so a general solution to the differential equation in (7) is

$$X(x) = C_1 e^{\sqrt{-\lambda}x} + C_2 e^{-\sqrt{-\lambda}x} .$$

To determine C_1 and C_2, we appeal to the boundary conditions:

$$X(0) = C_1 + C_2 = 0 ,$$

$$X(L) = C_1 e^{\sqrt{-\lambda}L} + C_2 e^{-\sqrt{-\lambda}L} = 0 .$$

From the first equation, we see that $C_2 = -C_1$. The second equation can then be written as $C_1(e^{\sqrt{-\lambda}L} - e^{-\sqrt{-\lambda}L}) = 0$ or $C_1(e^{2\sqrt{-\lambda}L} - 1) = 0$. Since $-\lambda > 0$, it follows that $(e^{2\sqrt{-\lambda}L} - 1) > 0$. Therefore C_1, and hence C_2, is zero. Consequently, there is *no* nontrivial solution to (7) for $\lambda < 0$.

Case 2. $\lambda = 0$. Here $r = 0$ is a repeated root to the auxiliary equation, and a general solution to the differential equation is

$$X(x) = C_1 + C_2 x .$$

The boundary conditions in (7) yield $C_1 = 0$ and $C_1 + C_2 L = 0$, which imply that $C_1 = C_2 = 0$. Thus, for $\lambda = 0$, there is *no* nontrivial solution to (7).

Case 3. $\lambda > 0$. In this case the roots of the auxiliary equation are $\pm i\sqrt{\lambda}$. Thus a general solution to $X'' + \lambda X = 0$ is

(8) $X(x) = C_1 \cos \sqrt{\lambda}x + C_2 \sin \sqrt{\lambda}x .$

This time the boundary conditions $X(0) = X(L) = 0$ give the system

$$C_1 = 0,$$

$$C_1 \cos \sqrt{\lambda} L + C_2 \sin \sqrt{\lambda} L = 0.$$

Because $C_1 = 0$, the system reduces to solving $C_2 \sin \sqrt{\lambda} L = 0$. Hence, either $\sin \sqrt{\lambda} L = 0$ or $C_2 = 0$. Now $\sin \sqrt{\lambda} L = 0$ only when $\sqrt{\lambda} L = n\pi$, where n is an integer. Therefore, (7) has a nontrivial solution $(C_2 \neq 0)$ when $\sqrt{\lambda} L = n\pi$ or $\lambda = (n\pi/L)^2$, $n = 1, 2, 3, \dots$ (we exclude $n = 0$, since it makes $\lambda = 0$). Furthermore, the nontrivial solutions (eigenfunctions) X_n corresponding to the eigenvalue $\lambda = (n\pi/L)^2$ are given by [cf. (8)]

$$(9) \qquad X_n(x) = a_n \sin\left(\frac{n\pi x}{L}\right),$$

where the a_n's are arbitrary nonzero constants.

The eigenvalues and eigenfunctions for this example have many features that are common to all of the separation of variables solutions that we will study. Take note of these features of Figure 10.2.

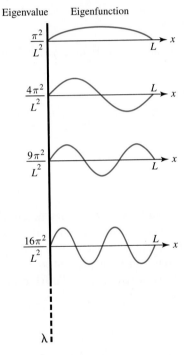

Eigenvalue Eigenfunction

$$\frac{\pi^2}{L^2}$$

$$\frac{4\pi^2}{L^2}$$

$$\frac{9\pi^2}{L^2}$$

$$\frac{16\pi^2}{L^2}$$

λ

Figure 10.2 Eigenfunctions

SEPARATION OF VARIABLES: EIGENFUNCTION PROPERTIES

(*i*) The eigenfunctions are solutions to a second-order ordinary differential equation (7) containing a parameter λ called the eigenvalue.

(*ii*) Each eigenfunction satisfies a homogeneous boundary condition at each end (7).

(*iii*) The only values of λ that admit nontrivial solutions, i.e., the eigenvalues, form an infinite set accumulating at ∞.

(*iv*) The eigenfunctions oscillate faster as λ increases; the first contains no interior zeros, and each subsequent eigenfunction contains one more zero than its predecessor.

(*v*) If any two distinct eigenfunctions are multiplied together, the resulting function is zero on the average; specifically,

$$\sin\frac{m\pi x}{L} \sin\frac{n\pi x}{L} = \frac{1}{2}\left[\cos\frac{(m-n)\pi x}{L} - \cos\frac{(m+n)\pi x}{L}\right]$$

and both cosines integrate to 0 over $[0, L]$.

Having determined that $\lambda = (n\pi/L)^2$ for any positive integer n, let's consider the second equation in (5):

$$T'(t) + \beta\left(\frac{n\pi}{L}\right)^2 T(t) = 0.$$

For each $n = 1, 2, 3, \dots$, a general solution to this linear first-order equation is

$$T_n(t) = b_n e^{-\beta(n\pi/L)^2 t}.$$

(Note that the time factor has none of the eigenfunction properties.) Combining this with equation (9) we obtain, for each $n = 1, 2, 3, \ldots$, the functions

(10) $u_n(x, t) := X_n(x)T_n(t) = a_n \sin(n\pi x/L)\, b_n e^{-\beta(n\pi/L)^2 t}$

$$= c_n e^{-\beta(n\pi/L)^2 t} \sin(n\pi x/L),$$

where c_n is also an arbitrary constant.

It is easy to see that each $u_n(x, t)$ is a solution to (1)–(2). A simple computation also shows that if u_n and u_m are solutions to (1)–(2), then so is any linear combination $au_n + bu_m$. (This is a consequence of the facts that the operator $\mathcal{L} := \partial/\partial t - \beta \partial^2/\partial x^2$ is a *linear* operator and the boundary conditions in (2) are *linear homogeneous*.)

This enables us to solve the following example.

Example 1 Find the solution to the heat flow problem

(11) $\dfrac{\partial u}{\partial t} = 7\dfrac{\partial^2 u}{\partial x^2},$ $0 < x < \pi,$ $t > 0,$

(12) $u(0, t) = u(\pi, t) = 0,$ $t > 0,$

(13) $u(x, 0) = 3\sin 2x - 6\sin 5x,$ $0 < x < \pi.$

Solution Comparing equation (11) with (1), we see that $\beta = 7$ and $L = \pi$. Hence, we need only find a combination of terms like (10) that satisfies the initial condition (13):

$$u(x, 0) = \sum c_n e^0 \sin nx = 3\sin 2x - 6\sin 5x.$$

For these data the task is simple; $c_2 = 3$ and $c_5 = -6$. The solution to the heat flow problem (11)–(13) is

$$u(x, t) = c_2 e^{-\beta(2\pi/L)^2 t}\sin(2\pi x/L) + c_5 e^{-\beta(5\pi/L)^2 t}\sin(5\pi x/L)$$

$$= 3e^{-28t}\sin 2x - 6e^{-175t}\sin 5x. \;\blacklozenge$$

What would we do if the initial condition (13) had been an *arbitrary* function, rather than a simple combination of a few of the eigenfunctions we found? Possibly the most beautiful property of the eigenfunctions—one that we did not list on page 565—is *completeness*; virtually any function f likely to arise in applications can be expressed as a convergent series of eigenfunctions! For the sines we have been working with, the **Fourier sine series** looks like

(14) $f(x) = \displaystyle\sum_{n=1}^{\infty} c_n \sin\left(\dfrac{n\pi x}{L}\right)$ for $0 < x < L.$

This enables the complete solution to the generic problem given by (1)–(3):

(15) $u(x, t) = \displaystyle\sum_{n=1}^{\infty} c_n e^{-\beta(n\pi/L)^2 t} \sin\left(\dfrac{n\pi x}{L}\right),$

provided this expansion and its first two derivatives converge.

The convergence of the sine series (among others), as well as the procedure for finding the coefficients c_n, will be discussed in the next two sections (and generalized in Chapter 11). For now, let us turn to the mathematical description of a vibrating string, another situation in which the separation of variables approach applies. This concerns the transverse vibrations of a string stretched between two points, such as a guitar string or piano wire. The goal is to find a function $u(x, t)$ that gives the displacement (deflection) of the string at any point

Figure 10.3 Displacement of string at time t

x $(0 \leq x \leq L)$ and any time $t \geq 0$ (see Figure 10.3). In developing the mathematical model, we assume that the string is perfectly flexible and has constant linear density, the tension on the string is constant, gravity is negligible, and no other forces are acting on the string. Under these conditions and the additional assumption that the displacements $u(x, t)$ are small in comparison to the length of the string, it turns out that the motion of the string is governed by the following initial-boundary value problem.[†]

$$(16) \qquad \frac{\partial^2 u}{\partial t^2} = \alpha^2 \frac{\partial^2 u}{\partial x^2}, \qquad 0 < x < L, \qquad t > 0,$$

$$(17) \qquad u(0, t) = u(L, t) = 0, \qquad t \geq 0,$$

$$(18) \qquad u(x, 0) = f(x), \qquad 0 \leq x \leq L,$$

$$(19) \qquad \frac{\partial u}{\partial t}(x, 0) = g(x), \qquad 0 \leq x \leq L.$$

Equation (16) essentially states Newton's third law, $F = ma$, for the string. The second (time) derivative on the left is the acceleration, and the second (spatial) derivative on the right arises because the restoring force is produced by the string's *curvature*. The constant α^2 is strictly positive and equals the ratio of the tension to the (linear) density of the string. The physical significance of α, which has units of velocity, will be revealed in Section 10.6. The boundary conditions in (17) reflect the fact that the string is held fixed at the two endpoints $x = 0$ and $x = L$. Equations (18) and (19), respectively, specify the initial displacement of the string and the initial velocity of each point on the string. Recall that these are the typical initial data for all mechanical systems. For the initial and boundary conditions to be consistent, we assume $f(0) = f(L) = 0$ and $g(0) = g(L) = 0$.

Let's apply the method of separation of variables to the initial-boundary value problem for the vibrating string (16)–(19). Thus, we begin by assuming equation (16) has a solution of the form

$$u(x, t) = X(x)T(t),$$

where X is a function of x alone and T is a function of t alone. Differentiating u, we obtain

$$\frac{\partial^2 u}{\partial t^2} = X(x)T''(t), \qquad \frac{\partial^2 u}{\partial x^2} = X''(x)T(t).$$

Substituting these expressions into (16), we have

$$X(x)T''(t) = \alpha^2 X''(x)T(t),$$

[†]For a derivation of this mathematical model, see *Partial Differential Equations: Sources and Solutions*, by Arthur D. Snider (Dover Publications, N.Y., 2006).

and separating variables gives

$$\frac{T''(t)}{\alpha^2 T(t)} = \frac{X''(x)}{X(x)}.$$

Just as before, these ratios must equal some constant $-\lambda$:

(20) $\qquad \dfrac{X''(x)}{X(x)} = -\lambda \qquad$ and $\qquad \dfrac{T''(t)}{\alpha^2 T(t)} = -\lambda.$

Furthermore, with $u(x, t) = X(x)T(t)$, the boundary conditions in (17) give

$$X(0)T(t) = 0, \qquad X(L)T(t) = 0, \qquad t \geq 0.$$

In order for these equations to hold for all $t \geq 0$, either $T(t) \equiv 0$, which implies that $u(x, t) \equiv 0$, or

$$X(0) = X(L) = 0.$$

Ignoring the trivial solution, we combine these boundary conditions with the differential equation for X in (20) and obtain the boundary value problem

(21) $\qquad X''(x) + \lambda X(x) = 0, \qquad X(0) = X(L) = 0,$

where λ can be any constant.

This is the same boundary value problem that we encountered earlier while solving the heat equation. There we found that the suitable values for λ are

$$\lambda = \left(\frac{n\pi}{L}\right)^2, \qquad n = 1, 2, 3, \ldots,$$

with corresponding eigenfunctions (nontrivial solutions)

(22) $\qquad X_n(x) = c_n \sin\left(\dfrac{n\pi x}{L}\right),$

where the c_n's are arbitrary nonzero constants. Recall Figure 10.2, page 565.

Having determined that $\lambda = (n\pi/L)^2$ for some positive integer n, let's consider the second equation in (20) for such λ:

$$T''(t) + \frac{\alpha^2 n^2 \pi^2}{L^2} T(t) = 0.$$

For each $n = 1, 2, 3, \ldots$, a general solution is

$$T_n(t) = c_{n,1} \cos\frac{n\pi\alpha}{L}t + c_{n,2} \sin\frac{n\pi\alpha}{L}t.$$

Combining this with equation (22), we obtain, for each $n = 1, 2, 3, \ldots$, the function

$$u_n(x, t) = X_n(x)T_n(t) = \left(c_n \sin\frac{n\pi x}{L}\right)\left(c_{n,1} \cos\frac{n\pi\alpha}{L}t + c_{n,2} \sin\frac{n\pi\alpha}{L}t\right),$$

or, reassembling the constants,

(23) $\qquad u_n(x, t) = \left(a_n \cos\dfrac{n\pi\alpha}{L}t + b_n \sin\dfrac{n\pi\alpha}{L}t\right)\sin\dfrac{n\pi x}{L}.$

Using the fact that linear combinations of solutions to (16)–(17) are again solutions, we consider a superposition of the functions in (23):

$$(24) \qquad u(x,t) = \sum_{n=1}^{\infty} \left[a_n \cos\frac{n\pi\alpha}{L}t + b_n \sin\frac{n\pi\alpha}{L}t \right] \sin\frac{n\pi x}{L} \,.$$

For a solution of the form (24), substitution into the initial conditions (18)–(19) gives

$$(25) \qquad u(x,0) = \sum_{n=1}^{\infty} a_n \sin\frac{n\pi x}{L} = f(x) \,, \qquad 0 \le x \le L \,,$$

$$(26) \qquad \frac{\partial u}{\partial t}(x,0) = \sum_{n=1}^{\infty} \frac{n\pi\alpha}{L} b_n \sin\frac{n\pi x}{L} = g(x) \,, \qquad 0 \le x \le L \,.$$

We have now reduced the vibrating string problem (16)–(19) to the problem of determining the Fourier sine series expansions for $f(x)$ and $g(x)$:

$$(27) \qquad f(x) = \sum_{n=1}^{\infty} a_n \sin\frac{n\pi x}{L} \,, \qquad g(x) = \sum_{n=1}^{\infty} B_n \sin\frac{n\pi x}{L} \,,$$

where $B_n = (n\pi\alpha/L)b_n$. If we choose the a_n's and b_n's so that the equations in (25) and (26) hold, then the expansion for $u(x,t)$ in (24) is a **formal solution** to the vibrating string problem (16)–(19). If this expansion is finite, or converges to a function with continuous second partial derivatives, then the formal solution is an actual (genuine) solution.

Example 2 Find the solution to the vibrating string problem

$$(28) \qquad \frac{\partial^2 u}{\partial t^2} = 4\frac{\partial^2 u}{\partial x^2} \,, \qquad 0 < x < \pi \,, \qquad t > 0 \,,$$

$$(29) \qquad u(0,t) = u(\pi,t) = 0 \,, \qquad t \ge 0 \,,$$

$$(30) \qquad u(x,0) = \sin 3x - 4\sin 10x \,, \qquad 0 \le x \le \pi \,,$$

$$(31) \qquad \frac{\partial u}{\partial t}(x,0) = 2\sin 4x + \sin 6x \,, \qquad 0 \le x \le \pi \,.$$

Solution Comparing equation (28) with equation (16), we see that $\alpha = 2$ and $L = \pi$. Hence, we need only determine the values of the coefficients a_n and b_n in formula (24). The a_n's are chosen so that equation (25) holds; that is,

$$u(x,0) = \sin 3x - 4\sin 10x = \sum_{n=1}^{\infty} a_n \sin nx \,.$$

Equating coefficients of like terms, we see that

$$a_3 = 1 \,, \qquad a_{10} = -4 \,,$$

and the remaining a_n's are zero. Similarly, referring to equation (26), we must choose the b_n's so that

$$\frac{\partial u}{\partial t}(x,0) = 2\sin 4x + \sin 6x = \sum_{n=1}^{\infty} n2b_n \sin nx \,.$$

Comparing coefficients, we find

$$2 = (4)(2)b_4 \quad \text{or} \quad b_4 = \frac{1}{4} \,,$$

$$1 = (6)(2)b_6 \quad \text{or} \quad b_6 = \frac{1}{12} \,,$$

and the remaining b_n's are zero. Hence, from formula (24), the solution to the vibrating string problem (28)–(31) is

$$(32) \qquad u(x,t) = \cos 6t \sin 3x + \frac{1}{4}\sin 8t \sin 4x + \frac{1}{12}\sin 12t \sin 6x - 4\cos 20t \sin 10x . \quad \blacklozenge$$

In later sections the method of separation of variables is used to study a wide variety of problems for the heat, wave, and Laplace's equations. However, to use the method effectively, one must be able to compute trigonometric series (or, more generally, eigenfunction expansions) such as the Fourier sine series that we encountered here. These expansions are discussed in the next two sections.

10.2 EXERCISES

In Problems 1–8, determine all the solutions, if any, to the given boundary value problem by first finding a general solution to the differential equation.

1. $y'' - y = 0$; $\quad 0 < x < 1$,
 $y(0) = 0$, $\quad y(1) = -4$
2. $y'' - 6y' + 5y = 0$; $\quad 0 < x < 2$,
 $y(0) = 1$, $\quad y(2) = 1$
3. $y'' + 4y = 0$; $\quad 0 < x < \pi$,
 $y(0) = 0$, $\quad y'(\pi) = 0$
4. $y'' + 9y = 0$; $\quad 0 < x < \pi$,
 $y(0) = 0$, $\quad y'(\pi) = -6$
5. $y'' - y = 1 - 2x$; $\quad 0 < x < 1$,
 $y(0) = 0$, $\quad y(1) = 1 + e$
6. $y'' + y = 0$; $\quad 0 < x < 2\pi$,
 $y(0) = 0$, $\quad y(2\pi) = 1$
7. $y'' + y = 0$; $\quad 0 < x < 2\pi$,
 $y(0) = 1$, $\quad y(2\pi) = 1$
8. $y'' - 2y' + y = 0$; $\quad -1 < x < 1$,
 $y(-1) = 0$, $\quad y(1) = 2$

In Problems 9–14, find the values of λ (eigenvalues) for which the given problem has a nontrivial solution. Also determine the corresponding nontrivial solutions (eigenfunctions).

9. $y'' + \lambda y = 0$; $\quad 0 < x < \pi$,
 $y(0) = 0$, $\quad y'(\pi) = 0$
10. $y'' + \lambda y = 0$; $\quad 0 < x < \pi$,
 $y'(0) = 0$, $\quad y(\pi) = 0$
11. $y'' + \lambda y = 0$; $\quad 0 < x < 2\pi$,
 $y(0) = y(2\pi)$, $\quad y'(0) = y'(2\pi)$
12. $y'' + \lambda y = 0$; $\quad 0 < x < \pi/2$,
 $y'(0) = 0$, $\quad y'(\pi/2) = 0$
13. $y'' + \lambda y = 0$; $\quad 0 < x < \pi$,
 $y(0) - y'(0) = 0$, $\quad y(\pi) = 0$
14. $y'' - 2y' + \lambda y = 0$; $\quad 0 < x < \pi$,
 $y(0) = 0$, $\quad y(\pi) = 0$

In Problems 15–18, solve the heat flow problem (1)–(3) with $\beta = 3, L = \pi$, and the given function $f(x)$.

15. $f(x) = \sin x - 6\sin 4x$
16. $f(x) = \sin 3x + 5\sin 7x - 2\sin 13x$
17. $f(x) = \sin x - 7\sin 3x + \sin 5x$
18. $f(x) = \sin 4x + 3\sin 6x - \sin 10x$

In Problems 19–22, solve the vibrating string problem (16)–(19) with $\alpha = 3, L = \pi$, and the given initial functions $f(x)$ and $g(x)$.

19. $f(x) = 3\sin 2x + 12\sin 13x$, $\quad g(x) \equiv 0$
20. $f(x) \equiv 0$,
 $g(x) = -2\sin 3x + 9\sin 7x - \sin 10x$
21. $f(x) = 6\sin 2x + 2\sin 6x$,
 $g(x) = 11\sin 9x - 14\sin 15x$
22. $f(x) = \sin x - \sin 2x + \sin 3x$,
 $g(x) = 6\sin 3x - 7\sin 5x$

23. Find the formal solution to the heat flow problem (1)–(3) with $\beta = 2$ and $L = 1$ if
 $$f(x) = \sum_{n=1}^{\infty} \frac{1}{n^2}\sin n\pi x .$$

24. Find the formal solution to the vibrating string problem (16)–(19) with $\alpha = 4, L = \pi$, and
 $$f(x) = \sum_{n=1}^{\infty} \frac{1}{n^2}\sin nx ,$$
 $$g(x) = \sum_{n=1}^{\infty} \frac{(-1)^{n+1}}{n}\sin nx .$$

25. By considering the behavior of the solutions of the equation
 $$T'(t) = -\lambda\beta T(t) , \qquad t > 0 ,$$
 give an argument that is based on physical grounds to rule out the case where $\lambda < 0$ in equation (5).

26. Verify that $u_n(x,t)$ given in equation (10) satisfies equation (1) and the boundary conditions in (2) by substituting $u_n(x,t)$ directly into the equations involved.

In Problems 27–30, a partial differential equation (PDE) is given along with the form of a solution having separated variables. Show that such a solution must satisfy the indicated set of ordinary differential equations.

27. $\dfrac{\partial^2 u}{\partial r^2} + \dfrac{1}{r}\dfrac{\partial u}{\partial r} + \dfrac{1}{r^2}\dfrac{\partial^2 u}{\partial \theta^2} = 0$

with $u(r,\theta) = R(r)\Theta(\theta)$ yields

$r^2 R''(r) + rR'(r) - \lambda R(r) = 0$,

$\Theta''(\theta) + \lambda\Theta(\theta) = 0$,

where λ is a constant.

28. $\dfrac{\partial^2 u}{\partial t^2} + \dfrac{\partial u}{\partial t} + u = \alpha^2 \dfrac{\partial^2 u}{\partial x^2}$

with $u(x,t) = X(x)T(t)$ yields

$X''(x) + \lambda X(x) = 0$,

$T''(t) + T'(t) + (1 + \lambda\alpha^2)T(t) = 0$,

where λ is a constant.

29. $\dfrac{\partial u}{\partial t} = \beta\left\{\dfrac{\partial^2 u}{\partial x^2} + \dfrac{\partial^2 u}{\partial y^2}\right\}$

with $u(x,y,t) = X(x)Y(y)T(t)$ yields

$T'(t) + \beta\lambda T(t) = 0$,

$X''(x) + \mu X(x) = 0$,

$Y''(y) + (\lambda - \mu)Y(y) = 0$,

where λ, μ are constants.

30. $\dfrac{\partial^2 u}{\partial r^2} + \dfrac{1}{r}\dfrac{\partial u}{\partial r} + \dfrac{1}{r^2}\dfrac{\partial^2 u}{\partial \theta^2} + \dfrac{\partial^2 u}{\partial z^2} = 0$

with $u(r,\theta,z) = R(r)\Theta(\theta)Z(z)$ yields

$\Theta''(\theta) + \mu\Theta(\theta) = 0$,

$Z''(z) + \lambda Z(z) = 0$,

$r^2 R''(r) + rR'(r) - (r^2\lambda + \mu)R(r) = 0$,

where μ, λ are constants.

31. For the PDE in Problem 27, assume that the following boundary conditions are imposed:

$u(r,0) = u(r,\pi) = 0$,

$u(r,\theta)$ remains bounded as $r \to 0^+$.

Show that a nontrivial solution of the form $u(r,\theta) = R(r)\Theta(\theta)$ must satisfy the boundary conditions

$\Theta(0) = \Theta(\pi) = 0$,

$R(r)$ remains bounded as $r \to 0^+$.

32. For the PDE in Problem 29, assume that the following boundary conditions are imposed:

$u(0,y,t) = u(a,y,t) = 0; \quad 0 \le y \le b, \quad t \ge 0$,

$\dfrac{\partial u}{\partial y}(x,0,t) = \dfrac{\partial u}{\partial y}(x,b,t) = 0; \quad 0 \le x \le a, \quad t \ge 0$.

Show that a nontrivial solution of the form $u(x,y,t) = X(x)Y(y)T(t)$ must satisfy the boundary conditions

$X(0) = X(a) = 0$,

$Y'(0) = Y'(b) = 0$.

33. When the temperature in a wire reaches a steady state, that is, when u depends only on x, then $u(x)$ satisfies Laplace's equation $\partial^2 u/\partial x^2 = 0$.

(a) Find the steady-state solution when the ends of the wire are kept at a constant temperature of 50°C, that is, when $u(0) = u(L) = 50$.

(b) Find the steady-state solution when one end of the wire is kept at 10°C, while the other is kept at 40°C, that is, when $u(0) = 10$ and $u(L) = 40$.

10.3 Fourier Series

While analyzing heat flow and vibrating strings in the previous section, we encountered the problem of expressing a function in a trigonometric series [compare equations (12) and (25) in Section 10.2]. In the next two sections, we discuss the theory of Fourier series, which deals with trigonometric series expansions. First, however, we review some function properties that are particularly relevant to this study: piecewise continuity, periodicity, and even and odd symmetry.

In Section 7.2 we defined a **piecewise continuous** function on $[a,b]$ as a function f that is continuous at every point in $[a,b]$, except possibly for a finite number of points at which f has a *jump discontinuity*. Such functions are necessarily integrable over any finite interval on which they are piecewise continuous.

Recall also that a function is **periodic of period** T if $f(x + T) = f(x)$ for all x in the domain of f. The smallest positive value of T is called the **fundamental period**. The trigonometric functions $\sin x$ and $\cos x$ are examples of periodic functions with fundamental period 2π and $\tan x$ is periodic with fundamental period π. A constant function is a periodic function with arbitrary period T.

Two symmetry properties of functions will be useful in the study of Fourier series. A function f that satisfies $f(-x) = f(x)$ for all x in the domain of f has a graph that is symmetric with respect to the y-axis [see Figure 10.4(a)]. We say that such a function is **even**. A function f that satisfies $f(-x) = -f(x)$ for all x in the domain of f has a graph that is symmetric with respect to the origin [see Figure 10.4(b)]. It is said to be an **odd** function. The functions $1, x^2, x^4, \ldots$ are examples of even functions, while the functions $x, x^3, x^5, \ldots$ are odd. The trigonometric functions $\sin x$ and $\tan x$ are odd functions and $\cos x$ is an even function.

Example 1 Determine whether the given function is even, odd, or neither.

(a) $f(x) = \sqrt{1 + x^2}$ (b) $g(x) = x^{1/3} - \sin x$ (c) $h(x) = e^x$

Solution (a) Since $f(-x) = \sqrt{1 + (-x)^2} = \sqrt{1 + x^2} = f(x)$, then $f(x)$ is an even function.

(b) Because $g(-x) = (-x)^{1/3} - \sin(-x) = -x^{1/3} + \sin x = -(x^{1/3} - \sin x) = -g(x)$, it follows that $g(x)$ is an odd function.

(c) Here $h(-x) = e^{-x}$. Since $e^{-x} = e^x$ only when $x = 0$ and e^{-x} is never equal to $-e^x$, then $h(x)$ is neither an even nor an odd function. ◆

Knowing that a function is even or odd can be useful in evaluating definite integrals. The following result, illustrated in Figure 10.4, is a straightforward consequence of the definition of the definite integral.

Properties of Symmetric Functions

Theorem 1. If f is an even piecewise continuous function on $[-a, a]$, then

(1) $$\int_{-a}^{a} f(x)\, dx = 2 \int_{0}^{a} f(x)\, dx .$$

If f is an odd piecewise continuous function on $[-a, a]$, then

(2) $$\int_{-a}^{a} f(x)\, dx = 0 .$$

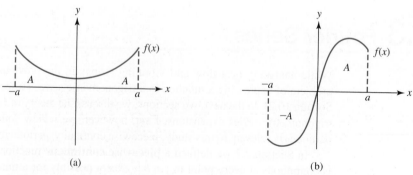

Figure 10.4 (a) Even function $\int_{-a}^{a} f = A + A = 2\int_{0}^{a} f$ (b) Odd function $\int_{-a}^{a} f = A - A = 0$

The next example deals with certain integrals that are crucial in Fourier series.

Example 2 Evaluate the following integrals when m and n are nonnegative integers:

(a) $\displaystyle\int_{-L}^{L} \sin\frac{m\pi x}{L} \cos\frac{n\pi x}{L}\, dx$. (b) $\displaystyle\int_{-L}^{L} \sin\frac{m\pi x}{L} \sin\frac{n\pi x}{L}\, dx$.

(c) $\displaystyle\int_{-L}^{L} \cos\frac{m\pi x}{L} \cos\frac{n\pi x}{L}\, dx$.

Solution The even and odd functions occurring in the integrands are sketched in Figure 10.5.
The given integrals are easily evaluated by invoking the trigonometric formula for products of sines and cosines:

$$\sin\frac{m\pi x}{L} \cos\frac{n\pi x}{L} = \frac{1}{2}\sin\frac{(m-n)\pi x}{L} + \frac{1}{2}\sin\frac{(m+n)\pi x}{L},$$

$$\sin\frac{m\pi x}{L} \sin\frac{n\pi x}{L} = \frac{1}{2}\cos\frac{(m-n)\pi x}{L} - \frac{1}{2}\cos\frac{(m+n)\pi x}{L},$$

$$\cos\frac{m\pi x}{L} \cos\frac{n\pi x}{L} = \frac{1}{2}\cos\frac{(m-n)\pi x}{L} + \frac{1}{2}\cos\frac{(m+n)\pi x}{L}.$$

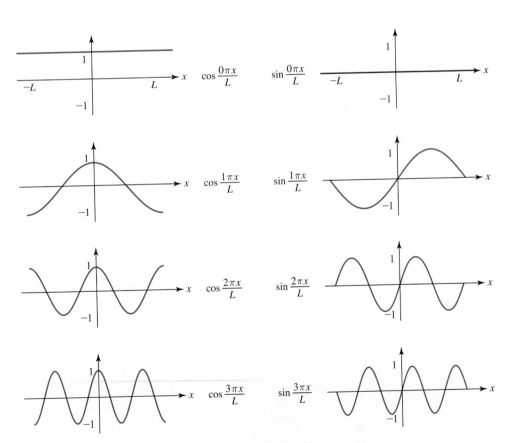

Figure 10.5 The sinusoids

Thus if $m \neq n$, each of the integrals calls for the (signed) area under an oscillating sinusoid over a whole number of periods, and is therefore zero. In fact, the only way to avoid these "oscillators" is to take $m = n$; whence $\cos[(m-n)\pi x/L] = \cos 0 = 1$ and subtends an area of $2L$, while $\sin[(m-n)\pi x/L] = \sin 0 = 0$ subtending zero area (again). But note that if m and n are *both* zero, then $\cos[(m+n)\pi x/L] = \cos 0$ also subtends area $2L$. Restoring the factors of $1/2$, we summarize with

$$(3) \qquad \int_{-L}^{L} \sin\frac{m\pi x}{L} \cos\frac{n\pi x}{L} \, dx = 0 \,,$$

$$(4) \qquad \int_{-L}^{L} \sin\frac{m\pi x}{L} \sin\frac{n\pi x}{L} \, dx = \begin{cases} 0\,, & m \neq n\,, \\ L\,, & m = n\,, \end{cases}$$

$$(5) \qquad \int_{-L}^{L} \cos\frac{m\pi x}{L} \cos\frac{n\pi x}{L} \, dx = \begin{cases} 0\,, & m \neq n\,, \\ L\,, & m = n \neq 0\,, \quad \blacklozenge \\ 2L\,, & m = n = 0\,. \end{cases}$$

Equations (3)–(5) express an **orthogonality condition**[†] satisfied by the set of trigonometric functions $\{1 = \cos 0x,\ \cos x,\ \sin x,\ \cos 2x,\ \sin 2x,\ \dots\}$, where $L = \pi$. We will say more about this later in this section.

It is easy to verify that if each of the functions $f_1, \dots, f_n$ is periodic of period T, then so is any linear combination

$$c_1 f_1(x) + \cdots + c_n f_n(x) \,.$$

For example, the sum $7 + 3\cos\pi x - 8\sin\pi x + 4\cos 2\pi x - 6\sin 2\pi x$ has period 2, since each term has period 2. Furthermore, if the infinite series

$$\frac{a_0}{2} + \sum_{n=1}^{\infty} \left(a_n \cos\frac{n\pi x}{L} + b_n \sin\frac{n\pi x}{L}\right)$$

consisting of $2L$-periodic functions converges for all x, then the function to which it converges will be periodic of period $2L$.

Just as we can associate a Taylor series with a function that has derivatives of all orders at a fixed point, we can identify a particular trigonometric series with a piecewise continuous function. To illustrate this, let's assume that $f(x)$ has the series expansion[‡]

$$(6) \qquad f(x) = \frac{a_0}{2} + \sum_{n=1}^{\infty} \left\{a_n \cos\frac{n\pi x}{L} + b_n \sin\frac{n\pi x}{L}\right\},$$

where the a_n's and b_n's are constants. (Necessarily, f has period $2L$.)

To determine the coefficients $a_0, a_1, b_1, a_2, b_2, \dots$, we proceed as follows. Let's integrate $f(x)$ from $-L$ to L, assuming that we can integrate term by term:

$$\int_{-L}^{L} f(x)\,dx = \int_{-L}^{L} \frac{a_0}{2}\,dx + \sum_{n=1}^{\infty} a_n \int_{-L}^{L} \cos\frac{n\pi x}{L}\,dx + \sum_{n=1}^{\infty} b_n \int_{-L}^{L} \sin\frac{n\pi x}{L}\,dx \,.$$

[†]This nomenclature is suggested by the fact that the formulas for the Riemann sums approximating the integrals (3)–(5) look like dot products (Section 9.1, page 496) of higher-dimensional vectors. In most calculus texts, it is shown that the dot product of orthogonal vectors in two dimensions is zero.

[‡]The choice of constant $a_0/2$ instead of just a_0 will be motivated shortly.

The (signed) area under the "oscillators" is zero. Hence,

$$\int_{-L}^{L} f(x)\, dx = \int_{-L}^{L} \frac{a_0}{2}\, dx = a_0 L \,,$$

and so

$$a_0 = \frac{1}{L} \int_{-L}^{L} f(x)\, dx \,.$$

(Notice that $a_0/2$ is the average value of f over one period $2L$.) Next, to find the coefficient a_m when $m \geq 1$, we multiply (6) by $\cos(m\pi x/L)$ and integrate:

$$(7) \qquad \int_{-L}^{L} f(x) \cos\frac{m\pi x}{L}\, dx = \frac{a_0}{2} \int_{-L}^{L} \cos\frac{m\pi x}{L}\, dx + \sum_{n=1}^{\infty} a_n \int_{-L}^{L} \cos\frac{n\pi x}{L} \cos\frac{m\pi x}{L}\, dx$$

$$+ \sum_{n=1}^{\infty} b_n \int_{-L}^{L} \sin\frac{n\pi x}{L} \cos\frac{m\pi x}{L}\, dx \,.$$

The orthogonality conditions (3)–(5) render the integrals on the right-hand side quite immediately. We have already observed that

$$\int_{-L}^{L} \cos\frac{m\pi x}{L}\, dx = 0 \,, \qquad m \geq 1 \,,$$

and, by formulas (3) and (5), we have

$$\int_{-L}^{L} \sin\frac{n\pi x}{L} \cos\frac{m\pi x}{L}\, dx = 0 \,,$$

$$\int_{-L}^{L} \cos\frac{n\pi x}{L} \cos\frac{m\pi x}{L}\, dx = \begin{cases} 0, & n \neq m \,, \\ L, & n = m \,. \end{cases}$$

Hence, in (7) we see that only one term on the right-hand side survives the integration:

$$\int_{-L}^{L} f(x) \cos\frac{m\pi x}{L}\, dx = a_m L \,.$$

Thus, we have a formula for the coefficient a_m:

$$a_m = \frac{1}{L} \int_{-L}^{L} f(x) \cos\frac{m\pi x}{L}\, dx \,.$$

Similarly, multiplying (6) by $\sin(m\pi x/L)$ and integrating yields

$$\int_{-L}^{L} f(x) \sin\frac{m\pi x}{L}\, dx = b_m L$$

so that the formula for b_m is

$$b_m = \frac{1}{L} \int_{-L}^{L} f(x) \sin\frac{m\pi x}{L}\, dx \,.$$

Motivated by the above computations, we now make the following definition.

Fourier Series

Definition 1. Let f be a piecewise continuous function on the interval $[-L, L]$. The **Fourier series**[†] of f is the trigonometric series

$$(8) \qquad f(x) \sim \frac{a_0}{2} + \sum_{n=1}^{\infty} \left\{ a_n \cos\frac{n\pi x}{L} + b_n \sin\frac{n\pi x}{L} \right\},$$

where the a_n's and b_n's are given by the formulas[‡]

$$(9) \qquad a_n = \frac{1}{L} \int_{-L}^{L} f(x) \cos\frac{n\pi x}{L}\, dx\,, \qquad n = 0, 1, 2, \ldots,$$

$$(10) \qquad b_n = \frac{1}{L} \int_{-L}^{L} f(x) \sin\frac{n\pi x}{L}\, dx\,, \qquad n = 1, 2, 3, \ldots.$$

Formulas (9) and (10) are called the **Euler-Fourier formulas.** We use the symbol $\sim$ in (8) to remind us that this series is associated with $f(x)$ but may not converge to $f(x)$. We will return to the question of convergence later in this section. Let's first consider a few examples of Fourier series.

Example 3 Compute the Fourier series for

$$f(x) = \begin{cases} 0\,, & -\pi < x < 0\,, \\ x\,, & 0 < x < \pi\,. \end{cases}$$

Solution Here $L = \pi$. Using formulas (9) and (10), we have

$$a_0 = \frac{1}{\pi} \int_{-\pi}^{\pi} f(x)\, dx = \frac{1}{\pi} \int_{0}^{\pi} x\, dx = \frac{x^2}{2\pi}\bigg|_0^\pi = \frac{\pi}{2}\,,$$

$$a_n = \frac{1}{\pi} \int_{-\pi}^{\pi} f(x) \cos nx\, dx = \frac{1}{\pi} \int_{0}^{\pi} x \cos nx\, dx$$

$$= \frac{1}{\pi n^2} \int_{0}^{\pi n} u \cos u\, du = \frac{1}{\pi n^2} \big[\cos u + u \sin u\big]\bigg|_0^{\pi n}$$

$$= \frac{1}{\pi n^2}(\cos n\pi - 1) = \frac{1}{\pi n^2}\big[(-1)^n - 1\big]\,, \qquad n = 1, 2, 3, \ldots,$$

$$b_n = \frac{1}{\pi} \int_{-\pi}^{\pi} f(x) \sin nx\, dx = \frac{1}{\pi} \int_{0}^{\pi} x \sin nx\, dx$$

$$= \frac{1}{\pi n^2} \int_{0}^{\pi n} u \sin u\, du = \frac{1}{\pi n^2} \big[\sin u - u \cos u\big]\bigg|_0^{\pi n}$$

$$= \frac{-\cos n\pi}{n} = \frac{(-1)^{n+1}}{n}\,, \qquad n = 1, 2, 3, \ldots.$$

[†]*Historical Footnote:* Joseph B. J. Fourier (1768–1830) developed his series for solving heat flow problems. Lagrange expressed doubts about the validity of the representation, but Dirichlet devised conditions that ensured its convergence. Note that the notational choice of $a_0/2$ in equation (6) made it unnecessary to insert an extra formula for a_0 in (9).

[‡]Notice that $f(x)$ need not be defined for every x in $[-L, L]$; we need only that the integrals in (9) and (10) exist.

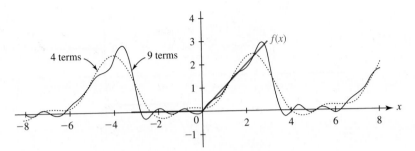

Figure 10.6 Partial sums of Fourier series in Example 3

Therefore,

$$
(11) \qquad f(x) \sim \frac{\pi}{4} + \sum_{n=1}^{\infty} \left\{ \frac{1}{\pi n^2}[(-1)^n - 1]\cos nx + \frac{(-1)^{n+1}}{n}\sin nx \right\}
$$

$$
= \frac{\pi}{4} - \frac{2}{\pi}\left\{ \cos x + \frac{1}{9}\cos 3x + \frac{1}{25}\cos 5x + \cdots \right\}
$$

$$
+ \left\{ \sin x - \frac{1}{2}\sin 2x + \frac{1}{3}\sin 3x + \cdots \right\}.
$$

Some partial sums of this series are displayed in Figure 10.6. ◆

Example 4 Compute the Fourier series for

$$
f(x) = \begin{cases} -1, & -\pi < x < 0, \\ 1, & 0 < x < \pi. \end{cases}
$$

Solution Again, $L = \pi$. Notice that f is an odd function. Since the product of an odd function and an even function is odd (see Problem 7), $f(x)\cos nx$ is also an odd function. Thus,

$$
a_n = \frac{1}{\pi} \int_{-\pi}^{\pi} f(x)\cos nx\, dx = 0, \qquad n = 0, 1, 2, \ldots .
$$

Furthermore, $f(x)\sin nx$ is the product of two odd functions and therefore is even, so

$$
b_n = \frac{1}{\pi} \int_{-\pi}^{\pi} f(x)\sin nx\, dx = \frac{2}{\pi} \int_{0}^{\pi} \sin nx\, dx
$$

$$
= \frac{2}{\pi}\left[\frac{-\cos nx}{n} \right]\Big|_{0}^{\pi} = \frac{2}{\pi}\left[\frac{1}{n} - \frac{(-1)^n}{n} \right], \qquad n = 1, 2, 3, \ldots ,
$$

$$
= \begin{cases} 0, & n \text{ even}, \\ \dfrac{4}{\pi n}, & n \text{ odd}. \end{cases}
$$

Thus,

$$
(12) \qquad f(x) \sim \frac{2}{\pi} \sum_{n=1}^{\infty} \frac{[1 - (-1)^n]}{n}\sin nx = \frac{4}{\pi}\left[\sin x + \frac{1}{3}\sin 3x + \frac{1}{5}\sin 5x + \cdots \right].
$$

Some partial sums of (12) are sketched in Figure 10.7 on page 578. ◆

Figure 10.7 Partial sums of Fourier series in Example 4

In Example 4 the odd function f has a Fourier series consisting only of sine functions. It is easy to see that, in general, if f is any odd function, then its Fourier series consists only of sine terms.

Example 5 Compute the Fourier series for $f(x) = |x|, -1 < x < 1$.

Solution Here $L = 1$. Since f is an even function, $f(x)\sin(n\pi x)$ is an odd function. Therefore,

$$b_n = \int_{-1}^{1} f(x)\sin(n\pi x)\,dx = 0\,, \qquad n = 1, 2, 3, \dots \,.$$

Since $f(x)\cos(n\pi x)$ is an even function, we have

$$a_0 = \int_{-1}^{1} f(x)\,dx = 2\int_{0}^{1} x\,dx = x^2 \Big|_0^1 = 1\,,$$

$$a_n = \int_{-1}^{1} f(x)\cos(n\pi x)\,dx = 2\int_0^1 x\cos(n\pi x)\,dx$$

$$= \frac{2}{\pi^2 n^2} \int_0^{\pi n} u\cos u\,du = \frac{2}{\pi^2 n^2}\left[\cos u + u\sin u\right]\Big|_0^{\pi n} = \frac{2}{\pi^2 n^2}(\cos n\pi - 1)$$

$$= \frac{2}{\pi^2 n^2}\left[(-1)^n - 1\right]\,, \qquad n = 1, 2, 3, \dots \,.$$

Therefore,

$$(13) \qquad f(x) \sim \frac{1}{2} + \sum_{n=1}^{\infty} \frac{2}{\pi^2 n^2}\left[(-1)^n - 1\right]\cos(n\pi x)$$

$$= \frac{1}{2} - \frac{4}{\pi^2}\left\{\cos(\pi x) + \frac{1}{9}\cos(3\pi x) + \frac{1}{25}\cos(5\pi x) + \cdots\right\}\,.$$

Partial sums for (13) are displayed in Figure 10.8. ◆

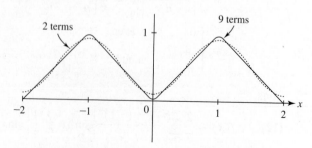

Figure 10.8 Partial sums of Fourier series in Example 5

Notice that the even function f of Example 5 has a Fourier series consisting only of cosine functions and the constant function $1 = \cos(0\pi x)$. In general, if f is an even function, then its Fourier series consists only of cosine functions [including $\cos(0\pi x)$].

Orthogonal Expansions

Fourier series are examples of orthogonal expansions.[†] A set of functions $\{f_n(x)\}_{n=1}^{\infty}$ is said to be an **orthogonal system** or just **orthogonal** with respect to the nonnegative weight function $w(x)$ on the interval $[a, b]$ if

$$(14) \qquad \int_a^b f_m(x)f_n(x)w(x)\,dx = 0, \qquad \text{whenever } m \neq n.$$

As we have seen, the set of trigonometric functions

$$(15) \qquad \{1, \cos x, \sin x, \cos 2x, \sin 2x, \ldots\}$$

is orthogonal on $[-\pi, \pi]$ with respect to the weight function $w(x) \equiv 1$. If we define the **norm** of f as

$$(16) \qquad \|f\| := \left[\int_a^b f^2(x)w(x)\,dx\right]^{1/2},$$

then we say that a set of functions $\{f_n(x)\}_{n=1}^{\infty}$ (or $\{f_n(x)\}_{n=1}^{N}$) is an **orthonormal system with respect to** $w(x)$ if (14) holds and also $\|f_n\| = 1$ for each n. Equivalently, we say the set is an orthonormal system if

$$(17) \qquad \int_a^b f_m(x)f_n(x)w(x)\,dx = \begin{cases} 0, & m \neq n, \\ 1, & m = n. \end{cases}$$

We can always obtain an orthonormal system from an orthogonal system just by dividing each function by its norm. In particular, since

$$\int_{-\pi}^{\pi} \cos^2 nx\,dx = \int_{-\pi}^{\pi} \sin^2 nx\,dx = \pi, \qquad n = 1, 2, 3, \ldots$$

and

$$\int_{-\pi}^{\pi} 1\,dx = 2\pi,$$

then the orthogonal system (15) gives rise on $[-\pi, \pi]$ to the orthonormal system

$$\left\{ (2\pi)^{-1/2}, \; \pi^{-1/2}\cos x, \; \pi^{-1/2}\sin x, \; \pi^{-1/2}\cos 2x, \; \pi^{-1/2}\sin 2x, \; \ldots \right\}.$$

If $\{f_n(x)\}_{n=1}^{\infty}$ is an orthogonal system with respect to $w(x)$ on $[a, b]$, we might ask if we can expand a function $f(x)$ in terms of these functions; that is, can we express f in the form

$$(18) \qquad f(x) = c_1 f_1(x) + c_2 f_2(x) + c_3 f_3(x) + \cdots$$

for a suitable choice of constants $c_1, c_2, \ldots$? Such an expansion is called an **orthogonal expansion,** or a **generalized Fourier series.**

[†]Orthogonality is also discussed in Section 8.8 on page 474.

To determine the constants in (18), we can proceed as we did in deriving Euler's formulas for the coefficients of a Fourier series, this time using the orthogonality of the system. Presuming the representation (18) is valid, we multiply by $f_m(x)w(x)$ and integrate to obtain

$$(19) \quad \int_a^b f(x)f_m(x)\,w(x)\,dx = c_1 \int_a^b f_1(x)f_m(x)\,w(x)\,dx + c_2 \int_a^b f_2(x)f_m(x)\,w(x)\,dx + \cdots$$

$$= \sum_{n=1}^{\infty} c_n \int_a^b f_n(x)f_m(x)\,w(x)\,dx\,.$$

(Here we have also assumed that we can integrate term by term.) Because the system is orthogonal with respect to $w(x)$, every integral on the right-hand side of (19) is zero except when $n = m$. Solving for c_m gives

$$(20) \quad c_m = \frac{\displaystyle\int_a^b f(x)f_m(x)\,w(x)\,dx}{\displaystyle\int_a^b f_m^2(x)\,w(x)\,dx} = \frac{\displaystyle\int_c^b f(x)f_m(x)\,w(x)\,dx}{\|f_m\|^2}\,, \qquad n = 1, 2, 3, \ldots\,.$$

The derivation of the formula for c_m was only *formal*, since the question of the convergence of the expansion in (18) was not answered. If the series $\sum_{n=1}^{\infty} c_n f_n(x)$ converges *uniformly* to $f(x)$ on $[a, b]$, then each step can be justified, and indeed, the coefficients are given by formula (20). The notion of uniform convergence is discussed in the next subsection and in Section 13.2.[†]

Convergence of Fourier Series

Let's turn to the question of the convergence of a Fourier series. For Example 5 it is possible to use a comparison or limit comparison test to show that the series is absolutely dominated by a *p*-series of the form $\sum_{n=1}^{\infty} 1/n^2$, which converges. However, this is much harder to do in Example 4, since the terms go to zero like $1/n$. Matters can be even worse, since there exist Fourier series that *diverge*.[‡] We state two theorems that deal with the convergence of a Fourier series and two dealing with the properties of termwise differentiation and integration. For proofs of these results, see *Partial Differential Equations of Mathematical Physics,* 2nd ed., by Tyn Myint-U (Elsevier North Holland, Inc., New York, 1983), Chapter 5; *Advanced Calculus with Applications,* by N. J. DeLillo (Macmillan, New York, 1982), Chapter 9; or an advanced text on the theory of Fourier series.

Before proceeding, we need a notation for the left- and right-hand limits of a function. Let

$$f(x^+) := \lim_{h \to 0^+} f(x+h) \quad \text{and} \quad f(x^-) := \lim_{h \to 0^+} f(x-h)\,.$$

We now present the fundamental pointwise convergence theorem for Fourier series. While reading it, keep in mind Figures 10.6-10.8 .

[†]All references to Chapters 11–13 refer to the expanded text, *Fundamentals of Differential Equations and Boundary Value Problems,* 7th ed.

[‡]In fact, there are trigonometric series that converge but are not Fourier series; an example is $\displaystyle\sum_{n=1}^{\infty} \frac{\sin nx}{\ln(n+1)}$.

Pointwise Convergence of Fourier Series

Theorem 2. If f and f' are piecewise continuous on $[-L, L]$, then for any x in $(-L, L)$

$$(21) \qquad \frac{a_0}{2} + \sum_{n=1}^{\infty} \left\{ a_n \cos\frac{n\pi x}{L} + b_n \sin\frac{n\pi x}{L} \right\} = \frac{1}{2}[f(x^+) + f(x^-)],$$

where the a_n's and b_n's are given by the Euler-Fourier formulas (9) and (10). For $x = \pm L$, the series converges to $\frac{1}{2}[f(-L^+) + f(L^-)]$.[†]

In other words, when f and f' are piecewise continuous on $[-L, L]$, the Fourier series converges to $f(x)$ whenever f is continuous at x and converges to the average of the left- and right-hand limits at points where f is discontinuous.

Observe that the left-hand side of (21) is periodic of period $2L$. This means that if we extend $f(x)$ from the interval $(-L, L)$ to the entire real line using $2L$-periodicity, then equation (21) holds for all x for the $2L$-periodic extension of $f(x)$.

Example 6 To which function does the Fourier series for

$$f(x) = \begin{cases} -1, & -\pi < x < 0, \\ 1, & 0 < x < \pi, \end{cases}$$

converge?

Solution In Example 4 we found that the Fourier series for $f(x)$ is given by (12), and in Figure 10.7, we sketched the graphs of two of its partial sums. Now $f(x)$ and $f'(x)$ are piecewise continuous in $[-\pi, \pi]$. Moreover, f is continuous except at $x = 0$. Thus, by Theorem 2, the Fourier series of f in (12) converges to the 2π-periodic function $g(x)$, where $g(x) = f(x) = -1$ for $-\pi < x < 0$, $g(x) = f(x) = 1$ for $0 < x < \pi$, $g(0) = [f(0^+) + f(0^-)]/2 = 0$, and at $\pm\pi$ we have $g(\pm\pi) = [f(-\pi^+) + f(\pi^-)]/2 = (-1+1)/2 = 0$. The graph of $g(x)$ is given in Figure 10.9. ◆

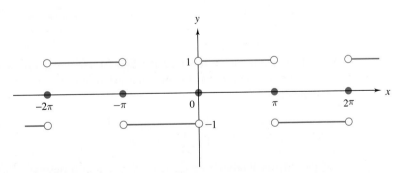

Figure 10.9 The limit function of the Fourier series for $f(x) = \begin{cases} -1, & -\pi < x < 0, \\ 1, & 0 < x < \pi \end{cases}$

[†]From formula (21), we see that it doesn't matter how we define $f(x)$ at its points of discontinuity, since only the left- and right-hand limits are involved. The derivative $f'(x)$, of course, is undefined at such points.

When f is a $2L$-periodic function that is continuous on $(-\infty, \infty)$ and has a piecewise continuous derivative, its Fourier series not only converges at each point—it converges **uniformly on** $(-\infty, \infty)$. This means that for any prescribed tolerance $\varepsilon > 0$, the graph of the partial sum

$$s_N(x) := \frac{a_0}{2} + \sum_{n=1}^{N} \left\{ a_n \cos\frac{n\pi x}{L} + b_n \sin\frac{n\pi x}{L} \right\}$$

will, for all N sufficiently large, lie in an ε-corridor about the graph of f on $(-\infty, \infty)$ (see Figure 10.10). The property of uniform convergence of Fourier series is particularly helpful when one needs to verify that a formal solution to a partial differential equation is an actual (genuine) solution.

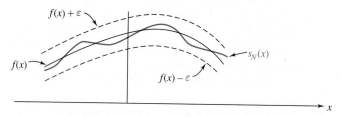

Figure 10.10 An ε-corridor about f

Uniform Convergence of Fourier Series

Theorem 3. Let f be a continuous function on $(-\infty, \infty)$ and periodic of period $2L$. If f' is piecewise continuous on $[-L, L]$, then the Fourier series for f converges uniformly to f on $[-L, L]$ and hence on any interval. That is, for each $\varepsilon > 0$, there exists an integer N_0 (that depends on ε) such that

$$\left| f(x) - \left[\frac{a_0}{2} + \sum_{n=1}^{N} \left\{ a_n \cos\frac{n\pi x}{L} + b_n \sin\frac{n\pi x}{L} \right\} \right] \right| < \varepsilon ,$$

for all $N \geq N_0$, and all $x \in (-\infty, \infty)$.

In Example 5 we obtained the Fourier series expansion given in (13) for $f(x) = |x|$, $-1 < x < 1$. Since $g(x)$, the periodic extension of $f(x)$ (see Figure 10.11 on page 583) is continuous on $(-\infty, \infty)$ and

$$f'(x) = \begin{cases} -1, & -1 < x < 0, \\ 1, & 0 < x < 1, \end{cases}$$

is piecewise continuous on $[-1, 1]$, the Fourier series expansion (13) converges uniformly to $|x|$ on $[-1, 1]$. Compare Figure 10.8 on page 578.

The term-by-term differentiation of a Fourier series is not always permissible. For example, the Fourier series for $f(x) = x$, $-\pi < x < \pi$ (see Problem 9), is

$$(22) \qquad f(x) \sim 2 \sum_{n=1}^{\infty} (-1)^{n+1} \frac{\sin nx}{n} ,$$

which converges for all x, whereas its derived series

$$2 \sum_{n=1}^{\infty} (-1)^{n+1} \cos nx$$

diverges for every x. The following theorem gives sufficient conditions for using termwise differentiation.

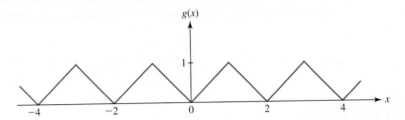

Figure 10.11 Periodic extension of $f(x) = |x|, -1 < x < 1$

Differentiation of Fourier Series

Theorem 4. Let $f(x)$ be continuous on $(-\infty, \infty)$ and $2L$-periodic. Let $f'(x)$ and $f''(x)$ be piecewise continuous on $[-L, L]$. Then, the Fourier series of $f'(x)$ can be obtained from the Fourier series for $f(x)$ by termwise differentiation. In particular, if

$$f(x) = \frac{a_0}{2} + \sum_{n=1}^{\infty} \left\{ a_n \cos \frac{n\pi x}{L} + b_n \sin \frac{n\pi x}{L} \right\},$$

then

$$f'(x) \sim \sum_{n=1}^{\infty} \frac{\pi n}{L} \left\{ -a_n \sin \frac{n\pi x}{L} + b_n \cos \frac{n\pi x}{L} \right\}.$$

Notice that Theorem 4 does not apply to the example function $f(x)$ and its Fourier series expansion shown in (22), since the 2π-periodic extension of this $f(x)$ fails to be continuous on $(-\infty, \infty)$.

Termwise integration of a Fourier series is permissible under much weaker conditions.

Integration of Fourier Series

Theorem 5. Let $f(x)$ be piecewise continuous on $[-L, L]$ with Fourier series

$$f(x) \sim \frac{a_0}{2} + \sum_{n=1}^{\infty} \left\{ a_n \cos \frac{n\pi x}{L} + b_n \sin \frac{n\pi x}{L} \right\}.$$

Then, for any x in $[-L, L]$,

$$\int_{-L}^{x} f(t) \, dt = \int_{-L}^{x} \frac{a_0}{2} \, dt + \sum_{n=1}^{\infty} \int_{-L}^{x} \left\{ a_n \cos \frac{n\pi t}{L} + b_n \sin \frac{n\pi t}{L} \right\} dt .$$

A final note on the convergence issue: The question as to under what conditions the Fourier series converges has led to a tremendous amount of beautiful mathematics. As a practical matter, however, time and economics will often dictate that you can afford to compute only a few terms of an eigenfunction series for your particular partial differential equation. Convergence fades into the background, and you need to know how to "do the best with what you've got." So it is extremely gratifying to know that even if you're going to use only part of an eigenfunction expansion, the Fourier coefficients (9)–(10) are still the best choice; Problem 37 demonstrates that every partial sum of a Fourier series outperforms any comparable superposition of the trigometric functions, in terms of mean-square approximation.

10.3 EXERCISES

In Problems 1–6, determine whether the given function is even, odd, or neither.

1. $f(x) = x^3 + \sin 2x$

2. $f(x) = \sin^2 x$

3. $f(x) = (1 - x^2)^{-1/2}$

4. $f(x) = \sin(x + 1)$

5. $f(x) = e^{-x} \cos 3x$

6. $f(x) = x^{1/5} \cos x^2$

7. Prove the following properties:

(a) If f and g are even functions, then so is the product fg.

(b) If f and g are odd functions, then fg is an even function.

(c) If f is an even function and g is an odd function, then fg is an odd function.

8. Verify formula (5). [*Hint:* Use the identity $2\cos A \cos B = \cos(A + B) + \cos(A - B).$]

In Problems 9–16, compute the Fourier series for the given function f on the specified interval. Use a computer or graphing calculator to plot a few partial sums of the Fourier series.

9. $f(x) = x, \quad -\pi < x < \pi$

10. $f(x) = |x|, \quad -\pi < x < \pi$

11. $f(x) = \begin{cases} 1, & -2 < x < 0, \\ x, & 0 < x < 2 \end{cases}$

12. $f(x) = \begin{cases} 0, & -\pi < x < 0, \\ x^2, & 0 < x < \pi \end{cases}$

13. $f(x) = x^2, \quad -1 < x < 1$

14. $f(x) = \begin{cases} x, & 0 < x < \pi, \\ x + \pi, & -\pi < x < 0 \end{cases}$

15. $f(x) = e^x, \quad -\pi < x < \pi$

16. $f(x) = \begin{cases} 0, & -\pi < x < -\pi/2, \\ -1, & -\pi/2 < x < 0, \\ 1, & 0 < x < \pi/2, \\ 0, & \pi/2 < x < \pi \end{cases}$

In Problems 17–24, determine the function to which the Fourier series for $f(x)$, given in the indicated problem, converges.

17. Problem 9

18. Problem 10

19. Problem 11

20. Problem 12

21. Problem 13

22. Problem 14

23. Problem 15

24. Problem 16

25. Find the functions represented by the series obtained by the termwise integration of the given series from $-\pi$ to x.

(a) $2 \sum_{n=1}^{\infty} \dfrac{(-1)^{n+1}}{n} \sin nx \sim x, \quad -\pi < x < \pi$

(b) $\dfrac{4}{\pi} \sum_{n=0}^{\infty} \dfrac{\sin(2n+1)x}{(2n+1)} \sim f(x),$

$f(x) = \begin{cases} -1, & -\pi < x < 0, \\ 1, & 0 < x < \pi \end{cases}$

26. Show that the set of functions

$\left\{ \cos\dfrac{\pi}{2}x, \ \sin\dfrac{\pi}{2}x, \ \cos\dfrac{3\pi}{2}x, \ \sin\dfrac{3\pi}{2}x, \ \dots, \right.$

$\left. \cos\dfrac{(2n-1)\pi}{2}x, \ \sin\dfrac{(2n-1)\pi}{2}x, \ \dots \right\}$

is an orthonormal system on $[-1, 1]$ with respect to the weight function $w(x) \equiv 1$.

27. Find the orthogonal expansion (generalized Fourier series) for

$f(x) = \begin{cases} 0, & -1 < x < 0, \\ 1, & 0 < x < 1, \end{cases}$

in terms of the orthonormal system of Problem 26.

28. (a) Show that the function $f(x) = x^2$ has the Fourier series, on $-\pi < x < \pi$,

$f(x) \sim \dfrac{\pi^2}{3} + 4 \sum_{n=1}^{\infty} \dfrac{(-1)^n}{n^2} \cos nx.$

(b) Use the result of part (a) and Theorem 2 to show that

$$\sum_{n=1}^{\infty} \frac{(-1)^{n+1}}{n^2} = \frac{\pi^2}{12}.$$

(c) Use the result of part (a) and Theorem 2 to show that

$$\sum_{n=1}^{\infty} \frac{1}{n^2} = \frac{\pi^2}{6}.$$

29. In Section 8.8, it was shown that the Legendre polynomials $P_n(x)$ are orthogonal on the interval $[-1, 1]$ with respect to the weight function $w(x) \equiv 1$. Using the fact that the first three Legendre polynomials are

$$P_0(x) \equiv 1, \ P_1(x) = x, \ P_2(x) = (3/2)x^2 - (1/2),$$

find the first three coefficients in the expansion

$$f(x) = c_0 P_0(x) + c_1 P_1(x) + c_2 P_2(x) + \cdots,$$

where $f(x)$ is the function

$$f(x) := \begin{cases} -1, & -1 < x < 0, \\ 1, & 0 < x < 1. \end{cases}$$

30. As in Problem 29, find the first three coefficients in the expansion

$$f(x) = c_0 P_0(x) + c_1 P_1(x) + c_2 P_2(x) + \cdots,$$

when $f(x) = |x|, -1 < x < 1$.

31. The Hermite polynomials $H_n(x)$ are orthogonal on the interval $(-\infty, \infty)$ with respect to the weight function $W(x) = e^{-x^2}$. Verify this fact for the first three Hermite polynomials:

$$H_0(x) \equiv 1, \ H_1(x) = 2x, \ H_2(x) = 4x^2 - 2.$$

32. The Chebyshev (Tchebichef) polynomials $T_n(x)$ are orthogonal on the interval $[-1, 1]$ with respect to the weight function $w(x) = (1 - x^2)^{-1/2}$. Verify this fact for the first three Chebyshev polynomials:

$$T_0(x) \equiv 1, \ T_1(x) = x, \ T_2(x) = 2x^2 - 1.$$

33. Let $\{f_n(x)\}$ be an orthogonal set of functions on the interval $[a, b]$ with respect to the weight function $w(x)$. Show that they satisfy the **Pythagorean property**

$$\|f_m + f_n\|^2 = \|f_m\|^2 + \|f_n\|^2$$

if $m \neq n$.

34. Norm. The norm of a function $\|f\|$ is like the length of a vector in R^n. In particular, show that the norm defined in (16) satisfies the following properties associated with length (assume f and g are continuous and $w(x) > 0$ on $[a, b]$):

(a) $\|f\| \geq 0$, and $\|f\| = 0$ if and only if $f \equiv 0$.

(b) $\|cf\| = |c| \ \|f\|$, where c is any real number.

(c) $\|f + g\| \leq \|f\| + \|g\|$.

35. Inner Product. The integral in the orthogonality condition (14) is like the dot product of two vectors in R^n. In

particular, show that the **inner product** of two functions defined by

$$(23) \quad \langle f, g \rangle := \int_a^b f(x)g(x)w(x)\,dx,$$

where $w(x)$ is a positive weight function, satisfies the following properties associated with the dot product (assume f, g, and h are continuous on $[a, b]$):

(a) $\langle f + g, h \rangle = \langle f, h \rangle + \langle g, h \rangle$.

(b) $\langle cf, h \rangle = c\langle f, h \rangle$, where c is any real number.

(c) $\langle f, g \rangle = \langle g, f \rangle$.

36. Complex Form of the Fourier Series.

(a) Using the Euler formula $e^{i\theta} = \cos\theta + i\sin\theta$, $i = \sqrt{-1}$, prove that

$$\cos nx = \frac{e^{inx} + e^{-inx}}{2} \quad \text{and} \quad \sin nx = \frac{e^{inx} - e^{-inx}}{2i}.$$

(b) Show that the Fourier series

$$f(x) \sim \frac{a_0}{2} + \sum_{n=1}^{\infty} \{a_n \cos nx + b_n \sin nx\}$$

$$= c_0 + \sum_{n=1}^{\infty} \{c_n e^{inx} + c_{-n} e^{-inx}\},$$

where

$$c_0 = \frac{a_0}{2}, \quad c_n = \frac{a_n - ib_n}{2}, \quad c_{-n} = \frac{a_n + ib_n}{2}.$$

(c) Finally, use the results of part (b) to show that

$$f(x) \sim \sum_{n=-\infty}^{\infty} c_n e^{inx}, \quad -\pi < x < \pi,$$

where

$$c_n = \frac{1}{2\pi} \int_{-\pi}^{\pi} f(x) e^{-inx}\,dx.$$

37. Least-Squares Approximation Property. The Nth partial sum of the Fourier series

$$f(x) \sim \frac{a_0}{2} + \sum_{n=1}^{\infty} \{a_n \cos nx + b_n \sin nx\}$$

gives the best mean-square approximation of f by a trigonometric polynomial. To prove this, let $F_N(x)$ denote an arbitrary trigonometric polynomial of degree N:

$$F_N(x) = \frac{\alpha_0}{2} + \sum_{n=1}^{N} \{\alpha_n \cos nx + \beta_n \sin nx\},$$

and define

$$E := \int_{-\pi}^{\pi} [f(x) - F_N(x)]^2\,dx,$$

which is the *total square error.* Expanding the integrand, we get

$$E = \int_{-\pi}^{\pi} f^2(x)\,dx - 2\int_{-\pi}^{\pi} f(x)F_N(x)\,dx$$
$$+ \int_{-\pi}^{\pi} F_N^2(x)\,dx\,.$$

(a) Use the orthogonality of the functions $\{1, \cos x, \sin x, \cos 2x, \dots\}$ to show that

$$\int_{-\pi}^{\pi} F_N^2(x)\,dx = \pi\left(\frac{\alpha_0^2}{2} + \alpha_1^2 + \cdots + \alpha_N^2\right.$$
$$\left. + \beta_1^2 + \cdots + \beta_N^2\right)$$

and

$$\int_{-\pi}^{\pi} f(x)F_N(x)\,dx = \pi\left(\frac{\alpha_0 a_0}{2} + \alpha_1 a_1 + \cdots + \alpha_N a_N\right.$$
$$\left. + \beta_1 b_1 + \cdots + \beta_N b_N\right).$$

(b) Let E^* be the error when we approximate f by the Nth partial sum of its Fourier series, that is, when we choose $\alpha_n = a_n$ and $\beta_n = b_n$. Show that

$$E^* = \int_{-\pi}^{\pi} f^2(x)\,dx - \pi\left(\frac{a_0^2}{2} + a_1^2 + \cdots + a_N^2\right.$$
$$\left. + b_1^2 + \cdots + b_N^2\right).$$

(c) Using the results of parts (a) and (b), show that $E - E^* \geq 0$, that is, $E \geq E^*$, by proving that

$$E - E^* = \pi\left\{\frac{(\alpha_0 - a_0)^2}{2} + (\alpha_1 - a_1)^2\right.$$
$$+ \cdots + (\alpha_N - a_N)^2 + (\beta_1 - b_1)^2$$
$$\left. + \cdots + (\beta_N - b_N)^2\right\}.$$

Hence, the Nth partial sum of the Fourier series gives the least total square error, since $E \geq E^*$.

38. Bessel's Inequality. Use the fact that E^*, defined in part (b) of Problem 37, is nonnegative to prove **Bessel's inequality**

$$(24) \qquad \frac{a_0^2}{2} + \sum_{n=1}^{\infty}\{a_n^2 + b_n^2\} \leq \frac{1}{\pi}\int_{-\pi}^{\pi} f^2(x)\,dx\,.$$

(If f is piecewise continuous on $[-\pi, \pi]$, then we have equality in (24). This result is called **Parseval's identity.**)

39. Gibbs Phenomenon.[†] Josiah Willard Gibbs, who was awarded the first American doctorate in engineering (Yale, 1863), observed that near points of discontinuity of f, the partial sums of the Fourier series for f may overshoot by approximately 9% of the jump, regardless of the number of terms. This is illustrated in Figure 10.12 for the function

$$f(x) = \begin{cases} -1, & -\pi < x < 0, \\ 1, & 0 < x < \pi, \end{cases}$$

(a) Graph of $f_{11}(x)$

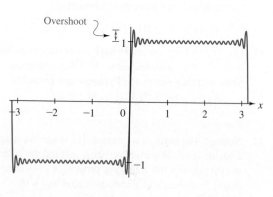

(b) Graph of $f_{51}(x)$

Figure 10.12 Gibbs phenomenon for partial sums of Fourier series

[†]*Historical Footnote:* Actually, H. Wilbraham discovered this phenomenon some 50 years earlier than Gibbs did. It is more appropriately called the *Gibbs–Wilbraham* phenomenon.

whose Fourier series has the partial sums

$$f_{2n-1}(x) = \frac{4}{\pi}\left[\sin x + \frac{1}{3}\sin 3x\right.$$
$$\left. + \cdots + \frac{\sin(2n-1)x}{(2n-1)}\right].$$

To verify this for $f(x)$, proceed as follows:

(a) Show that

$$\pi(\sin x)f'_{2n-1}(x) = 4\sin x[\cos x + \cos 3x$$
$$+ \cdots + \cos(2n-1)x]$$
$$= 2\sin 2nx.$$

(b) Infer from part (a) and the figure that the maximum occurs at $x = \pi/(2n)$ and has the value

$$f_{2n-1}\left(\frac{\pi}{2n}\right) = \frac{4}{\pi}\left[\sin\frac{\pi}{2n} + \frac{1}{3}\sin\frac{3\pi}{2n}\right.$$
$$\left. + \cdots + \frac{1}{2n-1}\sin\frac{(2n-1)\pi}{2n}\right].$$

(c) Show that if one approximates

$$\int_0^\pi \frac{\sin x}{x}dx$$

using the partition $x_k := (2k-1)(\pi/2n)$, $k = 1$, $2, \ldots, n$, $\Delta x_k = \pi/n$ and choosing the midpoint of

each interval as the place to evaluate the integrand, then

$$\int_0^\pi \frac{\sin x}{x}dx \approx \frac{\sin(\pi/2n)}{\pi/2n}\frac{\pi}{n} + \cdots$$
$$+ \frac{\sin[(2n-1)\pi/2n]}{(2n-1)\pi/2n}\frac{\pi}{n}$$
$$= \frac{\pi}{2}f_{2n-1}\left(\frac{\pi}{2n}\right).$$

(d) Use the result of part (c) to show that the overshoot satisfies

$$\lim_{n\to\infty} f_{2n-1}\left(\frac{\pi}{2n}\right) = \frac{2}{\pi}\int_0^\pi \frac{\sin x}{x}dx.$$

(e) Using the result of part (d) and a numerical integration algorithm (such as Simpson's rule, Appendix C) for the **sine integral function**

$$Si(z) := \int_0^z \frac{\sin x}{x}dx,$$

show that $\lim_{n\to\infty} f_{2n-1}(\pi/(2n)) \approx 1.18$. Thus, the approximations overshoot the true value of $f(0^+) = 1$ by 0.18, or 9% of the jump from $f(0^-)$ to $f(0^+)$.

10.4 Fourier Cosine and Sine Series

A typical problem encountered in using separation of variables to solve a partial differential equation is the problem of representing a function defined on some finite interval by a trigonometric series consisting of only sine functions or only cosine functions. For example, in Section 10.2, equation (25), page 569, we needed to express the initial values $u(x, 0) = f(x)$, $0 < x < L$, of the solution to the initial-boundary value problem associated with the problem of a vibrating string as a trigonometric series of the form

$$(1) \qquad f(x) = \sum_{n=1}^{\infty} a_n \sin\left(\frac{n\pi x}{L}\right).$$

Recalling that the Fourier series for an odd function defined on $[-L, L]$ consists entirely of sine terms, we might try to achieve (1) by artificially extending the function $f(x), 0 < x < L$, to the interval $(-L, L)$ in such a way that the extended function is odd. This is accomplished by defining the function

$$f_o(x) := \begin{cases} f(x), & 0 < x < L, \\ -f(-x), & -L < x < 0, \end{cases}$$

and extending $f_o(x)$ to all x using $2L$-periodicity.[†] Since $f_o(x)$ is an odd function, it has a Fourier series consisting entirely of sine terms. Moreover, $f_o(x)$ is an extension of $f(x)$,

[†]Strictly speaking, we have extended $f_o(x)$ to all x other than the integer multiples of L. Continuity considerations often suggest appropriate values for the extended functions at some of these points, as well. Figure 10.13 on page 589 illustrates this.

since $f_o(x) = f(x)$ on $(0, L)$. This extension is called the **odd 2L-periodic extension** of $f(x)$. The resulting Fourier series expansion is called a half-range expansion for $f(x)$, since it represents the function $f(x)$ on $(0, L)$, which is half of the interval $(-L, L)$ where it represents $f_o(x)$.

In a similar fashion, we can define the **even 2L-periodic extension** of $f(x)$ as the function

$$f_e(x) := \begin{cases} f(x), & 0 < x < L, \\ f(-x), & -L < x < 0, \end{cases}$$

with $f_e(x + 2L) = f_e(x)$.

To illustrate the various extensions, let's consider the function $f(x) = x, 0 < x < \pi$. If we extend $f(x)$ to the interval $(-\pi, \pi)$ using π-periodicity, then the extension $\tilde{f}$ is given by

$$\tilde{f}(x) = \begin{cases} x, & 0 < x < \pi, \\ x + \pi, & -\pi < x < 0, \end{cases}$$

with $\tilde{f}(x + 2\pi) = \tilde{f}(x)$. In Problem 14 of Exercises 10.3, the Fourier series for $\tilde{f}(x)$ was found to be

$$\tilde{f}(x) \sim \frac{\pi}{2} - \sum_{n=1}^{\infty} \frac{1}{n} \sin 2nx,$$

which consists of both odd functions (the sine terms) and even functions (the constant term), since the π-periodic extension $\tilde{f}(x)$ is neither an even nor an odd function. The odd 2π-periodic extension of $f(x)$ is just $f_o(x) = x, -\pi < x < \pi$, which has the Fourier series expansion

$$(2) \qquad f_o(x) \sim 2 \sum_{n=1}^{\infty} \frac{(-1)^{n+1}}{n} \sin nx$$

(see Problem 9 in Exercises 10.3). Because $f_o(x) = f(x)$ on the interval $(0, \pi)$, the expansion in (2) is a half-range expansion for $f(x)$. The even 2π-periodic extension of $f(x)$ is the function $f_e(x) = |x|, -\pi < x < \pi$, which has the Fourier series expansion

$$(3) \qquad f_e(x) = \frac{\pi}{2} - \frac{4}{\pi} \sum_{n=0}^{\infty} \frac{1}{(2n+1)^2} \cos(2n+1)x$$

(see Problem 10 in Exercises 10.3).

The preceding three extensions, the π-periodic function $\tilde{f}(x)$, the odd 2π-periodic function $f_o(x)$, and the even 2π-periodic function $f_e(x)$, are natural extensions of $f(x)$. There are many other ways of extending $f(x)$. For example, the function

$$g(x) = \begin{cases} x, & 0 < x < \pi, \\ 0, & -\pi < x < 0, \end{cases} \qquad g(x + 2\pi) = g(x),$$

which we studied in Example 3 of Section 10.3, is also an extension of $f(x)$. However, its Fourier series contains both sine and cosine terms and hence is not as useful as previous extensions. The graphs of these extensions of $f(x)$ are given in Figure 10.13 on page 589.

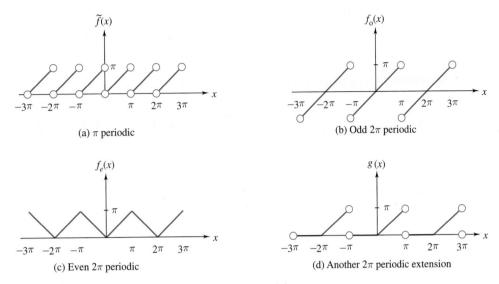

(a) π periodic

(b) Odd 2π periodic

(c) Even 2π periodic

(d) Another 2π periodic extension

Figure 10.13 Extensions of $f(x) = x, 0 < x < \pi$

The Fourier series expansions for $f_o(x)$ and $f_e(x)$ given in (2) and (3) represent $f(x)$ on the interval $(0, \pi)$ $\big($ actually, they equal $f(x)$ on $(0, \pi)$ $\big)$. This motivates the following definitions.

Fourier Cosine and Sine Series

Definition 2. Let $f(x)$ be piecewise continuous on the interval $[0, L]$. The **Fourier cosine series** of $f(x)$ on $[0, L]$ is

(4) $$\frac{a_0}{2} + \sum_{n=1}^{\infty} a_n \cos \frac{n\pi x}{L} ,$$

where

(5) $$a_n = \frac{2}{L} \int_0^L f(x) \cos \frac{n\pi x}{L} \, dx , \qquad n = 0, 1, \dots .$$

The **Fourier sine series** of $f(x)$ on $[0, L]$ is

(6) $$\sum_{n=1}^{\infty} b_n \sin \frac{n\pi x}{L} ,$$

where

(7) $$b_n = \frac{2}{L} \int_0^L f(x) \sin \frac{n\pi x}{L} \, dx , \qquad n = 1, 2, \dots .$$

The trigonometric series in (4) is just the Fourier series for $f_e(x)$, the even $2L$-periodic extension of $f(x)$, and that in (6) is the Fourier series for $f_o(x)$, the odd $2L$-periodic extension of $f(x)$. These are called **half-range expansions** for $f(x)$.

Example 1 Compute the Fourier sine series for

$$f(x) = \begin{cases} x, & 0 \le x \le \pi/2, \\ \pi - x, & \pi/2 \le x \le \pi. \end{cases}$$

Solution Using formula (7) with $L = \pi$, we find

$$b_n = \frac{2}{\pi} \int_0^{\pi} f(x) \sin nx\, dx = \frac{2}{\pi} \int_0^{\pi/2} x \sin nx\, dx + \frac{2}{\pi} \int_{\pi/2}^{\pi} (\pi - x) \sin nx\, dx$$

$$= \frac{2}{\pi n^2} \int_0^{\pi n/2} u \sin u\, du + 2 \int_{\pi/2}^{\pi} \sin nx\, dx - \frac{2}{\pi n^2} \int_{\pi n/2}^{\pi n} u \sin u\, du$$

$$= \frac{2}{\pi n^2} \left[\sin u - u \cos u \right] \Big|_0^{\pi n/2} - \frac{2}{n} \left[\cos \pi n - \cos \frac{n\pi}{2} \right]$$

$$- \frac{2}{\pi n^2} \left[\sin u - u \cos u \right] \Big|_{\pi n/2}^{\pi n}$$

$$= \frac{4}{\pi n^2} \sin \frac{n\pi}{2} = \begin{cases} 0, & n \text{ even}, \\ \dfrac{4(-1)^{(n-1)/2}}{\pi n^2}, & n \text{ odd}. \end{cases}$$

So on letting $n = 2k + 1$, we find the Fourier sine series for $f(x)$ to be

(8) $$\frac{4}{\pi} \sum_{k=0}^{\infty} \frac{(-1)^k}{(2k+1)^2} \sin(2k+1)x = \frac{4}{\pi} \left\{ \sin x - \frac{1}{9}\sin 3x + \frac{1}{25}\sin 5x + \cdots \right\}. \quad \blacklozenge$$

Since, in Example 1, the function $f(x)$ is continuous and $f'(x)$ is piecewise continuous on $(0, \pi)$, it follows from Theorem 2 on pointwise convergence of Fourier series that

$$f(x) = \frac{4}{\pi} \left\{ \sin x - \frac{1}{9}\sin 3x + \frac{1}{25}\sin 5x - \frac{1}{49}\sin 7x + \cdots \right\}$$

for all x in $[0, \pi]$.

Let's return to the problem of heat flow in one dimension.

Example 2 Find the solution to the heat flow problem

(9) $$\frac{\partial u}{\partial t} = 2\frac{\partial^2 u}{\partial x^2}, \qquad 0 < x < \pi, \qquad t > 0,$$

(10) $$u(0, t) = u(\pi, t) = 0, \qquad t > 0,$$

(11) $$u(x, 0) = \begin{cases} x, & 0 < x \le \pi/2, \\ \pi - x, & \pi/2 \le x < \pi. \end{cases}$$

Solution Comparing equation (9) with equation (1) in Section 10.2 (page 563), we see that $\beta = 2$ and $L = \pi$. Hence, we need only represent $u(x, 0) = f(x)$ in a Fourier sine series of the form

$$\sum_{n=1}^{\infty} c_n \sin nx.$$

In Example 1 we obtained this expansion and showed that

$$c_n = \frac{4}{\pi n^2} \sin \frac{n\pi}{2} = \begin{cases} 0, & n \text{ even}, \\ \dfrac{4(-1)^{(n-1)/2}}{\pi n^2}, & n \text{ odd}. \end{cases}$$

Hence, from equation (15) on page 566, the solution to the heat flow problem (9)–(11) is

$$
\begin{aligned}
u(x, t) &= \sum_{n=1}^{\infty} c_n e^{-2n^2 t} \sin nx \\
&= \frac{4}{\pi} \sum_{k=0}^{\infty} \frac{(-1)^k}{(2k+1)^2} e^{-2(2k+1)^2 t} \sin(2k+1)x \\
&= \frac{4}{\pi} \left\{ e^{-2t} \sin x - \frac{1}{9} e^{-18t} \sin 3x + \frac{1}{25} e^{-50t} \sin 5x + \cdots \right\}.
\end{aligned}
$$

A sketch of a partial sum for $u(x, t)$ is displayed in Figure 10.14. ◆

Figure 10.14 Partial sum for $u(x, t)$ in Example 2

10.4 EXERCISES

In Problems 1–4, determine (a) the π-periodic extension $\tilde{f}$, (b) the odd 2π-periodic extension f_o, and (c) the even 2π-periodic extension f_e for the given function f and sketch their graphs.

1. $f(x) = x^2$, $0 < x < \pi$

2. $f(x) = \sin 2x$, $0 < x < \pi$

3. $f(x) = \begin{cases} 0, & 0 < x < \pi/2, \\ 1, & \pi/2 < x < \pi \end{cases}$

4. $f(x) = \pi - x$, $0 < x < \pi$

In Problems 5–10, compute the Fourier sine series for the given function.

5. $f(x) = -1$, $0 < x < 1$

6. $f(x) = \cos x$, $0 < x < \pi$

7. $f(x) = x^2$, $0 < x < \pi$

8. $f(x) = \pi - x$, $0 < x < \pi$

9. $f(x) = x - x^2$, $0 < x < 1$

10. $f(x) = e^x$, $0 < x < 1$

In Problems 11–16, compute the Fourier cosine series for the given function.

11. $f(x) = \pi - x$, $0 < x < \pi$

12. $f(x) = 1 + x$, $0 < x < \pi$

13. $f(x) = e^x$, $0 < x < 1$

14. $f(x) = e^{-x}$, $0 < x < 1$

15. $f(x) = \sin x$, $0 < x < \pi$

16. $f(x) = x - x^2$, $0 < x < 1$

In Problems 17–19, for the given $f(x)$, find the solution to the heat flow problem.

$$\frac{\partial u}{\partial t} = 5\frac{\partial^2 u}{\partial x^2}, \qquad 0 < x < \pi, \qquad t > 0,$$

$$u(0, t) = u(\pi, t) = 0, \qquad t > 0,$$

$$u(x, 0) = f(x), \qquad 0 < x < \pi,$$

17. $f(x) = 1 - \cos 2x$

18. $f(x) = x(\pi - x)$

19. $f(x) = \begin{cases} -x, & 0 < x \leq \pi/2, \\ x - \pi, & \pi/2 \leq x < \pi \end{cases}$

10.5 The Heat Equation

In Section 10.1 we developed a model for heat flow in an insulated uniform wire whose ends are kept at the constant temperature 0°C. In particular, we found that the temperature $u(x, t)$ in the wire is governed by the initial-boundary value problem

(1) $$\frac{\partial u}{\partial t} = \beta\frac{\partial^2 u}{\partial x^2}, \qquad 0 < x < L, \qquad t > 0,$$

(2) $$u(0, t) = u(L, t) = 0, \qquad t > 0,$$

(3) $$u(x, 0) = f(x), \qquad 0 < x < L$$

[see equations (7)–(9) in Section 10.1, page 562]. Here equation (2) specifies that the temperature at the ends of the wire is zero, whereas equation (3) specifies the initial temperature distribution.

In Section 10.2 we also derived a formal solution to (1)–(3) using separation of variables. There we found the solution to (1)–(3) to have the form

(4) $$u(x, t) = \sum_{n=1}^{\infty} c_n e^{-\beta(n\pi/L)^2 t} \sin\frac{n\pi x}{L},$$

where the c_n's are the coefficients in the Fourier sine series for $f(x)$:

(5) $$f(x) = \sum_{n=1}^{\infty} c_n \sin\frac{n\pi x}{L}.$$

In other words, solving (1)–(3) reduces to computing the Fourier sine series for the initial value function $f(x)$.

In this section we discuss heat flow problems where the ends of the wire are insulated or kept at a constant, but nonzero, temperature. (The latter involves nonhomogeneous boundary conditions.) We will also discuss the problem in which a heat source is adding heat to the wire. (This results in a nonhomogeneous partial differential equation.) The problem of heat flow in a rectangular plate is also discussed and leads to the topic of double Fourier series. We conclude this section with a discussion of the existence and uniqueness of solutions to the heat flow problem.

In the model of heat flow in a uniform wire, let's replace the assumption that the ends of the wire are kept at a constant temperature zero and instead assume that the ends of the wire are *insulated*—that is, no heat flows out (or in) at the ends of the wire. It follows from the principle of heat conduction (see Section 10.1) that the temperature gradient must be zero at these endpoints, that is,

$$\frac{\partial u}{\partial x}(0, t) = \frac{\partial u}{\partial x}(L, t) = 0, \qquad t > 0.$$

In the next example we obtain the formal solution to the heat flow problem with these boundary conditions.

Example 1 Find a formal solution to the heat flow problem governed by the initial-boundary value problem

(6)
$$\frac{\partial u}{\partial t} = \beta \frac{\partial^2 u}{\partial x^2}, \quad 0 < x < L, \quad t > 0,$$

(7)
$$\frac{\partial u}{\partial x}(0, t) = \frac{\partial u}{\partial x}(L, t) = 0, \quad t > 0,$$

(8)
$$u(x, 0) = f(x), \quad 0 < x < L.$$

Solution Using the method of separation of variables, we first assume that

$$u(x, t) = X(x)T(t).$$

Substituting into equation (6) and separating variables, as was done in Section 10.2 [compare equation (5) on page 564], we get the two equations

(9)
$$X''(x) + \lambda X(x) = 0,$$

(10)
$$T'(t) + \beta \lambda T(t) = 0,$$

where λ is some constant. The boundary conditions in (7) become

$$X'(0)T(t) = 0 \quad \text{and} \quad X'(L)T(t) = 0.$$

For these equations to hold for all $t > 0$, either $T(t) \equiv 0$, which implies that $u(x, t) \equiv 0$, or

(11)
$$X'(0) = X'(L) = 0.$$

Combining the boundary conditions in (11) with equation (9) gives the boundary value problem

(12)
$$X''(x) + \lambda X(x) = 0; \quad X'(0) = X'(L) = 0,$$

where λ can be any constant.

To solve for the nontrivial solutions to (12), we propose $X(x) = e^{rx}$ and form the auxiliary equation $r^2 + \lambda = 0$. When $\lambda < 0$, arguments similar to those used in Section 10.2 show that there are no nontrivial solutions to (12).

When $\lambda = 0$, the auxiliary equation has the repeated root 0 and a general solution to the differential equation is

$$X(x) = A + Bx.$$

The boundary conditions in (12) reduce to $B = 0$ with A arbitrary. Thus, for $\lambda = 0$, the nontrivial solutions to (12) are of the form

$$X(x) = c_0,$$

where $c_0 = A$ is an arbitrary nonzero constant.

When $\lambda > 0$, the auxiliary equation has the roots $r = \pm i\sqrt{\lambda}$. Thus, a general solution to the differential equation in (12) is

$$X(x) = C_1 \cos \sqrt{\lambda} x + C_2 \sin \sqrt{\lambda} x.$$

The boundary conditions in (12) lead to the system

$$\sqrt{\lambda} C_2 = 0,$$
$$-\sqrt{\lambda} C_1 \sin \sqrt{\lambda} L + \sqrt{\lambda} C_2 \cos \sqrt{\lambda} L = 0.$$

Hence, $C_2 = 0$ and the system reduces to solving $C_1 \sin \sqrt{\lambda} L = 0$. Since $\sin \sqrt{\lambda} L = 0$ only when $\sqrt{\lambda} L = n\pi$, where n is an integer, we obtain a nontrivial solution only when $\sqrt{\lambda} = n\pi/L$ or $\lambda = (n\pi/L)^2$, $n = 1, 2, 3, \ldots$. Furthermore, the nontrivial solutions (eigenfunctions) X_n corresponding to the eigenvalue $\lambda = (n\pi/L)^2$ are given by

$$(13) \qquad X_n(x) = c_n \cos \frac{n\pi x}{L},$$

where the c_n's are arbitrary nonzero constants. In fact, formula (13) also holds for $n = 0$, since $\lambda = 0$ has the eigenfunctions $X_0(x) = c_0$.

Note that these eigenvalues and eigenfunctions (see Figure 10.15) have the same properties as those on page 565 (which we now recognize as the components of the sine series): homogeneous boundary conditions, eigenvalues clustering at infinity, increasingly oscillatory, orthogonal.

Having determined that $\lambda = (n\pi/L)^2$, $n = 0, 1, 2, \ldots$, let's consider equation (10) for such λ:

$$T'(t) + \beta(n\pi/L)^2 T(t) = 0.$$

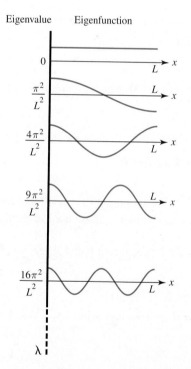

Eigenvalue Eigenfunction

0

$\dfrac{\pi^2}{L^2}$

$\dfrac{4\pi^2}{L^2}$

$\dfrac{9\pi^2}{L^2}$

$\dfrac{16\pi^2}{L^2}$

λ

Figure 10.15 Cosine eigenfunctions

For $n = 0, 1, 2, \ldots$, the general solution is

$$T_n(t) = b_n e^{-\beta(n\pi/L)^2 t},$$

where the b_n's are arbitrary constants. Combining this with equation (13), we obtain the functions

$$u_n(x, t) = X_n(x) T_n(t) = \left[c_n \cos \frac{n\pi x}{L} \right] \left[b_n e^{-\beta(n\pi/L)^2 t} \right],$$

$$u_n(x, t) = a_n e^{-\beta(n\pi/L)^2 t} \cos \frac{n\pi x}{L},$$

where $a_n = b_n c_n$ is, again, an arbitrary constant.

If we take an infinite series of these functions, we obtain

$$(14) \qquad u(x, t) = \frac{a_0}{2} + \sum_{n=1}^{\infty} a_n e^{-\beta(n\pi/L)^2 t} \cos \frac{n\pi x}{L},$$

which will be a solution to (6)–(7) provided the series has the proper convergence behavior. Notice that in (14) we have altered the constant term and written it as $a_0/2$, thus producing the standard form for cosine expansions.

Assuming a solution to (6)–(7) is given by the series in (14) and substituting into the initial condition (8), we get

$$(15) \qquad u(x, 0) = \frac{a_0}{2} + \sum_{n=1}^{\infty} a_n \cos \frac{n\pi x}{L} = f(x), \quad 0 < x < L.$$

This means that if we choose the a_n's as the coefficients in the Fourier cosine series for f,

$$a_n = \frac{2}{L} \int_0^L f(x) \cos \frac{n\pi x}{L} dx, \quad n = 0, 1, 2, \ldots ,$$

then $u(x, t)$ given in (14) will be a **formal solution** to the heat flow problem (6)–(8). Again, if this expansion converges to a continuous function with continuous second partial derivatives, then the formal solution is an actual solution. ◆

Note that the individual terms in the series (4) and (14) are genuine solutions to the heat equation and the associated boundary conditions (but not the initial condition, of course). These solutions are called the **modes** of the system. Observe that the time factor $e^{-\beta(n\pi/L)^2 t}$ decays faster for the more oscillatory (higher n) modes. This makes sense physically; a temperature pattern of many alternating hot and cold strips arranged closely together will equilibrate faster than a more uniform pattern. See Figure 10.16.

A good way of expressing how separation of variables "works" is as follows: The initial temperature profile $f(x)$ is decomposed into modes through Fourier analysis—equations (5) or (15). Then each mode is matched with a time decay factor—equations (4) or (14)—and it evolves in time accordingly.

When the ends of the wire are kept at 0°C or when the ends are insulated, the boundary conditions are said to be **homogeneous**. But when the ends of the wire are kept at constant temperatures different from zero, that is,

(16) $u(0, t) = U_1$ and $u(L, t) = U_2, \quad t > 0,$

then the boundary conditions are called **nonhomogeneous.**

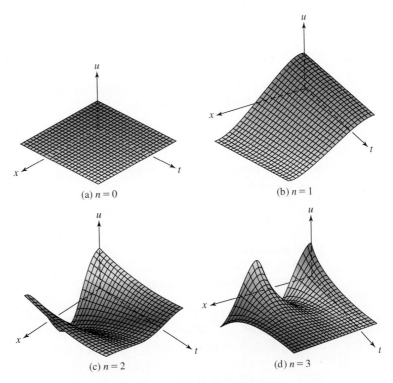

(a) $n = 0$ (b) $n = 1$

(c) $n = 2$ (d) $n = 3$

Figure 10.16 Modes for equation (14)

From our experience with vibration problems in Section 4.9 we expect that the solution to the heat flow problem with nonhomogeneous boundary conditions will consist of a **steady-state solution** $v(x)$ that satisfies the nonhomogeneous boundary conditions in (16) plus a **transient solution** $w(x, t)$. That is,

$$u(x, t) = v(x) + w(x, t),$$

where $w(x, t)$ and its partial derivatives tend to zero as $t \to \infty$. The function $w(x, t)$ will then satisfy homogeneous boundary conditions, as illustrated in the next example.

Example 2 Find a formal solution to the heat flow problem governed by the initial-boundary value problem

$$(17) \qquad \frac{\partial u}{\partial t} = \beta \frac{\partial^2 u}{\partial x^2}, \qquad 0 < x < L, \qquad t > 0,$$

$$(18) \qquad u(0, t) = U_1, \quad u(L, t) = U_2, \qquad t > 0,$$

$$(19) \qquad u(x, 0) = f(x), \quad 0 < x < L.$$

Solution Let's assume that the solution $u(x, t)$ consists of a steady-state solution $v(x)$ and a transient solution $w(x, t)$—that is,

$$(20) \qquad u(x, t) = v(x) + w(x, t).$$

Substituting for $u(x, t)$ in equations (17)–(19) leads to

$$(21) \qquad \frac{\partial u}{\partial t} = \frac{\partial w}{\partial t} = \beta v''(x) + \beta \frac{\partial^2 w}{\partial x^2}, \quad 0 < x < L, \qquad t > 0,$$

$$(22) \qquad v(0) + w(0, t) = U_1, \quad v(L) + w(L, t) = U_2, \qquad t > 0,$$

$$(23) \qquad v(x) + w(x, 0) = f(x), \quad 0 < x < L.$$

If we allow $t \to \infty$ in (21)–(22), assuming that $w(x, t)$ is a transient solution, we obtain the steady-state boundary value problem

$$v''(x) = 0, \quad 0 < x < L,$$
$$v(0) = U_1, \quad v(L) = U_2.$$

Solving for v, we obtain $v(x) = Ax + B$, and choosing A and B so that the boundary conditions are satisfied yields

$$(24) \qquad v(x) = U_1 + \frac{(U_2 - U_1)x}{L}$$

as the steady-state solution.

With this choice for $v(x)$, the initial-boundary value problem (21)–(23) reduces to the following initial-boundary value problem for $w(x, t)$:

$$(25) \qquad \frac{\partial w}{\partial t} = \beta \frac{\partial^2 w}{\partial x^2}, \quad 0 < x < L, \qquad t > 0,$$

$$(26) \qquad w(0, t) = w(L, t) = 0, \qquad t > 0,$$

$$(27) \qquad w(x, 0) = f(x) - U_1 - \frac{(U_2 - U_1)x}{L}, \qquad 0 < x < L.$$

Recall that a formal solution to (25)–(27) is given by equation (4). Hence,

$$w(x, t) = \sum_{n=1}^{\infty} c_n e^{-\beta(n\pi/L)^2 t} \sin \frac{n\pi x}{L} ,$$

where the c_n's are the coefficients of the Fourier sine series expansion

$$f(x) - U_1 - \frac{(U_2 - U_1)x}{L} = \sum_{n=1}^{\infty} c_n \sin \frac{n\pi x}{L} .$$

Therefore, the formal solution to (17)–(19) is

$$(28) \qquad u(x, t) = U_1 + \frac{(U_2 - U_1)x}{L} + \sum_{n=1}^{\infty} c_n e^{-\beta(n\pi/L)^2 t} \sin \frac{n\pi x}{L} ,$$

with

$$c_n = \frac{2}{L} \int_0^L \left[f(x) - U_1 - \frac{(U_2 - U_1)x}{L} \right] \sin \frac{n\pi x}{L} dx . \; \blacklozenge$$

In the next example, we consider the heat flow problem when a heat source is present but is independent of time. (Recall the derivation in Section 10.1.)

Example 3 Find a formal solution to the heat flow problem governed by the initial-boundary value problem

$$(29) \qquad \frac{\partial u}{\partial t} = \beta \frac{\partial^2 u}{\partial x^2} + P(x) , \qquad 0 < x < L , \qquad t > 0 ,$$

$$(30) \qquad u(0, t) = U_1 , \qquad u(L, t) = U_2 , \qquad t > 0 ,$$

$$(31) \qquad u(x, 0) = f(x) , \qquad 0 < x < L .$$

Solution We begin by assuming that the solution consists of a steady-state solution $v(x)$ and a transient solution $w(x, t)$, namely,

$$u(x, t) = v(x) + w(x, t) ,$$

where $w(x, t)$ and its partial derivatives tend to zero as $t \to \infty$. Substituting for $u(x, t)$ in (29)–(31) yields

$$(32) \qquad \frac{\partial u}{\partial t} = \frac{\partial w}{\partial t} = \beta v''(x) + \beta \frac{\partial^2 w}{\partial x^2} + P(x) , \qquad 0 < x < L , \qquad t > 0 ,$$

$$(33) \qquad v(0) + w(0, t) = U_1 , \qquad v(L) + w(L, t) = U_2 , \qquad t > 0 ,$$

$$(34) \qquad v(x) + w(x, 0) = f(x) , \qquad 0 < x < L .$$

Letting $t \to \infty$ in (32)–(33), we obtain the steady-state boundary value problem

$$v''(x) = -\frac{1}{\beta} P(x) , \qquad 0 < x < L ,$$

$$v(0) = U_1 , \qquad v(L) = U_2 .$$

The solution to this boundary value problem can be obtained by two integrations using the boundary conditions to determine the constants of integration. The reader can verify that the solution is given by the formula

$$(35) \qquad v(x) = \left[U_2 - U_1 + \int_0^L \left(\int_0^z \frac{1}{\beta} P(s) \, ds \right) dz \right] \frac{x}{L} + U_1 - \int_0^x \left(\int_0^z \frac{1}{\beta} P(s) \, ds \right) dz .$$

With this choice for $v(x)$, we find that the initial-boundary value problem (32)–(34) reduces to the following initial-boundary value problem for $w(x, t)$:

$$(36) \qquad \frac{\partial w}{\partial t} = \beta \frac{\partial^2 w}{\partial x^2}, \qquad 0 < x < L, \qquad t > 0,$$

$$(37) \qquad w(0, t) = w(L, t) = 0, \qquad t > 0,$$

$$(38) \qquad w(x, 0) = f(x) - v(x), \qquad 0 < x < L,$$

where $v(x)$ is given by formula (35). As before, the solution to this initial-boundary value problem is

$$(39) \qquad w(x, t) = \sum_{n=1}^{\infty} c_n e^{-\beta(n\pi/L)^2 t} \sin \frac{n\pi x}{L},$$

where the c_n's are determined from the Fourier sine series expansion of $f(x) - v(x)$:

$$(40) \qquad f(x) - v(x) = \sum_{n=1}^{\infty} c_n \sin \frac{n\pi x}{L} .$$

Thus the formal solution to (29)–(31) is given by

$$u(x, t) = v(x) + w(x, t) ,$$

where $v(x)$ is given in (35) and $w(x, t)$ is prescribed by (39)–(40). ◆

The method of separation of variables is also applicable to problems in higher dimensions. For example, consider the problem of heat flow in a rectangular plate with sides $x = 0$, $x = L$, $y = 0$, and $y = W$. If the two sides $y = 0$, $y = W$ are kept at a constant temperature of 0°C and the two sides $x = 0$, $x = L$ are perfectly insulated, then heat flow is governed by the initial-boundary value problem in the following example (see Figure 10.17).

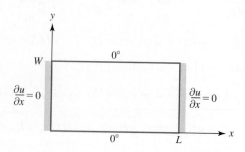

Figure 10.17 Plate with insulated sides

Example 4 Find a formal solution $u(x, y, t)$ to the initial-boundary value problem

(41) $\dfrac{\partial u}{\partial t} = \beta\left\{\dfrac{\partial^2 u}{\partial x^2} + \dfrac{\partial^2 u}{\partial y^2}\right\},\qquad 0 < x < L,\qquad 0 < y < W,\qquad t > 0,$

(42) $\dfrac{\partial u}{\partial x}(0, y, t) = \dfrac{\partial u}{\partial x}(L, y, t) = 0,\qquad 0 < y < W,\qquad t > 0,$

(43) $u(x, 0, t) = u(x, W, t) = 0,\qquad 0 < x < L,\qquad t > 0,$

(44) $u(x, y, 0) = f(x, y),\qquad 0 < x < L,\qquad 0 < y < W.$

Solution If we assume a solution of the form $u(x, y, t) = V(x, y)T(t)$, then equation (41) separates into the two equations

(45) $T'(t) + \beta\lambda T(t) = 0,$

(46) $\dfrac{\partial^2 V}{\partial x^2}(x, y) + \dfrac{\partial^2 V}{\partial y^2}(x, y) + \lambda V(x, y) = 0,$

where λ can be any constant. To solve equation (46), we again use separation of variables. Here we assume $V(x, y) = X(x)Y(y)$. This allows us to separate equation (46) into the two equations

(47) $X''(x) + \mu X(x) = 0,$

(48) $Y''(y) + (\lambda - \mu)Y(y) = 0,$

where μ can be any constant (see Problem 29 in Exercises 10.2 on page 571).

To solve for $X(x)$, we observe that the boundary conditions in (42), in terms of the separated variables, become

$X'(0)Y(y)T(t) = X'(L)Y(y)T(t) = 0;\qquad 0 < y < W,\qquad t > 0.$

Hence, in order to get a nontrivial solution, we must have

(49) $X'(0) = X'(L) = 0.$

The boundary value problem for X given in equations (47) and (49) was solved in Example 1 [compare equations (12) and (13)]. Here $\mu = (m\pi/L)^2, m = 0, 1, 2, \ldots$, and

$X_m(x) = c_m \cos\dfrac{m\pi x}{L},$

where the c_m's are arbitrary. (We forgo the "$c_0/2$" convention here.)

To solve for $Y(y)$, we first observe that the boundary conditions in (43) become

(50) $Y(0) = Y(W) = 0.$

Next, substituting $\mu = (m\pi/L)^2$ into equation (48) yields

$Y''(y) + \left(\lambda - (m\pi/L)^2\right)Y(y) = 0,$

which we can rewrite as

(51) $Y''(y) + EY(y) = 0,$

where $E = \lambda - (m\pi/L)^2$. The boundary value problem for Y consisting of (50)–(51) has also been solved before. In Section 10.2 [compare equations (7) and (9), pages 564 and 565] we showed that $E = (n\pi/W)^2, n = 1, 2, 3, \ldots$, and the nontrivial solutions are given by

$Y_n(y) = a_n \sin\dfrac{n\pi y}{W},$

where the a_n's are arbitrary.

Since $\lambda = E + (m\pi/L)^2$, we have

$$\lambda = (n\pi/W)^2 + (m\pi/L)^2 ; \qquad m = 0, 1, 2, \dots , \qquad n = 1, 2, 3, \dots .$$

Substituting λ into equation (45), we can solve for $T(t)$ and obtain

$$T_{mn}(t) = b_{mn}e^{-(m^2/L^2+n^2/W^2)\beta\pi^2 t} .$$

Substituting in for X_m, Y_n, and T_{mn}, we get

$$u_{mn}(x, y, t) = \left(c_m \cos\frac{m\pi x}{L} \right)\left(a_n \sin\frac{n\pi y}{W} \right)\left(b_{mn}e^{-(m^2/L^2+n^2/W^2)\beta\pi^2 t} \right) ,$$

$$u_{mn}(x, y, t) = a_{mn}e^{-(m^2/L^2+n^2/W^2)\beta\pi^2 t} \cos\frac{m\pi x}{L} \sin\frac{n\pi y}{W} ,$$

where $a_{mn} := a_n b_{mn} c_m$ $(m = 0, 1, 2, \dots , n = 1, 2, 3, \dots)$ are arbitrary constants.

If we now take a doubly infinite series of such functions, then we obtain the formal series

$$(52) \qquad u(x, y, t) = \sum_{m=0}^{\infty} \sum_{n=1}^{\infty} a_{mn}e^{-(m^2/L^2+n^2/W^2)\beta\pi^2 t} \cos\frac{m\pi x}{L} \sin\frac{n\pi y}{W} .$$

We are now ready to apply the initial conditions (44). Setting $t = 0$, we obtain

$$(53) \qquad u(x, y, 0) = f(x, y) = \sum_{m=0}^{\infty} \sum_{n=1}^{\infty} a_{mn} \cos\frac{m\pi x}{L} \sin\frac{n\pi y}{W} .$$

This is a **double Fourier series.**[†] The formulas for the coefficients a_{mn} are obtained by exploiting the orthogonality conditions twice. Presuming (53) is valid and permits term-by-term integration, we multiply each side by $\cos(p\pi x/L) \sin(q\pi y/W)$ and integrate over x and y:

$$\int_0^L \int_0^W f(x, y) \cos\frac{p\pi x}{L} \sin\frac{q\pi y}{W} dy\, dx$$

$$= \sum_{m=0}^{\infty} \sum_{n=1}^{\infty} a_{mn} \int_0^L \int_0^W \cos\frac{m\pi x}{L} \sin\frac{n\pi y}{W} \cos\frac{p\pi x}{L} \sin\frac{q\pi y}{W} dy\, dx .$$

According to the orthogonality conditions, each integral on the right is zero, except when $m = p$ and $n = q$; thus,

$$\int_0^L \int_0^W f(x, y) \cos\frac{p\pi x}{L} \sin\frac{q\pi y}{W} dy\, dx$$

$$= a_{pq} \int_0^L \cos^2\frac{p\pi x}{L} dx \int_0^W \sin^2\frac{q\pi y}{W} dy = \begin{cases} \dfrac{LW}{4}a_{pq} , & p \neq 0 , \\[2mm] \dfrac{LW}{2}a_{pq} , & p = 0 . \end{cases}$$

Hence,

$$(54) \qquad a_{0q} = \frac{2}{LW} \int_0^L \int_0^W f(x, y)\sin\frac{q\pi y}{W} dy\, dx , \qquad q = 1, 2, 3, \dots ,$$

[†]For a discussion of double Fourier series, see *Partial Differential Equations of Mathematical Physics*, 2nd ed. by Tyn Myint-U (Elsevier North Holland, New York, 1983), Sec. 5.14.

and for $p \geq 1, q \geq 1$,

$$(55) \quad a_{pq} = \frac{4}{LW} \int_0^L \int_0^W f(x, y) \cos \frac{p\pi x}{L} \sin \frac{q\pi y}{W} \, dy \, dx \, .$$

Finally, the solution to the initial-boundary value problem (41)–(44) is given by equation (52), where the coefficients are prescribed by equations (54) and (55). ◆

We derived formulas (54) and (55) under the assumption that the double series (53) truly converged (uniformly). How can we justify the assumption that a double Fourier series converges? A rough argument goes as follows. If we *fix y* in $f(x, y)$, then $f(x, y)$ presumably has a convergent cosine series in x: $f(x, y) = \sum_{m=0}^{\infty} d_m \cos(m\pi x/L)$. But the coefficients d_m depend on the value of the y that we fixed: $d_m = d_m(y)$. So presumably each *function* $d_m(y)$ has a convergent sine series in y: $d_m(y) = \sum_{n=1}^{\infty} a_{mn} \sin(n\pi y/W)$. Assembling all this, we get $f(x, y) = \sum_{m=0}^{\infty} \sum_{n=1}^{\infty} a_{mn} \cos(m\pi x/L) \sin(n\pi y/W)$.

Existence and Uniqueness of Solutions

In the examples that we have studied in this section and in Section 10.2, we were able to obtain formal solutions in the sense that we could express the solution in terms of a series expansion consisting of exponentials, sines, and cosines. To prove that these series converge to actual solutions requires results on the convergence of Fourier series and results from real analysis on uniform convergence. We will not go into these details here, but refer the reader to Section 6.5 of the text by Tyn Myint-U (see footnote on page 600) for a proof of the existence of a solution to the heat flow problem discussed in Sections 10.1 and 10.2. (A proof of uniqueness is also given there.)

As might be expected, by using Fourier series and the method of separation of variables one can also obtain "solutions" when the initial data are discontinuous, since the formal solutions require only the existence of a convergent Fourier series. This allows one to study idealized problems in which the initial conditions do not agree with the boundary conditions or the initial conditions involve a jump discontinuity. For example, we may assume that initially one half of the wire is at one temperature, whereas the other half is at a different temperature, that is,

$$f(x) = \begin{cases} U_1, & 0 < x < L/2, \\ U_2, & L/2 < x < L. \end{cases}$$

The formal solution that we obtain will make sense for $0 < x < L, t > 0$, but near the points of discontinuity $x = 0, L/2$, and L, we have to expect behavior like that exhibited in Figure 10.7 of Section 10.3, page 578.

The question of the uniqueness of the solution to the heat flow problem can be answered in various ways. One is tempted to argue that the method of separation of variables yields *formulas* for the solutions and therefore a unique solution. However, this does *not* exclude the possibility of solutions existing that cannot be obtained by the method of separation of variables.

From physical considerations, we know that the *peak* temperature along a wire does not increase spontaneously if there are no heat sources to drive it up. Indeed, if hot objects could draw heat from colder objects without external intervention, we would have violations of the second law of thermodynamics. Thus no point on an unheated wire will ever reach a

higher temperature than the initial peak temperature. Such statements are called **maximum principles** and can be proved mathematically for the heat equation and Laplace's equation. One such result is the following.[†]

Maximum Principle for the Heat Equation

Theorem 6. Let $u(x, t)$ be a continuously differentiable function that satisfies the heat equation

(56) $\dfrac{\partial u}{\partial t} = \beta \dfrac{\partial^2 u}{\partial x^2}$, $0 < x < L$, $t > 0$,

and the boundary conditions

(57) $u(0, t) = u(L, t) = 0$, $t \geq 0$.

Then $u(x, t)$ attains its maximum value at $t = 0$, for some x in $[0, L]$—that is,

$$\max_{\substack{t \geq 0 \\ 0 \leq x \leq L}} u(x, t) = \max_{0 \leq x \leq L} u(x, 0).$$

We can use the maximum principle to show that the heat flow problem has a unique solution.

Uniqueness of Solution

Theorem 7. The initial-boundary value problem

(58) $\dfrac{\partial u}{\partial t} = \beta \dfrac{\partial^2 u}{\partial x^2}$, $0 < x < L$, $t > 0$,

(59) $u(0, t) = u(L, t) = 0$, $t \geq 0$,

(60) $u(x, 0) = f(x)$, $0 < x < L$,

has at most one continuously differentiable solution.

Proof. Assume $u(x, t)$ and $v(x, t)$ are continuously differentiable functions that satisfy the initial-boundary value problem (58)–(60). Let $w = u - v$. Now w is a continuously differentiable solution to the boundary value problem (56)–(57). By the maximum principle, w must attain its maximum at $t = 0$, and since

$$w(x, 0) = u(x, 0) - v(x, 0) = f(x) - f(x) = 0,$$

we have $w(x, t) \leq 0$. Hence, $u(x, t) \leq v(x, t)$ for all $0 \leq x \leq L, t \geq 0$. A similar argument using $\hat{w} = v - u$ yields $v(x, t) \leq u(x, t)$. Therefore we have $u(x, t) = v(x, t)$ for all $0 \leq x \leq L$, $t \geq 0$. Thus, there is at most one continuously differentiable solution to the problem (58)–(60). ◆

[†]For a discussion of maximum principles and their applications, see *Maximum Principles in Differential Equations,* by M. H. Protter and H. F. Weinberger (Springer-Verlag, New York, 1984).

10.5 EXERCISES

In Problems 1–10, find a formal solution to the given initial-boundary value problem.

1. $\dfrac{\partial u}{\partial t} = 5\dfrac{\partial^2 u}{\partial x^2}$, $\quad 0 < x < 1$, $\quad t > 0$,

$u(0, t) = u(1, t) = 0$, $\quad t > 0$,

$u(x, 0) = (1 - x)x^2$, $\quad 0 < x < 1$

2. $\dfrac{\partial u}{\partial t} = \dfrac{\partial^2 u}{\partial x^2}$, $\quad 0 < x < \pi$, $\quad t > 0$,

$u(0, t) = u(\pi, t) = 0$, $\quad t > 0$,

$u(x, 0) = x^2$, $\quad 0 < x < \pi$

3. $\dfrac{\partial u}{\partial t} = 3\dfrac{\partial^2 u}{\partial x^2}$, $\quad 0 < x < \pi$, $\quad t > 0$,

$\dfrac{\partial u}{\partial x}(0, t) = \dfrac{\partial u}{\partial x}(\pi, t) = 0$, $\quad t > 0$,

$u(x, 0) = x$, $\quad 0 < x < \pi$

4. $\dfrac{\partial u}{\partial t} = 2\dfrac{\partial^2 u}{\partial x^2}$, $\quad 0 < x < 1$, $\quad t > 0$,

$\dfrac{\partial u}{\partial x}(0, t) = \dfrac{\partial u}{\partial x}(1, t) = 0$, $\quad t > 0$,

$u(x, 0) = x(1 - x)$, $\quad 0 < x < 1$

5. $\dfrac{\partial u}{\partial t} = \dfrac{\partial^2 u}{\partial x^2}$, $\quad 0 < x < \pi$, $\quad t > 0$,

$\dfrac{\partial u}{\partial x}(0, t) = \dfrac{\partial u}{\partial x}(\pi, t) = 0$, $\quad t > 0$,

$u(x, 0) = e^x$, $\quad 0 < x < \pi$

6. $\dfrac{\partial u}{\partial t} = 7\dfrac{\partial^2 u}{\partial x^2}$, $\quad 0 < x < \pi$, $\quad t > 0$,

$\dfrac{\partial u}{\partial x}(0, t) = \dfrac{\partial u}{\partial x}(\pi, t) = 0$, $\quad t > 0$,

$u(x, 0) = 1 - \sin x$, $\quad 0 < x < \pi$

7. $\dfrac{\partial u}{\partial t} = 2\dfrac{\partial^2 u}{\partial x^2}$, $\quad 0 < x < \pi$, $\quad t > 0$,

$u(0, t) = 5$, $\quad u(\pi, t) = 10$, $\quad t > 0$,

$u(x, 0) = \sin 3x - \sin 5x$, $\quad 0 < x < \pi$

8. $\dfrac{\partial u}{\partial t} = \dfrac{\partial^2 u}{\partial x^2}$, $\quad 0 < x < \pi$, $\quad t > 0$,

$u(0, t) = 0$, $\quad u(\pi, t) = 3\pi$, $\quad t > 0$,

$u(x, 0) = 0$, $\quad 0 < x < \pi$

9. $\dfrac{\partial u}{\partial t} = \dfrac{\partial^2 u}{\partial x^2} + e^{-x}$, $\quad 0 < x < \pi$, $\quad t > 0$,

$u(0, t) = u(\pi, t) = 0$, $\quad t > 0$,

$u(x, 0) = \sin 2x$, $\quad 0 < x < \pi$

10. $\dfrac{\partial u}{\partial t} = 3\dfrac{\partial^2 u}{\partial x^2} + x$, $\quad 0 < x < \pi$, $\quad t > 0$,

$u(0, t) = u(\pi, t) = 0$, $\quad t > 0$,

$u(x, 0) = \sin x$, $\quad 0 < x < \pi$

11. Find a formal solution to the initial-boundary value problem

$$\dfrac{\partial u}{\partial t} = 4\dfrac{\partial^2 u}{\partial x^2}, \quad 0 < x < \pi, \quad t > 0,$$

$$\dfrac{\partial u}{\partial x}(0, t) = 0, \quad u(\pi, t) = 0, \quad t > 0,$$

$$u(x, 0) = f(x), \quad 0 < x < \pi.$$

12. Find a formal solution to the initial-boundary value problem

$$\dfrac{\partial u}{\partial t} = \dfrac{\partial^2 u}{\partial x^2}, \quad 0 < x < \pi, \quad t > 0,$$

$$u(0, t) = 0, \quad u(\pi, t) + \dfrac{\partial u}{\partial x}(\pi, t) = 0, \quad t > 0,$$

$$u(x, 0) = f(x), \quad 0 < x < \pi.$$

13. Find a formal solution to the initial-boundary value problem

$$\dfrac{\partial u}{\partial t} = 2\dfrac{\partial^2 u}{\partial x^2} + 4x, \quad 0 < x < \pi, \quad t > 0,$$

$$u(0, t) = u(\pi, t) = 0, \quad t > 0,$$

$$u(x, 0) = \sin x, \quad 0 < x < \pi.$$

14. Find a formal solution to the initial-boundary value problem

$$\dfrac{\partial u}{\partial t} = 3\dfrac{\partial^2 u}{\partial x^2} + 5, \quad 0 < x < \pi, \quad t > 0,$$

$$u(0, t) = u(\pi, t) = 1, \quad t > 0,$$

$$u(x, 0) = 1, \quad 0 < x < \pi.$$

In Problems 15–18, find a formal solution to the initial-boundary value problem.

$$\dfrac{\partial u}{\partial t} = \dfrac{\partial^2 u}{\partial x^2} + \dfrac{\partial^2 u}{\partial y^2}, \quad 0 < x < \pi, \quad 0 < y < \pi, \quad t > 0,$$

$$\dfrac{\partial u}{\partial x}(0, y, t) = \dfrac{\partial u}{\partial x}(\pi, y, t) = 0, \quad 0 < y < \pi, \quad t > 0,$$

$$u(x, 0, t) = u(x, \pi, t) = 0, \quad 0 < x < \pi, \quad t > 0,$$

$$u(x, y, 0) = f(x, y), \quad 0 < x < \pi, \quad 0 < y < \pi,$$

for the given function $f(x, y)$.

15. $f(x, y) = \cos 6x \sin 4y - 3 \cos x \sin 11y$

16. $f(x, y) = \cos x \sin y + 4 \cos 2x \sin y - 3 \cos 3x \sin 4y$

17. $f(x, y) = y$

18. $f(x, y) = x \sin y$

19. Chemical Diffusion. Chemical diffusion through a thin layer is governed by the equation

$$\frac{\partial C}{\partial t} = k \frac{\partial^2 C}{\partial x^2} - LC \, ,$$

where $C(x, t)$ is the concentration in moles/cm^3, the diffusivity k is a positive constant with units cm^2/sec, and

$L > 0$ is a consumption rate with units sec^{-1}. Assume the boundary conditions are

$$C(0, t) = C(a, t) = 0 \, , \qquad t > 0 \, ,$$

and the initial concentration is given by

$$C(x, 0) = f(x) \, , \qquad 0 < x < a \, .$$

Use the method of separation of variables to solve formally for the concentration $C(x, t)$. What happens to the concentration as $t \to +\infty$?

10.6 The Wave Equation

In Section 10.2 we presented a model for the motion of a vibrating string. If $u(x, t)$ represents the displacement (deflection) of the string and the ends of the string are held fixed, then the motion of the string is governed by the initial-boundary value problem

$$(1) \qquad \frac{\partial^2 u}{\partial t^2} = \alpha^2 \frac{\partial^2 u}{\partial x^2} \, , \qquad 0 < x < L \, , \qquad t > 0 \, ,$$

$$(2) \qquad u(0, t) = u(L, t) = 0 \, , \qquad t > 0 \, ,$$

$$(3) \qquad u(x, 0) = f(x) \, , \qquad 0 < x < L \, ,$$

$$(4) \qquad \frac{\partial u}{\partial t}(x, 0) = g(x) \, , \qquad 0 < x < L \, .$$

Equation (1) is called the **wave equation.**

The constant α^2 appearing in (1) is strictly positive and depends on the linear density and tension of the string. The boundary conditions in (2) reflect the fact that the string is held fixed at the two endpoints $x = 0$ and $x = L$.

Equations (3) and (4) specify, respectively, the initial displacement and the initial velocity of each point on the string. For the initial and boundary conditions to be consistent, we assume $f(0) = f(L) = 0$ and $g(0) = g(L) = 0$.

Using the method of separation of variables, we found in Section 10.2 that a formal solution to (1)–(4) is given by [compare equations (24)–(26) on page 569]

$$(5) \qquad u(x, t) = \sum_{n=1}^{\infty} \left[a_n \cos \frac{n \pi \alpha}{L} t + b_n \sin \frac{n \pi \alpha}{L} t \right] \sin \frac{n \pi x}{L} \, ,$$

where the a_n's and b_n's are determined from the Fourier sine series

$$(6) \qquad f(x) = \sum_{n=1}^{\infty} a_n \sin \frac{n \pi x}{L} \, ,$$

$$(7) \qquad g(x) = \sum_{n=1}^{\infty} b_n \left(\frac{n \pi \alpha}{L} \right) \sin \frac{n \pi x}{L} \, .$$

See Figure 10.18 on page 605 for a sketch of partial sums for (5).

Figure 10.18 Partial sums for equation (5) with $f(x) = \begin{cases} x, & 0 < x \le \pi/2, \\ \pi - x, & \pi/2 \le x < \pi, \end{cases}$ and $g(x) = 0$

Each term (or mode) in expansion (5) can be viewed as a **standing wave** (a wave that vibrates in place without lateral motion along the string). For example, the first term,

$$\left(a_1 \cos\frac{\pi\alpha}{L}t + b_1 \sin\frac{\pi\alpha}{L}t \right) \sin\frac{\pi x}{L},$$

consists of a sinusoidal shape function $\sin(\pi x/L)$ multiplied by a time-varying amplitude. The second term is also a sinusoid $\sin(2\pi x/L)$ with a time-varying amplitude. In the latter case, there is a *node* in the middle at $x = L/2$ that never moves. For the nth term, we have a sinusoid $\sin(n\pi x/L)$ with a time-varying amplitude and $(n-1)$ nodes. This is illustrated in Figure 10.19. Thus, separation of variables decomposes the initial data into sinusoids or modes [equations (6)–(7)], assigns each mode a frequency at which to vibrate [equation (5)], and represents the solution as the superposition of infinitely many standing waves. Modes with more nodes vibrate at higher frequencies. By choosing the initial configuration of the string to have the same shape as one of the individual terms in the solution, we can activate only that mode.

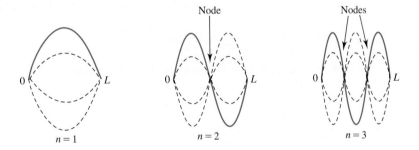

Figure 10.19 Standing waves. Time-varying amplitudes are shown by dashed curves

The different modes of vibration of a guitar string are distinguishable to the human ear by the frequency of the sound they generate; different frequencies are discerned as different "pitches." The *fundamental* is the frequency of the lowest mode, and integer multiples of the fundamental frequency are called "harmonics." A cello and a trombone sound different even if they are playing the same fundamental, say F$^{\#}$, because of the difference in the intensities of the harmonics they generate.

Many engineering devices operate better in some modes than in others. For example, certain modes propagate down a waveguide or an optical fiber with less attenuation than the others. In such cases, engineers design the systems to suppress the undesirable modes, either by shaping the initial excitation or by introducing devices (like resistor cards in a waveguide) which damp out certain modes preferentially.

The fundamental mode on a guitar string can be suppressed by holding the finger lightly in contact with the midpoint, thereby creating a stationary point or node there. According to equation (5), the pitch of the note then doubles, thus producing an "octave." This style of fingering, called "playing harmonics," is used frequently in musical performance.

An important characteristic of an engineering device is the set of angular frequencies supported by its eigenmodes—its "eigenfrequencies." According to equation (5), the eigenfrequencies of the vibrating string are the harmonics $\omega_n = n\pi\alpha/L, n = 1, 2, 3, \ldots$. These blend together pleasantly to the human ear, and we enjoy melodies played on string instruments. The eigenfrequencies of a drum are not harmonics (see Problem 21). Therefore, drums are used for rhythm, not for melody.

As we have seen in the preceding section, the method of separation of variables can be used to solve problems with nonhomogeneous boundary conditions and nonhomogeneous equations where the forcing term is time independent. In the next example, we will consider a problem with a time-dependent forcing term $h(x, t)$.

Example 1 For given functions f, g, and h, find a formal solution to the initial-boundary value problem

(8) $$\frac{\partial^2 u}{\partial t^2} = \alpha^2 \frac{\partial^2 u}{\partial x^2} + h(x, t), \quad 0 < x < L, \quad t > 0,$$

(9) $$u(0, t) = u(L, t) = 0, \quad t > 0,$$

(10) $$u(x, 0) = f(x), \quad 0 < x < L,$$

(11) $$\frac{\partial u}{\partial t}(x, 0) = g(x), \quad 0 < x < L.$$

Solution The boundary conditions in (9) certainly require that the solution be zero for $x = 0$ and $x = L$. Motivated by the fact that the solution to the corresponding homogeneous system (1)–(4) consists of a superposition of standing waves, let's try to find a solution to (8)–(11) of the form

(12) $$u(x, t) = \sum_{n=1}^{\infty} u_n(t) \sin\frac{n\pi x}{L},$$

where the $u_n(t)$'s are functions of t to be determined.

For each fixed t, we can compute a Fourier sine series for $h(x, t)$. If we assume that the series is convergent to $h(x, t)$, then

(13) $$h(x, t) = \sum_{n=1}^{\infty} h_n(t) \sin\frac{n\pi x}{L},$$

where the coefficient $h_n(t)$ is given by [recall equation (7) on page 589]

$$h_n(t) = \frac{2}{L} \int_0^L h(x, t) \sin\frac{n\pi x}{L} dx, \quad n = 1, 2, \ldots.$$

If the series in (13) has the proper convergence properties, then we can substitute (12) and (13) into equation (8) and obtain

$$\sum_{n=1}^{\infty} \left[u_n''(t) + \left(\frac{n\pi\alpha}{L}\right)^2 u_n(t) \right] \sin\frac{n\pi x}{L} = \sum_{n=1}^{\infty} h_n(t) \sin\frac{n\pi x}{L}.$$

Equating the coefficients in each series (why?), we have

$$u_n''(t) + \left(\frac{n\pi\alpha}{L}\right)^2 u_n(t) = h_n(t) \,.$$

This is a nonhomogeneous, constant-coefficient equation that can be solved using variation of parameters. You should verify that

$$u_n(t) = a_n \cos\frac{n\pi\alpha}{L}t + b_n \sin\frac{n\pi\alpha}{L}t + \frac{L}{n\pi\alpha}\int_0^t h_n(s)\sin\left[\frac{n\pi\alpha}{L}(t-s)\right]ds$$

[compare Problem 20 of Exercises 4.6]. Hence, with this choice of $u_n(t)$, the series in (12) is a formal solution to the partial differential equation (8).

Since[†]

$$u_n(0) = a_n \quad \text{and} \quad u_n'(0) = b_n\left(\frac{n\pi\alpha}{L}\right),$$

substituting (12) into the initial conditions (10)–(11) yields

$$(14) \qquad u(x,0) = f(x) = \sum_{n=1}^{\infty} a_n \sin\frac{n\pi x}{L}\,,$$

$$(15) \qquad \frac{\partial u}{\partial t}(x,0) = g(x) = \sum_{n=1}^{\infty} b_n\left(\frac{n\pi\alpha}{L}\right)\sin\frac{n\pi x}{L}\,.$$

Thus, if we choose the a_n's and b_n's so that equations (14) and (15) are satisfied, a formal solution to (8)–(11) is given by

$$(16) \qquad u(x,t) = \sum_{n=1}^{\infty}\left\{a_n\cos\frac{n\pi\alpha}{L}t + b_n\sin\frac{n\pi\alpha}{L}t\right.$$
$$\left. + \frac{L}{n\pi\alpha}\int_0^t h_n(s)\sin\left[\frac{n\pi\alpha}{L}(t-s)\right]ds\right\}\sin\frac{n\pi x}{L}\,. \ \blacklozenge$$

The method of separation of variables can also be used to solve initial-boundary value problems for the wave equation in higher dimensions. For example, a vibrating rectangular membrane of length L and width W (see Figure 10.20 on page 608) is governed by the following initial-boundary value problem for $u(x, y, t)$:

$$(17) \qquad \frac{\partial^2 u}{\partial t^2} = \alpha^2\left(\frac{\partial^2 u}{\partial x^2} + \frac{\partial^2 u}{\partial y^2}\right), \quad 0<x<L, \quad 0<y<W, \quad t>0,$$

$$(18) \qquad u(0,y,t) = u(L,y,t) = 0, \quad 0<y<W, \quad t>0,$$

$$(19) \qquad u(x,0,t) = u(x,W,t) = 0, \quad 0<x<L, \quad t>0,$$

$$(20) \qquad u(x,y,0) = f(x,y), \quad 0<x<L, \quad 0<y<W,$$

$$(21) \qquad \frac{\partial u}{\partial t}(x,y,0) = g(x,y), \quad 0<x<L, \quad 0<y<W.$$

[†]To compute $u_n'(0)$, we use the fact that $\dfrac{d}{dt}\displaystyle\int_0^t G(s,t)\,ds = G(t,t) + \int_0^t \dfrac{\partial G}{\partial t}(s,t)\,ds.$

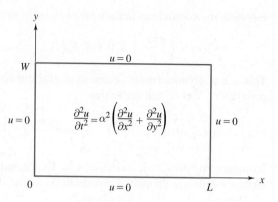

Figure 10.20 Vibrating membrane

Using an argument similar to the one given for the problem of heat flow in a rectangular plate (Example 4 in Section 10.5), we find that the initial-boundary value problem (17)–(21) has a formal solution

$$(22) \qquad u(x, y, t) = \sum_{m=1}^{\infty} \sum_{n=1}^{\infty} \left\{ a_{mn} \cos \left(\sqrt{\frac{m^2}{L^2} + \frac{n^2}{W^2}}\, \alpha \pi t \right) \right.$$

$$\left. + b_{mn} \sin \left(\sqrt{\frac{m^2}{L^2} + \frac{n^2}{W^2}}\, \alpha \pi t \right) \right\} \sin \frac{m\pi x}{L} \sin \frac{n\pi y}{W} ,$$

where the constants a_{mn} and b_{mn} are determined from the double Fourier series

$$f(x, y) = \sum_{m=1}^{\infty} \sum_{n=1}^{\infty} a_{mn} \sin \frac{m\pi x}{L} \sin \frac{n\pi y}{W} ,$$

$$g(x, y) = \sum_{m=1}^{\infty} \sum_{n=1}^{\infty} \alpha \pi \sqrt{\frac{m^2}{L^2} + \frac{n^2}{W^2}}\, b_{mn} \sin \frac{m\pi x}{L} \sin \frac{n\pi y}{W} .$$

In particular,

$$(23) \qquad a_{mn} = \frac{4}{LW} \int_0^L \int_0^W f(x, y) \sin \frac{m\pi x}{L} \sin \frac{n\pi y}{W}\, dy\, dx ,$$

$$(24) \qquad b_{mn} = \frac{4}{LW\pi\alpha \sqrt{\dfrac{m^2}{L^2} + \dfrac{n^2}{W^2}}} \int_0^L \int_0^W g(x, y) \sin \frac{m\pi x}{L} \sin \frac{n\pi y}{W}\, dy\, dx .$$

We leave the derivation of this solution as an exercise (see Problem 19).

We mentioned earlier that the solution to the vibrating string problem (1)–(4) consisted of a superposition of standing waves. There are also "traveling waves" associated with the wave equation. Traveling waves arise naturally out of d'Alembert's solution to the wave equation for an "infinite" string.

To obtain d'Alembert's solution to the wave equation

$$\frac{\partial^2 u}{\partial t^2} = \alpha^2 \frac{\partial^2 u}{\partial x^2} ,$$

we use the change of variables

$$\psi = x + \alpha t, \qquad \eta = x - \alpha t.$$

If u has continuous second partial derivatives, then $\partial u / \partial x = \partial u / \partial \psi + \partial u / \partial \eta$ and $\partial u / \partial t = \alpha(\partial u / \partial \psi - \partial u / \partial \eta)$, from which we obtain

$$\frac{\partial^2 u}{\partial x^2} = \frac{\partial^2 u}{\partial \psi^2} + 2\frac{\partial^2 u}{\partial \psi \partial \eta} + \frac{\partial^2 u}{\partial \eta^2},$$

$$\frac{\partial^2 u}{\partial t^2} = \alpha^2 \left\{ \frac{\partial^2 u}{\partial \psi^2} - 2\frac{\partial^2 u}{\partial \psi \partial \eta} + \frac{\partial^2 u}{\partial \eta^2} \right\}.$$

Substituting these expressions into the wave equation and simplifying yields

$$\frac{\partial^2 u}{\partial \psi \partial \eta} = 0.$$

We can solve this equation directly by first integrating with respect to ψ to obtain

$$\frac{\partial u}{\partial \eta} = b(\eta),$$

where $b(\eta)$ is an arbitrary function of η, and then integrating with respect to η to find

$$u(\psi, \eta) = A(\psi) + B(\eta),$$

where $A(\psi)$ and $B(\eta) = \int b(\eta)\, d\eta$ are arbitrary functions. Substituting the original variables x and t gives **d'Alembert's solution**

$$(25) \qquad u(x,t) = A(x + \alpha t) + B(x - \alpha t).$$

It is easy to check by direct substitution that $u(x, t)$, defined by formula (25), is indeed a solution to the wave equation, provided A and B are twice-differentiable functions.

Example 2 Using d'Alembert's formula (25), find a solution to the initial value problem

$$(26) \qquad \frac{\partial^2 u}{\partial t^2} = \alpha^2 \frac{\partial^2 u}{\partial x^2} \qquad -\infty < x < \infty, \qquad t > 0,$$

$$(27) \qquad u(x, 0) = f(x) \qquad -\infty < x < \infty,$$

$$(28) \qquad \frac{\partial u}{\partial t}(x, 0) = g(x) \qquad -\infty < x < \infty.$$

Solution A solution to (26) is given by formula (25), so we need only choose the functions A and B so that the initial conditions (27)–(28) are satisfied. For this we need

$$(29) \qquad u(x, 0) = A(x) + B(x) = f(x),$$

$$(30) \qquad \frac{\partial u}{\partial t}(x, 0) = \alpha A'(x) - \alpha B'(x) = g(x).$$

Integrating equation (30) from x_0 to x (x_0 arbitrary) and dividing by α gives

$$(31) \qquad A(x) - B(x) = \frac{1}{\alpha} \int_{x_0}^{x} g(s)\, ds + C,$$

where C is also arbitrary. Solving the system (29) and (31), we obtain

$$A(x) = \frac{1}{2}f(x) + \frac{1}{2\alpha}\int_{x_0}^{x} g(s)\,ds + \frac{C}{2},$$

$$B(x) = \frac{1}{2}f(x) - \frac{1}{2\alpha}\int_{x_0}^{x} g(s)\,ds - \frac{C}{2}.$$

Using these functions in formula (25) gives

$$u(x,t) = \frac{1}{2}[f(x+\alpha t)+f(x-\alpha t)] + \frac{1}{2\alpha}\left[\int_{x_0}^{x+\alpha t} g(s)ds - \int_{x_0}^{x-\alpha t} g(s)ds\right],$$

which simplifies to

(32) $$u(x,t) = \frac{1}{2}[f(x+\alpha t)+f(x-\alpha t)] + \frac{1}{2\alpha}\int_{x-\alpha t}^{x+\alpha t} g(s)\,ds . \quad \blacklozenge$$

Example 3 Find the solution to the initial value problem

(33) $$\frac{\partial^2 u}{\partial t^2} = 4\frac{\partial^2 u}{\partial x^2}, \qquad -\infty < x < \infty, \qquad t > 0,$$

(34) $$u(x,0) = \sin x, \qquad -\infty < x < \infty,$$

(35) $$\frac{\partial u}{\partial t}(x,0) = 1, \qquad -\infty < x < \infty.$$

Solution This is just a special case of the preceding example where $\alpha = 2$, $f(x) = \sin x$, and $g(x) = 1$. Substituting into (32), we obtain the solution

(36) $$u(x,t) = \frac{1}{2}[\sin(x+2t)+\sin(x-2t)] + \frac{1}{4}\int_{x-2t}^{x+2t} ds$$

$$= \sin x \cos 2t + t . \quad \blacklozenge$$

We now use d'Alembert's formula to show that the solution to the "infinite" string problem consists of traveling waves.

Let $h(x)$ be a function defined on $(-\infty, \infty)$. The function $h(x+a)$, where $a > 0$, is a translation of the function $h(x)$ in the sense that its "shape" is the same as $h(x)$, but its position has been shifted to the left by an amount a. This is illustrated in Figure 10.21 for a function $h(x)$ whose graph consists of a triangular "bump." If we let $t \geq 0$ be a parameter (say, time), then the functions $h(x+\alpha t)$ represent a family of functions with the same shape but shifted farther and farther to the left as $t \to \infty$ (Figure 10.22 on page 611). We say that $h(x+\alpha t)$ is a **traveling wave** moving to the left with speed α. In a similar fashion, $h(x-\alpha t)$ is a traveling wave moving to the right with speed α.

Figure 10.21 Graphs of $h(x)$ and $h(x+a)$

Figure 10.22 Traveling wave

If we refer to formula (25), we find that the solution to $\partial^2 u/\partial t^2 = \alpha^2 \partial^2 u/\partial x^2$ consists of traveling waves $A(x + \alpha t)$ moving to the left with speed α and $B(x - \alpha t)$ moving to the right at the same speed.

In the special case when the initial velocity $g(x) \equiv 0$, we have

$$u(x, t) = \frac{1}{2}[f(x + \alpha t) + f(x - \alpha t)] .$$

Hence, $u(x, t)$ is the sum of the traveling waves

$$\frac{1}{2}f(x + \alpha t) \quad \text{and} \quad \frac{1}{2}f(x - \alpha t) .$$

These waves are initially superimposed, since

$$u(x, 0) = \frac{1}{2}f(x) + \frac{1}{2}f(x) = f(x) .$$

As t increases, the two waves move away from each other with speed 2α. This is illustrated in Figure 10.23 for a triangular wave.

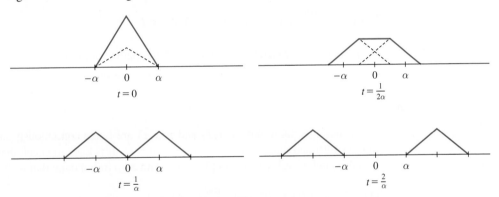

Figure 10.23 Decomposition of initial displacement into traveling waves

Example 4 Express the standing wave $\cos\left(\dfrac{\pi \alpha}{L}t\right)\sin\left(\dfrac{\pi x}{L}\right)$ as a superposition of traveling waves.

Solution By a familiar trigonometric identity,

$$\cos\left(\frac{\pi \alpha}{L}t\right)\sin\left(\frac{\pi x}{L}\right) = \frac{1}{2}\sin\frac{\pi}{L}(x - \alpha t) + \frac{1}{2}\sin\frac{\pi}{L}(x + \alpha t)$$
$$= h(x - \alpha t) + h(x + \alpha t) ,$$

where

$$h(x) = \frac{1}{2}\sin\frac{\pi x}{L} .$$

This standing wave results from adding a wave traveling to the right to the same waveform traveling to the left. ◆

Existence and Uniqueness of Solutions

In Example 1 the method of separation of variables was used to derive a formal solution to the given initial-boundary value problem. To show that these series converge to an actual solution requires results from real analysis, just as was the case for the formal solutions to the heat equation in Section 10.5. In Examples 2 and 3, we can establish the validity of d'Alembert's solution by direct substitution into the initial value problem, assuming sufficient differentiability of the initial functions. We leave it as an exercise for you to show that if f has a continuous second derivative and g has a continuous first derivative, then d'Alembert's solution is a true solution (see Problem 12).

The question of the uniqueness of the solution to the initial-boundary value problem (1)–(4) can be answered using an **energy argument.**

Uniqueness of the Solution to the Vibrating String Problem

Theorem 8. The initial-boundary value problem

(37)
$$\frac{\partial^2 u}{\partial t^2} = \alpha^2 \frac{\partial^2 u}{\partial x^2}, \qquad 0 < x < L, \qquad t > 0,$$

(38) $u(0, t) = u(L, t) = 0, \qquad t > 0,$

(39) $u(x, 0) = f(x), \qquad 0 < x < L,$

(40) $\dfrac{\partial u}{\partial t}(x, 0) = g(x), \qquad 0 < x < L,$

has at most one twice continuously differentiable solution.

Proof. Assume both $u(x, t)$ and $v(x, t)$ are twice continuously differentiable solutions to (37)–(40) and let $w(x, t) := u(x, t) - v(x, t)$. It is easy to check that $w(x, t)$ satisfies the initial-boundary value problem (37)–(40) with zero initial data; that is, for $0 \le x \le L$,

(41) $w(x, 0) = 0 \quad \text{and} \quad \dfrac{\partial w}{\partial t}(x, 0) = 0.$

We now show that $w(x, t) \equiv 0$ for $0 \le x \le L, t \ge 0$.

If $w(x, t)$ is the displacement of the vibrating string at location x for time t, then with the appropriate units, the total energy $E(t)$ of the vibrating string at time t equals the integral

(42) $E(t) := \dfrac{1}{2} \displaystyle\int_0^L \left[\alpha^2 \left(\dfrac{\partial w}{\partial x} \right)^2 + \left(\dfrac{\partial w}{\partial t} \right)^2 \right] dx .$

(The first term in the integrand relates to the stretching of the string at location x and represents the potential energy. The second term is the square of the velocity of the vibrating string at x and represents the kinetic energy.)

We now consider the derivative of $E(t)$:

$$\frac{dE}{dt} = \frac{d}{dt} \left\{ \frac{1}{2} \int_0^L \left[\alpha^2 \left(\frac{\partial w}{\partial x} \right)^2 + \left(\frac{\partial w}{\partial t} \right)^2 \right] dx \right\} .$$

Since w has continuous second partial derivatives (because u and v do), we can interchange the order of integration and differentiation. This gives

$$(43) \qquad \frac{dE}{dt} = \int_0^L \left[\alpha^2 \frac{\partial w}{\partial x} \frac{\partial^2 w}{\partial t \, \partial x} + \frac{\partial w}{\partial t} \frac{\partial^2 w}{\partial t^2} \right] dx .$$

Again the continuity of the second partials of w guarantees that the mixed partials are equal; that is,

$$\frac{\partial^2 w}{\partial t \, \partial x} = \frac{\partial^2 w}{\partial x \, \partial t} .$$

Combining this fact with integration by parts, we obtain

$$(44) \qquad \int_0^L \alpha^2 \frac{\partial w}{\partial x} \frac{\partial^2 w}{\partial t \, \partial x} dx = \int_0^L \alpha^2 \frac{\partial w}{\partial x} \frac{\partial^2 w}{\partial x \, \partial t} dx$$

$$= \alpha^2 \frac{\partial w}{\partial x}(L, t) \frac{\partial w}{\partial t}(L, t) - \alpha^2 \frac{\partial w}{\partial x}(0, t) \frac{\partial w}{\partial t}(0, t)$$

$$- \int_0^L \alpha^2 \frac{\partial^2 w}{\partial x^2} \frac{\partial w}{\partial t} dx .$$

The boundary conditions $w(0, t) = w(L, t) = 0$, $t \geq 0$, imply that $(\partial w / \partial t)(0, t) = (\partial w / \partial t)(L, t) = 0$, $t \geq 0$. This reduces equation (44) to

$$\int_0^L \alpha^2 \frac{\partial w}{\partial x} \frac{\partial^2 w}{\partial t \, \partial x} dx = - \int_0^L \alpha^2 \frac{\partial w}{\partial t} \frac{\partial^2 w}{\partial x^2} dx .$$

Substituting this in for the first integrand in (43), we find

$$\frac{dE}{dt} = \int_0^L \frac{\partial w}{\partial t} \left[\frac{\partial^2 w}{\partial t^2} - \alpha^2 \frac{\partial^2 w}{\partial x^2} \right] dx .$$

Since w satisfies equation (37), the integrand is zero for all x. Thus, $dE/dt = 0$, and so $E(t) \equiv C$, where C is a constant. This means that the total energy is conserved within the vibrating string.

The first boundary condition in (41) states that $w(x, 0) = 0$ for $0 \leq x \leq L$. Hence, $(\partial w / \partial x)(x, 0) = 0$ for $0 < x < L$. Combining this with the second boundary condition in (41), we find that when $t = 0$, the integrand in (42) is zero for $0 < x < L$. Therefore, $E(0) = 0$. Since $E(t) \equiv C$, we must have $C = 0$. Hence,

$$(45) \qquad E(t) = \frac{1}{2} \int_0^L \left[\alpha^2 \left(\frac{\partial w}{\partial x} \right)^2 + \left(\frac{\partial w}{\partial t} \right)^2 \right] dx \equiv 0 .$$

That is, the total energy of this system is zero.

Because the integrand in (45) is nonnegative and continuous and the integral is zero, the integrand must be zero for $0 \leq x \leq L$. Moreover, the integrand is the sum of two squares and so each term must be zero. Hence,

$$\frac{\partial w}{\partial x}(x, t) = 0 \quad \text{and} \quad \frac{\partial w}{\partial t}(x, t) = 0$$

for all $0 \leq x \leq L, t \geq 0$. Thus $w(x, t) = K$, where K is a constant. Physically, this says that

there is no motion in the string.

Finally, since w is constant and w is zero when $t = 0$, then $w(x, t) \equiv 0$. Consequently, $u(x, t) = v(x, t)$ and the initial-boundary value problem has at most one solution. ◆

10.6 EXERCISES

In Problems 1–4, find a formal solution to the vibrating string problem governed by the given initial-boundary value problem.

1. $\dfrac{\partial^2 u}{\partial t^2} = \dfrac{\partial^2 u}{\partial x^2}$, $\quad 0 < x < 1$, $\quad t > 0$,

 $u(0, t) = u(1, t) = 0$, $\quad t > 0$,

 $u(x, 0) = x(1 - x)$, $\quad 0 < x < 1$,

 $\dfrac{\partial u}{\partial t}(x, 0) = \sin 7\pi x$, $\quad 0 < x < 1$

2. $\dfrac{\partial^2 u}{\partial t^2} = 16\dfrac{\partial^2 u}{\partial x^2}$, $\quad 0 < x < \pi$, $\quad t > 0$,

 $u(0, t) = u(\pi, t) = 0$, $\quad t > 0$,

 $u(x, 0) = \sin^2 x$, $\quad 0 < x < \pi$,

 $\dfrac{\partial u}{\partial t}(x, 0) = 1 - \cos x$, $\quad 0 < x < \pi$

3. $\dfrac{\partial^2 u}{\partial t^2} = 4\dfrac{\partial^2 u}{\partial x^2}$, $\quad 0 < x < \pi$, $\quad t > 0$,

 $u(0, t) = u(\pi, t) = 0$, $\quad t > 0$,

 $u(x, 0) = x^2(\pi - x)$, $\quad 0 < x < \pi$,

 $\dfrac{\partial u}{\partial t}(x, 0) = 0$, $\quad 0 < x < \pi$

4. $\dfrac{\partial^2 u}{\partial t^2} = 9\dfrac{\partial^2 u}{\partial x^2}$, $\quad 0 < x < \pi$, $\quad t > 0$,

 $u(0, t) = u(\pi, t) = 0$, $\quad t > 0$,

 $u(x, 0) = \sin 4x + 7\sin 5x$, $\quad 0 < x < \pi$,

 $\dfrac{\partial u}{\partial t}(x, 0) = \begin{cases} x, & 0 < x < \pi/2, \\ \pi - x, & \pi/2 < x < \pi \end{cases}$

5. **The Plucked String.** A vibrating string is governed by the initial-boundary value problem (1)–(4). If the string is lifted to a height h_0 at $x = a$ and released, then the initial conditions are

 $f(x) = \begin{cases} h_0 x/a, & 0 < x \le a, \\ h_0(L - x)/(L - a), & a < x < L, \end{cases}$

 and $g(x) \equiv 0$. Find a formal solution.

6. A vibrating string is governed by the initial-boundary value problem (1)–(4). The initial conditions are given by $f(x) \equiv 0$ and

 $g(x) = \begin{cases} v_0 x/a, & 0 < x \le a, \\ v_0(L - x)/(L - a) & a < x < L, \end{cases}$

 where v_0 is a constant. Find a formal solution.

In Problems 7 and 8, find a formal solution to the vibrating string problem governed by the given nonhomogeneous initial-boundary value problem.

7. $\dfrac{\partial^2 u}{\partial t^2} = \dfrac{\partial^2 u}{\partial x^2} + tx$, $\quad 0 < x < \pi$, $\quad t > 0$,

 $u(0, t) = u(\pi, t) = 0$, $\quad t > 0$,

 $u(x, 0) = \sin x$, $\quad 0 < x < \pi$,

 $\dfrac{\partial u}{\partial t}(x, 0) = 5\sin 2x - 3\sin 5x$, $\quad 0 < x < \pi$

8. $\dfrac{\partial^2 u}{\partial t^2} = \dfrac{\partial^2 u}{\partial x^2} + x\sin t$, $\quad 0 < x < \pi$, $\quad t > 0$,

 $u(0, t) = u(\pi, t) = 0$, $\quad t > 0$,

 $u(x, 0) = 0$, $\quad 0 < x < \pi$,

 $\dfrac{\partial u}{\partial t}(x, 0) = 0$, $\quad 0 < x < \pi$

9. If one end of a string is held fixed while the other is free, then the motion of the string is governed by the initial-boundary value problem

 $\dfrac{\partial^2 u}{\partial t^2} = \alpha^2\dfrac{\partial^2 u}{\partial x^2}$, $\quad 0 < x < L$, $\quad t > 0$,

 $u(0, t) = 0$ and $\dfrac{\partial u}{\partial x}(L, t) = 0$, $\quad t > 0$,

 $u(x, 0) = f(x)$, $\quad 0 < x < L$,

 $\dfrac{\partial u}{\partial t}(x, 0) = g(x)$, $\quad 0 < x < L$.

 Derive a formula for a formal solution.

10. Derive a formula for the solution to the following initial-boundary value problem involving nonhomogeneous boundary conditions

$$\frac{\partial^2 u}{\partial t^2} = \alpha^2 \frac{\partial^2 u}{\partial x^2}, \qquad 0 < x < L, \qquad t > 0,$$

$$u(0, t) = U_1, \qquad u(L, t) = U_2, \qquad t > 0,$$

$$u(x, 0) = f(x), \qquad 0 < x < L,$$

$$\frac{\partial u}{\partial t}(x, 0) = g(x), \qquad 0 < x < L,$$

where U_1 and U_2 are constants.

11. The Telegraph Problem.[†] Use the method of separation of variables to derive a formal solution to the telegraph problem

$$\frac{\partial^2 u}{\partial t^2} + \frac{\partial u}{\partial t} + u = \alpha^2 \frac{\partial^2 u}{\partial x^2}, \quad 0 < x < L, \quad t > 0,$$

$$u(0, t) = u(L, t) = 0, \qquad t > 0,$$

$$u(x, 0) = f(x), \qquad 0 < x < L,$$

$$\frac{\partial u}{\partial t}(x, 0) = 0, \qquad 0 < x < L.$$

12. Verify d'Alembert's solution (32) to the initial value problem (26)–(28) when $f(x)$ has a continuous second derivative and $g(x)$ has a continuous first derivative by substituting it directly into the equations.

In Problems 13–18, find the solution to the initial value problem.

$$\frac{\partial^2 u}{\partial t^2} = \alpha^2 \frac{\partial^2 u}{\partial x^2}, \qquad -\infty < x < \infty, \qquad t > 0,$$

$$u(x, 0) = f(x), \qquad -\infty < x < \infty,$$

$$\frac{\partial u}{\partial t}(x, 0) = g(x), \qquad -\infty < x < \infty,$$

for the given functions $f(x)$ and $g(x)$.

13. $f(x) \equiv 0$, $\qquad g(x) = \cos x$

14. $f(x) = x^2$, $\qquad g(x) \equiv 0$

15. $f(x) = x$, $\qquad g(x) = x$

16. $f(x) = \sin 3x$, $\qquad g(x) \equiv 1$

17. $f(x) = e^{-x^2}$, $\qquad g(x) = \sin x$

18. $f(x) = \cos 2x$, $\qquad g(x) = 1 - x$

19. Derive the formal solution given in equations (22)–(24) to the vibrating membrane problem governed by the initial-boundary value problem (17)–(21).

20. Long Water Waves. The motion of long water waves in a channel of constant depth is governed by the **linearized Korteweg and de Vries (KdV) equation**

$$(46) \qquad u_t + \alpha u_x + \beta u_{xxx} = 0,$$

where $u(x, t)$ is the displacement of the water from its equilibrium depth at location x and at time t, and α and β are positive constants.

(a) Show that equation (46) has a solution of the form

$$(47) \qquad u(x, t) = V(z),$$
$$z = kx - w(k)t,$$

where k is a fixed constant and $w(k)$ is a function of k, provided V satisfies

$$(48) \qquad -w\frac{dV}{dz} + \alpha k \frac{dV}{dz} + \beta k^3 \frac{d^3 V}{dz^3} = 0.$$

These solutions, defined by (47), are called **uniform waves.**

(b) Physically, we are interested only in solutions $V(z)$ that are bounded and nonconstant on the infinite interval $(-\infty, \infty)$. Show that such solutions exist only if $\alpha k - w > 0$.

(c) Let $\lambda^2 = (\alpha k - w)/(\beta k^3)$. Show that the solutions from part (b) can be expressed in the form

$$V(x, t) = A \sin\left[\lambda k x - (\alpha\lambda k - \beta\lambda^3 k^3)t + B\right],$$

where A and B are arbitrary constants. [*Hint:* Solve for w in terms of λ and k and use (47).]

(d) Since both λ and k can be chosen arbitrarily and they always appear together as the product λk, we can set $\lambda = 1$ without loss of generality. Hence, we have

$$V(x, t) = A \sin\left[kx - (\alpha k - \beta k^3)t + B\right]$$

as a uniform wave solution to (46). The defining relation $w(k) = \alpha k - \beta k^3$ is called the **dispersion relation**, the ratio $w(k)/k = \alpha - \beta k^2$ is called the **phase velocity**, and the derivative $dw/dk = \alpha - 3\beta k^2$ is called the **group velocity**. When the group velocity is not constant, the waves are called **dispersive**. Show that the standard wave equation $u_{tt} = \alpha^2 u_{xx}$ has only nondispersive waves.

21. Vibrating Drum. A vibrating circular membrane of unit radius whose edges are held fixed in a plane and

[†]For a discussion of the telegraph problem, see Project E at the end of this chapter.

whose displacement $u(r, t)$ depends only on the radial distance r from the center and on the time t is governed by the initial-boundary value problem.

$$\frac{\partial^2 u}{\partial t^2} = \alpha^2 \left(\frac{\partial^2 u}{\partial r^2} + \frac{1}{r} \frac{\partial u}{\partial r} \right),$$

$$0 < r < 1, \qquad t > 0,$$

$$u(1, t) = 0, \qquad t > 0,$$

$$u(r, t) \quad \text{remains finite as } r \to 0^+.$$

Show that there is a family of solutions of the form

$$u_n(r, t) = [a_n \cos(k_n \alpha t) + b_n \sin(k_n \alpha t)] J_0(k_n r),$$

where J_0 is the Bessel function of the first kind of order zero (see page 478) and $0 < k_1 < k_2 < \cdots < k_n < \cdots$ are the positive zeros of J_0. See Figure 10.24.

(a) (b) (c)

Figure 10.24 Mode shapes for vibrating drum. (a) $J_0(2.405r)$, (b) $J_0(5.520r)$, (c) $J_0(8.654r)$

10.7 Laplace's Equation

In Section 10.1 we showed how Laplace's equation,

$$\Delta u = \frac{\partial^2 u}{\partial x^2} + \frac{\partial^2 u}{\partial y^2} = 0,$$

arises in the study of steady-state or time-independent solutions to the heat equation. Because these solutions do not depend on time, initial conditions are irrelevant and only boundary conditions are specified. Other applications include the static displacement $u(x, y)$ of a stretched membrane fastened in space along the boundary of a region (here u must satisfy Laplace's equation inside the region); the electrostatic and gravitational potentials in certain force fields (here u must satisfy Laplace's equation in any region that is free of electrical charges or mass); and, in fluid mechanics for an idealized fluid, the stream function $u(x, y)$ whose level curves (stream lines), $u(x, y) = $ constant, represent the path of particles in the fluid (again u satisfies Laplace's equation in the flow region).

There are two basic types of boundary conditions that are usually associated with Laplace's equation: **Dirichlet boundary conditions,** where the solution $u(x, y)$ to Laplace's equation in a domain D is required to satisfy

$$u(x, y) = f(x, y) \qquad \text{on } \partial D,$$

with $f(x, y)$ a specified function defined on the boundary ∂D of D; and **Neumann boundary conditions,** where the directional derivative $\partial u / \partial n$ along the outward normal to the boundary is required to satisfy

$$\frac{\partial u}{\partial n}(x, y) = g(x, y) \qquad \text{on } \partial D,$$

with $g(x, y)$ a specified function defined on ∂D.[†] We say that the boundary conditions are **mixed** if the solution is required to satisfy $u(x, y) = f(x, y)$ on part of the boundary and $(\partial u / \partial n)(x, y) = g(x, y)$ on the remaining portion of the boundary.

In this section we use the method of separation of variables to find solutions to Laplace's equation with various boundary conditions for rectangular, circular, and cylindrical domains. We also discuss the existence and uniqueness of such solutions.

Example 1 Find a solution to the following mixed boundary value problem for a rectangle (see Figure 10.25):

(1) $$\frac{\partial^2 u}{\partial x^2} + \frac{\partial^2 u}{\partial y^2} = 0, \qquad 0 < x < a, \qquad 0 < y < b,$$

(2) $$\frac{\partial u}{\partial x}(0, y) = \frac{\partial u}{\partial x}(a, y) = 0, \qquad 0 \le y \le b,$$

(3) $$u(x, b) = 0, \qquad 0 \le x \le a,$$

(4) $$u(x, 0) = f(x), \qquad 0 \le x \le a.$$

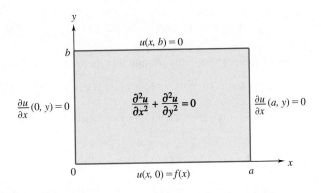

Figure 10.25 Mixed boundary value problem

Solution Separating variables, we first let $u(x, y) = X(x)Y(y)$. Substituting into equation (1), we have

$$X''(x)Y(y) + X(x)Y''(y) = 0,$$

which separates into

$$\frac{X''(x)}{X(x)} = -\frac{Y''(y)}{Y(y)} = -\lambda,$$

where λ is some constant. This leads to the two ordinary differential equations

(5) $$X''(x) + \lambda X(x) = 0,$$

(6) $$Y''(y) - \lambda Y(y) = 0.$$

From the boundary condition (2), we observe that

(7) $$X'(0) = X'(a) = 0.$$

[†]These generalize the boundary conditions (5) in Section 10.1 and (42) in Section 10.5.

We have encountered the eigenvalue problem in (5) and (7) previously (see Example 1 in Section 10.5). The eigenvalues are $\lambda = \lambda_n = (n\pi/a)^2$, $n = 0, 1, 2, \ldots$, with corresponding solutions

$$\text{(8)} \qquad X_n(x) = a_n \cos\left(\frac{n\pi x}{a}\right),$$

where the a_n's are arbitrary constants.

Setting $\lambda = \lambda_n = (n\pi/a)^2$ in equation (6) and solving for Y gives[†]

$$\text{(9)} \qquad Y_0(y) = A_0 + B_0 y,$$

$$Y_n(y) = A_n \cosh\left(\frac{n\pi y}{a}\right) + B_n \sinh\left(\frac{n\pi y}{a}\right), \qquad n = 1, 2, \ldots$$

Now the boundary condition $u(x, b) = 0$ in (3) will be satisfied if $Y(b) = 0$. Setting $y = b$ in (9), we see that we want $A_0 = -bB_0$ and

$$A_n = -B_n \tanh\left(\frac{n\pi b}{a}\right), \quad \text{or equivalently,}$$

$$Y_n(y) = C_n \sinh\left[\frac{n\pi}{a}(y - b)\right], \qquad n = 1, 2, \ldots$$

(see Problem 18). Thus we find that there are solutions to (1)–(3) of the form

$$u_0(x, y) = X_0(x)Y_0(y) = a_0 B_0(y - b) = E_0(y - b),$$

$$u_n(x, y) = X_n(x)Y_n(y) = a_n \cos\left(\frac{n\pi x}{a}\right) C_n \sinh\left[\frac{n\pi}{a}(y - b)\right]$$

$$= E_n \cos\left(\frac{n\pi x}{a}\right) \sinh\left[\frac{n\pi}{a}(y - b)\right], \qquad n = 1, 2, \ldots,$$

where the E_n's are constants. In fact, by the superposition principle,

$$\text{(10)} \qquad u(x, y) = E_0(y - b) + \sum_{n=1}^{\infty} E_n \cos\left(\frac{n\pi x}{a}\right) \sinh\left[\frac{n\pi}{a}(y - b)\right]$$

is a formal solution to (1)–(3).

Applying the remaining nonhomogeneous boundary condition in (4), we have

$$u(x, 0) = f(x) = -E_0 b + \sum_{n=1}^{\infty} E_n \sinh\left(-\frac{n\pi b}{a}\right) \cos\left(\frac{n\pi x}{a}\right).$$

This is a Fourier cosine series for $f(x)$ and hence the coefficients are given by the formulas

$$\text{(11)} \qquad E_0 = \frac{1}{(-ba)} \int_0^a f(x)\, dx,$$

$$E_n = \frac{2}{a \sinh\left(-\dfrac{n\pi b}{a}\right)} \int_0^a f(x) \cos\left(\frac{n\pi x}{a}\right) dx, \qquad n = 1, 2, \ldots.$$

Thus a formal solution is given by (10) with the constants E_n given by (11). See Figure 10.26 on page 619 for a sketch of partial sums for (10). ◆

[†]We usually express $Y_n(y) = a_n e^{n\pi y/a} + b_n e^{-n\pi y/a}$. However, computation is simplified in this case by using the hyperbolic functions $\cosh z = (e^z + e^{-z})/2$ and $\sinh z = (e^z - e^{-z})/2$.

Figure 10.26 Partial sums for Example 1 with $f(x) = \begin{cases} x, & 0 < x \leq \pi/2, \\ \pi - x, & \pi/2 \leq x < \pi \end{cases}$

In Example 1 the boundary conditions were homogeneous on three sides of the rectangle and nonhomogeneous on the fourth side, $\{(x, y): y = 0, 0 \leq x \leq a\}$. It is important to note that the method used in Example 1 can also be used to solve problems for which the boundary conditions are nonhomogeneous on all sides. This is accomplished by solving four separate boundary value problems in which three sides have homogeneous boundary conditions and only one side is nonhomogeneous. The solution is then obtained by summing these four solutions (see Problem 5).

For problems involving circular domains, it is usually more convenient to use polar coordinates. In rectangular coordinates the Laplacian has the form

$$\Delta u = \frac{\partial^2 u}{\partial x^2} + \frac{\partial^2 u}{\partial y^2}.$$

In polar coordinates (r, θ), we let

$$x = r \cos \theta, \qquad y = r \sin \theta$$

so that

$$r = \sqrt{x^2 + y^2}, \qquad \tan \theta = y/x.$$

With patience and a little care in applying the chain rule, one can show that the Laplacian in polar coordinates is

(12) $$\Delta u = \frac{\partial^2 u}{\partial r^2} + \frac{1}{r}\frac{\partial u}{\partial r} + \frac{1}{r^2}\frac{\partial^2 u}{\partial \theta^2}$$

(see Problem 6). In the next example we obtain a solution to the **Dirichlet problem** in a disk of radius a.

Example 2 A circular metal disk of radius a has its top and bottom insulated. The edge $(r = a)$ of the disk is kept at a specified temperature that depends on its location (varies with θ). The steady-state temperature inside the disk satisfies Laplace's equation. Determine the temperature distribution $u(r, \theta)$ inside the disk by finding the solution to the following Dirichlet boundary value problem, depicted in Figure 10.27 on page 620:

(13) $$\frac{\partial^2 u}{\partial r^2} + \frac{1}{r}\frac{\partial u}{\partial r} + \frac{1}{r^2}\frac{\partial^2 u}{\partial \theta^2} = 0, \qquad 0 \leq r < a, \qquad -\pi \leq \theta \leq \pi,$$

(14) $$u(a, \theta) = f(\theta), \qquad -\pi \leq \theta \leq \pi.$$

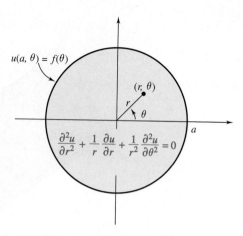

Figure 10.27 Steady-state temperature distribution in a disk

Solution To use the method of separation of variables, we first set

$$u(r, \theta) = R(r)\Theta(\theta),$$

where $0 \le r < a$ and $-\pi \le \theta \le \pi$. Substituting into (13) and separating variables give

$$\frac{r^2 R''(r) + r R'(r)}{R(r)} = -\frac{\Theta''(\theta)}{\Theta(\theta)} = \lambda,$$

where λ is any constant. This leads to the two ordinary differential equations

(15) $r^2 R''(r) + r R'(r) - \lambda R(r) = 0,$

(16) $\Theta''(\theta) + \lambda \Theta(\theta) = 0.$

For $u(r, \theta)$ to be continuous in the disk $0 \le r < a$, we need $\Theta(\theta)$ to be 2π-periodic; in particular, we require

(17) $\Theta(-\pi) = \Theta(\pi)$ and $\Theta'(-\pi) = \Theta'(\pi).$

Therefore, we seek nontrivial solutions to the eigenvalue problem (16)–(17).

When $\lambda < 0$, the general solution to (16) is the sum of two exponentials. Hence we have only trivial 2π-periodic solutions.

When $\lambda = 0$, we find $\Theta(\theta) = A\theta + B$ to be the solution to (16). This linear function is periodic only when $A = 0$, that is, $\Theta_0(\theta) = B$ is the only 2π-periodic solution corresponding to $\lambda = 0$.

When $\lambda > 0$, the general solution to (16) is

$$\Theta(\theta) = A \cos \sqrt{\lambda}\theta + B \sin \sqrt{\lambda}\theta.$$

Here we get a nontrivial 2π-periodic solution only when $\sqrt{\lambda} = n, n = 1, 2, \ldots$. [You can check this using (17).] Hence, we obtain the nontrivial 2π-periodic solutions

(18) $\Theta_n(\theta) = A_n \cos n\theta + B_n \sin n\theta$

corresponding to $\sqrt{\lambda} = n, n = 1, 2, \ldots$. Notice that the eigenfunctions are the terms of the Fourier series discussed in Section 10.3 (recall Figure 10.5, page 573).

Now for $\lambda = n^2, n = 0, 1, 2, \ldots$, equation (15) is the Cauchy–Euler equation

(19) $r^2 R''(r) + r R'(r) - n^2 R(r) = 0$

(see Section 4.7, page 192). When $n = 0$, the general solution is

$$R_0(r) = C + D \ln r \,.$$

Since $\ln r \to -\infty$ as $r \to 0^+$, this solution is unbounded near $r = 0$ when $D \neq 0$. Therefore, we must choose $D = 0$ if $u(r, \theta)$ is to be continuous at $r = 0$. We now have $R_0(r) = C$ and so $u_0(r, \theta) = R_0(r)\Theta(\theta) = CB$, which for convenience we write in the form

(20) $u_0(r, \theta) = \dfrac{A_0}{2} \,,$

where A_0 is an arbitrary constant.

When $\lambda = n^2, n = 1, 2, \ldots$, you should verify that equation (19) has the general solution

$$R_n(r) = C_n r^n + D_n r^{-n} \,.$$

Since $r^{-n} \to \infty$ as $r \to 0^+$, we must set $D_n = 0$ in order for $u(r, \theta)$ to be bounded at $r = 0$. Thus,

$$R_n(r) = C_n r^n \,.$$

Now for each $n = 1, 2, \ldots$, we have the solutions

(21) $u_n(r, \theta) = R_n(r)\Theta_n(\theta) = C_n r^n (A_n \cos n\theta + B_n \sin n\theta) \,,$

and by forming an infinite series from the solutions in (20) and (21), we get the following formal solution to (13):

$$u(r, \theta) = \dfrac{A_0}{2} + \sum_{n=1}^{\infty} C_n r^n (A_n \cos n\theta + B_n \sin n\theta) \,.$$

It is more convenient to write this series in the equivalent form

(22) $u(r, \theta) = \dfrac{a_0}{2} + \sum_{n=1}^{\infty} \left(\dfrac{r}{a}\right)^n (a_n \cos n\theta + b_n \sin n\theta) \,,$

where the a_n's and b_n's are constants. These constants can be determined from the boundary condition; indeed, with $r = a$ in (22), condition (14) becomes

$$f(\theta) = \dfrac{a_0}{2} + \sum_{n=1}^{\infty} (a_n \cos n\theta + b_n \sin n\theta) \,.$$

Hence, we recognize that a_n, b_n are Fourier coefficients for $f(\theta)$. Thus,

(23) $a_n = \dfrac{1}{\pi} \displaystyle\int_{-\pi}^{\pi} f(\theta)\cos n\theta \, d\theta \,, \qquad n = 0, 1, \ldots \,,$

(24) $b_n = \dfrac{1}{\pi} \displaystyle\int_{-\pi}^{\pi} f(\theta)\sin n\theta \, d\theta \,, \qquad n = 1, 2, \ldots \,.$

To summarize, if a_n and b_n are defined by formulas (23) and (24), then $u(r, \theta)$ given in (22) is a formal solution to the Dirichlet problem (13)–(14). ◆

The procedure in Example 2 can also be used to study the **Neumann problem** in a disk:

$$(25) \qquad \Delta u = 0, \qquad 0 \le r < a, \qquad -\pi \le \theta \le \pi,$$

$$(26) \qquad \frac{\partial u}{\partial r}(a, \theta) = f(\theta), \qquad -\pi \le \theta \le \pi.$$

For this problem there is no unique solution, since if u is a solution, then the function u plus a constant is also a solution. Moreover, f must satisfy the **consistency condition**

$$(27) \qquad \int_{-\pi}^{\pi} f(\theta)\, d\theta = 0.$$

If we interpret the solution $u(r, \theta)$ of equation (25) as the steady-state temperature distribution inside a circular disk that does not contain either a heat source or heat sink, then equation (26) specifies the flow of heat across the boundary of the disk. Here the consistency condition (27) is simply the requirement that the net flow of heat across the boundary is zero. (If we keep pumping heat in, the temperature won't reach equilibrium!) We leave the solution of the Neumann problem and the derivation of the consistency condition for the exercises.

The technique used in Example 2 also applies to annular domains, $\{(r, \theta): 0 < a < r < b\}$, and to exterior domains, $\{(r, \theta): a < r\}$. We also leave these applications as exercises.

Laplace's equation in cylindrical coordinates arises in the study of steady-state temperature distributions in a solid cylinder and in determining the electric potential inside a cylinder. In cylindrical coordinates,

$$x = r\cos\theta, \qquad y = r\sin\theta, \qquad z = z,$$

Laplace's equation becomes

$$(28) \qquad \Delta u = \frac{\partial^2 u}{\partial r^2} + \frac{1}{r}\frac{\partial u}{\partial r} + \frac{1}{r^2}\frac{\partial^2 u}{\partial \theta^2} + \frac{\partial^2 u}{\partial z^2} = 0.$$

The Dirichlet problem for the cylinder $\{(r, \theta, z): 0 \le r \le a, 0 \le z \le b\}$ has the boundary conditions

$$(29) \qquad u(a, \theta, z) = f(\theta, z), \qquad -\pi \le \theta \le \pi, \qquad 0 \le z \le b,$$

$$(30) \qquad u(r, \theta, 0) = g(r, \theta), \qquad 0 \le r \le a, \qquad -\pi \le \theta \le \pi,$$

$$(31) \qquad u(r, \theta, b) = h(r, \theta), \qquad 0 \le r \le a, \qquad -\pi \le \theta \le \pi.$$

To solve the Dirichlet boundary value problem (28)–(31), we first solve the three boundary value problems corresponding to: (i) $g \equiv 0$ and $h \equiv 0$; (ii) $f \equiv 0$ and $h \equiv 0$; and (iii) $f \equiv 0$ and $g \equiv 0$. Then by the superposition principle, the solution to (28)–(31) will be the sum of these three solutions. This is the same method that was discussed in dealing with Dirichlet problems on rectangular domains. (See the remarks following Example 1.) In the next example we solve the Dirichlet problem when $g \equiv 0$ and $h \equiv 0$.

Example 3 The base $(z = 0)$ and the top $(z = b)$ of a charge-free cylinder are grounded and therefore are at zero potential. The potential on the lateral surface $(r = a)$ of the cylinder is given by $u(a, \theta, z) = f(\theta, z)$ where $f(\theta, 0) = f(\theta, b) = 0$. Inside the cylinder, the potential $u(r, \theta, z)$

satisfies Laplace's equation. Determine the potential u inside the cylinder by finding a solution to the Dirichlet boundary value problem

$$(32) \quad \frac{\partial^2 u}{\partial r^2} + \frac{1}{r}\frac{\partial u}{\partial r} + \frac{1}{r^2}\frac{\partial^2 u}{\partial \theta^2} + \frac{\partial^2 u}{\partial z^2} = 0, \quad 0 \le r < a, \quad -\pi \le \theta \le \pi, \quad 0 < z < b,$$

$$(33) \quad u(a, \theta, z) = f(\theta, z), \quad -\pi \le \theta \le \pi, \quad 0 \le z \le b,$$

$$(34) \quad u(r, \theta, 0) = u(r, \theta, b) = 0, \quad 0 \le r < a, \quad -\pi \le \theta \le \pi.$$

Solution Using the method of separation of variables, we first assume that

$$u(r, \theta, z) = R(r)\Theta(\theta)Z(z).$$

Substituting into equation (32) and separating out the Z's, we find

$$\frac{R''(r) + (1/r)R'(r)}{R(r)} + \frac{1}{r^2}\frac{\Theta''(\theta)}{\Theta(\theta)} = -\frac{Z''(z)}{Z(z)} = \lambda,$$

where λ can be any constant. Separating further the R's and Θ's gives

$$\frac{r^2 R''(r) + rR'(r)}{R(r)} - r^2\lambda = -\frac{\Theta''(\theta)}{\Theta(\theta)} = \mu,$$

where μ can also be any constant. We now have the three ordinary differential equations:

$$(35) \quad r^2 R''(r) + rR'(r) - (r^2\lambda + \mu)R(r) = 0,$$

$$(36) \quad \Theta''(\theta) + \mu\Theta(\theta) = 0,$$

$$(37) \quad Z''(z) + \lambda Z(z) = 0.$$

For u to be continuous in the cylinder, $\Theta(\theta)$ must be 2π-periodic. Thus, let's begin with the eigenvalue problem

$$\Theta''(\theta) + \mu\Theta(\theta) = 0, \quad -\pi < \theta < \pi,$$

$$\Theta(-\pi) = \Theta(\pi) \quad \text{and} \quad \Theta'(-\pi) = \Theta'(\pi).$$

In Example 2 we showed that this problem has nontrivial solutions for $\mu = n^2, n = 0, 1, 2, \ldots$, that are given by the Fourier series components

$$(38) \quad \Theta_n(\theta) = A_n \cos n\theta + B_n \sin n\theta,$$

where the A_n's and B_n's are arbitrary constants.

The boundary conditions in (34) imply that $Z(0) = Z(b) = 0$. Therefore, Z must satisfy the eigenvalue problem

$$Z''(z) + \lambda Z(z) = 0, \quad 0 < z < b, \quad Z(0) = Z(b) = 0.$$

We have seen this eigenvalue problem several times before. Nontrivial solutions exist for $\lambda = (m\pi/b)^2, m = 1, 2, 3, \ldots$ and are given by the sine series components

$$(39) \quad Z_m(z) = C_m \sin\left(\frac{m\pi z}{b}\right),$$

where the C_m's are arbitrary constants.

Substituting for μ and λ in equation (35) gives

$$r^2 R''(r) + rR'(r) - \left(r^2\left(\frac{m\pi}{b}\right)^2 + n^2\right)R(r) = 0, \quad 0 \le r < a.$$

You should verify that the change of variables $s = (m\pi r/b)$ transforms this equation into the **modified Bessel's equation of order n** (page 483)[†]

$$(40) \quad s^2R''(s) + sR'(s) - (s^2 + n^2)R(s) = 0, \qquad 0 \le s < \frac{m\pi a}{b}.$$

The modified Bessel's equation of order n has two linearly independent solutions—the **modified Bessel function of the first kind:**

$$I_n(s) = \sum_{k=0}^{\infty} \frac{(s/2)^{2k+n}}{k!\,\Gamma(k+n+1)},$$

which remains bounded near zero, and the **modified Bessel function of the second kind:**

$$K_n(s) = \lim_{v \to n} \frac{\pi}{2} \frac{I_{-v}(s) - I_v(s)}{\sin v\pi},$$

which becomes unbounded as $s \to 0$. (Recall that Γ is the gamma function discussed in Section 7.6.) A general solution to (40) has the form $CK_n + DI_n$, where C and D are constants. Since u must remain bounded near $s = 0$, we must take $C = 0$. Thus, the desired solutions to (40) have the form

$$(41) \quad R_{mn}(r) = D_{mn}I_n\left(\frac{m\pi r}{b}\right); \qquad n = 0, 1, \dots, \qquad m = 1, 2, \dots,$$

where the D_{mn}'s are arbitrary constants.

If we multiply the functions in (38), (39), and (41) and then sum over m and n, we obtain the following series solution to (32) and (34):

$$(42) \quad u(r, \theta, z) = \sum_{m=1}^{\infty} a_{m0}I_0\left(\frac{m\pi r}{b}\right)\sin\left(\frac{m\pi z}{b}\right)$$

$$+ \sum_{n=1}^{\infty}\sum_{m=1}^{\infty}(a_{mn}\cos n\theta + b_{mn}\sin n\theta)I_n\left(\frac{m\pi r}{b}\right)\sin\left(\frac{m\pi z}{b}\right),$$

where the a_{mn}'s and b_{mn}'s are arbitrary constants.

The constants in (42) can be obtained by imposing the boundary condition (33). Setting $r = a$ and rearranging terms, we have

$$(43) \quad f(\theta, z) = \sum_{m=1}^{\infty} a_{m0}I_0\left(\frac{m\pi a}{b}\right)\sin\left(\frac{m\pi z}{b}\right)$$

$$+ \sum_{n=1}^{\infty}\left[\sum_{m=1}^{\infty} a_{mn}I_n\left(\frac{m\pi a}{b}\right)\sin\left(\frac{m\pi z}{b}\right)\right]\cos n\theta$$

$$+ \sum_{n=1}^{\infty}\left[\sum_{m=1}^{\infty} b_{mn}I_n\left(\frac{m\pi a}{b}\right)\sin\left(\frac{m\pi z}{b}\right)\right]\sin n\theta.$$

[†]The modified Bessel's equation of order n arises in many applications and has been studied extensively. We refer the reader to the text *Special Functions* by E. D. Rainville (Chelsea Publishing, New York, 1972), for details about its solution.

Treating this double Fourier series like the one in Section 10.5 (page 600), we find

$$(44) \qquad a_{m0} = \int_0^b \int_{-\pi}^{\pi} f(\theta, z) \sin\left(\frac{m\pi z}{b}\right) d\theta\, dz \Big/ \pi b I_0\left(\frac{m\pi a}{b}\right),$$

$$(45) \qquad a_{mn} = 2 \int_0^b \int_{-\pi}^{\pi} f(\theta, z) \sin\left(\frac{m\pi z}{b}\right) \cos n\theta\, d\theta\, dz \Big/ \pi b I_n\left(\frac{m\pi a}{b}\right), \qquad n \geq 1,$$

$$(46) \qquad b_{mn} = 2 \int_0^b \int_{-\pi}^{\pi} f(\theta, z) \sin\left(\frac{m\pi z}{b}\right) \sin n\theta\, d\theta\, dz \Big/ \pi b I_n\left(\frac{m\pi a}{b}\right), \qquad n \geq 1.$$

Consequently, a formal solution to (32)–(34) is given by equation (42) with the constants a_{mn} and b_{mn} determined by equations (44)–(46). ◆

Existence and Uniqueness of Solutions

The existence of solutions to the boundary value problems for Laplace's equation can be established by studying the convergence of the formal solutions that we obtained using the method of separation of variables.

To answer the question of the uniqueness of the solution to a Dirichlet boundary value problem for Laplace's equation, recall that Laplace's equation arises in the search for steady-state solutions to the heat equation. Just as there are maximum principles for the heat equation, there are also maximum principles for Laplace's equation. We state one such result here.[†]

Maximum Principle for Laplace's Equation

Theorem 9. Let $u(x, y)$ be a solution to Laplace's equation in a bounded domain D with $u(x, y)$ continuous in $\overline{D}$, the closure of D. ($\overline{D} = D \cup \partial D$, where ∂D denotes the boundary of D.) Then $u(x, y)$ attains its maximum value on ∂D.

The uniqueness of the solution to the Dirichlet boundary value problem follows from the maximum principle. We state this result in the next theorem but leave its proof as an exercise (see Problem 19).

Uniqueness of Solution

Theorem 10. Let D be a bounded domain. If there is a continuous solution to the Dirichlet boundary value problem

$$\Delta u(x, y) = 0 \quad \text{in } D,$$
$$u(x, y) = f(x, y) \quad \text{on } \partial D,$$

then the solution is unique.

[†]A proof can be found in Section 8.2 of the text, *Partial Differential Equations of Mathematical Physics*, 2nd ed., by Tyn Myint-U (Elsevier North Holland, New York, 1983).

Solutions to Laplace's equation in two variables are called **harmonic** functions. These functions arise naturally in the study of analytic functions of a single complex variable. Moreover, complex analysis provides many useful results about harmonic functions. For a discussion of this interaction, we refer the reader to an introductory text on complex analysis such as *Fundamentals of Complex Analysis,* 3rd ed., by E. B. Saff and A. D. Snider (Prentice Hall, Englewood Cliffs, N.J., 2003).

10.7 EXERCISES

In Problems 1–5, find a formal solution to the given boundary value problem.

1. $\dfrac{\partial^2 u}{\partial x^2} + \dfrac{\partial^2 u}{\partial y^2} = 0$, $\quad 0 < x < \pi$, $\quad 0 < y < 1$,

$\dfrac{\partial u}{\partial x}(0, y) = \dfrac{\partial u}{\partial x}(\pi, y) = 0$, $\quad 0 \le y \le 1$,

$u(x, 0) = 4\cos 6x + \cos 7x$, $\quad 0 \le x \le \pi$,

$u(x, 1) = 0$, $\quad 0 \le x \le \pi$

2. $\dfrac{\partial^2 u}{\partial x^2} + \dfrac{\partial^2 u}{\partial y^2} = 0$, $\quad 0 < x < \pi$, $\quad 0 < y < \pi$,

$\dfrac{\partial u}{\partial x}(0, y) = \dfrac{\partial u}{\partial x}(\pi, y) = 0$, $\quad 0 \le y \le \pi$,

$u(x, 0) = \cos x - 2\cos 4x$, $\quad 0 \le x \le \pi$,

$u(x, \pi) = 0$, $\quad 0 \le x \le \pi$

3. $\dfrac{\partial^2 u}{\partial x^2} + \dfrac{\partial^2 u}{\partial y^2} = 0$, $\quad 0 < x < \pi$, $\quad 0 < y < \pi$,

$u(0, y) = u(\pi, y) = 0$, $\quad 0 \le y \le \pi$,

$u(x, 0) = f(x)$, $\quad 0 \le x \le \pi$,

$u(x, \pi) = 0$, $\quad 0 \le x \le \pi$

4. $\dfrac{\partial^2 u}{\partial x^2} + \dfrac{\partial^2 u}{\partial y^2} = 0$, $\quad 0 < x < \pi$, $\quad 0 < y < \pi$,

$u(0, y) = u(\pi, y) = 0$, $\quad 0 \le y \le \pi$,

$u(x, 0) = \sin x + \sin 4x$, $\quad 0 \le x \le \pi$,

$u(x, \pi) = 0$, $\quad 0 \le x \le \pi$

5. $\dfrac{\partial^2 u}{\partial x^2} + \dfrac{\partial^2 u}{\partial y^2} = 0$, $\quad 0 < x < \pi$, $\quad 0 < y < 1$,

$\dfrac{\partial u}{\partial x}(0, y) = \dfrac{\partial u}{\partial x}(\pi, y) = 0$, $\quad 0 \le y \le 1$,

$u(x, 0) = \cos x - \cos 3x$, $\quad 0 \le x \le \pi$,

$u(x, 1) = \cos 2x$, $\quad 0 \le x \le \pi$

6. Derive the polar coordinate form of the Laplacian given in equation (12).

In Problems 7 and 8, find a solution to the Dirichlet boundary value problem for a disk:

$$\frac{\partial^2 u}{\partial r^2} + \frac{1}{r}\frac{\partial u}{\partial r} + \frac{1}{r^2}\frac{\partial^2 u}{\partial \theta^2} = 0, \qquad 0 \le r < 2, \qquad -\pi \le \theta \le \pi,$$

$$u(2, \theta) = f(\theta), \qquad -\pi \le \theta \le \pi$$

for the given function $f(\theta)$.

7. $f(\theta) = |\theta|$, $\quad -\pi \le \theta \le \pi$

8. $f(\theta) = \cos^2 \theta$, $\quad -\pi \le \theta \le \pi$

9. Find a solution to the Neumann boundary value problem for a disk:

$$\frac{\partial^2 u}{\partial r^2} + \frac{1}{r}\frac{\partial u}{\partial r} + \frac{1}{r^2}\frac{\partial^2 u}{\partial \theta^2} = 0, \qquad 0 \le r < a, \qquad -\pi \le \theta \le \pi,$$

$$\frac{\partial u}{\partial r}(a, \theta) = f(\theta), \qquad -\pi \le \theta \le \pi.$$

10. A solution to the Neumann problem (25)–(26) must also satisfy the consistency condition in (27). To show this, use **Green's second formula**

$$\iint_D (v\Delta u - u\Delta v)\, dx\, dy = \int_{\partial D} \left(v\frac{\partial u}{\partial n} - u\frac{\partial v}{\partial n} \right) ds,$$

where $\partial/\partial n$ is the outward normal derivative and ds is the differential of arc length. [*Hint:* Take $v \equiv 1$ and observe that $\partial u / \partial n = \partial u / \partial r$.]

11. Find a solution to the following Dirichlet problem for an annulus:

$$\frac{\partial^2 u}{\partial r^2} + \frac{1}{r}\frac{\partial u}{\partial r} + \frac{1}{r^2}\frac{\partial^2 u}{\partial \theta^2} = 0, \qquad 1 < r < 2, \qquad -\pi \le \theta \le \pi,$$

$$u(1, \theta) = \sin 4\theta - \cos \theta, \qquad -\pi \le \theta \le \pi,$$

$$u(2, \theta) = \sin \theta, \qquad -\pi \le \theta \le \pi.$$

12. Find a solution to the following Dirichlet problem for an annulus:

$$\frac{\partial^2 u}{\partial r^2} + \frac{1}{r}\frac{\partial u}{\partial r} + \frac{1}{r^2}\frac{\partial^2 u}{\partial \theta^2} = 0, \qquad 1 < r < 3, \qquad -\pi \le \theta \le \pi,$$

$$u(1, \theta) = 0, \qquad -\pi \le \theta \le \pi,$$

$$u(3, \theta) = \cos 3\theta + \sin 5\theta, \qquad -\pi \le \theta \le \pi.$$

13. Find a solution to the following Dirichlet problem for an exterior domain:

$$\frac{\partial^2 u}{\partial r^2} + \frac{1}{r}\frac{\partial u}{\partial r} + \frac{1}{r^2}\frac{\partial^2 u}{\partial \theta^2} = 0, \qquad 1 < r, \qquad -\pi \le \theta \le \pi,$$

$$u(1, \theta) = f(\theta) \qquad -\pi \le \theta \le \pi,$$

$$u(r, \theta) \quad \text{remains bounded as } r \to \infty.$$

14. Find a solution to the following Neumann problem for an exterior domain:

$$\frac{\partial^2 u}{\partial r^2} + \frac{1}{r}\frac{\partial u}{\partial r} + \frac{1}{r^2}\frac{\partial^2 u}{\partial \theta^2} = 0, \qquad 1 < r, \qquad -\pi \le \theta \le \pi,$$

$$\frac{\partial u}{\partial r}(1, \theta) = f(\theta), \qquad -\pi \le \theta \le \pi,$$

$$u(r, \theta) \quad \text{remains bounded as } r \to \infty.$$

15. Find a solution to the following Dirichlet problem for a half disk:

$$\frac{\partial^2 u}{\partial r^2} + \frac{1}{r}\frac{\partial u}{\partial r} + \frac{1}{r^2}\frac{\partial^2 u}{\partial \theta^2} = 0, \qquad 0 < r < 1, \qquad 0 < \theta < \pi,$$

$$u(r, 0) = 0, \qquad 0 \le r \le 1,$$

$$u(r, \pi) = 0, \qquad 0 \le r \le 1,$$

$$u(1, \theta) = \sin 3\theta, \qquad 0 \le \theta \le \pi,$$

$$u(0, \theta) \quad \text{bounded.}$$

16. Find a solution to the following Dirichlet problem for a half annulus:

$$\frac{\partial^2 u}{\partial r^2} + \frac{1}{r}\frac{\partial u}{\partial r} + \frac{1}{r^2}\frac{\partial^2 u}{\partial \theta^2} = 0, \qquad \pi < r < 2\pi, \qquad 0 < \theta < \pi,$$

$$u(r, 0) = \sin r, \qquad \pi \le r \le 2\pi,$$

$$u(r, \pi) = 0, \qquad \pi \le r \le 2\pi,$$

$$u(\pi, \theta) = u(2\pi, \theta) = 0, \qquad 0 \le \theta \le \pi.$$

17. Find a solution to the mixed boundary value problem

$$\frac{\partial^2 u}{\partial r^2} + \frac{1}{r}\frac{\partial u}{\partial r} + \frac{1}{r^2}\frac{\partial^2 u}{\partial \theta^2} = 0, \qquad 1 < r < 3, \qquad -\pi \le \theta \le \pi,$$

$$u(1, \theta) = f(\theta), \qquad -\pi \le \theta \le \pi,$$

$$\frac{\partial u}{\partial r}(3, \theta) = g(\theta), \qquad -\pi \le \theta \le \pi.$$

18. Show that

$$-B_n \tanh\left(\frac{n\pi b}{a}\right)\cosh\left(\frac{n\pi y}{a}\right) + B_n \sinh\left(\frac{n\pi y}{a}\right)$$

$$= C_n \sinh\left[\frac{n\pi}{a}(y - b)\right],$$

where

$$C_n = B_n / \cosh\left(\frac{n\pi b}{a}\right).$$

19. Prove Theorem 10 on the uniqueness of the solution to the Dirichlet problem.

20. Stability. Use the maximum principle to prove the following theorem on the continuous dependence of the solution on the boundary conditions:

Theorem. Let f_1 and f_2 be continuous functions on ∂D, where D is a bounded domain. For $i = 1$ and 2, let u_i be the solution to the Dirichlet problem

$$\Delta u = 0, \quad \text{in } D,$$

$$u = f_i, \quad \text{on } \partial D.$$

If the boundary values satisfy

$$|f_1(x, y) - f_2(x, y)| \le \varepsilon \qquad \text{for all } (x, y) \text{ on } \partial D,$$

where $\varepsilon > 0$ is some constant, then

$$|u_1(x, y) - u_2(x, y)| \le \varepsilon \qquad \text{for all } (x, y) \text{ in } D.$$

21. For the Dirichlet problem described in Example 3, let $a = b = \pi$ and assume the potential on the lateral side $(r = \pi)$ of the cylinder is $f(\theta, z) = \sin z$. Use equations (44)–(46) to compute the solution given by equation (42).

22. Invariance of Laplace's Equation. A complex-valued function $f(z)$ of the complex variable $z = x + iy$ can be written in the form $f(z) = u(x, y) + iv(x, y)$, where u and v are real-valued functions. If $f(z)$ is analytic in a planar region D, then its real and imaginary parts satisfy the Cauchy–Riemann equations in D; that is, in D

$$\frac{\partial u}{\partial x} = \frac{\partial v}{\partial y}, \qquad \frac{\partial u}{\partial y} = -\frac{\partial v}{\partial x}.$$

Let f be analytic and one-to-one in D and assume its inverse f^{-1} is analytic in D', where D' is the image of D under f. Then

$$\frac{\partial x}{\partial u} = \frac{\partial y}{\partial v}, \qquad \frac{\partial x}{\partial v} = -\frac{\partial y}{\partial u}$$

are just the Cauchy–Riemann equations for f^{-1}. Show that if $\phi(x, y)$ satisfies Laplace's equation for (x, y) in D, then $\psi(u, v) := \phi\big(x(u, v), y(u, v)\big)$ satisfies Laplace's equation for (u, v) in D'.

23. Fluid Flow Around a Corner. The stream lines that describe the fluid flow around a corner (see Figure 10.28) are given by $\phi(x, y) = k$, where k is a constant and ϕ, the stream function, satisfies the boundary value problem

$$\frac{\partial^2 \phi}{\partial x^2} + \frac{\partial^2 \phi}{\partial y^2} = 0, \quad x > 0, \quad y > 0,$$

$$\phi(x, 0) = 0, \quad 0 \le x,$$

$$\phi(0, y) = 0, \quad 0 \le y.$$

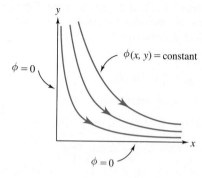

Figure 10.28 Flow around a corner

(a) Using the results of Problem 22, show that this problem can be reduced to finding the flow above a flat plate (see Figure 10.29). That is, show that the problem reduces to finding the solution to

$$\frac{\partial^2 \psi}{\partial u^2} + \frac{\partial^2 \psi}{\partial v^2} = 0, \quad v > 0, \quad -\infty < u < \infty,$$

$$\psi(u, 0) = 0, \quad -\infty < u < \infty,$$

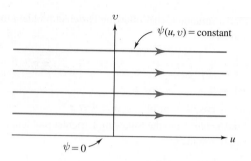

Figure 10.29 Flow above a flat surface

where ψ and ϕ are related as follows: $\psi(u, v) = \phi\big(x(u, v), y(u, v)\big)$ with the mapping between (u, v) and (x, y) given by the analytic function $f(z) = z^2$.

(b) Verify that a nonconstant solution to the problem in part (a) is given by $\psi(u, v) = v$.

(c) Using the result of part (b), find a stream function $\phi(x, y)$ for the original problem.

24. Unbounded Domain. Using separation of variables, find a solution of Laplace's equation in the infinite rectangle $0 < x < \pi, 0 < y < \infty$ that is zero on the sides $x = 0$ and $x = \pi$, approaches zero as $y \to \infty$, and equals $f(x)$ for $y = 0$.

Chapter 10 Summary

Separation of Variables. A classical technique that is effective in solving boundary value problems for partial differential equations is the method of separation of variables. Briefly, the idea is to assume first that there exists a solution that can be written with the variables separated: e.g., $u(x, t) = X(x)T(t)$. Substituting $u = XT$ into the partial differential equation and then imposing the boundary conditions leads to the problem of finding the eigenvalues and eigenfunctions for a boundary value problem for an ordinary differential equation. Solving for the eigenvalues and eigenfunctions, one eventually obtains solutions $u_n(x, t) = X_n(x)T_n(t)$, $n = 1, 2, 3, \ldots$ that solve the partial differential equation and the homogeneous boundary conditions. Taking infinite linear combinations of the $u_n(x, t)$ yields complete solutions of the form

$$u(x, t) = \sum_{n=1}^{\infty} c_n u_n(x, t).$$

The coefficients c_n are then obtained using the initial conditions or nonhomogeneous conditions.

Fourier Series. Let $f(x)$ be a piecewise continuous function on the interval $[-L, L]$. The Fourier series of f is the trigonometric series

$$f(x) \sim \frac{a_0}{2} + \sum_{n=1}^{\infty}\left\{a_n \cos\left(\frac{n\pi x}{L}\right) + b_n \sin\left(\frac{n\pi x}{L}\right)\right\},$$

where the a_n's and b_n's are given by **Euler's formulas:**

$$a_n = \frac{1}{L}\int_{-L}^{L} f(x)\cos\left(\frac{n\pi x}{L}\right)dx, \qquad n = 0, 1, 2, \ldots,$$

$$b_n = \frac{1}{L}\int_{-L}^{L} f(x)\sin\left(\frac{n\pi x}{L}\right)dx, \qquad n = 1, 2, 3, \ldots.$$

Let f be piecewise continuous on the interval $[0, L]$. The **Fourier cosine series** of $f(x)$ on $[0, L]$ is

$$f(x) \sim \frac{a_0}{2} + \sum_{n=1}^{\infty} a_n \cos\left(\frac{n\pi x}{L}\right),$$

where

$$a_n = \frac{2}{L}\int_{0}^{L} f(x)\cos\left(\frac{n\pi x}{L}\right)dx.$$

The **Fourier sine series** of $f(x)$ on $[0, L]$ is

$$f(x) \sim \sum_{n=1}^{\infty} b_n \sin\left(\frac{n\pi x}{L}\right),$$

where

$$b_n = \frac{2}{L}\int_{0}^{L} f(x)\sin\left(\frac{n\pi x}{L}\right)dx.$$

Fourier series and the method of separation of variables are used to solve boundary value problems and initial-boundary value problems for the three classical equations:

Heat equation $\qquad \dfrac{\partial u}{\partial t} = \beta\dfrac{\partial^2 u}{\partial x^2}.$

Wave equation $\qquad \dfrac{\partial^2 u}{\partial t^2} = \alpha^2\dfrac{\partial^2 u}{\partial x^2}.$

Laplace's equation $\qquad \dfrac{\partial^2 u}{\partial x^2} + \dfrac{\partial^2 u}{\partial y^2} = 0.$

For the heat equation, there is a unique solution $u(x, t)$ when the boundary values at the ends of a conducting wire are specified along with the initial temperature distribution. The wave equation yields the displacement $u(x, t)$ of a vibrating string when the ends are held fixed and the initial displacement and initial velocity at each point of the string are given. In such a case, separation of variables yields a solution that is the sum of standing waves. For an infinite string with specified initial displacement and velocity, d'Alembert's solution yields traveling waves. Laplace's equation arises in the study of steady-state solutions to the heat and wave equations.

TECHNICAL WRITING EXERCISES FOR CHAPTER 10

1. The method of separation of variables is an important technique in solving initial-boundary value problems and boundary value problems for *linear* partial differential equations. Explain where the linearity of the differential equation plays a crucial role in the method of separation of variables.

2. In applying the method of separation of variables, we have encountered a variety of *special functions,* such as sines, cosines, Bessel functions, and modified Bessel functions. Describe three or four examples of partial differential equations that involve other special functions, such as Legendre polynomials, Hermite polynomials, and Laguerre polynomials. (Some exploring in the library may be needed; start with the table on page 483.)

3. A constant-coefficient second-order partial differential equation of the form

$$a\frac{\partial^2 u}{\partial x^2} + b\frac{\partial^2 u}{\partial x \partial y} + c\frac{\partial^2 u}{\partial y^2} = 0$$

can be classified using the discriminant $D := b^2 - 4ac$. In particular, the equation is called

 hyperbolic if $D > 0$, and

 elliptic if $D < 0$.

Verify that the wave equation is hyperbolic and Laplace's equation is elliptic. It can be shown that such hyperbolic (elliptic) equations can be transformed by a linear change of variables into the wave (Laplace's) equation. Based on your knowledge of the latter equations, describe which types of problems (initial value, boundary value, etc.) are appropriate for hyperbolic equations and elliptic equations.

A Steady-State Temperature Distribution in a Circular Cylinder

When the temperature u inside a circular cylinder reaches a steady state, it satisfies Laplace's equation $\Delta u = 0$. If the temperature on the lateral surface $(r = a)$ is kept at zero, the temperature on the top $(z = b)$ is kept at zero, and the temperature on the bottom $(z = 0)$ is given by $u(r, \theta, 0) = f(r, \theta)$, then the steady-state temperature satisfies the boundary value problem

$$\frac{\partial^2 u}{\partial r^2} + \frac{1}{r}\frac{\partial u}{\partial r} + \frac{1}{r^2}\frac{\partial^2 u}{\partial \theta^2} + \frac{\partial^2 u}{\partial z^2} = 0, \quad 0 \le r < a, \quad -\pi \le \theta \le \pi, \quad 0 < z < b,$$

$$u(a, \theta, z) = 0, \quad -\pi \le \theta \le \pi, \quad 0 \le z \le b,$$

$$u(r, \theta, b) = 0, \quad 0 \le r < a, \quad -\pi \le \theta \le \pi,$$

$$u(r, \theta, 0) = f(r, \theta), \quad 0 \le r < a, \quad -\pi \le \theta \le \pi,$$

where $f(a, \theta) = 0$ for $-\pi \le \theta \le \pi$ (see Figure 10.30). To find a solution to this boundary value problem, proceed as follows:

(a) Let $u(r, \theta, z) = R(r)\Theta(\theta)Z(z)$. Show that R, Θ, and Z must satisfy the three ordinary differential equations

$$r^2 R'' + rR' - (r^2\lambda + \mu)R = 0,$$

$$\Theta'' + \mu\Theta = 0,$$

$$Z'' + \lambda Z = 0.$$

(b) Show that $\Theta(\theta)$ has the form

$$\Theta(\theta) = A\cos n\theta + B\sin n\theta$$

for $\mu = n^2$, $n = 0, 1, 2, \ldots$.

(c) Show that $Z(z)$ has the form

$$Z(z) = C\sinh\left[\beta(b - z)\right]$$

for $\lambda = -\beta^2$, where $\beta > 0$.

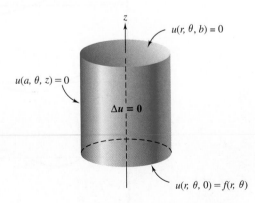

Figure 10.30 Dirichlet problem for cylinder

(d) Show that $R(r)$ has the form, for each n,

$$R_n(r) = DJ_n(\beta r),$$

where J_n is the Bessel function of the first kind.

(e) Show that the boundary conditions require $R(a) = 0$, and so

$$J_n(\beta a) = 0.$$

Hence, for each n, if $0 < \alpha_{n1} < \alpha_{n2} < \cdots < \alpha_{nm} < \cdots$ are the zeros of J_n, then

$$\beta_{nm} = \alpha_{nm}/a.$$

Moreover,

$$R_n(r) = DJ_n(\alpha_{nm}r/a).$$

Some typical eigenfunctions are displayed in Figure 10.31.

(f) Use the preceding results to show that $u(r, \theta, z)$ has the form

$$u(r, \theta, z) = \sum_{n=0}^{\infty} \sum_{m=1}^{\infty} J_n\left(\frac{\alpha_{nm}r}{a}\right)(a_{nm}\cos n\theta + b_{nm}\sin n\theta)\sinh\left(\frac{\alpha_{nm}(b-z)}{a}\right),$$

where a_{nm} and b_{nm} are constants.

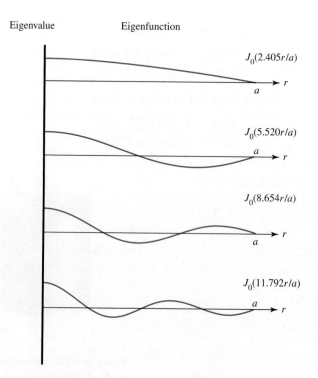

Figure 10.31 Bessel eigenfunctions. See also Figure 10.24, page 616.

(g) Use the final boundary condition and a double orthogonal expansion involving Bessel functions and trigonometric functions to derive the formulas

$$a_{0m} = \frac{1}{\pi a^2 \sinh(\alpha_{0m}b/a)[J_1(\alpha_{0m})]^2}\int_0^a\int_0^{2\pi} f(r,\theta)J_0\left(\frac{\alpha_{0m}r}{a}\right)r\,dr\,d\theta\,,$$

for $m = 1, 2, 3, \ldots$, and for $n, m = 1, 2, 3, \ldots$,

$$a_{nm} = \frac{2}{\pi a^2 \sinh(\alpha_{nm}b/a)[J_{n+1}(\alpha_{nm})]^2}\int_0^a\int_0^{2\pi} f(r,\theta)J_n\left(\frac{\alpha_{nm}r}{a}\right)\cos(n\theta)\,r\,dr\,d\theta\,,$$

$$b_{nm} = \frac{2}{\pi a^2 \sinh(\alpha_{nm}b/a)[J_{n+1}(\alpha_{nm})]^2}\int_0^a\int_0^{2\pi} f(r,\theta)J_n\left(\frac{\alpha_{nm}r}{a}\right)\sin(n\theta)\,r\,dr\,d\theta\,.$$

(The orthogonality relations for Bessel function expansions are described by equations (21), (22) in Section 11.7, page 689.)*

B Laplace Transform Solution of the Wave Equation

Laplace transforms can be used to solve certain partial differential equations. To illustrate this technique, consider the initial-boundary value problem

(1) $\dfrac{\partial^2 u}{\partial t^2} = \alpha^2 \dfrac{\partial^2 u}{\partial x^2}\,,\qquad 0 < x < \infty\,,\qquad t > 0\,,$

(2) $u(0,t) = h(t)\,,\qquad t > 0\,,$

(3) $u(x,0) = 0\,,\qquad 0 < x < \infty\,,$

(4) $\dfrac{\partial u}{\partial t}(x,0) = 0\,,\qquad 0 < x < \infty\,,$

(5) $\displaystyle\lim_{x\to\infty} u(x,t) = 0\,,\qquad t \geq 0\,.$

This problem arises in studying a semi-infinite string that is initially horizontal and at rest and where one end is being moved vertically. Let $u(x,t)$ be the solution to (1)–(5). For each x, let

$$U(x,s) := \mathcal{L}\{u(x,t)\}(x,s) = \int_0^\infty e^{-st}u(x,t)\,dt\,.$$

(a) Using the fact that

$$\mathcal{L}\left\{\frac{\partial^2 u}{\partial x^2}\right\} = \frac{\partial^2}{\partial x^2}\mathcal{L}\{u\}\,,$$

show that $U(x,s)$ satisfies the equation

(6) $s^2 U(x,s) = \alpha^2 \dfrac{\partial^2 U}{\partial x^2}\,,\qquad 0 < x < \infty\,.$

(b) Show that the general solution to (6) is

$$U(x,s) = A(s)e^{-sx/\alpha} + B(s)e^{sx/\alpha}\,,$$

where $A(s)$ and $B(s)$ are arbitrary functions of s.

†All references to Chapters 11–13 refer to the expanded text, *Fundamentals of Differential Equations and Boundary Value Problems*, 7th ed.

(c) Since $u(x, t) \to 0$ as $x \to \infty$ for all $0 \le t < \infty$, we have $U(x, s) \to 0$ as $x \to \infty$. Use this fact to show that the $B(s)$ in part (b) must be zero.

(d) Using equation (2), show that

$$A(s) = H(s) = \mathscr{L}\{h\}(s),$$

where $A(s)$ is given in part (b).

(e) Use the results of parts (b), (c), and (d) to obtain a formal solution to (1)–(5).

C Green's Function

Let Ω be a region in the xy-plane having a smooth boundary $\partial\Omega$. Associated with Ω is a **Green's function** $G(x, y; \xi, \eta)$ defined for pairs of distinct points (x, y), (ξ, η) in Ω. The function $G(x, y; \xi, \eta)$ has the following property.

Let $\Delta u := \partial^2 u/\partial x^2 + \partial^2 u/\partial y^2$ denote the Laplacian operator on Ω; let $h(x, y)$ be a given continuous function on Ω; and let $f(x, y)$ be a given continuous function on $\partial\Omega$. Then, a continuous solution to the Dirichlet boundary value problem

$$\Delta u(x, y) = h(x, y), \qquad \text{in } \Omega,$$

$$u(x, y) = f(x, y), \qquad \text{on } \partial\Omega,$$

is given by

$$(7) \qquad u(x, y) = \iint_\Omega G(x, y; \xi, \eta)\, h(\xi, \eta)\, d\xi\, d\eta$$

$$+ \int_{\partial\Omega} f(\xi, \eta) \frac{\partial G}{\partial n}(x, y; \xi, \eta)\, d\sigma(\xi, \eta),$$

where n is the outward normal to the boundary $\partial\Omega$ of Ω and the second integral is the line integral, with respect to arc length, around the boundary of Ω with the interior of Ω on the left as the boundary is traversed. In (7) we assume that $\partial\Omega$ is sufficiently smooth so that the integrands and integrals exist.

When Ω is the upper half-plane, the Green's function is[†]

$$G(x, y; \xi, \eta) = \frac{1}{4\pi} \ln\left[\frac{(x - \xi)^2 + (y - \eta)^2}{(x - \xi)^2 + (y + \eta)^2}\right].$$

(a) Using (7), show that a solution to

$$(8) \qquad \Delta u(x, y) = h(x, y), \qquad -\infty < x < \infty, \qquad 0 < y,$$

$$(9) \qquad u(x, 0) = f(x), \qquad -\infty < x < \infty,$$

is given by

$$u(x, y) = \frac{y}{\pi} \int_{-\infty}^{\infty} \frac{f(\xi)}{(x - \xi)^2 + y^2}\, d\xi$$

$$+ \frac{1}{4\pi} \int_0^\infty \int_{-\infty}^\infty \ln\left[\frac{(x - \xi)^2 + (y - \eta)^2}{(x - \xi)^2 + (y + \eta)^2}\right] h(\xi, \eta)\, d\xi\, d\eta.$$

[†]Techniques for determining Green's functions can be found in texts on partial differential equations such as *Partial Differential Equations of Mathematical Physics,* 2nd ed., by Tyn Myint-U (Elsevier North Holland, New York, 1983), Chapter 10.

(b) Use the result of part (a) to determine a solution to (8)–(9) when $h \equiv 0$ and $f \equiv 1$.

(c) When Ω is the interior of the unit circle $x^2 + y^2 = 1$, the Green's function, in polar coordinates (r, θ) and (ρ, ϕ), is given by

$$G(r, \theta; \rho, \phi) = \frac{1}{4\pi} \ln\left[r^2 + \rho^2 - 2r\rho \cos(\phi - \theta)\right]$$

$$- \frac{1}{4\pi} \ln[r^2 + \rho^{-2} - 2r\rho^{-1} \cos(\phi - \theta)] - \frac{1}{4\pi} \ln\rho \, .$$

Using (7), show that the solution to

$$\frac{\partial^2 u}{\partial r^2} + \frac{1}{r}\frac{\partial u}{\partial r} + \frac{1}{r^2}\frac{\partial^2 u}{\partial \theta^2} = 0, \qquad 0 \le r < 1, \qquad 0 \le \theta \le 2\pi,$$

$$u(1, \theta) = f(\theta), \qquad 0 \le \theta \le 2\pi,$$

is given by

$$u(r, \theta) = \frac{1}{2\pi} \int_0^{2\pi} \frac{1 - r^2}{1 + r^2 - 2r \cos(\phi - \theta)} f(\phi) \, d\phi \, .$$

This is known as **Poisson's integral formula.**

(d) Use Poisson's integral formula to derive the following mean value property for solutions to Laplace's equation.

Mean Value Property

Theorem 11. If $u(x, y)$ satisfies $\Delta u = 0$ in a bounded domain Ω in R^2 and (x_0, y_0) lies in Ω, then,

(10) $$u(x_0, y_0) = \frac{1}{2\pi} \int_0^{2\pi} u(x_0 + r\cos\theta, y_0 + r\sin\theta) \, d\theta$$

for all $r > 0$ for which the disk $\overline{B}(x_0, y_0; r) = \{(x, y): (x - x_0)^2 + (y - y_0)^2 \le r^2\}$ lies entirely in Ω.

[*Hint:* Use a change of variables that maps the disk $\overline{B}(x_0, y_0; r)$ to the unit disk $\overline{B}(0, 0; 1)$.]

D Numerical Method for $\Delta u = f$ on a Rectangle

Let R denote the open rectangle

$$R = \{(x, y): a < x < b, \ c < y < d\}$$

and ∂R be its boundary. Here we describe a numerical technique for solving the generalized Dirichlet problem

(11) $$\frac{\partial^2 u}{\partial x^2} + \frac{\partial^2 u}{\partial y^2} = f(x, y), \qquad \text{for } (x, y) \text{ in } R,$$

$$u(x, y) = g(x, y), \qquad \text{for } (x, y) \text{ on } \partial R.$$

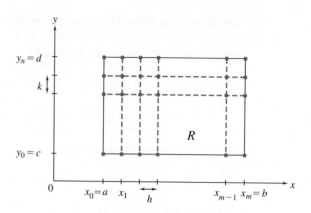

Figure 10.32 Rectangular grid

The method is similar to the finite-difference technique discussed in Chapter 4. We begin by selecting positive integers m and n and step sizes h and k so that $b - a = hm$ and $d - c = kn$. The interval $[a, b]$ is now partitioned into m equal subintervals and $[c, d]$ into n equal subintervals. The partition points are

$$x_i = a + ih, \qquad 0 \le i \le m,$$
$$y_j = c + jk, \qquad 0 \le j \le n$$

(see Figure 10.32). The (dashed) lines $x = x_i$ and $y = y_j$ are called **grid lines** and their intersections (x_i, y_j) are the **mesh points** of the partition. Our goal is to obtain approximations to the solution $u(x, y)$ of problem (11) at each *interior* mesh point, i.e., at (x_i, y_j) where $1 \le i \le m - 1$, $1 \le j \le n - 1$. [Of course, from (11), we are given the values of $u(x, y)$ at the boundary mesh points, e.g., $u(x_0, y_j) = g(x_0, y_j), 0 \le j \le n$.]

The next step is to approximate the partial derivatives $\partial^2 u / \partial x^2$ and $\partial^2 u / \partial y^2$ using the *centered-difference approximations*

$$(12) \qquad \frac{\partial^2 u}{\partial x^2}(x_i, y_j) \approx \frac{1}{h^2} \left[u(x_{i+1}, y_j) - 2u(x_i, y_j) + u(x_{i-1}, y_j) \right] \qquad \text{(for } 1 \le i \le m - 1\text{)},$$

and

$$(13) \qquad \frac{\partial^2 u}{\partial y^2}(x_i, y_j) \approx \frac{1}{k^2} \left[u(x_i, y_{j+1}) - 2u(x_i, y_j) + u(x_i, y_{j-1}) \right] \qquad \text{(for } 1 \le i \le n - 1\text{)}.$$

These approximations are based on the generic formula

$$y''(x) = [y(x + h) - 2y(x) + y(x - h)]/h^2 + [\text{terms involving } h^2, h^3, \dots],$$

which follows from the Taylor series expansion for $y(x)$.

(a) Show that substituting the approximations (12) and (13) into the Dirichlet problem (11) yields the following system:

$$(14) \qquad 2\left[\left(\frac{h}{k}\right)^2 + 1 \right] u_{i,j} - (u_{i+1,j} + u_{i-1,j}) - \left(\frac{h}{k}\right)^2 (u_{i,j+1} + u_{i,j-1}) = -h^2 f(x_i, y_j)$$

for $i = 1, 2, \dots, m - 1$, and $j = 1, 2, \dots, n - 1$;

$$u_{0,j} = g(a, y_j), \qquad u_{m,j} = g(b, y_j), \qquad j = 0, 1, \dots, n,$$
$$u_{i,0} = g(x_i, c), \qquad u_{i,n} = g(x_i, d), \qquad i = 1, 2, \dots, m - 1,$$

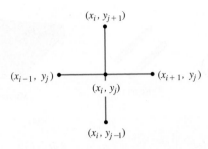

Figure 10.33 Approximate solution $u_{i,j}$ at (x_i, y_j) is obtained from values at the ends of the cross

where $u_{i,j}$ approximates $u(x_i, y_j)$. Notice that each equation in (14) involves approximations to the solution that appear in a cross centered at a mesh point (see Figure 10.33).

(b) Show that the system in part (a) is a linear system of $(m-1)(n-1)$ unknowns in $(m-1)(n-1)$ equations.

(c) For Laplace's equation where $f(x, y) \equiv 0$, show that when $h = k$, equation (14) yields

$$u_{i,j} = \frac{1}{4}\left(u_{i-1,j} + u_{i+1,j} + u_{i,j-1} + u_{i,j+1}\right).$$

Compare this averaging formula with the mean value property of Theorem 11 (Project C).

(d) For the square plate

$$R = \{(x, y) : 0 \le x \le 0.4, 0 \le y \le 0.4\},$$

the boundary is maintained at the following temperatures:

$$u(0, y) = 0, \qquad u(0.4, y) = 10y, \qquad 0 \le y \le 0.4.$$
$$u(x, 0) = 0, \qquad u(x, 0.4) = 10x, \qquad 0 \le x \le 0.4.$$

Using the system in part (a) with $f(x, y) \equiv 0$, $m = n = 4$, and $h = k = 0.1$, find approximations to the steady-state temperatures at the mesh points of the plate. [*Hint:* It is helpful to label these grid points with a single index, say $p_1, p_2, \ldots, p_q$, choosing the ordering in a book-reading sequence.]

E The Telegrapher's Equation and the Cable Equation

Electrical engineers often employ transmission lines to guide electromagnetic waves from a source to a device. The transmission line may take any of several forms, such as depicted in Figure 10.34(a) on page 638. They have common characteristics: uniformity in one direction (z) and uniform capacitance, inductance, longitudinal resistance, and transversal resistance *per unit length* in the z direction (if they are operated at low frequency). Thus any short length dz of the line can be modeled as an *RLC* circuit as depicted in Figure 10.34(b).

In the figure, the circuit parameters are interpreted *per unit length*; thus the incremental inductance is $L\,dz$, the incremental capacitance is $C\,dz$, the (very low) incremental longitudinal

Figure 10.34 Transmission lines (a) common forms (b) circuit model

resistance is $R\, dz$, and the (very low) incremental transversal conductance (reciprocal of resistance) is $G\, dz$.

(a) Apply the circuit analysis of Section 5.7 to the length dz of the transmission line to derive the equations

$$(15) \qquad V(z+dz,t) = V(z,t) - (R\,dz)I(z,t) - (L\,dz)\frac{\partial I(z,t)}{\partial t},$$

$$(16) \qquad I(z+dz,t) = I(z,t) - (G\,dz)V(z,t) - (C\,dz)\frac{\partial V(z,t)}{\partial t},$$

(b) Divide these equations by dz and take limits to obtain the system

$$(17) \qquad \frac{\partial V(z,t)}{\partial z} = -RI(z,t) - L\frac{\partial I(z,t)}{\partial t},$$

$$\frac{\partial I(z,t)}{\partial z} = -GV(z,t) - C\frac{\partial V(z,t)}{\partial t}.$$

(c) Eliminate $I(z,t)$ from the system (17) to derive

$$(18) \qquad \frac{\partial^2 V(z,t)}{\partial z^2} = LC\frac{\partial^2 V(z,t)}{\partial t^2} + [RC+LG]\frac{\partial V(z,t)}{\partial t} + RGV(z,t).$$

[*Hint:* Differentiate the first equation with respect to z and the second with respect to t, and observe the equality of the mixed partials of $I(z,t)$.]

Equation (18) is known as the *telegrapher's equation* for the transmission line; (17) is also known as the *system of telegrapher's equations.*

(d) Make the substitution $V(z,t) = T(t)v(z,t)$ in the telegrapher's equation. For what choices of $T(t)$ is the term proportional to $\partial v/\partial t$ eliminated? For what choices of $T(t)$ is the term proportional to v eliminated?

(e) Show that the telegrapher's equation for a *lossless line*, where the longitudinal resistance R and the transversal conductance G of the line are zero, reduces to the wave equation of Section 10.6. What is the speed of the voltage waves on a lossless line?

(f) Naturally engineers are interested in oscillatory waves on transmission lines, and thanks to Euler's formula (Section 4.3, page 166), solutions to the system (17) can be constructed from the real and imaginary parts of solutions of the form $V(z, t) = V(z)e^{i\omega t}$ and $I(z, t) = I(z)e^{i\omega t}$. (Elaboration of this procedure is given in Project B, Chapter 7, page 418.) Derive a first-order matrix system of the form

(19)
$$\begin{bmatrix} V(z) \\ I(z) \end{bmatrix}' = \mathbf{A} \begin{bmatrix} V(z) \\ I(z) \end{bmatrix}$$

for such solutions, where $\mathbf{A}$ is a constant matrix.

(g) For the lossless case $(R = G = 0)$, construct the matrix exponential $e^{\mathbf{A}t}$ for the matrix in (19) by direct calculation of the series expansion in equation (2) of Section 9.8 (page 545). Express a general solution to the oscillatory transmission line system (19) for the lossless line.

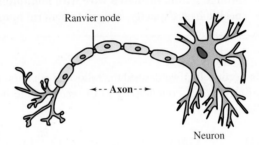

Ranvier node

Axon

Neuron

Figure 10.35 Axon model

(h) In the neurons of an invertebrate's central nervous system, electric signals are propagated between *Ranvier nodes* along *axons*, which are fibers containing an axoplasm liquid and insulated by a myelin sheaf; see Figure 10.35. Their electrical properties are quite similar to those of the coaxial cable in Figure 10.34 on page 638. However, the inductance is negligible. Therefore the voltages in a myelinated axon propagate in accordance with the *cable equation*, which is the transmission line equation (18) with L omitted:

(20)
$$\frac{\partial^2 V(z, t)}{\partial z^2} = RC\frac{\partial V(z, t)}{\partial t} + RGV(z, t) .$$

(i) Use the strategy of part (d) to find $T(t)$ such that the substitution $V(z, t) = T(t)v(z, t)$ results in an analog to the heat equation (Section 10.5)

(21)
$$\frac{\partial^2 v(z, t)}{\partial z^2} = RC\frac{\partial v(z, t)}{\partial t} .$$

(Therefore propagation along an axon proceeds not like a wavefront, but rather like a warm front.)

Eigenvalue Problems and Sturm–Liouville Equations

11.1 Introduction: Heat Flow in a Nonuniform Wire

> **Formulate the differential equation and boundary conditions for the flow of heat through a thin, insulated wire with nonuniform density, thermal conductivity, and specific heat capacity, subject to general boundary conditions at the ends.**

In Section 10.1 we derived the following equation for temperature $u(x, t)$ in a wire with density $\rho(x)$, thermal conductivity $k(x)$, and specific heat capacity $c(x)$ [compare equation (3) in Section 10.1]:

$$(1) \qquad c(x)\rho(x)\frac{\partial u}{\partial t}(x, t) = \frac{\partial}{\partial x}\left[k(x)\frac{\partial u}{\partial x}(x, t) \right] + Q(x, t) \,.$$

We analyzed this equation for the case where c, ρ, and k were independent of position x; now we no longer have this luxury.

If the ends of the wire are held at 0°C, $u(x, t)$ is required to satisfy the boundary conditions

$$(2) \qquad u(0, t) = u(L, t) = 0 \,, \qquad t > 0 \,,$$

where L is the length of the wire. With $f(x)$ denoting the initial temperature distribution in the wire, then $u(x, t)$ must also satisfy the initial condition

$$(3) \qquad u(x, 0) = f(x) \,, \qquad 0 < x < L \,.$$

Thus we see that a mathematical model for the flow of heat through a nonuniform wire is given by the initial-boundary value problem consisting of equations (1), (2), and (3).

In certain cases, such as heat given off by a chemical reaction, the internal source/sink expression $Q(x, t)$ can be modeled by a term proportional to the temperature, $q(x)u(x, t)$. Most other types of sources are independent of temperature and are modeled by an explicit (known) function $h(x, t)$. Hence, a very general equation for heat flow in a nonuniform wire is given by

$$(4) \qquad c(x)\rho(x)\frac{\partial u}{\partial t}(x, t) = \frac{\partial}{\partial x}\left[k(x)\frac{\partial u}{\partial x}(x, t) \right] + q(x)u(x, t) + h(x, t) \,.$$

In the original problem the temperatures at the ends of the wire were held constant. More generally, the ends of the wire could be insulated, $\partial u/\partial x = 0$, or just partially insulated, $\partial u/\partial x + cu = 0$. To generalize, then, we assume $u(x, t)$ satisfies boundary conditions at $x = a$ and $x = b$ of the form

$$(5) \qquad a_1 u(a, t) + a_2\frac{\partial u}{\partial x}(a, t) = 0 \,, \qquad b_1 u(b, t) + b_2\frac{\partial u}{\partial x}(b, t) = 0$$

for all $t > 0$. These are called **separated boundary conditions,** since the condition imposed on $u(x, t)$ at $x = a$ does not depend on the values of $u(x, t)$ at $x = b$ and vice versa.

From (4), (5), and (3), we get the following general model for heat flow in a nonuniform wire:

(6) $$r(x)\frac{\partial u}{\partial t}(x, t) = \frac{\partial}{\partial x}\left[p(x)\frac{\partial u}{\partial x}(x, t)\right] + q(x)u(x, t) + h(x, t), \quad a < x < b, \quad t > 0,$$

(7) $$a_1 u(a, t) + a_2\frac{\partial u}{\partial x}(a, t) = 0, \qquad b_1 u(b, t) + b_2\frac{\partial u}{\partial x}(b, t) = 0, \qquad t > 0,$$

(8) $$u(x, 0) = f(x), \qquad a < x < b,$$

where $r(x) = c(x)\rho(x)$ and $p(x) = k(x)$.

In Chapter 10 we used the method of separation of variables to solve these types of problems. In particular, we discussed techniques for handling the nonhomogeneous term $h(x, t)$ (see Example 3 in Section 10.5 and Example 1 in Section 10.6). For now, we will assume that $h(x, t) \equiv 0$. Applying the method of separation of variables to equations (6) and (7)—that is, setting $u(x, t) = X(x)T(t)$—leads to the **eigenvalue problem for $X(x) := y(x)$:**

(9) $$\frac{d}{dx}\left[p(x)\frac{dy}{dx}\right] + q(x)y + \lambda r(x)y = 0, \qquad a < x < b,$$

(10) $$a_1 y(a) + a_2 y'(a) = 0, \qquad b_1 y(b) + b_2 y'(b) = 0,$$

where λ is a number to be determined (you should verify this). The system (9)–(10) is called a **Sturm–Liouville boundary value problem.**[†] We exclude the trivial boundary conditions where $a_1 = a_2 = 0$ or $b_1 = b_2 = 0$. When $p(x), q(x)$, and $r(x)$ are continuous on $[a, b]$ and $p(x) > 0$ and $r(x) > 0$ on $[a, b]$, we refer to this system as a **regular** Sturm–Liouville boundary value problem.

Most of the eigenvalue problems encountered in Chapter 10 were regular. However, we say that the equation is **singular** when the assumptions on p, p', q, and r hold only on the open interval (a, b) and not the closed, and either p is zero at a or b; $p(x), q(x)$, or $r(x)$ becomes unbounded as x approaches either a or b; or the interval (a, b) is unbounded. An example of a singular Sturm–Liouville boundary value problem is the problem of finding nontrivial solutions for Bessel's equation

(11) $$\frac{d}{dx}\left[x\frac{dy}{dx}\right] + \lambda xy = 0, \qquad 0 < x < b,$$

that satisfy the boundary conditions

(12) $$\lim_{x \to 0^+} y(x) \text{ and } \lim_{x \to 0^+} y'(x) \text{ exist and are finite and } y(b) = 0.$$

[†]*Historical Footnote:* These systems are also called **Sturm–Liouville systems** or **Sturm–Liouville problems.** They are named after the Swiss mathematician Jacques Charles François Sturm (1803–1855) and the French mathematician Joseph Liouville (1809–1882), who studied this class of boundary value problems in the 1830s.

This problem arises in the study of heat flow in a cylinder. Here $p(x) = r(x) = x$, which is zero at $x = 0$. The boundary conditions in (12) are typical of singular Sturm–Liouville boundary value problems.

Another important class of boundary conditions is the class of periodic conditions that we previously encountered in solving Laplace's equation in a disk (see Section 10.7). If we seek P-periodic solutions to equation (9), then we require the solution $y(x)$ to satisfy the **periodic boundary conditions**

$$(13) \qquad y(0) = y(P), \qquad y'(0) = y'(P).$$

As we will see, the theory of periodic Sturm–Liouville problems is closely related to the theory of regular Sturm–Liouville boundary value problems.

The type of boundary value problem that arises from the application of the method of separation of variables depends on the geometry of the region. Of course, a coordinate system must be chosen that allows the variables to be separated. In Chapter 10, we made use of rectangular and polar coordinates.[†]

In this chapter we study the properties of Sturm–Liouville boundary value problems. We begin, in Section 11.2, with a discussion of eigenvalues and eigenfunctions. In Section 11.3 we analyze regular Sturm–Liouville boundary value problems. Nonhomogeneous boundary value problems and the Fredholm alternative are discussed in Section 11.4. We give two methods for solving nonhomogeneous problems: the method of eigenfunction expansions (Section 11.5) and Green's functions (Section 11.6). In Section 11.7 we discuss three examples of singular Sturm–Liouville boundary value problems: Bessel's equation, Legendre's equation, and Hermite's equation. Finally, in Section 11.8, we describe the location of the zeros for solutions to Sturm–Liouville equations as given by the Sturm separation theorem, the Sturm fundamental theorem, and the Picone comparison theorem.

11.2 Eigenvalues and Eigenfunctions

The problem of finding a solution to a second-order linear differential equation

$$(1) \qquad y'' + p(x)y' + q(x)y = f(x), \qquad a < x < b,$$

that satisfies the boundary conditions

$$(2) \qquad \begin{aligned} a_{11}y(a) + a_{12}y'(a) + b_{11}y(b) + b_{12}y'(b) &= c_1, \\ a_{21}y(a) + a_{22}y'(a) + b_{21}y(b) + b_{22}y'(b) &= c_2, \end{aligned}$$

is called a **two-point boundary value problem.** The conditions in (2) are examples of linear boundary conditions. When $c_1 = c_2 = 0$, we say that the boundary conditions are **homogeneous;** otherwise, we refer to them as **nonhomogeneous.**

[†]Other coordinate systems that are used for special geometric regions are discussed in the text by G. Arfken, H. J. Weber, and F. E. Harris, *Mathematical Methods for Physicists,* 7th ed. (Elsevier Academic Press, Boston, 2012), Chapter 2.

There are special types of boundary conditions that occur frequently in applications. They are

Separated: $a_1 y(a) + a_2 y'(a) = c_1$,
$b_1 y(b) + b_2 y'(b) = c_2$.

Dirichlet:[†] $y(a) = c_1$, $y(b) = c_2$.

Neumann:[‡] $y'(a) = c_1$, $y'(b) = c_2$.

Periodic: $y(-T) = y(T)$, $y'(-T) = y'(T)$,
or
$y(0) = y(2T)$, $y'(0) = y'(2T)$,

where the period is $2T$.

Notice that the Dirichlet and Neumann conditions are special types of separated boundary conditions. If neither coefficient is zero in a separated boundary condition, it is called a **Robin** condition.

For boundary value problems involving third- or higher-order equations, one usually specifies the same number of boundary conditions as the order of the equation. When the order is even, these boundary conditions are often separated with half the conditions at the endpoint $x = a$ and half at $x = b$. We give an example of a fourth-order problem later in this section.

For the homogeneous equation with homogeneous boundary conditions

(3) $y'' + p(x)y' + q(x)y = 0$, $a < x < b$;

(4)
$a_{11}y(a) + a_{12}y'(a) + b_{11}y(b) + b_{12}y'(b) = 0$,
$a_{21}y(a) + a_{22}y'(a) + b_{21}y(b) + b_{22}y'(b) = 0$,

it is easy to verify that if $\phi(x)$ is a nontrivial solution to (3)–(4), then so is $A\phi$ for any choice of the constant A. This gives a *one-parameter family* of solutions. If (3)–(4) has two linearly independent solutions $\phi_1(x)$ and $\phi_2(x)$, then $A_1\phi_1(x) + A_2\phi_2(x)$ is also a solution for any choice of A_1 and A_2; hence we have a *two-parameter family* of solutions. The remaining possibility is that $\phi(x) \equiv 0$ is the only solution to (3)–(4); that is, there is a unique solution. Consequently, three situations can arise: The boundary value problem (3)–(4) has a unique solution, a one-parameter family of solutions, or a two-parameter family of solutions (see Problem 31).

The situation for the nonhomogeneous problem (1)–(2) is similar. However, there is an additional possibility: *The nonhomogeneous problem may have no solutions.* The following examples illustrate the various situations.

Example 1 Find all the solutions to the boundary value problem

(5) $y'' + 2y' + 5y = 0$;

(6) $y(0) = 2$, $y(\pi/4) = 1$.

Solution The auxiliary equation for (5) is $r^2 + 2r + 5 = 0$, which has roots $r = -1 \pm 2i$. Hence, a general solution for (5) is $y(x) = c_1 e^{-x} \cos 2x + c_2 e^{-x} \sin 2x$. We will now try to determine c_1 and c_2 so that the boundary conditions in (6) are satisfied. Setting $x = 0$ and $\pi/4$, we find

$$y(0) = c_1 = 2, \qquad y(\pi/4) = c_2 e^{-\pi/4} = 1.$$

Hence, $c_1 = 2$ and $c_2 = e^{\pi/4}$, and so problem (5)–(6) has the unique solution

$$y(x) = 2e^{-x} \cos 2x + e^{\pi/4} e^{-x} \sin 2x. \quad \blacklozenge$$

Example 2 Find all the solutions to the boundary value problem

(7) $y'' + y = \cos 2x$;

(8) $y'(0) = 0, \qquad y'(\pi) = 0$.

Solution The auxiliary equation for (7) is $r^2 + 1 = 0$, so a general solution for the corresponding homogeneous equation is $y_h(x) = c_1 \cos x + c_2 \sin x$. As we know from the method of undetermined coefficients, a particular solution to (7) has the form $y_p(x) = A \cos 2x + B \sin 2x$. Substituting y_p into (7) and solving for A and B, we find $A = -1/3$ and $B = 0$. Hence, $y_p(x) = -(1/3)\cos 2x$. Therefore, a general solution for (7) is

(9) $y(x) = c_1 \cos x + c_2 \sin x - (1/3) \cos 2x$.

To determine the values c_1 and c_2, we substitute the general solution in (9) into (8) and obtain

$$y'(0) = c_2 = 0, \qquad y'(\pi) = -c_2 = 0.$$

Thus, $c_2 = 0$ and c_1 is arbitrary. So problem (7)–(8) has a one-parameter family of solutions, namely,

$$y(x) = c_1 \cos x - (1/3) \cos 2x,$$

where c_1 is any real number. $\blacklozenge$

Example 3 Find the solutions to

(10) $y'' + 4y = 0$;

(11) $y(-\pi) = y(\pi), \qquad y'(-\pi) = y'(\pi)$.

Solution The auxiliary equation for (10) is $r^2 + 4 = 0$, so a general solution is

(12) $y(x) = c_1 \cos 2x + c_2 \sin 2x$.

Since $\cos 2x$ and $\sin 2x$ are π- (and hence 2π-) periodic functions, it follows that $y(-\pi) = y(\pi)$. Moreover, $y'(x) = -2c_1 \sin 2x + 2c_2 \cos 2x$ is also π-periodic and so $y'(-\pi) = y'(\pi)$. Thus, (10)–(11) has a two-parameter family of solutions given by (12), where c_1 and c_2 are the parameters. $\blacklozenge$

Let's see what happens if, in the preceding example, we replace equation (10) by the non-homogeneous equation

(13) $y'' + 4y = 4x$.

A general solution for (13) is

(14) $y(x) = x + c_1 \cos 2x + c_2 \sin 2x$.

Since $y(-\pi) = -\pi + c_1$ and $y(\pi) = \pi + c_1$, we find that there are no solutions to (13) that satisfy $y(-\pi) = y(\pi)$. Thus, the nonhomogeneous boundary value problem (13)–(11) has *no solutions*.

The Sturm–Liouville boundary value problems introduced in the preceding section are examples of two-point boundary problems that involve a parameter λ. Our goal is to determine for which values of λ the boundary value problem

$$(15) \qquad \frac{d}{dx}\left(p(x)\frac{dy}{dx}\right) + q(x)y + \lambda r(x)y = 0, \qquad a < x < b,$$

$$(16) \qquad a_1 y(a) + a_2 y'(a) = 0, \qquad b_1 y(b) + b_2 y'(b) = 0,$$

has nontrivial solutions. Such problems are called **eigenvalue problems.** The nontrivial solutions are called **eigenfunctions** and the corresponding number λ an **eigenvalue.**[†]

One reason eigenvalue problems are important is that they arise in using the method of separation of variables to solve partial differential equations (see Section 10.2). It is therefore worthwhile to keep in mind some simple examples of these problems that were solved in Chapter 10.

Dirichlet boundary conditions:

$$(17) \qquad y'' + \lambda y = 0; \qquad y(0) = y(L) = 0,$$

which has eigenvalues $\lambda_n = (n\pi/L)^2, n = 1, 2, 3, \ldots$, and corresponding eigenfunctions

$$\phi_n(x) = a_n \sin\left(\frac{n\pi x}{L}\right),$$

where the a_n's are arbitrary nonzero constants (see Figure 10.2, Section 10.2, page 565).

Neumann boundary conditions:

$$(18) \qquad y'' + \lambda y = 0; \qquad y'(0) = y'(L) = 0,$$

which has eigenvalues $\lambda_n = (n\pi/L)^2, n = 0, 1, 2, \ldots$, and corresponding eigenfunctions

$$\phi_n(x) = c_n \cos\left(\frac{n\pi x}{L}\right),$$

where the c_n's are arbitrary nonzero constants (see Figure 10.15, Section 10.5, page 594).

Periodic boundary conditions:

$$(19) \qquad y'' + \lambda y = 0; \qquad y(-\pi) = y(\pi), \qquad y'(-\pi) = y'(\pi),$$

which has eigenvalues $\lambda_n = n^2, n = 0, 1, 2, \ldots$, and corresponding eigenfunctions

$$\phi_0(x) = A_0, \qquad \phi_n(x) = A_n \cos nx + B_n \sin nx, \qquad n \geq 1,$$

where the A_n's and B_n's are arbitrary constants, not both zero (see Figure 10.5, Section 10.3, page 573). Notice that for the eigenvalue problem (19) involving periodic boundary conditions, there are two linearly independent eigenfunctions for each nonzero eigenvalue λ_n, $n = 1, 2, 3, \ldots$.

[†]Readers familiar with linear algebra will recognize these eigenvalues and eigenfunctions as the eigenvalues and eigenvectors for the linear operator defined by $\mathcal{L}[y] := -(1/r)(py')' - (q/r)y$, where the domain of $\mathcal{L}$ consists of those twice-differentiable functions satisfying (16).

In the next example we illustrate that even for the simple equation $y'' + \lambda y = 0$, eigenvalues may be difficult to determine exactly when one has a Robin boundary condition.

Example 4 Find all the real eigenvalues and eigenfunctions for

(20) $y'' + \lambda y = 0$;

(21) $y(0) = 0$, $3y(\pi) - y'(\pi) = 0$.

Solution The auxiliary equation for (20) is $r^2 + \lambda = 0$. The nature of the roots varies for λ negative, zero, and positive, so we will consider these cases separately.

Case 1. $\lambda < 0$. Let $\lambda = -\mu^2$ for $\mu > 0$. The roots of the auxiliary equation are $r = \pm \mu$, so a general solution to (20) is $y(x) = c_1 e^{\mu x} + c_2 e^{-\mu x}$. It is more convenient to express the general solution in terms of hyperbolic functions. Thus, we write $y(x)$ in the equivalent form $y(x) = C_1 \cosh \mu x + C_2 \sinh \mu x$. Substituting into the boundary conditions in (21) yields

$$C_1 = 0, \qquad 3(C_1 \cosh \mu\pi + C_2 \sinh \mu\pi) - (\mu C_1 \sinh \mu\pi + \mu C_2 \cosh \mu\pi) = 0.$$

With $C_1 = 0$, the last equation becomes

$$C_2(3 \sinh \mu\pi - \mu \cosh \mu\pi) = 0.$$

To obtain a nontrivial solution, we must have $C_2 \neq 0$, so $3 \sinh \mu\pi - \mu \cosh \mu\pi = 0$; that is, μ must satisfy

(22) $\tanh \mu\pi = \dfrac{1}{3}\mu$.

It is easy to see that in the μy-plane, the line $y = \mu/3$ intersects the curve $y = \tanh \mu\pi$ only once for $\mu > 0$ (see Figure 11.1). Therefore, there is only one positive solution to (22), which we denote by μ_0.

To summarize, the boundary value problem (20)–(21) has one negative eigenvalue $\lambda_0 = -\mu_0^2$, where $\tanh \mu_0\pi = \mu_0/3$, and the corresponding eigenfunctions are $y_0(x) = c_0 \sinh \mu_0 x$, with $c_0 \neq 0$ arbitrary.

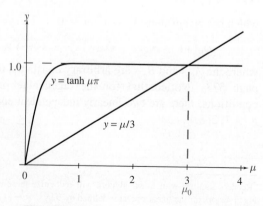

Figure 11.1 Graphs of $y = \mu/3$ and $y = \tanh \mu\pi$

Case 2. $\lambda = 0$. Here zero is a double root of the auxiliary equation, so a general solution to (20) is $y(x) = c_1 x + c_2$. Substituting into (21) gives

$$c_2 = 0 \,, \qquad 3\pi c_1 + 3c_2 - c_1 = 0 \,,$$

which has the solution $c_1 = c_2 = 0$. Thus, we have only the trivial solution (and, in particular, no eigenfunctions).

Case 3. $\lambda > 0$. Let $\lambda = \mu^2$ for $\mu > 0$. The roots of the auxiliary equation are $r = \pm\mu i$, and a general solution to (20) is $y(x) = c_1 \cos\mu x + c_2 \sin\mu x$. Substituting into (21) gives

$$c_1 = 0 \,, \qquad 3(c_1 \cos\mu\pi + c_2 \sin\mu\pi) - (-\mu c_1 \sin\mu\pi + \mu c_2 \cos\mu\pi) = 0 \,.$$

Setting $c_1 = 0$ in the last equation, we get

$$c_2(3 \sin\mu\pi - \mu \cos\mu\pi) = 0 \,.$$

Hence, either c_2 is zero or $3 \sin\mu\pi - \mu \cos\mu\pi = 0$. That is, for nontrivial solutions to exist, μ must satisfy the equation

$$(23) \qquad \tan\mu\pi = \frac{1}{3}\mu \,.$$

To see that there exist values for $\mu > 0$ that satisfy (23), we examine the graphs of $y = \mu/3$ and $y = \tan\mu\pi$. For those values of $\mu > 0$ where the two graphs intersect, we have a solution to (23). From Figure 11.2(a), we see that there are infinitely many such values of μ. Let μ_1 denote the smallest value, μ_2 the next larger, and so on. Now for each μ_n, we get that $y_n(x) = c_2 \sin\mu_n x$ with $c_2 \neq 0$ is a nontrivial solution to (20)–(21) with $\lambda = \mu_n^2$, $n = 1, 2, 3, \ldots$

From the three cases above, we see that the eigenvalues of (20)–(21) are $\lambda_0 = -\mu_0^2$ where $\tanh\mu_0\pi = \mu_0/3$ and $\lambda_n = \mu_n^2$, $n = 1, 2, 3, \ldots$, where $\tan\mu_n\pi = \mu_n/3$. The corresponding eigenfunctions are $y_0 = c_0 \sinh\mu_0 x$ and $y_n = c_n \sin\mu_n x$, $n = 1, 2, 3, \ldots$, with the c_n's being arbitrary nonzero constants. It is interesting to note from Figure 11.2(a) that as n becomes large, μ_n approaches an odd multiple of $1/2$. That is, $\lambda_n \approx (2n+1)^2/4$ when n is large. Using a numerical scheme such as Newton's method to solve $\tanh\mu_0\pi = \mu_0/3$ and $\tan\mu_1\pi = \mu_1/3$, we can show that $\lambda_0 \approx -9.00000$ and $\lambda_1 \approx 1.23897$. ◆

Note the similar features of the eigenfunctions in Figure 11.2(b) on page 653 and those in Figures 10.2 and 10.15 (pages 565 and 594).

In the above example we dealt only with the *real* eigenvalues. In the next section we show that the eigenvalues for regular Sturm–Liouville boundary value problems are necessarily real, not complex, numbers.

Example 5 Determine all real eigenvalues and eigenfunctions for

$$(24) \qquad (xy')' + \frac{\lambda}{x}y = 0 \,, \qquad 1 < x < e \,,$$

$$(25) \qquad y(1) = 0 \,, \qquad y(e) = 0 \,.$$

Solution Equation (24) can be rewritten as the Cauchy–Euler equation

$$(26) \qquad x^2 y'' + xy' + \lambda y = 0 \,.$$

(Cauchy–Euler equations were studied in Sections 4.7 and 8.5.) Substituting $y = x^r$ gives $r^2 + \lambda = 0$ as the auxiliary equation for (26). We again consider the three cases $\lambda < 0$, $\lambda = 0$, and $\lambda > 0$.

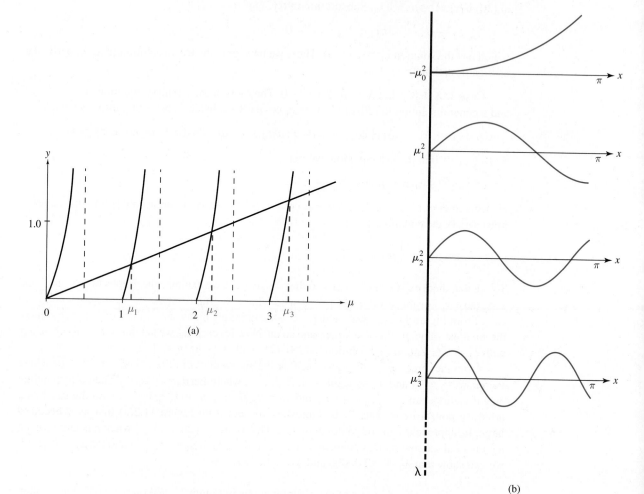

Figure 11.2 (a) Graphs of $y = \mu/3$ and $y = \tan\mu\pi$ (b) Eigenvalues and eigenfunctions

Case 1. $\lambda < 0$. Let $\lambda = -\mu^2$ for $\mu > 0$. The roots of the auxiliary equation are $r = \pm\mu$, so a general solution to (26) is $y(x) = c_1 x^\mu + c_2 x^{-\mu}$. Substituting into the boundary conditions (25) gives

$$c_1 + c_2 = 0\,,$$
$$c_1 e^\mu + c_2 e^{-\mu} = 0\,,$$

whose only solution for $\mu > 0$ is $c_1 = c_2 = 0$. Thus, there are no eigenvalues $\lambda < 0$.

Case 2. $\lambda = 0$. Here zero is a double root of the auxiliary equation, and a general solution to (26) is $y(x) = c_1 + c_2 \ln x$. Substituting into (25), we have

$$c_1 = 0\,, \qquad c_1 + c_2 = 0\,.$$

Again the solution is $c_1 = c_2 = 0$, so $\lambda = 0$ is not an eigenvalue.

Case 3. $\lambda > 0$. Let $\lambda = \mu^2$ for $\mu > 0$. The roots of the auxiliary equation are now $r = \pm \mu i$, and a general solution to (26) is $y(x) = c_1 \cos(\mu \ln x) + c_2 \sin(\mu \ln x)$. Substituting into (25) yields

$$c_1 = 0, \qquad c_1 \cos\mu + c_2 \sin\mu = 0 .$$

With $c_1 = 0$, the last equation becomes $c_2 \sin\mu = 0$. For $\mu = n\pi$, $n = 1, 2, 3, \ldots$, $\sin\mu = \sin n\pi = 0$ and we get the nontrivial solutions $y = c_2 \sin(n\pi \ln x)$ to (24)–(25).

To summarize, the problem (24)–(25) has eigenvalues $\lambda_n = (n\pi)^2, n = 1, 2, 3, \ldots$, with corresponding eigenfunctions $y_n = c_n \sin(n\pi \ln x)$, where c_n is an arbitrary nonzero constant. See Figure 11.3. ◆

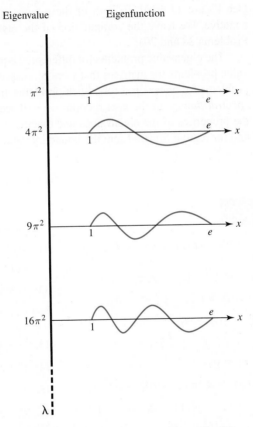

Figure 11.3 Eigenfunctions for equations (24)–(25)

Eigenvalue problems for higher-order linear equations occur in continuum mechanics and, in particular, in vibration and bending problems for beams. One such equation is

$$(27) \qquad \frac{d^4y}{dx^4} + \lambda y = 0 , \qquad 0 < x < L ,$$

where y is the transverse displacement of a uniform elastic beam of length L. If the beam is simply supported at its ends, then one imposes the boundary conditions

$$y(0) = y''(0) = 0 , \qquad y(L) = y''(L) = 0$$

Figure 11.4 Simply supported beam (a) and a cantilever beam (b)

[see Figure 11.4(a)]. If we have a cantilever beam where one end is clamped while the other is free to move, then the boundary conditions are

$$y(0) = y'(0) = 0, \qquad y''(L) = y'''(L) = 0$$

[see Figure 11.4(b)]. In both of these cases, it can be shown that the eigenvalues must be negative. We leave the computation of the eigenvalues and eigenfunctions as exercises (see Problems 28 and 29).

The eigenvalue problems for differential equations that we have studied here and the eigenvalue problems for matrices that were discussed in Chapter 9 are both examples of eigenvalue problems for linear operators. Such problems are studied in the field of mathematics known as *spectral theory*. In the next section, we will see that there is a very close connection between the properties of the eigenvalues and eigenvectors for real symmetric (or Hermitian) matrices and for regular Sturm–Liouville boundary value problems.

11.2 EXERCISES

In Problems 1–12, determine the solutions, if any, to the given boundary value problem.

1. $y'' + 2y' + 26y = 0$; $y(0) = 1$, $y(\pi) = -e^{-\pi}$

2. $y'' + y = x + 1$; $y(0) = 0$, $y'(\pi) = 0$

3. $y'' - 4y' + 13y = 0$; $y(0) = 0$, $y(\pi) = 0$

4. $y'' + 4y' + 8y = 0$;
$y'(0) = 0$, $y'(\pi) = 0$

5. $y'' + y = \sin 2x$; $y(0) = y(2\pi)$, $y'(0) = y'(2\pi)$

6. $y'' - y = x$; $y(0) = 3$, $y'(1) = 2e - e^{-1} - 1$

7. $y'' + y = e^x$;
$y(0) = 0$, $y(\pi) + y'(\pi) = 0$

8. $y'' + 4y = 4$; $y(-\pi) = y(\pi)$, $y'(-\pi) = y'(\pi)$

9. $y'' + 4y = \sin 2x$;
$y(-\pi) = y(\pi)$, $y'(-\pi) = y'(\pi)$

10. $y'' + 2y' + 2y = e^x \cos x$; $y(0) = 0$, $y(\pi) = 0$

11. $x^2 y'' + xy' + y = 4x^{-1}$;
$y(1) = 2$, $y(e^\pi) = 2e^{-\pi}$

12. $y'' + y = x$; $y'(0) = 0$, $y'(\pi) = 0$

In Problems 13–20, find all the real eigenvalues and eigenfunctions for the given eigenvalue problem.

13. $y'' + \lambda y = 0$; $y(0) = 0$, $y'(1) = 0$

14. $y'' + \lambda y = 0$; $y(-3) = y(3)$, $y'(-3) = y'(3)$

15. $y'' + 3y + \lambda y = 0$; $y'(0) = 0$, $y'(\pi) = 0$

16. $y'' - 2y + \lambda y = 0$; $y(0) = 0$, $y'(\pi) = 0$

17. $y'' + \lambda y = 0$; $2y(0) + y'(0) = 0$, $y(\pi) = 0$

18. $y'' + \lambda y = 0$; $y(0) = 0$, $y(\pi) - 2y'(\pi) = 0$

19. $(xy')' + \lambda x^{-1} y = 0$; $y'(1) = 0$, $y(e^\pi) = 0$

20. $(xy')' + \lambda x^{-1} y = 0$; $y'(1) = 0$, $y'(e) = 0$

In Problems 21 and 22, use Newton's method (Appendix B) to determine to three decimal places the first three eigenvalues for the given eigenvalue problem.

21. $y'' + \lambda y = 0$; $y(0) + 2y'(0) = 0$, $y(\pi) = 0$

22. $y'' + \lambda y = 0$; $y(0) = 0$, $y(1) + y'(1) = 0$

In Problems 23–26, find all real values of λ for which the given boundary value problem has a nontrivial solution. Also find the corresponding solutions.

23. $y'' + y' + \lambda(y' + y) = 0$; $y'(0) = 0$, $y(\pi) = 0$

24. $x^2 y'' + \lambda(xy' - y) = 0$; $y(1) = 0$, $y'(e) = 0$

25. $y'' + \lambda y = 0$; $y(0) = 0$, $y(1 + \lambda^2) = 0$

26. $y'' + \lambda y = 0$; $y(0) = 0$, $\lambda y(\pi) - y'(\pi) = 0$

In Problems 27–29, find all the nonpositive real eigenvalues and eigenfunctions of the given fourth-order boundary value problems. [Hint: Let $\lambda = -\mu^4$, $\mu > 0$ or $\lambda = 0$.]

27. $y^{(4)} + \lambda y = 0$;
$y(0) = y'(0) = 0$, $y(L) = y'(L) = 0$

28. $y^{(4)} + \lambda y = 0$;
$y(0) = y'(0) = 0$, $y''(L) = y'''(L) = 0$

29. $y^{(4)} + \lambda y = 0$;
$y(0) = y''(0) = 0$, $y(L) = y''(L) = 0$

30. Prove that the nonhomogeneous boundary value problem (1)–(2) has at most one solution for each continuous function f and numbers c_1 and c_2 if and only if the corresponding homogeneous boundary value problem, with $f(x) \equiv 0$ and $c_1 = c_2 = 0$, has only the trivial solution $y(x) \equiv 0$.

31. Prove that the homogeneous boundary value problem (3)–(4) has either the trivial solution, a one-parameter family of solutions, or a two-parameter family of solutions.

32. Prove that the nonhomogeneous boundary value problem (1)–(2) has either no solution, a unique solution, a one-parameter family of solutions, or a two-parameter family of solutions. [*Hint:* If y_p is a solution to (1)–(2) and y is another solution to (1)–(2), use the superposition principle

to show that $y - y_p$ is a solution to the homogeneous problem (3)–(4). The conclusion follows from the result in Problem 31.]

33. Partial Differential Equation. Assume that the problem

$$u_{tt} = u_{xx} + 2u_x; 0 < x < \pi, t > 0,$$
$$u(0, t) = 0, t > 0,$$
$$u(\pi, t) = 0, t > 0$$

has a solution of the form $u(x, t) = X(x)T(t)$.

(a) Use the method of separation of variables discussed in Section 10.2 to derive the following eigenvalue problem for $X(x)$:

$$X''(x) + 2X'(x) + \lambda X(x) = 0;$$
$$X(0) = X(\pi) = 0.$$

(b) Find the eigenvalues λ and the corresponding eigenfunctions for the problem given in part (a).

34. Assume that the problem

$$u_t = u_{xx} + u; 0 < x < 1, t > 0,$$
$$u_x(0, t) = 0, t > 0,$$
$$u_x(1, t) = 0, t > 0$$

has a solution of the form $u(x, t) = X(x)T(t)$.

(a) Use the method of separation of variables discussed in Section 10.2 to derive the eigenvalue problem

$$X''(x) + (1 + \lambda)X(x) = 0;$$
$$X'(0) = X'(1) = 0.$$

(b) Find the eigenvalues λ and the corresponding eigenfunctions for the problem given in part (a).

35. Helmholtz Equation. In solving the vibrating membrane problem, one encounters the Helmholtz equation

$$u_{xx} + u_{yy} + \lambda u = 0,$$
$$0 < x < a, 0 < y < b,$$

with boundary conditions

$$u(0, y) = u(a, y) = 0, 0 < y < b,$$
$$u(x, 0) = u(x, b) = 0, 0 < x < a.$$

Use the method of separation of variables discussed in Section 10.2 to find the eigenvalues λ and corresponding eigenfunctions.

11.3 Regular Sturm–Liouville Boundary Value Problems

In this section we analyze regular Sturm–Liouville boundary value problems—that is, eigenvalue problems of the form

(1) $(p(x)y'(x))' + q(x)y(x) + \lambda r(x)y(x) = 0, \qquad a < x < b;$

(2) $a_1 y(a) + a_2 y'(a) = 0,$
 $b_1 y(b) + b_2 y'(b) = 0,$

where $p(x), p'(x), q(x),$ and $r(x)$ are real-valued continuous functions on $[a, b]$ and $p(x) > 0$ and $r(x) > 0$ on $[a, b]$. We exclude those cases when $a_1 = a_2 = 0$ or $b_1 = b_2 = 0$.

The following question immediately comes to mind. Can a general second-order homogeneous linear differential equation be written in the form of a Sturm–Liouville equation? That is, can we convert the equation

(3) $A_2(x)y''(x) + A_1(x)y'(x) + A_0(x)y(x) + \lambda p(x)y(x) = 0$

into an equation of the same type as (1)? To do this, we would like to determine p so that $(py')' = py'' + p'y' = A_2 y'' + A_1 y'$. This means $p = A_2$ and $p' = A_1$. But, in general, $A_2' \neq A_1$, so this straightforward approach is not always possible. We can remedy this situation by finding an *integrating factor* $\mu(x)$ such that when (3) is multiplied by μ, we do get $(\mu A_2)' = \mu A_1$. Setting $p = \mu A_2$, the last equation becomes $p' = \mu A_1 = pA_1/A_2$. When we solve this separable equation for p, we find

$$p(x) = Ce^{\int A_1(x)/A_2(x)\,dx},$$

where C is arbitrary. Taking $C = 1$ and recalling that $\mu = p/A_2$, we get

(4) $\mu(x) = [1/A_2(x)]e^{\int A_1(x)/A_2(x)\,dx}.$

Multiplying (3) by this choice for μ gives

$$\mu A_2 y'' + \mu A_1 y' + \mu A_0 y + \lambda \mu p y = 0,$$

which simplifies to

(5) $(py')' + qy + \lambda ry = 0,$

where $p = \mu A_2, q = \mu A_0,$ and $r = \mu p$. Notice that we need the coefficients continuous and $A_2(x)$ not equal to zero. This procedure is illustrated in the following example.

Example 1 Convert the following equation into the form of a Sturm–Liouville equation:

(6) $3x^2 y''(x) + 4xy'(x) + 6y(x) + \lambda y(x) = 0, \qquad x > 0.$

Solution Here $A_2(x) = 3x^2$ and $A_1(x) = 4x$. Using formula (4), we find

$$\mu(x) = \frac{1}{3x^2}e^{\int A_1(x)/A_2(x)\,dx} = \frac{1}{3x^2}e^{\int (4x)/(3x^2)\,dx}$$

$$= \frac{1}{3x^2}e^{(4/3)\int x^{-1}\,dx} = \frac{1}{3x^2}e^{(4/3)\ln x} = \frac{x^{4/3}}{3x^2} = \frac{1}{3x^{2/3}}.$$

Multiplying equation (6) by $\mu(x) = 1/(3x^{2/3})$, we get

$$x^{4/3}y''(x) + \frac{4}{3}x^{1/3}y'(x) + 2x^{-2/3}y(x) + \lambda(3x^{2/3})^{-1}y(x) = 0 \,,$$

(7) $[x^{4/3}y'(x)]' + 2x^{-2/3}y(x) + \lambda(3x^{2/3})^{-1}y(x) = 0 \,.$ ◆

Throughout this section we set

(8) $L[y](x) := \big(p(x)y'(x)\big)' + q(x)y(x) \,,$

so that equation (1) can be written in the compact form

$$L[y](x) + \lambda r(x)y(x) = 0 \,.$$

We'll use this notation to express the various identities that arise in our study of regular Sturm–Liouville boundary value problems.

Lagrange's Identity for $L[y] = (py')' + qy$

Theorem 1. If u and v are functions having continuous second derivatives on the interval $[a, b]$, then

(9) $uL[v] - vL[u] = \dfrac{d}{dx}\big(pW[u, v]\big) \,,$

where $W[u, v] = uv' - vu'$ is the Wronskian of u and v.

Proof. Using the product rule and adding and subtracting $pu'v'$, we have

$$\begin{aligned}
uL[v] - vL[u] &= u[(pv')' + qv] - v[(pu')' + qu] \\
&= u(pv')' + quv - v(pu')' - quv \\
&= u(pv')' + u'(pv') - v'(pu') - v(pu')' \\
&= [u(pv')]' - [v(pu')]' \\
&= [p(uv' - vu')]' \\
&= \frac{d}{dx}\big(pW[u, v]\big) \,. \quad ◆
\end{aligned}$$

By integrating Lagrange's identity, we obtain the following equation called **Green's formula.**

Green's Formula for $L[y] = (py')' + qy$

Corollary 1. If u and v are functions having continuous second derivatives on the interval $[a, b]$, then

(10) $\displaystyle\int_a^b \big(uL[v] - vL[u]\big)(x)\, dx = \big(pW[u, v]\big)(x)\Big|_a^b \,.$

If u and v also satisfy the boundary conditions in (2), then Green's formula simplifies to

(11) $\qquad \displaystyle\int_a^b \left(uL[v] - vL[u] \right)(x)\, dx = 0\,.$

To see this, observe that when $a_2 \neq 0$ and $b_2 \neq 0$ in (2), then

$$u'(a) = -(a_1/a_2)u(a)\,, \qquad u'(b) = -(b_1/b_2)u(b)\,,$$
$$v'(a) = -(a_1/a_2)v(a)\,, \qquad v'(b) = -(b_1/b_2)v(b)\,.$$

Substituting these values into the right-hand side of (10) gives

$$p(b)W[u,v](b) - p(a)W[u,v](a)$$
$$= p(b)\left[u(b)v'(b) - u'(b)v(b) \right] - p(a)\left[u(a)v'(a) - u'(a)v(a) \right]$$
$$= p(b)\left[u(b)(-b_1/b_2)v(b) - (-b_1/b_2)u(b)v(b) \right]$$
$$\quad - p(a)\left[u(a)(-a_1/a_2)v(a) - (-a_1/a_2)u(a)v(a) \right]$$
$$= 0\,.$$

When $a_2 = 0$ (so that $a_1 \neq 0$), it follows from (2) that both $u(a)$ and $v(a)$ are zero. Hence, $[u(a)v'(a) - u'(a)v(a)] = 0$. Similarly when $b_2 = 0$, we get $[u(b)v'(b) - u'(b)v(b)] = 0$. Therefore, if u and v satisfy any instance of the boundary conditions in (2), then (11) holds.

Equation (11) takes a particularly useful and elegant form when expressed in terms of an inner product. If we set

(12) $\qquad (f, g) = \displaystyle\int_a^b f(x)g(x)\, dx\,,$

then $(\,\cdot\,,\cdot\,)$ is an **inner product** on the set of continuous real-valued functions on $[a, b]$.[†] Using this notation, we can rewrite (11) as

(13) $\qquad (u, L[v]) = (L[u], v)\,.$

A linear differential operator L that satisfies (13) for all functions u and v in its domain is called a **selfadjoint operator.** We have just shown that if L is defined by (8) and the domain of L is the set of functions that have continuous second derivatives on $[a, b]$ and satisfy the boundary conditions in (2), then L is a selfadjoint operator. Selfadjoint operators are like symmetric matrices in that their eigenvalues are real-valued. We now prove this for regular Sturm–Liouville boundary value problems.

Eigenvalues and Eigenfunctions are Real-Valued

Theorem 2. The eigenvalues for the regular Sturm–Liouville boundary value problem (1)–(2) are real and have real-valued eigenfunctions.

[†]For further discussion on inner products, see *Fundamentals of Matrix Analysis with Applications*, by Edward Barry Saff and Arthur David Snider (John Wiley & Sons, Hoboken, New Jersey, 2016).

Proof. Assume λ, possibly a complex number, is an eigenvalue for (1)–(2) with eigenfunction $\phi(x)$, which may be complex-valued. That is,

(14) $L[\phi](x) + \lambda r(x)\phi(x) = 0$,

and $\phi(\not\equiv 0)$ satisfies the boundary conditions in (2). Taking the complex conjugate of both sides of (14) and using the fact that p, q, and r are real-valued, we obtain

(15) $\overline{L[\phi](x) + \lambda r(x)\phi(x)} = L[\overline{\phi}](x) + \overline{\lambda}r(x)\overline{\phi}(x) = 0$.

Further, since a_1, a_2, b_1, and b_2 are real, a similar argument shows that $\overline{\phi}$ also satisfies the boundary conditions in (2). Hence, $\overline{\lambda}$ is an eigenvalue with eigenfunction $\overline{\phi}$.

From equation (14), we get

(16) $\displaystyle\int_a^b L[\phi](x)\overline{\phi}(x)\,dx = -\lambda\int_a^b r(x)\phi(x)\overline{\phi}(x)\,dx = -\lambda\int_a^b r(x)|\phi(x)|^2\,dx$.

In addition, using identity (11) and the fact that $\overline{\lambda}$ is an eigenvalue with eigenfunction $\overline{\phi}$, we find

(17) $\displaystyle\int_a^b L[\phi](x)\,\overline{\phi}(x)\,dx = \int_a^b \phi(x)L[\overline{\phi}](x)\,dx = -\overline{\lambda}\int_a^b r(x)|\phi(x)|^2\,dx$.

But the left-hand sides of (16) and (17) are the same, so equating the right-hand sides yields

$$-\lambda\int_a^b r(x)|\phi(x)|^2\,dx = -\overline{\lambda}\int_a^b r(x)|\phi(x)|^2\,dx.$$

Since $r(x) > 0$ and ϕ is a nontrivial solution to (1)–(2), then $\int_a^b r(x)|\phi(x)|^2\,dx > 0$. Thus, we can divide out this integral to obtain $\lambda = \overline{\lambda}$, which means that λ is a real number.

If ϕ itself is not real-valued, we can obtain real-valued eigenfunctions corresponding to λ simply by taking the real or imaginary parts of ϕ (see Problem 16). This completes the proof of Theorem 2. ◆

If all the eigenfunctions associated with a particular eigenvalue are just scalar multiples of each other, then the eigenvalue is called **simple.**

Eigenvalues Are Simple

Theorem 3. All the eigenvalues of the regular Sturm–Liouville boundary value problem (1)–(2) are simple.

Proof. Assume $\phi(x)$ and $\psi(x)$ are two eigenfunctions corresponding to the eigenvalue λ (thus, ϕ and ψ are solutions to the same differential equation). They satisfy (2), so we have

$$a_1\phi(a) + a_2\phi'(a) = 0, \qquad a_1\psi(a) + a_2\psi'(a) = 0.$$

Let's assume that $a_2 \neq 0$. (A similar argument applies if $a_2 = 0$ and $a_1 \neq 0$.) Solving for $\phi'(a)$ and $\psi'(a)$ gives

$$\phi'(a) = (-a_1/a_2)\phi(a), \qquad \psi'(a) = (-a_1/a_2)\psi(a).$$

If we now compute the Wronskian of ϕ and ψ at $x = a$, we find

(18) $W[\phi, \psi](a) = \phi(a)\psi'(a) - \phi'(a)\psi(a)$

$$= \phi(a)(-a_1/a_2)\psi(a) - (-a_1/a_2)\phi(a)\psi(a) = 0.$$

Recall that in Section 6.1, we proved that if the Wronskian of two solutions to a second-order linear homogeneous equation is zero at a point, then the two solutions are linearly dependent. Consequently, ϕ and ψ are linearly dependent, so one must be a multiple of the other. Thus, λ is a simple eigenvalue. ◆

We now show that the eigenfunctions associated with different eigenvalues enjoy the property of orthogonality. Recall that in the discussion of orthogonal expansions in Section 10.3, we said that two real-valued functions f and g are **orthogonal** with respect to a positive weight function $w(x)$ on the interval $[a, b]$ if

(19) $$\int_a^b f(x)g(x)w(x)dx = 0.$$

Orthogonality of Eigenfunctions

Theorem 4. Eigenfunctions that correspond to distinct eigenvalues of the regular Sturm–Liouville boundary value problem (1)–(2) are orthogonal with respect to the weight function $r(x)$ on $[a, b]$.

Proof. Let λ and μ be distinct eigenvalues with corresponding eigenfunctions ϕ and ψ, respectively; that is, $L[\phi] = -\lambda r\phi$ and $L[\psi] = -\mu r\psi$. We can assume, by Theorem 2, that ϕ and ψ are real-valued functions. Setting $u = \phi$ and $v = \psi$ in (11) and substituting for $L[\phi]$ and $L[\psi]$, we find

$$0 = \int_a^b (\phi L[\psi] - \psi L[\phi])(x)dx$$

$$= \int_a^b (-\phi\mu r\psi + \psi\lambda r\phi)(x)dx$$

$$= (\lambda - \mu)\int_a^b \phi(x)\psi(x)r(x)dx.$$

Since $\lambda \neq \mu$, we see that

$$\int_a^b \phi(x)\psi(x)r(x)dx = 0.$$

Therefore the eigenfunctions ϕ and ψ are orthogonal with respect to $r(x)$ on $[a, b]$. ◆

To illustrate the properties of regular Sturm–Liouville boundary value problems, let's revisit the equations in Example 5 of Section 11.2 (page 647):

(20) $(xy')' + \dfrac{\lambda}{x}y = 0$; $y(1) = y(e) = 0$.

In the example, we showed that the eigenvalues of (20) are $\lambda_n = n^2\pi^2$, $n = 1, 2, 3, \ldots$, with corresponding eigenfunctions $c_n \sin(n\pi \ln x)$. Notice that the eigenvalues are real and simple, as predicted by Theorems 2 and 3. Comparing (20) with (1), we see that the weight function $r(x)$ is $1/x$ for this problem, so the orthogonality relation (19) takes the form

$$(21) \qquad \int_1^e \sin(n\pi \ln x) \sin(m\pi \ln x) \frac{1}{x} \, dx = 0, \qquad \text{for } m \neq n.$$

Indeed, equation (21) can be confirmed directly, since the substitution $\xi = \pi \ln x$ renders it identical with the orthogonality relation for the sine series, $\int_0^\pi \sin n\xi \sin m\xi \, d\xi = 0 \, (m \neq n)$, that we have been exploiting since Chapter 10 (cf. Section 10.7).

Notice further that the eigenvalues for problem (20) form an increasing sequence

$$\lambda_1 = 1^2 < \lambda_2 = 2^2 < \lambda_3 = 3^2 < \cdots,$$

with $\lim_{n \to \infty} \lambda_n = \lim_{n \to \infty} n^2 = +\infty$. A similar ordering is true for any regular Sturm–Liouville boundary value problem, as one would expect from the examples we have encountered in this and the preceding chapter.

Sequence of Eigenvalues

Theorem 5. The eigenvalues for the regular Sturm–Liouville boundary value problem (1)–(2) form a countable, increasing sequence

$$(22) \qquad \lambda_1 < \lambda_2 < \lambda_3 < \cdots,$$

with $\lim_{n \to \infty} \lambda_n = +\infty$.

This theorem tells us that a regular Sturm–Liouville boundary value problem has infinitely many eigenvalues and that they can be arranged in a sequence that increases to infinity. The proof of this fact is beyond the scope of this book, so we refer you to more advanced texts.[†]

Eigenfunction Expansions

Having in hand a sequence of orthogonal functions, we can use these functions to construct series expansions for a general class of functions. For example, the eigenfunctions $\phi_n(x) = \sin nx$ for problem (20) lead to the Fourier sine series expansions discussed in Section 10.4:

$$f(x) \sim \sum_{n=1}^{\infty} b_n \sin nx.$$

In fact, if f is continuous on $[0, \pi]$, f' is piecewise continuous on $[0, \pi]$, and $f(0) = f(\pi) = 0$, then

$$f(x) = \sum_{n=1}^{\infty} b_n \sin nx, \qquad 0 \leq x \leq \pi,$$

[†]See, for example, the texts by G. Birkhoff and G. Rota, *Ordinary Differential Equations*, 4th ed. (Wiley, New York, 1989), Chapter 10, or E. Coddington and N. Levinson, *Theory of Ordinary Differential Equations* (Krieger, New York, 1984), Chapter 8.

where

$$b_n = \frac{2}{\pi} \int_0^\pi f(x) \sin nx\, dx, \qquad n = 1, 2, 3, \ldots.$$

General orthogonal expansions were discussed in Section 10.3. There we showed that if the set $\{\phi_n(x)\}_{n=1}^\infty$ is orthonormal with respect to a positive weight function $w(x)$ on $[a, b]$, that is,

$$(23) \qquad \int_a^b \phi_n(x)\phi_m(x)w(x)\, dx = \begin{cases} 0, & m \neq n, \\ 1, & m = n, \end{cases}$$

then with any piecewise continuous function f, we can identify an orthogonal expansion

$$(24) \qquad f(x) \sim \sum_{n=1}^\infty c_n \phi_n(x),$$

where

$$(25) \qquad c_n = \int_a^b f(x)\phi_n(x)w(x)\, dx.$$

The eigenfunctions corresponding to distinct eigenvalues are orthogonal with respect to $r(x)$ for a regular Sturm–Liouville boundary value problem. Hence, we can use the eigenfunctions to form an *orthonormal* system with respect to $r(x)$ simply by normalizing (scaling) each eigenfunction ϕ_n so that

$$(26) \qquad \int_a^b \phi_n^2(x)r(x)\, dx = 1.$$

Example 2 Construct an orthonormal system of eigenfunctions corresponding to the regular Sturm–Liouville boundary value problem

$$(27) \qquad y'' + \lambda y = 0; \qquad y(0) = 0, \qquad y'(\pi) = 0.$$

Solution In Exercises 10.2, Problem 9, we found that the eigenvalues of (27) are $\lambda_n = (2n - 1)^2/4$, $n = 1, 2, 3, \ldots$, with corresponding eigenfunctions

$$\phi_n(x) = a_n \sin\left(\frac{2n - 1}{2}\right)x.$$

The eigenfunctions are orthogonal with respect to $r(x) \equiv 1$ on $[0, \pi]$, so we need only choose the a_n's so that (26) holds. We compute

$$\int_0^\pi a_n^2 \sin^2\left(\frac{2n - 1}{2}x\right) dx = a_n^2 \int_0^\pi \left[\frac{1}{2} - \frac{1}{2}\cos(2n - 1)x\right] dx$$

$$= a_n^2 \left[\frac{x}{2} - \frac{1}{2(2n - 1)}\sin(2n - 1)x\right]\Bigg|_0^\pi$$

$$= a_n^2(\pi/2).$$

Hence, we can take $a_n = \sqrt{2/\pi}$, which gives

$$(28) \qquad \left\{\sqrt{2/\pi}\,\sin\left(\frac{2n - 1}{2}x\right)\right\}_{n=1}^\infty$$

as an orthonormal system of eigenfunctions. Figure 11.5 on page 659 displays a pattern of eigenfunction features that should be familiar by now. ◆

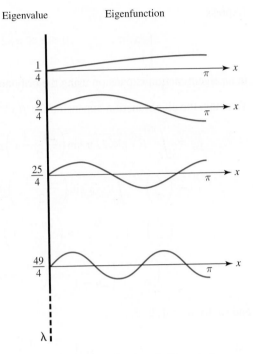

Figure 11.5 Eigenfunctions for Example 2

If $\{\phi_n\}_{n=1}^{\infty}$ is an orthonormal system of eigenfunctions, then we can use (24) and (25) to obtain an eigenfunction expansion for a function f. But, does the eigenfunction expansion converge? If so, to what function does it converge? These questions are partially answered in the next theorem.

Uniform Convergence of Eigenfunction Expansions

Theorem 6. Let $\{\phi_n\}_{n=1}^{\infty}$ be an orthonormal system of eigenfunctions for the regular Sturm–Liouville boundary value problem (1)–(2). Let f be a continuous function on $[a, b]$ with f' piecewise continuous[†] on $[a, b]$. If f satisfies the boundary conditions in (2), then

$$(29) \qquad f(x) = \sum_{n=1}^{\infty} c_n \phi_n(x) \,, \qquad a \le x \le b \,,$$

where c_n is given by formula (25) with $w(x) = r(x)$. Moreover, the eigenfunction expansion in (29) converges uniformly on $[a, b]$.

The proof of Theorem 6 can be found in more advanced texts.[‡]

[†]A function $g(x)$ is piecewise continuous on $[a, b]$ if it is continuous except for a finite number of points at which g has jump discontinuities (see Section 7.2).

[‡]See, for example, the text by E. Coddington and N. Levinson mentioned in the footnote on page 657.

Example 3 Express

$$f(x) = \begin{cases} 2x/\pi, & 0 \le x \le \pi/2, \\ 1, & \pi/2 \le x \le \pi, \end{cases}$$

in an eigenfunction expansion using the orthonormal system of eigenfunctions given in (28).

Solution To obtain the eigenfunction expansion for $f(x)$, we use formula (25). Thus,

$$c_n = \int_0^\pi f(x)\sqrt{2/\pi}\,\sin\left(\frac{2n-1}{2}x\right)dx$$

$$= \left(\frac{2}{\pi}\right)^{3/2}\int_0^{\pi/2} x\sin\left(\frac{2n-1}{2}x\right)dx + \left(\frac{2}{\pi}\right)^{1/2}\int_{\pi/2}^\pi \sin\left(\frac{2n-1}{2}x\right)dx$$

$$= \left(\frac{2}{\pi}\right)^{3/2}\left[\frac{-2x}{2n-1}\cos\left(\frac{2n-1}{2}x\right) + \frac{4}{(2n-1)^2}\sin\left(\frac{2n-1}{2}x\right)\right]\Bigg|_0^{\pi/2}$$

$$- \left(\frac{2}{\pi}\right)^{1/2}\left[\frac{2}{2n-1}\cos\left(\frac{2n-1}{2}x\right)\right]\Bigg|_{\pi/2}^\pi,$$

and so, for $n = 1, 2, 3, \ldots,$

$$(30) \qquad c_n = \frac{2^{7/2}}{\pi^{3/2}(2n-1)^2}\sin\left(\frac{n\pi}{2} - \frac{\pi}{4}\right).$$

Hence,

$$(31)\quad f(x) = \sum_{n=1}^\infty c_n\sqrt{2/\pi}\,\sin\left(\frac{2n-1}{2}x\right)$$

$$= \frac{2^{7/2}}{\pi^2}\sin(x/2) + \frac{2^{7/2}}{9\pi^2}\sin(3x/2) - \frac{2^{7/2}}{25\pi^2}\sin(5x/2) - \frac{2^{7/2}}{49\pi^2}\sin(7x/2) + \cdots,$$

where the c_n's are as given in equation (30). Since $f(0) = 0$ and $f'(\pi) = 0$, then f satisfies the boundary conditions in (27). Moreover, f is continuous and f' piecewise continuous on $[0, \pi]$. Therefore, by Theorem 6, the series in (31) converges uniformly to f on $[0, \pi]$. ◆

There are further results concerning the pointwise convergence of the eigenfunction expansions that are similar to the results for Fourier series (compare Theorem 2, Section 10.3). We state without proof one such result.

Pointwise Convergence of Eigenfunction Expansions

Theorem 7. Let $\{\phi_n\}_{n=1}^\infty$ be an orthonormal sequence of eigenfunctions for the regular Sturm–Liouville boundary value problem (1)–(2). Let f and f' be piecewise continuous on $[a, b]$. Then, for any x in (a, b),

$$(32) \qquad \sum_{n=1}^\infty c_n\phi_n(x) = \frac{1}{2}[f(x^+) + f(x^-)],$$

where the c_n's are given by formula (25) with $w(x) = r(x)$.

Periodic Sturm–Liouville Problems

Periodic Sturm–Liouville problems have properties similar to those of regular Sturm–Liouville boundary value problems. For example, the periodic Sturm–Liouville problem

$$(33) \qquad y'' + \lambda y = 0 ; \qquad y(0) = y(2\pi) , \qquad y'(0) = y'(2\pi) ,$$

has eigenvalues $\lambda_n = n^2$, $n = 0, 1, 2, \dots$, and eigenfunctions

$$\phi_0 = a_0/2 , \qquad \phi_{n,1}(x) = a_n \cos nx , \qquad \phi_{n,2}(x) = b_n \sin nx$$

for $n = 1, 2, 3, \dots$.

The set of eigenfunctions

$$\{ 1/2, \cos x, \sin x, \cos 2x, \sin 2x, \dots \}$$

form an orthogonal set. The usual Fourier series expansions discussed in Section 10.3 are just eigenfunction expansions using this set; that is,

$$f(x) \sim a_0/2 + \sum_{n=1}^{\infty} \left[a_n \cos nx + b_n \sin nx \right] .$$

The main distinction between the properties of regular and periodic Sturm–Liouville problems is the fact that the eigenvalues for the periodic Sturm–Liouville problem do not have to be simple. For example, both cos nx and sin nx are eigenfunctions corresponding to the eigenvalue $\lambda_n = n^2$ for the periodic problem (33). In the following theorem we give, without proof, the properties of the **periodic Sturm–Liouville problem:**

$$(34) \qquad (py')' + qy + \lambda ry = 0 ; \qquad y(0) = y(2T) , \qquad y'(0) = y'(2T) ,$$

where p, p', q, and r are continuous $2T$-periodic functions with p and r positive.

Properties of Periodic Sturm–Liouville Problems

Theorem 8. For the periodic Sturm–Liouville problem (34):

1. The eigenvalues are real and form a countable, increasing sequence
 $\lambda_1 < \lambda_2 < \lambda_3 \dots$ with $\lim_{n \to \infty} \lambda_n = +\infty$.
2. The eigenfunctions can be chosen to be real-valued.
3. The eigenfunctions corresponding to distinct eigenvalues are orthogonal with respect to $r(x)$ on $[0, 2T]$.
4. Let $\{ \phi_n \}_{n=1}^{\infty}$ be an orthonormal sequence of all the eigenfunctions and let f be a $2T$-periodic function such that f is continuous and f' is piecewise continuous. Then,

$$(35) \qquad f(x) = \sum_{n=1}^{\infty} a_n \phi_n(x) ,$$

where

$$(36) \qquad a_n = \int_{0}^{2T} f(x) \phi_n(x) r(x) \, dx .$$

Moreover, the series in (35) converges uniformly on $[0, 2T]$ [and hence on $(-\infty, \infty)$].

11.3 EXERCISES

In Problems 1–6, convert the given equation into the form of a Sturm–Liouville equation.

1. $y'' + 6y' + \lambda y = 0$

2. $xy'' + \lambda y = 0$, $\quad x > 0$

3. $x(1-x)y'' - 2xy' + \lambda y = 0$, $\quad 0 < x < 1$

4. $x^2 y'' + xy' + (x^2 - \lambda)y = 0$, $\quad x > 0$

5. $x^2 y'' + xy' + \lambda y = 0$, $\quad x > 0$

6. $(1 - x^2)y'' - 2xy' + \lambda y = 0$, $\quad -1 < x < 1$

In Problems 7–11, determine whether the given linear differential operator $L[y]$, whose domain consists of all functions that have continuous second derivatives on the interval $[0, \pi]$ and satisfy the given boundary conditions, is selfadjoint.

7. $L[y] := y'' + 2y' + 5y$;
$\quad y(0) = 0$, $\quad y(\pi) = 0$

8. $L[y] := y'' + \lambda y$;
$\quad y(0) = y(\pi)$, $\quad y'(0) = y'(\pi)$

9. $L[y] := y'' + \lambda y$;
$\quad y(0) = -y(\pi)$, $\quad y'(0) = -y'(\pi)$

10. $L[y] := y'' + \lambda y$;
$\quad y(0) + y'(\pi) = 0$, $\quad y(\pi) + y'(0) = 0$

11. $L[y] := (1 + x^2)y'' + 2xy' + \lambda ry$;
$\quad y'(0) = 0$, $\quad y(\pi) - (1 + \pi^2)y'(\pi) = 0$

12. Inner Product. The notion of inner product $(\cdot, \cdot)$ is important in the study of general vector spaces and generalizes the dot product of vectors in $\mathbf{R}^n$. Recall we encountered inner products in our study of Fourier series and orthogonal expansions in Section 10.3. Verify that (f, g) defined in (12) is an inner product for f and g continuous on $[a, b]$; that is, show for f, g, and h continuous on $[a, b]$ and c a real number that (f, g) satisfies the following:

(a) $(f, f) \geq 0$ and equals zero only when $f \equiv 0$.

(b) $(f, g) = (g, f)$.

(c) $(f + h, g) = (f, g) + (h, g)$.

(d) $(cf, g) = c(f, g)$.

13. Let $L[y] := y^{(4)}$. Let the domain of L be the set of functions that have four continuous derivatives on $[0, \pi]$ and satisfy

$$y(0) = y'(0) = 0 \text{ and } y(\pi) = y'(\pi) = 0.$$

(a) Show that L is selfadjoint.

(b) Use the result of part (a) to show that eigenfunctions corresponding to distinct eigenvalues of the problem

$$(37) \qquad y^{(4)} + \lambda y = 0; \quad y(0) = y'(0) = 0,$$
$$y(\pi) = y'(\pi) = 0$$

are orthogonal with respect to $w(x) \equiv 1$ on $[0, \pi]$.

14. Let λ be an eigenvalue and ϕ a corresponding eigenfunction for the regular Sturm–Liouville boundary value problem (1)–(2).

(a) Show that

$$(38) \quad \lambda \int_a^b r(x)\phi^2(x)\,dx$$
$$= \int_a^b \{p(x)[\phi'(x)]^2 - q(x)\phi^2(x)\}\,dx$$
$$- [p(x)\phi(x)\phi'(x)]\Big|_a^b.$$

(b) Use equation (38) to show that if $q(x) \leq 0$ and $a_1 = b_1 = 0$ in (2), then λ is nonnegative.

(c) Show that if $q(x) \leq 0$ and $a_2 = b_2 = 0$ in (2), then λ is nonnegative.

(d) Show that if $q(x) \leq 0$, $a_1/a_2 < 0$, and $b_1/b_2 > 0$, then λ is nonnegative.

15. Show that the eigenvalues of the eigenvalue problem (37) in Problem 13 are nonpositive. [*Hint:* Derive an equation similar to equation (38) in Problem 14.]

16. Show that if $\phi = u + iv$ is an eigenfunction corresponding to λ for the regular Sturm–Liouville problem (1)–(2), then the real part u and the imaginary part v each satisfy (1)–(2).

In Problems 17–24, (a) determine the normalized eigenfunctions for the given problem and (b) use the results of part (a) to express $f(x) = x$ in an eigenfunction expansion. (These are the same equations as in Problems 13 through 20 of Exercises 11.2.)

17. $y'' + \lambda y = 0$; $y(0) = 0$, $y'(1) = 0$

18. $y'' + \lambda y = 0$;
$y(-3) = y(3)$, $y'(-3) = y'(3)$

19. $y'' + 3y + \lambda y = 0$;
$y'(0) = 0$, $y'(\pi) = 0$

20. $y'' - 2y + \lambda y = 0$;
$y(0) = 0$, $y'(\pi) = 0$

21. $y'' + \lambda y = 0$;
$2y(0) + y'(0) = 0$, $y(\pi) = 0$

22. $y'' + \lambda y = 0$;
$y(0) = 0$, $y(\pi) - 2y'(\pi) = 0$

23. $(xy')' + \lambda x^{-1}y = 0$;
$y'(1) = 0$, $y(e^{\pi}) = 0$

24. $(xy')' + \lambda x^{-1}y = 0$; $y'(1) = 0$, $y'(e) = 0$

25. Bessel's Inequality. Let $\{\phi_n\}_{n=1}^{\infty}$ be an orthonormal system with respect to a positive weight function $w(x)$ on $[a, b]$ and let f be continuous on $[a, b]$. Show that if

$$f(x) \sim \sum_{n=1}^{\infty} c_n\phi_n(x),$$

then

(39) $$\sum_{n=1}^{\infty} c_n^2 \leq \int_a^b f^2(x)w(x)\,dx.$$

[*Hint:* Model your proof after that of Bessel's inequality for Fourier series (compare Problem 38 in Exercises 10.3, page 579).]

26. Prove that the linear differential operator associated with the periodic Sturm–Liouville problem (34) is selfadjoint.

11.4 Nonhomogeneous Boundary Value Problems and the Fredholm Alternative

Unlike homogeneous boundary value problems (which always have $y \equiv 0$ as a solution), nonhomogeneous problems need not have any solution. In this section our goal is to develop a test to determine whether the nonhomogeneous problem

(1) $L[y](x) := A_2(x)y''(x) + A_1(x)y'(x) + A_0(x)y(x) = h(x)$, $a < x < b$;

(2) $B_1[y] := a_{11}y(a) + a_{12}y'(a) + b_{11}y(b) + b_{12}y'(b) = 0$,
$B_2[y] := a_{21}y(a) + a_{22}y'(a) + b_{21}y(b) + b_{22}y'(b) = 0$

has a solution. Before we proceed, we need to introduce the concept of an adjoint.

Formal Adjoint

Definition 1. Let L be the differential operator given by the left-hand side of equation (1), i.e., $L[y] = A_2y'' + A_1y' + A_0y$. The **formal adjoint** (or Lagrange adjoint) of L is the differential operator L^+ defined by

(3) $L^+[y] := (A_2y)'' - (A_1y)' + A_0y$,

provided A_2'', A_1', and A_0 are continuous on $[a, b]$.

Although at this stage the peculiar definition of L^+ is unmotivated, we will soon uncover its relationship to the operator L. You should take a minute to verify that if $A_1 = A_2'$ (i.e., if equation (1) is in Sturm–Liouville form), then $L^+ = L$. In the following examples, we compute the formal adjoint for an operator with constant coefficients and for a Cauchy–Euler type operator.

Example 1 Determine the formal adjoint of

(4) $L[y] = y'' + 2y' + 2y$.

Solution Here $A_0 = A_1 = 2$ and $A_2 = 1$. Substituting into formula (3) gives

(5) $L^+[y] = y'' - 2y' + 2y$. ◆

Example 2 Determine the formal adjoint of

(6) $L[y](x) = x^2 y''(x) + xy'(x) + 4y(x)$.

Solution Here $A_2(x) = x^2$, $A_1(x) = x$, and $A_0(x) = 4$. Substituting into formula (3) and simplifying gives

(7) $L^+[y](x) = [x^2 y(x)]'' - [xy(x)]' + 4y(x)$
$= x^2 y''(x) + 3xy'(x) + 5y(x)$. ◆

The differential operator L and its formal adjoint are related by a Lagrange identity. This is a generalization of the Lagrange identity we proved for Sturm–Liouville operators (see page 653).

Lagrange's Identity

Theorem 9. Let L be the differential operator defined in (1) and let L^+ be its formal adjoint. Then,

(8) $L[u]v - uL^+[v] = \dfrac{d}{dx}[P(u, v)]$,

where $P(u, v)$, the **bilinear concomitant** associated with L, is defined as

(9) $P(u, v) := uA_1 v - u(A_2 v)' + u' A_2 v$.

Proof. Expanding the left-hand side of (8), we get

(10) $L[u]v - uL^+[v] = (A_2 u'' + A_1 u' + A_0 u)v - u[(A_2 v)'' - (A_1 v)' + A_0 v]$
$= A_2 u''v + A_1 u'v - A_2 uv'' - 2A_2' uv' - A_2'' uv + A_1' uv + A_1 uv'$.

From the definition of $P(u, v)$ in (9), we compute

(11) $\dfrac{d}{dx}[P(u, v)] = (u' A_1 v + u A_1' v + u A_1 v') - (u A_2 v' + u A_2' v)'$
$+ (u'' A_2 v + u' A_2' v + u' A_2 v')$
$= A_2 u''v + A_1 u'v - A_2 uv'' - 2A_2' uv' - A_2'' uv + A_1' uv + A_1 uv'$,

which is the same as the right-hand side of (10). ◆

Integrating Lagrange's identity leads to Green's formula.

> ### Green's[†] Formula
>
> **Corollary 2.** If L is the differential operator defined in (1) and L^+ is its formal adjoint, then
>
> $$(12) \qquad \int_a^b (L[u]v - uL^+[v])(x)\,dx = P(u,v)(x)\Big|_a^b .$$

In Section 11.3 we defined an inner product for continuous real-valued functions on $[a, b]$ by

$$(13) \qquad (f, g) = \int_a^b f(x)g(x)\,dx .$$

With this notation, Green's formula becomes

$$(14) \qquad (L[u], v) = (u, L^+[v]) + P(u, v)\Big|_a^b .$$

The elegance of formula (14) serves as a justification for the peculiar definitions of L^+ and $P(u, v)$. Indeed, it shows us how to transfer the operator L from the function u to the other function v, when computing inner products.

While the operators L and L^+ are meaningful when applied to any function y having a continuous second derivative on $[a, b]$ (i.e., for y in $C^2[a, b]$), in applications we need to further restrict the collections of functions y on which L and L^+ act. For example, in solving boundary value problems, we are interested in applying L only to those functions in $C^2[a, b]$ that satisfy the given boundary conditions. Therefore, when we consider the operators L and L^+, we will hereafter assume their domains $D(L)$ and $D(L^+)$, respectively, to be certain subspaces of $C^2[a, b]$.

Given the linear differential operator L with domain $D(L)$, we shall see that it is advantageous to restrict the formal adjoint L^+ to a domain $D(L^+)$ so that

$$(15) \qquad (L[u], v) = (u, L^+[v])$$

holds for all u in $D(L)$ and all v in $D(L^+)$. Of course, simply taking $D(L^+)$ to consist of the single function $v \equiv 0$ will do the job, but this is of no interest. What we would really like is to choose $D(L^+)$ as the largest possible set of functions in $C^2[a, b]$ for which (15) holds. With this optimal choice for $D(L^+)$, we say that L^+ is the **adjoint operator**[‡] of L (i.e., we drop the adjective "formal"). For those special cases when $L^+ = L$ and (15) holds with $D(L^+) = D(L)$, we say that L is **selfadjoint.** For example, in the preceding section, we showed that for a regular Sturm–Liouville boundary value problem, the associated operator L is selfadjoint (the domain of L consisting of those functions in $C^2[a, b]$ that satisfy the boundary conditions).

[†]*Historical Footnote:* George Green (1793–1841), an English mathematician, made the first attempt at a mathematical theory of electromagnetism (1828).

[‡]The adjoint of an operator is usually studied in connection with a Hilbert space. For a further discussion of Hilbert spaces and adjoint operators, see a text on functional analysis such as *Functional Analysis,* 2nd ed., by W. Rudin (McGraw-Hill, New York, 2007).

Thanks to Green's formula, the adjoint will simply be L^+ restricted to a suitable class of functions v so that

$$P(u, v)\Big|_a^b = 0, \quad \text{for } u \text{ in } D(L), \quad v \text{ in } D(L^+).$$

Example 3 Find the adjoint operator for

(16) $L[y] = y'' + 2y' + 2y$,

where the domain of L consists of those functions that have continuous second derivatives on $[0, \pi]$ and satisfy the boundary conditions

(17) $y(0) = y(\pi) = 0$.

Solution In Example 1 we found that $L^+[v] = v'' - 2v' + 2v$. We must now determine the domain of L^+. We first observe that v'' must be continuous on $[0, \pi]$. Further, in order for equation (15) to hold, we must have

(18) $P(u, v)(x)\Big|_0^\pi = 0$

for all u in $D(L)$ and v in $D(L^+)$. From the formula for $P(u, v)$ in (9) with $A_1 = 2$ and $A_2 = 1$, condition (18) becomes

$$\{2u(\pi)v(\pi) - u(\pi)v'(\pi) + u'(\pi)v(\pi)\}$$
$$- \{2u(0)v(0) - u(0)v'(0) + u'(0)v(0)\} = 0.$$

Since u is in $D(L)$, we know that $u(0) = u(\pi) = 0$. Thus, the above equation simplifies to

(19) $u'(\pi)v(\pi) - u'(0)v(0) = 0$.

Now observe that $u'(\pi)$ and $u'(0)$ can take on any value, so we must have $v(0) = v(\pi) = 0$ in order for equation (19) to hold for all u in $D(L)$. Hence, the domain of L^+ consists of those functions v that have continuous second derivatives on $[0, \pi]$ and that satisfy the boundary conditions

(20) $v(0) = v(\pi) = 0$. ◆

Adjoint Boundary Value Problem

Definition 2. Let $B[u] = 0$ represent the boundary conditions required for a function to be in the domain of L and let $B^+[u] = 0$ be the boundary conditions required for a function to be in the domain of its adjoint operator L^+. Then, the **adjoint boundary value problem** for

(21) $L[u] = 0, \quad B[u] = 0$

is the boundary value problem

(22) $L^+[u] = 0, \quad B^+[u] = 0$.

For example, as seen from Example 3, the adjoint boundary value problem for

(23) $y'' + 2y' + 2y = 0$; $y(0) = y(\pi) = 0$,

is the boundary value problem

(24) $y'' - 2y' + 2y = 0$; $y(0) = y(\pi) = 0$.

Two other examples follow.

Example 4 Find the adjoint boundary value problem for

(25) $x^2 y''(x) + xy'(x) + 4y(x) = 0$; $y(1) = y(e^\pi) = 0$.

Solution In Example 2 we found that the formal adjoint for $L[y](x) = x^2 y''(x) + xy'(x) + 4y(x)$ is

(26) $L^+[y](x) = x^2 y''(x) + 3xy'(x) + 5y(x)$.

To determine the adjoint boundary conditions $B^+(y) = 0$, we use Green's formula and require that

$$P(u, v)(x)\Big|_1^{e^\pi} = 0,$$

for u in $D(L)$ and v in $D(L^+)$. Since $A_2(x) = x^2$ and $A_1(x) = x$, this condition becomes

$$\left[xu(x)v(x) - x^2 u(x)v'(x) - 2xu(x)v(x) + x^2 u'(x)v(x) \right]\Big|_1^{e^\pi} = 0.$$

That is,

$$e^\pi u(e^\pi)v(e^\pi) - e^{2\pi}u(e^\pi)v'(e^\pi) - 2e^\pi u(e^\pi)v(e^\pi) + e^{2\pi}u'(e^\pi)v(e^\pi)$$
$$- u(1)v(1) + u(1)v'(1) + 2u(1)v(1) - u'(1)v(1) = 0.$$

For u in $D(L)$, we have $u(1) = u(e^\pi) = 0$, so the above equation simplifies to

(27) $e^{2\pi}u'(e^\pi)v(e^\pi) - u'(1)v(1) = 0$.

Because $u'(e^\pi)$ and $u'(1)$ can take on any values, we must have $v(e^\pi) = v(1) = 0$ in order for equation (27) to hold for all u in $D(L)$. Therefore, the adjoint boundary value problem is

(28) $x^2 y''(x) + 3xy'(x) + 5y(x) = 0$; $y(1) = y(e^\pi) = 0$. ◆

Example 5 Find the adjoint boundary value problem for

(29) $y'' + y = 0$; $y(0) + y(\pi) = 0$, $y'(0) = 0$.

Solution Here $L^+[y] = L[y] = y'' + y$. To determine the adjoint boundary value conditions, we again require

$$P(u, v)\Big|_0^\pi = 0$$

for u in $D(L)$ and v in $D(L^+)$. Since $A_1 = 0$ and $A_2 = 1$, we want

$$u'(\pi)v(\pi) - u(\pi)v'(\pi) - u'(0)v(0) + u(0)v'(0) = 0.$$

Now for u in $D(L)$, we have $u(\pi) = -u(0)$ and $u'(0) = 0$. Thus, the above equation reduces to

(30) $u'(\pi)v(\pi) + u(0)[v'(\pi) + v'(0)] = 0$.

Since $u'(\pi)$ and $u(0)$ are arbitrary, we require that $v(\pi) = 0$ and $v'(0) + v'(\pi) = 0$ so that equation (30) holds for all u in $D(L)$. Therefore, the adjoint boundary value problem is

(31) $y'' + y = 0$; $y'(0) + y'(\pi) = 0$, $y(\pi) = 0$. ◆

Let's return to the question of when does the nonhomogeneous boundary value problem

(32) $L[y] = h$, $B[y] = 0$,

have a solution. If it does have a solution u, then

$$(L[u], v) = (h, v)$$

for any v. Choosing v to be a solution to the (homogeneous) adjoint problem

(33) $L^+[v] = 0$, $B^+[v] = 0$,

we find that

$$(L[u], v) = (u, L^+[v]) = (u, 0) = 0 .$$

Hence, a necessary condition for the existence of a solution to (32) is that

$$(h, v) = 0 ,$$

for all v that satisfy (33). In other words, for the nonhomogeneous problem to have a solution, the nonhomogeneous term h must be orthogonal to all solutions of the adjoint problem. It turns out that this condition is also sufficient for the existence of solutions.[†] This test is known as the Fredholm alternative.[‡]

Fredholm Alternative

Theorem 10. Let L be a linear differential operator and let B represent a set of linear boundary conditions. The nonhomogeneous boundary value problem

(34) $L[y](x) = h(x)$, $a < x < b$,
 $B[y] = 0$

has a solution if and only if

(35) $\displaystyle\int_a^b h(x)z(x)\,dx = 0$

for every solution z of the adjoint boundary value problem

(36) $L^+[z](x) = 0$, $a < x < b$,
 $B^+[z] = 0$.

[†]For a proof that the condition is sufficient, we refer you to Chapter 11 of the text by E. Coddington and N. Levinson referenced in the footnote on page 657.

[‡]*Historical Footnote:* I. Fredholm (1866–1927), in his study of linear integral equations (circa 1900), realized that an analogy exists between linear integral equations and systems of n equations in n unknowns. This analogy also extends to differential equations and, more generally, to linear operators.

Remark. According to Theorem 10, if the adjoint boundary value problem (36) has only the trivial solution, then the nonhomogeneous boundary value problem (34) has a solution for each h. In such a case, if the boundary conditions in (2) are either separated or periodic, then it can be shown that there is a *unique* solution for each h. For more general boundary conditions, it is possible to have a one-parameter family of solutions (see Problem 28).

In the following examples, we illustrate how the Fredholm alternative is used to determine whether a nonhomogeneous boundary value problem has a solution.

Example 6 Determine conditions on h that guarantee that the problem

(37) $y'' + 2y' + 2y = h$; $y(0) = y(\pi) = 0$,

has a solution.

Solution The operator $L[y] = y'' + 2y' + 2y$ is the same as that in Example 3. Hence [see (23) and (24)], the adjoint problem is

(38) $y'' - 2y' + 2y = 0$; $y(0) = y(\pi) = 0$.

The auxiliary equation for (38) is $r^2 - 2r + 2 = 0$, whose roots are $r = 1 \pm i$. Hence, a general solution to the differential equation in (38) is

$$y(x) = c_1 e^x \cos x + c_2 e^x \sin x.$$

Using the boundary conditions in (38) to determine c_1 and c_2, we find

$$y(0) = c_1 = 0 \text{ and } y(\pi) = -c_1 e^\pi = 0.$$

Thus, $c_1 = 0$ and c_2 is arbitrary. Therefore, every solution to the adjoint problem (38) has the form

$$z(x) = c_2 e^x \sin x.$$

It now follows from the Fredholm alternative that if h is continuous, then the nonhomogeneous problem (37) has a solution if and only if

$$\int_0^\pi h(x) e^x \sin x \, dx = 0. \;\blacklozenge$$

Recall from Section 11.2 that if (37) does have a solution, say when $h(x) = e^{-x} \cos x$, then it has a one-parameter family of solutions, since the corresponding homogeneous problem [(37) with $h = 0$] has a one-parameter family of solutions, $y(x) = ce^{-x} \sin x$.

Example 7 Determine conditions on h that guarantee that the problem

(39) $x^2 y''(x) + xy'(x) + 4y(x) = h(x)$; $y(1) = y(e^\pi) = 0$,

has a solution.

Solution In Example 4 we determined that the corresponding adjoint problem is

(40) $x^2 y''(x) + 3xy'(x) + 5y(x) = 0$; $y(1) = y(e^\pi) = 0$.

Now the auxiliary equation for the Cauchy–Euler equation in (40) is $r(r-1) + 3r + 5 = 0$, which has roots $r = -1 \pm 2i$. Hence, a general solution to the differential equation in (40) is

$$y(x) = c_1 x^{-1} \cos(2 \ln x) + c_2 x^{-1} \sin(2 \ln x).$$

Using the boundary conditions in (40) to determine c_1 and c_2, we find

$$y(1) = c_1 = 0 \quad \text{and} \quad y(e^\pi) = e^{-\pi} c_1 = 0.$$

Hence, $c_1 = 0$ and c_2 is arbitrary. Therefore, any solution to the adjoint problem (40) has the form

$$z(x) = c_2 x^{-1} \sin(2 \ln x).$$

It now follows from the Fredholm alternative that for h continuous, the nonhomogeneous problem (39) has a solution if and only if

(41) $$\int_1^{e^\pi} h(x) x^{-1} \sin(2 \ln x)\, dx = 0. \quad \blacklozenge$$

Applying the Fredholm alternative to a nonhomogeneous regular Sturm–Liouville boundary value problem is easy because the operator is selfadjoint.

Example 8 Give conditions on h that guarantee that the nonhomogeneous regular Sturm–Liouville boundary value problem

(42) $$y'' + \mu y = h; \qquad y(0) = 0, \qquad y'(\pi) = 0,$$

has a solution.

Solution Since (42) is selfadjoint, we must have h orthogonal to every solution to the corresponding homogeneous problem. From Example 2 of Section 11.3 (page 658), we know that this homogeneous problem has nontrivial solutions when and only when $\mu = \lambda_n = (2n-1)^2/4$ for some $n = 1, 2, 3, \ldots$; that is, μ is an eigenvalue. Therefore, we must consider two cases.

 Case 1. If $\mu \neq (2n-1)^2/4$ for $n = 1, 2, 3, \ldots$, then the only solution to the homogeneous problem is the trivial solution. It then follows from the Fredholm alternative that the nonhomogeneous problem (42) has a unique solution for every h.

 Case 2. If $\mu = (2n-1)^2/4$ for some $n = 1, 2, 3, \ldots$, then the solutions of the homogeneous problem (eigenfunctions) are of the form $\phi_n(x) = b_n \sin[(2n-1)x/2]$, where b_n is arbitrary. The Fredholm alternative tells us that the nonhomogeneous problem (42) has a solution if and only if

(43) $$\int_0^\pi h(x) \sin\left(\frac{2n-1}{2} x\right) dx = 0.$$

Moreover, if this condition is satisfied, there is a one-parameter family of solutions, since we can always add a multiple of $\sin[(2n-1)x/2]$ to an existing solution to obtain a new solution. $\blacklozenge$

In the next two sections we give procedures for solving nonhomogeneous boundary value problems. The first procedure produces an eigenfunction expansion for the solution. The second procedure involves Green's functions and expresses the solution as a definite integral.

11.4 EXERCISES

In Problems 1–6, compute the formal adjoint of the given linear differential operator L.

1. $L[y](x) = xy''(x) + (2 - x^2)y'(x) + 3y(x)$

2. $L[y](x) = x^2y''(x) + (\sin x)y'(x) + 2xy(x)$

3. $L[y](x) = (1 + x^2)y''(x) + 2xy'(x) + y(x)$

4. $L[y](x) = x^2y''(x) - 2xy'(x) + 3y(x)$

5. $L[y](x) = 2x^2y''(x) + xy'(x) - 5y(x)$

6. $L[y](x) = (\sin x)y''(x) - e^x y'(x) + y(x)$

In Problems 7–10, find the adjoint operator and its domain for the given linear differential operator L whose domain consists of functions that have continuous second derivatives and satisfy the indicated boundary conditions.

7. $L[y] = y'' - 2y' + 10y$;
 $y(0) = 0, \qquad y(\pi) = 0$

8. $L[y] = y'' - 4y' + 5y$;
 $y(0) = 0, \qquad y(2\pi) = 0$

9. $L[y](x) = 2x^2y''(x) + 3xy'(x) - y(x)$;
 $y(1) = 0, \qquad y(4) = 0$

10. $L[y](x) = x^2y''(x) + 2xy'(x) + (5/4)y(x)$;
 $y(1) = 0, \qquad y(e^\pi) = 0$

In Problems 11–16, find the adjoint boundary value problem for the given boundary value problem.

11. $y'' + 6y' + 10y = 0$;
 $y'(0) = 0, \qquad y'(\pi) = 0$

12. $y'' + y' + y = 0$;
 $y(0) = y(\pi), \qquad y'(0) = y'(\pi)$

13. $y'' + 4y = 0$;
 $y(0) = y(\pi), \qquad y'(0) = y'(\pi)$

14. $y'' = 0$;
 $y(0) = y(\pi/2), \qquad y'(0) = y'(\pi/2)$

15. $x^2y''(x) + 3xy'(x) + 2y(x) = 0$;
 $y'(1) = 0, \qquad y'(e^\pi) = 0$

16. $x^2y''(x) + 2xy'(x) = 0$;
 $y(1) = y(2), \qquad y'(1) = y'(2)$

In Problems 17–26, determine conditions on h that guarantee that the given nonhomogeneous boundary value problem has a solution. (These are nonhomogeneous problems associated with the homogeneous Problems 7–16.)

17. $y'' - 2y' + 10y = h$;
 $y(0) = 0, \qquad y(\pi) = 0$

18. $y'' - 4y' + 5y = h$;
 $y(0) = 0, \qquad y(2\pi) = 0$

19. $2x^2y''(x) + 3xy'(x) - y(x) = h(x)$;
 $y(1) = 0, \qquad y(4) = 0$

20. $x^2y''(x) + 2xy'(x) + (5/4)y(x) = h(x)$;
 $y(1) = 0, \qquad y(e^\pi) = 0$

21. $y'' + 6y' + 10y = h$;
 $y'(0) = 0, \qquad y'(\pi) = 0$

22. $y'' + y' + y = h$;
 $y(0) = y(\pi), \qquad y'(0) = y'(\pi)$

23. $y'' + 4y = h$;
 $y(0) = y(\pi), \qquad y'(0) = y'(\pi)$

24. $y'' = h$;
 $y(0) = y(\pi/2), \qquad y'(0) = y'(\pi/2)$

25. $x^2y''(x) + 3xy'(x) + 2y(x) = h(x)$;
 $y'(1) = 0, \qquad y'(e^\pi) = 0$

26. $x^2y''(x) + 2xy'(x) = h(x)$;
 $y(1) = y(2), \qquad y'(1) = y'(2)$

27. Prove Green's formula by starting with

 $$(L[u], v) = \int_a^b \left(A_2 u'' + A_1 u' + A_0 u\right) v \, dx$$

 and then, performing integration by parts, obtain

 $$(L[u], v) = (u, L^\dagger[v]) + P(u, v)\Big|_a^b .$$

28. Show that the "boundary value problem"

 $$y'' + y = h(x), \qquad 0 < x < \pi, \qquad y(0) = 0,$$

 has a one-parameter family of solutions for each h continuous on $[0, \pi]$. [*Hint:* Show that the homogeneous equation has a one-parameter family of solutions, while the adjoint problem has only the trivial solution.]

29. Show that the boundary value problem

 $$y'' + y = h(x) ;$$
 $$y(0) = 0, \qquad y(\pi) = 0, \qquad y'(\pi) = 0,$$

 has a unique solution for each h that is continuous on $[0, \pi]$ and satisfies

 $$\int_0^\pi h(x) \sin x \, dx = 0 .$$

 [*Hint:* Show that the homogeneous equation has only the trivial solution.]

11.5 Solution by Eigenfunction Expansion

In the previous section we found that the Fredholm alternative provides a test to determine whether a given nonhomogeneous boundary value problem has a solution. But how do we find such a solution? Of course, if we can find a general solution to the nonhomogeneous equation, then we can simply substitute into the boundary conditions and determine a solution (see Example 2 in Section 11.2). However, most regular Sturm–Liouville boundary value problems are not constant-coefficient equations. Therefore it may be difficult to find such a general solution.

In this section, we take advantage of the fact that the eigenfunctions associated with a regular Sturm–Liouville boundary value problem form an orthogonal system. Our goal is to find an **eigenfunction expansion** for a solution to the nonhomogeneous regular Sturm–Liouville boundary value problem

(1) $L[y] + \mu r y = f$;

(2) $a_1 y(a) + a_2 y'(a) = 0$, $b_1 y(b) + b_2 y'(b) = 0$,

where μ is a fixed real number and

(3) $L[y] := (py')' + qy$.

Let $\{\lambda_n\}_{n=1}^{\infty}$ be the eigenvalues with corresponding eigenfunctions $\{\phi_n\}_{n=1}^{\infty}$ for the homogeneous eigenvalue problem associated with (1)–(2), with μ replaced by the parameter λ; that is, $L[\phi_n] + \lambda_n r \phi_n = 0$ and ϕ_n satisfies (2). We know (recall Theorem 4 in Section 11.3) that $\{\phi_n\}_{n=1}^{\infty}$ forms an orthogonal system with respect to the weight function $r(x)$ on $[a, b]$.

Let

(4) $\phi(x) = \displaystyle\sum_{n=1}^{\infty} c_n \phi_n(x)$.

Our aim is to determine the coefficients c_n so that ϕ is a solution to (1)–(2).

For the moment, we'll ignore the important issues of convergence and differentiability of this series; that is, we will treat ϕ as a *formal* series expansion. First, notice that since each ϕ_n satisfies the boundary conditions in (2), so does ϕ. Further, substituting the expansion for ϕ into the left-hand side of (1) and using the fact that $L[\phi_n] = -\lambda_n r \phi_n$, we find

(5) $L[\phi] + \mu r \phi = L\left[\displaystyle\sum_{n=1}^{\infty} c_n \phi_n \right] + \mu r \displaystyle\sum_{n=1}^{\infty} c_n \phi_n$

$= \displaystyle\sum_{n=1}^{\infty} c_n L[\phi_n] + \mu r \displaystyle\sum_{n=1}^{\infty} c_n \phi_n$

$= \displaystyle\sum_{n=1}^{\infty} c_n (-\lambda_n r \phi_n) + \mu r \displaystyle\sum_{n=1}^{\infty} c_n \phi_n$

$= r \displaystyle\sum_{n=1}^{\infty} (\mu - \lambda_n) c_n \phi_n$.

If we express f as the product $r(f/r)$ and expand f/r in an eigenfunction expansion

$$ f/r = \sum_{n=1}^{\infty} \gamma_n \phi_n , \qquad \gamma_n = \frac{\displaystyle\int_a^b (f/r) \phi_n r \, dx}{\displaystyle\int_a^b \phi_n^2 \, r \, dx} = \frac{\displaystyle\int_a^b f \phi_n \, dx}{\displaystyle\int_a^b \phi_n^2 \, r \, dx} , $$

then the right-hand side of (1) becomes

(6) $$r \sum_{n=1}^{\infty} \gamma_n \phi_n \, .$$

Equating the left- and right-hand sides of (1) given in (5) and (6), we find

(7) $$r \sum_{n=1}^{\infty} (\mu - \lambda_n) c_n \phi_n = r \sum_{n=1}^{\infty} \gamma_n \phi_n \, .$$

Since $r > 0$, we can simplify (7) to

$$\sum_{n=1}^{\infty} [(\mu - \lambda_n) c_n - \gamma_n] \phi_n = 0 \, .$$

This expansion sums to zero if and only if the coefficients are all zero. Therefore, we set

(8) $$(\mu - \lambda_n) c_n - \gamma_n = 0 \, , \qquad n = 1, 2, 3, \dots \, .$$

We now consider two cases.

Case 1. $\mu \neq \lambda_n$ for every $n = 1, 2, 3, \dots$. In this case, we can solve (8) for c_n:

$$c_n = \frac{\gamma_n}{\mu - \lambda_n} \, .$$

The unique solution to (1)–(2) is thus given formally by

(9) $$\phi = \sum_{n=1}^{\infty} \frac{\gamma_n}{\mu - \lambda_n} \phi_n \, .$$

We have referred to ϕ as the unique solution, since, if $\mu \neq \lambda_n$ for every $n = 1, 2, 3, \dots$, the corresponding homogeneous equation has only the trivial solution and hence the nonhomogeneous boundary value problem has a unique solution.

Case 2. $\mu = \lambda_N$ for some N. In the case when $n = N$, equation (8) reduces to

$$\gamma_N = 0 \, .$$

There are now two possibilities. If $\gamma_N \neq 0$, we have a contradiction and so the nonhomogeneous equation does not have a solution expressible as an eigenfunction expansion. If $\gamma_N = 0$, then there is no restriction on c_N. Thus, we get a one-parameter family of solutions to the nonhomogeneous boundary value problem (1)–(2), with c_N being the parameter.

But what does it mean for $\gamma_N = 0$? From the definition of γ_N, we see that since

$$\gamma_N = \int_a^b f(x) \phi_N(x) \, dx \bigg/ \int_a^b \phi_N^2(x) r(x) \, dx \, ,$$

the function f must be orthogonal to ϕ_N when $\gamma_N = 0$. But this is just the Fredholm alternative condition for the nonhomogeneous problem (1)–(2).

Method of Eigenfunction Expansions

To obtain an eigenfunction expansion for the solution ϕ to the nonhomogeneous regular Sturm–Liouville boundary value problem

$$L[y] + \mu ry = f; \qquad B[y] = 0:$$

(a) Find an orthogonal system of eigenfunctions $\{\phi_n\}_{n=1}^{\infty}$ and corresponding eigenvalues $\{\lambda_n\}_{n=1}^{\infty}$ for

$$L[y] + \lambda ry = 0, \qquad B[y] = 0.$$

(b) Compute the eigenfunction expansion for f/r; that is, determine the coefficients γ_n so that

$$f/r = \sum_{n=1}^{\infty} \gamma_n \phi_n.$$

(c) If $\mu \neq \lambda_n$ for $n = 1, 2, 3, \ldots$, then the solution ϕ is given by

$$\phi = \sum_{n=1}^{\infty} \frac{\gamma_n}{\mu - \lambda_n} \phi_n.$$

(d) If $\mu = \lambda_N$ for some N and $\gamma_N \neq 0$, then there is no solution.

(e) If $\mu = \lambda_N$ for some N and $\gamma_N = 0$, then

$$\phi = c_N \phi_N + \sum_{\substack{n=1 \\ n \neq N}}^{\infty} \frac{\gamma_n}{\mu - \lambda_n} \phi_n$$

is a solution for any choice of the parameter c_N.

In the following examples we illustrate this procedure.

Example 1 Find an eigenfunction expansion for the solution to the regular Sturm–Liouville boundary value problem

$$(10) \qquad y'' + 3y = 5\sin 3x + 2\sin 8x; \qquad y(0) = 0, \qquad y(\pi) = 0.$$

Solution Recall from Section 11.2 that the boundary value problem

$$y'' + \lambda y = 0; \qquad y(0) = 0, \qquad y(\pi) = 0,$$

has eigenvalues $\lambda_n = n^2$, $n = 1, 2, \ldots$, with corresponding eigenfunctions

$$\phi_n(x) = \sin nx, \qquad n = 1, 2, \ldots.$$

Here $r(x) \equiv 1$, so we need to determine coefficients γ_n such that

$$f(x)/r(x) = f(x) = \sum_{n=1}^{\infty} \gamma_n \sin nx$$
$$= 5\sin 3x + 2\sin 8x.$$

Clearly, $\gamma_3 = 5$, $\gamma_8 = 2$, and the remaining γ_n's are zero. Since $\mu = 3 \neq n^2 = \lambda_n$ for $n = 1, 2, \ldots,$ it follows that the unique solution to (10) is

$$
(11) \qquad \phi(x) = \sum_{n=1}^{\infty} \frac{\gamma_n}{\mu - \lambda_n} \phi_n(x) = \frac{5}{3-9}\sin 3x + \frac{2}{3-64}\sin 8x
$$

$$
\qquad\qquad = -\frac{5}{6}\sin 3x - \frac{2}{61}\sin 8x . \quad \blacklozenge
$$

As a check on our computations and as a verification of the theory, you are invited to show directly by substitution that ϕ given in (11) is the solution to problem (10).

Example 2 Find a formal eigenfunction expansion for the solution to the regular Sturm–Liouville boundary value problem

$$
(12) \qquad y'' + y = f; \qquad y(0) = y'(\pi) = 0 ,
$$

where

$$
(13) \qquad f(x) = \begin{cases} 2x/\pi, & 0 \le x \le \pi/2 , \\ 1, & \pi/2 \le x \le \pi . \end{cases}
$$

Solution In Examples 2 and 3 in Section 11.3, we determined that the sequence

$$
\left\{ \sqrt{\frac{2}{\pi}}\sin\left(\frac{2n-1}{2}x\right) \right\}_{n=1}^{\infty}
$$

is an orthonormal system of eigenfunctions for the boundary value problem

$$
y'' + \lambda y = 0 ; \qquad y(0) = y'(\pi) = 0 ,
$$

where the associated eigenvalues are $\lambda_n = (2n-1)^2/4$. We also found that $f(x)\, (= f(x)/r(x)$, since $r(x) \equiv 1)$ given in (13) has the eigenfunction expansion

$$
(14) \qquad f(x) = \sum_{n=1}^{\infty} \gamma_n \sqrt{\frac{2}{\pi}}\sin\left(\frac{2n-1}{2}x\right) ,
$$

where

$$
(15) \qquad \gamma_n = \frac{2^{7/2}}{\pi^{3/2}(2n-1)^2}\sin\left(\frac{n\pi}{2} - \frac{\pi}{4}\right) , \qquad n = 1, 2, 3, \ldots .
$$

Since $\mu = 1$ is *not* an eigenvalue, the nonhomogeneous problem (12) has a unique solution ϕ given formally by the expansion

$$
(16) \qquad \phi(x) = \sum_{n=1}^{\infty} \frac{\gamma_n}{1 - \lambda_n}\phi_n(x) = \sum_{n=1}^{\infty} \frac{\gamma_n}{1 - \dfrac{(2n-1)^2}{4}} \sqrt{\frac{2}{\pi}}\sin\left(\frac{2n-1}{2}x\right) ,
$$

where the γ_n's are given by equation (15). $\blacklozenge$

A solution to the nonhomogeneous boundary value problem in (12) can also be obtained by finding two general solutions to the differential equation in (12), one valid on the interval $[0, \pi/2]$ and the other on $[\pi/2, \pi]$. The *four* arbitrary constants in these general solutions are selected to ensure the boundary conditions in (12) are satisfied and that the solution and its

derivative are continuous on $[0, \pi]$. We leave it to the reader to carry out this procedure and obtain the solution

$$(17) \qquad \phi(x) = \begin{cases} \dfrac{2}{\pi}x, & 0 \leq x \leq \pi/2, \\ 1 - \dfrac{2}{\pi}\cos x, & \pi/2 \leq x \leq \pi. \end{cases}$$

In advanced texts it is shown that if f is continuous on $[a, b]$, then the formal solution in (9) converges uniformly to the genuine solution. For example, the expansion in (16) converges uniformly to the function ϕ in (17).

If one has an eigenfunction expansion for the solution to a nonhomogeneous boundary value problem, then one can obtain an *approximate solution* by truncating the expansion after N terms. This truncated expansion is an example of a **Galerkin approximation** for the solution. Galerkin's method and **collocation**[†] are two widely used procedures for approximating the solution to a boundary value problem.

11.5 EXERCISES

In Problems 1–8, find a formal eigenfunction expansion for the solution to the given nonhomogeneous boundary value problem, if it exists.

1. $y'' + 2y = 5\sin x - 7\sin 3x$;
 $y(0) = 0$, $y(\pi) = 0$

2. $y'' - y = \sin 4x - \sin 11x$;
 $y(0) = 0$, $y(\pi) = 0$

3. $y'' + 4y = \sin 2x + \sin 8x$;
 $y(0) = 0$, $y(\pi) = 0$

4. $y'' + \pi y = \cos 7x + 5\cos 10x$;
 $y'(0) = 0$, $y'(\pi) = 0$

5. $y'' + y = \cos 4x + \cos 7x$;
 $y'(0) = 0$, $y'(\pi) = 0$

6. $y'' + 6y = \cos 4x - \cos 5x$;
 $y'(0) = 0$, $y'(\pi) = 0$

7. $y'' + y = x^2 - 2\pi x$;
 $y(0) = 0$, $y'(\pi) = 0$

8. $y'' + 2y = \pi x - x^2$;
 $y(0) = 0$, $y(\pi) = 0$

In Problems 9–14, find a formal eigenfunction expansion for the solution to the given nonhomogeneous boundary value problem. Assume f satisfies the conditions of the Fredholm alternative.

9. $y'' + y = f$; $y'(0) = 0$, $y(\pi) = 0$

10. $y'' + 7y = f$; $y(0) = 0$, $y'(\pi) = 0$

11. $(xy')' + 5x^{-1}y = f$;
 $y'(1) = 0$, $y(e^\pi) = 0$

12. $xy'' + y' + x^{-1}y = f$;
 $y(1) = 0$, $y(e^\pi) = 0$

13. $y'' + 6y = f$;
 $y(0) = 0$, $3y(\pi) - y'(\pi) = 0$

 [*Hint:* See Example 4 in Section 11.2.]

14. $(xy')' - x^{-1}y = f$;
 $y'(1) = 0$, $y'(e) = 0$

15. Derive the solution to Problem (12) given in equation (17).

[†]For more information on numerical methods for solving boundary value problems, see *Numerical Methods for Two-Point Boundary Value Problems*, by H. B. Keller (Dover Publications, New York, 1993). The **shooting method,** in particular, is illustrated in Project D.

11.6 Green's Functions

In this section we derive an integral representation for the solution to a nonhomogeneous boundary value problem. Namely, we show that the solution can be expressed in the form

$$(1) \qquad y(x) = \int_a^b G(x, s)f(s)\,ds,$$

where the kernel function, $G(x, s)$, is called a *Green's function*, or an *influence function*. This representation is known to exist for very general problems.[†] However, we will consider only the regular Sturm–Liouville boundary value problems

$$(2) \qquad L[y](x) = -f(x), \qquad a < x < b,$$
$$(3) \qquad a_1 y(a) + a_2 y'(a) = 0,$$
$$(4) \qquad b_1 y(b) + b_2 y'(b) = 0,$$

where

$$(5) \qquad L[y] := \frac{d}{dx}\left(p\frac{dy}{dx}\right) + qy$$

and p, p', and q are continuous on $[a, b]$ with $p(x) > 0$ on $[a, b]$. We will assume f is continuous on $[a, b]$. While it is not necessary, it is customary in the present context to write $-f(x)$, rather than $f(x)$, as the nonhomogeneous term in equation (2).

Throughout this section **we assume that the corresponding homogeneous problem** ($f(x) \equiv 0$) **has only the trivial solution,** or equivalently, we assume that $\lambda = 0$ is not an eigenvalue. It then follows from the Fredholm alternative that the nonhomogeneous problem (2)–(4) has a solution. This solution is unique, since the corresponding homogeneous problem has a unique solution.

We will use the method of variation of parameters to obtain an integral representation for the solution to (2)–(4) (see Section 4.7). To begin, choose two nontrivial solutions z_1 and z_2 to the homogeneous equation $L[y] = 0$. Pick z_1 so that it satisfies (3) and z_2 so that it satisfies (4). The existence of z_1 and z_2 follows from the existence and uniqueness theorem for initial value problems (see Section 4.7), since one can choose initial conditions at a (respectively b) so that the boundary condition (3) [respectively (4)] is satisfied and the solution is nontrivial.

Before we can use z_1 and z_2 with the method of variation of parameters, we must verify that z_1 and z_2 are linearly independent. Note that if z_1 and z_2 were linearly *dependent,* then one would be a constant multiple of the other, say $z_1 = cz_2$. Since z_2 satisfies (4), it follows that z_1 also satisfies (4). Recall that z_1 also satisfies (3). Thus, z_1 satisfies $L[z_1] = 0$ and (3)–(4). But we have assumed that the corresponding homogeneous problem has *only* the trivial solution. This contradiction implies that z_1 and z_2 are linearly independent.

To apply the formulas from the variation of parameters method, we first write the nonhomogeneous equation (2) in the standard form

$$y'' + \frac{p'}{p}y' + \frac{q}{p}y = -\frac{f}{p}.$$

[†]For a discussion of Green's functions for boundary value problems for nth-order linear ordinary differential equations, see the text by E. Coddington and N. Levinson, *Theory of Ordinary Differential Equations* (Krieger, New York, 1984), Chapter 7.

A particular solution to this equation is then given by (compare Section 4.7).

(6) $y(x) = c_1(x)z_1(x) + c_2(x)z_2(x)$,

where

(7) $c_1'(x) = \dfrac{f(x)z_2(x)}{p(x)W[z_1, z_2](x)}$, $c_2'(x) = \dfrac{-f(x)z_1(x)}{p(x)W[z_1, z_2](x)}$,

and $W[z_1, z_2] = z_1 z_2' - z_1' z_2$ is the Wronskian of z_1 and z_2. Since we are free to pick the constants in the antiderivatives for c_1' and c_2', it will turn out to be convenient to choose

(8) $c_1(x) = \displaystyle\int_x^b \dfrac{-z_2(s)f(s)}{p(s)W[z_1, z_2](s)}\, ds$

and

(9) $c_2(x) = \displaystyle\int_a^x \dfrac{-z_1(s)f(s)}{p(s)W[z_1, z_2](s)}\, ds$.

Substituting into (6), we obtain the following solution to (2):

(10) $y(x) = z_1(x) \displaystyle\int_x^b \dfrac{-z_2(s)f(s)}{p(s)W[z_1, z_2](s)}\, ds + z_2(x) \displaystyle\int_a^x \dfrac{-z_1(s)f(s)}{p(s)W[z_1, z_2](s)}\, ds$

$\qquad = \displaystyle\int_a^b G(x, s)f(s)\, ds$,

where

$$G(x, s) := \begin{cases} \dfrac{-z_1(s)z_2(x)}{p(s)W[z_1, z_2](s)}, & a \le s \le x, \\[3ex] \dfrac{-z_1(x)z_2(s)}{p(s)W[z_1, z_2](s)}, & x \le s \le b. \end{cases}$$

Using the fact that z_1 and z_2 satisfy the equation $L[y] = 0$, one can show that

(11) $p(x)W[z_1, z_2](x) = C$, x in $[a, b]$,

where C is a constant (see Problem 31). Hence, $G(x, s)$ has the simpler form

(12) $G(x, s) = \begin{cases} -z_1(s)z_2(x)/C, & a \le s \le x, \\ -z_1(x)z_2(s)/C, & x \le s \le b, \end{cases}$

where C is given by equation (11). The function $G(x, s)$ is called the **Green's function** for the Problem (2)–(4).

Integral Representation of Solution

Theorem 11. If $G(x, s)$ is the Green's function for a regular Sturm–Liouville boundary value problem, then the function

(13) $y(x) = \displaystyle\int_a^b G(x, s)f(s)\, ds$

is the solution to the regular Sturm–Liouville boundary value problem with $L[y](x) = -f(x)$ for any function f that is continuous on $[a, b]$.

Proof. We used the method of variation of parameters to derive $y(x)$, so we know that it satisfies (2). Hence, we need only show that $y(x)$ given in (13) satisfies the boundary conditions (3) and (4). We first compute $y'(x)$ from equation (10) (with $pW[z_1, z_2] = C$):

$$(14)\qquad y'(x) = z_1'(x)\int_x^b \frac{-z_2(s)f(s)}{C}\,ds + \frac{z_1(x)z_2(x)f(x)}{C}$$

$$+ z_2'(x)\int_a^x \frac{-z_1(s)f(s)}{C}\,ds - \frac{z_1(x)z_2(x)f(x)}{C}$$

$$= z_1'(x)\int_x^b [-z_2(s)f(s)/C]\,ds + z_2'(x)\int_a^x [-z_1(s)f(s)/C]\,ds .$$

Using equations (10) and (14), we have

$$a_1 y(a) + a_2 y'(a) = a_1 z_1(a)\int_a^b [-z_2(s)f(s)/C]\,ds + a_2 z_1'(a)\int_a^b [-z_2(s)f(s)/C]\,ds$$

$$= [a_1 z_1(a) + a_2 z_1'(a)]\int_a^b [-z_2(s)f(s)/C]\,ds$$

$$= 0 ,$$

since z_1 satisfies (3). Thus, y given in (13) satisfies (3). A similar computation shows that y also satisfies (4). Hence, y solves the problem (2)–(4). ◆

We now summarize the above procedure for solving the nonhomogeneous boundary value problem (2)–(4) when the homogeneous problem has only the trivial solution. Here $B_1[y] = 0$ imposes a boundary condition at $x = a$, and B_2 imposes one at $x = b$.

Solution Using Green's Functions

To obtain the solution to

$$L[y] = -f; \qquad B_1[y] = 0, \qquad B_2[y] = 0,$$

where $L[y] = (py')' + qy$ and f is continuous on $[a, b]$:

(a) Find two independent solutions z_1 and z_2 to $L[y] = 0$ that satisfy $B_1[z_1] = 0$ and $B_2[z_2] = 0$.

(b) Compute $C = p(x)W[z_1, z_2](x)$, where $W[z_1, z_2] = z_1 z_2' - z_1' z_2$.

(c) Construct the Green's function:

$$(15)\qquad G(x, s) = \begin{cases} -z_1(s)z_2(x)/C, & a \le s \le x, \\ -z_1(x)z_2(s)/C, & x \le s \le b. \end{cases}$$

The desired solution is given by

$$(16)\qquad y(x) = \int_a^b G(x, s)f(s)\,ds .$$

Example 1 Find the Green's function $G(x, s)$ for the boundary value problem

$$(17)\qquad y''(x) = -f(x); \qquad y(0) = 0, \qquad y(\pi) = 0 .$$

Use the Green's function to obtain the solution when $f(x) = x$.

Solution A general solution to the homogeneous problem $y'' = 0$ is $y_h(x) = Ax + B$, so $z_1(x)$ and $z_2(x)$ must be of this form. To get $z_1(x)$, we want to choose A and B so that $z_1(0) = B = 0$. Since A is arbitrary, we can set it equal to 1. Hence, we take $z_1(x) = x$. To get $z_2(x)$, we want to choose A and B so that $z_2(\pi) = A\pi + B = 0$. Thus, $B = -A\pi$. Taking $A = -1$, we get $z_2(x) = \pi - x$.

We now compute

$$C = p(x)W[z_1, z_2](x) = (1)[(x)(-1) - (1)(\pi - x)] = -\pi.$$

(This step also serves as a check on our work, since all appearances of the variable x must cancel to obtain a constant.) Thus, the Green's function is

(18)
$$G(x, s) = \begin{cases} -z_1(s)z_2(x)/C, & 0 \le s \le x, \\ -z_1(x)z_2(s)/C, & x \le s \le \pi, \end{cases}$$
$$= \begin{cases} s(\pi - x)/\pi, & 0 \le s \le x, \\ x(\pi - s)/\pi, & x \le s \le \pi. \end{cases}$$

When $f(x) = x$, the solution to (17) is given by equation (16). Substituting for $f(x)$ and $G(x, s)$, we find

$$y(x) = \int_a^b G(x, s)f(s)\, ds$$
$$= \int_0^x [s(\pi - x)s/\pi]\, ds + \int_x^\pi [x(\pi - s)s/\pi]\, ds$$
$$= \frac{(\pi - x)}{\pi}\left[\frac{s^3}{3}\right]\Big|_0^x + \frac{x}{\pi}\left[\frac{\pi s^2}{2} - \frac{s^3}{3}\right]\Big|_x^\pi$$
$$= x(\pi^2 - x^2)/6. \quad \blacklozenge$$

In the next theorem we give some basic properties of Green's functions.

Properties of Green's Functions

Theorem 12. Let $G(x, s)$ be the Green's function defined by equation (12) for the regular Sturm–Liouville boundary value problem (2)–(4). Then

(a) $G(x, s)$ is continuous on the square $[a, b] \times [a, b]$. For each fixed s, the partial derivatives $(\partial G/\partial x)(x, s)$ and $(\partial^2 G/\partial x^2)(x, s)$ are continuous functions of x for $x \ne s$.

(b) At $x = s$, the partial derivative $\partial G/\partial x$ has a jump discontinuity:

(19)
$$\lim_{x \to s^+} \frac{\partial G}{\partial x}(x, s) - \lim_{x \to s^-} \frac{\partial G}{\partial x}(x, s) = \frac{-1}{p(s)}.$$

(c) For each fixed s, the function $G(x, s)$ satisfies the corresponding homogeneous problem for $x \ne s$; that is,[†]

(20)
$$L[G(\cdot, s)](x) = 0, \quad \text{for } x \ne s;$$

(21)
$$a_1 G(a, s) + a_2 \frac{\partial G}{\partial x}(a, s) = 0, \quad b_1 G(b, s) + b_2 \frac{\partial G}{\partial x}(b, s) = 0.$$

(d) There is only one function satisfying properties (a)–(c).

(e) $G(x, s)$ is symmetric; that is,

$$G(x, s) = G(s, x).$$

[†]For $L[y] = (py')' + qy$, the notation $L[G(\cdot, s)](x)$ stands for $\frac{\partial}{\partial x}\left[p(x)\frac{\partial G}{\partial x}(x, s)\right] + q(x)G(x, s)$.

Proof.

(a) Since z_1 and z_2 are both solutions to $L[y] = 0$ on $[a, b]$, they are continuous and have continuous first and second derivatives. Hence, it follows from the formula for $G(x, s)$ that $G(x, s)$ is continuous on the square $[a, b] \times [a, b]$. Furthermore, for fixed s, it follows that $\partial G/\partial x$ and $\partial^2 G/\partial x^2$ are continuous functions of x for $x \neq s$.

(b) Using the fact that $C = p(s)W[z_1, z_2](s)$, we find

$$\lim_{x \to s^+} \frac{\partial G}{\partial x}(x, s) - \lim_{x \to s^-} \frac{\partial G}{\partial x}(x, s) = \frac{-z_1(s)z_2'(s)}{C} - \frac{-z_1'(s)z_2(s)}{C}$$

$$= -\frac{z_1(s)z_2'(s) - z_1'(s)z_2(s)}{C}$$

$$= -\frac{W[z_1, z_2](s)}{C} = \frac{-1}{p(s)}.$$

(c) For fixed s, the function $G(x, s)$ is a constant multiple of $z_1(x)$ when $x < s$ and a constant multiple of $z_2(x)$ when $s < x$. Since both z_1 and z_2 satisfy $L[y] = 0$, then so does $G(\cdot, s)$ for $x \neq s$. Next, we verify (21). When $x < s$, we have $G(x, s) = (-z_2(s)/C)z_1(x)$. Since $z_1(x)$ satisfies (3), we have

$$a_1 G(a, s) + a_2 \frac{\partial G}{\partial x}(a, s) = (-z_2(s)/C)[a_1 z_1(a) + a_2 z_1'(a)] = 0.$$

Furthermore, when $s < x$, we have $G(x, s) = (-z_1(s)/C)z_2(x)$. Since $z_2(x)$ satisfies (4), we find

$$b_1 G(b, s) + b_2 \frac{\partial G}{\partial x}(b, s) = (-z_1(s)/C)[b_1 z_2(b) + b_2 z_2'(b)] = 0.$$

Thus, for each fixed s, $G(\cdot, s)$ satisfies (20)–(21).

(d) Assume that both $G(x, s)$ and $H(x, s)$ satisfy properties (a), (b), and (c). We will show that

$$K(x, s) := G(x, s) - H(x, s)$$

is identically zero on $[a, b] \times [a, b]$. For this purpose, fix $s = s_0$. By property (a), $K(x, s_0)$ is continuous on $[a, b]$, and $(\partial K/\partial x)(x, s_0)$ and $(\partial^2 K/\partial x^2)(x, s_0)$ are continuous for $x \neq s_0$. Moreover, since property (b) says that $(\partial G/\partial x)(x, s_0)$ and $(\partial H/\partial x)(x, s_0)$ have equal jumps at $x = s_0$, we see that $(\partial K/\partial x)(x, s_0) = (\partial G/\partial x)(x, s_0) - (\partial H/\partial x)(x, s_0)$ exists and is continuous (has no jump) at $x = s_0$. Hence $K(x, s_0)$ and $(\partial K/\partial x)(x, s_0)$ are continuous on $[a, b]$. From property (c), equation (20), it follows that for $x \neq s_0$, $L[K(\cdot, s_0)] = 0$. Solving this equation for $(\partial^2 K/\partial x^2)(x, s_0)$, we find that $(\partial^2 K/\partial x^2)(x, s_0)$ is equal to a function that is continuous at $x = s_0$. Consequently $(\partial^2 K/\partial x^2)(x, s_0)$ is continuous on $[a, b]$. Thus, $K(x, s_0)$ is a solution to the homogeneous boundary value problem on the whole interval $[a, b]$. But this problem has *only* the trivial solution, so $K(x, s_0) \equiv 0$. Finally, since s_0 is arbitrary, $K(x, s) \equiv 0$ on $[a, b] \times [a, b]$, that is, the Green's function is unique.

(e) The symmetry is obvious from the definition of $G(x, s)$ in (12); simply interchange x and s in the formulas. ◆

Green's functions satisfying properties (a)–(d) also exist for general nonselfadjoint boundary value problems. Furthermore, it can be shown that the existence of a *symmetric* Green's function is equivalent to the linear operator L being selfadjoint.†

Because of the uniqueness of the Green's function, we could have defined $G(x, s)$ to be the function that satisfies properties (a)–(c) in Theorem 12 and then used these properties to derive formula (12). We leave this derivation as an exercise (see Problem 32).

The Green's function can also be characterized using the Dirac delta function (see Section 7.9, page 405). If we assume that there exists a function $G(x, s)$ such that

$$y(x) = \int_a^b G(x, s)f(s) \, ds$$

is the solution to (2)–(4), then operating on both sides with L, and assuming we can interchange differentiation and integration, gives

$$L[y](x) = \int_a^b L[G(\cdot, s)](x) f(s) \, ds = -f(x) \,.$$

That is, $L[G(\cdot, s)](x)$ must act like the Dirac delta $-\delta(x - s)$. Moreover, $G(\cdot, s)$ must satisfy the boundary conditions (3) and (4). If we define $G(x, s)$ as the solution to

$$(22) \qquad L[G(\cdot, s)](x) = -\delta(x - s) \,,$$

$$(23) \qquad a_1 G(a, s) + a_2 \frac{\partial G}{\partial x}(a, s) = 0 \,,$$

$$(24) \qquad b_1 G(b, s) + b_2 \frac{\partial G}{\partial x}(b, s) = 0 \,,$$

then, using the properties of the Dirac delta function, one can formally show that $G(x, s)$ satisfies properties (a)–(c) in Theorem 12 (see Problem 33).

The characterization of the Green's function given in (22)–(24) supports the following physical interpretation. We view $G(x, s)$ as the deflection of a string (beam, etc.) at a point x as a result of the action of a point force of unit intensity at some other point s. That is, $G(x, s)$ represents the *influence* of the force at s on the string at the point x—hence the name, **influence function.** Furthermore, $G(x, s)f(s)$ is the influence of a point force of intensity $f(s)$. Taking a superposition of *all* the point forces for s between a and b, we obtain

$$y(x) = \int_a^b G(x, s)f(s) \, ds$$

as the resultant deflection of a string under an external force of $-f(x)$.

†See E. Coddington and N. Levinson, ibid.

In Problems 1–10, find the Green's function $G(x, s)$ for the given boundary value problem.

1. $y'' = -f(x)$; $y(0) = 0$, $y'(\pi) = 0$

2. $y'' - y = -f(x)$; $y(0) = 0$, $y(1) = 0$

3. $y'' = -f(x)$; $y(0) = 0$, $y(\pi) + y'(\pi) = 0$

4. $y'' + y = -f(x)$; $y'(0) = 0$, $y(\pi) = 0$

5. $y'' + 4y = -f(x)$; $y(0) = 0$, $y'(\pi) = 0$

6. $y'' + y = -f(x)$; $y(0) + y'(0) = 0$, $y(\pi) = 0$

7. $(x^2 y')' - 2y = -f(x)$; $y(1) = 0$, $y'(2) = 0$

8. $(xy')' - (4/x)y = -f(x)$; $y'(1) = 0$, $y(2) = 0$

9. $(e^{2x} y')' + e^{2x} y = -f(x)$; $y(0) = 0$, $y(1) = 0$

10. $(e^{-4x} y')' - 5e^{-4x} y = -f(x)$;
 $y(0) = 0$, $y'(1) = 0$

In Problems 11–20, use Green's functions to solve the given boundary value problem.

11. $y'' = -e^x$; $y(0) = 0$, $y(\pi) = 0$

12. $y'' = x^2$; $y(0) = 0$, $y(\pi) = 0$

13. $y'' = -x$; $y(0) = 0$, $y(\pi) + y'(\pi) = 0$

14. $y'' = -x^2$; $y(0) = 0$, $y'(\pi) = 0$

15. $y'' + y = 1$; $y(0) = 0$, $y(1) = 0$

16. $y'' + y = -12$; $y(0) = 0$, $y(2) = 0$

17. $y'' - y = x$; $y(0) = 0$, $y(1) = 0$

18. $y'' - y = 24$; $y'(0) = 0$, $y'(1) = 0$

19. $(xy')' - (4/x)y = -x$; $y(1) = 0$, $y(2) = 0$

20. $(x^2 y')' = 1$; $y(1) = 0$, $y(2) = 0$

21. Derive a formula using a Green's function for the solution to the following nonhomogeneous equation with nonhomogeneous boundary conditions

 (25) $y''(x) = -f(x)$; $y(0) = \alpha$, $y(1) = \beta$.

[*Hint:* Let $y = u + v$, where $u'' = 0$ and u satisfies the nonhomogeneous boundary conditions in (25), while $v'' = -f$ and v satisfies the corresponding homogeneous boundary conditions where $\alpha = \beta = 0$.]

22. Nonselfadjoint Equations. Assume the homogeneous problem associated with the boundary value problem

$$A_2(x)y''(x) + A_1(x)y'(x) + A_0(x)y(x) = -f(x),$$
$$a < x < b,$$

$$a_1 y(a) + a_2 y'(a) = 0 \quad \text{and} \quad b_1 y(b) + b_2 y'(b) = 0$$

has only the trivial solution. Using the method of variation of parameters and proceeding in a manner similar to that used in this section, show that the solution to this problem is given by the formula

$$y(x) = \int_a^b K(x, s) f(s)\, ds,$$

where

$$K(x, s) = \begin{cases} \dfrac{-z_1(s)z_2(x)}{W[z_1, z_2](s)A_2(s)}, & a \leq s \leq x, \\[2ex] \dfrac{-z_1(x)z_2(s)}{W[z_1, z_2](s)A_2(s)}, & x \leq s \leq b. \end{cases}$$

The function $K(x, s)$ is called the Green's function for the nonselfadjoint problem.

In Problems 23–26, use the results of Problem 22 to (a) find the Green's function $K(x, s)$ for the given (nonselfadjoint) boundary value problem and (b) solve the given boundary value problem using the Green's function.

23. $y'' + 3y' + 2y = -x$; $y(0) = 0$, $y(1) = 0$

24. $y'' - 2y' + y = 1$; $y(0) = 0$, $y(1) = 0$

25. $x^2 y'' - 2xy' + 2y = x$; $y(1) = 0$, $y(2) = 0$

26. $x^2 y'' + 3xy' - 3y = x$; $y(1) = 0$, $y'(2) = 0$

In Problems 27–30, find the Green's function $H(x, s)$ for the fourth-order equation

$$y^{(4)}(x) = -f(x),$$

with given boundary conditions. [*Hint: Solve the boundary value problem for* $f(x) = \delta(x - s)$. *Observe that the jump condition is*

$$\lim_{x \to s^+} \frac{\partial^3 H(x, s)}{\partial x^3} - \lim_{x \to s^-} \frac{\partial^3 H(x, s)}{\partial x^3} = -1.]$$

27. $y(0) = y'(0) = 0,$ $y(\pi) = y'(\pi) = 0$

28. $y(0) = y''(0) = 0,$ $y(\pi) = y''(\pi) = 0$

29. $y(0) = y'(0) = 0,$ $y''(\pi) = y'''(\pi) = 0$

30. $y(0) = y'(0) = 0,$ $y(\pi) = y''(\pi) = 0$

31. Let y_1 and y_2 be solutions to

$$y'' + \frac{p'}{p} y' + \frac{q}{p} y = 0,$$

on (a, b). Use Abel's identity (equation (26) Section 6.1, page 324) to show that

$$p(x) W[y_1, y_2](x) = C$$

for all x in $[a, b]$, where C is a constant.

32. Define $G(x, s)$ to be the (unique) function that satisfies properties (a)–(c) in Theorem 12. To show that $G(x, s)$ is given by formula (12), proceed as follows:

(i) Using property (c), show that $G(x, s)$ has the form

$$G(x, s) = \begin{cases} u(s) z_1(x), & a \le x < s, \\ v(s) z_2(x), & s < x \le b, \end{cases}$$

where $L[z_1] = 0$ and z_1 satisfies the boundary conditions in (21) at $x = a$, while $L[z_2] = 0$ and z_2 satisfies the boundary conditions in (21) at $x = b$.

(ii) Imposing conditions (a) and (b) on $G(x, s)$, show that u and v must satisfy the system

$$v(s) z_2(s) - u(s) z_1(s) = 0,$$

$$v(s) z_2'(s) - u(s) z_1'(s) = -\frac{1}{p(s)}.$$

(iii) Solving this system for u and v, derive formula (12) for $G(x, s)$.

33. Define $G(x, s)$ to be the solution to equations (22)–(24). To formally show that $G(x, s)$ has properties (a)–(c) of Theorem 12, proceed as follows:

(i) Use the fact that $\delta(x - s) = 0$ for $x \ne s$ to show that $G(x, s)$ satisfies property (c). Moreover, for each fixed s, the partial derivatives

$(\partial G / \partial x)(x, s)$ and $(\partial^2 G / \partial x^2)(x, s)$ are continuous functions of x for $x \ne s$.

(ii) To show that $G(x, s)$ is continuous on $[a, b] \times [a, b]$ and complete the proof of property (a), use the fact that

$$y(x) = \int_a^b G(x, s) f(s) \, ds,$$

the solution to (2)–(4), is continuous and hence, for any f,

$$0 = \lim_{\varepsilon \to 0^+} [y(s + \varepsilon) - y(s - \varepsilon)]$$

$$= \int_a^b \lim_{\varepsilon \to 0^+} [G(s + \varepsilon, s) - G(s - \varepsilon, s)] f(s) \, ds.$$

(iii) Fix s and let $x_1 < s < x_2$. Formally integrate both sides of (22) with respect to x from $x = x_1$ to $x = x_2$, to obtain

$$p(x) \frac{\partial G}{\partial x}(x, s) \Big|_{x_1}^{x_2} + \int_{x_1}^{x_2} q(x) G(x, s) \, dx$$

$$= -\int_{x_1}^{x_2} \delta(x - s) \, dx.$$

From this equation, use the continuity of p, q, and G and the property of the Dirac delta function that

$$\int_{-\infty}^{\infty} \delta(x - s) \, dx = 1$$

to derive property (b).

34. Let $G(x, s)$ be the Green's function for the regular Sturm–Liouville boundary value problem

$$L[y] + \mu y = -f; \qquad B[y] = 0.$$

Let $f(x) = \delta(x - s)$. Using the method of eigenfunction expansions discussed in Section 11.5, derive the eigenfunction formula for $G(x, s)$,

$$G(x, s) = \sum_{n=1}^{\infty} \frac{\phi_n(x) \phi_n(s)}{\lambda_n - \mu},$$

where $\{\phi_n\}_{n=1}^{\infty}$ is an orthonormal system of eigenfunctions with corresponding eigenvalues $\{\lambda_n\}_{n=1}^{\infty}$ for

$$L[y] + \lambda y = 0; \qquad B[y] = 0.$$

Assume that μ is *not* an eigenvalue.

35. Show that property (b) in Theorem 12 is equivalent to

$$\lim_{s \to x^+} \frac{\partial G}{\partial x}(x, s) - \lim_{s \to x^-} \frac{\partial G}{\partial x}(x, s) = \frac{1}{p(x)}.$$

11.7 Singular Sturm–Liouville Boundary Value Problems

In applications to heat flow on a cylinder, one encounters Sturm–Liouville boundary value problems that do not satisfy *all* the conditions required for the problem to be regular (see Section 11.3).

Let

(1) $$L[y](x) := \frac{d}{dx}\left[p(x)\frac{dy}{dx}\right] + q(x)y$$

and consider the equation

(2) $$L[y](x) + \lambda r(x)y(x) = 0\,,\qquad a < x < b\,.$$

We assume that $p(x), p'(x), q(x)$, and $r(x)$ are real-valued continuous functions on (a,b) and $p(x)$ and $r(x)$ are positive on (a,b). Notice that our assumptions apply only on the *open* interval (a,b) and *not* the closed interval $[a,b]$, which is the case for a regular Sturm–Liouville boundary value problem. We will call equation (2) a **singular Sturm–Liouville equation**[†] if one or more of the following situations hold:

(a) $\lim_{x\to a^+}p(x) = 0$ or $\lim_{x\to b^-}p(x) = 0$.

(b) $p(x), q(x)$, or $r(x)$ becomes unbounded as x approaches either a or b.

(c) The interval (a,b) is unbounded (e.g., (a,∞), $(-\infty, b)$, or $(-\infty,\infty)$).

Three important examples of singular Sturm–Liouville equations are Bessel's equation, Legendre's equation, and Hermite's equation (see page 485). Bessel's equation governs the radial variable when the heat, wave, or Laplace equation is separated in cylindrical coordinates; the coefficient p becomes zero as the radial coordinate approaches zero, near the z-axis. Legendre's equation governs the latitudinal angle variable when these partial differential equations are separated in spherical coordinates; its coefficient p becomes zero as the angle approaches 0 or π (near the north and south poles). Hermite's equation arises from separation of the Schrödinger equation for a quantum mechanical mass–spring oscillator; the independent variable x runs from $-\infty$ to $+\infty$.

Bessel's Equation of Order v

(3) $$t^2 y''(t) + ty'(t) + (t^2 - v^2)y = 0\,,\qquad 0 < t < \sqrt{\lambda}\,b$$

can be transformed by substituting $t = \sqrt{\lambda}x$ and dividing by x into the singular Sturm–Liouville equation

(4) $$\frac{d}{dx}\left[x\frac{dy}{dx}\right] - \frac{v^2}{x}y + \lambda xy = 0\,,\qquad 0 < x < b\,.$$

Here $p(x) = x$ approaches zero and $q(x) = -v^2/x$ becomes unbounded as $x \to 0^+$.

[†]For a more rigorous treatment of general singular Sturm–Liouville problems, see "Orthogonal Polynomials and Singular Sturm–Liouville Systems, I" by Lance L. Littlejohn and Allan M. Krall, *Rocky Mountain Journal of Mathematics*, Vol. 16, No. 3 (1986): 435–479.

Legendre's Equation

$$(5) \qquad (1 - x^2)y'' - 2xy' + n(n+1)y = 0, \qquad -1 < x < 1$$

can be written as the singular Sturm–Liouville equation

$$(6) \qquad \frac{d}{dx}\left[(1 - x^2)\frac{dy}{dx}\right] + \lambda y = 0, \qquad -1 < x < 1,$$

where $\lambda = n(n+1)$. Here $p(x) = 1 - x^2$ is zero at both endpoints ± 1.

Hermite's Equation†

$$(7) \qquad y'' - 2xy' + 2ny = 0, \qquad -\infty < x < \infty,$$

when multiplied by e^{-x^2}, becomes the singular Sturm–Liouville equation

$$(8) \qquad \frac{d}{dx}\left[e^{-x^2}\frac{dy}{dx}\right] + \lambda e^{-x^2}y = 0, \qquad -\infty < x < \infty,$$

where $\lambda = 2n$. Here the interval $(-\infty, \infty)$ is unbounded.

In Section 11.3 we observed that the linear operator associated with a regular Sturm–Liouville boundary value problem was a *selfadjoint* operator. Since many of the properties of regular Sturm–Liouville boundary value problems are consequences of this fact, it is useful to know which boundary conditions make a singular Sturm–Liouville equation into a selfadjoint problem. To investigate this question, we return to Green's formula [see equation (10) in Section 11.3]. Since p and q are continuous on (a, b), the only possible change in Green's formula for a singular Sturm–Liouville equation is that the integral may be improper at one or both endpoints. In this case, we have

$$(9) \qquad \int_a^b (uL[v] - vL[u])(x)dx = \lim_{x \to b^-} p(x)W[u, v](x) - \lim_{x \to a^+} p(x)W[u, v](x).$$

Now the operator L will be selfadjoint if the right-hand side of (9) is zero [compare equation (13) in Section 11.3]. This would be the case if each limit, taken separately, were zero. So let's consider just one of the limits on the right-hand side of (9) and ask what boundary conditions on u and v ensure that

$$(10) \qquad \lim_{x \to a^+} p(x)[u(x)v'(x) - u'(x)v(x)] = 0.$$

One type of boundary condition for which (10) holds is the separated boundary condition (page 641). This is the type of boundary condition we imposed on a regular Sturm–Liouville boundary value problem. Furthermore, it occurs naturally in such applications as the flow of heat in a wire (see Section 11.1). Another type of boundary condition that we encountered in our study of Laplace's equation is the requirement that both $y(x)$ and $y'(x)$ remain bounded as x approaches the endpoint (see Examples 2 and 3 in Section 10.7). This condition is more general than the separated boundary condition. A third condition arising in applications is the requirement that $\sqrt{p(x)}\,y(x)$ and $\sqrt{p(x)}\,y'(x)$ go to zero as x approaches the boundary (see Project A). These three types of boundary conditions, with suitable assumptions on $p(x)$, guarantee that equation (10) holds at the endpoint a, as we now show.

†*Historical Footnote:* Named after the French mathematician Charles Hermite (1822–1901), who contributed significantly to algebra and analysis.

Singular Boundary Conditions

Lemma 1. Any one of the following three conditions guarantees that equation (10) holds:

(i) $\lim_{x \to a^+} p(x) = p(a)$ exists and u, v satisfy the boundary condition

(11) $a_1 y(a) + a_2 y'(a) = 0$, a_1, a_2 not both zero.

(ii) $\lim_{x \to a^+} p(x) = 0$ and u, v satisfy the boundary condition

(12) $y(x), y'(x)$ remain bounded as $x \to a^+$.

(iii) The functions u, v satisfy the boundary condition

(13) $\lim_{x \to a^+} \sqrt{p(x)}\, y(x) = 0$ and $\lim_{x \to a^+} \sqrt{p(x)}\, y'(x) = 0$.

Proof.

(i) If $a_1 = 0$, $u'(a) = v'(a) = 0$ and (10) is immediate. Otherwise, from (11) we have $u(a) = -(a_2/a_1)u'(a)$ and $v(a) = -(a_2/a_1)v'(a)$. Substituting into the left-hand side of (10) yields

(14) $p(a)[u(a)v'(a) - u'(a)v(a)]$
$$= p(a)[-(a_2/a_1)u'(a)v'(a) + (a_2/a_1)u'(a)v'(a)] = 0.$$

(ii) If u, u', v, and v' remain bounded as x approaches a from the right, then the factor in brackets on the left-hand side of (10) remains bounded. Since $\lim_{x \to a^+} p(x) = 0$, the left-hand side is just a bounded factor times a factor that goes to zero. Hence, the left-hand side goes to zero and (10) holds.

(iii) If u and v satisfy (13), then substituting into the left-hand side of (10) gives

$$\lim_{x \to a^+} p(x)[u(x)v'(x) - u'(x)v(x)] = \left(\lim_{x \to a^+} \sqrt{p(x)}\, u(x) \right)\left(\lim_{x \to a^+} \sqrt{p(x)}\, v'(x) \right)$$
$$- \left(\lim_{x \to a^+} \sqrt{p(x)}\, u'(x) \right)\left(\lim_{x \to a^+} \sqrt{p(x)}\, v(x) \right)$$
$$= 0 - 0 = 0.$$

Hence, equation (10) is satisfied. ◆

Although Lemma 1 deals with the left endpoint a, an analogous lemma can be stated for the right endpoint b that ensures

(15) $\lim_{x \to b^-} p(x)[u(x)v'(x) - u'(x)v(x)] = 0$.

When both (10) and (15) hold, the operator L will be selfadjoint [recall equation (9)]. For example, if condition (i) of Lemma 1 holds for a and the analog of condition (ii) holds for b, then L is selfadjoint.

Even when boundary conditions imposed on the singular Sturm–Liouville equation give a linear operator that is selfadjoint, other difficulties can arise. For example, the eigenvalues may *not* be discrete (isolated). Indeed, there may be an entire interval I such that if λ is any number in I, then the problem has a nontrivial solution (see Project B on page 707). In such a case, the problem

Figure 11.6 Set of eigenvalues containing discrete points as well as an interval

is said to have a **continuous spectrum.** It is also possible to have both discrete eigenvalues and a continuous spectrum (see Figure 11.6). When the singular Sturm–Liouville boundary value problem has no continuous spectra but has an infinite number of discrete eigenvalues, the arguments given in Section 11.3 for regular Sturm–Liouville boundary value problems can be modified to show that the eigenvalues must be real and that eigenfunctions corresponding to distinct eigenvalues are orthogonal with respect to $r(x)$ on (a, b) (see Problem 7). Moreover, the formulas for the coefficients of an eigenfunction expansion remain the same. However, the convergence of the eigenfunction expansion must be examined on an individual basis. For an in-depth discussion of singular Sturm–Liouville boundary value problems, we refer the reader to an advanced text on the subject.[†]

To illustrate the behavior of solutions to singular Sturm–Liouville boundary value problems, let's study separately the boundary value problems associated with Bessel's equation, Legendre's equation, and Hermite's equation.

Bessel's Equation

A typical singular Sturm–Liouville boundary value problem associated with Bessel's equation of order v is

(16) $$\frac{d}{dx}\left[x\frac{dy}{dx}\right] - \frac{v^2}{x}y + \lambda xy = 0, \qquad 0 < x < 1,$$

(17) $y(x), y'(x)$ remain bounded as $x \to 0^+$,

(18) $y(1) = 0$.

This problem is selfadjoint, since condition (ii) of Lemma 1 holds at $a = 0$ and the analog of condition (i) of Lemma 1 holds at $b = 1$.

If $\lambda > 0$, the substitution $t = \sqrt{\lambda}x$ transforms (16) into the usual Bessel's equation

(19) $$t^2z'' + tz' + (t^2 - v^2)z = 0$$

with $z(t) := y(t/\sqrt{\lambda})$. Using power series methods, we found in Section 8.8 that a general solution to (19) is

$$z(t) = c_1J_v(t) + c_2Y_v(t),$$

or

$$y(x) = c_1J_v(\sqrt{\lambda}x) + c_2Y_v(\sqrt{\lambda}x).$$

It follows from the power series expansion for $J_v(\sqrt{\lambda}x)$ that when $v = 0$ or $v \geq 1$, both $J_v(\sqrt{\lambda}x)$ and $\sqrt{\lambda}J'_v(\sqrt{\lambda}x)$ remain bounded as $x \to 0^+$ (see Problem 9). However, the function $Y_v(\sqrt{\lambda}x)$ does not. Hence, we must have $c_2 = 0$ for (17) to be satisfied. (The case when $0 < v < 1$ is discussed in Problem 10.)

[†]See, for example, the book by K. Yosida, *Lectures on Differential and Integral Equations* (Dover Publications, New York, 1991).

For (18) to be satisfied for $y(x) = c_1 J_\nu(\sqrt{\lambda} x)$, we need $c_1 = 0$ or

$$J_\nu(\sqrt{\lambda}) = 0 .$$

That is, $\sqrt{\lambda}$ must be a zero of J_ν in order for $J_\nu(\sqrt{\lambda} x)$ to be a nontrivial solution to (16)–(18). It is known that the Bessel function J_ν has an increasing sequence of real zeros (see Figure 10.31, page 632):

$$0 < \alpha_{\nu 1} < \alpha_{\nu 2} < \alpha_{\nu 3} < \cdots .$$

Hence, $\lambda_n = \alpha_{\nu n}^2$, $n = 1, 2, 3, \ldots$ are eigenvalues of (16)–(18) with corresponding eigenfunctions

(20) $y_n(x) = c_1 J_\nu(\alpha_{\nu n} x) .$

These are the only positive eigenvalues. Moreover, one can show that there are no nonpositive eigenvalues of (16)–(18) (see Problem 11).

The eigenfunctions in (20) are orthogonal with respect to the weight function $r(x) = x$:

(21) $\displaystyle \int_0^1 J_\nu(\alpha_{\nu n} x) J_\nu(\alpha_{\nu m} x) x \, dx = 0 , \qquad n \neq m .$

If $f(x)$ is a given function, then an eigenfunction expansion associated with f is

(22) $\displaystyle f \sim \sum_{n=1}^\infty a_n J_\nu(\alpha_{\nu n} x) ,$

where

$$a_n = \frac{\displaystyle \int_0^1 f(x) J_\nu(\alpha_{\nu n} x) x \, dx}{\displaystyle \int_0^1 J_\nu^2(\alpha_{\nu n} x) x \, dx} , \qquad n = 1, 2, 3, \ldots .$$

When f and f' are piecewise continuous on $[0, 1]$, the expansion in (22) converges for x in $(0, 1)$ to $[f(x^+) + f(x^-)]/2$.

Legendre's Equation

A typical singular Sturm–Liouville boundary value problem associated with Legendre's equation is

(23) $\displaystyle \frac{d}{dx}\left[(1 - x^2) \frac{dy}{dx} \right] + \lambda y = 0 , \qquad -1 < x < 1 ,$

(24) $y(x), y'(x)$ remain bounded as $x \to \pm 1$.

This problem is selfadjoint, since condition (ii) of Lemma 1 holds at each endpoint. In Section 8.8 we found that (23) has polynomial solutions for $\lambda = n(n + 1)$, $n = 0, 1, 2, \ldots$. These solutions are just constant multiples of the Legendre polynomials

(25) $\displaystyle P_n(x) = 2^{-n} \sum_{m=0}^{[n/2]} \frac{(-1)^m (2n - 2m)!}{(n - m)! \, m! \, (n - 2m)!} x^{n-2m} ,$

where $\lceil n/2 \rceil$ is the greatest integer less than or equal to $n/2$ [compare (43) on page 481, Section 8.8]. The first four Legendre polynomials are

$$P_0(x) = 1, \qquad P_1(x) = x, \qquad P_2(x) = \frac{3}{2}x^2 - \frac{1}{2}, \qquad P_3(x) = \frac{5}{2}x^3 - \frac{3}{2}x.$$

It is left to you to show that the only eigenvalues for (23)–(24) are $\lambda = n(n+1)$, $n = 0, 1, 2, \ldots$, with corresponding eigenfunctions $P_n(x)$, the Legendre polynomials (see Problem 12). That the Legendre polynomials are (up to constant multiples) the only eigenfunctions corresponding to these eigenvalues follows from the fact that a second linear independent solution to (23) involves a $\ln(1 - x)$ term that becomes unbounded as $x \to 1^-$ (see Problem 13).

The Legendre polynomials $P_n(x)$ are orthogonal on $[-1, 1]$ with respect to $r(x) = 1$:

$$(26) \qquad \int_{-1}^{1} P_n(x) P_m(x) \, dx = 0, \qquad n \neq m.$$

If $f(x)$ is a given function, then an eigenfunction expansion for $f(x)$ in terms of Legendre polynomials is

$$(27) \qquad f \sim \sum_{n=0}^{\infty} a_n \, P_n(x),$$

where

$$a_n = \frac{\displaystyle\int_{-1}^{1} f(x) P_n(x) \, dx}{\displaystyle\int_{-1}^{1} P_n^2(x) \, dx}, \qquad n = 0, 1, 2, \ldots.$$

Convergence results similar to those for Fourier series hold for the expansion in (27) (compare Section 10.3).

Hermite's Equation

A singular Sturm–Liouville boundary value problem associated with Hermite's equation is

$$(28) \qquad \frac{d}{dx}\left[e^{-x^2} \frac{dy}{dx} \right] + \lambda e^{-x^2} y = 0, \qquad -\infty < x < \infty,$$

$$(29) \qquad \lim_{x \to \pm\infty} e^{-x^2/2} y(x) = 0 \quad \text{and} \quad \lim_{x \to \pm\infty} e^{-x^2/2} y'(x) = 0.$$

By condition (iii) of Lemma 1, the linear operator associated with (28)–(29) is selfadjoint.

In Section 8.8 we observed that Hermite's equation

$$(30) \qquad y'' - 2xy' + 2ny = 0$$

[which is equivalent to (28)] has polynomial solutions that are constant multiples of the Hermite polynomials $H_n(x)$, $n = 0, 1, 2, \ldots$. Since polynomial solutions satisfy the boundary conditions (29), we find that the singular Sturm–Liouville boundary value problem (28)–(29) has eigenvalues $\lambda_n = 2n$, $n = 0, 1, 2, \ldots$, with corresponding eigenfunctions $H_n(x)$. In fact, it can be shown that these are the only eigenvalues. The first four Hermite polynomials are

$$H_0(x) = 1, \qquad H_1(x) = 2x, \qquad H_2(x) = 4x^2 - 2, \qquad H_3(x) = 8x^3 - 12x$$

(see Exercises 8.8, Problem 37).

The Hermite polynomials $H_n(x)$ are orthogonal on $(-\infty, \infty)$ with respect to the weight function $r(x) = e^{-x^2}$:

(31) $$\int_{-\infty}^{\infty} H_n(x)H_m(x)e^{-x^2}\,dx = 0\,, \qquad n \neq m\,.$$

If f is a given function, then an eigenfunction expansion associated with f is given by

(32) $$f \sim \sum_{n=0}^{\infty} a_n H_n(x)\,,$$

where

$$a_n = \frac{\displaystyle\int_{-\infty}^{\infty} f(x)H_n(x)e^{-x^2}\,dx}{\displaystyle\int_{-\infty}^{\infty} H_n^2(x)e^{-x^2}\,dx}\,, \qquad n = 0, 1, 2, \dots\,.$$

A detailed study of the eigenvalues and eigenfunctions for Hermite's boundary value Problem (28)–(29), along with conditions for the convergence of the eigenfunction expansion in (32), is beyond the scope of this book. We refer you to an advanced book on these topics.[†]

11.7 EXERCISES

In Problems 1–6, first verify that the corresponding homogeneous problem is selfadjoint. Then, find a formal eigenfunction expansion for the solution to the given nonhomogeneous boundary value problem where f is a continuous function and μ is not an eigenvalue of the corresponding homogeneous problem.

1. $(xy')' - \dfrac{4}{x}y + \mu xy = f(x)\,, \qquad 0 < x < 1\,,$
 $y(x), y'(x)$ remain bounded as $x \to 0^+, y(1) = 0$

2. $(xy')' - \dfrac{9}{x}y + \mu xy = f(x)\,, \qquad 0 < x < 1\,,$
 $y(x), y'(x)$ remain bounded as $x \to 0^+, y(1) = 0$

3. $(xy')' + \mu xy = f(x)\,, \qquad 0 < x < 1\,,$
 $y(x), y'(x)$ remain bounded as $x \to 0^+, y'(1) = 0$

4. $[(1 - x^2)y']' + \mu y = f(x)\,, \qquad -1 < x < 1\,,$
 $y(x), y'(x)$ remain bounded as $x \to \pm 1$

5. $[(1 - x^2)y']' + \mu y = f(x)\,, \qquad 0 < x < 1\,,$
 $y(0) = 0, y(x), y'(x)$ remain bounded as $x \to 1^-$
 [*Hint:* Eigenfunctions are odd-degree Legendre polynomials.]

6. $[(1 - x^2)y']' + \mu y = f(x)\,, \qquad 0 < x < 1\,,$
 $y'(0) = 0, y(x), y'(x)$ remain bounded as $x \to 1^-$
 [*Hint:* Eigenfunctions are even-degree Legendre polynomials.]

7. Let (2) be a singular Sturm–Liouville equation. Let $B[y] = 0$ represent linear homogeneous boundary conditions such that, together with (1), they give rise to a selfadjoint operator L. Assume this singular Sturm–Liouville boundary value problem has no continuous spectra but has an infinite number of discrete eigenvalues $\lambda_n, n = 1, 2, 3, \dots$. Using arguments similar to those in Section 11.3 for regular Sturm–Liouville boundary value problems:
 (a) Show that the eigenvalues must be real.
 (b) Show that the eigenfunctions corresponding to distinct eigenvalues are orthogonal with respect to $r(x)$ on (a, b).

8. Verify that the change of variables $t = \sqrt{\lambda}\,x$ transforms equation (3) into equation (4).

[†]See, for example, the books by F. V. Atkinson, *Discrete and Continuous Boundary Value Problems* (Academic Press, Orlando, FL., 1964), and E. C. Titchmarsh, *Eigenfunction Expansions Associated with Second-Order Differential Equations*, Vols. I and II (Camp Press, La Vergne, TN, 2007).

9. The power series expansion for $J_\nu(\sqrt{\lambda}x)$ has the form $(\sqrt{\lambda}x)^\nu g(\sqrt{\lambda}x)$, where g is analytic about $x = 0$. Show that for $\nu = 0$ or $\nu \geq 1$, both $J_\nu(\sqrt{\lambda}x)$ and $J'_\nu(\sqrt{\lambda}x)$ remain bounded as $x \to 0^+$.

10. For Bessel's equation of order ν where $0 < \nu < 1$, we replace the boundary condition (17) by

$$(33) \qquad \lim_{x \to 0^+} y(x) = 0, \qquad \lim_{x \to 0^+} xy'(x) = 0.$$

(a) Show that if u and v satisfy the boundary conditions in (33), then equation (10) holds at $a = 0$ and hence the singular Sturm–Liouville boundary value problem (16), (18), and (33) gives rise to a selfadjoint operator.

(b) Using the representation for $J_\nu(\sqrt{\lambda}x)$ given in Problem 9, show that $J_\nu(\sqrt{\lambda}x)$ satisfies the boundary conditions in (33).

11. To show that the eigenvalues for Bessel's equation must be positive, proceed as follows:

(a) Let $\phi(x)$ be an eigenfunction for (16)–(18) with corresponding eigenvalue λ. Substitute $\phi(x)$ into (16). Now multiply both sides of (16) by $\phi(x)$ and integrate from 0 to 1. Using integration by parts, obtain

$$(34) \qquad -\int_0^1 x[\phi'(x)]^2\, dx - \nu^2 \int_0^1 x^{-1}[\phi(x)]^2\, dx$$
$$+ \lambda \int_0^1 x[\phi(x)]^2\, dx = 0.$$

(b) For $\nu > 0$, use equation (34) to conclude that $\lambda > 0$.

(c) For $\nu = 0$, use equation (34) to conclude that $\lambda \geq 0$. To prove $\lambda = 0$ is not an eigenvalue for $\nu = 0$, find a general solution to $xy'' + y' = 0$ and show that only the trivial solution satisfies the boundary conditions in (17)–(18).

12. (a) Show that the only nonnegative eigenvalues for (23)–(24) are $\lambda_n = n(n+1)$, $n = 0, 1, 2, \ldots$. [*Hint:* Find a power series expansion for a general solution to (23).]

(b) Show that (23)–(24) does not have any negative eigenvalues. [*Hint:* Use a procedure similar to the one outlined in Problem 11.]

13. Show that the only eigenfunctions of (23)–(24) corresponding to the eigenvalues $\lambda_n = n(n+1)$, $n = 0, 1, 2, \ldots$, are multiples of the Legendre polynomials $P_n(x)$. [*Hint:* In Section 8.8, we showed that exponents of the singularity $x_0 = 1$ for Legendre's equation were both zero. Use Theorem 7 in Section 8.7

to show that any other solution has a logarithmic term that becomes unbounded as $x \to 1^-$.]

14. (a) Use formula (25) to show that $P_n(x)$ is an odd function when n is odd and is an even function when n is even.

(b) Show that the odd degree Legendre polynomials are orthogonal on $(0, 1)$; that is,

$$\int_0^1 P_{2m+1}(x) P_{2n+1}(x)\, dx = 0, \qquad m \neq n.$$

(c) Show that the even degree Legendre polynomials are orthogonal on $(0, 1)$; that is,

$$\int_0^1 P_{2m}(x) P_{2n}(x)\, dx = 0, \qquad m \neq n.$$

15. **Chebyshev's (Tchebichef's) Equation.**

(a) Show that Chebyshev's equation,

$$(1 - x^2)y'' - xy' + \lambda y = 0,$$
$$-1 < x < 1,$$

can be written in the Sturm–Liouville form

$$(35) \qquad [(1 - x^2)^{1/2}y']' + \lambda(1 - x^2)^{-1/2}y = 0,$$
$$-1 < x < 1.$$

(b) It can be shown that Chebyshev's equation (35) with the boundary conditions

$$y(x), y'(x) \text{ remain bounded as } x \to \pm 1$$

has eigenvalues $\lambda_n = n^2$, $n = 0, 1, 2, \ldots$. The corresponding eigenfunctions are the Chebyshev polynomials $T_n(x)$. The first four are

$$T_0(x) = 1, \qquad T_1(x) = x,$$
$$T_2(x) = 2x^2 - 1, \; T_3(x) = 4x^3 - 3x.$$

Show that

$$\int_{-1}^1 \frac{T_m(x) T_n(x)}{(1 - x^2)^{1/2}}\, dx = 0, \qquad m \neq n.$$

(c) Find a formal eigenfunction expansion for the solution to the nonhomogeneous boundary value problem

$$[(1 - x^2)^{1/2}y']' + \mu(1 - x^2)^{-1/2}y = f(x),$$
$$-1 < x < 1,$$
$$y(x), y'(x) \text{ remain bounded as } x \to \pm 1,$$

where f is continuous on $[-1, 1]$ and μ is not an eigenvalue of the corresponding homogeneous problem.

16. Laguerre's Equation.

(a) Show that Laguerre's equation

$$xy'' + (1-x)y' + \lambda y = 0,$$

$$0 < x < \infty$$

can be written in the Sturm–Liouville form,

$$(36) \quad (xe^{-x}y')' + \lambda e^{-x}y = 0,$$

$$0 < x < \infty .$$

(b) It can be shown that Laguerre's equation (36) with boundary conditions

$y(x), y'(x)$ remain bounded as $x \to 0^+$,

$x^{1/2}e^{-x/2}y(x)$ and $x^{1/2}e^{-x/2}y'(x) \to 0$ as $x \to \infty$

has eigenvalues $\lambda_n = n$, $n = 0, 1, 2, \ldots$. The corresponding eigenfunctions are the Laguerre polynomials $L_n(x)$. The first four are

$L_0(x) = 1$, $L_1(x) = 1 - x$,

$L_2(x) = (2 - 4x + x^2)/2$,

$L_3(x) = (6 - 18x + 9x^2 - x^3)/6$.

Show that

$$\int_0^\infty L_m(x)L_n(x)e^{-x}\,dx = 0, \qquad m \neq n .$$

(c) Find a formal eigenfunction expansion for the solution to the nonhomogeneous equation

$$(xe^{-x}y')' + \mu e^{-x}y = f(x), \qquad 0 < x < \infty ,$$

satisfying the boundary conditions in part (b), where μ is not an eigenvalue of the corresponding homogeneous problem and f is continuous on $[0, \infty)$ and satisfies

$$\lim_{x \to \infty} f(x)e^{-x/2} = 0.$$

11.8 Oscillation and Comparison Theory

In Chapter 4 we developed a model for a simple mass–spring system involving a second-order linear differential equation. We found that when the damping constant is sufficiently small, the mass vibrates back and forth or **oscillates.** In fact, the solution was a product of a damping factor $Ae^{-(b/2m)t}$ and a sine function $\sin(\beta t + \theta)$ [compare equation (17) in Section 4.9].

Functions that oscillate cannot generally be expressed in such a nice fashion. In fact, the concept of oscillation is difficult to capture mathematically. Clearly (nonconstant) periodic functions oscillate, but many other types of functions also exhibit what we would call oscillatory behavior. For our purposes, we will identify oscillatory behavior of $y = f(x)$ with repeated contact with the x-axis. Thus, the more frequently the graph of a function intersects the x-axis, the more it oscillates. If a function has an *infinite* number of zeros on an interval $[a, \infty)$, we say that the function is **oscillatory.** Hence, studying the oscillatory behavior of a function means we are investigating the number and locations of its zeros.

In this section we investigate the zeros of solutions to the Sturm–Liouville equation, which we write as

$$(1) \quad \frac{d}{dx}\left(p(x)\frac{dy}{dx}\right) + Q(x)y = p(x)y'' + p'(x)y' + Q(x)y = 0, \quad a < x < b .$$

If the equation expresses an eigenvalue problem, then $Q(x) = q(x) + \lambda r(x)$; for the present we do not need to make a distinction. The functions p, p', Q, q, and r are assumed continuous on $[a,b]$, and p and r are positive on $[a,b]$.[†]

[†]For an unbounded interval such as $(0, \infty)$, we assume that the conditions on p and Q hold on the open interval $(0, \infty)$.

To get a rough feeling for how the factor Q affects the solutions, note first of all that in the mass–spring analogy (see Section 4.1, on page 152), p plays the role of the inertia, p' is the damping coefficient, and Q is the spring stiffness. Because these are typically nonconstant functions, all kinds of solution behaviors can be expected. But since the "inertia" is positive, we can reasonably expect that as the "stiffness" increases, the oscillation "frequency" also increases; if $Q_2(x)$ is larger than $Q_1(x)$, then solutions for Q_2 should oscillate faster than solutions for Q_1. On the other hand, if the stiffness Q is large negative, the solutions won't oscillate at all—recall the discussion of the negative spring in Example 5 of Section 4.3. (In fact the speculations of Section 4.8 will be quantified in this section.)

In particular, for the Sturm–Liouville eigenvalue problem, the stiffness $Q(x) = q(x) + \lambda r(x)$ varies directly with λ (because the factor $r(x)$ is also presumed positive). So, as we have surmised from the eigenfunction graphs displayed in this and the preceding chapter, Sturm–Liouville eigenfunctions oscillate faster as the eigenvalue λ increases, and they stop oscillating altogether as λ decreases.

Let's get started with the analysis. The first question one might ask is: Can a solution have an infinite number of zeros in (a, b)? The following theorem shows that only the trivial solution may have an infinite number of zeros.

Finite Number of Zeros

Theorem 13. Let ϕ be a solution to (1) on (a, b). If ϕ has an infinite number of zeros in *any* closed, bounded interval $[a', b'] \subset (a, b)$, then $\phi(x) \equiv 0$ in (a, b).

Proof. To prove this theorem, we need a result from real analysis called the *Bolzano–Weierstrass* theorem.[†] It states that a bounded infinite set has at least one limit point. That is, there is a sequence of distinct points $\{x_n\}_{n=1}^{\infty}$ in the set and a point x_0 such that $\lim_{n \to \infty} x_n = x_0$. Here we are assuming an infinite number of zeros of ϕ lie in $[a', b']$, a bounded interval. By the Bolzano–Weierstrass theorem, there is a sequence $\{x_n\}_{n=1}^{\infty}$ of (distinct) zeros of ϕ that converges to some point x_0. Since $[a', b']$ is a closed interval, the point x_0 must be in $[a', b']$. We claim

The point x_0 is a zero of ϕ. Since x_n is a zero of ϕ (that is, $\phi(x_n) = 0, n = 1, 2, \dots$), it follows from the continuity of ϕ that

$$\phi(x_0) = \lim_{n \to \infty} \phi(x_n) = \lim_{n \to \infty} 0 = 0 .$$

The point x_0 is a zero of ϕ'. By assumption, ϕ' exists for x in $[a', b']$, and we have

$$\phi'(x_0) = \lim_{x \to x_0} \frac{\phi(x) - \phi(x_0)}{x - x_0} = \lim_{x \to x_0} \frac{\phi(x)}{x - x_0} .$$

Since this limit must hold for *any* sequence of distinct points tending to x_0, we can choose the sequence $\{x_n\}_{n=1}^{\infty}$ of zeros of ϕ. Thus,

$$\phi'(x_0) = \lim_{n \to \infty} \frac{\phi(x_n)}{x_n - x_0} = 0 .$$

Finally, since $\phi(x_0) = 0$ and $\phi'(x_0) = 0$, it follows from the existence and uniqueness theorem for second-order linear equations that $\phi(x) \equiv 0$ on (a, b) (see Section 6.1). ◆

[†]See, for example, the book by W. Rudin, *Principles of Mathematical Analysis*, 3rd ed. (Example Product Manufacturer, 2013).

Remark. Theorem 13 shows that solutions with infinitely many zeros can occur only on an unbounded interval if p is everywhere positive.

Another question one might ask is: Can two solutions to equation (1) have a common zero on $[a, b]$? The answer is yes, but if they do, one function must be a constant times the other.

Common Zeros

> **Theorem 14.** If ϕ and ψ are two solutions to (1) and $\phi(x_0) = \psi(x_0) = 0$ for some x_0 in $[a, b]$, then ϕ and ψ are linearly dependent on $[a, b]$.

Proof. Computing the Wronskian of ϕ and ψ and using the fact that $\phi(x_0) = \psi(x_0) = 0$, we have

$$(2) \qquad W[\phi, \psi](x_0) = \phi(x_0)\psi'(x_0) - \phi'(x_0)\psi(x_0) = 0 \,.$$

But when the Wronskian of two solutions vanishes at some point in $[a, b]$, these solutions must be linearly dependent on $[a, b]$ according to Section 4.7. Thus, ϕ and ψ are linearly dependent. ◆

An immediate consequence of Theorem 14 is that if ϕ and ψ are two linearly independent solutions, then they must have *different zeros*. But, is there any relationship between their respective zeros? One answer is given by the Sturm separation theorem.

Sturm Separation Theorem

> **Theorem 15.** Let ϕ and ψ be two linearly independent solutions to (1). Then, between any two consecutive zeros of ϕ in $[a, b]$, there is precisely one zero of ψ (see Figure 11.7).

Proof. Let x_1 and x_2 be two consecutive zeros of ϕ in $[a, b]$. Since ϕ and ψ are linearly independent, it follows from the preceding theorem that $\psi(x_1) \neq 0$ and $\psi(x_2) \neq 0$.

Assume that the conclusion is false; that is, $\psi(x) \neq 0$ for x in $[x_1, x_2]$. Then the function $f(x) := \phi(x)/\psi(x)$ is well defined and continuous on $[x_1, x_2]$ and continuously differentiable on (x_1, x_2). Since $\phi(x_1) = \phi(x_2) = 0$, we have $f(x_1) = f(x_2) = 0$. It now follows from Rolle's theorem that there exists a number z in (x_1, x_2) such that $f'(z) = 0$.

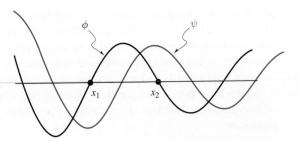

Figure 11.7 Interlacing of zeros of linearly independent solutions

Computing $f'(z)$, we find

$$(3) \qquad 0 = f'(z) = \frac{\phi'(z)\psi(z) - \phi(z)\psi'(z)}{[\psi(z)]^2} = \frac{W[\psi, \phi](z)}{[\psi(z)]^2}.$$

But this means that the Wronskian of ϕ and ψ vanishes at z, which contradicts the fact that ϕ and ψ are linearly independent. Thus $\psi(x)$ must have a zero in (x_1, x_2).

By interchanging the roles of ϕ and ψ, we also see that between any two consecutive zeros of ψ is a zero of ϕ. Hence, there cannot be more than one zero of ψ between two consecutive zeros of ϕ. ◆

(For example, the zeros of $\sin x$ and $\cos x$, two independent solutions of $y'' + y = 0$, alternate.)

A consequence of this theorem is the fact that between two consecutive zeros of a Bessel function of the first kind J_ν, there is a zero of the Bessel function of the second kind Y_ν (and vice versa). This is because these two functions are linearly independent solutions to the Sturm–Liouville equation

$$(4) \qquad \frac{d}{dx}\left(x\frac{dy}{dx}\right) + \left(x - \frac{\nu^2}{x}\right)y = 0, \qquad x > 0.$$

Since J_ν has infinitely many zeros on $(0, \infty)$, we conclude from this interlacing property of the zeros that Y_ν also has infinitely many zeros. This illustrates a general phenomenon.

Oscillatory Solutions

Corollary 3. Let the interval in Theorem 15 be the interval $[a, \infty)$. If one (nontrivial) solution to (1) is oscillatory, then all solutions to (1) are oscillatory.

Proof. Let ϕ be a nontrivial solution to (1) that has an infinite number of zeros on $[a, \infty)$. By the Sturm separation theorem, any other solution must have a zero between each successive pair of zeros of ϕ and hence must also have an infinite number of zeros on $[a, \infty)$. ◆

When we apply Theorem 15 to Hermite's equation in Sturm–Liouville form

$$(5) \qquad \frac{d}{dx}\left(e^{-x^2}\frac{dy}{dx}\right) + 2ne^{-x^2}y = 0, \qquad -\infty < x < \infty,$$

where n is a nonnegative integer, we find that since one solution is the Hermite polynomial $H_n(x)$ of degree n which is known to have n real zeros, then the other solutions must have only one zero between consecutive zeros of $H_n(x)$ for a minimum of $n - 1$ zeros. Moreover, any other (nontrivial) solution cannot have more than $n + 1$ zeros because if the contrary were true, it would imply that $H_n(x)$ has more zeros than its degree permits.

Thus far, we have compared the zeros of two solutions to the *same* equation. For solutions of two different equations, it still may be possible to relate their zeros. For example, for the equations

$$y'' + m^2 y = 0 \quad \text{and} \quad y'' + n^2 y = 0,$$

we see that the first has a general solution of the form $\phi_1(x) = A_1 \sin[m(x - \theta_1)]$ and the second, $\phi_2(x) = A_2 \sin[n(x - \theta_2)]$. The distance between the consecutive zeros of the first is π/m and the second is π/n. Therefore, when $m < n$, the distance between the zeros of ϕ_1 is greater than the distance between the zeros of ϕ_2. In other words, for $n^2 > m^2$, between any two consecutive zeros of $\phi_1(x)$ there is a zero of $\phi_2(x)$. A similar result holds when the constants m^2 and n^2 in the differential equations are replaced by functions of x. The next result is a version of the Sturm fundamental theorem.

Sturm Fundamental Theorem

Theorem 16. Let ϕ_1 be a nontrivial solution to

$$(6) \qquad y'' + Q_1(x)y = 0, \qquad a < x < b,$$

and let ϕ_2 be a nontrivial solution to

$$(7) \qquad y'' + Q_2(x)y = 0, \qquad a < x < b.$$

Assume $Q_2(x) \geq Q_1(x)$ for x in (a, b). If x_1 and x_2 are two consecutive zeros of $\phi_1(x)$ on (a, b), then there exists a zero of ϕ_2 between x_1 and x_2 unless $Q_2(x) = Q_1(x)$ on $[x_1, x_2]$ and ϕ_1 and ϕ_2 are linearly dependent on $[x_1, x_2]$.

Notice that the conclusion tells us that either ϕ_2 has a zero between x_1 and x_2 or $\phi_2(x_1) = \phi_2(x_2) = 0$. See Figure 11.8.

 Proof. Let x_1 and x_2 be two successive zeros of ϕ_1. Assume that $\phi_2(x) \neq 0$ in (x_1, x_2). Our goal is to show that $Q_2 = Q_1$ and that ϕ_1, ϕ_2 are linearly dependent on $[x_1, x_2]$. Without loss of generality, we further assume that $\phi_1(x) > 0$ and $\phi_2(x) > 0$ in (x_1, x_2). Combining this with the fact that $Q_2(x) \geq Q_1(x)$, we find

$$(8) \qquad \frac{d}{dx}(W[\phi_2, \phi_1]) = \frac{d}{dx}(\phi_2\phi_1' - \phi_2'\phi_1) = \phi_2'\phi_1' + \phi_2\phi_1'' - \phi_2'\phi_1' - \phi_1\phi_2''$$

$$= \phi_2\phi_1'' - \phi_2''\phi_1 = \phi_2(-Q_1\phi_1) - (-Q_2\phi_2)\phi_1$$

$$= \phi_1\phi_2(Q_2 - Q_1) \geq 0,$$

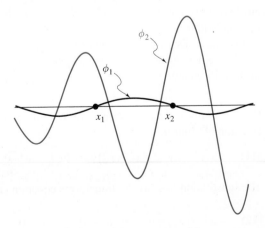

Figure 11.8 Sturm fundamental theorem: zeros of ϕ_2 separate zeros of ϕ_1

for x in (x_1, x_2). Hence, $W[\phi_2, \phi_1](x)$ is nondecreasing on (x_1, x_2). However, since $\phi_1(x_1) = \phi_1(x_2) = 0$ and ϕ_1 is positive on (x_1, x_2), we must have $\phi_1'(x_1) \geq 0$ and $\phi_1'(x_2) \leq 0$. Therefore, since $\phi_2(x) \geq 0$,

$$W[\phi_2, \phi_1](x_1) = \phi_2(x_1)\phi_1'(x_1) \geq 0$$

and

$$W[\phi_2, \phi_1](x_2) = \phi_2(x_2)\phi_1'(x_2) \leq 0 \,.$$

Because $W[\phi_2, \phi_1]$ is nondecreasing, the only way for it to be nonnegative at x_1 and nonpositive at x_2 is for

(9) $W[\phi_2, \phi_1](x) \equiv 0 \,,$ x in $[x_1, x_2]$.

Clearly equation (9) implies that $\frac{d}{dx} W[\phi_2, \phi_1](x) = 0$ on $[x_1, x_2]$. From equation (8), we deduce that $Q_2 = Q_1$ on $[x_1, x_2]$. But then ϕ_1, ϕ_2 satisfy the same differential equation on $[x_1, x_2]$ and their Wronskian vanishes on this interval. Therefore, ϕ_1 and ϕ_2 are linearly dependent on $[x_1, x_2]$. ◆

One immediate consequence of the Sturm fundamental theorem is the following.

Nonoscillating Solutions

Corollary 4. If $Q(x) \leq 0$ on $[a, b]$, then a nontrivial solution ϕ of the equation

(10) $y'' + Q(x)y = 0 \,,$ $a < x < b \,,$

can have at most one zero in $[a, b]$.

Proof. Since $\psi(x) \equiv 1$ is a solution to the equation $y'' + 0 \cdot y = 0$ and $Q(x) \leq 0$, it follows from the Sturm fundamental theorem (with $Q_1 = Q, Q_2 \equiv 0$) that if ϕ has two or more zeros in $[a, b]$, then ψ must have a zero between them. Since ψ is never zero, ϕ can have at most one zero in $[a, b]$. ◆

The Sturm fundamental theorem tells us that the oscillatory behavior of a solution to equation (10) is determined by the sign and magnitude of $Q(x)$. For example, when $Q(x) \leq 0$, a solution can have at most one zero or, when $Q(x) > m^2$, a solution must have a zero between any two zeros of a solution to $y'' + m^2 y = 0$. In the latter case, the zeros of the solutions to (10) must lie within a distance π/m of each other because $y'' + m^2 y = 0$ has a solution with zeros that are π/m distance apart (see Problem 12).

In the next example, we consider the distance between consecutive zeros of the solutions to Bessel's equation.

Example 1 For $a > 0$, discuss the number of zeros in the interval $[a, a + \pi)$ of the solutions to Bessel's equation of order v:

(11) $x^2 y'' + xy' + (x^2 - v^2)y = 0 \,,$ $x > 0 \,.$

Solution The substitution $y = ux^{-1/2}$ transforms equation (11) into

(12) $u'' + \left[1 - \dfrac{4v^2 - 1}{4x^2}\right]u = 0 \,,$ $x > 0$

(see Problem 13). We will now compare the solutions of (12) with those of $u'' + u = 0$. Observe that $\phi(x) = A \sin(x - a)$ is a solution to $u'' + u = 0$ and has zeros at a and $a + \pi$.

Case 1. $\nu > 1/2$**.** In this case, $4\nu^2 - 1 > 0$, so

$$1 - \frac{4\nu^2 - 1}{4x^2} < 1$$

for x in $[a, a + \pi)$. Thus, by the Sturm fundamental theorem, a solution to (12) cannot have more than one zero in $[a, a + \pi)$ because ϕ does not have a zero in $(a, a + \pi)$. Since $y = ux^{-1/2}$, we have shown that when $\nu > 1/2$ any solution to Bessel's equation (11) can have *at most one zero* in $[a, a + \pi)$.

Case 2. $0 \leq \nu < 1/2$ **.** In this case, $4\nu^2 - 1 < 0$, so

$$1 - \frac{4\nu^2 - 1}{4x^2} > 1$$

for x in $[a, a + \pi)$. Again, by the Sturm fundamental theorem, a solution to (12) must have a zero in $(a, a + \pi)$, since a and $a + \pi$ are consecutive zeros of $\phi(x) = A \sin(x - a)$. Because $y = ux^{-1/2}$, we have shown that when $0 \leq \nu < 1/2$, every solution to Bessel's equation (11) *must have a zero* in $[a, a + \pi)$.

Case 3. $\nu = 1/2$ **.** In this case, (12) reduces to $u'' + u = 0$, which has the general solution $u(x) = c_1 \cos x + c_2 \sin x$. Thus,

$$y(x) = c_1 x^{-1/2} \cos x + c_2 x^{-1/2} \sin x = A x^{-1/2} \sin(x - \theta)$$

for a suitable constant θ. Consequently any solution to Bessel's equation (11) with $\nu = 1/2$ has *exactly one zero* in $[a, a + \pi)$.

Summarizing, we have shown that a solution to Bessel's equation (11) has

(a) at most one zero in $[a, a + \pi)$ when $\nu > 1/2$;

(b) at least one zero in $[a, a + \pi)$ when $0 \leq \nu < 1/2$; and

(c) exactly one zero in $[a, a + \pi)$ when $\nu = 1/2$. ◆

The Sturm fundamental theorem was extended to more general Sturm–Liouville equations by M. Picone.

Picone Comparison Theorem

Theorem 17. Let $\phi_1(x)$ be a nontrivial solution to the Sturm–Liouville equation

(13) $$\frac{d}{dx}\left(p_1 \frac{dy}{dx}\right) + Q_1 y = 0 \,, \qquad a < x < b \,,$$

and let $\phi_2(x)$ be a nontrivial solution to

(14) $$\frac{d}{dx}\left(p_2 \frac{dy}{dx}\right) + Q_2 y = 0 \,, \qquad a < x < b \,.$$

Assume that $p_1(x) \geq p_2(x) > 0$ and $Q_1(x) \leq Q_2(x)$ for x in $[a, b]$. Then, between any two consecutive zeros x_1 and x_2 of ϕ_1 in $[a, b]$, there is a zero of ϕ_2, unless ϕ_1 and ϕ_2 are linearly dependent on $[x_1, x_2]$, in which case $Q_1(x) \equiv Q_2(x)$ on $[x_1, x_2]$.

A proof of the Picone comparison theorem using the **Picone identity** is outlined in Project C. It can also be proved using the Prüfer substitution.[†]

If we apply the Picone comparison theorem to a Sturm–Liouville eigenvalue equation, we can obtain estimates on the distance between consecutive zeros of its solutions. In the next corollary, let f_M and f_m denote, respectively, the maximum and minimum values of f over the interval $[a, b]$.

Distance Between Zeros

Corollary 5. Let

(15) $$\frac{d}{dx}\left(p\,\frac{dy}{dx}\right) + qy + \lambda ry = 0\,, \qquad a < x < b\,,$$

be a Sturm–Liouville equation, where p, p', q, and r are continuous on $[a, b]$ and p and r are positive on $[a, b]$. If ϕ is a nontrivial solution to (15) with consecutive zeros x_1 and x_2 in $[a, b]$ and

$$\lambda > \max\left\{\frac{-q_M}{r_M},\ \frac{-q_m}{r_m},\ 0\right\},$$

then $x_2 - x_1$, the distance between the zeros of ϕ, is bounded by

(16) $$\pi\sqrt{\frac{p_m}{q_M + \lambda r_M}} \le x_2 - x_1 \le \pi\sqrt{\frac{p_M}{q_m + \lambda r_m}}\,.$$

Proof. The constant-coefficient equation

(17) $$p_M y'' + (q_m + \lambda r_m)y = 0$$

has the solution $\psi(x) = \sin[K(x - x_1)]$, where $K^2 = (q_m + \lambda r_m)/p_M$ (here we have used the hypothesis $\lambda > -q_m/r_m$ so that $q_m + \lambda r_m > 0$). Moreover, $\psi(x)$ has a zero at $x = x_1$ and its next zero is at $x_1 + (\pi/K)$; that is, the distance to its next zero is

$$\pi\sqrt{\frac{p_M}{q_m + \lambda r_m}}\,.$$

Since $p_M \ge p(x)$, $\lambda > 0$, and $q_m + \lambda r_m \le q(x) + \lambda r(x)$ for all x in (a, b), it follows from the Picone comparison theorem that ϕ must have a zero between any two successive zeros of $\psi(x)$ [or $\phi(x_1 + (\pi/K)) = 0$ if ϕ and ψ are linearly dependent]. Therefore, $x_2 \le x_1 + (\pi/K)$, so

$$x_2 - x_1 \le \pi\sqrt{\frac{p_M}{q_m + \lambda r_m}}$$

(see Figure 11.9 on the next page).

[†]For a discussion of the Prüfer substitution and a proof of the Picone comparison theorem, also referred to as the Sturm comparison theorem, see the text by G. Birkhoff and G. C. Rota, *Ordinary Differential Equations,* 4th ed. (Wiley, New York, 1989), Chapter 10.

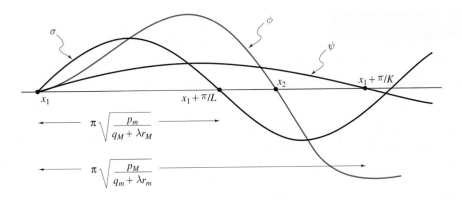

Figure 11.9 Estimating the distance between consecutive zeros of ϕ

To obtain a lower estimate for the distance between consecutive zeros, we compare the given equation with the constant-coefficient equation

(18) $p_m y'' + (q_M + \lambda r_M)y = 0$,

which has the solution $\sigma(x) = \sin[L(x - x_1)]$, where $L^2 = (q_M + \lambda r_M)/p_m$. Notice that $\sigma(x)$ has consecutive zeros x_1 and $x_1 + (\pi/L)$. Since $p(x) \geq p_m$ and $q(x) + \lambda r(x) \leq q_M + \lambda r_M$ for x in (a, b), it follows from the Picone comparison theorem that $\sigma(x)$ must have a zero between any two consecutive zeros of ϕ or have the same zeros (see Figure 11.9). Thus, $x_1 + (\pi/L) \leq x_2$, so

$$\pi \sqrt{\frac{p_m}{q_M + \lambda r_M}} \leq x_2 - x_1 . \quad \blacklozenge$$

An immediate consequence of Corollary 5 is that as $\lambda \to +\infty$, the distance between the zeros of a solution to a Sturm–Liouville equation approaches zero. This means that since the eigenvalues λ_n associated with a regular Sturm–Liouville boundary value problem increase to infinity, the number of zeros in (a, b) of an eigenfunction ϕ_n increases as n increases. In fact, one can show using the Prüfer substitution that if a regular Sturm–Liouville boundary value problem has eigenvalues $\lambda_1 < \lambda_2 < \cdots$, then the eigenfunction ϕ_n belonging to λ_n has *exactly* $n - 1$ zeros in the open interval (a, b).[†]

[†]For a proof of this result, see the text by G. Birkhoff and G. C. Rota, ibid.

11.8 EXERCISES

1. Let $\phi(x)$ be a solution to the equation

$$\frac{d}{dx}\left(e^x\frac{dy}{dx}\right) + xy = 0$$

on $(-\infty, \infty)$. Given that $\phi(1/n) = 0$ for $n = 1, 2, \ldots$, determine $\phi(x)$.

2. The function $\phi(x) = \sin x$ is a solution to $y'' + y = 0$ on $(-\infty, \infty)$ and has an infinite number of zeros. Does this contradict Theorem 13? Explain.

3. Let $\phi(x)$ be a solution on $(-\infty, \infty)$ to

$$\frac{d}{dx}\left[(x^2 + 1)\frac{dy}{dx}\right] - 2y = 0 \,.$$

If $\phi(0) = 0$, prove that $\phi(x) = cx$ for some constant c.

4. Can the function $\phi(x) = x^4\sin(1/x)$ be a solution on $(-1, 1)$ to some differential equation of the form (1), where p, p', and q are continuous on $[-1, 1]$ and p is positive on $[-1, 1]$? Explain.

5. Use the Sturm fundamental theorem to prove that any nontrivial solution to $y'' + (1 - e^x)y = 0$ has at most one zero in $(0, \infty)$ but has infinitely many zeros in $(-\infty, 0)$.

6. Use the Sturm fundamental theorem to prove that any nontrivial solution to $y'' + (x^2 - 1)y = 0$ has at most one zero in $(-1, 1)$ but has infinitely many zeros on $(-\infty, -1)$ and on $(1, \infty)$.

7. The Legendre polynomial $P_n(x)$ of degree n satisfies the equation

$$\frac{d}{dx}\left[(1 - x^2)\frac{dy}{dx}\right] + n(n+1)y = 0 \,,$$

$$-1 < x < 1 \,,$$

and has n distinct zeros in $(-1, 1)$. Use the Picone comparison theorem to show that between any two zeros of $P_{m-1}(x)$, there is a zero of $P_m(x)$ for $m > 2$.

8. Let $\phi(x)$ be a nontrivial solution to

$$y'' + 9y + e^{-3x}y = 0$$

on $(0, \infty)$. Use the Sturm fundamental theorem to estimate the distance between consecutive zeros of $\phi(x)$ when x is large.

9. Let $\phi(x)$ be a nontrivial solution to

$$y'' + x^{-2}y' + (4 - e^{-x})y = 0$$

on $(0, \infty)$. Use the Picone comparison theorem to estimate the distance between consecutive zeros of $\phi(x)$ when x is large. [*Hint:* First express the equation in Sturm–Liouville form.]

10. Use Corollary 5 to estimate the distance between consecutive zeros in $[0, 5]$ of a nontrivial solution to

$$\frac{d}{dx}\left[e^{-x^2}\frac{dy}{dx}\right] + (\sin x)y + \lambda(1 + x^2)^{-1}y = 0 \,.$$

11. Use Corollary 5 to estimate the distance between consecutive zeros in $[0, 5]$ of a nontrivial solution to

$$\frac{d}{dx}\left[(1 + x)\frac{dy}{dx}\right] + e^{-x}y + \lambda y = 0 \,.$$

12. In equation (10), assume $Q(x) \geq m^2$ on $[a, b]$. Prove that the distance between two consecutive zeros of a solution ϕ to (10) is less than or equal to π/m. [*Hint:* If x_1 is a zero of ϕ, compare ϕ with the function $\psi(x) = \sin[m(x - x_1)]$, which is a solution to $y'' + m^2y = 0$.]

13. Show that the substitution $y = ux^{-1/2}$ transforms equation (11) into equation (12).

14. Show that if $Q(x) \geq m^2 > 0$ on $[a, \infty)$, then every solution to $y'' + Q(x)y = 0$ is oscillatory.

15. Assume $0 < Q_m \leq Q(x) \leq Q_M$ on $[a, b]$. Prove the following:
 (a) If $Q_m > k^2\pi^2/(b - a)^2$, k a positive integer, then any solution to equation (10) has at least k zeros in $[a, b]$.
 (b) If $Q_M < k^2\pi^2/(b - a)^2$, then any nontrivial solution to equation (10) has at most k zeros in $[a, b]$.

16. **Sturm Convexity Theorem.** Prove that if $Q(x)$ is a continuous, increasing function on $[0, \infty)$ and $\phi(x)$ is a nontrivial solution to

$$y'' + Q(x)y = 0 \,, \qquad 0 < x < \infty \,,$$

then $t_k - t_{k-1} > t_{k+1} - t_k$, where $\{t_n\}_{n=1}^{\infty}$ is the sequence of consecutive zeros in $(0, \infty)$ of the solution $\phi(x)$.

17. Show that if $w(x)$ and $Q(x)$ are continuous on $[a, b]$ and $Q(x) < 0$ on $[a, b]$, then a nontrivial solution ϕ of the equation

$$y'' + w(x)y' + Q(x)y = 0$$

can have at most one zero in $[a, b]$.

Chapter **11** Summary

In this chapter we discussed the theory of regular Sturm–Liouville boundary value problems, the Fredholm alternative and nonhomogeneous boundary value problems, examples of singular Sturm–Liouville boundary value problems, and the oscillation and comparison theory for Sturm–Liouville equations. The important properties and concepts are given below.

Eigenvalue Problems

The problem of finding those values of λ for which the boundary value problem

$$(1) \quad \begin{aligned} (py')' + qy + \lambda ry &= 0, & a &< x < b; \\ a_1 y(a) + a_2 y'(a) &= 0, & b_1 y(b) + b_2 y'(b) &= 0, \end{aligned}$$

has nontrivial solutions is called an **eigenvalue problem.** The number λ is an **eigenvalue,** and the corresponding nontrivial solutions are **eigenfunctions.**

Sturm–Liouville Equation

A second-order linear differential equation of the form

$$\left(p(x)y'(x) \right)' + Q(x)y(x) = 0, \quad a < x < b,$$

is called a **Sturm–Liouville equation.** Any homogeneous second-order linear equation can be rewritten as a Sturm–Liouville equation.

Regular Sturm–Liouville Boundary Value Problems

An eigenvalue problem of the form (1) where $p(x), p'(x), q(x)$, and $r(x)$ are real-valued continuous functions on $[a, b]$ and $p(x) > 0$ and $r(x) > 0$ on $[a, b]$ is called a **regular Sturm–Liouville boundary value problem.** These boundary value problems have the following properties:

1. The eigenvalues are real, simple, and form a countable, increasing sequence $\lambda_1 < \lambda_2 < \cdots$ with $\lim_{n \to \infty} \lambda_n = +\infty$.

2. The eigenfunctions can be chosen to be real-valued.

3. The eigenfunctions corresponding to distinct eigenvalues are orthogonal with respect to $r(x)$ on $[a, b]$.

4. Let $\{\phi_n\}_{n=1}^{\infty}$ be an orthonormal sequence of eigenfunctions and let f be a continuous function on $[a, b]$ with f' piecewise continuous on $[a, b]$. If f satisfies the two boundary conditions, then

$$f(x) = \sum_{n=1}^{\infty} a_n \phi_n(x), \quad a \le x \le b,$$

where

$$a_n = \int_a^b f(x)\phi_n(x)r(x)\,dx.$$

Moreover, the series converges uniformly on $[a, b]$.

Periodic Sturm–Liouville problems have similar properties.

Fredholm Alternative

Let $L[y](x) := A_2(x)y''(x) + A_1(x)y'(x) + A_0(x)y(x)$ and let $B[y] = 0$ represent a set of separated boundary conditions. Let $L^+[y](x) = (A_2(x)y(x))'' - (A_1(x)y(x))' + A_0(x)y(x)$ be the adjoint of L with adjoint boundary conditions $B^+[y] = 0$. (The adjoint boundary conditions are chosen so that $(Lu, v) = (u, L^+v)$ for all u and v satisfying $B[u] = 0$ and $B^+[v] = 0$.) The nonhomogeneous boundary value problem

$$L[y](x) = h(x), \qquad a < x < b; \qquad B[y] = 0,$$

has a solution if and only if

$$\int_a^b h(x)z(x)\, dx = 0$$

for every solution z of the adjoint boundary value problem

$$L^+[z](x) = 0, \qquad a < x < b; \qquad B^+[z] = 0.$$

Solution by Eigenfunction Expansion

Let $L[y] := (py')' + qy$ and assume $L[y] + \mu ry = 0, B[y] = 0$ is a regular Sturm–Liouville boundary value problem with eigenvalues $\{\lambda_n\}_{n=1}^\infty$ and corresponding eigenfunctions $\{\phi_n\}_{n=1}^\infty$.

If $\mu \neq \lambda_n$ for $n = 1, 2, \ldots$, then the solution to the nonhomogeneous problem

$$L[y] + \mu ry = h; \qquad B[y] = 0,$$

is given by

$$\phi = \sum_{n=1}^\infty \frac{\gamma_n}{\mu - \lambda_n}\phi_n,$$

where the constants γ_n are determined from the eigenfunction expansion of h/r:

$$\frac{h}{r} = \sum_{n=1}^\infty \gamma_n\phi_n.$$

If $\mu = \lambda_N$ for some N and $\gamma_N \neq 0$, then there is no solution to the nonhomogeneous problem.

If $\mu = \lambda_N$ for some N and $\gamma_N = 0$, then

$$\phi = c_N\phi_N + \sum_{\substack{n=1 \\ n \neq N}}^\infty \frac{\gamma_n}{\mu - \lambda_n}\phi_n$$

is a solution to the nonhomogeneous problem for any choice of the parameter c_N.

Green's Functions

The solution to

$$L[y] = -f; \qquad B_1[y] = 0, \qquad B_2[y] = 0,$$

where we assume that $L[y] := (py')' + qy$, the function f is continuous on $[a, b]$, and the corresponding homogeneous problem has only the trivial solution, is given by

$$\phi(x) = \int_a^b G(x, s)f(s)\, ds,$$

where $G(x, s)$ is the **Green's function**

$$G(x, s) = \begin{cases} -z_1(s)z_2(x)/C, & a \le s \le x, \\ -z_1(x)z_2(s)/C, & x \le s \le b. \end{cases}$$

Here z_1 and z_2 are two nontrivial solutions to $L[y] = 0$ satisfying $B_1[z_1] = 0$ and $B_2[z_2] = 0$. The constant C is given by $C = p(x)W[z_1, z_2](x)$ for any $x \in [a, b]$.

Singular Sturm–Liouville Problems

Certain *non*regular Sturm–Liouville boundary value problems, called **singular** Sturm–Liouville boundary value problems, arise in solving Bessel's, Legendre's, and Hermite's equations. These singular Sturm–Liouville problems also have increasing sequences of simple eigenvalues. Moreover, their corresponding sequences of eigenfunctions are orthogonal and can be used to form eigenfunction expansions with convergence properties similar to those associated with regular Sturm–Liouville problems.

Oscillation and Comparison Theory

Several results are known about the location of the zeros of solutions to Sturm–Liouville equations. The **Sturm separation theorem** says that the zeros of two linearly independent solutions are intertwined. The **Sturm fundamental theorem** and the **Picone comparison theorem** allow one to compare the location of the zeros of a solution to one Sturm–Liouville equation with those of a related equation.

REVIEW PROBLEMS FOR CHAPTER 11

1. Find all the real eigenvalues and eigenfunctions for the given eigenvalue problem.
 (a) $y'' + 6y' + \lambda y = 0$;
 $$y(0) = 0, \quad y(1) = 0$$
 (b) $y'' + \lambda y = 0$;
 $$y(0) = 0, \quad y(\pi) + 2y'(\pi) = 0$$

2. Convert the given equation into the form of a Sturm–Liouville equation.
 (a) $y'' + 7y' + \lambda y = 0$
 (b) $y'' - 3xy' + \lambda y = 0$
 (c) $xy'' + (1 - x)y' + \lambda y = 0$, $\quad x > 0$

3. (a) Determine the eigenfunctions, which are normalized with respect to the weight function $r(x) \equiv 1$ on $[0, \pi]$, for the problem
 $$y'' + \lambda y = 0;$$
 $$y'(0) = 0, \quad y(\pi) = 0.$$
 (b) Express $f(x) = 1 - x$ in an eigenfunction expansion using the result of part (a).

4. Find the adjoint boundary value problem for the given boundary value problem.
 (a) $y'' + xy' + y = 0$;
 $$y'(0) = 0, \quad y(1) + y'(1) = 0$$
 (b) $x^2y'' + 2xy' - 3y = 0$;
 $$y(1) = 0, \quad y'(e) = 0$$

5. Use the Fredholm alternative to determine conditions on h that guarantee that the given nonhomogeneous boundary value problem has a solution.
 (a) $y'' - 6y' + 13y = h$;
 $$y(0) = 0, \quad y(2\pi) = 0$$
 (b) $xy'' + y' = h$;
 $$y'(1) = 0, \quad y'(2) = 0$$

6. Find a formal eigenfunction expansion for the solution to the nonhomogeneous boundary value problem
 $$y'' + 3y = x - 4\cos 2x;$$
 $$y'(0) = 0, \quad y'(\pi) = 0.$$

7. Find the Green's function $G(x, s)$ and use it to solve the given boundary value problem.

 (a) $y'' = x + 1$; $y(0) = 0$, $y'(1) = 0$

 (b) $y'' - y = 4$; $y'(0) = 0$, $y(1) = 0$

8. Find a formal eigenfunction expansion for the solution to the given nonhomogeneous boundary value problem where f is a continuous function and μ is not an eigenvalue of the corresponding homogeneous problem.

 (a) $(xy')' - (49/x)y + \mu xy = f(x)$, $0 < x < 1$,
 $y(x), y'(x)$ remain bounded as $x \to 0^+$, $y(1) = 0$

 (b) $[(1 - x^2)y']' + \mu y = f(x)$, $-1 < x < 1$,
 $y(x), y'(x)$ remain bounded as $x \to \pm 1$

9. Let $\phi(x)$ be a nontrivial solution to

$$[(1 - e^{-x})y']' + \left(\frac{x+1}{x}\right)y = 0$$

on $(0, \infty)$. Use the Picone comparison theorem to estimate the distance between consecutive zeros of $\phi(x)$ when x is large.

10. Use Corollary 5 in Section 11.8 to estimate the distance between consecutive zeros in $[0, \pi]$ of a nontrivial solution to

$$[(2 + \cos x)y']' + (\sin x)y + 5y = 0.$$

TECHNICAL WRITING EXERCISES CHAPTER 11

1. The method of solving a nonhomogeneous regular Sturm–Liouville boundary value problem using an eigenfunction expansion as discussed in Section 11.5 can be extended to periodic Sturm–Liouville problems. Describe this extension and illustrate it with an example.

2. Green's functions also exist for certain periodic Sturm–Liouville problems. Explain how to construct a Green's function for the problem

$$y''(x) + 2y(x) = -f(x);$$
$$y(0) = y(2\pi), y'(0) = y'(2\pi).$$

3. Formulate a Fredholm alternative for the system $\mathbf{Ax} = \mathbf{b}$, where $\mathbf{A}$ is an $n \times n$ real-valued constant matrix, $\mathbf{b}$ is an $n \times 1$ constant vector, and $\mathbf{x}$ is the $n \times 1$ unknown vector. Illustrate your formulation by examples.

4. Discuss the mass–spring oscillator interpretation of the topics in Section 11.8 (recall Section 4.8).

Projects for Chapter 11

A Hermite Polynomials and the Harmonic Oscillator

In quantum mechanics the motion of a particle is described by Schrödinger's equation. In one-dimensional space, the equation has the form

$$\text{(1)} \qquad \psi'' + (2m/\hbar^2)[E - V(x)]\psi = 0 ,$$

where the function $V(x)$ represents the potential energy, m is the mass of the particle, $\hbar$ is Planck's constant h divided by 2π, the constant E represents the energy, and ψ is the **wave function.** The wave function ψ may be real or complex valued. The quantity $\psi(x)\overline{\psi(x)}\, dx = |\psi(x)|^2\, dx$ gives the probability that the particle will be found in the interval $(x, x + dx)$. Physicists are interested in the solutions to (1) that satisfy the boundary conditions

$$\text{(2)} \qquad \psi(x) \quad \text{and} \quad \psi'(x) \qquad \text{remain bounded as } x \to \pm\infty .$$

The eigenvalues $E = E_n$ of (1)–(2) are called the **energy levels** of the system.

(a) For the **harmonic oscillator,** the potential energy is $V(x) = \alpha x^2$, where $\alpha > 0$ is a constant. Show that for the harmonic oscillator, the boundary value problem (1)–(2) simplifies to

$$\text{(3)} \qquad y''(t) + (\lambda - t^2)y(t) = 0 ;$$

$$\text{(4)} \qquad y(t) \text{ and } y'(t) \text{ remain bounded as } t \to \pm\infty ,$$

where $t = \beta x$, $\beta^4 = 2m\alpha/\hbar^2$, $y(t) = \psi(x)$, and $\lambda = 2mE/(\beta^2\hbar^2)$.

(b) Verify that the change of variables $y(t) = e^{-t^2/2}u(t)$ transforms (3)–(4) into

$$\text{(5)} \qquad (e^{-t^2}u')' + (\lambda - 1)e^{-t^2}u = 0 ;$$

$$\text{(6)} \qquad e^{-t^2/2}u(t) \quad \text{and} \quad e^{-t^2/2}u'(t) - te^{-t^2/2}u(t) \quad \text{remain bounded as } t \to \pm\infty .$$

(c) Use the result of part (b) to show that $\lambda = 2n + 1$, $n = 0, 1, 2, \ldots$, are eigenvalues of the singular Sturm–Liouville boundary value problem (3)–(4) with corresponding eigenfunctions $y_n(t) = e^{-t^2/2}H_n(t)$, where H_n is the **Hermite polynomial** of degree n (compare Section 11.7 or Section 8.8).

(d) Sketch the first three wave functions $y_0(t)$, $y_1(t)$, and $y_2(t)$.

B Continuous and Mixed Spectra

Free Particles. When the potential energy $V(x)$ in Schrödinger's equation (1) is identically zero, the particle is said to be **free.** Thus the motion of a **free particle** is governed by the singular Sturm–Liouville boundary value problem

$$\text{(7)} \qquad \psi'' + \lambda\psi = 0 ;$$

$$\text{(8)} \qquad \psi(x) \quad \text{and} \quad \psi'(x) \quad \text{remain bounded as } x \to \pm\infty ,$$

where $\lambda = 2mE/\hbar^2$.

(a) Show that every nonnegative number λ is an eigenvalue of (7)–(8) and that there are no negative eigenvalues. Hence the spectrum of the free particle is continuous and consists of the interval $[0, \infty)$.

Square-Well Potential. A **square-well** potential is a function of the form

$$V(x) = \begin{cases} -\alpha^2, & |x| < a, \\ 0, & |x| > a, \end{cases}$$

where α and a are positive constants. For such a potential, Schrödinger's equation has the form

(9) $\psi'' + \lambda\psi = \begin{cases} -\beta^2\psi, & |x| < a, \\ 0, & |x| > a, \end{cases}$

where $\lambda = 2mE/\hbar^2$ and $\beta^2 = 2m\alpha^2/\hbar^2$. We also impose the boundary conditions

(10) $\psi(x)$ and $\psi'(x)$ remain bounded as $x \to \pm \infty$.

Equation (9) can be solved in the sense that we can find solutions on the three intervals $(-\infty, -a)$, $(-a, a)$, and (a, ∞) and then patch them together to get a function ψ such that ψ and ψ' are continuous on $(-\infty, \infty)$.

(b) Show that each $\lambda > 0$ is an eigenvalue for the singular Sturm–Liouville boundary value problem (9)–(10) and has two linearly independent eigenfunctions associated with it.

(c) Show that $\lambda = 0$ is an eigenvalue for the boundary value problem (9)–(10) if and only if $\beta a = k\pi/a$ for some integer k. Moreover, when $\beta a = k\pi/a$, there is only one linearly independent eigenfunction associated with $\lambda = 0$.

(d) Show that for $\lambda < -\beta^2$ the only solution to (9)–(10) is the trivial solution and hence there are no eigenvalues less than $-\beta^2$.

(e) Show that for $-\beta^2 < \lambda < 0$, there can exist a finite number of discrete eigenvalues λ_n of (9)–(10). [*Hint:* Let $\gamma = \sqrt{\beta^2 + \lambda}$. The function $A \cos(\gamma x)$ is part of an eigenfunction for $x > a$ if and only if $\gamma \tan(\gamma a) = \sqrt{-\lambda}$. Further, $B \sin(\gamma x)$ is part of an eigenfunction for $x > a$ if and only if $\gamma \cot(\gamma a) = -\sqrt{-\lambda}$.]

Consequently, the spectrum of a particle under the influence of a square-well potential may have a discrete spectrum made up of a finite number of negative eigenvalues in addition to its continuous spectrum $[0, \infty)$.

C Picone Comparison Theorem

The purpose of this project is to give a proof of the Picone comparison theorem (see Theorem 17, page 699). Using the notation of that theorem, we proceed as follows.

Let x_1 and x_2 be two consecutive zeros of ϕ_1 and assume $\phi_2 \neq 0$ in $[x_1, x_2]$.

(a) Show that

(11) $(Q_2 - Q_1)\phi_1^2\phi_2^2 + (p_2\phi_2')'\phi_1^2\phi_2 - (p_1\phi_1')'\phi_1\phi_2^2 = 0$.

(b) Integrating equation (11) from x_1 to x_2 and using integration by parts, obtain

(12) $\displaystyle\int_{x_1}^{x_2}(Q_2 - Q_1)\phi_1^2\,dx + \int_{x_1}^{x_2}p_1(\phi_1')^2\,dx - \int_{x_1}^{x_2}\frac{2p_2\phi_2'\phi_1\phi_1'}{\phi_2}\,dx + \int_{x_1}^{x_2}\frac{p_2(\phi_2')^2\phi_1^2}{\phi_2^2}\,dx = 0$.

(c) From equation (12), derive the **Picone identity**

$$(13) \qquad \int_{x_1}^{x_2} (Q_2 - Q_1)\phi_1^2 \, dx + \int_{x_1}^{x_2} (p_1 - p_2)(\phi_1')^2 \, dx + \int_{x_1}^{x_2} p_2 \left[\phi_1' - \frac{\phi_1 \phi_2'}{\phi_2} \right]^2 dx = 0 \, .$$

(d) If $\phi_2(x) \neq 0$ for x in (x_1, x_2), but $\phi_2(x)$ is zero at either $x = x_1$ or $x = x_2$, then some of the preceding integrals must be treated as improper integrals. Verify that the Picone identity still holds. [*Hint:* If $\phi_2(x_1) = 0$, then $\phi_2'(x_1) \neq 0$. Use this fact and L'Hôpital's rule to show that the boundary terms, which arise from using integration by parts, are still zero.]

(e) Use the Picone identity and the hypotheses of Theorem 17 to show that if $Q_1(x) \neq Q_2(x)$ on $[x_1, x_2]$, then ϕ_2 must have a zero in (x_1, x_2).

(f) Use the Picone identity and the hypotheses of Theorem 17 to show that if $Q_1(x) \equiv Q_2(x)$ on $[x_1, x_2]$, then ϕ_1 and ϕ_2 are linearly dependent on $[x_1, x_2]$. [*Hint:* On the intervals where $p_1 \equiv p_2$, show that ϕ_1 and ϕ_2 satisfy the same equation and their Wronskian is zero. On the intervals where $p_1(x) \neq p_2(x)$, show that $\phi_1' = 0$ and $\phi_2' = 0$. Combine to conclude ϕ_1 and ϕ_2 are linearly dependent.]

This completes the proof of the Picone comparison theorem.

D Shooting Method

The shooting method for approximating the solution to the second-order boundary value problem

$$(14) \qquad y''(x) = f(x, y, y'), \qquad a \leq x \leq b; \qquad y(a) = \alpha, \qquad y(b) = \beta,$$

involves solving a sequence of initial value problems of the form

$$(15) \qquad y''(x) = f(x, y, y'), \qquad a \leq x \leq b; \qquad y(a) = \alpha, \qquad y'(a) = t,$$

where t is a parameter. The idea is to "shoot" from a starting location, $y(a) = \alpha$, with an initial "elevation," $y'(a) = t$, at a target located at $y(b) = \beta$. (See Figure 11.10.) If we miss on the first attempt, we adjust the "elevation" t and try again.

The shooting method generates a sequence of values t_k that (we hope) converges to a value $\hat{t}$ for which the solution to the initial value problem (15) with $y'(a) = \hat{t}$ satisfies $y(b) = \beta$. If we let $y(x, t)$ denote the solution to (15) with $y'(a) = t$, then our problem is to find t such that $y(b, t) = \beta$ or

$$(16) \qquad y(b, t) - \beta = 0 \, .$$

We can use a root-finding procedure such as the **bisection method** to generate a sequence t_k that converges to the solution to (16). To begin, we need two guesses t_0 and t_1 such that $y(b, t_0) - \beta$ and $y(b, t_1) - \beta$ are of opposite signs (one positive and one negative). We then pick

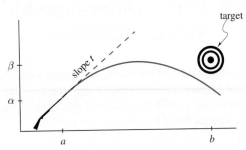

Figure 11.10 Shooting method

$t_2 = (t_0 + t_1)/2$. If $y(b, t_2) - \beta = 0$, we are done. Otherwise, $y(b, t_2) - \beta$ will have the opposite sign of either $y(b, t_0) - \beta$ or $y(b, t_1) - \beta$. If $y(b, t_2) - \beta$ and $y(b, t_0) - \beta$ have opposite signs, then we pick $t_3 = (t_0 + t_2)/2$. Otherwise we pick $t_3 = (t_1 + t_2)/2$. We continue until t_n and t_{n-1} are sufficiently close together (that is, within a prescribed tolerance) and recall that the desired root of (16) is located somewhere between them. Notice that this procedure requires solving an initial value problem for each t_k (see Section 5.3).

(a) Use the shooting method to approximate the solution to the nonlinear boundary value problem

$$y''(x) = \frac{1}{2}y^3(x), \qquad 1 \le x \le 3; \qquad y(1) = 2/3, \qquad y(3) = 2/5.$$

Compare your approximate solution with the actual solution $y(x) = 2/(x + 2)$.

(b) For linear boundary value problems of the form

(17) $y''(x) = -p(x)y'(x) - q(x)y(x) + g(x)$, $a \le x \le b$; $y(a) = \alpha$, $y(b) = \beta$,

the shooting method requires only two "shots." Using the superposition principle, show that the solution to (17) is given by

$$y(x) = y_1(x) + \frac{\beta - y_1(b)}{y_2(b)}y_2(x),$$

where $y_1(x)$ is the solution to the initial value problem

$$y''(x) = -p(x)y'(x) - q(x)y(x) + g(x), \quad a \le x \le b; \quad y(a) = \alpha, \quad y'(a) = 0,$$

and $y_2(x)$ is the solution to the initial value problem

$$y''(x) = -p(x)y'(x) - q(x)y(x), \quad a \le x \le b; \quad y(a) = 0, \quad y'(a) = 1.$$

(We assume that $y_2(b) \ne 0$; otherwise, the boundary value problem (17) does not have a unique solution.)

(c) Use the result of part (b) to approximate the solution to the linear boundary value problem

$$y''(x) = -\frac{11}{3x}y'(x) + \frac{1}{x^2}y(x) + \frac{8}{3x^2} - \frac{\ln x}{x^2}, \qquad 1 \le x \le 4;$$

$$y(1) = 1, \qquad y(4) = 4^{1/3} + \ln 4 \approx 2.973695413.$$

Compare your approximate solution with the actual solution

$$y(x) = x^{1/3} + \ln x.$$

E Finite-Difference Method for Boundary Value Problems

A classical method for numerically approximating a solution to a boundary value problem for a linear second-order differential equation consists of replacing each derivative in the differential equation by a difference quotient that approximates that derivative. To illustrate this procedure, consider the linear boundary value problem

(18) $y''(x) + p(x)y'(x) + q(x)y(x) = g(x)$, $a < x < b$;

(19) $y(a) = \alpha$, $y(b) = \beta$,

where p, q, and g are continuous on $[a, b]$ and α, β are given constants.

We begin by partitioning the interval $[a, b]$ into $N + 1$ subintervals of equal length h. That is, let $h := (b - a)/(N + 1)$ and for $k = 0, 1, \ldots, N + 1$, set

$$x_k := a + kh .$$

Then $a = x_0 < x_1 < \cdots < x_N < x_{N+1} = b$.

Assuming a solution $y(x)$ to (1)–(2) exists, our aim is to approximate $y(x)$ at the interior mesh points x_k, $k = 1, 2, \ldots, N$. At these points, we know that

(20) $y''(x_k) + p(x_k)y'(x_k) + q(x_k)y(x_k) = g(x_k) .$

The essence of the finite-difference method is to find difference quotients that closely approximate the derivatives $y'(x_k)$ and $y''(x_k)$ when h is small.

(a) Assuming y has a Taylor series representation of the form

(21) $y(x) = y(x_k) + y'(x_k)(x - x_k) + \dfrac{y''(x_k)}{2}(x - x_k)^2 + \dfrac{y'''(x_k)}{3!}(x - x_k)^3 + \cdots$

$$= \sum_{j=0}^{\infty} \frac{y^{(j)}(x_k)}{j!}(x - x_k)^j ,$$

derive the following **centered-difference formula** for $y'(x_k)$:

(22) $y'(x_k) = \dfrac{1}{2h}[y(x_{k+1}) - y(x_{k-1})] + [\text{terms involving } h^2, h^3, \ldots] .$

[*Hint:* Work with the Taylor representations for $y(x_{k+1})$ and $y(x_{k-1})$.]

(b) Again using (21), derive the following centered-difference formula for $y''(x_k)$:

(23) $y''(x_k) = \dfrac{1}{h^2}[y(x_{k+1}) - 2y(x_k) + y(x_{k-1})] + [\text{terms involving } h^2, h^3, \ldots] .$

(c) If we let w_k denote the approximation (to be computed) for $y(x_k)$, $k = 1, 2, \ldots, N$, and set $w_0 = y(x_0) = \alpha$, $w_{N+1} = y(x_{N+1}) = \beta$, then from (23) and (22) we obtain the approximations

(24) $y''(x_k) \approx \dfrac{1}{h^2}(w_{k+1} - 2w_k + w_{k-1}) ,$

(25) $y'(x_k) \approx \dfrac{1}{2h}(w_{k+1} - w_{k-1}) .$

Using the centered-difference approximations (24) and (25) in place of $y''(x_k)$ and $y'(x_k)$ in (20) yields the system of equations

(26) $\left(\dfrac{w_{k+1} - 2w_k + w_{k-1}}{h^2}\right) + p(x_k)\left(\dfrac{w_{k+1} - w_{k-1}}{2h}\right) + q(x_k)w_k = g(x_k) ,$

$k = 1, 2, \ldots, N$. The equation (26) represents N linear equations in the N unknowns $w_1, \ldots, w_N$. Thus, on solving (26), we obtain approximations for $y(x)$ at the interior mesh points x_k, $k = 1, \ldots, N$.

Use this method to approximate the solution to the boundary value problem

$$y''(x) - 9y(x) = 0 ; \quad y(0) = 1 , \quad y(1) = e^{-3} ,$$

taking $h = 1/4$ $(N = 3)$.

(d) Compare the approximations obtained in part (c) with the actual solution $y = e^{-3x}$.

(e) Let V denote the electrostatic potential between two concentric metal spheres of radii r_1 and r_2 $(r_1 < r_2)$, where the potential of the inner sphere is kept constant at 100 volts and the potential of the outer sphere at 0 volts. The potential V will then depend only on the distance r from the center of the spheres and must satisfy Laplace's equation. This leads to the following boundary value problem for V:

$$\frac{d^2V}{dr^2} + \frac{2}{r}\frac{dV}{dr} = 0; \qquad V(r_1) = 100, \qquad V(r_2) = 0,$$

where $r_1 < r < r_2$.

For $r_1 = 1$ and $r_2 = 3$, obtain finite-difference approximations for $V(2)$ using $h = 1/2$ $(N = 3)$ and $h = 1/4$ $(N = 7)$.

(f) Compare your approximations in part (e) with the actual solution $V(r) = 150r^{-1} - 50$.

CHAPTER 12

Stability of Autonomous Systems*

12.1 Introduction: Competing Species

> The populations $x(t)$ and $y(t)$ at time t of two species that compete for the same food supply are governed by the system
>
> (1) $\qquad \dfrac{dx}{dt} = x(a_1 - b_1 x - c_1 y)$,
>
> (2) $\qquad \dfrac{dy}{dt} = y(a_2 - b_2 y - c_2 x)$,
>
> where $a_1, a_2, b_1, b_2, c_1,$ and c_2 are positive constants that depend on the species and environment under investigation. Describe the long-term behavior of these populations.

That (1)–(2) is a plausible model for competing species can be seen as follows. Using the logistic model of Section 3.2 for the growth of a single species, let's assume that in the absence of species y, population x is governed by the equation of the form

$$\frac{dx}{dt} = a_1 x - b_1 x^2 = x(a_1 - b_1 x)$$

and, similarly, that in the absence of species x, population y is governed by

$$\frac{dy}{dt} = a_2 y - b_2 y^2 = y(a_2 - b_2 y)\,,$$

where $a_1, a_2, b_1,$ and b_2 are positive constants. When both species are present and competing for the same food, the effect is to decrease their rate of growth. Because this effect depends on the size of both populations, it is reasonable to assume it is proportional to the product xy. Consequently, we replace the growth rate for species x by $a_1 x - b_1 x^2 - c_1 xy$ and for species y by $a_2 y - b_2 y^2 - c_2 xy$, which yields the model (1)–(2).

Generally, we cannot explicitly solve the nonlinear system (1)–(2) for x and y in terms of t. Nor can we find explicit analytic formulas for the trajectories in the phase plane, as defined in Section 5.4. However, in the present chapter we will extend the qualitative methods introduced in that section to obtain useful information about certain trajectories—specifically, we will try to determine how the solutions behave near a critical point. Recall that a **critical point**

*Before proceeding with this chapter, the reader is urged to read the introductory Section 5.4 on phase plane analysis and particularly to be familiarized with the notions of trajectories, phase plane equations, and critical points of autonomous systems.

of system (1)–(2) is a point (x_0, y_0) where both right-hand sides are zero. If (x_0, y_0) is a critical point, then the constant functions $x(t) \equiv x_0, y(t) \equiv y_0$ form a solution to (1)–(2). This is called an **equilibrium solution,** since the functions do not change with t.

Let's first determine the critical points for the system (1)–(2). Setting the right-hand sides equal to zero, we have

$$x(a_1 - b_1 x - c_1 y) = 0 \,,$$

$$y(a_2 - b_2 y - c_2 x) = 0 \,.$$

If we assume $b_1 b_2 - c_1 c_2 \neq 0$, this system has (at most) four solutions: $(0, 0)$, $(0, a_2/b_2)$, $(a_1/b_1, 0)$, and the point of intersection of the two lines $a_1 - b_1 x - c_1 y = 0$, $a_2 - b_2 y - c_2 x = 0$, which is the point

$$(3) \qquad P = \left(\frac{a_1 b_2 - a_2 c_1}{b_1 b_2 - c_1 c_2}, \frac{a_2 b_1 - a_1 c_2}{b_1 b_2 - c_1 c_2} \right) .$$

Notice that the first three equilibrium solutions lie on the nonnegative axes that border the first quadrant in the xy-plane. Because populations must be nonnegative, we will confine the rest of our analysis to the first quadrant.

To determine the flow of trajectories in the first quadrant, let's start with equation (1) involving dx/dt:

$$\frac{dx}{dt} = x(a_1 - b_1 x - c_1 y) \,.$$

From this equation, we see that in the first quadrant, the sign of dx/dt is positive when $a_1 - b_1 x - c_1 y > 0$ and negative when $a_1 - b_1 x - c_1 y < 0$. Consequently, $x(t)$ increases along a trajectory that lies below the line $a_1 - b_1 x - c_1 y = 0$ and decreases along trajectories above this line [see Figure 12.1(a)]. Similarly, $y(t)$ increases along a trajectory that lies below the line $a_2 - b_2 y - c_2 x = 0$ and decreases when it is above this line [see Figure 12.1(b)].

It turns out that there are four cases to analyze, depending on the relative positions of the two lines and whether they intersect in the first quadrant. These four cases are illustrated in Figure 12.2 on page 715. The arrow systems (e.g., ↘) indicate whether the variables x and y are increasing or decreasing and are based on the analysis illustrated in Figure 12.1. The colored dots are the critical points in the first quadrant. The black lines denote where $dx/dt = 0$ and the colored lines where $dy/dt = 0$. Typical trajectories are also shown in color.

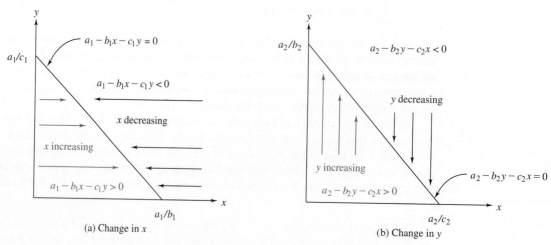

(a) Change in x — (b) Change in y

Figure 12.1 Regions where the populations are increasing or decreasing

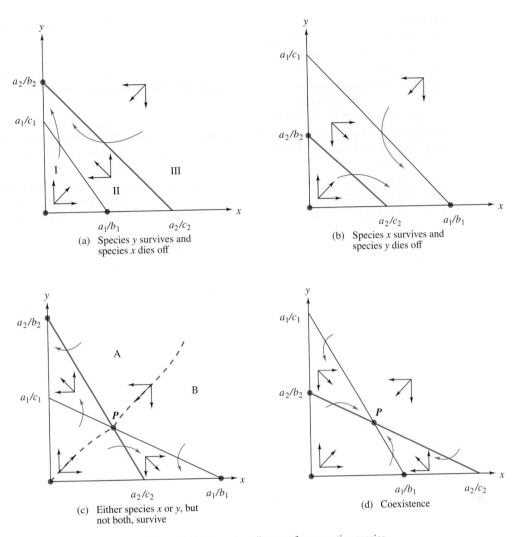

(a) Species y survives and species x dies off

(b) Species x survives and species y dies off

(c) Either species x or y, but not both, survive

(d) Coexistence

Figure 12.2 Phase plane diagrams for competing species

Let's begin with the case illustrated in Figure 12.2(a). If we start in region I, that is, $(x(0), y(0)) = (x_0, y_0)$ lies in region I, then populations x and y are increasing ($\nearrow$). At some point in time, the trajectory $(x(t), y(t))$ will move into region II. In region II, the population $x(t)$ decreases while $y(t)$ continues to increase. Observe that the trajectory *cannot* cross the colored line bounding region II because this would require either $x(t)$ to be increasing, which it is not in region II, or $y(t)$ to increase *across* the colored line, which is impossible because $dy/dt = 0$ on that line. Consequently, the trajectories in region II must move up and to the left ($\nwarrow$). Since $dx/dt = 0$ on the y-axis, the trajectory cannot cross it and so must approach the critical point $(0, a_2/b_2)$ in the upper corner of region II. When the populations start in region III, both $x(t)$ and $y(t)$ are decreasing ($\swarrow$). Since the trajectory cannot leave the first quadrant, the trajectory must cross the colored line and enter region II, where it then moves up toward the critical point $(0, a_2/b_2)$. To summarize, we have shown that when the initial populations $x(0)$ and $y(0)$ are positive, the species x dies off $[x(t) \to 0]$, while the population of species y approaches the limiting population a_2/b_2. Graphically, all trajectories beginning in the first

quadrant approach the critical point $(0, a_2/b_2)$ as $t \to +\infty$. Typical trajectories are illustrated by colored curves in Figure 12.2(a).

A similar analysis of the case illustrated in Figure 12.2(b) shows that species y dies off, while the population of species x approaches the limiting population a_1/b_1. Graphically, all trajectories beginning in the first quadrant approach the critical point $(a_1/b_1, 0)$ as $t \to +\infty$. Typical trajectories are again illustrated by colored curves.

The analysis of the case illustrated in Figure 12.2(c) yields a different result. There exists a dividing curve called a *separatrix* (illustrated by a dashed colored curve) that separates the first quadrant into two regions A and B. Trajectories beginning in region A approach the critical point $(0, a_2/b_2)$ as $t \to +\infty$, and those beginning in region B approach the critical point $(a_1/b_1, 0)$. In other words, for some initial populations, species x dies off, while the population of species y approaches a_2/b_2. For other initial populations, species y dies off, while the population of species x approaches a_1/b_1. Unfortunately, the separatrix is usually difficult to determine.

Finally, an analysis of the case illustrated in Figure 12.2(d) shows that every trajectory in the first quadrant heads toward the critical point P given in (3). That is, the populations $x(t)$ and $y(t)$ approach the equilibrium populations

$$x = \frac{a_1 b_2 - a_2 c_1}{b_1 b_2 - c_1 c_2}, \qquad y = \frac{a_2 b_1 - a_1 c_2}{b_1 b_2 - c_1 c_2}.$$

This is the only case that allows for the coexistence of the two species. The critical point P is an example of an *asymptotically stable* critical point, since all the nearby trajectories approach P as $t \to +\infty$. By comparison, the origin is an *unstable* critical point because nearby trajectories move away from it.

In Section 12.2 we study the stability of critical points for general linear systems. Namely, we analyze the behavior of trajectories near the critical point of a linear system in the plane and classify the type of behavior as $t \to +\infty$. In Section 12.3 we discuss how the stability of a critical point for an "almost linear system" can be determined by studying the corresponding linear system. In Section 12.4, we introduce the energy method and use it to study conservative systems. Section 12.5 describes a generalization of the energy method known as Lyapunov's direct method, which is used to analyze the stability of a critical point. In Section 12.6 we present results of Poincaré and Bendixson, which concern the existence of periodic solutions for autonomous systems in the plane. Finally, in Section 12.7 we discuss stability for higher-dimensional systems using results from Chapter 9.

12.2 Linear Systems in the Plane

A linear autonomous system in the plane has the form

(1) $x'(t) = a_{11}x + a_{12}y + b_1$,

(2) $y'(t) = a_{21}x + a_{22}y + b_2$,

where the a_{ij}'s and the b_i's are constants. In this section, we study the issue of stability of critical points for linear systems; later in the chapter we will turn to nonlinear systems.

By a simple translation, we can always transform a given linear system into one of the form

(3) $x'(t) = ax + by$,

(4) $y'(t) = cx + dy$,

 or $\begin{bmatrix} x' \\ y' \end{bmatrix} = \mathbf{A}\begin{bmatrix} x \\ y \end{bmatrix}$ with $\mathbf{A} := \begin{bmatrix} a & b \\ c & d \end{bmatrix}$,

where the origin $(0, 0)$ is now the critical point (see Problem 21). So, without loss of generality, we analyze the linear system (3)–(4), but under the additional assumption that $ad - bc \neq 0$, which implies that $(0, 0)$ is the only critical point.

System (3)–(4) can be solved using the elimination method discussed in Section 5.2 or by the matrix methods of Sections 9.4 and 9.5. Either approach yields solutions of the form $x(t) = ue^{rt}, y(t) = ve^{rt}$, where u, v, and r are constants to be determined. Substituting these expressions for x and y into system (3)–(4), we have

$$rue^{rt} = aue^{rt} + bve^{rt},$$
$$rve^{rt} = cue^{rt} + dve^{rt},$$

which simplifies to

$$(r - a)u - bv = 0,$$
$$-cu + (r - d)v = 0.$$

This pair of equations has a nontrivial solution for u, v if and only if the determinant of the coefficient matrix is zero; that is,

(5) $(r - a)(r - d) - bc = r^2 - (a + d)r + (ad - bc) = 0.$

This is called the **characteristic equation** for the system (3)–(4) and is also the condition for r to be an eigenvalue for the matrix **A**. (Recall Section 9.5.)

The asymptotic (long-term) behavior of the solutions to (3)–(4) is linked to the nature of the roots r_1 and r_2 of the characteristic equation—namely, whether these roots (eigenvalues) are both real or both complex, both positive, both negative, of opposite signs, etc. We consider each case in detail by first discussing a special example and then indicating what happens in general. These general results (whose proofs we omit) can be obtained using a linear change of variables[†] (see Problem 23). For your convenience, at the end of this section we summarize all these results in Theorem 1.

Case 1. r_1, r_2 real, distinct, and positive. In this case, the general solution to system (3)–(4) has the form

(6)
$$x(t) = A_1 e^{r_1 t} + A_2 e^{r_2 t},$$
$$y(t) = B_1 e^{r_1 t} + B_2 e^{r_2 t},$$

where the A_i's and B_i's are interrelated. As a simple example, we discuss the system

$$\frac{dx}{dt} = x, \qquad \frac{dy}{dt} = 3y, \qquad \text{or} \qquad \begin{bmatrix} x' \\ y' \end{bmatrix} = \begin{bmatrix} 1 & 0 \\ 0 & 3 \end{bmatrix} \begin{bmatrix} x \\ y \end{bmatrix}.$$

Here $r_1 = 1$ and $r_2 = 3$ and a general solution is

$$x(t) = Ae^{t}, \qquad y(t) = Be^{3t}.$$

Consequently, the trajectories of the system lie on the curves $y = B(e^t)^3 = (B/A^3)x^3$ in the phase plane. A phase plane diagram for this system is given in Figure 12.3(a) on page 718. Notice that the trajectories move *away* from the origin and are tangent to the x-axis at the origin, except for the two along the y-axis. In this case, we say that the origin is an **unstable improper node.**

[†]Such change of variables procedures are studied in linear algebra as "change of coordinates," where it is shown that the structure of the eigenvalues and eigenspaces remains unchanged when a matrix is diagonalized. See *Fundamentals of Matrix Analysis with Applications*, by Edward Barry Saff and Arthur David Snider (John Wiley & Sons, Hoboken, New Jersey, 2016), Chapter 3.

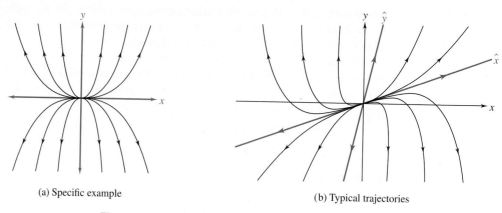

(a) Specific example (b) Typical trajectories

Figure 12.3 Unstable improper node; roots are real, distinct, and positive

It is unstable because the trajectories move *away* from the origin, and it is an improper node because (almost all) trajectories have the same tangent line at the origin. (A proper node occurs in Case 4; see Figures 12.3 and 12.8 for a comparison.)

In general, when the roots are distinct, real, and positive, the origin is an *unstable improper node* whose phase plane diagram near the origin resembles Figure 12.3(b). The colored lines labeled $\hat{x}$ and $\hat{y}$ represent the images of the x- and y-axes under a change of variables. We make two key observations. First, (most) trajectories in Figures 12.3(a) and 12.3(b) are becoming parallel to the y- and $\hat{y}$-axes, respectively, as they move away from the origin. Second, (most) trajectories are leaving the origin along directions that are tangent to the x- or $\hat{x}$-axes. We refer to the $\hat{x}$- and $\hat{y}$-axes as the **transformed axes**, and they are parallel to the eigenvectors for the matrix $\mathbf{A} = \begin{bmatrix} a & b \\ c & d \end{bmatrix}$. (Recall Section 9.5, and see Problems 25 and 26.) Observe that these transformed axes support trajectories that are simply rays.[†]

Case 2. r_1, r_2 real, distinct, and negative. The general solutions to (3)–(4) have the same form as in (6). A simple example is the system

$$\frac{dx}{dt} = -2x, \qquad \frac{dy}{dt} = -y, \qquad \text{or} \qquad \begin{bmatrix} x' \\ y' \end{bmatrix} = \begin{bmatrix} -2 & 0 \\ 0 & -1 \end{bmatrix} \begin{bmatrix} x \\ y \end{bmatrix},$$

for which $r_1 = -2, r_2 = -1$ and

$$x(t) = Ae^{-2t}, \qquad y(t) = Be^{-t}$$

is a general solution. The reader should verify that the trajectories for this example travel along curves of the form $x = (A/B^2)y^2$. A phase plane diagram for this system is given in Figure 12.4(a) on page 719. Notice that in this example, the trajectories are moving *toward* the origin. For this reason, the origin is called an **asymptotically stable** critical point. Again, since (most) trajectories approach the origin along curves that are tangent to the y-axis, the origin is an improper node. Thus the critical point at the origin is an asymptotically stable improper node.

[†]For example, in Figure 12.3(a) the solution (ray) $x(t) = e^t, y(t) = 0$ lies on the positive x-axis, and the solution $x(t) = 0, y(t) = -e^{3t}$ lies on the negative y-axis. And Figure 12.3(b) depicts "ray-solutions" lying on the positive and negative $\hat{x}$- and $\hat{y}$-axes.

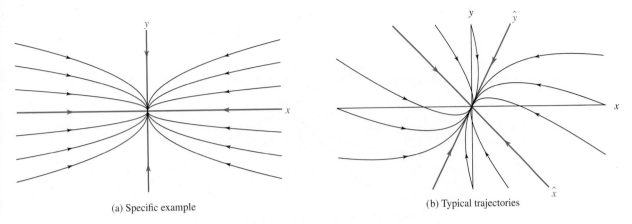

(a) Specific example (b) Typical trajectories

Figure 12.4 Stable improper node; roots are real, distinct, and negative

In general, when the roots are distinct, real, and negative, the origin is an *asymptotically stable improper node* whose phase plane diagram near the origin resembles that in Figure 12.4(b). Cases 1 and 2 are quite similar. They differ mainly in that when the roots are both positive, the trajectories move away from the origin (the origin is unstable), while when the roots are both negative, the trajectories approach the origin (the origin is stable).

Example 1 Classify the critical point at the origin and sketch a phase diagram for the system

(7) $$\frac{dx}{dt} = -5x + 2y\,,$$

$$\text{or}\quad \begin{bmatrix} x' \\ y' \end{bmatrix} = \begin{bmatrix} -5 & 2 \\ 1 & -4 \end{bmatrix} \begin{bmatrix} x \\ y \end{bmatrix}.$$

(8) $$\frac{dy}{dt} = x - 4y\,,$$

Solution The characteristic equation for this system is $r^2 + 9r + 18 = (r+6)(r+3) = 0$, which has roots $r_1 = -6$, $r_2 = -3$. Thus the origin is an asymptotically stable improper node. To sketch the phase plane diagram, we must first determine the two lines passing through the origin that correspond to the transformed axes. For this purpose we can solve for the eigenvectors corresponding to the eigenvalues -6 and -3. Alternatively, since these lines must support trajectories, we can utilize the phase plane equation

(9) $$\frac{dy}{dx} = \frac{dy/dt}{dx/dt} = \frac{x - 4y}{-5x + 2y}\,.$$

Substituting $y = mx$ (for a line through the origin) gives

$$m = \frac{x - 4mx}{-5x + 2mx} = \frac{1 - 4m}{-5 + 2m}\,.$$

Solving for m yields $2m^2 - m - 1 = (m-1)(2m+1) = 0$, so $m = 1, -1/2$. Hence, the two axes are $y = x$ and $y = -(1/2)x$. As we know, the trajectories approach one of these lines tangentially. To determine which, let's graph a few of the isoclines for (9). For example, when $dy/dx = 0$, we get $x - 4y = 0$ or $y = (1/4)x$. So, when a trajectory crosses the line $y = (1/4)x$, its slope is zero. Moreover, when dy/dx is infinite or undefined, then $-5x + 2y = 0$. Hence, when a trajectory crosses the line $y = (5/2)x$, the tangent line to the

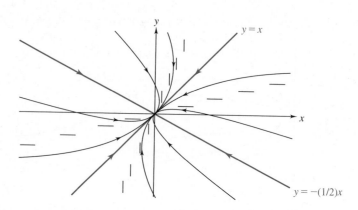

Figure 12.5 Phase plane diagram for $dx/dt = -5x + 2y$; $dy/dt = x - 4y$

trajectory must be vertical. These isoclines are represented in Figure 12.5 by horizontal and vertical hash marks, respectively. It is now clear that the trajectories must be tangent to the line $y = x$ as they approach the origin. ◆

Case 3. r_1, r_2, real opposite signs. Again, a general solution to system (3)–(4) has the same form as in (6). A simple example is the system

$$\frac{dx}{dt} = x, \qquad \frac{dy}{dt} = -y,$$

which has a general solution of the form

$$x(t) = Ae^t, \qquad y(t) = Be^{-t},$$

with $r_1 = 1$, $r_2 = -1$. The trajectories lie on the curves $y = ABx^{-1}$ [see Figure 12.6(a)].

In this case the trajectories come in along the positive and negative y-axis and go out along the positive and negative x-axis. The origin is unstable, since there are trajectories that pass arbitrarily near to the origin, but then eventually move away. (The two trajectories on the y-axis do approach the origin, but they are the only ones.) For apt reasons,[†] the origin is said to be a **saddle point.** Hence, when the roots are real and of opposite signs, the origin is an *unstable*

(a) Specific example (b) Typical trajectories

Figure 12.6 Unstable saddle point; roots are real and of opposite signs

[†]Clearly xy is constant along any trajectory. A three-dimensional graph of xy versus x and y looks like a saddle.

saddle point. In general, the trajectories come in alongside one of the transformed axes[†] and go out alongside the other. This is illustrated in Figure 12.6(b). Observe that the trajectories have both of these lines as asymptotes and that both lines themselves support ray trajectories for the system.

Example 2 Classify the critical point at the origin and sketch a phase plane diagram for the system

(10) $$\frac{dx}{dt} = 5x - 3y,$$

(11) $$\frac{dy}{dt} = 4x - 3y.$$

Solution The characteristic equation for this system is $r^2 - 2r - 3 = (r+1)(r-3) = 0$, which has roots $r_1 = -1, r_2 = 3$. Thus the origin is an unstable saddle point. To sketch the phase plane diagram, we must first determine the two lines corresponding to the transformed axes. As in the preceding case we can utilize the phase plane equation

$$\frac{dy}{dx} = \frac{dy/dt}{dx/dt} = \frac{4x - 3y}{5x - 3y}.$$

Next we substitute $y = mx$ and find

$$m = \frac{4x - 3mx}{5x - 3mx} = \frac{4 - 3m}{5 - 3m}.$$

Solving for m gives $3m^2 - 8m + 4 = (3m - 2)(m - 2) = 0$ so that $m = 2, 2/3$. Hence, the two transformed axes are $y = 2x$ and $y = (2/3)x$. Since the origin is a saddle point, the trajectories are moving in toward the origin alongside one of the transformed axes and they are moving out alongside the other. On the line $y = 2x$ itself, we find that

$$\frac{dx}{dt} = 5x - 3y = 5x - 3(2x) = -x.$$

So, when $x > 0$, this trajectory moves toward the origin, since $dx/dt < 0$. The same is true for $x < 0$, since $dx/dt > 0$. Consequently, on the other line, $y = (2/3)x$, the trajectories must be moving away from the origin. Indeed, a quick check shows that on $y = (2/3)x$

$$\frac{dx}{dt} = 5x - 3y = 5x - 3(2/3)x = 3x,$$

so the trajectories do, in fact, recede from the origin. The phase plane diagram is shown in Figure 12.7 on page 722. ◆

We shall see that the "transformed axes" concept is not relevant in the remaining cases.

[†]The phase plane diagram can be generated using one of the computer software packages described in Appendix G.; however, knowing the transformed axes is still helpful.

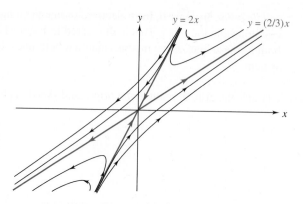

Figure 12.7 Phase plane diagram for $dx/dt = 5x - 3y$; $dy/dt = 4x - 3y$

Case 4. Equal roots, $r_1 = r_2$. In this case, a general solution may or may not[†] involve a factor of t times the exponential e^{rt}. As we shall see, the presence of the factor of t alters the phase plane diagram, but not the stability. We will discuss the two possibilities.

A simple example where the factor of t is not present is

$$\frac{dx}{dt} = rx, \qquad \frac{dy}{dt} = ry,$$

which has the general solution

$$x(t) = Ae^{rt}, \qquad y(t) = Be^{rt}.$$

Here the trajectories lie on the integral curves $y = (B/A)x$. When $r > 0$, these trajectories move away from the origin, so the origin is unstable [see Figure 12.8(a)]. When $r < 0$, the trajectories approach the origin and the origin is asymptotically stable [see Figure 12.8(b)]. In either case all the trajectories lie on lines passing through the origin. Because every direction through the origin defines a trajectory, the origin is called a **proper node.**

In general, when the roots are equal and a factor of t is not present, the origin is a *proper node* that is *unstable* when the root $r > 0$ and *asympototically stable* when $r < 0$.

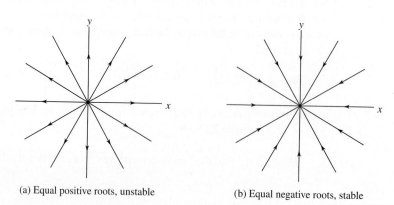

(a) Equal positive roots, unstable (b) Equal negative roots, stable

Figure 12.8 Proper nodes; equal roots with no factor of t

[†]In Section 9.8 we found that when we have a repeated eigenvalue and there are not two linearly independent eigenvectors, then the factor of t is present.

An example where the factor of t arises is

$$\frac{dx}{dt} = rx, \qquad \frac{dy}{dt} = x + ry,$$

which has the general solution

$$x(t) = Ae^{rt}, \qquad y(t) = (At + B)e^{rt}.$$

To find the trajectories, we first solve for t in terms of x to obtain

$$t = \frac{1}{r}\ln(x/A).$$

Substituting for t in the expression for $y(t)$ and simplifying, we find

$$y = \frac{B}{A}x + \frac{x}{r}\ln(x/A).$$

(Alternatively, we can deduce this formula by solving the phase plane equation.) The integral curves are sketched in Figure 12.9(a). Observe that dy/dx is undefined for $x = 0$. When $r > 0$, the trajectories move away from the origin, so it is unstable. Since the trajectories lie only on curves that are tangent to the y-axis at the origin, we have an improper node [see Figure 12.9(a)]. When $r < 0$, the trajectories approach the origin, so the origin is an asymptotically stable point.

In summary, when the roots are equal and a factor of t is present in the general solution, the origin is an *improper node* that is *unstable* when $r > 0$ [see Figure 12.9(b)] and *asymptotically stable* when $r < 0$ [see Figure 12.9(c)].

Case 5. Complex roots, $r = \alpha \pm i\beta$. We now consider the case of complex roots $r = \alpha \pm i\beta$, where $\alpha \neq 0$ and $\beta \neq 0$. (Pure imaginary roots, $r = \pm i\beta$, are discussed in Case 6.) For complex roots, a general solution to system (3)–(4) is given by

$$x(t) = e^{\alpha t}[A_1 \cos \beta t + B_1 \sin \beta t],$$
$$y(t) = e^{\alpha t}[A_2 \cos \beta t + B_2 \sin \beta t].$$

(a) Specific example;
$r > 0$ unstable

(b) Typical trajectory;
$r > 0$ unstable

(c) Typical trajectory;
$r < 0$ asymptotically stable

Figure 12.9 Improper node; equal roots with a factor of t

It is clear that when $\alpha > 0$, the trajectories will travel far away from the origin. Hence, the origin is unstable. But, when $\alpha < 0$, the trajectories approach the origin and the origin is asymptotically stable.

To better describe the behavior of trajectories near the origin, we analyze the specific system

$$(12) \qquad \frac{dx}{dt} = \alpha x - \beta y,$$

$$(13) \qquad \frac{dy}{dt} = \beta x + \alpha y,$$

which has characteristic roots $r = \alpha \pm i\beta$. If we set $z(t) = x(t) + iy(t)$ to form a complex-valued function, then

$$\frac{dz}{dt} = \frac{dx}{dt} + i\frac{dy}{dt} = (\alpha x - \beta y) + i(\beta x + \alpha y)$$
$$= \alpha(x + iy) + i\beta(x + iy) = (\alpha + i\beta)z.$$

Using the polar coordinate representation $z(t) = \rho(t)e^{i\theta(t)}$, we obtain

$$\frac{dz}{dt} = \frac{d\rho}{dt}e^{i\theta} + i\rho e^{i\theta}\frac{d\theta}{dt}.$$

Setting the two expressions for dz/dt equal yields

$$\frac{d\rho}{dt}e^{i\theta} + i\rho e^{i\theta}\frac{d\theta}{dt} = (\alpha + i\beta)\rho e^{i\theta},$$

and simplifying, we have

$$\frac{d\rho}{dt} + i\rho\frac{d\theta}{dt} = \alpha\rho + i\beta\rho.$$

By equating the real and imaginary parts, we obtain the simple system

$$\frac{d\rho}{dt} = \alpha\rho, \qquad \frac{d\theta}{dt} = \beta,$$

which has the general solution

$$\rho(t) = \rho_0 e^{\alpha t}, \qquad \theta(t) = \beta t + \theta_0,$$

where ρ_0 and θ_0 are real-valued constants. Again, we see that if $\alpha > 0$, the solution moves away from the origin, since $\rho \to +\infty$ as $t \to +\infty$. But, because θ is a linear function of t, the trajectories spiral around the origin. Similarly, when $\alpha < 0$, the solutions spiral in toward the origin. We therefore refer to the origin as a **spiral point.** We can determine the direction of rotation of the spiral from the original system. For example, setting $y = 0$ in equation (13), we find that $dy/dt = \beta x$. Now, if $\beta < 0$, then as the trajectory crosses the positive x-axis $(y = 0)$, we have $dy/dt < 0$. Consequently, the trajectory is spiraling clockwise about the origin [see Figure 12.10(a) and (b) on page 725].

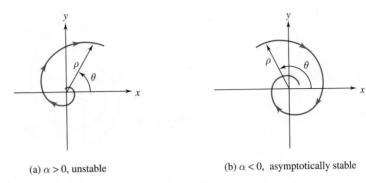

(a) $\alpha > 0$, unstable (b) $\alpha < 0$, asymptotically stable

Figure 12.10 Spiral point; $r = \alpha \pm i\beta$ with clockwise rotation

In the general case, the spirals may be distorted, but the stability criterion persists; when the roots $r = \alpha \pm i\beta$ are complex numbers, the origin is a *spiral point* that is *unstable* when $\alpha > 0$ [see Figure 12.11(a)] and *asymptotically stable* when $\alpha < 0$ [see Figure 12.11(b)].

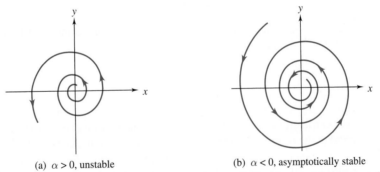

(a) $\alpha > 0$, unstable (b) $\alpha < 0$, asymptotically stable

Figure 12.11 General spiral point; $r = \alpha \pm i\beta$ with counterclockwise rotation

Example 3 Classify the critical point at the origin and sketch the phase plane diagram for the system

(14) $$\frac{dx}{dt} = x - 4y ,$$

(15) $$\frac{dy}{dt} = 4x + y .$$

Solution The characteristic equation for this system is $(r - 1)^2 + 16 = 0$, which has complex roots $r = 1 \pm 4i$. Thus, the origin is an unstable spiral point, since $\alpha = 1 > 0$. Setting $y = 0$ in the equation (15), we obtain $dy/dt = 4x$. Hence $dy/dt > 0$ when a trajectory crosses the positive x-axis. Thus, the trajectories are spiraling with a counterclockwise rotation about the origin. See Figure 12.12 on page 726. ◆

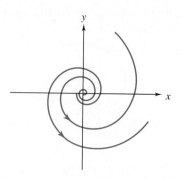

Figure 12.12 Trajectories for $dx/dt = x - 4y$; $dy/dt = 4x + y$

Case 6. Pure imaginary roots, $r = \pm i\beta$. When the roots are pure imaginary, $r = \pm i\beta$, a general solution has the form

$$x(t) = A_1 \cos \beta t + B_1 \sin \beta t,$$
$$y(t) = A_2 \cos \beta t + B_2 \sin \beta t.$$

These solutions all have period $2\pi/\beta$ and remain bounded. A simple example is the system

$$\frac{dx}{dt} = -\beta y, \qquad \frac{dy}{dt} = \beta x,$$

which is just system (12)–(13) with $\alpha = 0$. Hence, taking $\alpha = 0$ in the analysis of Case 5, we find that, in polar coordinates,

$$\frac{d\rho}{dt} = 0, \qquad \frac{d\theta}{dt} = \beta.$$

Hence, $\rho(t) \equiv \rho_0$ and $\theta(t) = \beta t + \theta_0$. Since ρ is constant (but arbitrary), the trajectories are concentric circles about the origin. Thus, the motion is a periodic rotation around a circle centered at the origin. Appropriately the origin is called a **center** and is a stable critical point.

In general, when the roots are pure imaginary, the origin is a *stable center*. The trajectories are ellipses centered at the origin (see Problem 24). A typical phase diagram is shown in Figure 12.13.

We now summarize the analysis of the behavior of trajectories of a linear system near a critical point at the origin.

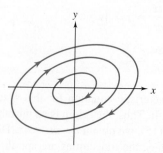

Figure 12.13 Stable center; pure imaginary roots, clockwise rotation

Stability of Linear Systems

Theorem 1. Assume the origin $(0, 0)$ is the only critical point for the linear system

$$\begin{aligned} x'(t) &= ax + by \\ y'(t) &= cx + dy \,, \end{aligned} \quad \text{or} \quad \begin{bmatrix} x' \\ y' \end{bmatrix} = \begin{bmatrix} a & b \\ c & d \end{bmatrix} \begin{bmatrix} x \\ y \end{bmatrix},$$

where a, b, c, and d are real, and let r_1 and r_2 be the eigenvalues of the coefficient matrix or, equivalently, the roots of the characteristic equation

$$r^2 - (a+d)r + (ad - bc) = 0 \,.$$

The stability of the origin and the classification of the origin as a critical point depends on the roots r_1 and r_2 as follows:

Roots (eigenvalues)	Type of Critical Point	Stability
distinct, positive	improper node	unstable
distinct, negative	improper node	asymptotically stable
opposite signs	saddle point	unstable
equal, positive	proper or improper node	unstable
equal, negative	proper or improper node	asymptotically stable
complex-valued:		
positive real part	spiral point	unstable
negative real part	spiral point	asymptotically stable
pure imaginary	center	stable

By a simple change of variables, Theorem 1 can be used to classify critical points that are not located at the origin.

Example 4 Find and classify the critical point of the linear system

(16) $$\dfrac{dx}{dt} = 2x + y + 3 \,,$$

(17) $$\dfrac{dy}{dt} = -3x - 2y - 4 \,.$$

Solution The critical point is the solution to the system

$$\begin{aligned} 2x + y + 3 &= 0 \,, \\ -3x - 2y - 4 &= 0 \,. \end{aligned}$$

Solving this system, we obtain the critical point $(-2, 1)$. Now we will use the change of variables

$$x = u - 2 \,, \qquad y = v + 1 \,,$$

to translate the critical point $(-2, 1)$ to the origin $(0, 0)$. Substituting into system (16)–(17) and simplifying, we obtain a system of differential equations in u and v:

$$\frac{du}{dt} = \frac{dx}{dt} = 2(u - 2) + (v + 1) + 3 = 2u + v \,,$$

$$\frac{dv}{dt} = \frac{dy}{dt} = -3(u - 2) - 2(v + 1) - 4 = -3u - 2v \,.$$

The characteristic equation for this linear system is $r^2 - 1 = 0$, which has roots $r = \pm 1$. Hence, the origin is an unstable saddle point for the new system. Therefore, the critical point $(-2, 1)$ is an unstable saddle point for the original system (16)–(17). ◆

12.2 EXERCISES

In Problems 1–6, classify the critical point at the origin of the given linear system.

1. $\dfrac{dx}{dt} = 5x + 6y$,

$\dfrac{dy}{dt} = -5x - 8y$

2. $\dfrac{dx}{dt} = 3x$,

$\dfrac{dy}{dt} = 3y$

3. $\dfrac{dx}{dt} = 3x + 5y$,

$\dfrac{dy}{dt} = -5x - 5y$

4. $\dfrac{dx}{dt} = 6x - y$,

$\dfrac{dy}{dt} = 8y$

5. $\dfrac{dx}{dt} = -6x + 3y$,

$\dfrac{dy}{dt} = x - 4y$

6. $\dfrac{dx}{dt} = 2x + 13y$,

$\dfrac{dy}{dt} = -x - 2y$

In Problems 7–12, find and classify the critical point of the given linear system.

7. $\dfrac{dx}{dt} = -4x + 2y + 8$,

$\dfrac{dy}{dt} = x - 2y + 1$

8. $\dfrac{dx}{dt} = -3x - y - 4$,

$\dfrac{dy}{dt} = 29x + y + 30$

9. $\dfrac{dx}{dt} = 2x + y + 9$,

$\dfrac{dy}{dt} = -5x - 2y - 22$

10. $\dfrac{dx}{dt} = 2x + 4y + 4$,

$\dfrac{dy}{dt} = 3x + 5y + 4$

11. $\dfrac{dx}{dt} = -x + y + 8$,

$\dfrac{dy}{dt} = -x - 4y - 17$

12. $\dfrac{dx}{dt} = -x + y + 4$,

$\dfrac{dy}{dt} = -y + 1$

In Problems 13–20, classify the critical point at the origin and, either by hand or using a computer software package, sketch a phase plane diagram for the given system.

13. $\dfrac{dx}{dt} = x - 6y$,

$\dfrac{dy}{dt} = 2x - 7y$

14. $\dfrac{dx}{dt} = 5x$,

$\dfrac{dy}{dt} = 5y$

15. $\dfrac{dx}{dt} = x + 2y$,

$\dfrac{dy}{dt} = 5x - 2y$

16. $\dfrac{dx}{dt} = -3x - 5y$,

$\dfrac{dy}{dt} = 2x - y$

17. $\dfrac{dx}{dt} = -8x + y$,

$\dfrac{dy}{dt} = 2x + 2y$

18. $\dfrac{dx}{dt} = 3x - y$,

$\dfrac{dy}{dt} = -2x + 4y$

19. $\dfrac{dx}{dt} = -2x + y$,

$\dfrac{dy}{dt} = -2y$

20. $\dfrac{dx}{dt} = -3x - y$,

$\dfrac{dy}{dt} = 13x + 3y$

21. Show that when the system

$$x'(t) = ax + by + p,$$
$$y'(t) = cx + dy + q$$

has a critical point at (h, k), then the change of variables $x = u + h$, $y = v + k$ yields the new system

$$u'(t) = au + bv,$$
$$v'(t) = cu + dv,$$

with the origin as the critical point.

22. Show that if $ad - bc = 0$, then there exist critical points for the system (3)–(4) that are arbitrarily close to $(0, 0)$. [*Hint:* If a or b is not zero, then the points on the line $ax + by = 0$ are all critical points.]

23. Show that the change of variables

$$x = 2u + v, \qquad y = 3u + v$$

transforms the linear system

$$\frac{dx}{dt} = x - 2y, \qquad \frac{dy}{dt} = 3x - 4y$$

into the linear system

$$\frac{du}{dt} = -2u, \qquad \frac{dv}{dt} = -v,$$

which is the example studied in Case 2. Verify that the u-axis maps to the line $y = (3/2)x$ and the v-axis maps to the line $y = x$.

24. Show that the set of points

$$C := \{(x, y) : x^2 + y^2 = r^2\}$$

is mapped by the change of variables

$$x = au + bv, \qquad y = cu + dv,$$

where $ad - bc \neq 0$, into the set of points

$$E := \{ (u, v): (a^2 + c^2)u^2 + (2ab + 2cd)uv$$
$$+ (b^2 + d^2)v^2 = r^2 \} .$$

Verify that E is an ellipse in the uv-plane.

25. Show that the roots of the characteristic equation (5) are the eigenvalues of the matrix

$$A = \begin{bmatrix} a & b \\ c & d \end{bmatrix} .$$

26. Show that when the roots of the characteristic equation (5) are real and distinct, the transformed axes are parallel to the eigenvectors of the matrix

$$A = \begin{bmatrix} a & b \\ c & d \end{bmatrix} .$$

27. Compute the eigenvectors of the matrix for the system described by equations (10), (11). Show that they are parallel to the lines $y = 2x$ and $y = 2x/3$. Identify the directions taken by trajectories along these rays.

12.3 Almost Linear Systems

Bearing in mind the discussion of the preceding section concerning stability for linear autonomous systems, we now turn to the general situation of a *nonlinear* system and ask: Exactly what is meant by stability of a critical point? Let's say we are dealing with the autonomous system

(1) $$\frac{dx}{dt} = f(x, y) ,$$

(2) $$\frac{dy}{dt} = g(x, y) ,$$

where f and g are real-valued functions that do not depend explicitly on t.

Stability of a Critical Point

Definition 1. A critical point (x_0, y_0) of the autonomous system (1)–(2) is **stable** if, given any $\varepsilon > 0$, there exists a $\delta > 0$ such that every solution $x = \phi(t), y = \psi(t)$ of the system that satisfies

$$\sqrt{(\phi(0) - x_0)^2 + (\psi(0) - y_0)^2} < \delta$$

at $t = 0$ also satisfies

$$\sqrt{(\phi(t) - x_0)^2 + (\psi(t) - y_0)^2} < \varepsilon$$

for all $t \geq 0$.

 If (x_0, y_0) is stable and there exists an $\eta > 0$ such that every solution $x = \phi(t), y = \psi(t)$ that satisfies

$$\sqrt{(\phi(0) - x_0)^2 + (\psi(0) - y_0)^2} < \eta$$

at $t = 0$ also satisfies

$$\lim_{t \to +\infty} \phi(t) = x_0 , \qquad \lim_{t \to +\infty} \psi(t) = y_0 ,$$

then the critical point is **asymptotically stable**.
 A critical point that is not stable is called **unstable**.

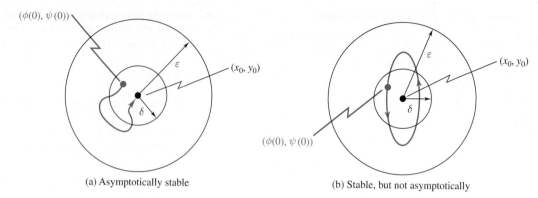

(a) Asymptotically stable (b) Stable, but not asymptotically

Figure 12.14 Stability of a critical point

In other words, a critical point (x_0, y_0) is stable if any trajectory that begins near (within δ of) the point (x_0, y_0) remains near (within ε of) this point. This definition is illustrated in Figure 12.14(a) and (b). When all the trajectories that start close to a stable critical point actually approach it as $t \to +\infty$, then the critical point is asymptotically stable [see Figure 12.14(a)]. As we have seen in the case of a linear system whose characteristic equation has pure imaginary roots, it is possible for trajectories to remain close, but not actually approach a critical point [see Figure 12.14(b)]; that is, there are indeed critical points that are stable but not asymptotically stable.

A remark is in order. The essential property of asymptotic stability, that $\phi(t) \to x_0$ and $\psi(t) \to y_0$ as $t \to +\infty$, is *not* sufficient for a critical point to be stable. There are in fact peculiar systems with the property that trajectories starting arbitrarily close to a critical point first travel away from the critical point before turning back.[†] This is shown schematically in Figure 12.15. For such a system, (x_0, y_0) is not stable and hence not asymptotically stable.

Our analysis in the preceding section concerning the behavior of trajectories near a critical point for a linear system suggests that our classification of the critical point as stable, asymptotically stable, and unstable is consistent with the definitions just given. To rigorously verify this, we would have to explain, for example, how to select the δ's given the ε's. However, we will not pursue such details here.

Our intuition leads us to believe that if we were to make a small perturbation of a linear system near a critical point, then the stability properties of the critical point for the perturbed system would be the same as those for the original linear system. In many cases this is true, but there are certain exceptions. These exceptions can be observed even when the perturbed system is linear. Since the classification of a critical point depends only on the roots of the characteristic

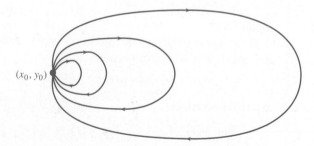

Figure 12.15 A peculiar unstable critical point

equation [see equation (5) in Section 12.2], this classification can be determined by answering the question: How do the roots of the characteristic equation change under small perturbations of the coefficients? If the roots are real, distinct, and both positive (negative), then under a small perturbation they would still be positive (negative). Consequently, the nature of the critical point would remain the same. This is also true if the roots are of opposite signs or are complex (not pure imaginary). However, there are two cases where small changes can substantially alter the nature of the roots and hence possibly alter the classification of the critical point.

The first is the case of pure imaginary roots. A small perturbation of such a system could result in any of three possible situations: pure imaginary roots, complex conjugate roots with positive real parts $(\alpha > 0)$, or complex conjugate roots with negative real parts $(\alpha < 0)$. For example, the system

$$\frac{dx}{dt} = \alpha x - (1 + \gamma)y \,,$$

$$\frac{dy}{dt} = (1 + \gamma)x + \alpha y$$

has the characteristic equation

$$r^2 - 2\alpha r + \left[\alpha^2 + (1 + \gamma)^2\right] = 0 \,,$$

which has roots $r = \alpha \pm (1 + \gamma)i$. Let the unperturbed system be the system with $\alpha = \gamma = 0$, for which the characteristic equation has pure imaginary roots $r = \pm i$. If we perturb this system, keeping α and γ small, we can have

(a) pure imaginary roots $r = \pm(1 + \gamma)i$ for $\alpha = 0$, or
(b) complex roots $r = \alpha \pm (1 + \gamma)i$ with positive real parts when $\alpha > 0$ and negative real parts when $\alpha < 0$.

Consequently, if the critical point for the unperturbed system is a center, then the critical point for the perturbed linear system can be a center, an unstable spiral point, or an asymptotically stable spiral point. For sufficiently small perturbations, these are the only possibilities.

A second exceptional situation is that of real, equal roots. Here an arbitrarily small perturbation may result in distinct real roots or complex conjugate roots (depending on whether the double root separates horizontally or vertically in the complex plane; see Problem 18). If the equal roots were positive (negative), then under small perturbations, we would have real positive (negative) roots or complex conjugate roots whose real parts are positive (negative). In either case the stability is not affected.

The situation is similar when the perturbed system is "almost linear."

Almost Linear Systems

Definition 2. Let the origin $(0, 0)$ be a critical point of the autonomous system

(3) $x'(t) = ax + by + F(x, y) \,,$

(4) $y'(t) = cx + dy + G(x, y) \,,$

where a, b, c, d are constants and F, G are continuous in a neighborhood (disk) about the origin. Assume that $ad - bc \neq 0$ so that the origin is the only critical point for the **corresponding linear system,** which we obtain by setting $F = G = 0$ in (3) and (4). Then, system (3)–(4) is said to be **almost linear** near the origin if

(5) $\dfrac{F(x, y)}{\sqrt{x^2 + y^2}} \to 0$ and $\dfrac{G(x, y)}{\sqrt{x^2 + y^2}} \to 0,$ as $\sqrt{x^2 + y^2} \to 0 \,.$

One implication of conditions (5) is that the first partial derivatives of F and G must vanish at the critical point $(0, 0)$. Thus on writing the system (3)–(4) in the standard form $x' = f(x, y)$, $y' = g(x, y)$, we note that $f_x(0, 0) = a$, $f_y(0, 0) = b$, $g_x(0, 0) = c$, and $g_y(0, 0) = d$. Therefore, the corresponding linear system (or *linearization*) can be written in matrix form as $\begin{bmatrix} x' \\ y' \end{bmatrix} = \mathbf{A} \begin{bmatrix} x \\ y \end{bmatrix}$, where the coefficient matrix $\mathbf{A} = \begin{bmatrix} f_x(0,0) & f_y(0,0) \\ g_x(0,0) & g_y(0,0) \end{bmatrix}$ is the familiar *Jacobian* for the change of variables $u = f(x, y)$, $v = g(x, y)$, evaluated at the critical point $(0, 0)$.

Notice also that conditions (5) can be written in a simpler fashion using the notation $r = \sqrt{x^2 + y^2}$ for the distance from (x, y) to the origin. The requirement now becomes $F(x, y)/r \to 0$ and $G(x, y)/r \to 0$ as $r \to 0^+$.

Almost linear systems are often encountered in applications where F and G involve higher-order powers or products of x and y. For example, the system

$$\frac{dx}{dt} = 3x + 2y - xy + y^3,$$

$$\frac{dy}{dt} = -x + y + x^2 - y^2$$

is almost linear since $ad - bc = (3)(1) - (2)(-1) \neq 0$, and $F(x, y) = -xy + y^3$ and $G(x, y) = x^2 - y^2$ involve only higher-order terms in x and y. In fact, if the function $f(x, y)$ in equation (1) can be expanded in a Taylor series in x and y about $(0, 0)$ with $f(0, 0) = 0$, and if $g(x, y)$ in equation (2) can also be expanded in a Taylor series about $(0, 0)$ with $g(0, 0) = 0$, then system (1)–(2) will be almost linear, provided the origin is the only critical point for the corresponding linear system. (The linear system is given by the matrix equation $\mathbf{x}' = \mathbf{Ax}$, where $\mathbf{A}$ is (as above) the Jacobian matrix at $(0, 0)$.)

We now state a classical result due to Poincaré that relates the stability of the origin for an almost linear system with the stability of the origin in the corresponding linear system. The situation turns out to be the same as we found for perturbations that result in a linear system.[†]

Stability of Almost Linear Systems

Theorem 2. Let r_1 and r_2 be the roots of the characteristic equation

(6) $r^2 - (a + d)r + (ad - bc) = 0$

(equivalently, the eigenvalues) for the linear system corresponding to the almost linear system (3)–(4). Then, the stability properties of the critical point at the origin for the almost linear system are the same as the stability properties of the origin for the corresponding linear system with one exception: When the roots are pure imaginary, the stability properties for the almost linear system cannot be deduced from the linear system (a different approach is required).

In addition to the stability, the behavior of the trajectories near the origin is essentially the same for the almost linear system and its corresponding linear system. The exceptions discussed for linear perturbations are the same as in the almost linear case. These are summarized in Table 12.1 on page 733.

[†]We do not give a proof here but refer the reader to the book by D. W. Jordan and P. Smith, *Nonlinear Ordinary Differential Equations*, 4th ed. (Clarendon Press, Oxford, 2007), Chapter 10, where a detailed analysis using Lyapunov functions is given.

TABLE 12.1	**Stability of Almost Linear and Linear Systems**				
Roots (eigenvalues)	**Linear Systems**			**Almost Linear Systems**	
r_1 and r_2	Type	Stability		Type	Stability
distinct, positive	improper node	unstable		improper node	unstable
distinct, negative	improper node	asymptotically stable		improper node	asymptotically stable
opposite signs	saddle point	unstable		saddle point	unstable
equal, positive	improper or proper node	unstable		improper or proper node or spiral point	unstable
equal, negative	improper or proper node	asymptotically stable		improper or proper node or spiral point	asymptotically stable
complex; real part					
positive	spiral point	unstable		spiral point	unstable
negative	spiral point	asymptotically stable		spiral point	asymptotically stable
pure imaginary	center	stable		center or spiral point	indeterminate

Here r_1 and r_2 denote the roots of the characteristic equation (6) for the linear system corresponding to the almost linear system (3)–(4).

Competing Species

In Section 12.1 we presented a model for two competing species with populations $x(t)$ and $y(t)$. These populations were assumed to satisfy the system

$$(7) \qquad \frac{dx}{dt} = x(a_1 - b_1 x - c_1 y),$$

$$(8) \qquad \frac{dy}{dt} = y(a_2 - b_2 y - c_2 x),$$

where the positive constants $a_1, a_2, b_1, b_2, c_1, c_2$ depend on the species and environment under investigation. Using Theorem 2 and Table 12.1, we present a detailed analysis of the phase plane diagram and discuss the stability for a particular competing species problem. The analytical methods presented here combined with the use of a computer can produce a more detailed phase plane diagram (see Project C).

Assume the populations satisfy the almost linear system

$$(9) \qquad \frac{dx}{dt} = x(7 - x - 2y) = 7x - x^2 - 2xy,$$

$$(10) \qquad \frac{dy}{dt} = y(5 - y - x) = 5y - y^2 - xy.$$

Then the critical points for this system are solutions to the pair of equations

$$x(7 - x - 2y) = 0,$$
$$y(5 - y - x) = 0.$$

One solution is $(0, 0)$. Setting $x = 0$, we also obtain $y = 5$ and setting $y = 0$, we get $x = 7$. So three of the critical points are $(0, 0)$, $(0, 5)$, and $(7, 0)$. While these points are not strictly inside the first quadrant, they do lie on its border. Therefore, knowing the behavior of trajectories near these points will help us to obtain a phase diagram for the first quadrant. A fourth critical point occurs at the intersection of the lines

$$7 - x - 2y = 0 ,$$
$$5 - y - x = 0 ,$$

which is the point $(3, 2)$. We begin with an analysis of the trajectories near the origin $(0, 0)$.

Critical point $(0, 0)$. We get the corresponding linear system by deleting the second-order terms in (9)–(10). This yields

$$\frac{dx}{dt} = 7x , \qquad \frac{dy}{dt} = 5y .$$

The characteristic equation for this system is $(r - 7)(r - 5) = 0$, which has roots $r_1 = 7$, $r_2 = 5$. Thus the origin is an unstable improper node. Furthermore, this system is easy to solve (the variables are uncoupled!). We find that $x(t) = Ae^{7t}$ and $y(t) = Be^{5t}$, so we can directly compute the trajectories. They are along the curves $y = B(e^t)^5 = (B/A^{5/7})x^{5/7}$. Observe that the trajectories are tangent to the y-axis and move away from the origin. In Figure 12.16, we have sketched the phase plane diagram in the first quadrant near the origin.

Critical point $(0, 5)$. We begin by translating the point $(0, 5)$ to the origin by making the change of variables $x = u, y = v + 5$. Substituting into system (9)–(10) and simplifying, we obtain the almost linear system

$$\frac{du}{dt} = -3u - u^2 - 2uv ,$$

$$\frac{dv}{dt} = -5u - 5v - v^2 - uv .$$

The corresponding linear system is

$$\frac{du}{dt} = -3u , \qquad \frac{dv}{dt} = -5u - 5v ,$$

which has the characteristic equation $(r + 3)(r + 5) = 0$ with roots (eigenvalues) $r_1 = -3, r_2 = -5$. Hence, the point $(0, 5)$ is an asymptotically stable improper node. To find the transformed axes $\hat{u}$ and $\hat{v}$, we set $v = mu$. Substituting into

$$\frac{dv}{du} = \frac{dv/dt}{du/dt} = \frac{-5u - 5v}{-3u} ,$$

Figure 12.16 Phase plane diagram near $(0, 0)$

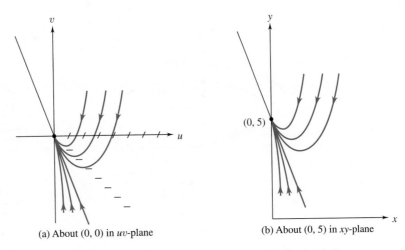

(a) About $(0, 0)$ in uv-plane (b) About $(0, 5)$ in xy-plane

Figure 12.17 Phase plane diagram near the critical point $(0, 5)$

we obtain

$$m = \frac{-5 - 5m}{-3},$$

and solving for m, we find $m = -5/2$. Thus, one transformed axis is $v = -(5/2)u$. What about the other axis? By substituting $v = mu$, we allow for all lines through the origin except the vertical line $u = 0$. Thus we suspect that the v-axis $(u = 0)$ is the missing transformed axis. We can confirm this by solving for du/dv:

$$\frac{du}{dv} = \frac{-3u}{-5u - 5v},$$

which is indeed satisfied by $u = 0$. Therefore, the transformed axes are $v = -(5/2)u$ and $u = 0$. If we check the direction field along the u-axis, we see that $dv/du = 5/3$. Moreover, $dv/du = 0$ along the line $v = -u$. Consequently, the trajectories approach the transformed axis $v = -(5/2)u$ tangentially. The phase plane diagram for the linearized system is sketched in Figure 12.17(a), and the diagram near the critical point $(0, 5)$ of the original system is shown in Figure 12.17(b).

Critical point (7, 0). To translate the point $(7, 0)$ to the origin, we set $x = u + 7$, $y = v$. Substituting into (9)–(10) and simplifying yields the almost linear system

$$\frac{du}{dt} = -7u - 14v - u^2 - 2uv,$$

$$\frac{dv}{dt} = -2v - v^2 - uv.$$

The corresponding linear system

$$\frac{du}{dt} = -7u - 14v, \qquad \frac{dv}{dt} = -2v$$

(a) About $(0, 0)$ in uv-plane (b) About $(7, 0)$ in xy-plane

Figure 12.18 Phase plane diagram near the critical point $(7, 0)$

has the characteristic equation $(r + 7)(r + 2) = 0$ with roots $r_1 = -7, r_2 = -2$. Thus, the point $(7, 0)$ is an asymptotically stable improper node. To get the transformed axes, we set $v = mu$. For the linear system, substituting into

$$\frac{dv}{du} = \frac{dv/dt}{du/dt} = \frac{-2v}{-7u - 14v},$$

we obtain

$$m = \frac{-2m}{-7 - 14m},$$

which yields $m = 0, -5/14$. Therefore, the two transformed axes are $v = 0$ and $v = -(5/14)u$. We now check the direction field along a few lines. When $u = 0, dv/du = 1/7$, and when $v = -u, dv/du = 2/7$. Moreover, dv/du is undefined when $v = -(1/2)u$. Consequently, the trajectories approach the transformed axis $v = -(5/14)u$ tangentially. The phase plane diagrams in the uv- and xy-planes are sketched in Figure 12.18(a) and (b).

Critical point $(3, 2)$. Setting $x = u + 3, y = v + 2$ and substituting into system (9)–(10) we obtain the almost linear system

$$\frac{du}{dt} = -3u - 6v - u^2 - 2uv,$$

$$\frac{dv}{dt} = -2u - 2v - v^2 - uv.$$

The corresponding linear system has the characteristic equation $r^2 + 5r - 6 = (r + 6)(r - 1) = 0$ with roots $r = -6, 1$. Hence, the point $(3, 2)$ is an unstable saddle point. To obtain the transformed axes, we set $v = mu$. For the linear system, substituting into

$$\frac{dv}{du} = \frac{dv/dt}{du/dt} = \frac{-2u - 2v}{-3u - 6v},$$

we obtain

$$m = \frac{-2 - 2m}{-3 - 6m}.$$

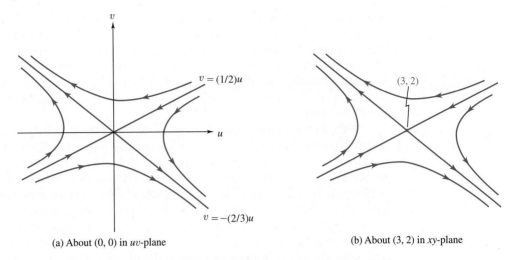

(a) About $(0, 0)$ in uv-plane (b) About $(3, 2)$ in xy-plane

Figure 12.19 Phase plane diagram near the critical point $(3, 2)$

Solving for m yields $m = 1/2, -2/3$. Thus, the two transformed axes are $v = (1/2)u$ and $v = -(2/3)u$. For a saddle point, the trajectories move toward the critical point alongside one transformed axis and away alongside the other. If we set $v = -(2/3)u$, then for the linear system, we have $du/dt = -3u - 6(-2/3)u = u$. The trajectories along this transformed axis are moving away from the origin because $du/dt > 0$ if $u > 0$ and $du/dt < 0$ for $u < 0$. Checking the line $v = (1/2)u$, we see that $du/dt = -3u - 6(1/2)u = -6u$, so the trajectories are moving toward the origin. The phase plane diagrams in the uv- and xy-plane are sketched in Figure 12.19.

In Figure 12.20, we have combined the results of each of the four critical points and sketched the phase plane diagram in the first quadrant for the system (9)–(10). We have included four trajectories that separate the first quadrant into four regions labeled I, II, III, and IV. (These trajectories are difficult to find and the curves sketched represent only rough approximations of

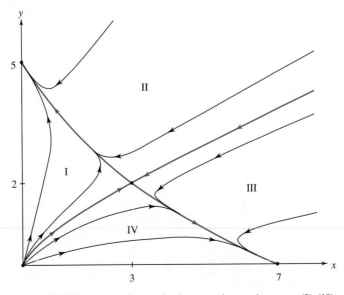

Figure 12.20 Phase plane diagram for the competing species system (9)–(10)

the true curves; see Project C.) These curves are called *separatrices*. Observe that a trajectory in region I remains in that region and approaches the critical point $(0, 5)$ as $t \rightarrow +\infty$. Similarly, trajectories in region II also approach $(0, 5)$ as $t \rightarrow +\infty$. The trajectories in regions III and IV approach the critical point $(7, 0)$ as $t \rightarrow +\infty$. Regions I and II form part of the **regions of asymptotic stability for the critical point** $(0, 5)$; that is, a trajectory beginning in that region is attracted to $(0, 5)$. Similarly, regions III and IV are part of the regions of asymptotic stability for the critical point $(7, 0)$.[†]

What does the phase plane diagram tell us about the competing species problem? If the initial population vector $(x(0), y(0))$ lies in either region I or II, then species x dies off and species y approaches the limiting population of 5. (Units may be thousands, millions, etc.) If, however, $(x(0), y(0))$ lies in region III or IV, then species y dies off and the population x will approach the limiting value 7. Mathematically, there can be coexistence only if the initial population vector $(x(0), y(0))$ lies on either separatrix that approaches the critical point $(3, 2)$ or lies precisely at the critical point $(3, 2)$. However, this situation does not arise in practice because the model, no matter how good, is just an approximation and any error can move the true trajectory off a separatrix. Another point: the time scale that the model (9)–(10) predicts for these "extinctions" to occur may be very large, possibly exceeding the limits of credibility for the model itself.

The Pendulum

The pendulum equation (see (21), Section 4.8 page 208 with $g = l$) $\theta'' = -\sin\theta$ has the equivalent normal form (Section 5.3, equations (3), (4), page 253)

$$x' = y,$$
$$y' = -\sin x,$$

with the interpretation $x = \theta$, $y = \theta'$. Clearly it has critical points at $y = 0$ and $x = n\pi$, $n = 0, \pm 1, \pm 2, \ldots$. At these points $\sin x$ has the Taylor expansions

$$\sin x = 0 + (\cos n\pi)(x - n\pi) - (0) - (\cos n\pi)\frac{(x - n\pi)^3}{3!} + \cdots.$$

Translating these to the origin with the change of variable $u = x - n\pi$, $v = y$, we derive the almost linear system

$$u' = v + F(u, v),$$
$$v' = -(-1)^n u + G(u, v)$$

with $F \equiv 0$ and $G = (-1)^n \dfrac{u^3}{3!} + \cdots$ clearly satisfying conditions (5).

The corresponding linear system,

$$u' = v,$$
$$v' = -(-1)^n u,$$

has a critical point at the origin with characteristic equation $r^2 + (-1)^n = 0$. Thus if n is odd, $r_1 = 1, r_2 = -1$ and Table 12.1 tells us the critical point is an unstable saddle point, both for the linear and almost linear systems.

[†]Regions of asymptotic stability are also called "basins of attraction." When *every* trajectory approaches a critical point, the critical point is said to be *globally stable*. This happens in the case of an asymptotically stable critical point for an autonomous *linear* system. But, as we see here, an asymptotically stable critical point need not be globally stable.

If n is even, $r_1 = i$, $r_2 = -i$ and the critical point is a (stable) center for the linear system. Table 12.1 is inconclusive regarding its stability for the almost linear system. The phase plane diagram for the pendulum can be seen in Section 12.4 (Figure 12.25 on page 746), where by reexamining it using energy methods we find it remains a center.

12.3 EXERCISES

In Problems 1–8, show that the given system is almost linear near the origin and discuss the type and stability of the critical point at the origin.

1. $\dfrac{dx}{dt} = 3x + 2y - y^2$,

$\dfrac{dy}{dt} = -2x - 2y + xy$

2. $\dfrac{dx}{dt} = -2x + 2xy$,

$\dfrac{dy}{dt} = x - y + x^2$

3. $\dfrac{dx}{dt} = x - y$,

$\dfrac{dy}{dt} = 3x + 5y - x^3$

4. $\dfrac{dx}{dt} = -x + 4y + y^2$,

$\dfrac{dy}{dt} = -x - y - xy$

5. $\dfrac{dx}{dt} = x + 5y - y^2$,

$\dfrac{dy}{dt} = -x - y - y^2$

6. $\dfrac{dx}{dt} = 4x + 5y + xy$,

$\dfrac{dy}{dt} = x + x^2 - y^2$

7. $\dfrac{dx}{dt} = e^{x+y} - \cos x$,

$\dfrac{dy}{dt} = \cos y + x - 1$

8. $\dfrac{dx}{dt} = \sin(y - 3x)$,

$\dfrac{dy}{dt} = \cos x - e^y$

In Problems 9–12, find all the critical points for the given system and discuss the type and stability of the critical points.

9. $\dfrac{dx}{dt} = 16 - xy$,

$\dfrac{dy}{dt} = x - y^3$

10. $\dfrac{dx}{dt} = x^2 - y$,

$\dfrac{dy}{dt} = x + y^2$

11. $\dfrac{dx}{dt} = x + y$,

$\dfrac{dy}{dt} = 5y - xy + 6$

12. $\dfrac{dx}{dt} = 4 - xy$,

$\dfrac{dy}{dt} = x - y$

In Problems 13–16, convert the second-order differential equation to an equivalent first-order system, find its critical points, and discuss the type and stability of each critical point.

13. $x'' = \cos x$ [*Hint:* What would be the form of the pendulum equation if its angle were measured from the horizontal?]

14. $x'' = -bx' - \sin x$ (The damped pendulum: consider all positive values of b.)

15. $x'' = [1 - x^2]x' - x$ (**van der Pol equation**) [*Hint:* Confirm your answer by examining Example 7 of Section 4.8, page 208.]

16. $x'' = [1 - (x')^2]x' - x$ (**Rayleigh equation**) [*Hint:* Confirm your answer by examining Problem 11 of Exercises 4.8, page 211.]

In Problems 17–20, find all the critical points for the given system, discuss the type and stability of each critical point, and—either by hand or using a computer software package—sketch the phase plane diagrams near each of the critical points.

17. $\dfrac{dx}{dt} = 1 - xy$,

$\dfrac{dy}{dt} = x - y^3$

18. $\dfrac{dx}{dt} = -x + xy - 6$,

$\dfrac{dy}{dt} = x - y$

19. $\dfrac{dx}{dt} = 6x - 2x^2 - xy$,

$\dfrac{dy}{dt} = 6y - 2y^2 - xy$

20. $\dfrac{dx}{dt} = x + y^2$,

$\dfrac{dy}{dt} = x - 2y$

21. van der Pol's Equation.

(a) Show that van der Pol's equation

$$\frac{d^2x}{dt^2} + \varepsilon(x^2 - 1)\frac{dx}{dt} + x = 0$$

is equivalent to the system

$$\frac{dx}{dt} = y,$$

$$\frac{dy}{dt} = -x + \varepsilon y - \varepsilon x^2 y,$$

where $y := dx/dt$.

(b) Describe how the stability of the zero solution of van der Pol's equation (equivalently, the origin of the system) depends on the parameter ε.

22. Consider the system

$$\frac{dx}{dt} = (\lambda + \alpha)x + \beta y,$$

$$\frac{dy}{dt} = -\beta x + (\lambda + \gamma)y,$$

where α, β, γ, and λ are real numbers. The characteristic equation for the unperturbed system, where

$\alpha = \beta = \gamma = 0$, has a repeated root λ. Show that if we perturb the system, keeping α, β, and λ small, then the characteristic equation for the perturbed system can have the following:

(a) Equal roots for $\beta = 0$ and $\alpha = \gamma$.
(b) Distinct real roots for $\beta = 0$ and $\alpha \neq \gamma$.
(c) Complex roots for $\beta \neq 0$ and $\alpha = \gamma = 0$.
(d) What does this suggest about the type of critical point for an almost linear system if the characteristic equation for the corresponding linear system has equal real roots?

23. Let $x(t)$ and $y(t)$ equal the populations at time t of two species of insects that compete with each other for food. Assume their populations are governed by the system (7)–(8) [see equations (1) and (2) in Section 12.1].

(a) Show that when $y \equiv 0$, species x has a stable equilibrium population of a_1/b_1. This is called the *carrying capacity* for x.
(b) What is the carrying capacity for y?
(c) Let A and B denote the carrying capacities for species x and y, respectively. Show that the system can be rewritten in the equivalent form

$$\frac{dx}{dt} = a_1 x\left(1 - \frac{x}{A} - \frac{d_1 y}{A}\right),$$
$$\frac{dy}{dt} = a_2 y\left(1 - \frac{y}{B} - \frac{d_2 x}{B}\right),$$

where $d_i = c_i/b_i$, $i = 1, 2$.

(d) In Section 12.1 we observed that when $a_1/c_1 > a_2/b_2$ and $a_1/b_1 < a_2/c_2$, the two species can coexist. Assume this is initially the case but now an insecticide is being applied that kills only species x. Describe what happens to the competing populations if the application of the insecticide is increased (that is, a_1 is decreased).

24. Show that coexistence occurs in the competing species model governed by system (7)–(8) if and only if there is an asymptotically stable critical point in the first quadrant.

25. When one of the populations in a competing species model is being harvested, the phase plane diagram is completely changed. For example, consider the competition model

$$\frac{dx}{dt} = x(1 - 4x - y) - h,$$
$$\frac{dy}{dt} = y(1 - 2y - 5x),$$

where h is the constant harvest rate. For $h = 0, 1/32$, and $5/32$, determine the critical points, discuss the type and stability of the critical points, and sketch the phase plane diagrams. In each case discuss the implications of the competition model.

12.4 Energy Methods

Many differential equations arise from problems in mechanics, and for these equations it is natural to study the effect that energy in the system has on the solutions to the system. By analyzing the energy, we can often determine the stability of a critical point (even in the more elusive case when it is a center of the corresponding linear equation). As we will see in the next section, these ideas can be generalized to systems that are unrelated to problems in mechanics.

Conservative Systems

In Chapter 4 we considered some mechanical systems that were governed by Newton's second law

$$F = ma = m\frac{d^2x}{dt^2}$$

(*force equals mass times acceleration*). When the force $F = F(t, x, x')$ depends only on x, we call the system **conservative**. In such a case we introduce the notation $-U(x)$ for the antiderivative of $F(x)$,

$$F = F(x) = -\frac{dU(x)}{dx}, \quad \text{or} \quad U(x) = -\int F(x)\,dx + K,$$

and express the second law as

(1) $\quad m\dfrac{d^2x}{dt^2} + \dfrac{dU(x)}{dx} = 0 .$

Under this condition, the quantity

(2) $\quad \dfrac{1}{2}m\left(\dfrac{dx}{dt}\right)^2 + U(x)$

is constant during the motion, because

$$\dfrac{d}{dt}\left\{\dfrac{1}{2}m\left(\dfrac{dx}{dt}\right)^2 + U(x)\right\} = \dfrac{2}{2}m\dfrac{dx}{dt}\dfrac{d^2x}{dt^2} + \dfrac{dU}{dx}\dfrac{dx}{dt} = \dfrac{dx}{dt}\left\{m\dfrac{d^2x}{dt^2} - F\right\} = 0 .$$

Since $(m/2)(dx/dt)^2$ is the *kinetic energy* of the system, it is customary to refer to $U(x)$ as the *potential energy* and $(m/2)(dx/dt)^2 + U(x)$ as the *total energy* of the system. We have thus shown that the total energy is constant. This "energy conservation" principle appeared in a different context in Section 4.8, where it was called the "energy integral lemma."

By dividing the equations throughout by the bothersome constant m and setting $g(x) := U'(x)/m$, we obtain the *standard form of the differential equation for a conservative system*

(3) $\quad \dfrac{d^2x}{dt^2} + g(x) = 0 ;$

the equivalent *phase plane system*

(4) $\quad \dfrac{dx}{dt} = v ,$

(5) $\quad \dfrac{dv}{dt} = -g(x) ;$

the *potential function*

$$G(x) := \int g(x)\,dx + C ;$$

and the *energy function*

$$E(x, v) := v^2/2 + G(x) .$$

The fact that the total energy is constant means that the *level curves*

(6) $\quad E(x, v) = k , \quad k$ a constant,

contain the phase plane trajectories of the system (4)–(5).

Example 1 Find the energy function $E(x, v)$ for the following equations. Select E so that the energy at the critical point $(0, 0)$ is zero, that is, $E(0, 0) = 0$.

(a) $\dfrac{d^2x}{dt^2} + x - x^3 = 0$

(b) $\dfrac{d^2x}{dt^2} + \sin x = 0$ (pendulum equation)

(c) $\dfrac{d^2x}{dt^2} + x - x^4 = 0$

Solution

(a) Here $g(x) = x - x^3$. By integrating g, we obtain the potential function $G(x) = (1/2)x^2 - (1/4)x^4 + C$. Thus, $E(x, v) = (1/2)v^2 + (1/2)x^2 - (1/4)x^4 + C$. Now $E(0, 0) = 0$ implies that $C = 0$. Hence, the energy function we seek is $E(x, v) = (1/2)v^2 + (1/2)x^2 - (1/4)x^4$.

(b) Since $g(x) = \sin x$, the potential function is $G(x) = -\cos x + C$, which gives $E(x, v) = (1/2)v^2 - \cos x + C$. Setting $E(0, 0) = 0$, we find $C = 1$. Thus, the energy function is $E(x, v) = (1/2)v^2 - \cos x + 1$.

(c) Here $g(x) = x - x^4$ and the potential function is $G(x) = (1/2)x^2 - (1/5)x^5 + C$. Setting $E(0, 0) = 0$, we get $C = 0$, so the energy function is $E(x, v) = (1/2)v^2 + (1/2)x^2 - (1/5)x^5$. ◆

Notice that the pendulum equation in part (b) also has a critical point at $(\pi, 0)$. Setting the energy to be zero at $(\pi, 0)$—that is, $E(\pi, 0) = 0$—we arrive at the energy function $E(x, v) = (1/2)v^2 - \cos x - 1$.

For conservative systems, much useful information about the phase plane can be gained directly from the potential function $G(x)$. A convenient way of doing this is to place the graph of the potential function $z = G(x)$ (the potential plane) directly above the phase plane diagram with the z-axis aligned with the v-axis (see Figure 12.21). This places the relative extrema for $G(x)$ directly over the critical points for the system, since $G'(x_0) = g(x_0) = 0$ implies that $(x_0, 0)$ is a critical point for (4)–(5).

To understand the relationship between the potential plane and the phase plane, let's look at the level curves for the energy function. In general, we have

$$\frac{v^2}{2} + G(x) = k,$$

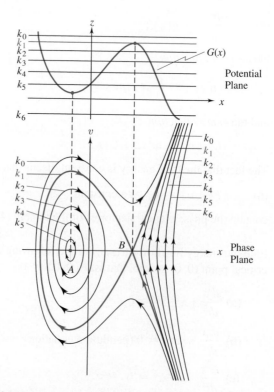

Figure 12.21 Relationship between potential plane (above) and phase plane (below)

where $G(x)$ is the potential function with $G'(x) = g(x)$. Solving for v gives

$$v = \pm\sqrt{2[k - G(x)]}.$$

Thus, the velocity v exists (is real-valued) only when $k - G(x) \geq 0$. This generally results in three types of behavior.

When $G(x)$ has a strict local minimum at x_0 (above the point A) as illustrated in Figure 12.21, the level curves $E(x, v) = k$, where k is slightly greater than $G(x_0)$, are closed curves encircling the critical point $A = (x_0, 0)$. To see this, observe that there is an interval (a, b) containing x_0 with $G(a) = G(b) = k_5$ and $G(x) < k_5$ for x in (a, b). Since $v = \sqrt{2[k_5 - G(x)]}$ is defined on (a, b) with $v = 0$ at $x = a$ and b, the two curves $v = \pm\sqrt{2[k_5 - G(x)]}$ join up at a and b to produce the closed curve. As this occurs for any k satisfying $G(x_0) < k \leq k_5$, the critical point A is encircled by closed trajectories corresponding to periodic solutions. Thus, A is a center.

When $G(x)$ has a strict local maximum at x_1 (above the point B), the level curves $E(x, v) = k$ are *not* closed curves encircling the critical point $B = (x_1, 0)$. For $k = G(x_1)$, the curves touch the x-axis at x_1, but for $k > G(x_1)$, $v = \sqrt{2[k - G(x)]}$ is strictly positive and the level curves do not touch the x-axis. Finally, when $k < G(x_1)$, there is an interval about x_1 above and below in which there are *no* points of the curve $E(x, v) = k$ (the quantity under the radical in the formula for v is negative). Hence, the critical point B is a saddle point, and the level curves near B resemble hyperbolas. This is illustrated in Figure 12.21 by the level curves corresponding to k_0, k_1, and k_2 near B.

Finally, away from the extreme points of $G(x)$, the level curves may be part of a closed trajectory, such as the level curve corresponding to $k = k_2$, or may be unbounded, such as the level curves corresponding to $k = k_1$ or k_6.

Example 2 Use the potential plane to help sketch the phase plane diagram for the equations:

(a) $\dfrac{d^2x}{dt^2} + x - x^3 = 0$. (b) $\dfrac{d^2x}{dt^2} + \sin x = 0$. (c) $\dfrac{d^2x}{dt^2} + x - x^4 = 0$.

Solution In Example 1 we found that potential functions for these equations are

(a) $G(x) = (1/2)x^2 - (1/4)x^4$; (b) $G(x) = 1 - \cos x$; (c) $G(x) = (1/2)x^2 - (1/5)x^5$.

(a) The local maxima and minima of $G(x)$ occur when $G'(x) = g(x) = x - x^3 = 0$; namely, at $x = 0, -1, 1$. Consequently, the phase plane diagram for equation (a) has critical points at $(0, 0)$, $(-1, 0)$, and $(1, 0)$. Since $G(x)$ has a strict local minimum at $x = 0$, the critical point $(0, 0)$ is a center. Furthermore, $G(x)$ has local maxima at $x = \pm 1$, so $(-1, 0)$ and $(1, 0)$ are saddle points. The potential plane and the phase plane are displayed in Figure 12.22 on page 744.

(b) The local maxima and minima of $G(x) = 1 - \cos x$ occur when $G'(x) = g(x) = \sin x = 0$. Here $G(x)$ has local minima for $x = 0, \pm 2\pi, \pm 4\pi, \dots$ and local maxima for $x = \pm\pi, \pm 3\pi, \pm 5\pi, \dots$. Consequently, the critical points $(2n\pi, 0)$, where n is an integer, are centers, while the critical points $((2n + 1)\pi, 0)$ are saddle points. The potential and phase planes are given in Figure 12.23 on page 744.

(c) The extreme values for $G(x) = (1/2)x^2 - (1/5)x^5$ occur when $G'(x) = g(x) = x - x^4 = 0$; that is, for $x = 0$ and 1. Here G has a minimum at $x = 0$ and a maximum at $x = 1$. Hence, the critical point $(0, 0)$ is a center, and the critical point $(1, 0)$ is a saddle point (see Figure 12.24 on page 745). ◆

You may have already observed that the three conservative systems in Example 2 have the *same* corresponding linear equation $d^2x/dt^2 + x = 0$. As we know, the origin is a stable center

Figure 12.22 Potential and phase plane for $d^2x/dt^2 + x - x^3 = 0$

for this linear equation. Thus, as mentioned in the discussion of the pendulum equation in the preceding section, Theorem 2 gives us no information about the stability of the origin for the three almost linear equations. However, using the energy method approach, we were able to determine that for these three conservative systems the origin is indeed a stable center.

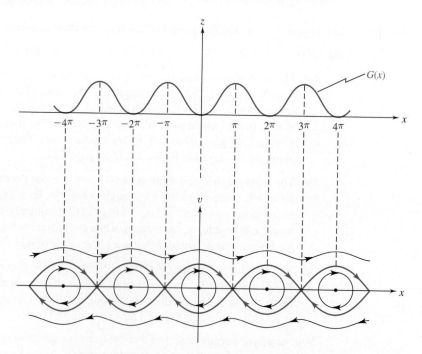

Figure 12.23 Potential and phase plane for $d^2x/dt^2 + \sin x = 0$

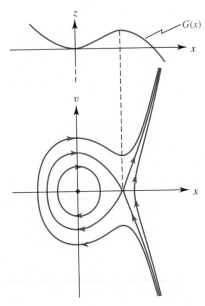

Figure 12.24 Potential and phase plane for $d^2x/dt^2 + x - x^4 = 0$

Nonconservative Systems

For nonconservative systems the energy of the system does *not* remain constant along its trajectories in the Poincaré phase plane. Consequently, the level curves of the energy function are *not* integral curves for the system. However, if we know that the energy is decreasing along a trajectory, then it follows that the trajectory is moving toward a state that has a lower total energy. This state may correspond to a critical point or a limit cycle (defined in Section 12.6). We will demonstrate this behavior in the following examples.

A nonconservative system with nonlinear damping has the form

$$\frac{d^2x}{dt^2} + h(x, dx/dt) + g(x) = 0 \, ,$$

where $G(x) := \int g(x)\,dx$ corresponds to the potential energy and the velocity-dependent term $h(x, dx/dt)$ represents damping. Its equivalent phase plane formulation generalizes (4)–(5):

$$(7) \qquad \frac{dx}{dt} = v \, ,$$

$$(8) \qquad \frac{dv}{dt} = -h(x, dx/dt) - g(x).$$

Let

$$E(x, dx/dt) := \frac{1}{2}\left(\frac{dx}{dt}\right)^2 + G(x) \, ,$$

as in the case of a conservative system ($h \equiv 0$). If we now compute dE/dt, we find

$$\frac{dE}{dt} = \frac{dx}{dt}\frac{d^2x}{dt^2} + G'(x)\frac{dx}{dt} = \frac{dx}{dt}\left[\frac{d^2x}{dt^2} + g(x)\right]$$

$$= -\frac{dx}{dt}h(x, dx/dt) \, ,$$

where we have used $d^2x/dt^2 + g(x) = -h(x, dx/dt)$. Consequently, the *energy E* is decreasing when $vh(x, v) > 0$ and is increasing when $vh(x, v) < 0$.

Example 3 Sketch the phase plane diagram for the damped pendulum equation

$$\frac{d^2x}{dt^2} + b\frac{dx}{dt} + \sin x = 0 ,$$

where b is a positive constant.

Solution As already mentioned, the energy $E(x, v)$ is the same as that for the (conservative, undamped) simple pendulum; i.e., $E(x, v) = (1/2)v^2 + 1 - \cos x$, whose level curves $E(x, v) = k$ are plotted in Figure 12.23. It is important to note that these level curves are symmetric about the x-axis and shrink to points that are even-integer multiples of π on the x-axis as k decreases. For the damped pendulum, we have $vh(x, v) = v(bv) = bv^2 > 0$ for $v \neq 0$, and so its trajectories no longer lie along the energy level curves. Rather, the energy is continually decreasing along a (nontrivial) trajectory and, based on the aforementioned properties of the energy level curves, such a trajectory will have to cross these curves heading in the direction of the x-axis. Moreover, it is not difficult to verify that the critical points for the damped pendulum equation are the same as for the undamped equation (although the Jacobian's eigenvalues are different). Using the method of almost linear systems, the reader is requested to verify in Problem 21 that all those points along the x-axis located at even-integer multiples of π are indeed asymptotically stable, while those at odd-integer multiples are unstable saddle points. The resulting phase plane diagram is sketched in Figure 12.25; the colored curves (separatrices) divide the plane into regions inside of which all the trajectories approach the same critical point. Exactly how these trajectories approach the stable critical points depends upon the size of the damping coefficient b. For example, when $0 < b < 2$, underdamping occurs and the stable critical points are spirals as illustrated in Figure 12.25 (see Problem 21). ◆

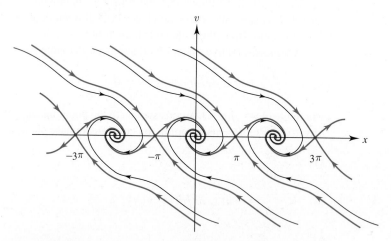

Figure 12.25 Phase plane diagram for damped pendulum of Example 3 with $0 < b < 2$. The stable trajectories are spirals.

In the next example, the trajectories approach a limit cycle.

Example 4 Sketch the phase plane diagram for the equation

$$\frac{d^2x}{dt^2} + \left[x^2 + \left(\frac{dx}{dt}\right)^2 - 1\right]\frac{dx}{dt} + x = 0 .$$

Solution Here $h(x, v) = (x^2 + v^2 - 1)v$. Hence, $vh(x, v) = v^2(x^2 + v^2 - 1)$, which is zero on the unit circle $x^2 + v^2 = 1$, positive outside the unit circle, and negative inside. Consequently, other than the unstable critical point at the origin, trajectories beginning inside the unit circle spiral out and trajectories beginning outside the circle spiral in (see Figure 12.26). The unit circle (in color) is the trajectory for the solution $x(t) = \cos t$ and is called a **limit cycle**. ◆

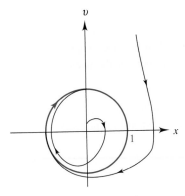

Figure 12.26 Phase plane diagram for $d^2x/dt^2 + [x^2 + (dx/dt)^2 - 1]dx/dt + x = 0$

12.4 EXERCISES

In Problems 1–6, find the potential energy function $G(x)$ and the energy function $E(x, v)$ for the given equations. Select E so that $E(0, 0) = 0$.

1. $\dfrac{d^2x}{dt^2} + x^2 - 3x + 1 = 0$ **2.** $\dfrac{d^2x}{dt^2} + \cos x = 0$

3. $\dfrac{d^2x}{dt^2} + \dfrac{x^2}{x-1} = 0$ **4.** $\dfrac{d^2x}{dt^2} + x - \dfrac{x^3}{6} + \dfrac{x^5}{120} = 0$

5. $\dfrac{d^2x}{dt^2} + x^3 + x^2 - x = 0$ **6.** $\dfrac{d^2x}{dt^2} + e^x - 1 = 0$

In Problems 7–12, use the potential plane to help sketch the phase plane diagrams for the given equations.

7. $\dfrac{d^2x}{dt^2} + x^3 - x = 0$ **8.** $\dfrac{d^2x}{dt^2} + 9x = 0$

9. $\dfrac{d^2x}{dt^2} + 2x^2 + x - 1 = 0$ **10.** $\dfrac{d^2x}{dt^2} - \sin x = 0$

11. $\dfrac{d^2x}{dt^2} + \dfrac{x}{x-2} = 0$ **12.** $\dfrac{d^2x}{dt^2} + (x-1)^3 = 0$

In Problems 13–16, use the energy function, almost linear system methods, and software, if necessary, to assist in sketching the phase plane diagrams for the given nonconservative systems.

13. $\dfrac{d^2x}{dt^2} + \dfrac{dx}{dt} + x - x^3 = 0$

14. $\dfrac{d^2x}{dt^2} + \dfrac{dx}{dt} + x - x^4 = 0$

15. $\dfrac{d^2x}{dt^2} + \dfrac{dx}{dt} + \dfrac{x}{x-2} = 0$

16. $\dfrac{d^2x}{dt^2} + \dfrac{dx}{dt} + 2x^2 + x - 1 = 0$

17. Nonlinear Spring (Duffing's Equation). The general nonlinear spring equation

$$\frac{d^2x}{dt^2} + \alpha x + \beta x^3 = 0,$$

where $\alpha > 0$ and β are parameters, is used to model a variety of physical phenomena. Holding $\alpha > 0$ fixed, sketch the potential function and the phase plane diagram for $\beta > 0$ and also for $\beta < 0$. Describe how the behavior of solutions to the equation differs in these cases.

18. Cusps. We have observed that where the potential energy function $G(x)$ for a conservative system has a strict maximum (minimum), the corresponding critical point is a saddle point (center). At a critical point $(x_0, 0)$ for which $G(x)$ also has a point of inflection (e.g., $G'(x_0) = g(x_0) = 0$, $G''(x_0) = g'(x_0) = 0$, and $G'''(x_0) = g''(x_0) \neq 0$), the curve in the phase plane

has a **cusp.** Demonstrate this by sketching the potential plane and the phase plane diagrams for the equation

$$\frac{d^2x}{dt^2} + x^2 = 0.$$

19. General Relativity. In studying the relativistic motion of a particle moving in the gravitational field of a larger body, one encounters the equation

$$\frac{d^2u}{d\theta^2} + u - \lambda u^2 - 1 = 0,$$

where u is inversely proportional to the distance of the particle from the body, θ is an angle in the plane of motion, and λ is a parameter with $0 < \lambda < 1$.

(a) Sketch the phase plane diagram for $0 < \lambda < 1/4$.
(b) Sketch the phase plane diagram for $1/4 < \lambda < 1$.
(c) What observations can be made about the motion of the particle in these two cases?

20. Verify that the method of almost linear systems applied to part (b) of Example 2 again confirms that the critical points $\big((2n+1)\pi, 0\big)$, where n is an integer, are indeed saddle points.

21. For the damped pendulum equation of Example 3,
(a) Use the method of almost linear systems to verify that the points $(2n\pi, 0)$, where n is an integer, are asymptotically stable, while those at $\big((2n+1)\pi, 0\big)$ are saddle points.
(b) Verify that for $b > 2$ the stable critical points are improper nodes, while for $0 < b < 2$, these become spiral points.
(c) Describe what can happen when $b = 2$.

12.5 Lyapunov's Direct Method

In the preceding section, we used the energy of the system, $E(x, v)$, to determine the qualitative behavior of the trajectories of solutions of conservative as well as nonconservative systems. An advantage of that method is that we do not actually have to solve the equation but can deduce the behavior directly. As we will see, this approach can be extended to the study of the stability of an *isolated* critical point for a general autonomous system; i.e., there is a disk about the critical point that contains no other critical points. The technique is called **Lyapunov's direct method** (also known as **Lyapunov's second method**).[†]

Before we present the results that make up Lyapunov's direct method, let's analyze the system

$$(1) \qquad \frac{dx}{dt} = y - xy^2 - x^3,$$

$$(2) \qquad \frac{dy}{dt} = -x - x^2y - y^3.$$

This is almost linear with the origin as its only critical point. The corresponding linear system

$$(3) \qquad \frac{dx}{dt} = y, \qquad \frac{dy}{dt} = -x$$

has a *center* at the origin, and therefore the results of Section 12.3 do not tell us whether the origin is stable, asymptotically stable, or unstable for the almost linear system.

Let's try to apply the energy method. We begin by observing that the linear system (3) is equivalent to the second-order equation

(4) $$\frac{d^2x}{dt^2} + x = 0 \, ,$$

where $y = x'$. This suggests that we investigate the "energy" function

$$V(x, y) = (1/2)x^2 + (1/2)y^2$$

that is associated with the mechanical system governed by equation (4). The function $V(x, y)$ is positive except at the origin, where $V(0, 0) = 0$. Moreover, the level curves $V(x, y) = k$ are concentric circles about the origin. Again, following the energy method, we now consider how $V(x, y)$ changes along trajectories of the almost linear system. Let $x = \phi(t)$ and $y = \psi(t)$ be a solution to system (1)–(2). The partial derivatives V_x, V_y are just $V_x = x, V_y = y$, so we can compute dV/dt using the chain rule:

$$\frac{d}{dt} V\bigl(\phi(t), \psi(t)\bigr) = V_x\bigl(\phi(t), \psi(t)\bigr) \frac{d\phi}{dt} + V_y\bigl(\phi(t), \psi(t)\bigr) \frac{d\psi}{dt}$$

$$= \phi(t) \frac{d\phi}{dt} + \psi(t) \frac{d\psi}{dt}$$

$$= \phi(t)[\psi(t) - \phi(t)\psi^2(t) - \phi^3(t)] + \psi(t)[-\phi(t) - \phi^2(t)\psi(t) - \psi^3(t)]$$

$$= -[\phi^2(t) + \psi^2(t)]^2 \, .$$

Hence, dV/dt is negative everywhere except at the origin, where it is zero. Consequently, $V(\phi(t), \psi(t))$ is decreasing along any trajectory of (1)–(2). Geometrically this means that as a trajectory intersects the level curve $V(x, y) = k$ (a circle), it then has to move into the region enclosed by that level curve (see Figure 12.27). Therefore, every trajectory must approach the origin as $t \to +\infty$, which means the origin is asymptotically stable.

In this example, the function $V(x, y)$ was selected based on our knowledge of the total energy for the mass–spring system that was governed by the corresponding linear system. Actually, all that is needed is a function $V(x, y)$ whose level curves encircle the critical point and whose values decrease along the trajectories of the almost linear system. To help describe such a function, we introduce the following definitions.

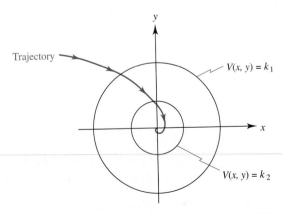

Figure 12.27 Trajectory entering regions bounded by level curves $V(x, y) = k$

Definite and Semidefinite Functions

Definition 3. Let $V(x, y)$ be continuous on some domain D containing the origin and assume that $V(0, 0) = 0$. We say $V(x, y)$ is **positive definite** on D if $V(x, y) > 0$ for all (x, y) in D except the origin.

If $V(x, y) \geq 0$ for all (x, y) in D, we say that $V(x, y)$ is **positive semidefinite** on D.

Similarly, we define **negative definite** and **negative semidefinite** by reversing the appropriate signs in the above inequalities.

The energy functions

$$V(x, y) = \frac{y^2}{2} + \frac{x^2}{2} - \frac{x^4}{4},$$

(5)
$$V(x, y) = \frac{y^2}{2} + 1 - \cos x,$$

$$V(x, y) = \frac{y^2}{2} + \frac{x^2}{2} - \frac{x^5}{5},$$

which we encountered in Example 1 of Section 12.4, are all positive definite near the origin (see Problem 16).

To study the effect on $V(x, y)$ as we move along a trajectory, we introduce the notion of the derivative of $V(x, y)$ with respect to a system.

The Derivative $\dot{V}(x, y)$

Definition 4. Suppose the first-order partial derivatives of $V(x, y)$ are continuous in a domain D containing the origin. The **derivative of V with respect to the system**

$$\frac{dx}{dt} = f(x, y), \qquad \frac{dy}{dt} = g(x, y)$$

is the function

$$\dot{V}(x, y) := V_x(x, y)f(x, y) + V_y(x, y)g(x, y).$$

Remark. As we saw in the preceding example, the derivative of $V(x, y)$ with respect to the system, when evaluated at a solution $x = \phi(t), y = \psi(t)$ of the system, is just the total derivative of $V(\phi(t), \psi(t))$ with respect to t; that is,

$$\frac{dV}{dt}(\phi, \psi) = V_x(\phi, \psi)\frac{d\phi}{dt} + V_y(\phi, \psi)\frac{d\psi}{dt}$$

$$= V_x(\phi, \psi)f(\phi, \psi) + V_y(\phi, \psi)g(\phi, \psi)$$

$$= \dot{V}(\phi, \psi).$$

We now state and prove Lyapunov's theorem, which gives sufficient conditions for the origin to be stable.

Lyapunov's Stability Theorem

Theorem 3. Let $V(x, y)$ be a positive definite function on a domain D containing the origin. Suppose the origin is an isolated critical point for the system

$$(6) \qquad \frac{dx}{dt} = f(x, y),$$

$$(7) \qquad \frac{dy}{dt} = g(x, y)$$

and let $\dot{V}(x, y)$ be the derivative of $V(x, y)$ with respect to this system.

(a) If $\dot{V}(x, y)$ is negative semidefinite on D, then the origin is stable.

(b) If $\dot{V}(x, y)$ is negative definite on D, then the origin is asymptotically stable.

Proof.

(a) Let B_r denote the open disk

$$B_r := \{(x, y): x^2 + y^2 < r^2\}$$

and $\bar{B}_r$ be the closed disk of radius r. For a given $\varepsilon > 0$, we need to determine $\delta > 0$ such that for any solution $x = \phi(t), y = \psi(t)$ of (6)–(7) satisfying $\big(\phi(t_0), \psi(t_0)\big) \in B_\delta$, we have $\big(\phi(t), \psi(t)\big) \in B_\varepsilon$ for all $t \geq t_0$. To specify δ we proceed as follows.

 Choose $\alpha > 0$ so that $\bar{B}_\alpha \subset D$ and, for $\varepsilon < \alpha$, let $\bar{A}_{\varepsilon, \alpha}$ denote the closed annulus

$$\bar{A}_{\varepsilon, \alpha} := \{(x, y): \varepsilon^2 \leq x^2 + y^2 \leq \alpha^2\}$$

with inner radius ε and outer radius α. Let μ denote the minimum value of V on $\bar{A}_{\varepsilon, \alpha}$. Since V is continuous and is positive everywhere except at the origin, we have $\mu > 0$. Also, since V is zero at the origin, we can select a $\delta > 0$ such that $\delta < \varepsilon$ and $V(x, y) < \mu$ for all (x, y) in B_δ. We now show that for this δ, the stability condition is satisfied.

 Let $0 < \varepsilon < \alpha$ and $x = \phi(t), y = \psi(t)$ be a nontrivial solution to (6)–(7) such that $\big(\phi(t_0), \psi(t_0)\big) \in B_\delta$. If $\big(\phi(t), \psi(t)\big) \in B_\varepsilon$ for all $t \geq t_0$, then we are done. If this is not the case, then there is a first value of t, say $t_1 > t_0$, where the trajectory touches the boundary of B_ε (see Figure 12.28 on page 752). Then on the interval $[t_0, t_1]$, we have $\big(\phi(t), \psi(t)\big) \in \bar{B}_\varepsilon \subset B_\alpha$. Since $\dot{V}$ is negative semidefinite, we have for $t \in [t_0, t_1]$,

$$V\big(\phi(t), \psi(t)\big) - V\big(\phi(t_0), \psi(t_0)\big) = \int_{t_0}^{t} \frac{dV}{ds}\big(\phi(s), \psi(s)\big)\, ds$$

$$= \int_{t_0}^{t} \dot{V}\big(\phi(s), \psi(s)\big)\, ds$$

$$\leq 0 .$$

Thus,

$$(8) \qquad V\big(\phi(t), \psi(t)\big) \leq V\big(\phi(t_0), \psi(t_0)\big), \quad \text{for } t \in [t_0, t_1].$$

But $V\big(\phi(t_0), \psi(t_0)\big) < \mu$ because $\big(\phi(t_0), \psi(t_0)\big) \in B_\delta$, and $V\big(\phi(t_1), \psi(t_1)\big) \geq \mu$, since $\big(\phi(t_1), \psi(t_1)\big) \in \bar{A}_{\varepsilon, \alpha}$. Consequently, from (8),

$$\mu \leq V\big(\phi(t_1), \psi(t_1)\big) \leq V\big(\phi(t_0), \psi(t_0)\big) < \mu ,$$

which is a contradiction. Therefore $\big(\phi(t), \psi(t)\big) \in B_\varepsilon$ for all $t \geq t_0$. Hence, the stability is established.

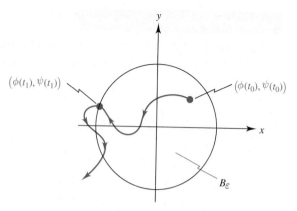

Figure 12.28 A trajectory leaving B_ε at $t = t_1$

(b) Here $\dot{V}$ is negative definite on D, so we already know that the origin is stable from part (a). It remains to be shown that there exists an $\eta > 0$ such that if $x = \phi(t)$, $y = \psi(t)$ is a solution to (6)–(7) for which $\big(\phi(t_0), \psi(t_0)\big)$ lies in B_η, then

$$\lim_{t \to +\infty} \phi(t) = 0, \qquad \lim_{t \to +\infty} \psi(t) = 0.$$

Now if we can prove that $V\big(\phi(t), \psi(t)\big) \to 0$ as $t \to +\infty$, then it will follow from the positive definiteness of V that $\phi(t) \to 0$ and $\psi(t) \to 0$ as $t \to +\infty$.

To specify the constant η, take $\varepsilon = \alpha/2$ and for this ε choose δ as in the proof of part (a).[†] We will show that with $\eta = \delta$ the asymptotic stability criterion is satisfied.

Suppose to the contrary that there exists a solution $x = \phi(t), y = \psi(t)$ to (6)–(7) such that $\big(\phi(t_0), \psi(t_0)\big)$ lies in B_η but $\big(\phi(t), \psi(t)\big)$ does not approach the origin as $t \to +\infty$. Since $V\big(\phi(t), \psi(t)\big)$ is decreasing for $t \geq t_0$ (see (8)), it follows that there is an $L > 0$ such that

$$V\big(\phi(t), \psi(t)\big) \geq L > 0 \quad \text{for all} \quad t \geq t_0.$$

Also, since V is continuous and $V(0,0) = 0$, there exists an $h > 0$ such that

$$V(x, y) < L \quad \text{for all} \quad (x, y) \in B_h.$$

Hence, the trajectory $\big(\phi(t), \psi(t)\big)$ must lie in the annulus $\bar{A}_{h, \alpha/2}$ for $t \geq t_0$ (see Figure 12.29 on page 753).

Let $\gamma\,(<0)$ be the maximum value of $\dot{V}$ on the annulus $\bar{A}_{h, \alpha/2}$. Then, since the trajectory $\big(\phi(t), \psi(t)\big)$ lies in $\bar{A}_{h, \alpha/2}$, we have

$$\frac{dV}{dt}\big(\phi(t), \psi(t)\big) \leq \gamma.$$

So, integrating we find

$$V\big(\phi(t), \psi(t)\big) \leq V\big(\phi(t_0), \psi(t_0)\big) + \gamma(t - t_0) \quad \text{for all} \quad t \geq t_0.$$

Since $\gamma < 0$, it now follows that, for t sufficiently large, $V\big(\phi(t), \psi(t)\big) < 0$, which contradicts the fact that V is positive definite. Hence, we must have $V\big(\phi(t), \psi(t)\big) \to 0$ as $t \to +\infty$. ◆

[†]There is nothing special about $\alpha/2$. Any fixed choice of ε less than α would work as well.

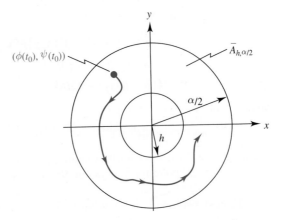

Figure 12.29 The trajectory lies in the annulus $\bar{A}_{h,\alpha/2}$

The function $V(x, y)$ in Theorem 3 is called a **Lyapunov function.** It is clear from the theorem that Lyapunov functions play a key role in determining the stability of an isolated critical point. *Unfortunately, there is no clear strategy available for choosing $V(x, y)$. However, for those systems that can be identified with a mechanical problem, the total energy function is a good candidate and, for many other systems, expressions of the form*

$$V(x, y) = ax^2 + by^2 , \qquad V(x, y) = ax^2 + by^4 ,$$
$$V(x, y) = ax^4 + by^2 , \qquad V(x, y) = ax^4 + by^4 ,$$

may work for an appropriate choice of the constants a, b.[†]

In the next example, we use Lyapunov's direct method to determine the stability of the origin for an autonomous system whose corresponding linear system does *not* have the origin as an isolated critical point; such a system cannot be analyzed using the results of Section 12.3.

Example 1 Use Lyapunov's direct method to determine the stability of the origin for the system

(9) $$\frac{dx}{dt} = -2y^3 ,$$

(10) $$\frac{dy}{dt} = x - 3y^3 .$$

Solution The origin is an isolated critical point for this system, since the pair of equations

$$-2y^3 = 0 , \qquad x - 3y^3 = 0$$

has only $x = y = 0$ as a solution. (Note, however, that for the corresponding linear system $dx/dt = 0, dy/dt = x$ the origin is *not* an isolated critical point $(ad - bc = 0)$. Hence, (9)–(10) is not an almost linear system.)

Since $f(x, y) = -2y^3$ and $g(x, y) = x - 3y^3$, let's try to find a Lyapunov function of the form $V(x, y) = ax^2 + by^4$, with $a, b > 0$. Clearly, V is positive definite on any domain containing the origin.

[†]For further discussion, see the book by J. LaSalle and S. Lefschetz, *Stability by Liapunov's Direct Method with Applications* (Academic Press, New York, 1961).

Next, computing $\dot{V}(x, y)$, we find

$$\dot{V}(x, y) = V_x(x, y)f(x, y) + V_y(x, y)g(x, y)$$
$$= (2ax)(-2y^3) + (4by^3)(x - 3y^3)$$
$$= 4(b - a)xy^3 - 12by^6.$$

So, if we take $a = b = 1$, then

$$\dot{V}(x, y) = -12y^6 \leq 0,$$

and we get that $\dot{V}(x, y)$ is negative semidefinite. It now follows from Theorem 3 that the origin is a stable critical point for the system (9)–(10). ◆

The next result will give conditions for the *instability* of a system.

Lyapunov's Instability Theorem

Theorem 4. Suppose the origin is an isolated critical point for the autonomous system (6)–(7) and let $V(x, y)$ with $V(0, 0) = 0$ be defined and continuous on a domain D containing the origin. Suppose further that $\dot{V}(x, y)$ is positive definite on D. If, in every disk centered at the origin, there exists some point (a, b) such that $V(a, b)$ is positive, then the origin is unstable.

Proof. Fix $t_0 > 0$ and let $\alpha > 0$ be chosen so that $\bar{B}_\alpha \subset D$, where $\bar{B}_\alpha$ is the closed disk with center $(0, 0)$ and radius α. Since V is continuous, there exists an $L > 0$ such that $|V(x, y)| \leq L$ for all (x, y) in $\bar{B}_\alpha$. We will show that no matter how small we take $\varepsilon > 0$, there is a solution $x = \phi(t), y = \psi(t)$ with $(\phi(t_0), \psi(t_0))$ in B_ε such that the trajectory $(\phi(t), \psi(t))$ eventually leaves the disk B_α. Thereby, it follows that the origin is unstable.

Now, by hypothesis, given ε, with $0 < \varepsilon < \alpha$, there exists a point (a, b) in B_ε for which $V(a, b) > 0$. Let $x = \phi(t), y = \psi(t)$ be the solution to (6)–(7) satisfying the initial condition

$$\phi(t_0) = a, \qquad \psi(t_0) = b.$$

Also, let t_1 be the first value of $t > t_0$ for which the trajectory $(\phi(t), \psi(t))$ leaves B_α [$t_1 = +\infty$ means $(\phi(t), \psi(t))$ remains in B_α]. Since $\dot{V}$ is positive definite on B_α, we have

$$V(\phi(t), \psi(t)) - V(\phi(t_0), \psi(t_0)) = \int_{t_0}^{t} \frac{dV}{ds}(\phi(s), \psi(s))\, ds$$
$$\geq 0$$

for all t in $[t_0, t_1)$. If we let V_0 denote $V(\phi(t_0), \psi(t_0))$, then $V_0 > 0$ and, for t in $[t_0, t_1)$,

$$V(\phi(t), \psi(t)) \geq V_0.$$

Now since V is continuous, there exists a $\lambda > 0$ such that $|V(x, y)| < V_0$ for (x, y) in B_λ. Therefore the trajectory $(\phi(t), \psi(t))$ must lie in the annulus $\bar{A}_{\lambda, \alpha}$ for $t \in [t_0, t_1)$. Let $\mu > 0$ be the minimum value of $\dot{V}$ on $\bar{A}_{\lambda, \alpha}$. Then, for t in $[t_0, t_1)$, we have

$$V(\phi(t), \psi(t)) - V(\phi(t_0), \psi(t_0)) = \int_{t_0}^{t} \frac{dV}{ds}(\phi(s), \psi(s))\, ds$$
$$\geq \mu(t - t_0).$$

That is,

$$V\big(\phi(t), \psi(t)\big) \geq V_0 + \mu(t - t_0) \quad \text{for } t \in [t_0, t_1] \, .$$

If $t_1 = +\infty$, then it follows that $V\big(\phi(t), \psi(t)\big) \to +\infty$ as $t \to +\infty$. But this is absurd, since $|V(x, y)| \leq L$ for (x, y) in B_α. Hence, $t_1 < \infty$; that is, the trajectory $\big(\phi(t), \psi(t)\big)$ leaves the disk B_α. ◆

Example 2 Determine the stability of the critical point at the origin for the system

(11) $$\frac{dx}{dt} = -y^3 \, ,$$

(12) $$\frac{dy}{dt} = -x^3 \, .$$

Solution The origin is not an isolated critical point for the corresponding linear system, so we cannot use the comparison approach of Section 12.3. Suspecting that the origin is unstable, we will attempt to apply Theorem 4.

Notice that $f(x, y) = -y^3$ is a function only of y and $g(x, y) = -x^3$ a function only of x. So let's try $V(x, y) = -xy$. Then V is continuous, $V(0, 0) = 0$, and in every disk centered at the origin, V is positive at some point (namely, those points of the disk lying in the second and fourth quadrants). Moreover,

$$\dot{V}(x, y) = (-y)(-y^3) + (-x)(-x^3) = y^4 + x^4 \, ,$$

which is positive definite in any domain containing the origin. Hence, by Lyapunov's instability theorem, the system (11)–(12) is unstable. ◆

12.5 EXERCISES

In Problems 1–8, use Lyapunov's direct method to determine the stability of the origin for the given system.

1. $$\frac{dx}{dt} = -3x^3 + 2xy^2 \, ,$$

$$\frac{dy}{dt} = -y^3$$

[*Hint:* Try $V(x, y) = x^2 + y^2$.]

2. $$\frac{dx}{dt} = -y - x^3 - xy^2 \, ,$$

$$\frac{dy}{dt} = x - x^2y - y^3$$

[*Hint:* Try $V(x, y) = x^2 + y^2$.]

3. $$\frac{dx}{dt} = y^2 + xy^2 - x^3 \, ,$$

$$\frac{dy}{dt} = -xy + x^2y - y^3$$

[*Hint:* Try $V(x, y) = x^2 + y^2$.]

4. $$\frac{dx}{dt} = x^3y + x^2y^3 - x^5 \, ,$$

$$\frac{dy}{dt} = -2x^4 - 6x^3y^2 - 2y^5$$

[*Hint:* Try $V(x, y) = 2x^2 + y^2$.]

5. $$\frac{dx}{dt} = 2x^3 \, ,$$

$$\frac{dy}{dt} = 2x^2y - y^3$$

[*Hint:* Try $V(x, y) = x^2 - y^2$.]

6. $$\frac{dx}{dt} = x + y + y^3 \, ,$$

$$\frac{dy}{dt} = x + y + x^5$$

[*Hint:* Try $V(x, y) = xy$.]

7. $\dfrac{dx}{dt} = 2y - x^3$,

$\dfrac{dy}{dt} = -x^3 - y^5$

[*Hint:* Try $V(x, y) = ax^4 + by^2$.]

8. $\dfrac{dx}{dt} = y^3 - 2x^3$,

$\dfrac{dy}{dt} = -3x - y^3$

[*Hint:* Try $V(x, y) = ax^2 + by^4$.]

In Problems 9–14, use Lyapunov's direct method to determine the stability of the zero solution for the given equation.

9. $\dfrac{d^2x}{dt^2} + |x|\dfrac{dx}{dt} + x^3 = 0$

10. $\dfrac{d^2x}{dt^2} + (1 - x^2)\dfrac{dx}{dt} + x = 0$

11. $\dfrac{d^2x}{dt^2} + \left[1 - \left(\dfrac{dx}{dt}\right)^2\right]\dfrac{dx}{dt} + x = 0$

12. $\dfrac{d^2x}{dt^2} + x^2\dfrac{dx}{dt} + x^5 = 0$

13. $\dfrac{d^2x}{dt^2} + \dfrac{dx}{dt} + \sin x = 0$

[*Hint:* Use $V(x, y) = (1/2)y^2 - \cos x$, where $y = dx/dt$.]

14. $\dfrac{d^2x}{dt^2} + \cos x\dfrac{dx}{dt} + \sin x = 0$

15. Let f be continuous and have continuous first partial derivatives in some domain D containing the origin. Show that the system

$\dfrac{dx}{dt} = -y + xf(x, y)$,

$\dfrac{dy}{dt} = x + yf(x, y)$

(a) is asymptotically stable if $f(x, y) < 0$ in D.
(b) is unstable if $f(x, y) > 0$ in D.

16. Verify that the three energy functions given in (5) are positive definite in some domain containing the origin.

17. Prove that the zero solution for a conservative system

$\dfrac{d^2x}{dt^2} + g(x) = 0$,

where g is continuous in an interval containing zero, is stable provided $g(0) = 0$ and, for some $k > 0$, $sg(s) > 0$ for $0 < |s| < k$.

18. Assume $V(x, y)$ is positive definite in a domain containing the origin and the level curves $V(x, y) = k$ are simple closed curves that expand out from the origin such that the outward normal $\mathbf{N}$ to the level curve at the point (a, b) is given by

$$\mathbf{N} = \nabla V(a, b) = V_x(a, b)\mathbf{i} + V_y(a, b)\mathbf{j},$$

where $\mathbf{i}$ and $\mathbf{j}$ are unit vectors in the positive x and y directions, respectively. Let $\big(\phi(t), \psi(t)\big), t \to +\infty$, be a trajectory for the solution to system (6)–(7) that passes through (a, b) at time $t = t_0$. Recall that the tangent $\mathbf{T}$ to the trajectory at (a, b) is

$$\mathbf{T} = \dfrac{d\phi}{dt}(t_0)\mathbf{i} + \dfrac{d\psi}{dt}(t_0)\mathbf{j}.$$

(a) Verify that

$$\dot{V}(\phi, \psi) = \mathbf{N} \cdot \mathbf{T} = |\mathbf{N}|\,|\mathbf{T}|\cos\theta,$$

where θ is the angle between $\mathbf{N}$ and $\mathbf{T}$ (see Figure 12.30).
(b) Show that if $\dot{V}$ is negative definite, then a trajectory $\big(\phi(t), \psi(t)\big)$ that touches the level curve $V(x, y) = k$ must enter the region bounded by this level curve.
(c) Show that if $\dot{V}$ is negative semidefinite, then a trajectory that enters the region bounded by $V(x, y) = k$ cannot leave this region.

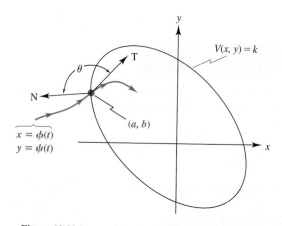

Figure 12.30 Intersection of a trajectory with a level curve

(d) Show that if $\dot{V}$ is positive definite, then a trajectory $\big(\phi(t), \psi(t)\big)$ that touches the level curve $V(x, y) = k$ must leave the region bounded by this level curve.

12.6 Limit Cycles and Periodic Solutions

In Example 4 in Section 12.4, we used the energy method to study a nonconservative system involving nonlinear damping. Our analysis led us to the existence of a closed trajectory that encircled the unstable spiral point at the origin. What was interesting about this closed trajectory was that all trajectories beginning inside of it spiraled out toward it, while those beginning outside spiraled into it. Since closed trajectories correspond to periodic solutions, we have in hand an example of when all solutions approach the periodic solution as $t \to \infty$ (except for the zero solution corresponding to the critical point at the origin). This closed trajectory is an example of a **limit cycle.** In this section we discuss the criteria for the existence and nonexistence of a limit cycle in the plane.

> **Limit Cycle**
>
> **Definition 5.** A nontrivial[†] closed trajectory with at least one other trajectory spiraling into it (as time approaches plus or minus infinity) is called a **limit cycle.**

The next example illustrates the behavior of trajectories near a limit cycle. Let $r^2 = x^2 + y^2$ and consider the system

(1) $$\frac{dx}{dt} = x - y - x(3r - r^2 - 1) ,$$

(2) $$\frac{dy}{dt} = x + y - y(3r - r^2 - 1) ,$$

for which the origin is the only critical point. It is helpful to use polar coordinates $x = r\cos\theta, y = r\sin\theta$ to study this system. We need two identities

$$x\frac{dx}{dt} + y\frac{dy}{dt} = r\frac{dr}{dt} ,$$

$$x\frac{dy}{dt} - y\frac{dx}{dt} = r^2\frac{d\theta}{dt} ,$$

which follow from the polar equations (see Problem 2). Now multiplying equation (1) by x and equation (2) by y and adding, we find

$$r\frac{dr}{dt} = x\frac{dx}{dt} + y\frac{dy}{dt} = x\left(x - y - x(3r - r^2 - 1)\right) + y\left(x + y - y(3r - r^2 - 1)\right)$$
$$= (x^2 + y^2) - (x^2 + y^2)(3r - r^2 - 1)$$
$$= r^2(r - 1)(r - 2) .$$

Thus, we have

(3) $$\frac{dr}{dt} = r(r - 1)(r - 2) .$$

[†]Critical points are technically closed trajectories, but we exclude these trivial ones.

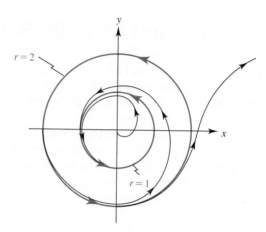

Figure 12.31 Phase plane diagram with limit cycles and spiraling trajectories

We now multiply equation (2) by x and equation (1) by y and subtract the latter from the former to get

$$r^2 \frac{d\theta}{dt} = x \frac{dy}{dt} - y \frac{dx}{dt} = x\left(x + y - y(3r - r^2 - 1)\right) - y\left(x - y - x(3r - r^2 - 1)\right)$$
$$= x^2 + y^2 = r^2 \, .$$

Hence, $d\theta/dt = 1$, which tells us that the trajectories revolve counterclockwise about the origin. From equation (3), we observe that dr/dt is zero when $r = 0$, 1, and 2. The critical point $(0, 0)$ is represented by $r = 0$. When $r = 1$ or 2, the corresponding trajectories are just circles of radius 1 or 2, respectively; hence, these circles are limit cycles. Now when r lies in $(0, 1)$, we see from (3) that $dr/dt > 0$, so a trajectory in this region spirals out toward the limit cycle $r = 1$. When r lies in $(1, 2)$, $dr/dt < 0$ and the trajectories spiral in toward the limit cycle $r = 1$ and away from the limit cycle $r = 2$. Finally, when $r > 2$, $dr/dt > 0$ and the trajectories spiral out away from the limit cycle $r = 2$ (see Figure 12.31).

When nearby trajectories approach a limit cycle, we say that the limit cycle is **stable.** When they recede, the limit cycle is **unstable.** It is possible for trajectories to approach one side of the cycle while receding from the other. Such a limit cycle is called **semistable** (see Problem 1). In the previous example, the limit cycle $r = 2$ is unstable while the limit cycle $r = 1$ is stable. Clearly the origin is an unstable spiral point.

In applications when limit cycles do exist, they are seldom circles and, unfortunately, finding them may be difficult. In this regard, certain theoretical results, which we now present, may be helpful. The first theorem certainly makes sense intuitively; however, its proof is beyond the scope of this text.[†]

Limit Cycles Surround Critical Points

Theorem 5. A limit cycle in the plane must enclose at least one critical point. Moreover, any critical point enclosed by a limit cycle cannot be a saddle point.

[†]For a proof, see the book by P. W. Jordan and P. Smith, *Nonlinear Ordinary Differential Equations*, 4th ed. (Clarendon Press, Oxford, 2007), Section 3.4.

Observe that in the previous example, both limit cycles encircled the origin, which was an unstable spiral point.

The next result is due to I. Bendixson. It tells us when a limit cycle does not exist.

Bendixson Negative Criterion

Theorem 6. Let $f(x, y)$ and $g(x, y)$ have continuous first partial derivatives in the simply connected domain[†] D and assume that

$$f_x(x, y) + g_y(x, y)$$

is of one sign in D. Then there are no (nonconstant) periodic solutions to

(4) $\dfrac{dx}{dt} = f(x, y)$,

(5) $\dfrac{dy}{dt} = g(x, y)$

that lie entirely in D.

Proof. Assume that $x = \phi(t), y = \psi(t)$ is a (nonconstant) T-periodic solution to system (4)–(5) whose trajectory $\Gamma(t) := \big(\phi(t), \psi(t)\big)$ lies entirely in D. Let Ω denote the region enclosed by Γ. Green's theorem in the plane says that[‡]

(6) $\displaystyle \oint_\Gamma [f(x, y)\, dy - g(x, y)\, dx] = \int\int_\Omega [f_x(x, y) + g_y(x, y)]\, dx\, dy$.

The assumption that $f_x + g_y$ is of one sign in D implies the integral on the right-hand side is not zero. Using the parametric representation $\Gamma(t) = \big(\phi(t), \psi(t)\big), 0 \le t \le T$, and the fact that ϕ, ψ are solutions to (4)–(5), we find the integral on the left-hand side of (6) to be equal to

$$\int_0^T \left[f\big(\phi(t), \psi(t)\big) \frac{d\psi}{dt} - g\big(\phi(t), \psi(t)\big) \frac{d\phi}{dt} \right] dt = \int_0^T \left[\frac{d\phi}{dt}\frac{d\psi}{dt} - \frac{d\psi}{dt}\frac{d\phi}{dt} \right] dt = 0 \, .$$

This yields a contradiction, since the right-hand side of (6) is not zero. Thus, there is no (nonconstant) periodic solution whose trajectory lies entirely in D. ◆

We use the Bendixson criterion in the next example to show that limit cycles do not exist for the given systems.

Example 1 Show that the system

(7) $\dfrac{dx}{dt} = \alpha(y)x + \beta(y)$,

(8) $\dfrac{dy}{dt} = \gamma(x)y + \delta(x)$,

where β, δ are continuous and α, γ are both continuous and positive (negative) in some simply connected domain, has no nonconstant periodic solution.

[†]A simply connected domain is a connected region that does not have any "holes" in it.

[‡]If ϕ, ψ is a nonconstant periodic solution of (4)–(5), then the assumptions on f and g will guarantee that Γ and Ω satisfy the conditions of Green's theorem.

Solution Here $f(x, y) = \alpha(y)x + \beta(y)$ and $g(x, y) = \gamma(x)y + \delta(x)$. Now computing $f_x + g_y$, we find

$$f_x + g_y = \alpha(y) + \gamma(x) ,$$

which is always positive (negative). Hence, by the Bendixson negative criterion, there can be no nonconstant periodic solutions. ◆

A mechanical system with positive (negative) damping is governed by a second-order equation of the form

$$(9) \qquad \frac{d^2x}{dt^2} + \gamma(x)\frac{dx}{dt} + \delta(x) = 0 ,$$

where $\gamma(x)$ is positive (negative). This equation can be written in the form of system (7)–(8). It then follows from the Bendixson negative criterion that this mechanical system has no nonconstant periodic solution (see Problem 23). This conclusion is consistent with the results of the energy method that show that positive damping causes trajectories to move in toward the origin and negative damping causes them to move away from the origin.

What happens if there is negative damping near the origin and positive damping away from the origin? Is there a nonconstant periodic solution? The following result gives sufficient conditions for the existence of a nonconstant periodic solution. Its proof is left for more advanced texts.[†]

Poincaré–Bendixson Theorem

Theorem 7. If f and g have continuous first partial derivatives on the closed bounded region R and the system

$$\frac{dx}{dt} = f(x, y) , \qquad \frac{dy}{dt} = g(x, y)$$

has no critical points in R, then any solution $x = \phi(t), y = \psi(t)$ to the system that exists for all $t \geq t_0$ is either a periodic solution or approaches a limit cycle. In either case, the system has a nonconstant periodic solution.

To apply this theorem, we must find a closed bounded region R that contains a trajectory Γ, but not a critical point of the system. If we are searching for limit cycles, then this region R must have a hole in it, since Theorem 5 asserts that limit cycles encircle critical points.

A typical strategy for determining R is as follows. First, find a curve C_1 encircling a critical point and show that all trajectories crossing C_1 must head out. Second, find a curve C_2 encircling C_1 and show that all trajectories crossing C_2 must head in. If R is the region between C_1 and C_2, including C_1 and C_2 as its boundary, then any trajectory beginning in R must remain in R. Otherwise, it must cross either C_1 headed in or C_2 headed out. Both situations yield a contradiction. We illustrate this approach in the next example.

[†]For a proof, see the book by J. K. Hale, *Ordinary Differential Equations,* 2nd ed. (Dover, New York, 2009), Chapter 2.

Example 2 Show that the equation

$$(10) \qquad \frac{d^2x}{dt^2} + \left[4x^2 + \left(\frac{dx}{dt} \right)^2 - 4 \right] \frac{dx}{dt} + x^3 = 0$$

has a nonconstant periodic solution.

Solution Let $y = dx/dt$ and rewrite equation (10) as the equivalent system

$$\frac{dx}{dt} = y \,,$$

$$\frac{dy}{dt} = -x^3 - (4x^2 + y^2 - 4)y \,.$$

Notice that the origin is the only critical point for this system.

Motivated by the energy method, let's consider the Lyapunov function

$$V(x, y) := \frac{1}{2}y^2 + \frac{1}{4}x^4 \,.$$

The level curves $V(x, y) = k, k > 0$, consist of simple closed curves encircling the origin. Moreover,

$$\dot{V}(x, y) = V_x \frac{dx}{dt} + V_y \frac{dy}{dt}$$

$$= x^3y + y[-x^3 - y(4x^2 + y^2 - 4)]$$

$$= -y^2(4x^2 + y^2 - 4) \,.$$

Now if (x, y) lies inside the ellipse $4x^2 + y^2 = 4$, then $\dot{V}(x, y) \geq 0$, and if it lies outside, then $\dot{V}(x, y) \leq 0$.

To apply the Poincaré–Bendixson theorem, we must determine a region R that contains a trajectory but no critical point. To define R, select k_1 sufficiently small so that the level curve $V(x, y) = k_1$ lies inside the ellipse $4x^2 + y^2 = 4$. On this level curve, $\dot{V}(x, y) \geq 0$, so trajectories cannot cross into the region enclosed by $V(x, y) = k_1$. Next select k_2 sufficiently large so that the level curve $V(x, y) = k_2$ encircles the ellipse $4x^2 + y^2 = 4$. Here, on this level curve, $\dot{V}(x, y) \leq 0$, so trajectories cannot leave the region enclosed by $V(x, y) = k_2$. Let R be the region between $V(x, y) = k_1$ and $V(x, y) = k_2$ (see Figure 12.32). It now follows from the Poincaré–Bendixson theorem that there exists a nonconstant periodic solution inside R. ◆

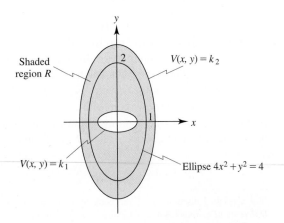

Figure 12.32 Region R where periodic solution exists

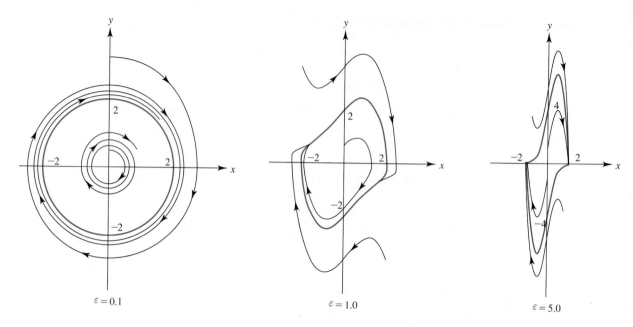

Figure 12.33 Limit cycles for van der Pol equations

In the above example (page 761), the ellipse $4x^2 + y^2 = 4$ is *not* a solution curve for the associated phase plane equation and hence does not support a trajectory, much less a limit cycle. However, there is a limit cycle and it must be intertwined with the ellipse. Indeed, while the limit cycle lies inside the ellipse, its "energy" V is increasing, and while it is outside, its "energy" is decreasing.

Example 2 is a situation where the damping is negative near the origin and positive away from the origin. In the analysis of the triode vacuum tube, a slightly different behavior arises. The model governing the tube is the van der Pol equation[†] (discussed qualitatively in Section 4.8, page 201).

$$(11) \qquad \frac{d^2x}{dt^2} + \varepsilon(x^2 - 1)\frac{dx}{dt} + x = 0 \,,$$

where ε is a positive constant. Here the damping "coefficient" $\varepsilon(x^2 - 1)$ is positive for $|x| > 1$ but negative for $|x| < 1$. Therefore, this damping coefficient has to be negative somewhere, on every curve encircling the origin. While it is not obvious, it turns out that the van der Pol equation has a unique periodic solution for any $\varepsilon > 0$. The limit cycles corresponding to these periodic solutions are illustrated in Figure 12.33 for $\varepsilon = 0.1$, 1, and 5.

We conclude this section with a statement of a theorem on **Lienard equations** due to N. Levinson and O. K. Smith,[‡] which can be used to show that van der Pol's equation has a unique nonconstant periodic solution (see Problem 24).

[†]For a detailed discussion, see the book by L. Cesari, *Asymptotic Behavior and Stability Problems in Ordinary Differential Equations*, 3rd ed. (Springer-Verlag, New York, 1971), pages 154–156.
[‡]See *Duke Mathematics Journal*, Vol. 9, 1942, pages 382–403.

Levinson and Smith's Theorem

Theorem 8. Let $f(x)$ and $g(x)$ be continuous functions and let

$$F(x) := \int_0^x f(s)\,ds\,, \qquad G(x) := \int_0^x g(s)\,ds\,.$$

The Lienard equation

$$\frac{d^2x}{dt^2} + f(x)\frac{dx}{dt} + g(x) = 0$$

has a unique nonconstant periodic solution whenever all of the following conditions hold:

(a) $f(x)$ is even.

(b) $F(x) < 0$, for $0 < x < a$;
 $F(x) > 0$, for $x > a$, for some a.

(c) $F(x) \to +\infty$ as $x \to +\infty$, monotonically for $x > a$.

(d) $g(x)$ is an odd function with $g(x) > 0$ for $x > 0$.

(e) $G(x) \to +\infty$ as $x \to +\infty$.

12.6 EXERCISES

1. **Semistable Limit Cycle.** For the system

$$\frac{dx}{dt} = x - y - x(2r - r^2)\,,$$

$$\frac{dy}{dt} = x + y - y(2r - r^2)\,,$$

where $x = r\cos\theta$ and $y = r\sin\theta$:

(a) Sketch the phase plane diagram and identify each limit cycle and critical point.

(b) Explain why the limit cycle $r = 1$ is semistable.

2. For $x = r\cos\theta$ and $y = r\sin\theta$, prove that

$$r\frac{dr}{dt} = x\frac{dx}{dt} + y\frac{dy}{dt}\,,$$

$$r^2\frac{d\theta}{dt} = x\frac{dy}{dt} - y\frac{dx}{dt}\,.$$

[*Hint:* Use the facts that $r^2 = x^2 + y^2$ and $\tan\theta = y/x$.]

3. For systems of the form

$$\frac{dx}{dt} = x - y + xf(r)/r\,,$$

$$\frac{dy}{dt} = x + y + yf(r)/r\,,$$

where $x = r\cos\theta$ and $y = r\sin\theta$:

(a) Show that the system has periodic solutions corresponding to the values of r such that $r + f(r) = 0$.

(b) Determine the direction of rotation for the trajectories of this system.

4. For systems of the form

$$\frac{dx}{dt} = y + xf(r)/r\,,$$

$$\frac{dy}{dt} = -x + yf(r)/r\,,$$

where $x = r\cos\theta$ and $y = r\sin\theta$:

(a) Show that the system has periodic solutions corresponding to the zeros of $f(r)$.

(b) Determine the direction of rotation for the trajectories of this system.

In Problems 5–12, either by hand or using a computer software package, sketch the phase plane diagrams for the given system. Discuss the stability of each limit cycle and critical point. Here $x = r\cos\theta$ and $y = r\sin\theta$.

5. $\dfrac{dx}{dt} = x - y + x(r^3 - 4r^2 + 5r - 3)\,,$

 $\dfrac{dy}{dt} = x + y + y(r^3 - 4r^2 + 5r - 3)$

6. $\dfrac{dx}{dt} = y + x(r^2 - 7r + 10)$,

$\dfrac{dy}{dt} = -x + y(r^2 - 7r + 10)$

7. $\dfrac{dx}{dt} = x - y + x(r^3 - r - 1)$,

$\dfrac{dy}{dt} = x + y + y(r^3 - r - 1)$

8. $\dfrac{dx}{dt} = x - y + x(x^2 + y^2)$,

$\dfrac{dy}{dt} = x + y + y(x^2 + y^2)$

9. $\dfrac{dx}{dt} = x + y + x(r^2 - 3r + 1)$,

$\dfrac{dy}{dt} = -x + y + y(r^2 - 3r + 1)$

10. $\dfrac{dx}{dt} = -y + x(r^2 - 5r + 6)$,

$\dfrac{dy}{dt} = x + y(r^2 - 5r + 6)$

11. $\dfrac{dx}{dt} = y + x\sin(1/r)$,

$\dfrac{dy}{dt} = -x + y\sin(1/r)$

12. $\dfrac{dx}{dt} = y + \dfrac{x}{r}\sin r$, $\quad \dfrac{dy}{dt} = -x + \dfrac{y}{r}\sin r$

In Problems 13–18, show that the given system or equation has only constant periodic solutions in the specified domain D.

13. $\dfrac{dx}{dt} = x - y + x^2 y$,

$\dfrac{dy}{dt} = x + x^2 y + \dfrac{1}{3}y^3$

(*D* is the entire plane.)

14. $\dfrac{dx}{dt} = 2x + y - xy^2$,

$\dfrac{dy}{dt} = x - 3y - x^2 y$

(*D* is the entire plane.)

15. $\dfrac{dx}{dt} = -8x + y + x^3$,

$\dfrac{dy}{dt} = 2x - 7y + y^3$

(*D* is the disk $x^2 + y^2 < 5$.)

16. $\dfrac{dx}{dt} = 3x + 2y - xy^2$,

$\dfrac{dy}{dt} = x + 4y - 2x^2 y$

(*D* is the interior of the ellipse $2x^2 + y^2 = 7$.)

17. $\dfrac{d^2x}{dt^2} + (1 - x^2)\dfrac{dx}{dt} + x^3 = 0$

(*D* is the strip $-1 < x < 1$ in the *xv*-plane.)

18. $\dfrac{d^2x}{dt^2} + \left[3 - \left(\dfrac{dx}{dt}\right)^2\right]\dfrac{dx}{dt} + x = 0$

(*D* is the slab $-1 < v < 1$ in the *xv*-plane.)

In Problems 19–22, show that the given system or equation has a nonconstant periodic solution.

19. $\dfrac{dx}{dt} = 2x - y - 2x^3 - 3xy^2$,

$\dfrac{dy}{dt} = 2x + 4y - 4y^3 - 2x^2 y$

[*Hint:* Use the Lyapunov function $V(x, y) = 2x^2 + y^2$.]

20. $\dfrac{dx}{dt} = 8x - 2y - 4x^3 - 2xy^2$,

$\dfrac{dy}{dt} = x + 4y - 2y^3 - 3x^2 y$

[*Hint:* Use the Lyapunov function $V(x, y) = x^2 + 2y^2$.]

21. $\dfrac{d^2x}{dt^2} + \left(x^4 + \left(\dfrac{dx}{dt}\right)^2 - 1\right)\dfrac{dx}{dt} + x = 0$

22. $\dfrac{d^2x}{dt^2} + \ln\left(x^2 + \left(\dfrac{dx}{dt}\right)^2 + \dfrac{1}{2}\right)\dfrac{dx}{dt} + x^3 = 0$

23. Strictly Positive Damping. Use the Bendixson negative criterion to show that a mechanical system governed by equation (9) with γ and δ continuous on $(-\infty, \infty)$ and γ strictly positive (negative) has only constant periodic solutions.

24. Show that van der Pol's equation (11) satisfies the conditions of Levinson–Smith's theorem and therefore has a unique nonconstant periodic solution.

In Problems 25–28, show that the given Lienard equations have unique nonconstant periodic solutions.

25. $\dfrac{d^2x}{dt^2} + x^2(x^2 - 1)\dfrac{dx}{dt} + x = 0$

26. $\dfrac{d^2x}{dt^2} + (x^2 - 4)\dfrac{dx}{dt} + x^3 = 0$

27. $\dfrac{d^2x}{dt^2} + (x^2 - \cos x - 1)\dfrac{dx}{dt} + x + \sin x = 0$

28. $\dfrac{d^2x}{dt^2} + (x^4 - 1)\dfrac{dx}{dt} + x^5 = 0$

12.7 Stability of Higher-Dimensional Systems

Many of the results we have obtained on the stability of linear and almost linear systems in the plane can be extended to higher-dimensional systems (a notable exception is the Poincaré–Bendixson theorem discussed in Section 12.6). To give these extensions, we must first introduce some definitions and notation. We assume that you are familiar with the material in Chapter 9.

Let $\mathbf{z} = \text{col}(z_1, z_2, \ldots, z_n)$ be a vector in $\mathbf{R}^n$. We define the **norm**[†] of $\mathbf{z}$ to be

$$\|\mathbf{z}\| := \max_{i=1,\ldots,n} |z_i| \, .$$

In a similar fashion, if $\mathbf{A}$ is an $n \times n$ matrix with entries a_{ij}, $i, j = 1, \ldots, n$, we define the norm of $\mathbf{A}$ to be

$$\|\mathbf{A}\| := \max_{i,j=1,\ldots,n} |a_{ij}| \, .$$

A simple analysis using the triangle inequality shows that

$$
\begin{aligned}
\|\mathbf{A}\mathbf{z}\| &= \max_{i=1,\ldots,n} \left| \sum_{k=1}^{n} a_{ik} z_k \right| \\
&\leq \max_{i=1,\ldots,n} \left(\sum_{k=1}^{n} |a_{ik}| |z_k| \right) \\
&\leq \sum_{k=1}^{n} \|\mathbf{A}\| |z_k| \\
&\leq \|\mathbf{A}\| \sum_{k=1}^{n} \|\mathbf{z}\| \, ,
\end{aligned}
$$

which reduces to the inequality

$$(1) \qquad \|\mathbf{A}\mathbf{z}\| \leq n \|\mathbf{A}\| \, \|\mathbf{z}\| \, .$$

We can now define stability for a solution to the system of differential equations

$$(2) \qquad \mathbf{x}'(t) = \mathbf{f}(t, \mathbf{x}) \, ,$$

where $\mathbf{x}'(t) = \text{col}\big(x_1'(t), \ldots, x_n'(t)\big)$ and $\mathbf{f}(t, \mathbf{x}) = \text{col}\big(f_1(t, \mathbf{x}), \ldots, f_n(t, \mathbf{x})\big)$. We assume that each $f_i(t, \mathbf{x})$, $i = 1, \ldots, n$, is a continuous function of $t, x_1, \ldots, x_n$ with continuous partial derivatives with respect to $x_1, \ldots, x_n$.

Lyapunov Stability

Definition 6. A solution $\boldsymbol{\phi}(t)$ to system (2) is **stable** (also called **Lyapunov stable**) for $t \geq t_0$ if for any $\varepsilon > 0$ there exists $\delta(t_0, \varepsilon) > 0$ such that whenever $\|\mathbf{x}(t_0) - \boldsymbol{\phi}(t_0)\| < \delta$, where $\mathbf{x}(t)$ is any solution to (2), then $\|\mathbf{x}(t) - \boldsymbol{\phi}(t)\| < \varepsilon$ for all $t \geq t_0$. In addition, if for any such $\mathbf{x}$ we have $\mathbf{x}(t) - \boldsymbol{\phi}(t) \to \mathbf{0}$ as $t \to +\infty$, then $\boldsymbol{\phi}$ is **asymptotically stable**.

If $\boldsymbol{\phi}$ is not stable, then it is **unstable**.

[†]There are other norms that can be used in $\mathbf{R}^n$, such as the Euclidean norm $\sqrt{z_1^2 + z_2^2 + \cdots + z_n^2}$. In most of the results that we give, the choice of norm is not essential.

The relationship between Lyapunov stability of a solution and the stability of a critical point is the following. A solution $\boldsymbol{\phi}$ of (2) is stable if and only if the equilibrium solution $\mathbf{z}(t) \equiv \mathbf{0}$ is a stable solution to the system

$$(3) \qquad \mathbf{z}'(t) = \mathbf{g}(t, \mathbf{z}) \,,$$

where

$$\mathbf{g}(t, \mathbf{z}) := \mathbf{f}(t, \mathbf{z} + \boldsymbol{\phi}) - \mathbf{f}(t, \boldsymbol{\phi}) \,.$$

This in turn is equivalent to the origin $\mathbf{0} = \mathrm{col}(0, 0, \ldots, 0)$ being a stable critical point for (3). Indeed, if $\mathbf{x}$ and $\boldsymbol{\phi}$ are solutions to (2), then $\mathbf{z} = \mathbf{x} - \boldsymbol{\phi}$ is a solution to (3). So if $\boldsymbol{\phi}$ is stable and $\|\mathbf{x}(t) - \boldsymbol{\phi}(t)\| < \varepsilon$, then $\|\mathbf{z}(t)\| < \varepsilon$ and the zero solution is stable. The converse is also true.

For linear systems it is sufficient to study the stability of the zero solution to the homogeneous system.

Stability and the Zero Solution

Theorem 9. A solution $\boldsymbol{\phi}(t)$ to the linear system

$$(4) \qquad \mathbf{x}'(t) = \mathbf{A}(t)\mathbf{x}(t) + \mathbf{h}(t) \,,$$

where $\mathbf{A}(t)$ is an $n \times n$ matrix whose entries are functions of t and $\mathbf{h}(t) = \mathrm{col}\big(h_1(t), \ldots, h_n(t)\big)$ is stable (asymptotically stable) if and only if the zero solution $\mathbf{z}(t) \equiv \mathbf{0}$ is a stable (an asymptotically stable) solution to $\mathbf{x}'(t) = \mathbf{A}(t)\mathbf{x}(t)$.

We leave the proof as an exercise (Problem 7).

The next result relates the stability of a linear system to properties of a fundamental matrix for the system.

Fundamental Matrix and Stability

Theorem 10. Let $\mathbf{A}(t)$ be an $n \times n$ matrix-valued function whose entries are continuous functions for $t \geq t_0$ and let $\mathbf{X}(t)$ be a fundamental matrix for the homogeneous system

$$(5) \qquad \mathbf{x}'(t) = \mathbf{A}(t)\mathbf{x}(t) \,, \qquad t \geq t_0 \,.$$

If there exists a constant $K > 0$ such that $\|\mathbf{X}(t)\| \leq K$ for $t \geq t_0$, then the zero solution $\mathbf{x}(t) = \mathbf{0}$ is stable. Moreover, if $\lim_{t \to +\infty} \|\mathbf{X}(t)\| = 0$, then the zero solution is asymptotically stable.

Proof. Without loss of generality, assume that $\mathbf{X}(t_0) = \mathbf{I}$. [Otherwise, use the fundamental matrix $\mathbf{Y}(t) = \mathbf{X}(t)\mathbf{X}^{-1}(t_0)$.] If $\mathbf{x}(t)$ is any solution to (5), then we can express $\mathbf{x}(t)$ in the form

$$\mathbf{x}(t) = \mathbf{X}(t)\mathbf{x}(t_0)$$

(see Theorem 3 in Section 9.4, page 518, and the subsequent discussion). Taking the norm of both sides and using inequality (1), we find

$$(6) \qquad \|\mathbf{x}(t)\| = \|\mathbf{X}(t)\mathbf{x}(t_0)\| \leq n\|\mathbf{X}(t)\|\|\mathbf{x}(t_0)\| \,.$$

If $\||\mathbf{X}(t)\|| \le K$ for $t \ge t_0$, then we have

$$\|\mathbf{x}(t)\| \le nK\|\mathbf{x}(t_0)\| .$$

So, given $\varepsilon > 0$, we select $\delta = \varepsilon/(nK)$. Then, whenever $\|\mathbf{x}(t_0)\| < \delta = \varepsilon/(nK)$, it follows that

$$\|\mathbf{x}(t)\| < nK\delta = \varepsilon \quad \text{for} \quad t \ge t_0 ,$$

and the origin is stable.

Moreover, if $\lim_{t \to +\infty} \||\mathbf{X}(t)\|| = 0$, then it follows from (6) that $\mathbf{x}(t) \to \mathbf{0}$ as $t \to \infty$, and so the zero solution is asymptotically stable. ◆

As a corollary, we have the following result for homogeneous linear systems with constant coefficients. The case of 2×2 matrices $\mathbf{A}$ was treated in Theorem 1 of Section 12.2, page 727.

Homogeneous Systems with Constant Coefficients

Corollary 1. Let $\mathbf{A}$ be an $n \times n$ constant matrix. If all eigenvalues of $\mathbf{A}$ have negative real parts, then for any t_0 the zero solution $\mathbf{x}(t) \equiv \mathbf{0}$ to

$$\mathbf{x}'(t) = \mathbf{A}\mathbf{x}(t) , \qquad t \ge t_0 ,$$

is asymptotically stable.

However, if at least *one* eigenvalue of $\mathbf{A}$ has positive real part, then the zero solution is unstable.

Proof. Recall from Chapter 9 that the entries of a fundamental matrix consist of functions of the form

(7) $e^{\alpha t}[p(t)\cos\beta t + q(t)\sin\beta t] ,$

where $r = \alpha + i\beta$ is an eigenvalue of $\mathbf{A}$ and $p(t)$ and $q(t)$ are polynomials in t. [*Note:* If r is real, then $\beta = 0$ and the function reduces to $e^{\alpha t}p(t)$.] Now if all eigenvalues have negative real parts, that is, all alpha's, satisfy $\alpha < 0$, then each of the entries in the fundamental matrix $\mathbf{X}(t)$ is not only bounded, but also goes to zero as $t \to +\infty$, since

$$\lim_{t \to +\infty} e^{\alpha t}p(t) = 0 , \quad \text{for} \quad \alpha < 0 .$$

Hence, the zero solution is asymptotically stable.

If some eigenvalue $r = \alpha + i\beta$ has a positive real part, $\alpha > 0$, then the function in (7) becomes unbounded as $t \to +\infty$, since

$$\lim_{t \to +\infty} |e^{\alpha t}p(t)| = +\infty , \quad \text{for} \quad \alpha > 0 .$$

We can multiply a solution to a linear system by an arbitrarily small constant and still have a solution, so we conclude that starting arbitrarily close to the zero solution is a solution that becomes unbounded. Thus, the zero solution is unstable. ◆

Notice that, in the corollary, we omitted the case when one of the eigenvalues has zero real part. In this case, the zero solution may be stable or unstable, depending on whether the dimension of the eigenspace is the same as the multiplicity of the eigenvalue (see Problem 8).

Corollary 1 gives us an easy test for stability.

Example 1 Determine the stability of the zero solution to the system

$$(8) \qquad \mathbf{x}'(t) = \begin{bmatrix} -1 & 0 & 0 \\ 0 & -1 & 4 \\ 0 & -1 & -1 \end{bmatrix} \mathbf{x}(t), \qquad t \ge 0.$$

Solution From the characteristic equation

$$(r+1)\big[(r+1)^2 + 4\big] = 0,$$

we find that the eigenvalues are $-1, -1 \pm 2i$. Since all the eigenvalues have negative real parts, the zero solution is asymptotically stable. ◆

Example 2 Determine the stability of the zero solution to the system

$$(9) \qquad \mathbf{x}'(t) = \begin{bmatrix} 1 & -1 & 0 \\ 3 & 0 & 0 \\ 0 & 0 & -2 \end{bmatrix} \mathbf{x}(t), \qquad t \ge 0.$$

Solution From the characteristic equation

$$(r+2)(r^2 - r + 3) = 0,$$

we find that the eigenvalues are $-2, (1/2) \pm (\sqrt{11}/2)i$. Because at least one eigenvalue, $(1/2) + (\sqrt{11}/2)i$, has positive real part, the zero solution is unstable. ◆

Example 3 Determine the stability of the zero solution to the system

$$(10) \qquad \mathbf{x}'(t) = \begin{bmatrix} -2 & 1 & 0 \\ 0 & -2 & 1 \\ 0 & 0 & -2 \end{bmatrix} \mathbf{x}(t), \qquad t \ge 0.$$

Solution Here there is just one eigenvalue -2, which has multiplicity 3. Even though the eigenspace is one dimensional, the zero solution is asymptotically stable, since the eigenvalue is negative. ◆

Certain results for almost linear systems in the plane can be extended to higher-dimensional systems and to nonautonomous systems.

Almost Linear Systems

Definition 7. Let $\mathbf{A}$ be an $n \times n$ constant matrix whose determinant is nonzero. Let $\mathbf{f}(t, \mathbf{x})$ and $(\partial \mathbf{f}/\partial x_i)(t, \mathbf{x})$, $i = 1, \dots, n$, be continuous functions for $t \ge 0$ and for $\|\mathbf{x}\| < K$. Assume that $\mathbf{f}(t, \mathbf{0}) = \mathbf{0}$ for $t \ge 0$. If

$$\lim_{\|\mathbf{z}\| \to 0} \frac{\|\mathbf{f}(t, \mathbf{z})\|}{\|\mathbf{z}\|} = 0,$$

uniformly[†] in t, then we say that

$$(11) \qquad \mathbf{x}'(t) = \mathbf{A}\mathbf{x}(t) + \mathbf{f}(t, \mathbf{x})$$

is an **almost linear system.**

[†]That is, for every $\varepsilon > 0$, there exists a $\delta > 0$ such that whenever $0 < \|\mathbf{z}\| < \delta$, then $\|\mathbf{f}(t, \mathbf{z})\|/\|\mathbf{z}\| < \varepsilon$ for all $t \ge 0$.

We conclude with the following result due to Poincaré and Perron, which gives conditions on **A** for the zero solution to the almost linear system to be asymptotically stable.[†]

Poincaré and Perron's Theorem

Theorem 11. If all the eigenvalues of **A** have negative real parts, then the zero solution of the almost linear system (11) is asymptotically stable. If at least one eigenvalue of **A** has positive real part, then the zero solution of (11) is unstable.

12.7 EXERCISES

In Problems 1–6, determine the stability of the zero solution to the system $\mathbf{x}' = \mathbf{A}\mathbf{x}$ *for the given matrix* **A**.

1. $\begin{bmatrix} -1 & 2 & 1 \\ 0 & -3 & 2 \\ 0 & 0 & -4 \end{bmatrix}$

2. $\begin{bmatrix} -3 & 0 & 0 \\ 0 & -2 & -1 \\ 0 & 1 & -2 \end{bmatrix}$

3. $\begin{bmatrix} -1 & -2 & 1 \\ -1 & 0 & -1 \\ -4 & 4 & -5 \end{bmatrix}$

4. $\begin{bmatrix} 1 & 1 & 0 \\ 0 & 1 & 0 \\ 0 & 0 & 2 \end{bmatrix}$

5. $\begin{bmatrix} 0 & 1 & 0 \\ 0 & 0 & 1 \\ 1 & -1 & 1 \end{bmatrix}$

6. $\begin{bmatrix} 2 & 1 & -1 \\ -3 & -1 & 1 \\ 9 & 3 & -4 \end{bmatrix}$

7. Prove Theorem 9.

8. Eigenvalues with Zero Real Part. To illustrate the relationship between stability and the degeneracy of the eigenspace for an eigenvalue with zero real part, consider the system

$$\mathbf{x}' = \begin{bmatrix} 0 & \alpha \\ 0 & 0 \end{bmatrix} \mathbf{x} ,$$

where α is a constant.

(a) Show that when $\alpha = 0$, zero is an eigenvalue of multiplicity two and has two linearly independent eigenvectors. Find the general solution and show directly that the zero solution is stable.

(b) Show that when $\alpha = 1$, zero is an eigenvalue of multiplicity two and has only one linearly independent eigenvector. Find the general solution and show directly that the zero solution is unstable.

(c) When an eigenspace is degenerate, then one of the polynomials, $p(t)$ or $q(t)$, in (7) is at least of order one. Use this fact to prove that if an eigenvalue for the matrix **A** has zero real part and is degenerate, then the zero solution to $\mathbf{x}' = \mathbf{A}\mathbf{x}$ is unstable.

(d) When an eigenspace is nondegenerate, then the polynomials $p(t)$ and $q(t)$ in (7) are constant. Use this fact to prove that if all the eigenvalues of the matrix **A** have nonpositive real parts and those with zero real parts are nondegenerate, then the zero solution to $\mathbf{x}' = \mathbf{A}\mathbf{x}$ is stable.

In Problems 9 and 10, use the results of Problem 8 to determine the stability of the zero solution to the system $\mathbf{x}' = \mathbf{A}\mathbf{x}$ *for the given matrix* **A**.

9. $\begin{bmatrix} 0 & 1 & 1 & 0 \\ -1 & 0 & 0 & 1 \\ 0 & 0 & 0 & 1 \\ 0 & 0 & -1 & 0 \end{bmatrix}$

10. $\begin{bmatrix} 0 & 4 & 0 & 0 \\ -4 & 0 & 0 & 0 \\ 0 & 0 & 0 & 4 \\ 0 & 0 & -4 & 0 \end{bmatrix}$

In Problems 11–14, verify that the given system is almost linear and discuss the stability of the zero solution.

11. $\mathbf{x}' = \begin{bmatrix} 0 & 1 & 0 \\ 0 & 0 & 1 \\ -2 & -5 & -4 \end{bmatrix} \mathbf{x} + \begin{bmatrix} x_1^3 - x_2 x_3 \\ x_3^2 \\ -x_1 x_2 x_3 \end{bmatrix}$

12. $\mathbf{x}' = \begin{bmatrix} -1 & -1 & 0 \\ 2 & -1 & 1 \\ 0 & 1 & -1 \end{bmatrix} \mathbf{x} + (2 - \cos t) \begin{bmatrix} x_1^2 \\ x_2^2 \\ x_3^2 \end{bmatrix}$

13. $\dfrac{dx_1}{dt} = e^{-x_1} + \cos x_2 - 2 ,$

$\dfrac{dx_2}{dt} = -x_2 + \sin x_3 ,$

$\dfrac{dx_3}{dt} = 1 - e^{x_2 + x_3}$

14. $\dfrac{dx_1}{dt} = -x_1 + x_2 + e^{x_3} - 1 ,$

$\dfrac{dx_2}{dt} = \sin x_3 - x_2 ,$

$\dfrac{dx_3}{dt} = -2x_3$

[†]See the book by F. Brauer and J. A. Nohel, *The Qualitative Theory of Ordinary Differential Equations* (Dover Publications, New York, 1989), Chapter 4, for a proof.

In Problems 15–18, find all the critical points for the given system and discuss the stability of the corresponding equilibrium solutions.

15. $\dfrac{dx_1}{dt} = -x_1 + 1$,

$\dfrac{dx_2}{dt} = -2x_1 - x_2 + 2x_3 - 4$,

$\dfrac{dx_3}{dt} = -3x_1 - 2x_2 - x_3 + 1$

16. $\dfrac{dx_1}{dt} = x_2 + x_3 - 1$,

$\dfrac{dx_2}{dt} = x_1 + x_3 + 2$,

$\dfrac{dx_3}{dt} = x_1 + x_2 + 3$

17. $\dfrac{dx_1}{dt} = x_1\left(1 - x_1 - \dfrac{1}{2}x_2 - x_3\right)$,

$\dfrac{dx_2}{dt} = x_2\left(1 - x_1 - x_2 - \dfrac{1}{2}x_3\right)$,

$\dfrac{dx_3}{dt} = x_3\left(1 - \dfrac{1}{2}x_1 - x_2 - x_3\right)$

18. $\dfrac{dx_1}{dt} = x_3 - x_3^2$,

$\dfrac{dx_2}{dt} = -x_1$,

$\dfrac{dx_3}{dt} = x_2$

12.8 Neurons and the FitzHugh–Nagumo Equations

The central nervous system is an intricate biological network that, basically, carries electrical signals between the brain and the muscles. The network is comprised of *neuron* cells connected by electrical carriers called *axons*, which are comprised of "Schwann cell" segments similar to coaxial transmission lines, interrupted by *Ranvier nodes* wherein the voltage signal may be regenerated. See Figure 12.34.

The propagation through the Schwann cell is governed by the cable equation, studied in Project E, Chapter 10 (page 637). The voltage signal diffuses from one node to the next, like a meteorological warm front. In a healthy axon, the arriving voltage has sufficient strength to

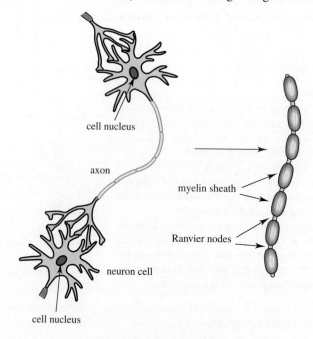

cell nucleus

axon

myelin sheath

Ranvier nodes

neuron cell

cell nucleus

Figure 12.34 Neuron structure

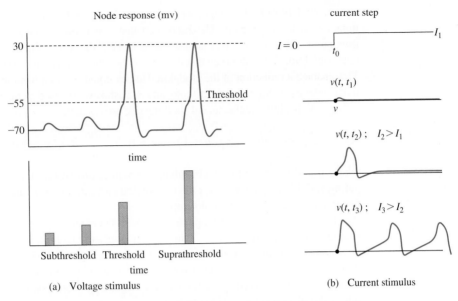

Figure 12.35 The action potential: Ranvier membrane responses

stimulate the Ranvier node to regenerate (and propagate) a new signal. Such regenerated signals are known as *action potentials*, and they always take the same form. If the arriving signal has deteriorated below a certain level, the Ranvier node will not regenerate the signal. This is often due to a depletion of the *myelin* insulation around the Schwann cell and is one of the symptoms of multiple sclerosis. The all-or-nothing "threshhold" phenomenon is depicted in Figure 12.35(a).

In 1939 Hodgkin, Huxley, and others were able to insert wire probes into the giant axon of the squid, which has a diameter of around 0.5 mm. (It is used by the squid to control its propulsion "jets.") This permitted extensive experimentation whereby the threshold behavior could be artificially stimulated by current pulses and steps. It was found that short current pulses induced node responses like those in Figure 12.35(a): feeble responses for subthreshold pulse intensities, and identical action-potential profiles for suprathreshhold intensities. Current *steps* induce the behavior in Figure 12.35(b): weak response, one single action potential, and a string of action potentials—all settling on elevated equilibrium voltage levels.

We have not encountered such threshold phenomena in any of the differential equations so far, and the compilation of a differential equation model is very difficult. In 1952 Hodgkin and Huxley[†] published an *RC* circuit model (Section 3.5) that reproduced the threshold behavior, and in 1963 they won a Nobel prize.

The analysis of the Hodgkins–Huxley circuit is quite complicated, because the values of the resistances are not constant. Briefly, they are expressed as functions of three probabilistic (nonphysical) "gating variables" that are found to describe the levels of sodium and potassium ions that have diffused through the wall of the Ranvier node. Each gating variable satisfies a differential equation of the form $du/dt = \alpha(1 - u) - \beta u$, which we studied in Problem 40, Exercises 2.3 (page 57), *but with coefficients α, β that depend on the node's voltage.*

The gating equations and the circuit equation comprise a system of four coupled nonlinear equations in four unknowns, and defy analytic solution. Nonetheless, numerical solutions demonstrate that they accurately trace the behavior of the Ranvier node. Still, scientists were frustrated by their inability to *visualize* the workings of the model.

[†]A. L. Hodgkin, A. F. Huxley, and B. Katz in "Measurement of Current-Voltage Relations in the Membrane of the Giant Axon of Loligo", *Journal of Physiology* (London), Vol. 116 (1952): 424–48.

Around 1960 R. FitzHugh, working at the National Institutes of Health, simulated the equations on analog computers. He discovered that one gating variable evolves quite rapidly to the asymptotic value $\alpha/(\alpha + \beta)$ that it would achieve if the Ranvier node voltage were held at its current level. Further, he observed that the sum of the other two gating variables stayed nearly constant. This enabled a reduction of the Hodgkin–Huxley equations to a *two*-dimensional system. Working independently, J. Nagumo, S. Arimoto, and S. Yoshizawa devised a circuit based on the same observations. With a little mathematical shuffling one arrives at the *FitzHugh–Nagumo* equations[†]

(1) $$\frac{dv}{dt} = -v(v-a)(v-1) - w + I(t), \quad \frac{dw}{dt} = \varepsilon(v - \gamma w),$$

where v and w are shifted and scaled approximations to the node voltage and the potassium gating variable respectively, $I(t)$ is the applied current, and α, ε, and γ are adjustable constants depending on the experimental setup.

If $I(t)$ is a either a current pulse or step, it is constant for $t > 0$, and the FitzHugh–Nagumo equations become autonomous. Thus we can apply the techniques of phase plane analysis.

First we consider current pulse excitations, as in Figure 12.35(a) on page 771. A current pulse into a capacitor simply induces a voltage across the plates instantaneously, and $I(t) = 0$ thereafter. Thus we are considering trajectories originating from points on the positive v-axis. They satisfy the system (1) with $I(t)$ omitted:

(2) $$\frac{dv}{dt} = -v(v-a)(v-1) - w, \quad \frac{dw}{dt} = \varepsilon(v - \gamma w).$$

A plot of the direction field for specific values of the parameters is displayed in Figure 12.36. Also shown are the *nullclines*—isoclines where $\dfrac{dv}{dt} = 0$ $\left(\text{a cubic, } w = -v(v-a)(v-1)\right)$ and $\dfrac{dw}{dt} = 0$ (a straight line, $w = v/\gamma$). These intersect in the critical point $(0, 0)$.

Figure 12.36 Phase plane and solutions v vs. t for equation (2): $I = 0$, $a = 0.3$, $\varepsilon = 0.01$, $\gamma = 3.0$

[†]R. FitzHugh, "Mathematical Models of Threshold Phenomena in the Nerve Membrane," *Bulletin of Mathematical Biophysics*, Vol. 17 (1955): 257–278; J. Nagumo, S. Arimoto, and S. Yoshizawa, "An active pulse transmission line simulating nerve axon," *Proc. IRE*, Vol. 50 (1962): 2061–2070.

The critical point appears to be a stable improper node; recall Figure 12.3, page 718. In fact for the data $a = 0.3$, $\varepsilon = 0.01$, $\gamma = 3.0$, the eigenvalues of the corresponding linear system turn out to be approximately -0.26 and -0.07. A few trajectories, starting from successive points to the right along the v-axis (corresponding to ever more intense current pulses) are displayed in the figure, together with the numerically computed graphs of $v(t)$.

We see that low-intensity current pulses (a, b) produce weak voltage "bumps" that return gently to the (stable) critical point, but that pulses driving the voltage beyond the rising portion of the v-nullcline (c, d) are ushered alongside that nullcline to its remote descending portion, resulting in much more expansive excursions before returning to the origin. Indeed, not only is the thresholding phenomena of Figure 12.35(a) explained, but also the negative overshoot of the action potential, and the approximate independence of its shape on the intensity of the suprathreshold excitation!

Now let's turn to the curious responses of the neuron to current steps, Figure 12.35(b) on page 771. For this situation the system (1) becomes

$$(3) \qquad \frac{dv}{dt} = -v(v-a)(v-1) - w + I_0 \,, \qquad \frac{dw}{dt} = \varepsilon(v - \gamma w) \,,$$

and we consider trajectories originating at the origin (which is no longer a critical point). Note that the mathematical effect of the constant current I_0 is to shift the v nullcline up the w-axis $w = -v(v-a)(v-1) + I_0$; the direction field plots in Figure 12.37 are shifted replicas of Figure 12.36. Typical phase plane trajectories launched from the origin appear in the figure.

a. $I_0 = 0.02$ b. $I_0 = 0.04$ c. $I_0 = 0.10$

Figure 12.37 Trajectories and solutions v vs. t stimulated by current steps: $a = 0.3$, $\varepsilon = 0.01$, $\gamma = 3.0$

The critical points in the first two graphs are stable spirals. For $I_0 = 0.02$, Figure 12.37(a) on 773 displays a spiral at approximately $(0.03, 0.01)$ with approximate Jacobian eigenvalues given by $-0.12 \pm 0.04i$. For $I_0 = 0.04$ we see the action potential in Figure 12.37(b). The spiral appears at $(0.07, 0.02)$ with Jacobian eigenvalues $-0.08 \pm 0.09i$.

For $I_0 = 0.10$ the critical point in Figure 12.37(c) is an unstable improper node at $(0.3, 0.1)$, with Jacobian eigenvalues 0.16 and 0.02. Indeed, the Poincaré–Bendixson theorem of Section 12.6 (page 760) can be applied to the direction field diagram; in the detailed Figure 12.38 the region R is trapped between an outer polygon and an inner triangle which allow no exit trajectories, and it does not contain the critical point; R meets the conditions of the theorem. Thus we identify the Ranvier node's train of action potentials in Figure 12.35(b) as a limit cycle. And note the importance of the FitzHugh–Nagumo reformulation of the Hodgkin–Huxley equations; there is no Poincaré–Bendixson theorem in four dimensions.

Figure 12.38 The region R for the Poincaré–Bendixson theorem

Some other interesting configurations of critical points are demonstrated in Figures 12.39 and 12.40. They will be analyzed in the exercises.

Figure 12.39 Phase plane for Problem 7: $a = 0.01$, $\varepsilon = 0.01$, $\gamma = 4.8$, $I_0 = 0$

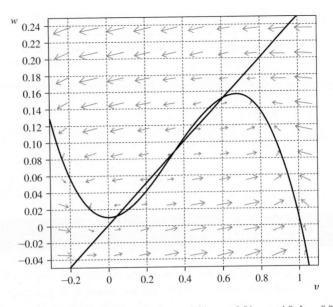

Figure 12.40 Phase plane for Problem 8: $a = 0.01$, $\varepsilon = 0.01$, $\gamma = 4.0$, $I_0 = 0.01$

12.8 EXERCISES

1. Calculate the Jacobian eigenvalues at the critical point $(0, 0)$ for equation (2) with the data of Figure 12.36. Confirm the values cited in the text, and verify that there are no other critical points.

2. Calculate the critical point and Jacobian eigenvalues for equation (3) with the data of Figure 12.37(a). Confirm the values cited in the text.

3. Calculate the critical point and Jacobian eigenvalues for equation (3) with the data of Figure 12.37(b). Confirm the values cited in the text.

4. Calculate the critical point and Jacobian eigenvalues for equation (3) with the data of Figure 12.37(c). Confirm the values cited in the text.

5. Experiment with software to determine the approximate threshold value of the initial voltage v for which the action potential is created in Figure 12.36.

6. Experiment with software to determine the approximate threshold values of the current step I_0 for which the single action potential, and the limit cycle, are created in Figures 12.37(b,c).

7. Calculate the critical points and eigenvalues for the phase plane shown in Figure 12.39. Identify which are stable and which are unstable. [*Hint*: The configuration is known as "bistable."] Inside each circle indicated in the figure, experiment with software to find points launching trajectories converging to different stable critical points.

8. Calculate the critical points and eigenvalues for the phase planes shown in Figure 12.40. Identify which are stable and which are unstable. Experiment with software to see if all trajectories converge to the same limit cycle.

9. (a) Calculate the Jacobian matrix at the critical points for the generic FitzHugh–Nagumo equations (1).

 (b) Show that the sum of the eigenvalues of this matrix is given by $w' - \varepsilon\gamma$, where $w' = dw/dv$ is the slope of the v-nullcline.

 (c) Show that the product of the eigenvalues is given by $\varepsilon(1 - \gamma w')$.

10. One way of characterizing the eigenvalues for the FitzHugh–Nagumo equations is displayed in the top row of Table 12.2 on page 776. Fill in the missing the entries of the table.

11. Using Table 12.2, argue that if a critical point lies on a descending portion of the v-nullcline, it is stable. Confirm this in Figures 12.36 and 12.37. If you worked Problem 7, confirm it in Figure 12.39.

12. Using Table 12.2, argue that if a critical point lies on the ascending portion of the v-nullcline, it is unstable if ε is sufficiently small. Confirm this in Figure 12.37. If you worked Problem 8, confirm it in Figure 12.40.

TABLE 12.2	Eigenvalues λ_1, λ_2 of the FitzHugh–Nagumo Jacobian				
	Both negative	**Both nonnegative**	**One negative, one nonnegative**	**Complex conjugate, real part negative**	**Complex conjugate, real part nonnegative**
$\lambda_1 + \lambda_2 = w' - \varepsilon\gamma$	Negative	?	Indeterminant	?	Nonnegative
$\lambda_1\lambda_2 = \varepsilon(1 - \gamma w')$	Positive	?	?	?	Nonnegative
Guaranteed stable?	Yes	?	?	?	No

Chapter 12 Summary

In this chapter we discussed the stability of critical points, limit cycles, and solutions for autonomous systems. While the discussion dealt primarily with systems in the plane, many of the results can be extended to higher-order systems. The techniques can loosely be grouped into two categories. First, there is an analytic method dating back to Poincaré that uses information about a related linear system to deduce information about the nonlinear system. Second, there is a geometric method mainly due to Lyapunov that uses a generalization of the energy of a physical system.

Linear Systems in the Plane

If the origin is an isolated critical point, then the stability of the origin and the classification of the origin as a critical point depends on the roots of the characteristic equation. The origin may be stable, asymptotically stable, or unstable and may be a node, saddle point, spiral point, or center. The results are summarized in Theorem 1, page 732, and again in Table 12.1, page 733.

Almost Linear Systems

If the origin is an isolated critical point for the corresponding linear system, then the stability and classification of the critical point at the origin for the linear and nonlinear systems are closely related, as shown in Table 12.1. In particular, the stability of the origin for the linear system and for the nonlinear system is the same, except when the origin of the associated linear system is a center. In this case, the nonlinear system may have either a center or spiral point at the origin and the stability must be decided by a different method.

Conservative and Nonconservative Systems

For second-order differential equations that arise from problems in mechanics, the **energy method** is very helpful in sketching the phase plane diagram for the equation. For a conservative system governed by an equation of the form

$$\frac{d^2x}{dt^2} + g(x) = 0,$$

the function $G(x) = \int g(x)\,dx$ is called a **potential function** and $E(x, v) = (1/2)v^2 + G(x)$ is an **energy function.** The level curves for the energy function contain the trajectories for the system. Moreover, the critical points for the system are located at the critical values of the potential function. Graphing the potential function above the phase plane diagram can be very helpful in sketching the phase plane diagram.

Lyapunov's Direct Method

The energy method generalizes to arbitrary systems. This generalization is called Lyapunov's direct method. If $V(x, y)$ is positive definite and

$$\dot{V}(x, y) = V_x(x, y)f(x, y) + V_y(x, y)g(x, y)$$

is the derivative of $V(x, y)$ with respect to the system

$$\frac{dx}{dt} = f(x, y), \qquad \frac{dy}{dt} = g(x, y),$$

then the origin is stable if $\dot{V}$ is negative semidefinite, asymptotically stable if $\dot{V}$ is negative definite, and unstable if $\dot{V}$ is positive definite. The function $V(x, y)$ is called a **Lyapunov function.**

Limit Cycles and Periodic Solutions

A **limit cycle** is a nontrivial closed trajectory with at least one trajectory spiraling into it. Such trajectories correspond to special types of nonconstant periodic solutions. The Bendixson negative criterion is a simple test to show that no limit cycles exist in a certain region of the plane, while the Poincaré–Bendixson theorem gives sufficient conditions for the existence of a nonconstant periodic solution.

Higher-Dimensional Systems

The concept of stability extends to higher-dimensional systems. For linear systems the stability of the zero solution (the origin) follows from properties of a fundamental matrix for the system. If the norm of the fundamental matrix $\|\mathbf{X}(t)\|$ is bounded, then the zero solution is stable. If $\|\mathbf{X}(t)\| \to 0$ as $t \to +\infty$, then the zero solution is asymptotically stable. When the linear system has constant coefficients, the stability is determined by the eigenvalues of the coefficient matrix $\mathbf{A}$. If all eigenvalues have negative real parts, then the zero solution is asymptotically stable. If at least one eigenvalue has positive real part, then the zero solution is unstable.

For almost linear systems, if the eigenvalues of the corresponding linear system all have negative real parts, then the zero solution of the almost linear system is asymptotically stable.

REVIEW PROBLEMS FOR CHAPTER 12

In Problems 1–6, find all the critical points for the given system, discuss the type and stability of each critical point, and sketch the phase plane diagrams near each of the critical points.

1. $\dfrac{dx}{dt} = -x - 4y,$

$\dfrac{dy}{dt} = -2x - 7y$

2. $\dfrac{dx}{dt} = -x + 3y,$

$\dfrac{dy}{dt} = 5x - 3y$

3. $\dfrac{dx}{dt} = x + 5y + xy,$

$\dfrac{dy}{dt} = -5x - 5y - xy$

4. $\dfrac{dx}{dt} = x + 13y,$

$\dfrac{dy}{dt} = -2x - y$

5. $\dfrac{dx}{dt} = x - 3x^2 - xy,$

$\dfrac{dy}{dt} = 2y - y^2 - 4xy$

6. $\dfrac{dx}{dt} = 3x + 2y - y^3,$

$\dfrac{dy}{dt} = x + 2y + y^3$

In Problems 7 and 8, use the potential plane to help sketch the phase plane diagrams for the given equations.

7. $\dfrac{d^2x}{dt^2} + x^3 - 2x^2 - x + 2 = 0$

8. $\dfrac{d^2x}{dt^2} + x^3 - x^2 = 0$

In Problems 9–12, use Lyapunov's direct method to determine the stability for the origin of the given system.

9. $\dfrac{dx}{dt} = -x^3 - xy^2 + xy^5,$

$\dfrac{dy}{dt} = -y^3 + x^2y - 3x^2y^4$

[*Hint:* Try $V(x, y) = 3x^2 + y^2.$]

10. $\dfrac{dx}{dt} = x^2 y + y^3$,

$\dfrac{dy}{dt} = -xy^2 + x^3$

[*Hint:* Try $V(x, y) = xy$.]

11. $\dfrac{dx}{dt} = x^3 + xy^2$,

$\dfrac{dy}{dt} = -y^3 - x^2 y$

[*Hint:* Try $V(x, y) = x^2 - y^2$.]

12. $\dfrac{dx}{dt} = -x^3 + y^3$,

$\dfrac{dy}{dt} = -y^5$

[*Hint:* Try $V(x, y) = x^4 + y^2$.]

In Problems 13 and 14, sketch the phase plane diagrams for the given system. Discuss the stability of each limit cycle and critical point. Here $x = r\cos\theta$ and $y = r\sin\theta$.

13. $\dfrac{dx}{dt} = -y + x(r^3 - 5r^2 + 8r - 4)$,

$\dfrac{dy}{dt} = x + y(r^3 - 5r^2 + 8r - 4)$

14. $\dfrac{dx}{dt} = x - y + x(r^3 - 9r^2 + 26r - 25)$,

$\dfrac{dy}{dt} = x + y + y(r^3 - 9r^2 + 26r - 25)$

In Problems 15 and 16, determine whether the given system has a nonconstant periodic solution in the entire plane.

15. $\dfrac{dx}{dt} = 2x - 8x^3 - 3xy^2$,

$\dfrac{dy}{dt} = \dfrac{1}{2}y - \dfrac{1}{2}y^3 - x^2 y$

[*Hint:* Use the Lyapunov function $V(x, y) = x^2 + y^2$.]

16. $\dfrac{dx}{dt} = x - x^3 + 2x^3 y$,

$\dfrac{dy}{dt} = -2y - y^3 - 3x^2 y^2$

In Problems 17 and 18, determine the stability of the zero solution to the given system.

17. $\mathbf{x}' = \begin{bmatrix} -1 & 1 & 0 \\ 0 & -1 & 0 \\ 0 & 0 & -2 \end{bmatrix} \mathbf{x} + \begin{bmatrix} x_1^2 x_2 \\ x_2^3 \\ x_3^3 \end{bmatrix}$

18. $\mathbf{x}' = \begin{bmatrix} -3 & 0 & 0 \\ 0 & -1 & 2 \\ 0 & -2 & -1 \end{bmatrix} \mathbf{x} + \begin{bmatrix} -x_1 x_2^2 \\ x_1 x_3 \\ -x_2 x_3 \end{bmatrix}$

TECHNICAL WRITING EXERCISES FOR CHAPTER 12

1. Discuss the relationship between the stability of the almost linear system and the stability of the corresponding linear system.

2. What are the advantages and disadvantages of the two general approaches (analytic vs. geometric) in determining the stability of a critical point?

3. Periodic solutions of autonomous systems in the plane and limit cycles are closely identified. Describe this relationship and discuss the existence and stability of limit cycles.

4. Much of the chapter deals with almost linear systems in the plane. Discuss the extension of the results to almost linear systems in higher dimensions.

Projects for Chapter 12

A | Solitons and Korteweg–de Vries Equation

The Korteweg–de Vries (KdV) equation

$$\frac{\partial u}{\partial t} + \frac{\partial^3 u}{\partial x^3} + \frac{1}{2}\frac{\partial}{\partial x}(u^2) = 0,$$

where $u(x, t)$ is the vertical displacement at position x and time t, is used to model long water waves. Of particular interest are **traveling wave** solutions, which are solutions of the form $u(x, t) = v(x - ct)$, where c is the constant wave speed.

 (a) Show that $v(x - ct)$ is a solution to the KdV equation provided $v(z)$, $z = x - ct$, satisfies

$$-c\frac{dv}{dz} + \frac{d^3 v}{dz^3} + \frac{1}{2}\frac{d}{dz}(v^2) = 0.$$

 (b) Of the traveling waves, we are most interested in **solitons,** traveling waves that are positive and decay to zero at $\pm\infty$. Therefore, integrate once and set the constant of integration equal to zero to obtain

$$\frac{d^2 v}{dz^2} + \frac{1}{2}v^2 - cv = 0.$$

 (c) Sketch the phase plane diagram for this conservative system.
 (d) Use the results of parts (a) and (b) to argue that there exist solitons for the KdV equation. [*Hint:* The existence of a positive function $v(z)$ that decays to zero as $z \to \pm\infty$ corresponds to a trajectory in the vv'-plane that begins at the origin and returns to the origin.]
 (e) Does increasing the speed of the soliton produce a shorter or taller wave?

B | Burger's Equation

A simple model for diffusive waves in fluid mechanics that incorporates nonlinearity and diffusion is **Burger's equation:**

$$\frac{\partial u}{\partial t} + u\frac{\partial u}{\partial x} = c\frac{\partial^2 u}{\partial x^2}.$$

The existence of shock waves is associated with the existence of traveling waves $u(x, t) = v(x - ct)$.

 (a) Using an approach similar to that used in Project A, find a second-order equation for $v(z)$, $z = x - ct$, that corresponds to a traveling wave solution to Burger's equation.
 (b) Sketch the phase plane diagram for the equation obtained in part (a).
 (c) Describe the type of traveling waves that exist for Burger's equation.

C Computing Phase Plane Diagrams

In this chapter our emphasis was on analytic techniques for determining the stability of a critical point or a limit cycle. The phase plane diagrams that we constructed, although somewhat crude, provided us with a geometric representation for the behavior of the trajectories. Naturally, having a detailed phase plane diagram can improve our understanding of the behavior of the trajectories.

Using a numerical method such as Runge–Kutta for systems discussed in Section 5.3 (or a computer software package) and a little experimentation—*keeping in mind the results of the analytical methods*—one can obtain a detailed phase plane diagram. In addition to knowing the type and location of each critical point and where limit cycles might be located, we offer a few practical suggestions.

(i) In computing trajectories near an unstable critical point, replace the step size h by $-h$. This has the effect of reversing "time" and makes unstable nodes and spiral points into stable nodes and spiral points. It can also transform an unstable limit cycle into a stable limit cycle.

(ii) In general there is no way of finding an initial point lying on a limit cycle. Hence, it is customary to locate limit cycles by following two trajectories—one approaching from the outside and one from the inside—until they appear to converge together and jointly map out the limit cycle. (The limit cycle lies between these two trajectories.) If the limit cycle is not stable, then the trajectories are computed using $-h$.

(a) Using a numerical method such as Runge–Kutta for systems or a computer software package, generate a detailed phase plane diagram for the system

$$\frac{dx}{dt} = x(7 - x - 2y) \,,$$
$$\frac{dy}{dt} = y(5 - y - x) \,.$$

Compare with Figure 12.20 on page 737.

(b) Generate a detailed phase plane diagram for the equation

$$\frac{d^2x}{dt^2} + x - x^3 = 0 \,.$$

Compare with Figure 12.22 on page 744.

(c) Generate a detailed phase plane diagram for the van der Pol equation

$$\frac{d^2x}{dt^2} + 5(x^2 - 1)\frac{dx}{dt} + x = 0 \,.$$

Compare with Figure 12.33 on page 762.

D Ecosystem on Planet GLIA-2

Courtesy of Alfred Clark, Jr., University of Rochester

You are the chief mathematical biologist on a scientific expedition to the starsystem Glia. For the last week, your spaceship has been orbiting the small planet Glia-2, gathering data on what appears to be a unique biological system. Although most of the planet is barren, a small portion

contains a thriving ecosystem consisting of just three species. There is a flowering plant, Jancor; a small mouselike mammal, Murrot, which feeds on Jancor; and a ferretlike predator, Vekton, which feeds on Murrot. After studying the data brought back by landing parties, you have concluded that this system can be modeled by the following equations for the Murrot population M and the Vekton population V, both in units of millions:

$$\frac{dM}{dt} = rM\left(1 - \frac{M}{A}\right) - \frac{\beta MV}{M + B}, \qquad \frac{dV}{dt} = \frac{bVM}{M + B} - cV.$$

Thus, the Murrots obey a logistic law with a net birthrate r and a maximum sustainable population A. They suffer a loss to the Vektons, represented by the term $-\beta MV/(M + B)$. The Vektons have a natural death rate c—hence the loss term cV—and a birthrate $bVM/(M + B)$, which is strongly dependent on the Murrot population. From the landing party observations, the following parameter values are established:

$r = 12 \ \text{yr}^{-1}$

$A = 14 \ \text{(units of 1 million)}$

$\beta = 20 \ \text{yr}^{-1}$

$B = 4 \ \text{(units of 1 million)}$

$b = 4 \ \text{yr}^{-1}$

$c = 1.6 \ \text{yr}^{-1}$

In the following calculations, you will need to make use of a software package in order to integrate the equations.

(a) Determine whether the population is in stable equilibrium or in a limit cycle. You may assume that the system has been going for some time with the above parameter values.

(b) Consider the possibility of intervening. If the population of either species drops below 1 million, either in equilibrium or during a limit cycle, the ecosystem will be susceptible to destruction by random natural events, such as extreme weather. Show that the system with the above parameter values is, by this criterion, in danger. Hence, intervention—even though contrary to general Federation guidelines—is warranted. The only intervention tool you have is the ability to clear out some of the Jancor, thereby reducing the maximum sustainable population A. Assume you have the power to make A take on any value ≤ 14. Is it actually feasible to increase the minimum population by restricting the Murrot food supply (that is, reducing A)?

E Spread of Staph Infections in Hospitals—Part II

Courtesy of Joanna Wares, University of Richmond, and Glenn Webb, Vanderbilt University

In Project B in Chapter 5, page 308, we used a vector field to study the transmission dynamics of two strains of methicillin-resistant *Staphylococcus aureus* (MRSA) in the hospital setting. Our goal was to predict whether or not CA-MRSA, the newer and deadlier strain, will overtake HA-MRSA. In that project, we used particular data from one hospital. But for the model to be

useful, we should be able to say something about the results in terms of the parameters, which differ for different hospitals. Here we study the full system in terms of parameters.

(1) $$\frac{dS}{dt} = \underbrace{\Lambda}_{\text{entrance rate}} - \underbrace{\frac{\beta_H S(t) H(t)}{N}}_{\text{acquire HA-MRSA}} - \underbrace{\frac{\beta_C S(t) C(t)}{N}}_{\text{acquire CA-MRSA}}$$

$$+ \underbrace{\alpha_H H(t)}_{\text{HA-MRSA decolonized}} + \underbrace{\alpha_C C(t)}_{\text{CA-MRSA decolonized}} - \underbrace{\delta_S S(t)}_{\text{exit hospital}}$$

(2) $$\frac{dH}{dt} = \underbrace{\frac{\beta_H S(t) H(t)}{N}}_{\text{from S}} - \underbrace{\alpha_H H(t)}_{\text{decolonized}} - \underbrace{\delta_H H(t)}_{\text{exit hospital}}$$

(3) $$\frac{dC}{dt} = \underbrace{\frac{\beta_C S(t) C(t)}{N}}_{\text{from S}} - \underbrace{\alpha_C C(t)}_{\text{decolonized}} - \underbrace{\delta_C C(t)}_{\text{exit hospital}} .$$

This model classifies all patients in the hospital into three groups:

- $H(t) = $ patients colonized with the traditional hospital strain, HA-MRSA.
- $C(t) = $ patients colonized with the community strain, CA-MRSA.
- $S(t) = $ susceptible patients, those not colonized with either strain.

The *parameters* of the model are defined in Project B, Chapter 5.
Let

$$R_C := \frac{\Lambda \beta_C}{\delta_S N (\alpha_C + \delta_C)} \quad \text{and} \quad R_H := \frac{\Lambda \beta_H}{\delta_S N (\alpha_H + \delta_H)} .$$

(a) Find all equilibria (critical points) of the system in terms of the parameters.
(b) Under what conditions on R_C and R_H does each equilibrium in part (a) exist and under what conditions is each stable?
(c) Under what conditions on R_C and R_H will MRSA be completely eliminated from the hospital over time?
(d) It is believed that the most common mode of transmission between patients is through health care workers. Contaminated health care workers transit from one patient to the next and possibly transmit the bacteria from a patient who is colonized with MRSA to one who is not. If health care workers properly follow hand-hygiene measures, the problem is greatly reduced. However, because these measures take time and must be done before and after every patient visit, health care workers sometimes do not fully comply with these practices.

 Health care worker compliance with hand-hygiene measures can be incorporated into this model by multiplying each of the transition rates by $(1 - \eta)$, where $0 < \eta < 1$ and 100η represents the percentage of hand-hygiene compliance. How will this inclusion of hand-hygiene affect the stability conditions of the equilibria in part (a)?
(e) Under what conditions on R_C and R_H would you expect CA-MRSA to overtake HA-MRSA in the hospital? Why?

CHAPTER

13

Existence and Uniqueness Theory

13.1 Introduction: Successive Approximations

In nearly every chapter of this book, the theory, application, or procedure discussed depended on a fundamental theorem concerning the existence and uniqueness of the solution to an initial value problem. In this chapter, we provide the proofs of these basic theorems.

How does one know that a solution exists? When possible, the simplest way is actually to exhibit the solution in an explicit form. While we were able to do this for the entire class of linear first-order differential equations (see Section 2.3) and also for homogeneous equations (or homogeneous normal systems) that have constant coefficients (see Chapters 4, 6, 9), most differential equations do not fall into these categories.

Another approach to proving the existence of a solution is to construct a sequence of **approximate solutions** to the initial value problem and then show that these approximations converge to the desired solution. This is the strategy we adopt in this chapter.

Some methods for constructing approximations are already familiar to us. For example, we have discussed numerical procedures such as Euler's method (see Sections 1.4 and 3.6) and analytic techniques such as Taylor's method for generating the partial sums of a power series solution (see Section 8.1). For the purpose of handling a very general class of differential equations, our procedure will be to construct a sequence of **successive approximations (iterations)** and show that these converge to the desired solution.

To describe the procedure, we begin by expressing the initial value problem

$$(1) \qquad y'(x) = f\big(x, y(x)\big) , \qquad y(x_0) = y_0$$

as an **integral equation.** This is accomplished by integrating both sides of equation (1) from x_0 to x:

$$(2) \qquad \int_{x_0}^{x} y'(t)\, dt = y(x) - y(x_0) = \int_{x_0}^{x} f\big(t, y(t)\big)\, dt .$$

Setting $y(x_0) = y_0$ and solving for $y(x)$, we have

$$(3) \qquad y(x) = y_0 + \int_{x_0}^{x} f\big(t, y(t)\big)\, dt .$$

Our derivation shows that a solution to (1) must also be a solution to (3). Conversely, using the fundamental theorem of calculus, we can show that if f and y are continuous, then a solution to (3) satisfies the differential equation in (1) (see Problem 17). Furthermore, such a solution must satisfy the initial condition $y(x_0) = y_0$, as can be seen by substituting $x = x_0$ in (3). Therefore, with these continuity assumptions, the initial value problem (1) and the integral equation (3) are equivalent in the sense that they have the same solutions.

We can view the right-hand side of (3) as defining an operator that maps a function y to a function $T[y]$, where

$$(4) \qquad T[y](x) := y_0 + \int_{x_0}^{x} f\big(t, y(t)\big)\, dt \ .$$

(In general, T is a nonlinear operator.) With this notation, we can express equation (3) in operator form as

$$(5) \qquad y = T[y] \ .$$

In (5), equality is meant in the sense that the functions y and $T[y]$ are equal on some interval I. A solution to (5) is called a **fixed point** of T because for equation (5) to be satisfied, T must map y to itself, and y therefore remains fixed. Of course, the term "point" refers to the *function* $y(t)$, in this context.

To gain some insight into the procedure for determining approximate solutions to (5), let's consider the simpler but analogous problem of finding the fixed points of a function $g(x)$; that is, we seek the roots of the equation

$$(6) \qquad x = g(x) \ .$$

Geometrically, solving equation (6) is equivalent to finding the points where the graph of $y = g(x)$ intersects the line $y = x$. If we consider the particular graphs in Figure 13.1, then we can devise the following geometric procedure that leads to the point of intersection P. Starting with some point P_0 on the curve $y = g(x)$, we follow a horizontal line until we reach a point Q_0 on the line $y = x$. From Q_0, we follow a vertical line down to a point P_1 on the graph of $y = g(x)$. Repeating the procedure, we follow a horizontal line to the point Q_1 on $y = x$, then a vertical line to the point P_2. Continuing in this staircase fashion, we draw closer and closer to the point of intersection P.

To describe this procedure algebraically, let (x_0, y_0) be the coordinates of the point P_0, (x_1, y_1) be the coordinates of the point P_1, and so on. Each point P_n lies on the graph of the curve $y = g(x)$, so we have $y_n = g(x_n)$. Recall that the points P_0 and Q_0 lie on the same horizontal line; thus the coordinates of Q_0 are (y_0, y_0). Also the points P_1 and Q_0 are on the same

Figure 13.1 Method of successive substitutions

vertical line, so $x_1 = y_0$. Since $y_0 = g(x_0)$, we conclude that $x_1 = g(x_0)$. By the same reasoning, we find that $x_2 = g(x_1)$, $x_3 = g(x_2)$, and so forth. In general, we have

$$(7) \qquad x_{n+1} = g(x_n), \qquad n = 0, 1, 2, \dots .$$

To summarize, we start with an initial approximation (a guess) x_0, substitute it into (7), and obtain a new approximation x_1. Continuing this iteration, we obtain a sequence of approximations $\{x_n\}$ that we hope will converge to a solution of (6). This procedure is called the method of **successive substitutions.**

Example 1 Use the method of successive substitutions to approximate a solution to the equation

$$(8) \qquad x = \frac{x}{2} + \frac{1}{x} .$$

Solution The graph of the function $g(x) = x/2 + 1/x$ is shown in Figure 13.1. Using formula (7), the recurrence relation is

$$(9) \qquad x_{n+1} = \frac{x_n}{2} + \frac{1}{x_n} .$$

With $x_0 = 1$ as an initial approximation, we find

$$x_1 = \frac{1}{2} + \frac{1}{1} = 1.5 , \qquad x_2 = \frac{1.5}{2} + \frac{1}{1.5} = 1.4166667 ,$$

and so on. In Table 13.1 we give x_n for $n = 0, 1, \dots, 5$ as computed on a pocket calculator. We stopped iterating after x_5 because x_4 and x_5 agree to seven decimal places, which is the accuracy of our calculator.[†] Hence,

$$x_4 \approx x_5 = \frac{x_4}{2} + \frac{1}{x_4} ,$$

and 1.4142136 is the desired approximation for a solution to (8). ◆

TABLE 13.1	Approximations for a Solution of $x = x/2 + 1/x$	
$x_0 = 1$		$x_3 = 1.4142157$
$x_1 = 1.5$		$x_4 = 1.4142136$
$x_2 = 1.4166667$		$x_5 = 1.4142136$

If the value of x_5 seems familiar, that is because x_5 agrees with the number $\sqrt{2}$ to eight digits. Indeed, one can verify that equation (8) is equivalent to the equation $x^2 = 2$.

While the method of successive substitutions worked quite well (converged) for the problem of Example 1, there are situations in which it will not converge to a fixed point (see Problem 19). The following theorem[‡] gives a set of conditions that ensure convergence.

[†]Stopping procedures for numerical approximations are discussed in Section 3.6.

[‡]*Historical Footnote:* This theorem is a special case of a theorem proved by Stephen Banach in 1922.

Banach Fixed Point Theorem

Theorem 1. Suppose g is a differentiable function on the closed interval I and $g(x)$ lies in I for all x in I. If there exists a constant K such that $|g'(x)| \le K < 1$ for all x in I, then the equation $x = g(x)$ has a unique solution in I. Moreover, the sequence of **successive approximations** defined by

$$(10) \qquad x_{n+1} = g(x_n), \qquad n = 0, 1, 2, \ldots,$$

converges to this solution for any choice of the starting value x_0 in I.

Proof. While it is not difficult to prove directly that g has a fixed point in I, we will establish this fact by analyzing the sequence of successive approximations. We divide the argument into four parts.

The Approximations Are Well Defined. Choose an x_0 in I. Since x_0 lies in the domain of g, then $x_1 = g(x_0)$ is well defined. Moreover, x_1 is a point of I, so $x_2 = g(x_1)$ is well defined. By induction, we see that the entire sequence $\{x_n\}$ is well defined.

The Approximations Converge to an Element of I. To establish convergence, observe that we can write

$$(11) \qquad x_n = x_0 + (x_1 - x_0) + (x_2 - x_1) + \cdots + (x_n - x_{n-1}).$$

That is, x_n is the $(n+1)$th partial sum of the infinite series

$$(12) \qquad x_0 + \sum_{j=0}^{\infty} (x_{j+1} - x_j).$$

Therefore, to prove that the sequence $\{x_n\}$ converges, we just have to show that the series in (12) converges. We can prove this using the comparison test, but first we must obtain an estimate on the absolute value of each term in the series.

Let x and y be two numbers in I. By the mean value theorem, there exists some number z between x and y such that

$$g(x) - g(y) = g'(z)(x - y).$$

Using the assumption that $|g'(z)| \le K$ for all z in I, we have

$$(13) \qquad |g(x) - g(y)| \le K|x - y|, \qquad (K < 1)$$

for all x, y in I. In particular, from formula (10), we have

$$|x_2 - x_1| = |g(x_1) - g(x_0)| \le K|x_1 - x_0|.$$

In a similar manner, we find

$$|x_3 - x_2| \le K|x_2 - x_1| \le K^2|x_1 - x_0|$$

and, in general,

$$(14) \qquad |x_{j+1} - x_j| \le K^j|x_1 - x_0|, \qquad j = 0, 1, 2, \ldots.$$

Since $K < 1$, the geometric series

$$\sum_{j=0}^{\infty} K^j |x_1 - x_0| = |x_1 - x_0| \sum_{j=0}^{\infty} K^j$$

converges. Thus, by the comparison test, the series in (12) converges absolutely, which means that $\{x_n\}$ converges. Let $x_\infty := \lim x_n$. Since $\{x_n\}$ lies in I and I is a closed interval, then x_∞ is in I.

x_∞ *Is a Fixed Point.* To see that x_∞ is a solution of the equation $x = g(x)$, we take the limit as $n \to \infty$ of both sides of the equation $x_{n+1} = g(x_n)$. Since g is continuous, we have

$$x_\infty = \lim_{n \to \infty} x_{n+1} = \lim_{n \to \infty} g(x_n) = g\left(\lim_{n \to \infty} x_n\right) = g(x_\infty) .$$

Hence x_∞ is a fixed point of g.

Uniqueness. To show that x_∞ is the unique fixed point in I, assume z is some other fixed point in I. Then, by inequality (13),

$$|x_\infty - z| = |g(x_\infty) - g(z)| \le K|x_\infty - z| .$$

Simplifying, we find $(1 - K)|x_\infty - z| \le 0$, and since $0 \le K < 1$, we must have $x_\infty = z$. Therefore there is only one fixed point of g in I, and it is the limit of the successive approximations defined in (10) for any starting value x_0 in I. ◆

We leave it as an exercise to verify that the function $g(x)$ in Example 1 satisfies the hypotheses of Theorem 1 on the interval $[1, 2]$ (see Problem 20).

Several times in the proof of Theorem 1 we used inequality (13). Functions that satisfy such an inequality (with $K < 1$) for all x, y in an interval I are called **contractions.** Contractions map two points x, y into two points $g(x), g(y)$ that are closer together than were the original points. If we view the interval I as pliable, then we can visualize the Banach fixed point theorem as stating that if g contracts the interval I into a subset of itself, then some point remains fixed. Not surprisingly, Theorem 1 is also called the **contraction mapping principle.**

In the next section, we use another form of the Banach fixed point theorem to prove the existence of a unique solution to the initial value problem (1). For now, let's apply the method of successive substitutions to the integral equation (3). That is, we define the sequence of functions $\{y_n\}_{n=0}^{\infty}$ by

$$(15) \qquad y_{n+1}(x) := T[y_n](x) = y_0 + \int_{x_0}^{x} f\big(t, y_n(t)\big) dt ,$$

where we must specify the starting function $y_0(x)$. The simplest choice for $y_0(x)$ is to let $y_0(x) \equiv y_0$. The successive approximations y_n of the solution to the integral equation (3) are called the **Picard iterations.**[†]

[†]*Historical Footnote:* The method can be traced back to J. Liouville, who used it in 1838 in his study of linear second-order equations. It was also used by J. Caque in 1864, L. Fuchs in 1870, and G. Peano in 1888. But it was E. Picard whose name became associated with the method because of his work on successive approximations in 1890 and 1896.

Example 2 Compute the Picard iterations for the initial value problem

(16) $y'(x) = 2y(x), \qquad y(0) = 1,$

starting with $y_0(x) \equiv 1$.

Solution The initial value problem (16) is equivalent to the integral equation

(17) $y(x) = 1 + \displaystyle\int_0^x 2y(t)\,dt$.

Computing the Picard iterations via (15), we find

$$y_1(x) = 1 + \int_0^x 2y_0(t)\,dt = 1 + \int_0^x 2\,dt = 1 + 2x,$$

$$y_2(x) = 1 + \int_0^x 2(1 + 2t)\,dt = 1 + 2x + \frac{(2x)^2}{2!},$$

$$y_3(x) = 1 + \int_0^x 2\left[1 + 2t + \frac{(2t)^2}{2!}\right]dt = 1 + 2x + \frac{(2x)^2}{2!} + \frac{(2x)^3}{3!},$$

and so forth. The pattern seems clear: The nth iteration is given by

(18) $y_n(x) = 1 + 2x + \dfrac{(2x)^2}{2!} + \cdots + \dfrac{(2x)^n}{n!}$.

Here we recognize $y_n(x)$ as the partial sum of the Maclaurin series for e^{2x}:

$$e^{2x} = 1 + 2x + \frac{(2x)^2}{2!} + \cdots + \frac{(2x)^n}{n!} + \cdots.$$

Hence, we see that the sequence of Picard iterations converges to e^{2x}, which is the solution to the initial value problem (16). ◆

In Example 2, we were very fortunate to be able to decipher the general form of the sequence of Picard iterations. This enabled us to determine that the sequence converged to the solution to the initial value problem. In most cases, it is impossible to determine a simple form for the Picard iterations. As we show in the next section, however, there are conditions on f that guarantee that the sequence of iterations converges to a solution.

13.1 EXERCISES

In Problems 1–4, express the given initial value problem as an integral equation.

1. $y'(x) = x^2 - y(x), \qquad y(1) = 4$

2. $y'(x) = \sin(x + y(x)), \qquad y(\pi) = 0$

3. $y'(x) = [y(x) - x]^2, \qquad y(1) = -3$

4. $y'(x) = e^{y(x)}, \qquad y(0) = 1$

In Problems 5–10, use the method of successive substitutions to approximate a solution of the given equation, starting with the given value for x_0.

5. $x = \dfrac{1}{2}\left(x + \dfrac{3}{x}\right), \qquad x_0 = 3$

6. $x = \dfrac{3}{8 - x^3}, \qquad x_0 = 0$

7. $x = \dfrac{1}{x^2 + 4}$, $\quad x_0 = 0.5$

8. $x = \dfrac{1}{2}\left(x + \dfrac{5}{x}\right)$, $\quad x_0 = 2$

9. $x = \sqrt[4]{\dfrac{5 - x}{3}}$, $\quad x_0 = 1$

10. $x = 1 + \sin x$, $\quad x_0 = 2$

In Problems 11–16, compute the Picard iterations $y_1(x)$ *and* $y_2(x)$ *for the given initial value problem, starting with* $y_0(x) \equiv y_0$, *where* y_0 *is the initial value for y.*

11. $y'(x) = -y(x)$, $\quad y(0) = 2$

12. $y'(x) = y^2(x)$, $\quad y(0) = 1$

13. $y'(x) = 3x^2$, $\quad y(1) = 2$

14. $y'(x) = \cos x$, $\quad y(\pi) = 0$

15. $y'(x) = y(x) - e^x$, $\quad y(0) = 0$

16. $y'(x) = x - y(x)$, $\quad y(1) = 1$

17. Assuming f and y in equation (3) are continuous, use the fundamental theorem of calculus to show

that a solution to (3) is a solution to the initial value problem (1).

18. Show that equation (8) is equivalent to $x^2 - 2 = 0$ in the sense that a solution to one is a solution to the other.

19. Show that the method of successive substitutions for the equation $x = (x^2 + 1)/2$ converges to the solution $x = 1$ when $x_0 = 0$ but diverges when x_0 is chosen to be 2 or even 1.1. [*Hint:* Sketch the curve $y = (x^2 + 1)/2$.]

20. To verify that $g(x) = x/2 + 1/x$ satisfies the hypotheses of Theorem 1 on the interval $[1, 2]$, proceed as follows:

(a) Show that for x in $[1, 2]$,
$$-\frac{1}{2} \leq g'(x) \leq \frac{1}{4},$$
so $|g'(x)| \leq 1/2$.

(b) Show that for x in $[1, 2]$,
$$\sqrt{2} \leq g(x) \leq \frac{3}{2},$$
so $g(x)$ lies in $[1, 2]$.

13.2 Picard's Existence and Uniqueness Theorem

In this section we show that, under mild assumptions, the successive approximations obtained using Picard's method converge to the solution to the initial value problem.

What do we mean when we say a sequence of functions $\{y_n\}$ converges to a function y on an interval $[a, b]$? Intuitively, we want the functions y_n to get close to the function y as n approaches ∞. We therefore need to define what is meant by *close*.

Let $C[a, b]$ denote the set of all functions that are continuous on the interval $[a, b]$. For y in $C[a, b]$, we define the **norm** of y to be

$$(1) \qquad \|y\| := \max_{x \in [a,b]} |y(x)|.$$

When we wish to emphasize the interval I over which the norm is taken, we write $\|y\|_I$. The norm of y can be thought of as the distance from the function y to the identically zero function. This is like saying that $|x|$ is the distance from x to the origin. Analogous to the formula for the distance between two real numbers, we define the **distance between two functions** y and z in $C[a, b]$ to be $\|y - z\|$.

Using the properties of the absolute value, it can be shown that the norm satisfies the following three properties:

$$(2) \qquad \|y - z\| = 0 \quad \text{if and only if } y(x) = z(x) \text{ for all } x \text{ in } [a, b].$$

$$(3) \qquad \|cy\| = |c|\,\|y\|, \quad \text{where } c \text{ is any real number.}$$

$$(4) \qquad \|y + z\| \leq \|y\| + \|z\| \quad \text{(triangle inequality)}.$$

Here we have assumed that y and z are elements of $C[a, b]$.

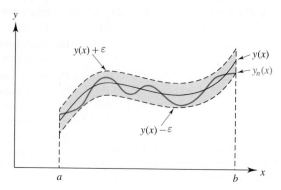

Figure 13.2 An ε-corridor about $y(x)$

With this concept of distance, we say a sequence of functions $\{y_n\}$ **converges uniformly** to y, denoted $y_n \xrightarrow{\text{unif}} y$, if

(5) $$\lim_{n \to \infty} \|y_n - y\| = 0 .$$

That is, the distance between y_n and y goes to zero as n approaches ∞. Geometrically we can think of uniform convergence as requiring that the sequence $\{y_n\}$ eventually gets inside an arbitrarily small ε-**corridor** about $y(x)$ (see Figure 13.2).

Example 1 Show that the sequence of functions $y_n(x) = x^2 + x/n$, $n = 1, 2, 3, \ldots$, converges uniformly on $[0, 1]$ to the function $y(x) = x^2$.

Solution We first compute

$$
\begin{aligned}
\|y_n - y\| &= \max_{x \in [0, 1]} |y_n(x) - y(x)| \\
&= \max_{x \in [0,1]} \left| x^2 + \frac{x}{n} - x^2 \right| \\
&= \max_{x \in [0,1]} \left| \frac{x}{n} \right| = \frac{1}{n} .
\end{aligned}
$$

Thus

$$\lim_{n \to \infty} \|y_n - y\| = \lim_{n \to \infty} \frac{1}{n} = 0 ,$$

and so y_n converges uniformly on $[0, 1]$ to y. ◆

You are probably familiar with a related type of convergence for functions that is called pointwise convergence. We say that a sequence of functions $\{y_n\}$ **converges pointwise** on $[a, b]$ to the function y if for every x in $[a, b]$,

(6) $$\lim_{n \to \infty} y_n(x) = y(x) .$$

Uniform convergence is stronger than pointwise convergence because if $\|y_n - y\| \to 0$, then certainly

$$\lim_{n \to \infty} [y_n(x) - y(x)] = 0$$

for every x in $[a, b]$; hence, y_n converges pointwise to y. However, pointwise convergence does *not* imply uniform convergence (see Problem 7).

In the previous section, we proved that under suitable conditions, the sequence of approximations obtained by the method of successive substitutions converged to the fixed point of the function. This was a consequence of the Banach fixed point theorem for functions defined on a closed interval I. To prove that the Picard iterations converge, we need an extension of this theorem that applies to *operators* defined on a closed subset of functions in $C[a, b]$.

Banach Fixed Point Theorem for Operators

Theorem 2. Let S denote the set of continuous functions on $[a, b]$ that lie within a distance α of a given function y_0; that is,

$$S = \{y \in C[a, b]: \|y - y_0\| \leq \alpha\} .$$

Suppose G is an operator mapping S into S and that G is a **contraction** on S in the sense that there exists a constant K, $0 \leq K < 1$, such that

(7) $\|G[w] - G[z]\| \leq K\|w - z\|$ for all w, z in S.

Then the equation $y = G[y]$ has a unique solution in S. Moreover, the sequence of successive approximations defined by

(8) $y_{n+1} := G[y_n], \qquad n = 0, 1, 2, \dots ,$

converges uniformly to this solution for any choice of the starting function y_0 in S.

Proof. The proof of Theorem 2 is essentially the same as the proof of Theorem 1 in Section 13.1, so we only outline the argument.

The Approximations Are Well Defined. As in the proof of Theorem 1, the y_n's are well defined because of the hypothesis that G maps S into S.

The Approximations Converge Uniformly to an Element of S. Using the fact that G is a contraction on S, we can derive the inequalities

(9) $\|y_{j+1} - y_j\| \leq K^j\|y_1 - y_0\| , \qquad j = 0, 1, 2, \dots ,$

in the same way that we established inequalities (14) in Section 13.1. Furthermore, y_n is the $(n + 1)$th partial sum of the series

$$y_0 + \sum_{j=0}^{\infty} (y_{j+1} - y_j) ,$$

and from (9), we know that the norm of each term in this series is bounded by the constant $M_j := K^j\|y_1 - y_0\|$. Since the series ΣM_j is convergent (because $K < 1$), we can appeal to an extension of the comparison test, called the **Weierstrass M test,**[†] to conclude that the sequence $\{y_n\}$ converges uniformly to a function y_∞ in S.

[†]This test is also called the **Weierstrass comparison test** and is discussed in most advanced calculus or real analysis texts; for example, *Advanced Calculus,* 5th ed., by Wilfred Kaplan (Pearson, Boston, Mass., 2003), or *Principles of Mathematical Analysis,* 3rd ed., by Walter Rudin (McGraw-Hill, New York, 1976).

y_∞ Is a Fixed Point. Because G is a contraction, $\|G[y_\infty] - G[y_n]\| \leq K\|y_\infty - y_n\|$ for all n. We already know that $\|y_\infty - y_n\| \to 0$. Thus, $\|G[y_\infty] - G[y_n]\| \to 0$ as $n \to \infty$; that is, $G[y_n] \xrightarrow{\text{unif}} G[y_\infty]$. But $y_{n+1} = G[y_n]$ and $y_{n+1} \xrightarrow{\text{unif}} y_\infty$. Therefore, by the uniqueness of the limit, we have $y_\infty = G[y_\infty]$. Hence y_∞ is a fixed point of G.

Uniqueness. That y_∞ is the unique fixed point of G again follows from the fact that G is a contraction, just as in the proof of Theorem 1. ◆

In Section 13.1 we showed that when f and y are continuous functions, the initial value problem

(10) $y'(x) = f(x, y(x))$, $y(x_0) = y_0$,

is equivalent to the operator equation

(11) $y = T[y]$,

where

(12) $T[y](x) := y_0 + \displaystyle\int_{x_0}^{x} f(t, y(t)) \, dt$.

Our goal is to apply the Banach fixed point theorem to this operator T and thereby show (with certain conditions on f) that equation (11) has a unique solution, which is the limit of the **Picard iterations**

(13) $y_{n+1} := T[y_n]$, $y_0(x) \equiv y_0$.

Picard's Existence and Uniqueness Theorem

Theorem 3. If f and $\partial f / \partial y$ are continuous functions in a rectangle $R = \{(x, y): a < x < b, c < y < d\}$ that contains the point (x_0, y_0), then the initial value problem (10) has a unique solution in some interval $[x_0 - h, x_0 + h]$, where h is a positive number. Moreover, the Picard iterations defined in (13) converge uniformly on $[x_0 - h, x_0 + h]$ to this solution.

Proof. To apply Theorem 2 to the operator T in (12), we must show that there exist an interval $I := [x_0 - h, x_0 + h]$ and a positive constant α such that T maps the set

$$S := \{g \in C(I): \|g - y_0\|_I \leq \alpha\}$$

into itself and T is a contraction on S.

First we choose positive constants h_1 and α_1 so that

$$R_1 := \{(x, y): |x - x_0| \leq h_1, \ |y - y_0| \leq \alpha_1\} \subset R.$$

Since R_1 is a closed rectangle and f as well as $\partial f / \partial y$ are continuous on R_1, there exist constants $M > 0$ and $L > 0$ such that for all (x, y) in R_1,

$$|f(x, y)| \leq M \quad \text{and} \quad \left|\frac{\partial f}{\partial y}(x, y)\right| \leq L.$$

Let g be a continuous function on $I_1 := [x_0 - h_1, x_0 + h_1]$ that satisfies $|g(x) - y_0| \leq \alpha_1$ for x in I_1. For such a g, we have

$$T[g](x) = y_0 + \int_{x_0}^{x} f(t, g(t)) \, dt.$$

Hence, for $|x - x_0| \leq h_1$,

$$(14) \quad |T[g](x) - y_0| = \left| \int_{x_0}^{x} f(t, g(t)) \, dt \right| \leq \left| \int_{x_0}^{x} |f(t, g(t))| \, dt \right|$$

$$\leq M \left| \int_{x_0}^{x} dt \right| = M|x - x_0| \, .$$

From a geometric point of view, we have shown that if the function g satisfies $|g(x) - y_0| \leq \alpha_1$ for x in I_1, then the graph of the function $T[g]$ must lie in the sectors bounded by the lines $y = y_0 \pm M(x - x_0)$ (see Figure 13.3). To keep $T[g]$ in the rectangle R_1, we must choose h less than $h_2 := \min\{h_1, \alpha_1/M\}$. We will see later that we also need h to be less than $1/L$ in order for T to be a contraction. Therefore we choose h so that

$$0 < h < \min\{h_1, \alpha_1/M, 1/L\} \, .$$

Now let $S := \{g \in C(I): \|g - y_0\|_I \leq \alpha\}$, where $\alpha = \alpha_1$ and $I = [x_0 - h, x_0 + h]$.

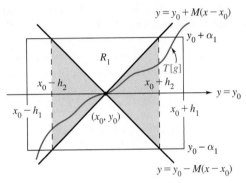

Figure 13.3 Graph of $T[g]$

T Maps S into S. Clearly $T[g](x)$ is a continuous function on I. For g in S, inequality (14) holds, so for x in I, we have

$$|T[g](x) - y_0| \leq M|x - x_0| \leq Mh < M\left(\frac{\alpha_1}{M}\right) = \alpha_1 \, .$$

Hence, $\|T[g] - y_0\|_I < \alpha_1$ (see the shaded region in Figure 13.3). Therefore, if g lies in S, then $T[g]$ does also. Thus, T maps S into S.

T Is a Contraction on S. Let u and v be functions in S. On R_1, we have $|\partial f/\partial y| \leq L$. Thus, using the mean value theorem,

$$|T[u](x) - T[v](x)| = \left| \int_{x_0}^{x} \{f(t, u(t)) - f(t, v(t))\} \, dt \right|$$

$$= \left| \int_{x_0}^{x} \frac{\partial f}{\partial y}(t, z(t)) [u(t) - v(t)] \, dt \right|$$

$$\leq L \left| \int_{x_0}^{x} |u(t) - v(t)| \, dt \right|$$

$$\leq L\|u - v\|_I |x - x_0| \leq Lh\|u - v\|_I \, ,$$

where $z(t)$ is a function that lies between $u(t)$ and $v(t)$. Setting $K := Lh$, we have shown that

$$\|T[u] - T[v]\|_I \le K\|u - v\|_I \,,$$

and since (by assumption) $h < 1/L$, we have $K < 1$. Hence, T is a contraction.

It now follows from the Banach fixed point theorem (page 786) that T has a unique fixed point in S, so the initial value problem (10) has a unique solution in S. Moreover, this solution is the uniform limit of the sequence of successive approximations defined in (13).

To complete the proof, we need to resolve one subtle but important point. That is, although we have shown that problem (10) has a unique solution in S, we have not shown that every solution to (10) on the interval $[x_0 - h, x_0 + h]$ must lie in S. To see this, suppose $u(x)$ is a solution to (10) on $[x_0 - h, x_0 + h]$ and recall that f is bounded by M on the rectangle R_1 of Figure 13.3. Since $u(x_0) = y_0$, the graph of $u(x)$ must lie in R_1 for x near x_0. For such values of x, we have $|f(x, u(x))| \le M$. Thus $|u'(x)| = |f(x, u(x))| \le M$. Therefore, for x near x_0, the graph of $u = T[u]$ must be in the shaded sector shown in Figure 13.3. In fact, the graph of u cannot escape from this sector on the interval $|x - x_0| \le h \,(\le h_2)$ because this would require that $|u'(x)| = |f(x, u(x))| > M$ at some point of the sector. Hence $|u(x) - y_0| \le \alpha_1$ for all $|x - x_0| \le h$; that is, $u \in S$. ◆

The preceding discussion carries over to first-order systems if we define the distance between two vector functions properly. To be specific, let $C_n[a, b]$ denote the set of all vector functions $\mathbf{y} = (y_1, y_2, \ldots, y_n)$ with n components that are continuous on $[a, b]$. For $\mathbf{y}$ in $C_n[a, b]$, define

$$\|\mathbf{y}\| := \max_{x \in [a,b]} |\mathbf{y}(x)| \,,$$

where $|\mathbf{y}(x)| := [|y_1(x)|^2 + |y_2(x)|^2 + \cdots + |y_n(x)|^2]^{1/2}$.

By extending the Banach fixed point theorem to operators defined on closed subsets of $C_n[a, b]$, we obtain the following version of Picard's existence and uniqueness theorem.

Picard's Theorem for Systems

Theorem 4. If $\mathbf{f}$ and $\partial \mathbf{f}/\partial x_i$, $i = 1, \ldots, n$, are continuous functions in a 'rectangle'

$$R = \{(t, x_1, \ldots, x_n): a < t < b, \quad c_i < x_i < d_i, \quad i = 1, \ldots, n\}$$

that contains the point $(t_0, \mathbf{x}_0)$, then the initial value problem

$$\mathbf{x}'(t) = \mathbf{f}(t, \mathbf{x}), \qquad \mathbf{x}(t_0) = \mathbf{x}_0,$$

where $\mathbf{x}_0 = \mathrm{col}(x_{1,0}, \ldots, x_{n,0})$, has a unique solution in some interval $[t_0 - h, t_0 + h]$, where h is a positive constant.

Example 2 Find an interval $[-h, h]$ over which the initial value problem

$$y' = \frac{1}{y} + x^2, \qquad y(0) = 1$$

is guaranteed to have a unique solution by the Picard theorem.

Solution We trace the steps of the proof of Theorem 3. Note that we are not requested to find the *largest* such interval—any one will do. So first we pick a rectangle R centered about the initial point $(x_0, y_0) = (0, 1)$, such that $f := 1/y + x^2$ and $\partial f/\partial y = -1/y^2$ are continuous inside. Clearly, $R := \{(x, y): -1 < x < 1 \text{ and } 0 < y < 2\}$ suffices.

Next we choose a *closed* rectangle contained in R; for definiteness, we choose $R_1 := \{(x, y): -0.9 \le x \le 0.9 \text{ and } 0.1 \le y \le 1.9\}$. Inside R_1 we easily derive the upper bounds

$$|f(x, y)| \le \frac{1}{0.1} + 0.9^2 = 10.81$$

and

$$\left|\frac{\partial f}{\partial y}(x, y)\right| \le \frac{1}{0.1^2} = 100.$$

Thus, the parameters h_1, α_1, M, and L in the theorem can be assigned the values

$$h_1 = 0.9, \quad \alpha_1 = 0.9, \quad M = 10.81, \quad \text{and} \quad L = 100.$$

The choice $h = \min\{0.9, 0.9/10.81, 1/100\} = 1/100$ then suffices; a unique solution is guaranteed to exist for $-0.01 \le x \le 0.01$. ◆

13.2 EXERCISES

*In Problems 1–6, determine whether the given sequence of functions $\{y_n\}$ converges **uniformly** to the given function y on the interval specified.*

1. $y_n(x) = 1 - \dfrac{x}{n}$ and $y(x) \equiv 1$ on $[-1, 1]$

2. $y_n(x) = \left(\dfrac{x}{2}\right)^n$ and $y(x) \equiv 0$ on $[0, 2]$

3. $y_n(x) = \dfrac{nx}{1 + n^2 x^2}$ and $y(x) \equiv 0$ on $[0, 1]$

4. $y_n(x) = \dfrac{1}{2x + n}$ and $y(x) \equiv 0$ on $[0, 1]$

5. $y_n(x) = \displaystyle\sum_{k=0}^{n} x^k$ and $y(x) = \dfrac{1}{1 - x}$ on $\left[0, \dfrac{1}{2}\right]$

6. $y_n(x) = x + \dfrac{x^2}{n}$ and $y(x) = x$ on $[0, 1]$

7. Show that the sequence of functions

$$y_n(x) = \begin{cases} n^2 x, & 0 \le x \le 1/n, \\ 2n - n^2 x, & 1/n < x < 2/n, \\ 0, & 2/n \le x \le 1, \end{cases}$$

$$n = 2, 3, 4, \ldots$$

converges pointwise on $[0, 1]$ to $y(x) \equiv 0$ but does not converge uniformly to y on $[0, 1]$. [*Hint:* Show that $\|y_n - y\| = n$ for all n.]

8. Prove inequality (9) in the proof of Theorem 2.

9. In the proof of Theorem 3, h is chosen to be strictly less than $\min\{h_1, \alpha_1/M, 1/L\}$, where h_1, α_1, M, and L are defined in the proof. Find an h for which the initial value problem

$$y'(x) = y^2(x) - x, \qquad y(1) = 0$$

has a unique solution on $[1 - h, 1 + h]$ if

$$R_1 = \{(x, y): |x - 1| \le 1, |y| \le 1\};$$

that is, $h_1 = 1$, $\alpha_1 = 1$.

10. The initial value problem

(15) $\begin{aligned} y''(x) + y(x) &= 6x; \\ y(0) &= 1, \quad y'(0) = -1 \end{aligned}$

is equivalent to the integral equation

(16) $y(x) = x^3 - x + 1 + \displaystyle\int_0^x (u - x) y(u)\, du.$

To prove this, proceed as follows:

(a) Integrate equation (15) from 0 to x using the initial condition $y'(0) = -1$ to obtain

$$(17)\quad y'(x) = 3x^2 - 1 - \int_0^x y(u)\,du\ .$$

(b) Integrate equation (17) from 0 to x using the initial condition $y(0) = 1$ to get

$$(18)\quad y(x) = x^3 - x + 1 - \int_0^x \left[\int_0^t y(u)\,du\right]dt\ .$$

(c) Inverting the order of integration and integrating with respect to t, obtain equation (16).

11. Let

$$T[y](x) := x^3 - x + 1 + \int_0^x (u - x)y(u)\,du\ .$$

Equation (16) in Problem 10 can now be written $y = T[y]$, and we can therefore try to approximate the solution to the initial value problem (15) using the Picard iterations $y_{n+1} = T[y_n]$. Setting $y_0(x) = x^3 - x + 1$, compute the approximations $y_1(x)$ and $y_2(x)$.

12. Let T be the operator defined in Problem 11. We can use the Banach fixed point theorem for operators to show that T has a fixed point in the set

$$S = \{g \in C[0, 1]: \|g - y_0\| \le 2\}\ ,$$

where $y_0(x) = x^3 - x + 1$ and the norm is taken over $[0, 1]$.

(a) Show that if g, f lie in S, then

$$\|T[g] - T[f]\| \le \frac{1}{2}\|g - f\|\ ,$$

and hence T is a contraction on S.

(b) Show that $\|y_0\| = 1$.

(c) Show that if $\|g - y_0\| \le 2$, then $\|g\| \le 2 + \|y_0\| = 3$.

(d) Using the results of parts (a), (b), and (c), show that if g is in S, then

$$\|T[g] - y_0\| \le \frac{1}{2}\|g\| \le \frac{3}{2}\ .$$

Hence, T maps S into S.

(e) Apply the Banach fixed point theorem for operators to prove that T has a fixed point in S.

(f) Why is $y_0(x) = x^3 - x + 1$ a good starting point for the Picard iterations in Problem 11?

13. Let $\{y_n\}$ be a sequence of functions in $C[a, b]$ that converge uniformly to a function y in $C[a, b]$. Prove that

$$\lim_{n \to \infty} \int_a^b y_n(x)\,dx = \int_a^b y(x)\,dx\ .$$

14. Is the assertion of Problem 13 true if we replace uniform convergence by pointwise convergence? [*Hint:* See Problem 7.]

15. Picard Iterations for Systems.

(a) Show that the initial value problem

$$\begin{aligned}
x'(t) &= -y^2(t)\ , & x(0) &= 0\ ; \\
y'(t) &= z(t)\ , & y(0) &= 1\ ; \\
z'(t) &= x(t)y(t)\ , & z(0) &= 1
\end{aligned}$$

is equivalent to the system of integral equations

$$x(t) = -\int_0^t y^2(s)\,ds\ ,$$

$$y(t) = 1 + \int_0^t z(s)\,ds\ ,$$

$$z(t) = 1 + \int_0^t x(s)y(s)\,ds\ .$$

(b) The Picard iterations for this system consist of the three sequences $\{x_n\}$, $\{y_n\}$, and $\{z_n\}$ defined by

$$x_{n+1}(t) := -\int_0^t y_n^2(s)\,ds\ , \qquad x_0(t) \equiv 0\ ;$$

$$y_{n+1}(t) := 1 + \int_0^t z_n(s)\,ds\ , \qquad y_0(t) \equiv 1\ ;$$

$$z_{n+1}(t) := 1 + \int_0^t x_n(s)y_n(s)\,ds\ , \qquad z_0(t) \equiv 1\ .$$

Compute the first two iterations: $x_1(t), y_1(t), z_1(t)$ and $x_2(t), y_2(t), z_2(t)$.

13.3 Existence of Solutions of Linear Equations

In Chapter 2 we established an important property of first-order *linear* differential equations. Namely, if the functions $P(t)$ and $Q(t)$ are continuous in an interval I containing t_0, then there exists a solution to

(1) $y'(t) + P(t)y(t) = Q(t) , \qquad y(t_0) = y_0 ,$

on the *entire* interval I. In the preceding section the best we could say was that if f and $\partial f / \partial y$ are continuous in a rectangle

$$R = \{ (t, y) : a < t < b, c < y < d \}$$

containing (t_0, y_0), then the solution to

$$y'(t) = f\big(t, y(t)\big) , \qquad y(t_0) = y_0 ,$$

exists on *some* subinterval $[t_0 - h, t_0 + h] \subset [a, b]$.

For example, the seemingly simple nonlinear problem

(2) $y'(t) = y^2(t) , \qquad y(0) = a (\neq 0) ,$

has the solution $y(t) = 1 / (a^{-1} - t)$, which is undefined at $t = 1/a$ even though $f(t, y) = y^2$ is very well behaved.

In this section we show that for *linear* systems in normal form, a solution exists over the entire interval. This fact is a consequence of the following theorem.

Continuation of Solution

Theorem 5. Let $\mathbf{x} = (x_1, \dots, x_n)$ and $\mathbf{f}(t, \mathbf{x})$ denote the vector function $\mathbf{f}(t, \mathbf{x}) = \big(f_1(t, x_1, \dots, x_n), \dots, f_n(t, x_1, \dots, x_n)\big)$. Suppose $\mathbf{f}$ and $\partial \mathbf{f} / \partial x_i$, $i = 1, \dots, n$, are continuous on the strip

$$R = \{ (t, \mathbf{x}) \in \mathbf{R}^{n+1} : a \leq t \leq b, \mathbf{x} \text{ arbitrary} \}$$

containing the point $(t_0, \mathbf{x}_0)$. Assume further that there exists a positive constant L such that, for $i = 1, \dots, n$,

(3) $\left| \dfrac{\partial \mathbf{f}}{\partial x_i}(t, \mathbf{x}) \right| \leq L$

for all $(t, \mathbf{x})$ in R. Then the initial value problem

(4) $\mathbf{x}'(t) = \mathbf{f}(t, \mathbf{x}) , \qquad \mathbf{x}(t_0) = \mathbf{x}_0 ,$

has a unique solution on the entire interval $a \leq t \leq b$.

Proof. We prove Theorem 5 for scalar equations ($n = 1$). The proof for systems is essentially the same.

The proof when $n = 1$ proceeds along the same line of reasoning as the proofs of the Banach fixed point theorems. We show that the Picard iterations $\{x_n\}$, where

(5) $x_{n+1}(t) = T[x_n](t) := x_0 + \displaystyle\int_{t_0}^{t} f\big(s, x_n(s)\big) ds , \qquad x_0(t) \equiv x_0 ,$

converges uniformly to a unique solution on the interval $[a, b]$.

The Approximations Are Well Defined. Since $x_0(t) \equiv x_0$ is continuous on $[a, b]$ and f is continuous on the strip R, it follows that $x_1(t) = T[x_0](t)$ is well defined and continuous on $[a, b]$. By induction, we find that $x_n(t)$ is well defined and continuous on $[a, b]$.

The Approximations Converge Uniformly on $[a, b]$. Recall that the sequence of successive approximations $\{x_n\}$ consists of the partial sums of the series

$$(6) \qquad x_0 + \sum_{j=0}^{\infty} \left(x_{j+1}(t) - x_j(t) \right).$$

Thus, if this series converges uniformly on $[a, b]$, then so does the sequence. But to show convergence on the entire interval $[a, b]$, we need more precise estimates for the terms in (6) than we obtained in the proofs of the Banach fixed point theorems.

Since $f(t, x_0)$ is continuous on $[a, b]$, it is bounded; that is, there is some constant M such that $|f(t, x_0)| \leq M$ on $[a, b]$. To simplify the computations, let's consider t in the interval $[t_0, b]$. (The case when $t \in [a, t_0]$ can be handled in a similar fashion.) For such t,

$$(7) \qquad |x_1(t) - x_0| = \left| \int_{t_0}^{t} f(s, x_0(s)) \, ds \right| \leq \int_{t_0}^{t} |f(s, x_0)| \, ds \leq M(t - t_0).$$

Let $w(t)$ and $v(t)$ be two continuous functions on $[t_0, b]$. Using the mean value theorem and the hypothesis that $|\partial f / \partial x|$ is bounded by L on R, we find

$$(8) \qquad \left| f(t, w(t)) - f(t, v(t)) \right| = \left| \frac{\partial f}{\partial x}(t, z(t)) \left[w(t) - v(t) \right] \right| \leq L |w(t) - v(t)|$$

for all t in $[t_0, b]$, where $z(t)$ is some function between $w(t)$ and $v(t)$. Using inequality (8) and then (7), we find that $x_2 = T[x_1]$ satisfies

$$|x_2(t) - x_1(t)| \leq \int_{t_0}^{t} \left| f(s, x_1(s)) - f(s, x_0(s)) \right| \, ds$$

$$\leq L \int_{t_0}^{t} \left| x_1(s) - x_0(s) \right| \, ds$$

$$\leq LM \int_{t_0}^{t} (s - t_0) \, ds = LM \frac{(t - t_0)^2}{2!}$$

for t in $[t_0, b]$. Similarly, for $n = 1, 2, 3, \ldots$, we find by induction that

$$|x_{n+1}(t) - x_n(t)| \leq \int_{t_0}^{t} \left| f(s, x_n(s)) - f(s, x_{n-1}(s)) \right| \, ds$$

$$(9) \qquad\qquad \leq L \int_{t_0}^{t} \left| x_n(s) - x_{n-1}(s) \right| \, ds$$

$$\leq L^n M \int_{t_0}^{t} \frac{(s - t_0)^n}{n!} \, ds = L^n M \frac{(t - t_0)^{n+1}}{(n+1)!}.$$

From inequalities (9) we see that the terms in (6) are bounded in absolute value by the terms in the positive series

$$\left(\frac{M}{L} \right) \sum_{j=0}^{\infty} \frac{\left[L(b - t_0) \right]^{j+1}}{(j+1)!},$$

which converges to $(M/L)(e^{L(b-t_0)} - 1)$. Therefore, by the Weierstrass M test,[†] the series in (6) converges uniformly on $[t_0, b]$ to a continuous function $x_\infty(t)$. Hence, the sequence $\{x_n\}$ converges uniformly on $[t_0, b]$ to x_∞.

A similar argument establishes convergence on $[a, t_0]$, so $\{x_n\}$ converges uniformly on $[a, b]$ to x_∞.

x_∞ *Is a Solution.* Since $\{x_n\}$ converges uniformly to x_∞, we can take the limit inside the integral (see Problem 13 in Exercises 13.2). Thus, using equation (5), we obtain

$$x_\infty(t) = \lim_{n \to \infty} x_{n+1}(t) = \lim_{n \to \infty} \left[x_0 + \int_{t_0}^t f(s, x_n(s)) \, ds \right]$$

$$= x_0 + \int_{t_0}^t f(s, x_\infty(s)) \, ds \, .$$

Since x_∞ satisfies the equivalent integral equation, it must satisfy the initial value problem (4).

Uniqueness. Let z be a solution to (4) on $[a, b]$. Then z must satisfy the integral equation

(10) $$z(t) = x_0 + \int_{t_0}^t f(s, z(s)) \, ds \, ,$$

just as x_∞ does. Therefore, for t in $[t_0, b]$,

(11) $$|z(t) - x_\infty(t)| \le \int_{t_0}^t \left| f(s, z(s)) - f(s, x_\infty(s)) \right| ds$$

$$\le L \int_{t_0}^t |z(s) - x_\infty(s)| \, ds$$

$$\le L \|z - x_\infty\| \int_{t_0}^t ds = L \|z - x_\infty\| (t - t_0) \, ,$$

where the norm is taken over $[t_0, b]$. Using (11), we next show by a similar string of inequalities that

$$|z(t) - x_\infty(t)| \le \int_{t_0}^t \left| f(s, z(s)) - f(s, x_\infty(s)) \right| ds$$

$$\le L \int_{t_0}^t |z(s) - x_\infty(s)| \, ds$$

$$\le L^2 \|z - x_\infty\| \int_{t_0}^t (s - t_0) \, ds = L^2 \|z - x_\infty\| \frac{(t - t_0)^2}{2!} \, .$$

By induction, we find

$$|z(t) - x_\infty(t)| \le L^n \frac{(t - t_0)^n}{n!} \|z - x_\infty\| \, , \qquad n = 1, 2, 3, \ldots ,$$

[†]See footnote on page 791.

for t in $[t_0, b]$. Taking the limit as $n \to \infty$, we deduce that $|z(t) - x_\infty(t)| \equiv 0$. That is, z and x_∞ agree on $[t_0, b]$. A similar argument shows that they agree on $[a, t_0]$, so the initial value problem has a unique solution on the interval $[a, b]$. ◆

Theorem 5 is called a **continuation theorem** because it specifies conditions that guarantee that the solution to initial value problem (4) can be **continued** from the small interval $[t_0 - h, t_0 + h]$ to the entire interval $[a, b]$. As we now show, these conditions are satisfied by linear systems.

Existence and Uniqueness Theorem for Linear Systems

Theorem 6. If the $n \times n$ matrix function $\mathbf{A}(t) = \left[a_{ij}(t) \right]$ and the vector function $\mathbf{f}(t)$ are continuous on an open interval (a, b) that contains the point t_0, then, for any choice of the initial vector $\mathbf{x}_0$, there exists a unique solution on the whole interval (a, b) to the initial value problem

$$(12) \qquad \mathbf{x}'(t) = \mathbf{A}(t)\mathbf{x}(t) + \mathbf{f}(t), \qquad \mathbf{x}(t_0) = \mathbf{x}_0.$$

Proof. Let $[\tilde{a}, \tilde{b}]$ be a closed subinterval of (a, b) containing t_0. For system (12) we let $\mathbf{g}(t, \mathbf{x}) := \mathbf{A}(t)\mathbf{x} + \mathbf{f}(t)$, which is continuous on $\tilde{R} := \{ (t, \mathbf{x}): \tilde{a} \le t \le \tilde{b}, \mathbf{x} \text{ arbitrary} \}$. Furthermore,

$$\frac{\partial \mathbf{g}}{\partial x_i} = \text{col}\left[a_{1i}(t), a_{2i}(t), \ldots, a_{ni}(t) \right]$$

for $i = 1, \ldots, n$. Since each a_{ij} is continuous on the closed interval $[\tilde{a}, \tilde{b}]$, we can find a positive constant L such that $|\partial \mathbf{g}/\partial x_i| \le L$ in $\tilde{R}$. Therefore, by Theorem 5, there exists a unique solution to (12) on the interval $[\tilde{a}, \tilde{b}]$. Since $[\tilde{a}, \tilde{b}]$ is an arbitrary interval in (a, b), we must have a unique solution on the whole interval (a, b). ◆

Example 1 Over what interval are we assured that there is a unique solution to

$$(13) \qquad \mathbf{x}'(t) = \begin{bmatrix} t^2 & \tan t \\ 1 & \sin t \end{bmatrix} \mathbf{x}(t) + \begin{bmatrix} \sqrt{1-t} \\ 0 \end{bmatrix}, \qquad \mathbf{x}(0) = \begin{bmatrix} -1 \\ 1 \end{bmatrix} ?$$

Solution The matrix $\mathbf{A}(t)$ is continuous on the interval $(-\pi/2, \pi/2)$, but on no larger interval containing 0 because $a_{12}(t) = \tan t$ has vertical asymptotes at $t = \pm\pi/2$. Here $\mathbf{f}(t)$ is continuous on the interval $(-\infty, 1]$ but not for $t > 1$, since $f_1(t) = \sqrt{1-t}$ is not defined for $t > 1$. Hence, $\mathbf{A}(t)$ and $\mathbf{f}(t)$ are simultaneously continuous on the interval $(-\pi/2, 1)$, which contains 0. Therefore, by Theorem 6, the initial value problem (13) has a unique solution on the interval $(-\pi/2, 1)$. ◆

A careful examination of the proof of Theorem 6 reveals that the initial value problem (13) actually has a (unique) solution on the half-closed interval $(-\pi/2, 1]$. We leave the verification as an exercise.

In Sections 5.3 and 9.4 we demonstrated how a higher-order linear equation can be reduced to a first-order system in normal form. Using this reduction, we prove the following.

Existence and Uniqueness Theorem for Linear Equations

Theorem 7. If $p_1(t), \ldots, p_n(t)$ and $g(t)$ are continuous on an interval (a, b) containing the point t_0, then, for every choice of the initial values $y_0, y_1, \ldots, y_{n-1}$, there exists a unique solution on the whole interval (a, b) to the initial value problem

(14)
$$y^{(n)}(t) + p_1(t)y^{(n-1)}(t) + \cdots + p_n(t)y(t) = g(t) ;$$

$$y(t_0) = y_0 , \qquad y'(t_0) = y_1 , \qquad \ldots, \qquad y^{(n-1)}(t_0) = y_{n-1} .$$

Proof. Let $x_1 = y$, $x_2 = y'$, $\ldots$, $x_n = y^{(n-1)}$. Then the initial value problem (14) is reduced to the system

(15)
$$x_1'(t) = x_2(t) , \qquad\qquad\qquad x_1(t_0) = y_0 ,$$
$$\vdots \qquad\qquad\qquad\qquad \vdots$$
$$x_{n-1}'(t) = x_n(t) , \qquad\qquad x_{n-1}(t_0) = y_{n-2} ,$$
$$x_n'(t) = -p_n(t)x_1(t) - \cdots - p_1(t)x_n(t) + g(t) , \qquad x_n(t_0) = y_{n-1} .$$

When we express (15) in the matrix form $\mathbf{x}'(t) = \mathbf{A}(t)\mathbf{x}(t) + \mathbf{f}(t)$, we find that the entries in $\mathbf{A}(t)$ are either $-p_1(t), \ldots, -p_n(t), 0$, or 1 and the entries in $\mathbf{f}(t)$ are either 0 or $g(t)$. Hence, if $p_1(t), \ldots, p_n(t)$ and $g(t)$ are continuous on an interval (a, b), then so are $\mathbf{A}(t)$ and $\mathbf{f}(t)$. Therefore, by Theorem 6, the initial value problem in (15) has a unique solution on (a, b). But since (15) is equivalent to (14), there is a unique solution to (14) on (a, b). ◆

Example 2 Over what interval are we assured that there is a unique solution to

(16) $$y'''(t) + t^2 y''(t) + (\ln t)y'(t) - (\cos t)y(t) = \frac{1}{2-t} ;$$

$$y(1) = 3 , \qquad y'(1) = -2 , \qquad y''(1) = -6 ?$$

Solution Here $p_1(t) = t^2$ is continuous everywhere, $p_2(t) = \ln t$ is continuous on $(0, \infty)$, $p_3(t) = -\cos t$ is continuous everywhere, and $g(t) = (2-t)^{-1}$ is continuous for $t \neq 2$. Thus, $p_1(t)$, $p_2(t), p_3(t)$, and $g(t)$ are simultaneously continuous on $(0, 2)$, which contains $t = 1$. Hence, by Theorem 7, there is a unique solution to (16) on $(0, 2)$. ◆

There are many other theorems dealing with the existence and uniqueness of solutions to initial value problems. For these and other results, we refer the reader to an advanced text that studies the theory of ordinary differential equations.[†]

[†]For example, *Ordinary Differential Equations*, 4th ed., by G. Birkhoff and G. C. Rota (Wiley, New York, 1989).

13.3 EXERCISES

In Problems 1–6, determine over what interval we are assured that there is a unique solution to the given initial value problem.

1. $\mathbf{x}'(t) = \begin{bmatrix} \cos t & \sqrt{t} \\ t^3 & -1 \end{bmatrix} \mathbf{x}(t) + \begin{bmatrix} \tan t \\ e^t \end{bmatrix}$,

$\mathbf{x}(2) = \begin{bmatrix} 0 \\ 4 \end{bmatrix}$

2. $\mathbf{x}'(t) = \begin{bmatrix} e^t & 5 \\ 0 & \sqrt{2+t} \end{bmatrix} \mathbf{x}(t) + \begin{bmatrix} (1-t)^{-1} \\ t \end{bmatrix}$,

$\mathbf{x}(0) = \begin{bmatrix} 1 \\ 2 \end{bmatrix}$

3. $y'''(t) - (\ln t)y''(t) + (\tan t)y(t) = e^{2t}$;

$y(1) = 1, \qquad y'(1) = -2, \qquad y''(1) = 0$

4. $ty^{(4)}(t) - y(t) = \sqrt{3 - t}$;

$y(1) = y'(1) = y''(1) = 0, \qquad y'''(1) = -2$

5. $\mathbf{x}'(t) = \begin{bmatrix} x_1'(t) \\ x_2'(t) \end{bmatrix} = \begin{bmatrix} \sin x_2(t) \\ 3x_1(t) \end{bmatrix}$, $\qquad \mathbf{x}(0) = \begin{bmatrix} 1 \\ 2 \end{bmatrix}$

6. $x'(t) = \arctan(x(t))$, $\qquad x(0) = 2$

7. Find the unique solution to the initial value problem

$y'''(t) - (\sin t)y'(t) + e^{-t}y(t) = 0$;

$y(0) = y'(0) = y''(0) = 0$

8. Prove that the initial value problem (13) in Example 1 actually has a (unique) solution on the half-closed interval $(-\pi/2, 1]$.

13.4 Continuous Dependence of Solutions

In the preceding sections we discussed the existence and uniqueness of solutions to the initial value problem

(1) $\qquad \dfrac{dy}{dx} = f(x, y), \qquad y(x_0) = y_0$.

In this section we investigate properties of the solution. Namely, we give conditions for the solution to be a continuous function of the initial value y_0 or of the function f. We state and prove these results for a scalar first-order equation. Similar results hold for systems.[†]

To show that the solution to equation (1) is a continuous function of the initial value y_0, we need the following inequality due to T. H. Gronwall.

Gronwall's Inequality

Theorem 8. Let u and v be continuous functions satisfying $u(x) > 0$ and $v(x) \geq 0$ on $[a, b]$. Let $c \geq 0$ be a constant. If

(2) $\qquad v(x) \leq c + \displaystyle\int_a^x u(t)v(t)\,dt, \qquad a \leq x \leq b$,

then

(3) $\qquad v(x) \leq c \exp\left\{ \displaystyle\int_a^x u(t)\,dt \right\}, \qquad a \leq x \leq b$.

[†]For a discussion of the results of this section for systems, see an advanced text such as *Ordinary Differential Equations,* 4th ed., by G. Birkhoff and G. C. Rota (Wiley, New York, 1989), Chapter 6.

Proof. First assume that $c > 0$ and define

$$V(x) := c + \int_a^x u(t)v(t)\,dt\,, \qquad a \le x \le b\,.$$

Then $V(x) \ge v(x)$ by (2), and since u and v are nonnegative, $V(x) \ge V(a) = c > 0$ on $[a, b]$. Moreover, $V'(x) = u(x)v(x) \le u(x)V(x)$. Dividing by $V(x)$, we get

(4) $$u(x) \ge \frac{V'(x)}{V(x)}\,, \qquad a \le x \le b\,.$$

When we integrate both sides of (4) from a to x, we find

(5) $$\int_a^x u(s)\,ds \ge \int_a^x \frac{V'(s)}{V(s)}\,ds = \ln V(s)\Big|_a^x = \ln V(x) - \ln V(a) = \ln\big(V(x)/c\big)\,.$$

Solving for $V(x)$ (recall that the exponential function is increasing), we obtain

(6) $$V(x) \le c \exp\left\{ \int_a^x u(s)\,ds \right\}, \qquad a \le x \le b\,.$$

Since $v(x) \le V(x)$, inequality (3) follows from (6).

 If inequality (2) holds with $c = 0$, then it certainly holds for all $c > 0$. Hence, as we have shown in the preceding discussion, inequality (3) is valid for all $c > 0$. Taking the limit as $c \to 0^+$, we get $v(x) \le 0$, which is the same as (3) with $c = 0$. ◆

 As we have just seen, when $c = 0$ in inequality (2), it follows that $v(x) \le 0$. But since $v(x)$ is nonnegative, we must have $v(x) \equiv 0$. This observation can be used to prove that an initial value problem has a unique solution (see Problem 12).

 The next theorem is of practical importance. Suppose we need to solve the initial value Problem (1), but we know the value of y_0 only to within $\pm 10^{-3}$ (for example, if y_0 is obtained by some measurement, its value will be known only up to the accuracy of the measuring instrument). Let's denote the approximate value of y_0 by $\tilde{y}_0$ so that $|y_0 - \tilde{y}_0| \le 10^{-3}$. When we solve

(7) $$y' = f(x, y)\,, \qquad y(x_0) = \tilde{y}_0\,,$$

we obtain a solution $\phi(x, \tilde{y}_0)$, where we have displayed the initial value for emphasis. The question is: Will $\phi(x, \tilde{y}_0)$ be close to the solution $\phi(x, y_0)$ of (1) for x in an interval about x_0? In other words, if we slightly perturb the initial value y_0, will the solution change significantly? As we now show, under suitable assumptions on f, small changes in y_0 lead to small changes in the solution.

Continuous Dependence on Initial Value

Theorem 9. Let f and $\partial f/\partial y$ be continuous functions on an open rectangle

$$R = \{(x, y): a < x < b, c < y < d\}$$

containing the point (x_0, y_0). Assume that for all $\tilde{y}_0$ sufficiently close to y_0, the solution $\phi(x, \tilde{y}_0)$ to (7) exists on the interval $[x_0 - h, x_0 + h]$ and its graph lies within a fixed closed rectangle $R_0 \subset R$. Then, for $|x - x_0| \le h$,

(8) $$|\phi(x, y_0) - \phi(x, \tilde{y}_0)| \le |y_0 - \tilde{y}_0|e^{Lh}\,,$$

where L is any positive constant such that $|(\partial f/\partial y)(x, y)| \le L$ for all (x, y) in R_0. Moreover, as $\tilde{y}_0$ approaches y_0, the solution $\phi(x, \tilde{y}_0)$ approaches $\phi(x, y_0)$ uniformly on $[x_0 - h, x_0 + h]$.

Notice that (8) implies that, for each $\varepsilon > 0$,

$$|\phi(x, y_0) - \phi(x, \tilde{y}_0)| \le \varepsilon \quad \text{on} \quad [x_0 - h, x_0 + h],$$

whenever $|y_0 - \tilde{y}_0| \le \varepsilon e^{-Lh}$.

Proof of Theorem 9. We know that the solution $\phi(x, y_0)$ to the initial value problem (1) satisfies the integral equation

$$(9) \qquad \phi(x, y_0) = y_0 + \int_{x_0}^{x} f\left(s, \phi(s, y_0)\right) ds, \qquad |x - x_0| \le h.$$

Similarly, for $\tilde{y}$ close enough to y_0,

$$(10) \qquad \phi(x, \tilde{y}_0) = \tilde{y}_0 + \int_{x_0}^{x} f\left(s, \phi(s, \tilde{y}_0)\right) ds, \qquad |x - x_0| \le h.$$

Subtracting the second of these equations from the first gives

$$(11) \qquad \phi(x, y_0) - \phi(x, \tilde{y}_0) = y_0 - \tilde{y}_0 + \int_{x_0}^{x} [f\left(s, \phi(s, y_0)\right) - f\left(s, \phi(s, \tilde{y}_0)\right)] ds.$$

Assume first that $x \ge x_0$. Taking the absolute value of both sides of (11) and using the triangle inequality, we obtain

$$(12) \qquad |\phi(x, y_0) - \phi(x, \tilde{y}_0)| \le |y_0 - \tilde{y}_0| + \int_{x_0}^{x} \left| f\left(s, \phi(s, y_0)\right) - f\left(s, \phi(s, \tilde{y}_0)\right) \right| ds.$$

Since $|\partial f/\partial y|$ is bounded by L on R_0, it follows from the mean value theorem that

$$(13) \qquad |f(s, w) - f(s, z)| \le L|w - z|$$

for (s, w) and (s, z) in R_0. Using this last inequality in (12), we get

$$(14) \qquad |\phi(x, y_0) - \phi(x, \tilde{y}_0)| \le |y_0 - \tilde{y}_0| + \int_{x_0}^{x} L|\phi(s, y_0) - \phi(s, \tilde{y}_0)| ds.$$

Now we can apply Gronwall's inequality $\left(\text{with } v(x) = |\phi(x, y_0) - \phi(x, \tilde{y}_0)|\right)$ to (14) and obtain

$$|\phi(x, y_0) - \phi(x, \tilde{y}_0)| \le |y_0 - \tilde{y}_0| \exp\left\{ \int_{x_0}^{x} L\, dt \right\}$$

$$= |y_0 - \tilde{y}_0| e^{L(x - x_0)}$$

$$\le |y_0 - \tilde{y}_0| e^{Lh}, \qquad x_0 \le x \le x_0 + h.$$

For $x \le x_0$, we can use the substitution $x = x_0 - t$ to show that inequality (8) continues to hold in this range (see Problem 13). ◆

Geometrically, Theorem 9 says that for initial values that are close together, the graphs of the solutions passing through them will remain close, at least on the interval $|x - x_0| \le h$.

In the next theorem we show that solutions to different equations, but with the same initial conditions, remain close if the differential equations are nearly the same.

Continuous Dependence on f(x, y)

Theorem 10. Let f and $\partial f / \partial y$ be continuous functions on an open rectangle $R = \{(x, y): a < x < b, c < y < d\}$ containing the point (x_0, y_0). Let F be continuous on R and assume that

(15) $|F(x, y) - f(x, y)| \leq \varepsilon$ for (x, y) in R.

If ϕ is the solution to the initial value problem

(16) $\dfrac{dy}{dx} = f(x, y),$ $y(x_0) = y_0,$

and ψ is a solution to

(17) $\dfrac{dy}{dx} = F(x, y),$ $y(x_0) = y_0,$

and both solutions exist on $[x_0 - h, x_0 + h]$ and their graphs lie in a closed rectangle $R_0 \subset R$, then, for $|x - x_0| \leq h$,

(18) $|\phi(x) - \psi(x)| \leq \varepsilon h e^{Lh},$

where L is any positive constant such that $|(\partial f / \partial y)(x, y)| \leq L$ for all (x, y) in R_0.
 In particular, as F approaches f uniformly on R—that is, as $\varepsilon \to 0^+$ in (15)—the solution $\psi(x)$ approaches $\phi(x)$ uniformly on $[x_0 - h, x_0 + h]$.

Proof. We begin with the integral representations for $\phi(x)$ and $\psi(x)$; namely,

$$\phi(x) = y_0 + \int_{x_0}^{x} f\big(s, \phi(s)\big) ds, \qquad \psi(x) = y_0 + \int_{x_0}^{x} F\big(s, \psi(s)\big) ds.$$

Subtracting these equations gives

$$\phi(x) - \psi(x) = \int_{x_0}^{x} \big[f\big(s, \phi(s)\big) - F\big(s, \psi(s)\big) \big] ds$$

and inserting $0 = -f\big(s, \psi(s)\big) + f\big(s, \psi(s)\big)$ in the integrand yields

(19)
$$\phi(x) - \psi(x) = \int_{x_0}^{x} \big[f\big(s, \phi(s)\big) - f\big(s, \psi(s)\big) \big] ds + \int_{x_0}^{x} \big[f\big(s, \psi(s)\big) - F\big(s, \psi(s)\big) \big] ds.$$

For $x_0 \leq x \leq x_0 + h$, the triangle inequality applied to (19) gives

(20)
$$|\phi(x) - \psi(x)| \leq \int_{x_0}^{x} \big| f\big(s, \phi(s)\big) - f\big(s, \psi(s)\big) \big| ds + \int_{x_0}^{x} \big| f\big(s, \psi(s)\big) - F\big(s, \psi(s)\big) \big| ds.$$

The first integral in (20) can be estimated just as in the proof of Theorem 9. That is, since

$$\left| \frac{\partial f}{\partial y}(s, t) \right| \leq L \quad \text{for } (s, t) \text{ in } R_0,$$

we have

$$\left| f\big(s, \phi(s)\big) - f\big(s, \psi(s)\big) \right| \leq L |\phi(s) - \psi(s)|.$$

Hence, for $x_0 \leq x \leq x_0 + h$,

$$(21) \quad \int_{x_0}^{x} \left| f(s, \phi(s)) - f(s, \psi(s)) \right| ds \leq \int_{x_0}^{x} L|\phi(s) - \psi(s)| ds .$$

To estimate the second integral in (20), we use the hypothesis (15). This gives

$$(22) \quad \int_{x_0}^{x} \left| f(s, \psi(s)) - F(s, \psi(s)) \right| ds \leq \varepsilon(x - x_0) \leq \varepsilon h .$$

Using the estimates (21) and (22) in inequality (20), we obtain

$$|\phi(x) - \psi(x)| \leq \varepsilon h + \int_{x_0}^{x} L|\phi(s) - \psi(s)| ds .$$

Now we are in position to apply Gronwall's inequality (Theorem 8). It yields

$$|\phi(x) - \psi(x)| \leq \varepsilon h \exp\left\{ \int_{x_0}^{x} L ds \right\}$$

$$\leq \varepsilon h e^{Lh} \quad \text{for} \quad x_0 \leq x \leq x_0 + h .$$

Since the last inequality can also be verified for $x_0 - h \leq x \leq x_0$, the conclusion (18) of the theorem is proved.

Finally, by letting $\varepsilon \to 0^+$ in (18), we see that $\psi(x)$ approaches $\phi(x)$ uniformly on $[x_0 - h, x_0 + h]$. That is, as F comes uniformly close to f on R, then ψ comes uniformly close to ϕ on $[x_0 - h, x_0 + h]$. ◆

In the next example, we use inequality (18) to bound the difference between the solutions to two different equations that are nearly the same.

Example 1 Let $\phi(x)$ be the solution to

$$(23) \quad y' = e^y , \qquad y(0) = y_0 ,$$

and let $\psi(x)$ be the solution to

$$(24) \quad y' = 1 + y , \qquad y(0) = y_0 .$$

Assume the solutions exist on $[-1/2, 1/2]$ and their graphs remain in the rectangle $R_0 = \{ (x, y) : -1/2 \leq x \leq 1/2, -2 \leq y \leq 2 \}$. Use Theorem 10 to find a bound on the difference between $\phi(x)$ and $\psi(x)$ for x in $[-1/2, 1/2]$.

Solution We can use inequality (18) in Theorem 10 to obtain the bound, but first we must determine the constant L and the constant ε. Here $x_0 = 0$ and $h = 1/2$.

Since e^y and $1 + y$ are both continuous and have continuous partial derivatives with respect to y in R_0, we will use the one with the smaller bound. If we set $F(x, y) := e^y$, then $(\partial F/\partial y)(x, y) = e^y$. Hence $|(\partial F/\partial y)(x, y)| \leq e^2$ for (x, y) in R_0. For $f(x, y) = 1 + y$, we find that $|(\partial f/\partial y)(x, y)| = 1$ for (x, y) in R_0. Since $1 \leq e^2$, we will let $f(x, y) = 1 + y$ in Theorem 10 and take $L = 1$.

We must now measure the distance between $1 + y$ and e^y on R_0. Recall that Taylor's theorem with remainder gives the representation

$$(25) \quad g(y) = g(0) + g'(0)y + \frac{g''(0)y^2}{2!} + \cdots + \frac{g^{(k)}(0)y^k}{k!} + \frac{g^{(k+1)}(\xi)y^{k+1}}{(k+1)!} ,$$

where ξ, which depends on y, lies between 0 and y. For $g(y) = e^y$ and $k = 1$, we have

$$e^y = 1 + y + \frac{e^{\xi}y^2}{2}, \qquad \xi \text{ between } 0 \text{ and } y.$$

Thus,

$$\left|e^y - (1+y)\right| = \left|\frac{e^{\xi}y^2}{2}\right| \le \frac{e^2(2)^2}{2} = 2e^2$$

for y in $[-2, 2]$. Consequently,

$$|F(x, y) - f(x, y)| = \left|e^y - (1+y)\right| \le 2e^2 =: \varepsilon$$

for (x, y) in R_0.

It now follows from inequality (18) in Theorem 10 that

$$(26) \qquad |\phi(x) - \psi(x)| \le (2e^2)\left(\frac{1}{2}\right)e^{(1)(1/2)} = e^{2.5}$$

for x in $[-1/2, 1/2]$. ◆

13.4 EXERCISES

In Problems 1–6, let $\phi(x, y_0)$ be the solution to the given differential equation that satisfies $y(0) = y_0$ and let $\phi(x, \tilde{y}_0)$ be the solution satisfying $y(0) = \tilde{y}_0$. Assume the solutions exist on $[-1, 1]$ and their graphs remain in the rectangle $R_0 = \{(x, y): -1 \le x \le 1, -4 \le y \le 4\}$. Use Theorem 9 to find a bound on the difference between $\phi(x, y_0)$ and $\phi(x, \tilde{y}_0)$ for x in $[-1, 1]$ if $|y_0 - \tilde{y}_0| < 10^{-2}$.

1. $y' = \sin(x + y)$ **2.** $y' = x + y$

3. $y' = e^{\cos y} + x^2$ **4.** $y' = e^{-y^2}$

5. $y' = \dfrac{x^2 - y^2}{1 + x^2 + y^2}$ **6.** $y' = \arctan(xy)$

In Problems 7–10, let $\phi(x)$ be the solution to $y' = f(x, y)$, $y(0) = y_0$ and let $\psi(x)$ be the solution to $y' = F(x, y)$, $y(0) = y_0$, where f and F are given functions. Assume the solutions exist on $[-1, 1]$ and their graphs remain in the rectangle $R_0 = \{(x, y): -1 \le x \le 1, -1 \le y \le 1\}$. Use Theorem 10 to find a bound on the difference between $\phi(x)$ and $\psi(x)$ for x in $[-1, 1]$.

7. $f(x, y) = e^{-y}$, $F(x, y) = 1 - y + y^2/2 - y^3/6$

8. $f(x, y) = x - \cos y$, $F(x, y) = x - 1 + y^2/2$

9. $f(x, y) = \sin x + \dfrac{1}{1 + y^2}$, $F(x, y) = x + 1 - y^2$

10. $f(x, y) = \sin(xy)$, $F(x, y) = xy$

11. Let $f(x, y) = y^2$.

 (a) Solve explicitly for $\phi(x, 1)$, the solution to $y' = f(x, y)$, $y(0) = 1$, and for $\phi(x, 1 + \varepsilon)$, the solution to $y' = f(x, y)$, $y(0) = 1 + \varepsilon$, where $\varepsilon > 0$.

 (b) Show that $|\partial f/\partial y| \le 4$ on the rectangle $R = \{(x, y): -1 < x < 1, -2 < y < 2\}$.

 (c) Using the result of part (b) in inequality (8) in Theorem 9, it appears that

$$|\phi(x, 1) - \phi(x, 1 + \varepsilon)| \le \varepsilon e^4, \qquad x \text{ in } (-1, 1).$$

Show that this is *not* true (even for $\varepsilon > 0$ small) by using the results of part (a) to verify that $\phi(x, 1) \to 1 + \varepsilon^{-1}$ as $x \to 1/(1 + \varepsilon)$, while $|\phi(x, 1 + \varepsilon)| \to +\infty$ as $x \to 1/(1 + \varepsilon)$.

 (d) Which assumption in Theorem 9 is *not* satisfied so that inequality (8) is not applicable to this situation?

12. Let f and $\partial f/\partial y$ be continuous functions on an open rectangle R containing the point (x_0, y_0). Let $\phi(x)$ and $\psi(x)$ be two solutions to $y' = f(x, y)$, $y(x_0) = y_0$, on the interval $[x_0 - h, x_0 + h]$. Assume their graphs lie in a fixed closed rectangle $R_0 \subset R$. Use Gronwall's inequality to show that $\phi(x) \equiv \psi(x)$ for x in $[x_0 - h, x_0 + h]$. [*Hint:* Proceed as in the proof of Theorem 9.]

13. To verify that inequality (8) holds for $x \le x_0$, let $x = x_0 - t$ and define $\psi(t, y_0) := \phi(x_0 - t, y_0)$ and $\psi(t, \tilde{y}_0) := \phi(x_0 - t, \tilde{y}_0)$.

(a) Show that $\psi(t, y_0)$ is a solution to the initial value problem

$$\frac{dz}{dt} = g(t, z), \qquad z(0) = y_0,$$

where $g(t, z) := -f(x_0 - t, z)$.

(b) Use inequality (8) to show that for $0 \le t \le h$,

$$\left| \psi(t, y_0) - \psi(t, \tilde{y}_0) \right| \le \left| y_0 - \tilde{y}_0 \right| e^{Lh}.$$

(c) Use the result of part (b) to show that for $x_0 - h \le x \le x_0$,

$$\left| \phi(x, y_0) - \phi(x, \tilde{y}_0) \right| \le \left| y_0 - \tilde{y}_0 \right| e^{Lh}.$$

14. Assume that the constant L in Theorem 9 is replaced by a positive, continuous function $\ell(x)$, that is

$$\left| \frac{\partial f}{\partial y}(x, y) \right| \le \ell(x) \quad \text{for } (x, y) \text{ in } R_0.$$

Show that Theorem 9 remains valid if inequality (8) is replaced by

$$\left| \phi(x, y_0) - \phi(x, \tilde{y}_0) \right| \le \left| y_0 - \tilde{y}_0 \right| \exp\left\{ \left| \int_{x_0}^{x} \ell(x)\, dx \right| \right\}.$$

15. Let $f(x, y, \lambda)$ and $(\partial f/\partial y)(x, y, \lambda)$ be continuous for $|x - x_0| < A$, $|y - y_0| < B$, $|\lambda - \lambda_0| < C$. Let $\phi(x)$ be the solution to

$$dy/dx = f(x, y, \lambda_0), \qquad y(x_0) = y_0$$

and let $\psi(x)$ be the solution to

$$dy/dx = f(x, y, \lambda), \qquad y(x_0) = y_0.$$

Assume that both solutions $\phi(x)$ and $\psi(x)$ exist on $[x_0 - h, x_0 + h]$ and their graphs lie in a closed rectangle

$$R_0 \subset R = \left\{ (x, y) : |x - x_0| < A, |y - y_0| < B \right\}.$$

Show that for each $\varepsilon > 0$, there exists a $\delta > 0$ such that whenever $|\lambda - \lambda_0| < \delta$, the solutions $\phi(x)$ and $\psi(x)$ satisfy

$$\left| \phi(x) - \psi(x) \right| < \varepsilon, \qquad x_0 - h \le x \le x_0 + h.$$

16. Prove the following combination of Theorems 9 and 10: Let f and $\partial f/\partial y$ be continuous functions on an open rectangle $R = \left\{ (x, y) : a < x < b, c < y < d \right\}$ containing the point (x_0, y_0). Let F be continuous on R and assume that

$$\left| F(x, y) - f(x, y) \right| \le \varepsilon \quad \text{for } (x, y) \text{ in } R.$$

Let ϕ be the solution to the initial value problem

$$dy/dx = f(x, y), \qquad y(x_0) = y_0$$

and let ψ be a solution to

$$dy/dx = F(x, y), \qquad y(x_0) = \tilde{y}_0.$$

Assume that both solutions $\phi(x)$, $\psi(x)$ exist on $[x_0 - h, x_0 + h]$ and their graphs lie in a closed rectangle $R_0 \subset R$. Then, for $|x - x_0| \le h$,

$$\left| \phi(x) - \psi(x) \right| \le \left\{ |y_0 - \tilde{y}_0| + \varepsilon h \right\} e^{Lh},$$

where L is a positive constant such that $|(\partial f/\partial y)(x, y)| \le L$ for (x, y) in R_0.

Chapter **13** Summary

In this chapter we discussed the existence and uniqueness of solutions to initial value problems. Our approach was to construct a sequence of **successive approximations** that converge uniformly to the desired solution.

For first-order equations, we showed that the initial value problem

$$(1) \qquad y'(x) = f\big(x, y(x)\big) , \qquad y(x_0) = y_0$$

is equivalent to the **integral equation**

$$(2) \qquad y(x) = y_0 + \int_{x_0}^{x} f\big(t, y(t)\big) dt .$$

If we let the right-hand side of (2) define the operator

$$(3) \qquad T[y](x) := y_0 + \int_{x_0}^{x} f\big(t, y(t)\big) dt ,$$

then the integral equation (2) is equivalent to the **operator equation**

$$(4) \qquad y = T[y] .$$

Under certain restrictions on f, we used the **Banach fixed point theorem** to show that the operator equation (4) has a unique solution and consequently the initial value problem (1) has a unique solution on some interval $[x_0 - h, x_0 + h]$, where h is some positive constant. Moreover, this solution can be obtained as the uniform limit of the **Picard iterations** $\{y_n\}$, where

$$(5) \qquad y_{n+1}(x) := T[y_n](x) = y_0 + \int_{x_0}^{x} f\big(t, y_n(t)\big) dt , \qquad y_0(x) \equiv y_0 .$$

These results have natural generalizations to first-order systems in normal form.

For linear systems and linear higher-order equations, we showed that the solutions can be extended beyond the (possibly) small interval $[x_0 - h, x_0 + h]$. For example, if $\mathbf{A}(t)$ and $\mathbf{f}(t)$ are continuous on the interval (a, b) containing t_0, then there exists a unique solution on the *entire* interval (a, b) to the initial value problem

$$(6) \qquad \mathbf{x}'(t) = \mathbf{A}(t)\mathbf{x}(t) + \mathbf{f}(t) , \qquad \mathbf{x}(t_0) = \mathbf{x}_0 .$$

When $f(x, y)$ and $f_y(x, y)$ are continuous functions on an open rectangle containing the point (x_0, y_0), we proved that the solution to the initial value problem (1) depends continuously on the initial value y_0 and on the function f. That is, if y_0 or f are "slightly changed," then the solution to the perturbed initial value problem will be close to the solution of the original problem. Both of these properties follow from **Gronwall's inequality.**

REVIEW PROBLEMS FOR CHAPTER 13

In Problems 1 and 2, use the method of successive substitutions to approximate a solution of the given equation starting with the given value for x_0.

1. $x = \dfrac{2x^3 + 1}{3x^2 - 1}$, $x_0 = 1$

2. $x = \cos x$, $x_0 = 0$

In Problems 3 and 4, express the given initial value problem as an integral equation.

3. $y'(x) = y^2(x) + \sin x$, $y(-1) = -2$

4. $y'(x) = x^2 y^3(x) - y^2(x)$, $y(0) = 9$

In Problems 5 and 6, compute the Picard iterations $y_1(x)$ and $y_2(x)$ for the given initial value problem, starting with $y_0(x) \equiv y_0$, where y_0 is the initial value for y.

5. $y'(x) = y^3(x)$, $y(0) = 1$

6. $y'(x) = -2y(x)$, $y(0) = -1$

In Problems 7 and 8, determine whether the given sequence of functions $\{y_n\}_{n=1}^{\infty}$ converges uniformly to the given limit function y on the specified interval.

7. $y_n(x) = \dfrac{\sin x}{1 + nx}$ and $y(x) \equiv 0$ on $[0, 1]$

 [*Hint:* $|\sin x| \le 2x/\pi$ on $[0, 1]$.]

8. $y_n(x) = \dfrac{1}{1 + nx^2}$ and $y(x) \equiv 0$ on $[-1, 1]$

In Problems 9 and 10, determine over what interval we are assured that there exists a unique solution to the given initial value problem.

9. $\begin{bmatrix} x_1'(t) \\ x_2'(t) \\ x_3'(t) \end{bmatrix} = \begin{bmatrix} \exp(-x_2^2(t)) \\ (\sin t)x_3(t) \\ x_1(t) - x_3(t) \end{bmatrix}$, $\mathbf{x}(0) = \begin{bmatrix} 0 \\ 1 \\ 1 \end{bmatrix}$

10. $y'''(t) + (\tan t)y'(t) - e^t y(t) = t^2 - t + 1$;
 $y(1) = 0$, $y'(1) = 1$, $y''(1) = 5$

11. Let $\phi(x, y_0)$ be the solution to $y' = x \tan y$ satisfying $y(0) = y_0$ and let $\phi(x, \tilde{y}_0)$ be the solution satisfying $y(0) = \tilde{y}_0$. Assume these solutions exist on $[0, 2]$ and their graphs remain in the rectangle $R_0 = \{(x, y): 0 \le x \le 2, -\pi/4 \le y \le \pi/4\}$. Find a bound on the difference between $\phi(x, y_0)$ and $\phi(x, \tilde{y}_0)$ for x on $[0, 2]$ if $|y_0 - \tilde{y}_0| < 10^{-3}$.

12. Let $\phi(x)$ be the solution to $y' = x \sin y, y(0) = y_0$, and let $\psi(x)$ be the solution to $y' = xy, y(0) = y_0$. Assume the solutions exist on $[-1, 1]$ and their graphs remain in the rectangle $R_0 = \{(x, y): -1 \le x \le 1, -\pi \le y \le \pi\}$. Find a bound on the difference between $\phi(x)$ and $\psi(x)$ for x in $[-1, 1]$.

TECHNICAL WRITING EXERCISES FOR CHAPTER 13

1. The Banach fixed point theorem for operators (Theorem 2, page 791) gives conditions for an operator mapping a set of continuous functions into itself to have a fixed point. A similar result holds for an operator mapping a set of functions analytic in a disk into itself. What implications does such a result have on the existence of a power series solution to an initial value problem (see Sections 8.3 and 8.4)?

2. Given the initial value problem

 $$y' = f(x, y), y(x_0) = y_0,$$

 suppose you know that f and $\partial f/\partial y$ are continuous in an open rectangle $R = \{(x, y): a < x < b, c < y < d\}$ that contains the point (x_0, y_0). Give a step-by-step procedure (an algorithm) that can be used to determine a constant h so that the initial value problem has a unique solution on $[x_0 - h, x_0 + h]$. Pretend that your algorithm will be converted to a computer program and discuss how you intend to implement each step.

3. The results of Theorems 9 and 10 on the continuous dependence of solutions discussed in Section 13.4 are stated only for scalar first-order equations. How should these results be stated for first-order systems?

Appendices

Review of Integration Techniques

In solving differential equations, we cannot overemphasize the importance of integration techniques. Indeed, we sometimes refer to the problem of solving a differential equation by asking, "How can we integrate this equation?"

This section is devoted to reviewing three of the standard techniques that will be useful for integrating many of the functions encountered in this text. Of course, one can always search in a table of integrals (a brief table appears on the inside back cover of this text), but familiarity with tricks of integration is a worthwhile investment of time for several reasons: (i) we may not find the particular integral we need in the table; (ii) using one of the techniques may be quicker; and (iii) we might be curious to know how the various integrals in the table were obtained.

So what techniques can we use for the following integrals?

$$\textbf{(a)} \quad \int x\sqrt{2x+1}\,dx \qquad\qquad \textbf{(b)} \quad \int \frac{x^2}{\sqrt{9-x^2}}\,dx \qquad\qquad \textbf{(c)} \quad \int x\ln x\,dx$$

$$\textbf{(d)} \quad \int \frac{7x^2+10x-1}{x^3+3x^2-x-3}\,dx \qquad \textbf{(e)} \quad \int e^{x^2}dx$$

For **(a)** we can use *algebraic substitution*, while **(b)** can be handled with *trigonometric substitution*. For **(c)** we have available *integration by parts*, and the integral in **(d)** will yield to *partial fractions*. But try as we may, no trick is sufficient to give **(e)** in a finite number of terms. By this we mean that if we have available the elementary functions (rational functions, trigonometric functions, inverse trigonometric functions, exponential functions, and logarithmic functions) and permit use of the standard operations—addition, subtraction, multiplication, division, taking roots, and composite functions—a finite number of times, we cannot find a formula for the antiderivative of e^{x^2}. A similar conundrum arises with integration of $\sqrt{1+\cos^2 x}$. So we can add a fourth reason for honing our skills with integration techniques: They will help us to recognize those integrals that *can* be expressed in a finite number of terms.

In this context, we are speaking of integration in closed form—analytically—as opposed to numerical integration. Thus integration is the inverse of differentiation; it requires finding an anti*derivative*. As a warmup, then, we begin this appendix by jogging your recollection of rules for *derivatives*. We hope you will find our approach a little different and refreshing.

Differentiation

Of course you remember that the derivative of a constant equals zero, and the derivative of a sum equals the sum of the derivatives. So let's jump in with the basic power rule.

Derivative of a power function $\quad \dfrac{d}{dx}x^p = px^{p-1}.$

This holds for *any* constant exponent p, be it integer, fraction, or negative.

Examples: $\dfrac{d}{dx}x^{100} = 100x^{99}, \quad \dfrac{d}{dx}x^{7/2} = \dfrac{7}{2}x^{5/2}, \quad \dfrac{d}{dx}x^{1/3} = \dfrac{1}{3}x^{-2/3}, \quad \dfrac{d}{dx}x^{-e} = -ex^{-e-1}$

Derivative of the exponential $\dfrac{d}{dx}e^{x} = e^{x}.$

This is the essential feature of the exponential function.

Derivative of trig functions $\dfrac{d}{dx}\sin x = \cos x, \quad \dfrac{d}{dx}\cos x = -\sin x.$

Remember which one gets the minus sign.

Derivative of a product $\dfrac{d}{dx}(fg) = f\dfrac{d}{dx}g + g\dfrac{d}{dx}f.$

Example: $\dfrac{d}{dx}(x^{2}\sin x) = x^{2}\cos x + (\sin x)(2x)$

(*Bonus*: If we rewrite this rule as $f\dfrac{d}{dx}g = \dfrac{d}{dx}(fg) - g\dfrac{d}{dx}f$, we begin to see how to build an antiderivative (fg) for $f\dfrac{d}{dx}g$, by compensating with $-g\dfrac{d}{dx}f$. This is the essence of *integration by parts*.)

Derivative of a quotient $\dfrac{d}{dx}\left(\dfrac{f}{g}\right) = \dfrac{g\dfrac{d}{dx}f - f\dfrac{d}{dx}g}{g^{2}}.$

(If you forget this, you can just apply the product rule to fg^{-1})

Examples:

$$\dfrac{d}{dx}\tan x = \dfrac{d}{dx}\dfrac{\sin x}{\cos x} = \dfrac{(\cos x)(\cos x) - (\sin x)(-\sin x)}{\cos^{2}x} = \dfrac{\cos^{2}x + \sin^{2}x}{\cos^{2}x}$$

$$= \dfrac{1}{\cos^{2}x} = \sec^{2}x = 1 + \tan^{2}x$$

(*Bonus*: if you reverse the roles, so $\tan x = y$ and $x = \arctan y$, and invert, you get a new rule: $\dfrac{d}{dy}\arctan y = \dfrac{1}{1+y^{2}}.$)

$$\dfrac{d}{dx}\sec x = \dfrac{d}{dx}\dfrac{1}{\cos x} = \dfrac{(\cos x)(0) - 1(-\sin x)}{\cos^{2}x} = \dfrac{\sin x}{\cos^{2}x} = (\sec x)(\tan x)$$

Chain rule $\dfrac{d}{dx}f(u) = \left[\dfrac{d}{du}f(u)\right]\left[\dfrac{du}{dx}\right]. \left(\textbf{Think } \dfrac{df(u)}{du}\dfrac{du}{dx}.\right)$

Examples: $\dfrac{d}{dx}(7x^{2}+x-2)^{-1/4} = \left[-\dfrac{1}{4}(7x^{2}+x-2)^{-5/4}\right]\left[\dfrac{d}{dx}(7x^{2}+x-2)\right]$

$$= -\dfrac{1}{4}\dfrac{14x+1}{(7x^{2}+x-2)^{5/4}}.$$

$$\dfrac{d}{dx}e^{x^{2}} = [e^{(x^{2})}]\left[\dfrac{d}{dx}(x^{2})\right] = 2xe^{x^{2}}.$$

Now we begin our review of integration with the *method of substitution* (also called *change of variable*), which is derived from the chain rule.

Method of Substitution

From the formula $\dfrac{d}{dx}f(g(x)) = f'(g(x))g'(x)$ we obtain

$$\int f'(g(x))g'(x)\,dx = f(g(x)) + C.$$

Thus if we can recognize an integrand as being of the form $f'(g(x))g'(x)$, then integration is immediate via the substitution $u = g(x)$. Here are some simple illustrations:

- $\displaystyle\int \sin^3 2x \cos 2x\,dx$

 Since $d(\sin 2x) = 2\cos 2x\,dx$, the substitution $u = \sin 2x$ reduces the integration to that of $u^3/2\,du$.

- $\displaystyle\int x^3 e^{5x^4}\,dx$

 Observe that x^3 is essentially (except for a constant factor) the derivative of $5x^4$, so we set $u = 5x^4$ and integrate $e^u/20\,du$.

- $\displaystyle\int \frac{6\ln(x+5)}{x+5}\,dx$

 Recognizing that $1/(x+5)$ is the derivative of $\ln(x+5)$, we put $u = \ln(x+5)$ and integrate $6u\,du$.

- $\displaystyle\int \tan^3 x \sec x\,dx$

 Here it is convenient to use the identity $\tan^2 x = \sec^2 x - 1$ and write the integrand as $(\sec^2 x - 1)\sec x \tan x$. Recalling that $(\sec x)' = \sec x \tan x$, the substitution $u = \sec x$ reduces the problem to integrating $(u^2 - 1)\,du$.

When a portion of the integrand involves some fractional power of an expression, an algebraic substitution can sometimes simplify the integration process. The technique is amply illu-strated in the following example.

Example 1 Find

(1) $\displaystyle I_1 := \int x\sqrt{2x+1}\,dx$.

Solution The difficulty here arises from the square root factor, so let's try to rid ourselves of this problem term by making the substitution $u = \sqrt{2x+1}$. Here are the details:

(2) $u^2 = 2x+1, \qquad x = \dfrac{u^2 - 1}{2}, \qquad dx = u\,du$.

If we use these expressions in (1), we find that in terms of the new variable u,

(3)
$$\int x\sqrt{2x+1}\,dx = \int \left(\frac{u^2-1}{2}\right)(u)\,(u\,du) = \frac{1}{2}\int (u^4 - u^2)\,du$$

$$= \frac{1}{2}\left(\frac{u^5}{5} - \frac{u^3}{3}\right) + C$$

$$= \frac{u^3}{30}(3u^2 - 5) + C.$$

Returning to the original variable x, via $u = \sqrt{2x+1}$, we deduce that

$$I_1 = \frac{(2x+1)^{3/2}}{30}[3(2x+1)-5] + C = \frac{(2x+1)^{3/2}}{15}(3x-1) + C. \quad \blacklozenge$$

We remark that the substitution $u = \sqrt{2x+1}$ is not the only substitution that will allow us to compute the integral in (1). Another substitution that works just as well is $u = 2x+1$.
In a similar manner we can compute the integrals

$$\int \frac{x^3}{(1+x^2)^{3/2}}\,dx \qquad \text{and} \qquad \int \frac{e^{x^{1/3}+1}}{x^{2/3}}\,dx$$

by making the substitution $u = 1 + x^2$ in the first and $u = x^{1/3} + 1$ in the second.

The integration of certain algebraic expressions can sometimes be simplified by utilizing *trigonometric substitutions*. The effectiveness of these substitutions in handling quadratic expressions under a radical derives from the familiar identities

$$1 - \sin^2\theta = \cos^2\theta, \quad 1 + \tan^2\theta = \sec^2\theta, \quad \sec^2\theta - 1 = \tan^2\theta.$$

Accordingly, if the integrand involves:

(I)	$\sqrt{a^2 - x^2}$,	set $x = a\sin\theta$,
(II)	$\sqrt{a^2 + x^2}$,	set $x = a\tan\theta$,
(III)	$\sqrt{x^2 - a^2}$,	set $x = a\sec\theta$.

These substitutions are easy to remember if we associate with each of the cases a right triangle whose sides are a, x, and a suitable radical.[†] The labeling of the triangle depends on the particular case at hand. For example, in (II) it is clear that the hypotenuse must be $\sqrt{a^2 + x^2}$, while in (I) the hypotenuse must be a, and in (III) the hypotenuse is x.

Example 2 Find the indefinite integral $I_2 := \int \dfrac{x^2\,dx}{\sqrt{9-x^2}}$.

Solution This integral is of type (I) with $a = 3$. We first construct a triangle as shown in Figure A.1.

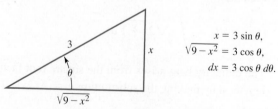

$$x = 3\sin\theta,$$
$$\sqrt{9-x^2} = 3\cos\theta,$$
$$dx = 3\cos\theta\,d\theta.$$

Figure A.1 Triangle for Example 2

[†]The radical sign may not be visible. For example, if the integrand is $1/(a^2 + x^2)^5$, this comes under case (II), because $1/(a^2 + x^2)^5 = 1/(\sqrt{a^2 + x^2})^{10}$.

From this triangle we see immediately that

$$I_2 = \int \frac{x^2\,dx}{\sqrt{9-x^2}} = \int \frac{(3\sin\theta)^2\,3\cos\theta\,d\theta}{3\cos\theta} = 9\int \sin^2\theta\,d\theta.$$

To integrate $\sin^2\theta$ we use the trigonometric identity

$$\sin^2\theta = \frac{1-\cos 2\theta}{2}$$

and obtain

$$I_2 = \frac{9}{2}\int (1-\cos 2\theta)\,d\theta = \frac{9}{2}\theta - \frac{9}{4}\sin 2\theta + C$$

$$= \frac{9}{2}\theta - \frac{9}{2}\sin\theta\cos\theta + C.$$

Returning to the original variable x, we can derive the expressions for θ, $\sin\theta$, and $\cos\theta$ directly from the triangle in Figure A.1. We find that

$$\theta = \arcsin\frac{x}{3}, \quad \sin\theta = \frac{x}{3}, \quad \cos\theta = \frac{\sqrt{9-x^2}}{3}.$$

Hence

$$I_2 = \frac{9}{2}\arcsin\frac{x}{3} - \frac{9}{2}\frac{x}{3}\frac{\sqrt{9-x^2}}{3} + C = \frac{9}{2}\arcsin\frac{x}{3} - \frac{1}{2}x\sqrt{9-x^2} + C. \; \blacklozenge$$

Integration by Parts

The method of integration by parts is of both practical and theoretical importance. It is based on the product rule for differentiation which we now write in the format

$$\frac{d}{dx}(uv) = u\frac{dv}{dx} + v\frac{du}{dx} = uv' + vu'.$$

Integrating both sides of this equation and transposing, we obtain

(4) $$\int uv'\,dx = uv - \int vu'\,dx.$$

Since $dv = v'\,dx$ and $du = u'\,dx$, equation (4) can be put in the more compact form

(5) $$\int u\,dv = uv - \int v\,du.$$

Equation (5) is the basic equation for integration by parts. The idea is that the integrand on the left side may be very complicated, but if we select the part for u and the part for dv quite carefully, the integral on the right side may be much easier. (Some experimentation may be necessary to find the right selection.)

Example 3 Find the indefinite integral $I_3 := \int x \ln x \, dx$.

Solution There are two natural ways of selecting u and dv so that the product is $x \ln x \, dx$:

First possibility:	let $u = x$,	$dv = \ln x \, dx$.
Second possibility:	let $u = \ln x$,	$dv = x \, dx$.

With the first selection we cannot easily find v, the integral of $\ln x$, so we come to a dead end. In the second case we can find v. We arrange the work as follows. Let

$$u = \ln x, \quad dv = x \, dx .$$

Then

$$du = \frac{1}{x} dx, \quad v = \frac{x^2}{2} .$$

Making these substitutions in (5), we have for the left side

$$\int u \, dv = \int (\ln x)(x \, dx) = \int x \ln x \, dx .$$

and for the right side

(6) $\qquad uv - \int v \, du = \frac{x^2}{2} \ln x - \int \frac{x^2}{2} \frac{1}{x} dx = \frac{x^2}{2} \ln x - \frac{1}{2} \int x \, dx .$

Therefore, by (5),

(7) $\qquad I_3 = \int x \ln x \, dx = \frac{x^2}{2} \ln x - \frac{1}{4} x^2 + C . \quad \blacklozenge$

Note that we used $v = x^2/2$ rather than $v = x^2/2 + C$. It is clear that equation (5) is correct for every value of C, and in most cases it is simpler to set $C = 0$. The arbitrary constant that we need can be added in the last step, as we have done in going from equation (6) to (7).

It is sometimes necessary to apply integration by parts more than once in order to evaluate a particular integral, such as

$$I_4 := \int x^2 e^x \, dx .$$

In the first application, we set $u = x^2$, $dv = e^x \, dx$ to deduce that

(8) $\qquad I_4 = x^2 e^x - 2 \int x e^x \, dx$

and to evaluate the last integral in (8) we perform a second application of integration by parts with $u = x$, $dv = e^x \, dx$.

There are some integrals that may appear at first not to be split into two parts, but we need to keep in mind that "1" can be considered as a factor of any expression. For example, to compute the integrals

$$I_5 := \int \arctan(3x) \, dx \qquad \text{and} \qquad I_6 := \int \ln x \, dx .$$

an effective splitting for evaluating I_5 is $u = \arctan(3x)$, $dv = 1 \, dx$, while for I_6 an appropriate choice is $u = \ln x$, $dv = 1 \, dx$. The reader should verify that integration by parts then reduces these integrals to routine problems.

Partial Fractions

Any rational function, that is, any function of the form $R(x) = P(x)/Q(x)$, where $P(x)$ and $Q(x)$ are polynomials, can be integrated in finite terms. The first step in the process is to compare the degree of $P(x)$ with that of $Q(x)$. If deg $P \geq$ deg Q, then division gives

$$(9) \qquad \frac{P(x)}{Q(x)} = s(x) + \frac{p(x)}{Q(x)},$$

where $s(x)$ and $p(x)$ are polynomials and deg $p <$ deg Q. Since the integration of the polynomial $s(x)$ is a trivial matter, we need only focus on the integration of the remainder term $p(x)/Q(x)$.

The idea of the method of partial fractions is to express $p(x)/Q(x)$ as the sum of simpler fractions that are easy to integrate. For example, knowing that

$$(10) \qquad R(x) = \frac{7x^2 + 10x - 1}{x^3 + 3x^2 - x - 3}$$

can be written as

$$\frac{2}{x-1} + \frac{1}{x+1} + \frac{4}{x+3}$$

leads immediately to the evaluation

$$\int R(x)\, dx = 2 \ln|x-1| + \ln|x+1| + 4 \ln|x+3| + C.$$

The key to finding such a simple fraction representation is to first factor the denominator polynomial $Q(x)$. Indeed, in (10) the denominator is just the product $(x-1)(x+1)(x+3)$. In general, we recall that any polynomial $Q(x)$ with real coefficients can be factored into a product of linear and irreducible quadratic factors with real coefficients. There are four cases that may arise in the factorization: **(i)** nonrepeated linear factors; **(ii)** repeated linear factors; **(iii)** nonrepeated quadratic factors; and **(iv)** repeated quadratic factors.

In the case when a nonrepeated linear factor $(x-a)$ occurs in the factorization of $Q(x)$, we associate the partial fraction $A/(x-a)$, where A is a constant to be determined. If a linear factor $(x-a)$ is repeated m times in the factorization of $Q(x)$, we associate the sum of m fractions

$$\sum_{k=1}^{m} \frac{A_k}{(x-a)^k},$$

where the m constants A_k are to be determined. In the case of a nonrepeated quadratic factor $ax^2 + bx + c$ (where $b^2 - 4ac < 0$) we associate the single fraction

$$\frac{Ax + B}{ax^2 + bx + c},$$

where we now have two constants in the numerator to be determined, while if this quadratic factor is repeated m times, then we associate the sum

$$\sum_{k=1}^{m} \frac{A_k x + B_k}{(ax^2 + bx + c)^k},$$

which involves the determination of $2m$ constants A_k and B_k.

We illustrate the approach in the following example. Further discussion of partial fractions appears in Section 7.4, page 370.

Example 4 Find the indefinite integral $I_7 = \displaystyle\int \frac{3x^3 + 3x^2 + 3x + 2}{x^3(x+1)}\,dx.$

Solution Since the factor x is repeated three times in the denominator, while $(x+1)$ is a nonrepeated factor, our partial fraction decomposition must be

(11) $\quad \dfrac{3x^3 + 3x^2 + 3x + 2}{x^3(x+1)} = \dfrac{A}{x+1} + \dfrac{B}{x} + \dfrac{C}{x^2} + \dfrac{D}{x^3},$

where A, B, C, and D are unknowns to be determined. Multiplying both sides of (11) by $x^3(x+1)$, we have

(12) $\quad 3x^3 + 3x^2 + 3x + 2 = Ax^3 + Bx^2(x+1) + Cx(x+1) + D(x+1).$

In order for (12) to be true for all x, the corresponding coefficients must be the same:

The coefficients of x^3 yield $\qquad A + B \qquad = 3.$
The coefficients of x^2 yield $\qquad\qquad B + C \quad = 3.$
The coefficients of x^1 yield $\qquad\qquad\qquad C + D = 3.$
The coefficients of x^0 yield $\qquad\qquad\qquad\qquad D = 2.$

Solving this system of four linear equations in four unknowns gives $A = 1$, $B = 2$, $C = 1$, and $D = 2$. Hence

$$I_7 = \int \left(\frac{1}{x+1} + \frac{2}{x} + \frac{1}{x^2} + \frac{2}{x^3} \right) dx = \ln|(x+1)x^2| - \frac{1}{x} - \frac{1}{x^2} + C. \; \blacklozenge$$

We remark that the coefficients A and D in the representation (11) can be determined directly. Indeed, if we set $x = 0$ in (12), we obtain $2 = D$, and if we set $x = -1$ in (12), we obtain $-3 + 3 - 3 + 2 = A(-1)$ or $A = 1$. But B and C cannot be obtained so readily. If we are fortunate to have a denominator with only nonrepeated linear factors, then this simple substitution method will work to quickly determine all the unknown coefficients.

In dealing with the integration of rational functions that have a quadratic factor in the denominator, the familiar formula

$$\int \frac{1}{1+x^2}\,dx = \arctan x + C$$

plays a central role. For example, by completing the square and making a simple change of variables, we find

$$\int \frac{3}{x^2 + 4x + 5}\,dx = \int \frac{3}{(x+2)^2 + 1}\,dx = 3\arctan(x+2) + C.$$

A EXERCISES

In Problems 1–26, find the indicated indefinite integral.

1. $\displaystyle\int x^3(1+x^2)^{3/2}\,dx$

2. $\displaystyle\int x\cos(2x)\,dx$

3. $\displaystyle\int 4t^4 e^{6t^5}\,dt$

4. $\displaystyle\int x^3 e^{3x}\,dx$

5. $\displaystyle\int \cos^2(\pi\theta)\,d\theta$

6. $\displaystyle\int \frac{x^{1/3}}{x^{1/3}+1}\,dx$

7. $\displaystyle\int \frac{t+1}{t^2+4}\,dt$

8. $\displaystyle\int 10\sec^2(5x)\,dx$

9. $\displaystyle\int \frac{1}{\sqrt{1-2x^2}}\,dx$

10. $\displaystyle\int \frac{1}{2x^2+4x+4}\,dx$

11. $\displaystyle\int \frac{4}{(x-1)(x-2)(x-3)}\,dx$

12. $\displaystyle\int \sec(2x)\tan(2x)\,dx$

13. $\displaystyle\int \frac{\sqrt{9x^2-1}}{x}\,dx$

14. $\displaystyle\int \frac{2x+41}{x^2+5x-14}\,dx$

15. $\displaystyle\int \frac{x^3+2x^2+8}{x^2(x^2+4)}\,dx$

16. $\displaystyle\int \frac{x^3+4x+1}{x^3+3x^2+4x+2}\,dx$

17. $\displaystyle\int \frac{1}{x\ln x}\,dx$

18. $\displaystyle\int \frac{x^2}{x^2-1}\,dx$

19. $\displaystyle\int y\sinh y\,dy$

20. $\displaystyle\int \sec(3\theta+1)\,d\theta$

21. $\displaystyle\int \frac{1}{(x^2+4)^{3/2}}\,dx$

22. $\displaystyle\int 36x^5\exp(2x^3)\,dx$

23. $\displaystyle\int t\ln(t+3)\,dt$

24. $\displaystyle\int \sin\sqrt{x}\,dx$

25. $\displaystyle\int \frac{\sin t}{1+\cos^2 t}\,dt$

26. $\displaystyle\int \frac{e^{3y}}{1+e^{6y}}\,dy$

In Problems 27–34, use an appropriate trigonometric identity to help find the indicated integral.

27. $\displaystyle\int \sin^{2/5}x\cos^3 x\,dx$

28. $\displaystyle\int \tan^3 x\sqrt{\sec x}\,dx$

29. $\displaystyle\int \cos(3x)\cos(7x)\,dx$

30. $\displaystyle\int \cot^3\theta\,d\theta$

31. $\displaystyle\int \tan^4\theta\sec^6\theta\,d\theta$

32. $\displaystyle\int \cos^{1/3}x\sin^3 x\,dx$

33. $\displaystyle\int \sin^2(3x)\cos^2(3x)\,dx$

34. $\displaystyle\int \cot^2 x\csc^4 x\,dx$

APPENDIX B

Newton's Method

To solve an equation $g(x)=0$, we must find the point or points where the graph of $y=g(x)$ meets the *x*-axis. One procedure for approximating a solution is **Newton's method.**

To motivate Newton's method geometrically, we let $\tilde{x}$ be a root of $g(x)=0$ and let x_1 be our guess at the value of $\tilde{x}$. If $g(x_1)=0$, we are done. If $g(x_1)\neq 0$, then we are off by some amount that we call dy (see Figure B.1). Then

$$\frac{dy}{dx}=g'(x_1)\,,$$

and so

$$(1)\qquad dx=\frac{dy}{g'(x_1)}\,.$$

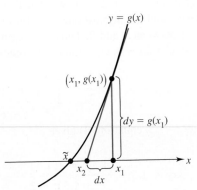

Figure B.1 Tangent line approximation of root

Now $dx = x_1 - x_2$ or $x_2 = x_1 - dx$, where x_2 is at the point where the tangent line through $\left(x_1, g(x_1)\right)$ intersects the x-axis (Figure B.1). Using equation (1) and the fact that $dy = g(x_1)$, we obtain

$$x_2 = x_1 - \frac{dy}{g'(x_1)} = x_1 - \frac{g(x_1)}{g'(x_1)},$$

which we use as the next approximation to the root $\widetilde{x}$.

Repeating this process with x_2 in place of x_1, we obtain the next approximation x_3 to the root $\widetilde{x}$. In general, we find the next approximation x_{n+1} by the formula

(2) $$x_{n+1} = x_n - \frac{g(x_n)}{g'(x_n)}, \quad n = 1, 2, \ldots .$$

The process is illustrated in Figure B.2.

If the initial guess x_1 is sufficiently close to a root $\widetilde{x}$, then the sequence of iterations $\{x_n\}_{n=1}^{\infty}$ usually converges to the root $\widetilde{x}$. However, if we make a bad guess for x_1, then the process may lead away from $\widetilde{x}$.

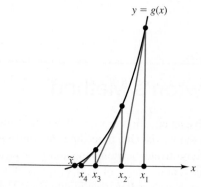

Figure B.2 Sequence of iterations converging to root

Example 1 Find a root to four decimal places of the equation

(3) $$x^3 + 2x - 4 = 0 .$$

Solution Setting $g(x) = x^3 + 2x - 4$, we find that $g'(x) = 3x^2 + 2$ is positive for all x. Hence, g is increasing and has at most one zero. Furthermore, since $g(1) = -1$ and $g(2) = 8$, this zero must lie between 1 and 2. Thus, we begin the procedure with the initial guess $x_1 = 1.5$. For $g(x) = x^3 + 2x - 4$, equation (2) becomes

(4) $$x_{n+1} = x_n - \frac{x_n^3 + 2x_n - 4}{3x_n^2 + 2}, \quad n = 1, 2, \ldots .$$

With $x_1 = 1.5$, equation (4) gives

$$x_2 = 1.5 - \frac{(1.5)^3 + 2(1.5) - 4}{3(1.5)^2 + 2} = 1.5 - \frac{2.375}{8.75} \approx 1.22857 .$$

Using x_2 to compute x_3 and so on, we find

$$x_3 = 1.18085 \,,$$

$$x_4 = 1.17951 \,,$$

$$x_5 = 1.17951 \,,$$

where we have rounded off the computations to five decimal places. Since x_4 and x_5 agree to four decimal places and we are uncertain of the fifth decimal place because of roundoff, we surmise that the root $\widetilde{x}$ of (3) agrees with 1.1795 to four decimal places. Indeed,

$$g(1.1795) = -0.00005 \dots \quad \text{and} \quad g(1.1796) = 0.00056 \dots \,,$$

so $1.1795 < \widetilde{x} < 1.1796$. Consequently, $\widetilde{x} = 1.1795 \dots$ ◆

Observe that Newton's method transforms the problem of finding a root to the equation $g(x) = 0$ into the problem of finding a fixed point for the function $h(x) = x - g(x)/g'(x)$; that is, finding a number x such that $x = h(x)$. [See equation (2).]

Several theorems give conditions that guarantee that the sequence of iterations $\{x_n\}_{n=1}^{\infty}$ defined by (2) will converge to a zero of $g(x)$. We mention one such result.

Convergence of Newton's Method

Theorem 1. Suppose a zero $\widetilde{x}$ of $g(x)$ lies in the interval (a, b) and that in this interval

$$g'(x) > 0 \quad \text{and} \quad g''(x) > 0 \,.$$

If we select x_1 so that $\widetilde{x} < x_1 < b$, then the sequence of iterations defined by (2) will decrease to $\widetilde{x}$.

We do not give a proof of this theorem, but refer the reader to an introductory numerical analysis text such as *Numerical Analysis,* 10th ed., by R. Burden, J. Faires, and A. Burden (Cengage Learning, 2016).

APPENDIX C

Simpson's Rule

A useful procedure for approximating the value of a definite integral is **Simpson's rule.**

Let the interval $[a, b]$ be divided into $2n$ equal parts and let $x_0, x_1, \dots, x_{2n}$ be the points of the partition, that is,

$$x_k := a + kh \,, \quad k = 0, 1, \dots, 2n \,,$$

where $h := (b - a)/(2n)$. If

$$y_k := f(x_k) \,, \quad k = 0, 1, \dots, 2n \,,$$

then the Simpson's rule approximation I_S for the value of the definite integral

$$\int_a^b f(x)\,dx$$

is given by

(1) $I_S = \dfrac{h}{3}[y_0 + 4y_1 + 2y_2 + 4y_3 + \cdots + 2y_{2n-2} + 4y_{2n-1} + y_{2n}]$

$$= \dfrac{h}{3}\sum_{k=1}^n (y_{2k-2} + 4y_{2k-1} + y_{2k})\,.$$

If

$$E := \int_a^b f(x)\,dx - I_S$$

is the error that results from using Simpson's rule to approximate the value of the definite integral, then

(2) $|E| \le \dfrac{(b-a)}{180}h^4 M\,,$

where $M := \max|f^{(4)}(x)|$ for all x in $[a, b]$.

Example 1 Use Simpson's rule with $n = 4$ to approximate the value of the definite integral

(3) $\displaystyle\int_0^1 \dfrac{1}{1+x^2}\,dx\,.$

Solution Here $h = 1/8$, $x_k = k/8$, $k = 0, 1, \ldots, 8$, and

$$y_k = (1 + x_k^2)^{-1} = \dfrac{1}{1 + \dfrac{k^2}{64}} = \dfrac{64}{64 + k^2}\,.$$

By Simpson's rule (1), we find

$$I_S = \dfrac{\left(\frac{1}{8}\right)}{3}\left[1 + 4\left(\dfrac{64}{64+1}\right) + 2\left(\dfrac{64}{64+4}\right) + 4\left(\dfrac{64}{64+9}\right)\right.$$

$$+\, 2\left(\dfrac{64}{64+16}\right) + 4\left(\dfrac{64}{64+25}\right) + 2\left(\dfrac{64}{64+36}\right)$$

$$\left. +\, 4\left(\dfrac{64}{64+49}\right) + \left(\dfrac{64}{64+64}\right)\right] = 0.7854\,.$$

Hence, the value of the definite integral in (3) is approximately $I_S = 0.7854$. ◆

For a more detailed discussion of Simpson's rule, we refer the reader to a numerical analysis book such as *Numerical Analysis,* 10th ed., by R. Burden, J. Faires, and A. Burden (Cengage Learning, 2016).

APPENDIX D
Cramer's Rule

When a system of n linear equations in n unknowns has a unique solution, determinants can be used to obtain a formula for the unknowns. This procedure is called **Cramer's rule.** When n is small, these formulas provide a simple procedure for solving the system.

Suppose that for a system of n linear equations in n unknowns,

$$
(1) \quad
\begin{aligned}
a_{11}x_1 + a_{12}x_2 + \cdots + a_{1n}x_n &= b_1, \\
a_{21}x_1 + a_{22}x_2 + \cdots + a_{2n}x_n &= b_2, \\
\vdots \qquad \vdots \qquad\qquad \vdots \quad\ \ &\ \ \vdots \\
a_{n1}x_1 + a_{n2}x_2 + \cdots + a_{nn}x_n &= b_n,
\end{aligned}
$$

the coefficient matrix

$$
(2) \quad \mathbf{A} :=
\begin{bmatrix}
a_{11} & a_{12} & \cdots & a_{1n} \\
a_{21} & a_{22} & \cdots & a_{2n} \\
\vdots & \vdots & & \vdots \\
a_{n1} & a_{n2} & \cdots & a_{nn}
\end{bmatrix}
$$

has a nonzero determinant. Then Cramer's rule gives the solutions

$$
(3) \quad x_i = \frac{\det \mathbf{A}_i}{\det \mathbf{A}}, \qquad i = 1, 2, \ldots, n,
$$

where $\mathbf{A}_i$ is the matrix obtained from $\mathbf{A}$ by replacing the ith column of $\mathbf{A}$ by the column vector

$$
\begin{bmatrix}
b_1 \\
b_2 \\
\vdots \\
b_n
\end{bmatrix}
$$

consisting of the constants on the right-hand side of system (1).

Example 1 Use Cramer's rule to solve the system

$$
\begin{aligned}
x_1 + 2x_2 - x_3 &= 0, \\
2x_1 + x_2 + x_3 &= 9, \\
x_1 - x_2 - 2x_3 &= 1.
\end{aligned}
$$

Solution We first compute the determinant of the coefficient matrix:

$$
\det
\begin{bmatrix}
1 & 2 & -1 \\
2 & 1 & 1 \\
1 & -1 & -2
\end{bmatrix}
= 12.
$$

Using formula (3), we find

$$
x_1 = \frac{1}{12} \det
\begin{bmatrix}
0 & 2 & -1 \\
9 & 1 & 1 \\
1 & -1 & -2
\end{bmatrix}
= \frac{48}{12} = 4,
$$

$$
x_2 = \frac{1}{12} \det
\begin{bmatrix}
1 & 0 & -1 \\
2 & 9 & 1 \\
1 & 1 & -2
\end{bmatrix}
= \frac{-12}{12} = -1,
$$

$$
x_3 = \frac{1}{12} \det
\begin{bmatrix}
1 & 2 & 0 \\
2 & 1 & 9 \\
1 & -1 & 1
\end{bmatrix}
= \frac{24}{12} = 2. \ \blacklozenge
$$

For a more detailed discussion of Cramer's rule, we refer the reader to *Fundamentals of Matrix Analysis with Applications*, by Edward Barry Saff and Arthur David Snider (John Wiley & Sons, Hoboken, New Jersey, 2016).

APPENDIX E

Method of Least Squares

The **method of least squares** is a procedure for fitting a straight line to a set of measured data. Consider the *scatter diagram* in Figure E.1, consisting of an x, y-plot of some data points $\{(x_i, y_i): i = 1, 2, \ldots, N\}$. It is desired to construct the straight line $y = \alpha + \beta x$ that best fits the data points, in the sense that the sum of the squares of the vertical deviations from the points to the line is minimized.

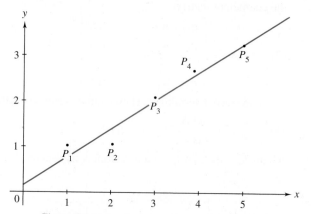

Figure E.1 Scatter diagram and least-squares linear fit

If the y-axis intercept α and the slope β of the line are known, then the y-value on the line corresponding to the measured x-value x_i is given by $\alpha + \beta x_i$. The corresponding measured y-value, y_i, thus deviates from the line by $[y_i - (\alpha + \beta x_i)]$, and the total sum of squares of deviations is

$$S := \sum_{i=1}^{N} [y_i - (\alpha + \beta x_i)]^2 .$$

The interesting feature of the function S is that the symbols x_i and y_i are constants, while α and β are the (unknown) variables. The values of α and β that minimize S will force its partial derivatives to be zero:

$$0 = \frac{\partial S}{\partial \alpha} = \sum_{i=1}^{N} 2[y_i - (\alpha + \beta x_i)](-1) ,$$

$$0 = \frac{\partial S}{\partial \beta} = \sum_{i=1}^{N} 2[y_i - (\alpha + \beta x_i)](-x_i) .$$

Displaying these conditions in a form that emphasizes the roles of α and β, we rewrite them after a little algebra as

$$\alpha N + \beta \sum_{i=1}^{N} x_i = \sum_{i=1}^{N} y_i ,$$

$$\alpha \sum_{i=1}^{N} x_i + \beta \sum_{i=1}^{N} x_i^2 = \sum_{i=1}^{N} x_i y_i$$

and obtain the formulas for the optimal values of intercept and slope:

$$\alpha = \frac{\left(\sum_{i=1}^{N} x_i^2\right)\left(\sum_{i=1}^{N} y_i\right) - \left(\sum_{i=1}^{N} x_i\right)\left(\sum_{i=1}^{N} x_i y_i\right)}{N\sum_{i=1}^{N} x_i^2 - \left(\sum_{i=1}^{N} x_i\right)^2},$$

$$\beta = \frac{N\sum_{i=1}^{N} x_i y_i - \left(\sum_{i=1}^{N} x_i\right)\left(\sum_{i=1}^{N} y_i\right)}{N\sum_{i=1}^{N} x_i^2 - \left(\sum_{i=1}^{N} x_i\right)^2}.$$

Example 1 Find the least-squares linear fit to the data points $P_1(1, 1), P_2(2, 1), P_3(3, 2), P_4(4, 2.5)$, and $P_5(5, 3.1)$, which are plotted in Figure E.1.

Solution Arranging the data as in Table E.1 yields

$$\alpha = \frac{55(9.6) - 15(34.5)}{5(55) - (15)^2} = \frac{10.5}{50} = 0.21,$$

$$\beta = \frac{5(34.5) - 15(9.6)}{5(55) - (15)^2} = \frac{28.5}{50} = 0.57.$$

TABLE E.1				
i	x_i	y_i	$x_i y_i$	x_i^2
1	1	1	1	1
2	2	1	2	4
3	3	2	6	9
4	4	2.5	10	16
5	5	3.1	15.5	25
Sums	15	9.6	34.5	55

Thus, $y = 0.21 + 0.57x$ is the equation for the best-fitting line, which is graphed in Figure E.1. ◆

APPENDIX F

Runge–Kutta Procedure for *n* Equations

Below are program outlines for the classical fourth-order Runge–Kutta algorithm for systems discussed in Section 5.3. The first of these, called the "subroutine," provides approximations to the solution functions of a system of *n* first-order ordinary differential equations over an interval. The second outline on page A-17 utilizes the method of halving the step size in order to obtain approximations (to within a prescribed tolerance) to the solution functions at a single given point.

Classical Fourth-Order Runge–Kutta Subroutine (*n* Equations)

Purpose To approximate the solution to the initial value problem

$$x_i' = f_i(t, x_1, \ldots, x_n) ;$$
$$x_i(t_0) = a_i , \qquad i = 1, 2, \ldots, n$$

for $t_0 \leq t \leq c$.

INPUT $n, t_0, a_1, \ldots, a_n, c, N$ (number of steps), PRNTR ($= 1$ to print table).

Step 1 Set step size

$$h = (c - t_0)/N , \quad t = t_0 , \quad x_1 = a_1 , \ldots , \quad x_n = a_n .$$

Step 2 For $j = 1$ to N do Steps 3–5.

Step 3 Set

$$k_{i,1} = hf_i(t, x_1, \ldots, x_n) ,$$
$$i = 1, \ldots, n ;$$

$$k_{i,2} = hf_i\left(t + \frac{h}{2}, x_1 + \frac{1}{2}k_{1,1}, \ldots, x_n + \frac{1}{2}k_{n,1} \right) ,$$
$$i = 1, \ldots, n ;$$

$$k_{i,3} = hf_i\left(t + \frac{h}{2}, x_1 + \frac{1}{2}k_{1,2}, \ldots, x_n + \frac{1}{2}k_{n,2} \right) ,$$
$$i = 1, \ldots, n ;$$

$$k_{i,4} = hf_i(t + h, x_1 + k_{1,3}, \ldots, x_n + k_{n,3}) ,$$
$$i = 1, \ldots, n .$$

Step 4 Set

$$t = t + h ;$$

$$x_i = x_i + \frac{1}{6}(k_{i,1} + 2k_{i,2} + 2k_{i,3} + k_{i,4}) , \quad i = 1, \ldots, n .$$

Step 5 If PRNTR $= 1$, print $t, x_1, x_2, \ldots, x_n$.

Classical Fourth-Order Runge–Kutta Algorithm with Tolerance (n Equations)

Purpose To approximate the solution to the initial value problem

$$x_i' = f_i(t, x_1, \ldots, x_n) \, ;$$
$$x_i(t_0) = a_i, \quad i = 1, 2, \ldots, n$$

at $t = c$, with tolerance ε.

INPUT $n, t_0, a_1, \ldots, a_n, c$
ε (tolerance)
M (maximum number of iterations)

Step 1 Set $z_i = a_i, \quad i = 1, 2, \ldots, n$; set PRNTR $= 0$.
Step 2 For $m = 0$ to M do Steps 3–7 (or, to save time, start with $m > 0$).
Step 3 Set $N = 2^m$.
Step 4 Call FOURTH-ORDER RUNGE–KUTTA SUBROUTINE (n EQUATIONS).
Step 5 Print $h, x_1, x_2, \ldots, x_n$.
Step 6 If $|z_i - x_i| < \varepsilon$ for $i = 1, \ldots, n$, go to Step 10.
Step 7 Set $z_i = x_i, \quad i = 1, \ldots, n$.
Step 8 Print "$x_i(c)$ is approximately"; x_i (for $i = 1, \ldots, n$); "but may not be within the tolerance"; ε.
Step 9 Go to Step 11.
Step 10 Print "$x_i(c)$ is approximately"; x_i (for $i = 1, \ldots, n$); "with tolerance"; ε.
Step 11 STOP.
OUTPUT Approximations of the solution to the initial value problem at $t = c$ using 2^m steps.

APPENDIX G

Software for Analyzing Differential Equations

In this section we shall list some commercial software and some freeware that we have found to be useful in the various aspects of analyzing systems of differential equations numerically: direction field plotting, Euler and Runge–Kutta codes for solutions, and phase plane experimentation. These items are available at the time of printing this edition, but the accessibility of freeware, in particular, is notoriously tenuous. Each of the web sites mentioned was successfully accessed on July 6, 2016.

Some *commercial products* that are helpful:

MATLAB® (Mathworks, 1 Apple Hill Drive, Natick, MA 01760-2098 USA)
(http://www.mathworks.com/products/matlab/):
 "The MATLAB ODE Suite" at
 http://www.mathworks.com/videos/solving-odes-in-matlab-9-the-matlab-ode-suite-117653.html?s_tid=srchtitle
 "Choose an ODE Solver" at
 http://www.mathworks.com/help/matlab/math/choose-an-ode-solver.html
 "Vector Fields" at
 http://www.mathworks.com/help/matlab/vector-fields.html .

MATHEMATICA® (Wolfram Research, 100 Trade Center Dr, Champaign, IL 61820 USA)
(https://www.wolfram.com/mathematica/):
> **"How to Solve a Differential Equation"** at
> https://reference.wolfram.com/language/howto/SolveADifferentialEquation.html
> **"How to Plot a Direction Field"** at
> https://reference.wolfram.com/language/howto/PlotAVectorField.html .

MAPLE® (Maplesoft, 615 Kumpf Drive, Waterloo, ON N2V 1K8 CAN)
(http://www.maplesoft.com/).
> **"dsolve"** at
> https://www.maplesoft.com/support/help/maple/view.aspx?path=dsolve
> **"dfieldplot"** at
> https://www.maplesoft.com/support/help/maple/view.aspx?path=DEtools%2Fdfieldplot .

PTC Mathcad® (Parametric Technology Corporation, 140 Kendrick Street, Needham, MA
02494 USA)
(http://www.ptc.com/engineering-math-software/mathcad).
> **"Differential Equations worksheet"** in
> PTC Mathcad Worksheet Library - Applied Math.

Some *freeware sources* that are helpful:

Casio Computer Co. Ltd® (6-2, Hon-machi 1-chome, Shibuya-ku, Tokyo 151-8543, Japan)
(http://world.casio.com/)
> **"Euler's Method Calculator"** at
> http://keisan.casio.com/exec/system/1392171850
> **"Runge–Kutta method (2nd-order) Calculator"** at
> http://keisan.casio.com/exec/system/1392171606
> **"Runge–Kutta method (4th-order) Calculator"** at
> http://keisan.casio.com/exec/system/1222997077

The codes *dfield* and *pplane* are graciously provided by our differential equations colleague
John C. Polking of the Department of Mathematics at Rice University. He holds a copyright
and they are not in the public domain; however, they are being made available free for use in
educational institutions.
> **"dfield and pplane: The Java Versions"** (Polking, J. C. and Castellanos, J.) at
> http://math.rice.edu/~dfield/dfpp.html

The Dynamic Web Tools site provides codes dedicated to the FitzHugh-Nagumo equation:
> **"FitzHugh-Nagumo System of Differential Equations"** (Martin, M.) at
> http://math.jccc.edu:8180/webMathematica/JSP/mmartin/fitznag.jsp

The MIT Interactive Mathematics Site provides the following dedicated codes:
> **"Vector Fields"** (Miller, H., MIT Mathlets) at
> http://mathlets.org/mathlets/vector-fields/
> **"Linear Phase Portraits: Matrix Entry"** (Miller, H., MIT Mathlets) at
> http://mathlets.org/mathlets/linear-phase-portraits-matrix-entry/

Answers to Odd-Numbered Problems

CHAPTER 1

Exercises 1.1, page 5

1. ODE, 2nd-order, ind. var. t, dep. var. x, linear
3. ODE, 1st-order, ind. var. x, dep. var. y, nonlinear
5. ODE, 1st-order, ind. var. x, dep. var. y, nonlinear
7. ODE, 1st-order, ind. var. t, dep. var. p, nonlinear
9. ODE, 2nd-order, ind. var. x, dep. var. y, linear
11. PDE, 2nd-order, ind. var. t, r, dep. var. N
13. $dp/dt = kp$, where k is the proportionality constant
15. $dT/dt = k(M - T)$, where k is the proportionality
 constant
17. Kevin wins by $6\sqrt{3} - 4\sqrt{6} \approx 0.594$ sec.

Exercises 1.2, page 13

3. Yes 5. No 7. Yes 9. No 11. Yes
13. Yes 19. The left-hand side is always ≥ 4.
21. (a) $1/3, -3$ (b) $1 \pm \sqrt{6}$ 23. Yes 25. Yes
27. No 29. (b) Yes 31. (a) No
 (c) $y = 2x$ and $y = -2x$

Exercises 1.3, page 21

1. (b)

(c)

Figure B.1
Solution to Problem 1(b)

Figure B.2
Solution to Problem 1(c)

(d) The solutions in parts (b) and (c) both become infinite
 and have the line $y = 2x$ as an asymptote as $x \to \infty$. As
 $x \to -\infty$, the solution in part (b) becomes infinite and
 has $y = -2x$ as an asymptote, but the solution in part (c)
 does not even exist for x negative.
3. All solutions have limiting value 8 as $t \to +\infty$.

Figure B.3 Solutions to Problem 3 satisfying
$v(0) = 5$, $v(0) = 8$, and $v(0) = 15$

5. (a)

Figure B.4 Direction field for Problem 5(a)

(b) $\lim_{t \to \infty} p(t) = 3/2$ (c) $\lim_{t \to \infty} p(t) = 3/2$
(d) No

7. (a)

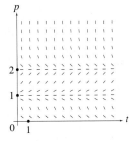

Figure B.5 Direction field for Problem 7(a)

(b) 2 (c) 2 (d) 0 (e) No

9. (d) It increases and asymptotically approaches the line
 $y = x - 1$.
 (f) and (g)

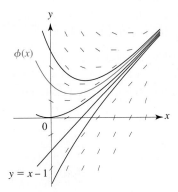

Figure B.6 Direction field and sketch of $\phi(x)$
for Problems 9 (f), (g)

11.

Figure B.7 Solution to Problem 11

13.

Figure B.8 Solution to Problem 13

15.

Figure B.9 Solution to Problem 15

17. It approaches 3.

Exercises 1.4, page 28

1. (rounded to three decimal places)

x_n	0.1	0.2	0.3	0.4	0.5
y_n	4.000	3.998	3.992	3.985	3.975

3. (rounded to three decimal places)

x_n	0.1	0.2	0.3	0.4	0.5
y_n	1.100	1.220	1.362	1.528	1.721

5. (rounded to three decimal places)

x_n	1.1	1.2	1.3	1.4	1.5
y_n	0.100	0.209	0.325	0.444	0.564

7. (rounded to three decimal places)

N	h	y_N
1	π	3.142
2	$\pi/2$	1.571
4	$\pi/4$	1.207
8	$\pi/8$	1.148

9.

x_n	y_n
1.1	−0.9
1.2	−0.81654
1.3	−0.74572
1.4	−0.68480
1.5	−0.63176
1.6	−0.58511
1.7	−0.54371
1.8	−0.50669
1.9	−0.47335
2.0	−0.44314

11. $x(1)$ stabilizes at 1.56 ± 0.01 for h below 2^{-10}; $[\tan(1) = 1.5574\ldots] \; x(t_0) = 1$ for some t_0 in $(0.78, 0.80)$; $[\arctan(1) = 0.785\ldots]$.

15. **(a)** $T(30) = 311.7$ **(b)** $T(60) = 298.2$

Review Problems, page 29

1. ind. var. t, dep. var. x, nonlinear
3. ind. var. y, dep. var. x, linear
5. ind. var. t, dep. var. v, linear
7. True **9.** False **11.** True **13.** True
15. $y_n = \left(1 - \dfrac{1}{n}\right)^n$, $y(1) = e^{-1} \approx 0.3679$
17. **(a)** 500 **(b)** No **(c)** 500

CHAPTER 2

Exercises 2.2, page 46

1. No **3.** Yes **5.** Yes
7. $y = \pm\sqrt[4]{\ln(x^4) + C}$
9. $e^{2x}(2x - 1) + 4e^{-t}(t + 1) = C$
11. $v = \pm\dfrac{\sqrt{1 - Cx^{-8/3}}}{2}$
13. $\dfrac{y}{\sqrt{1 + y^2}} - x^3 = C$
15. $y = \pm\sqrt{C\exp\left[\exp(-x^2)\right] - 1}$
17. $y = \tan\left(-\ln|\cos x| + \dfrac{\pi}{3}\right)$
19. $y = (\sin x + 1)^2 - 1$
21. $y^2/2 + \ln y = \sin\theta - \theta\cos\theta + 1/2 - \pi$
23. $y = \arctan(t^2 + 1)$
25. $y = 4e^{x^3/3} - 1$
27. **(a)** $y(x) = \displaystyle\int_0^x e^{t^2}\, dt$

(b) $y(x) = \left(1 + 3\displaystyle\int_0^x e^{t^2}\, dt\right)^{1/3}$

(c) $y(x) = \tan\left(\displaystyle\int_0^x \sqrt{1 + \sin t}\, dt + \pi/4\right)$

(d) $y(0.5) \approx 1.381$

29. **(d)** $\partial f/\partial y$ is not continuous at $(0, 0)$.
31. **(a)** $x^2 + y^{-2} = C$

(b) $\dfrac{1}{\sqrt{1 - x^2}}$; $\dfrac{1}{\sqrt{4 - x^2}}$; $\dfrac{1}{\sqrt{1/4 - x^2}}$

(c) $-1 < x < 1$; $-2 < x < 2$; $-\dfrac{1}{2} < x < \dfrac{1}{2}$

(d) $|x| < \dfrac{1}{a}$ is domain.

33. 28.1 kg
35. **(a)** 82.2 min **(b)** 31.8 min
(c) Never attains desired temperature

37. (a) $1105.17 **(b)** 27.73 years
(c) $4427.59
39. *A* wins.

Exercises 2.3, page 54

1. Linear **3.** Both **5.** Neither
7. $y = (1/2)e^{3x} + Ce^x$
9. $r = \sin\theta + C\cos\theta$
11. $y = -t - 2 + Ce^t$
13. $x = y^3 + Cy^{-2}$
15. $y = 1 + C(x^2 + 1)^{-1/2}$
17. $y = xe^x - x$
19. $x = \dfrac{t^3}{6}\ln t - \dfrac{t^3}{36} + \dfrac{1}{2t} - \dfrac{17}{36t^3}$
21. $y = x^2\cos x - \pi^2\cos x$
23. $y(t) = (38/3)e^{-5t} - (8/3)e^{-20t}$
25. (b) $y(3) \approx 0.183$
27. (b) 0.9960 **(c)** 0.9486 , 0.9729
29. $x = e^{4y}/2 + Ce^{2y}$
31. (a) $y = x - 1 + Ce^{-x}$
(b) $y = x - 1 + 2e^{-x}$
(c) $y = x/3 - 1/9 + Ce^{-3x}$
(d) $y = \begin{cases} x - 1 + 2e^{-x}, & \text{if } 0 \le x \le 2 \\ x/3 - 1/9 + (4e^6/9 + 2e^4)e^{-3x}, & \text{if } 2 < x \end{cases}$
(e)

Figure B.10 Solution to Problem 31(e)

33. (a) $y = x$ is only solution in a neighborhood of $x = 0$.
(b) $y = -3x + Cx^2$ satisfies $y(0) = 0$ for any C.
35. (a) 0.0281 kg/L **(b)** 0.0598 kg/L
37. $x(t) = \dfrac{1}{2} - \dfrac{2\cos(\pi t/12)}{4 + (\pi/12)^2} - \dfrac{(\pi/12)\sin(\pi t/12)}{4 + (\pi/12)^2}$
$\qquad + \left(\dfrac{19}{2} + \dfrac{2}{4 + (\pi/12)^2}\right)e^{-2t}$
39. 71.8°F at noon; 26.9°F at 5 P.M.

Exercises 2.4, page 64

1. Linear with y as dep. var.
3. Separable
5. Separable, also linear with x as dep. var.
7. Exact
9. $y = (C - 3x)/(x^2 - 1)$
11. $e^x \sin y - x^3 + \sqrt[5]{y} = C$
13. $y = \dfrac{(t-1)e^t + C}{1 + e^t}$
15. $r = (C - e^\theta)\sec\theta$
17. Not exact
19. $x^2 - y^2 + \arctan(xy) = C$

21. $\ln x + x^2y^2 - \sin y = \pi^2$
23. $y = -2/(te^t + 2)$
25. $\sin x - x \cos x = \ln y + 1/y + \pi - 1$ (equation is separable, not exact)
27. (a) $-\ln|y| + f(x)$
(b) $\cos x \sin y - y^2/2 + f(x)$, where f is a function of x only
29. (c) $y = x^2/(C - x)$ **(d)** Yes, $y \equiv 0$
33. (a) $x = cy^2$; $x \equiv 0$; $y \equiv 0$
(b) $x^2 + 4y^2 = c$
(c) $2y^2\ln y - y^2 + 2x^2 = c$
(d) $2x^2 + y^2 = c$

Exercises 2.5, page 69

1. Integrating factor depending on x alone
3. Exact
5. Linear with x as dep. var., has integrating factor depending on x alone
7. $\mu = y^{-4}$; $x^2y^{-3} - y^{-1} = C$ and $y \equiv 0$
9. $\mu = x^{-2}$; $y = x^4/3 - x \ln|x| + Cx$ and $x \equiv 0$
11. $\mu = y^{-2}$; $x^2y^{-1} + x = C$ and $y \equiv 0$
13. $\mu = xy$; $x^2y^3 - 2x^3y^2 = C$
15. (a) $\mu(z) = \exp\left[\int H(z)\,dz\right]$; $z = xy$
(b) $\ln(x^3y) + 2xy = 2$
17. (a) $\dfrac{\partial N/\partial x - \partial M/\partial y}{x^2M - 2xyN}$ depends only on x^2y .
(b) $\exp(x^2y)[x^2 + 2xy] = C$
19. (b) $\mu = e^y$; $x = y - 1 + Ce^{-y}$

Exercises 2.6, page 76

1. Homogeneous and Bernoulli **3.** Bernoulli
5. Homogeneous and Bernoulli
7. $y' = G(ax + by)$
9. $y = -x/(\ln|x| + C)$ and $y \equiv 0$
11. $y = x/(\ln|x| + C)$ and $y \equiv 0, x \equiv 0$
13. $\sqrt{1 + x^2/t^2} = \ln|t| + C$
15. $(x^2 - 4y^2)^3x^2 = C$
17. $y = (x + C)^2/4 - x$ and $y = -x$
19. $y = x + (6 + 4Ce^{2x})/(1 + Ce^{2x})$ and $y = x + 4$
21. $y = 2/(Cx - x^3)$ and $y \equiv 0$
23. $y = 5x^2/(x^5 + C)$ and $y \equiv 0$
25. $x^{-2} = 2t^2\ln|t| + Ct^2$ and $x \equiv 0$
27. $r = \theta^2/(C - \theta)$ and $r \equiv 0$
29. $2\arctan\left(\dfrac{y-3}{x+2}\right) - \ln\left[(x+2)^2 + (y-3)^2\right] = C$.
31. $(2x + 2y - 3)^3 = C(2x + y - 2)^2$
33. $x^2 + t^2 - Ct = 0$ and $t \equiv 0$
35. $y = -3/(x^4 + Cx)$ and $y \equiv 0$
37. $y = (\theta/4)(\ln|\theta| + C)^2$ and $\theta \equiv 0$
39. $\sin(x + y) + \cos(x + y) = Ce^{x-y}$
41. $(x - y + 2)^2 = Ce^{2x} + 1$
45. $(y - 4x)^2(y + x)^3 = C$
47. (a) $v' + [2\,Pu + Q]v = -P$
(b) $y = x + 5x/(C - x^5)$

Review Problems, page 79

1. $e^x + ye^{-y} = C$

3. $x^2y - x^3 + y^{-2} = C$

5. $y + x\sin(xy) = C$

7. $t = C\exp(1/(7y^7))$

9. $(x^2 + 4y^2)^3 x^2 = C$

11. $\tan(t - x) + t = C$

13. $y = -(x^2/2)\cos(2x) + (x/4)\sin(2x) + Cx$

15. $y = 2x + 3 - (x + C)^2/4$

17. $y = 2/(1 + Ce^{2\theta})$ and $y \equiv 0$

19. $y^2 = x^2 + Cx^3$ and $x \equiv 0$

21. $xy - x^2 - x + y^2/2 - 4y = C$

23. $y^2 + 2xy - x^2 = C$

25. $x^2y^{-2} - 2xy^{-1} - 4xy^{-2} = C$ and $y \equiv 0$

27. $\left[(y-4)^2 - 3(x-3)^2\right]\left[\sqrt{3}(x-3) + (y-4)\right]$
$$/\left[\sqrt{3}(x-3) - (y-4)\right]\Big|^{1/\sqrt{3}} = C$$

29. $x^4y^3 - 3x^3y^2 + x^4y^2 = C$ **31.** $y = -x^3/2 + 7x/2$

33. $x = -t - 2 + 3e^{-t}$ **35.** $y = -2x\sqrt{2x^2 - 1}$

37. $\ln\left[(y-2)^2 + 2(x-1)^2\right]$
$$+ \sqrt{2}\arctan\left[\frac{y-2}{\sqrt{2}(x-1)}\right] = \ln 2$$

39. $y = \sqrt{(19x^4 - 1)/2}$

41. $y(t) = e^{-t}\int_2^t \frac{e^r}{1 + r^2}\,dr + 3e^{-(t-2)},\quad y(3) \approx 1.1883\dots$

CHAPTER 3

Exercises 3.2, page 100

1. $5 - 4.5e^{-2t/25}$ kg; 5.07 min

3. $(0.4)(100 - t) - (3.9 \times 10^{-7})(100 - t)^4$ L; 19.96 min

5. 0.0097%; 73.24 h **7.** 20 min later; $1/e$ times as salty

9. 110,868 **13.** 5970; 6000 **15.** 6572; 6693

17. (a)

Year	$\dfrac{1}{p}\dfrac{dp}{dt}$	Logistic (Least Squares)
1790	0.0351	3.29
1800	0.0363	4.51
1810	0.0331	6.16
1820	0.0335	8.41
1830	0.0326	11.45
1840	0.0359	15.53
1850	0.0356	20.97
1860	0.0267	28.14
1870	0.0260	37.45
1880	0.0255	49.31
1890	0.0210	64.07

Year	$\dfrac{1}{p}\dfrac{dp}{dt}$	Logistic (Least Squares)
1900	0.0210	81.89
1910	0.0150	102.65
1920	0.0162	125.87
1930	0.0073	150.64
1940	0.0145	175.80
1950	0.0185	200.10
1960	0.0134	222.47
1970	0.0114	242.16
1980	0.0098	258.81
1990	0.0132	272.44
2000	0.0097	283.28
2010		291.73

(b) $p_1 = 316.920$, $A = 0.00010050$

(c) $p_0 = 3.28780$ (using all data, including 2010)

(d) See table in part (a).

19. $(1/2)\ln(15) \approx 1.354$ yr; 14 million tons per yr

21. 1 hr; 2 h **23.** 11.7% **25.** 31,606 yr

27. e^{-2t} kg of Hh, $2e^{-t} - 2e^{-2t}$ kg of It, and $1 - 2e^{-t} + e^{-2t}$ kg of Bu

Exercises 3.3, page 107

1. 20.7 min **3.** 22.6 min **5.** 9:08 A.M.

7. 28.3°C; 32.5°C; 1:16 P.M.

9. 16.3°C; 19.1°C; 31.7°C; 28.9°C **11.** 39.5 min

13. 148.6°F

15. $T - M = C(T + M)\exp[2\arctan(T/M) - 4M^3kt]$;
for T near M, $M^4 - T^4 \approx 4M^3(M - T)$, and so
$dT/dt \approx k_1(M - T)$, where $k_1 = 4M^3k$

Exercises 3.4, page 115

1. $(0.981)t + (0.0981)e^{-10t} - 0.0981$ m; 1019 sec

3. 18.6 sec **5.** $4.91t + 22.55 - 22.55e^{-2t}$ m; 97.3 sec

7. 241 sec

9. $95.65t + 956.5e^{-t/10} - 956.5$ m; 13.2 sec

11. $e^{bv}|bv - mg|^{mg} = e^{v_0 b}|bv_0 - mg|^{mg}e^{-b^2x/m}$

13. 2.69 sec; 101.19 m **15.** $(\omega_0 - T/k)e^{-kt/I} + T/k$

17. 300 sec

19. $2636e^{-t/20} + 131.8t - 2636$ m; 1.768 sec

21. $5e^{-2t}/2 + 6t - 5/2$; 6 m/sec **23.** Sailboat B

25. (e) 11.18 km/sec **(f)** 2.38 km/sec

Exercises 3.5, page 121

1. $I = \{1.44e^{-100t} + \cos 120t + 1.2\sin 120t\}/2.44$;
$E_L = (-7.2e^{-100t} - 6\sin 120t + 7.2\cos 120t)/2.44$

3. $-(\ln .4) \times 10^{-10} \approx 9.2 \times 10^{-11}$ sec

5. $VI = IRI = I^2R$; $VI = L\dfrac{dI}{dt}I = \dfrac{d}{dt}\dfrac{LI^2}{2}$;
$$VI = E_C\frac{dCE_C}{dt} = \frac{d}{dt}\frac{CE_C^2}{2}$$

7. $-(10\ln .1)/3 \approx 7.68$ sec

2okokokok.kook.okI'll transcribe this page.

Exercises 3.6, page 129

3.

h	"e"
1	3
0.1	2.72055
0.01	2.71830
0.001	2.71828
0.0001	2.71828

7.

x_n	y_n
1.1	0.10450
1.2	0.21668
1.3	0.33382
1.4	0.45300
1.5	0.57135

9.

x_n	y_n
0.2	0.61784
0.4	1.23864
0.6	1.73653
0.8	1.98111
1.0	1.99705
1.2	1.88461
1.4	1.72447
1.6	1.56184
1.8	1.41732
2.0	1.29779

11. $\phi(1) \approx x(1; 2^{-3}) = 1.25494$

13. $\phi(1) \approx y(1; 2^{-3}) = 0.71698$

15. $x = 1.27$

17.

x_n	$y_n(h = 0.2)$	$y_n(h = 0.1)$	$y_n(h = 0.025)$
0.1		-1	0.06250
0.2	-3	1	0.00391
0.3		-1	0.00024
0.4	9	1	0.00002
0.5		-1	0.00000
0.6	-27	1	0.00000
0.7		-1	0.00000
0.8	81	1	0.00000
0.9		-1	0.00000
1.0	-243	1	0.00000

We conclude that step size can dramatically affect convergence.

19.

Time	$K = 0.2$	$K = 0.4$	$K = 0.6$
	T_n		
Midnight	65.0000	65.0000	65.0000
4 A.M.	69.1639	68.5644	68.1300
8 A.M.	71.4836	72.6669	73.6678
Noon	72.9089	75.1605	76.9783
4 P.M.	72.0714	73.5977	74.7853
8 P.M.	69.8095	69.5425	69.2831
Midnight	68.3852	67.0500	65.9740

Exercises 3.7, page 139

1. $y_{n+1} = y_n + h \cos(x_n + y_n)$
$$- \frac{h^2}{2} \sin(x_n + y_n)[1 + \cos(x_n + y_n)]$$

3. $y_{n+1} = y_n + h(x_n - y_n) + \frac{h^2}{2}(1 - x_n + y_n)$
$$- \frac{h^3}{6}(1 - x_n + y_n) + \frac{h^4}{24}(1 - x_n + y_n)$$

5. Order 2, $\phi(1) \approx 1.3725$; order 4, $\phi(1) \approx 1.3679$

7. -11.7679 **9.** 1.36789 **11.** $x = 1.41$

13. $x = 0.50$

15.

x_n	y_n
0.5	0.21462
1.0	0.13890
1.5	-0.02668
2.0	-0.81879
2.5	-1.69491
3.0	-2.99510

19. $v(3) \approx 0.24193$ with $h = 0.0625$

21. $z(1) \approx 2.87083$ with $h = 0.03125$

CHAPTER 4

Exercises 4.1, page 156

3. Both approach zero. **5.** 0

7. $y(t) = -(30/61)\cos 3t - (25/61)\sin 3t$

9. $y(t) = -2\cos 2t + (3/2)\sin 2t$

Exercises 4.2, page 164

1. $c_1 e^{t/2} + c_2 e^{-4t}$

3. $c_1 e^{-3t} + c_2 e^{-2t}$

5. $c_1 e^{-4t} + c_2 t e^{-4t}$

7. $c_1 e^{t/2} + c_2 e^{-2t/3}$

9. $c_1 e^{t/2} + c_2 t e^{t/2}$

11. $c_1 e^{-5t/2} + c_2 t e^{-5t/2}$

13. $3e^{-4t}$ **15.** $\frac{4}{3}e^t - \frac{1}{3}e^{3t}$

17. $\left(\frac{7t}{3} + 2\right)e^{3t}$. **19.** $e^{-t} - 2te^{-t}$

21. (a) $ar + b = 0$ (b) $ce^{-bt/a}$

23. $ce^{-4t/5}$ **25.** $ce^{13t/6}$

27. Lin. dep. **29.** Lin. indep. **31.** Lin. dep.

33. If $c_1 \neq 0$, then $y_1 = -(c_2/c_1)y_2$.

35. (a) Lin. indep. (b) Lin. dep. (c) Lin. indep.
(d) Lin. dep.

37. $c_1 e^t + c_2 e^{(-1-\sqrt{5})t} + c_3 e^{(-1+\sqrt{5})t}$

39. $c_1 e^{-2t} + c_2 t e^{-2t} + c_3 e^{2t}$

41. $c_1 e^{-3t} + c_2 e^{-2t} + c_3 e^{2t}$

43. $3 + e^t - 2e^{-t}$

45. (a) $c_1 e^{r_1 t} + c_2 e^{r_2 t} + c_3 e^{r_3 t}$ (where $r_1 = -4.832$,
$r_2 = -1.869$, and $r_3 = 0.701$)
(b) $c_1 e^{r_1 t} + c_2 e^{-r_1 t} + c_3 e^{r_2 t} + c_4 e^{-r_2 t}$
(where $r_1 = 1.176$, $r_2 = 1.902$)
(c) $c_1 e^{-t} + c_2 e^t + c_3 e^{-2t} + c_4 e^{2t} + c_5 e^{3t}$

Exercises 4.3, page 172

1. $c_1 \cos 3t + c_2 \sin 3t$

3. $c_1 e^{3t} \cos t + c_2 e^{3t} \sin t$

5. $c_1 e^{-2t} \cos \sqrt{2}t + c_2 e^{-2t} \sin \sqrt{2}t$

7. $c_1 e^{-t/2} \cos(\sqrt{5}t/2) + c_2 e^{-t/2} \sin(\sqrt{5}t/2)$

9. $c_1 e^t + c_2 e^{7t}$

11. $c_1 e^{-5t} + c_2 t e^{-5t}$

13. $c_1 e^t \cos 5t + c_2 e^t \sin 5t$

15. $c_1 e^{(3+\sqrt{53})t/2} + c_2 e^{(3-\sqrt{53})t/2}$.

17. $c_1 e^{t/2} \cos(3\sqrt{3}t/2) + c_2 e^{t/2} \sin(3\sqrt{3}t/2)$

19. $c_1 e^t + c_2 e^{-t} \cos 2t + c_3 e^{-t} \sin 2t$

21. $2e^{-t} \cos t + 3e^{-t} \sin t$

23. $(\sqrt{2}/4)\left[e^{(2+\sqrt{2})t} - e^{(2-\sqrt{2})t}\right]$

25. $e^t \sin t - e^t \cos t$

27. $e^{2t} - \sqrt{2}\, e^t \sin \sqrt{2}t$

29. **(a)** $c_1 e^{-t} + c_2 e^t \cos \sqrt{2}t + c_3 e^t \sin \sqrt{2}t$
(b) $c_1 e^{2t} + c_2 e^{-2t} \cos 3t + c_3 e^{-2t} \sin 3t$
(c) $c_1 \cos 2t + c_2 \sin 2t + c_3 \cos 3t + c_4 \sin 3t$

31. **(a)** Oscillatory **(b)** Tends to zero
(c) Tends to $-\infty$ **(d)** Tends to $-\infty$
(e) Tends to $+\infty$

33. **(a)** $y(t) = 0.3e^{-3t} \cos 4t + 0.2e^{-3t} \sin 4t$
(b) $[\pi + \arctan(-1.5)]/4$
(c) $2/\pi$
(d) Decreases the frequency of oscillation, introduces the factor e^{-3t}, causing the solution to decay to zero

35. $b \geq 2\sqrt{Ik}$

37. **(a)** $c_1 \cos t + c_2 \sin t + c_3 t \cos t + c_4 t \sin t$
(b) $(c_1 + c_2 t)e^{-t} \cos(\sqrt{3}t) + (c_3 + c_4 t)e^{-t} \sin(\sqrt{3}t)$

Exercises 4.4, page 180

1. No **3.** Yes **5.** No **7.** Yes

9. $y_p \equiv -3$

11. $y_p(x) = \left[(\ln 2)^2 + 1\right]^{-1} 2^x$

13. $\cos 3t$

15. $xe^x/2 + 3e^x/4$

17. $-2t \cos 2t$

19. $\left(\dfrac{t}{13} + \dfrac{8}{169}\right)te^{-3t}$ **21.** $t^3 e^{2t}/6$

23. $-\dfrac{1}{21}\theta^3 - \dfrac{1}{49}\theta^2 - \dfrac{2}{343}\theta$

25. $e^{2t}(\cos 3t + 6 \sin 3t)$

27. $(A_3 t^4 + A_2 t^3 + A_1 t^2 + A_0 t) \cos 3t$
$\quad + (B_3 t^4 + B_2 t^3 + B_1 t^2 + B_0 t) \sin 3t$

29. $e^t(A_4 t^4 + A_3 t^3 + A_2 t^2 + A_1 t + A_0)$

31. $(A_3 t^4 + A_2 t^3 + A_1 t^2 + A_0 t)e^{-t} \cos t$
$\quad + (B_3 t^4 + B_2 t^3 + B_1 t^2 + B_0 t)e^{-t} \sin t$

33. $(1/5) \cos t + (2/5) \sin t$

35. $\left(\dfrac{1}{10}t^2 - \dfrac{4}{25}t\right)e^t$

Exercises 4.5, page 185

1. **(a)** $5 \cos t$ **(b)** $\cos t - e^{2t}$ **(c)** $4 \cos t + 6e^{2t}$

3. $y = -t + c_1 e^t + c_2 e^{-t}$

5. $\theta = t - 1 + c_1 e^{2t} + c_2 e^{-t}$

7. $y = \tan x + c_1 e^{\sqrt{2}x} + c_2 e^{-\sqrt{2}x}$

9. Yes **11.** No **13.** Yes **15.** No

17. $y(t) = -t^2 + \dfrac{4t}{3} + \dfrac{1}{9} + c_1 e^{3t} + c_2 e^{-t}$.

19. $y = (\cos x - \sin x)e^x/2 + c_1 e^x + c_2 e^{2x}$

21. $y = (1/2)\theta e^{-\theta} \sin \theta + (c_1 \cos \theta + c_2 \sin \theta)e^{-\theta}$

23. $y = e^t - 1$

25. $z = e^{-x} - \cos x + \sin x$

27. $y = -(3/10) \cos x - (1/10) \sin x - (1/20) \cos 2x$
$\quad + (3/20) \sin 2x$

29. $y = -(1/2) \sin \theta - (1/3)e^{2\theta} + (3/4)e^\theta + (7/12)e^{-\theta}$

31. $y_p = (A_1 t + A_0)t \cos t + (B_1 t + B_0)t \sin t + C \cdot 10^t$

33. $x_p(t) = (A \cos t + B \sin t)e^t + C_2 t^2 + C_1 t + C_0$
$\quad + D_1 \cos 3t + D_2 \sin 3t + E_1 \cos t + E_2 \sin t$

35. $y_p = (A_1 t + A_0) \cos 3t + (B_1 t + B_0) \sin 3t + Ce^{5t}$

37. $y_p = t^2 + 3t - 1$

39. $y_p = (t/10 - 4/25)te^t - 1/2$

41. **(a)** $y_1 = -(2\cos 2t + \sin 2t)e^{-t} + 2$
for $0 \leq t \leq 3\pi/2$
(b) $y_2 = y_h = (c_1\cos 2t + c_2\sin 2t)e^{-t}$
for $t > 3\pi/2$
(c) $c_1 = -2(e^{3\pi/2} + 1)$, $c_2 = -(e^{3\pi/2} + 1)$

43. $y = -\cos t + (1/2) \sin t - (1/2)e^{-3t} + 2e^{-t}$

45. **(a)** $y(t) = \dfrac{2V\cos(\pi/2V)}{V^2 - 1} \sin t$ for $V \neq 1$;

$\quad y(t) = \dfrac{\pi}{2} \sin t$ for $V = 1$

(b) $V \approx 0.73$

47. **(a)** $2 \sin 3t - \cos 6t$ **(b)** No solution
(c) $c \sin 3t - \cos 6t$, where c is any constant

Exercises 4.6, page 191

1. $c_1 \cos 2t + c_2 \sin 2t$
$\quad - (1/4)(\cos 2t) \ln|\sec 2t + \tan 2t|$

3. $c_1 e^t + c_2 te^t + te^t \ln|t|$

5. $c_1 \cos 4\theta + c_2 \sin 4\theta + (\theta/4) \sin 4\theta$
$\quad + (1/16)(\cos 4\theta) \ln|\cos 4\theta|$

7. $(2\ln t - 3)t^2 e^{-2t}/4 + c_1 e^{-2t} + c_2 te^{-2t}$

9. $y_p = -2t - 4$

11. $y = -(\cos t) \ln|\sec t + \tan t| + (1/10)e^{3t}$
$\qquad\qquad -1 + c_1 \cos t + c_2 \sin t$

13. $c_1 \cos 2t + c_2 \sin 2t + (1/24)\sec^2 2t - 1/8$
$\quad + (1/8)(\sin 2t)\ln|\sec 2t + \tan 2t|$

15. $c_1 \cos t + c_2 \sin t - t^2 + 3 + 3t \sin t$
$\quad + 3(\cos t)\ln|\cos t|$

17. $c_1 \cos 2t + c_2 \sin 2t - e^t/5$
$\quad - (1/2)(\cos 2t)\ln|\sec 2t + \tan 2t|$

19. $y = e^{1-t} - e^{t-1} + \dfrac{e^t}{2}\displaystyle\int_1^t \dfrac{e^{-u}}{u}du - \dfrac{e^{-t}}{2}\displaystyle\int_1^t \dfrac{e^u}{u}du$

$\quad [y(2) \approx -1.93]$

21. 0.3785

23. $c_1 e^t + c_2(t + 1) - t^2$

25. $c_1(5t - 1) + c_2 e^{-5t} - t^2 e^{-5t}/10$

Exercises 4.7, page 199

1. Unique solution on $(-\pi/2, \pi/2)$

3. Unique solution on $(0, \infty)$

5. Does not apply; $t = 0$ is a point of discontinuity

7. Does not apply; not an initial value problem

9. $c_1 t + c_2 t^{-7}$

11. $c_1 t^{-2} + c_2 t^{-2} \ln t$

13. $c_1 t^{-1/3} + c_2 t^{-1/3} \ln t$

15. $c_1 t \cos\left[2\ln(-t)\right] + c_2 t \sin\left[2\ln(-t)\right]$

17. $t^{-4}\left\{c_1 \cos\left[\ln(-t)\right] + c_2 \sin\left[\ln(-t)\right]\right\}$

19. $t - 3t^4$

21. $c_1(t-2) + c_2(t-2)^7$

23. (c) $t^\alpha \cos(\beta\ln|t|), t^\alpha \sin(\beta\ln|t|); t^r, t^r \ln|t|$

25. (a) True **(b)** False

27. (e) No, because the coefficient of y'' vanishes at $t = 0$ and the equation cannot be written in standard form.

29. Otherwise their Wronskian would be zero at t_0, contradicting linear independence.

31. (a) Yes **(b)** No **(c)** Yes **(d)** Yes

33. Cte^{-t}

35. $1 + 2t - t^2$

37. $c_1 \cos(3\ln t) + c_2 \sin(3\ln t)$
 $+ (1/9)\cos(3\ln t)\ln\left|\sec(3\ln t) + \tan(3\ln t)\right|$

39. $c_1 t + c_2 t \ln t + (1/2)t(\ln t)^2 + 3t(\ln t)\left[\ln|\ln t|\right]$

41. t^4

43. $t + 1$

45. $(t-1)e^{2t}/2$

47. (a) $(1 - 2t^2)\int (1 - 2t^2)^{-2} e^{t^2} dt$

 (b) $(3t - 2t^3)\int (3t - 2t^3)^{-2} e^{t^2} dt$

49. $tw'' + 2tw' + (t+1)w = 0$

51. (a) $\phi'(t_0) = \lim\limits_{n\to\infty} \dfrac{\phi(t_n) - \phi(t_0)}{t_n - t_0} = \lim\limits_{n\to\infty} \dfrac{0 - 0}{t_n - t_0} = 0$

Exercises 4.8, page 210

1. Let $Y(t) = y(-t)$. Then
 $Y'(t) = -y'(-t), Y''(t) = y''(-t)$. But
 $y''(s) - sy(s) = 0$, so $y''(-t) + ty(-t) = 0$ or
 $Y''(t) + tY(t) = 0$.

3. The spring stiffness is $(-6y)$, so it opposes negative displacements $(y < 0)$ and reinforces positive displacements $(y > 0)$. Initially $y < 0$ and $y' < 0$, so the (positive) stiffness reverses the negative velocity and restores y to 0. Thereafter, $y > 0$ and the negative stiffness drives y to $+\infty$.

5. (a) $y'' = 2y^3 = \dfrac{d}{dy}(y^4/2)$. Thus, by setting $K = 0$ and choosing the $(-)$ sign in equation (11), we get
 $t = -\int \dfrac{dy}{\sqrt{2y^4/2}} + c = \dfrac{1}{y} + c$, or $y = 1/(t - c)$.

 (b) Linear dependence would imply
 $\dfrac{y_1(t)}{y_2(t)} = \dfrac{1/(t - c_1)}{1/(t - c_2)} = \dfrac{t - c_2}{t - c_1} \equiv$ constant
 in a neighborhood of 0, which is false if $c_1 \neq c_2$.

 (c) If $y(t) = 1/(t - c)$, then $y(0) = -1/c$,
 $y'(0) = -1/c^2 = -y(0)^2$, which is false for the given data.

7. (a) The velocity, which is always perpendicular to the lever arm, is $\ell\, d\theta/dt$. Thus, (lever arm) times (perpendicular momentum) $= \ell m\ell\, d\theta/dt = m\ell^2 d\theta/dt$.

 (b) The component of the gravitational force perpendicular to lever arm is $mg\sin\theta$, and is directed toward decreasing θ. Thus, torque $= -\ell mg\sin\theta$.

 (c) Torque $= \dfrac{d}{dt}$ (angular momentum) or
 $-\ell mg\sin\theta = (m\ell^2\theta')' = m\ell^2\theta''$.

9. 2 or -2

11. The sign of the damping coefficient $(y')^2 - 1$ indicates that low velocities are boosted by negative damping but that high velocities are slowed. Hence, one expects a limit cycle.

13. (a) Airy **(b)** Duffing **(c)** van der Pol

15. (a) Yes $(t^2 =$ positive stiffness)

 (b) No $(-t^2 =$ negative stiffness)

 (c) Yes $(y^4 =$ positive stiffness)

 (d) No $(y^5 =$ negative stiffness for $y < 0)$

 (e) Yes $(4 + 2\cos t =$ positive stiffness)

 (f) Yes (positive stiffness and damping)

 (g) No (negative stiffness and damping)

17. $1/4\sqrt{2}$

Exercises 4.9, page 220

1. $y(t) = -(1/4)\cos 5t - (1/5)\sin 5t$
 amplitude $= \sqrt{41}/20$; period $= 2\pi/5$;
 frequency $= 5/2\pi$; $\left[\pi - \arctan(5/4)\right]/5$ sec

3. $b = 0$: $y(t) = \cos 4t$

Figure B.11 $b = 0$

$b = 6$: $y(t) = e^{-3t}\cos\sqrt{7}t + (3/\sqrt{7})e^{-3t}\sin\sqrt{7}t$
 $= (4/\sqrt{7})e^{-3t}\sin(\sqrt{7}t + \phi)$, where
 $\phi = \arctan\sqrt{7}/3 \approx 0.723$

Figure B.12 $b = 6$

$b = 8$: $y(t) = (1 + 4t)e^{-4t}$

Figure B.13 $b = 8$

$b = 10$: $y(t) = (4/3)e^{-2t} - (1/3)e^{-8t}$

Figure B.14 $b = 10$

5. $k = 20$: $y(t) = [(1 + \sqrt{5})/2]e^{(-5+\sqrt{5})t}$
 $+ [(1 - \sqrt{5})/2]e^{(-5-\sqrt{5})t}$

Figure B.15 $k = 20$

$k = 25$: $y(t) = (1 + 5t)e^{-5t}$

Figure B.16 $k = 25$

$k = 30$: $y(t) = e^{-5t}\cos\sqrt{5}t + \sqrt{5}e^{-5t}\sin\sqrt{5}t$
 $= \sqrt{6}e^{-5t}\sin(\sqrt{5}t + \phi)$, where
 $\phi = \arctan(1/\sqrt{5}) \approx 0.421$

Figure B.17 $k = 30$

7. $y(t) = (-3/4)e^{-8t}\cos 8t - e^{-8t}\sin 8t$
 $= (5/4)e^{-8t}\sin(8t + \phi)$, where
 $\phi = \pi + \arctan(3/4) \approx 3.785$;
 damp. factor $= (5/4)e^{-8t}$;
 quasiperiod $= \pi/4$; quasifreq. $= 4/\pi$
9. 0.242 m
11. $(10/\sqrt{9999})\arctan(\sqrt{9999}) \approx 0.156$ sec
13. Relative extrema at
 $t = [\pi/3 + n\pi - \arctan(\sqrt{3}/2)]/(2\sqrt{3})$
 for $n = 0, 1, 2, \ldots$; but touches curves $\pm\sqrt{7/12}e^{-2t}$
 at $t = [\pi/2 + m\pi - \arctan(\sqrt{3}/2)]/(2\sqrt{3})$
 for $m = 0, 1, 2, \ldots$.
15. First measure half the quasiperiod P as the time between
 two successive zero crossings. Then compute the ratio
 $y(t + P)/y(t) = e^{-(b/2m)P}$.

Exercises 4.10, page 227

1. $M(\gamma) = 1/\sqrt{(1 - 4\gamma^2)^2 + 4\gamma^2}$

Figure B.18

3. $y(t) = \cos 3t + (1/3)t\sin 3t$

Figure B.19

5. (a) $y(t) = -[F_0/(k - m\gamma^2)]\cos(\sqrt{k/m}\,t)$
 $+ [F_0/(k - m\gamma^2)]\cos\gamma t$
 $= (F_0/[m(\omega^2 - \gamma^2)])(\cos\gamma t - \cos\omega t)$
(c) $y(t) = \sin 8t \,\sin t$

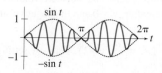

Figure B.20

7. $y(t) = c_1 e^{r_1 t} + c_2 e^{r_2 t} + \dfrac{F_0\sin(\gamma t + \theta)}{\sqrt{(k - m\gamma^2)^2 + b^2\gamma^2}}$,
 where $r_1, r_2 = -(b/2m) \pm (1/2m)\sqrt{b^2 - 4mk}$
 and $\tan\theta = (k - m\gamma^2)/(b\gamma)$ as in equation (7)
9. $y_p(t) = (0.08)\cos 2t + (0.06)\sin 2t$
 $= (0.1)\sin(2t + \theta)$, where
 $\theta = \arctan(4/3) \approx 0.927$

11. $y(t) = -(18/85)e^{-2t}\cos 6t - (22/255)e^{-2t}\sin 6t$
$+ (2/\sqrt{85})\sin(2t + \theta)$, where
$\theta = \arctan(9/2) \approx 1.352$;
res. freq. $= 2\sqrt{2}/\pi$ cycles/sec

13. $y_p(t) = (3/185)(8\sin 4t - 11\cos 4t)$

15. Amp $= \sqrt{\left(-\dfrac{11}{986}\right)^2 + \left(-\dfrac{3}{1972}\right)^2} \approx 0.01\,(\mathrm{m})$,

freq. $= 2/\pi$

Review Problems, page 231

1. $c_1 e^{-9t} + c_2 e^t$

3. $c_1 e^{t/2}\cos(3t/2) + c_2 e^{t/2}\sin(3t/2)$

5. $c_1 e^{3t/2} + c_2 e^{t/3}$

7. $c_1 e^{-t/3}\cos(t/6) + c_2 e^{-t/3}\sin(t/6)$

9. $c_1 e^{7t/4} + c_2 t e^{7t/4}$

11. $t^{1/2}\{c_1\cos\left[(\sqrt{19}/2)\ln t\right] + c_2\sin\left[(\sqrt{19}/2)\ln t\right]\}$

13. $c_1\cos 4t + c_2\sin 4t + (1/17)te^t - (2/289)e^t$

15. $c_1 e^{-2t} + c_2 e^{-t} + c_3 e^{-t/3}$

17. $c_1 e^t + c_2 e^{-t/2}\cos(\sqrt{43}t/2) + c_3 e^{-t/2}\sin(\sqrt{43}t/2)$

19. $c_1 e^{-3t} + c_2 e^{t/2} + c_3 t e^{t/2}$

21. $c_1 e^{3t/2}\cos(\sqrt{19}t/2) + c_2 e^{3t/2}\sin(\sqrt{19}t/2) - e^t/5$
$+ t^2 + 6t/7 + 4/49$

23. $c_1\cos 4\theta + c_2\sin 4\theta$
$- (1/16)(\cos 4\theta)\ln|\sec 4\theta + \tan 4\theta|$

25. $c_1 e^{3t/2} + c_2 t e^{3t/2} + e^{3t}/9 + e^{5t}/49$

27. $c_1 x + c_2 x^{-2} - 2x^{-2}\ln x + x\ln x$

29. $e^{-2t}\cos(\sqrt{3}t)$

31. $2e^t\cos 3t - (7/3)e^t\sin 3t - \sin 3t$

33. $-e^{-t} - 3e^{5t} + e^{8t}$

35. $\cos\theta + 2\sin\theta + \theta\sin\theta + (\cos\theta)\ln|\cos\theta|$

37. **(a)**, **(c)**, **(e)**, and **(f)** have all solutions bounded as
$t \to +\infty$

39. $y_p(t) = (1/4)\sin 8t$; $\sqrt{62}/2\pi$

CHAPTER 5

Exercises 5.2, page 249

1. **(a)** $-t^3 + 3t^2 + 8$
(b) $-2t^3 + 3t^2 + 6t + 16$
(c) $2t^3 + 3t^2 - 16$
(d) $-2t^3 + 3t^2 + 6t + 16$
(e) $-2t^3 + 3t^2 + 6t + 16$

3. $x = c_1 + c_2 e^{-2t}$; $y = c_2 e^{-2t}$

5. $x \equiv -5$; $y \equiv 1$

7. $u = c_1 - (1/2)c_2 e^{-t} + (1/2)e^t + (5/3)t$;
$v = c_1 + c_2 e^{-t} + (5/3)t$

9. $x = c_1 e^t + (1/4)\cos t - (1/4)\sin t$;
$y = -3c_1 e^t - (3/4)\cos t - (1/4)\sin t$

11. $u = c_1\cos 2t + c_2\sin 2t + c_3 e^{\sqrt{3}t} + c_4 e^{-\sqrt{3}t}$
$- (3/10)e^t$;
$v = c_1\cos 2t + c_2\sin 2t - (2/5)c_3 e^{\sqrt{3}t}$
$- (2/5)c_4 e^{-\sqrt{3}t} + (1/5)e^t$

13. $x = 2c_2 e^t\cos 2t - 2c_1 e^t\sin 2t$;
$y = c_1 e^t\cos 2t + c_2 e^t\sin 2t$

15. $w = -(2/3)c_1 e^{2t} + c_2 e^{7t} + t + 1$;
$z = c_1 e^{2t} + c_2 e^{7t} - 5t - 2$

17. $x = c_1\cos t + c_2\sin t - 4c_3\cos(\sqrt{6}t)$
$- 4c_4\sin(\sqrt{6}t)$;
$y = c_1\cos t + c_2\sin t + c_3\cos(\sqrt{6}t) + c_4\sin(\sqrt{6}t)$

19. $x = 2e^{3t} - e^{2t}$;
$y = -2e^{3t} + 2e^{2t}$

21. $x = e^t + e^{-t} + \cos t + \sin t$;
$y = e^t + e^{-t} - \cos t - \sin t$

23. Infinitely many solutions satisfying
$x + y = e^t + e^{-2t}$

25. $x(t) = -c_1 e^t - 2c_2 e^{2t} - c_3 e^{3t}$,
$y(t) = c_1 e^t + c_2 e^{2t} + c_3 e^{3t}$,
$z(t) = 2c_1 e^t + 4c_2 e^{2t} + 4c_3 e^{3t}$

27. $x(t) = c_1 e^{8t} + c_2 e^{4t} + c_3$,
$y(t) = \dfrac{1}{2}(c_1 e^{8t} - c_2 e^{4t} + c_3)$,
$z(t) = -c_1 e^{8t} + c_3$

29. $-3 \le \lambda \le -1$

31. $x(t) = -\left(10 + \dfrac{20}{\sqrt{7}}\right)e^{r_1 t} - \left(10 - \dfrac{20}{\sqrt{7}}\right)e^{r_2 t}$
$+ 20$ kg ,
$y(t) = \dfrac{30}{\sqrt{7}}e^{r_1 t} - \dfrac{30}{\sqrt{7}}e^{r_2 t} + 20$ kg, where
$r_1 = \dfrac{-5 - \sqrt{7}}{100} \approx -0.0765$,
$r_2 = \dfrac{-5 + \sqrt{7}}{100} \approx -0.0235$

33. $x = \left[\dfrac{20 - 10\sqrt{5}}{\sqrt{5}}\right]e^{(-3+\sqrt{5})t/100}$
$- \left[\dfrac{20 + 10\sqrt{5}}{\sqrt{5}}\right]e^{(-3-\sqrt{5})t/100} + 20$;
$y = -\left(\dfrac{10}{\sqrt{5}}\right)e^{(-3+\sqrt{5})t/100} + \left(\dfrac{10}{\sqrt{5}}\right)e^{(-3-\sqrt{5})t/100} + 20$

35. $90.4°\mathrm{F}$ **37.** $460/11 \approx 41.8°\mathrm{F}$

Exercises 5.3, page 259

1. $x_1' = x_2$, $x_2' = 3x_1 - tx_2 + t^2$;
$x_1(0) = 3$, $x_2(0) = -6$

3. $x_1' = x_2$, $x_2' = x_3$, $x_3' = x_4$,
$x_4' = x_4 - 7x_1 + \cos t$; $x_1(0) = x_2(0) = 1$,
$x_3(0) = 0$, $x_4(0) = 2$

5. $x_1' = x_2$, $x_2' = x_2 - x_3 + 2t$,
$x_3' = x_4$, $x_4' = x_1 - x_3 - 1$;
$x_1(3) = 5$, $x_2(3) = 2$, $x_3(3) = 1$, $x_4(3) = -1$

7. $x_1' = x_2$, $x_2' = x_3$, $x_3' = x_4 + t$, $x_4' = x_5$,
$x_5' = (2x_4 - 2x_3 + 1)/5$; $x_1(0) = x_2(0) = x_3(0) = 4$,
$x_4(0) = x_5(0) = 1$

9. $t_{n+1} = t_n + h, \quad n = 0, 1, 2, \ldots;$

$$x_{i,n+1} = x_{i,n} + \frac{h}{2}\left[f_i(t_n, x_{1,n}, \ldots, x_{m,n}) + \right.$$
$$f_i(t_n + h, x_{1,n} + hf_1(t_n, x_{1,n}, \ldots, x_{m,n}), \ldots, $$
$$\left. x_{m,n} + hf_m(t_n, x_{1,n}, \ldots, x_{m,n}))\right]$$
$$i = 1, 2, \ldots, m$$

11.

i	t_i	$y(t_i)$
1	0.250	0.750000
2	0.500	0.625000
3	0.750	0.573529
4	1.000	0.563603

13.

i	t_i	$y(t_i)$
1	0.250	0.25000
2	0.500	0.50000
3	0.750	0.75000
4	1.000	1.00000

15. $y(1) \approx x_1(1, 2^{-2}) = 1.69, y'(1) \approx 1.82$

17. $u(1; 2^{-2}) = v(1; 2^{-2}) = 0.36789$

19.

i	t_i	Part (a) $x(t_i)$	$y(t_i)$	Part (b) $x(t_i)$	$y(t_i)$	Part (c) $x(t_i)$	$y(t_i)$
1	0.5	1.95247	2.25065	1.48118	2.42311	0.91390	2.79704
2	1.0	3.34588	1.83601	2.66294	1.45358	1.63657	1.13415
3	1.5	4.53662	3.36527	5.19629	2.40348	4.49334	1.07811
4	2.0	2.47788	4.32906	3.10706	4.64923	5.96115	5.47788
5	2.5	1.96093	2.71900	1.92574	3.32426	1.51830	5.93110
6	3.0	2.86412	1.96166	2.34143	2.05910	0.95601	2.18079
7	3.5	4.28449	2.77457	3.90106	2.18977	2.06006	0.98131
8	4.0	3.00965	4.11886	3.83241	3.89043	5.62642	1.38072
9	4.5	2.18643	3.14344	2.32171	3.79362	5.10594	5.10462
10	5.0	2.63187	2.25824	2.21926	2.49307	1.74187	5.02491

21.

i	t_i	$x_1(t_i) \approx H(t_i)$
1	0.5	0.09573
2	1.0	0.37389
3	1.5	0.81045
4	2.0	1.37361
5	2.5	2.03111
6	3.0	2.75497
7	3.5	3.52322
8	4.0	4.31970
9	4.5	5.13307
10	5.0	5.95554

23. Yes, yes

25. $y(1) \approx x_1(1; 2^{-3}) = 1.25958$

27. $y(0.1) \approx 0.00647, \ldots, y(2.0) \approx 1.60009$

29. (a) $P_1(10) \approx 0.567, P_2(10) \approx 0.463, P_3(10) \approx 0.463$

(b) $P_1(10) \approx 0.463, P_2(10) \approx 0.567, P_3(10) \approx 0.463$

(c) $P_1(10) \approx 0.463, P_2(10) \approx 0.463, P_3(10) \approx 0.567$

All populations approach 0.5.

Exercises 5.4, page 271

1. $x = y^3, \quad y > 0$

Figure B.21

3. $\left(1/\sqrt{2}, 1/\sqrt{2}\right), \quad \left(-1/\sqrt{2}, -1/\sqrt{2}\right)$ **5.** $(0, 0)$

7. $e^x + ye^{-y} = c$ **9.** $e^x + xy - y^2 = c$

11. $y^2 - x^2 = c$ **13.** $(x - 1)^2 + (y - 1)^2 = c$

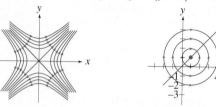

Figure B.22 Figure B.23

15. $(-2, 1)$ is a saddle point (unstable).

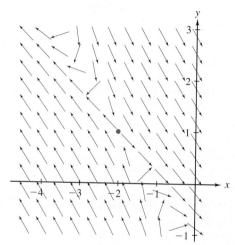

Figure B.24

17. $(0, 0)$ is a center (stable).

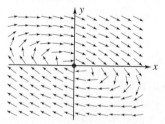

Figure B.25

19. $(0, 0)$ is a saddle point (unstable).

Figure B.26

21. $(0,0)$ is a center (stable).

Figure B.27

23. $(0,0)$ is a center (stable); $(1,0)$ is a saddle point (unstable).

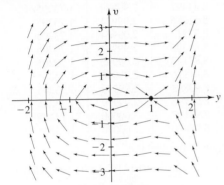

Figure B.28

25. (a) Periodic **(b)** Not periodic
 (c) Critical point (periodic)

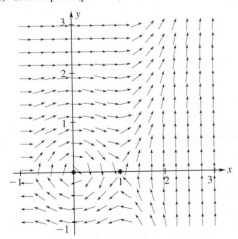

Figure B.29

27. $(x(t), y(t))$ approaches $(0,0)$.

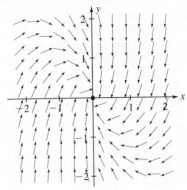

Figure B.30

29. Critical points are $(1,0)$ and $(x,1)$ with x arbitrary. Phase plane solutions are $x = 1$, and $y = C(x-1)$. Solution starting at $(3,2)$ goes to infinity. The others converge to $(1,0)$.

31. (a) $y' = v, \quad v' = f(y)$

 (b) $\dfrac{dv}{dy} = \dfrac{f(y)}{v}$ implies $\displaystyle\int v\,dv = \int f(y)\,dy +$ constant, or
 $v^2/2 = F(y) + K$

33. (a) $x' = v, v' = -x + 1/(\lambda - x)$

 (b) $\dfrac{dv}{dx} = \dfrac{-x + 1/(\lambda - x)}{v}$ implies

 $\displaystyle\int v\,dv = \int \left(-x + \dfrac{1}{\lambda - x}\right)dx +$ constant, or

 $v^2/2 = -x^2/2 - \ln(\lambda - x) +$ constant. The solution equation follows with the choice $C/2$ for the constant.

 (c) At a critical point $v = 0$ and $-x + 1/(\lambda - x) = 0$.

 Solutions for the latter are $x = \dfrac{\lambda \pm \sqrt{\lambda^2 - 4}}{2}$, and

 are real only for $\lambda \geq 2$.

 (d)

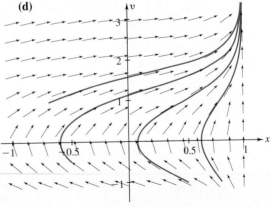

Figure B.31 $\lambda = 1$

(e) See Figures B.31 and B.32. When $\lambda = 3$, one critical point is a center and the other is a saddle. For $\lambda = 1$, the bar is attracted to the magnet. For $\lambda = 3$, the bar may oscillate periodically, or (rarely) come to rest at the saddle critical point.

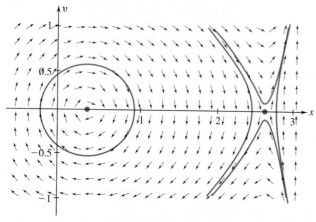

Figure B.32 $\lambda = 3$

35. (c) For upper half-plane, center is at $v = 0$, $y = -1$; for lower half-plane, center is at $v = 0$, $y = 1$.
(d) All points on the segment $v = 0$, $-1 \le y \le 1$
(e) $y = -0.5$; see Figure B.33.

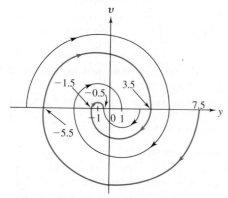

Figure B.33

Exercises 5.5, page 281

1. Runge–Kutta approximations

t_n	$p_n(r = 1.5)$	$p_n(r = 2)$	$p_n(r = 3)$
0.25	1.59600	1.54249	1.43911
0.5	2.37945	2.07410	1.64557
0.75	3.30824	2.47727	1.70801
1.0	4.30243	2.72769	1.72545
1.25	5.27054	2.86458	1.73024
1.5	6.13869	2.93427	1.73156
1.75	6.86600	2.96848	1.73192
2.0	7.44350	2.98498	1.73201
2.25	7.88372	2.99286	1.73204
2.5	8.20933	2.99661	1.73205
2.75	8.44497	2.99839	1.73205
3.0	8.61286	2.99924	1.73205
3.25	8.73117	2.99964	1.73205
3.5	8.81392	2.99983	1.73205
3.75	8.87147	2.99992	1.73205
4.0	8.91136	2.99996	1.73205
4.25	8.93893	2.99998	1.73205
4.5	8.95796	2.99999	1.73205
4.75	8.97107	3.00000	1.73205
5.0	8.98010	3.00000	1.73205

Limiting population is $3^{1/(r-1)}$.

3. $x(t) = \dfrac{1}{2} - \dfrac{2\cos(\pi t/12)}{4 + (\pi/12)^2} - \dfrac{(\pi/12)\sin(\pi t/12)}{4 + (\pi/12)^2}$
$\qquad + \left(\dfrac{19}{2} + \dfrac{2}{4 + (\pi/12)^2} \right) e^{-2t}$

5. (a)

Figure B.34

(b) 295 persons

7. $P(t) = \exp\left\{ \dfrac{c[1 - \exp(-bt)]}{b} - bt \right\}$,
$\qquad Q(t) = \exp\left\{ \dfrac{c[1 - \exp(-bt)]}{b} - bt \right\}[\exp(bt) - 1]$

9. $N(t) = \dfrac{N_0(c + s)}{N_0 s + (c + s - N_0 s)e^{-(c+s)t}}$, $N_0 := N(0)$

11. Roughly, 2

Exercises 5.6, page 287

1. $m_1 x'' = k_1(y - x)$,
$\quad m_2 y'' = -k_1(y - x) - k_2 y$;
$\quad x(0) = -1, \ x'(0) = 0, \ y(0) = 0, \ y'(0) = 0$
$\quad x(t) = -(8/17)\cos t - (9/17)\cos\left(\sqrt{20/3}\,t\right)$,
$\quad y(t) = -(6/17)\cos t + (6/17)\cos\left(\sqrt{20/3}\,t\right)$

3. $mx'' = -kx + k(y - x)$,
$\quad my'' = -k(y - x) + k(z - y)$,
$\quad mz'' = -k(z - y) - kz$;

The normal frequency $(1/2\pi)\sqrt{\left(2 + \sqrt{2}\right)(k/m)}$ has the mode $x(t) = z(t) = -\left(1/\sqrt{2}\right)y(t)$; the normal frequency $(1/2\pi)\sqrt{\left(2 - \sqrt{2}\right)(k/m)}$ has the mode $x(t) = z(t) = \left(1/\sqrt{2}\right)y(t)$; and the normal frequency $(1/2\pi)\sqrt{2k/m}$ has the mode $x(t) = -z(t), y(t) \equiv 0$.

5. $x(t) = -e^{-t} - te^{-t} - \cos t$; $y(t) = e^{-t} + te^{-t} - \cos t$

7. $x(t) = (2/5)\cos t + (4/5)\sin t - (2/5)\cos\sqrt{6}t$
$\qquad + \left(\sqrt{6}/5\right)\sin\sqrt{6}t - \sin 2t$;
$\quad y(t) = (4/5)\cos t + (8/5)\sin t + (1/5)\cos\sqrt{6}t$
$\qquad - \left(\sqrt{6}/10\right)\sin\sqrt{6}t - (1/2)\sin 2t$

9. $(1/2\pi)\sqrt{g/l}$; $(1/2\pi)\sqrt{(g/l) + (2k/m)}$

Exercises 5.7, page 294

1. $I(t) = \left(19/\sqrt{21}\right)\left[e^{(-25 - 5\sqrt{21})t/2} - e^{(-25 + 5\sqrt{21})t/2}\right]$

3. $I_p(t) = (4/51)\cos 20t - (1/51)\sin 20t$; resonance frequency is $5/\pi$.

5. $M(\gamma) = 1/\sqrt{(100 - 4\gamma^2)^2 + 100\gamma^2}$

Figure B.35

7. $L = 35$ H, $R = 10\ \Omega$, $C = 1/15$ F, and $E(t) = 50\cos 10t$ V

11. $I_1 = (3/5)e^{-3t/2} - (8/5)e^{-2t/3} + 1$,
$\quad I_2 = (1/5)e^{-3t/2} - (6/5)e^{-2t/3} + 1$,
$\quad I_3 = (2/5)e^{-3t/2} - (2/5)e^{-2t/3}$

13. $(1/2)I_1' + 2q_3 = \cos 3t$ (where $I_3 = q_3'$) ,
$\quad (1/2)I_1' + I_2 = 0, I_1 = I_2 + I_3$:
$\quad I_1(0) = I_2(0) = I_3(0) = 0$;
$\quad I_1 = -(36/61)e^{-t}\cos\sqrt{3}t$
$\qquad - \left(42\sqrt{3}/61\right)e^{-t}\sin\sqrt{3}t + (36/61)\cos 3t$
$\qquad + (30/61)\sin 3t$,
$\quad I_2 = (45/61)e^{-t}\cos\sqrt{3}t - \left(39\sqrt{3}/61\right)e^{-t}\sin\sqrt{3}t$
$\qquad - (45/61)\cos 3t + (54/61)\sin 3t$,
$\quad I_3 = -(81/61)e^{-t}\cos\sqrt{3}t$
$\qquad - \left(3\sqrt{3}/61\right)e^{-t}\sin\sqrt{3}t + (81/61)\cos 3t$
$\qquad - (24/61)\sin 3t$

Exercises 5.8, page 303

1. For $\omega = 3/2$: The Poincaré map alternates between the points $(0.8, 1.5)$ and $(0.8, -1.5)$. There is a subharmonic solution of period 4π. For $\omega = 3/5$: The Poincaré map cycles through the points $(-1.5625, 0.6)$, $(-2.1503, -0.4854)$, $(-0.6114, 0.1854)$, $(-2.5136, 0.1854)$, and $(-0.9747, -0.4854)$. There is a subharmonic solution of period 10π.

3. The points become unbounded.

5. The attractor is the point $(-1.0601, 0.2624)$.

9. (a) $\{1/7, 2/7, 4/7, 1/7, \dots\}$,
$\qquad \{3/7, 6/7, 5/7, 3/7, \dots\}$
$\quad$ **(b)** $\{1/15, 2/15, 4/15, 8/15, 1/15, \dots\}$,
$\qquad \{1/5, 2/5, 4/5, 3/5, 1/5, \dots\}$,
$\qquad \{1/3, 2/3, 1/3, \dots\}$,
$\qquad \{7/15, 14/15, 13/15, 11/15, 7/15, \dots\}$
$\quad$ **(c)** $x_n = 0$ for $n \geq j$

13.

Figure B.36

Review Problems, page 306

1. $x(t) = -(c_1/3)t^3 - (c_2/2)t^2 - (c_3 + 2c_1)t + c_4$,
$\quad y(t) = c_1 t^2 + c_2 t + c_3$

3. $x(t) = c_1\cos 3t + c_2\sin 3t + e^t/10$,
$\quad y(t) = (3/2)(c_1 + c_2)\cos 3t - (3/2)(c_1 - c_2)\sin 3t$
$\qquad - (11/20)e^t - (1/4)e^{-t}$

5. $x(t) = 2\sin t, \ y(t) = e^t - \cos t + \sin t$,
$\quad z(t) = e^t + \cos t + \sin t$

7. $x(t) = -(13.9/4)e^{-t/6} - (4.9/4)e^{-t/2} + 4.8$,
$\quad y(t) = -(13.9/2)e^{-t/6} + (4.9/2)e^{-t/2} + 4.8$

9. $x_1' = x_2, \ x_2' = x_3, \ x_3' = (1/3)(5 + e^t x_1 - 2x_2)$

11. $x_1' = x_2, \ x_2' = x_3, \ x_3' = t - x_5 - x_6$,
$\quad x_4' = x_5, \ x_5' = x_6, \ x_6' = x_2 - x_3$

13. $x^2 - (y - 1)^2 = c$; critical point $(0, 1)$ is saddle (unstable).

15. Asymptotically stable spiral point

17. $I_1 + I_2 + I_3 = 0$, $q/C = R_2 I_2$,
$\quad R_2 I_2 = R_1 I_3 + L\, dI_3/dt$, where q is the charge on the capacitor $(I_1 = dq/dt)$;
$\quad I_3 = e^{-t}(A\cos t + B\sin t)$,
$\quad I_2 = e^{-t}(B\cos t - A\sin t)$,
$\quad I_1 = e^{-t}[(A - B)\sin t - (A + B)\cos t]$.

CHAPTER 6

Exercises 6.1, page 326

1. $(-\infty, 0)$ **3.** $(3\pi/2, 5\pi/2)$ **5.** $(0, \infty)$
7. Lin. indep. **9.** Lin. dep.
11. Lin. indep. **13.** Lin. indep.
15. $c_1 e^{3x} + c_2 e^{-x} + c_3 e^{-4x}$
17. $c_1 x + c_2 x^2 + c_3 x^3$
19. (a) $c_1 e^x + c_2 e^{-x} \cos 2x + c_3 e^{-x} \sin 2x + x^2$
 (b) $-e^x + e^{-x} \sin 2x + x^2$
21. (a) $c_1 x + c_2 x \ln x + c_3 x (\ln x)^2 + \ln x$
 (b) $3x - x \ln x + x(\ln x)^2 + \ln x$
23. (a) $2 \sin x - x$
 (b) $4x - 6 \sin x$
29. (b) Let $f_1(x) = |x-1|$ and $f_2(x) = x - 1$
33. e^{2x}, $(\sin x - 2 \cos x)/5$, $-(2 \sin x + \cos x)/5$
35. $xy''' - y'' + xy' - y = 0$

Exercises 6.2, page 332

1. $c_1 + c_2 e^{2x} + c_3 e^{-4x}$
3. $c_1 e^{-x} + c_2 e^{-2x/3} + c_3 e^{x/2}$
5. $c_1 e^{-x} + c_2 e^{-x} \cos 5x + c_3 e^{-x} \sin 5x$
7. $c_1 e^{-x} + c_2 e^{(3+\sqrt{65})x/4} + c_3 e^{(3-\sqrt{65})x/4}$
9. $c_1 e^{3x} + c_2 x e^{3x} + c_3 x^2 e^{3x}$
11. $c_1 e^{-x} + c_2 x e^{-x} + c_3 x^2 e^{-x} + c_4 x^3 e^{-x}$
13. $c_1 \cos \sqrt{2}x + c_2 x \cos \sqrt{2}x + c_3 \sin \sqrt{2}x + c_4 x \sin \sqrt{2}x$
15. $c_1 e^x + c_2 x e^x + c_3 e^{-3x} + (c_4 + c_5 x) e^{-x} \cos 2x$
 $+ (c_6 + c_7 x) e^{-x} \sin 2x$
17. $c_1 e^{-4x} + c_2 e^{3x} + (c_3 + c_4 x + c_5 x^2) e^{-2x}$
 $+ (c_6 + c_7 x) e^{-2x} \cos x + (c_8 + c_9 x) e^{-2x} \sin x + c_{10}$
 $+ c_{11} x + c_{12} x^2 + c_{13} x^3 + c_{14} x^4$
19. $e^x - 2e^{-2x} - 3e^{2x}$
21. $e^{2x} - \sqrt{2} e^x \sin \sqrt{2}x$
23. $x(t) = c_1 + c_2 t + c_3 e^t$, $y(t) = c_1 - c_2 + c_2 t$
27. $c_1 e^{1.120x} + c_2 e^{0.296x} + c_3 e^{-0.520x} + c_4 e^{-2.896x}$
29. $c_1 e^{-0.5x} \cos(0.866x) + c_2 e^{-0.5x} \sin(0.866x)$
 $+ c_3 e^{-0.5x} \cos(1.323x) + c_4 e^{-0.5x} \sin(1.323x)$
31. (a) $\{x, x^{-1}, x^2\}$
 (b) $\{x, x^2, x^{-1}, x^{-2}\}$
 (c) $\{x, x^2 \cos(3 \ln x), x^2 \sin(3 \ln x)\}$
33. (b) $x(t) = c_1 \cos t + c_2 \sin t + c_3 \cos \sqrt{6}t$
 $+ c_4 \sin \sqrt{6}t$
 (c) $y(t) = 2c_1 \cos t + 2c_2 \sin t - (c_3/2) \cos \sqrt{6}t$
 $- (c_4/2) \sin \sqrt{6}t$
 (d) $x(t) = (3/5)\cos t + (2/5)\cos \sqrt{6}t$,
 $y(t) = (6/5)\cos t - (1/5)\cos \sqrt{6}t$
35. $c_1 \cosh rx + c_2 \sinh rx + c_3 \cos rx + c_4 \sin rx$, where
 $r^4 = k/(EI)$

Exercises 6.3, page 337

1. $c_1 x e^x + c_2 + c_3 x + c_4 x^2$
3. $c_1 x^2 e^{-2x}$
5. $c_1 e^x + c_2 e^{3x} + c_3 e^{-2x} - (1/6)x e^x + (1/6)x^2$
 $+ (5/18)x + 37/108$

7. $c_1 e^x + c_2 e^{-2x} + c_3 x e^{-2x} - (1/6)x^2 e^{-2x}$
9. $c_1 e^x + c_2 x e^x + c_3 x^2 e^x + (1/6)x^3 e^x$
11. D^5
13. $D + 7$
15. $(D-2)(D-1)$
17. $[(D+1)^2 + 4]^3$
19. $(D+2)^2[(D+5)^2 + 9]^2$
21. $c_3 \cos 2x + c_4 \sin 2x + c_5$
23. $c_3 x e^{3x} + c_4 x^2 + c_5 x + c_6$
25. $c_3 + c_4 x + c_5 \cos 2x + c_6 \sin 2x$
27. $c_3 x e^{-x} \cos x + c_4 x e^{-x} \sin x + c_5 x^2 + c_6 x + c_7$
29. $c_2 x + c_3 x^2 + c_6 x^2 e^x$
31. $-2e^{3x} + e^{-2x} + x^2 - 1$
33. $x^2 e^{-2x} - x^2 + 3$
39. $x(t) = -(1/63)e^{3t} + c_1 + c_2 t - c_3 e^{\sqrt{2}t} - c_4 e^{-\sqrt{2}t}$,
 $y(t) = (8/63)e^{3t} + c_1 + c_2 t + c_3 e^{\sqrt{2}t} + c_4 e^{-\sqrt{2}t}$

Exercises 6.4, page 341

1. $(1/6)x^2 e^{2x}$
3. $e^{2x}/16$
5. $\ln(\sec x) - (\sin x)\ln(\sec x + \tan x)$
7. $c_1 x + c_2 x^2 + c_3 x^3 - (1/24)x^{-1}$
9. $-(1/2)e^x \int e^{-x} g(x)\,dx + (1/6)e^{-x} \int e^x g(x)\,dx$
 $+ (1/3)e^{2x} \int e^{-2x} g(x)\,dx$
11. $c_1 x + c_2 x^{-1} + c_3 x^3 - x \sin x - 3 \cos x + 3x^{-1} \sin x$

Review Problems, page 343

1. (a) $(0, \infty)$
 (b) $(-4, -1)$, $(-1, 1)$, $(1, \infty)$
5. (a) $e^{-5x}(c_1 + c_2 x) + e^{2x}(c_3 + c_4 x + c_5 x^2)$
 $+ (\cos x)(c_6 + c_7 x) + (\sin x)(c_8 + c_9 x)$
 (b) $c_1 + c_2 x + c_3 x^2 + c_4 x^3 + e^x(c_5 + c_6 x)$
 $+ (e^{-x} \cos \sqrt{3}x)(c_7 + c_8 x)$
 $+ (e^{-x} \sin \sqrt{3}x)(c_9 + c_{10}x)$
7. (a) D^3 (b) $D^2(D-3)$ (c) $[D^2 + 4]^2$
 (d) $[(D+2)^2 + 9]^3$
 (e) $D^3(D+1)^2(D^2 + 4)(D^2 + 9)$
9. $c_1 x + c_2 x^5 + c_3 x^{-1} - (1/21)x^{-2}$

CHAPTER 7

Exercises 7.2, page 360

1. $\dfrac{1}{s^2}$, $s > 0$ **3.** $\dfrac{1}{s-6}$, $s > 6$

5. $\dfrac{s}{s^2 + 4}$, $s > 0$

7. $\dfrac{s-2}{(s-2)^2 + 9}$, $s > 2$

9. $e^{-2s}\left(\dfrac{2s+1}{s^2}\right)$, $s > 0$ **11.** $\dfrac{e^{-\pi s} + 1}{s^2 + 1}$, all s

13. $\dfrac{6}{s+3} - \dfrac{2}{s^3} + \dfrac{2}{s^2} - \dfrac{8}{s}$, $s > 0$

15. $\dfrac{6}{s^4} - \dfrac{1}{(s-1)^2} + \dfrac{s-4}{(s-4)^2+1}$, $s > 4$

17. $\dfrac{6}{(s-3)^2+36} - \dfrac{6}{s^4} + \dfrac{1}{s-1}$, $s > 3$

19. $\dfrac{24}{(s-5)^5} - \dfrac{s-1}{(s-1)^2+7}$, $s > 5$

21. Continuous (hence piecewise continuous)

23. Piecewise continuous

25. Continuous (hence piecewise continuous)

27. Neither

29. All but functions (c), (e), and (h)

Exercises 7.3, page 365

1. $\dfrac{2}{s^3} + \dfrac{2}{(s-1)^2+4}$

3. $\dfrac{s+1}{(s+1)^2+9} + \dfrac{1}{s-6} - \dfrac{1}{s}$

5. $\dfrac{4}{(s+1)^3} - \dfrac{1}{s^2} + \dfrac{s}{s^2+16}$

7. $\dfrac{24}{s^5} - \dfrac{24}{s^4} + \dfrac{12}{s^3} - \dfrac{4}{s^2} + \dfrac{1}{s}$

9. $\dfrac{4(s+1)}{[(s+1)^2+4]^2}$ **11.** $\dfrac{s}{s^2-b^2}$

13. $\dfrac{1}{2s} - \dfrac{s}{2(s^2+4)}$ **15.** $\dfrac{3s}{4(s^2+1)} + \dfrac{s}{4(s^2+9)}$

17. $\dfrac{s}{2(s^2+9)} - \dfrac{s}{2(s^2+49)}$

19. $\dfrac{n+m}{2[s^2+(n+m)^2]} + \dfrac{m-n}{2[s^2+(m-n)^2]}$

21. $\dfrac{s-a}{(s-a)^2+b^2}$

25. (a) $\dfrac{s^2-b^2}{(s^2+b^2)^2}$ **(b)** $\dfrac{2s^3-6sb^2}{(s^2+b^2)^3}$

29. $\dfrac{1}{s^2+6s+10}$ **33.** e^{-s}/s^2

35. $e^{-(\pi/2)s}/(s^2+1)$

Exercises 7.4, page 374

1. $e^t t^3$ **3.** $e^{-t}\cos 3t$ **5.** $(1/2)e^{-2t}\sin 2t$

7. $2e^{-2t}\cos 3t + 4e^{-2t}\sin 3t$

9. $(3/2)e^t\cos 2t - 3e^t\sin 2t$

11. $\dfrac{6}{s+5} - \dfrac{1}{s+2} - \dfrac{4}{s-1}$ **13.** $\dfrac{1}{(s+1)^2} - \dfrac{2}{s}$

15. $-\dfrac{3}{s+1} + \dfrac{(s-1)+2}{(s-1)^2+4}$

17. $-\dfrac{5}{6s} + \dfrac{11}{10(s-2)} - \dfrac{4}{15(s+3)}$

19. $\dfrac{1}{17}\left[\dfrac{1}{s-3} - \dfrac{s+1}{(s+1)^2+1} - \dfrac{4}{(s+1)^2+1} \right]$

21. $(1/3) + e^t + (14/3)e^{6t}$

23. $-e^{-3t} + 2te^{-3t} + 6e^{-t}$

25. $8e^{2t} - e^{-t}\cos 2t + 3e^{-t}\sin 2t$

27. $-(5/3)e^{-t} + (5/12)e^{2t} + (5/4)e^{-2t}$

29. $3e^{-2t} + 7e^t\cos t + 11e^t\sin t$

31. $F_1(s) = F_2(s) = F_3(s) = 1/s^2$,
$\mathcal{L}^{-1}\{1/s^2\}(t) = f_3(t) = t$

33. $e^{5t}/t - e^{-2t}/t$

35. $2(\cos t - \cos 3t)/t$

39. $\dfrac{A}{s} + \dfrac{B}{s-1} + \dfrac{C}{s+2}$, where

$A = \dfrac{2s+1}{(s-1)(s+2)}\bigg|_{s=0} = \dfrac{-1}{2}$,

$B = \dfrac{2s+1}{s(s+2)}\bigg|_{s=1} = 1$,

$C = \dfrac{2s+1}{s(s-1)}\bigg|_{s=-2} = \dfrac{-1}{2}$

41. $2e^{-t} - 4e^{3t} + 5e^{2t}$

43. $\dfrac{4}{s+2} + \dfrac{2(s-1)+2(3)}{(s-1)^2+2^2}$

Exercises 7.5, page 382

1. $2e^t\cos 2t + e^t\sin 2t$ **3.** $-e^{-3t} + 3te^{-3t}$

5. $t^2 + \cos t - \sin t$ **7.** $\cos t - 4e^{5t} + 8e^{2t}$

9. $3(t-1)e^{t-1} - e^{-6(t-1)}$

11. $2 - t + e^{2-t} + 2e^{t-2}$

13. $(7/5)\sin t + (11/5)\cos t + (3/5)e^{2t-\pi} - e^{-(t-\pi/2)}$

15. $\dfrac{-s^2+s-1}{(s^2+1)(s-1)(s-2)}$

17. $\dfrac{s^5+s^4+6}{s^4(s^2+s-1)}$ **19.** $\dfrac{s^3+5s^2-6s+1}{s(s-1)(s^2+5s-1)}$

21. $\dfrac{s^3+s^2+2s}{(s^2+1)(s-1)^2}$ **23.** $\dfrac{-s^3+1+3se^{-2s}-e^{-2s}}{s^2(s^2+4)}$

25. $2e^t - \cos t - \sin t$ **27.** $(t^2-4)e^{-t}$

29. $(3a-b)e^t/2 + (b-a)e^{3t}/2$

31. $5/2 + (a-5/2)e^{-t}\cos t + (a+b-5/2)e^{-t}\sin t$

35. $t^2/2$

37. $\cos t + t\sin t + c(\sin t - t\cos t)$, (c arbitrary)

39. $e(t) = -a\cos\left(\sqrt{k/I}\,t\right)$

41. $e(t) = \left(-2aI/\sqrt{4Ik-\mu^2}\right)e^{-\mu t/(2I)}\sin\left(\sqrt{4Ik-\mu^2}\,t/(2I)\right)$

Exercises 7.6, page 390

1. $2e^{-s}/s^3$ **3.** $e^{-2s}(4s^2+4s+2)/s^3$

5. $(2e^{-s} - e^{-2s} + 2e^{-3s})/s$

7. $[(e^{-s} - e^{-2s})(s+1)]/s^2$

9. $(e^{-s} - 2e^{-2s} + e^{-3s})/s^2$

11. $e^{t-2}u(t-2)$

13. $e^{-2(t-2)}u(t-2) - 3e^{-2(t-4)}u(t-4)$

15. $e^{-2(t-3)}[\cos(t-3) - 2\sin(t-3)]u(t-3)$

17. $(7e^{6-2t} - 6e^{3-t})u(t-3)$

19. $10 - 10u(t-3\pi)[1 + e^{-(t-3\pi)}(\cos t + \sin t)]$
$\qquad + 10u(t-4\pi)[1 - e^{-(t-4\pi)}(\cos t + \sin t)]$

21. $\sin t + [1 - \cos(t-3)]u(t-3)$

Figure B.37

23. $t + [4 - t + \sin(t-2) - 2\cos(t-2)]u(t-2)$

Figure B.38

25. $e^{-t}\cos t + 2e^{-t}\sin t$
$+ (1/2)[1 - e^{2\pi - t}(\cos t + \sin t)]u(t - 2\pi)$
$- (1/2)[1 - e^{4\pi - t}(\cos t + \sin t)]u(t - 4\pi)$
27. $e^{-t} + e^{-2t} + (1/2)[e^{-3t} - 2e^{-2(t+1)} + e^{-(t+4)}]u(t-2)$
29. $\cos 2t + (1/3)[1 - u(t - 2\pi)]\sin t$
$+ (1/6)[8 + u(t - 2\pi)]\sin 2t$
31. $2e^{-2t} - 2e^{-3t}$
$+ [1/36 + (1/6)(t-1) - (1/4)e^{-2(t-1)}$
$+ (2/9)e^{-3(t-1)}]u(t-1)$
$- [19/36 + (1/6)(t-5)$
$- (7/4)e^{-2(t-5)} + (11/9)e^{-3(t-5)}]u(t-5)$
33. $0.04 - 0.02\,e^{-3t/125} + 0.02\,u(t-10)[1 - e^{-3(t-10)/125}]$
35. The resulting differential equation has polynomial
coefficients, so the Laplace transform method will result
in a differential equation for the transform.

Exercises 7.7, page 396

1. $\dfrac{1 - 2se^{-2s} - e^{-2s}}{s^2(1 - e^{-2s})}$

$f(t)$

Figure B.39

3. $\dfrac{1}{1 - e^{-2s}}\left[\dfrac{1 - e^{-s-1}}{s+1} + \dfrac{e^{-s} - e^{-2s}}{s}\right]$

$f(t)$

Figure B.40

5. $\dfrac{1}{s(1 + e^{-as})}$ **7.** $\dfrac{1 - e^{-as}}{as^2(1 + e^{-as})}$

13. $\dfrac{e^{t-n}}{6} - \dfrac{e^{-t}}{2}(1 + e - e^{n+1})$
$+ \dfrac{e^{-2t}}{3}\left[\dfrac{1 + e + e^2 - e^{2n+2}}{e+1}\right]$,
for $n < t < n + 1$

15. $\displaystyle\sum_{n=1}^{\infty} \dfrac{1}{s^n} = \dfrac{1}{s-1}$

17. $\displaystyle\sum_{n=1}^{\infty} \dfrac{(-1)^{n+1}}{2ns^{2n}} = \dfrac{1}{2}\ln\left(1 + \dfrac{1}{s^2}\right)$

19. (a) $\sqrt{\pi/s}$
 (b) $105\sqrt{\pi}/(16s^{9/2})$
27. (a) and (b), but not (c)

Exercises 7.8, page 404

1. $2te^t - e^t + \displaystyle\int_0^t e^{t-v}(t-v)g(v)dv$

3. $\displaystyle\int_0^t g(v)e^{2v-2t}\sin(t-v)dv + e^{-2t}\cos t + 3e^{-2t}\sin t$

5. $1 - \cos t$ **7.** $2e^{5t} - 2e^{-2t}$ **9.** $(t/2)\sin t$
11. $(2/3)e^{-2t} + (1/3)e^t$ **13.** $s^{-2}(s-3)^{-1}$
15. $t/4 + (3/8)\sin 2t$ **17.** $\cos t$ **19.** 3
21. $e^{-t/2}\cos\left(\sqrt{3}t/2\right) - \left(1/\sqrt{3}\right)e^{-t/2}\sin\left(\sqrt{3}t/2\right)$
23. $H(s) = (s^2 + 9)^{-1}$; $h(t) = (1/3)\sin 3t$;
$y_k(t) = 2\cos 3t - \sin 3t$;
$y(t) = (1/3)\displaystyle\int_0^t [\sin 3(t-v)]g(v)dv + 2\cos 3t - \sin 3t$
25. $H(s) = (s^2 - s - 6)^{-1}$; $h(t) = (e^{3t} - e^{-2t})/5$;
$y_k(t) = 2e^{3t} - e^{-2t}$;
$y(t) = (1/5)\displaystyle\int_0^t [e^{3(t-v)} - e^{-2(t-v)}]g(v)dv + 2e^{3t} - e^{-2t}$
27. $H(s) = (s^2 - 2s + 5)^{-1}$; $h(t) = (1/2)e^t\sin 2t$;
$y_k(t) = e^t\sin 2t$;
$y(t) = (1/2)\displaystyle\int_0^t e^{(t-v)}[\sin 2(t-v)]g(v)dv + e^t\sin 2t$
29. $(1/30)\displaystyle\int_0^t e^{-2(t-v)}[\sin 6(t-v)]e(v)dv - e^{-2t}\cos 6t$
$+ e^{-2t}\sin 6t$
31. $t^2/2$

Exercises 7.9, page 410

1. -1 **3.** -1 **5.** e^{-2} **7.** $e^{-s} - e^{-3s}$
9. e^{-s} **11.** 0 **13.** $-(\sin t)u(t - \pi)$
15. $e^t + e^{-3t} + (1/4)(e^{t-1} - e^{3-3t})u(t-1)$
$- (1/4)(e^{t-2} - e^{6-3t})u(t-2)$
17. $2(e^{t-2} - e^{-(t-2)})u(t-2) + 2e^t - t^2 - 2$
19. $e^{-t} - e^{-5t} + (e/4)(e^{1-t} - e^{5-5t})u(t-1)$

21. $\sin t + (\sin t)u(t - 2\pi)$; see Fig. B.41.

Figure B.41

23. $\sin t + (\sin t)u(t - \pi) + (\sin t)u(t - 2\pi)$

Figure B.42

25. $(1/2)e^{-2t}\sin 2t$ **27.** $(1/2)e^t\sin 2t$

29. The mass remains stopped at $x(t) \equiv 0$, $t > \pi/2$

35. $\dfrac{L}{6EI}[3\lambda x^2 - x^3 + (x - \lambda)^3 u(x - \lambda)]$

Exercises 7.10, page 413

1. $x = e^t$; $y = e^t$ **3.** $z = e^t$; $w \equiv 0$

5. $x = (7/4)e^t + (7/4)e^{-t} - (3/2)\cos t$;
 $y = (7/4)e^t - (7/4)e^{-t} + (1/2)\sin t$

7. $x = -(150/17)e^{5t/2}\cos\left(\sqrt{15}t/2\right)$
 $- \left(334\sqrt{15}/85\right)e^{5t/2}\sin\left(\sqrt{15}t/2\right) - (3/17)e^{-3t}$;
 $y = (46/17)e^{5t/2}\cos\left(\sqrt{15}t/2\right)$
 $- \left(146\sqrt{15}/85\right)e^{5t/2}\sin\left(\sqrt{15}t/2\right) + (22/17)e^{-3t}$

9. $x = 4e^{-2t} - e^{-t} - \cos t$; $y = 5e^{-2t} - e^{-t}$

11. $x = (e^t - e^{-t})/2 - (1/2)[e^{t-2} - e^{-(t-2)}]u(t - 2)$;
 $y = 1 - (e^t + e^{-t})/2$
 $- [1 - (e^{t-2} + e^{-(t-2)})/2]u(t - 2)$

13. $x = e^{-t} - (1/2)[e^{-(t-\pi)} + \cos t - \sin t]u(t - \pi)$;
 $y = e^{-t} + [1 - (1/2)e^{-(t-\pi)} + (1/2)\cos t$
 $+ (1/2)\sin t]u(t - \pi)$

15. $x = t^2$; $y = t - 1$

17. $x = (t - 2)e^{t-2}$; $y = e^{t-2}$

19. $x = -7e^{-t} + e^t$; $y = 2e^{-t}$; $z = -13e^{-t} + e^t$

21. $x = 0.01(-e^{-t/2} + e^{-t/6}) + [0.48 - 0.36e^{-(t-5)/6}$
 $- 0.12e^{-(t-5)/2}]\,u(t - 5)$;
 $y = 0.02(e^{-t/2} + e^{-t/6}) + [0.48 - 0.72e^{-(t-5)/6}$
 $+ 0.24e^{-(t-5)/2}]\,u(t - 5)$

23. $2I_1 + (0.1)I_3' + (0.2)I_1' = 6$, $(0.1)I_3' - I_2 = 0$,
 $I_1 = I_2 + I_3$; $I_1(0) = I_2(0) = I_3(0) = 0$;
 $I_1 = -e^{-20t} - 2e^{-5t} + 3$,
 $I_2 = -2e^{-20t} + 2e^{-5t}$,
 $I_3 = e^{-20t} - 4e^{-5t} + 3$

Review Problems, page 415

1. $\dfrac{3}{s} + e^{-2s}\left[\dfrac{1}{s} - \dfrac{1}{s^2}\right]$ **3.** $\dfrac{2}{(s+9)^3}$

5. $\dfrac{1}{s-2} - \dfrac{6}{s^4} + \dfrac{2}{s^3} - \dfrac{5}{s^2+25}$ **7.** $\dfrac{s^2-36}{(s^2+36)^2}$

9. $2e^{-4s}\left[\dfrac{1}{s^3} + \dfrac{4}{s^2} + \dfrac{8}{s}\right]$ **11.** $(7/2)t^2e^{-3t}$

13. $2e^t + 2e^{-2t}\cos 3t + e^{-2t}\sin 3t$

15. $e^{-2t} + e^{-t} - 2te^{-t}$

17. $[2e^{t-2} + 2e^{4-2t}]u(t - 2)$

19. $e^{2t} - e^{5t}$

21. $-(3/2) + t + t^2/2 + (3/2)e^{-t}\cos t - (1/2)e^{-t}\sin t$

23. $(2/\sqrt{7})e^{-3t/2}\sin\left(\sqrt{7}t/2\right)$
 $+ \{(1/4) - [3/(4\sqrt{7})]e^{-3(t-1)/2}\sin[\sqrt{7}(t-1)/2]$
 $- (1/4)e^{-3(t-1)/2}\cos[\sqrt{7}(t-1)/2]\}u(t-1)$

25. $c[t + te^{-2t} + e^{-2t} - 1]$

27. $(9/10)e^{-3t} + (1/10)\cos t - (3/10)\sin t$

29. $(s^2 - 5s + 6)^{-1}$; $e^{3t} - e^{2t}$

31. $x = 1 - (e^t + e^{-t})/2$
 $- [1 - (e^{t-2} + e^{-(t-2)})/2]u(t-2)$,
 $y = (e^t - e^{-t})/2 - (1/2)[e^{t-2} - e^{-(t-2)}]u(t-2)$

CHAPTER 8

Exercises 8.1, page 425

1. $1 + x + x^2 + \cdots$

3. $x + x^2 + (1/2)x^3 + \cdots$

5. $1 - (1/6)t^3 + (1/180)t^6 + \cdots$

7. $(1/6)\theta^3 - (1/120)\theta^5 + (1/5040)\theta^7 + \cdots$

9. (a) $p_3(x) = (x-1) - \dfrac{1}{2}(x-1)^2 + \dfrac{1}{3}(x-1)^3$
 (b) $|\varepsilon_3(1.5)| = |f^{(4)}(\xi)(1.5-1)^4/4!| \le 6(0.5)^4/4!$
 (c) $|\ln(1.5) - p_3(1.5)| = 0.011202...$
 (d)

Figure B.43

13. $t + (1/2)t^2 - (1/6)t^3 + \cdots$ **15.** $1 - x^2/6$

Exercises 8.2, page 433

1. $[-1, 3)$ **3.** $(-4, 0)$ **5.** $[1, 3]$

9. $\displaystyle\sum_{n=0}^{\infty}\left[\dfrac{1}{n+1} + 2^{-n-1}\right]x^n$

11. $x + x^2 + (1/3)x^3 + \cdots$

13. $1 - 2x + (5/2)x^2 + \cdots$

15. (c) $1 - (1/2)x + (1/4)x^2 - (1/24)x^3 + \cdots$

17. $\displaystyle\sum_{n=1}^{\infty} (-1)^n n x^{n-1}$ **19.** $\displaystyle\sum_{k=1}^{\infty} a_k 2k x^{2k-1}$

21. $\ln(x+1) = \displaystyle\sum_{n=0}^{\infty} \frac{(-1)^n}{n+1} x^{n+1}$

23. $\displaystyle\sum_{k=0}^{\infty} (k+1) a_{k+1} x^k$

25. $\displaystyle\sum_{k=1}^{\infty} a_{k-1} x^k$ **29.** $\displaystyle\sum_{n=0}^{\infty} \frac{(-1)^{n+1}(x-\pi)^{2n}}{(2n)!}$

31. $1 + \displaystyle\sum_{n=1}^{\infty} 2x^n$

33. $6(x-1) + 3(x-1)^2 + (x-1)^3$

35. (a) $\displaystyle\sum_{n=0}^{\infty} (-1)^n (x-1)^n$ **(b)** $\displaystyle\sum_{n=1}^{\infty} \frac{(-1)^{n-1}}{n}(x-1)^n$

Exercises 8.3, page 443

1. -1 **3.** $\pm\sqrt{2}$ **5.** $-1, 2$

7. $x = n\pi$, n an integer

9. $\theta \le 0$ and $\theta = n\pi$, $n = 1, 2, 3, \ldots$

11. $y = a_0(1 - 2x + (3/2)x^2 - x^3/3 + \cdots)$

13. $a_0(1 + x^4/12 + \cdots) + a_1(x + x^5/20 + \cdots)$

15. $a_0(1 - x^2/2 - x^3/6 + \cdots)$
$\quad + a_1(x + x^2/2 - x^3/6 + \cdots)$

17. $a_0(1 - x^2/2 + \cdots) + a_1(x - x^3/6 + \cdots)$

19. $a_0 \displaystyle\sum_{n=0}^{\infty} \frac{1}{n!} x^{2n} = a_0 e^{x^2}$

21. $a_0(1 - 2x^2 + x^4/3)$
$\quad + a_1\left[x + \displaystyle\sum_{k=1}^{\infty} \frac{(-3)(-1)\cdots(2k-5)}{(2k+1)!} x^{2k+1} \right]$

23. $a_{3k+2} = 0$, $k = 0, 1, \ldots$
$$a_0\left(1 + \sum_{k=1}^{\infty} \frac{[1\cdot4\cdot7\cdots(3k-2)]^2}{(3k)!} x^{3k}\right)$$
$$+ a_1\left(x + \sum_{k=1}^{\infty} \frac{[2\cdot5\cdot8\cdots(3k-1)]^2}{(3k+1)!} x^{3k+1}\right)$$

25. $2 + x^2 - (5/12)x^4 + (11/72)x^6 + \cdots$

27. $x + (1/6)x^3 - (1/12)x^4 + (7/120)x^5 + \cdots$

29. $1 - 2x + x^2 - (1/6)x^3$

31. $1 + 2x - (3/4)x^2 - (5/6)x^3$

35. $10 - 25t^2 + (250/3)t^3 - (775/4)t^4 + \cdots$

Exercises 8.4, page 449

1. 2 **3.** $\sqrt{3}$ **5.** $\pi/2$

7. $a_0[1 - (x-1)^2 + (1/2)(x-1)^4$
$\quad - (1/6)(x-1)^6 + \cdots]$

9. $a_0[1 + (x-1)^2 + \cdots]$
$\quad + a_1[(x-1) + (1/3)(x-1)^3 + \cdots]$

11. $a_0[1 - (1/8)(x-2)^2 + (1/32)(x-2)^3 + \cdots]$
$\quad + a_1[(x-2) + (1/8)(x-2)^2$
$\qquad - (7/96)(x-2)^3 + \cdots]$

13. $1 - (1/2)t^2 + (1/6)t^4 - (31/720)t^6 + \cdots$

15. $1 + x + (1/24)x^4 + (1/60)x^5 + \cdots$

17. $1 - (1/6)(x-\pi)^3 + (1/120)(x-\pi)^5$
$\quad + (1/180)(x-\pi)^6 + \cdots$

19. $-1 + x + x^2 + (1/2)x^3 + \cdots$

21. $a_0[1 + (1/2)x^2 + (1/8)x^4 + (1/48)x^6 + \cdots]$
$\quad + [(1/2)x^2 + (1/12)x^4 + (11/720)x^6 + \cdots]$

23. $a_0[1 - (1/2)x^2 + \cdots] + a_1[x - (1/3)x^3 + \cdots]$
$\quad + [(1/2)x^2 + (1/3)x^3 + \cdots]$

25. $a_0[1 - (1/2)x^2 + \cdots] + a_1[x + \cdots]$
$\quad + [(1/2)x^2 - (1/6)x^3 + \cdots]$

27. $a_0[1 - (1/2)x^2 - (1/6)x^3 + \cdots]$
$\quad + a_1[x + (1/2)x^2 + \cdots] + [(1/6)x^3 + \cdots]$

29. (a) $a_0\left[1 + \displaystyle\sum_{k=1}^{\infty} (-1)^k \frac{n(n-2)(n-4)\cdots(n-2k+2)(n+1)(n+3)\cdots(n+2k-1)}{(2k)!} x^{2k}\right]$
$\quad + a_1\left[x + \displaystyle\sum_{k=1}^{\infty} (-1)^k \frac{(n-1)(n-3)\cdots(n-2k+1)(n+2)(n+4)\cdots(n+2k)}{(2k+1)!} x^{2k+1}\right]$

(c) $P_0(x) = 1$, $P_1(x) = x$, $P_2(x) = (1/2)(3x^2 - 1)$

31. (a) $1 - (1/2)t^2 + (\eta/6)t^3 + [(1-\eta^2)/24]t^4 + \cdots$ **(b)** Yes

Exercises 8.5, page 453

1. $c_1 x^{-2} + c_2 x^{-3}$

3. $c_1 x \cos(4 \ln x) + c_2 x \sin(4 \ln x)$

5. $c_1 x^3 \cos(2 \ln x) + c_2 x^3 \sin(2 \ln x)$

7. $c_1 x + c_2 x^{-1} \cos(3 \ln x) + c_3 x^{-1} \sin(3 \ln x)$

9. $c_1 x + c_2 x^{-1/2} \cos\left[\left(\sqrt{19}/2 \right) \ln x \right]$
$\qquad + c_3 x^{-1/2} \sin\left[\left(\sqrt{19}/2 \right) \ln x \right]$

11. $c_1 (x-3)^{-2} + c_2 (x-3)^{1/2}$

13. $c_1 x + c_2 x^2 + (4/15) x^{-1/2}$

15. $2 t^4 + t^{-3}$

17. $(31/17) x + (3/17) x^{-2} \cos(5 \ln x)$
$\qquad - (76/85) x^{-2} \sin(5 \ln x)$

Exercises 8.6, page 464

1. ± 1 are regular.

3. $\pm i$ are regular.

5. 1 is regular, -1 is irregular.

7. 2 is regular, -1 is irregular.

9. -4 is regular and 2 is irregular.

11. $r^2 - 3r - 10 = 0$; $\quad r_1 = 5$, $\quad r_2 = -2$

13. $r^2 - 5r/9 - 4/3 = 0$; $\quad r_1 = \left(5 + \sqrt{457} \right)/18$,
$\qquad r_2 = \left(5 - \sqrt{457} \right)/18$

15. $r^2 - 1 = 0$; $\quad r_1 = 1$, $\quad r_2 = -1$

17. $r^2 + r - 12 = 0$; $\quad r_1 = 3$, $\quad r_2 = -4$

19. $a_0 \left[x^{2/3} - (1/2) x^{5/3} + (5/28) x^{8/3} - (1/21) x^{11/3} + \cdots \right]$

21. $a_0 \left[1 - (1/4) x^2 + (1/64) x^4 - (1/2304) x^6 + \cdots \right]$

23. $a_0 \left[x - (1/3) x^2 + (1/12) x^3 - (1/60) x^4 + \cdots \right]$

25. $a_0 \displaystyle\sum_{n=0}^{\infty} \frac{(-1)^n x^{n+(3/2)}}{2^{n-1}(n+2)!}$

27. $a_0 \displaystyle\sum_{n=0}^{\infty} \frac{x^{2n+2}}{2^{2n}(n+1)! \, n!}$

29. $a_0 x^2$

31. $a_0 \displaystyle\sum_{n=0}^{\infty} \frac{x^n}{n!} = a_0 e^x$; $\quad$ yes, $\quad a_0 < 0$

33. $a_0 \left[x^{1/3} + \displaystyle\sum_{n=1}^{\infty} \frac{2^{n-2}(3n+4) x^{n+(1/3)}}{3^n n!} \right]$; $\quad$ yes, $\quad a_0 < 0$

35. $a_0 \left[x^{5/6} - (1/11) x^{11/6} + (1/374) x^{17/6} \right.$
$\qquad \left. - (1/25{,}806) x^{23/6} + \cdots \right]$

37. $a_0 \left[x^{4/3} + (1/17) x^{7/3} + (1/782) x^{10/3} \right.$
$\qquad \left. + (1/68{,}034) x^{13/3} + \cdots \right]$

39. The expansion $\displaystyle\sum_{n=0}^{\infty} n! x^n$ diverges for $x \neq 0$.

41. The transformed equation is
$z \, d^2 y/dz^2 + 3 \, dy/dz - y = 0$. Also $zp(z) = 3$ and
$z^2 q(z) = -z$ are analytic at $z = 0$; hence, $z = 0$ is a
regular singular point.
$y_1(x) = a_0 \left[1 + (1/3) x^{-1} + (1/24) x^{-2} \right.$
$\qquad\qquad \left. + (1/360) x^{-3} + \cdots \right]$

45. $a_0 \left[x^{-3} - (1/2) x^{-1} - (1/8) x - (1/144) x^3 + \cdots \right]$
$\qquad + a_3 \left[1 + (1/10) x^2 + (1/280) x^4 \right.$
$\qquad\qquad \left. + (1/15{,}120) x^6 + \cdots \right]$

Exercises 8.7, page 473

1. $c_1 y_1(x) + c_2 y_2(x)$, $\quad$ where
$y_1(x) = x^{2/3} - (1/2) x^{5/3} + (5/28) x^{8/3} + \cdots \quad$ and
$y_2(x) = x^{1/3} - (1/2) x^{4/3} + (1/5) x^{7/3} + \cdots$

3. $c_1 y_1(x) + c_2 y_2(x)$, $\quad$ where
$y_1(x) = 1 - (1/4) x^2 + (1/64) x^4 + \cdots \quad$ and
$y_2(x) = y_1(x) \ln x + (1/4) x^2 - (3/128) x^4$
$\qquad + (11/13{,}824) x^6 + \cdots$

5. $c_1 \left[x - (1/3) x^2 + (1/12) x^3 + \cdots \right] + c_2 \left[x^{-1} - 1 \right]$

7. $c_1 y_1(x) + c_2 y_2(x)$, $\quad$ where
$y_1(x) = x^{3/2} - (1/6) x^{5/2} + (1/48) x^{7/2} + \cdots \quad$ and
$y_2(x) = x^{-1/2} - (1/2) x^{1/2}$

9. $c_1 w_1 + c_2 w_2$, $\quad$ where
$w_1(x) = x^2 + (1/8) x^4 + (1/192) x^6 + \cdots \quad$ and
$w_2(x) = w_1(x) \ln x + 2 - (3/32) x^4 - (7/1152) x^6 + \cdots$

11. $c_1 y_1(x) + c_2 y_2(x)$, $\quad$ where $y_1(x) = x^2 \quad$ and
$y_2(x) = x^2 \ln x - 1 + 2x - (1/3) x^3 + (1/24) x^4 + \cdots$

13. $c_1 y_1(x) + c_2 y_2(x)$, $\quad$ where
$y_1(x) = 1 + x + (1/2) x^2 + \cdots \quad$ and
$y_2(x) = y_1(x) \ln x - \left[x + (3/4) x^2 + (11/36) x^3 + \cdots \right]$

15. $c_1 y_1(x) + c_2 y_2(x)$, $\quad$ where
$y_1(x) = x^{1/3} + (7/6) x^{4/3} + (5/9) x^{7/3} + \cdots \quad$ and
$y_2(x) = 1 + 2x + (6/5) x^2 + \cdots$; $\quad$ all solutions are
bounded near the origin.

17. $c_1 y_1(x) + c_2 y_2(x) + c_3 y_3(x)$, $\quad$ where
$y_1(x) = x^{5/6} - (1/11) x^{11/6} + (1/374) x^{17/6} + \cdots$,
$y_2(x) = 1 - x + (1/14) x^2 + \cdots$, $\quad$ and
$y_3(x) = y_2(x) \ln x + 7x - (117/196) x^2$
$\qquad + (4997/298116) x^3 + \cdots$

19. $c_1 y_1(x) + c_2 y_2(x) + c_3 y_3(x)$, $\quad$ where
$y_1(x) = x^{4/3} + (1/17) x^{7/3} + (1/782) x^{10/3} + \cdots$,
$y_2(x) = 1 - (1/3) x - (1/30) x^2 + \cdots$, $\quad$ and
$y_3(x) = x^{-1/2} - (1/5) x^{1/2} - (1/10) x^{3/2} + \cdots$

21. **(b)** $c_1 y_1(t) + c_2 y_2(t)$, $\quad$ where
$y_1(t) = 1 - (1/6)(\alpha t)^2 + (1/120)(\alpha t)^4 + \cdots$,
and
$y_2(t) = t^{-1} \left[1 - (1/2)(\alpha t)^2 + (1/24)(\alpha t)^4 + \cdots \right]$
(c) $c_1 y_1(x) + c_2 y_2(x)$, $\quad$ where
$y_1(x) = 1 - (1/6)(\alpha/x)^2 + (1/120)(\alpha/x)^4 + \cdots$,
and
$y_2(x) = x \left[1 - (1/2)(\alpha/x)^2 + (1/24)(\alpha/x)^4 + \cdots \right]$

23. $c_1 y_1(x) + c_2 y_2(x)$, $\quad$ where
$y_1(x) = 1 + 2x + 2x^2 \quad$ and
$y_2(x) = -(1/6) x^{-1} - (1/24) x^{-2} - (1/120) x^{-3} + \cdots$

25. For $n = 0$,
$c_1 + c_2 \left[\ln x + x + (1/4) x^2 + (1/18) x^3 + \cdots \right]$;
and for $n = 1$,
$c_1 (1 - x) + c_2 \left[(1 - x) \ln x + 3x - (1/4) x^2 \right.$
$\qquad\qquad \left. - (1/36) x^3 + \cdots \right]$

Exercises 8.8, page 485

1. $c_1 F\left(1, 2; \dfrac{1}{2}; x \right) + c_2 x^{1/2} F\left(\dfrac{3}{2}, \dfrac{5}{2}; \dfrac{3}{2}; x \right)$

3. $c_1 F\left(1, 1; \dfrac{1}{2}; x \right) + c_2 x^{1/2} F\left(\dfrac{3}{2}, \dfrac{3}{2}; \dfrac{3}{2}; x \right)$

5. $F(1, 1; 2; x) = \sum\limits_{n=0}^{\infty} \dfrac{1}{(n+1)} x^n = -x^{-1} \ln(1-x)$

7. $F\left(\dfrac{1}{2}, 1; \dfrac{3}{2}; x^2\right) = \sum\limits_{n=0}^{\infty} (1/2)_n x^{2n}/(3/2)_n$

$\qquad = \sum\limits_{n=0}^{\infty} x^{2n}/(2n+1)$

$\qquad = \dfrac{1}{2} x^{-1} \ln\left(\dfrac{1+x}{1-x}\right)$

9. $(1-x)^{-1}, \quad (1-x)^{-1} \ln x$

13. $c_1 J_{1/2}(x) + c_2 J_{-1/2}(x)$ **15.** $c_1 J_1(x) + c_2 Y_1(x)$

17. $c_1 J_{2/3}(x) + c_2 J_{-2/3}(x)$

19. $J_1(x) \ln x - x^{-1} + (3/64)x^3 - (7/2304)x^5 + \cdots$

21. $x^{3/2} J_{3/2}(x)$

27. $J_0(x) \ln x + \sum\limits_{n=1}^{\infty} \dfrac{(-1)^{n+1}}{2^{2n}(n!)^2}\left[1 + \dfrac{1}{2} + \cdots + \dfrac{1}{n}\right] x^{2n}$

29. $1, \quad x, \quad (3x^2 - 1)/2,$
$(5x^3 - 3x)/2, \quad (35x^4 - 30x^2 + 3)/8$

37. $1, \quad 2x, \quad 4x^2 - 2, \quad 8x^3 - 12x$

39. $1, \quad 1 - x, \quad (2 - 4x + x^2)/2,$
$(6 - 18x + 9x^2 - x^3)/6$

41. (b) $z = c_1 \cos x + c_2 \sin x$, so for x large,
$y(x) \approx c_1 x^{-1/2} \cos x + c_2 x^{-1/2} \sin x$

Review Problems, page 489

1. (a) $1 - x + (3/2)x^2 - (5/3)x^3 + \cdots$
(b) $-1 + x - (1/6)x^3 + (1/6)x^4 + \cdots$

3. (a) $a_0[1 + x^2 + \cdots] + a_1[x + (1/3)x^3 + \cdots]$
(b) $a_0[1 + (1/2)x^2 - (1/6)x^3 + \cdots]$
$\quad + a_1[x - (1/2)x^2 + (1/2)x^3 + \cdots]$

5. $a_0[1 + (1/2)(x-2)^2 - (1/24)(x-2)^4 + \cdots]$
$\quad + a_1(x-2)$

7. (a) $a_0[x^3 + x^4 + (1/4)x^5 + (1/36)x^6 + \cdots]$
(b) $a_0[x - (1/4)x^2 + (1/20)x^3 - (1/120)x^4 + \cdots]$

9. (a) $c_1 y_1(x) + c_2 y_2(x)$, where
$y_1(x) = x + x^2 + (1/2)x^3 + \cdots = xe^x$ and
$y_2(x) = y_1(x)\ln x - x^2 - (3/4)x^3 - (11/36)x^4 + \cdots$
(b) $c_1 y_1(x) + c_2 y_2(x)$, where
$y_1(x) = 1 + 2x + x^2 + \cdots$ and
$y_2(x) = y_1(x)\ln x - [4x + 3x^2 + (22/27)x^3 + \cdots]$
(c) $c_1 y_1(x) + c_2 y_2(x)$, where
$y_1(x) = 1 - (1/6)x + (1/96)x^2 + \cdots$ and
$y_2(x) = y_1(x)\ln x - 8x^{-2} - 4x^{-1} + (29/36) + \cdots$
(d) $c_1 y_1(x) + c_2 y_2(x)$, where
$y_1(x) = x^2 - (1/4)x^3 + (1/40)x^4 + \cdots$ and
$y_2(x) = x^{-1} + (1/2) + (1/4)x + \cdots$

CHAPTER 9

Exercises 9.1, page 500

1. $\begin{bmatrix} x \\ y \end{bmatrix}' = \begin{bmatrix} 7 & 2 \\ 3 & -2 \end{bmatrix} \begin{bmatrix} x \\ y \end{bmatrix}$

3. $\begin{bmatrix} x \\ y \\ z \end{bmatrix}' = \begin{bmatrix} 1 & 1 & 1 \\ -1 & 0 & 2 \\ 0 & 4 & 0 \end{bmatrix} \begin{bmatrix} x \\ y \\ z \end{bmatrix}$

5. $\begin{bmatrix} x \\ y \end{bmatrix}' = \begin{bmatrix} \sin t & e^t \\ \cos t & a + bt^3 \end{bmatrix} \begin{bmatrix} x \\ y \end{bmatrix}$

7. $\begin{bmatrix} y \\ v \end{bmatrix}' = \begin{bmatrix} 0 & 1 \\ -k/m & -b/m \end{bmatrix} \begin{bmatrix} y \\ v \end{bmatrix}$

9. $\begin{bmatrix} x_1 \\ x_2 \end{bmatrix}' = \begin{bmatrix} 0 & 1 \\ t & 0 \end{bmatrix} \begin{bmatrix} x_1 \\ x_2 \end{bmatrix}; \quad x_1 = y, \quad x_2 = y'$

11. $\begin{bmatrix} x_1 \\ x_2 \\ x_3 \\ x_4 \end{bmatrix}' = \begin{bmatrix} 0 & 1 & 0 & 0 \\ -3 & 0 & -2 & 0 \\ 0 & 0 & 0 & 1 \\ 2 & 0 & 0 & 0 \end{bmatrix} \begin{bmatrix} x_1 \\ x_2 \\ x_3 \\ x_4 \end{bmatrix};$
$x_1 = x, \quad x_2 = x', \quad x_3 = y, \quad x_4 = y'$

13. $\begin{bmatrix} x_1 \\ x_2 \\ x_3 \\ x_4 \\ x_5 \end{bmatrix}' = \begin{bmatrix} 0 & 1 & 0 & 0 & 0 \\ \cos t & 3 & -t^2 & 0 & 0 \\ 0 & 0 & 0 & 1 & 0 \\ 0 & 0 & 0 & 0 & 1 \\ -e^t & t & 0 & -1 & -1 \end{bmatrix} \begin{bmatrix} x_1 \\ x_2 \\ x_3 \\ x_4 \\ x_5 \end{bmatrix};$
$x_1 = x, \quad x_2 = x', \quad x_3 = y, \quad x_4 = y', \quad x_5 = y''$

Exercises 9.2, page 504

1. $x_1 = 2, \quad x_2 = 1, \quad x_3 = 1$

3. $x_1 = -s + t, \quad x_2 = s, \quad x_3 = t$

5. $x_1 = 0, \quad x_2 = 0$

7. $x_1 = 3s, \quad x_2 = s \quad (-\infty < s < \infty)$

9. $x_1 = 2s, \quad x_2 = (-1 + i)s$, with s any complex number

11. $x_1 = -(s+1)/2, \quad x_2 = -(11s+1)/2, \quad x_3 = s$
$(-\infty < s < \infty)$

13. For $r = 2$, unique solution is $x_1 = x_2 = 0$.
For $r = 1$, solutions are $x_1 = 3s, x_2 = s, -\infty < s < \infty$.

Exercises 9.3, page 513

1. (a) $\begin{bmatrix} 1 & 1 \\ 5 & 2 \end{bmatrix}$ **(b)** $\begin{bmatrix} 7 & 3 \\ 7 & 18 \end{bmatrix}$

3. (a) $\begin{bmatrix} 18 & 14 \\ 4 & 5 \end{bmatrix}$ **(b)** $\begin{bmatrix} 8 & 12 \\ 3 & 5 \end{bmatrix}$ **(c)** $\begin{bmatrix} 16 & 3 \\ 5 & 19 \end{bmatrix}$

5. (a) $\begin{bmatrix} -1 & -2 \\ -1 & -3 \end{bmatrix}$ **(b)** $\begin{bmatrix} -5 & -1 \\ -8 & -1 \end{bmatrix}$ **(c)** $\begin{bmatrix} -6 & -3 \\ -9 & -4 \end{bmatrix}$

7. (c) Yes **(d)** Yes **9.** $\begin{bmatrix} 4/9 & -1/9 \\ 1/9 & 2/9 \end{bmatrix}$ **11.** Doesn't exist

13. $\begin{bmatrix} 1 & 0 & 1 \\ -2 & 1 & -2 \\ 1 & 1 & 0 \end{bmatrix}$ **17.** $\begin{bmatrix} (4/3)e^{-t} & -(1/3)e^{-t} \\ -(1/3)e^{-4t} & (1/3)e^{-4t} \end{bmatrix}$

19. $\begin{bmatrix} e^{-t} & (1/2)e^{-t} & -(1/2)e^{-t} \\ (1/3)e^t & -(1/2)e^t & (1/6)e^t \\ -(1/3)e^{-2t} & 0 & (1/3)e^{-2t} \end{bmatrix}$

21. 11 **23.** -12 **25.** 25 **27.** 2, 3

29. 0, 1, 1 **31.** $\begin{bmatrix} 3e^{3t} \\ 6e^{3t} \\ -3e^{3t} \end{bmatrix}$ **33.** $\begin{bmatrix} 5e^{5t} & 6e^{2t} \\ -10e^{5t} & -2e^{2t} \end{bmatrix}$

39. (a) $\begin{bmatrix} (1/2)t^2 + c_1 & e^t + c_2 \\ t + c_3 & e^t + c_4 \end{bmatrix}$

(b) $\begin{bmatrix} \sin 1 & -1 + \cos 1 \\ 1 - \cos 1 & \sin 1 \end{bmatrix}$

(c) $\begin{bmatrix} (1 + e^t)\cos t + (e^t - t)\sin t & (e^t - t)\cos t - (e^t + 1)\sin t \\ (e^t - 1)\sin t + e^t \cos t & (e^t - 1)\cos t - e^t \sin t \end{bmatrix}$

Exercises 9.4, page 521

1. $\begin{bmatrix} x'(t) \\ y'(t) \end{bmatrix} = \begin{bmatrix} 3 & -1 \\ -1 & 2 \end{bmatrix} \begin{bmatrix} x(t) \\ y(t) \end{bmatrix} + \begin{bmatrix} t^2 \\ e^t \end{bmatrix}$

3. $\begin{bmatrix} dx/dt \\ dy/dt \\ dz/dt \end{bmatrix} = \begin{bmatrix} t^2 & -1 & -1 \\ 0 & 0 & e^t \\ t & -1 & 3 \end{bmatrix} \begin{bmatrix} x \\ y \\ z \end{bmatrix} + \begin{bmatrix} t \\ 5 \\ -e^t \end{bmatrix}$

5. $\begin{bmatrix} x_1'(t) \\ x_2'(t) \end{bmatrix} = \begin{bmatrix} 0 & 1 \\ 10 & 3 \end{bmatrix} \begin{bmatrix} x_1(t) \\ x_2(t) \end{bmatrix} + \begin{bmatrix} 0 \\ \sin t \end{bmatrix};$
$x_1 = y, \quad x_2 = y'$

7. $\begin{bmatrix} x_1'(t) \\ x_2'(t) \\ x_3'(t) \\ x_4'(t) \end{bmatrix} = \begin{bmatrix} 0 & 1 & 0 & 0 \\ 0 & 0 & 1 & 0 \\ 0 & 0 & 0 & 1 \\ -1 & 0 & 0 & 0 \end{bmatrix} \begin{bmatrix} x_1(t) \\ x_2(t) \\ x_3(t) \\ x_4(t) \end{bmatrix} + \begin{bmatrix} 0 \\ 0 \\ 0 \\ t^2 \end{bmatrix};$
$x_1 = w, \quad x_2 = w', \quad x_3 = w'', \quad x_4 = w'''$

9. $x_1'(t) = 5x_1(t) + 2e^{-2t};$
$x_2'(t) = -2x_1(t) + 4x_2(t) - 3e^{-2t}$

11. $x_1'(t) = x_1(t) + x_3(t) + e^t;$
$x_2'(t) = -x_1(t) + 2x_2(t) + 5x_3(t) + t;$
$x_3'(t) = 5x_2(t) + x_3(t)$

13. LI **15.** LD **17.** LI **19.** LI

21. Not a fundamental solution set

23. Yes; $\begin{bmatrix} e^{-t} & e^t & e^{3t} \\ 2e^{-t} & 0 & -e^{3t} \\ e^{-t} & e^t & 2e^{3t} \end{bmatrix}$;

$c_1 \begin{bmatrix} e^{-t} \\ 2e^{-t} \\ e^{-t} \end{bmatrix} + c_2 \begin{bmatrix} e^t \\ 0 \\ e^t \end{bmatrix} + c_3 \begin{bmatrix} e^{3t} \\ -e^{3t} \\ 2e^{3t} \end{bmatrix}$

25. $\dfrac{3}{2} \begin{bmatrix} te^t \\ te^t \end{bmatrix} - \dfrac{1}{4} \begin{bmatrix} e^t \\ 3e^t \end{bmatrix} + \begin{bmatrix} t \\ 2t \end{bmatrix} - \begin{bmatrix} 0 \\ 1 \end{bmatrix} + c_1 \begin{bmatrix} e^t \\ e^t \end{bmatrix} + c_2 \begin{bmatrix} e^{-t} \\ 3e^{-t} \end{bmatrix}$

29. $\mathbf{X}^{-1}(t) = \begin{bmatrix} 0 & (1/4)e^t & -(1/4)e^t \\ -(1/5)e^{2t} & (4/5)e^{2t} & -(2/5)e^{2t} \\ (1/5)e^{-3t} & (9/20)e^{-3t} & (3/20)e^{-3t} \end{bmatrix};$

$\mathbf{x}(t) = \begin{bmatrix} -(3/2)e^{-t} + (3/5)e^{-2t} - (1/10)e^{3t} \\ (1/4)e^{-t} - (1/5)e^{-2t} - (1/20)e^{3t} \\ (5/4)e^{-t} - (1/5)e^{-2t} - (1/20)e^{3t} \end{bmatrix}$

37. (b) $2e^{6t}$ **(e)** $2e^{6t}$; same

Exercises 9.5, page 531

1. Eigenvalues are $r_1 = 0$ and $r_2 = -5$ with associated eigenvectors $\mathbf{u}_1 = \begin{bmatrix} s \\ 2s \end{bmatrix}$ and $\mathbf{u}_2 = \begin{bmatrix} 2s \\ -s \end{bmatrix}$.

3. Eigenvalues are $r_1 = 2$ and $r_2 = 3$ with associated eigenvectors $\mathbf{u}_1 = s\begin{bmatrix} 1 \\ -1 \end{bmatrix}$ and $\mathbf{u}_2 = s\begin{bmatrix} 1 \\ -2 \end{bmatrix}$.

5. Eigenvalues are $r_1 = 1$, $r_2 = 2$, and $r_3 = -2$ with associated eigenvectors $\mathbf{u}_1 = s\begin{bmatrix} 1 \\ 0 \\ 0 \end{bmatrix}$, $\mathbf{u}_2 = s\begin{bmatrix} 0 \\ 1 \\ 1 \end{bmatrix}$, and $\mathbf{u}_3 = s\begin{bmatrix} 0 \\ -1 \\ 1 \end{bmatrix}$.

7. Eigenvalues are $r_1 = 1$, $r_2 = 2$, and $r_3 = 5$ with associated eigenvectors $\mathbf{u}_1 = s\begin{bmatrix} 2 \\ -3 \\ 2 \end{bmatrix}$, $\mathbf{u}_2 = s\begin{bmatrix} 0 \\ -1 \\ 1 \end{bmatrix}$, and $\mathbf{u}_3 = s\begin{bmatrix} 0 \\ 1 \\ 2 \end{bmatrix}$.

9. Eigenvalues are $r_1 = i$ and $r_2 = -i$, with associated eigenvectors $\mathbf{u}_1 = s\begin{bmatrix} i \\ 1 \end{bmatrix}$, and $\mathbf{u}_2 = s\begin{bmatrix} -i \\ 1 \end{bmatrix}$,

11. $c_1 e^{3t/2} \begin{bmatrix} 3 \\ 10 \end{bmatrix} + c_2 e^{t/2} \begin{bmatrix} 1 \\ 2 \end{bmatrix}$

13. $c_1 e^{-t} \begin{bmatrix} -2 \\ 1 \\ 1 \end{bmatrix} + c_2 e^{-3t} \begin{bmatrix} 0 \\ -1 \\ 1 \end{bmatrix} + c_3 e^{-5t} \begin{bmatrix} 1 \\ 1 \\ 1 \end{bmatrix}$

15. $c_1 e^{-t} \begin{bmatrix} 3 \\ 0 \\ -2 \end{bmatrix} + c_2 e^t \begin{bmatrix} 1 \\ -6 \\ 4 \end{bmatrix} + c_3 e^{4t} \begin{bmatrix} 1 \\ 0 \\ 1 \end{bmatrix}$

17. (b) **(c)**

Figure B.43 **Figure B.44**

(d)

Figure B.45

19. $\begin{bmatrix} e^{3t} & e^{-3t} \\ 4e^{3t} & -2e^{-3t} \end{bmatrix}$

21. $\begin{bmatrix} e^t & e^{2t} & e^{4t} \\ e^t & 2e^{2t} & 4e^{4t} \\ e^t & 4e^{2t} & 16e^{4t} \end{bmatrix}$

23. $\begin{bmatrix} e^{2t} & e^{-t} & e^{3t} & -e^{7t} \\ 0 & -3e^{-t} & 0 & e^{7t} \\ 0 & 0 & e^{3t} & 2e^{7t} \\ 0 & 0 & 0 & 8e^{7t} \end{bmatrix}$

25. $x = -c_1 e^t - 2c_2 e^{2t} - c_3 e^{3t}$,
$y = c_1 e^t + c_2 e^{2t} + c_3 e^{3t}$,
$z = 2c_1 e^t + 4c_2 e^{2t} + 4c_3 e^{3t}$

(d) $x_3(t) = \dfrac{t^2}{2} e^{2t} \begin{bmatrix} 1 \\ 0 \\ 0 \end{bmatrix} + te^{2t} \begin{bmatrix} 0 \\ 1 \\ 0 \end{bmatrix} + e^{2t} \begin{bmatrix} 0 \\ -6/5 \\ 1/5 \end{bmatrix}$

27. $\begin{bmatrix} e^{-0.3473t} & e^{0.5237t} & 0.0286e^{-7.0764t} \\ -0.3157e^{-0.3473t} & 0.4761e^{0.5237t} & -0.1837e^{-7.0764t} \\ 0.0844e^{-0.3473t} & 0.1918e^{0.5237t} & e^{-7.0764t} \end{bmatrix}$

39. (b) $x_1(t) = e^t \begin{bmatrix} -1 \\ 1 \\ 0 \end{bmatrix}$; $\quad x_2(t) = e^t \begin{bmatrix} -1 \\ 0 \\ 1 \end{bmatrix}$

29. $\begin{bmatrix} 0.0251e^{3.4142t} & -0.2361e^{-1.6180t} & e^{0.6180t} & e^{0.5858t} \\ 0.0858e^{3.4142t} & 0.3820e^{-1.6180t} & 0.6180e^{0.6180t} & 0.5858e^{0.5858t} \\ 0.2929e^{3.4142t} & -0.6180e^{-1.6180t} & 0.3820e^{0.6180t} & 0.3431e^{0.5858t} \\ e^{3.4142t} & e^{-1.6180t} & 0.2361e^{0.6180t} & 0.2010e^{0.5858t} \end{bmatrix}$

(c) $x_3(t) = te^t \begin{bmatrix} 1 \\ 1 \\ -2 \end{bmatrix} + e^t \begin{bmatrix} 1 \\ 0 \\ 0 \end{bmatrix}$ **(d) 0**

31. $\begin{bmatrix} 2e^{4t} + e^{-2t} \\ 2e^{4t} - e^{-2t} \end{bmatrix}$ **33.** $\begin{bmatrix} -3e^{-t} + e^{5t} \\ -2e^{-t} - e^{5t} \\ e^{-t} + e^{5t} \end{bmatrix}$

43. $c_1 \begin{bmatrix} 3t^2 \\ t^2 \end{bmatrix} + c_2 \begin{bmatrix} t^4 \\ t^4 \end{bmatrix}$

45. $x_1(t) = (2.5/2)(e^{-3t/25} + e^{-t/25})$,
$x_2(t) = (2.5)(e^{-t/25} - e^{-3t/25})$

35. (b) $x_1(t) = e^{-t} \begin{bmatrix} 1 \\ 2 \end{bmatrix}$

47. (a) $r_1 = 2.39091$, $r_2 = -2.94338$, $r_3 = 3.55247$

(c) $x_2(t) = te^{-t} \begin{bmatrix} 1 \\ 2 \end{bmatrix} + e^{-t} \begin{bmatrix} 1 \\ 1 \end{bmatrix}$

(b) $u_1 = \begin{bmatrix} 1 \\ -2.64178 \\ -9.31625 \end{bmatrix}$; $\quad u_2 = \begin{bmatrix} 1 \\ -1.16825 \\ 0.43862 \end{bmatrix}$;

37. (b) $x_1(t) = e^{2t} \begin{bmatrix} 1 \\ 0 \\ 0 \end{bmatrix}$

$u_3 = \begin{bmatrix} 1 \\ 0.81004 \\ -0.12236 \end{bmatrix}$

(c) $x_2(t) = te^{2t} \begin{bmatrix} 1 \\ 0 \\ 0 \end{bmatrix} + e^{2t} \begin{bmatrix} 0 \\ 1 \\ 0 \end{bmatrix}$

(c) $c_1 e^{r_1 t} u_1 + c_2 e^{r_2 t} u_2 + c_3 e^{r_3 t} u_3$, where the r_i's and the u_i's are given in parts (a) and (b).

Exercises 9.6, page 537

1. $c_1 \begin{bmatrix} 2\cos 2t \\ \cos 2t + \sin 2t \end{bmatrix} + c_2 \begin{bmatrix} 2\sin 2t \\ \sin 2t - \cos 2t \end{bmatrix}$

3. $c_1 e^t \cos t \begin{bmatrix} -1 \\ 1 \\ 0 \end{bmatrix} - c_1 e^t \sin t \begin{bmatrix} -2 \\ 0 \\ 1 \end{bmatrix} + c_2 e^t \sin t \begin{bmatrix} -1 \\ 1 \\ 0 \end{bmatrix} + c_2 e^t \cos t \begin{bmatrix} -2 \\ 0 \\ 1 \end{bmatrix} + c_3 e^t \begin{bmatrix} 1 \\ 0 \\ 0 \end{bmatrix}$

Note: Equivalent answer is obtained if col $(-1, 1, 0)$ and col $(-2, 0, 1)$ are replaced by col $(-5, 1, 2)$ and col $(0, -2, 1)$, respectively.

5. $\begin{bmatrix} e^{-t}\cos 4t & e^{-t}\sin 4t \\ 2e^{-t}\sin 4t & -2e^{-t}\cos 4t \end{bmatrix}$ **7.** $\begin{bmatrix} 1 & \cos t & \sin t \\ 0 & -\cos t & -\sin t \\ 0 & -\sin t & \cos t \end{bmatrix}$

9. $\begin{bmatrix} \cos t & \sin t & 0 & 0 \\ -\sin t & \cos t & 0 & 0 \\ 0 & 0 & \cos t & \sin t \\ 0 & 0 & -\sin t & \cos t \end{bmatrix}$

11. $\begin{bmatrix} e^t\sin t - e^t\cos t & -e^t\sin t - e^t\cos t & -\cos t & -\sin t \\ 2e^t\sin t & -2e^t\cos t & \sin t & -\cos t \\ 2e^t\sin t + 2e^t\cos t & 2e^t\sin t - 2e^t\cos t & \cos t & \sin t \\ 4e^t\cos t & 4e^t\sin t & -\sin t & \cos t \end{bmatrix}$

13. (a) $\begin{bmatrix} e^{-2t}(\sin t - \cos t) \\ -2e^{-2t}\sin t \end{bmatrix}$ **(b)** $\begin{bmatrix} -e^{-2(t-\pi)}\cos t \\ e^{-2(t-\pi)}(\cos t - \sin t) \end{bmatrix}$

(c) $\begin{bmatrix} e^{-2(t+2\pi)}(2\cos t - 3\sin t) \\ e^{-2(t+2\pi)}(\cos t + 5\sin t) \end{bmatrix}$ **(d)** $\begin{bmatrix} e^{\pi-2t}\cos t \\ e^{\pi-2t}(\sin t - \cos t) \end{bmatrix}$

17. $c_1 \begin{bmatrix} t^{-1} \\ 0 \\ -2t^{-1} \end{bmatrix} + c_2 \begin{bmatrix} t^{-1}\cos(\ln t) \\ t^{-1}\sin(\ln t) \\ -t^{-1}\cos(\ln t) \end{bmatrix} + c_3 \begin{bmatrix} t^{-1}\sin(\ln t) \\ -t^{-1}\cos(\ln t) \\ -t^{-1}\sin(\ln t) \end{bmatrix}$

19. $\dfrac{\sqrt{9 - \sqrt{17}}}{2\sqrt{2\pi}} \approx 0.249$; $\dfrac{\sqrt{9 + \sqrt{17}}}{2\sqrt{2\pi}} \approx 0.408$

21. $I_1 = (5/6)e^{-2t}\sin 3t$,
$I_2 = -(10/13)e^{-2t}\cos 3t + (25/78)e^{-2t}\sin 3t$,
$I_3 = (10/13)e^{-2t}\cos 3t + (20/39)e^{-2t}\sin 3t$

Exercises 9.7, page 542

1. $c_1 e^{7t} \begin{bmatrix} 1 \\ 1 \end{bmatrix} + c_2 e^{2t} \begin{bmatrix} 1 \\ -4 \end{bmatrix} + \begin{bmatrix} 2 \\ -1 \end{bmatrix}$

3. $c_1 e^{-3t} \begin{bmatrix} 1 \\ 1 \\ -1 \end{bmatrix} + c_2 e^{3t} \begin{bmatrix} 1 \\ 0 \\ 1 \end{bmatrix} + c_3 e^{3t} \begin{bmatrix} -1 \\ 1 \\ 0 \end{bmatrix} + e^{t} \begin{bmatrix} 1 \\ 0 \\ -1 \end{bmatrix}$

5. $t \begin{bmatrix} 0 \\ 0 \\ -1 \end{bmatrix} + \begin{bmatrix} 0 \\ 0 \\ -1 \end{bmatrix} + \dfrac{e^{2t}}{3} \begin{bmatrix} 2 \\ -1 \\ 0 \end{bmatrix} + \dfrac{\sin t}{2} \begin{bmatrix} 1 \\ 0 \\ 0 \end{bmatrix} + \dfrac{\cos t}{2} \begin{bmatrix} 0 \\ -1 \\ 0 \end{bmatrix}$

$\qquad + c_1 e^{-t} \begin{bmatrix} 1 \\ 1 \\ 0 \end{bmatrix} + c_2 e^{t} \begin{bmatrix} -1 \\ 1 \\ 0 \end{bmatrix} + c_3 e^{t} \begin{bmatrix} 0 \\ 0 \\ 1 \end{bmatrix}$

7. $\mathbf{x}_p = t\mathbf{a} + \mathbf{b} + \sin 3t\,\mathbf{c} + \cos 3t\,\mathbf{d}$

9. $\mathbf{x}(t) = t\mathbf{a} + \mathbf{b} + e^{2t}\mathbf{c} + (\sin t)\mathbf{d} + (\cos t)\mathbf{e}$

11. $c_1 \begin{bmatrix} \sin t \\ \cos t \end{bmatrix} + c_2 \begin{bmatrix} \cos t \\ -\sin t \end{bmatrix} + \begin{bmatrix} 0 \\ -1 \end{bmatrix}$

13. $c_1 \begin{bmatrix} 1 \\ 2 \end{bmatrix} + c_2 e^{6t} \begin{bmatrix} 2 \\ 1 \end{bmatrix} - \begin{bmatrix} t^{-1}/2 \\ t^{-1} \end{bmatrix}$

15. $c_1 \begin{bmatrix} 1 \\ 2 \end{bmatrix} + c_2 e^{-5t} \begin{bmatrix} -2 \\ 1 \end{bmatrix} + \begin{bmatrix} \ln|t| + (8/5)t - 8/25 \\ 2\ln|t| + (16/5)t + 4/25 \end{bmatrix}$

17. $c_1 e^{2t} \begin{bmatrix} 1 \\ 1 \\ 1 \end{bmatrix} + c_2 e^{-t} \begin{bmatrix} -1 \\ 0 \\ 1 \end{bmatrix} + c_3 e^{-t} \begin{bmatrix} -1 \\ 1 \\ 0 \end{bmatrix} + e^{t} \begin{bmatrix} 1 \\ -1 \\ -1 \end{bmatrix}$

19. $\begin{bmatrix} c_1\cos t + c_2\sin t + 1 \\ -c_1\sin t + c_2\cos t - t \\ c_3 e^{t} + c_4 e^{-t} - (1/4)e^{-t} - t + (1/2)te^{-t} \\ c_3 e^{t} - c_4 e^{-t} - (1/4)e^{-t} - 1 - (1/2)te^{-t} \end{bmatrix}$

21. (a) $\begin{bmatrix} 4te^{t} + 5e^{t} \\ 2te^{t} + 4e^{t} \end{bmatrix}$

(b) $\begin{bmatrix} -2e^{t-1} + 2e^{2(t-1)} - 3e^{2t-1} + (4t-1)e^{t} \\ -e^{t-1} + 2e^{2(t-1)} - 3e^{2t-1} + (2t+1)e^{t} \end{bmatrix}$

(c) $\begin{bmatrix} (-20 + 2e^{-5})e^{t} + (-3e^{-5} - e^{-10})e^{2t} + (4t+3)e^{t} \\ (-10 + e^{-5})e^{t} + (-3e^{-5} - e^{-10})e^{2t} + (2t+3)e^{t} \end{bmatrix}$

(d) $\begin{bmatrix} -18e^{t+1} + 14e^{2t+2} - 3e^{2t+1} + (4t+7)e^{t} \\ -9e^{t+1} + 14e^{2t+2} - 3e^{2t+1} + (2t+5)e^{t} \end{bmatrix}$

23. $x = (c_1 + c_2 t)e^{t} + c_3 e^{-t} - t^2 - 4t - 6$;

$y = -c_2 e^{t} - 2c_3 e^{-t} - t^2 - 2t - 3$

25. (a) $\left\{ \begin{bmatrix} e^{t} \\ e^{t} \end{bmatrix}, \begin{bmatrix} e^{2t} \\ 2e^{2t} \end{bmatrix} \right\}$ (c) $te^{t} \begin{bmatrix} 2 \\ 2 \end{bmatrix} + e^{t} \begin{bmatrix} 0 \\ 1 \end{bmatrix}$

(d) $c_1 e^{t} \begin{bmatrix} 1 \\ 1 \end{bmatrix} + c_2 e^{2t} \begin{bmatrix} 1 \\ 2 \end{bmatrix} + te^{t} \begin{bmatrix} 2 \\ 2 \end{bmatrix} + e^{t} \begin{bmatrix} 0 \\ 1 \end{bmatrix}$

27. $c_1 e^{2t} \begin{bmatrix} 1 \\ 1 \\ 1 \end{bmatrix} + c_2 e^{-t} \begin{bmatrix} -1 \\ 0 \\ 1 \end{bmatrix} + c_3 e^{-t} \begin{bmatrix} -1 \\ 1 \\ 0 \end{bmatrix}$

$\qquad + te^{-t} \begin{bmatrix} 1 \\ 0 \\ -1 \end{bmatrix} + e^{-t} \begin{bmatrix} 1 \\ 0 \\ 0 \end{bmatrix} + \begin{bmatrix} 0 \\ 0 \\ 1 \end{bmatrix}$

29. $c_1 t \begin{bmatrix} 1 \\ 1 \end{bmatrix} + c_2 t^{-1} \begin{bmatrix} 1 \\ 3 \end{bmatrix}$

$\qquad + \begin{bmatrix} -(3/4)t^{-1} - (1/2)t^{-1}\ln t + 1 \\ -(3/4)t^{-1} - (3/2)t^{-1}\ln t + 2 \end{bmatrix}$

33. $I_1 = I_2 + 3I_5$, $I_3 = 3I_5$, $I_4 = 2I_5$, where

$I_2 = \left(1/\left(20\sqrt{817} \right) \right) \left[\left(13 - \sqrt{817} \right) e^{-\left(31+\sqrt{817} \right)5t/2} \right.$

$\qquad \left. - \left(13 + \sqrt{817} \right) e^{\left(-31+\sqrt{817} \right)5t/2} \right] + 1/10$,

$I_5 = \left(3/\left(40\sqrt{817} \right) \right) \left[\left(31 - \sqrt{817} \right) e^{-\left(31+\sqrt{817} \right)5t/2} \right.$

$\qquad \left. - \left(31 + \sqrt{817} \right) e^{\left(-31+\sqrt{817} \right)5t/2} \right] + 3/20$

Exercises 9.8, page 551

1. (a) $r = 3$; $k = 2$ (b) $e^{3t} \begin{bmatrix} 1 & -2t \\ 0 & 1 \end{bmatrix}$

3. (a) $r = -1$; $k = 3$ (b) $e^{-t} \begin{bmatrix} 1 + 3t - (3/2)t^2 & t & -t + (1/2)t^2 \\ -3t & 1 & t \\ 9t - (9/2)t^2 & 3t & 1 - 3t + (3/2)t^2 \end{bmatrix}$

5. (a) $r = -2$; $k = 2$ (b) $e^{-2t} \begin{bmatrix} 1 & 0 & 0 \\ 4t & 1 & 0 \\ t & 0 & 1 \end{bmatrix}$ **7.** $\begin{bmatrix} \cos t & \sin t \\ -\sin t & \cos t \end{bmatrix}$

9. $\dfrac{1}{2} \begin{bmatrix} e^{t} + \cos t - \sin t & 2\sin t & e^{t} - \cos t - \sin t \\ e^{t} - \sin t - \cos t & 2\cos t & e^{t} + \sin t - \cos t \\ e^{t} - \cos t + \sin t & -2\sin t & e^{t} + \cos t + \sin t \end{bmatrix}$

11. $\dfrac{1}{25} \begin{bmatrix} -4 + 29e^{5t} - 20te^{5t} & 20 - 20e^{5t} & -8 + 8e^{5t} - 40te^{5t} \\ -5 + 5e^{5t} & 25 & -10 + 10e^{5t} \\ 2 - 2e^{5t} + 10te^{5t} & -10 + 10e^{5t} & 4 + 21e^{5t} + 20te^{5t} \end{bmatrix}$

13. $\begin{bmatrix} (1 - t + t^2/2)e^{t} & (t - t^2)e^{t} & (t^2/2)e^{t} & 0 & 0 \\ (t^2/2)e^{t} & (1 - t + t^2)e^{t} & (t + t^2/2)e^{t} & 0 & 0 \\ (t + t^2/2)e^{t} & (-3t - t^2)e^{t} & (1 + 2t + t^2/2)e^{t} & 0 & 0 \\ 0 & 0 & 0 & \cos t & \sin t \\ 0 & 0 & 0 & -\sin t & \cos t \end{bmatrix}$

15.
$$\begin{bmatrix} (1+t+t^2/2)e^{-t} & (t+t^2)e^{-t} & (t^2/2)e^{-t} & 0 & 0 \\ (-t^2/2)e^{-t} & (1+t-t^2)e^{-t} & (t-t^2/2)e^{-t} & 0 & 0 \\ (-t+t^2/2)e^{-t} & (-3t+t^2)e^{-t} & (1-2t+t^2/2)e^{-t} & 0 & 0 \\ 0 & 0 & 0 & (1+2t)e^{-2t} & te^{-2t} \\ 0 & 0 & 0 & -4te^{-2t} & (1-2t)e^{-2t} \end{bmatrix}$$

17. $c_1 e^{-t}\begin{bmatrix} 1 \\ -1 \\ 1 \end{bmatrix} + c_2 e^{-t}\begin{bmatrix} -t \\ -1+t \\ 2-t \end{bmatrix} + c_3 e^{-2t}\begin{bmatrix} 1 \\ -2 \\ 4 \end{bmatrix}$ **19.** $c_1 e^{2t}\begin{bmatrix} 3 \\ 6 \\ 1 \\ 1 \end{bmatrix} + c_2 e^{t}\begin{bmatrix} 0 \\ 1 \\ 0 \\ 0 \end{bmatrix} + c_3 e^{t}\begin{bmatrix} 1 \\ t \\ 0 \\ 0 \end{bmatrix} + c_4 e^{t}\begin{bmatrix} 2t \\ t+t^2 \\ 0 \\ 1 \end{bmatrix}$

21. $\begin{bmatrix} e^{-2t} \\ 4te^{-2t}+e^{-2t} \\ te^{-2t}-e^{-2t} \end{bmatrix}$. **23.** $e^{-t}\begin{bmatrix} 3t \\ 3 \\ 9t \end{bmatrix} + e^{At}\begin{bmatrix} 2-e^{t}(t^2-2t+2) \\ 1+e^{t}(t-1) \\ 6-3e^{t}(t^2-2t+2) \end{bmatrix}$, where e^{At} is the matrix in the answer to Problem 3.

29. $\mathbf{x}(t) = \begin{bmatrix} (4/13)e^{-3t}+6te^{3t}+(2/13)(3\sin 2t-2\cos 2t) \\ e^{3t} \\ (8/13)e^{-3t}+(3t-1)e^{3t}+(4/13)(3\sin 2t-2\cos 2t) \end{bmatrix}$.

Review Problems, page 555

1. $c_1 e^{3t}\begin{bmatrix} 1 \\ 1 \end{bmatrix} + c_2 e^{4t}\begin{bmatrix} 3 \\ 2 \end{bmatrix}$ **3.** $c_1 e^{-t}\begin{bmatrix} 1 \\ -1 \\ 0 \\ 0 \end{bmatrix} + c_2 e^{-t}\begin{bmatrix} 0 \\ 0 \\ 1 \\ -1 \end{bmatrix} + c_3 e^{3t}\begin{bmatrix} 1 \\ 1 \\ 0 \\ 0 \end{bmatrix} + c_4 e^{3t}\begin{bmatrix} 0 \\ 0 \\ 1 \\ 1 \end{bmatrix}$ **5.** $\begin{bmatrix} e^{2t} & e^{3t} \\ -e^{2t} & -2e^{3t} \end{bmatrix}$

7. $c_1 e^{-t}\begin{bmatrix} 1 \\ -2 \end{bmatrix} + c_2 e^{3t}\begin{bmatrix} 1 \\ 2 \end{bmatrix} + \begin{bmatrix} -1/3 \\ -14/3 \end{bmatrix}$ **9.** $c_1 e^{-t}\begin{bmatrix} 1 \\ 0 \\ 3 \end{bmatrix} + c_2 e^{-t}\begin{bmatrix} t \\ 1 \\ 3t \end{bmatrix} + c_3 e^{-t}\begin{bmatrix} -t+(1/2)t^2 \\ t \\ 1-3t+(3/2)t^2 \end{bmatrix} + \begin{bmatrix} 2+t \\ 7-3t \\ 10 \end{bmatrix}$

11. $\begin{bmatrix} 3e^{t}-2e^{2t} \\ 3e^{t}-4e^{2t} \end{bmatrix}$ **13.** $c_1 t^{-1}\begin{bmatrix} -3 \\ 1 \\ 0 \end{bmatrix} + c_2 t^{-1}\begin{bmatrix} -1 \\ 0 \\ 1 \end{bmatrix} + c_3 t^{4}\begin{bmatrix} 1 \\ 1 \\ 1 \end{bmatrix}$

15. $\begin{bmatrix} (1/2)e^{t}+(2/3)e^{5t}-(1/6)e^{-t} & -e^{t}+(2/3)e^{5t}+(1/3)e^{-t} & -(1/2)e^{t}+(2/3)e^{5t}-(1/6)e^{-t} \\ (1/3)e^{5t}-(1/3)e^{-t} & (1/3)e^{5t}+(2/3)e^{-t} & (1/3)e^{5t}-(1/3)e^{-t} \\ -(1/2)e^{t}+(1/2)e^{-t} & e^{t}-e^{-t} & (1/2)e^{t}+(1/2)e^{-t} \end{bmatrix}$

17. (a) false **(b)** false **(c)** false **(d)** true **(e)** true **(f)** false **(g)** true

CHAPTER 10

Exercises 10.2, page 570

1. $y = [4/(e-e^{-1})](e^{-x}-e^{x})$ **3.** $y \equiv 0$

5. $y = e^{x}+2x-1$ **7.** $y = \cos x + c\sin x$; c arbitrary

9. $\lambda_n = (2n-1)^2/4$ and $y_n = c_n \sin[(2n-1)x/2]$, where $n = 1, 2, 3, \dots$ and the c_n's are arbitrary.

11. $\lambda_n = n^2, n = 0, 1, 2, \dots$; $y_0 = a_0$ and $y_n = a_n \cos nx + b_n \sin nx, n = 1, 2, 3, \dots$, where $a_0, a_n,$ and b_n are arbitrary.

13. The eigenvalues are the roots of $\tan(\sqrt{\lambda_n}\,\pi) + \sqrt{\lambda_n} = 0$, where $\lambda_n > 0$. For n large, $\lambda_n \approx (2n-1)^2/4$, n is a positive integer. The eigenfunctions are $y_n = c_n[\sin(\sqrt{\lambda_n}x) + \sqrt{\lambda_n}\cos(\sqrt{\lambda_n}x)]$, where the c_n's are arbitrary.

15. $u(x,t) = e^{-3t}\sin x - 6e^{-48t}\sin 4x$

17. $u(x,t) = e^{-3t}\sin x - 7e^{-27t}\sin 3x + e^{-75t}\sin 5x$

19. $u(x,t) = 3\cos 6t\sin 2x + 12\cos 39t\sin 13x$

21. $u(x,t) = 6\cos 6t\sin 2x + 2\cos 18t\sin 6x$
$\qquad\qquad + (11/27)\sin 27t\sin 9x$
$\qquad\qquad - (14/45)\sin 45t\sin 15x$

23. $u(x,t) = \displaystyle\sum_{n=1}^{\infty} n^{-2}e^{-2\pi^2 n^2 t}\sin n\pi x$

25. If $K > 0$, then $T(t)$ becomes unbounded as $t \to \infty$, and so the temperature $u(x,t) = X(x)T(t)$ becomes unbounded at each position x. Since the temperature must remain bounded for all time, $K \not> 0$.

33. (a) $u(x) \equiv 50$ **(b)** $u(x) = 30x/L + 10$

Exercises 10.3, page 584

1. Odd **3.** Even **5.** Neither

9. $f(x) \sim \sum_{n=1}^{\infty} \frac{2(-1)^{n+1}}{n} \sin nx$

11. $f(x) \sim 1 + \sum_{n=1}^{\infty} \left[\frac{2}{\pi^2 n^2} \left(-1 + (-1)^n \right) \cos \frac{n\pi x}{2} \right.$
$\left. + \frac{1}{\pi n} \left((-1)^{n+1} - 1 \right) \sin \frac{n\pi x}{2} \right]$

13. $f(x) \sim \frac{1}{3} + \sum_{n=1}^{\infty} \frac{4(-1)^n}{n^2 \pi^2} \cos n\pi x$

15. $f(x) \sim \left[(\sinh \pi) / \pi \right] \left(1 + 2 \sum_{n=1}^{\infty} \left[\frac{(-1)^n}{1+n^2} \cos nx \right. \right.$
$\left. \left. + \frac{(-1)^{n+1} n}{1+n^2} \sin nx \right] \right)$

17. The 2π-periodic function $g(x)$, where
$$g(x) = \begin{cases} x & (-\pi < x < \pi), \\ 0 & (x = \pm \pi) \end{cases}$$

19. The 4-periodic function $g(x)$, where
$$g(x) = \begin{cases} 1 & (-2 < x < 0), \\ x & (0 < x < 2), \\ 1/2 & (x = 0), \\ 3/2 & (x = \pm 2) \end{cases}$$

21. The 2-periodic function $g(x)$, where
$g(x) = x^2 \quad (-1 \le x \le 1)$

23. The 2π-periodic function $g(x)$, where
$$g(x) = \begin{cases} e^x & (-\pi < x < \pi), \\ (e^{\pi} + e^{-\pi})/2 & (x = \pm \pi) \end{cases}$$

25. (a) $F(x) = (x^2 - \pi^2)/2$ **(b)** $F(x) = |x| - \pi$

27. $f(x) \sim \sum_{n=1}^{\infty} \frac{2}{(2n-1)\pi} \left[(-1)^{n+1} \cos \left(\frac{(2n-1)\pi x}{2} \right) \right.$
$\left. + \sin \left(\frac{(2n-1)\pi x}{2} \right) \right]$

29. $c_0 = 0$; $c_1 = 3/2$; $c_2 = 0$

Exercises 10.4, page 591

1. (a) The π-periodic function $\tilde{f}(x)$, where
$\tilde{f}(x) = x^2 \quad (0 < x < \pi)$
(b) The 2π-periodic function $f_o(x)$, where
$$f_o(x) = \begin{cases} x^2 & (0 < x < \pi), \\ -x^2 & (-\pi < x < 0) \end{cases}$$
(c) The 2π-periodic function $f_e(x)$, where
$$f_e(x) = \begin{cases} x^2 & (0 < x < \pi), \\ x^2 & (-\pi < x < 0) \end{cases}$$

3. (a) The π-periodic function $\tilde{f}(x)$, where
$$\tilde{f}(x) = \begin{cases} 0 & (0 < x < \pi/2), \\ 1 & (\pi/2 < x < \pi) \end{cases}$$

(b) The 2π-periodic function $f_o(x)$, where
$$f_o(x) = \begin{cases} -1 & (-\pi < x < -\pi/2), \\ 0 & (-\pi/2 < x < 0), \\ 0 & (0 < x < \pi/2), \\ 1 & (\pi/2 < x < \pi) \end{cases}$$

(c) The 2π-periodic function $f_e(x)$, where
$$f_e(x) = \begin{cases} 1 & (-\pi < x < -\pi/2), \\ 0 & (-\pi/2 < x < 0), \\ 0 & (0 < x < \pi/2), \\ 1 & (\pi/2 < x < \pi) \end{cases}$$

5. $f(x) \sim -\frac{4}{\pi} \sum_{k=1}^{\infty} \frac{1}{2k-1} \sin(2k-1)\pi x$

7. $f(x) \sim \sum_{n=1}^{\infty} \left[\frac{2\pi(-1)^{n+1}}{n} + \frac{4}{\pi n^3} \left((-1)^n - 1 \right) \right] \sin nx$

9. $f(x) \sim \sum_{k=0}^{\infty} \frac{8}{(2k+1)^3 \pi^3} \sin(2k+1)\pi x$

11. $f(x) \sim \frac{\pi}{2} + \frac{4}{\pi} \sum_{k=1}^{\infty} \frac{1}{(2k-1)^2} \cos(2k-1)x$

13. $f(x) \sim e - 1 + 2 \sum_{n=1}^{\infty} \frac{(-1)^n e - 1}{1 + \pi^2 n^2} \cos n\pi x$

15. $f(x) \sim \frac{2}{\pi} + \frac{2}{\pi} \sum_{k=1}^{\infty} \left(\frac{1}{2k+1} - \frac{1}{2k-1} \right) \cos 2kx$

17. $u(x,t) =$
$\frac{2}{\pi} \sum_{k=1}^{\infty} \left[\frac{2}{2k-1} - \frac{1}{2k+1} - \frac{1}{2k-3} \right] e^{-5(2k-1)^2 t} \sin(2k-1)x$

19. $u(x,t) = \frac{4}{\pi} \sum_{k=0}^{\infty} \frac{(-1)^{k+1}}{(2k+1)^2} e^{-5(2k+1)^2 t} \sin(2k+1)x$

Exercises 10.5, page 603

1. $u(x,t) = \sum_{n=1}^{\infty} \frac{8(-1)^{n+1} - 4}{\pi^3 n^3} e^{-5\pi^2 n^2 t} \sin n\pi x$

3. $u(x,t) = \frac{\pi}{2} - \sum_{k=0}^{\infty} \frac{4}{\pi(2k+1)^2} e^{-3(2k+1)^2 t} \cos(2k+1)x$

5. $u(x,t) = \frac{2(e^{\pi} - 1)}{\pi} + \sum_{n=1}^{\infty} \frac{2e^{\pi}(-1)^n - 2}{\pi(1+n^2)} e^{-n^2 t} \cos nx$

7. $u(x,t) = 5 + \frac{5}{\pi} x - \frac{30}{\pi} e^{-2t} \sin x + \frac{5}{\pi} e^{-8t} \sin 2x$
$+ \left(1 - \frac{10}{\pi} \right) e^{-18t} \sin 3x$
$+ \frac{5}{2\pi} e^{-32t} \sin 4x - \left(1 + \frac{6}{\pi} \right) e^{-50t} \sin 5x$
$+ \sum_{n=6}^{\infty} \frac{10}{\pi n} [2(-1)^n - 1] e^{-2n^2 t} \sin nx$

9. $u(x, t) = \dfrac{e^{-\pi} - 1}{\pi} x - e^{-x} + 1 + \sum\limits_{n=1}^{\infty} c_n e^{-n^2 t} \sin nx$,

where

$$c_n = \begin{cases} \dfrac{2e^{-\pi} - 2}{\pi n}(-1)^n + \dfrac{2n}{\pi(1 + n^2)}[(-1)^{n+1} e^{-\pi} + 1] \\[2mm] \quad + \dfrac{2}{\pi n}[(-1)^n - 1] \quad (n \neq 2), \\[3mm] \dfrac{e^{-\pi} - 1}{\pi} + \dfrac{4}{5\pi}(1 - e^{-\pi}) + 1 \quad (n = 2) \end{cases}$$

11. $u(x, t) = \sum\limits_{n=0}^{\infty} a_n e^{-4(n+1/2)^2 t} \cos(n + 1/2)x$, where

$f(x) = \sum\limits_{n=0}^{\infty} a_n \cos(n + 1/2)x$

13. $u(x, t) = \dfrac{\pi^2}{3} x - \dfrac{1}{3} x^3 - 3e^{-2t} \sin x$

$\qquad + \sum\limits_{n=2}^{\infty} \dfrac{4(-1)^n}{n^3} e^{-2n^2 t} \sin nx$

15. $u(x, y, t) = e^{-52t} \cos 6x \sin 4y - 3e^{-122t} \cos x \sin 11y$

17. $u(x, y, t) = 2 \sum\limits_{n=1}^{\infty} \dfrac{(-1)^{n+1}}{n} e^{-n^2 t} \sin ny$

19. $C(x, t) = \sum\limits_{n=1}^{\infty} c_n e^{-[L + kn^2 \pi^2/a^2] t} \sin\left(\dfrac{n\pi x}{a}\right)$, where

$c_n = \dfrac{2}{a} \int_0^a f(x) \sin\left(\dfrac{n\pi x}{a}\right) dx$

Concentration goes to zero as $t \to +\infty$.

Exercises 10.6, page 614

1. $u(x, t) = \dfrac{1}{7\pi} \sin 7\pi t \sin 7\pi x$

$\qquad + \sum\limits_{k=0}^{\infty} \dfrac{8}{((2k+1)\pi)^3} \cos(2k+1)\pi t \sin(2k+1)\pi x$

3. $u(x, t) = \sum\limits_{n=1}^{\infty} \dfrac{4}{n^3} [2(-1)^{n+1} - 1] \cos 2nt \sin nx$

5. $u(x, t) = \dfrac{2h_0 L^2}{\pi^2 a(L - a)} \sum\limits_{n=1}^{\infty} \dfrac{1}{n^2} \sin\dfrac{n\pi a}{L} \sin\dfrac{n\pi x}{L} \cos\dfrac{n\pi at}{L}$

7. $u(x, t) = \cos t \sin x + \dfrac{5}{2} \sin 2t \sin 2x - \dfrac{3}{5} \sin 5t \sin 5x$

$\qquad + 2 \sum\limits_{n=1}^{\infty} \dfrac{(-1)^{n+1}}{n^3} \left[t - \dfrac{\sin nt}{n}\right] \sin nx$

9. $u(x, t) = \sum\limits_{n=0}^{\infty} \left[a_n \cos\dfrac{(2n+1)\pi at}{2L} \right.$

$\qquad \left. + b_n \sin\dfrac{(2n+1)\pi at}{2L} \right] \sin\dfrac{(2n+1)\pi x}{2L}$, where

$f(x) = \sum\limits_{n=0}^{\infty} a_n \sin\dfrac{(2n+1)\pi x}{2L}$ and

$g(x) = \sum\limits_{n=0}^{\infty} b_n \dfrac{(2n+1)\pi a}{2L} \sin\dfrac{(2n+1)\pi x}{2L}$

11. $u(x, t) = \sum\limits_{n=1}^{\infty} a_n T_n(t) \sin\dfrac{n\pi x}{L}$, where

$a_n = \dfrac{2}{L} \int_0^L f(x) \sin\dfrac{n\pi x}{L} dx$, and

$T_n(t) = e^{-t/2}\left(\cos\beta_n t + \dfrac{1}{2\beta_n} \sin\beta_n t \right)$, where

$\beta_n = \dfrac{1}{2L} \sqrt{3L^2 + 4\alpha^2 \pi^2 n^2}$

13. $u(x, t) = \dfrac{1}{2\alpha}[\sin(x + \alpha t) - \sin(x - \alpha t)]$

$\qquad = \dfrac{1}{\alpha} \sin\alpha t \cos x$

15. $u(x, t) = x + tx$

17. $u(x, t) = \dfrac{1}{2}\left[e^{-(x+\alpha t)^2} + e^{-(x-\alpha t)^2} \right.$

$\qquad\qquad \left. + \dfrac{\cos(x - \alpha t) - \cos(x + \alpha t)}{\alpha} \right]$

21. $u(r, t) = \sum\limits_{n=1}^{\infty} [a_n \cos(k_n \alpha t) + b_n \sin(k_n \alpha t)] J_0(k_n r)$,

where

$a_n = \dfrac{1}{c_n} \int_0^1 f(r) J_0(k_n r) r \, dr$, and

$b_n = \dfrac{1}{\alpha k_n c_n} \int_0^1 g(r) J_0(k_n r) r \, dr$, with

$c_n = \int_0^1 J_0^2(k_n r) r \, dr$

Exercises 10.7, page 626

1. $u(x, y) = \dfrac{4 \cos 6x \sinh[6(y - 1)]}{\sinh(-6)}$

$\qquad + \dfrac{\cos 7x \sinh[7(y - 1)]}{\sinh(-7)}$

3. $u(x, y) = \sum\limits_{n=1}^{\infty} A_n \sin nx \sinh(ny - n\pi)$, where

$A_n = \dfrac{2}{\pi \sinh(-n\pi)} \int_0^\pi f(x) \sin nx \, dx$

5. $u(x, y) = \dfrac{\cos x \sinh(y - 1)}{\sinh(-1)} - \dfrac{\cos 3x \sinh(3y - 3)}{\sinh(-3)}$

$\qquad + \dfrac{\cos 2x \sinh 2y}{\sinh(2)}$

7. $u(r, \theta) = \dfrac{\pi}{2} - \sum\limits_{k=0}^{\infty} \dfrac{r^{2k+1}}{(2k+1)^2 \pi 2^{2k-1}} \cos(2k + 1)\theta$

9. $u(r, \theta) = \dfrac{a_0}{2} + \sum\limits_{n=1}^{\infty} \left(\dfrac{r}{a}\right)^n (a_n \cos n\theta + b_n \sin n\theta)$, where

a_0 is arbitrary, and for $n = 1, 2, 3, \ldots$

$a_n = \dfrac{a}{\pi n} \int_{-\pi}^{\pi} f(\theta) \cos n\theta \, d\theta$,

$b_n = \dfrac{a}{n\pi} \int_{-\pi}^{\pi} f(\theta) \sin n\theta \, d\theta$

11. $u(r, \theta) = \left(\dfrac{1}{3}r - \dfrac{4}{3}r^{-1}\right)\cos\theta + \left(\dfrac{2}{3}r - \dfrac{2}{3}r^{-1}\right)\sin\theta$
$$+ \left(-\dfrac{1}{255}r^4 + \dfrac{256}{255}r^{-4}\right)\sin 4\theta$$

13. $u(r, \theta) = \dfrac{a_0}{2} + \displaystyle\sum_{n=1}^{\infty} r^{-n}(a_n \cos n\theta + b_n \sin n\theta)$,

where

$$a_n = \dfrac{1}{\pi}\int_{-\pi}^{\pi} f(\theta)\cos n\theta\, d\theta \qquad (n = 0, 1, 2, \ldots\,),$$

and

$$b_n = \dfrac{1}{\pi}\int_{-\pi}^{\pi} f(\theta)\sin n\theta\, d\theta \qquad (n = 1, 2, 3, \ldots\,)$$

15. $u(r, \theta) = r^3 \sin 3\theta$

17. $u(r, \theta) = a_0 + b_0 \ln r + \displaystyle\sum_{n=1}^{\infty} \left[(a_n r^n + b_n r^{-n})\cos n\theta \right.$
$$\left. + (c_n r^n + d_n r^{-n})\sin n\theta\right],$$

where

$$a_0 = \dfrac{1}{2\pi}\int_{-\pi}^{\pi} f(\theta)\, d\theta, \qquad b_0 = \dfrac{3}{2\pi}\int_{-\pi}^{\pi} g(\theta)\, d\theta,$$

and for $n = 1, 2, 3, \ldots$

$$a_n + b_n = \dfrac{1}{\pi}\int_{-\pi}^{\pi} f(\theta)\cos n\theta\, d\theta,$$

$$n3^{n-1}a_n - n3^{-n-1}b_n = \dfrac{1}{\pi}\int_{-\pi}^{\pi} g(\theta)\cos n\theta\, d\theta,$$

and

$$c_n + d_n = \dfrac{1}{\pi}\int_{-\pi}^{\pi} f(\theta)\sin n\theta\, d\theta,$$

$$n3^{n-1}c_n - n3^{-n-1}d_n = \dfrac{1}{\pi}\int_{-\pi}^{\pi} g(\theta)\sin n\theta\, d\theta$$

21. $u(r, \theta, z) = [I_0(r)/I_0(\pi)]\sin z$

23. **(c)** $\phi(x, y) = 2xy$

CHAPTER 11

Exercises 11.2, page 650

1. $y = e^{-x}\cos 5x + ce^{-x}\sin 5x$, where c is any real number.

3. $y = ce^{2x}\sin 3x$, where c is any real number.

5. $y = c_1 \cos x + c_2 \sin x - (1/3)\sin 2x$, where c_1 and c_2 are real numbers.

7. $y = -(1/2)\cos x + (1/2 + e^{\pi})\sin x + (1/2)e^x$

9. No solution

11. $y = c\sin(\ln x) + 2x^{-1}$, where c is any real number.

13. $\lambda_n = (n + 1/2)^2 \pi^2$, $n = 0, 1, 2, \ldots$; and
$y_n = c_n \sin[(n + 1/2)\pi x]$, $n = 0, 1, 2, \ldots$, where the c_n's are arbitrary.

15. $\lambda_n = n^2 - 3$, $n = 0, 1, 2, \ldots$; and
$y_n = c_n \cos nx$, $n = 0, 1, 2, \ldots$,
where the c_n's are arbitrary.

17. $\lambda_0 = -\mu_0^2$, where $\tanh(\mu_0\pi) = \mu_0/2$; and
$y_0 = c_0\big(\cosh(\mu_0 x) - (2/\mu_0)\sinh(\mu_0 x)\big)$ and
$\lambda_n = \mu_n^2$,
where $\tan(\mu_n\pi) = \mu_n/2$, $n = 1, 2, 3, \ldots$; and
$y_n = c_n\big(\cos(\mu_n x) - (2/\mu_n)\sin(\mu_n x)\big)$, where the c_n's are arbitrary.

19. $\lambda_n = (n + 1/2)^2$, $n = 0, 1, 2, \ldots$; and
$y_n = c_n \cos[(n + 1/2)\ln x]$, $n = 0, 1, 2, \ldots$, where the c_n's are arbitrary.

21. $\lambda_1 = -0.195$, $\lambda_2 = 1.932$, $\lambda_3 = 5.932$

23. No nontrivial solutions

25. λ_n, $n = 1, 2, 3, \ldots$ such that $(\lambda_n^2 + 1)\sqrt{\lambda_n} = n\pi$; and
$y_n = c_n \sin(\sqrt{\lambda_n}x)$, $n = 1, 2, 3, \ldots$, where the c_n's are arbitrary.

27. $\lambda_n = -\mu_n^4$, $\mu_n > 0$ and $\cos\mu_n L \cdot \cosh\mu_n L = 1$;
$y_n = c_n[(\cos\mu_n L - \cosh\mu_n L)(\sin\mu_n x - \sinh\mu_n x)$
$\quad - (\sin\mu_n L - \sinh\mu_n L)(\cos\mu_n x - \cosh\mu_n x)]$,
$n = 1, 2, \ldots$, where the c_n's are arbitrary.

29. $\lambda_n = -(n\pi/L)^4$, $n = 1, 2, 3, \ldots$, and
$y_n = c_n \sin(n\pi x/L)$, $n = 1, 2, 3, \ldots$, where the c_n's are arbitrary.

33. **(b)** $\lambda_n = n^2 + 1$, $n = 1, 2, \ldots$, and
$y_n = c_n e^{-x}\sin nx$, where the c_n's are arbitrary.

35. $U_{m, n} = c_{n, m}\left(\sin\dfrac{\pi n}{a}x\right)\left(\sin\dfrac{\pi m}{b}y\right)$,
$$\lambda_{m, n} = \left(\dfrac{\pi n}{a}\right)^2 + \left(\dfrac{\pi m}{b}\right)^2, \quad n = 1, 2, \ldots,$$
$m = 1, 2, \ldots$, where the $c_{n, m}$'s are arbitrary.

Exercises 11.3, page 662

1. $(e^{6x}y')' + \lambda e^{6x}y = 0$

3. $((1 - x)^2 y')' + \lambda[(1 - x)/x]y = 0$

5. $(xy')' + \lambda yx^{-1} = 0$ **7.** Not selfadjoint

9. Selfadjoint **11.** Selfadjoint

17. $\phi_n = \sqrt{2}\sin[(n + 1/2)\pi x]$, $n = 0, 1, 2, \ldots$;
$$x = \dfrac{8}{\pi^2}\sum_{n=0}^{\infty}\dfrac{(-1)^n}{(2n + 1)^2}\sin[(n + 1/2)\pi x]$$

19. $\phi_o = 1/\sqrt{\pi}$; $\phi_n = \sqrt{\dfrac{2}{\pi}}\cos nx$, $n = 1, 2, \ldots$;
$$\dfrac{\pi}{2} + \dfrac{2}{\pi}\sum_{n=1}^{\infty}\dfrac{(-1)^n - 1}{n^2}\cos nx$$

21. $\phi_0 = \gamma_0^{-1}[\cosh(\mu_0 x) - (2/\mu_0)\sinh(\mu_0 x)]$ and
$\phi_n = \gamma_n^{-1}[-(\mu_n/2)\cos(\mu_n x) + \sin(\mu_n x)]$,
$n = 1, 2, 3, \ldots$, where $\mu_0 = 2\tanh(\mu_0 x)$,
$\mu_n = 2\tan(\mu_n\pi)$,
$\gamma_0^2 = (4\mu_0^{-2} - 1)[(4\mu_0)^{-1}\sinh(2\mu_0\pi) - \pi/2]$, and
$\gamma_n^2 = (\mu_n^2/4)[\pi/2 + (4\mu_n)^{-1}\sin(2\mu_n\pi)]$
$\quad - (1/2)\sin^2(\mu_n\pi) + \pi/2 - (4\mu_n)^{-1}\sin(2\mu_n\pi)$;
$\gamma_0^{-2}[(\pi/\mu_0)\sinh(\mu_0\pi)(1 - 4/\mu_0) + \mu_0^{-2}][\cosh(\mu_0 x)$
$\quad - (2/\mu_0)\sinh(\mu_0 x)] + \sum_{n=1}^{\infty} \gamma_n^{-2}[(\mu_n^{-2}$
$\quad - \pi/2)\sin(\mu_n\pi) + \mu_n^{-1}(\pi - 1/2)\cos(\mu_n\pi)$
$\quad + (2/\mu_n)^{-1}][-(\mu_n/2)\cos(\mu_n x) + \sin(\mu_n x)]$

23. $\phi_n = \sqrt{2/\pi}\cos[(n + 1/2)\ln x]$, $n = 0, 1, 2, \ldots$;
$\frac{2}{\pi}\sum_{n=0}^{\infty} \frac{(-1)^n e^\pi(n + 1/2) - 1}{1 + (n + 1/2)^2}\cos[(n + 1/2)\ln x]$

Exercises 11.4, page 671

1. $L^+[y] = xy'' + x^2 y' + (2x + 3)y$
3. $L^+[y] = (1 + x^2)y'' + 2xy' + y$
5. $L^+[y] = 2x^2 y'' + 7xy' - 2y$
7. $L^+[y] = y'' + 2y' + 10y$, where
$D(L^+) = \{v \text{ in } C^2[0, \pi]: v(0) = v(\pi) = 0\}$
9. $L^+[y] = 2x^2 y'' + 5xy'$, where
$D(L^+) = \{v \text{ in } C^2[1, 4]: v(1) = v(4) = 0\}$
11. $y'' - 6y' + 10y = 0$; $6y(0) - y'(0) = 0$,
$6y(\pi) - y'(\pi) = 0$
13. $y'' + 4y = 0$; $y(0) = y(\pi)$, $y'(0) = y'(\pi)$
15. $x^2 y'' + xy' + y = 0$; $y(1) - y'(1) = 0$,
$y(e^\pi) - e^\pi y'(e^\pi) = 0$
17. $\int_0^\pi h(x)e^{-x}\sin 3x \, dx = 0$
19. Unique solution for each h
21. $\int_0^\pi h(x)e^{3x}(\cos x + 3\sin x)\, dx = 0$
23. $\int_0^\pi h(x)\cos 2x \, dx = 0$ and
$\int_0^\pi h(x)\sin 2x \, dx = 0$
25. $\int_1^{e^\pi} h(x)[\cos(\ln x) + \sin(\ln x)]\, dx = 0$

Exercises 11.5, page 676

1. $5\sin x + \sin 3x$ **3.** No solution
5. $C_1\cos x - (1/15)\cos 4x - (1/48)\cos 7x$,
where C_1 is arbitrary
7. $\sum_{n=1}^{\infty} \frac{-32}{\pi(2n-1)^3[1 - (n - 1/2)^2]}\sin[(n - 1/2)x]$

9. $\sum_{n=0}^{\infty} \frac{\gamma_n}{[1 - (n + 1/2)^2]}\sqrt{\frac{2}{\pi}}\cos[(n + 1/2)x]$, where
$\gamma_n = \sqrt{\frac{2}{\pi}}\int_0^\pi f(x)\cos[(n + 1/2)x]\, dx$

11. $\sum_{n=0}^{\infty} \frac{\gamma_n}{[5 - (n + 1/2)^2]}\sqrt{\frac{2}{\pi}}\cos[(n + 1/2)\ln x]$,
where $\gamma_n = \sqrt{\frac{2}{\pi}}\int_1^{e^\pi} f(x)\cos[(n + 1/2)\ln x]\, dx$

13. $\frac{\gamma_0}{6 - \lambda_0}\sinh(\mu_0 x) + \sum_{n=1}^{\infty} \frac{\gamma_n}{6 - \lambda_n}\sin(\mu_n x)$,
where $\gamma_0 = \dfrac{\int_0^\pi f(x)\sinh(\mu_0 x)\, dx}{\int_0^\pi \sinh^2(\mu_0 x)\, dx}$ and
$\gamma_n = \dfrac{\int_0^\pi f(x)\sin(\mu_n x)\, dx}{\int_0^\pi \sin^2(\mu_n x)\, dx}$;
and μ_0 and μ_n satisfy
$\tanh\mu_0\pi = \mu_0/3$, $\lambda_0 = -\mu_0^2$,
and
$\tan\mu_n\pi = \mu_n/3$, $\lambda_n = \mu_n^2$, $n = 1, 2, 3, \ldots$.

Exercises 11.6, page 683

1. $G(x, s) = \begin{cases} s, & 0 \le s \le x, \\ x, & x \le s \le \pi \end{cases}$

3. $G(x, s) = \begin{cases} -\dfrac{s(x - 1 - \pi)}{1 + \pi}, & 0 \le s \le x, \\ -\dfrac{x(s - 1 - \pi)}{1 + \pi}, & x \le s \le \pi \end{cases}$

5. $G(x, s) = \begin{cases} (\sin 2s\cos 2x)/2, & 0 \le s \le x, \\ (\sin 2x\cos 2s)/2, & x \le s \le \pi \end{cases}$

7. $G(x, s) = \begin{cases} -(s^{-2} - s)(4x^{-2} + x)/15, & 1 \le s \le x, \\ -(x^{-2} - x)(4s^{-2} + s)/15, & x \le s \le 2 \end{cases}$

9. $G(x, s) = \begin{cases} -se^{-s}(xe^{-x} - e^{-x}), & 0 \le s \le x, \\ -xe^{-x}(se^{-s} - e^{-s}), & x \le s \le 1 \end{cases}$

11. $G(x, s) = \begin{cases} -s(x - \pi)/\pi, & 0 \le s \le x, \\ -x(s - \pi)/\pi, & x \le s \le \pi, \end{cases}$ and
$y = 1 - e^x - x/\pi + e^\pi x/\pi$

13. $G(x, s) = \begin{cases} -s(x - 1 - \pi)/(1 + \pi), & 0 \le s \le x, \\ -x(s - 1 - \pi)/(1 + \pi), & x \le s \le \pi, \end{cases}$
and $y = -x^3/6 + (3\pi^2 + \pi^3)x/(6 + 6\pi)$

15. $G(x, s) = \begin{cases} -\sin s\sin(x - 1)/(\sin 1), & 0 \le s \le x, \\ -\sin x\sin(s - 1)/(\sin 1), & x \le s \le 1, \end{cases}$
and $y = [\sin 1 + \sin(x - 1) - \sin x]/(\sin 1)$

17. $G(x, s) =$
$\begin{cases} (e^s - e^{-s})(e^x - e^{2-x})/(2 - 2e^2), & 0 \le s \le x, \\ (e^x - e^{-x})(e^s - e^{2-s})/(2 - 2e^2), & x \le s \le 1, \end{cases}$
and
$y = -x + \dfrac{\sinh x}{\sinh 1} = -x + \dfrac{e^{x+1} - e^{1-x}}{e^2 - 1}$

19. $G(x, s) =$
$$\begin{cases} -(s^2 - s^{-2})(x^2 - 16x^{-2})/60, & 1 \le s \le x, \\ -(x^2 - x^{-2})(s^2 - 16s^{-2})/60, & x \le s \le 2, \end{cases}$$
and
$$y = [16x^2 \ln 2 - 16x^{-2} \ln 2 - 15x^2 \ln x]/60$$

21. $G(x, s) = \begin{cases} -s(x-1), & 0 \le s \le x, \\ -x(s-1), & x \le s \le 1, \end{cases}$ and
$$y = \alpha + (\beta - \alpha)x - x\int_0^1 sf(s)\,ds + \int_0^x sf(s)\,ds$$
$$+ x\int_x^1 f(s)\,ds$$

23. (a) $K(x, s) =$
$$\begin{cases} -(e^{-s} - e^{-2s})(e^{-x} - e^{1-2x})/[(e-1)e^{-3s}], \\ \qquad 0 \le s \le x, \\ -(e^{-x} - e^{-2x})(e^{-s} - e^{1-2s})/[(e-1)e^{-3s}], \\ \qquad x \le s \le 1 \end{cases}$$

(b) $y = (1/4)(3 - e^2)e^{-x}/(e-1)$
$$+ (1/4)e(e-3)e^{-2x}/(e-1)$$
$$- (1/2)x + (3/4)$$

25. (a) $K(x, s) = \begin{cases} -x(2-x)(s^{-3} - s^{-2}), & 1 \le s \le x, \\ -x(1-x)(2s^{-3} - s^{-2}), & x \le s \le 2 \end{cases}$

(b) $y = x^2 \ln 2 - x \ln x - x \ln 2$

27. $H(x, s) =$
$$\begin{cases} -[(x^3/2 - \pi x^2)/\pi^2 + x/2]s^2 \\ \quad - [(-x^3/3 + \pi x^2/2)/\pi^3 - 1/6]s^3, \\ \qquad 0 \le s \le x, \\ -[(s^3/2 - \pi s^2)/\pi^2 + s/2]x^2 \\ \quad - [(-s^3/3 + \pi s^2/2)/\pi^3 - 1/6]x^3, \\ \qquad x \le s \le \pi \end{cases}$$

29. $H(x, s) = \begin{cases} s^2(s - 3x)/6, & 0 \le s \le x, \\ x^2(x - 3s)/6, & x \le s \le \pi \end{cases}$

Exercises 11.7, page 691

1. $\displaystyle\sum_{n=1}^{\infty} b_n J_2(\alpha_{2n}x)$, where $\{\alpha_{2n}\}$ is the increasing sequence of real zeros of J_2 and
$$b_n = \frac{\int_0^1 f(x)J_2(\alpha_{2n}x)\,dx}{(\mu - \alpha_{2n}^2)\int_0^1 J_2^2(\alpha_{2n}x)x\,dx}.$$

3. $\displaystyle\sum_{n=1}^{\infty} a_n J_0(\alpha_{1n}x)$, where $\{\alpha_{1n}\}$ is the increasing sequence of real zeros of J_0', which are also the zeros of J_1, and
$$a_n = \frac{\int_0^1 f(x)J_0(\alpha_{1n}x)\,dx}{(\mu - \alpha_{1n}^2)\int_0^1 J_0^2(\alpha_{1n}x)x\,dx}.$$

5. $\displaystyle\sum_{n=0}^{\infty} a_{2n+1} P_{2n+1}(x)$, where
$$a_{2n+1} = \frac{\int_0^1 f(x)P_{2n+1}(x)\,dx}{[\mu - (2n+1)(2n+2)]\int_0^1 P_{2n+1}^2(x)\,dx}.$$

15. (c) $\displaystyle\sum_{n=0}^{\infty} a_n T_n(x)$, where
$$a_n = \frac{\int_{-1}^1 f(x)T_n(x)\,dx}{(\mu - n^2)\int_{-1}^1 T_n^2(x)(1-x^2)^{-1/2}\,dx}.$$

Exercises 11.8, page 702

1. $\phi(x) \equiv 0$ **9.** $\pi/2$

11. Between $\pi\sqrt{1/(\lambda + 1)}$ and $\pi\sqrt{6/(\lambda + e^{-5})}$

Review Problems, page 705

1. (a) $\lambda_n = 9 + n^2\pi^2$, $n = 1, 2, 3, \ldots$; and
$y_n = c_n e^{-3x} \sin(n\pi x)$, $n = 1, 2, 3, \ldots$,
where c_n is any real number.

(b) $\lambda_n = \mu_n^2$, $n = 1, 2, 3, \ldots$, where
$\tan \mu_n \pi = -2\mu_n$; and $y_n = c_n \sin \mu_n x$,
$n = 1, 2, 3, \ldots$, where c_n is any real number.

3. (a) $\lambda_n = (n - 1/2)^2$, $n = 1, 2, 3, \ldots$, and
$y_n = \sqrt{2/\pi}\cos[(n - 1/2)x]$, $n = 1, 2, 3, \ldots$

(b) $\displaystyle\frac{2}{\pi}\sum_{n=1}^{\infty}\left[\frac{(-1)^{n-1}(1-\pi)}{(n-1/2)} + \frac{1}{(n-1/2)^2}\right]$
$\times \cos[(n-1/2)x]$

5. (a) $\displaystyle\int_0^{2\pi} h(x)e^{-3x}\sin 2x\,dx = 0$

(b) $\displaystyle\int_1^2 h(x)\,dx = 0$

7. (a) $G(x, s) = \begin{cases} s, & 0 \le s \le x, \\ x, & x \le s \le 1; \end{cases}$
$y = x^3/6 + x^2/2 - 3x/2$

(b) $G(x, s) =$
$$\begin{cases} -(e^s + e^{-s})(e^x - e^{2-x})/(2e^2 + 2), \\ \qquad 0 \le s \le x, \\ -(e^x + e^{-x})(e^s - e^{2-s})/(2e^2 + 2), \\ \qquad x \le s \le 1; \end{cases}$$
$y = 4(e^{1+x} + e^{1-x} - 1 - e^2)/(e^2 + 1)$

9. π

CHAPTER 12

Exercises 12.2, page 728

1. Unstable saddle point

3. Asymptotically stable spiral point

5. Asymptotically stable improper node

7. $(3, 2)$ is an asymptotically stable improper node.

9. $(-4, -1)$ is a stable center.

11. $(3, -5)$ is an asymptotically stable improper node.

13. Asymptotically stable improper node

Figure B.46

15. Unstable saddle point

Figure B.47

17. Unstable saddle point

Figure B.48

19. Asymptotically stable improper node

Figure B.49

27. Eigenvalues 3, -1 correspond to eigenvectors $[\,3\ 2\,]^T$ and $[\,1\ 2\,]^T$ respectively. Solution rays on $y = 2x/3$ go to infinity; solution rays on $y = 2x$ approach the origin.

Exercises 12.3, page 739

1. Unstable saddle point **3.** Unstable improper node
5. Center or spiral point of indeterminate stability
7. Unstable saddle point
9. $(8, 2)$ is an asymptotically stable improper node; $(-8, -2)$ is an unstable saddle point.
11. $(2, -2)$ is an unstable improper node; $(3, -3)$ is an unstable saddle point.
13. Critical points $x = (2n + 1)\pi/2$, $x' = 0$ are unstable saddles for all odd n, inconclusive for even n.
15. $x = x' = 0$ is an unstable spiral.
17. $(1, 1)$ is an asymptotically stable improper node, proper node, or spiral point; $(-1, -1)$ is an unstable saddle point. (See Figure B.50.)

Figure B.50

19. $(0, 0)$ is an unstable improper node, proper node, or spiral point; $(0, 3)$ and $(3, 0)$ are unstable saddle points; and $(2, 2)$ is an asymptotically stable improper node. (See Figure B.51.)

Figure B.51

21. (b) $\varepsilon < -2$, asymptotically stable improper node;
 $\varepsilon = -2$, asymptotically stable improper node,
 proper node, or spiral point;
 $-2 < \varepsilon < 0$, asymptotically stable spiral point;
 $\varepsilon = 0$, stable center;
 $0 < \varepsilon < 2$, unstable spiral point;
 $\varepsilon = 2$, unstable improper node, proper node,
 or spiral point;
 $\varepsilon > 2$, unstable improper node

23. (b) a_2/b_2 **(d)** As a_1 is decreased, the population of
species x approaches zero (reaches zero when
$a_1/c_1 = a_2/b_2$) and the population of species y
approaches a_2/b_2.

25. $h = 0$: $(0, 0)$ is an unstable improper node. See Fig. B.52;
 $(0, 1/2)$ is an unstable saddle point;
 $(1/4, 0)$ is an asymptotically stable
 improper node;
 $(1/3, -1/3)$ is an unstable saddle point, but not
 of interest.

Species x survives while
species y dies off.

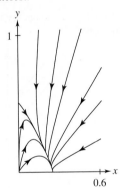

Figure B.52

$h = 1/32$: $\left((2 - \sqrt{2})/16, 0\right)$ is an unstable improper
 node;
 $\left((2 + \sqrt{2})/16, 0\right)$ is an asymptotically
 stable improper node;
 $(1/12, 7/24)$ is an unstable saddle point;
 $(1/4, -1/8)$ is an unstable saddle point,
 but not of interest.

Competitive exclusion: One
survives, but the other dies off.
(See Figure B.53.)

Figure B.53

$h = 5/32$: There are no critical points. Species y
 survives while species x dies off.
 (See Figure B.54.)

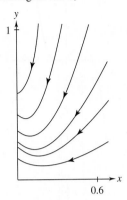

Figure B.54

Exercises 12.4, page 747

1. $G(x) = (1/3)x^3 - (3/2)x^2 + x + C$;
 $E(x, v) = (1/2)v^2 + (1/3)x^3 - (3/2)x^2 + x$

3. $G(x) = (1/2)x^2 + x + \ln|x - 1| + C$;
 $E(x, v) = (1/2)v^2 + (1/2)x^2 + x + \ln|1 - x|$

5. $G(x) = (1/4)x^4 + (1/3)x^3 - (1/2)x^2 + C$;
 $E(x, v) = (1/2)v^2 + (1/4)x^4 + (1/3)x^3 - (1/2)x^2$

7.

Figure B.55

9.

Figure B.56

11.

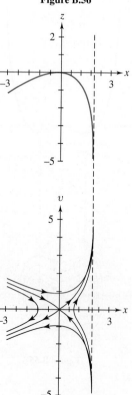

Figure B.57

13. $vh(x, v) = v^2$, so energy is decreasing along a trajectory.

Figure B.58

15. $vh(x, v) = v^2$, so energy is decreasing along a trajectory.

Figure B.59

17. When $\beta > 0$, all trajectories are closed and encircle the origin. Hence, all solutions are periodic. When $\beta < 0$, only trajectories near the origin are closed. Hence, periodic solutions exist, but only for initial values near the origin.

Figure B.60

Figure B.61

19. (a)

Figure B.62

(b)

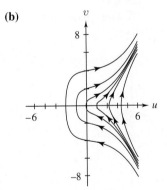

Figure B.63

 (c) When $0 < \lambda < 1/4$, closed trajectories exist and hence there can be periodic motion.
 When $1/4 < \lambda < 1$, there cannot be periodic motion.

21. The Jacobian's eigenvalues at $(2n\pi, 0)$ are $\left[-b \pm \sqrt{b^2 - 4}\right]/2$, and at $([2n + 1]\pi, 0)$ they are $\left[-b \pm \sqrt{b^2 + 4}\right]/2$. The stability properties for all cases can be read from Table 12.1.

Exercises 12.5, page 755

 1. Asymptotically stable **3.** Stable **5.** Unstable
 7. Asymptotically stable **9.** Stable **11.** Stable
13. Stable

Exercises 12.6, page 763

1. (a)

Figure B.64

(b) Trajectories inside the unit circle spiral out and approach the unit circle as a limit cycle. Trajectories outside the unit circle also spiral out, away from the unit circle.

3. (b) Counterclockwise

5. $r = 0$ is an asymptotically stable spiral point.
$r = 1$ is a semistable limit cycle.
$r = 2$ is an unstable limit cycle.
(See Figure B.65.)

Figure B.65

7. $r = 0$ is an asymptotically stable spiral point.
$r = 1$ is an unstable limit cycle.
(See Figure B.66.)

Figure B.66

9. $r = 0$ is an unstable spiral point.
$r = 1$ is a stable limit cycle.
$r = 2$ is an unstable limit cycle.
(See Figure B.67.)

Figure B.67

11. $r = 0$ is not an isolated critical point, but it is stable. $r = (n\pi)^{-1}, n = 1, 2, \ldots$, is a limit cycle that is stable for n even and unstable for n odd. (See Figure B.68.)

Figure B.68

Exercises 12.7, page 769

1. Asymptotically stable **3.** Asymptotically stable
5. Unstable **9.** Unstable
11. Asymptotically stable
13. Asymptotically stable
15. The equilibrium solution corresponding to the critical point $(1, -2, 2)$ is asymptotically stable.
17. The equilibrium solutions corresponding to the critical points $(0, 0, 0)$, $(1, 0, 0)$, $(0, 1, 0)$, and $(0, 0, 1)$ are unstable. The one corresponding to $(2/5, 2/5, 2/5)$ is asymptotically stable.

Exercises 12.8, page 775

5. About 0.37.
7. $(0, 0)$ is a stable spiral with Jacobian eigenvalues $-0.029 \pm 0.098i$.
$(0.313, 0.065)$ is a saddle point with Jacobian eigenvalues 0.30 and -0.019.
$(0.697, 0.145)$ is a stable spiral with Jacobian eigenvalues $-0.053 \pm 0.100i$.

Figure B.69

9. If we express the FitzHugh-Nagumo system as

$$\frac{dv}{dt} = F \equiv -v(v-a)(v-1) - w + I,$$

$$\frac{dw}{dt} = G \equiv \varepsilon(v - \gamma w), \text{ then } \frac{\partial F}{\partial v} \text{ is the same as the}$$

slope $\frac{dw}{dv} = w'$ of the v-nullcline $F = 0$. Therefore

the Jacobian $\begin{bmatrix} \dfrac{\partial F}{\partial v} & \dfrac{\partial F}{\partial w} \\ \dfrac{\partial G}{\partial v} & \dfrac{\partial G}{\partial v} \end{bmatrix}$ at the critical point equals

$\begin{bmatrix} w' & -1 \\ \varepsilon & -\varepsilon\gamma \end{bmatrix}$.

The eigenvalue sum is its trace; the eigenvalue product is its determinant.

11. If $w' < 0$, then the sum of the eigenvalues of the Jacobian is negative and the product is positive; they must both be negative.

Review Problems, page 777

1. $(0,0)$ is an unstable saddle point.

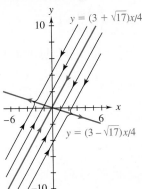

Figure B.70

3. $(0,0)$ is an asymptotically stable spiral point. (See Figure B.79.)

Figure B.71

5. $(0,0)$ is an unstable improper node;
$(0,2)$ is an asymptotically stable improper node;
$(1/3,0)$ is an unstable saddle point;
$(1,-2)$ is an asymptotically stable spiral point.
(See Figure B.72.)

Figure B.72

7.

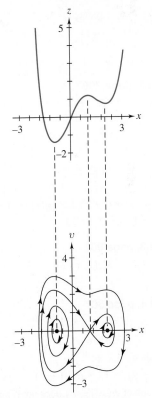

Figure B.73

9. Asymptotically stable **11.** Unstable
13. $r = 0$ is an asymptotically stable spiral point.
$r = 1$ is an unstable limit cycle.
$r = 2$ is a semistable limit cycle.
(See Figure B.74 on page B-36.)

Figure B.74

15. Yes **17.** Asymptotically stable

CHAPTER 13

Exercises 13.1, page 788

1. $y(x) = 4 + \int_1^x [t^2 - y(t)] \, dt$

3. $y(x) = -3 + \int_1^x [y(t) - t]^2 \, dt$ **5.** 1.7320508

7. 0.2462662 **9.** 1.0698479

11. $y_1(x) = 2 - 2x$, $y_2(x) = 2 - 2x + x^2$

13. $y_1(x) = y_2(x) = 1 + x^3$

15. $y_1(x) = 1 - e^x$, $y_2(x) = 2 + x - 2e^x$

Exercises 13.2, page 795

1. Yes **3.** No **5.** Yes

9. Any h satisfying $0 < h < 1/2$ will work.

11. $y_1(x) = -(1/20)x^5 + (7/6)x^3 - (1/2)x^2 - x + 1$;

$y_2(x) = (1/840)x^7 - (7/120)x^5 + (1/24)x^4$
$+ (7/6)x^3 - (1/2)x^2 - x + 1$

15. (b) $x_1(t) = -t$, $y_1(t) = 1 + t$, $z_1(t) = 1$;

$x_2(t) = -t - t^2 - t^3/3$, $y_2(t) = 1 + t$,
$z_2(t) = 1 - t^2/2 - t^3/3$

Exercises 13.3, page 802

1. $(\pi/2, 3\pi/2)$ **3.** $(0, \pi/2)$ **5.** $(-\infty, \infty)$

7. $y(t) \equiv 0$

Exercises 13.4, page 807

1. $10^{-2}e$ **3.** $10^{-2}e^e$ **5.** $10^{-2}e^{16}$

7. $(1/24)e^e$ **9.** $(73/6)e^{3\sqrt{3}/8}$

11. (a) $\phi(x, 1) = (1 - x)^{-1}$;

$\phi(x, 1 + \varepsilon) = (1 + \varepsilon)\big(1 - (1 + \varepsilon)x\big)^{-1}$

(d) The graph of $\phi(x, 1 + \varepsilon)$ does not remain in any closed rectangle $R_0 \subset R$.

Review Problems, page 810

1. 1.3247180

3. $y(x) = -2 + \int_{-1}^x [y^2(t) + \sin t] \, dt$

5. $y_1(x) = 1 + x$;

$y_2(x) = 1 + x + (3/2)x^2 + x^3 + (1/4)x^4$

7. Yes **9.** $(-\infty, \infty)$ **11.** $10^{-3}e^4$

APPENDIX A

Exercises A, page A-1

1. $(x^2 + 1)^{5/2}(5x^2 - 2)/35 + C$ **3.** $\dfrac{2}{15}e^{6t^5} + C$

5. $\dfrac{1}{2}\theta + \dfrac{1}{4\pi}\sin(2\pi\theta) + C$

7. $\dfrac{1}{2}\ln(t^2 + 4) + \dfrac{1}{2}\arctan(t/2) + C$

9. $(1/\sqrt{2})\arcsin\sqrt{2}x + C$

11. $2\ln|x - 1| - 4\ln|x - 2| + 2\ln|x - 3| + C$

13. $\sqrt{9x^2 - 1} - \arccos\left(\dfrac{1}{3x}\right) + C$

15. $\dfrac{1}{2}\ln(x^2 + 4) - \dfrac{2}{x} + C$ **17.** $\ln|\ln x| + C$

19. $y\cosh y - \sinh y + C$ **21.** $\dfrac{x}{4\sqrt{x^2 + 4}} + C$

23. $\dfrac{1}{2}(t^2 - 9)\ln(t + 3) - \dfrac{t^2}{4} + \dfrac{3t}{2} + C$

25. $-\arctan(\cos t) + C$

27. $\dfrac{5}{7}\sin^{7/5}x - \dfrac{5}{17}\sin^{17/5}x + C$

29. $\dfrac{\sin(10x)}{20} + \dfrac{\sin(4x)}{8} + C$

31. $\dfrac{1}{9}\tan^9\theta + \dfrac{2}{7}\tan^7\theta + \dfrac{1}{5}\tan^5\theta + C$

33. $\dfrac{1}{8}x - \dfrac{1}{96}\sin(12x) + C$

Index

Note: *n* indicates footnote.

A BRIEF TABLE OF INTEGRALS*

$$\int (f(u) + g(u))\,du = \int f(u)\,du + \int g(u)\,du. \qquad \int cf(u)\,du = c\int f(u)\,du.$$

$$\int u\,dv = uv - \int v\,du. \qquad \int u^n\,du = \frac{u^{n+1}}{n+1}, \qquad n \neq -1. \qquad \int \frac{du}{u} = \ln|u|.$$

$$\int e^u\,du = e^u. \qquad \int u e^u\,du = (u-1)e^u. \qquad \int u^n e^u\,du = u^n e^u - n\int u^{n-1} e^u\,du.$$

$$\int a^u\,du = \frac{a^u}{\ln a}, \qquad a > 0, a \neq 1. \qquad \int \ln u\,du = u\ln u - u. \qquad \int \frac{du}{u\ln u} = \ln|\ln u|.$$

$$\int u^n \ln u\,du = u^{n+1}\left(\frac{\ln u}{n+1} - \frac{1}{(n+1)^2}\right), \qquad n \neq -1. \qquad \int \frac{du}{a^2 + u^2} = \frac{1}{a}\arctan\frac{u}{a}.$$

$$\int \frac{du}{u^2 - a^2} = \frac{1}{2a}\ln\left|\frac{u-a}{u+a}\right|. \qquad \int \frac{du}{a^2 - u^2} = \frac{1}{2a}\ln\left|\frac{a+u}{a-u}\right|.$$

$$\int \frac{du}{(a+bu)(\alpha+\beta u)} = \frac{1}{a\beta - \alpha b}\ln\left|\frac{\alpha+\beta u}{a+bu}\right|$$

$$\int \frac{u\,du}{(a+bu)(\alpha+\beta u)} = \frac{1}{a\beta - \alpha b}\left[\frac{a}{b}\ln|a+bu| - \frac{\alpha}{\beta}\ln|\alpha+\beta u|\right].$$

$$\int \frac{du}{\sqrt{u^2 + a^2}} = \ln\left|u + \sqrt{u^2 + a^2}\right|. \qquad \int \frac{du}{\sqrt{u^2 - a^2}} = \ln\left|u + \sqrt{u^2 - a^2}\right|, \qquad u^2 \geq a^2.$$

$$\int \frac{du}{\sqrt{a^2 - u^2}} = \arcsin\frac{u}{a}, \qquad a^2 \geq u^2. \qquad \int \frac{du}{u\sqrt{u^2 - a^2}} = \frac{1}{a}\arccos\frac{a}{u}, \qquad u > a > 0.$$

$$\int \sin u\,du = -\cos u. \qquad \int \cos u\,du = \sin u. \qquad \int \tan u\,du = -\ln|\cos u|.$$

$$\int \cot u\,du = \ln|\sin u|. \qquad\qquad\qquad \int \sec u\,du = \ln|\sec u + \tan u|.$$
$$\int \csc u\,du = -\ln|\csc u + \cot u| = \ln|\csc u - \cot u|.$$

$$\int \sec^2 u\,du = \tan u. \qquad \int \csc^2 u\,du = -\cot u. \qquad \int \sec u\tan u\,du = \sec u.$$

$$\int \sin^2 u\,du = \frac{1}{2}u - \frac{1}{4}\sin 2u. \qquad \int \cos^2 u\,du = \frac{1}{2}u + \frac{1}{4}\sin 2u. \qquad \int \tan^2 u\,du = \tan u - u.$$

$$\int \sin^n u\,du = -\frac{\sin^{n-1} u \cos u}{n} + \frac{n-1}{n}\int \sin^{n-2} u\,du.$$

$$\int \cos^n u\,du = \frac{\cos^{n-1} u \sin u}{n} + \frac{n-1}{n}\int \cos^{n-2} u\,du.$$

Note: An arbitrary constant is to be added to each indefinite integral.

$$\int u \sin u \, du = \sin u - u \cos u.$$

$$\int u^n \sin u \, du = -u^n \cos u + n \int u^{n-1} \cos u \, du.$$

$$\int u \cos u \, du = \cos u + u \sin u.$$

$$\int u^n \cos u \, du = u^n \sin u - n \int u^{n-1} \sin u \, du.$$

$$\int e^{au} \sin nu \, du = \frac{e^{au}(a \sin nu - n \cos nu)}{a^2 + n^2}.$$

$$\int e^{au} \cos nu \, du = \frac{e^{au}(a \cos nu + n \sin nu)}{a^2 + n^2}.$$

$$\int \sin au \sin bu \, du = -\frac{\sin(a+b)u}{2(a+b)} + \frac{\sin(a-b)u}{2(a-b)}, \qquad a^2 \neq b^2.$$

$$\int \cos au \cos bu \, du = \frac{\sin(a+b)u}{2(a+b)} + \frac{\sin(a-b)u}{2(a-b)}, \qquad a^2 \neq b^2.$$

$$\int \sin au \cos bu \, du = -\frac{\cos(a+b)u}{2(a+b)} - \frac{\cos(a-b)u}{2(a-b)}, \qquad a^2 \neq b^2.$$

$$\int \sinh u \, du = \cosh u.$$

$$\int \cosh u \, du = \sinh u.$$

$$\Gamma(t) = \int_0^\infty e^{-u} u^{t-1} du, \quad t > 0; \qquad \Gamma\left(\tfrac{1}{2}\right) = \sqrt{\pi}; \quad \text{and} \quad \Gamma(n+1) = n!, \text{ if } n \text{ is a positive integer.}$$

SOME POWER SERIES EXPANSIONS

$$f(x) = f(a) + f'(a)(x-a) + \frac{f''(a)}{2!}(x-a)^2 + \cdots + \frac{f^{(n)}(a)}{n!}(x-a)^n + \cdots \quad (\text{Taylor series})$$

$$e^x = \sum_{n=0}^\infty \frac{x^n}{n!} \qquad \sin x = \sum_{n=0}^\infty \frac{(-1)^n x^{2n+1}}{(2n+1)!} \qquad \cos x = \sum_{n=0}^\infty \frac{(-1)^n x^{2n}}{(2n)!}$$

$$(1-x)^{-1} = \sum_{n=0}^\infty x^n \qquad (1-x)^{-2} = \sum_{n=0}^\infty (n+1)x^n \qquad \ln(1-x) = -\sum_{n=1}^\infty \frac{x^n}{n}$$

$$\tan x = x + \tfrac{1}{3}x^3 + \tfrac{2}{15}x^5 + \tfrac{17}{315}x^7 + \tfrac{62}{2835}x^9 + \cdots$$

$$\arctan x = \sum_{n=0}^\infty (-1)^n \frac{x^{2n+1}}{2n+1}$$

$$\arcsin x = x + \frac{x^3}{2\cdot3} + \frac{1\cdot3}{2\cdot4\cdot5}x^5 + \frac{1\cdot3\cdot5}{2\cdot4\cdot6\cdot7}x^7 + \cdots$$

$$J_0(x) = \sum_{k=0}^\infty \frac{(-1)^k x^{2k}}{(k!)^2 2^{2k}} \qquad J_1(x) = \sum_{k=0}^\infty \frac{(-1)^k x^{2k+1}}{k!(k+1)!2^{2k+1}} \qquad J_n(x) = \sum_{k=0}^\infty \frac{(-1)^k x^{2k+n}}{k!\Gamma(n+k+1)2^{2k+n}}$$

Note: An arbitrary constant is to be added to each indefinite integral.

LINEAR FIRST-ORDER EQUATIONS

The solution to $y' + P(x)y = Q(x)$ taking the value $y(x_0)$ at $x = x_0$ is

$$y(x) = e^{-\int_{x_0}^{x} P(\xi)d\xi}\left[\int_{x_0}^{x} e^{\int_{x_0}^{\xi} P(\zeta)d\zeta} Q(\xi)d\xi + y(x_0)\right].$$

METHOD OF UNDETERMINED COEFFICIENTS

To find a particular solution to the constant-coefficient differential equation

$$ay'' + by' + cy = P_m(t)e^{rt},$$

where $P_m(t)$ is a polynomial of degree m, use the form

$$y_p(t) = t^s(A_m t^m + \cdots + A_1 t + A_0)e^{rt};$$

if r is not a root of the associated auxiliary equation, take $s = 0$; if r is a simple root of the associated auxiliary equation, take $s = 1$; and if r is a double root of the associated auxiliary equation, take $s = 2$.

To find a particular solution to the differential equation

$$ay'' + by' + cy = P_m(t)e^{\alpha t}\cos\beta t + Q_n(t)e^{\alpha t}\sin\beta t, \quad \beta \neq 0,$$

where $P_m(t)$ is a polynomial of degree m and $Q_n(t)$ is a polynomial of degree n, use the form

$$y_p(t) = t^s(A_k t^k + \cdots + A_1 t + A_0)e^{\alpha t}\cos\beta t$$
$$+ t^s(B_k t^k + \cdots + B_1 t + B_0)e^{\alpha t}\sin\beta t,$$

where k is the larger of m and n. If $\alpha + i\beta$ is not a root of the associated auxiliary equation, take $s = 0$; if $\alpha + i\beta$ is a root of the associated auxiliary equation, take $s = 1$.

VARIATION OF PARAMETERS FORMULA

If y_1 and y_2 are two linearly independent solutions to $ay'' + by' + cy = 0$, then a particular solution to $ay'' + by' + cy = f$ is $y = v_1 y_1 + v_2 y_2$, where

$$v_1(t) = \int \frac{-f(t)y_2(t)}{aW[y_1, y_2](t)}\,dt, \qquad v_2(t) = \int \frac{f(t)y_1(t)}{aW[y_1, y_2](t)}\,dt,$$

and $W[y_1, y_2](t) = y_1(t)y_2'(t) - y_1'(t)y_2(t).$

$f(t)$	$F(s) = \mathcal{L}\{f\}(s)$	$f(t)$	$F(s) = \mathcal{L}\{f\}(s)$
1. $f(at)$	$\dfrac{1}{a}F\!\left(\dfrac{s}{a}\right)$	**20.** $\dfrac{1}{\sqrt{t}}$	$\dfrac{\sqrt{\pi}}{\sqrt{s}}$
2. $e^{at}f(t)$	$F(s-a)$	**21.** $\sqrt{t}$	$\dfrac{\sqrt{\pi}}{2s^{3/2}}$
3. $f'(t)$	$sF(s) - f(0)$	**22.** $t^{n-(1/2)},\quad n = 1, 2, \ldots$	$\dfrac{1\cdot 3\cdot 5\cdots(2n-1)\sqrt{\pi}}{2^{n}s^{n+(1/2)}}$
4. $f^{(n)}(t)$	$s^{n}F(s) - s^{n-1}f(0) - s^{n-2}f'(0)$	**23.** $t^{r},\quad r > -1$	$\dfrac{\Gamma(r+1)}{s^{r+1}}$
	$-\cdots - sf^{(n-2)}(0) - f^{(n-1)}(0)$	**24.** $\sin bt$	$\dfrac{b}{s^2 + b^2}$
5. $t^{n}f(t)$	$(-1)^{n}F^{(n)}(s)$	**25.** $\cos bt$	$\dfrac{s}{s^2 + b^2}$
6. $\dfrac{1}{t}f(t)$	$\displaystyle\int_{s}^{\infty} F(u)\,du$	**26.** $e^{at}\sin bt$	$\dfrac{b}{(s-a)^2 + b^2}$
7. $\int_{0}^{t} f(v)\,dv$	$\dfrac{F(s)}{s}$	**27.** $e^{at}\cos bt$	$\dfrac{s-a}{(s-a)^2 + b^2}$
8. $(f * g)(t)$	$F(s)\,G(s)$	**28.** $\sinh bt$	$\dfrac{b}{s^2 - b^2}$
9. $f(t+T) = f(t)$	$\dfrac{\int_{0}^{T} e^{-st}f(t)\,dt}{1 - e^{-sT}}$	**29.** $\cosh bt$	$\dfrac{s}{s^2 - b^2}$
10. $f(t-a)u(t-a),\ a \geq 0$	$e^{-as}F(s)$	**30.** $\sin bt - bt\cos bt$	$\dfrac{2b^3}{(s^2 + b^2)^2}$
11. $g(t)u(t-a),\quad a \geq 0$	$e^{-as}\mathcal{L}\{g(t+a)\}(s)$	**31.** $t\sin bt$	$\dfrac{2bs}{(s^2 + b^2)^2}$
12. $u(t-a),\quad a \geq 0$	$\dfrac{e^{-as}}{s}$	**32.** $\sin bt + bt\cos bt$	$\dfrac{2bs^2}{(s^2 + b^2)^2}$
13. $\Pi_{a,b}(t),\quad 0 < a < b$	$\dfrac{e^{-sa} - e^{-sb}}{s}$	**33.** $t\cos bt$	$\dfrac{s^2 - b^2}{(s^2 + b^2)^2}$
14. $\delta(t-a),\quad a \geq 0$	e^{-as}	**34.** $\sin bt\cosh bt - \cos bt\sinh bt$	$\dfrac{4b^3}{s^4 + 4b^4}$
15. e^{at}	$\dfrac{1}{s-a}$	**35.** $\sin bt\sinh bt$	$\dfrac{2b^2 s}{s^4 + 4b^4}$
16. $t^{n},\quad n = 1, 2, \ldots$	$\dfrac{n!}{s^{n+1}}$	**36.** $\sinh bt - \sin bt$	$\dfrac{2b^3}{s^4 - b^4}$
17. $e^{at}t^{n},\quad n = 1, 2, \ldots$	$\dfrac{n!}{(s-a)^{n+1}}$	**37.** $\cosh bt - \cos bt$	$\dfrac{2b^2 s}{s^4 - b^4}$
18. $e^{at} - e^{bt}$	$\dfrac{(a-b)}{(s-a)(s-b)}$	**38.** $J_{v}(bt), v > -1$	$\dfrac{(\sqrt{s^2 + b^2} - s)^{v}}{b^{v}\sqrt{s^2 + b^2}}$
19. $ae^{at} - be^{bt}$	$\dfrac{(a-b)s}{(s-a)(s-b)}$		